W9-CDI-987

Introducing the Interactive Companion Web site for *Human Sexuality in a World of Diversity, 5/e*

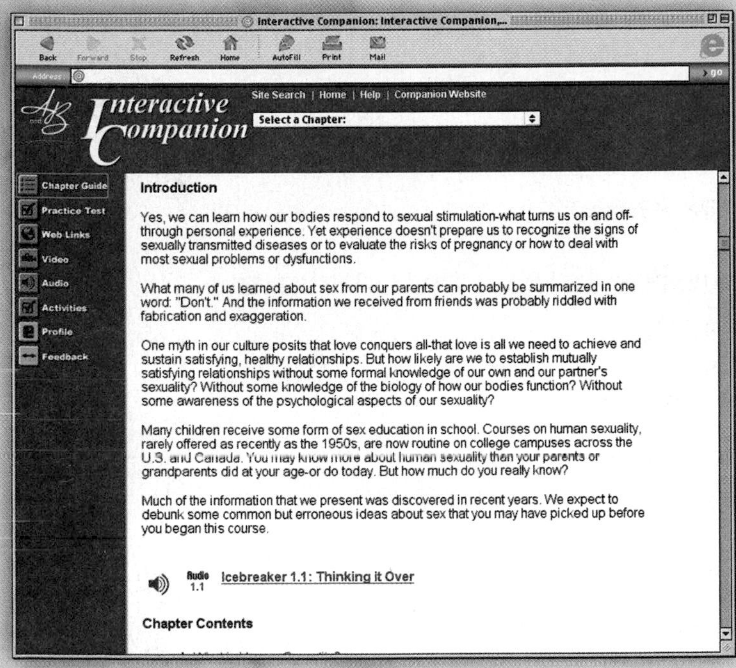

Learning has reached a whole new level with the Interactive Companion Web site – *by encouraging you to apply what you've learned through audio and video clips, Web sites, activities and practice tests.*

With the purchase of *Human Sexuality in a World of Diversity*, 5/e you'll get access to this PIN-code protected Web site FREE of charge. The Allyn & Bacon Interactive Companion Web site represents an exciting new study tool that uses the latest in multimedia to review, enrich, and expand upon key concepts presented in its companion textbook.

Using chapter highlights as its organizing structure, the Interactive Companion helps you apply what you've learned by presenting you with hundreds of links to audio and video clips, Web sites, activities, and practice tests. These links are annotated with brief descriptions that help you understand the value and purpose of each type of media in the context of the chapter.

Everything you need is right here at your fingertips!

ACTIVATE YOUR PIN.

How will you benefit from using the Interactive Companion Web site?

- ✓ **Provides you with frequent feedback on your learning progress to perform better on tests.**
- ✓ **Offers highly interactive ways for you to engage with the textbook content.**
- ✓ **Adds variety to course materials and helps you study more effectively.**
- ✓ **Helps you to think critically about the information presented to you in the textbook and on the Web site.**
- ✓ **Gives you access to the latest information related to the textbook topics via the Web.**

> *"A major advantage of the Web site is that it allows students to connect with a wealth of learning support any time and any place. Professors can create a vast array of assignments and projects, knowing that all students – those in dorms and those commuting from home miles away, those with easy access to libraries and those who are more isolated – have the resources to complete the assignments and projects."*
>
> ***Anita Woolfolk***
> ***Educational Psychology Professor***
> ***The Ohio State University***

Easy navigation that lets you study the way you want!

1. Each chapter begins with an attention-grabbing opener. **Click on the associated audio icon and you'll hear questions or issues framed around the chapter opener;** this allows you to hear as well as read about new concepts.
2. **Chapter learning objectives are linked to the various topic areas**, allowing you to go directly from what you need to learn in the chapter to the media assets that will help you learn this new information.
3. **Each topic area ends with an activity** that is a review of key terms. These are created as flash card exercises – allowing you to reinforce the information that you've learned from that section of the book.
4. **Every chapter ends with three items:**

   a. **Concept Check Activity** which asks you to match terms and concepts in the chapter with their definitions.

   b. **"Closing Thoughts"** in which you revisit the opening scenario and consider the situation in light of what you've learned in the chapter.

   c. **Practice Test,** which promotes self-regulation and self-monitoring, two key elements of successful learning.

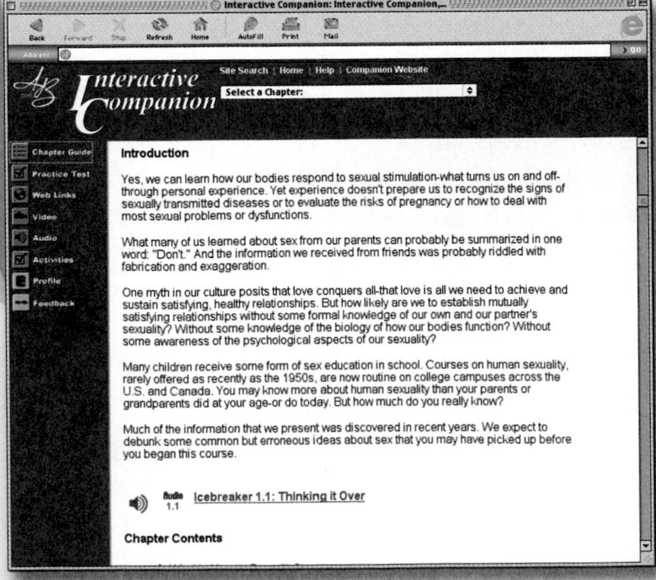

Everything you need is right here at your fingertips!

ACTIVATE YOUR PIN.

Interactive ways to prepare for that in-class exam and research paper!

Practice Test

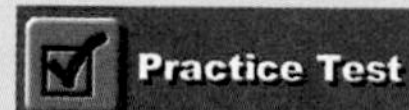

Click on a **"Practice Test"** icon and you'll be able to test your understanding of the chapter material by completing a self-scoring practice test. You'll receive immediate results and feedback from your test, allowing you to review your weak areas in preparation for the actual in-class exam.

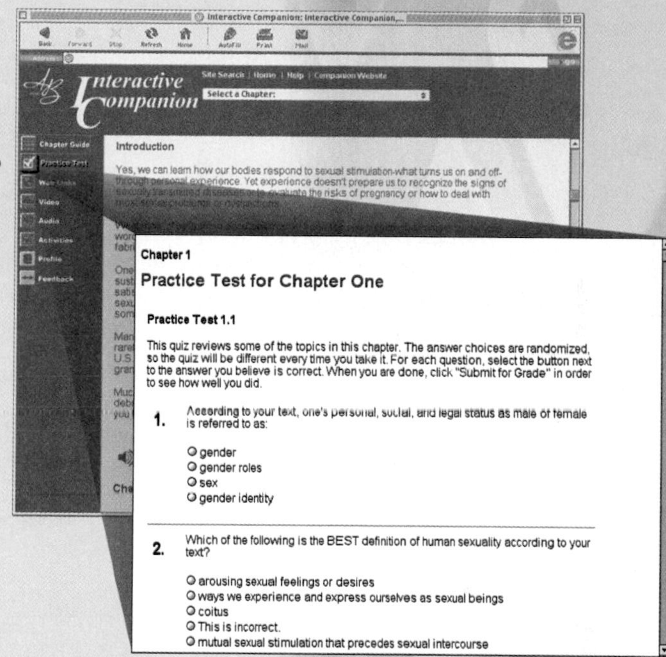

> *I think it's great. You can take practice tests so that when you really get tested, you already know what to expect. I think it really contributes to your learning.*
> ***Gloria, age 27***

Web links

With the benefit of an Internet connection, by clicking on the **"Web links"** icon you'll jump to current Web sites that provide you with additional information about the specific topics you're studying. Web links are continuously monitored and updated by Allyn & Bacon, so you'll always have the most current sites to access. This is a great resource for you to utilize when writing a research paper!

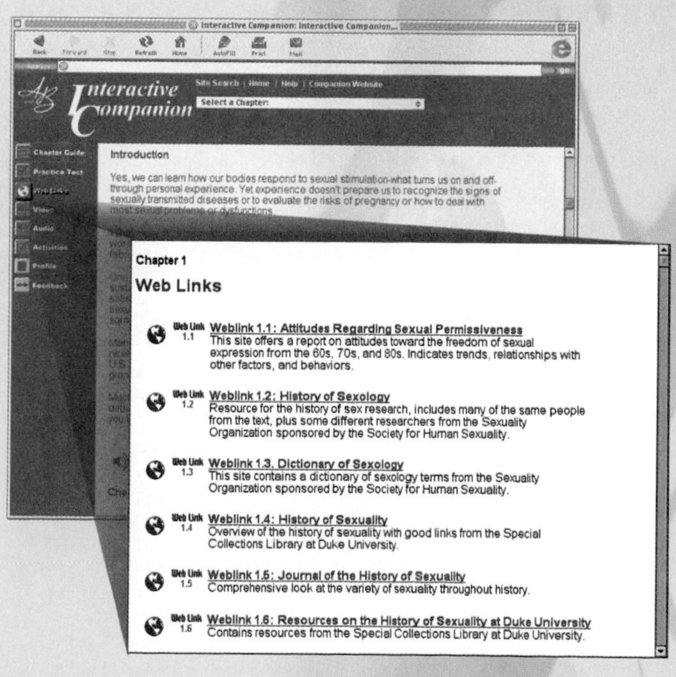

> *These Web links help me find quality Internet resources for the types of assignments required for class.*
> ***Debra, age 22***

Information comes alive when you see it and hear it!

Video

Click on a **"Video"** icon and you'll be captivated by the **sights and sounds of video segments directly related to the material you just read.** For some textbooks, these segments come from leading news sources. For others, they are part of custom videos developed to demonstrate concepts in the discipline.

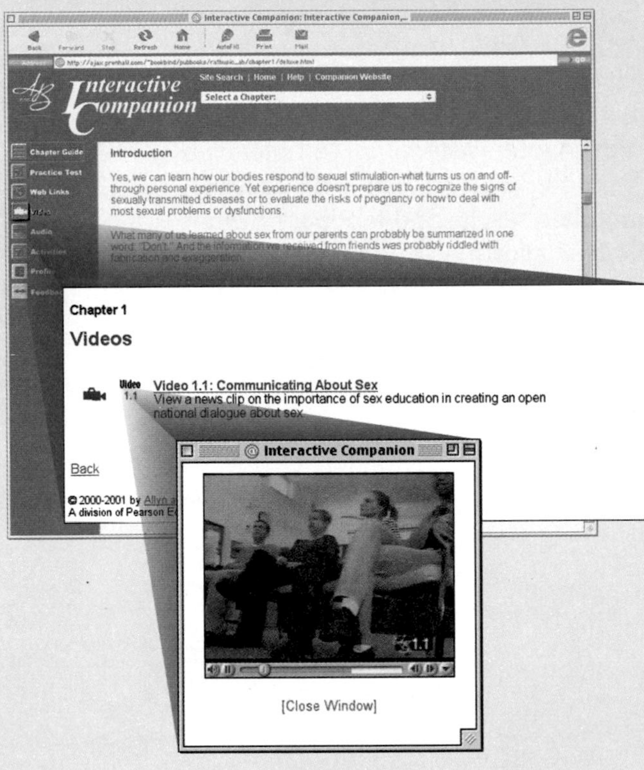

> *The audio, feedback, and videos help me understand each chapter.*
> ***Janel, age 40***

Audio

Click on an **"Audio"** icon and you'll hear either the author of the textbook or a specialist in the field speaking directly about concepts in the book. Often **the "voice" will add background information or give examples – material that enhances and extends the chapter material.**

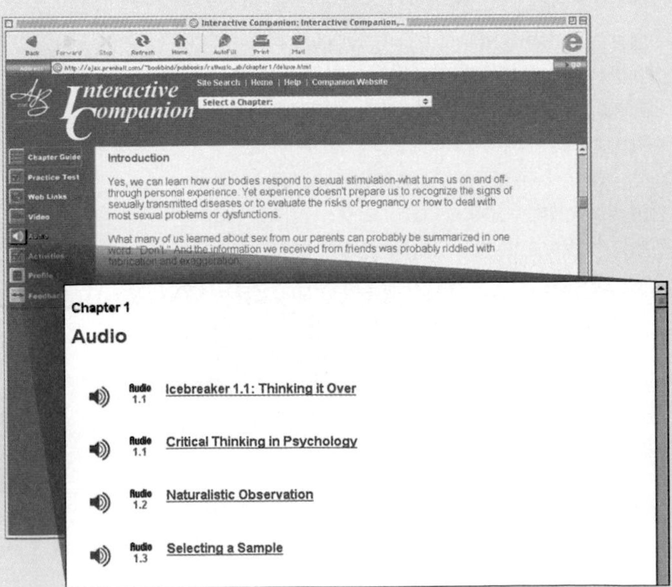

Everything you need is right here at your fingertips!

ACTIVATE YOUR PIN.

Activities give you more opportunities to test your level of understanding!

Activities

Click on an **"Activities"** icon and you can complete interesting activities directly related to the information presented in the textbook. **You'll be asked to research, discuss, think critically, and more!**

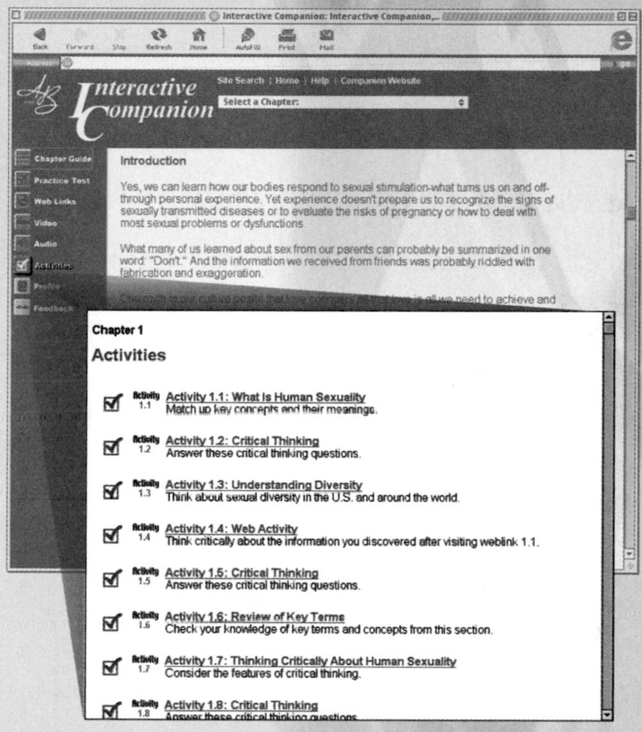

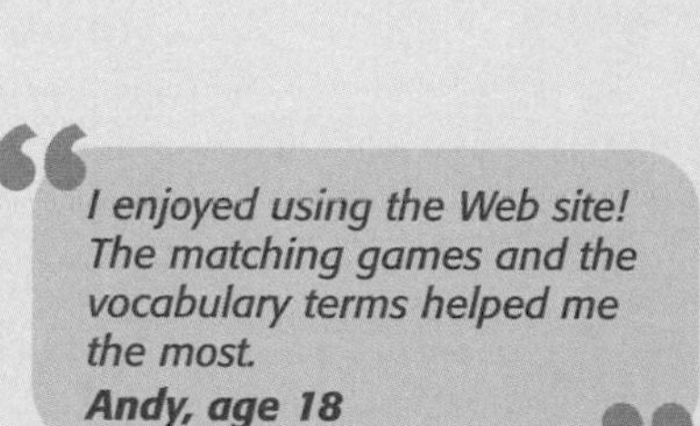

Profile

By entering your profile on the site, you can avoid retyping it every time you need to submit homework. Once saved, the information will appear automatically whenever it's required.

Feedback

We are always looking to improve and expand upon the information and technology we offer you. If you have any suggestions, questions or technical difficulties, please contact us through the feedback feature on the site.

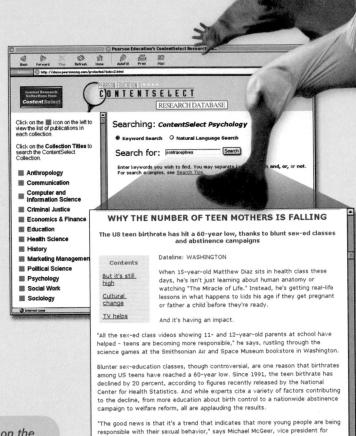

fifth edition

Human Sexuality in a World of Diversity

Spencer A. Rathus
Montclair State University

Jeffrey S. Nevid
St. John's University

Lois Fichner-Rathus
The College of New Jersey

Allyn and Bacon
Boston ■ London ■ Toronto ■ Sydney ■ Tokyo ■ Singapore

Executive Editor: *Rebecca Pascal*
Senior Marketing Manager: *Caroline Croley*
Production Administrator: *Susan Brown*
Editorial/Production Services: *Kathleen Deselle*
Text Designer: *Glenna Collett*
Formatting/Page Layout: *Omegatype Typography, Inc.*
Cover Administrator: *Linda Knowles*
Composition Buyer: *Linda Cox*
Manufacturing Buyer: *Megan Cochran*
Photo Researcher: *Helane Prottas*

A Pearson Education Company
75 Arlington St.
Boston, MA 02116
Internet: www.abacon.com

Library of Congress Cataloging-in-Publication Data

Rathus, Spencer A.
Human sexuality in a world of diversity/Spencer A. Rathus, Jeffrey S. Nevid, Lois Fichner-Rathus. 5th ed.
p. cm.
Includes bibliographical references and index.
ISBN 0-205-33517-9
1. Sex. I. Nevid, Jeffrey, S. II. Fichner-Rathus, Lois 1953– III. Title.
HQ21.R23 2002
306.7—dc21 2001022871

Printed in the United States of America
10 9 8 7 6 5 4 3 2 1 RRDW 04 03 02 01

Photo Credits: **Chapter 1:** p. 7, David Austen/Stock Boston; p. 10, Ali Meyer/The Bridgeman Art Library, Ltd.; p. 14, © Bettmann/Corbis; p. 19, Kent Meireis/The Image Works; p. 21, © Hulton Getty; p. 22, © Birkhead/Animals Animals; p. 26, Jerome Delay/AP/ Wide World Photos; p. 31, Elizabeth Crews/The Image Works. **Chapter 2:** p. 41, Courtesy of Farrall Instruments; p. 45, © Hulton Getty; p. 50, Rick Browne/Stock Boston; p. 52, L. Ka Tai/Woodfin Camp & Associates. **Chapter 3:** p. 67, (all) © Susan Lerner 1999/Joel Gordon Photography 2001; p. 71, Catherine Leroy/SIPA Press; p. 79, John Giannicchi/Science Source/Photo Researchers; p. 82, (left) © Susan Werner/Joel Gordon Photography 2001, (middle) Custom Medical Stock Photo, (right) Joel Gordon/©Joel Gordon Photography 2001; p. 83, Charles Gupton/Stock Boston; p. 97, Ed Honowitz/© Stone; p. 98, George Gardner/The Image Works; p. 102, © Mark Richards/PhotoEdit; p. 105, Dan Bosler/© Stone. **Chapter 4:** p. 113, (left and right) Joel Gordon/© Joel Gordon Photography 2001, (middle) Custom Medical Stock; p.116, © David Young-Wolff/PhotoEdit; p. 118, Cleo/PhotoEdit; p. 119, Professor P. Motta/Dept. of Anatomy/Rome University/SPL/Science Source/Photo Researchers; p. 120, © Jack Clark/Animals Animals; p. 124, Laurent Rebours/AP/Wide World Photos; p. 126, Robert Mecea/AP/Wide World Photos.

Photo credits continue on page xii, which constitutes a continuation of the copyright page.

Dedicated with love to our children
Taylor Lane Rathus and Michael Zev Nevid,
who were born at the time the first edition
of this book was written

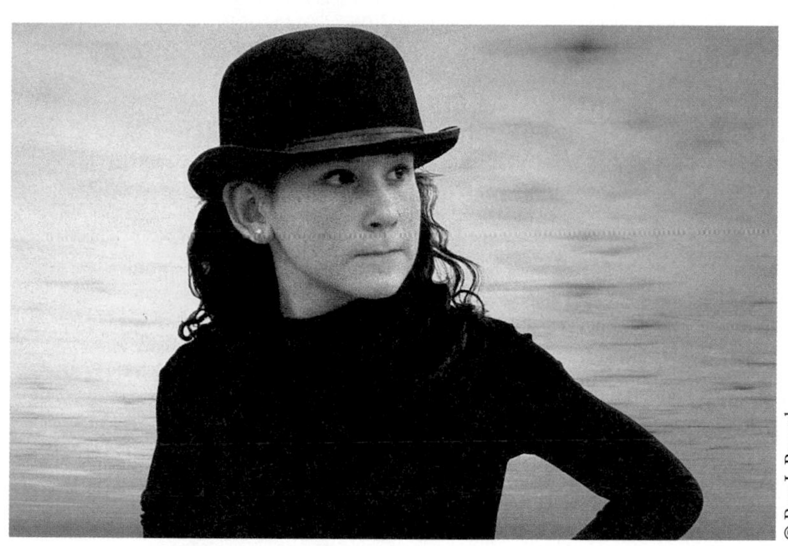

Taylor Lane Rathus
November 3, 1990

Michael Zev Nevid
January 17, 1991

Photo Credits: *(continued from page x)* **Chapter 5:** p. 144, © D. Thompson/Animals Animals; p. 145, Luc Beziat/Stone; p. 146, © CAP Features/The Image Works; p. 148, © Ralph Reinhold/ Animals Animals; p. 150, Bruce Lee Smith/Gamma Liaison; p. 157, © M. K. Denny/PhotoEdit. **Chapter 6:** pp. 177 and 178, from John Money, *Sex Errors of the Body and Related Syndromes: A Guide for Counseling Children, Adolescents and Their Families*, 1994, pp. 37 and 46. Published by Paul H. Brookes Publishing Co., P. O. Box 10624, Baltimore, MD 21285–0624. Used with permission of the author; p. 184, Courtesy of Jayne Thomas; p. 187, Brian Smith; p. 190, Loren Santow/ © Stone; p. 193, Christopher Arnesen/ © Stone; p. 196, © Tony Freeman/PhotoEdit. **Chapter 7:** p. 208, (top left) Adam Wolfitt/Woodfin Camp & Associates, (top middle) Reuters/Fred Prouser/Archive Photos, (top right) Claus Meyer/Black Star, (bottom left) Mitch Kezar/Black Star, (bottom right) Nicholas DeVore III/ Photographers Aspen; p. 210, ©Spencer Grant/PhotoEdit; p. 211, (both) D. Perrett, K. May, and S. Yoshikawa, University of St. Andrews/SPL/Science Source/Photo Researchers; p. 215, Frank Siteman/ Stock Boston; p. 220, Chris Weeks/© Liaison Agency. **Chapter 8:** p. 233, Brian Smith; p. 237, Michael Newman/PhotoEdit; p. 240, Paul Edmondson/© Stone; p. 243, © Hulton Getty; p. 246, Joel Gordon/ © Joel Gordon Photography 2001; p. 252, Robert Harbison. **Chapter 9:** p. 274, Costa Manos/Magnum Photos. **Chapter 10:** p. 291, Joel Gordon/© Joel Gordon Photography 2001; p. 301, Jonathan Nourok/ PhotoEdit; p. 305, Rommel Pecson/Impact Visuals; p. 319, Chuck Nacke/Woodfin Camp & Associates; p. 321, Karen Beard/© Stone. **Chapter 11:** p. 329, Francis LeRoy/Bio Cosmos/SPL/Science Source/ Photo Researchers; p. 331, Anna Zuckerman/PhotoEdit; p. 335, © Zig Leszcznski/Animals Animals; p. 337, Mehau Kulyk/SPL/Science Source/Photo Researchers; p. 342, (all) Petit Format/Nestle/Science Source/Photo Researchers; p. 352, Will Hart; p. 358, SIU/Science Source/Photo Researchers; p. 365, © A. Ramey/Stock Boston. **Chapter 12:** p. 377, Joel Gordon/© Joel Gordon Photography 2001; p. 383, © Will & Deni McIntyre/Photo Researchers; p. 385, Joel Gordon/© Joel Gordon Photography; p. 386, (left) Coutesy of Alza, (right) PhotoTake. **Chapter 13:** p. 423, Rick Smolen/ Stock, Boston; p. 428, © Michael Newman/PhotoEdit; p. 431, David Harry Stewart/Stone; p. 435, Bob Torrez/Stone; p. 443, Mark Lewis/© Stone; p. 445, Yvonne Hemsey/Gamma Liaison. **Chapter 14:** p. 456, Frank Herboldt/ Stone; p. 458, Frank Siteman/Stone; p. 463, B. D. V./© Corbis; p. 468, (left) B. Petry/Gamma Liaison, (right) AP/Wide World Photos; p. 473, AP/Wide World Photos; p. 478, Laurence Monneret/Stone. **Chapter 15:** p. 492, © Michael Newman/PhotoEdit; p. 497, Paul Gauguin. *And the gold of their bodies.* Photo: B. Hatala. Musée d'Orsay, Paris, France. Courtesy of Réunion des Musées Nationaux/Art Resource/NY; p. 502, Robert A. Mitchell/© Stone; p. 506, © Tom McCarthy/PhotoEdit; p. 511, G. Thomas Bishop/Custom Medical Stock Photo; p. 513, © David Young-Wolff/PhotoEdit; p. 515, © CLANG/CMP. **Chapter 16:** p, 529, Jim Corwin/Stone (This photograph is used for illustrative purposes only.); p. 536, Courtesy of Dr. Nicholas J. Fiumara; p. 537, Dr. David Chase/PhotoTake; p. 538, Custom Medical Stock Photo; p. 540, Brian Palmer/Impact Visuals; p. 543, National Medical Slide/Custom Medical Stock Photo; p. 547, Bill Longcone/Science Source/Photo Researchers; p. 549, Douglas Mason/Woodfin Camp & Associates; p. 553, Jana Birchum/Impact Visuals; p. 554, © F. Martinez/ PhotoEdit; p. 557, Dr. John Wilson/Science Source/Photo Researchers; p. 559, BioPhoto Assoc./Science Source/Photo Researchers; p. 561, E. Gray/SPL/Science Source/Photo Researchers. **Chapter 17:** p. 575, Herb Snitzer/Stock Boston; p. 576, Donna Binder/Impact Visuals; p. 585, Photo: David Appleby/© Fox Searchlight/PhotoFest; p. 588, © David Young-Wolff/PhotoEdit; p. 596, Bob Daemmrich/The Image Works. **Chapter 18:** p. 607, © Richard Pasley/Stock Boston; p. 613, Michael Nichols/Magnum Photos; p. 617, Carolina Kroon/Impact Visuals; p. 622, Courtesy of LaPorte County Child Abuse Prevention Council; p. 627, Charles Gupton/Stone; p. 632, John Coletti. **Chapter 19:** p. 643, © A. Woodfitt/ Woodfin Camp & Associates; p. 645, Douglas Mason/Woodfin Camp & Associates; p. 648, Ian Jackson/© Stone; p. 652, Paul Chesley/© Stone; p. 656, (left) Reuters/Jeff Christensen/Archive Photos, (right) *Patrice, 1977.* Copyright © 1977 The Estate of Robert Mapplethorpe. Used with permission.

brief contents

1 What Is Human Sexuality? *2*

2 Research Methods *36*

3 Female Sexual Anatomy and Physiology *64*

4 Male Sexual Anatomy and Physiology *108*

5 Sexual Arousal and Response *140*

6 Gender Identity and Gender Roles *170*

7 Attraction and Love *204*

8 Relationships, Intimacy, and Communication *227*

9 Sexual Techniques and Behavior Patterns *259*

10 Sexual Orientation *288*

11 Conception, Pregnancy, and Childbirth *326*

12 Contraception and Abortion *372*

13 Sexuality in Childhood and Adolescence *418*

14 Sexuality in Adulthood *452*

15 Sexual Dysfunctions *486*

16 Sexually Transmitted Infections *526*

17 Atypical Sexual Variations *570*

18 Sexual Coercion *600*

19 Commercial Sex *638*

20 Making Responsible Sexual Decisions—An Epilogue *660*

contents

Preface *xxvi*

1 What Is Human Sexuality? *2*

Outline *3*

Truth or Fiction? *3*

What Is Human Sexuality? *5*
The Study of Human Sexuality *5*

A World of Diversity Around the World in Eighty Ways—A Preview *6*

Sexuality and Values *6*

Thinking Critically About Human Sexuality *8*
Principles of Critical Thinking *8*

Perspectives on Human Sexuality *9*
The Historical Perspective *9*
The Biological Perspective *19*
The Evolutionary Perspective *20*

A Closer Look Females Throughout the Animal Kingdom Are Monogamous—*Not!* *22*

The Cross-Species Perspective *24*
The Cross-Cultural Perspective *24*

A World of Diversity Love Finds a Way in Iran: "Temporary Marriage" *26*

Psychological Perspectives *28*
Sociocultural Perspectives *31*
Multiple Perspectives on Human Sexuality *33*

Human Sexuality Online Web Sites on Human Sexuality *34*

Summing Up *34*
Questions for Critical Thinking *35*

2 Research Methods *36*

Outline *37*

Truth or Fiction? *37*

A Scientific Approach to Human Sexuality *38*
The Scientific Method *38*
Goals and Methods of the Science of Human Sexuality *39*
Operational Definitions *40*

A Closer Look Physiological Measures of Sexual Arousal *41*

Populations and Samples: Representing the World of Diversity *42*
Sampling Methods *42*

Methods of Observation *44*
The Case-Study Method *44*
The Survey Method *45*

A Closer Look With a Horse? *47*

A World of Diversity Surveys of the Sexual Behavior of Diverse Populations *50*

The Naturalistic-Observation Method *53*
The Ethnographic-Observation Method *54*
The Participant-Observation Method *54*
The Laboratory-Observation Method *54*

The Correlational Method *56*
Limitations of the Correlational Method *56*

The Experimental Method *57*
Aspects of the Experimental Method *57*

Ethics in Sex Research *59*
Pain and Stress *59*
Confidentiality *60*
Informed Consent *60*
The Use of Deception *60*

Human Sexuality Online Web Sites for Use in Research on Human Sexuality *61*

Summing Up *61*
Questions for Critical Thinking *62*

3 Female Sexual Anatomy and Physiology *64*

Outline *65*

Truth or Fiction? *65*

External Sex Organs *66*
The Mons Veneris *67*
The Labia Majora *67*
The Labia Minora *68*
The Clitoris *68*
The Vestibule *69*

The Urethral Opening *69*

A World of Diversity Clitoridectomy—Ritual Genital Mutilation *70*

The Vaginal Opening *72*
The Perineum *73*
Structures That Underlie the External Sex Organs *73*

A Closer Look Kegels *74*

Internal Sex Organs *75*
The Vagina *75*
The Cervix *76*
The Uterus *77*
The Fallopian Tubes *78*
The Ovaries *78*
The Pelvic Examination *79*

The Breasts *80*
Breast Cancer *82*

A World of Diversity African American Women and Breast Cancer *83*

A Closer Look Breast Self-Examination *86*

The Menstrual Cycle *88*
Menstruation Versus Estrus *88*
Regulation of the Menstrual Cycle *89*
Phases of the Menstrual Cycle *90*

A World of Diversity Historical and Cross-Cultural Perspectives on Menstruation *94*

Coitus During Menstruation *95*
Menopause *95*

A World of Diversity Symptoms of Menopause, Ethnicity, Socioeconomic Status, and Lifestyle *96*

A World of Diversity Osteoporosis—Not for Women Only *98*

A Closer Look Myths About Menopause *99*

Menstrual Problems *100*
Dysmenorrhea *100*
Amenorrhea *101*
Premenstrual Syndrome (PMS) *101*

A Closer Look "Lifting the Curse"—Should Monthly Periods Be Optional? *102*

How to Handle Menstrual Discomfort *104*

Human Sexuality Online Web Sites Related to Female Sexual Anatomy and Physiology *105*

Summing Up *106*

Questions for Critical Thinking *107*

4 Male Sexual Anatomy and Physiology *108*

Outline *109*

Truth or Fiction? *109*

External Sex Organs *111*
The Penis *111*

A Closer Look On Penis Size and Sexual Performance *115*

The Scrotum *115*

Human Sexuality Online Another Sort of Gender Equality: Up-Jock Pics Now Compete with Up-Skirt Photos *116*

Internal Sex Organs *116*
The Testes *117*

A Closer Look Is There a *Man*opause? *118*

The Vas Deferens *119*

A Closer Look Big *120*

The Seminal Vesicles *120*
The Prostate Gland *121*
Cowper's Glands *121*
Semen *121*

Diseases of the Urogenital System *122*
Urethritis *122*
Cancer of the Testes *122*
Disorders of the Prostate *123*

A Closer Look Lance Armstrong—Winner of Two Contests *124*

A Closer Look More Options, and Decisions, for Men with Prostate Cancer *126*

Male Sexual Functions *130*
Erection *130*
Spinal Reflexes and Sexual Response *131*
Ejaculation *135*

Human Sexuality Online Web Sites Related to Male Sexual Anatomy and Physiology *137*

Summing Up *137*

Questions for Critical Thinking *139*

5 Sexual Arousal and Response *140*

Outline *141*

Truth or Fiction? *141*

Making Sense of Sex: The Role of the Senses in Sexual Arousal *142*
Vision: The Better to See You With *142*
Smell: Does the Nose Know Best? *142*

A Closer Look The Buds and the Bees *144*

The Skin Senses: Sex as a Touching Experience *144*
Taste: On Savory Sex *145*

Human Sexuality Online www.seduction.net *146*

Hearing: The Better to Hear You With *146*

Aphrodisiacs: Of Spanish Flies and Rhino Horns *146*

A Closer Look Sex Toys for Pandas? 148

Anaphrodisiacs 149
Psychoactive Drugs 149

Sexual Response and the Brain: Cerebral Sex? 151
The Geography of the Brain 151
Brain Mechanisms in Sexual Functioning 153
On Pushing the Right Buttons: Are There Pleasure Centers in the Brain? 153

Sex Hormones: Do They "Goad" Us into Sex? 154
Sex Hormones and Sexual Behavior: Organizing and Activating Influences 154
Sex Hormones and Male Sexual Behavior 155
Sex Hormones and Female Sexual Behavior 156

The Sexual Response Cycle 157
Excitement Phase 158
Plateau Phase 159
Orgasmic Phase 160
Resolution Phase 162
Kaplan's Three Stages of Sexual Response: An Alternative Model 163

Controversies About Orgasm 163
Multiple Orgasms 164
How Many Kinds of Orgasms Do Women Have? One, Two, or Three? 165
The G-Spot 166

Human Sexuality Online Web Sites Related to Sexual Arousal and Response 168

Summing Up 168

Questions for Critical Thinking 169

6 Gender Identity and Gender Roles 170

Outline 171

Truth or Fiction? 171

Prenatal Sexual Differentiation 172
Genetic Factors in Sexual Differentiation 174
The Role of Sex Hormones in Sexual Differentiation 175
Descent of the Testes and the Ovaries 175
Sex-Chromosomal Abnormalities 176
Prenatal Sexual Differentiation of the Brain 176

Gender Identity 176
Nature and Nurture in Gender Identity 176
Hermaphroditism 177
Transsexualism 179

A Closer Look Boys Who Are Reared as Girls 180

A World of Diversity Using the Back Door in Iran: Can a Transsexual Change Her Mind? 183

A Closer Look Jayne Thomas, Ph.D.—In Her Own Words 184

Gender Roles and Stereotypes 186
Sexism 188
Gender Differences: Vive la Différence or Vive la Similarité? 189
Differences in Cognitive Abilities 189
Differences in Personality 190

A World of Diversity Men and Women Iron Their Way to Domestic Equality 191

On Becoming a Man or a Woman: Gender Typing 191
Biological Perspectives 191
Cross-Cultural Perspectives 192
Psychological Perspectives 194

Gender Roles and Sexual Behavior 198
Men as Sexually Aggressive, Women as Sexually Passive 198
Men as Overaroused, Women as Underaroused 199

Psychological Androgyny: The More Traits, the Merrier? 199
Psychological Androgyny, Psychological Well-Being, and Personal Development 200
Psychological Androgyny and Sexual Behavior 201

Human Sexuality Online Web Sites Related to Gender Identity and Gender Roles 201

Summing Up 202

Questions for Critical Thinking 203

7 Attraction and Love 204

Outline 205

Truth or Fiction? 205

Attraction 207
Physical Attractiveness: How Important Is Looking Good? 207

Human Sexuality Online Hospital Holds 'Net Auction for Breast Reduction 210

A World of Diversity Wide-Eyed with . . . Beauty? 211

A World of Diversity Gender Differences in Preferences in Mates Across 37 Cultures 214

The Matching Hypothesis: Who Is "Right" for You? 215
Similarity in Attitudes: Do Opposites Attract? 215
Reciprocity: If You Like Me, You Must Have Excellent Judgment 216

Love 216
The Greek Heritage 217
Romantic Love in Contemporary Western Culture 217

Questionnaire Are You a Romantic or a Realist? The Love Attitudes Scale *218*

Contemporary Models of Love: Dare Science Intrude? *220*

Questionnaire Sternberg's Triangular Love Scale *224*

Human Sexuality Online Web Sites Related to Attraction and Love *226*

Summing Up *227*

Questions for Critical Thinking *227*

8 Relationships, Intimacy, and Communication *228*

Outline *229*

Truth or Fiction? *229*

The ABC(DE)'s of Romantic Relationships *230*

A World of Diversity Don't Flirt (with Anyone!) in Qir *231*

The A's—Attraction *231*

Human Sexuality Online Modem Matchmaking *232*

The B's—Building *232*

A Closer Look How to Improve Date-Seeking Skills *235*

Human Sexuality Online In Modern E-Mail Romance, "Trash" Is Just a Click Away *236*

The C's—Continuation *238*
The D's—Deterioration *239*

Human Sexuality Online Long-Distance Romance, Web Enabled *240*

The E's—Ending *240*

Human Sexuality Online Advice for Young Women: How to Prepare for Married Life *242*

Loneliness: "All the Lonely People, Where Do They All Come From?" *244*
Causes of Loneliness *244*
Coping with Loneliness *245*

Intimacy *246*
Knowing and Liking Yourself *246*
Trusting and Caring *246*

A Closer Look Heaven Sent: The Blind Date Who Is Your Destiny *247*

Being Honest *248*
Making a Commitment *248*
Maintaining Individuality When the *I* Becomes *We* *248*
The Michelangelo Phenomenon *248*
Communicating *249*

Communication Skills for Enhancing Relationships and Sexual Relations *250*
Common Difficulties in Sexual Communication *250*
Getting Started *251*
Listening to the Other Side *252*
Learning About Your Partner's Needs *252*
Providing Information *253*
Making Requests *254*
Delivering Criticism *254*
Receiving Criticism *255*
When Communication Is Not Enough: Handling Impasses *257*

Human Sexuality Online Web Sites Related to Relationships, Intimacy, and Communication *258*

Summing Up *258*

Questions for Critical Thinking *259*

9 Sexual Techniques and Behavior Patterns *260*

Outline *261*

Truth or Fiction? *621*

Solitary Sexual Behavior *262*
Masturbation *262*

A Closer Look The Technology of Orgasm: "Hysteria," the Vibrator, and Women's Sexual Satisfaction *265*

Use of Fantasy *270*

Sexual Behavior with Others *272*
Foreplay *272*
Kissing *273*
Touching *273*
Stimulation of the Breasts *275*
Oral–Genital Stimulation *275*

A World of Diversity Sociocultural Factors and Oral Sex *276*

Sexual Intercourse: Positions and Techniques *278*

A World of Diversity Are African Americans Sexually Permissive? Cultural Myths Versus Data *284*

Human Sexuality Online Web Sites Related to Sexual Techniques and Behavior Patterns *286*

Summing Up *287*

Questions for Critical Thinking *287*

10 Sexual Orientation *288*

Outline *289*

Truth or Fiction? *289*

Sexual Orientation *291*
Coming to Terms with Terms *291*
Sexual Orientation and Gender Identity *292*
Classification of Sexual Orientation *292*
Bisexuality *297*

Perspectives on Gay Male and Lesbian Sexual Orientations *298*
Historical Perspectives *298*
Cross-Cultural Perspectives *299*
A World of Diversity Ethnicity and Sexual Orientation: A Matter of Belonging *300*
Cross-Species Perspectives *300*
Attitudes Toward Sexual Orientation in Contemporary Society *302*
Biological Perspectives *307*
Psychological Perspectives *309*
Gender Nonconformity *312*
Adjustment of Gay Males and Lesbians *313*
Treatment of Gay Male and Lesbian Sexual Orientations *315*
Coming Out: Coming to Terms with Being Gay *316*
Coming Out to Oneself *316*
Coming Out to Others *317*
Human Sexuality Online Out Into the "Gay Global Village" of Cyberspace *318*
Patterns of Sexual Activity Among Gay Males and Lesbians *318*
Sexual Techniques *318*
Gay Lifestyles *319*
A Closer Look "Don't Ask, Don't Tell"—Far from Clear as a Bell *321*
Lifestyle Differences Between Gay Males and Lesbians *322*
Variations in Gay Lifestyles *322*
Human Sexuality Online Web Sites Related to Sexual Orientation *323*
Summing Up *324*
Questions for Critical Thinking *325*

11 Conception, Pregnancy, and Childbirth *326*

Outline *327*
Truth or Fiction? *327*
Conception: Against All Odds *328*
Optimizing the Chances of Conception *329*
A Closer Look Selecting the Gender of Your Child *330*
Infertility and Alternative Ways of Becoming Parents *333*
Male Fertility Problems *333*
Female Fertility Problems *334*
A Closer Look Sex and the Single Lizard *335*
Pregnancy *337*
Questionnaire Should You Have a Child? *338*
Early Signs of Pregnancy *338*
Pregnancy Tests *338*
Early Effects of Pregnancy *340*
Miscarriage (Spontaneous Abortion) *340*
Sex During Pregnancy *340*
Psychological Changes During Pregnancy *341*
Prenatal Development *342*
The Germinal Stage *343*
The Embryonic Stage *344*
The Fetal Stage *346*
Environmental Influences on Prenatal Development *346*
A Closer Look Spacing Children the Goldilocks Way *349*
Chromosomal and Genetic Abnormalities *352*
Childbirth *355*
The Stages of Childbirth *355*
Methods of Childbirth *357*
Laboring Through the Birthing Options: Where Should a Child Be Born? *360*
Human Sexuality Online Childbirth Advice Is Just Keystrokes Away *362*
A World of Diversity A Racial Gap in Infant Deaths, and a Search for Reasons *364*
Birth Problems *364*
Anoxia *365*
Preterm and Low-Birth-Weight Children *366*
The Postpartum Period *366*
Maternal Depression *366*
Breast-Feeding Versus Bottle-Feeding *367*
Human Sexuality Online Where to Get Help Breast-Feeding *368*
Resumption of Ovulation and Menstruation *368*
Resumption of Sexual Activity *368*
Human Sexuality Online Web Sites Related to Conception, Pregnancy, and Childbirth *369*
Summing Up *369*
Questions for Critical Thinking *370*

12 Contraception and Abortion *372*

Outline *373*
Truth or Fiction? *373*
Contraception *375*
Contraception in the United States: The Legal Battle *375*
A Closer Look Talking with Your Partner About Contraception *376*

Selecting a Method of Contraception 377
Methods of Contraception 380
Oral Contraceptives ("the Pill") 380
A World of Diversity African Americans and Birth Control—The Intersection of Suspicion and Hope 382
Norplant 384
A Closer Look Congratulations, Wheat Ridge, It's a . . . ! 385
Intrauterine Devices (IUDs) 386
The Diaphragm 388
Spermicides 390
The Cervical Cap 391
Condoms 392
Douching 395
Withdrawal (Coitus Interruptus) 395
Fertility Awareness Methods (Rhythm Methods) 395
Sterilization 398
The Search Goes On 401
Abortion 404
When Does Human Life Begin? 405
Historical and Legal Perspectives on Abortion 405
Questionnaire Pro-Choice or Pro-Life? Where Do You Stand? 406
Methods of Abortion 410
A Closer Look Partial-Birth Abortion 411
A Closer Look RU-486 in Europe 412
Psychological Consequences of Abortion 413
Human Sexuality Online Web Sites Related to Contraception and Abortion 415
Summing Up 416
Questions for Critical Thinking 417

13 Sexuality in Childhood and Adolescence 418

Outline 419
Truth or Fiction? 419
Infancy (0 to 2 Years): The Search for the Origins of Human Sexuality 420
The Infant's Capacity for Sexual Response 420
A World of Diversity Cross-Cultural Perspectives on Childhood Sexuality 421
Masturbation 421
Genital Play 422
Early Childhood (3 to 8 Years) 422
Masturbation 422
Male–Female Sexual Behavior 423
Male–Male and Female–Female Sexual Behavior 423
A Closer Look How Should Parents React to Children Who Masturbate? 424
Preadolescence (9 to 13 Years) 425
Masturbation 425
Male–Female Sexual Behavior 425
Male–Male and Female–Female Sexual Behavior 425
Sex Education and Miseducation 425
A Closer Look Talking with Your Children About Sex 426
A Closer Look What Parents Want from Sex Education Courses 428
Adolescence 430
Puberty 431
A Closer Look Teenage Lingo: The Case of "Hooking Up" 434
Masturbation 436
Male–Female Sexual Behavior 436
A Closer Look "School Days, School Days," Third Millennium 438
A Closer Look The First Time 440
A World of Diversity Ethnicity, Teenagers, and HIV 442
Male–Male and Female–Female Sexual Behavior 442
Teenage Pregnancy 443
Contraceptive Use Among Sexually Active Teens 445
Teenagers and Sex: Where Do We Stand? 447
Human Sexuality Online Web Sites Related to Sexuality in Childhood and Adolescence 449
Summing Up 450
Questions for Critical Thinking 451

14 Sexuality in Adulthood 452

Outline 453
Truth or Fiction? 453
Singlehood 454
Cohabitation: Darling, Would You Be My POSSLQ? 456
Some Facts About Cohabitation 457
Reasons for Cohabitation 457
Cohabitation First and Marriage Later: Benefit or Risk? 458
Marriage 459
Historical Perspectives 459
A World of Diversity Russian Region Stops Men from Marrying Four Times 461
Why Do People Marry? 461
A World of Diversity Housewives Urged to Strike 462
Types of Marriage 462
Whom We Marry: Are Marriages Made in Heaven or in the Neighborhood? 462

Marital Sexuality *463*
The Sexual Revolution Hits Home *464*
Sexual Satisfaction *466*
Extramarital Sex *468*
Patterns of Extramarital Sex *469*
Attitudes Toward Extramarital Sex *469*
Human Sexuality Online "Someone to Watch Over Me"? Yes—Snoopware *470*
Effects of Extramarital Sex *472*
Swinging *472*
Domestic Violence *472*
Divorce *474*
The Cost of Divorce *474*
Alternative Lifestyles *476*
Open Marriage *476*
Group Marriage *476*
Sex in the Later Years *477*
Physical Changes *477*
Patterns of Sexual Activity *479*
Sex and Disability *480*
Physical Disabilities *481*
Human Sexuality Online Web Sites Related to Sexuality in Adulthood *483*
Psychological Disabilities *483*
Summing Up *484*
Questions for Critical Thinking *485*

15 Sexual Dysfunctions *486*

Outline *487*
Truth or Fiction? *487*
Types of Sexual Dysfunctions *489*
Sexual Desire Disorders *490*
Sexual Arousal Disorders *491*
Orgasmic Disorders *493*
Sexual Pain Disorders *495*
A World of Diversity Inis Beag and Mangaia—Worlds Apart *496*
Origins of Sexual Dysfunctions *498*
Organic Causes *498*
Psychosocial Causes *500*
A World of Diversity An Odd Couple: Koro and Dhat Syndromes *501*
Treatment of Sexual Dysfunctions *504*
The Masters-and-Johnson Approach *504*
The Helen Singer Kaplan Approach *505*
Sexual Desire Disorders *505*
A Closer Look "Sexercise" Credited with Boosting Performance *506*
Sexual Arousal Disorders *508*
Human Sexuality Online Buying Viagra and Other Drugs Online *512*
Orgasmic Disorders *515*
A Closer Look The "Orgasm Pill" for Women *519*
Sexual Pain Disorders *520*
Evaluation of Sex Therapy *521*
How Do You Find a Qualified Sex Therapist? *522*
Human Sexuality Online Web Sites Related to Sexual Dysfunctions *523*
Summing Up *524*
Questions for Critical Thinking *525*

16 Sexually Transmitted Infections *526*

Outline *527*
Truth or Fiction? *527*
An Epidemic *529*
A Closer Look Talking with Your Partner About STIs *530*
Questionnaire STI Attitude Scale *531*
Bacterial Infections *532*
Gonorrhea *532*
Syphilis *536*
Chlamydia *539*
Other Bacterial Diseases *541*
Vaginal Infections *542*
Bacterial Vaginosis *542*
Candidiasis *542*
Trichomoniasis *543*
Viral Infections *544*
HIV Infection and AIDS *544*
A World of Diversity HIV/AIDS: A Global Plague *546*
A World of Diversity Gender Is a Crucial Issue in the Fight Against HIV/AIDS *554*
Genital Herpes *555*
Viral Hepatitis *558*
Genital Warts *559*
Molluscum Contagiosum *560*
Ectoparasitic Infestations *560*
Pediculosis *560*
Scabies *561*
Prevention of STIs: More Than Safer Sex *561*
Human Sexuality Online Web Sites Related to Sexually Transmitted Infections *566*
Sources of Help *567*
Summing Up *567*
Questions for Critical Thinking *569*

17 Atypical Sexual Variations 570

Outline 571
Truth or Fiction? 571

Normal Versus Deviant Sexual Behavior 573
The Paraphilias 574
Fetishism 575
Transvestism 576
A Closer Look A Case of Transvestism 577
Exhibitionism 578
A Closer Look How to Respond to an Exhibitionist 579
Obscene Telephone Calling 580
Voyeurism 581
A Closer Look Shooting Private Parts in Public Places 582
Sexual Masochism 582
Sexual Sadism 584
Frotteurism 586
Other Paraphilias 587
Human Sexuality Online "Cybersex Addiction"—A New Psychological Disorder? 588
A Closer Look Sexual Addiction and Compulsivity 590
Theoretical Perspectives 590
Biological Perspectives 590
Psychoanalytic Perspectives 591
Learning Perspectives 592
Sociological Perspectives 593
An Integrated Perspective: The "Lovemap" 594
Treatment of the Paraphilias 594
Psychotherapy 595
Behavior Therapy 595
Biochemical Approaches 596
Human Sexuality Online Web Sites Related to Atypical Sexual Variations 597
Summing Up 598
Questions for Critical Thinking 599

18 Sexual Coercion 600

Outline 601
Truth or Fiction? 601

Rape 602
A World of Diversity Not All Sex Offenders Are Men 603
Incidence of Rape 604
Types of Rapes 605
A Closer Look Anatomy of a Date Rape: Ann and Jim 607
Social Attitudes and Myths That Encourage Rape 609
Sociocultural Factors in Rape 609
A Closer Look Feral Fowl Females Reject Rapists 610
Psychological Characteristics of Rapists 611
Adjustment of Rape Survivors 612
A World of Diversity Blaming the Victim to an Extreme 616
Treatment of Rape Survivors 616
Rape Prevention 617
A Closer Look Rape Prevention 618
Verbal Sexual Coercion 619
A Closer Look Handling Sexual Pressure Lines 620
Sexual Abuse of Children 621
What Is Sexual Abuse of Children? 621
Patterns of Abuse 621
Pedophilia 623
Incest 624
Effects of Sexual Abuse of Children 626
A Closer Look New Methods for Getting to the Truth in Cases of Child Abuse 627
Prevention of Sexual Abuse of Children 628
Human Sexuality Online Online Registries of Sex Offenders: Boon or Boondoggle? 628
Treatment of Survivors of Sexual Abuse 629
Treatment of Rapists and Child Molesters 629
Sexual Harassment 630
Sexual Harassment in the Workplace 631
Sexual Harassment on Campus 633
Sexual Harassment in the Schools 633
How to Resist Sexual Harassment 634
Human Sexuality Online Web Sites Related to Sexual Coercion 635
Summing Up 636
Questions for Critical Thinking 637

19 Commercial Sex 638

Outline 639
Truth or Fiction? 639

The World of Commercial Sex: A Disneyland for Adults 640
Prostitution 640
Incidence of Prostitution in the United States Today 641
Types of Female Prostitutes 641
A World of Diversity The Joy of Amsterdam's Red-Light District: A Police Guide 642
A World of Diversity From Russia, with . . . Sex? 644
Human Sexuality Online Online Sex Workers Have Rights Too 648

Characteristics of Female Prostitutes *648*

A World of Diversity Sinking to a New Low in Romania *650*

Customers of Female Prostitutes *650*

A World of Diversity Schoolgirls as Sex Toys *652*

Male Prostitution *652*
HIV, AIDS, and Prostitution *654*

Pornography and Obscenity *655*
What Is Pornographic? *655*
Pornography and the Law *656*

Human Sexuality Online Surfing for Sex *658*

Prevalence and Use of Erotica and Pornography *659*

Human Sexuality Online Filtering Sex Sites: Tough Work, But . . . *660*

Pornography and Sexual Coercion *660*

Human Sexuality Online Web Sites Related to Commercial Sex *664*

Summing Up *664*

Questions for Critical Thinking *665*

20 Making Responsible Sexual Decisions—An Epilogue *666*

Outline *667*

Choices, Information, and Decision Making *668*
Conflict *668*
Decisions, Decisions, Decisions . . . *669*

Value Systems *669*
Legalism *669*

Human Sexuality Online What's Wrong with Virtual Adultery? A Sin Is a Sin Is a Sin *670*

Situational Ethics *671*
Ethical Relativism *671*
Hedonism *672*
Asceticism *672*
Utilitarianism *672*
Rationalism *672*

The Balance Sheet for Decision Making *673*

Human Sexuality Online Web Sites Related to Making Responsible Sexual Decisions *675*

Questions for Critical Thinking *676*

Appendix A Scoring Keys for Self-Scoring Questionnaires *A-1*

References *R-1*

Name Index *NI-1*

Subject Index *SI-1*

feature box contents

A Racial Gap in Infant Deaths, and a Search for Reasons *364*
African American Women and Breast Cancer *83*
African Americans and Birth Control—The Intersection of Suspicion and Hope *382*
An Odd Couple: Koro and Dhat Syndromes *501*
Are African Americans Sexually Permissive? Cultural Myths Versus Data *284*
Around the World in Eighty Ways—A Preview *6*
Blaming the Victim to an Extreme *616*
Clitoridectomy—Ritual Genital Mutilation *70*
Cross-Cultural Perspectives on Childhood Sexuality *421*
Don't Flirt (with Anyone!) in Qir *231*
Ethnicity and Sexual Orientation: A Matter of Belonging *300*
Ethnicity, Teenagers, and HIV *422*
From Russia, with . . . Sex? *644*
Gender Differences in Preferences in Mates Across 37 Cultures *214*
Gender Is a Crucial Issue in the Fight Against HIV/AIDS *554*
Historical and Cross-Cultural Perspectives on Menstruation *94*
HIV/AIDS: A Global Plague *546*
Housewives Urged to Strike *462*
Inis Beag and Mangaia—Worlds Apart *496*
Love Finds a Way in Iran: "Temporary Marriage" *26*
Men and Women Iron Their Way to Domestic Equality *191*
Not All Sex Offenders Are Men *603*
Osteoporosis—Not for Women Only *98*
Russian Region Stops Men from Marrying Four Times *461*
Schoolgirls as Sex Toys *652*
Sinking to a New Low in Romania *650*
Sociocultural Factors and Oral Sex *276*
Surveys of the Sexual Behavior of Diverse Populations *50*
Symptoms of Menopause, Ethnicity, Socioeconomic Status, and Lifestyle *96*
The Joy of Amsterdam's Red-Light District: A Police Guide *642*
Using the Back Door in Iran: Can a Transsexual Change Her Mind? *183*
Wide-Eyed with . . . Beauty? *211*

a CLOSER look

A Case of Transvestism *577*
Big *120*
Anatomy of a Date Rape: Ann and Jim *607*
Boys Who Are Reared as Girls *180*
Breast Self-Examination *86*
Congratulations, Wheat Ridge, It's a . . . ! *385*
"Don't Ask, Don't Tell"—Far from Clear as a Bell *321*
Females Throughout the Animal Kingdom Are Monogamous—*Not!* 22
Feral Fowl Females Reject Rapists *610*
Handling Sexual Pressure Lines *620*
Heaven Sent: The Blind Date Who Is Your Destiny *247*
How Should Parents React to Children Who Masturbate? *424*
How to Improve Date-Seeking Skills *235*
How to Respond to an Exhibitionist *579*
Is There a *Man*opause? *118*
Jayne Thomas, Ph.D.—In Her Own Words *184*
Kegels *74*
Lance Armstrong—Winner of Two Contests *124*
"Lifting the Curse"—Should Monthly Periods Be Optional? *102*
More Options, and Decisions, for Men with Prostate Cancer *126*
Myths About Menopause *99*
New Methods for Getting to the Truth in Cases of Child Abuse *627*
On Penis Size and Sexual Performance *115*
Partial-Birth Abortion *411*
Physiological Measures of Sexual Arousal *41*
Rape Prevention *618*
RU-486 in Europe *412*
"School Days, School Days," Third Millennium *438*
Selecting the Gender of Your Child *330*
Sex and the Single Lizard *335*
Sex Toys for Pandas? *148*
"Sexercise" Credited with Boosting Performance *506*
Sexual Addiction and Compulsivity *590*
Shooting Private Parts in Public Places *582*
Spacing Children the Goldilocks Way *349*
Talking with Your Children About Sex *426*
Talking with Your Partner About Contraception *376*
Talking with Your Partner About STIs *530*
Teenage Lingo: The Case of "Hooking Up" *434*
The "Orgasm Pill" for Women *519*
The Buds and the Bees *144*
The First Time *440*
The Technology of Orgasm: "Hysteria," the Vibrator, and Women's Sexual Satisfaction *265*
What Parents Want from Sex Education Courses *428*
With a Horse? *47*

Human Sexuality ONLINE //

Advice for Young Women: How to Prepare for Married Life *242*
Another Sort of Gender Equality: Up-Jock Pics Now Compete with Up-Skirt Photos *116*
Buying Viagra and Other Drugs Online *512*
Childbirth Advice Is Just Keystrokes Away *362*
"Cybersex Addiction"—A New Psychological Disorder? *588*
Filtering Sex Sites: Tough Work, But . . . *660*
Hospital Holds 'Net Auction for Breast Reduction *210*
In Modern E-Mail Romance, "Trash" Is Just a Click Away *236*
Long-Distance Romance, Web Enabled *240*
Modem Matchmaking *232*
Online Registries of Sex Offenders: Boon or Boondoggle? *628*
Online Sex Workers Have Rights Too *648*
Out Into the "Gay Global Village" of Cyberspace *318*
"Someone to Watch Over Me"? Yes—Snoopware *470*
Surfing for Sex *658*
Web Sites for Use in Research on Human Sexuality *61*
Web Sites on Human Sexuality *34*
Web Sites Related to Attraction and Love *226*
Web Sites Related to Atypical Sexual Variations *597*
Web Sites Related to Commercial Sex *664*
Web Sites Related to Contraception and Abortion *415*
Web Sites Related to Conception, Pregnancy, and Childbirth *369*
Web Sites Related to Female Sexual Anatomy and Physiology *105*
Web Sites Related to Gender Identity and Gender Roles *201*
Web Sites Related to Making Responsible Sexual Decisions *675*
Web Sites Related to Male Sexual Anatomy and Physiology *137*
Web Sites Related to Relationships, Intimacy, and Communication *258*
Web Sites Related to Sexual Arousal and Response *168*
Web Sites Related to Sexual Coercion *635*
Web Sites Related to Sexual Dysfunctions *523*
Web Sites Related to Sexual Orientation *323*
Web Sites Related to Sexual Techniques and Behavior Patterns *286*
Web Sites Related to Sexuality in Adulthood *483*
Web Sites Related to Sexuality in Childhood and Adolescence *449*
Web Sites Related to Sexually Transmitted Infections *566*
What's Wrong with Virtual Adultery? A Sin Is a Sin Is a Sin *670*
Where to Get Help Breast-Feeding *368*
www.seduction.net *146*

Are You a Romantic or a Realist? The Love Attitudes Scale *218*
Pro-Choice or Pro-Life? Where Do You Stand? *406*
Should You Have a Child? *338*
Sternberg's Triangular Love Scale *224*
STI Attitude Scale *531*

preface

> There are more things in heaven and earth, Horatio,
> Than are dreamt of in your philosophy.
>
> —Shakespeare, *Hamlet*

There are indeed more kinds of people in this world, and more ways in which people experience their sexuality, than most of us might imagine. Human sexuality may be intimately related to human biology, but it is embedded within the sociocultural fabric of human society. The approach that has separated *Human Sexuality in a World of Diversity* from other human sexuality textbooks is its full embrace of the richness of human diversity.

Let us first consider what has changed in the fifth edition, which is, in many ways, a very new book. Then we will consider the themes of the text, which represent continuing concerns in the realm of human sexuality. We will also discuss the features that give the book its vitality and motivational force and those that enhance its service as a pedagogical tool.

What's New in the Fifth Edition

Although previous editions of *Human Sexuality in a World of Diversity* have been well received, we are always aware of the instructor's concern, which can be expressed as "What have you done for my students lately?" The answer is: *Much.* The fifth edition of *Human Sexuality in a World of Diversity* contains many exciting changes—changes that reflect our entrance into the largely unknown universe of the new millennium. The key new elements in the fifth edition include some of the very latest research.

A GENERAL UPDATING OF INFORMATION Because of the rapid developments in the behavioral and social sciences, and in biology and medicine, there are literally hundreds of new references from the 2000s throughout the text. No part of the text has been left untouched by change. Just two handfuls of examples of updatings within chapters include:

Chapter 3: New coverage on the debate over whether menstruation should be optional
Chapter 4: New information concerning treatment options for prostate cancer
Chapter 6: New research concerning the SRY (sex-determining region Y) gene
Chapter 6: New research concerning gender differences in response to stress
Chapter 10: New research and theory concerning the origins of sexual orientation
Chapter 11: New information concerning developments in birthing children
Chapter 12: The latest developments in the search for safe and effective contraceptives
Chapter 15: New research on biological treatments of sexual dysfunctions
Chapter 16: The latest information on the worldwide epidemic of HIV/AIDS and its treatment
Chapter 17: The latest information on biological treatments for paraphilias

ENHANCED EMPHASIS ON THE EVOLUTIONARY PERSPECTIVE The behavioral and biological sciences today recognize the influence of evolution not only on physical traits but also on social behavior, including sexual behavior. The process of natural selection has led to some fascinating adaptations among species throughout the animal kingdom, including humans. As humans and their ancestors evolved over millions of years, their fitness for survival gained prominence in terms of physical and psychological traits that permitted them to reach sexual maturity and bear and rear children—not only their brawn, their sharpness of eye, and their fleetness of foot but also their ability to acquire skills, their language ability, and their tendency to form bonds of attachment with their mates and their children.

The fifth edition of *Human Sexuality in a World of Diversity* demonstrates recognition of the importance of the evolutionary perspective in a new section in Chapter 1 called "The Evolutionary Perspective." However, the overlaps between evolution and sexuality are illustrated in nearly every chapter, as noted in the following examples:

Chapter 1: New discussion of evolution and "erotic plasticity" and of (the lack of) female monogamy throughout much of the animal kingdom
Chapter 4: Discussion of the evolution of the penis
Chapter 5: Discussion of the evolution of sexual attractants (for bees, if not for birds) in plants
Chapter 6: Discussion of (possible) evolutionary contributors to gender-typed behavior patterns
Chapter 7: Discussion of (possible) evolutionary contributors to gender differences in preferences for mates
Chapter 8: Discussion of evolution and (possible) gender differences in jealousy
Chapter 10: Discussion of the possible evolution of differences in sexual orientation
Chapter 11: Discussion of the evolutionary advantages (we can't think of any others!) in "morning sickness"
Chapter 18: Discussion of the controversial issue as to whether sexual aggression is "natural" in males (but we show that even if it is, sexual aggression remains completely unacceptable)

HUMAN SEXUALITY ONLINE Perhaps the single most exciting change in the fifth edition is the inclusion of a *new* boxed feature entitled Human Sexuality Online. This feature highlights the ways in which human social and sexual experience is being transformed as we connect with, and spend more time dwelling on, the "information superhighway." People today not only go online to arrange for travel to foreign countries but also to arrange for sexual liaisons around the world. People go online not only to obtain information about dieting and wills and pop stars but also about pregnancy, breast feeding, and cancers of the sexual organs. People go online not only for photos and films of Paris and Buenos Aires but also to interact with pornographic sites. A few years ago, American Express advertised its credit card with the slogan "Don't leave home without it." Today, many people in search of experience no longer leave home at all. Instead, they surf for sex on their personal computers or Web TVs. Yet the World Wide Web has done much more than bring titillation into the home; it has brought instant access to the great libraries and archives and research institutes of the world. And much of the information is free.

Each chapter of the text has a closing Human Sexuality Online feature that lists and annotates Web sites related to the material in that chapter. For example, Chapter 11, "Conception, Pregnancy, and Childbirth," has a closing Human Sexuality Online feature titled "Web Sites Related to Conception, Pregnancy, and Childbirth." The Web sites in the feature include the American College of Obstetricians and Gynecologists, the American College of Nurse-Midwives, the Doulas of North America, "The Universe of Women's Health," Childbirth.org, and La Leche League. The annotation for La Leche League explains that "La Leche League [is] an 'international, nonprofit, nonsectarian organization dedicated to providing education, information, support, and encouragement to women who want to breast-feed.' There are meetings for breast-feeding mothers, as well as future breast-feeding mothers, and information about how to obtain a consultant near you. The league also updates health care professionals on research on lactation management."

Human Sexuality Online features are also found throughout the text of the chapters. Here is a sampling of the new Human Sexuality Online features that are found in the fifth edition:

Chapter 4: Another Sort of Gender Equality: Up-Jock Pics Now Compete with Up-Skirt Photos
Chapter 5: www.seduction.net
Chapter 7: Hospital Holds 'Net Auction for Breast Reduction
Chapter 8: In Modern E-Mail Romance, "Trash" Is Just a Click Away
Chapter 8: Long-Distance Romance, Web Enabled
Chapter 11: Childbirth Advice Is Just Keystrokes Away
Chapter 14: "Someone to Watch Over Me"? Yes—Snoopware
Chapter 15: Buying Viagra and Other Drugs Online
Chapter 17: "Cybersex Addiction"—A New Psychological Disorder?
Chapter 18: Online Registries of Sex Offenders: Boon or Boondoggle?
Chapter 19: Online Sex Workers Have Rights Too
Chapter 19: Filtering Sex Sites: Tough Work, But . . .
Chapter 20: What's Wrong with Virtual Adultery? A Sin Is a Sin Is a Sin

QUESTIONS FOR CRITICAL THINKING Each chapter also has a closing section titled Questions for Critical Thinking. These questions prompt active consideration of key issues in human sexuality—many of which are quite controversial. They may also serve as the basis of class discussions.

The Themes of Human Sexuality in a World of Diversity

The fifth edition of *Human Sexuality in a World of Diversity* builds upon the strong themes for which it has come to be known. Four themes thread their way through the text:

- The rich diversity found in gender roles, sexual attitudes, and sexual behaviors and customs
- Critical thinking
- Making responsible sexual decisions
- Sexual health

THEME 1: HUMAN DIVERSITY Colleges and universities are undertaking the mission of broadening students' perspectives so that they will appreciate and tolerate human diversity.

The United States is a nation of hundreds of different ethnic and religious groups, many of whom endorse culturally distinct beliefs about appropriate gender roles for men and women and sexual practices and customs. Diversity is yet greater within the global village of the world's nearly 200 nations and those nations' own distinctive subcultures. *Human Sexuality in a World of Diversity* incorporates a multicultural, multi-ethnic perspective that reflects the diversity of sexual experience in our society and around the world. Our book thereby broadens the student's understanding of the range of cultural differences in sexual attitudes and behavior around the world and within our own society. Discussion of diversity encourages respect for people who hold diverse beliefs and attitudes. We also encourage students to question what is appropriate for women and men in terms of social roles and sexual conduct in light of cultural traditions and standards.

THEME 2: CRITICAL THINKING Colleges and universities in the new millennium are also encouraging students to become critical thinkers. Today's students are so inundated with information about gender and sexuality that it can be difficult to sort out truth from fiction. Not only do politicians, theologians, and community leaders influence our gender- and sex-related attitudes and behaviors, but newspapers, TV programs, and other media also brim with features about gender roles and issues concerning human sexuality.

Critical thinking means being skeptical of information that is presented in print or uttered by authority figures or celebrities. Critical thinking requires thoughtful analysis and probing of the claims and arguments of others in light of evidence. Moreover, it requires a willingness to challenge conventional wisdom and common knowledge that many of us take for granted. It means scrutinizing definitions of terms, evaluating the premises or assumptions that underlie arguments, and examining the logic of arguments.

Throughout the book we raise issues that demand critical thinking. These issues are intended to stimulate student interest in analyzing and evaluating their beliefs and attitudes toward gender roles and sexuality in light of the accumulated scientific evidence.

Moreover, the fifth edition fosters critical thinking through the sections titled Questions for Critical Thinking, which are found at the conclusion of every chapter.

THEME 3: RESPONSIBLE SEXUAL DECISION MAKING We also encourage students to make responsible sexual decisions. There are psychological and physical dangers in "going with the flow" or being passive about our sexuality. Of course, we do not encourage students to be sexually active (such a decision is personal). On the other hand, *we do encourage students to actively make their own sexual decisions on the basis of accurate information.*

Decision making is deeply intertwined with our sexual experiences. For example, we need to decide

- Whom to date, and how and when to become sexually intimate
- Whether to practice contraception and which methods to use
- How we will protect ourselves against HIV/AIDS and other sexually transmitted infections (STIs)

Responsible sexual decision making is based not only on accurate information but also on carefully evaluating this information in the light of one's own moral values. We encourage students to actively consider their own values, needs, and interests, rather than go along with the crowd or accede to the demands of their partners.

Throughout the text we provide students with the information they need to make responsible decisions about their physical health, the gender roles they will enact, sexual practices, birth control, and prevention of sexually transmitted diseases.

Moreover, Chapter 20, "Making Responsible Sexual Decisions," explains the major ethical systems that are used in moral reasoning and provides a decision-making model that will help students weigh the pluses and minuses of the choices they face.

THEME 4: SEXUAL HEALTH *Human Sexuality in a World of Diversity* places a strong emphasis on issues relating to sexual health, including extensive coverage of such topics as HIV/AIDS and other STIs, innovations in contraception and reproductive technologies, breast cancer, menstrual distress, sex and disabilities, and diseases that affect the reproductive tract. The text encourages students to take an active—in fact, *pro*active—role in health promotion. For example, it includes exercises and features that will help students examine their breasts and testes for abnormalities, reduce the risk of HIV infection, and cope with menstrual discomfort.

Features

The fifth edition of *Human Sexuality in a World of Diversity,* as earlier editions, contains various features that stimulate student interest and enhance understanding.

HUMAN SEXUALITY ONLINE New Human Sexuality Online sections highlight the relationships between human sexuality and the technological and societal transformations that people are experiencing as they enter the new millennium. Developments in electronics not only gave birth to the "information superhighway," they also made surfing the Internet a way of finding dates, seeking sexual stimulation and gratification, and learning about biomedical developments, like innovations that aid in the detection and treatment of diseases of the sex organs and STIs.

WORLD OF DIVERSITY World of Diversity boxes highlight the rich variety of human sexual customs and practices in our own society and around the world. Viewing human sexuality in a multicultural context helps students better understand how cultural beliefs, values, and attitudes can influence the expression of sexuality. Students may come to understand that their partners, who may not share the same ethnic or religious heritages as themselves, may feel differently than they do about sexual intimacy. Students will learn about cultural differences that relate to gender roles, sexual orientation, sexual jealousy, and premarital and extramarital sexual patterns.

CLOSER LOOK Closer Look sections provide in-depth discussions of scientific techniques (e.g., "Physiological Measures of Sexual Arousal"), self-scoring questionnaires (e.g., "The Triangular Love Scale"), skill-building exercises (e.g., "Breast Self-Examination," "Self-Examination of the Testes," "What to Do If You Suspect You Have Contracted an STD," "How to Respond to an Exhibitionist"), even humor (e.g., "Big" and "Sex and the Single Lizard").

THE WRITING STYLE Despite the stimulating role that sexuality plays in people's lives, most human sexuality textbooks manage to convert that which is energizing into just another dull textbook. *Human Sexuality in a World of Diversity*, by contrast, conveys the excitement of the science of human sexuality in all its dimensions. The text is written in a style intended to capture student interest in the material and at a level that students will find compelling and accessible, but not oversimplified, prurient, or patronizing. The selective use of humor and personal asides build interest and enliven descriptions.

Learning Aids

A textbook is more than a "good read." It is also a teaching tool—a device for presenting material in a way that stimulates learning and critical thinking. *Human Sexuality in a World of Diversity* was designed to maximize this goal by means of several pedagogical aids.

CHAPTER OUTLINES Each chapter begins with an outline that organizes the subject matter for the student. Headings were created to be succinct and to promote student interest in the topics they address.

TRUTH OR FICTION? AND TRUTH OR FICTION REVISITED SECTIONS Many students assume that they are experts on sexual matters. After all, they have seen the soap operas and the TV talk shows and have "been around" for a while.

Are they? Truth or Fiction? sections allow students to find out just how expert they are. These unique chapter-opening devices motivate students by challenging common sense (which is just as often common *non*sense), stereotypes, and folklore.

Truth or Fiction Revisited sections are interspersed throughout each chapter and provide feedback to students regarding the accuracy of their assumptions in the light of the evidence presented in the chapter.

RUNNING GLOSSARY Research shows that most students do not make use of glossaries at the ends of books. Searching for the meanings of terms is a difficult task and distracts them from the subject matter. *Human Sexuality in a World of Diversity* therefore has a running glossary. Key terms are boldfaced in the text and defined in the margins where they appear. Students can readily find the meanings of key terms and also maintain their concentration on the flow of the material.

HUMAN SEXUALITY ONLINE Human Sexuality Online sections at the conclusion of each chapter point students in the direction of supplementary online information—information that they can use to enhance their general knowledge of human sexuality, and information that they can apply to help them cope with sexually related issues in their own lives.

SUMMING UP SECTIONS To prompt students' memories, the Summing Up sections at the end of each chapter organize and review the subject matter according to the heads within the chapter.

QUESTIONS FOR CRITICAL THINKING These chapter-concluding sections follow the summaries and pose challenging questions for critical thinking and possible class discussion.

The Package

Human Sexuality in a World of Diversity, Fifth Edition, is supported by a wide range of ancillaries and teaching aids. Every copy of the text is packaged with a unique, Pin-protected **Interactive Companion Website.** This Website offers a wide variety of teaching and learning aids organized by each major section of each chapter of the text, including video and audio clips, Web activities, matching and identification activities, essay questions, annotated Web links, multiple-choice practice tests, and much more. Please see the Pincode package inside the front cover of your textbook for access information.

Additional teaching and learning aids include:

- *Instructor's Resource Manual*, Dr. Luis Montesinos, Montclair State University. Each chapter contains a Chapter-at-a-Glance table that keys material within the manual to the corresponding material within the text; learning objectives; teaching tips; discussion questions; recommended readings, videos, and Websites; and a rich collection of activities, handouts, and additional lecture material, complete with notes on how to incorporate them into your lesson plans.
- *Test Bank*, David Libhart. A separate test item file containing more than 2000 test items provides teachers with an array of items to use for assessing students' knowledge and retention of basic concepts in *Human Sexuality.* The test items are available as both a print supplement and in computerized format.
- *Allyn and Bacon Test Manager—Computerized Test Bank.* (Available for Windows, CD-ROM and disk, and Macintosh disk; DOS disk available upon request.) Allyn and Bacon Test Manager is an integrated suite of testing and assessment tools for Windows and Macintosh. You can use Test Manager to create professional-looking exams in just minutes by building tests from the existing database of questions, editing questions, or adding your own. Course management features include a class roster, gradebook, and item analysis. Test Manager also has everything you need to create and administer online tests. For first-time users, there is a guided tour of the entire Test Manager system and screen wizards to walk you through each area.
- *PowerPoint Presentation CD-ROM with Instructor's Manual.* New to the Fifth Edition, this book-specific presentation provides detailed outlines of key points for each chapter, supported by charts, graphs, diagrams, and other visuals from the textbook. Resources from the Rathus *Instructor's Resource Manual* and Companion Website are also integrated into each chapter of this presentation, for ultimate customization. This material is also available in a Web format accessible at www.abacon.com/ppt.
- *Allyn and Bacon Transparencies for Human Sexuality.* A full set of color acetate transparencies is available to enhance classroom lectures and discussion. These images come from a wide range of sources to support and extend teaching and learning.
- *Custom Videotapes.* Two custom videos are available upon adoption of the Fifth Edition of *Human Sexuality in a World of Diversity.* Each includes a wide variety of brief video clips and is designed to stimulate critical thinking and support the main topics covered in the textbook. Ask your local Allyn and Bacon representative for more information about the Allyn and Bacon Interactive Video and Sexuality in Society Videotape.
- *Contraceptive Kit.* In conjunction with Planned Parenthood, Allyn and Bacon makes available this kit containing a wide range of sample contraceptives for classroom presentation. Ask your local Allyn and Bacon representative for more information regarding availability.
- *Practice Tests*, Janet Snoyer, Cornell University. New to the Fifth Edition of *Human Sexuality*, this practice test booklet contains twenty-five all new multiple-choice questions per chapter that will aid students in preparing for course exams. An answer key appears in the back of the booklet; answer justifications have been provided to give students an even deeper understanding of the material.
- *Study Guide with Practice Tests*, Janet Snoyer, Cornell University. The student *Study Guide* contains chapter summaries, learning objectives, and a variety of exercises geared toward helping students retain the material from the text: fill-in-the-blank, matching, vocabulary, and labeling activities; short answer questions; and twenty-five multiple-choice questions per chapter, complete with appropriate page references and learning objectives.

- *On the Net: Human Sexuality, 2001.* Updated to reflect the most current URLs related to the study of human sexuality, this easy-to-read guide helps point you and your students in the right direction when looking at the tremendous array of information related to psychology on the Internet. This guide also includes an overview about doing research on the Internet and a variety of Internet-based activities for additional exploration.
- *Companion Website.* The companion Website for *Human Sexuality in a World of Diversity,* which can be accessed at www.ablongman.com/rathus offers a wide range of resources for both instructors and students. Each chapter contains learning objectives; annotated Web links; Web activities; interactive online quizzes; labeling activities; and multiple-choice practice tests.

Acknowledgments

The authors owe a great debt of gratitude to the many researchers and scholars whose contributions to the body of knowledge in the field of human sexuality is represented in these pages. Underscoring the interdisciplinary nature of the field, we have drawn upon the work of scholars in such fields as psychology, sociology, medicine, anthropology, theology, and philosophy, to name a few. We are also indebted to the many researchers who have generously allowed us to quote from their work and reprint tabular material representing their findings. We also wish to thank our professional colleagues who reviewed this text at various stages in its development:

Reviewers for the fifth edition are Sharon M. Ballard, East Carolina University; Keith E. Davis, University of South Carolina; Lyndall Ellingson, California State University, Chico; Richard Ellsworth, Chapman University; Jerry Marshall, Green River Community College; Jason Moses, University of the Virgin Islands; Suzanne L. Osman, Salisbury State University; and John Wiebe, University of Texas at El Paso.

Reviewers of previous editions of *Human Sexuality in a World of Diversity* are Richard Archer, Southwest Texas State University; Elaine Baker, Marshall University; Judith Baker, Texas Woman's University; Kenneth Beausang, Black Hawk College; Larry Bell, Northeastern University; Thomas Billimek, San Antonio College; Barry R. Burkhart, Auburn University; Jean Byrne, Kent State University; Willie A. Campbell, South Suburban College; Jerome Cerny, Indiana State University Michael Cleary, Slippery Rock University; Mona Coates, Orange Coast College; Clive Davis, Syracuse University; Laurie Desiderato, Bemidji State University; Linda DeVillers, Chaffey College; Beverly Drinnin, Des Moines Area Community College; Vera Dunwoody, Chaffey College; Katherine Ellison, Montclair College; Basil Fiorito, California Polytechnic State University; Heinz Fischer, Long Beach City College; Julio R. Garcia, Southwestern College; Michael Gonzalez, University Psychological Associates; Elizabeth A. Guillette, University of Florida; Ralph Hammond, University of Arkansas at Little Rock; Richard J. Hardy, Central Michigan University; Robert Holdsambeck, Allan Hancock College; Karen Huffman, Palomar College; G. David Johnson, University of South Alabama; Vicki Krenz, California State University, Fresno Herbert H. Laube, University of Minnesota; R. Martin Lobdell, Pierce College; John T. Long, San Antonio College; Deborah McDonald, New Mexico State College; Gilbert Meyer, Illinois Valley Community College; Gregory D. Murrow, Edinboro University of Pennsylvania; Robert Pollack, University of Georgia; Daphne Long Rankin, Virginia Commonwealth University; Jane Ellen Smith, University of New Mexico; Tom Springer, Louisiana Tech University; Marlene Tufts, Clackamus Community College; Mary Ann Watson, Metro State College of Denver; Charles Weichert, San Antonio College; and Paul Zelhart, East Texas State University.

Spencer A. Rathus
Montclair State University

Jeffrey S. Nevid
St. John's University

Lois Fichner-Rathus
The College of New Jersey

Human Sexuality in a World of Diversity

chapter

1

What Is Human Sexuality?

Pablo Picasso. *La Joie de vivre, 1946.* © Copyright ARS, NY. Musée Picasso, Antibes, France. Credit: Giraudon/Art Resource, NY. Credit: © 2002 Estate of Pablo Picasso/Artists Rights Society (ARS), New York.

outline

Truth or Fiction?

WHAT IS HUMAN SEXUALITY?

The Study of Human Sexuality

A World of Diversity Around the World in Eighty Ways—A Preview

Sexuality and Values

THINKING CRITICALLY ABOUT HUMAN SEXUALITY

Principles of Critical Thinking

PERSPECTIVES ON HUMAN SEXUALITY

The Historical Perspective

The Biological Perspective

The Evolutionary Perspective

A Closer Look Females Throughout the Animal Kingdom Are Monogamous—*Not!*

The Cross-Species Perspective

The Cross-Cultural Perspective

A World of Diversity Love Finds a Way in Iran: "Temporary Marriage"

Psychological Perspectives

Sociocultural Perspectives

Multiple Perspectives on Human Sexuality

Human Sexuality Online Web Sites on Human Sexuality

SUMMING UP

QUESTIONS FOR CRITICAL THINKING

Truth or Fiction?

____ Ancient civilizations worshiped women's ability to bear children and perpetuate the species.

____ In ancient Greece, a mature man would take a sexual interest in an adolescent boy, often with the blessing of the boy's parents.

____ The production of illustrated sex manuals originated in modern times.

____ The graham cracker came into being as a means of helping young men control their sexual appetites.

____ Female redback spiders eat their mates after the female has been inseminated.

____ Trobrianders consider their children old enough to engage in sexual intercourse when they are . . . old enough.

____ In Iran's conservative Islamic Republic, many couples get around the prohibitions against premarital sex by forming officially sanctioned temporary marriages—called *sigheh*—that can last from a few minutes to 99 years.

____ In our dreams, airplanes, bullets, snakes, sticks, and similar objects symbolize the male genitals.

We are about to embark on the study of human sexuality. But why, you may wonder, do we need to *study* human sexuality? Isn't sex something to *do* rather than to *talk about?* Isn't sex a natural function? Don't we learn what we need to know from personal experience or from our parents or our friends?

Yes, we can learn how our bodies respond to sexual stimulation—what turns us on and what turns us off—through personal experience. Personal experience teaches us little, however, about the biological processes that bring about sexual response and orgasm. Nor does experience inform us about the variations in sexual behavior that exist around the world, or in the neighborhood. Experience does not prepare us to recognize the signs of sexually transmitted diseases or to evaluate the risks of pregnancy. Nor does experience help us deal with most sexual problems, or dysfunctions. What of our parents? What many of us learned about sex from our parents can probably be summarized in a single word: "Don't." The information we received from our friends was probably riddled with fabrication, exaggeration, and folklore. Yes, many young people today do receive accurate information through sex education courses in the schools, which is all the more important now that the specter of HIV/AIDS hangs over every sexual decision.

There is also something of a myth in our culture that love conquers all—that love is all we need to achieve and sustain satisfying and healthy relationships. Yet how likely are we to establish healthy and mutually satisfying relationships without some formal knowledge of our own and our partner's sexuality? Without some knowledge of the biology of how our bodies function? Without some awareness of the psychological aspects of our sexuality? In a scientific vacuum?

Concerns about AIDS and unwanted teenage pregnancies have focused greater attention today on the importance of sex education. Many children receive some form of sex education as early as elementary school. Courses on human sexuality, which were rarely offered as recently as the 1950s, are now routine on college campuses across the United States and Canada. You may know more about human sexuality than your parents or grandparents did at your age—perhaps more than they do today. But how much do you really know? What, for example, happens inside your body when you are sexually stimulated? What causes erection or vaginal lubrication? Can people who are paralyzed from the neck down become erect or lubricated? What do we know of the factors that determine a person's sexual orientation? What are the causes of sexual dysfunctions? How do our sexual responsiveness and interests change as we age? Why does the United States have the highest incidence of rape in the industrialized world? Can you contract a sexually transmitted disease and not know that you have it until you wind up sterile? Can you infect others without having any symptoms yourself?

These are just a few of the issues we will explore in this book. Much of the information we present was discovered in recent years. It is almost as new to us as it may be to you. We also expect to debunk some common but erroneous ideas about sex that you may have picked up before you began this course. Before we proceed further, let us define our subject.

WHAT IS HUMAN SEXUALITY?

What *is* human sexuality? This is not a trick question. Consider the meaning, or rather meanings, of the word *sex*. The word derives from Latin roots that mean "to cut or divide," signifying the division of organisms into male and female genders. One use of the term *sex*, then, refers to our **gender,** or state of being male or female. The word *sex* (or *sexual*) is also used to refer to those anatomic structures, called sex (or sexual) organs, that play a role in reproduction or sexual pleasure. We may also speak of sex when referring to physical activities that involve our sex organs and are engaged in purposes of reproduction or pleasure: masturbation, hugging, kissing, **coitus,** and so on. Sex also relates to **erotic** feelings, experiences, or desires, such as sexual fantasies and thoughts, sexual urges, or feelings of sexual attraction to another person.

We usually make our usage of the term *sex* clear enough in our everyday speech. When we ask about the sex of a newborn, we are referring to anatomic sex. When we speak of "having sex" (a rather ugly phrase, which seems to imply that we engage in sexual activity much as we "have" a ham sandwich), we generally mean the physical expression of erotic feelings.

The term *sex organs* (or *sexual organs*) may be used in different ways. Sometimes the term is used to refer to those organs required for, and involved directly in, reproduction. These include the penis and testes in men, the vagina, uterus, and ovaries in women. The terms *sex organs* and *sexual organs* are also sometimes used to refer to organs or structures that are eroticized, even though they may play no direct role in reproduction (such as the clitoris or the breasts). Let us define sex (or sexual) organs as bodily structures that can be eroticized (such as the clitoris or the breasts) or that play a role in reproduction (such as the testes in the man and the uterus in the woman). They can also do both, of course, as do the penis in the man and the vagina in the woman. We will use the term *gender* in this text to refer to the state of being male or female, as in **gender identity** and **gender roles.**

The term *sexual behavior* refers to physical activities that involve the body in the expression of erotic or affectionate feelings. This description of sexual behavior includes, but is not limited to, behavior involving reproduction. *Masturbation*, for example, is sexual behavior that is performed for pleasure, not reproduction. Kissing, hugging, manual manipulation of the genitals, and oral–genital contact are all sexual behaviors that can provide sensual stimulation, even though they do not directly lead to reproduction. They may also be used as forms of **foreplay,** which leads to coitus, which can lead to reproduction.

We can now define **human sexuality** as the ways in which we experience and express ourselves as sexual beings. Our awareness of ourselves as females or males is part of our sexuality, as is the capacity we have for erotic experiences and responses. Our sexuality is an essential part of ourselves, whether or not we ever engage in sexual intercourse or sexual fantasy, and even if we lose sensation in our genitals because of injury.

The Study of Human Sexuality

The study of human sexuality is an interdisciplinary enterprise that draws upon the scientific expertise of anthropologists, biologists, medical researchers, sociologists, and psychologists, to name but a few of the professionals involved in the field. These disciplines all make contributions because sexual behavior reflects our biological capabilities, our psychological characteristics, and social and cultural influences. Biologists inform us about the physiological mechanisms of sexual arousal and response. Medical science teaches us about sexually transmitted diseases and the biological bases of sexual dysfunctions. Psychologists examine how our sexual behavior and attitudes are shaped by perception, learning, thought, motivation and emotion, and personality. Sociocultural theorists consider the sociocultural contexts of sexual behavior. For example, they examine relationships between sexual behavior and religion, race, and social class. Anthropologists focus on cross-cultural similarities and differences in sexual behavior. Scientists

Gender One's personal, social, and legal status as male or female.

Coitus (co-it-us or co-EET-us). Sexual intercourse.

Erotic Arousing sexual feelings or desires. (From the Greek *eros,* which means "love.")

Gender identity One's personal experience of being male or female.

Gender roles Complex clusters of ways in which males and females are expected to behave within a given culture.

Foreplay Mutual sexual stimulation that precedes sexual intercourse.

Human sexuality The ways in which we experience and express ourselves as sexual beings.

Around the World in Eighty Ways—A Preview

Like other aspects of human behavior, sexual beliefs and behaviors vary widely around the world. The United States alone is a nation of hundreds of different ethnic and religious groups, which vary in their sexual customs, attitudes, and beliefs. This diversity extends to the entire "global village" of the world's nearly 200 nations and to each nation's own distinctive subcultures.

The World of Diversity features that appear throughout this text explore the rich variety of sexual expression found worldwide. Seeing sexuality in contexts other than our own can help us understand the role of a culture's beliefs, values, and attitudes in our own and others' expressions of sexuality. It can help us understand why our partners, who may not share our ethnic or religious heritage, may have different beliefs about sexual intimacy. Exploring diversity can also help us understand cultural differences related to gender, sexual orientation, sexual attraction, sexual jealousy, premarital sex, teenage pregnancy, and risks of sexually transmitted diseases.

People in some societies believe, for instance, that a brother and sister who eat at the same table are engaging in a mildly erotic act. The practice is therefore forbidden (Davenport, 1977). In contemporary Islamic societies, female sexuality is often viewed as dangerous. If women's behavior and attire are not kept under strict control, they can be "fatal attractions" for men (Kammeyer et al., 1990). What is sexually arousing, too, varies from culture to culture. Among the Abkhasians in the southern part of what used to be the Soviet Union, men regard the female armpit as highly arousing. A woman's armpits are, therefore, a sight for her husband alone (Kammeyer et al., 1990).

Of course, one glaring reason for today's heightened interest in understanding sexuality in a broader perspective is the worldwide AIDS epidemic. Any effort to end this scourge requires that we open our eyes to cultural attitudes and traditions that may increase the risk of transmission of the disease.

If we take a quick tour of the world of diversity within our own borders and beyond, we find that

- Kissing is practiced nearly universally as a form of petting in the United States, but it is unpopular in Japan and unknown among some cultures in Africa and South America (see Chapter 9).
- Some societies encourage sexual experimentation and even sexual intercourse among children and adolescents, whereas others punish any form of childhood sexual play (see Chapter 13).
- Marital fidelity is a prominent value in Western culture, but among some people of the Arctic it is considered hospitable for a man to offer his wife to a visiting tribesman (see Chapter 14).
- In the United States, there remains a tendency to blame the victims of crimes—especially rape victims—rather than the perpetrators (see Chapter 18). In the strongly patriarchal society of Islamic Pakistan, however, the so-called Hudood Ordinance has actually resulted in *prison sentences* for some women who have brought rape charges against men. Hudood, you see, grants more credibility to the testimony of men than to that of women.

from many disciplines explore parallels between the sexual behavior of humans and that of other animals.

Science provides us with information, but it cannot make sexual decisions for us. Our **values** come into play in determining our sexual choices and behavior. The Declaration of Independence endorsed the fundamental values of "life, liberty, and the pursuit of happiness"—not a bad beginning. Religious traditions also play a prominent role in shaping values. Our study of human sexuality will thus consider how religious teachings shape sexual values.

Sexuality and Values

Our society is pluralistic. It embraces a wide range of sexual attitudes and values. Some readers may be liberal in their sexual views and behavior. Others may be conservative or traditional. Some will be staunchly pro-choice on abortion, others adamantly pro-life. Some will approve of premarital sex for couples who are dating casually. Others will hold the line at emotional commitment. Still others will believe that people should wait until marriage.

Because we encourage you, as you study this text, to explore your own values about the issues we discuss, let us reveal two values that guided *our* writing:

Values The qualities in life that are deemed important or unimportant, right or wrong, desirable or undesirable.

An accused man may claim that any sexual contact between himself and the woman making the accusations was consensual, and the court will be inclined to believe him. Women are also frequently prevented from testifying in court. As reported in the *Washington Post* (www.washingtonpost.com/), the result has been that women who bring charges of rape are sometimes prosecuted for adultery and jailed if found guilty, while their assailants go free (Schork, 1990).

- Sex was typically considered "indecent and unmentionable" in Russia; according to Kon (1995), it was "a subject only for the degenerate underground." However, since the collapse of the Soviet Union, sex has apparently been escaping into the mainstream of Russian life. In today's Russia, as reported in *The New York Times* (www.nytimes.com/), newspapers and television regularly cover prostitution and pornography as well as the traditional bus accidents (Schillinger, 1995). As sex has grabbed the Russian imagination, the incidences of premarital sex and rape are also apparently on the rise (Kon, 1995).
- Perhaps in response to feelings of guilt or remorse, many Japanese women who have abortions place miniature stone statues known as *mizuko-jizo* in Buddhist temples in memory of their aborted fetuses (see Chapter 12). They sometimes decorate the statues with bibs or tiny hats and surround them with little stuffed animals, baby food, and pacifiers—all in the belief that doing so keeps the souls of their aborted fetuses warm and entertained (Bumiller, 1990).
- The United States has its romantic Valentine's Day, but Japan has a blatantly erotic day: Christmas Eve. (Yes, Christmas Eve.) Whereas Christmas Eve is a time of religious devotion in many Western nations, it has become a time of sexual devotion in Japan. On Christmas Eve every single person must have a date—one that includes an overnight visit (Reid, 1990). The cost of a typical Christmas Eve date exceeds $1000. Where do Tokyo singles like to go before their overnight visits? Tokyo Disneyland.

We elaborate on such topics and issues concerning sexual diversity in subsequent chapters.

Kissing. Kissing is a nearly universal way of expressing affection or enhancing sexual arousal in the United States and Canada. But it is unpopular in Japan and unknown in many African and South American cultures.

1. *Sexual knowledge and critical thinking skills are of value because they allow us to make informed sexual decisions.* We hope that readers will confirm our belief. Having agreed on this much, your authors admit that they hold different values about a number of the issues we discuss. Therefore, we—your authors—do not try to persuade readers to adopt a particular stance concerning issues raised in the textbook. We present opposing points of view on controversial matters such as abortion and the distribution of condoms in schools. We hope that readers will critically consider their preconceptions and that the views they form will be their own.
2. *Students should take an active role in enhancing their health.* In the course of this text, we will urge you, for example, to examine your bodies for possible abnormalities, to see your physician when you have questions about painful menstruation or other physical complaints, to become sensitive to the signs of sexually transmitted infections, to get good prenatal care, and so forth.

People's sexual attitudes, experiences, and behaviors are shaped to a large extent by their cultural traditions and beliefs. Because our world consists of diverse peoples and cultures, the study of human sexuality is really the study of human sexual*ities*. In this book we highlight the many ways in which people experience their sexuality. We preview some of the findings of our review of sexuality in this chapter's World of Diversity feature.

THINKING CRITICALLY ABOUT HUMAN SEXUALITY

We are inundated with so much information about sex that it is difficult to separate truth from fiction. Newspapers, TV shows, and popular books and magazines contain one feature after another about sex. Many of these presentations contradict one another, contain half-truths, or draw misleading or unsubstantiated conclusions. A scientific approach to human sexuality encourages people to think critically about false claims and findings that are presented as truths.

Sad to say, most of us take certain "truths" for granted. We tend to assume that authority figures such as doctors and government officials provide us with factual information and are qualified to make decisions that affect our lives. When two doctors disagree on the need for a hysterectomy, however, or two officials disagree about whether condoms should be distributed in public schools, we wonder how both can be correct. Critical thinkers never say, "This is true because so-and-so says it is true."

To help students evaluate claims, arguments, and widely held beliefs, most colleges encourage *critical thinking*. Critical thinking has several features. One aspect of critical thinking is skepticism—not taking things for granted. It means being skeptical of things that are presented in print, uttered by authority figures or celebrities, or passed along by friends. Another aspect of critical thinking is the thoughtful analysis and probing of claims and arguments. Critical thinking requires willingness to challenge the conventional wisdom and common knowledge that many of us take for granted. It means scrutinizing definitions of terms and evaluating the premises of arguments and their logic. It also means finding *reasons* to support your beliefs, rather than relying on feelings. When people think critically, they maintain open minds. They suspend their beliefs until they have obtained and evaluated the evidence.

Throughout this book we raise issues that demand critical thinking. These issues may stimulate you to analyze and evaluate your beliefs and attitudes about sex in the light of scientific evidence. For example, upon reading Chapter 10 you may wish to reconsider your beliefs on whether gay males and lesbians *choose* their sexual orientation. Upon reading Chapter 4 you may reexamine folklore that suggests that sexual activity impairs a man's athletic performance on the following day. The discussion of abortion in Chapter 12 will encourage you to consider exactly when we begin to be *human*. When you read Chapters 13 and 17, you will face the question of whether schools should provide students with contraceptives.

Principles of Critical Thinking

Critical thinkers maintain a healthy skepticism. They examine definitions of terms, weigh premises, consider evidence, and decide whether arguments are valid and logical. Here are some suggestions for critical thinking:

1. *Be skeptical.* Politicians, religious leaders, and other authority figures attempt to convince you of their points of view. Even researchers and authors may hold certain biases. Resolve to adopt the attitude that you will accept nothing as true—including the comments of the authors of this text—until you have personally weighed the evidence.
2. *Examine definitions of terms.* Some statements are true when a term is defined in one way but are not true when it is defined in another. Consider the statement "Love is blind." If love is defined as head-over-heels infatuation, there may be substance to the statement. Infatuated people tend to idealize loved ones and overlook their faults. If love is defined as deep caring and commitment involving a more realistic (if still somewhat slanted) appraisal of the loved one, however, then love is not so much blind as a bit nearsighted.

3. *Examine the assumptions or premises of arguments.* Consider the statement "Abortion is murder." *Webster's New World Dictionary* defines *murder* as "the unlawful and malicious or premeditated killing of one human being by another." The statement can be true, according to this dictionary, only if the victim is held to be a human being (and if the act is unlawful and either malicious or premeditated). Pro-life advocates argue that embryos and fetuses are human beings. Pro-choice advocates claim that they are not, at least not until they are capable of surviving on their own. Hence the argument that abortion is murder would rest in part on the assumption that the embryo or fetus is a human being.
4. *Be cautious in drawing conclusions from evidence.* In Chapter 14 we shall discuss research findings that show that married people who cohabited before marriage are more likely to get divorced eventually than are those who didn't cohabit first. It may seem at first glance that cohabitation is a *cause* of divorce. However, married couples who cohabit before marriage may differ from those who do not in ways other than choosing cohabitation—which brings us to our next suggestion for critical thinking.
5. *Consider alternative interpretations of research evidence.* For example, cohabitors who later get married may be more likely to get divorced eventually because they are more liberal and less traditional than married couples who did not cohabit before marriage. Eventual divorce would then be *connected* with cohabitation but would not be *caused* by cohabitation.
6. *Consider the kinds of evidence on which conclusions are based.* Some conclusions, even seemingly "scientific" conclusions, are based on anecdotes and personal endorsements. They are not founded on sound research.
7. *Do not oversimplify.* Consider the statement "Homosexuality is inborn." There is some evidence that sexual orientation may involve inborn biological predispositions, such as genetic influences. However, biology is not destiny in human sexuality. Gay male, lesbian, and heterosexual sexual orientations appear to develop as the result of a complex interaction of biological and environmental factors.
8. *Do not overgeneralize.* Consider the belief that gay males are effeminate and lesbians are masculine. Yes, some gay males and lesbians fit these stereotypes. However, many do not. Overgeneralizing makes us vulnerable to accepting stereotypes.

PERSPECTIVES ON HUMAN SEXUALITY

Human sexuality is a complex topic. No single theory or perspective can capture all its nuances. In this book we explore human sexuality from many perspectives. In this section we introduce a number of perspectives—historical, biological, evolutionary, cross-species, cross-cultural, psychological, and sociocultural—that we will draw on in subsequent chapters.

The Historical Perspective

History places our sexual behavior in context. It informs us whether our sexual behavior reflects trends that have been with us through the millennia or the customs of a particular culture and era.

History shows little evidence of universal sexual trends. Attitudes and behaviors vary extensively from one time and place to another. Contemporary U.S. society may be permissive when compared to the Victorian and postwar eras. Yet it looks staid when compared to the sexual excesses of some ancient societies, most notably the ruling class of ancient Rome.

The Venus of Willendorf. Anthropologists believe that the Venus is an ancient fertility symbol.

History also involves the study of religious traditions. Religions explan natural events on the basis of the existence of a supreme being or beings or in terms of supernatural forces. Beliefs about the supernatural provide foundations for moral and ethical behavior. Religions have thus been major influences on sexual values and behavior. Yet the traditions and religions of diverse cultures have cast sexual behavior in different lights.

The rise of science since the Age of Enlightenment in the eighteenth century has also profoundly affected contemporary views of human sexuality. Let us trace some historical changes in attitudes toward sexuality. We begin by turning the clock back 20,000 or 30,000 years, to the days before written records were kept—that is, to *pre*history.

PREHISTORIC SEXUALITY: FROM FEMALE IDOLS TO PHALLIC WORSHIP Information about life among our Stone Age ancestors is drawn largely from cave drawings, stone artifacts, and the customs of modern-day preliterate peoples whose existence may have changed little over the millennia. From such sources, historians and anthropologists infer a prehistoric division of labor. By and large, men hunted for game. Women tended to remain close to home. They nurtured children and gathered edible plants and nuts, as well as crabs and other marine life that wandered along the shore or swam in shallow waters.

Art produced in the Stone Age, some 20,000 years ago, suggests the worship of women's ability to bear children and perpetuate the species (Fichner-Rathus, 2001). Primitive statues and cave drawings portray women with large, pendulous breasts, rounded hips, and prominent sex organs. Most theorists regard the figurines as fertility symbols. Emphasis on the female reproductive role may also have signified ignorance of the male's contribution to reproduction.

Truth or Fiction? **REVISITED**

It is true that ancient civilizations worshiped women's ability to bear children and perpetuate the species. Stone Age art suggests this. Primitive statues and cave drawings portraying women are regarded as fertility symbols.

As the ice sheets of the last ice age retreated (about 11,000 B.C.) and the climate warmed, human societies turned agrarian. Hunters and gatherers became farmers and herders. Villages sprang up around fields. Men tended the livestock. Women became farmers, supplementing their plant-gathering skills with expertise in cultivation. As people grew aware of the male role in reproduction, **phallic worship** sprang into being. Knowledge of paternity is believed to have developed around 9000 B.C., which is about the time that people shifted from being hunters and gatherers to being farmers and shepherds. The male's role in reproduction may seem obvious enough to us. However, uninformed people might not have connected childbirth to a sexual act that predated the most visible signs of pregnancy by months. Knowledge of paternity may have been a side benefit of the herding of livestock:

> For the first time, [people were] watching the same individual animals every day, all the year around, and [they] could scarcely fail to note the relatively constant length of the interval that elapsed between a ram servicing a ewe and the ewe dropping her lambs. (Tannahill, 1980, p. 46)

Of course, prehistoric people did not keep records that might help us to confirm this explanation. Had they done so, they would not have been *pre*historic.

In any event, the penis was glorified in art as a plough, an ax, or a sword. **Phallic symbols** figured in religious ceremonies in ancient Egypt. Ancient Greek art revered phalluses, rendering them sometimes as rings and sometimes as necklaces. Some phalluses were given wings, suggesting the power ascribed to them. In ancient Rome, a large phallus was carried like a float in a parade honoring Venus, the goddess of love.

Phallic worship Worship of the penis as a symbol of generative power.

Phallic symbols Images of the penis.

The **incest taboo** may have been the first human taboo (Tannahill, 1980). All human societies apparently have some form of incest taboo (Harris & Johnson, 2000; Whitten, 2001). Societies have varied in terms of the strictness of the taboo, however. Brother–sister marriages were permitted among the presumably divine rulers of ancient Egypt and among the royal families of the Incas and of Hawaii, even though they were generally prohibited among commoners. Father–daughter marriages were permitted among the aristocracy and royalty of ancient Egypt. Incestuous relationships in these royal blood lines may have ensured that wealth and power, as well as "divinity," would be kept in the family. Even today, many societies permit marriages between some blood relatives, such as first cousins.

THE ANCIENT HEBREWS The ancient Hebrews viewed sex, at least sex in marriage, as a fulfilling experience intended to fulfill the divine injunction to "be fruitful and multiply." The emphasis on the procreative function of sex led to some interesting social customs. For example, childlessness and the development of a repulsive abnormality, such as a boil, were grounds for divorce. After all, repulsiveness would dampen the mate's enthusiasm for reproducing. Male–male and female–female sexual behavior was strongly condemned, because it was believed to represent a threat to the perpetuation of the family. Adultery, too, was condemned—at least for a woman. Although the Hebrew Bible (called the Old Testament in the Christian faith) permitted **polygamy,** the vast majority of the Hebrews were **monogamous.**

The ancient Hebrews approved of sex within marriage not simply for procreation but also for mutual pleasure and fulfillment. They believed that the expression of sexual needs and desires helped strengthen marital bonds and solidify the family. Jewish law even legislated the minimum frequency of marital relations, which varied according to the man's profession and the amount of time he spent at home:

> Every day for those who have no occupation, twice a week for laborers, once a week for ass-drivers; once every thirty days for camel drivers; and once every six months for sailors. (Mishna Ketubot 5:6; Ketubot 62b-62b; quoted in Telushkin, 1991, p. 616)

What of the feminine gender role among the ancient Hebrews? Women were to be good wives and mothers. What *is* a "good wife"? According to the Book of Proverbs, a good wife rises before dawn to tend to her family's needs, brings home the food, instructs the servants, tends the vineyards, makes the clothes, keeps the ledger, helps the needy, and works well into the night. Even so, among the ancient Hebrews, a wife was considered the property of her husband. If she offended him, she could be divorced on a whim (although this almost never happened). A wife could be stoned to death for adultery. She might also have had to share her husband with his secondary wives and concubines. Men who committed adultery by consorting with the wives of other men were considered to have violated the property rights of those men. Although they were subject to harsh penalties for such violation of property rights, they were not put to death.

In case the notion that a woman is a man's property sounds ancient to you, we must note that in many cultures it remains current enough. For example, in the year 2000, Zambian judge Alfred Shilibwa ordered a hotel employee, Obert Siyankalanga, to pay a woman's husband $300 in compensation after he slipped his hand into the woman's blouse and fondled her breasts ("Man pays victim's husband," 2000). The woman, a hotel employee named Bertha Kosamu, had been ironing at the time. She explained the scars on Obert's face and head: "I clobbered him on the head with the iron." There is one contemporary ring to the story: Because Obert was Bertha's supervisor, the judge also found him guilty of sexual harassment.

THE ANCIENT GREEKS The classical or golden age of ancient Greece lasted about 200 years, from about 500 B.C. to 300 B.C. Within this relatively short span lived the

Incest taboo The prohibition against intercourse and reproduction among close blood relatives.

Polygamy The practice of having two or more spouses at the same time. (From the Greek roots meaning "many" [*poly-*] and "marriage" [*gamos*].)

Monogamy The practice of having one spouse. (From the Greek *mono-*, which means "single" or "alone.")

philosophers Socrates, Plato, and Aristotle; the playwrights Aristophanes, Aeschylus, and Sophocles; the natural scientist Archimedes; and the lawgiver Solon. Like the Hebrews, the Greeks valued family life. But the Greeks did not cement family ties by limiting sexual interests to marriage—at least not male sexual interests. The Greeks expressed sexual interests openly. They admired the well-developed male body and enjoyed nude wrestling among men in the arena. Erotic encounters and off-color jokes characterized the works of Aristophanes and other playwrights. The Greeks held that the healthy mind must dwell in a healthy body. They cultivated muscle and movement along with mind.

The Greeks viewed their gods—Zeus, god of gods, Apollo, who inspired art and music, Aphrodite, the goddess of carnal love whose name is the basis of the word *aphrodisiac*, and others—as voracious seekers of sexual variety. Not only were they believed to have sexual adventures among themselves, but they were also thought to have seduced mortals.

Three aspects of Greek sexuality are of particular interest to our study of sexual practices in the ancient world: male–male sexual behavior, pederasty, and prostitution. The Greeks viewed men and women as **bisexual.** One of their heroes was Heracles (Hercules). Heracles is said to have ravished 50 virgins in a night. Nevertheless, he also had affairs with men, including "sweet Hylas, he of the curling locks" (Tannahill, 1980, p. 85). Male–male *sex* was deemed normal. However, only a few male–male *relationships*, such as relationships between soldiers and between adolescents and older men, received the stamp of social approval. Male–male sexual behavior was tolerated so long as it did not threaten the institution of the family. Exclusive male–male sexual behavior was therefore discouraged. Some Greeks idealized romantic love between men—love of the sort that bound Achilles to Patroclus in Homer's *Iliad.* The warrior Achilles could not be moved to fight the Trojans by love of country or the pleas of his king. He sprang into action, however, when the enemy slew his lover Patroclus.

Pederasty means love of boys. Greek men might take on an adolescent male as a lover and pupil. Sex between men and prepubescent boys was illegal, however. Families were generally pleased if their adolescent sons attracted socially prominent mentors. Pederasty did not impede the boy's future male–female functioning, because the pederast himself was usually married, and the Greeks believed people equally capable of male–female and male–male sexual activity. Not all Greeks approved of pederasty, however. Aristotle, for one, considered it depraved.

Truth or Fiction? **REVISITED**

It is true that in ancient Greece, a mature man would take a sexual interest in an adolescent boy, often with the blessing of the boy's parents. The Greeks viewed people as naturally bisexual.

Prostitution flourished at every level of society. Prostitutes ranged from refined **courtesans** to **concubines,** who were usually slaves. Courtesans were similar to the geisha girls of Japan. They could play musical instruments, dance, engage in witty repartee, or discuss the latest political crisis. They were also skilled in the arts of love. No social stigma was attached to visiting a courtesan. Their clients included philosophers, playwrights, politicians, generals, and the very affluent. At the lower rungs of society were streetwalkers and prostitutes who lived in tawdry brothels. They were not hard to find. A wooden or painted penis invariably stood by the door.

Women in general held low social status. The women of Athens had no more legal or political rights than slaves. They were subject to the authority of their male next-of-kin before marriage and to that of their husbands afterwards. They received no formal education and were consigned most of the time to women's quarters in their homes. They were chaperoned when they ventured out-of-doors. A husband could divorce his wife without cause and was obligated to do so if she committed adultery. A wife, however, could divorce her husband only under extreme circumstances, which did not include adultery or pederasty. The legal and social rights of women in ancient Athens were similar to those of

Bisexual Sexually responsive to either gender. (From the Latin *bi-*, which means "two.")

Pederasty Sexual love of boys. (From the Greek *paidos,* which means "boy.")

Courtesan A prostitute—especially the mistress of a noble or wealthy man. (From Italian roots meaning "court lady.")

Concubine A secondary wife, usually of inferior legal and social status. (From Latin roots meaning "lying with.")

their contemporaries in Babylonia and Egypt and among the ancient Hebrews. All in all, the women of the ancient world were treated as *chattels*—property.

THE WORLD OF ANCIENT ROME Much is made of the sexual excesses of the Roman emperors and ruling families. Julius Caesar is reputed to have been bisexual—"a man to every woman and a woman to every man." Other emperors, such as Caligula, sponsored orgies at which guests engaged in a wide variety of sexual practices, including in some cases **bestiality** and **sadism.** These sexual excesses were found more often among the upper classes of palace society than among average Romans. Unlike their counterparts in ancient Greece, Romans viewed male–male sexual behavior as a threat to the integrity of the Roman family and to the position of the Roman woman. Thus, it was not held in favor.

Western society traces the roots of many of its sexual terms to Roman culture, as indicated by their Latin roots. **Fellatio,** for example, derives from the Latin *fellare*, which means "to suck." **Cunnilingus** derives from *cunnus*, which means "vulva," and *lingere*, "to lick." **Fornication** derives from *fornix*, which means "arch or vault." The term stems from the Roman streetwalkers' practice of serving their customers in the shadows of archways near public buildings such as stadiums and theaters.

The family was viewed as the source of strength of the Roman empire. Although Roman women were more likely than their Greek counterparts to share their husbands' social lives, they still were considered the property of their husbands.

THE EARLY CHRISTIANS Christianity emerged within the Roman Empire during the centuries following the death of Christ. According to the Christian Bible's New Testament, Jesus taught that love and tolerance are paramount in human relations and that God is forgiving. Little is known about Jesus's views on sex, however. Early Christian views on sexuality were largely shaped by Saint Paul and the church fathers in the first century and by Saint Augustine in the latter part of the fourth century. Adultery and fornication were rampant among the upper classes of Rome during this era. It was against this backdrop of sexual decadence that the early Christian leaders began to associate sexuality with sin.

In replacing the pagan values of Rome, the early Christians, like the Hebrews, sought to restrict sex to the marriage bed. They saw temptations of the flesh as distractions from spiritual devotion to God. Paul preached that celibacy was closer to the Christian ideal than marriage. He recognized that not everyone could achieve celibacy, however, so he said that it was "better to marry than to burn" (with passion, that is). Marriage was no sin, but it was spiritually inferior to celibacy.

Christians, like Jews before them, demanded virginity of brides. Masturbation and prostitution were condemned as sinful (Allen, 2000). Early Christians taught that men should love their wives with restraint, not passion. The goal of procreation should govern sexual behavior—the intellect should rule the flesh. Divorce was outlawed. Unhappiness with one's spouse might reflect sexual, and thus sinful, restlessness. Dissolving a marriage might also jeopardize the tight social structure that supported the church.

Over subsequent centuries, Christian leaders took an even more negative view of sexuality. Particularly influential were the ideas of Saint Augustine (A.D. 353–430), who associated sexual lust with the original sin of Adam and Eve in the Garden of Eden. According to Augustine, lust had transformed the innocent procreative instinct, instilled in humanity by God, into sin. Following their fall from grace, Adam and Eve cloaked their nakedness with fig leaves. Shame had entered the picture. To Augustine, lust and shame were passed down from Adam and Eve through the generations. Lust made any sexual expression, even intercourse in marriage, inherently evil and wicked. Only through celibacy, according to Augustine, could men and women attain a state of grace.

Nonprocreative sexual activity was deemed most sinful. Masturbation, male–male sexual behavior, female–female sexual behavior, oral–genital contact, anal intercourse—all

Bestiality Sexual relations between a person and an animal.

Sadism The practice of achieving sexual gratification through hurting or humiliating others.

Fellatio A sexual activity involving oral contact with the penis.

Cunnilingus A sexual activity involving oral contact with the female genitals.

Fornication Sexual intercourse between people who are not married to one another. (If one of the partners is married, the act may be labeled *adultery*.)

were viewed as abominations in the eyes of God. To the Jews, sex was a natural and pleasurable function, so long as it was practiced within marriage. There was no sin attached to sexual pleasure. To the early Christians, however, sexual pleasure, even within marriage, was stained by the original sin of Adam and Eve. Marital sex was deemed somewhat less sinful when practiced for procreation and without passion.

SEXUALITY AND THE EASTERN RELIGIONS An appreciation of the religious traditions of the Middle East and Far East can broaden our perspective on the history of sexual customs and practices. Islam, the dominant religion in the Middle East, was founded by the prophet Muhammad. Muhammad was born in Mecca, in what is now Saudi Arabia, in about A.D. 570. The Islamic tradition treasures marriage and sexual fulfillment in marriage. Premarital intercourse invites shame and social condemnation; in some fundamentalist Islamic states, it incurs the death penalty.

The family is the backbone of Islamic society. Celibacy is frowned upon (Ahmed, 1991). Muhammad decreed that marriage represents the only road to virtue (Minai, 1981). Islamic tradition permits a sexual double standard, however. Men may take up to four wives, but women are permitted only one husband. Public social interactions between men and women are severely restricted in Islamic societies. In traditional Islamic cultures, social dancing between the genders is shunned. Women in most traditional Islamic societies are expected to keep their heads and faces veiled in public and to avoid all contact, even a handshake, with men other than their husbands.

In the cultures of the Far East, sexuality was akin to spirituality. To the Taoist masters of China, who influenced Chinese culture for millennia, sex was anything but sinful. Rather, they taught that sex was a sacred duty. It was a form of worship that was believed to lead toward immortality. Sex was to be performed well and often if one was to achieve harmony with nature.

The Chinese culture was the first to produce a detailed sex manual, which came into use about 200 years before the birth of Jesus. This manual helped educate men and women in the art of lovemaking. The man was expected to extend intercourse as long as possible, thereby absorbing more of his wife's natural essence, or *yin*. Yin would enhance his own masculine essence, or *yang*. Moreover, he was expected to help bring his partner to orgasm so as to increase the flow of energy that he might absorb. (Her pleasure was incidental or secondary.)

Taoists believed that it was wasteful for a man to "spill his seed." Masturbation, though acceptable for women, was ruled out for men. Sexual practices such as anal intercourse and oral–genital contact (fellatio and cunnilingus) were permissible, so long as the man did not squander *yang* through wasteful ejaculation. Another parallel to Western cultures was the role accorded women in traditional Chinese society. Here the "good wife," like her Western counterparts, was limited largely to the domestic roles of child rearing and homemaking.

Perhaps no culture has cultivated sexual pleasure as a spiritual ideal to a greater extent than the ancient Hindus of India. From the fifth century onward, temples show sculptures of gods, heavenly nymphs, and ordinary people in erotic poses (Gupta, 1994). Hindu sexual practices were codified in a sex manual, the *Kama Sutra*. The *Kama Sutra* illustrates sexual positions, some of which would challenge a contortionist. It holds recipes for alleged aphrodisiacs, ways of making love, and so on. This manual remains the most influential sex manual ever produced. It is believed to have been written by the Hindu sage Vatsyayana sometime between the third and fifth centuries A.D., at about the time that Christianity was taking shape in the West as an organized religion.

An Illustration from the *Kama Sutra*. The *Kama Sutra*, an Indian sex manual believed to have been written sometime between the third and fifth centuries A.D., contained graphic illustrations of sexual techniques and practices.

It is not true that the production of illustrated sex manuals originated in modern times. Actually, the most influential sex manual ever produced was written and illustrated—profusely!—in ancient India. ■

In its graphic representations of sexual positions and practices, the *Kama Sutra* reflected the Hindu belief that sex was a religious duty, not a source of shame or guilt. In the Hindu doctrine of *karma* (the passage or transmigration of souls from one place to another), sexual fulfillment was regarded as one way to become reincarnated at a higher level of existence.

All in all, early Indian culture viewed sex as virtuous and natural. Indian society grew more restrictive toward sexuality after about A.D. 1000, however (Tannahill, 1980).

THE MIDDLE AGES The Middle Ages, sometimes called medieval times, span the millennium of Western history from about A.D. 476 to A.D. 1450. These years are sometimes termed the Dark Ages because some historians have depicted them as an era of cultural and intellectual decay and stagnation. The Roman Catholic Church continued to grow in influence. Its attitudes toward sexuality, largely unchanged since the time of Augustine, dominated medieval thought. Some crosscurrents of change crept across medieval Europe, however, especially in the social standing of women. The Roman Catholic Church had long regarded all women as being tainted by the sin of Eve, whom it blamed for humankind's downfall in the Garden of Eden. In the Eastern church of Constantinople, the cult of the Virgin Mary flourished. The ideal of womanhood was in the image of Mary: good, gracious, loving, and saintly.

Imported by the Crusaders and others who returned from the East, the cult of the Virgin Mary swept European Christendom and helped elevate the status of women. Hundreds of new abbeys were founded across Europe. Their monks were devoted to the Virgin. They erected chapels in their churches in her honor and wore white in tribute to her purity.

Two conflicting concepts of women thus dominated medieval thought: one, *woman as Eve*, the temptress; the other, *woman as Mary*, virtuous and pure. Contemporary Western images of women still reflect the schism between the good girl and the bad girl—the Madonna and the whore. (Part of the fascination of the rock star Madonna is that she combines the name of the Virgin Mary and crucifixes with an open display of undergarments and simulated lovemaking on the stage and in her videos.) Note also the Western double standard: An unmarried, sexually active woman runs the risk of being branded a "slut." An unmarried, sexually active man is likely to receive more complimentary labels, such as "stud," "playboy," or "ladies' man." Society has generally been more accepting of men's sexuality than of women's.

Among the upper classes of medieval times, a concept of courtly love also emerged. It was part chivalry and part romance novel. Troubadours of the twelfth and thirteenth centuries sang of pure and ennobling love that burned brightly but remained unconsummated. Their verses often depicted the chaste love between a married lady of the court and a handsome suitor of lower rank who sought her favor through heroic deeds.

THE PROTESTANT REFORMATION During the Reformation, Martin Luther (1483–1546) and other Christian reformers such as John Calvin (1509–1564) split off from the Roman Catholic Church and formed their own sects, which led to the development of the modern Protestant denominations of Western Europe (and later, the New World). Luther disputed many Roman Catholic doctrines on sexuality. He believed that priests should be allowed to marry and rear children. To Luther, marriage was as much a part of human nature as eating or drinking (Tannahill, 1980). Calvin rejected the Roman church's position that sex in marriage was permissible only for the purpose of procreation. To Calvin, sexual expression in marriage fulfilled other legitimate roles, such as strengthening the marriage bond and helping to relieve the stresses of everyday life. The Protestant Reformation encouraged a more accepting view of sexuality, although it strictly maintained that sex was permissible only in the context of marriage. Stern penalties were meted out for premarital and extramarital sex. The Puritans, one of the earliest Protestant sects to settle in America, subjected fornicators to flogging and sent parents to the pillory or the stocks if their children were born too soon after the wedding date. Adulterers, too, were flogged and sometimes branded. Remember the scarlet *A* that the adulteress Hester Prynne was compelled to wear in Nathaniel Hawthorne's novel *The Scarlet Letter.*

SHAPING THE PRESENT: FROM REPRESSION TO REVOLUTION TO REACTION Early settlers brought to the New World the religious teachings that had dominated Western thought and culture for centuries. Whatever their differences, these religions stressed the ideal of family life and viewed sex outside of marriage as immoral or sinful. A woman's place, by and large, was in the home and in the fields. Not until 1833, when Oberlin opened its doors to women, were women permitted to attend college in the United States. Not until the twentieth century did women gain the right to vote.

In this century, social change has swept through Western culture at a sometimes dizzying pace. Sexual behaviors and attitudes that would have been unthinkable a couple of generations ago—such as cohabitation ("living together")—have become commonplace. The social barriers that had restricted women's roles have largely broken down.

The middle and later parts of the nineteenth century are generally called the Victorian period after Queen Victoria of England. Victoria assumed the throne in 1837 and ruled until her death in 1901. Her name has become virtually synonymous with sexual repression. Victorian society in Europe and the United States, on the surface at least, was prim and proper. Sex was not discussed in polite society. Even the legs of pianos were draped with cloth for the sake of modesty. Many women viewed sex as a marital duty to be performed for procreation or to satisfy their husbands' cravings. Consider the following quotation:

> I am happy now that Charles calls on my bed chamber less frequently than of old. As it is, I now endure but two calls a week and when I hear his steps outside my door I lie down on my bed, close my eyes, open my legs and think of England.
>
> —Attributed to Alice, Lady Hillingdon, wife of the Second Baron Hillingdon

Women were assumed not to experience sexual desires or pleasures. "I would say," observed Dr. William Acton (1814–1875), an influential English physician, in 1857, "that the majority of women (happily for society) are not much troubled with sexual feeling of any kind." Women, thought Acton, were born with a sort of *sexual anesthesia.*

It was widely believed among medical authorities in England and the United States that sex drains the man of his natural vitality. Physicians thus recommended that intercourse be practiced infrequently, perhaps once a month or so. The Reverend Sylvester Graham (1794–1851) preached that ejaculation deprived men of the "vital fluids" they need to maintain health and vitality. Graham believed that the loss of an ounce of semen was equal to the loss of several ounces of blood. Each time a man ejaculated, in Graham's view, he risked his physical health. Graham preached against "wasting the seed" by masturbation or frequent marital intercourse. (How frequent was frequent? In Graham's view, intercourse more than once a month could dangerously deplete the man's vital energies.) Graham recommended that young men control their sexual appetites by a diet of simple foods based on whole-grain flours. To this day, his name is identified with a type of cracker he developed for this purpose in the 1830s, the graham cracker, derived from unbolted wheat (graham flour).

Truth or Fiction? REVISITED

It is true that the graham cracker came into being as a means of helping young men control their sexual appetites. Graham believed that a diet based on whole-grain flours was a key to self-control. ■

It appears, though, that the actual behavior of Victorians was not as repressed as advertised. Despite the belief in female sexual anesthesia, Victorian women, just like women before and after them, certainly did experience sexual pleasure and orgasm. One piece of evidence was provided by an early sex survey conducted in 1892 by a female physician, Celia Duel Mosher. Although her sample was small and nonrandom, 35 of the 44 women who responded admitted to desiring sexual intercourse. And 34 of them reported experiencing orgasm. Women's diaries of the time also contained accounts of passionate and sexually fulfilling love affairs (Gay, 1984).

Prostitution flourished during the Victorian era. Men apparently thought that they were doing their wives a favor by looking elsewhere. Accurate statistics are hard to come by, but there may have been as many as 1 prostitute for every 12 men in London during the nineteenth century; in Vienna, perhaps 1 for every 7 men.

THE FOUNDATIONS OF THE SCIENTIFIC STUDY OF SEXUALITY It was against this backdrop of sexual repression that scientists and scholars first began to approach sexuality as an area of legitimate scientific study. An important early contributor to the science of human sexuality was the English physician Havelock Ellis (1859–1939). Ellis compiled a veritable encyclopedia of sexuality: a series of volumes published between 1897 and 1910 entitled *Studies in the Psychology of Sex.* Ellis drew information from various sources, including case histories, anthropological findings, and medical knowledge. He challenged the prevailing view by arguing that sexual desires in women were natural and healthful. He promoted the view that many sexual problems had psychological rather than physical causes. He also promoted acceptance of the view that a gay male or lesbian sexual orientation was a naturally occurring variation within the spectrum of normal sexuality, not an aberration. Presaging some contemporary views, Ellis treated gay male and lesbian sexual orientations as inborn dispositions, not as vices or character flaws.

Another influential **sexologist,** the German psychiatrist Richard von Krafft-Ebing (1840–1902) described more than 200 case histories of individuals with various sexual deviations in his book *Psychopathia Sexualis.* His writings contain vivid descriptions of deviations ranging from sadomasochism (sexual gratification through inflicting or receiving pain) and bestiality (sex with animals) to yet more bizarre and frightening forms, such as necrophilia (intercourse with dead people). Krafft-Ebing viewed sexual deviations as mental diseases that could be studied and perhaps treated by medical science.

At about the same time, a Viennese physician, Sigmund Freud (1856–1939), was developing a theory of personality that has had an enormous influence on modern culture and science. Freud believed that the sex drive was our principal motivating force.

Alfred Kinsey (1894–1956), an Indiana University zoologist, conducted the first large-scale studies of sexual behavior in the 1930s and 1940s. It was then that sex research became recognized as a field of scientific study in its own right. In 1938 Kinsey had been asked to teach a course on marriage. When researching the course, Kinsey discovered that little was known about sexual practices in American society. He soon embarked on an ambitious research project. Detailed personal interviews with nearly 12,000 people across the United States were conducted. The results of his surveys were published in two volumes, *Sexual Behavior in the Human Male* (1948) and *Sexual Behavior in the Human Female* (1953). These books represent the first scientific attempts to provide a comprehensive picture of sexual behavior in the United States.

Kinsey's books made for rather dry reading and were filled with statistical tables rather than racy pictures or vignettes. Nevertheless, they became best sellers. They exploded on a public that had not yet learned to discuss sex openly. Their publication—especially that of the book on female sexuality—unleashed the dogs of criticism. Kinsey's work had some methodological flaws, but much of the criticism branded it immoral and obscene. *The New York Times* refused to run advertisements for the 1948 volume on male sexuality. Many newspapers refused to report the results of his survey on female sexuality. A congressional committee in the 1950s went so far as to claim that Kinsey's work undermined the moral fiber of the nation, rendering it more vulnerable to a Communist takeover (Gebhard, 1976).

Kinsey died in 1956. His death may have been hastened by the emotional strain he suffered because of the public's reaction to his work (Gagnon, 1990). Even so, Kinsey and

Sexologist A person who engages in the scientific study of sexual behavior.

his colleagues made sex research a scientifically respectable field of study. They helped lay the groundwork for greater openness in discussing sexual behavior.

THE PATH TO THE PRESENT: SEXUALITY IN THE TWENTIETH CENTURY Victorian attitudes toward sexuality dominated society's view of sexuality well into the middle of the twentieth century. The belief that women did not desire sex was something of a self-fulfilling prophecy. That is, couples who held this belief may not have sought ways of enhancing the woman's sexual pleasure. Women were expected to remain virgins until marriage, although society gave tacit permission to men to sow their wild oats. Men were permitted to seek sex with prostitutes or "loose" women, or at least people looked the other way when they did.

Although steamy passages were found in early twentieth-century writings such as D. H. Lawrence's *Lady Chatterley's Lover*, the most explicit sexual contact permitted in the films of the 1930s and 1940s was a discreet kiss. Open-mouth kissing was not allowed. No public forums featured sexual exotica or discussed sex openly. Nor was human sexuality widely taught in public schools, colleges, or even medical schools. By and large, sex remained shrouded in ignorance and secrecy.

THE SEXUAL REVOLUTION The period of the mid-1960s to the mid-1970s is often referred to as the *sexual revolution*. Dramatic changes occurred in U.S. sexual attitudes and practices during the "Swinging Sixties." When folksinger Bob Dylan sang "The Times They Are A-Changin'," our society was on the threshold of major social upheaval, not only in sexual behavior, but also in science, politics, fashion, music, art, and cinema. The so-called Woodstock generation, disheartened by commercialism and the Vietnam War, tuned in (to rock music on the radio), turned on (to drugs), and dropped out (of mainstream society). The heat was on between the hippies and the hardhats. Long hair became the mane of men. Bell-bottomed jeans flared out. Films became sexually explicit. Critics seriously contemplated whether the pornography classic *Deep Throat* had deep social implications. Hard rock music bellowed the message of rebellion and revolution.

No single event marked the onset of the sexual revolution. There was no charge up a sexual Bunker Hill. Social movements often gain momentum from a timely interplay of scientific, social, political, and economic forces. The war (in Vietnam), the bomb (fear of the nuclear bomb), the pill (the introduction of the birth-control pill), and the tube (TV, that is) were four such forces. The pill ended the risk of unwanted pregnancy for young people. It permitted them to engage in recreational or casual sex, rather than procreative sex (Asbell, 1995). Pop psychology movements, such as the Human Potential Movement of the 1960s and 1970s (the "Me Decade"), spread the message that people should get in touch with and express their genuine feelings, including their sexual feelings. "Doing your own thing" became one catchphrase. "If it feels right, go with it" became another. The lamp was rubbed. Out popped the sexual genie.

The sexual revolution was tied to social permissiveness and political liberalism. In part reflecting the times, in part acting the catalyst, the media dealt openly with sex. Popular books encouraged people to explore their sexuality. Film scenes of lovemaking became so commonplace that the movie rating system was introduced to alert parents. Protests against the Vietnam War and racial discrimination spilled over into broader protests against conventional morality and hypocrisy. Traditional prohibitions against drugs, casual sex, and even group sex crumbled suddenly.

Some of the alternative lifestyles that preoccupied the media in the 1960s and 1970s have fallen by the wayside. We no longer hear much about mate swapping (also called *swinging*) or "open marriages." Casual sex may also be on the wane—in part because of fear of AIDS, in part as a pendulum swing back toward commitment and romance.

Yet if the pendulum has swung partway back, the incidence of premarital sex among teenagers, especially younger teens, appears headed in the opposite direction. More teenagers are sexually active today, and at younger ages (Alan Guttmacher Institute, 1998). In addition to premarital sex, two other features of the sexual revolution have become permanent parts of our social fabric: the liberation of female sexuality and a greater

Are Today's Young People More or Less Liberal in the Expression of Their Sexuality Than People in Earlier Generations? Today the threat of AIDS hangs over every sexual encounter. Although many young people today are selective in their choice of partners and take precautions to make sex safer, more teenagers are engaging in sexual activity, and are doing so at younger ages, than in previous generations.

willingness to discuss sex openly. In 1998, TV networks broadcast President Bill Clinton's grand jury testimony, which included explicit references to oral sex with White House intern Monica Lewinsky, during daytime hours.

What, then, does history tell us about sex? Is there a universal standard for defining sexual values, or are there many standards? All societies have some form of an incest taboo. Most societies have placed a value on procreative sex within the context of an enduring relationship. The societal value of an enduring social, economic, and intimate relationship—usually in the form of marriage—lies in the role it plays in providing security for children, maintaining or increasing the population, and ensuring the orderly transfer of property from generation to generation.

Other sexual practices—masturbation, promiscuous sex, male–male sexual behavior, female–female sexual behavior, prostitution, polygamy, and so on—have been condemned in some societies, tolerated by others, and encouraged by others still. Some societies have looked with favor upon nonprocreative sex, at least within the context of marriage. Others have condemned it. Some historians argue that the pagan "degradations" of Rome led to its demise—that otherwise, a "purer" Rome might still bestride the earth. They warn that the "excesses" of our own sexual revolution may also bring us down. Rome, however, also suffered from the administrative difficulties of tending to a far-flung empire. Rome was also besieged by "barbarians" in the outposts and (eventually) at the city gates. In fact, the civilizations of ancient Greece and Rome maintained their prominence for hundreds of years. When we consider their contributions to Western art, philosophical thought, and the languages we speak, we may question whether they fell at all.

The Biological Perspective

The biological perspective focuses on the roles of genes, hormones, the nervous system, and other biological factors in human sexuality. Sex, after all, serves the biological function of reproduction. We are biologically endowed with anatomic structures and physiological capabilities that make sexual behavior possible and, for most people, pleasurable.

Study of the biology of sex acquaints us with the mechanisms of reproduction. It informs us of the physiological mechanisms of sexual arousal and response. By studying the biology of sex, we learn that erection occurs when the penis becomes engorged with blood. We learn that vaginal lubrication is the result of a "sweating" action of the vaginal walls. We learn that orgasm is a spinal reflex as well as a psychological event.

Biological researchers have made major strides in helping infertile couples conceive through laboratory-based methods of fertilization. "Test-tube babies" have been conceived in laboratory dishes and inserted into their mothers' uteruses, where they have developed to term. In some cases in which women cannot provide egg cells of their own, other women have donated egg cells that are fertilized by the father's sperm and then inserted into the mother's uterus.

Knowledge of biology has furthered our understanding of sexuality and our ability to overcome sexual problems. To what extent does biology govern sexual behavior? Is sex controlled by biological instincts? Or are psychosocial factors, such as culture, experience, and decision-making ability, more important? Although the sexuality of other species is largely governed by biological processes, culture and experience play essential roles—and in some cases, the more vital roles—in human sexuality. *Human* sexuality involves a complex interaction of biological and psychosocial factors. Biology indicates what is possible and, often, what is pleasurable or painful. Biology is not destiny, however. It does not imply what is proper and improper, nor does it determine the sexual decisions that we make. Religious tradition, cultural and personal values, and learning and experience guide these decisions.

The Evolutionary Perspective

Species vary not only in their physical characteristics but also in their social behavior, including their mating behavior. Scientists look to the process of **evolution** to help explain such variability. What is evolution? How might the sexual behavior of various species, including our own, be influenced by evolutionary forces?

The English naturalist Charles Darwin (1809–1882), the founder of the modern theory of evolution, believed that animal and plant species were not created independently but, rather, evolved from other life-forms. The mechanism by which species evolved was **natural selection,** or, in the vernacular, "survival of the fittest." In each species, some individuals are better adapted to their environment than others. Better-adapted members are more likely to survive to reproduce. Therefore, they are also more likely to transmit their traits to succeeding generations. As the generations pass, a greater proportion of the population of the species comes to carry the traits of the fittest members. Fitness, in the evolutionary sense, means reproductive success—the ability to produce surviving offspring. The fittest members of a species produce the greatest number of surviving offspring. (Note that these "fittest" individuals are not necessarily the strongest or fleetest of foot, although these traits may be adaptive in some environments and so enhance reproductive success.)

Over time, natural selection favors traits that contribute to survival and reproduction. When environmental conditions change, natural selection favors those members of a species who possess traits that help them adapt. These forms of the species proliferate, eventually replacing forms that fail to survive and reproduce. Species that lack forms that possess adaptive traits will eventually become extinct and be replaced by species that are better suited to the prevailing environmental conditions.

Darwin was ahead of his time—the technology of his day did not allow him to find the mechanisms by which traits are transmitted from generation to generation. Nor could he explain how a species evolved from one form to another. Not until the discovery of the principles of genetic inheritance by the Austrian monk Gregor Mendel (1822–1884) did the pieces of the puzzle of evolution begin to fall into place. Mendel discovered that traits are transmitted by units of heredity that we now call **genes.** Traits are determined by the combinations of genes that offspring inherit from their parents.

We now understand that genes are segments of **chromosomes,** which are composed of **DNA.** The chemical structure of genes provides genetic instructions. Each human cell normally contains a complement of 46 chromosomes, which are arranged in 23 pairs. Each human chromosome consists of more than 1,000 genes. A child normally inherits 1 member of each pair of chromosomes from each parent, so each offspring inherits 50% of his or

Evolution The development of a species to its present state, a process that is believed to involve adaptations to its environment.

Natural selection The evolutionary process by which adaptive traits enable members of a species to survive to reproductive age and transmit these traits to future generations.

Genes The basic units of heredity, which consist of chromosomal segments of DNA.

Chromosomes The rodlike structures that reside in the nuclei of every living cell and carry the genetic code in the form of genes.

DNA Deoxyribonucleic acid, the chemical substance whose molecules make up genes and chromosomes.

her genes from each parent. The particular combinations of genes that one inherits from one's parents determine whether one has blue eyes or brown eyes, light hair or dark hair, and a wide range of other characteristics.

New variations in species are introduced through random genetic changes called **mutations.** Mutations occur randomly but are subject to natural selection. That is, some mutations are adaptive in that they enhance reproductive success. As a result, these adaptive mutations are more likely to be retained and to proliferate in a species. More members of the species come to possess a variety of these adaptive traits, and the species as a whole changes in form. Most mutations are not adaptive, however, and quickly disappear from the genetic pool.

In recent years, some psychologists have suggested that there is a genetic basis to social behavior, including sexual behavior, among humans and other animals. This theory, which is called **evolutionary psychology,** proposes that dispositions toward *behavior patterns* that enhance reproductive success—as well as physical traits that do so—may be genetically transmitted. If so, we may carry behavioral traits that helped our prehistoric ancestors survive and reproduce successfully, even if these traits are no longer adaptive in modern culture. "Modern culture"—dating, say, from classical Greece—is but a moment in the lifetime of our species.

Charles Darwin

EVOLUTIONARY PERSPECTIVE AND EROTIC PLASTICITY Consider the concept of "erotic plasticity" (Baumeister, 2000), which addresses the fact that in response to various social and cultural forces, people show different levels of sex drive and express their sexual desires in a variety of ways. Roy Baumeister (2000) reports evidence that women show greater erotic plasticity than men do. For example, (1) individual women show greater variation than men in sexual behavior over time, (2) women seem to be more responsive than men to most specific cultural factors, such as cultural permissiveness or restraint, and (3) men's sexual behavior is more consistent with their sexual attitudes than is that of women. Baumeister concludes that evolutionary, biological forces may be an important factor in the greater female erotic plasticity.

There is a tendency to think of adaptive traits as somehow more "worthy," "good," or "admirable" than less adaptive traits. Evolution is not a moralistic enterprise, however. A trait either does or does not enhance reproductive success. It is not in itself good or bad. It is apparently adaptive for the female of one species of insect to eat the male after mating. "Dad" then literally nourishes his offspring during the period of gestation. In evolutionary terms, his personal sacrifice is adaptive if it increases the chances that the offspring will survive and carry his genes into the next generation. In other species, it may be adaptive for fathers to "love them and leave them"—that is, to mate with as many females as possible and abruptly abandon them to "plant their seed" elsewhere.

Evolutionary psychologists are interested in sexual behavior because it is so interwoven with reproductive success. They seek common sexual themes across cultures in the belief that common themes may represent traits that helped our ancestors survive and became part of the human genetic endowment. For example, there is considerable cross-cultural evidence that men are more promiscuous than women and have "spread their seed" widely.

Some evolutionary psychologists argue that men are naturally more promiscuous because they are the genetic heirs of ancestors whose reproductive success was related to the number of women they could impregnate (Bjorklund & Kipp, 1996; Buss, 1994). Women, by contrast, can produce only a few offspring in their lifetimes. Thus, the theory goes, they have to be more selective with respect to their mating partners. Women's reproductive success is enhanced by mating with the fittest males, not with any Tom, Dick, or Harry who happens by. From this perspective, the male's "roving eye" and the female's selectivity are embedded in their genes (Townsend, 1995). Evolutionary psychology theory is also sometimes invoked to explain why the incidences of infanticide and sexual abuse of children are higher in stepfamilies than in families where everyone is genetically related (Daly & Wilson, 1998).

Mutations Random changes in the molecular structure of DNA.

Evolutionary psychology The theory that dispositions toward behavior patterns that enhance reproductive success may be genetically transmitted.

a CLOSER look

Females Throughout the Animal Kingdom Are Monogamous—*Not!*

Many scientists and philosophers have looked to the behavior found throughout the animal kingdom in an effort to determine what kinds of behaviors are "natural" for people. Are there lessons about human behavior in the following examples? Perhaps, but read at your own risk. . . .

Contrary to what biologists once thought, females of most species are promiscuous—which explains a lot of weird behavior. For example, in the ordinary course of events, a male dung fly copulates with a female for a full 40 minutes, and even when he is finished delivering what he came to deliver and disengages, he hangs onto her for an additional 20 minutes. Otherwise, she is likely to—how to put this?—collect sperm from another suitor.

When a male ghost crab mates, his first move is to shoot into his beloved a bit of fluid that hardens into an epoxylike plug. The plug blocks any rival sperm that may have arrived earlier from swimming out of the tract where the female stores the wiggly gifts that males deposit, and so keeps the other guys' sperm from reaching her eggs. Only then does the latest crab, optimistic about his paternal chances, introduce his own sperm. The male redback spider, seconds after inseminating a female, does a somersault into her mouth so he becomes her postcoital meal; since matings followed by cannibalism last twice as long as those that don't, his sacrifice improves the chance that his own sperm will fertilize her eggs before someone else can have at her.

Truth or Fiction? REVISITED

It is true that female redback spiders eat their mates after they have been inseminated. Even so, female redbacks don't seem to choose their next . . . meal too carefully.

Are Females Throughout the Animal Kingdom Monogamous? Want a quick answer? It's no. Females in many species, including chimpanzees, are likely to have offspring that have been sired by more than one father.

More Bizarre Than the *Kama Sutra*

From behavior to physiology to anatomy, sex throughout the animal kingdom has always been, and will surely always be, more bizarre than the *Kama Sutra* lets on. But at least it's becoming less mysterious. Such previously inexplicable facts of life as weird genitalia and ludicrous copulatory practices (such as the 79 straight days that stick insects remain *in flagrante delicto*), biologists are now realizing, are adaptations to something they managed to overlook for a few millennia: female promiscuity. "Generations of reproductive biologists assumed females to be sexually monogamous," says biologist Tim Birkhead of the University of Sheffield in his book *Promiscuity*, which recounts scores of recent studies. "But it is now clear that this is wrong. Females of most species . . . routinely copulate with several different males."

How routinely? Once they started looking, biologists found promiscuous females in some 70% of the species they studied. A clutch of grasshopper eggs can have several fathers. Some 35% percent of baby indigo buntings (a pretty little songbird) are sired by a male other than the guy Mom came in with. So are 76% of Australian fairy wrens. In five hours, a female Scottish Soay sheep paired up with seven rams for a total of 163 encounters. Female chimps copulate

To some evolutionary psychologists, human beings are like marionettes on strings being tugged by invisible puppet masters—their genes. Genes govern the biological processes of sexual maturation and the production of sex hormones. Hormones, in turn, are largely responsible for regulating the sexual behavior of other animal species. Extending evolutionary psychology to human behavior sparks considerable controversy, however. Critics contend that learning, personal choice, and sociocultural factors may be more important determinants of human behavior than heredity (Hyde & Durik, 2000).

Critics also claim that evolutionary psychology is largely conjectural. No one, for example, has yet discovered genes for promiscuity. There is no direct evidence that either a total of 500 to 1,000 times for each pregnancy: A study using DNA to run paternity tests found that 54% of baby chimps were fathered by males other than Mom's supposed partner. A single clutch of goshawk eggs is inseminated some 500 times.

The Exception Rather Than the Rule

Contrary to the view that only men gain an evolutionary edge by spreading their seed widely, female monogamy is the exception rather than the rule. Surveys find that human females'—that is, women's—"ideal number" of lifetime sexual partners is less than men's and that they indeed have fewer partners than men. On the basis of this, "a lot of people want to simplify human mating and say that women are monogamous and men are promiscuous," says psychologist David Buss of the University of Texas. "But that's a gross oversimplification: both sexes pursue both strategies."

Female promiscuity triggers a war between the sexes. If a male is to have a fighting chance of fathering offspring and getting his genes into the next generation in the face of faithless females, he needs both crafty mating habits and seemingly outlandish mating equipment. One favored adaptation is a penis decked out with features much more elaborate than your basic sperm-meets-ovum job requires. Hence the tools of male damselflies and dragonflies: Both are covered with horns and hooks whose sole purpose, biologists have finally deduced, is to scoop out sperm that have arrived in the female's genital tract before theirs. The wild variety in the size of testes (relative to body size) throughout the animal kingdom suddenly makes sense in light of female promiscuity, too: The more promiscuous the females of a species, the larger the equipment a male grows so that his sperm will have a swimming chance at fatherhood. That's why gorillas' testes are small (faithful females) but chimps are . . . well, there's a reason why circus chimps are usually clothed.

Scientists aren't sure what influences the females of a species (or the males, for that matter) to remain monogamous. But in humans, female promiscuity is a response to tough living conditions. In Bhutan, many women practice polyandry because in the poor valleys of the Himalayas a lone husband cannot support a family. Spouse exchange among the Inuit improves their chance of survival in the unforgiving Arctic, because kin are morally obliged to provide for each other; the more in-laws, the more support.

Which brings us to the central question: What do females get out of promiscuity? (No, besides that.) One benefit of polyandry is increased resources and protection for a female and her brood. Most males are happy to trade food for sex; every male a female cricket copulates with brings her a protein-rich meal that is good for her eggs. And female Adelie penguins collect a stone, to build a nest, every time they offer themselves to a male. Females also gain by sowing the seeds of confusion, paternity-wise. Each Galapagos hawk that mates with a female helps rear her chicks, even though some are not the sires. But by trading sex for paternal care, the female increases the chance that her chicks will survive. Female red-winged blackbirds who copulate with multiple males are less likely to lose their chicks to predators: Each male attacks would-be predators. Male primates and lions commit infanticide against babies not their own; by confusing paternity, female chimps and lionesses keep more of their offspring out of harm's way.

Female promiscuity may bring not more offspring (as male promiscuity does) but better ones. Females of several species seem to have a mysterious detector built into their reproductive system that rejects "genetically incompatible" sperm but accepts sperm whose DNA complements that of their eggs, thus producing the most viable offspring. Several studies have found that females produce better-quality offspring by mating with several males; scientists are only starting to figure out how females discriminate duds from winners.[1]

[1]The first and second authors of this text point out that it is not necessarily advantageous for females to have the ability to discriminate duds from winners. The third author confesses that she wishes she had had that capacity years ago.

Source: From Sharon Begley with Erika Check (2000, August 5). "Sex and the Single Fly," *Newsweek*, 44–45.

male promiscuity or female selectivity is genetically determined. Critics also point to examples of cultural diversity as evidence that culture and experience, not genetics, play the pivotal role in human behavior. Nor is cross-cultural similarity in sexual practices necessarily proof of a common genetic factor. Different cultures may adopt similar customs because such customs serve a similar function. For example, marriage exists in some form in every human society. Perhaps marriage serves similar functions in various cultures, such as regulating the availability of sexual partners and furnishing an economic and social arrangement that provides for the care of offspring. All in all, the evidence that human social and sexual behaviors are direct products of our genes is far from clear or compelling.

The Cross-Species Perspective

The study of other animal species places human behavior in broader context. A surprising variety of sexual behaviors exists among nonhumans. There are animal examples, or **analogues,** of human male–male sexual behavior, female–female sexual behavior, oral–genital contact, and oral–oral behavior (that is, kissing). Foreplay is also well known in the animal world. Turtles massage their mates' heads with their claws. Male mice nibble at their partner's necks. Most mammals use only a rear-entry position for **copulation,** but some animals, such as apes, use a variety of coital positions.

We even find analogues of deviant forms of sexual behavior, such as rape, in the animal world. We should be careful about drawing a connection between human rape and animal analogues, however. Human (perhaps we should say *in*human) rape is often motivated by the desire to punish and humiliate the victim, a motive that may be entirely absent in the forced copulations among other animals that resemble rape among humans.

Cross-species research reveals an interesting pattern. Sexual behavior among "higher" mammals, such as primates, is less directly controlled by instinct than it is among the "lower" species, such as birds, fish, and lower mammals. Experience and learning play more important roles in sexuality as we travel up the evolutionary ladder.

The Cross-Cultural Perspective

The cross-cultural perspective, like the historical perspective, provides insight into the ways in which cultural beliefs affect sexual behavior and people's sense of morality. Unlike historians, who are limited in their sources to the eyewitness accounts of others and the shards of information that can be gleaned from fading relics, anthropologists can observe other cultures firsthand. Interest in the cross-cultural perspective on sexuality was spurred by the early twentieth-century work of the anthropologists Margaret Mead (1901–1978) and Bronislaw Malinowski (1884–1942).

In *Sex and Temperament in Three Primitive Societies* (1935), Mead laid the groundwork for recent psychological and sociological research challenging gender-role stereotypes. In most cultures characterized by a gender division of labor, men typically go to business or to the hunt, and—when necessary—to war. In such cultures, men are perceived as strong, active, independent, and logical. Women are viewed as passive, dependent, nurturant, and emotional. Mead concluded that these stereotypes are not inherent in our genetic heritage. Rather, they are acquired through cultural expectations and socialization. That is, men and women learn to behave in ways that are expected of them in their particular culture.

Malinowski lived on the Trobriand island of Boyawa in the South Pacific during World War I. There he gathered data on two societies of the South Pacific, the Trobrianders and the Amphett islanders. The Amphett islanders maintained strict sexual prohibitions, whereas the Trobrianders enjoyed greater freedom. Trobrianders, for example, encouraged their children to masturbate. Boys and girls were expected to begin to engage in intercourse when they were biologically old enough. According to custom, they would pair up, exchange a coconut, a bit of betel nut, a few beads, or some fruit from the bush. Then they would go off together and engage in intercourse (Malinowski, 1929, p. 488). Adolescents were expected to have multiple sex partners until they married.

It is true that Trobrianders consider their children old enough to engage in sexual intercourse when they are . . . old enough. In traditional Trobriand society, children were encouraged by their elders to engage in sexual intercourse when they were biologically mature enough to do so. ■

Malinowski found the Trobrianders less anxiety-ridden than the Amphett islanders. He attributed the difference to their sexual freedom, thus making an early plea to relax prohibitions in Western societies. Even sexually permissive cultures like that of the Trobrianders, however, placed limits on sexual freedom. They frowned on extramarital relationships, for example. Other cultures, however, like that of the Toda of southern

Analogue Something that is similar or comparable to something else.

Copulation Sexual intercourse. (From the Latin *copulare,* which means "to unite" or "to couple.")

India, consider it immoral for a husband to restrict his wife's extramarital relations (Howard, 1989). And Chukchee men of Arctic Siberia often exchange their wives with their friends.

CROSS-CULTURAL COMMONALITIES AND DIFFERENCES IN SEXUAL BEHAVIOR In 1951, Clellan Ford, an anthropologist, and Frank Beach, a psychologist, reviewed sexual behavior in preliterate societies around the world, as well as in other animals. They found great variety in sexual customs and beliefs among the almost 200 societies they studied. They also found some common threads, although there were exceptions to each. Ford and Beach's work is almost half a century old, but it remains a valuable source of information about cross-cultural and cross-species patterns in sexuality (Frayser, 1985). Throughout the text, our discussion of cross-cultural patterns in sexuality is guided by their work and by more recent cross-cultural studies conducted by Gwen Broude and Sarah Greene (1976) and by Suzanne Frayser (1985), among others.

Ford and Beach reported that kissing was quite common across the cultures they studied, though not universal. The Thonga of Africa were one society that did not practice kissing. Upon witnessing two European visitors kissing each other, members of the tribe commented that they could not understand why Europeans "ate" each other's saliva and dirt. The frequency of sexual intercourse also varies from culture to culture, but intercourse is relatively more frequent among young people everywhere.

Societies differ in their attitudes toward childhood masturbation. Some societies, such as the Hopi Native Americans of the southwest United States, ignore it. Trobrianders encourage children to stimulate themselves. Other societies condemn it. The people of the Pacific island of East Bay discourage children from touching their genitals and may subject those who do to ridicule or scolding (Davenport, 1965).

Although all cultures place some prohibitions on incestuous relationships, intercourse between brother and sister has been viewed as natural and desirable in some cultures, such as that of the Dahomey of West Africa (Stephens, 1982). The acceptability of incestuous relations has also varied in some cultures according to social class. In some societies, incestuous marriages were permitted among the ruling classes. Virtually all cultures have strict incest taboos, however. Ford and Beach reported that marriage between people who spoke the same dialect was even taboo in one Australian tribe—a prohibition that presumably did not enhance marital communication.

Eighty-four percent of Ford and Beach's (1951) preliterate cultures practiced polygamy. The researchers concluded that monogamy was relatively uncommon. More common is the form of polygamy called **polygyny,** in which men are permitted to have more than one wife. Similarly, Frayser (1985) found that polygyny was practiced by the great majority (82%) of societies in her cross-cultural sample. In many cultures, the number of wives a man has is an emblem of his wealth and status. The cultures studied in Ford and Beach's and Frayser's cross-cultural samples typically had few members, however. Monogamy, which is the custom in the more populous, technologically advanced cultures, is thus more prevalent worldwide. Even in polygynous cultures, more people are monogamously married than are married polygamously. Few societies have the oversupply of women that universal polygyny would entail (Harris & Johnson, 2000; Whitten, 2001). Rarer still is **polyandry,** wherein women may have more than one husband. Frayser (1985) found polyandry in only 2% of societies she studied. In fraternal polyandry, the most common form of polyandry, two or more brothers share a wife, and all live together in the same household. The wife visits each brother according to a schedule arranged by the men (Harris & Johnson, 2000; Whitten, 2001).

Polygyny has a long tradition in Western culture. King Solomon was reputed to have 700 wives. Polygyny was practiced in the nineteenth-century United States by an early leader of the Mormon church, Brigham Young, and by some of his followers. It has since been banned by the Mormon church as well as by state laws, although it is still practiced in some disenfranchised Mormon communities.

The cross-cultural perspective illustrates the importance of learning in human sexual behavior. Societies differ widely in their sexual attitudes, customs, and practices.

Polygyny A form of marriage in which a man has two or more wives. (From the Greek *gyne,* which means "woman.")

Polyandry A form of marriage in which a woman has two or more husbands. (From the Greek *andros,* which means "man" or "male.")

Love Finds a Way in Iran: "Temporary Marriage"

Even some of the most sexually repressive societies find ways for people to meet needs for intimacy and sex. Iran's conservative Islamic Republic may be no exception.

For five years, Maryam, the hairdresser, and Karim, the home appliance salesman, carried on a love affair, meeting secretly at the house where Karim lived with his parents. The young couple's relationship was officially sanctioned by Iran's Islamic Republic, even though unmarried couples who have sex or even date and hold hands can be arrested, fined, even flogged. That is because Maryam and Karim were married.

Sort of.

They had a valid contract of temporary marriage.

Iran is a country where rules are fluid, where people of all classes and degrees of religiosity pride themselves on finding loopholes in the Islamic system. Temporary marriage, or sigheh, is one of the oddest and biggest.

The practice of temporary marriage is said to have existed during the lifetime of Muhammad, who is believed to have recommended it to his companions and soldiers. The majority Sunni sect in Islam banned it; the minority Shiite sect did not. Historically, the practice was used most frequently in Iran by pilgrims in Shiite shrine cities like Meshed and Qum. Pilgrims who traveled had sexual needs, the argument went. Temporary marriage was a legal way to satisfy them.

Maryam and Karim chose temporary marriage for a practical reason. "We went out a lot together, and I didn't want to get into trouble," Maryam, 31, said. "We wanted to have documents so that if we were stopped on the street we could prove we weren't doing anything illegal."

Their "marriage" ritual was simple. Even though they could have sealed the contract privately, they went to a cleric in a marriage registry office in Tehran with their photographs and identity papers. Maryam had been forced into a loveless marriage at 15 to an opium-smoking, womanizing factory owner nearly two decades her senior who divorced her nine years later; so she brought along her divorce decree. If she had been a virgin, she would have needed her father's permission to marry.

The couple could have gotten married for as short a time as a few minutes or as long as 99 years. They could have specified whether and how much money Maryam would be paid as a kind of dowry, or how much time they would spend together. Instead, they decided on a straightforward contract of six months, which they renewed again and again.

What was unusual about Maryam's situation was her willingness to talk about it. Despite its religious imprimatur, temporary marriage has never been very popular in Iran. Tradition dictates that women be virgins when they marry; even when they're not, they should pretend to be. Many Iranians regard sigheh as little more than legalized prostitution, especially since it is an advertisement that a woman is not a virgin. In some circles, even illicit sex is considered better—as long as it can be kept secret.

But now an odd mix of feminists, clerics and officials have begun to discuss sigheh as a possible solution to the problems of Iran's youth. An extraordinarily large number of young people (about 65% of the population is under 25), combined with high unemployment, means that more couples are putting off marriage because they cannot afford it. Sigheh legally wraps premarital sex in an Islamic cloak.

"First, relations between young men and women will become a little bit freer," said Shahla Sherkat, editor of *Zanan*, a feminist monthly. "Second, they can satisfy their sexual needs. Third, sex will become depoliticized. Fourth, they will use up some of the energy they are putting into street demonstrations. Finally, our society's obsession with virginity will disappear."

A Legal Twist in Conservative Iran. Sex prior to or outside of marriage is strongly condemned in traditionalist Islamic Iran. However, the country permits temporary marriages called *sigheh* as a way of meeting social and sexual needs. One can theoretically have a *sigheh* that lasts a day.

Even conservatives like Muhammad Javad Larijani, a Berkeley-educated former legislator, favor temporary marriage. As Mr. Larijani put it: "What's wrong with temporary marriage? You've got a variation of it in California. It's called a partnership. Better to have it legal than have it done clandestinely in the streets."

Though most of Iran's reformist publications have closed in recent months, newspapers and magazines that remain have begun to discuss the issue. A recent front-page article in a weekly tabloid, *World of Medicine*, about a chador-wearing, AIDS-infected prostitute who took pleasure in infecting her clients included a recommendation on avoiding infection: take a temporary wife.

Advocates of temporary marriage also point out that children of such unions are legitimate and entitled to a share of the father's inheritance.

More rarely, unrelated couples have used nonsexual "temporary marriage" in order to live or work in close quarters.

But the popular response to such a sweeping societal solution has not been favorable. After *The Hope of Youth*, a weekly, ran an article in favor of sigheh, readers called and wrote in with scathing attacks.

"I am 23 years old," one unnamed young man told the paper. "If I temporarily marry a young woman for three years and then divorce her, would anyone be willing to marry her? It would be impossible that any man would want to have a family with this woman."

Another unidentified caller was quoted as saying: "Those who want to promote temporary marriage don't understand that they would be promoting prostitution. Who would be there to be a father for the children from temporary marriage?"

The paper wrote back: "The reality is that young men and women do have sexual relationships. If these relationships are defined within an Islamic framework, we will not have the danger of prostitution."

As for what to do about children of temporary marriages, the editor added, "It is not so complicated to use birth control anymore."

This is not the first time that people in the Islamic Republic have tried to promote sigheh. The first person to discuss it openly was none other than Ali Akbar Hashemi Rafsanjani when he was president. In a sermon in 1990, he called sexual desire a God-given trait. Don't be "promiscuous like the Westerners," he advocated, but use the God-given solution of temporary marriage.

That sermon brought thousands of protesters to Parliament, in part because a married man can have as many temporary wives as he wants, and up to four permanent ones, and can break the contract anytime he wants, whereas women cannot. Many secular Iranians are irked by what they perceive to be the hypocrisy of clerics, who have made ample use of temporary marriage over the years but are adamantly opposed to premarital or extramarital sex.

Clerics seldom talk about their experiences. But in the book "Law of Desire," Shahla Haeri, a Boston University cultural anthropologist and granddaughter of an ayatollah, cited interviews with clerics.

One proclaimed that because God banned alcohol, he allowed temporary marriage.

Ms. Haeri, who lectured on the subject in Iran, said that neither the clerics nor leading thinkers had begun to analyze its implications in a coherent way. "If they are really serious," she said, "they should study the matter in the context of sexuality, birth control, sexually transmitted diseases, morality, religion and gender relations."

But what of Maryam and Karim?

He gave her clothes and a little money from time to time during their "marriage," but not the gold coin he had promised her with each renewal of their contract. He told her she was beautiful, something her husband had never done. She cleaned his house occasionally and even met his brothers. He met her mother—who, twice divorced, had married (permanently) for the third time. They kept their temporary marriage a secret, even from her.

"She knew that I was with a man," Maryam said, "but would have preferred I was with him illegally than his sigheh."

In fact, Maryam and Karim are not the couple's real names. Maryam remains so ambivalent about what she did that she asked that not even their first names be used.

In the fifth year of their relationship, Karim began to call less frequently. Maryam went to a fortuneteller, who told her that Karim was to be married. When she confronted him, he said that it was over. After their contract ran out, he married a virgin chosen by his parents.

Because of her divorce, she said, "he told me right from the start that he couldn't marry me permanently. But he treated me so nicely that I thought things would change."

Maryam was so much in love that she even offered—half jokingly—to become Karim's temporary wife again after he was permanently married. He refused.

"I think sigheh is good, very good," she said, but added that she would not do it again. "I want to get married permanently now, as soon as possible."

Truth or Fiction? REVISITED

It is true that many couples in Iran's Islamic Republic get around the prohibitions against premarital sex by forming officially sanctioned temporary marriages. These unions—called *sigheh*—can last from a few minutes to 99 years. ■

Source: From Elaine Sciolino (2000, October 4). "Love Finds a Way in Iran: 'Temporary Marriage' " *The New York Times online.*

The members of all human societies share the same anatomic structures and physiological capacities for sexual pleasure, however. The same hormones flow through their arteries. Yet their sexual practices, and the pleasure they reap or fail to attain, may set them apart. Were human sexuality completely or predominantly determined by biology, we would not find such diversity.

Although sexual practices vary widely across cultures, there are some universal patterns. For one, all societies regulate sexual behavior in some fashion. No society allows unbounded sexual freedom. There is usually some societal control over the acceptability of sexual and marital partners, and extramarital relations. Nearly all societies have some sort of incest taboo. Nevertheless, cross-cultural comparisons show evidence of great variety. What is considered natural, normal, or moral in one culture may be deemed unnatural, abnormal, or immoral in another.

Yet people often apply an **ethnocentric** standard when judging other cultures. That is, they tend to treat the standards of their own cultures as the norm by which to judge other peoples. Cross-cultural information helps us to appreciate the relativity of the concept of normalcy and to recognize the cultural contexts of sexual behavior. When we consider the historical and cross-cultural perspectives, we come to appreciate that our behavior exists in the context of both our own culture and our own time.

Ethnocentric Adjectival form of the noun *ethnocentrism,* meaning the tendency to view other groups or cultures according to the standards of one's own. (From the Greek *ethnos,* which means "race," "culture," or "people," and *kentric,* which means "center.")

Psychoanalysis The theory of personality originated by Sigmund Freud, which proposes that human behavior represents the outcome of clashing inner forces.

Id In Freud's theory, the mental structure that is present at birth, embodies physiological drives, and is fully unconscious.

Ego In Freud's theory, the second mental structure to develop, which is characterized by self-awareness, planning, and delay of gratification.

Superego In Freud's theory, the third mental structure, which functions as a moral guardian and sets forth high standards for behavior.

Unconscious mind Those parts or contents of the mind that lie outside of conscious awareness.

Defense mechanisms In psychoanalytic theory, automatic processes that protect the ego from anxiety by disguising or ejecting unacceptable ideas and urges.

Repression The automatic ejection of anxiety-evoking ideas from consciousness.

Psychological Perspectives

Psychological perspectives focus on the many psychological influences—perception, learning, motivation, emotion, personality, and so on—that affect our sexual behavior and our experience of ourselves as female or male. Some psychological theorists, such as Sigmund Freud, focus on the motivational role of sex in human personality. Others focus on how our experiences and mental representations of the world affect our sexual behavior.

SIGMUND FREUD AND PSYCHOANALYTIC THEORY Sigmund Freud, a Viennese physician, formulated a grand theory of personality termed **psychoanalysis.** Freud believed that we are all born with biologically based sex drives. These drives must be channeled through socially approved outlets if family and social life are to carry on without undue conflict. He hypothesized that conflicts between sexuality and society become internalized in the form of an inner conflict between two opposing parts of the personality: the **id,** which is the repository of biologically based drives or "instincts" (such as hunger, thirst, elimination, sex, and aggression) and the **ego,** which represents reason and good sense. The ego seeks socially appropriate outlets for satisfying the basic drives that arise from the id. Your id, for example, prompts you to feel sexual urges. Your ego attempts to find ways of satisfying those urges without incurring condemnation from others or from your own moral conscience, which Freud called the **superego.** How these internal conflicts are resolved, in Freud's view, largely determines our ability to love, work, and lead well-adjusted lives.

Freud proposed that the mind operates on conscious and unconscious levels. The conscious level corresponds to our state of present awareness. The **unconscious mind** consists of the darker reaches of the mind that lie outside our direct awareness. The ego shields the conscious mind from awareness of our baser sexual and aggressive urges via **defense mechanisms** such as **repression,** the motivated forgetting of traumatic experiences.

Although many sexual ideas and impulses are banished to the unconscious, they continue to seek expression. One avenue of expression is the dream, through which sexual impulses may be perceived in disguised, or symbolic, form. The therapists and scholars who follow in the Freudian tradition are quite interested in analyzing dreams, and the dream objects listed in Table 1.1 are often considered sexual symbols. Freud himself maintained a bit of skepticism about the import of dream symbols, however. He once remarked, "Sometimes a cigar is just a cigar."

TABLE 1.1

Dream symbols in psychoanalytic theory*

Symbols for the Male Genital Organs				
airplanes	fish	neckties	tools	weapons
bullets	hands	poles	trains	
feet	hoses	snakes	trees	
fire	knives	sticks	umbrellas	
Symbols for the Female Genital Organs				
bottles	caves	doors	ovens	ships
boxes	chests	hats	pockets	tunnels
cases	closets	jars	pots	
Symbols for Sexual Intercourse				
climbing a ladder	flying in an airplane			
climbing a staircase	riding a horse			
crossing a bridge	riding an elevator			
driving an automobile	riding a roller coaster			
entering a room	walking into a tunnel or down a hall			
Symbols for the Breasts				
apples	peaches			

*Freud theorized that the content of dreams symbolized urges, wishes, and objects of fantasy that we would censor in the waking state.

Source: From the prepublication version of S. A. Rathus, *Psychology in the New Millennium*, 8th ed. to be published by Harcourt, Inc. Copyright © 2002 by Harcourt, Inc. Adapted by permission of the publisher.

To a psychoanalyst, dreams of airplanes, bullets, snakes, sticks, and similar objects may indeed symbolize the male genitals. Even Freud had to admit, however, that "sometimes a cigar is just a cigar"! ■

Freud introduced us to new and controversial ideas about ourselves as sexual beings. For example, he originated the concept of **erogenous zones**—the idea that many parts of the body, not just the genitals, are responsive to sexual stimulation.

One of Freud's most controversial beliefs was that children normally harbor erotic interests. He believed that the suckling of the infant in the oral stage was an erotic act. So too was anal bodily experimentation, through which children learn to experience pleasure in the control of their sphincter muscles and the processes of elimination. He theorized that it was normal for children to progress through stages of development in which the erotic interest shifts from one erogenous zone to another, as, for example, from the mouth or oral cavity to the anal cavity. According to his theory of **psychosexual development,** children undergo five stages of development: oral, anal, phallic, latency, and genital, which are named according to the predominant erogenous zones of each stage. Each stage gives rise to certain kinds of conflicts. Moreover, inadequate or excessive gratification in any stage can lead to **fixation** in that stage and to the development of traits and sexual preferences characteristic of that stage. (Parents who seek to rear their children according to psychoanalytic theory have been frustrated in that Freud never specified the proper amount of gratification in any stage or explained how to provide it.)

Erogenous zones Parts of the body, including but not limited to the sex organs, that are responsive to sexual stimulation.

Psychosexual development In psychoanalytic theory, the process by which sexual feelings shift from one erogenous zone to another.

Fixation In psychoanalytic theory, arrested development, which includes attachment to traits and sexual preferences that are characteristic of an earlier stage of psychosexual development.

Freud believed that it was normal for children to develop erotic feelings toward the parent of the other gender during the phallic stage. These incestuous urges lead to conflict with the parent of the same gender. In later chapters we shall see that these developments, which Freud termed the **Oedipus complex,** have profound implications for the assumption of gender roles and sexual orientation.

LEARNING THEORIES To what extent does sexual behavior reflect experience? Would you hold the same sexual attitudes and do the same things if you had been reared in another culture? We think not. Even within the same society, family and personal experiences can shape unique sexual attitudes and behaviors. Whereas psychoanalytic theory plumbs the depths of the unconscious, learning theorists focus on environmental factors that shape behavior.

Behaviorists such as John B. Watson (1878–1958) and B. F. Skinner (1904–1990) emphasized the importance of rewards and punishments in the learning process. In psychology, events (such as rewards) that increase the frequency or likelihood of behavior are termed reinforcements. Children left to explore their bodies without parental condemnation will learn what feels good and tend to repeat it. The Trobriand child who is rewarded for masturbation and premarital coitus through parental praise and encouragement will be more likely to repeat these behaviors (at least openly!) than the child in a more sexually restrictive culture, who is punished for the same behavior. When sexual behavior (such as masturbation) feels good but parents connect it with feelings of guilt and shame, the child is placed in conflict and may vacillate between masturbating and swearing off it.

Of course, parental punishment does not necessarily eliminate childhood masturbation. Nor has it stemmed the rising rate of teenage pregnancy in our society. Punishment tends to suppress behavior in circumstances in which it is expected to occur. People can learn to engage in prohibited behavior secretly, however. Still, if we as young children are severely punished for sexual exploration, we may come to associate sexual stimulation *in general* with feelings of guilt or anxiety. Such early learning experiences can set the stage for sexual dysfunctions in adulthood.

Social-learning theorists also use the concepts of reward and punishment, but they emphasize the importance of cognitive activity (anticipations, thoughts, plans, and so on) and of learning by observation. Observational learning, or **modeling,** refers to acquiring knowledge and skills by observing others. Observational learning involves more than direct observation of other people. It includes seeing models in films or on television, hearing about them, and reading about them. According to social-learning theory, children acquire the gender roles that are deemed appropriate in a society through reinforcement of gender-appropriate behavior and by observing the gender-role behavior of their parents, their peers, and other models on television, in films, and in books.

Psychological theories shed light on the ways in which sexuality is influenced by rewards, punishments, and mental processes such as fantasy, thoughts, attitudes, and expectations. Sigmund Freud helped bring sexuality within the province of scientific investigation. He also helped make it possible for people to recognize and talk about the importance of sexuality in their lives. Critics contend, however, that he may have placed too much emphasis on sexual motivation in determining behavior and on the role of unconscious processes.

The psychological perspective has much to offer to our understanding of human sexuality. Psychological factors affect every dimension of our sexuality. Those who harbor excessive guilt or anxiety over sexual activity may have difficulty enjoying sex or responding adequately to sexual stimulation. Even our basic gender identity is largely shaped by psychological factors, such as the experience, in a given society, of being reared as a girl or as a boy.

Oedipus complex In psychoanalytic theory, a conflict of the phallic stage in which the boy wishes to possess his mother sexually and perceives his father as a rival in love. (The analogous conflict for girls is the *Electra complex.*)

Behaviorists Learning theorists who argue that a scientific approach to understanding behavior must refer only to observable and measurable behaviors and who emphasize the importance of rewards and punishments in the learning process.

Social-learning theory A cognitively oriented learning theory in which observational learning, values, and expectations play key roles in determining behavior.

Modeling Acquiring knowledge and skills by observing others.

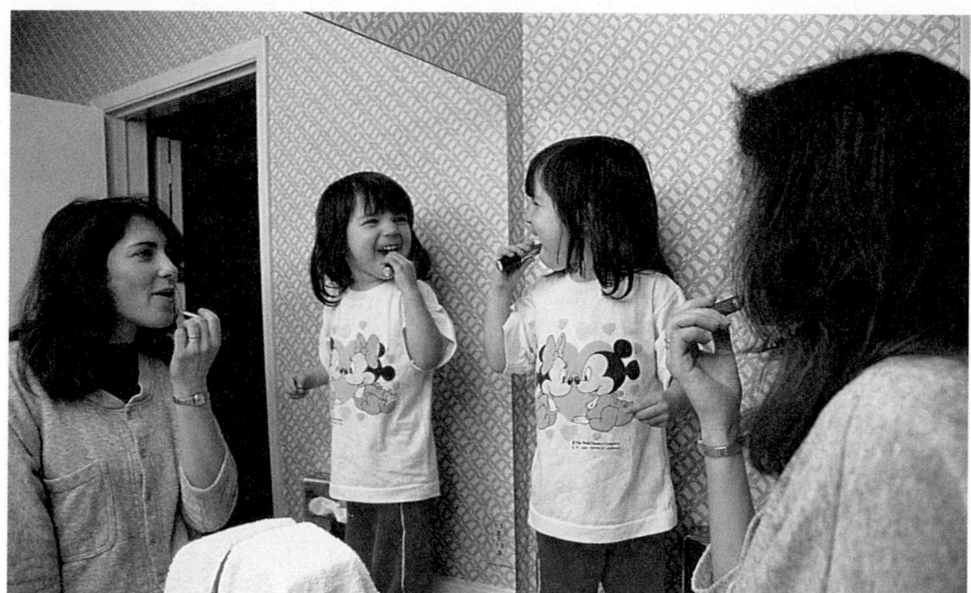

Acquisition of Gender Roles. According to social-learning theory, children learn gender roles that are considered appropriate in their society via the reinforcement of certain behavior patterns and by observing the gender-role behaviors of their parents, peers, and other role models in media such as TV, films, and books.

Sociocultural Perspectives

Sexual behavior is determined not only by biological and psychological factors, but also by social factors. Social factors contribute to the shaping of our sexual attitudes, beliefs, and behavior. Whereas anthropologists contribute to our understanding of cross-cultural variance in sexuality, sociocultural theorists focus on differences in sexuality among the subgroups of a society, as defined, for example, by differences in religion, race/ethnicity, country of origin, socioeconomic status, marital status, age, educational level, and gender. Such a society is the United States.

Consider the issue of the numbers of sex partners people have. Table 1.2 reports the results of a national survey concerning the number of sex partners people report having had since the age of 18. It considers the factors of gender, age, marital status, level of education, religion, and race/ethnicity (Laumann et al., 1994).

Consider gender. Males, according to the survey, report having had greater numbers of sex partners than females do. For example, 1 male in 3 (33%) reports having had 11 or more sex partners since the age of 18, as compared with fewer than 1 woman in 10 (9.2%). Throughout the text, we shall be focusing on gender differences and on why men seem generally more likely than women to seek a wide range of sexual experience.

Concerning age, the numbers of sex partners appears to rise with age into the 40s. As people gain in years, they have more opportunity to accumulate life experiences, including sexual experiences. But then the numbers of partners fall off among respondents in their 50s. Older respondents entered adulthood prior to the sexual revolution and were thus generally exposed to more conservative sexual attitudes. We shall find this sort of age difference, or age gradient, throughout the text as well.

Level of education is also connected with sexual behavior. Generally speaking, it would appear that education is something of a liberalizing influence. Therefore, it is not surprising that people with some college education, and people who have completed college, are likely to have had more sex partners than those who attended only grade school or high school.

If education is a liberating influence on sexuality, conservative religious experience would appear to be a restraining factor. In Table 1.2, those who report no religion and liberal Protestants (such as, Methodists, Lutherans, Presbyterians, Episcopalians, and

TABLE 1.2

Number of sex partners since age 18 as found in the NHSLS* study

Social Characteristics	NUMBER OF SEX PARTNERS (%)					
	0	**1**	**2–4**	**5–10**	**11–20**	**21+**
Gender						
Male	3.4	19.5	20.9	23.3	16.3	16.6
Female	2.5	31.5	36.4	20.4	6.0	3.2
Age						
18–24	7.8	32.1	34.1	15.4	7.8	2.8
25–29	2.2	25.3	31.3	22.2	9.9	9.0
30–34	3.1	21.3	29.3	25.2	10.8	10.3
35–39	1.7	18.9	29.7	24.9	14.0	10.8
40–44	0.7	21.9	27.6	24.2	13.7	12.0
45–49	2.0	25.7	23.8	25.1	9.6	13.9
50–54	2.4	33.9	27.8	18.0	9.0	9.0
55–59	1.3	40.0	28.3	15.2	8.3	7.0
Marital Status						
Never married (not cohabiting)	12.3	14.8	28.6	20.6	12.1	11.6
Never married (cohabiting)	0.0	24.6	37.3	15.7	9.7	12.7
Married	0.0	37.1	28.0	19.4	8.7	6.8
Education						
Less than high school	4.2	26.7	36.0	18.6	8.8	5.8
High school graduate	3.4	30.2	29.1	20.0	9.8	7.4
Some college	2.1	23.9	29.4	23.3	11.9	9.3
College graduate	2.1	24.1	25.8	23.9	11.1	13.0
Advanced degree	3.5	24.6	26.3	22.8	9.6	13.2
Religion						
None	2.6	16.2	29.0	20.3	15.9	15.9
Liberal, moderate Protestant	2.3	22.8	31.2	23.0	12.4	8.3
Conservative Protestant	2.9	29.8	30.4	20.4	9.5	7.0
Catholic	3.8	27.2	29.2	22.7	8.1	9.1
Jewish†	0.0	24.1	13.0	29.6	16.7	16.7
Race/Ethnicity						
European American	3.0	26.2	28.9	22.0	10.9	9.1
African American	2.2	18.0	34.2	24.1	11.0	10.5
Latino and Latina American	3.2	35.6	27.1	17.4	8.2	8.5
Asian American†	6.2	46.2	24.6	13.8	6.2	3.1
Native American†	5.0	27.5	35.0	22.5	5.0	5.0

*National Health and Social Life Survey, conducted by a research team centered at the University of Chicago.

†These sample sizes are quite small.

Source: From E. O. Laumann, J. H. Gagnon, R. T. Michael, and S. Michaels, *The Social Organization of Sexuality: Sexual Practices in the United States* (Chicago, IL: University of Chicago Press, (1994), Table 5.1C, p. 179. Copyright © 1994 by University of Chicago Press. Adapted by permission.

members of the United Church of Christ) report higher numbers of sex partners than do Catholics and conservative Protestants (such as members of Baptist churches, Pentecostal churches, churches of Christ, and Assemblies of God). Results for Jews are obscured by a small sample size. Research in general would suggests that Orthodox Judaism is a restraining factor on sexual behavior, compared to more liberal branches of Judaism.

The factor of ethnicity is also connected with sexual behavior. Throughout the text, our coverage of diversity will address differences among European Americans, African Americans, Latino and Latina Americans, Asian Americans, and Native Americans. The research findings listed in Table 1.2 suggest that European and African Americans have the highest numbers of sex partners. Latino and Latina Americans are mostly Catholic, and Catholicism, as we have noted, provides something of a restraint on sexual behavior. Asian Americans would appear to be the most sexually restrained ethnic group. However, as noted in the footnote to the table, the sample sizes for Asian Americans and Native Americans are relatively small.

The sociocultural perspective informs us of the relationship between sexuality and one's social group within a society. Sociocultural theorists view sexual behavior as occurring within a particular sociocultural system. They study the ways in which the values, beliefs, and norms of a group influence the sexual behavior of its members. To a certain extent, we share attitudes and behavior patterns with people from similar backgrounds—for example, people with the same ethnic identity. Even so, not all Protestants or all members of a given ethnic group act or think alike.

GENDER ROLES Sociocultural theorists also study gender roles. In Western cultures, men have traditionally been expected to be the breadwinners, whereas women have been expected to remain in the home and rear the children. Traditional gender roles also define sexual relations. Men, by this standard, are expected to be assertive, women compliant. Men are to initiate romantic overtures. Women are to perform a "gatekeeping" role and determine which advances they will accept. Today, many of these traditions are falling by the wayside. Most women today are members of the work force. Many are pursuing careers in traditionally male domains, such as law, medicine, and engineering. Some women command naval vessels. Others pilot military helicopters. Yet even women who become presidents and vice presidents of corporations are still burdened with the bulk of household chores (Lewin, 1995). Sexual practices are also changing to some degree. More women today initiate dates and sexual interactions than was the case in past generations.

Multiple Perspectives on Human Sexuality

Given the complexity and range of human sexual behavior, we need to consider multiple perspectives to understand sexuality. Each perspective—historical, biological, cross-species, cross-cultural, psychological, and sociocultural—has something to offer in this enterprise. Let us venture a few conclusions based on our overview of these perspectives. First, human sexuality appears to reflect a combination of biological, social, cultural, sociocultural, and psychological factors that interact in complex ways, perhaps in combinations that are unique for each individual. Second, there are few universal patterns of sexual behavior, and views on what is right and wrong show great diversity. Third, although our own cultural values and beliefs may be deeply meaningful to us, they may not indicate what is normal, natural, or moral in terms of sexual behavior. The complexity of human sexuality—complexity that causes it to remain somewhat baffling to scientists—adds to the wonder and richness of our sexual experience.

Human Sexuality ONLINE //

Web Sites on Human Sexuality

Many Human Sexuality Online features contain Web sites for readers to "check out." The lists are brief to help encourage actual exploration of the sites. That means that many important Web sites will be omitted. Your professor can suggest additional Web sites. Other Web sites of interest will frequently be noted within the chapters—as, for example, the more than a dozen Web sites of possible interest to pregnant women in Chapter 11. Here are some Web sites you might wish to explore to enhance your knowledge of some of the topics discussed in Chapter 1:

SSSS—the "Four S's"—that is, the Society for the Scientific Study of Sexuality: SSSS was founded in 1957 as an international organization dedicated to the advancement of knowledge about sexuality. It includes an interdisciplinary group of professionals who believe in the importance of both research and the clinical, educational, and social applications of research in sexuality. The society supports the study of sexuality as a valid area for scientific research. Members include biologists, physicians, nurses, therapists, psychologists, sociologists, anthropologists, historians, educators, and theologians.
www.ssc.wisc.edu/sss/

AASECT—The American Association of Sex Educators, Counselors and Therapists: AASECT promotes sexual health through the development and advancement of the fields of sex therapy, counseling, and education. The Association provides for the education and certification of sex educators, counselors, and therapists, as well as individuals who supervise sex therapists in training. AASECT also encourages research in sex education, counseling, and therapy. AASECT publishes the *The Journal of Sex Education and Therapy* and provides links that enable users to locate qualified sex therapists.
www.aasect.org/

The American Psychological Association: The APA provides a gateway to psychology journals, press releases, and other public information on issues such as aging, AIDS, children, depression, disabilities, sexual orientation, health, ethnic minority groups, parenting and the family, and women.
www.apa.org/

The American Sociological Association: This Web site contains information for sociologists and other social scientists, press releases, and other public information on issues such as affirmative action, families, the immigration experience, and violence.
www.asanet.org/

The *Journal of the American Medical Association:* This is the main journal of the AMA. Access is free, and users can search for articles on any medical topics, including those related to sexuality. Some of the articles are quite understandable to laypeople.
www.ama-assn.org/

summing up

WHAT IS HUMAN SEXUALITY?

The term *human sexuality* refers to matters of gender, sexual behavior, sexual feelings, and the biology of sex. Human sexuality comprises all the ways in which we experience and express ourselves as sexual beings.

■ The Study of Human Sexuality

The study of human sexuality draws on the expertise of anthropologists, biologists, medical researchers, sociologists, psychologists, and other scientists.

■ Sexuality and Values

Our pluralistic society embraces a wide range of sexual attitudes and values.

THINKING CRITICALLY ABOUT HUMAN SEXUALITY

The text encourages critical thinking to help students evaluate claims, arguments, and widely held beliefs. Critical thinking encourages the thoughtful analysis and probing of arguments, willingness to challenge conventional wisdom and common knowledge, and keeping on open mind.

■ Principles of Critical Thinking

The text enumerates several features of critical thinking. These include being skeptical, examining definitions of terms, examining the assumptions or premises of arguments, being cautious in drawing conclusions from evidence, considering alternative interpretations of evidence, and avoiding oversimplification and overgeneralization.

PERSPECTIVES ON HUMAN SEXUALITY

Human sexuality is a complex field, and no single theory or perspective can capture all its nuances. The text offers diverse perspectives on sexuality:

■ The Historical Perspective

History places our sexual behavior in the context of time. History shows little evidence of universal sexual trends. There is evidence of prehistoric worship of generative power in women and men. Jews and Christians have emphasized the role of sex as a means of propagation and have generally restricted sex to the context of family life. The ancient Greeks and Romans lived in male-oriented societies that viewed women as chattel. Some Eastern civilizations have equated sexual pleasure with religious experience and have developed sex manuals. Repressive Victorian sexual attitudes gave way to the sexual revolution of the 1960s and 1970s in the West.

■ The Biological Perspective

The biological perspective focuses on the role of biological processes, such as genetic, hormonal, vascular, and neural factors, in explaining human sexual behavior. Knowledge of biology helps us understand how our bodies respond to sexual stimulation and enhances aspects of our sexual health.

■ The Evolutionary Perspective

Darwin's theory of evolution proposes that organisms tend to inherit those traits that enabled their ancestors to reach the age of sexual maturity and reproduce. Evolutionary psychology proposes that dispositions toward behavior patterns that enhance reproductive success—as well as physical traits that do so—may be genetically transmitted.

■ The Cross-Species Perspective

The study of other animal species reveals a surprising variety of sexual behaviors among nonhumans. There are, for example, animal analogues of male–male sexual behavior, female–female sexual behavior, oral sex, foreplay, and rape. Still, we must be cautious in generalizing from lower animals to humans. We find that experience and learning play more important roles in sexuality as we travel up the evolutionary ladder.

■ The Cross-Cultural Perspective

This perspective, like the historical perspective, provides insight into the ways in which cultural beliefs affect sexual behavior and people's sense of morality. Anthropologists observe other cultures firsthand when possible. Cross-cultural evidence challenges the notion of the universality of gender-role stereotypes. All cultures apparently place some limits on sexual freedom (and nearly all cultures place some prohibitions on incestuous relationships), but some cultures are more sexually permissive than others.

■ Psychological Perspectives

Psychological perspectives focus on the psychological factors of perception, learning, motivation, emotion, personality, and so on that affect gender and sexual behavior in the individual. Sigmund Freud formulated the theory of psychoanalysis, which proposes that biologically based sex drives come into conflict with social codes. Erogenous zones shift through the process of psychosexual development, and defense mechanisms banish threatening ideas and impulses from conscious awareness. Learning theorists focus on the roles of rewards, punishments, and modeling on sexual behavior.

■ Sociocultural Perspectives

Sociocultural theorists focus on differences in sexuality among the groups within a society, as defined, for example, by differences in religion, race, country of origin, socioeconomic status, age, educational level, and gender.

■ Multiple Perspectives on Human Sexuality

Given the complexity and range of human sexual behavior, we need to consider multiple perspectives to understand human sexuality.

questions for critical thinking

1. What are the meanings of the word *sex?* What is the meaning of *gender?* Why is it important to differentiate between the two?
2. How does the study of human sexuality help you clarify your own values?
3. Why is critical thinking essential to the study of human sexuality?
4. A friend of yours insists that something is true because a TV personality, his minister, or this textbook claims that it is true. Would a critical thinker accept your friend's argument as proof? Why or why not?
5. How does the sociocultural perspective enhance our understanding of human sexuality?

chapter

2

Research Methods

Pablo Picasso. *Large Bather with a Book*, February 18, 1937. Oil, pastel and charcoal on canvas, 130 × 97.5. Photo: J. G. Berizzi. © Copyright ARS, NY. Musée Picasso, Paris, France. Credit: Réunion des Musées Nationaux/Art Resource, NY. Credit: © 2002 Estate of Pablo Picasso/Artists Rights Society (ARS), New York.

outline

Truth or Fiction?

A SCIENTIFIC APPROACH TO HUMAN SEXUALITY

The Scientific Method

Goals and Methods of the Science of Human Sexuality

Operational Definitions

A Closer Look Physiological Measures of Sexual Arousal

POPULATIONS AND SAMPLES: REPRESENTING THE WORLD OF DIVERSITY

Sampling Methods

METHODS OF OBSERVATION

The Case-Study Method

The Survey Method

A Closer Look With a Horse?

A World of Diversity Surveys of the Sexual Behavior of Diverse Populations

The Naturalistic-Observation Method

The Ethnographic-Observation Method

The Participant-Observation Method

The Laboratory-Observation Method

THE CORRELATIONAL METHOD

Limitations of the Correlational Method

THE EXPERIMENTAL METHOD

Aspects of the Experimental Method

ETHICS IN SEX RESEARCH

Pain and Stress

Confidentiality

Informed Consent

The Use of Deception

Human Sexuality Online Web Sites for Use in Research on Human Sexuality

SUMMING UP

QUESTIONS FOR CRITICAL THINKING

Truth or Fiction?

____ You could study the sexual behavior of millions of Americans and still not obtain an accurate picture of the sexual behavior of the general U.S. population.

____ Researchers can overcome the refusal of some people to participate in a study of sexual behavior by recruiting larger numbers of volunteers.

____ Case studies have been carried out on people who are dead.

____ Residents of China are more likely than people in the United States to approve of extramarital affairs.

____ Some sex researchers have engaged in “swinging” with the people they have studied.

____ Masters and Johnson created a transparent artificial penis containing photographic equipment to study female sexual response.

____ Regular churchgoers report higher levels of sexual satisfaction than others.

Have you ever wondered about questions such as these: Are my sexual interests and behavior patterns unique or shared by many others? Does alcohol stimulate or dampen sexual response? Why do people engage in male–male or female–female sexual behavior? How do people contract AIDS? Does pornography cause rape?

You may have thought of such questions. You may even have expressed opinions on them. But scientists insist that opinions about behavior, including sexual behavior, be supported by evidence. Evidence, in turn, must be based on careful observations in the laboratory or in the field.

In this chapter we explore the methods that scientists use to study human sexuality. We then focus on ethical issues in research on sex. Researchers in all behavioral sciences confront ethical problems, but the problems are heightened in sex research. After all, the field of human sexuality touches on some of the most personal and intimate experiences of our lives.

A SCIENTIFIC APPROACH TO HUMAN SEXUALITY

Scientists and researchers who study human sexuality take an **empirical** approach. That is, they base their knowledge on research evidence, rather than on intuition, faith, or superstition. Scientists' and other people's intuitions or religious beliefs may suggest topics to be studied scientifically, but once the topics are selected, answers are sought on the basis of the scientific method.

The Scientific Method

Critical thinking and the scientific approach share the hallmark of skepticism. As skeptics, scientists question prevailing assumptions and theories about sexual behavior. They are willing to dispute the assertions of authority figures such as political and religious leaders—and even other scientists. Scientists also recognize that they cannot gain perfect knowledge. One era's "truths" may become another era's ancient myths and fallacies. Scientists are involved in the continuous quest for truth, but they do not see themselves as experiencing revelations or defining final truths.

The *scientific method* is a systematic way of gathering scientific evidence and testing assumptions through research. It has a number of elements:

1. *Formulating a research question.* Does alcohol inspire or impair sexual response? Scientists formulate research questions on the basis of their observations of, or theories about, events or behavior. They then seek answers to such questions by conducting empirical research.
2. *Framing the research question in the form of a hypothesis.* Experiments are usually undertaken with a **hypothesis** in mind—a precise prediction about behavior that is often derived from theory. A hypothesis is tested through research. For instance, a scientist might theorize that alcohol enhances sexual responsiveness either by directly stimulating sexual response or by reducing feelings of guilt associated with sex. He or she might then hypothesize that an intervention (called, in experimental terms, a treatment), such as drinking alcohol in a laboratory setting, will lead to heightened sexual arousal in the presence of erotic stimuli (such as sexually explicit films). Hypotheses, then, can be considered educated guesses that anticipate the results of an experiment.

Empirical Derived from or based on observation and experimentation.

Hypothesis A precise prediction about behavior that is often derived from theory and is tested through research.

3. *Testing the hypothesis.* Scientists then test hypotheses through carefully controlled observation and experimentation. A specific hypothesis about alcohol and sexual arousal—that alcohol either increases or decreases sexual responsiveness—might be tested by administering a certain amount of alcohol to one group of people and then comparing their level of sexual arousal following specific types of sexual stimulation (such as exposure to sexually explicit films) to the level of sexual arousal of another group of people who were shown the films but not given any alcohol.
4. *Drawing conclusions.* Scientists then draw conclusions or inferences about the correctness of their hypotheses, on the basis of their analyses of the results of their studies. If the results of well-designed research studies fail to bear out certain hypotheses, scientists can revise the theories that served as the frameworks for the hypotheses. Research findings often lead scientists to modify their theories and, in turn, generate new hypotheses that can be tested in further research.

Some investigators include the publication of results in professional journals as part and parcel of the scientific method. Publication shares scientific knowledge with the public at large and also exposes research methods to evaluation by the scientific community. In this way, results can be interpreted in terms of potential flaws in methodology.

Goals and Methods of the Science of Human Sexuality

The goals of the science of human sexuality are congruent with those of other sciences: to describe, explain, predict, and control the events (in this case, the sexual behaviors) that are of interest. Let us discuss some of the general goals of science and how they are related to the study of human sexuality as a science.

Description is a basic objective of science. To understand sexual behavior, for example, we must first be able to describe it. Therefore, the description of behavior precedes understanding. Scientists attempt to be clear, unbiased, and precise in their descriptions of events and behavior. The scientific approach to human sexuality describes sexual behavior through techniques as varied as the field study, the survey, the individual case study, and the laboratory experiment.

To underscore the importance of the need for unbiased description, consider the name of a tropical fish that will be familiar to many readers: the kissing gourami. These small, flat fish—particularly the males—approach each other frontally and press their open mouths against one another. Yet the term *kissing gourami* may well be something of a misnomer if the word *kissing* is meant to imply affection. Prolonged observations of the fish suggest that it is more likely that kissing in gouramis is a test of strength. More powerful "kissers" apparently achieve positions of social dominance that are connected with privileges in feeding and mating.

The error in describing the behavior of gouramis as kissing involves confusing **inference** with description. It is, in fact, an **anthropomorphic** inference; it involves applying human standards to explain animal behavior. One of the great challenges to scientists is the separation of description from inference.

Inference, like description, is crucial to science. Inference allows us to move from observations or descriptions of particular events or behavior to general principles that can be woven into models and theories that help explain these events, such as the psychoanalytic and learning theories or models. Without a means for organizing our descriptions of events and behavior in terms of models and theories, we would be left with nothing more than a buzzing confusion of disconnected observations. Scientists need to distinguish between *description* and *inference*, however—to recognize when they jump from a description of events to an inference based on an interpretation of those events. For example, one does not *describe* a person's sexual behavior as "deviant." Rather, one *labels* or *classifies* it as deviant when one believes that it deviates from a certain norm. But what

Inference A conclusion or opinion.

Anthropomorphism The attributing of human characteristics to an animal.

is deviant in one culture may be normal in another. Incestuous relationships between brothers and sisters are permitted among the Dahomey of West Africa (Stephens, 1982). The same behavior is considered deviant in our own culture and virtually all others. Even within the same larger culture, behavior may be deviant by the standards of some subcultures and not by others.

Researchers attempt to relate their observations to other factors, or **variables,** that can help explain them. For example, researchers may attempt to explain variations in the frequency of coitus by relating—or *correlating*—coitus with **demographic** variables such as age, religious or social background, and cultural expectations. The variables that are commonly used to explain sexual behavior include biological (age, health), psychological (anxieties, skills), and sociological (educational level, socioeconomic status, ethnicity) variables. Explanations of behavior can involve reference to many variables, even those that cannot be measured directly, such as unconscious motivation. Relationships among variables may be tied together into theories of behavior. For example, psychoanalytic theory emphasizes the role of unconscious forces in determining behavior. Learning theories focus on how variables such as rewards, punishments, and expectations shape behavior.

Theories provide frameworks within which scientists can explain what they observe and can make predictions. It is not sufficient for theories to help us make sense of events that have already occurred. Theories must also enable us to make predictions. One test of the soundness of psychoanalytic and learning theories is whether they allow us to predict behavior. Prediction requires discovering variables that anticipate events. Geologists seek clues in the forces that affect the earth to forecast events such as earthquakes and volcanic eruptions. Sex researchers study factors that may predict various types of sexual behavior. Some researchers, for example, have examined childhood interests and behavior patterns that may predict the development of a gay male or lesbian sexual orientation. Others have explored factors, such as the age at which dating begins and the quality of the relationships between teens and their parents, that may predict the likelihood of premarital intercourse during adolescence.

The concept of "controlling" human behavior does not mean coercing people to do the bidding of others. Rather, it means drawing on scientific knowledge to help people create their own goals and marshal their resources to achieve these goals. Reputable scientists adhere to ethical and professional standards that ban the use of harmful techniques in research or practice and that safeguard the rights of participants in research.

The science of human sexuality does not tell people how they *ought* to behave. It does not attempt to limit or expand the variety of their sexual activities. Rather, it furnishes information that people may use to help themselves or others make decisions about their own behavior. For instance, the science of human sexuality provides information that increases the chances that a couple who are having difficulty becoming pregnant will be able to conceive. At the same time, it develops and evaluates means of birth control that can be used to help couples regulate their reproductive choices. The science of human sexuality also seeks to develop techniques that can help people overcome sexual dysfunctions and enhance the gratification they find in sexual relations. "Control" also takes the form of enabling couples to give and receive more sexual pleasure, enhancing fetal health, and preventing and curing sexually transmitted diseases.

Variables Quantities or qualities that vary or may vary.

Demographic Concerning the vital statistics (density, race, age, etc.) of human populations.

Penile strain gauge A device for measuring sexual arousal in men in terms of changes in the circumference of the penis.

Vaginal photoplethysmograph A tampon-shaped probe that is inserted in the vagina and suggests the level of vasocongestion by measuring the light reflected from the vaginal walls.

Operational Definitions

How do we study concepts such as sexual satisfaction or the even broader concept of marital satisfaction? One of the requirements in sex research, as in other types of research, is specifying the definitions of the concepts, or "constructs," of interest. Different investigators may define marital satisfaction or sexual satisfaction in different ways. They may be studying different events although they may use the same terms.

What, for example, is sexual arousal? Many studies seek to assess the effects of various stimuli (such as sexually explicit material) on sexual arousal, but sexual arousal can

a CLOSER look

Physiological Measures of Sexual Arousal

Scientific studies depend on the ability to measure the phenomena of interest. The phenomena of sexual arousal may be measured by different means, such as self-report and physiological measures. Self-report measures of sexual arousal are considered *subjective.* They ask people to give their impressions of the level of their sexual arousal at a given time, such as by circling their response on a ten-point scale that ranges from zero, "not at all aroused," to ten, "extremely aroused." Physiological devices measure the degree of vasocongestion that builds up in the genitals during sexual arousal. (Vasocongestion—that is, congestion with blood—leads to erection in men and to vaginal lubrication in women.) In men, vasocongestion is frequently measured by a **penile strain gauge.** This device is worn under the man's clothing. It is fitted around the penis and measures his erectile response by recording changes in the circumference of the penis. The device is sensitive to small changes in circumference that may not be noticed (and thus will not be reported) by the man.

Physiological measurement of sexual arousal in women is most often accomplished by means of a **vaginal photoplethysmograph.** The vaginal photoplethysmograph is a tampon-shaped probe with a light and a photocell in its tip. It is inserted in the vagina and indicates the level of blood congestion by measuring the amount of light reflected from the vaginal walls. The more light that is absorbed by the vaginal walls, the less that is reflected. Less reflected light indicates greater vasocongestion.

Sex researchers sometimes measure sexual arousal in response to stimuli such as erotic films or audiotaped dramatizations of erotic scenes. What happens when physiological devices give a different impression of sexual arousal than those offered by self-report? Objectively (physiologically) measured sexual arousal does not always agree with subjective feelings of sexual arousal, as measured by self-report. For example, a person may say that he or she is relatively unaroused at a time when the physiological measures suggest otherwise. Which is the *truer* measure of arousal, the person's subjective report or the level shown on the objective instruments?

Discrepancies across measures suggest that people may be sexually aroused (as measured by physiological indicators) but psychologically unprepared to recognize it or unwilling to admit it. In the real world of human relationships, sexual arousal has psychological as well as physiological aspects. The reflexes of erection and vaginal lubrication do not necessarily translate into "Yes."

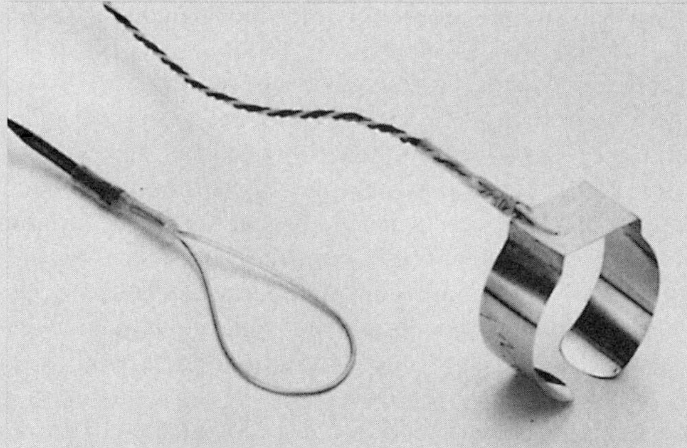

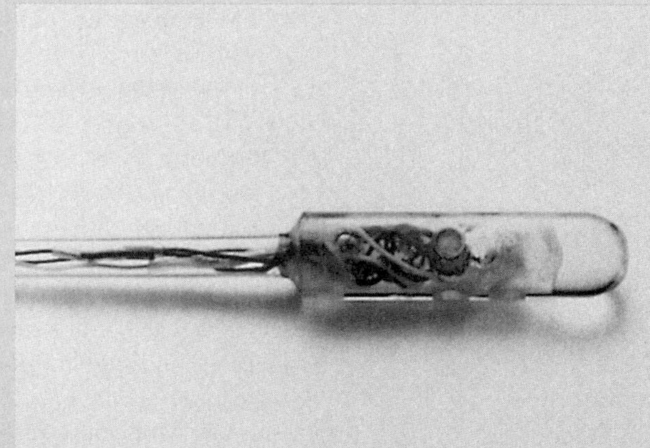

The Penile Strain Gauge and the Vaginal Photoplethysmograph. These devices measure vasocongestion in the genitals of men and women, providing an objective measure of the subjects' level of sexual arousal. Can you think of other aspects of sexual arousal?

mean several things. It can mean the *subjective* "feeling" that one is sexually aroused. It can mean one's self-report of genital sensations. It may refer to direct physiological measures of blood congestion **(vasocongestion)** in the genitals. We further explore the measurement of sexual arousal in the nearby Closer Look feature.

The **operational definition** of a construct links its meaning to the methods used to measure it. In some studies reported in this book, sexual arousal is operationally defined

Vasocongestion Congestion resulting from the flow of blood. (From the Latin *vas,* which means "vessel.")

Operational definition A definition of a construct or variable in terms of the methods used to measure it.

as vasocongestion. In others, it is operationally defined as self-report of sexual interest or genital sensations. In still others, a combination of measures is used. It is important to recognize that our ability to generalize the results of research is limited by the operational definitions of the variables. That is, we may be able to generalize only to instances that define terms in the same way.

Let us now examine the ways in which scientists study human sexuality. The first thing scientists do is to identify *whom* or *what* they will study, which brings us to the topic of sampling.

POPULATIONS AND SAMPLES: REPRESENTING THE WORLD OF DIVERSITY

Researchers undertake to learn about populations. **Populations** are complete groups of people or animals. Many researchers have attempted to learn about people in the United States, for example. (At the turn of the millennium, there were about 280 million of them—a sizable population.) Other researchers may identify American adults or American adolescents as their population. Still other researchers attempt to compare the sexual behavior of African Americans to that of Latino and Latina Americans and other Americans. These are termed the *populations of interest,* or *target populations.* These target populations are all sizable. It would be expensive, difficult, and all but impossible to study every individual in them.

Because of the impossibility of studying all members of a population, scientists select individuals from the population and study them. The individuals who participate in research are said to compose a **sample.** However, we cannot truly learn about the population of interest unless the sample *represents* that population. A *representative sample* is a research sample of participants who accurately represent the population of interest.

If you wished to learn about dating practices at your college or university, your population of interest would be the total student population. If you were to select your research sample from one dormitory or class, it would probably not be a representative sample (unless the dormitory or class contained a true cross section of the entire student body). If our samples do not represent the target populations, we cannot extend, or **generalize,** the results of our research to the populations of interest. If we wished to study the sexual behavior of Asian Americans, our population would consist of *all* Asian Americans. If we used only Asian American college students as our sample, we could not generalize our findings to all Asian Americans. Our sample would not *represent* all Asian Americans.

In a perfect world, research into populations would include every member of the target populations. Then we would know that our findings applied to these populations. Including all people in the United States in a study of sexual behavior would be impossible. We cannot even *find* all people in the United States when we conduct the census each decade. And incorporating sex research into the census would undoubtedly cause many more people to refuse to participate. Sampling a part of a target population makes research practical and possible—if imperfect.

Population A complete group of organisms or events.

Sample Part of a population.

Generalize To go from the particular to the general.

Sampling Methods

Because we cannot sample every member of a population, we need to select samples that represent that population. One way of acquiring a representative sample is through random sampling. A **random sample** is one in which every member of the target population has an equal chance of participating. A random sample is a type of **probability sample.** Statisticians use this term to describe samples in which we know the probability of any particular member's being included in the sample.

Now and then, magazine editors boast that they have surveyed samples of 20,000 or 30,000 readers, but size alone does not mean that a sample is representative. As an example, the magazine *Literary Digest* polled thousands of voters by telephone to predict the outcome of the 1936 presidential election. On the basis of the survey, the magazine predicted that Alfred Landon would defeat Franklin D. Roosevelt, but Roosevelt won by a landslide of nearly 11 million votes. The problem was that the election was held during the Great Depression, when only relatively affluent people could afford telephones. Those who could afford phones were also more likely to be Republicans and to vote for the Republican candidate, Landon. Thus, the *Digest* poll, although large, was biased. A sample of 30 million voters will not provide an accurate picture if it is biased.

It is true that you could study the sexual behavior of millions of people in the United States and still not obtain an accurate picture of the sexual behavior of the general U.S. population. A sample of many millions might not represent the U.S. population. Sample size alone does not guarantee that the sample's members have been selected in an unbiased manner.

Researchers overcome biased sampling by drawing *random* or *stratified random* samples of populations. In a random sample, every member of a population has an equal chance of participating. In a **stratified random sample,** known subgroups of a population are represented in proportion to their numbers in the population. For instance, about 12% of the U.S. population is African American. Researchers could therefore decide that 12% of their sample must be African American if the sample is to represent all people in the United States. The randomness of the sample would be preserved, because the members of the subgroups would be selected randomly from their particular subgroups. In practical terms, however, a reasonably large *random* sample with no prior stratification will turn out to be reasonably well stratified in the end. Put it another way: If you blindfold yourself, shake up a jar of jelly beans, and take out a scoopful, the proportions of the different colors of beans in the scoop are likely to approximate the proportions of those colors in the entire jar.

Random samples can be hard to come by, especially when the research involves asking people about their sexual attitudes or behavior. For instance, the Playboy Foundation (Hunt, 1974) sampled people listed in telephone directories in various cities in the United States in the early 1970s. Most people at the time had phones, yet the sample could not have included the very poor (who might not have been able to afford to keep a phone), those who had unlisted numbers, and college students without private phones.

Another problem is that sexual research is almost invariably conducted with people who volunteer to participate. Volunteers may differ from people who refuse to participate. For example, volunteers tend to be more open about their sexuality than the general population. They may even tend to exaggerate behaviors that others might consider deviant or abnormal.

The problem of a **volunteer bias** is a thorny one for sex researchers, because the refusal of people who have been randomly selected to participate in the survey can ruin the representativeness of the sample. It would be unethical to coerce people to participate in a study on sexual behavior (or in any other type of study), so researchers must use samples of volunteers, rather than true random samples. A low response rate to a voluntary survey is an indication that the responses do not represent the people to whom the survey was distributed.

Random sample A sample in which every member of a population has an equal chance of participating.

Probability sample A sample in which the probability that any particular member will be included in the sample is known.

Stratified random sample A random sample in which known subgroups in a population are represented in proportion to their numbers in the population.

Volunteer bias A slanting of research data that is caused by the characteristics of individuals who volunteer to participate, such as willingness to discuss intimate behavior.

Truth or Fiction? **REVISITED**

It is not true that researchers can overcome the refusal of some people to participate by recruiting larger numbers of volunteers. Volunteers differ from people who refuse to participate. Therefore, volunteers do not *represent* the entire population.

In some cases, as in the Janus and Janus (1993) survey, samples are samples of convenience. They consist of individuals who happen to be available to the researcher and who share some characteristics with the target population—perhaps religious background or sexual orientation. Still, they may not truly represent the target group. Convenience samples often consist of European American, middle-class college students who volunteer for studies conducted at their schools. They may not be (and probably are not) representative of students in general. They do not even represent the general student population at their own school. Random samples of gay males or lesbians are perhaps the most difficult to come by. The social stigma attached to their sexual orientations may discourage gay males and lesbians from disclosing them.

METHODS OF OBSERVATION

Once scientists have chosen those whom they will study, they observe them. In this section, we consider several methods of observation: the case-study method, the survey method, naturalistic observation, ethnographic observation, participant observation, and laboratory observation.

The Case-Study Method

A **case study** is a carefully drawn, in-depth biography of an individual or a small group. The focus is on understanding one or several individuals as fully as possible by unraveling the interplay of various factors in their backgrounds. In most case studies, the researcher comes to know the individual or group through interviews or other contacts conducted over a prolonged period of time. The interviewing pattern tends to build upon itself with a good deal of freedom, in contrast to the one-shot, standardized set of questions used in survey questionnaires.

Researchers may also conduct case studies by interviewing people who have known the individuals or by examining public records. Sigmund Freud, for example, drew on historical records in his case study of the Renaissance inventor and painter Leonardo da Vinci. Freud concluded that Leonardo's artistic productions represented the sublimating, or channeling, of male–male sexual impulses.

Truth or Fiction? **REVISITED**

It is true that case studies have been carried out on people who are dead. Such case studies rely on historical records rather than on interviews with the individuals themselves or their contemporaries.

Reports of innovative treatments for sexual dysfunctions usually appear as well-described case studies. A clinician typically reports the background of the client in depth, describes the treatment, reports the apparent outcomes, and suggests factors that might have contributed to the treatment's success or failure. In writing a treatment case study, the therapist tries to provide information that may be helpful to therapists who treat clients with similar problems. Case studies or "multiple case studies" (reports concerning a few individuals) that hold promise may be subjected to controlled investigation—ideally, to experimental studies involving treatment and control groups.

Case study A carefully drawn, in-depth biography of an individual or a small group of individuals that may be obtained through interviews, questionnaires, and historical records.

LIMITATIONS OF THE CASE-STUDY METHOD Despite the richness of material that may be derived from the case-study approach, it is not as rigorous a research design as an experiment. People often have gaps in memory, especially concerning childhood events. The potential for observer bias is also a prominent concern. Clinicians and interviewers may unintentionally guide people into saying what they expect to hear. Then, too, researchers may inadvertently color people's reports when they jot them down—shape them subtly in ways that reflect their own views.

Remember that clinicians who test a treatment method with a single client (or with a few clients) through a course of therapy are manipulating an independent variable (the treatment) with one or more people who are unlikely to represent the general population. For one thing, the clients have sought (or been placed in) professional treatment. In a case study, we cannot be certain whether treatment outcomes are due to the treatment methods, to giving clients hope that they will get better, to having clients talk about their problems, or to the simple passage of time. We will see that the experiment allows us to eliminate such rival explanations.

Alfred Kinsey. Kinsey and his colleagues conducted the first large-scale scientific study of sexual behavior in the United States.

The Survey Method

Surveys typically gather information about behavior through questionnaires or interviews. Researchers may interview or administer questionnaires to thousands of people from particular population groups to learn about their sexual behavior and attitudes. Interviews such as those used by Kinsey and his colleagues (1948, 1953) have the advantages of allowing face-to-face contact and giving the interviewer the opportunity to *probe*—that is, to follow up on answers that seem to lead toward useful information. A skilled interviewer may be able to set a respondent at ease and establish a sense of trust or *rapport* that encourages self-disclosure. On the other hand, unskilled interviewing may cause respondents to conceal information.

Questionnaires are inexpensive when compared to interviews. The major expenses in using questionnaires involve printing and distribution. Questionnaires can be administered to many people at once, and respondents can return them unsigned. (Anonymity may encourage respondents to disclose intimate information.) Questionnaires, of course, can be used only by people who can read and record their responses. Interviews can be used even with people who cannot read or write. But interviewers must be trained, sometimes extensively, and then paid for their time.

Some of the major surveys described in this book were conducted by Kinsey and his colleagues (1948, 1953), the Playboy Foundation (Hunt, 1974), Bell and his colleagues (1978, 1981), Coles and Stokes (1985), and the University of Chicago group (Laumann et al., 1994). The book also discusses surveys conducted by popular magazines such as *Redbook*. By and large, these surveys have reported the incidence and frequency of sexual activities among men and women both married and single; male–female, male–male, and female–female sexual activity; and sexual behaviors of adolescents, adults, and older people.

Many of these surveys have *something* to offer to our understanding of human sexuality, but some are more methodologically sound than others. None represents the U.S. population at large, however. Most people consider their sexuality to be among the most intimate, *private* aspects of their lives. People who willingly agree to be polled about their political preferences may resist participation in surveys concerning their sexual behavior. As a result, it is difficult, if not impossible, for researchers to recruit a truly representative sample of the population. Bear in mind, then, that survey results provide, at best, an approximation of the sexual attitudes, beliefs, and behaviors of the U.S. population.

Let us review the sampling techniques used in some of the major studies of human sexuality. Throughout the book we shall reconsider the findings of these surveys, especially those that shed light on the changes that have occurred between Kinsey's surveys and the present. But for now let us focus on the methods used by Kinsey and others to survey sexual behavior patterns in the United States.

Survey A detailed study of a sample obtained by means such as interviews and questionnaires.

THE KINSEY REPORTS Kinsey and his colleagues (1948, 1953) interviewed 5,300 males and 5,940 females in the United States between 1938 and 1949. They asked a wide array of questions on various types of sexual experiences, including masturbation, oral sex, and coitus that occurred before, during, and outside of marriage. The survey was a reasonable method for obtaining these data, because Kinsey was interested in studying the frequencies of various sexual behaviors, rather than their underlying causes. For obvious reasons, Kinsey could not use more direct observational methods, such as sending his researchers to peer through bedroom windows. Kinsey chose not to try to obtain a random sample. He believed that a high refusal rate would wreck the chances of accurately representing the general population. Instead, he adopted a *group sampling* approach. He recruited study participants from the organizations and community groups to which they belonged, such as college fraternities and sororities. He contacted representatives of groups in diverse communities and tried to persuade them to secure the cooperation of fellow group members. Kinsey hoped that if he showed these individuals that they would not be subjected to embarrassment or discomfort, they would persuade other members to participate. Kinsey understood that the groups he solicited were not necessarily representative of the general population. He believed, however, that his sampling approach was dictated by practical constraints. Even so, he made an attempt to sample as broadly as possible from the groups he solicited. In some cases, he obtained full participation. In other cases, he obtained a large enough proportion of the group membership to help ensure representativeness, at least of the group.

Still, Kinsey's samples did not represent the general population. People of color, people in rural areas, older people, the poor, and Catholics and Jews were all underrepresented in his samples. Statisticians who have reviewed Kinsey's methods have concluded that there were systematic biases in his sampling methods but that it would have been impossible to obtain a true probability sample from the general population (see, for example, Cochran et al., 1953). There is thus no way of knowing whether Kinsey's results accurately mirrored the U.S. population at the time. In Chapter 10 we shall see that his estimate that 37% of the male population had reached orgasm at least once through male–male sexual activity was probably too high. But the *relationships* Kinsey uncovered, such as the positive link between level of education and participation in oral sex, may be more generalizable.

To his credit, Kinsey took measures to encourage candor in the people he interviewed. For instance, study participants were assured of the confidentiality of their records. Kinsey's interviewers were also trained to conduct the interviews in an objective and matter-of-fact style. To reduce the tendency to slant responses in a socially desirable direction, participants were reassured that the interviewers were not passing judgment on them. Interviewers were trained not to show emotional reactions that the people they interviewed could interpret as signs of disapproval (they maintained a "calm and steady eye" and a constant tone of voice).

Kinsey also checked the **reliability** of his data by evaluating the consistency of the responses given by several hundred interviewees who were reexamined after at least 18 months. Their reports of the **incidence** of sexual activities (for example, whether or not they had ever engaged in premarital or extramarital coitus) were highly reliable. That is, participants tended to give the same answers on both occasions. But reports of the **frequency** of sexual activities (such as the number of times one has masturbated to orgasm and the frequency of coitus in marriage) were less consistent. People do tend to find it more difficult to compute the frequencies of their activities than to answer whether they have ever engaged in them.

Kinsey recognized that consistency of responses across time—or *retakes*, as he called them—did not guarantee their **validity.** That is, the retakes did not show whether the reported behaviors had some basis in fact. He could not validate self-reports directly, as one might validate reports that one is drug-free by means of urine analysis. He could not send his investigators to peer through bedroom windows, so he had to use indirect means to

Reliability The consistency or accuracy of a measure.

Incidence A measure of the occurrence or the degree of occurrence of an event.

Frequency The number of times an action is repeated within a given period.

Validity With respect to tests, the degree to which a particular test measures the constructs or traits that it purports to measure.

a CLOSER look

With a Horse?

Consider this account by Sarah Boxer (2000):

Wardell B. Pomeroy, one of Alfred Kinsey's associates in the 1940s, was interviewing a man about his first ejaculation. He asked, "When?" The man answered, "Fourteen." Pomeroy then asked, "How?" and was surprised to hear: "With a horse."

In his new biography, *Sex the Measure of All Things: A Life of Alfred C. Kinsey*, Jonathan Gathorne-Hardy records what happened next. Pomeroy asked the man, "How often were you having intercourse with animals at 14?"

The man looked confused and said, "Well, yes, it is true I had intercourse with a pony at 14." Pomeroy, it turned out, had misheard the man's previous answer. It was "with whores," not "with a horse." So the man was stunned that Pomeroy had had the insight to ask him the horse question out of the blue.

validate the data. One indirect measure was comparison of the reports of husbands and wives—for example, with respect to the *incidence* of oral–genital sex or the *frequency* of intercourse. There was a remarkable consistency in the reports of 706 pairs of spouses; this lends support to the view that their self-reports were accurate. (It is possible, but highly unlikely, that spouses colluded to misrepresent their behavior.)

THE NHSLS STUDY The National Health and Social Life Survey was intended to provide general information about sexual behavior in the United States and also specific information that might be used to predict and prevent the spread of AIDS. It was conducted by Edward O. Laumann of the University of Chicago and three colleagues—John H. Gagnon, Robert T. Michael, and Stuart Michaels—in the 1990s and published as *The Social Organization of Sexuality: Sexual Practices in the United States* in 1994. A companion volume authored by Michael, Gagnon, Laumann, and Gina Kolata (a *New York Times* science reporter)—was also published: *Sex in America: A Definitive Survey. Sex in America* is a bit less technical in presentation but offers some interesting data not found in the other version. The NHSLS study was originally to be supported by government funds, but Republican Senator Jesse Helms of North Carolina blocked federal financing on the grounds that it was inappropriate for the government to be supporting sex research (Bronner, 1998). The research team therefore obtained private funding but had to cut back the scope of the project.

The sample included 3,432 people. Of this number, 3,159 were drawn from English-speaking adults living in households (not dormitories, prisons, and so forth), ages 18 to 59. The other 273 were obtained by purposely oversampling African American and Latino and Latina American households, so that more information could be obtained about these ethnic groups. Although the sample probably represents the overall U.S. population quite well (or at least those of ages 18 to 59), there may be too few Asian Americans, Native Americans, and Jews to offer much information about these groups.

The researchers identified samples of households in geographic areas—by addresses, not names. They sent a letter to each household, describing the purpose and methods of the study, and an interviewer visited each household a week later. The people targeted were assured that the purposes of the study were important and that the identities of participants would be kept confidential. Incentives of up to $100 were offered for cooperating. A high completion rate of close to 80% was obtained in this way. All in all, the NHSLS study could be the only one since Kinsey's day that offers a reasonably accurate picture of the sexual practices of the general population of the United States.

THE PLAYBOY FOUNDATION SURVEY A survey of sexual practices in the 1970s was commissioned by the Playboy Foundation. The results were reported by Morton Hunt in his

1974 book *Sexual Behavior in the 1970's*. The Playboy survey, or Hunt survey, as it is sometimes called, sought to examine the changes that occurred in sexual behavior in the United States between Kinsey's time and the early 1970s.

The Playboy sample was drawn randomly from phone book listings in 24 U.S. cities. People were asked to participate in small-group discussions focusing on trends in sexual practices in the United States. They were not told that they would also be completing a questionnaire, and this omission has raised ethical concerns. That is, the participants were not fully informed about their role in the study when they agreed to participate. (Interestingly, though, none of 2,026 people who participated in the group meetings refused to complete the questionnaires.) The phone sampling method was supplemented by an additional sample of young people, who were likely to have been underrepresented in telephone directories. But even with these additional people, rural residents and inmates of prisons and mental hospitals remained underrepresented.

Because the Playboy Foundation did not sample rural people and selected only 24 urban centers, the study's participants do not represent the general U.S. population at the time. Hunt argues, perhaps with some justification, that the final sample of 2,026 is stratified properly as to the ages and races of urban residents across a diverse sample of U.S. cities. The major flaw in the method, however, is that 80% of the people contacted refused to participate. The final 2,026 were clearly willing to volunteer to participate in discussion groups and then to complete a sex questionnaire. We suspect that they were more open and frank about sexual issues than the population at large.

THE JANUS REPORT Another nationwide survey of sexual behavior in the United States was conducted from 1988 to 1992 by a husband-and-wife team, Samuel and Cynthia Janus (1993). *The Janus Report on Sexual Behavior* was based on a survey of 2,765 people of age 18 years or older (1,347 men and 1,418 women) who anonymously completed written questionnaires assessing a wide range of sexual behaviors and attitudes. In-depth interviews were also conducted with a subset of the larger sample.

The Janus sample, like the Kinsey sample before it, was not randomly selected from the general population. Rather, the Januses assembled a team of researchers from every region of the contiguous 48 states, who then made contact with groups of potential respondents. In all, the survey team distributed 4,550 questionnaires in various sites across the country. Satisfactorily completed questionnaires (those with few missing responses) were returned from 2,765 respondents, a return rate of only 61%. The surveyists attempted to construct a sample that represented a fairly typical cross section of the U.S. population with respect to characteristics such as age, gender, income, and educational background. Although the findings may offer some insights into contemporary sexuality in American society, we have no assurance that the sample truly represented the general population. People who respond to sex surveys may not only be more open about discussing their sexuality than the general population. They may also hold more permissive views about sex.

THE MAGAZINE SURVEYS Major readership surveys have also been conducted by popular magazines, such as *Psychology Today*, *Redbook*, *Ladies' Home Journal*, *McCall's*, *Cosmopolitan*, and even *Consumer Reports*. Although these surveys all offer some useful information and may be commended for obtaining large samples (ranging from 20,000 to 106,000!), their sampling techniques are inherently unscientific and biased. Each sample represents, at best, the readers of the magazine in which the questionnaire appeared. Moreover, we learn only about readers who volunteered to respond to the questionnaires. These volunteers probably differed in important ways from the majority of readers who failed to respond. Finally, readers of these magazines are more affluent than the public at large, and readers of *Cosmopolitan*, *Psychology Today*, and even *Redbook* tend to be more liberal. The samples may therefore represent only those readers who were willing to complete and mail in the surveys.

A nonacademic researcher, Shere Hite, published several popular—perhaps we should say notorious—books on sexual behavior in men and women. Her samples for her 1976 book on female sexuality, *The Hite Report*, and her 1981 book on male sexuality, *The Hite Report on Male Sexuality*, were gleaned from people who completed questionnaires received in direct mailings, printed in sexually explicit magazines like *Penthouse*, and made available through other outlets, including some churches. Her final samples of some 3,000 women, in the 1976 report, and 7,000 men, in the 1981 book, may seem large. However, they actually represent small return rates of 3% and 6%, respectively. For these reasons, among others, the "Hite reports" cannot be considered scientific studies.

Laumann and his colleagues (1994) are particularly harsh in their judgment of such surveys. They write that "such studies, in sum, produce junk statistics of no value whatsoever in making valid and reliable population projections" (p. 45).

SURVEYS OF SPECIFIC POPULATIONS The Kinsey and NHSLS studies were broad-based. They queried men and women from different localities, socioeconomic strata, and age groups. Magazine surveys tend to recruit people more narrowly, as defined by reader characteristics. In some cases, however, researchers have focused their efforts on particular populations, such as adolescents, older people, people from particular racial/ethnic groups, and gay men and lesbians.

In recent years, large-scale studies concerning sexual practices have been conducted in the United States and other countries to acquire information that might prove useful in the fight against AIDS (Adler, 1993). In the United States, for example, researchers from the Battelle Memorial Institute of Seattle interviewed a nationally representative sample of 3,321 men between the ages of 20 and 39 to determine the prevalence of unsafe sexual practices among young adult men (Billy et al., 1993; Tanfer et al., 1993).

In later chapters we shall discuss some of the findings from these various surveys. In this chapter's World of Diversity feature, we consider some survey findings concerning Native Americans, African Americans, and people in China.

THE KINSEY INSTITUTE REPORTS ON GAY PEOPLE: 1978 AND 1981 These reports by the Indiana University Institute for Sex Research, also called the Kinsey Institute, were based on a sample of 979 gay people from the San Francisco area and a reference group of 477 people matched for age, race, and educational and occupational achievements (Bell & Weinberg, 1978; Bell et al., 1981). In their 1978 book, *Homosexualities*, researchers Alan Bell and Martin Weinberg recognized that their findings could not necessarily be extended to gay people who lived in other cities or other sections of the country. Indeed, they acknowledged that their sample might not even represent gay people in San Francisco. It consisted of people who had "come out of the closet" to join gay-rights organizations, who attended gay bars and baths, and so forth. Nevertheless, these reports have provided wide-ranging information on parent–child relationships and sexual orientation and on the diversity of lifestyles among gay people (see Chapter 10).

LIMITATIONS OF THE SURVEY METHOD The Kinsey studies may be criticized because the interviewers were all men. Women respondents might have felt more free to open up to female interviewers. Gender of the interviewer can affect a respondent's willingness to disclose sensitive material. A recent investigation in the Asian country of Nepal, for example, showed that male interviewers generally elicited an underreporting of some sensitive aspects of sexual activity (Axinn, 1991). Problems in obtaining reliable estimates may also occur when interviewers and respondents are of different racial or socioeconomic backgrounds.

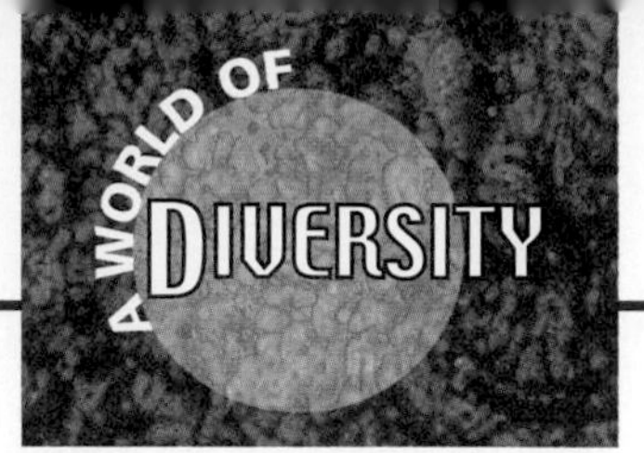

Surveys of the Sexual Behavior of Diverse Populations

Questions about the sexual behavior of Native Americans and African Americans abound. Kinsey did not survey Native Americans. Nor did Hunt, Wyatt, or other leading researchers. The 1982 government survey known as the National Survey of Family Growth (NSFG), which surveyed the reproductive behavior of more than 7,500 women nationwide, included only 83 Native American women in the sample, far too few to be a meaningful indicator of general patterns (Warren et al., 1990). The number of Native Americans in the NHSLS study was in the 40s (Laumann et al., 1994).

Kinsey obtained some data on the sexual behavior of African Americans but did not report it in his surveys because African Americans were clearly underrepresented in his samples. However, more recent studies by Gail Wyatt and the NHSLS group have yielded some useful information.

It is interesting to compare the results of these surveys with the results of similar surveys in different countries and cultures, such as China. China was never a particularly prudish country. China's tradition of concubines and prostitution predates the birth of Jesus. Nevertheless, after the Communists came into power in 1949, China was transformed into one of the world's more puritanical societies. Yet according to a survey by Shanghai-based sociologist Liu Dalin, since China's opening to the outside world in the 1970s, the times have been a-changin'. A little.

Let's briefly look at surveys of the sexual behavior of Native Americans, African Americans, and Chinese people.

The Billings Indian Health Service Survey

In one of the first efforts to survey the reproductive and health behavior of Native Americans, the Billings (Montana) Indian Health Service office surveyed 232 Native American women in the Billings area, ages 15 to 49. Half of the women lived on the Blackfoot Reservation. The other half lived off the reservation in nearby Great Falls (Warren et al., 1990). Surveys were conducted in 1987. The results were compared to data on European American and African American women obtained from the 1982 NSFG survey. Nearly 40% of the Native American households did not have telephones. Therefore, a face-to-face interview format was adopted. Virtually all of the women interviewed on the Blackfoot Reservation were Blackfeet, but among interviewees living in Great Falls, a variety of tribes were represented, including Chippewa-Cree, Little Shell, Assiniboine, and Chippewa, as well as Blackfeet.

Slightly more than half of the Native American women and African American women reported having engaged in sexual intercourse by the age of 17, as compared to 28% of the European

Native American Women. Research shows that the reproductive behavior of Native American women in urban areas is more similar to that of other U.S. minorities than to that of Native Americans living on reservations.

American women. Native American women bore an average of 3.4 to 4.0 children. This figure exceeded the average among European American women (2.7 children) but was similar to the average among African Americans (3.4). The percentage of last pregnancies that were unplanned among off-reservation Native American women (64%) was similar to that for African American women (61%) but higher than that for European American women (44.6%). Native American women living on the reservation showed an intermediate level (51.2%) of unplanned last pregnancies. Reservation women were more likely than European American women to use contraception (79% versus 69%). The percentage of off-reservation Native American women who used contraception was lower, 58%—a figure that was similar to the level of contraceptive use among African American women (60%). Although female sterilization was the most commonly used contraceptive method among all three groups, its incidence was higher among Native Americans (more than 30%) than among European Americans (16%) or African Americans (21%).

Overall, the survey underscores the similarities in reproductive patterns between Native American women living off the reservation and African American women in the larger community. Both groups tended to initiate intercourse at an early age, to have a high number of unplanned pregnancies, and to make only moderate use of contraception. The reproductive patterns they shared may reflect the plight of economically disadvantaged groups in the United States. Native American women living on the reservation showed reproductive patterns different from those of African Americans and European Americans, and even from those of the urbanized, off-reservation Native American sample. Perhaps their reproductive behavior adheres more closely to traditional cultural norms than does that of Native American women living off the reservation, who may be attempting to assimilate into the larger community.

The Wyatt Survey on African American and European American Women in Los Angeles

In the 1980s, UCLA researcher Gail Wyatt and her colleagues (Wyatt, 1985, 1989; Wyatt et al., 1988a, 1988b) examined the sexual behavior of a sample of 122 European American and 126 African American women in Los Angeles County. The women ranged in age from 18 to 36 years. Women in the study were sampled randomly from telephone listings. People who agreed to participate were selected to balance the sample with respect to demographic characteristics such as age, education, number of children, and marital status. One in three prospective participants refused to cooperate. The participants were interviewed in Kinsey-style, face-to-face interviews that lasted three to eight hours.

Because Wyatt and Kinsey used similar means to obtain data, direct comparisons can be made between their studies. Wyatt used statistical adjustments to control for differences in the sociodemographic characteristics (age, education, social class) between her samples and Kinsey's to make comparisons between the two possible.

Wyatt's work is important for several reasons. It addresses changes in sexual behavior that have taken place in U.S. society since Kinsey's day. It focuses on social issues that have become prominent, such as sexual abuse in childhood. It also is one of the few detailed studies of sexual behavior among African American women.

One of the striking differences between Kinsey's data and Wyatt's was that women in her 1980s sample—African American and European American—engaged in intercourse for the first time at earlier ages than the women in Kinsey's sample. Kinsey reported that by the age of 20, about one in five women had engaged in premarital coitus. By contrast, Wyatt reported that 98% of the people in her study (African American and European American) had experienced premarital intercourse by that age (Wyatt, 1989). When social-class differences were taken into consideration, the ages of first intercourse for African American and European American women in Wyatt's sample were quite similar.

Wyatt's research, of course, was limited to Los Angeles and may not represent the general U.S. population. Kinsey's sample was also geographically skewed. Kinsey overrepresented the northeastern United States, although he included respondents from other regions.

Let us also note the refusal rate in Wyatt's study. Wyatt was more successful in obtaining cooperation than the Playboy survey takers (67% of the people contacted agreed to participate, versus 20% for Playboy). Nonetheless, Wyatt's 33% refusal rate may have compromised representativeness. Also, to control for demographic differences between European American and African American women, Wyatt limited the pool of cooperating individuals to demographically comparable women. Her sample of African American women did match the demographic characteristics of the larger population of African American women in Los Angeles County. However, her European American sample did not match the population characteristics for European American women in the county. It contained a greater proportion of European American women from lower-income families. These kinds of trade-offs can occur when researchers attempt to control for demographic differences.

(continued)

Surveys of the Sexual Behavior of Diverse Populations *(continued)*

The Liu Report: Sexual Behavior in China

Liu interviewed 23,000 Chinese people over a period of 18 months (Southerland, 1990). China's "Kinsey Report" was the first nationwide sex survey ever conducted in China. It polled twice as many people as Kinsey and his colleagues did. Participants, including students, professionals, peasants, and even convicted sex offenders, were drawn from 3 major cities and 12 provinces. The researchers encountered problems in lack of funds, illiteracy, and reluctance to reveal intimate sexual information.

Here are some of Liu's findings:

- Chinese youth today are reaching sexual maturity about a year earlier than their grandparents did, partly because of improved nutrition.
- About 50% of the young people in cities and on farms reported that they engaged in premarital intercourse.
- About 14% of the women living in large cities said they engaged in extramarital sex. Overall, 69% of the respondents approved of such affairs. This is a larger approval rate than is found in the United States, where the majority of respondents to surveys disapprove of affairs (e.g., Blumstein & Schwartz, 1990; Hunt, 1974).
- Many Chinese couples use very little foreplay; 44% of urban wives and 37% of rural wives reported experiencing at least some pain during intercourse as a result of insufficient vaginal lubrication.
- Women are more likely than men to initiate divorces. Three out of five divorces are sought by women.

So go some survey findings about sexual attitudes and practices around the world.

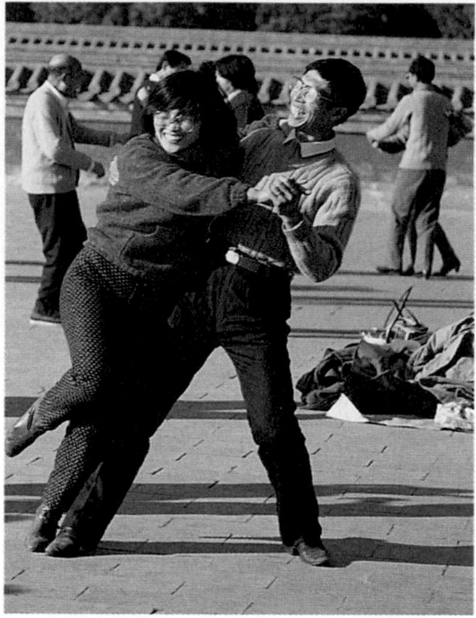

Sex in China. Historically speaking, China was not a prudish country. China had a tradition of concubines and prostitution that predated the birth of Jesus. But after the Communist Revolution in 1949, China was transformed into a puritanical society. However, a survey by sociologist Liu Dalin finds that China is not as controlled as had been thought. For example, about half of China's young people engage in premarital intercourse, and it would appear that more than half approve of extramarital affairs (though not nearly that many play around themselves).

Truth or Fiction? REVISITED

It does appear from the results of the Liu survey that the Chinese are more likely to approve of extramarital affairs than are people in the United States. ■

Yet another limitation of all surveys is that they require self-reports of respondents' behaviors. Self-reports are subject to inaccuracies or biases because of factors such as faulty memories of sexual behavior; tendencies to distort or conceal information because of embarrassment, shame, or guilt; or attempts to present a socially favorable image of oneself. Survey data may also be drawn from haphazard or nonrepresentative samples and thus may not represent the target population. Let us consider several weaknesses of the survey method.

VOLUNTEER BIAS Many people refuse to participate in surveys. Samples are thus biased by the inclusion of large numbers of volunteers. Volunteers are in general willing to take the time to participate. In the case of sex surveys, they also tend to be more sexually permissive and liberal-minded than nonvolunteers. Accordingly, the results of a survey based on a volunteer sample thus may not accurately reflect the population at large.

FAULTY ESTIMATION Respondents may recall their behavior inaccurately or may purposely misrepresent it. People may not recall the age at which they first engaged in petting or masturbated to orgasm. People may have difficulty remembering or calculating the frequencies of certain behaviors, such as the weekly frequency of marital intercourse

("Well, let's see. This week I think it was four times, but last week only two times, and I can't remember the week before that"). Kinsey and Hunt speculated that people who desire more frequent sex tend to underestimate the frequency of marital coitus, whereas people who want less frequent sex tend to overestimate it.

SOCIAL-DESIRABILITY RESPONSE BIAS Even people who consent to participate in surveys of sexual behavior may feel pressured to answer questions in the direction of **social desirability.** Some respondents, that is, try to ingratiate themselves with their interviewers by offering what they believe to be socially desirable answers. Although some respondents may readily divulge information concerning the frequency of marital coitus, they may deny experiences involving prohibited activities such as child molestation, voyeurism, or coerced sexual activity. (People who engage in these proscribed activities may also be more likely to decline to participate in sex surveys.)

Some people may not divulge sensitive information for fear of the interviewer's disapproval. Others may fear criminal prosecution. Even though interviewers may insist that they are nonjudgmental and that study participants will remain anonymous, respondents may fear that their identities will be uncovered someday.

EXAGGERATION For some respondents, the "socially desirable" response is exaggeration. In our culture, men may tend to exaggerate their sexual exploits. Women may tend to play them down (Harris & Johnson, 2000; Whitten, 2001). Males and "liberated" females may fear that the interviewer will think less of them for reporting only infrequent sexual contacts or for sounding too "straight." Some respondents falsify their attitudes and exaggerate the bizarreness of their behavior, perhaps to draw attention to themselves, perhaps to foul up the results.

DENIAL Another source of bias in sex survey research is denial. People may deny, even to themselves, sexual feelings or experiences that might elicit anxiety if they were acknowledged, such as male–male or female–female fantasies or feelings.

DIFFERENCES IN MEANINGS OF TERMS In 1998, when the White House sex scandal broke, Bill Clinton asserted that he had never had "sexual relations" with a young intern there. The tabloids feasted on the fact that President Clinton did not consider "sexual relations" to include fellatio.

Survey respondents can respond only to the questions posed by interviewers or included in questionnaires. A word or phrase may mean different things to different people, however. One person, when asked whether he or she has engaged in French kissing, may think of deep, open-mouthed tongue kissing (the generally accepted meaning of the term). To another, the term may denote a prolonged but gentle kiss on the lips. To some people, the term *sexual satisfaction* may denote an intense orgasm. To others, it signifies the pleasure of sharing physical intimacy with a loved one. To the extent that people interpret the same terms differently, they may differ in their responses to sex surveys because of differences in semantics, not because of differences in behavior.

You might think that these problems would render the survey method useless as a means of learning about human sexuality. Actually, carefully conceived and executed surveys offer many insights into sexual attitudes and practices. And in many cases they are the only available means. Before Kinsey, we had little information about the sexual practices of people in our society.

The Naturalistic-Observation Method

In **naturalistic observation,** also called the *field study,* scientists directly observe the behavior of animals and humans where it happens. Anthropologists, for example, have lived among preliterate societies and reported on their social and sexual customs. Other disci-

Social desirability A response bias to a questionnaire or interview in which the person provides a socially acceptable response.

Naturalistic observation A method in which organisms are observed in their natural environments.

plines, too, have adopted methods of naturalistic observation in their research on human sexuality. Sociologists have observed the street life of prostitutes. Psychologists have observed patterns of nonverbal communication and body language between couples in dating situations.

Scientists take precautions to keep their naturalistic observations *unobtrusive.* They try not to influence the behavior of the individuals they study. Over the years, naturalistic observers have been placed in ethical dilemmas. They have allowed sick or injured animals to die, rather than intervene, when medical assistance could have saved them. They have allowed substance abuse and illicit sexual behavior to go unreported to authorities. The ethical trade-off is that unobtrusive observation may yield data that will benefit large numbers of people—the greatest good for the greatest number.

The Ethnographic-Observation Method

Anthropologists have lived among societies of people in the four corners of the earth in order to observe and study human diversity. Margaret Mead (1935) reported on the social and sexual customs of various peoples of New Guinea. Bronislaw Malinowski (1929) studied the Trobriand islanders, among other peoples. Ford and Beach's (1951) account of sexual practices around the world and in nonhuman species remains a classic study of cross-cultural and cross-species comparisons. **Ethnographic** research has provided us with data concerning sexual behaviors and customs that occur widely across cultures and those that are limited to one or few cultures. Ethnographers are trained to be keen observers, but direct observation has its limits in the study of sexual behavior. Sexual activities are most commonly performed away from the watchful eyes of others, especially from those of visitors from other cultures. Ethnographers may thus have to rely on methods such as personal interviewing to learn more about sexual customs.

The Participant-Observation Method

In **participant observation,** the investigators learn about people's behavior by directly interacting with them. Participant observation has been used in studies of male–male sexual behavior and mate swapping. In effect, participation has been the "price of admission" for observation.

Investigators of mate swapping, or **swinging,** have contacted swinging couples through newsletter ads and other sources and presented themselves as "baby swingers" (novices) seeking sexual relations (Bartel, 1970; Palson & Palson, 1972). Thus, some investigators may have deceived the people they studied. In some cases, however, as in Bartel's study, individuals were informed that the investigator was conducting research after he was admitted to the party. In some cases, researchers have engaged in coitus with study participants during "swinging parties," which certainly raises questions about how far one should go "for the sake of science."

It is true that some sex researchers have engaged in "swinging" with the people they studied. But this is a rare occurrence, and the ethics of this research method have been questioned.

Sampling biases are another concern. There is no way of knowing whether the swingers who were contacted via newspaper ads were typical of swingers in general. The sampling methods also limited the investigations to active swingers and excluded people who had once "swung" but no longer did so.

The Laboratory-Observation Method

In *Human Sexual Response* (1966), William Masters and Virginia Johnson were among the first to report direct laboratory observations of individuals and couples engaged in sexual acts. In all, 694 people (312 men and 382 women) participated in the research. The

Ethnography The branch of anthropology that deals descriptively with specific cultures, especially preliterate societies.

Participant observation A method in which observers interact with the people they study as they collect data.

Swinging Partner or mate swapping.

women ranged from 18 to 78 in age, the men from 21 to 80. There were 276 married couples, 106 single women, and 36 single men. The married couples engaged in intercourse and other forms of mutual stimulation, such as manual and oral stimulation of the genitals. The unmarried people participated in studies that did not require intercourse, such as measurement of female sexual arousal in response to the insertion of a penis-shaped probe, and male ejaculation during masturbation. Masters and Johnson performed similar laboratory observations of sexual response among gay people for their 1979 book *Homosexuality in Perspective.*

Direct laboratory observation of biological processes was not invented by Masters and Johnson. However, they were confronting a society that was still unprepared to speak openly of sex, let alone to observe people engaged in sexual activity in the laboratory. Masters and Johnson were accused of immorality, voyeurism, and an assortment of other evils. Nevertheless, their methods offered the first reliable set of data on what happens to the body during sexual response. Their instruments permitted them to measure directly vasocongestion (blood flow to the genitals), myotonia (muscle tension), and other physiological responses. Perhaps their most controversial device was a "coition machine." This was a transparent artificial penis outfitted with photographic equipment. It enabled them to study changes in women's internal sexual organs as the women became sexually aroused. From these studies, Masters and Johnson observed that it is useful to divide sexual response into four stages (their "sexual response cycle").

Truth or Fiction? **REVISITED**

It is true that Masters and Johnson created a transparent artificial penis ("coition machine") containing photographic equipment to study female sexual response. ■

Researchers have since developed more sophisticated physiological methods of measuring sexual arousal and response. Masters and Johnson's laboratory method is now used, with some variations and less controversy, in research centers across the country (Rosen & Beck, 1988).

LIMITATIONS OF OBSERVATIONAL RESEARCH One of the basic problems with naturalistic observation is the possibility that the behavior under study may be *reactive* to the measurement itself. This source of bias is referred to as the **observer effect.**

The ethnographer who studies a particular culture or subgroup within a culture may unwittingly alter the behavior of the members of the group by focusing attention on some facets of their behavior. Falling prey to social desirability, some people may "clean up their act" while the ethnographer is present. Other people may try to impress the ethnographer by acting in ways that are more aggressive or sexually provocative than usual. In either case, such people supply distorted or biased information. Ethnographers must corroborate self-reported information by using multiple sources. They must also consider whether their own behavior is more obtrusive than they think.

Observer bias can distort researchers' perceptions of the behaviors they observe. Observers who hold rigid sexual attitudes may be relatively unwilling to examine sexual activities that they consider offensive or objectionable. They may unwittingly (or intentionally) slant interviews in a way that presents a "sanitized" view of the behavior of those they study, or they may exaggerate or "sensationalize" certain sexual practices to conform to their preconceptions of the sexual behavior of the people they study.

The method of laboratory observation used by Masters and Johnson may be even more subject to distortion. Unlike animals, which naturalists may observe unobtrusively from afar, people who participate in laboratory observation know that they are being observed and that their responses are being measured. The problem of volunteer bias, troublesome for sex surveys, is even thornier in laboratory observation. How many of us would assent to performing sexual activities in the laboratory while we were connected to physiological monitoring equipment in full view of researchers? Some of the women observed by Masters and Johnson were patients of Dr. Masters who felt indebted to him and

Observer effect A distortion of individuals' behavior caused by their awareness that they are being observed.

agreed to participate. Many were able to persuade their husbands to participate as well. Some were medical students and graduate students who may have been motivated to earn extra money (participants were paid for their time) as well as by scientific curiosity.

Another methodological concern of the Masters and Johnson approach is that observing people engaged in sexual activities may in itself alter their responses. People may not respond publicly in the same way they would in private. Perhaps sexual response in the laboratory bears little relationship to sexual response in the bedroom. The physiological monitoring equipment may also alter the subjects' natural responses. Given these constraints, it is perhaps remarkable that the people studied by Masters and Johnson were able to become sexually aroused and reach orgasm.

THE CORRELATIONAL METHOD

What are the relationships between age and frequency of coitus among married couples? What is the connection between socioeconomic status and teenage pregnancy? In each case, two variables are being related to one another: age and frequency of coitus, and socioeconomic status and rate of teenage pregnancy. Correlational research describes the relationship between variables such as these.

A **correlation** is a statistical measure of the relationship between two variables. In correlational studies, two or more variables are related, or linked, to one another by statistical means. The strength and direction (positive or negative) of the relationship between any two variables are expressed with a statistic called a **correlation coefficient.**

Research has shown relationships (correlations) between marital satisfaction and a host of variables: communication skills, shared values, flexibility, frequency of social interactions with friends, and churchgoing, to name a few. Although such research may give us an idea of the factors associated with marital satisfaction, the experimenters have not manipulated the variables of interest. For this reason we cannot say which, if any, of the factors is causally related to marital happiness. Consider the relationship between churchgoing and marital happiness. Couples who attend church more frequently report higher rates of marital satisfaction (Wilson & Filsinger, 1986). It is possible that churchgoing stabilizes marriages. It is also possible that people who attend church regularly are more stable and more committed to marriage in the first place.

Truth or Fiction? **REVISITED**

It is true that regular churchgoers report higher levels of sexual satisfaction than others. We cannot say, however, that churchgoing is causally related to marital satisfaction.

Correlations may be *positive* or *negative*. Two variables are positively correlated if one increases as the other increases. Frequency of intercourse, for example, has been found to be positively correlated with sexual satisfaction (Blumstein & Schwartz, 1983). That is, married couples who engage in coitus more frequently report higher levels of sexual satisfaction. However, the experimenters did not manipulate the variables. Therefore, we cannot conclude that sexual satisfaction *causes* high coital frequency. It could be, instead, that frequent coitus contributes to greater sexual satisfaction. It is also possible that there is no causal relationship between the variables. Perhaps both coital frequency and sexual satisfaction are affected by other factors, such as communication ability, general marital satisfaction, general health, and so on (see Figure 2.1). Similarly, height and weight are positively correlated but do not cause each other. Other factors that we label the growth process contribute to both.

Correlation A statistical measure of the relationship between two variables.

Correlation coefficient A statistic that expresses the strength and direction (positive or negative) of the relationship between two variables.

Limitations of the Correlational Method

Correlation is not causation. Many variables are correlated but are not causally related to one another.

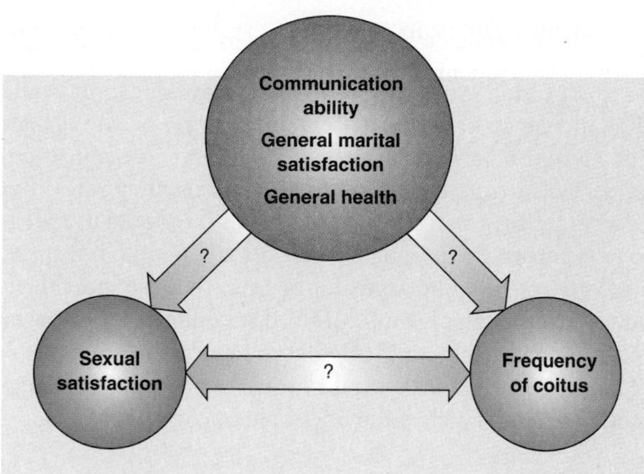

FIGURE 2.1

What Is the Relationship Between Frequency of Intercourse and Sexual Satisfaction? Married couples who engage in more frequent sexual relations report higher levels of sexual satisfaction, but why? Because researchers have not manipulated the variables, we cannot conclude that sexual satisfaction *causes* high coital frequency. Nor can we say that frequent coitus causes greater sexual satisfaction. Perhaps both variables are affected by other factors, such as communication ability, general marital satisfaction, and general health.

Although correlations do not show cause and effect, they can be used to make predictions. For example, we can predict that people who were sexually traumatized as children are more likely to encounter psychological problems later on. They find it more difficult to establish intimate relationships in adulthood. However, we cannot demonstrate that sexual trauma directly causes these problems. Although correlation does not provide causal knowledge, it permits us to anticipate the needs of victims of sexual assault for treatment services. Knowledge of correlational relationships may also lead to controlled experiments that more directly address questions of cause and effect.

THE EXPERIMENTAL METHOD

The best method (though not always a feasible method) for studying *cause-and-effect* relationships is the **experiment.** Experiments permit scientists to draw conclusions about cause-and-effect relationships because the experimenter is able to control or manipulate the factors or variables of interest directly and to observe their effects.

In an experiment on the effects of alcohol on sexual arousal, for example, a group of participants would receive an intervention, called a **treatment,** such as a dose of alcohol. (In other experiments, the intervention or treatment might involve the administration of a drug, exposure to violent pornography, or a program of sex education.) They would then be carefully observed to learn whether that treatment made a difference in their behavior—in this case, their sexual arousal.

Experiment A scientific method that seeks to confirm cause-and-effect relationships by manipulating independent variables and observing their effects on dependent variables.

Treatment In experiments, an intervention that is administered to participants (such as a test, a drug, or a sex education program) so that its effects can be observed.

Independent variable A condition in a scientific study that is manipulated so that its effects can be observed.

Dependent variables The measured results of an experiment, which are believed to be a function of the independent variables.

Aspects of the Experimental Method

INDEPENDENT AND DEPENDENT VARIABLES In an experiment, the variables (treatments) that are hypothesized to have a causal effect are manipulated or controlled by the researcher. Consider an experiment designed to determine whether alcohol stimulates sexual arousal. The design might involve giving one group of participants a certain dosage of alcohol and then measuring its effects. In such an experimental arrangement, the dosage of alcohol is considered an **independent variable,** whose presence and quantity are manipulated by the researchers. The measured results are called **dependent variables,** because changes in their values are believed to depend on the independent variable or variables. In this experiment, measures of sexual arousal would be the dependent variables. Dependent variables are outcomes; they are observed and measured by the researchers, but not manipulated. Sexual arousal might be measured by means such as physiological measurement (gauging the degree of penile erection in the male, for example) or self-report (asking participants to rate their sexual arousal on a rating scale).

In a study of the effects of sex education on teenage pregnancy, sex education would be the independent variable. The incidence of teenage pregnancy would be the dependent variable. Researchers would administer the experimental treatment (sex education) and track the participants for a period of time to determine their pregnancy rates. But how would experimenters know whether the treatment had made a difference in the pregnancy rate? One way would be to compare the pregnancy rate among people in experiments to rates found in public records. It could be argued, however, that the individuals in the experiment differed in other ways (besides having received sex education) from the people whose records were kept in the public documents, that the records were made at a different time, and so forth. Hence, experimenters would prefer to compare the pregnancy rates of people in experimental groups with those of people in control groups.

EXPERIMENTAL AND CONTROL GROUPS Well-designed experiments randomly assign people or animal-study participants to experimental and control groups. Participants in **experimental groups** receive the treatment. Participants in **control groups** do not. Every effort is made to hold all other conditions constant for both groups. By using random assignment and holding other conditions constant, researchers can be reasonably confident that the independent variable (treatment), and not extraneous factors (such as the temperature of the room in which the treatment was administered or differences between the participants in the experimental and control groups), brought about the results.

WHY IS RANDOM ASSIGNMENT IMPORTANT? Why do experimenters assign individuals at random to experimental and control groups whenever possible? Consider a study conducted to determine the effects of alcohol on sexual arousal in response to sexually explicit material, such as "XXX-rated" films. If we permitted study participants to choose whether they would drink alcohol, we might not know whether it was the alcohol itself that accounted for the results. Some other factor, called a **selection factor,** might discriminate between people who would and those who would not choose to drink alcohol. Perhaps people who chose to drink might also have more permissive attitudes toward sexually explicit material. Their permissiveness, rather than the alcohol, could affect their sexual responsiveness to these stimuli. If this were the case, experimental outcomes might reflect the effects of the selection factor rather than the alcohol.

EXPERIMENTATION WITH ANIMALS Researchers frequently undertake studies with other animal species that would be impractical or unethical with people. For example, it may be hypothesized that prenatal sexual hormones "feminize" or "masculinize" the brain to give rise to gender-typed behaviors or predispose the individual to a heterosexual, gay male, or lesbian sexual orientation (see Chapters 6 and 10). The ideal method for examining these hypotheses would be to alter the balance of sex hormones systematically at various stages of fetal development and then monitor the individuals' behavior in subsequent years. However, neither parents nor researchers would be willing to risk the well-being of children through such an experiment. Therefore, studies of this kind have been conducted on animals to try to shed some light on similar processes in people.

To the extent that we share physiological processes with animals, experiments with other species may inform us about ourselves. Because of their physiological similarities to people, rats and monkeys are commonly used in laboratory experiments. Unfortunately—or fortunately?—these animals are more similar to humans biologically than psychologically. Thus, we may make some generalizations from them to people concerning the effects of drugs and other biochemical treatments. We cannot, however, assume that their responses provide much of a clue to higher cognitive processes, such as thinking and reasoning. Nor can we claim that *pair bonding* between animals is equivalent to human mar-

Experimental group A group of study participants who receive a treatment.

Control group A group of study participants who do not receive the experimental treatment. However, other conditions are held comparable to those of individuals in the experimental group.

Selection factor A bias that may operate in research when people are allowed to determine whether they will receive a treatment.

riage or that (apparently) affectionate displays between animals are the equivalent of romantic love between people.

LIMITATIONS OF THE EXPERIMENTAL METHOD Although scientists agree that the experimental method provides the strongest evidence of cause-and-effect relationships, experimenters cannot manipulate many variables of interest directly. We may suspect that sexual trauma in the form of rape or sexual assault is a causal factor in the development of psychological or emotional problems (see Chapter 18). However, we would not expose people to sexual trauma in order to observe its effects. In the absence of a direct manipulation of the presumed causal factor (rape), we cannot be sure that adjustment problems in rape victims are directly attributable to rape itself, especially when these problems occur years later. Other factors may be involved, such as insensitive treatment of rape victims by their families or by representatives of the criminal justice system.

Nor can we conduct experiments to determine the effects of cohabitation on college students. We cannot assign an experimental group to cohabitation and a control group to separate living quarters. All we can do is compare groups of students who have chosen to cohabit to groups who have not. Our research may thus inform us that cohabitors have certain problems (for example, jealousy), but it cannot show that these problems are caused by cohabiting (see Chapter 14). Cohabitors and noncohabitors may differ on other factors that give rise to these differences in adjustment.

Similarly, we cannot conduct experiments to determine the effects of pornography on children and adolescents. Societal prohibitions and ethical standards preclude experimenters from exposing children or adolescents to erotic materials. Researchers must use other approaches, such as the correlational method, to study variables that cannot be manipulated.

ETHICS IN SEX RESEARCH

Sex researchers are required to protect the people being studied. This means that people will not be subjected to physical or psychological harm and will participate of their own free will.

In colleges, universities, hospitals, and research institutions, ethics review committees help researchers weigh the potential harm of administering the independent variables and review proposed studies in light of ethical guidelines. If the committee finds fault with a proposal, it may advise the researcher how to modify the research design to comply with ethical standards and will withhold approval until the proposal has been so modified.

Pain and Stress

Individuals may be harmed if they are exposed to pain or placed in stressful situations. For this reason, many potentially informative or useful studies have not been done. Researchers have not exposed children to erotic materials in order to determine the effects. Researchers have not exposed human fetuses to male or female sex hormones to learn whether they create fundamental predispositions toward tomboyishness, gay male and lesbian sexual orientations, and other variables of interest.

Sex researchers *have* studied children born to mothers who were given certain sex hormones to help maintain their pregnancies (see Chapter 6). But these treatments were standard medical procedures that were undertaken to help the women and their fetuses. The availability of these cases has made it possible for scientists to examine the effects of prenatal exposure to these hormones on later gender-role behavior and sexual behavior. But the investigators did not administer these treatments to the pregnant women for purposes of research.

When considering a study that poses a risk of harm, ethics review committees weigh the potential harm in light of the potential benefits to the participants, to science, and to society in general. Ethical standards require that research be conducted only when the expected benefits of the research outweigh the anticipated risks, when the experimenter attempts to minimize risks, and when the experimenter ensures that the pain and stress the participants experience is neither excessive nor lasting. Before the study begins, moreover, participants are given general information about the stress or pain they may encounter. Thus, they can decide whether they wish to participate and can prepare themselves.

Confidentiality

Sex researchers must keep the identities and responses of participants confidential to protect them from possible harm or embarrassment. Some participants may confide legally incriminating information to researchers, such as reports of illegal sexual activities. Some people might suffer harm or embarrassment if it were even known that they had participated in sex research.

Researchers can do many things to ensure the confidentiality of participants. They can make questionnaires anonymous. Interviewers may not be given the identities of interviewees. In reports of research, enough information about participants' backgrounds can be given to make the studies useful (size of city of origin, region of country, religion, age group, race, educational level, and so on) without divulging their identities. Once the need for follow-up has passed and the results have been fully analyzed, the names and addresses of participants and their records can be destroyed. If names and addresses are maintained, researchers can code them, keep them under lock and key, and spread the keys among several people, so that no one person has enough information to divulge an individual's sexual behavior.

Informed Consent

The principle of **informed consent** requires that people freely agree to participate after being given enough information about the procedures and purposes of the research, and its risks and benefits, to make an informed decision. Information must be presented in a manner that potential participants can understand. People may not be coerced into participating. Once the study has begun, they must be free to withdraw at any time without penalty. In some cases, people may be told that some information about the study must be withheld until it is completed. This is especially so when deception is involved.

The Use of Deception

Ethical conflicts may emerge when experiments require that participants not know all about their purposes and methods. For example, in experiments on the effects of violent pornography on aggression against women, participants may be misled into believing that they are administering electric shocks to women (who are actually confederates of the experimenter), even though no shocks are actually delivered. The experimenter seeks to determine participants' willingness to hurt women following exposure to aggressive erotic films. Such studies could not be carried out if participants knew that no shocks would actually be delivered. In studies that involve deception, researchers must demonstrate that the effects of the treatment are not seriously harmful or prolonged and that the benefits of the information obtained outweigh the risks.

Research is the backbone of human sexuality as a science. This textbook focuses on scientific findings that can illuminate our understanding of sexuality, enhance sexual experience, help prevent and treat sexually transmitted diseases, and build more rewarding relationships.

Informed consent The term used by researchers to indicate that people have agreed to participate in research after receiving information about the purposes and nature of the study and about its potential risks and benefits.

Human Sexuality ONLINE //

Web Sites for Use in Research on Human Sexuality

The Web sites listed for Chapter 2 are not highly technical sites that explain the minute details of research methods. Rather, they are key sources of information that you can use in your formal research (academically related) and informal research (checking into what's going on in the United States and the world from day to day). The government sites contain enormous data bases that you may be able to use in class projects. The newspaper and other public-information sites often report the results of important research in lay language. You can follow up by going to the Web sites of the journals and other sources mentioned in those articles.

The Centers for Disease Control and Prevention: The CDC is located in Atlanta, Georgia, and is an agency of the Department of Health and Human Services. The mission of the CDC is to promote health and quality of life by preventing and controlling disease, injury, and disability. Many of the issues covered by the CDC are sexual, as in its list of topics from A through Z. The Web site is an encyclopedia of research information about sexual behavior and related health issues in the United States.
www.cdc.gov/

The U.S. Census Bureau: The mission of the Bureau of the Census is to be "the preeminent collector and provider of timely, relevant, and quality data about the people and economy of the United States." The Bureau aims to provide timeliness, relevancy, quality, and cost for the research data it collects and disseminates. Exploration of this Web site will reveal that the Bureau does much more than simply tally numbers. It also analyses data and projects trends, many of them related to issues such as family composition—a key topic in human sexuality. An incredible data base.
www.census.gov/

The New York Times: The Web sites of *The New York Times* and other important newspapers across the country (such as, *The Washington Post, The Los Angeles Times*) and the world are found online. You can read the day's news, editorials, and so on without having to find (or purchase) the paper. You can also go to the specialized parts of the paper that are most relevant to human sexuality. When you reach *The New York Times* Web site, for example, you can click on Science/Health in the left-hand column. You can also search the archives of the various newspapers using keywords that may be relevant to class projects, such as *AIDS* or *prostitution.*
www.nytimes.com/

The Cable News Network: CNN is the original television provider of around-the-clock news. Other TV networks (e.g., ABC, MSNBC) also have Web sites with news. These Web sites, like the newspaper Web sites, include stories related to human sexuality, and you can usually click on Health to find them. The Web sites also typically have archives and search functions.
www.cnn.com/

Time magazine: Like newspapers and TV network Web sites, the Web sites of the news magazines (*Newsweek, U.S. News & World Report,* etc.) have areas such as Health, archives, and search functions. Newspapers, TV network news services, and news magazines frequently report the result of surveys on issues relevant to human sexuality.
www.time.com/time/

The Gallup Organization: The Gallup Organization of Princeton, NJ, conducts the Gallup polls. Click on "Gallup Poll" for current releases or for survey results topics A through Z.
www.gallup.com/

summing up

A SCIENTIFIC APPROACH TO HUMAN SEXUALITY

Scientists insist that assumptions about sexual behavior be supported by evidence. Evidence is based on careful observations in the laboratory or in the field.

■ The Scientific Method

The scientific method is a systematic way of gathering scientific evidence and testing assumptions through empirical research. It entails formulating a research question, framing a hypothesis, testing the hypothesis, and drawing conclusions about the hypothesis.

■ Goals and Methods of the Science of Human Sexuality

The goals of the science of human sexuality are to describe, explain, predict, and control sexual behaviors. People often confuse description with inference. Inferences are woven into theories, when possible.

Operational Definitions

The operational definition of a construct is linked to the methods used to measure it, enabling diverse researchers to understand what is being measured.

POPULATIONS AND SAMPLES: REPRESENTING THE WORLD OF DIVERSITY

Research samples should accurately represent the population of interest. Representative samples are usually obtained through random sampling.

METHODS OF OBSERVATION

The Case-Study Method

Case studies are carefully drawn biographies of individuals or small groups that focus on unraveling the interplay of various factors in individuals' backgrounds.

The Survey Method

Surveys typically gather information about behavior through interviews or questionnaires administered to large samples of people. The use of volunteers and the tendency of respondents to offer socially desirable responses are sources of bias in surveys.

The Naturalistic-Observation Method

In naturalistic observation, scientists directly observe the behavior of animals and humans where it happens—in the "field." The scientists remain unobtrusive.

The Ethnographic-Observation Method

Ethnographic research has provided us with data concerning sexual behaviors and customs that occur widely across cultures and those that are limited to one or few cultures.

The Participant-Observation Method

In participant observation, investigators learn about people's behavior by interacting with them.

The Laboratory-Observation Method

In the laboratory-observation method, people engage in the behavior under study in the laboratory setting. When methods of observation influence the behavior under study, that behavior may be distorted.

THE CORRELATIONAL METHOD

Correlational studies reveal the strength and direction of the relationships between variables. However, they do not show cause and effect.

THE EXPERIMENTAL METHOD

Experiments allow scientists to draw conclusions about cause-and-effect relationships because the scientists directly control or manipulate the variables of interest and observe their effects. Well-designed experiments randomly assign individuals to experimental and control groups.

ETHICS IN SEX RESEARCH

Ethics concerns the ways in which researchers protect participants in research studies from harm.

Pain and Stress

Ethical standards require that research be conducted only when the expected benefits of the research outweigh the anticipated risks to participants and when the experimenter attempts to minimize expected risks.

Confidentiality

Sex researchers keep the identities and responses of participants confidential to protect them from embarrassment and other potential sources of harm.

Informed Consent

The principle of informed consent requires that people agree to participate in research only after being given enough information about the purposes, procedures, risks, and benefits to make informed decisions.

The Use of Deception

Some research cannot be conducted without deceiving people as to its purposes and procedures. In such cases, the potential harm and benefits of the proposed research are weighed carefully.

questions for critical thinking

1. What is the scientific method? Why do scholars of human sexuality use it?
2. Does the science of human sexuality aim to tell people how they *ought* to behave? Why or why not?
3. Which strikes you as the *truer* measure of sexual arousal—the person's subjective reports of arousal or the levels shown on objective instruments such as the penile strain gauge or vaginal photoplethysmograph? Support your answer.
4. Would you participate in a survey on sexual behavior? Why or why not? Why do you think many people are reluctant to participate in such surveys?
5. What are your own feelings about the methods of laboratory observation used by Masters and Johnson? Why?

6. Couples who attend church more frequently report higher rates of marital satisfaction. Can researchers conclude that going to church enhances marital satisfaction? Why or why not?

7. To run a true experiment on the effect of a new drug on people with HIV/AIDS, some people with AIDS would have to be assigned to a control group. To control for the effects of expectations, they might be told they are being given the real drug when they are really receiving "sugar pills" that look like the real thing. Can researchers justify withholding the real drug from people in the control group? What do you think?

chapter

3

Female Sexual Anatomy and Physiology

Pablo Picasso. *Reclining Nude*, 1932. Oil. © Copyright ARS, NY. Photo: R. G. Ojeda. Musée Picasso, Paris, France. Credit: Réunion des Musées Nationaux/Art Resource, NY. Credit: © 2002 Estate of Pablo Picasso/Artists Rights Society (ARS), New York.

outline

Truth or Fiction?

EXTERNAL SEX ORGANS

The Mons Veneris

The Labia Majora

The Labia Minora

The Clitoris

The Vestibule

The Urethral Opening

A World of Diversity Clitoridectomy—Ritual Genital Mutilation

The Vaginal Opening

The Perineum

Structures That Underlie the External Sex Organs

A Closer Look Kegels

INTERNAL SEX ORGANS

The Vagina

The Cervix

The Uterus

The Fallopian Tubes

The Ovaries

The Pelvic Examination

THE BREASTS

Breast Cancer

A World of Diversity African American Women and Breast Cancer

A Closer Look Brest Self-Examination

THE MENSTRUAL CYCLE

Menstruation Versus Estrus

Regulation of the Menstrual Cycle

Phases of the Menstrual Cycle

A World of Diversity Historical and Cross-Cultural Perspectives on Menstruation

Coitus During Menstruation

Menopause

A World of Diversity Symptoms of Menopause, Ethnicity, Socioeconomic Status, and Lifestyle

A World of Diversity Osteoporosis—Not for Women Only

A Closer Look Myths About Menopause

MENSTRUAL PROBLEMS

Dysmenorrhea

Amenorrhea

Premenstrual Syndrome (PMS)

A Closer Look "Lifting the Curse"—Should Monthly Periods Be Optional?

How to Handle Menstrual Discomfort

Human Sexuality Online Web Sites Related to Female Sexual Anatomy and Physiology

SUMMING UP

QUESTIONS FOR CRITICAL THINKING

Truth or Fiction?

____ One name for the external female genitals is derived from Latin roots that mean "something to be ashamed of."

____ Women, but not men, have a sex organ whose only known function is the experiencing of sexual pleasure.

____ One may determine whether a woman is a virgin by examining the hymen.

____ Women with larger breasts produce more milk while nursing.

____ The incidence of breast cancer is on the rise in the United States.

____ Women who have had abortions are at greater risk of breast cancer.

____ The ancient Romans believed that menstrual blood soured wine and killed crops.

____ At menopause, women experience debilitating hot flashes.

____ Menopause signals an end to women's sexual appetite.

The French have a saying *Vive la différence!* ("Long live the difference!"). It celebrates the differences between men and women. The differences between the genders, at least their anatomic differences, have often been met with prejudice and misunderstanding, however. Men have historically exalted their own genitals. Too often, the less visible genitals of women have been deemed inferior. The derivation of the word **pudendum,** which refers to the external female genitals, speaks volumes about sexism in the ancient Mediterranean world. *Pudendum* derives from the Latin *pudendus*, which literally means "something to be ashamed of."

Truth or Fiction? REVISITED

It is true that one name for the external female genitals is derived from Latin roots that mean "something to be ashamed of." It is the Latin *pudendus.*

Even today, this cultural heritage may lead women to develop negative attitudes toward their genitals. Girls and boys are both sometimes reared to regard their genitals with shame or disgust. Both may be reprimanded for expressing normal curiosity about them. They may be reared with a "hands-off" attitude and warned to keep their "private parts" private, even to themselves. Touching them except for hygienic purposes may be discouraged. One woman recalls:

> When I was six years old I climbed up on the bathroom sink and looked at myself naked in the mirror. All of a sudden I realized I had three different holes. I was very excited about my discovery and ran down to the dinner table and announced it to everyone. "I have three holes!" Silence. "What are they for?" I asked. Silence even heavier than before. I sensed how uncomfortable everyone was and answered for myself. "I guess one is for pee-pee, the other for doo-doo and the third for ca-ca." A sigh of relief; no one had to answer my question. But I got the message—I wasn't supposed to ask "such" questions, though I didn't fully realize what "such" was about at that time. (Boston Women's Health Book Collective, 1992; Copyright © 1984, 1992 by The Boston Women's Health Book Collective. Reprinted with permission.)

In this chapter we tour the female sex organs. Even generally sophisticated students may fill in some gaps in their knowledge. Most of us know what a vagina is, but how many of us realize that only the female gender has an organ that is dedicated exclusively to pleasure? Or that a woman's passing of urine does not involve the vagina? How many of us know that a newborn girl already has all the **ova** she will ever produce?

As women readers encounter the features of their sexual anatomy in their reading, they may wish to examine their own genitals with a mirror. By following the text and the illustrations, students may discover some new anatomic features. They will see that their genitals can resemble those in the illustrations and yet also be unique.

EXTERNAL SEX ORGANS

Taken collectively, the external sexual structures of the female are termed the pudendum or the **vulva.** Because of its derivation, *pudendum*, may be a less desirable term than *vulva*. Vulva is a Latin word that means "wrapper" or "covering." The vulva consists of the *mons veneris*, the *labia majora* and *minora* (major and minor lips), the *clitoris*, and the vaginal opening (see Figure 3.1). Figure 3.2 shows variations in the appearance of women's genitals.

Pudendum The external female genitals.

Ova Egg cells. (Singular: ovum.)

Vulva The external sexual structures of the female.

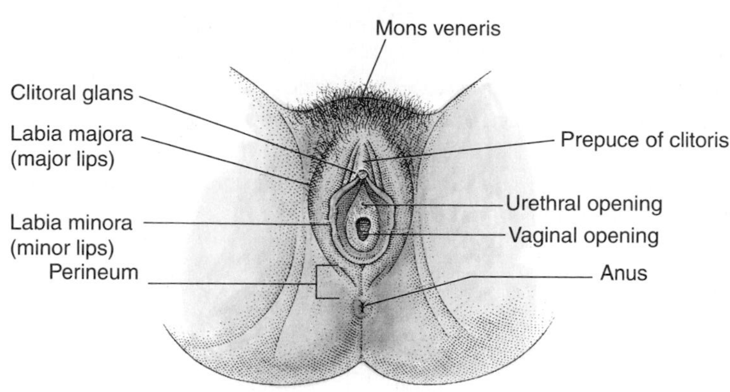

FIGURE 3.1

External Female Sex Organs. This figure shows the vulva with the labia opened to reveal the urethral and vaginal openings.

The Mons Veneris

The **mons veneris** consists of fatty tissue that covers the joint of the pubic bones in front of the body, below the abdomen and above the clitoris. At puberty the mons becomes covered with pubic hair that is often thick and curly but varies from person to person in waviness, texture, and color. The pubic hair captures the chemical secretions that exude from the vagina during sexual arousal. Despite the preoccupation in the United States with chemicals that mask odors, these secretions produce a scent that may allure lovers.

The mons cushions a woman's body during sexual intercourse, protecting her and her partner from the pressure against the pubic bone that stems from thrusting motions. There is an ample supply of nerve endings in the mons, so caresses of the area can produce pleasurable sexual sensations.

The Labia Majora

The **labia majora** are large folds of skin that run downward from the mons along the sides of the vulva. The labia majora of some women are thick and bulging. In other

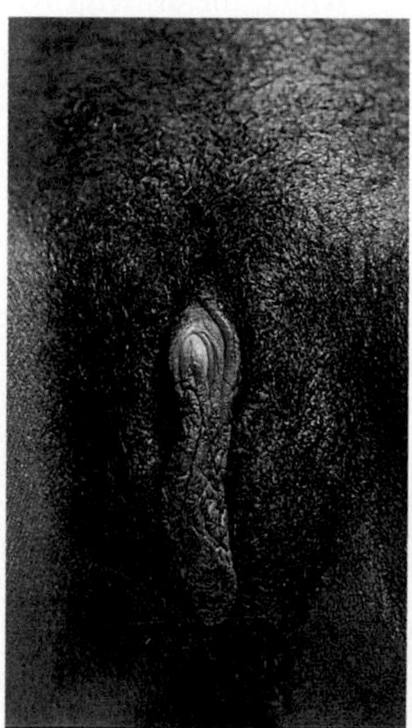

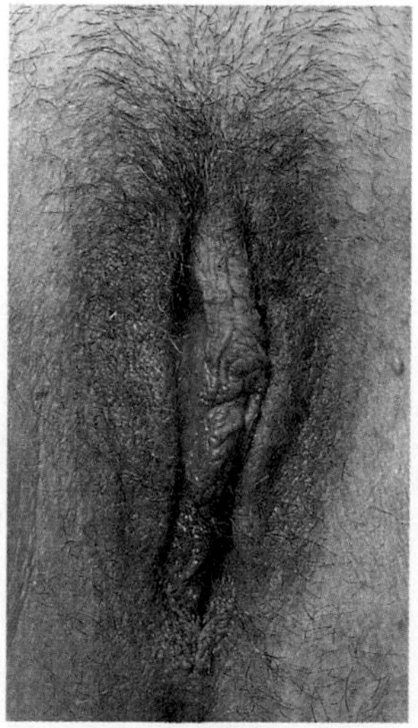

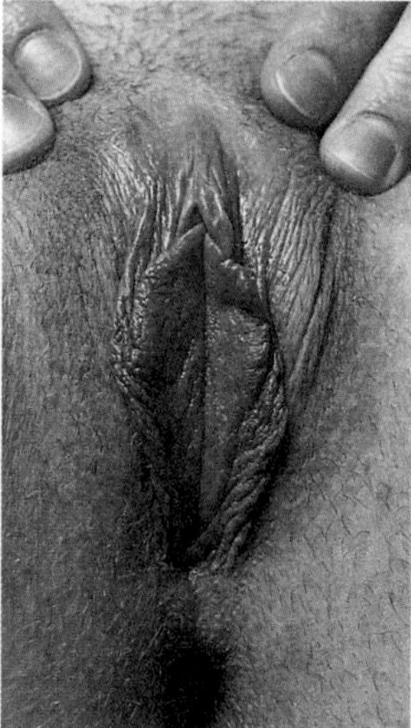

FIGURE 3.2

Normal Variations in the Vulva. The features of the vulva show a great deal of variation. A woman's attitude toward her genitals is likely to reflect her general self-concept and early childhood messages rather than the appearance of her vulva per se.

Mons veneris A mound of fatty tissue that covers the joint of the pubic bones in front of the body, below the abdomen and above the clitoris. (The name is a Latin phrase that means "hill or mount of Venus," the Roman goddess of love. Also known as the *mons pubis,* or simply the *mons.*)

Labia majora Large folds of skin that run downward from the mons along the sides of the vulva. (Latin for "large lips" or "major lips.")

women, they are thinner, flatter, and less noticeable. When close together, they hide the labia minora and the urethral and vaginal openings.

The outer surfaces of the labia majora, by the thighs, are covered with pubic hair and darker skin than that found on the thighs or labia minora. The inner surfaces of the labia majora are hairless and lighter in color. They are amply supplied with nerve endings that respond to stimulation and can produce sensations of sexual pleasure. The labia majora also shield the inner portion of the female genitals.

The Labia Minora

The **labia minora** are two hairless, light-colored membranes located between the major lips. They surround the urethral and vaginal openings. The outer surfaces of the labia minora merge with the major lips. At the top they join at the prepuce (hood) of the clitoris.

The labia minora differ markedly in appearance from woman to woman. The labia minora of some women form protruding flower shapes that are valued greatly in some cultures, such as that of the Hottentots of Africa. In fact, Hottentot women purposely elongate their labia minora by tugging at them.

Rich in blood vessels and nerve endings, the labia minora are highly sensitive to sexual stimulation. When stimulated they darken and swell, indicating engorgement with blood.

The Clitoris

> What's the matter, papa? please don't stall.
> Don't you know I love it and want it all?
> I'm wild about that thing. Just give my bell a ring.
> You pressed my button. I'm wild about that thing."
>
> —I'm Wild About That Thing," recorded by Bessie Smith, 1929

Worldwide, the clitoris is known by many names, from *bijou* (French for "jewel") to *pokhotnik* (Russian for "lust"). The Tuamotuan people of Polynesia have ten words for it, emblematic of their cultivated interest in female sexuality. By any name, however, the clitoris is the only sex organ whose only known function is the experiencing of pleasure.

Clitoris (Figure 3.1) derives from the Greek word *kleitoris*, which means "hill" or "slope." The clitoris receives its name from the manner in which it slopes upward in the shaft and forms a mound of spongy tissue at the glans. The body of the clitoris, termed the clitoral shaft, is about 1 inch long and ¼ inch wide. The clitoral shaft consists of erectile tissue that contains two spongy masses called **corpora cavernosa** ("cavernous bodies") that fill with blood (become engorged) and become erect in response to sexual stimulation. The stiffening of the clitoris is less apparent than the erection of the penis, because the clitoris does not swing free from the body as the penis does. The **prepuce** (meaning "before a swelling"), or hood, covers the clitoral shaft. It is a sheath of skin formed by the upper part of the labia minora. The clitoral glans is a smooth, round knob or lump of tissue. It resembles a button and is situated above the urethral opening. The clitoral glans may be covered by the clitoral hood but is readily revealed by gently separating the labia minora and retracting the hood. It is highly sensitive to touch because of its rich supply of nerve endings.

The clitoris is the female sex organ that is most sensitive to sexual sensation. The size of the clitoris varies from woman to woman, just as the size of the penis varies among men. There is no known connection between the size of the clitoris and sensitivity to sexual stimulation. The clitoral glans is highly sensitive to touch. Women thus usually prefer to be stroked or stimulated on the mons, or on the clitoral hood, rather than directly on the glans.

In some respects, the clitoris is the female counterpart of the penis. Both organs develop from the same embryonic tissue, which makes them similar in structure, or

Labia minora Hairless, light-colored membranes, located between the labia majora. (Latin for "small lips" or "minor lips.")

Clitoris A female sex organ consisting of a shaft and glans located above the urethral opening. It is extremely sensitive to sexual sensations.

Corpora cavernosa Masses of spongy tissue in the clitoral shaft that become engorged with blood and stiffen in response to sexual stimulation. (Latin for "cavernous bodies.")

Prepuce The fold of skin covering the glans of the clitoris (or penis). (From Latin roots that means "before a swelling.")

homologous. They are not fully similar in function, or **analogous,** however. Both organs receive and transmit sexual sensations, but the penis is directly involved in reproduction and excretion by serving as a conduit for sperm and urine, respectively. The clitoris, however, seems to be a unique sex organ. It serves no known purpose other than sexual pleasure. It is ironic that many cultures—including Victorian culture—have viewed women as unresponsive to sexual stimulation. It is ironic because women, not men, possess a sex organ that is apparently devoted solely to pleasurable sensations. The clitoris is the woman's most erotically charged organ, which is borne out by the fact that women most often masturbate through clitoral stimulation, not vaginal insertion.

Truth or Fiction? REVISITED

It is true that women, but not men, have a sex organ whose only known function is the experiencing of sexual pleasure. That organ is the clitoris. ■

Surgical removal of the clitoral hood is common among Moslems in the Near East and Africa. As we see in the nearby World of Diversity feature, this "rite of passage" to womanhood leaves many scars—physical and emotional.

The Vestibule

The word *vestibule*, which means "entranceway," refers to the area within the labia minora that contains the openings to the vagina and the urethra. The vestibule is richly supplied with nerve endings and is very sensitive to tactile or other sexual stimulation.

The Urethral Opening

Urine passes from the female's body through the **urethral opening** (see Figure 3.1), which is connected to the bladder by a short tube called the urethra (see Figure 3.3), where urine collects. The urethral opening lies below the clitoral glans and above the vaginal opening. The urethral opening, urethra, and bladder are unrelated to the reproductive system. Many males (and even some females), however, believe erroneously that for women, urination and coitus occur through the same bodily opening. The confusion may arise from the fact that urine and semen both pass through the penis of the male or because the urethral opening lies near the vaginal opening.

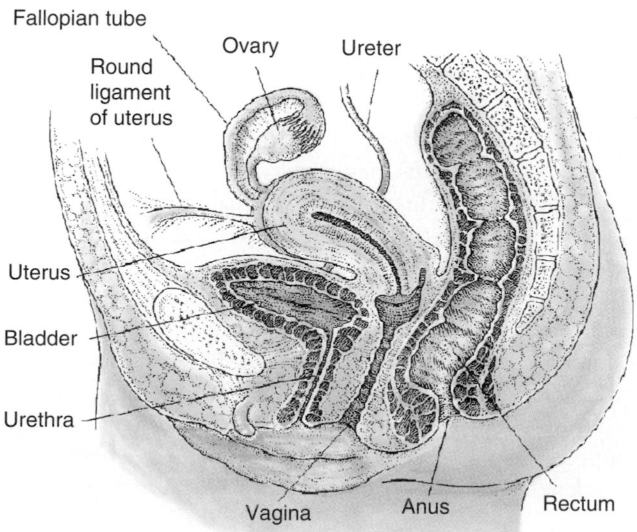

FIGURE 3.3

The Female Reproductive System. This cross section shows the location of many of the internal sex organs that compose the female reproductive system. Note that the uterus is normally tipped forward.

Homologous Similar in structure; developing from the same embryonic tissue.

Analogous Similar in function.

Urethral opening The opening through which urine passes from the female's body.

Clitoridectomy—Ritual Genital Mutilation

Despite hundreds of years of tradition, Hajia Zuwera Kassindja would not let it happen to her 17-year-old daughter, Fauziya. Hajia's own sister had died from it. So Hajia gave her daughter her inheritance from her deceased husband. the bequest amounted to only $3,500, but relinquishing it left Hajia a pauper. Fauziya used the money to buy a phony passport and flee from the African country of Togo to the United States (Dugger, 1996).

Upon arrival in the United States, Fauziya requested asylum from persecution. However, she was put into prison for more than a year. But then, in 1996, the Board of Immigration Appeals finally agreed that Fauziya was fleeing persecution, and she was allowed to remain in the United States.

From what had Hajia's sister died? From what was Fauziya escaping? *Ritual genital mutilation.*

Clitoridectomy

Cultures in some parts of Africa and the Middle East ritually mutilate or remove the clitoris, not just the clitoral hood. Removal of the clitoris, or **clitoridectomy,** is a rite of initiation into womanhood in many of these predominantly Islamic cultures. It is often performed as a puberty ritual in late childhood or early adolescence (not within a few days of birth, like male circumcision).

The clitoris gives rise to feelings of sexual pleasure in women. Its removal or mutilation represents an attempt to ensure the girl's chastity, because it is assumed that uncircumcised girls are consumed with sexual desires. Cairo physician Said M. Thabit says, "With circumcision we remove the external parts, so when a girl wears tight nylon underclothes she will not have any stimulation" (cited in MacFarquhar, 1996, p. A3). Some groups in rural Egypt and in the northern Sudan, however, perform clitoridectomies primarily because it is a social custom that has been maintained from ancient times by a sort of unspoken consensus (Missailidis & Gebre-Medhin, 2000). It is done by women to women (Nour, 2000). Some perceive it as part of their faith in Islam. However, the Koran—the Islamic bible—does not authorize it (Crossette, 1998; Nour, 2000). The typical young woman in this culture does not grasp that she is a victim. She assumes that clitoridectomy is part of being female. As one young woman told gynecologist Nawal M. Nour (2000), the clitoridectomy hurt but was a good thing, because now she was a woman.

Clitoridectomies are performed under unsanitary conditions without benezfit of anesthesia. Medical complications are common, including infections, bleeding, tissue scarring, painful menstruation, and obstructed labor. The procedure is psychologically traumatizing. An even more radical form of clitoridectomy, called *infibulation* or Pharaonic circumcision, is practiced widely in the Sudan. Pharaonic circumcision involves complete removal of the clitoris along with the labia minora and the inner layers of the labia majora. After removal of the skin tissue, the raw edges of the labia majora are sewn together. Only a tiny opening is left to allow passage of urine and menstrual discharge (Nour, 2000). The sewing together of the vulva is intended to ensure chastity until marriage (Crossette, 1998). Medical complications are common; they include menstrual and urinary problems and even death. After marriage, the opening is enlarged to permit intercourse. Enlargement is a gradual process that is often made difficult by scar tissue from the circumcision. Hemorrhaging and tearing of surrounding tissues are

---

The proximity of the urethral opening to the external sex organs may pose some hygienic problems for sexually active women. The urinary tract, which includes the urethra, bladder, and kidneys, may become infected from bacteria that are transmitted from the vagina or rectum. Infectious microscopic organisms may pass from the male's sex organs to the female's urethral opening during sexual intercourse. Manual stimulation of the vulva with dirty hands may also transmit bacteria through the urethral opening to the bladder. Anal intercourse followed by vaginal intercourse may transfer microscopic organisms from the rectum to the bladder and cause infection. For similar reasons, women should wipe first the vulva, then the anus, when using the bathroom.

Cystitis is a bladder inflammation that may stem from any of these sources. Its primary symptoms are burning and frequent urination (also called *urinary urgency*). Pus or a bloody discharge is common, and there may be an intermittent or persistent ache just above the pubic bone. These symptoms may disappear after several days, but consultation with a **gynecologist** is recommended, because untreated cystitis can lead to serious kidney infections.

So-called honeymoon cystitis is caused by the tugging on the bladder and urethral wall that occurs during vaginal intercourse. It may occur upon beginning coital activity

Clitoridectomy Surgical removal of the clitoris.

Cystitis An inflammation of the urinary bladder. (From the Greek *kystis,* which means "sac.")

Gynecologist A physician who treats women's diseases, especially of the reproductive tract. (From the Greek *gyne,* which means "woman.")

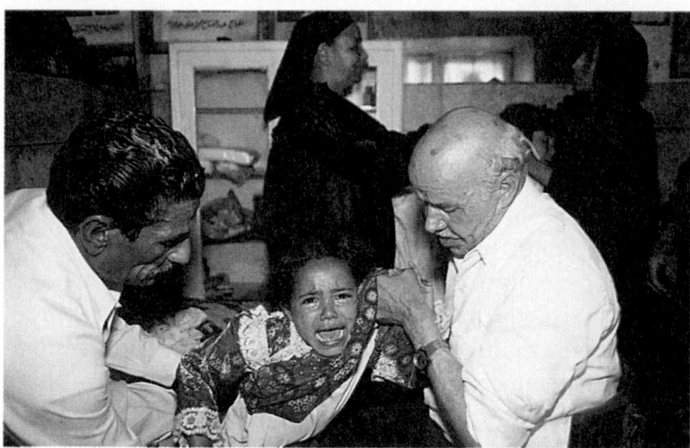

Ritual Genital Mutilation. Some predominantly Islamic cultures in Africa and the Middle East ritually mutilate or remove the clitoris as a rite of initiation into womanhood. Novelist Alice Walker drew attention to the practice in her novel *Possessing the Secret of Joy.* She has called for its abolition in her book and film, *Warrior Marks.*

common consequences. It may take three months or longer before the opening is large enough to allow penile penetration. Mutilation of the labia is now illegal in the Sudan, although the law continues to allow removal of the clitoris. Some African countries, including Egypt, have outlawed clitoridectomies, although such laws may not be enforced.

More than 100 million women in Africa and the Middle East have undergone removal of the clitoris and the labia minora. Clitoridectomies remain common or even universal in nearly 30 countries in Africa, in many countries in the Middle East, and in parts of Malaysia, Yemen, Oman, Indonesia, and the India–Pakistan subcontinent (Rosenthal, 1995). Thousands of African immigrant girls living in European countries and the United States have also been mutilated (Nour, 2000).

Do not confuse male circumcision with the maiming inflicted on girls in the name of circumcision. Nour (2000) maintains, rather, that the male equivalent of female genital mutilation would be cutting off the penis. *New York Times* columnist A. M. Rosenthal (1995) calls female genital mutilation the most widespread existing violation of human rights in the world. The Pulitzer Prize–winning African American novelist Alice Walker drew attention to the practice in her best-selling novel *Possessing the Secret of Joy* (1992). She also called for its abolition in her book and the movie based on it, *Warrior Marks.*

Outlawed

In 1996, the United States outlawed ritual genital mutilation within its borders. The government also directed U.S. representatives to world financial institutions to deny aid to countries that have not established educational programs to bring an end to the practice. Yet calls from Westerners to ban the practice in parts of Africa and the Middle East have sparked controversy on grounds of "cultural condescension"—that people in one culture cannot dictate the cultural traditions of another. Yet for Alice Walker and others of like mind, "torture is not culture." As the debate continues, some 2 million African girls are mutilated each year.

(though not necessarily on one's honeymoon) or upon resuming coital activity after lengthy abstinence. Figure 3.3 shows the close proximity of the urethra and vagina.

A few precautions may help women prevent serious inflammation of the bladder:

- Drinking two quarts of water a day to flush the bladder.
- Drinking orange or cranberry juice to maintain an acid environment that discourages growth of infectious organisms.
- Decreasing use of alcohol and caffeine (from coffee, tea, or cola drinks) that may irritate the bladder.
- Washing the hands prior to masturbation or self-examination.
- Washing one's partner's and one's own genitals before and after intercourse.
- Preventing objects that have touched the anus (fingers, penis, toilet tissue) from subsequently coming into contact with the vulva.
- Urinating soon after intercourse to help wash away bacteria.

These measures do not guarantee protection against transmission of infectious organisms, but they may help.

The Vaginal Opening

> When I was five or six, my mother told me about sex. I remember that I was confused about what my mother said, because somehow I couldn't conceptualize what the female vagina looked like. I was curious to see an actual vagina and not just how it looked diagramed in a book.
>
> —Morrison et al., 1980, p. 35

One does not see the entire vagina, but rather the vaginal opening, or **introitus,** when one parts the labia minora, or minor lips. The introitus lies below, and is larger than, the urethral opening. Its shape resembles that of the **hymen.**

The hymen is a fold of tissue across the vaginal opening that is usually present at birth and may remain at least partly intact until a woman engages in coitus. For this reason the hymen has been called the "maidenhead." Its presence has been taken as proof of virginity, and its absence as evidence of coitus. However, some women are born with incomplete hymens, and other women's hymens are torn accidentally, such as during horseback riding, strenuous exercise or gymnastics—or even when bicycle riding. A punctured hymen is therefore poor evidence of coital experience. A flexible hymen may also withstand many coital experiences, so its presence does not guarantee virginity. Some people believe incorrectly that virgins cannot insert tampons or fingers into their vaginas, but most hymens will accommodate these intrusions without great difficulty. Some hymens, however, are torn accidentally when inserting tampons.

Truth or Fiction? **REVISITED**

Contrary to myth, it is not true that one may determine whether a woman is a virgin by examining of the hymen. Examination of the hymen may yield false information.

Figure 3.4 illustrates various vaginal openings. The first three show hymen shapes that are frequently found among women who have not had coitus. The fifth drawing shows a *parous* ("passed through") vaginal opening, typical of a woman who has delivered a baby. Now and then the hymen consists of tough fibrous tissue and is closed, or *imperforate*, as in the fourth drawing. An imperforate hymen may not be discovered until after puberty, when menstrual discharges begin to accumulate in the vagina. In these rare cases, a simple surgical incision will perforate the hymen. A woman may also have a physician surgically perforate her hymen if she would rather forgo the tearing and discomfort that may accompany her initial coital experiences. This procedure is unnecessary, however, for the great majority of women. Despite old horror stories, most experience little pain or distress during initial coitus. A woman may also stretch the vaginal opening in preparation for intercourse by inserting a finger and gently pressing

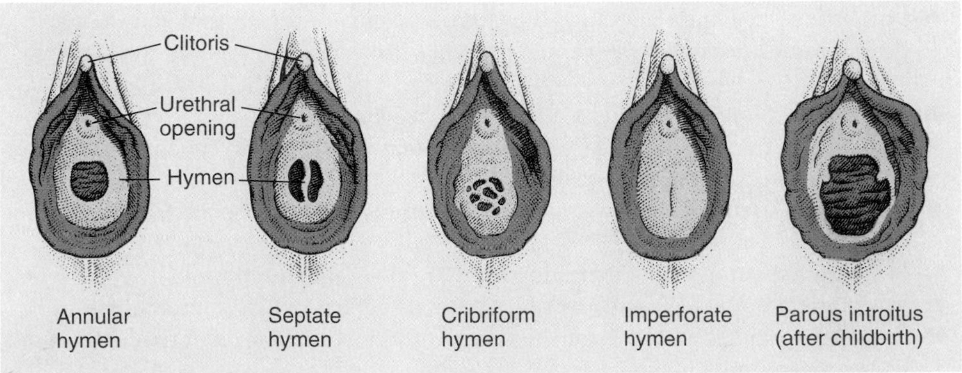

FIGURE 3.4

Appearance of Various Types of Hymens Before Coitus and the Introitus (at Right) As It Appears Following Delivery of a Baby.

Introitus The vaginal opening. (From the Latin for "entrance.")

Hymen A fold of tissue across the vaginal opening that is usually present at birth and remains at least partly intact until a woman engages in coitus. (Greek for "membrane.")

downward toward the anus. After several repetitions, she may insert two fingers and repeat the process, spreading the fingers slightly after insertion. This procedure is sometimes followed over several days or weeks.

The hymen is found only in female horses and humans. It is not present in animal species closest to humans on the evolutionary scale, such as chimps and gorillas. The hymen remains something of a biological mystery, because it serves no apparent biological function.

DEFLORATION The hymen has been of great cultural significance because of its (erroneous) association with virginity. In some societies its destruction, or **defloration,** is ritualized. For example, in some ancient religious rites, young girls were deflowered by sculpted phalluses in ceremonies conducted on the temple steps. During the Middle Ages, the lord of the manor held the *droit du seigneur* (French for "right of the lord") to deflower (by means of intercourse) a maiden on her wedding night, after which she became her husband's sexual property. Among the Yungar of Australia, older women deflowered maidens prior to marriage. Girls who were found to have ruptured hymens might be tortured or killed. Given that some girls are born with minimal hymens, many have probably been unjustly punished for purported sexual indiscretions.

The Perineum

The **perineum** consists of the skin and underlying tissue between the vaginal opening and the anus. The perineum is rich in nerve endings. Stimulation of the area may heighten sexual arousal. During labor, many physicians make a routine perineal incision, called an **episiotomy,** to facilitate childbirth.

Structures That Underlie the External Sex Organs

Figure 3.5 shows what lies beneath the skin of the vulva. The vestibular bulbs and Bartholin's glands are active during sexual arousal and are found on both sides (they are shown on the right in Figure 3.5). Muscular rings **(sphincters)** that constrict bodily openings, such as the vaginal and anal openings, are also found on both sides.

The clitoral **crura** are wing-shaped, leglike structures that attach the clitoris to the pubic bone beneath. The crura contain corpora cavernosa, which engorge with blood and stiffen during sexual arousal.

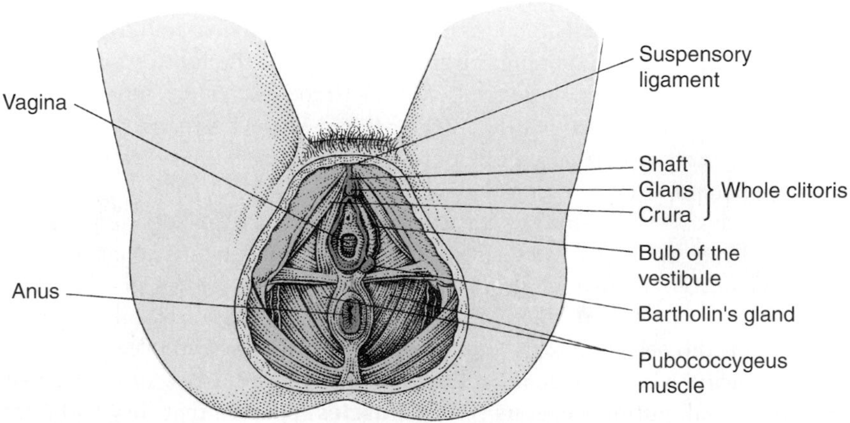

FIGURE 3.5

Structures That Underlie the Female External Sex Organs. If we could see beneath the vulva, we would find muscle fibers that constrict the various body openings, plus the crura ("legs") of the clitoris, the vestibular bulbs, and Bartholin's glands.

Defloration Destruction of the hymen (especially as a cultural ritual).

Perineum The skin and underlying tissue that lies between the vaginal opening and the anus. (From Greek roots that means "around" and "to empty out.")

Episiotomy A surgical incision in the perineum that may be made during childbirth to protect the vagina from tearing. (From the Greek roots *epision,* which means "pubic region," and *tome,* which means "cutting.")

Sphincters Ring-shaped muscles that surround body openings and open or close them by expanding or contracting. (From the Greek for "that which draws close.")

Crura Anatomic structures resembling legs that attach the clitoris to the pubic bone. (Singular: crus. A Latin word that means "leg" or "shank.")

a CLOSER look

Kegels

Kegel exercises are commonly used to help women heighten their awareness of vaginal sensations. They may also have a psychological benefit, because women who perform Kegel exercises assume a more active role in enhancing their genital sensations. Sex therapist Lonnie Barbach (1975) offers instructions for Kegel exercises:

1. Locate the P-C muscle by purposely stopping the flow of urine. The muscle you squeeze to stop the urine flow is the P-C muscle. (The P-C muscle acts as a sphincter for both the urethral and vaginal openings.)
2. In order to learn to focus consciously on contracting the P-C muscle, insert a finger into the vaginal opening and contract the muscle so that it can be felt to squeeze or contain the finger. (The P-C muscle can contract to contain objects as narrow as a finger. Thus men need not fear that a large penis is necessary to induce stimulating vaginal sensations in a woman.)
3. Remove your finger, squeeze the P-C muscle for three seconds, and then relax. Repeat several times. This part of the exercise may be performed while seated at a classroom or business desk. No one (except the woman herself) will be the wiser. Many women practice a series of Kegel exercises consisting of 10 contractions, 3 times a day.
4. The P-C muscle may also be tensed and relaxed in rapid sequence. Since this exercise may be more fatiguing than the above, women may choose to practice it perhaps 10 to 25 times, once a day.

The **vestibular bulbs** are attached to the clitoris at the top and extend downward along the sides of the vaginal opening. Blood congests them during sexual arousal, swelling the vulva and lengthening the vagina. This swelling contributes to coital sensations for both partners.

Bartholin's glands lie just inside the minor lips on each side of the vaginal opening. They secrete a couple of drops of lubrication just before orgasm. This lubrication is not essential for coitus. In fact, the fluid produced by Bartholin's glands has no known purpose. If these glands become infected and clogged, however, a woman may notice swelling and local irritation. It is wise to consult a gynecologist if these symptoms do not fade within a few days.

It was once believed that Bartholin's glands were the source of the vaginal lubrication, or "wetness," that women experience during sexual arousal. It is now known that engorgement of vaginal tissues during sexual excitement results in a form of "sweating" by the lining of the vaginal wall. During sexual arousal, the pressure from this engorgement causes moisture from the many small blood vessels that lie in the vaginal wall to be forced out and to pass through the vaginal lining, forming the basis of the lubrication. In less time than it takes to read this sentence (generally within 10 to 30 seconds), beads of vaginal lubrication, or "sweat," appear along the interior lining of the vagina in response to sexual stimulation, in much the same way that rising temperatures cause water to pass through the skin as perspiration.

Pelvic floor muscles permit women to constrict the vaginal and anal openings. They contract automatically, or involuntarily, during orgasm, and their tone may contribute to coital sensations. Gynecologist Arnold Kegel (1952) developed exercises to build pelvic muscle tone in women who had problems controlling urination after childbirth. The stress of childbirth on the pelvic muscles sometimes reduces muscle tone, leading to an involuntary loss of urine when a woman sneezes or coughs. Kegel also observed that women with thin or weak **pubococcygeus (P-C) muscles** reported that they had little or no vaginal sensations during coitus or experienced unpleasant vaginal sensations. Some women with weak P-C muscles complained, "I just don't feel anything" (during coitus) or "I don't like the feeling" (Kegel, 1952, p. 522). Kegel found that women who practiced his exercises improved their urinary control along with their genital sensations during coitus. He believed that many women could enhance vaginal sensations during coitus by exercising their P-C muscles through exercises that are now known as "Kegels."

Vestibular bulbs Cavernous structures that extend downward along the sides of the introitus and swell during sexual arousal.

Bartholin's glands Glands that lie just inside the minor lips and secrete fluid just before orgasm.

Pubococcygeus muscle The muscle that encircles the entrance to the vagina.

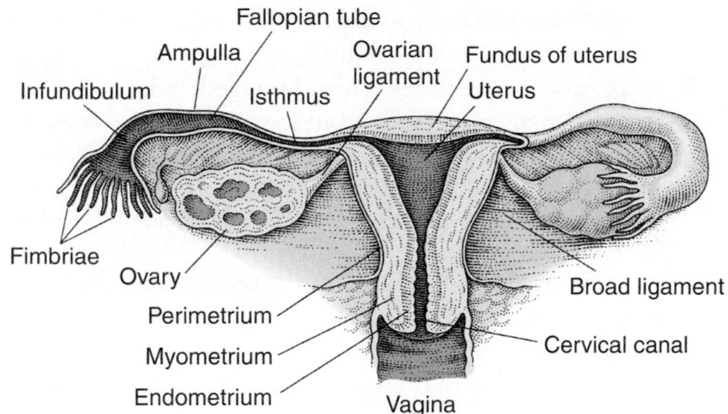

FIGURE 3.6

Female Internal Reproductive Organs. This drawing highlights the relationship of the uterus to the fallopian tubes and ovaries. Note the layers of the uterus, the ligaments that attach the ovaries to the uterus, and the relationship of the ovaries to the fimbriae of the fallopian tubes.

INTERNAL SEX ORGANS

The internal sex organs of the female include the innermost parts of the vagina, the cervix, the uterus, and two ovaries, each connected to the uterus by a fallopian tube (see Figures 3.3 and 3.6). These structures comprise the female reproductive system.

The Vagina

The **vagina** extends back and upward from the vaginal opening (see Figure 3.3). It is usually 3 to 5 inches long at rest. Menstrual flow and babies pass from the uterus to the outer world through the vagina. During coitus, the penis is contained within the vagina.

The vagina is commonly pictured as a canal or barrel, but when at rest, it is like a collapsed muscular tube. Its walls touch like the fingers of an empty glove. It vagina expands in length and width during sexual arousal. The vagina can also expand to allow the insertion of a tampon, as well as the passage of a baby's head and shoulders during childbirth.

The vaginal walls have three layers. The inner lining, or *vaginal mucosa*, is made visible by opening the labia minora. It is a mucous membrane similar to the skin that lines the inside of the mouth. It feels fleshy, soft, and corrugated. It may vary from very dry (especially if the female is anxious about something, such as examinations) to very wet, in which case fingers slide against it readily. The middle layer of the vaginal wall is muscular. The outer or deeper layer is a fibrous covering that connects the vagina to other pelvic structures.

The vaginal walls are rich with blood vessels but poorly supplied with nerve endings. Unlike the sensitive outer third of the vaginal barrel, the inner two thirds are so insensitive to touch that minor surgery can sometimes be performed on those portions without anesthesia. The entire vaginal barrel is sensitive to pressure, however, which may be experienced as sexually pleasurable.

The vaginal walls secrete substances that help maintain the vagina's normal acidity (pH 4.0 to 5.0). Normally the secretions taste salty. The odor and taste of these secretions may vary during the menstrual cycle. Although the evidence is not clear, the secretions are thought to contain substances that may act as sexual attractants. Women who frequently **douche** or use feminine deodorant sprays may thus remove or mask substances that may arouse sex partners. Douching or spraying may also alter the natural chemical balance of the vagina, which can increase the risk of vaginal infections. Feminine deodorant sprays can also irritate the vagina and evoke allergic reactions. The normal, healthy vagina cleanses itself through regular chemical secretions that are evidenced by a mild white or yellowish discharge.

Vaginitis is any vaginal inflammation, whether it is caused by an infection, an allergic reaction, or chemical irritation. Vaginitis may also stem from use of birth-control pills or antibiotics that alter the natural body chemistry, or from other factors, such as lowered resistance (from fatigue or poor diet). Changes in the natural body chemistry or lowered

Vagina The tubular female sex organ that contains the penis during sexual intercourse and through which a baby is born. (Latin for "sheath.")

Douche Application of a jet of liquid to the vagina as a rinse. (From the Italian *doccia,* which means "shower bath.")

Vaginitis Vaginal inflammation.

resistance may permit microscopic organisms normally found in the vagina to multiply to infectious levels. Vaginitis may be recognized by abnormal discharge, itching, burning of the vulva, and urinary urgency. Its causes and treatments are discussed in Chapter 16. Women with vaginitis are advised to seek medical attention, but let us note some suggestions that may help prevent vaginitis (Boston Women's Health Book Collective, 1992):

1. Wash your vulva and anus regularly with mild soap. Pat dry (taking care not to touch the vulva after dabbing the anus).
2. Wear cotton panties. Nylon underwear retains heat and moisture that cause harmful bacteria to flourish.
3. Avoid pants that are tight in the crotch.
4. Be certain that sex partners are well washed. Condoms may also reduce the spread of infections from one's sex partner.
5. Use a sterile, water-soluble jelly such as K-Y jelly if artificial lubrication is needed for intercourse. Do *not* use Vaseline. Birth-control jellies can also be used for lubrication.
6. Avoid intercourse that is painful or abrasive to the vagina.
7. Avoid diets high in sugar and refined carbohydrates because they alter the normal acidity of the vagina.
8. Women who are prone to vaginal infections may find it helpful to douche occasionally with plain water, a solution of 1 or 2 tablespoons of vinegar in a quart of warm water, or a solution of baking soda and water. Douches consisting of unpasteurized, plain (unflavored) yogurt may help replenish the "good" bacteria that are normally found in the vagina and that may be destroyed by the use of antibiotics. Be careful when douching, and do not douche when pregnant or when you suspect you may be pregnant. Consult your physician before deciding to douche or to apply any preparations to the vagina.
9. Watch your general health. Eating poorly or getting insufficient rest will reduce your resistance to infection.

The Cervix

> When someone first said to me two years ago, "You can feel the end of your own cervix with your finger," I was interested but flustered. I had hardly ever put my finger in my vagina at all, and felt squeamish about touching myself there, in that place "reserved" for lovers and doctors. It took me two months to get up nerve to try it, and then one afternoon, pretty nervously, I squatted down in the bathroom and put my finger in deep, back into my vagina. There it was, feeling slippery and rounded, with an indentation at the center through which, I realized, my menstrual flow came. It was both very exciting and beautifully ordinary at the same time. Last week I bought a plastic speculum so I can look at my cervix. Will it take as long this time? (Boston Women's Health Book Collective, 1992; Copyright © 1984, 1992 by The Boston Women's Health Book Collective. Reprinted with permission.)

The **cervix** is the lower end of the uterus. Its walls, like those of the vagina, produce secretions that contribute to the chemical balance of the vagina. The opening in the middle of the cervix, or **os,** is normally about the width of a straw, although it expands to permit passage of a baby from the uterus to the vagina during childbirth. Sperm pass from the vagina to the uterus through the cervical canal.

Cervical cancer is relatively uncommon in the United States, largely as a consequence of effective screening programs (Cannistra & Niloff, 1996). Nearly 16,000 new cases of cervical cancer are reported annually. Cervical cancer is more common among women who have had many sex partners, women who became sexually active at a relatively early age, women of lower socioeconomic status, and women who smoke. All women are at risk, however.

A **Pap test** examines a sample of cervical cells that are smeared on a slide to screen for cervical cancer and other abnormalities. The American Cancer Society (2000) recommends annual Pap tests, along with a pelvic examination, for women who are, or have been, sexually active or who have reached age 18. Most cases of cervical cancer can be

Cervix The lower end of the uterus. (Latin for "neck.")

Os The opening in the middle of the cervix. (Latin for "mouth.")

Pap test A test of a sample of cervical cells that screens for cervical cancer and other abnormalities. (Named after the originator of the technique, Dr. Papanicolaou.)

successfully treated by surgery and **radiotherapy** if they are detected early. The five-year survival rate for women with cervical cancer is 67%. For women diagnosed with localized cancer, the survival rate is 90% (National Cancer Institute, 2000). However, cervical cancer can also be prevented when precancerous changes are detected by a Pap test. There are nearly 5,000 deaths from cervical cancer a year. The mortality rate is more than twice as high for African American women as for European American women (National Cancer Institute, 2000).

The Uterus

The **uterus,** or womb (see Figures 3.3 and 3.6), is the organ in which a fertilized ovum implants and develops until birth. The uterus usually slants forward (is *antroverted*), although about 10% of women have uteruses that tip backward (are *retroverted*). In most instances a retroverted uterus causes no problems, but some women with retroverted uteruses find coitus in certain positions painful. (They quickly learn more comfortable positions by trial and error.) A retroverted uterus normally tips forward during pregnancy. This condition probably does not interfere with conception, although at one time it was thought that it might do so. The uterus is suspended in the pelvis by flexible ligaments. In a woman who has not given birth, it is about 3 inches long, 3 inches wide, and 1 inch thick near the top. The uterus expands to accommodate a fetus during pregnancy and shrinks after pregnancy, though not to its original size.

The uppermost part of the uterus is called the **fundus** (see Figure 3.6). The uterus is shaped like an inverted pear. If a ceramic model of a uterus were placed on a table, it would balance on the fundus. The central region of the uterus is called the body. The narrow lower region is the cervix, which leads downward to the vagina.

LAYERS OF THE UTERUS Like the vagina, the uterus has three layers (also shown in Figure 3.6). The innermost layer, or **endometrium,** is richly supplied with blood vessels and glands. Its structure varies according to a woman's age and the phase of the menstrual cycle. Endometrial tissue is discharged through the cervix and vagina at menstruation. For reasons not entirely understood, in some women endometrial tissue may also grow in the abdominal cavity or elsewhere in the reproductive system. This condition is called **endometriosis,** and the most common symptom is menstrual pain. If left untreated, endometriosis may lead to infertility.

Cancer of the endometrial lining is called endometrial cancer. Risk factors include obesity, a diet high in fats, early menarche or late menopause, history of failure to ovulate, and estrogen-replacement therapy (Grodstein et al., 1997). For women who obtain hormone replacement therapy (HRT), combining estrogen with progestin mitigates the risk of endometrial cancer (Grodstein et al., 1996; Rose, 1996). One of the symptoms of endometrial cancer is abnormal uterine staining or bleeding, especially after menopause. The most common treatment is surgery (Rose, 1996). The five-year survival rate for endometrial cancer is up to 95% if it is discovered early and limited to the endometrium (Rose, 1996). (Endometrial cancer is usually diagnosed early because women tend to report postmenopausal bleeding quickly to their doctors.) The survival rate drops when the cancer invades surrounding tissues or metastasizes.

The second layer of the uterus, the **myometrium,** is well muscled. It endows the uterus with flexibility and strength and creates the powerful contractions that propel a fetus outward during labor. The third or outermost layer, the **perimetrium,** provides an external cover.

HYSTERECTOMY One woman in three in the United States has a **hysterectomy** by the age of 60. Most women who obtain them do so between the ages of 35 and 45. The hysterectomy is the second most commonly performed operation on women in this country. (Cesarean sections are the most common). A hysterectomy may be performed when women develop cancer of the uterus, ovaries, or cervix or have other diseases that cause

Radiotherapy Treatment of a disease by X-rays or by emissions from a radioactive substance.

Uterus The hollow, muscular, pear-shaped organ in which a fertilized ovum implants and develops until birth.

Fundus The uppermost part of the uterus. (*Fundus* is a Latin word which means "base.")

Endometrium The innermost layer of the uterus. (From Latin and Greek roots that means "within the uterus.")

Endometriosis A condition caused by the growth of endometrial tissue in the abdominal cavity, or elsewhere outside the uterus, and characterized by menstrual pain.

Myometrium The middle, well-muscled layer of the uterus. (*Myo-* comes from the Greek *mys,* which means "muscle.")

Perimetrium The outer layer of the uterus. (From roots that mean "around the uterus.")

Hysterectomy Surgical removal of the uterus.

pain or excessive uterine bleeding. A hysterectomy may be partial or complete. A **complete hysterectomy** involves surgical removal of the ovaries, fallopian tubes, cervix, and uterus. It is usually performed to reduce the risk of cancer spreading throughout the reproductive system. A partial hysterectomy removes the uterus but not the ovaries and fallopian tubes. Sparing the ovaries allows the woman to continue to ovulate and produce adequate quantities of female sex hormones.

The hysterectomy has become steeped in controversy. It is generally accepted that the operation can relieve symptoms associated with various gynecological disorders and improve the quality of life for many women (Kjerulff et al., 2000). However, many gynecologists believe that hysterectomy is often recommended inappropriately, before necessary diagnostic steps are taken or when less radical medical interventions might successfully treat the problem (Broder et al., 2000). On the basis of a sample of more than 5,000 hysterectomies performed in seven health maintenance organizations (HMOs) across the country, a panel of experts in gynecology and obstetrics reported that 28% of these operations in younger women, and 16% in women overall, were inappropriate (Bernstein et al., 1993). We advise women whose physicians suggest a hysterectomy to seek a second opinion before proceeding.

Complete hysterectomy Surgical removal of the ovaries, fallopian tubes, cervix, and uterus.

Fallopian tubes Tubes that extend from the upper uterus toward the ovaries and conduct ova to the uterus. (After the Italian anatomist Gabriel Fallopio, who is credited with their discovery.)

Isthmus The segment of a fallopian tube closest to the uterus. (A Latin word that means "narrow passage.")

Ampulla The wide segment of a fallopian tube near the ovary. (A Latin word that means "bottle.")

Infundibulum The outer, funnel-shaped part of a fallopian tube. (A Latin word that means "funnel.")

Fimbriae Projections from a fallopian tube that extend toward an ovary. (Singular: fimbria. Latin for "fiber" or "fringe.")

Ectopic pregnancy A pregnancy in which the fertilized ovum implants outside the uterus, usually in the fallopian tube. (*Ectopic* derives from Greek roots that mean "out of place.")

Ovaries Almond-shaped organs that produce ova and the hormones estrogen and progesterone.

The Fallopian Tubes

Two uterine tubes, also called **fallopian tubes,** are about 4 inches in length and extend from the upper end of the uterus toward the ovaries (see Figure 3.6). The part of each tube nearest the uterus is the **isthmus,** which broadens into the **ampulla** as it approaches the ovary. The outer part, or **infundibulum,** has fringelike projections called **fimbriae** that extend toward, but are not attached to, the ovary.

Ova pass through the fallopian tubes on their way to the uterus. The fallopian tubes are not inert passageways. They help nourish and conduct ova. These tubes are lined with tiny hairlike projections termed cilia ("lashes") that help move ova through the tube. The exact mechanisms by which ova are guided are unknown, however. It is tempting to say that ova move at a snail's pace, but a snail would leave them far behind. They journey toward the uterus at a rate of about 1 inch per day. Because ova must be fertilized within a day or two after they are released from the ovaries, fertilization usually occurs in the infundibulum within a couple of inches of the ovaries. The form of sterilization called tubal ligation ties off the fallopian tubes so that ova cannot pass through them or become fertilized.

In an **ectopic pregnancy,** the fertilized ovum implants outside the uterus, most often in the fallopian tube where fertilization occurred. Ectopic pregnancies can eventually burst fallopian tubes, causing hemorrhaging and death. Ectopic pregnancies are thus terminated before the tube ruptures. They are not easily recognized, however, because their symptoms—missed menstrual period, abdominal pain, and irregular bleeding—suggest many conditions. Experiencing any these symptoms is an excellent reason for consulting a gynecologist. Women who have had pelvic inflammatory disease (PID), have undergone tubal surgery, or have used intrauterine devices (IUDs) are at increased risk of developing ectopic pregnancies (Marchbanks et al., 1988).

The Ovaries

The two **ovaries** are almond-shaped organs each about 1½ inches long. They lie on either side of the uterus, to which they are attached by ovarian ligaments. The ovaries produce ova (egg cells) and the female sex hormones **estrogen** and **progesterone.**

Estrogen is a generic term for several hormones (such as estradiol, estriol, and estrone) that promote the changes of puberty and regulate the menstrual cycle. Estrogen also helps older women maintain cognitive functioning and feelings of psychological well-being (Ross et al., 2000; Sourander, 1994). Progesterone too has multiple functions, including regulating the menstrual cycle and preparing the uterus for pregnancy by stimulating the development of the endometrium (uterine lining). Estrogen and progesterone levels vary with the phases of the menstrual cycle.

The human female is born with all the ova she will ever have (about 2 million), but they are immature in form. About 400,000 of these survive into puberty, each contained in the ovary within a thin capsule, or **follicle.** During a woman's reproductive years, from puberty to menopause, only 400 or so ripened ova, typically 1 per month, will be released by their rupturing follicles for possible fertilization. How these ova are selected is among the mysteries of nature.

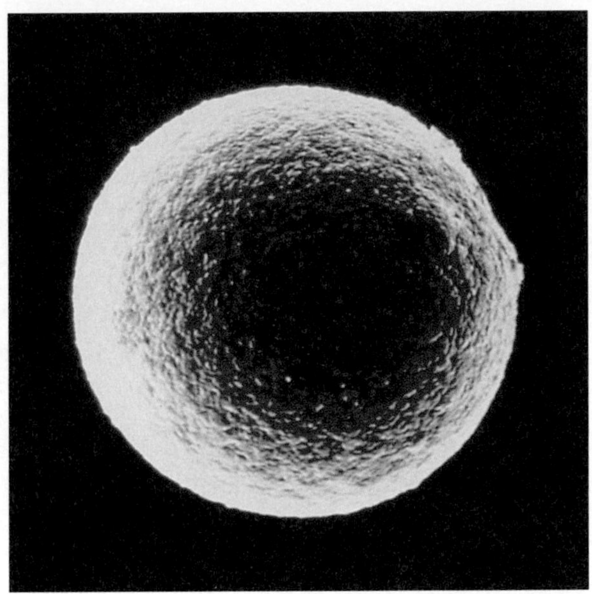

A Magnified Human Ovum (Egg Cell).

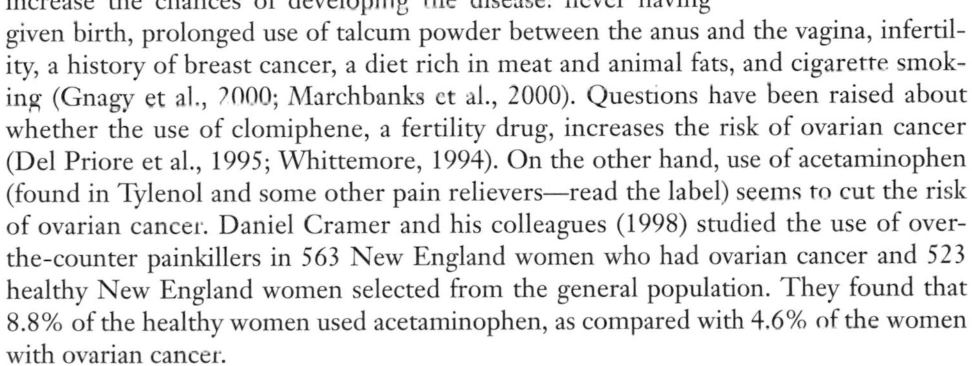

Each year some 30,000 women in the United States are diagnosed with ovarian cancer, and nearly half die from it (National Cancer Institute, 2000). It most often strikes women between the ages of 40 and 70 and ranks as the fourth leading cancer killer of women, behind lung cancer, breast cancer, and colon cancer. Women most at risk are those with blood relatives who had the disease, especially a first-degree relative (mother, sister, or daughter). Other risk factors are also important, because about 9 women in 10 who develop ovarian cancer do not have a family history of it. Researchers have identified several risk factors that increase the chances of developing the disease: never having given birth, prolonged use of talcum powder between the anus and the vagina, infertility, a history of breast cancer, a diet rich in meat and animal fats, and cigarette smoking (Gnagy et al., 2000; Marchbanks et al., 2000). Questions have been raised about whether the use of clomiphene, a fertility drug, increases the risk of ovarian cancer (Del Priore et al., 1995; Whittemore, 1994). On the other hand, use of acetaminophen (found in Tylenol and some other pain relievers—read the label) seems to cut the risk of ovarian cancer. Daniel Cramer and his colleagues (1998) studied the use of over-the-counter painkillers in 563 New England women who had ovarian cancer and 523 healthy New England women selected from the general population. They found that 8.8% of the healthy women used acetaminophen, as compared with 4.6% of the women with ovarian cancer.

Early detection is the key to fighting ovarian cancer. When it is detected before spreading beyond the ovary, 90% of victims survive. However, the overall five-year survival rate is about 50% (Altman, 1998c). Unfortunately, ovarian cancer is often "silent" in the early stages, showing no obvious signs or symptoms. The most common sign is enlargement of the abdomen, which is caused by the accumulation of fluid. Periodic, complete pelvic examinations are important. The Pap test, which is useful in detecting cervical cancer, does not reveal ovarian cancer. The American Cancer Society (2000) advises women over the age of 40 to have a cancer-related checkup every year.

Surgery, radiation therapy, and drug therapy are treatment options. Surgery usually includes the removal of one or both ovaries, the uterus, and the fallopian tubes.

The Pelvic Examination

Women are advised to have an internal (pelvic) examination at least once a year by the time they reach their late teens (or earlier if they become sexually active) and twice yearly if they are over age 35 or use birth-control pills. The physician (usually a gynecologist) first examines the woman externally for irritations, swellings, abnormal vaginal discharges, and clitoral adhesions. The physician normally inserts a speculum to help inspect the cervix and vaginal walls for discharges (which can be signs of infection), discoloration, lesions, or growths. This examination is typically followed by a Pap test to detect cervical cancer. A sample of vaginal discharge may also be taken to test for the sexually transmitted disease gonorrhea.

Estrogen A generic term for female sex hormones (including estradiol, estriol, estrone, and others) or synthetic compounds that promote the development of female sex characteristics and regulate the menstrual cycle. (From roots that mean "generating" (*-gen*) and "estrus.")

Progesterone A steroid hormone secreted by the corpus luteum or prepared synthetically that stimulates proliferation of the endometrium and is involved in regulation of the menstrual cycle. (From the root *pro-*, which means "promoting," and the words *gestation, steroid,* and *one.*)

Follicle A capsule within an ovary that contains an ovum. (From a Latin word that means "small bag.")

FIGURE 3.7

Use of the Speculum and Spatula During a Pelvic Examination. The speculum holds the vaginal walls apart, while the spatula is used to scrape cells gently from the cervix. The Pap test screens for cervical cancer and other abnormalities.

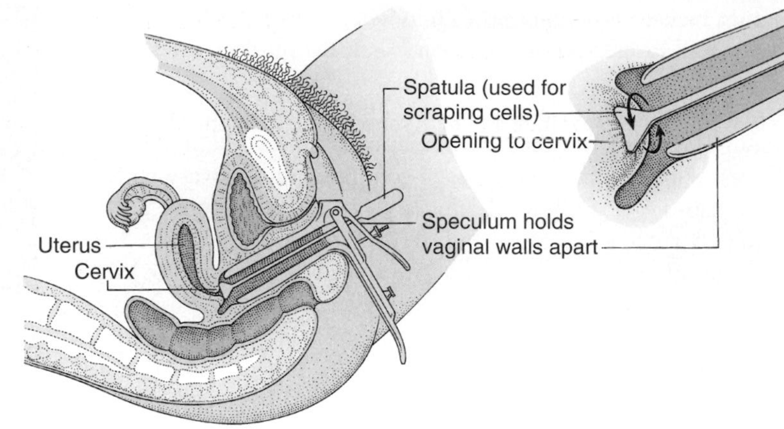

To take a Pap smear, the physician will hold the vaginal walls open with a plastic or (prewarmed!) metal speculum so that a sample of cells (a "smear") may be scraped from the cervix with a wooden spatula (see Figure 3.7). Women should not douche prior to Pap tests or schedule them during menstruation, because douches and blood confound analysis of the smear.

The speculum exam is normally followed by a bimanual vaginal exam in which the index and middle fingers of one hand are inserted into the vagina while the lower part of the abdomen is palpated (touched) by the other hand from the outside. The physician uses this technique to examine the location, shape, size, and movability of the internal sex organs, searching for abnormal growths and symptoms of other problems. Palpation may be somewhat uncomfortable, but severe pain is a sign that something is wrong. A woman should not try to be "brave" and hide such discomfort from the examiner. She may only be masking a symptom (that is, depriving the physician of useful information). Physical discomfort is usually mild, however, and psychological discomfort may often be relieved by discussing it frankly with the examiner.

Finally, the physician should do a recto-vaginal examination in which one finger is inserted into the rectum while the other is inserted into the vagina. This procedure provides additional information about the ligaments of the uterus, the ovaries, and the fallopian tubes. The procedure also helps the physician evaluate the health of the rectum.

Although it may be somewhat uncomfortable, the pelvic examination is not ordinarily painful. It is normal for a woman who has not had one, or who is visiting a new doctor, to be anxious about the exam. The doctor should be reassuring if the woman expresses concern. If the doctor is not reassuring, the woman should feel free to consult another doctor. She should not forgo the pelvic examination itself, however. It is essential for early detection of problems.

THE BREASTS

> The degree of attention which breasts receive, combined with the confusion about what the breast fetishists actually want, makes women unduly anxious about them. They can never be just right; they must always be too small, too big, the wrong shape, too flabby. The characteristics of the mammary stereotype are impossible to emulate because they are falsely simulated, but they must be faked somehow or another. Reality is either gross or scrawny.
>
> —Germaine Greer, *The Female Eunuch*

Some college women recall:

> I was very excited about my breast development. It was a big competition to see who was wearing a bra in elementary school. When I began wearing one, I also liked wearing see-through blouses so everyone would know. . . .

> My breasts were very late in developing. This brought me a lot of grief from my male peers. I just dreaded situations like going to the beach or showering in the locker room. . . .
>
> All through junior high and high school I felt unhappy about being "overendowed." I felt just too uncomfortable in sweaters—there was so much to reveal and I was always sure that the only reason boys liked me was because of my bustline. . . .
>
> By the time I was eleven I needed a bra. . . . The girls in my gym class in sixth grade laughed at me because my breasts were pretty big and I still didn't have a bra. I tried to cover myself up when I dressed and undressed. On my eleventh birthday my mom gave me a sailor blouse and inside was my first bra. . . . (It) was the best present I could have received. The bra made me feel a lot better about myself, but I was still unsure of my femininity for a long time. . . .
>
> —Morrison et al., 1980, pp. 66–70

In some cultures, the breasts are viewed merely as biological instruments for feeding infants. In our culture, however, breasts have taken on such erotic significance that a woman's self-esteem may become linked to her bustline.

The breasts are **secondary sex characteristics.** That is, like the rounding of the hips, they distinguish women from men but are not directly involved in reproduction. Each breast contains 15 to 20 clusters of milk-producing **mammary glands** (see Figure 3.8). Each gland opens at the nipple through its own duct. The mammary glands are separated by soft, fatty tissue. It is the amount of this fatty tissue, not the amount of glandular tissue, that largely determines the size of the breasts. Women vary little in their amount of glandular tissue, so breast size does not determine the quantity of milk that can be produced.

Truth or Fiction? **REVISITED**

It is not true that women with larger breasts produce more milk while nursing. It is the amount of fatty tissue, not the amount of glandular milk-producing tissue, that largely determines the size of the breasts. ■

The nipple, which lies in the center of the **areola,** contains smooth muscle fibers that make the nipple become erect when they contract. The areola, or area surrounding the nipple, darkens during pregnancy and remains darker after delivery. Oil-producing glands in the areola help lubricate the nipples during breast feeding. Milk ducts conduct milk from the mammary glands through the nipples. Nipples are richly endowed with nerve endings, so stimulation of the nipples heightens sexual arousal for many women. Male nipples are similar in sensitivity. Gay males often find nipple stimulation pleasurable. Male heterosexuals generally do not. Perhaps heterosexual men have learned to associate breast stimulation with the female sexual role. It is not that their nipples are less sensitive.

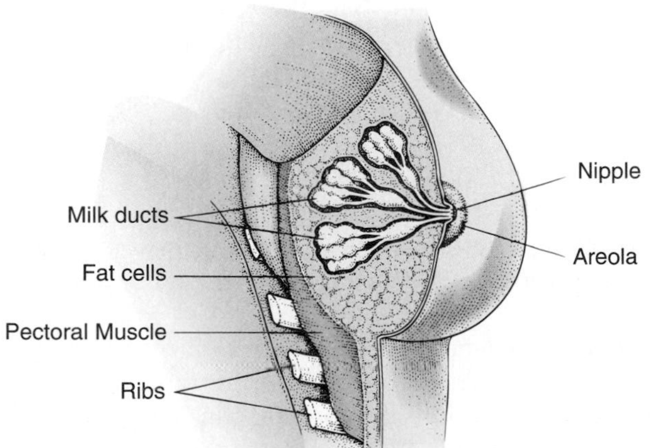

FIGURE 3.8

A Breast of an Adult Woman. This drawing reveals the structures underlying the breast, including milk ducts and fat cells.

Secondary sex characteristics Traits that distinguish women from men but are not directly involved in reproduction.

Mammary glands Milk-secreting glands. (From the Latin *mamma*, which means both "breast" and "mother.")

Areola The dark ring on the breast that encircles the nipple.

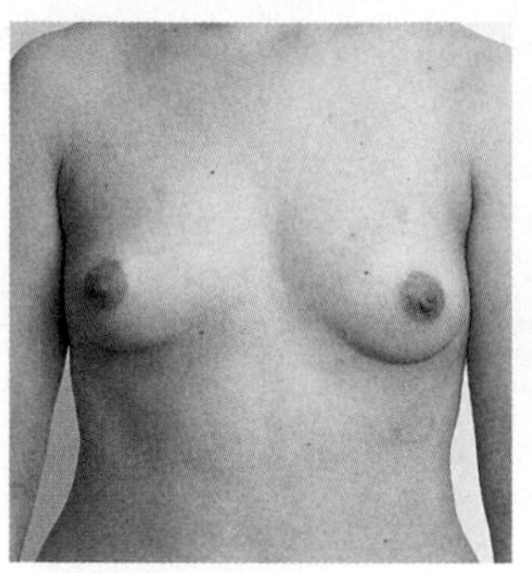
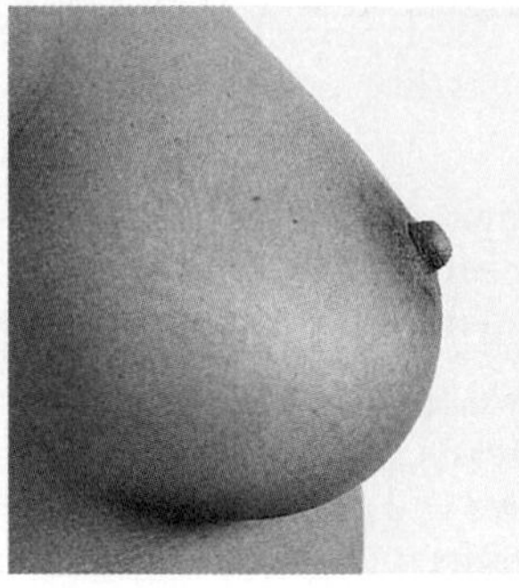
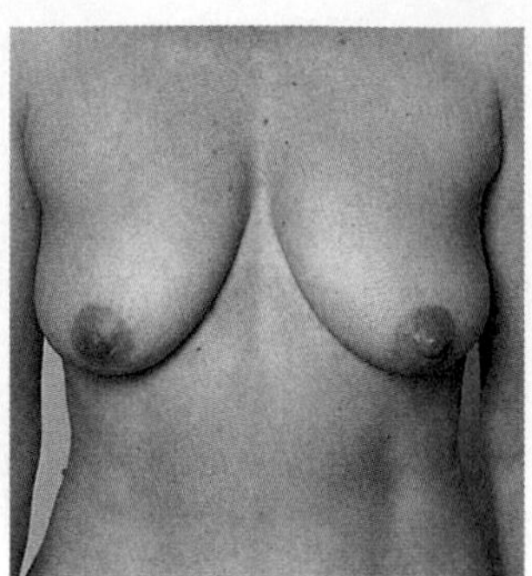

FIGURE 3.9

Normal Variations in the Size and Shape of the Breasts of Adult Women. The size and shape of the breasts have little bearing on ability to produce milk or on sensitivity to sexual stimulation. Breasts have become highly eroticized in our culture.

Figure 3.9 shows normal variations in the size and shape of the breasts of adult women. The sensitivity of the breasts to sexual stimulation is unrelated to their size. Small breasts may have as many nerve endings as large breasts, but they will be more densely packed.

Women can prompt their partners to provide breast stimulation by informing them that their breasts are sensitive to stimulation. They can also guide a partner's hands in ways that provide the type of stimulation they desire. The breasts vary in sensitivity with the phases of the menstrual cycle, and some women appear less responsive to breast stimulation than others. However, some less sensitive women may learn to enjoy breast stimulation by focusing on breast sensations during lovemaking in a relaxed atmosphere.

Breast Cancer

Breast cancer strikes nearly 180,000 women in the United States each year and takes nearly 44,000 lives (Armstrong et al., 2000). (An estimated 240 men also die of breast cancer annually.) The disease takes the lives of nearly 3 of 10 women who develop it. It is not cancer in the breast that causes death, but rather its spread to vital body parts, such as the brain, the bones, the lungs, or the liver.

Despite the popular impression to the contrary, rates of breast cancer in the United States are not on the rise (National Cancer Institute, 2000). Rather, more early cases of breast cancer are being detected because of an increased use of **mammography,** a kind of X-ray technique that detects cancerous lumps in the breast. Advances in early detection and treatment have led to increased rates of recovery. The five-year survival rate for women whose breast cancers have not metastasized—that is, spread beyond the breast—is near 93%, up from 78% in the 1940s (Miller et al., 1996). The five-year survival rate drops to 69% if the cancer has spread to the surrounding region and to 18% if it has spread to distant sites in the body. Overall, the five-year survival rate was about 85% in the 1990s as compared with 75% in the 1970s (Altman, 1998c).

Truth or Fiction? **REVISITED**

It is not true that the incidence of breast cancer is on the rise in the United States. Rather, breast cancer is being detected earlier through means such as mammography. ■

RISK FACTORS Breast cancer is rare in women under age 25. The risk increases sharply with age. About 4 of 5 cases develop in women over the age of 50 (Armstrong et al., 2000). The National Cancer Institute estimates that from birth to age 40, 1 in 217 women will develop breast cancer. By age 50, the risk rises to 1 in 50. By age 60, it rises to 1 in 24, and by age 70, to 1 in 14.

Mammography A special type of X-ray test that detects cancerous lumps in the breast.

African American Women and Breast Cancer

Overall, African Americans are more likely than European Americans to develop cancer. The case is somewhat different with breast cancer. As a group, African American women are somewhat less likely than European American women to develop breast cancer. However, when they do, they frequently do so at an earlier age. They tend to be diagnosed with the disease somewhat later, and they are also more likely to die from it (Jetter, 2000). Some aspects of the racial differences, such as the tendency to be diagnosed later, may reflect less access to health care; but genetic factors are also likely to be involved. It is usually estrogen that causes the proliferation of breast cancer cells, and thus some drugs, such as tamoxifen, treat breast cancer by suppressing the body's supply of estrogen. However, African American women are more likely to develop tumors that are "estrogen-receptor negative." That is, they develop rapidly even in the absence of estrogen. These tumors are highly aggressive—that is, they grow very rapidly—and are a major factor in the higher mortality rate for African American women (Jetter, 2000).

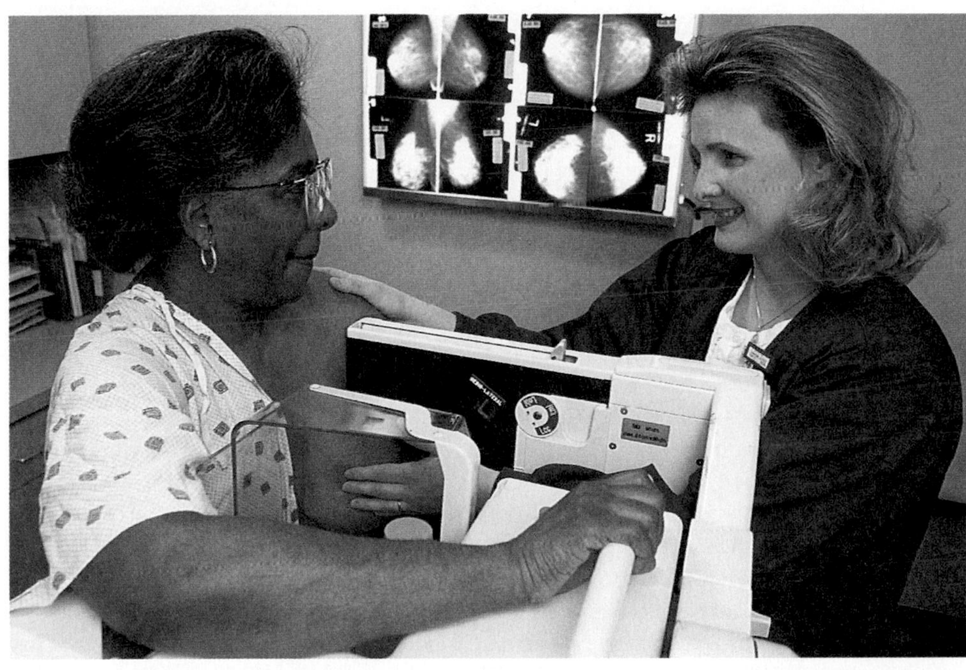

African American Women and Breast Cancer. African American women are generally less likely than European American Women to develop breast cancer. But when they do, the cancer tends to be more aggressive and deadly. The racial difference may be largely genetic.

Genetic factors are involved in breast cancer (Lichtenstein et al., 2000). The risk of breast cancer is higher among women with a family history of the disease (Armstrong et al., 2000). A study of more than 100,000 women nurses showed that those with mothers or sisters who had breast cancer had nearly twice the chance of developing the disease themselves (Colditz et al., 1993). Women who had both a mother and a sister with the disease had between two and three times greater risk. Still, about 80% of women with breast cancer have no family history of the disease. Only about 6% of the women with breast cancer in the nurses study had either a mother or a sister with the disease (Johnson & Williams, 1993).

A key risk factor in breast cancer is prolonged exposure to estrogen, which stimulates breast development in young women and also the proliferation of breast cancer cells (Brody, 1998f). The following all heighten the risk of breast cancer because they increase the woman's exposure to estrogen: early onset of menstruation (before age 14), late menopause (after age 55), delayed childbearing (after age 30), and never giving birth (Brody, 1998f). Despite the fact that birth-control pills contain estrogen, a review of 54 studies involving more than 150,000 women concluded that, generally speaking, there is no connection between using the pill and breast cancer (Gilbert, 1996). However, other research suggests that the pill may increase the risk of breast cancer in women with a family history of the disorder (Grabrick et al., 2000). Exercise, by the way, appears to reduce the risk of breast cancer, presumably by decreasing the amount of fatty tissue in the body (Dreyfuss, 1998). Fat is connected with higher levels of estrogen production.

Heavy drinking of alcohol also heightens the risk of breast cancer (Brody, 1998b; Fuchs, 1995). It has *not* been shown that a diet high in fats contributes to breast cancer (Hunter et al., 1996). Nevertheless, higher amounts of fatty tissue in the body appear to be connected with breast cancer (Dreyfuss, 1998).

Does abortion increase a woman's risk of breast cancer? Some researchers have speculated that pregnancy would cause breast cells to proliferate and that the abortion would prevent the cells from becoming specialized, allowing them to turn cancerous. However, carefully controlled studies do not find abortion to increase a woman's risk of breast cancer (Hartge, 1997; Melbye at al., 1997; Tang et al., 2000).

Truth or Fiction? **REVISITED**

It is *not* true that abortion increases a woman's risk of breast cancer. ■

DETECTION AND TREATMENT Women with breast cancer have lumps in the breast, *but most lumps in the breasts are not cancerous.* Most are either **cysts** or **benign** tumors called **fibroadenomas.** Breast cancer involves lumps in the breast that are **malignant.**

Early detection and treatment reduce the risk of mortality. The sooner cancer is detected, the less likely it is to have spread to critical organs.

Breast cancer may be detected in various ways, including breast self-examination, physical examination, and mammography. Through mammography, tiny, highly curable cancers can be detected—and treated—before they can be felt by touch (Brody, 1995a). By the time a malignant lump is large enough to be felt by touch, it already contains millions of cells. It may even have metastasized—that is, splintered off to form colonies elsewhere in the body. Mammography can detect tiny tumors before metastasis. One study found that 82% of women whose breast cancers were detected early by mammography survived for at least five years following surgery, as compared to 60% of those whose cancers were discovered later. Note that hormone replacement therapy can impair the accuracy of mammographic screening (Kavanagh et al., 2000).

We must note that a recent, large-scale Canadian study calls into question whether women with breast cancer that has been detected via mammography have a higher survival rate than women with breast cancer that is detected by physical examination. This study of nearly 40,000 women followed subjects for several years at a number of health centers (Miller et al., 2000). It was found that mammography is apparently no better than physical examination at reducing mortality from breast cancer (Miller et al., 2000). On the surface, it may appear that a woman can forgo mammography, but let us note several factors that give us pause. First of all, the women in the study all received careful regular physical examinations and self-screening. Moreover, the women receiving mammography in addition to physical examination detected cancerous growths some two years earlier than the women receiving physical examination alone. Therefore, they had a significantly longer period of time in which to undergo cancer treatment. It is true that this particular study did not find that the additional time was connected with a higher survival rate. However, it seems unwarranted to "throw away" the potential added benefits of the earlier warning. Moreover, this study, large-scale though it was, stands alone for the time being. We would want to see the results replicated.

Early detection may offer another benefit. Smaller cancerous lumps can often be removed by **lumpectomy,** which spares the breast. More advanced cancers are likely to be treated by **mastectomy.**

Many drugs are also used to treat breast cancer, and others are in the research pipeline. For example, *tamoxifen* locks into the estrogen receptors of breast cancer cells, thereby blocking the effects of estrogen that would otherwise stimulate the cells to grow

Cysts Saclike structures filled with fluid or diseased material.

Benign Doing little or no harm.

Fibroadenoma A benign, fibrous tumor.

Malignant Lethal; causing or likely to cause death.

Lumpectomy Surgical removal of a lump from the breast.

Mastectomy Surgical removal of the entire breast.

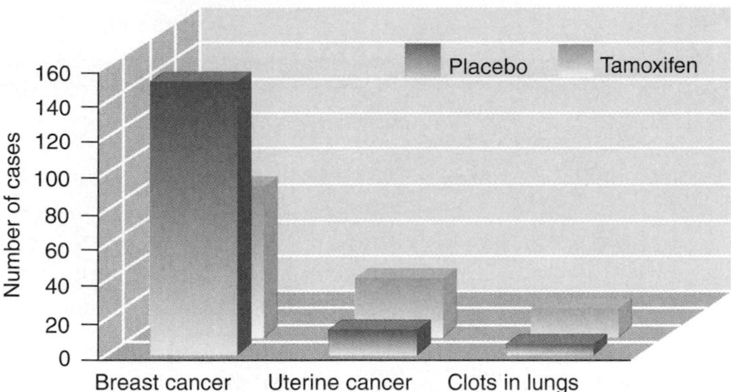

FIGURE 3.10

Effects of Tamoxifen in the National Cancer Institute Study. Over the 3½ years of the study, tamoxifen reduced the risk of breast cancer by about 45%. However, it increased the risks of uterine cancer and blood clots in the lungs. The study included 13,338 women, of whom about half were assigned to tamoxifen, and half to a placebo.

and proliferate. The National Cancer Institute conducted experimental research with 13,338 women at high risk—becuase of either family history or the appearance of precancerous lesions in the breast (Altman, 1998b). However, tamoxifen increases the risks of uterine cancer and of blood clots in the lungs, along with some other side effects (see Figure 3.10). The risks of these side effects were lowest among women below the age of 50. The drug raloxifene (sold under the brand name Evista), which is intended to treat osteoporosis and prevent heart problems in postmenopausal women (Marwick, 2000), has also been shown to reduce the risk of breast cancer dramatically (Cummings et al., 1999). Moreover, raloxifene does not appear to have the side effects associated with tamoxifen. Many other drugs, including Taxol (already approved by the Food and Drug Administration), and Herceptin, are also being studied for use against breast cancer. Ask your gynecologist for the latest research results and for advice about which drugs, if any, are right for you, given your risk of breast cancer, your age, and other factors.

Many women who have had mastectomies have had surgical breast implants to replace the tissue that has been removed. Other women have breast implants to augment their breast size. Articles published in *The New England Journal of Medicine* and *The Journal of the American Medical Association*, suggest that breast implants appear to have no effect on the probability of developing breast cancer (Bryant & Brasher, 1995), rheumatoid arthritis (Kolata, 1996a), and a number of other health problems (Sanchez-Guerrero et al., 1995), casting doubts on previous studies that had implicated them in the development of these problems. However, this issue remains quite controversial. Readers are advised to consult their gynecologists for the latest research information.

The American Cancer Society (2000) recommends that women have a breast exam every three years when they are between 20 and 39 years of age, every one or two years between ages 40 and 49, and annually thereafter. They also recommend that young women receive a baseline mammogram for comparison with later tests. The mortality rate from breast cancer may be cut by 30% or more if women follow these guidelines (Chlebowski, 2000). Mammography is not foolproof, however. Thus, the chances of early detection are optimized through a combination of monthly breast self-examinations (see the nearby "A Closer Look"), annual breast examinations by a physician, and regular mammograms.

Questions have been raised about the recommendation that women in their 40s obtain regular mammograms. The results of a controversial 1992 Canadian study showed that women in their 40s who had received mammograms had the same death rates from breast cancer as women who did not (Kolata, 1993a, 1993b). Yet analysis of data from eight major studies, including the Canadian study, suggests that mammography in younger women can reduce their death rate from breast cancer by about 14% (Smart et al., 1995). When the results of the Canadian study are left out, analysis suggests that women of ages 40 to 49 can reduce their death rate from breast cancer by 23% (Smart et al., 1995). Mammograms reduce the death rate from breast cancer in women age 50 and above by 25% to 30% (Chlebowski, 2000). Why the difference? It is possible that because

a CLOSER look

Breast Self-Examination

Regular breast self-examination and regular visits to a physician provide the best protection against breast cancer, because they may lead to early detection and treatment. The woman should conduct a breast self-examination at least once a month, preferably about a week after her period ends (when the breasts are least influenced by hormones), so that any changes can be reported promptly to a physician (see Figure 3.11).

The following instructions for breast self-examination are based on American Cancer Society guidelines. Additional material on breast self-examination may be obtained from the American Cancer Society at this toll-free number: (800) ACS-2345. However, women are advised to undertake their initial breast self-examinations with a physician in order to determine what degree of "lumpiness" seems normal for them and to learn the proper technique. Breast self-examinations can be anxiety-provoking experiences, and it appears that using the proper method is essential (Zuger, 1998).

1. *In the shower.* Examine your breasts during your bath or shower; hands glide more easily over wet skin. Keep your fingers flat and move gently over every part of each breast. Use the right hand to examine the left breast and the left hand for the right breast. Check for any lump, hard knot, or thickening.
2. *Before a mirror.* Inspect your breasts with your arms at your sides. Next, raise your arms high overhead. Look for any changes in the contour of each breast, a swelling, dimpling of skin, or changes in the nipple. Then rest your palms on your hips and press down firmly to flex your chest muscles. Your left and right breasts will not exactly match. Few women's breasts do. Regular inspection shows what is normal for you and will give you confidence in your examination.
3. *Lying down.* To examine your right breast, put a pillow or folded towel under your right shoulder. Place your right arm behind your head. This position distributes breast tissue more evenly on the chest. With your left hand, fingers flat, press gently with the finger pads (the top thirds of the fingers) of the three middle fingers in small circular motions around an imaginary clock face. Begin at the outermost top of your right breast for 12 o'clock, then move to 1 o'clock, and so on around the circle back to 12. A ridge of firm tissue in the lower curve of each breast is normal. Then move in 1 inch, toward the nipple. Keep circling to examine *every part of your breast,* including the nipple. This requires at least three more circles. Now slowly repeat the procedure on your left breast. Put the pillow beneath your left shoulder, place your left arm behind your head, and use the finger pads on your right hand.

After you examine your left breast fully, squeeze the nipple of each breast gently between your thumb and index finger. Any discharge, clear or bloody, should be reported to your doctor immediately.

younger women have denser breasts, mammograms in younger women are more difficult for doctors to read and so may be less reliable indicators of early cancers.

The decision whether a woman should have a mammogram rests with the woman and her doctor. We note, however, that only one study—the 1992 Canadian study—suggests that mammography is not of use in women under age 50. Moreover, the Canadian study has been criticized for relatively small sample size, relatively brief follow-ups of the women in the study, and failure to use the most up-to-date screening technology (Brody, 1995a). We therefore support the recommendations of the American Cancer Society that women of ages 40 to 50 have a mammogram every one to two years, and that women age 50 and above have an annual mammogram.

In sum, the National Cancer Institute estimates than 1 woman in 8 will still eventually develop breast cancer (Altman, 1998b). Breast cancer remains the second leading cancer killer among women, behind lung cancer. Yet there is good news—very good news—about breast cancer. The five-year survival rate for women who develop breast cancer is now 85% as compared with 75% in the 1970s (Altman, 1998c). And if the cancer is detected early—before it metastasizes—the five-year survival rate is about 93%, up from 78% in the 1940s (Miller et al., 1996).

More than ever is known about the development of breast cancer. Breast cancer that develops early, for example, is believed that be largely genetic, and genetic markers for early-onset breast cancer can indicate whether a particular woman is at risk (ask your gy-

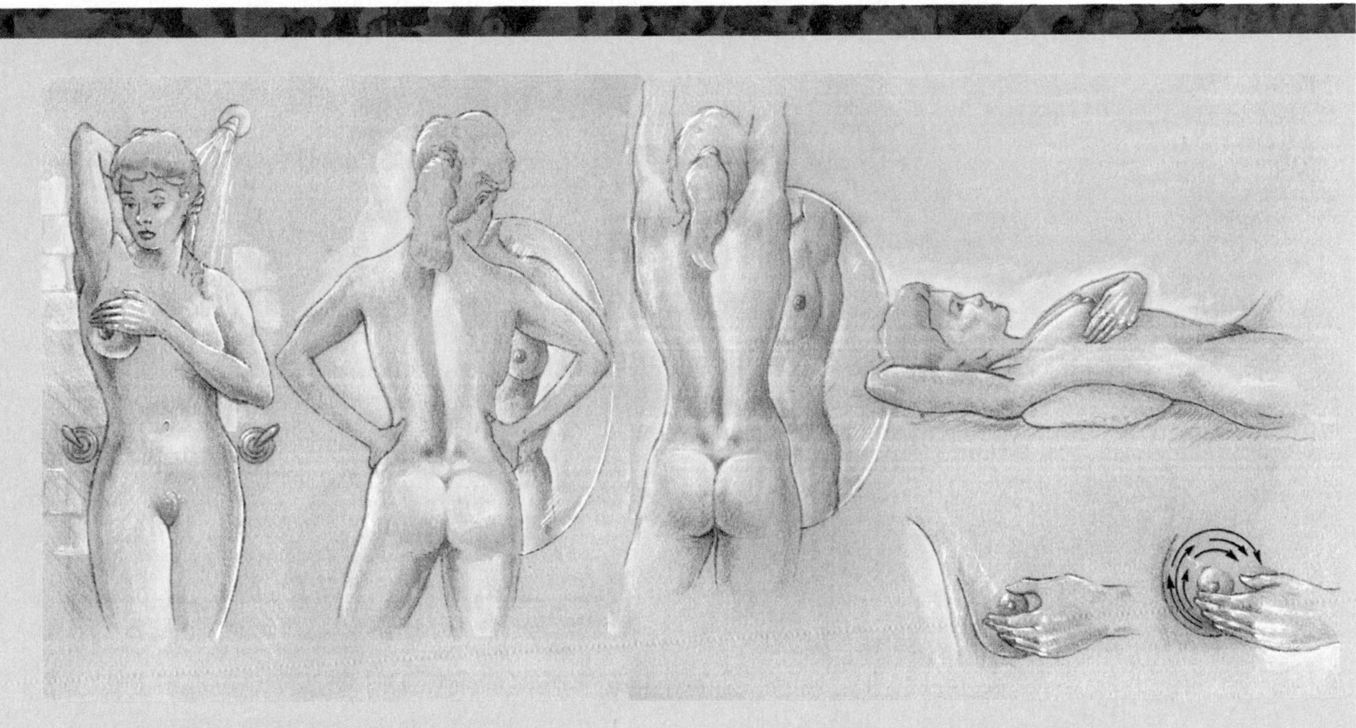

FIGURE 3.11

A Woman Examines Her Breast for Lumps.

Research shows that many women who know how to do breast self-examinations do not do them regularly. Why? There are many reasons, including fear of what one will find and doubts about whether self-examination will make a difference (Kash, 1998). But the most frequently mentioned reasons are being too busy and forgetting (Friedman et al., 1994).

necologist). Prolonged exposure to estrogen is a known risk factor because estrogen stimulates the proliferation of breast cancer cells as well as breast development in young women (Brody, 1998f). Exercise is believed to reduce the risk of breast cancer by decreasing the amount of fat in the body (Chlebowski, 2000). (Estrogen is produced by fat cells as well as by the ovaries.)

Some women who are at high risk for breast cancer choose to have their breasts removed prophylactically—that is, to prevent the development of breast cancer. They then typically have breast implants so that their breasts appear to be normal. Obviously, this preventive approach is radical. However, follow-up research suggests that it does have the positive effect of decreasing women's concern about breast cancer (Frost et al., 2000). On the other hand, there are the possibilities of problems with the implants and the reconstructive surgery, and some women are unhappy with the outcomes (Frost et al., 2000).

Mammography also allows for the detection of tiny, highly curable cancers before they can be felt by clinical examination or touch (Brody, 1995a). Early detection allows many women to have small cancerous lumps removed by lumpectomy (surgical removal of the lump that contains the mass of cancer cells) rather than by mastectomy (removal of the entire breast).

An increasing arsenal of anti-breast-cancer medicines are also available or under development. For example, the drug *tamoxifen* locks into the estrogen receptors of breast cancer cells, thus blocking the effects of estrogen. Tamoxifen apparently reduces the probability of developing breast cancer by about 45% (Chlebowski, 2000). *Raloxifene* is used to treat osteoporosis and prevent heart problems in postmenopausal women (Marwick, 2000), but it also reduces the risk of breast cancer (Cummings et al., 1999). Other drugs, including *Taxol* and *Herceptin*, are being studied for use against breast cancer (Altman, 1998c).

Your gynecologist should have the latest research results. Be proactive. Seize the opportunity to evaluate whether you have breast cancer. There are more treatment options than ever; the survival rates are higher than ever; and earlier detection is often connected with less radical treatments.

THE MENSTRUAL CYCLE

Menstruation is the cyclical bleeding that stems from the shedding of the uterine lining (endometrium.) Menstruation takes place when a reproductive cycle has not led to the fertilization of an ovum. The word *menstruation* derives from the Latin *mensis*, which means "month." The human menstrual cycle averages about 28 days in length.

The menstrual cycle is regulated by the hormones estrogen and progesterone and can be divided into four phases. During the first phase of the cycle, the *proliferative phase*, which follows menstruation, estrogen levels increase, causing the ripening of perhaps 10 to 20 ova (egg cells) within their follicles and the proliferation of endometrial tissue in the uterus. During the second phase of the cycle, estrogen reaches peak blood levels, and **ovulation** occurs. Normally only 1 ovum reaches maturity and is released by an ovary during ovulation. Then the third phase of the cycle—the *secretory*, or *luteal*, phase—begins right after ovulation and continues through the beginning of the next cycle.

The term *luteal phase* is derived from **corpus luteum,** the name given the follicle that releases an ovum. The corpus luteum functions as an **endocrine gland** and produces large amounts of progesterone and estrogen. Progesterone causes the endometrium to thicken, so that it will be able to support an embryo if fertilization occurs. If the ovum goes unfertilized, however, estrogen and progesterone levels plummet. These falloffs trigger the fourth phase, the *menstrual phase*, which leads to the beginning of a new cycle.

Ovulation may not occur in every menstrual cycle. Anovulatory ("without ovulation") cycles are most common in the years just after **menarche.** They may become frequent again in the years prior to menopause, but they may also occur irregularly among women in their 20s and 30s.

Although the menstrual cycle averages about 28 days, variations among women, and in the same woman from month to month, are quite common. Girls' cycles often are irregular for a few years after menarche but later assume reasonably regular patterns. Variations from cycle to cycle tend to occur during the proliferative phase that precedes ovulation. That is, menstruation tends to follow ovulation reliably by about 14 days. Variations of more than 2 days in the postovulation period are rare.

Although hormones regulate the menstrual cycle, psychological factors can influence the secretion of hormones. Stress can delay or halt menstruation. Anxiety that she may be pregnant and thus miss her period may also cause a woman to be late.

Menstruation Versus Estrus

The menstrual cycle is found only in women, female apes, and female monkeys. It differs from an ovarian cycle, or **estrous cycle,** which occurs in "lower" mammals such as rodents, cats, and dogs. **Estrus** is the periodic sexual excitement (otherwise referred to as being "in heat") when the female is most receptive to the advances of the male. Estrus occurs when the animal is ovulating and hence most likely to conceive offspring. Women (and other female primates), however, ovulate about halfway through the cycle and may be interested in sexual activity at any time during their cycles. Estrous cycles, moreover, may be characterized by little bleeding ("spotting") or no bleeding. Menstruation typically involves a heavier flow of blood.

Menstruation The cyclical bleeding that stems from the shedding of the uterine lining (endometrium).

Ovulation The release of an ovum from an ovary.

Corpus luteum The follicle that has released an ovum and then produces copious amounts of progesterone and estrogen during the luteal phase of a woman's cycle. (From Latin roots that mean "yellow body.")

Endocrine gland A ductless gland that releases its secretions directly into the bloodstream.

Menarche The first menstrual period.

Estrous cycle The female reproductive cycle of most mammals (other than primates), which is under hormonal control and includes a period of heat, followed by ovulation.

Estrus The periodic sexual excitement during which female mammals (other than primates) are most receptive to the sexual advances of males.

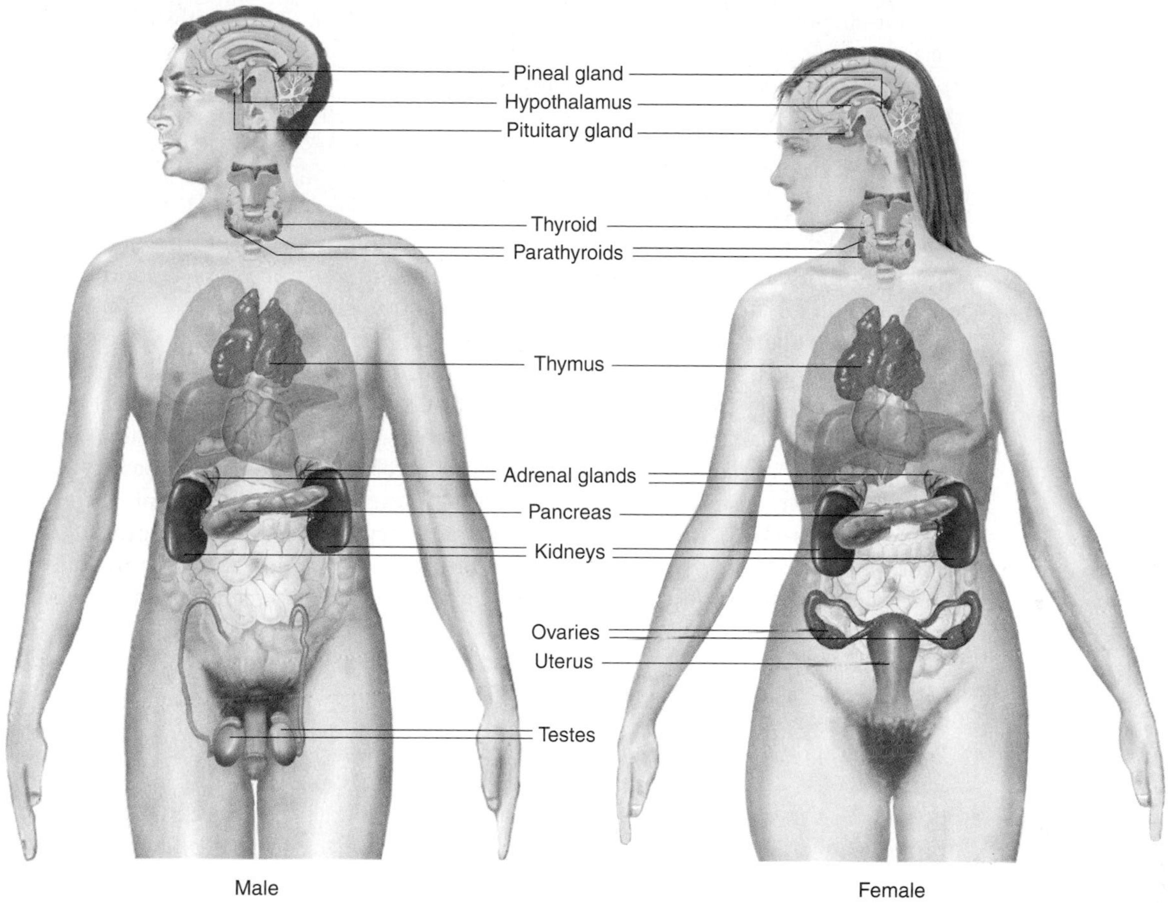

FIGURE 3.12

Major Glands of the Endocrine System. The endocrine system consists of glands that secrete chemicals called hormones directly into the bloodstream.

Regulation of the Menstrual Cycle

The menstrual cycle involves finely tuned relationships between structures in the brain—the **hypothalamus** and the **pituitary gland**—and the ovaries and uterus. All these structures are parts of the endocrine system, which means that they secrete chemicals directly into the bloodstream (see Figure 3.12). The ovaries and uterus are also reproductive organs. The chemicals secreted by endocrine glands are called **hormones.** (Other bodily secretions, such as milk, saliva, sweat, and tears, arrive at their destinations by passing through narrow, tubular structures in the body called ducts.)

Behavioral and social scientists are especially interested in hormones because of their behavioral effects. Hormones regulate bodily processes such as the metabolic rate, growth of bones and muscle, production of milk, metabolism of sugar, and storage of fats. Several hormones play important roles in sexual and reproductive functions.

The gonads—the **testes** (or testicles) in the male and the ovaries in the female—secrete sex hormones directly into the bloodstream. The female gonads, the ovaries, produce the sex hormones estrogen and progesterone. The male gonads, the testes, produce the male sex hormone **testosterone.** Males and females also produce sex hormones of the other gender, but in relatively small amounts.

The hypothalamus is a pea-sized structure in the front part of the brain. It weighs about 4 to 5 grams and lies above the pituitary gland and below (hence the prefix *hypo-*, for "under") the thalamus. Despite its small size, it is involved in regulating many states of motivation, including hunger, thirst, aggression, and sex. For example, when the rear part of a

Hypothalamus A bundle of neural cell bodies near the center of the brain that are involved in regulating body temperature, motivation, and emotion.

Pituitary gland The gland that secretes growth hormone, prolactin, oxytocin, and others.

Hormone A substance that is secreted by an endocrine gland and regulates various body functions. (From the Greek *horman,* which means "to stimulate" or "to goad.")

Testes The male gonads.

Testosterone The male sex hormone that fosters the development of male sex characteristics and is connected with the sex drive.

male rat's hypothalamus is stimulated by an electric probe, the rat runs through its courting and mating sequence. It nibbles at a female's ears and at the back of her neck. When she responds, they copulate. Human sexuality is not so stereotyped or mechanical, although in the cases of some people who are highly routinized in their behavior, it may appear so.

The pituitary gland, which is about the size of a pea, lies below the hypothalamus at the base of the brain. Because many pituitary secretions regulate other endocrine glands, the pituitary has also been called the *master gland.* Pituitary hormones regulate bone and muscle growth and urine production. Two pituitary hormones are active during pregnancy and motherhood: **prolactin,** which stimulates production of milk; and **oxytocin,** which stimulates uterine contractions in labor and the ejection of milk during nursing. The pituitary gland also produces **gonadotropins** (literally, "that which feeds the gonads") that stimulate the ovaries: **follicle-stimulating hormone (FSH)** and **luteinizing hormone (LH).** These hormones play central roles in regulating the menstrual cycle.

The hypothalamus receives information about bodily events through the nervous and circulatory systems. It monitors the blood levels of various hormones, including estrogen and progesterone, and releases a hormone called **gonadotropin-releasing hormone (Gn-RH),** which stimulates the pituitary to release gonadotropins. Gonadotropins, in turn, regulate the activity of the gonads. It was once thought that the pituitary gland ran the show, but it is now known that the pituitary gland is regulated by the hypothalamus. Even the "master gland" must serve another.

Phases of the Menstrual Cycle

We noted that the menstrual cycle has four stages or phases: the proliferative, ovulatory, secretory, and menstrual stages (see Figure 3.13). It might seem logical that a new cycle begins with the first day of the menstrual flow, this being the most clearly identifiable event of the cycle. Many women also count the days of the menstrual cycle beginning with the onset of menstruation. Biologically speaking, however, menstruation is really the culmination of the cycle. In fact, the cycle begins with the end of menstruation and the initiation of a series of biological events that lead to the maturation of an immature ovum in preparation for ovulation and possible fertilization.

THE PROLIFERATIVE PHASE The first phase, or the **proliferative phase,** begins with the end of menstruation and lasts about 9 or 10 days in an average 28-day cycle (see Figures 3.13 and 3.14). During this phase the endometrium develops, or "proliferates."

Prolactin A pituitary hormone that stimulates production of milk.

Oxytocin A pituitary hormone that stimulates uterine contractions in labor and the ejection of milk during nursing.

Gonadotropins Pituitary hormones that stimulate the gonads. (Literally, "that which feeds the gonads").

Follicle-stimulating hormone (FSH) A gonadotropin that stimulates the development of follicles in the ovaries.

Luteinizing hormone (LH) A gonadotropin that helps regulate the menstrual cycle by triggering ovulation.

Gonadotropin-releasing hormone (Gn-RH) A hormone that is secreted by the hypothalamus and stimulates the pituitary to release gonadotropins.

Proliferative phase The first phase of the menstrual cycle, which begins with the end of menstruation and lasts about 9 or 10 days. During this phase, the endometrium proliferates.

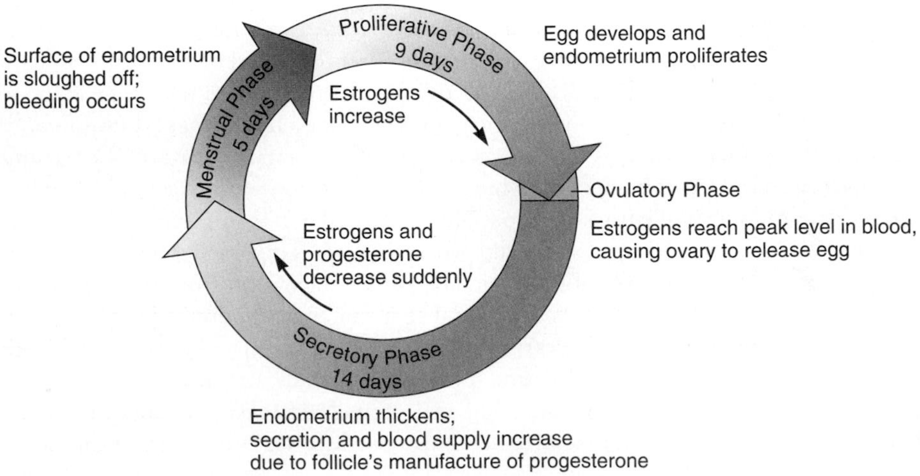

FIGURE 3.13

The Four Phases of the Menstrual Cycle. The menstrual cycle consists of the proliferative, ovulatory, secretory (luteal), and menstrual phases.

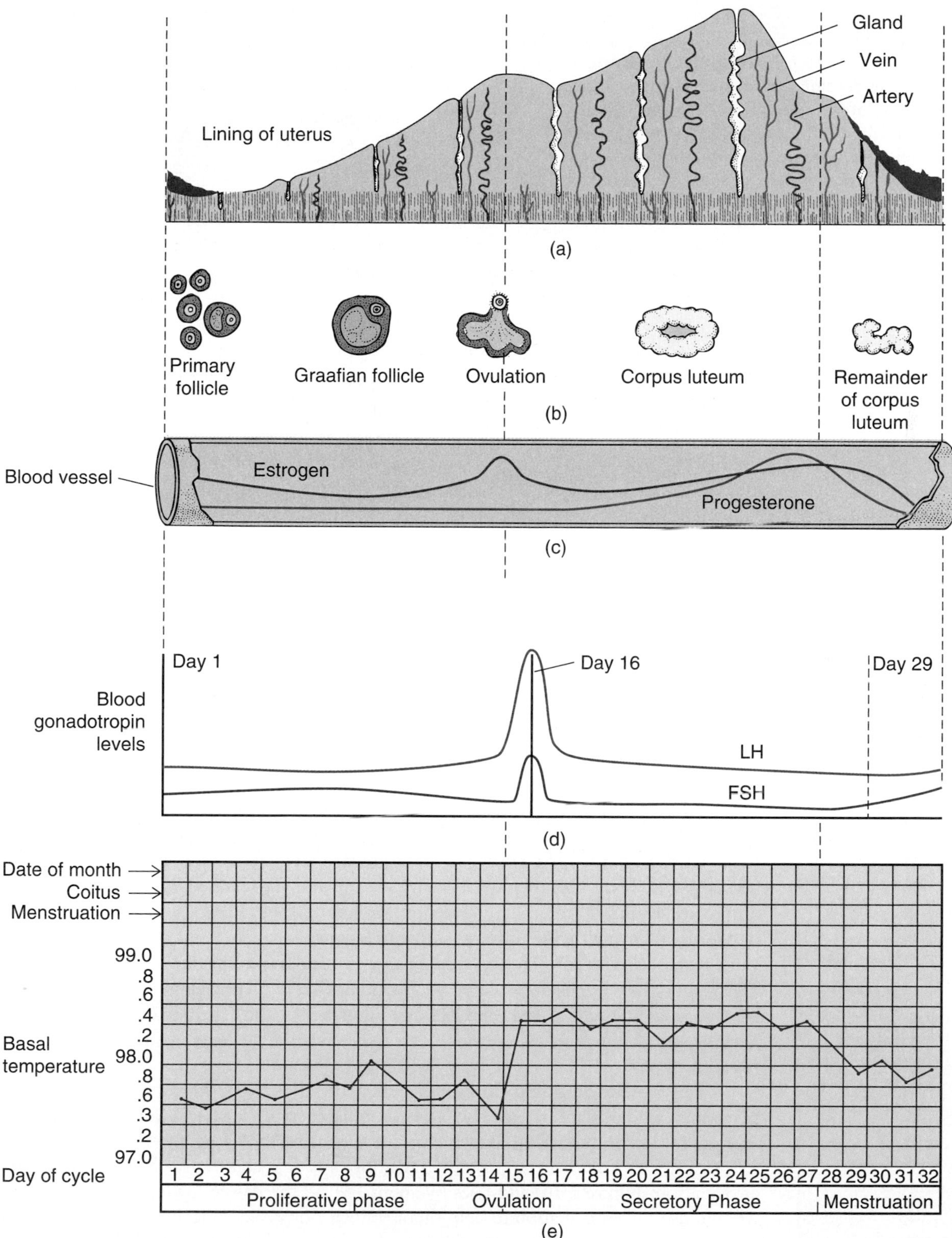

FIGURE 3.14

Changes That Occur During the Menstrual Cycle. This figure shows five categories of biological change: (a) changes in the development of the uterine lining (endometrium), (b) follicular changes, (c) changes in blood levels of ovarian hormones, (d) changes in blood levels of pituitary hormones, and (e) changes in basal temperature. Note the dip in temperature that is connected with ovulation.

FIGURE 3.15

Maturation and Eventual Decomposition of an Ovarian Follicle. Many follicles develop and produce estrogen during the proliferative phase of the menstrual cycle. Usually only one, the graafian follicle, ruptures and releases an ovum. The graafian follicle then develops into the corpus luteum, which produces copious quantities of estrogen and progesterone. When fertilization does not occur, the corpus luteum decomposes.

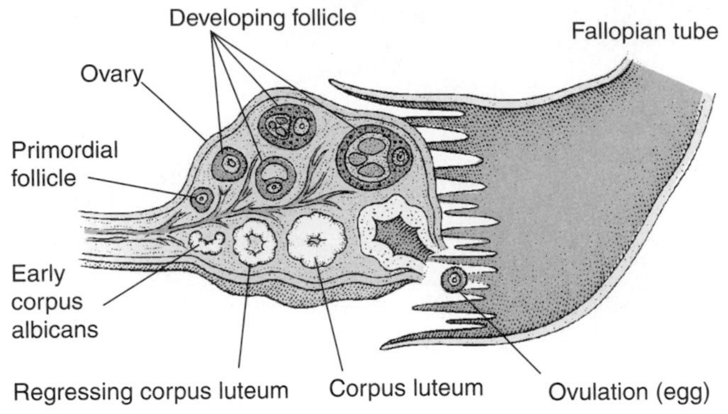

This phase is also known as the *preovulatory* or *follicular phase*, because certain ovarian follicles mature and the ovaries prepare for ovulation.

Low levels of estrogen and progesterone are circulating in the blood as menstruation draws to an end. When the hypothalamus senses a low level of estrogen in the blood, it increases its secretion of Gn-RH, which in turn triggers the pituitary gland to release FSH. When FSH reaches the ovaries, it stimulates some follicles (perhaps 10 to 20) to begin to mature. As the follicles ripen, they all start to produce estrogen. Normally, however, only one of them—called the *graafian follicle*—will reach full maturity in the days just preceding ovulation. As the graafian follicle matures, it moves toward the surface of the ovary, where it will eventually rupture and release a mature egg (see Figures 3.14 and 3.15).

Estrogen causes the endometrium in the uterus to thicken to about 1/8 inch. Glands develop that will eventually nourish the embryo if fertilization occurs. Estrogen also stimulates the appearance of a thin cervical mucus. This mucus is alkaline and provides a hospitable, nutritious medium for sperm. The chances are thus increased that sperm that enter the female reproductive system at the time of ovulation will remain viable.

THE OVULATORY PHASE During ovulation, or the **ovulatory phase,** the graafian follicle ruptures and releases a mature ovum *near* a fallopian tube, not actually *into* a fallopian tube (see Figure 3.15). The other ripening follicles degenerate and are harmlessly reabsorbed by the body. Occasionally, two ova mature and are released during ovulation, and if both are fertilized, fraternal (nonidentical) twins develop. Identical twins develop when one fertilized ovum divides into two separate **zygotes.**

Ovulation is set into motion when estrogen production reaches a critical level. The high level of estrogen is detected by the hypothalamus, which triggers the pituitary to release copious amounts of FSH and LH (see Figures 3.14 and 3.15). The surge of LH triggers ovulation, which usually begins 12 to 24 hours after the level of LH in the body has reached its peak.

The synthetic hormone **clomiphene** is chemically similar to LH and has been used by women who ovulate irregularly to induce reliable ovulation. The induction and accurate prediction of the timing of ovulation increase the chances of conceiving.

A woman's *basal body temperature*, taken by oral or rectal thermometer, dips slightly at ovulation (Figure 3.14) and rises by about 1°F on the day following ovulation. Many women use this information to help them conceive or avoid conceiving. Note, however, that Figure 3.15 is idealized. Many women show greater fluctuations in daily temperature or gradual rises in temperature for about two days after ovulation.

Some women have discomfort or cramping during ovulation, a condition termed **mittelschmerz.** Mittelschmerz is sometimes confused with appendicitis. Mittelschmerz, however, may occur on either the right or the left side of the abdomen, de-

Ovulatory phase The second stage of the menstrual cycle, during which a follicle ruptures and releases a mature ovum.

Zygote A fertilized ovum (egg cell).

Clomiphene A synthetic hormone that is chemically similar to LH and induces ovulation.

Mittelschmerz Pain that occurs during ovulation. (German for "middle pain," reflecting the fact that the pain occurs midway between menstrual periods.)

pending on which ovary is releasing an ovum. A ruptured appendix always causes pain on the right side.

THE SECRETORY PHASE The phase following ovulation is called the postovulatory or **secretory phase.** Some people refer to it as the *luteal phase,* which reflects the name given the ruptured (graafian) follicle—the *corpus luteum.* Figures 3.14 and 3.15 show the transformation of the graafian follicle into the corpus luteum.

Under the influence of LH, the corpus luteum, which has remained in the ovary, begins to produce large amounts of progesterone and estrogen. Levels of these hormones peak at around the 20th or 21st day of an average cycle (see Figure 3.14). These hormones cause the glands in the endometrium to secrete nutrients to sustain a fertilized ovum that becomes implanted in the uterine wall.

If implantation does not occur, the hypothalamus responds to the peak levels of progesterone in the blood by signaling the pituitary to stop producing LH and FSH. Although certainly more complex, this feedback process is similar to that of a thermostat in a house reacting to rising temperatures by shutting down the furnace. The levels of LH and FSH decline rapidly, leading the corpus luteum to decompose. After its decomposition, levels of estrogen and progesterone fall precipitously. In this sense, the corpus luteum sows the seeds of its own destruction: Its hormones signal the brain to shut down secretion of substances that maintain it.

THE MENSTRUAL PHASE: AN END AND A BEGINNING The **menstrual phase** is the sloughing off of the uterine lining (the endometrium) in the menstrual flow. Menstruation occurs when estrogen and progesterone levels decline to the point where they can no longer sustain the uterine lining. The lining then disintegrates and is discharged from the body along with the menstrual flow. Menstruation itself is the passing of the lining through the cervix and vagina.

The low estrogen levels of the menstrual phase signal the hypothalamus to release Gn-RH, which in turn stimulates the pituitary to secrete FSH. FSH prompts ovarian secretion of estrogen and the onset of another proliferative phase. Thus a new cycle begins. The menstrual phase is a beginning as well as an end.

Menstrual flow contains blood from the endometrium (uterine lining), endometrial tissue, and cervical and vaginal mucus. Although the flow can appear persistent and may last for five days or more, most women lose only a total of 2 or 3 ounces of blood (4 to 6 tablespoons). A typical blood donor, by contrast, donates 16 ounces of blood at a sitting. A woman's blood loss through menstruation is thus usually harmless. Extremely heavy or prolonged (over a week) menstrual bleeding may reflect health problems and should be discussed with a health care provider.

Prior to 1933, women generally used external sanitary napkins or pads to absorb the menstrual flow. In that year, however, **tampons** were introduced and altered the habits of millions of women. Women who use tampons can swim without concern while menstruating, can wear more revealing or comfortable apparel, and feel generally less burdened.

Tampons are inserted into the vagina and left in place to absorb menstrual fluid. In recent years, questions have arisen about whether or not tampons cause or exacerbate infections. For example, tampon use has been linked to toxic shock syndrome (TSS), an infection that is sometimes fatal. Signs of TSS include fever, headache, sore throat, vomiting, diarrhea, muscle aches, rash, and dizziness. Peeling skin, disorientation, and a plunge in blood pressure may follow.

TSS is caused by the *Staphylococcus aureus* ("staph") bacterium. In 1980, the peak year for cases of TSS, there were 344 recorded cases of TSS and 28 fatalities. Procter & Gamble removed its "extra absorbent" tampon Rely from the market when it was discovered that 71% of these TSS victims had used the product. Plugging the vagina

Secretory phase The third phase of the menstrual cycle, which follows ovulation. Also referred to as the *luteal phase,* after the *corpus luteum,* which begins to secrete large amounts of progesterone and estrogen following ovulation.

Menstrual phase The fourth phase of the menstrual cycle, during which the endometrium is sloughed off in the menstrual flow.

Tampon A cylindrical plug of cotton that is inserted into the vagina and left in place to absorb menstrual fluid. (A French word meaning a gun barrel "plug.")

Historical and Cross-Cultural Perspectives on Menstruation

In Peru, they speak of a "visit from Uncle Pepe," whereas in Samoa, menstruation is referred to as "the boogie man." One of the more common epithets given menstruation through the course of history is "the curse." The Fulani of Upper Volta in Africa use a term for it that translates as "to see dirt." Nationalism also rises to the call, with some nations blaming "the curse" on their historical enemies. In earlier times, the French referred to menstruation as "the English" and to its onset as "the English are coming."

It is a common folk belief that menstruating women are contaminated. Men thus avoid contact with menstruating women for fear of their lives. To prevent their contaminating others, menstruating women in tribal societies may be dispatched to special huts on the fringe of the village. In the traditional Navajo Indian culture, for instance, menstruating women were consigned to huts set apart from other living quarters. In many Islamic societies, a menstruating woman is considered polluted and is not permitted either to pray or to enter a mosque.

Women in industrialized nations are not consigned to special huts, but throughout the history of Western culture, menstruation has been seen as unclean, contaminating, and even magical. In A.D. 77, the Roman historian Pliny summed up Roman misbeliefs about menstrual blood:

> Contact with it turns new wine sour, crops touched by it become barren, grafts die, seeds in gardens are dried up, the fruit of trees falls off. . . . [The] edge of steel and the gleam of ivory are dulled, hives of bees die, even bronze and iron are at once seized by rust, and a horrible smell fills the air; to taste it drives dogs mad and infects their bites with an incurable poison.

Truth or Fiction? REVISITED

Yes, the ancient Romans did believe that menstrual blood soured wine and killed crops. Negative views of menstruation and menstrual blood have persisted throughout history. ■

The Old Testament (Leviticus 15:19) warns against any physical contact with a menstruating woman, including, of course, coitus:

> And if a woman have an issue, and her issue in her flesh be blood, she shall be put apart seven days; and whosoever toucheth her shall be unclean.

Orthodox Jews still abstain from coitus during menstruation and the week afterward. Prior to resuming sexual relations, the woman must attend a *mikvah*, a ritual cleansing in which the genitals are washed and all foreign substances are removed. Such practices are grounded in religious tradition, not science. The attitudes of people in ancient and present-day preliterate societies toward menstruation should be understood in terms of their limited grasp of bodily processes. Because they lack knowledge that menstruation is a part of the woman's natural menstrual cycle, they may regard the flow of blood from a woman's loins as a sign of filth or contamination and believe that it is healthier for others temporarily to avoid physical contact with her. Science teaches that there is no medical basis to these fears and that there are no dangers in menstrual coitus—despite the persistence of superstition.

Fears of contamination by menstruating women are nearly universal across cultures and remain quite current in some circles. As late as the 1950s, women were not allowed in some European breweries for fear that the presence of a menstruating woman would turn the beer sour. Some Indian castes still teach that a man who touches a woman during menses becomes contaminated and must be purified by a priest.

We might laugh off these misconceptions as folly and ignorance, if it were not for their profound effect on women. Women who believe the myths about menstruation may see themselves as sources of pollution and may endure anxiety, depression, and low self-esteem. Negative cultural beliefs concerning menstruation may also contribute to menstrual distress.

---

with a highly absorbent tampon that remains in place for 6 hours or more may create an ideal breeding ground for staph. The number of TSS cases has declined dramatically since then.

Some researchers believe that concern over TSS has been exaggerated. Nevertheless, many women now use regular rather than superabsorbent tampons to reduce the chance of creating a breeding ground for staph bacteria. Some women alternate tampons with sanitary napkins during each day of menstruation. Some change their tampons three or four times a day. These alternatives may also present problems, however. When more tampons are used, the greater number of insertions may increase the chances of transferring staph from the fingers or vaginal opening into the vagina. Other women have returned to external sanitary napkins. Still others use natural sponges. Women are encouraged to consult their health care providers about TSS.

Coitus During Menstruation

Many couples continue to engage in coitus during menstruation, but others abstain. One study found that men and women are less likely to initiate sexual activity during menstruation than during any other phase of the woman's cycle (Harvey, 1987). Some people abstain because of religious prohibitions. Others express concern about the "fuss" or the "mess" of the menstrual flow. Despite traditional attitudes that associate menstruation with uncleanliness, there is no evidence that coitus during menstruation is physically harmful to either partner. Ironically, menstrual coitus may be helpful to the woman. The uterine contractions that occur during orgasm may help relieve cramping by dispelling blood congestion. Orgasm achieved through masturbation may have the same effect.

Women may be sexually aroused at any time during the menstrual cycle. The preponderance of the research evidence, however, points to a peak in sexual desire in women around the time of ovulation (Kresin, 1993).

Human coital patterns during the phases of the menstrual cycle apparently reflect personal decisions, not hormone fluctuations. Some couples may decide to increase their frequency of coitus at ovulation in order to optimize the chances of conceiving; others may abstain during menstruation because of religious beliefs or beliefs linking menses with uncleanliness. Some may also increase their coital activity preceding menstruation to compensate for anticipated abstinence during menses or increase coital activity afterwards to make up for deprivation. In contrast, females of other species that are bound by the estrous cycle respond sexually only during estrus, except in relatively rare cases in which the female submits to sexual advances to fend off attacks from an aggressive male.

Menopause

Menopause, or the "change of life," is the cessation of menstruation. Menopause is a process that most commonly occurs between the ages of 46 and 50 and lasts for about two years. However, it may begin any time between the ages of 35 and 60. There is at least one case of a woman who became pregnant at age 61.

Menopause is a specific event in a long-term process known as the **climacteric** ("critical period"), the gradual decline in the reproductive capacity of the ovaries. The climacteric generally lasts for about 15 years, from ages 45 to 60 or so. After about the age of 35, the menstrual cycles of many women shorten, from an average of 28 days to 25 days at age 40 and to 23 days by the mid-40s. By the end of her 40s, a woman's cycles often become erratic, with some periods close together and others missed.

In menopause, the pituitary gland continues to pour normal levels of FSH and LH into the bloodstream, but for reasons that are not well understood, the ovaries gradually lose their capacity to respond. The ovaries no longer ripen egg cells or produce the sex hormones estrogen and progesterone.

The deficit in estrogen may lead to a number of unpleasant physical sensations, such as night sweats and hot flashes (suddenly feeling hot) and hot flushes (suddenly looking reddened) (Dennerstein et al., 2000). Hot flashes and flushes may alternate with cold sweats, in which a woman feels suddenly cold and clammy. Anyone who has experienced "cold feet" or hands from anxiety or fear will understand how dramatic the shifting patterns of blood flow can be. Hot flashes and flushes stem largely from "waves" of dilation of blood vessels across the face and upper body. All of these sensations reflect "vasomotor instability." That is, disruptions occur in the body mechanisms that dilate or constrict the blood vessels to maintain an even body temperature. Additional signs of estrogen deficiency include dizziness, headaches, pains in the joints, sensations of tingling in the

Menopause The cessation of menstruation.

Climacteric A long-term process, including menopause, that involves the gradual decline in the reproductive capacity of the ovaries.

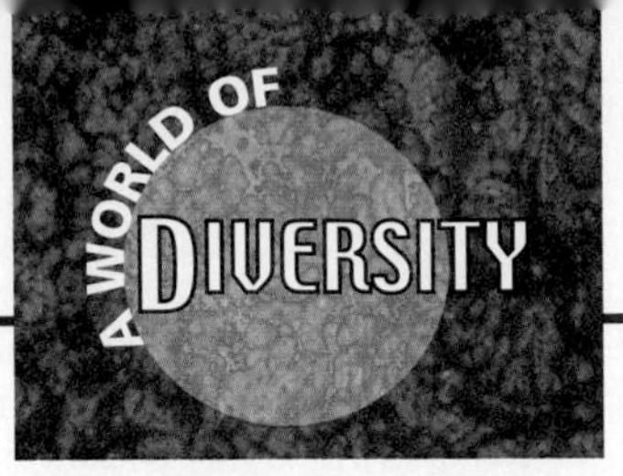

Symptoms of Menopause, Ethnicity, Socioeconomic Status, and Lifestyle

A study of more than 16,000 women aged 40 to 55 found that the symptoms of menopause differ according to ethnicity, socioeconomic status, and lifestyle. The research by Ellen P. Gold and her colleagues is taken from the first phase of the Study of Women's Health Across the Nation (funded by the National Institute on Aging and the National Institute on Nursing Research) and was published in the September 1, 2000, issue of the *American Journal of Epidemiology*. "This is one of the largest multi-ethnic studies of demographic and lifestyle factors associated with symptom reporting in midlife women, and the findings have potentially important preventive implications," Gold said in a University of California at Davis School of Medicine news release.

Women were questioned about seven symptoms of menopause: hot flashes or night sweats, urine leakage, vaginal dryness, stiffness or soreness in the joints, heart pounding or racing, forgetfulness, and difficulty sleeping. All symptoms except forgetfulness were reported less often by Japanese and Chinese women than by women from other ethnic groups. African American women were most likely to experience hot flashes, vaginal dryness, and forgetfulness; for example, 45% of African American women reported experiencing hot flashes or night sweats within the past two weeks, as compared with 31% of European American women. Latina American women were relatively more likely to report urine leakage, vaginal dryness, heart pounding, and forgetfulness.

Gold notes that the ethnic differences raise questions. "Are these just reporting differences, or are they related to factors in the women's lives, like diet or other lifestyle factors? Some literature suggests hormonal levels differ by ethnicity. Is this true and could it result in different symptoms [for women from different ethnic groups]?"

Gold and her colleagues also found that women were more likely to experience symptoms if they were older, had a lower educational level, had difficulty paying for basic needs, smoked, were more sedentary, or were heavier. Compared with women aged 40 to 43, women who entered menopause later were more likely to report hot flashes and night sweats, vaginal dryness, and forgetfulness. Women who had not completed high school were more likely to report forgetfulness, pounding of the heart, and hot flashes or night sweats. Women who reported that it was "somewhat hard" or "very hard" to pay for their basic needs were more likely to report all the symptoms.

Lifestyle also affected reporting of symptoms. As compared with women who had never smoked, women who had once smoked were still likely to report multiple symptoms, including hot flashes, urine leakage, stiffness or soreness of the joints, pounding of the heart, and forgetfulness. Current smokers of at least a pack a day were more likely to report difficulty sleeping, hot flashes, urine leakage, and stiffness or soreness of the joints. Women who believed they exercised less than other women were more likely to report all symptoms.

hands or feet, burning or itchy skin, and heart palpitations. The skin usually becomes drier. There is some loss of breast tissue and decreased vaginal lubrication during sexual arousal. Women may also encounter sleep problems, such as awakening more frequently at night and having difficulty going back to sleep.

Long-term estrogen deficiency has been linked to brittleness and porosity of the bones—a condition called **osteoporosis** (Delmas et al., 1997) Bones break more readily, and some women develop so-called dowager's hump. Osteoporosis can be severely handicapping, even life threatening. The increased brittleness of the bones increases the risk of serious fractures, especially of the hip, and many older women never recover from these fractures (Marwick, 2000).

Estrogen deficiency also has psychological effects. It can impair cognitive functioning and feelings of psychological well-being (Ross et al., 2000; Yaffe et al., 2000).

Osteoporosis A condition caused by estrogen deficiency and characterized by a decline in bone density, such that bones become porous and brittle. (From the Greek *osteon,* which means "bone," and the Latin *porus,* which means "pore.")

Hormone replacement therapy (HRT) Replacement of naturally occurring estrogen or estrogen and progesterone with synthetic equivalents, following menopause.

HORMONE REPLACEMENT THERAPY (HRT) Some women who experience severe physical symptoms have been helped by **hormone replacement therapy (HRT),** which typically consists of synthetic estrogen and progesterone. These synthetic hormones are used to offset the loss of their naturally occurring counterparts. HRT may help reduce the hot flushes and other symptoms brought about by hormonal deficiencies during

Do These Women Experience Menopause in Different Ways? Research shows that Asian women tend to have fewer symptom of menopause than women from other racial groups. African American women are most likely to have hot flashes. Why?

Compared with heavier women, slender women were more likely to report difficulty sleeping. Obese women were more likely to experience urine leakage, stiffness or soreness in the joints, and hot flashes or night sweats. However, in most studies, obesity has been associated with fewer, not more frequent, hot flashes.

"The theory is that women who are heavier produce more of a type of estrogen called estrone in their fat," Gold said in the news release. "We would expect that these women would have fewer hot flashes, which are thought to be caused by low estrogen levels. We didn't show that, but the increases in symptom reporting were not large. I think the relationship between estrogen levels and hot flashes is probably more complicated than we think."

menopause (den Tonkelaar & Oddens, 2000). It is especially helpful in preventing the development of osteoporosis (Davidson, 1995; Delmas et al., 1997). Drinking milk, which is high in calcium, increases bone density among girls and is likely to help prevent osteoporosis later in life ("More milk each day," 1997). Calcium supplements also seem to be helpful in decreasing the number of bone fractures among postmenopausal women (Eastell, 1998).

HRT is not without controversy. Although HRT has certainly been helpful to many menopausal women, exposure to estrogen increases the risk of some kinds of cancer, especially breast cancer (Brody, 1997c; Frank & Wenger, 1997). Use of a combination of estrogen and progestin appears to increase further the risk of breast cancer (Schairer et al., 2000). On the other hand, estrogen replacement lowers the woman's risk of osteoporosis (Davidson, 1995; Eastell, 1998), colon cancer (Newcomb & Storer, 1995), and perhaps Alzheimer's disease (Brody, 1997c). In fact, research is under way to determine whether estrogen replacement therapy may help women fight the cognitive effects (loss of memory, language problems, etc.) of Alzheimer's disease. To date, however, results are not promising (Henderson et al., 2000; Mulnard et al., 2000).

Other research shows that HRT raises levels of HDL ("good cholesterol") and lowers levels of LDL ("bad cholesterol") (Herrington et al., 2000; Nieto et al., 2000; Shlipak

Osteoporosis—Not for Women Only

Osteoporosis, a condition characterized by brittleness of the bones, is usually thought of as a health problem that afflicts older women (Orwoll et al., 2000). Yes, it does, but men make up about 2 million of the 10 million people with osteoporosis in the United States. Osteoporosis saps bone density, leading to crippling fractures.

Men are at less risk for osteoporosis because they have larger, stronger bones than women. Also, they do not experience the dramatic bone loss suffered by women following menopause. Osteoporosis is more likely to strike men who have low testosterone levels, use medications like steroids, smoke, drink heavily, and lead sedentary lives.

Not for Women Only. One person in five who have osteoporosis is a man.

et al., 2000). Because high levels of LDL are connected with cardiovascular disease, it was believed until recently that HRT may reduce the risk of such disease in postmenopausal women. However, recent research reports show mixed results. The women in the Hormone Replacement Therapy trial of the Women's Health Initiative apparently ran a slightly *greater* risk of heart attacks and strokes (Kolata, 2000a). Of the some 25,000 women in the study, nearly 1% of women using HRT developed such problems during the first year of HRT, but the risk tapered off rapidly during subsequent years. Dr. Jacques E. Rossouw (2000), director of the Hormone Replacement Therapy trial, considers the greater risk of cardiovascular problems to be "minuscule" and suggests that women still consider obtaining HRT because of its benefits. But a study of nearly 165,000 British women aged 50 to 74 from the General Practice Research Database found that HRT, in the form of either pills or skin patches, did have cardioprotective effects (Varas-Lorenzo et al., 2000). What can we conclude? Recall that (1) the director of the U.S. study describes as negligible the negative effects of HRT on the cardiovascular system that he (apparently) found and (2) other studies show that HRT may benefit the cardiovascular system. Therefore, women can probably be unconcerned about the effects of HRT on the cardiovascular system.

HRT is usually not recommended for women with a family history of breast cancer. On the other hand, researchers are attempting to develop estrogen compounds that will have the benefits of estrogen without the health risks (e.g., the risks of uterine and breast cancer) (Brody, 1997b; Fuleihan, 1997). Women are advised to explore the health benefits and risks of HRT with their gynecologists.

Overall, only about one postmenopausal woman in four in the United States chooses to use HRT, and of these, only half will remain on HRT for more than a few years (Brody, 1997c; Frank & Wenger, 1997). In what could be considered a vote of confidence for HRT, it turns out that nearly half (47%) of postmenopausal women doctors use HRT, and women gynecologists, who know the most about HRT, are the subgroup of doctors most

a CLOSER look

Myths About Menopause

Menopause is certainly a major life change for most women. For many women, menopause symbolizes the many midlife issues they face, including changes in appearance, sexuality, and health (Jones, 1994). Yet exactly what types of changes do we find? Many of us harbor misleading ideas about menopause—ideas that can be harmful to women. Consider the following myths and the realities. To which myths have you fallen prey?

Myth 1. *Menopause is abnormal.* Of course not. Menopause is a normal development in women's lives.

Myth 2. *The medical establishment considers menopause a disease.* No longer. Menopause is described as a "deficiency syndrome" today, to reflect the decline in secretion of estrogen and progesterone. Unfortunately, the term *deficiency* has negative meanings.

Myth 3. *After menopause, women need complete replacement of estrogen.* Not necessarily. Some estrogen continues to be produced by the adrenal glands, fatty tissue, and the brain (Guzick & Hoeger, 2000).

Myth 4. *Menopause is accompanied by depression and anxiety.* Not necessarily. Karen Matthews and her colleagues (1990) followed 541 healthy women through menopause and found that menopause was not significantly connected to depression, anxiety, stress, anger, or job dissatisfaction. (Outcomes may differ for women who have psychological problems prior to menopause.) Another group of researchers reported finding no overall relationship between mental health symptoms and menopausal status in a sample of 522 African American women (Jackson et al., 1991).

Much of a woman's response to menopause reflects its meaning to her, not physical changes (Jones, 1994). Women who adopt the commonly held belief that menopause signals the beginning of the end of life may develop a sense of hopelessness about the future, which in turn can set the stage for depression. Women whose entire lives have revolved around childbearing and child rearing are more likely to experience a sense of loss. Moreover, there is a cultural bias to explain depression and other problems of middle-aged women in terms of menopause, rather than exploring psychosocial factors.

Myth 5. *At menopause, women experience debilitating hot flashes.* Many women do not have hot flashes at all. Among those who do, the flashes are often relatively mild.

Truth or Fiction? REVISITED

It is not necessarily true that women at menopause experience debilitating hot flashes. Many women have none at all. For most of those who do, hot flashes are mild.

Myth 6. *A woman who has had a hysterectomy will not undergo menopause afterward.* It depends on whether or not the ovaries (the major producers of estrogen) were also removed. If they were not, menopause should proceed normally.

Myth 7. *Menopause signals an end to a woman's sexual appetite.* Not at all. Many women feel liberated by the severing of the ties between sex and reproduction.

Truth or Fiction? REVISITED

It is not true that menopause signals an end to women's sexual appetite. In fact, some women feel newly sexually liberated because of the separation of sex from reproduction.

Myth 8. *Menopause ends a woman's childbearing years.* Not necessarily! Postmenopausal women do not produce ova. However, ova from donors have been fertilized in laboratory dishes, and the developing embryos have been implanted in the uteruses of postmenopausal women and carried to term (Bohlen, 1995; Sauer et al., 1990).

Myth 9. *A woman's general level of activity is lower after menopause.* Many postmenopausal women become peppier and more assertive.

Myth 10. *Men are not affected by their wives' experience of menopause.* Many men are, of course. Men could become still more understanding if they learned about menopause and if their wives felt freer to talk to them about it.

likely to use it (Frank & Wenger, 1997). For women who do not receive HRT, other drugs are available to help them deal with menopausal problems, such as hot flashes.

MENSTRUAL PROBLEMS

Although menstruation is a natural biological process, 50% to 75% of women experience some discomfort prior to or during menstruation (Sommerfeld, 2000). Table 3.1 contains a list of commonly reported symptoms of menstrual problems. The problems we explore in this section include dysmenorrhea, mastalgia, menstrual migraine headaches, amenorrhea, and premenstrual syndrome (PMS).

Dysmenorrhea

Pain or discomfort during menstruation, or **dysmenorrhea,** is the most common type of menstrual problem. Most women at some time have at least mild menstrual pain or discomfort. Pelvic cramps are the most common manifestation of dysmenorrhea. They may be accompanied by headache, backache, nausea, or bloated feelings. Women who develop severe cases usually do so within a few years of menarche. **Primary dysmenorrhea** is menstrual pain or discomfort in the absence of known organic pathology. Women with **secondary dysmenorrhea** have identified organic problems that are believed to cause their menstrual problems. Their pain or discomfort is caused by, or *secondary to,* these problems. Endometriosis, pelvic inflammatory disease, and ovarian cysts are just a few of the organic disorders that can give rise to secondary dysmenorrhea. Yet evidence is accumulating that supposed primary dysmenorrhea is often *secondary* to hormonal changes, although the precise causes have not been delineated. For example, menstrual cramps sometimes decrease dramatically after childbirth, as a result of the massive hormonal changes that occur with pregnancy.

TABLE 3.1

Symptoms of menstrual problems

Physical Symptoms	Psychological Symptoms
Swelling of the breasts	Depressed mood, sudden tearfulness
Tenderness in the breasts	Loss of interest in usual social or recreational activities
Bloating	Anxiety, tension (feeling "on edge" or "keyed up")
Weight gain	Anger
Food cravings	Irritability
Abdominal discomfort	Changes in body image
Cramping	Concern over skipping routine activities, school, or work
Lack of energy	A sense of loss of control
Sleep disturbance, fatigue	A sense of loss of ability to cope
Migraine headache	
Pains in muscles and joints	
Aggravation of chronic disorders such as asthma and allergies	

Dysmenorrhea Pain or discomfort during menstruation.

Primary dysmenorrhea Menstrual pain or discomfort that occurs in the absence of known organic problems.

Secondary dysmenorrhea Menstrual pain or discomfort that is caused by identified organic problems.

Symptoms vary from person to person and also according to whether the woman has been pregnant. Women who have been pregnant report a lower incidence of sharp menstrual pain but a higher incidence of premenstrual symptoms and menstrual discomfort.

BIOLOGICAL ASPECTS OF DYSMENORRHEA Menstrual cramps appear to result from uterine spasms that may be brought about by copious secretion of hormones called **prostaglandins.** Prostaglandins apparently cause muscle fibers in the uterine wall to contract, as during labor. Most contractions go unnoticed, but powerful, persistent contractions are discomfiting in themselves and may temporarily deprive the uterus of oxygen, another source of distress. Women with more intense menstrual discomfort apparently produce higher quantities of prostaglandins. Prostaglandin-inhibiting drugs, such as ibuprofen, indomethacin, and aspirin are thus often helpful. Menstrual pain may also be secondary to endometriosis.

Pelvic pressure and bloating may be traced to pelvic edema (Greek for "swelling")—the congestion of fluid in the pelvic region. Fluid retention can lead to a gain of several pounds, sensations of heaviness, and **mastalgia**—a swelling of the breasts that sometimes causes premenstrual discomfort. Masters and Johnson (1966) noted that orgasm (through coitus or masturbation) can help relieve menstrual discomfort by reducing the pelvic congestion that spawns bloating and pressure. Orgasm may also increase the menstrual flow and shorten this phase of the cycle.

Headaches frequently accompany menstrual discomfort. Most headaches (in both sexes) stem from simple muscle tension, notably in the shoulders, the back of the neck, and the scalp. Pelvic discomfort may cause muscle contractions, thus contributing to the tension that produces headaches. Women who are tense about their menstrual flow are thus candidates for muscle tension headaches. Migraine headaches may arise from changes in the blood flow in the brain, however. Migraines are typically limited to one side of the head and are often accompanied by visual difficulties.

Amenorrhea

Amenorrhea, the absence of menstruation, is a primary sign of infertility. **Primary amenorrhea** is the absence of menstruation in a woman who has not menstruated at all by about the age of 16 or 17. **Secondary amenorrhea** is delayed or absent menstrual periods in women who have had regular periods in the past. Amenorrhea has various causes, including abnormalities in the structures of the reproductive system, hormonal abnormalities, growths such as cysts and tumors, and psychological problems, such as stress. Amenorrhea is normal during pregnancy and following menopause. Amenorrhea is also a symptom of **anorexia nervosa,** an eating disorder characterized by an intense fear of putting on weight and a refusal to eat enough to maintain a normal body weight, which often results in extreme (and sometimes life-threatening) weight loss. Hormonal changes that accompany emaciation are believed to be responsible for the cessation of menstruation. Amenorrhea may also occur in women who exercise strenuously, such as competitive long-distance runners. It is unclear whether the cessation of menstruation in female athletes is due to the effects of strenuous exercise itself, to related physical factors such as low body fat, to the stress of intensive training, or to a combination of factors.

Premenstrual Syndrome (PMS)

The term **premenstrual syndrome (PMS)** describes the combination of biological and psychological symptoms that may affect women during the four- to six-day interval that

Prostaglandins Hormones that cause muscle fibers in the uterine wall to contract, as during labor.

Mastalgia A swelling of the breasts that sometimes causes premenstrual discomfort.

Amenorrhea The absence of menstruation.

Primary amenorrhea Lack of menstruation in a woman who has never menstruated.

Secondary amenorrhea Lack of menstruation in a woman who has previously menstruated.

Anorexia nervosa A psychological disorder of eating characterized by intense fear of putting on weight and refusal to eat enough to maintain normal body weight.

Premenstrual syndrome (PMS) A combination of physical and psychological symptoms (such as anxiety, depression, irritability, weight gain from fluid retention, and abdominal discomfort) that regularly afflicts many women during the four- to six-day interval that precedes their menses each month.

a CLOSER look

Lifting the Curse—Should Monthly Periods Be Optional

Doctors have known for years that women could eliminate their monthly periods with the pill. But they are just now letting the general public in on what some experts call one of medicine's best-kept secrets.

In a recent essay in the international medical journal *The Lancet*, Charlotte Ellertson (2000), a reproductive health researcher at the Population Council, argues, "Health professionals and women ought to view menstruation as they would any other naturally occurring but frequently undesirable condition. This means providing those women who want it with safe and effective means to eliminate their menstrual cycles."

Continuous use of oral contraceptives—tossing out the 7 inactive pills typically packaged with 21 active pills and taking the active tablets for months on end—keeps hormone levels constant and eliminates monthly menstruation.

Although gynecologists sometimes let select patients in on this trick—such as those who want to avoid their periods on their honeymoons—they have generally not given women the option to suppress menstruation for long periods of time. Some doctors don't feel comfortable prescribing a relatively untested therapy and worry that it could increase the pill's risks.

But some gynecologists, endocrinologists, and contraceptive researchers are trying to spread the word about menstruation suppression. "For many women there is a menstruation-associated health problem month after month and there is no reason they have to put up with it," said Dr. Sheldon Segal, an endocrinologist at the Population Council and co-author, with Dr. Elsimar Coutinho, of the recently published book *Is Menstruation Obsolete?*

Health Benefits

According to the American College of Obstetricians and Gynecologists, 50% to 75% of women suffer some physical or emotional discomfort during or right before their periods. Up to 85% suffer from PMS and over 50% experience painful cramping. No period means no PMS and no cramps—and less monthly discomfort.

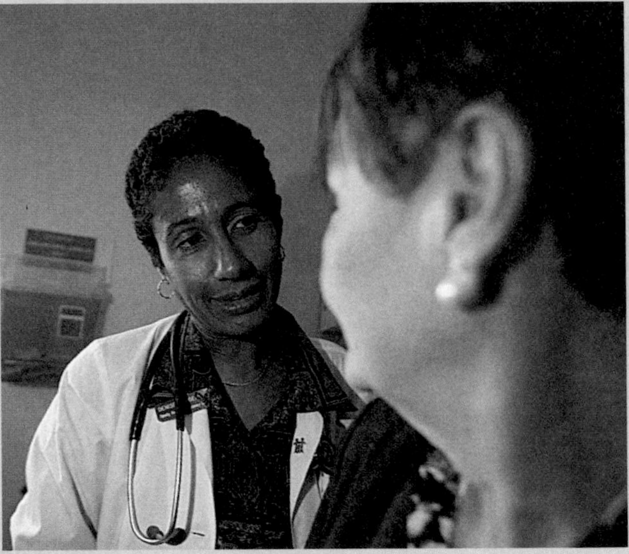

Is Menstruation Obsolete? What are the benefits and risks of eliminating menstruation through the use of hormones?

Women with endometriosis, which is caused by and made progressively worse by menstruation, may also benefit from skipping periods. The birth control pill has been shown to reduce the risk of endometrial cancer by 50% that of and ovarian cancer by 40%.

Segal advises the continuous use of oral contraceptives containing constant, low doses of both estrogen and progestin.

New Pill Promises Fewer Periods

There is no currently available oral contraceptive marketed for the purpose of suppressing menstruation, but Barr Laboratories has a patent on a four-periods-a-year pill, called Seasonale, that will probably enter clinical trials later this year and is expected to be on the market by about 2003.

Cultural Barriers and Misconceptions

Ellertson says common misconceptions about menstruation need to be debunked before women widely accept period suppression.

precedes their menses each month (see Table 3.2). For many women, premenstrual symptoms persist during menstruation.

Nearly three women in four experience some premenstrual symptoms (Brody, 1996). The great majority of cases involve mild to moderate levels of discomfort. Only about 2.5% of women report menstrual symptoms severe enough to impair their social, academic, or occupational functioning (Mortola, 1998), and that is when the term *premenstrual syndrome* is usually applied.

PMS is not unique to our culture. Researchers find premenstrual symptoms to be equally prevalent among women studied in the United States, in Italy, and in the Islamic nation of Bahrain (Brody, 1992b).

Today's menstrual pattern is far from natural, she says. Women living in industrialized countries begin menstruating at an earlier age, have fewer children, breast-feed for shorter periods of time, and experience menopause later than their foremothers. Hundreds of years ago, the average woman had about 160 periods in her lifetime; modern women have about 450.

And women on the traditional pill schedule who think the bleeding they experience each month is natural menstruation are being "duped," Ellertson says. This bleeding is artificially induced by the drop in hormones that results from going off the active pills. There is no ovulation and hardly any build-up of the uterine lining, so a period is not necessary, she says.

She adds that the placebo phase was built into the birth-control pill regimen by its developers to make it seem more natural to women—and in a failed attempt to make it more acceptable to the Catholic Church.

Why Didn't Anyone Tell Me?

Dr. Edward Levy, a gynecologist in private practice in St. Louis, tells some of his patients about this option, and about a dozen have taken him up on it, but he says most doctors aren't informing their patients.

"General gynecologists get lots of complaints about being on the pill—skin being a different texture, being headachy, weight gain—so they don't want to accentuate the side effects [and] so aren't likely to tell women to take more," said Dr. Michael Soules, a reproductive endocrinologist and professor of obstetrics and gynecology at the University of Washington in Seattle.

Soules also says that for some women there can be a downside to taking the pill continuously: It can result in small amounts of breakthrough bleeding. Spotting is already a side effect for half of women on the traditional pill regimen. This poses no danger, but some women find its unpredictable nature a nuisance.

Some doctors also have concerns about the safety of using the pill for extended periods of time, because it is not labeled for that purpose. Birth control pills "were designed to be used as cyclical agents. That's how they work and have been tested," said Dr. Gerson Weiss, chairman of the department of obstetrics and gynecology at New Jersey Medical School in Hackensack, NJ.

Although studies have found the birth-control pill safe for most women—with exceptions such as those who are over 35 and smoke or who have certain medical conditions such as heart disease, blood clots or breast cancer—they have suggested that pill users have a slightly higher chance of developing blood clots in the veins and lungs, stroke, and heart attack.

"If you take it continuously, that's effectively increasing the amount of hormones someone gets by 25%. That will probably up the risks by that percentage," Weiss said. "A potential increase in [blood clots] is probably the major [concern]."

Segal says the potential risks of continuous use of the pill are the same as with the three-weeks-on/one-week-off regimen. "Women who take the presently available oral contraceptive today are taking a much lower dose than in the past. The dose has been reduced dramatically. Even if you take it continuously, it's still less than the older products," he said.

A Period-Free Future?

Dr. Elsimar Coutinho, a gynecology professor at Federal University of Bahia in Brazil, predicts that in five years, women will be having periods only when they want to have them—such as when they want to get pregnant.

Ellerston also envisions a period-free future: "Pills are more frequently being used for reasons other than contraception, for instance to control acne, and menstruation suppression might grow to be just another use for pills," she said.

But Segal warns that continuous pill use is not OK for all women and that women should not attempt to suppress menstruation hormonally without first consulting with their doctor.

Source: From Julia Sommerfeld (2000, April 18). "Lifting the Curse—Should Monthly Periods by Optional?" www.msnbc.com/news/395750.asp. Reprinted with permission from MSNBC News Interactive L.L.C. *Note:* MSNBC Interactive News L.L.C. is not a sponsor of and does not endorse Prentice Hall or this textbook.

The causes of PMS are unclear, but evidence is accumulating for a biological basis. Researchers are looking to possible relationships between menstrual problems, including PMS, and chemical imbalances in the body. They have yet to find differences in levels of estrogen or progesterone between women with severe PMS and those with mild symptoms or no symptoms (Mortola, 1998; Rubinow & Schmidt, 1995). Research suggests that it is not the levels of these hormones themselves that contribute to PMS, but rather an abnormal response to the presence of the hormones (Schmidt et al., 1998). PMS also appears to be linked with imbalances in neurotransmitters such as serotonin (Mortola, 1998; Steiner et al., 1995). (Neurotransmitters are the chemical messengers in the nervous system.) Serotonin imbalances are also linked to changes in appetite. Women with PMS

TABLE 3.2

Symptoms of PMS*

Depression	Overeating or cravings for certain foods
Anxiety	Insomnia or too much sleeping
Mood swings	Feelings of being out of control or overwhelmed
Anger and irritability	Physical problems such as headaches, tenderness in the breasts, joint or muscle pain, weight gain or feeling bloated (both from fluid retention)
Loss of interest in usual activities	
Difficulty concentrating	
Lack of energy	

*Most women experience only a few of these symptoms, if they experience any at all.

Source: Jane E. Brody (1996, August 28). "PMS Need Not Be the Worry It Was Just Decades Ago." *The New York Times*, p. C9. Copyright © 1996 by the New York Times Co. Reprinted by permission.

show greater increases of appetite during the luteal phase than other women do. Another neurotransmitter, gamma-aminobutyric acid (GABA), also appears to be involved in PMS, because medicines that affect the levels of GABA help many women with PMS (Mortola, 1998). Premenstrual syndrome may well be caused by a complex interaction between ovarian hormones and neurotransmitters (Mortola, 1998).

Only a generation ago, PMS was seen as something a woman must put up with. No longer. Today there are many treatment options. These include exercise, dietary control (for example, eating several small meals a day rather than two or three large meals, limiting salt and sugar, and taking vitamin supplements), hormone treatments (usually progesterone), and medications that reduce anxiety or increase the amount of serotonin in the nervous system (Mortola, 1998). If you have PMS, get a clear idea which symptoms affect you most by using a PMS calendar or simply paying close attention to what happens. Then check with your physician about the most up-to-date treatment approaches.

How to Handle Menstrual Discomfort

Most women experience from some degree of menstrual discomfort. Women with persistent menstrual distress may profit from the suggestions listed below. Researchers are exploring the effectiveness of these techniques in controlled studies. For now, you might consider running a personal experiment. Adopt the techniques that sound right for you—all of them, if you wish. Try them out for a few months to see whether you reap any benefits. Such personal experiments are uncontrolled, and it is scientifically difficult to pin down the reasons for results. (You may feel better simply because you *expect* to or because of overall improvements in health.) Still, the methods may enhance your comfort, your health, and your outlook on menstruation—not bad outcomes at all!

1. Don't blame yourself! Menstrual problems were once erroneously attributed to women's "hysterical" nature. This is nonsense. Menstrual problems appear, in large part, to reflect hormonal variations or chemical fluctuations in the brain during the menstrual cycle. Researchers have not yet fully identified all the causal elements and patterns, but their lack of knowledge does *not* mean that women who have menstrual problems are hysterical.
2. Keep a menstrual calendar so that you can track your menstrual symptoms systematically and identify patterns.
3. Develop strategies for dealing with days when you experience the greatest distress—strategies that will help enhance your pleasure and minimize the stress affecting you on those days. Activities that distract you from your menstrual discomfort may be helpful. Go see a movie or get into that novel you've been meaning to read.

4. Ask yourself whether you harbor any self-defeating attitudes toward menstruation that might be compounding distress. Do close relatives or friends see menstruation as an illness, a time of "pollution," a "dirty thing"? Have you adopted any of these attitudes—if not verbally, then in ways that affect your behavior, such as by restricting your social activities during your period?
5. See a doctor about your concerns, especially if you have severe symptoms. Severe menstrual symptoms are often secondary to medical disorders such as endometriosis and pelvic inflammatory disease (PID). Check it out.
6. Develop nutritious eating habits and continue them throughout the entire cycle (that means always). Consider limiting your intake of alcohol, caffeine, fats, salt, and sweets, especially during the days preceding menstruation. Research suggests that a low-fat, vegetarian diet reduces the duration and intensity of menstrual pain and the duration of premenstrual symptoms (Barnard et al., 2000).
7. Eat several smaller meals (or nutritious snacks) throughout the day, rather than a few highly filling meals.
8. Some women find that vigorous exercise—jogging, swimming, bicycling, fast walking, dancing, skating, even jumping rope—helps relieve premenstrual and menstrual discomfort. Evidence suggests that exercise helps to relieve and possibly prevent menstrual discomfort (Choi, 1992). By the way, develop regular exercise habits. Don't seek to become solely a premenstrual athlete.
9. Check with your doctor about vitamin and mineral supplements (such as calcium and magnesium). Vitamin B6 appears to have helped some women.
10. Ibuprofen (brand names: Medipren, Advil, Motrin, etc.) and other medicines available over the counter may be helpful for cramping. Prescription drugs such as anti-anxiety drugs (such as alprazolam) and anti-depressant drugs (serotonin-reuptake inhibitors) may also be of help (Mortola, 1998). "Anti-depressants" affect levels of neurotransmitters in a way that can be helpful for women with PMS. Their benefits do not mean that women with PMS are depressed. Ask your doctor for a recommendation.
11. Remind yourself that menstrual problems are time-limited. Don't worry about getting through life or a career. Just get through the next couple of days.

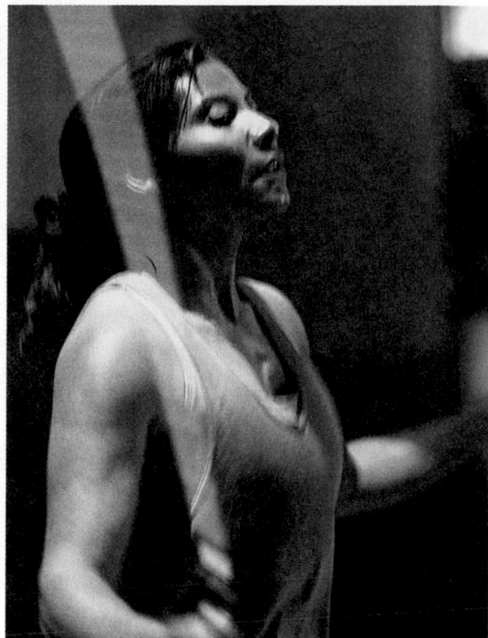

Exercise as a Strategy for Coping with Menstrual Discomfort. Some women find that vigorous exercise helps relieve menstrual discomfort.

In this chapter we have explored female sexual anatomy and physiology. In the following chapter, we turn our attention to the male.

Human Sexuality ONLINE //

Web Sites Related to Female Sexual Anatomy and Physiology

The American Cancer Society: This official web site of the American Cancer Society provides information on all types of cancer, cancer research, therapy, support groups, and community resources.
www.cancer.org/

Avon's Breast Cancer Awareness Campaign: This Web site provides women, particularly medically underserved women, with access to breast cancer education and early detection screening services.
www.avoncrusade.com/

Health tips and support for women.
www.thriveonline.com/

HealthGate: This is a Web site dedicated to facilitating improvements in patient care, biomedical research, and education for women.
www.healthgate.com/

A health guide for healthy and not-so-healthy people that provides detailed information about what makes the human body tick.
www.phys.com/

summing up

EXTERNAL SEX ORGANS

The female external sexual structures are collectively known as the vulva. They consist of the mons veneris, the labia majora and minora, the clitoris, the vestibule, and the vaginal opening.

■ The Mons Veneris

The mons veneris consists of fatty tissue that covers the joint of the pubic bones in front of the body.

■ The Labia Majora

The labia majora are large folds of skin that run downward from the mons along the sides of the vulva.

■ The Labia Minora

The labia minora are hairless, light-colored membranes that surround the urethral and vaginal openings.

■ The Clitoris

The clitoris is the female sex organ that is most sensitive to sexual sensation, but it is not directly involved in reproduction.

■ The Vestibule

The vestibule contains the openings to the vagina and the urethra.

■ The Urethral Opening

Urine passes from the female's body through the urethral opening.

■ The Vaginal Opening

The vaginal opening, or introitus, lies below the urethral opening.

■ The Perineum

The perineum is the area that lies between the vaginal opening and the anus.

■ Structures That Underlie the External Sexual Structures

These structures include the vestibular bulbs, Bartholin's glands, the sphincters, the clitoral crura, and the pubococcygeus (P-C) muscle.

INTERNAL SEX ORGANS

The internal female sex organs—or female reproductive system—include the innermost parts of the vagina, the cervix, the uterus, the ovaries, and the fallopian tubes.

■ The Vagina

Menstrual flow and babies pass from the uterus to the outer world through the vagina. During coitus, the vagina contains the penis.

■ The Cervix

The cervix is the lower end of the uterus.

■ The Uterus

The uterus or womb is the pear-shaped organ in which a fertilized ovum implants and develops until birth.

■ The Fallopian Tubes

Two fallopian tubes extend from the upper end of the uterus toward the ovaries. Ova pass through the fallopian tubes on their way to the uterus and are normally fertilized within these tubes.

■ The Ovaries

The ovaries lie on either side of the uterus and produce ova and the sex hormones estrogen and progesterone.

■ The Pelvic Examination

Regular pelvic examinations are essential for early detection of problems involving the reproductive tract.

THE BREASTS

In some cultures the breasts are viewed merely as biological instruments for feeding infants. In our culture, however, they have taken on erotic significance. The breasts are secondary sex characteristics that contain mammary glands.

■ Breast Cancer

Breast cancer is the second leading cancer killer in women, after lung cancer. Women with breast cancer will have lumps in the breast, but most lumps in the breasts are benign. Breast cancer may be detected in a number of ways, including breast self-examination, medical examinations, and mammography. Early detection yields the greatest chance of survival.

THE MENSTRUAL CYCLE

Menstruation is the cyclical bleeding that stems from the shedding of the endometrium when a reproductive cycle has not led to the fertilization of an ovum. The menstrual cycle is regulated by estrogen and progesterone.

■ Regulation of the Menstrual Cycle

The menstrual cycle involves finely tuned relationships among the hypothalamus, the pituitary gland, and the ovaries and uterus. Hormones produced by the hypothalamus regulate the pituitary, which in turn secretes hormones that regulate the secretions of the ovaries and uterus.

■ Phases of the Menstrual Cycle

The menstrual cycle has four stages or phases: the proliferative, ovulatory, secretory, and menstrual phases. During the first phase of the cycle, which follows menstruation, ova ripen within their follicles, and endometrial tissue proliferates. During the second phase, ovulation occurs. During the third phase, the corpus luteum produces copious amounts of progesterone and estrogen that cause the endometrium to thicken. If the ovum goes unfertilized, a plunge in estrogen and progesterone levels triggers the fourth, or menstrual, phase, which leads to the beginning of a new cycle.

■ **Coitus During Menstruation**
Couples are apparently less likely to initiate sexual activity during menstruation than during any other phase of the woman's cycle.

■ **Menopause**
Menopause, the cessation of menstruation, most commonly occurs between the ages of 46 and 50. Estrogen deficiency in menopause may give rise to night sweats, hot flashes, hot flushes, cold sweats, dry skin, loss of breast tissue, and decreased vaginal lubrication. Long-term estrogen deficiency has been linked to osteoporosis. Hormone replacement therapy can offset the loss of estrogen and progesterone but has been linked to a slightly increased risk of breast and endometrial cancers, though also to a reduction in the risk of osteoporosis. For most women, menopausal problems are mild. Psychological problems can reflect the meaning of menopause to the individual.

MENSTRUAL PROBLEMS

Most women experience some discomfort prior to or during menstruation. Common menstrual problems include dysmenorrhea, amenorrhea, and premenstrual syndrome (PMS).

■ **Dysmenorrhea**
Dysmenorrhea is the most common menstrual problem, and pelvic cramps are the most common symptom. Dysmenorrhea can be caused by problems such as endometriosis, pelvic inflammatory disease, and ovarian cysts.

■ **Amenorrhea**
Amenorrhea can be caused by problems such as abnormalities in the structures of the reproductive system, hormonal abnormalities, cysts, tumors, and stress.

■ **Premenstrual Syndrome (PMS)**
As many as three women in four have some form of PMS. The causes of PMS are unclear, but most researchers look to potential links between menstrual problems and hormone levels.

■ **How to Handle Menstrual Distress**
Women with persistent menstrual problems may benefit from a number of active coping strategies for handling menstrual distress.

questions for critical thinking

1. Do you believe that disapproval of female circumcision by Americans and other Westerners shows cultural "insensitivity"? Why or why not?
2. Are the female sex organs more complex than you had believed? If so, how?
3. If you are a woman, would you prefer a female or a male physician to conduct a pelvic exam? Why?
4. How important are the size and shape of your breasts (or of your partner's breasts) to you? Why?
5. How do you believe you would react if you (or your partner) had a mastectomy? Why?
6. Do most people from your sociocultural background hold any particular attitudes toward menstruation? What are they? Do you share these attitudes? Explain.
7. Would you engage in coitus during menstruation? Why or why not?
8. Do you (or your loved ones) experience PMS? What are you (or they) doing about it? Why?

chapter

4

Male Sexual Anatomy and Physiology

Pablo Picasso. *The Young Men*, 1906. Oil on canvas. © Copyright ARS, NY. Musée de l'Orangerie, Paris, France. Credit: Réunion des Musées Nationaux/Art Resource, NY. Credit: © 2002 Estate of Pablo Picasso/Artists Rights Society (ARS), New York.

outline

Truth or Fiction?

EXTERNAL SEX ORGANS

The Penis

A Closer Look On Penis Size and Sexual Performance

The Scrotum

Human Sexuality Online Another Sort of Gender Equality: Up-Jock Pics Now Compete with Up-Skirt Photos

INTERNAL SEX ORGANS

The Testes

A Closer Look Is There a *Man*opause?

The Vas Deferens

A Closer Look Big

The Seminal Vesicles

The Prostate Gland

Cowper's Glands

Semen

DISEASES OF THE UROGENITAL SYSTEM

Urethritis

Cancer of the Testes

Disorders of the Prostate

A Closer Look Lance Armstrong—Winner of Two Contests

A Closer Look More Options, and Decisions, for Men with Prostate Cancer

MALE SEXUAL FUNCTIONS

Erection

Spinal Reflexes and Sexual Response

Ejaculation

Human Sexuality Online Web Sites Related to Male Sexual Anatomy and Physiology

SUMMING UP

QUESTIONS FOR CRITICAL THINKING

Truth or Fiction?

____ The penis contains bone and muscle.

____ The father determines the baby's gender.

____ The sperm of the tiny fruit fly are longer than human sperm.

____ Morning erections reflect the need to urinate.

____ Men can will themselves to have erections.

____ The penis has a mind of its own.

____ Many men paralyzed below the waist can attain erection, engage in sexual intercourse, and ejaculate.

____ Men can have orgasms without ejaculating.

From the earliest foundations of Western civilization, male-dominated societies elevated men and exalted male genitalia. The ancient Greeks carried oversized images of fish as **phallic symbols** in their Dionysian processions, which celebrated the wilder and more frenzied aspects of human sexuality. In the murky predawn light of Western civilization, humankind engaged in phallic worship. Phallic symbols played roles in religious worship and became glorified in art in the form of ploughs, axes, and swords.

The tradition of phallic worship also achieved to higher aesthetic levels. The ancient Greeks adorned themselves with phallic rings and necklaces. In ancient Rome, celebrations were held to honor Venus, the goddess of love. The Romans outfitted a float in the shape of a large phallus and paraded it through the streets. There were no tributes to the female genitals. Even though Venus was being honored, no artisans devoted themselves to creating floats in the likeness of the vulva or the clitoris.

Men held their own genitals in such high esteem that it was common courtroom practice for them to swear to tell the truth with their hands on their genitals—as we swear to tell the truth in the name of God or by placing our hands on the Bible. The words **testes** and **testicles** derive from the same Latin word as *testify*. The Latin *testis* means "a witness."

Even today, we see evidence of pride in—indeed veneration of!—the male genitalia. Men with large genitals are accorded respect from their male peers and sometimes adoration from female admirers. In *The Sun Also Rises*, Ernest Hemingway describes how matadors stuffed the front of their trousers with fabric. With their "manhood" fully packed, they plunged the sword into the poor animal's head before their adoring public. U.S. slang describes men with large genitals as "well hung" or "hung like a bull" (or stallion).

Given these cultural attitudes, it is not surprising that young men (and some not-so-young men) belittle themselves if they feel, as many do, that they are little—that is, that their penises do not measure up to some ideal. Boys who mature late may be ridiculed by their peers for their small genitals. Their feelings of inadequacy may persist into adulthood. Adult men, too, may harbor doubts that their penises are large enough to satisfy their lovers. Or they may fear that their partners' earlier lovers had larger genitals.

In this chapter we examine male sexual anatomy and physiology, and we attempt to sort out truth from fiction. We see, for example, that despite his lingering doubts, a man's capabilities as a lover do not depend on the size of his penis (at least within broad limits). And even if size *did* matter, for a man to judge his sexual prowess on the basis of locker-room comparisons would make about as much sense as choosing a balloon by measuring it when it is deflated.

In our exploration of male sexual anatomy and physiology, as in our exploration of female sexual anatomy and physiology, we begin with the external genitalia and then move inward. Once inside, we focus on the route of sperm through the male reproductive system.

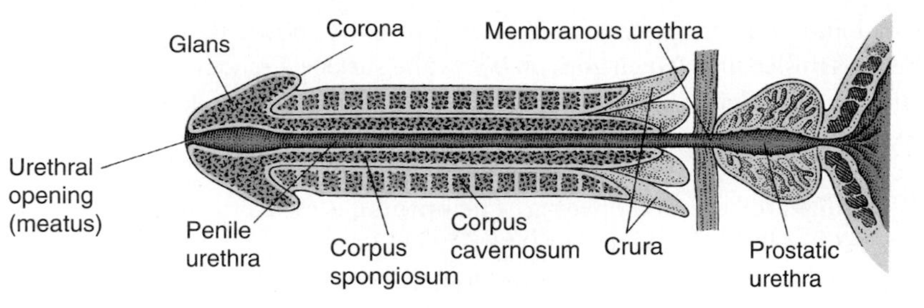

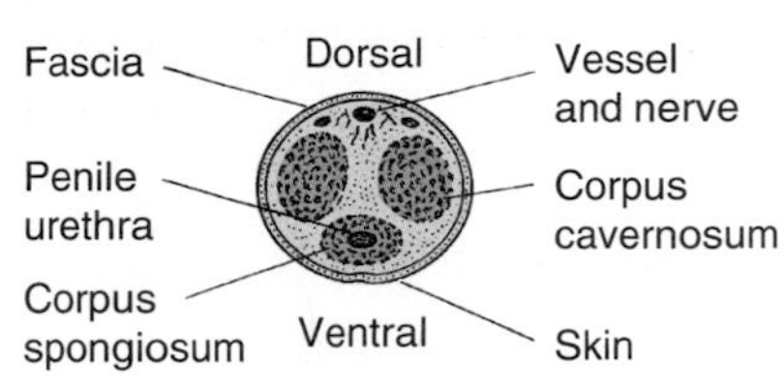

FIGURE 4.1

The Penis. During sexual arousal, the corpora cavernosa and corpus spongiosum become congested with blood, causing the penis to enlarge and stiffen.

EXTERNAL SEX ORGANS

The external male sex organs include the penis and the scrotum (see Figures 4.1 and 4.2).

The Penis

> The penis mightier than the sword.
>
> —Mark Twain
>
> Is that a gun in your pocket, or are you just glad to see me?
>
> —Mae West

At first glance, the **penis** may seem rather simple and obvious in its structures, particularly when compared to women's organs. This apparent simplicity may have contributed to cultural stereotypes that men are straightforward and aggressive, whereas women tend

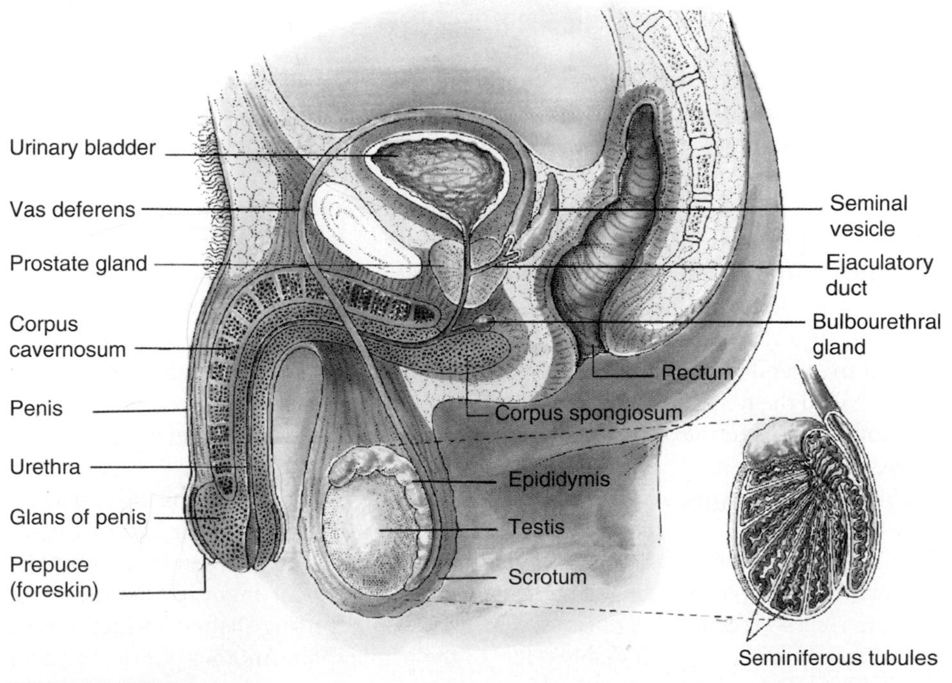

FIGURE 4.2

The Male Reproductive System. The external male sex organs include the penis and the scrotum.

Phallic symbols Images of the penis that are usually suggestive of generative power.

Testes The male sex glands, suspended in the scrotum, that produce sperm cells and male sex hormones. Singular: *testis.*

Testicles Testes.

Penis The male organ of sexual intercourse. (From the Latin for "tail.")

to be complicated and perhaps mysterious. Yet, as Figure 4.1 shows, the apparent simplicity of the penis is misleading. Much goes on below the surface. Gender stereotypes regarding anatomy are as misleading as those regarding personality (see Chapter 6).

The penis, like the vagina, is the sex organ used in sexual intercourse. Unlike the vagina, however, the penis also serves as a conduit for urine. Both semen and urine pass out of the penis through the urethral opening. The opening is called the urethral *meatus* (pronounced me-ATE-us), which means "passage."

The penis arrived on the evolutionary scene some 100 million years ago. It was first found in reptiles—the ancestors of present-day crocodiles and lizards and of dinosaurs. Earlier in the evolutionary process, both male and female animals had a genital opening called a **cloaca.** The cloaca functioned as both an excretory and a sex organ. It took some fancy bodywork for couples to align their cloacae so that the male's sperm could find their way into the female's cloacal cavity. The penis is a more efficient shape—a shaft—for funneling sperm to the female. It gives them a head start on their journey toward the ovum.

Many mammals, including dogs, have penile bones that stiffen the penis to facilitate copulation. Despite the slang term *boner*, the human penis contains no bones. Nor, despite another slang term, *muscle*, does the penis contain muscle tissue. However, muscles at the base of the penis, like the muscles surrounding the vaginal and urethral openings in women, are involved in controlling urination and ejaculation.

Despite the slang terms, there is no bone or muscle tissue in the penis.

Rather than bones or muscles, the penis contains three cylinders of spongy material that run its length. The larger two of these cylinders, the **corpora cavernosa** (see Figure 4.1), lie side by side and function like the cavernous bodies in the clitoris. These cylinders fill up with blood and stiffen during sexual arousal. In addition, a **corpus spongiosum** (spongy body) runs along the bottom, or ventral, surface of the penis. It contains the penile urethra that conducts urine through the penis to the urinary opening (urethral meatus) at the tip. At the tip of the penis, the spongy body enlarges to become the glans, or head, of the penis.

All three cylinders consist of spongy tissue that swells (becomes engorged) with blood during sexual arousal, resulting in erection. The urethra is connected to the bladder, which is unrelated to reproduction, and to those parts of the reproductive system that transport semen.

The glans of the penis, like the clitoral glans, is extremely sensitive to sexual stimulation. Direct, prolonged stimulation can become irritating, even painful. Men generally prefer to masturbate by stroking the shaft of the penis rather than the glans, although some prefer the latter. The **corona,** or coronal ridge, separates the glans from the body of the penis. It is also quite sensitive to sexual stimulation. After the glans, the parts of the penis that men tend to find most sensitive are the corona and an area on the underside of the penis called the frenulum. The **frenulum** is a thin strip of tissue that connects the underside of the glans to the shaft. Most men find the top part of the penis to be the least sensitive part.

The base of the penis, which is called the **root,** extends into the pelvis. It is attached to pelvic bones by leglike structures, called crura, that are like those that anchor the female's clitoris. The body of the penis is called the penile **shaft.** The penile shaft, unlike the clitoral shaft, is free-swinging. Thus, when sexual excitement engorges the penis with blood, the result—erection—is obvious. The expression "getting shafted," which means being taken advantage of, presumably refers to the penile shaft. Another phrase, "getting screwed," refers more directly to sexual intercourse.

The skin of the penis is hairless and loose, allowing expansion during erection. It is fixed to the penile shaft just behind the glans. Some of it, however, like the labia minora in the female, folds over to partially cover the glans. This covering is the prepuce, or **foreskin.** It covers part or all of the penile glans just as the clitoral prepuce (hood) covers the clitoral shaft. The prepuce consists of loose skin that freely moves over the glans.

Cloaca The cavity in birds, reptiles, and many fish into which the genitourinary and intestinal tracts empty. (From the Latin *cluere,* which means "to cleanse.")

Corpora cavernosa Cylinders of spongy tissue in the penis that become congested with blood and stiffen during sexual arousal.

Corpus spongiosum The spongy body that runs along the bottom of the penis, contains the penile urethra, and enlarges at the tip of the penis to form the glans.

Corona The ridge that separates the glans from the body of the penis. (From the Latin for "crown.")

Frenulum The sensitive strip of tissue that connects the underside of the penile glans to the shaft. (From the Latin *frenum,* which means "bridle.")

Root The base of the penis, which extends into the pelvis.

Shaft The body of the penis, which expands as a result of vasocongestion.

Foreskin The loose skin that covers the penile glans. Also referred to as the *prepuce.*

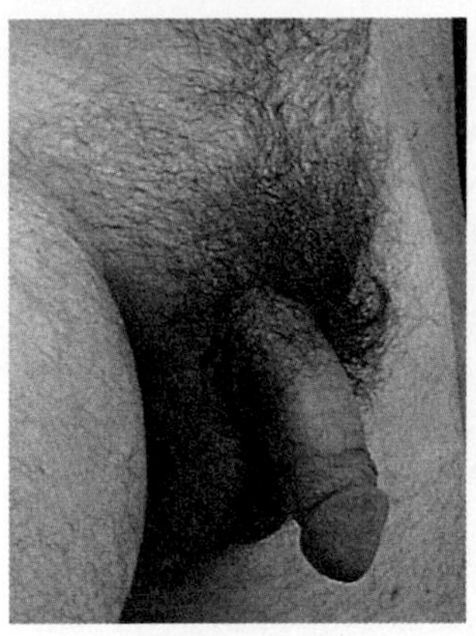
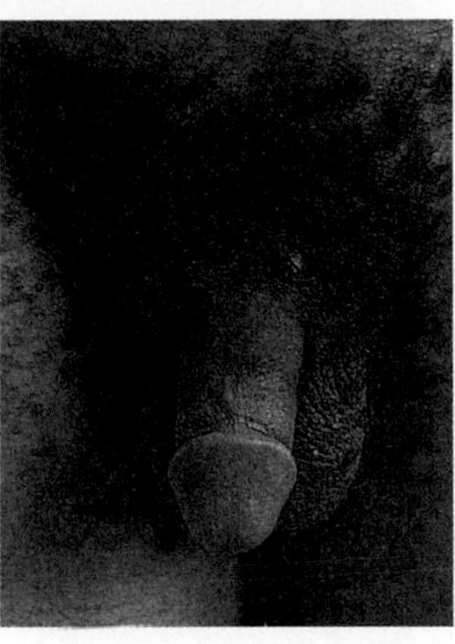
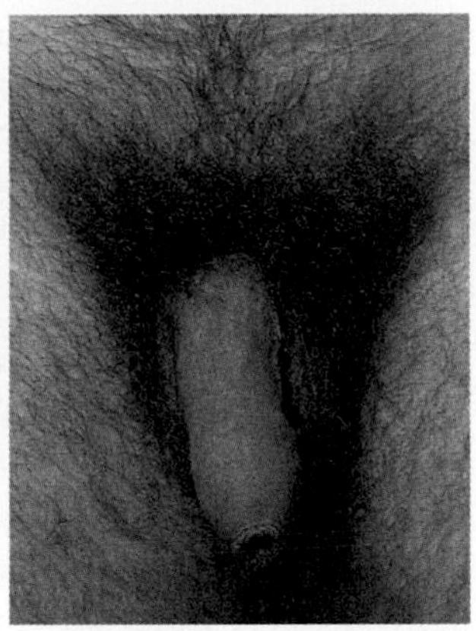

FIGURE 4.3

Normal Variations in the Male Genitals. The penis and scrotum vary a good deal in appearance from one man to another. The penis in the photo to the right is uncircumcised.

However, smegma—a cheeselike, foul-smelling secretion—may accumulate below the prepuce, causing the foreskin to adhere to the glans.

CIRCUMCISION **Circumcision** is the surgical removal of the prepuce (Figure 4.3). Advocates of circumcision believe that it enhances hygiene because it eliminates a site where smegma might accumulate and disease organisms might flourish. Opponents of circumcision believe that it is unnecessary because regular cleaning is sufficient to reduce the risk of these problems.

Male circumcision has a long history as a religious rite. Jews traditionally carry out male circumcision shortly after a baby is born. Circumcision is performed as a sign of the covenant between God and the people of Abraham. Muslims also have ritual circumcisions for religious reasons.

According to the NHSLS study of 1,410 men aged 18 to 59, circumcision is most common among European American men (81%) and men whose mothers graduated from college (87%) (Laumann et al., 1997). Figure 4.4 compares the circumcision rates among various ethnic groups in the United States. The groups have some overlap; the great majority of Jews—for whom circumcision is a religious rite—are also European Americans.

Circumcision became widespread in the United States because medical research suggested that it prevented the transmission of sexually transmitted infections (STIs). For

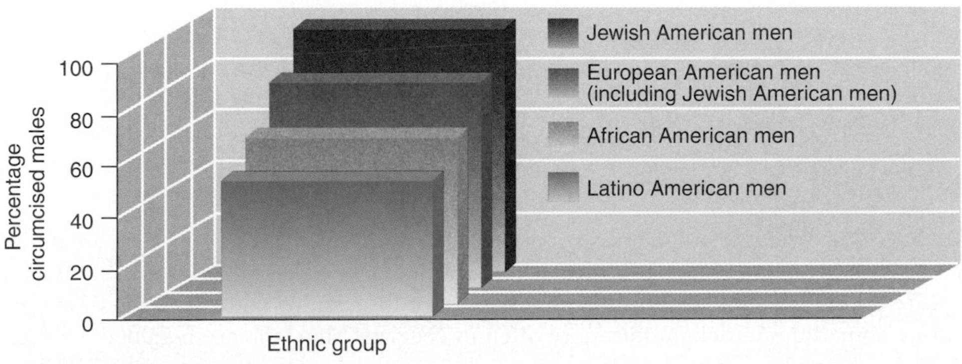

FIGURE 4.4

Who Is Circumcised? In the NHSLS study, circumcision was most common among Jewish American men, for whom it is a religious rite, and among European American men in general, as compared with African American men and Latino American men. [*Source:* From E. O. Laumann et al., (1997, April 2). "Circumcision in the U.S.: Prevalence, Prophylactic Effects, and Sexual Practice." *The Journal of the American Medical Association*, 227, pp. 1052–1057. Reprinted by permission of the American Medical Association.]

Circumcision Surgical removal of the foreskin of the penis. (From the Latin *circumcidere,* which means "to cut around.")

example, studies show that urinary tract infections are more common among uncircumcised male infants than among circumcised infants (Herzog, 1989). It also seems quite clear that uncircumcised men are at greater risk than circumcised men of becoming infected by the AIDS virus, at least during male–female sexual relations (Bailey, 2000; Halperin & Bailey, 1999; Cohen, 2000). For example, Ann Buve (2000) of the Belgian Institute of Tropical Medicine compared the incidence of HIV infection in two African cities with high rates and two with low rates. In Yaoundé, Cameroon, and Cotonou, Benin, the prevalence of HIV among sexually active men was about 4% to 5%, and 99% of the men were circumcised. In Kisumu, Kenya, and Ndola, Zambia, where circumcision rates were much lower, the rates of HIV infection among sexually active men were about 26% to 27%.

In a study carried out in Uganda, Quinn and his colleagues (2000) followed couples in whom one partner was infected with HIV and the other was not for 30 months. They found that more than 20% of the uncircumcised men became infected during the course of the study, as compared with none of the circumcised men. The protective effects of circumcision are likely to reflect a lower incidence of local inflammation and genital ulcers, both of which provide ports of entry for HIV, and the removal of cells in the foreskin—Langerhans cells—that are receptive to infection by HIV (Cohen, 2000; Szabo & Short, 2000).

Questions have also been raised about the *sexual* effects of circumcision. According to the NHSLS study, there is a substantial relationship between circumcision and sexual activity. The study found that circumcised men engaged in a wider variety of sexual practices, including masturbation and oral and anal sex. As an example, 81% of circumcised men reported having experienced heterosexual fellatio, as compared with 61% of uncircumcised men. Perhaps circumcised men are more likely to have liberal backgrounds.

The NHSLS study also found that circumcised men were less likely to have sexual dysfunctions such as lack of interest in sex, performance anxiety, or difficulty achieving or maintaining erections. Forty percent of the circumcised men reported sexual dysfunctions, as compared with 58% of the uncircumcised men.

Physicians once agreed that circumcision is the treatment of choice for **phimosis,** a condition in which it is difficult to retract the foreskin from the glans. But today, only a small minority of males with phimosis are circumcised for that reason (Rickwood et al., 2000).

Phimosis An abnormal condition in which the foreskin is so tight that it cannot be withdrawn from the glans. (From the Greek *phimos,* which means "muzzle.")

PENIS SIZE

> IRAS: Am I not an inch of fortune better than she?
>
> CHARMIAN: Well, if you were but an inch of fortune better than I, where would you choose it?
>
> IRAS: Not in my husband's nose.
>
> —William Shakespeare, *Antony and Cleopatra*

In our culture the size of the penis is sometimes seen as a measure of a man's masculinity and his ability to please his sex partner (see the nearby a Closer Look feature). Shakespeare and other writers inform us that men have looked down at themselves for centuries, sometimes in delight but more often in chagrin. Men who are heralded for their sexual or reproductive feats are presumed to have more prominent "testaments" to their manhood.

It may be of some comfort to note that even the smallest normal human penis is between three and four times the length of the phallus of the burly gorilla. Even so, the human penis does not merit comparison with the phallus of the blue whale. This ocean-dwelling mammal, the largest animal on earth, is about 100 feet from end to end and possesses a penis about 7 feet in length. When not in use, the penis is cached in the male's abdomen, which is a fortunate thing. If it were to trail down permanently, it might act as a rudder and muddle the animal's internal navigational system.

a CLOSER look

On Penis Size and Sexual Performance

Perhaps most men have had concerns about the size of their penises. As sex therapist Bernard Zilbergeld put it in his book *Male Sexuality*, "Women, we are given to believe, crave nothing so much as a penis that might be mistaken for a telephone pole" (1978, p. 27). Think critically about the belief that men with bigger penises make more effective lovers. What assumptions is it based on? Is there supportive evidence?

The belief that the size of the man's penis determines his sexual prowess is based upon the assumption that men with bigger penises are better equipped to satisfy a woman sexually. Zilbergeld and others point out, however, that women rarely mention penis size as an important element in their sexual satisfaction. Quite regularly they *do* mention ability to communicate with partners, the emotional atmosphere of the relationship, and sensitivity to employing sexual techniques that enhance their partner's pleasure.

Does your knowledge of female sexual physiology support or challenge the assumption that bigger is necessarily better? From your reading of Chapter 3, what do you know about how the vagina accommodates the penis? What is the role of the P-C muscle? Which part of the vagina is most sensitive to tactile stimulation? What, if any, might be the relationships between the clitoris—the woman's most erotically sensitive organ—and the size of the penis? Which seems more crucial: sexual technique and the quality of the relationship, or penis size? Why?

The diameter of the penis may have a greater bearing on a partner's sexual sensations than its length, because thicker penises may provide more clitoral stimulation during intercourse. Even though the inner vagina is relatively insensitive to touch, some women find the *pressure* of deeper penetration sexually pleasurable. Others, however, find deeper penetration to be uncomfortable or painful, especially if thrusting is too vigorous.

Masters and Johnson (1966) reported that the penises of the 312 male subjects they studied generally ranged in length from 3½ inches to a little more than 4 inches. The average erect penis ranges from 5 to 7 inches in length (Reinisch, 1990). Erect penises differ less in size than flaccid penises do (Jamison & Gebhard, 1988). Penises that are small when flaccid tend to gain more size when they become erect. Larger flaccid penises gain relatively less. Size differences in flaccid penises may thus be largely canceled out by erection. Nor is there a relationship between penis size and body weight, height, or build (Money et al., 1984).

Even when flaccid, the same penis can vary in size. Factors such as cold air or water and emotions of fear or anxiety can cause the penis (along with the scrotum and testicles) to draw closer to the body, reducing its size. The flaccid penis may also grow in size in warm water or when the man is relaxed.

The Scrotum

The **scrotum** is a pouch of loose skin that becomes covered lightly with hair at puberty. The scrotum consists of two compartments that hold the testes. Each testicle is held in place by a **spermatic cord,** a structure that contains the **vas deferens,** blood vessels and nerves, and the cremaster muscle. The **cremaster muscle** raises and lowers the testicle within the scrotum in response to temperature changes and sexual stimulation. (The testes are drawn closer to the body during sexual arousal.)

Sperm production is optimal at a temperature that is slightly cooler than the 98.6 degrees Fahrenheit that is desirable for most of the body. Typical scrotal temperature is 5 to 6 degrees lower than body temperature. The scrotum is loose-hanging and flexible. It permits the testes and nearby structures to escape the higher body heat, especially in warm weather. In the middle layer of the scrotum is the **dartos muscle,** which (like the cremaster) contracts and relaxes reflexively in response to temperature changes. In cold weather, or when a man jumps into a body of cold water, it contracts to bring the testes closer to the body. In warm weather it relaxes, allowing the testes to dangle farther from

Scrotum The pouch of loose skin that contains the testes. (From the same linguistic root as the word *shred,* which means "a long, narrow strip," probably referring to the long furrows on the scrotal sac.)

Spermatic cord The cord that suspends a testicle within the scrotum and contains a vas deferens, blood vessels, nerves, and the cremaster muscle.

Vas deferens A tube that conducts sperm from the testicle to the ejaculatory duct of the penis. (From Latin roots meaning "a vessel" that "carries down.")

Cremaster muscle The muscle that raises and lowers the testicle in response to temperature changes and sexual stimulation.

Dartos muscle The muscle in the middle layer of the scrotum that contracts and relaxes in response to temperature changes.

Human Sexuality ONLINE //

Another Sort of Gender Equality: Up-Jock Pics Now Compete with Up-Skirt Photos

There are countless "Up-Skirt" pay sex sites on the World Wide Web, which feature voyeuristic photos of women taken by people with hidden video cams. However, "Up-Jock" sites may not be far behind, as suggested by the covert filming of male college athletes in the buff. Female genitals are apparently not the only objects of fascination in the Net.

In 2000, a federal judge rejected a lawsuit against two Internet access providers, PSINet Inc. and GTE Corp. filed by dozens of Illinois State University athletes after their nude images were marketed on the World Wide Web. The ISU football players and other athletes were secretly videotaped in various states of undress by hidden cameras in restrooms, locker rooms, and showers. The videotapes were sold on various Web sites for which service was provided by the two companies. The video companies that made and sold the tapes are also facing lawsuits from the students.

Judge Charles Kocoras in the U.S. District Court for the Northern District of Illinois had previously granted motions by the two companies to dismiss the case after concluding that they could not be held liable as service providers under the Communications Decency Act of 1996. Kocoras said that he saw no evidence that PSINet or GTE provided any content for the Web sites in question.

Lawyers representing the companies hailed the ruling as a historic decision, describing it as "the first case to hold that Web hosting services are immune from suit under Section 230 of the Communications Decency Act." In his ruling, Kocoras said that 1996 law "creates a federal immunity to any cause of action that would make service providers liable for information originating with a third-party user of the service."

The Illinois State University athletes were among 200 students from more than 50 universities whose naked pictures were distributed on sexually oriented sites on the Internet. Without their knowledge, the athletes were videotaped at urinals or showers or weighing in naked at competitions. The tapes were made by employees or students working for video companies. These people posed as athletic trainers and slipped hidden cameras into locker rooms in gym bags. Taping occurred over the past decade at a host of other universities, including Northwestern University, Eastern Illinois University, Iowa State University, Michigan State, and the University of Pennsylvania.

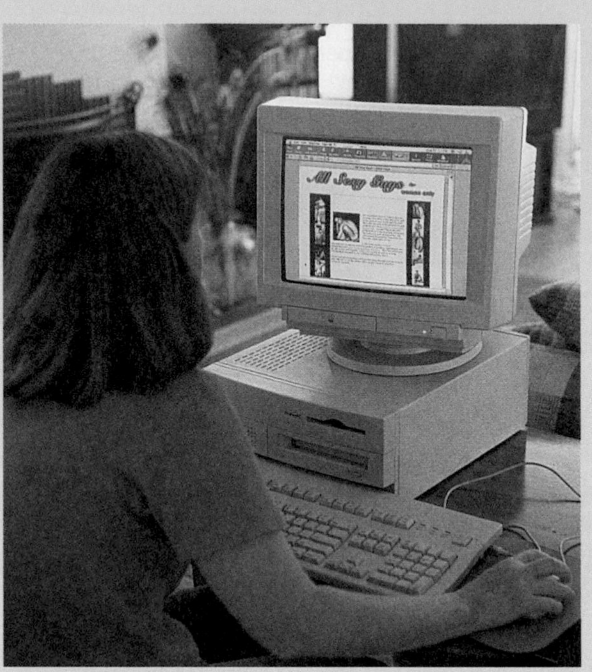

What Would You Do If Nude Photos of You Turned Up Online? In some cases, surprisingly enough, the law is less than perfectly clear.

Source: From "Athletes Rebuffed in Nude Photo Net Case." (2000, June 22), Reuters News Agency. Adapted by permission.

the body. The dartos muscle also increases or decreases the surface area of the scrotum in response to temperature changes. Smoothing allows greater dissipation of heat in hot weather. Tightening or constricting the skin surface helps retain heat and gives the scrotum a wrinkled appearance in the cold.

The scrotum is developed from the same embryonic tissue that becomes the labia majora of the female. Thus, like the labia majora, it is quite sensitive to sexual stimulation. It is somewhat more sensitive than the top side of the penis but less so than other areas of the penis.

INTERNAL SEX ORGANS

The male internal sex organs consist of the testes, the organs that manufacture sperm and the male sex hormone testosterone; the system of tubes and ducts that conduct sperm through the male reproductive system; and the organs that help nourish and activate sperm and neutralize some of the acidity that sperm encounter in the vagina.

The Testes

The testes are the male gonads (*gonad* derives from the Greek *gone*, which means "seed"). In slang the testes are frequently referred to as "balls" or "nuts." These terms are considered vulgar, but they are reasonably descriptive. They also make it easier for many people to refer to the testes in informal conversation.

The testes serve two functions analogous to those of the ovaries. They secrete sex hormones and produce mature **germ cells.** In the case of the testes, the germ cells are **sperm** and the sex hormones are **androgens.** The most important androgen is **testosterone.**

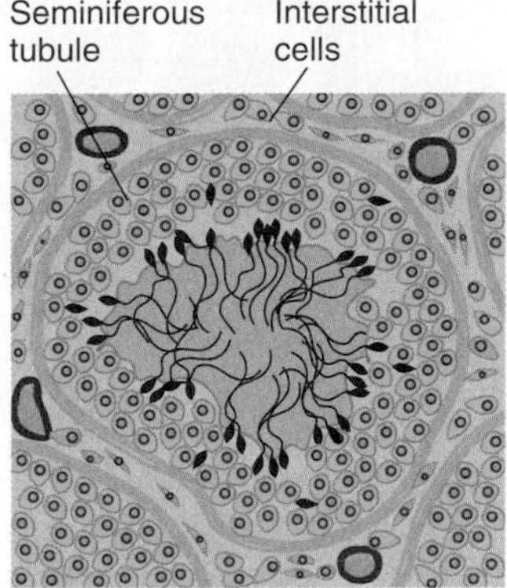

FIGURE 4.5

Interstitial Cells.
Testosterone is produced by the interstitial cells, which lie between the seminiferous tubules in each testis. Sperm (seen in the middle of the diagram) are produced within the seminiferous tubules.

TESTOSTERONE Testosterone is secreted by **interstitial cells,** which are also referred to as **Leydig's cells.** Interstitial cells lie between the seminiferous tubules and release testosterone directly into the bloodstream (see Figure 4.5). Testosterone stimulates the prenatal differentiation of male sex organs, sperm production, and the development of **secondary sex characteristics,** such as the beard, deep voice, and the growth of muscle mass.

In men, several endocrine glands—the hypothalamus, pituitary gland, and testes (see Figure 4.6)—keep blood testosterone levels at a more or less even level. This contrasts with the peaks and valleys in levels of female sex hormones during the phases of the menstrual cycle. Testosterone levels vary slightly with stress, time of day or month, and other factors, but a feedback loop among the endocrine glands keeps them relatively stable.

The same pituitary hormones, FSH and LH, that regulate the activity of the ovaries also regulate the activity of the testes. FSH regulates the production of sperm by the testes; LH stimulates secretion of testosterone by the interstitial cells. Low testosterone levels signal the hypothalamus to secrete a hormone called LH-releasing hormone (LH-RH). Like dominoes falling in a line, LH-RH causes the pituitary gland to secrete LH, which in turn stimulates the testes to release testosterone into the blood system. LH is also referred to as *interstitial-cell-stimulating hormone*, or ICSH.

When the level of testosterone in the blood system reaches a certain peak, the hypothalamus directs the pituitary gland *not* to secrete LH. This system for circling information around these three endocrine glands is called a *feedback loop.* This feedback loop is *negative.* That is, increases in hormone levels in one part of the system trigger another part to shut down, and decreases in those hormone levels activate that other part of the system.

The testes usually range between 1 and 1¾ inches in length. They are about half as wide and deep. The left testicle usually hangs lower, because the left spermatic cord tends to be somewhat longer.

Germ cell A cell from which a new organism develops. (From the Latin *germen,* which means "bud" or "sprout.")

Sperm The male germ cell. (From a Greek root that means "seed.")

Androgens Male sex hormones. (From the Greek *andros,* which means "man" or "males," and *-gene,* which means "born.")

Testosterone A male steroid sex hormone.

Interstitial cells Cells that lie between the seminiferous tubules and secrete testosterone. (*Interstitial* means "set between.")

Leydig's cells Another term for *interstitial cells.*

Secondary sex characteristics Traits that distinguish the genders but are not directly involved in reproduction.

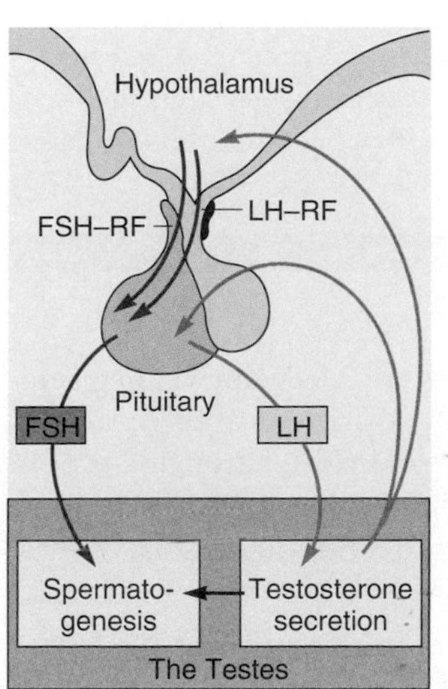

FIGURE 4.6

Hormonal Control of the Testes. Several endocrine glands—the hypothalamus, the pituitary gland, and the testes—keep blood testosterone levels at a more or less constant level. Low testosterone levels signal the hypothalamus to secrete LH-releasing hormone (LH-RH). Like dominoes falling in line, LH-RH causes the pituitary gland to secrete LH, which in turn stimulates the testes to release testosterone. Follicle-stimulating-hormone-releasing hormone (FSH-RH) from the hypothalamus causes the pituitary to secrete FSH which, in turn, causes the testes to produce sperm cells.

a CLOSER look

Is There a *Man*opause?

Men cannot undergo menopause; they have never menstruated. Yet one now and then hears of a so-called male menopause, occasionally referred to as "manopause." Some men during their later years are informally referred to as menopausal. Sadly, this description is usually meant to convey the negative, harmful stereotype of the aging person as crotchety and irritable. Such stereotypes of menopause are unfortunate reminders of sexism and ageism, and they are not necessarily consistent with the biology or psychology of aging.

The scientific jury is still out on the existence of the "male menopause." Women encounter relatively sudden age-related declines in sex hormones and fertility during menopause. Men experience a gradual decline in testosterone levels as they age, but nothing like the sharp plunge in estrogen levels that women experience during menopause (Sheehy, 1998). Testosterone levels begin to fall at about age 40 or 50 and may decline to one third or one half of their peak levels by age 80 (Brody, 1995c).

The drop in testosterone levels that occurs as men age may be connected to a variety of age-related symptoms, including reduced muscle mass and strength, accumulation of body fat, reduced energy levels, lowered fertility, and reduced erectile ability. However, despite a decline in testosterone levels, most men remain potent throughout their lives. Little is known about the critical levels of testosterone that are needed to maintain erectile ability. Certain age-related changes, such as reduced muscle mass and strength and increased body fat may be due not to declining testosterone production, but to other factors associated with aging, such as a gradual loss of *human growth hormone*, a hormone that helps maintain muscle strength and that may prevent fat buildup. Some experts believe that testosterone replacement may help avert erectile problems, bone loss, and frailty, in much the same way that estrogen replacement benefits postmenopausal women (Brody, 1995c). Others worry that excessive use of the hormone may increase the risks of prostate cancer and cardiovascular disease.

Although men as they age do experience a gradual decline in the number and motility of sperm, which reduces their fertility, some viable sperm continue to be produced even into late adulthood (Sheehy, 1997, 1998). It is not surprising, from a physical standpoint, to find a man in his 70s or older fathering a child.

Men can remain sexually active and father children at advanced ages. For both genders, marital satisfaction and attitudes toward aging can affect sexual behavior as profoundly as physical changes can.

Is There a *Man*opause? The term *manopause* is sometimes used to convey the negative, harmful stereotype of the aging person as crotchety and irritable. The scientific jury is still out on whether there is a "male menopause," but men do experience a gradual decline in testosterone levels as they age. The decline in testosterone is connected with age-related symptoms such as reduced muscle strength, accumulation of fat, reduced energy, lowered fertility, and reduced erectile ability. Nevertheless, at least half of men maintain erectile ability throughout their lives.

Seminiferous tubules Tiny, winding, sperm-producing tubes that are located within the lobes of the testes. (From Latin roots that mean "seed bearing.")

SPERM Each testicle is divided into many lobes. The lobes are filled with winding **seminiferous tubules** (see Figure 4.2). Although packed into a tiny space, these tubules, placed end to end, would span the length of several football fields. Through a process called **spermatogenesis,** these threadlike structures produce and store hundreds of billions of sperm through the course of a man's lifetime.

Sperm cells develop through several stages. It takes about 72 days for the testes to manufacture a mature sperm cell (Leary, 1990). In an early stage, sperm cells are called **spermatocytes.** Each one contains 46 chromosomes, including one X and one Y sex chromosome. Each spermatocyte divides into two **spermatids,** each of which has 23 chromosomes. Half the spermatids have X sex chromosomes, and the other half have Y sex chromosomes. Mature sperm cells, called **spermatozoa,** look something like tadpoles when examined under a microscope. Each has a head, a cone-shaped midpiece, and a tail. The head is about 5 microns (1/50,000 of an inch) long and contains the cell nucleus that houses the 23 chromosomes. The midpiece contains structures that provide the energy the tail needs to lash back and forth in a swimming motion. Each sperm cell is about 50 microns (1/5,000 of an inch) long, one of the smallest cells in the body (Thompson, 1993).

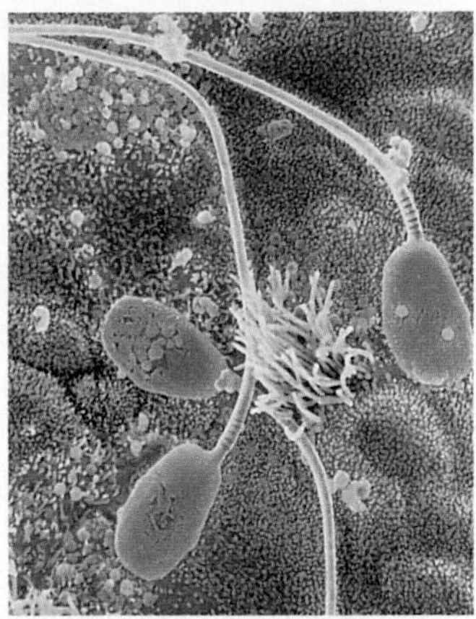

Human Sperm Cells Magnified Many Times.

During fertilization, the 23 chromosomes from the father's sperm cell combine with the 23 chromosomes from the mother's ovum, furnishing the standard ensemble of 46 chromosomes in the offspring. Among the 23 chromosomes borne by sperm cells is one sex chromosome—an X sex chromosome or a Y sex chromosome. Ova contain X sex chromosomes only. The union of an X sex chromosome and a Y sex chromosome leads to the development of male offspring. Two X sex chromosomes combine to yield female offspring. Thus, whether the father contributes an X or a Y sex chromosome determines the baby's gender.

It is true that the father determines the baby's gender through the presence of an X or a Y sex chromosome. If the fertilizing sperm has an X sex chromosome, the child will be a girl. If it has a Y sex chromosome, the child will be a boy. ■

The testes are veritable dynamos of manufacturing power, churning out about 1,000 sperm per second, or about 30 billion—yes, *billion*—per year. Mathematically speaking, 10 to 20 ejaculations hold enough sperm to populate the earth. (Men are always so taken with themselves, notes the third author.)

Belgian researchers have discovered that sperm cells possess the same kind of receptors that the nose uses to sense odors (Angier, 1992). This discovery suggests that sperm may find their way to an egg cell by detecting its scent. Researchers at the Texas Southwestern Medical Center in Dallas had earlier discovered that the fertile egg cells emit a compound that attracts the interest of sperm cells. These odor receptors may be the mechanism by which sperm recognize these attractants. If this turns out to be the case, it could lead to the development of contraceptives that prevent fertilization by blocking these receptors.

Sperm proceed from the seminiferous tubules through an intricate maze of ducts that converge in a single tube called the **epididymis.** The epididymis lies against the back wall of the testicle and serves as a storage facility for sperm. The epididymis, which is some 2 inches in length, consists of twisted passages that would be 10 to 20 feet in length if stretched end to end. Sperm are inactive when they enter the epididymis. They continue to mature as they slowly make their way through the epididymis for another two to four weeks.

Spermatogenesis The process by which sperm cells are produced and developed.

Spermatocyte An early stage in the development of sperm cells, in which each parent cell has 46 chromosomes, including one X and one Y sex chromosome.

Spermatids Cells formed by the division of spermatocytes. Each spermatid has 23 chromosomes.

Spermatozoa Mature sperm cells.

Epididymis A tube that lies against the back wall of each testicle and serves as a storage facility for sperm. (From Greek roots that mean "upon testicles.")

Vasectomy A sterilization procedure in which the vas deferens is severed, preventing sperm from reaching the ejaculatory duct.

The Vas Deferens

Each epididymis empties into a vas deferens (also called *ductus deferens*). The vas is a thin, cylindrical tube about 16 inches long that serves as a conduit for mature sperm. In the scrotum, the vas deferens lies near the skin surface within the spermatic cord. Therefore, a **vasectomy,** an operation in which the right and left vas deferens are severed, is a convenient means of sterilization. The tube leaves the scrotum and follows a circuitous path

a CLOSER look

Big

Don't be fooled by his overall size. The tiny fruit fly, *Drosophila bifurca*, is only a fraction of an inch long. You can crush it easily with the tip of your finger. But this little insect holds the distinction of being number 1 in the *Guinness Book of Records* for length of sperm. He produces coiled sperm that stretch out to nearly 2 ½ inches, or 20 times the length of his own body! How does the fruit fly accomplish this feat? By devoting about 11% of his body weight to his testes.

This fruit fly's sperm are 1,000 times as long as human sperm (Pitnick, 1995). Remember that a human sperm cell is about 1/5,000 of an inch long and is one of the smallest cells in the body (Thompson, 1993). In order to match the achievement of *Drosophila,* a man would have to produce sperm that are 100 to 120 feet long (Leary, 1995).

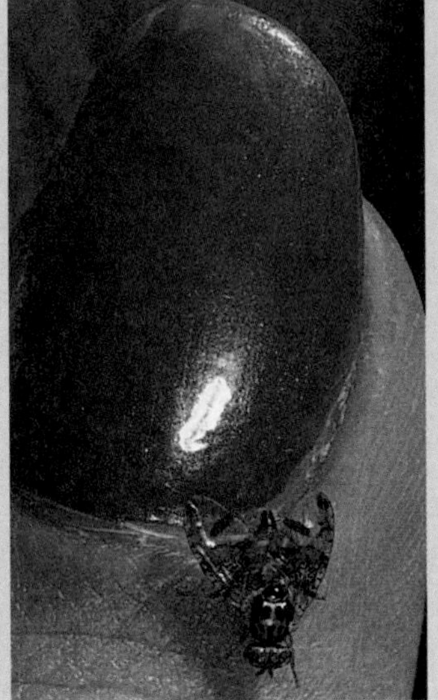

Drosophila bifurca. Although the fruit fly is only a fraction of an inch long, his sperm is 1,000 times longer than that of human males.

Truth or Fiction? **REVISITED**

It is true that the sperm of the tiny fruit fly are longer than human sperm. In fact, they are longer than the sperm of any mammal.

There may be many more fruit flies than people, but fruit flies are no match for people in terms of numbers of sperm. Men can ejaculate several hundred million sperm at a time, whereas the fruit fly releases only about 50 (Pitnick, 1995). Moreover, human sperm are strong swimmers, whereas the rolled-up fruit fly sperm slowly unwinds in the female and apparently relies on transportation mechanisms within the female to reach its goal.

Why, the third author wishes to ask, must men be so competitive—even with fruit flies?

up into the abdominal cavity. Then it loops back along the rear surface of the bladder (see Figure 4.7).

The Seminal Vesicles

The two **seminal vesicles** are small glands, each about 2 inches long. They lie behind the bladder and open into the **ejaculatory ducts,** where the fluids they secrete combine with sperm (see Figure 4.7). A vesicle is a small cavity or sac; the seminal vesicles were so named because they were mistakenly believed to be reservoirs for semen, rather than glands.

The fluid produced by the seminal vesicles is rich in fructose, a form of sugar, which nourishes sperm and helps them become active, or motile. Sperm motility is a major factor in male fertility. Before reaching the ejaculatory ducts, sperm are propelled along their journey by contractions of the epididymis and vas deferens and by **cilia** that line the walls of the vas deferens. Once they become motile, they propel themselves by whipping their tails.

At the base of the bladder, each vas deferens joins a seminal vesicle to form a short ejaculatory duct that runs through the middle of the prostate gland (see Figure 4.7). In the prostate, the ejaculatory duct opens into the urethra, which leads to the tip of the penis. The urethra carries sperm and urine out through the penis, but normally not at the same time.

Seminal vesicles Small glands that lie behind the bladder and secrete fluids that combine with sperm in the ejaculatory ducts.

Ejaculatory duct A duct, formed by the convergence of a vas deferens with a seminal vesicle, through which sperm pass through the prostate gland and into the urethra.

Cilia Hairlike projections from cells which beat rhythmically to produce locomotion or currents.

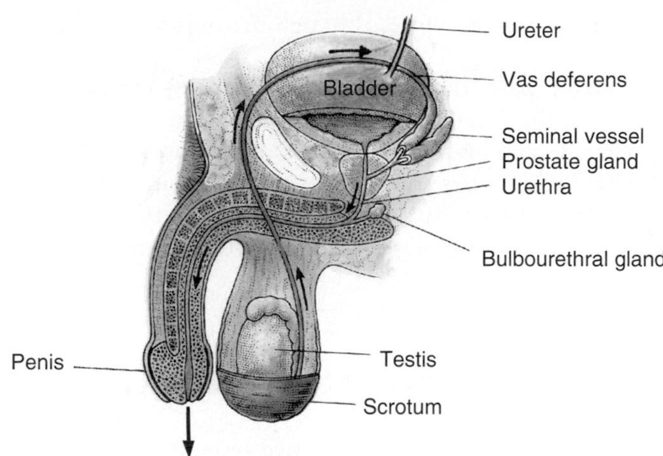

FIGURE 4.7

Passage of Spermatozoa. Each testicle is divided into lobes that contain threadlike seminiferous tubules. Through spermatogenesis, the tubules produce and store hundreds of billions of sperm over the course of a man's lifetime. During ejaculation, sperm cells travel through the vas deferens, up and over the bladder, into the ejaculatory duct, and then through the urethra. Secretions from the seminal vesicles and the bulbourethral glands join with sperm to compose semen.

The Prostate Gland

The **prostate gland** lies beneath the bladder and is approximately the shape and size of chestnut in (about ¾ inch in diameter). Note the spelling of the name of the gland—pros*tate*, not pros*trate*. (*Prostrate* means lying with one's face on the ground, as in some forms of prayer.) The prostate gland contains muscle fibers and glandular tissue that secrete prostatic fluid. Prostatic fluid is milky and alkaline. It provides the texture and odor characteristic of the seminal fluid. The alkalinity neutralizes some of the acidity of the vaginal tract, prolonging the life span of sperm as seminal fluid spreads through the female reproductive system. The prostate is continually active in mature males, but sexual arousal further stimulates secretions. Secretions are conveyed into the urethra by a sievelike duct system. There the secretions combine with sperm and fluid from the seminal vesicles.

A vasectomy prevents sperm from reaching the urethra but does not cut off fluids from the seminal vesicles or prostate gland. A man who has had a vasectomy thus emits an ejaculate that appears normal but contains no sperm.

Cowper's Glands

The **Cowper's glands** are also known as the **bulbourethral glands,** in recognition of their shape and location. These two structures lie below the prostate and empty their secretions into the urethra. During sexual arousal they secrete a drop or so of clear, slippery fluid that appears at the urethral opening. The functions of this fluid are not entirely understood. It may help buffer the acidity of the male's urethra and lubricate the urethral passageway to ease the passage of seminal fluid. The fluid is not produced in sufficient amounts to play a significant role in lubricating the vagina during intercourse.

Fluid from the Cowper's glands precedes the ejaculate and often contains viable sperm. Thus, coitus may lead to pregnancy even if the penis is withdrawn prior to ejaculation. This is one reason why people who practice the "withdrawal method" of birth control are frequently called "parents."

Semen

Sperm and the fluids contributed by the seminal vesicles, the prostate gland, and the Cowper's glands make up **semen,** the whitish seminal fluid that is expelled through the tip of the penis during ejaculation. The seminal vesicles secrete about 70% of the fluid that constitutes the ejaculate. The remaining 30% of seminal fluid consists of sperm and fluids produced by the prostate gland and the Cowper's glands. Sperm themselves account for only about 1% of the volume of semen. This is why men with vasectomies

Prostate gland The gland that lies beneath the bladder and secretes prostatic fluid, which gives semen its characteristic odor and texture.

Cowper's glands Structures that lie below the prostate and empty their secretions into the urethra during sexual arousal.

Bulbourethral glands Another term for *Cowper's glands.*

Semen The whitish fluid that constitutes the ejaculate, consisting of sperm and secretions from the seminal vesicles, prostate gland, and Cowper's glands.

continue to ejaculate about as much semen as before, although their ejaculates are devoid of sperm.

Semen is the medium that carries sperm through much of the male's reproductive system and the reproductive tract of the female. Semen contains water, mucus, sugar (fructose), acids, and bases. It activates and nourishes sperm, and the bases help shield sperm from vaginal acidity. The typical ejaculate contains between 200 and 400 million sperm and ranges between 3 and 5 milliliters in volume. (Five milliliters is equal to about 1 tablespoon.) The quantity of semen decreases with age and with frequency of ejaculation.

DISEASES OF THE UROGENITAL SYSTEM

Because the organs that make up the urinary and reproductive systems are near each other and share some "piping," they are referred to as the urinogenital or urogenital system. A number of diseases affect the urogenital system. The type of physician who specializes in their diagnosis and treatment is a **urologist.**

Urethritis

Men, like women, are subject to bladder and urethral inflammations, which are generally referred to as **urethritis.** The symptoms include frequent urination (urinary frequency), a strong need to urinate (urinary urgency), burning during urination, and a penile discharge. People with symptoms of urinary frequency and urinary urgency feel the pressing need to urinate repeatedly, even though they may have just done so and may have but another drop or two to expel. The discharge may dry on the urethral opening, in which case it may have to be peeled off or wiped away before it is possible to urinate. The urethra also may become constricted when it is inflamed, slowing or halting urination. It is a frightening sensation for a male to feel the urine rush from his bladder and then suddenly stop at the urethral opening!

Preventive measures for urethritis parallel those suggested for cystitis (bladder infection): drinking more water, drinking cranberry juice (4 ounces, two or three times a day), and lowering intake of alcohol and caffeine. Cranberry juice is highly acidic, and acid tends to eliminate many of the bacteria that can give rise to urethritis.

Cancer of the Testes

Cancer of the testicles remains a relatively rare form of cancer, accounting for about 6,000 new cases annually, or about 1% of all new cancers in men (National Cancer Institute, 2000). It is the most common form of solid-tumor cancer to strike men between the ages of 20 and 34, however (Vazi et al., 1989). It accounts for nearly 10% of all deaths from cancer among men in that age group.

There is no evidence that testicular cancer results from sexual overactivity or masturbation. Men who had **cryptorchidism** as children (a condition in which one or both testicles fail to descend from the abdomen into the scrotum) stand about a 40 times greater chance of contracting testicular cancer. Undescended testicles appear to occur more commonly in boys born to mothers who used the hormone diethylstilbestrol (DES) during pregnancy. In the 1940s and 1950s, pregnant women were often prescribed DES to help prevent miscarriages.

Although testicular cancer was generally fatal in earlier years, the prognosis today is quite favorable, especially for cases that are detected early. Treatments include surgical removal of the diseased testis, radiation, and chemotherapy. The survival rate among cases that are detected early, before the cancer has spread beyond the testes, is well above 90%

Urologist A physician who specializes in the diagnosis and treatment of diseases of the urogenital system.

Urethritis An inflammation of the bladder or urethra.

Cryptorchidism A condition in which one or both testicles fail to descend from the abdomen into the scrotum.

(National Cancer Institute, 2000). Delayed treatment markedly reduces the chances of survival, because survival is connected with the extent to which the cancer has spread.

The surgical removal of a testicle may have profound psychological implications. Some men who have lost a testicle feel less "manly." Fears related to sexual performance can engender sexual dysfunctions. From a physiological standpoint, sexual functioning should remain unimpaired, because adequate quantities of testosterone are produced by the remaining testis.

The early stages of testicular cancer usually produce no symptoms, other than the mass itself. Because early detection is crucial to survival, men are advised to examine themselves monthly following puberty (Reinisch, 1990) and to have regular medical checkups. Self-examination may also reveal evidence of sexually transmitted diseases and other problems.

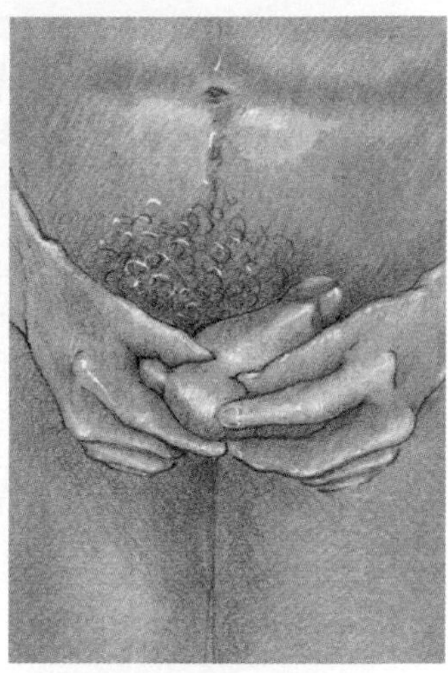

FIGURE 4.8

Self-Examination of the Testes.

SELF-EXAMINATION OF THE TESTES Self-examination (see Figure 4.8) is best performed shortly after a warm shower or bath, when the skin of the scrotum is most relaxed. The man should exam the scrotum for evidence of pea-sized lumps. Each testicle can be rolled gently between the thumb and the fingers. Lumps are generally found on the side or front of the testicle. The presence of a lump is not necessarily a sign of cancer, but it should be promptly reported to a physician for further evaluation. The American Cancer Society (2000) and the National Cancer Institute (2000) list these warning signals:

1. A slight enlargement of one of the testicles.
2. A change in the consistency of a testicle.
3. A dull ache in the lower abdomen or groin. (Pain may be absent in cancer of the testes, however).
4. Sensation of dragging and heaviness in a testicle.

Disorders of the Prostate

The prostate gland is tiny at birth and grows rapidly at puberty. It may shrink during adulthood but usually becomes enlarged past the age of 50.

BENIGN PROSTATIC HYPERPLASIA The prostate gland becomes enlarged in about one quarter of men past the age of 60 (Walsh, 1996). When enlargement is due to hormonal changes associated with aging rather than to other causes, such as inflammation from sexually transmitted diseases, it is referred to as **benign prostatic hyperplasia.** Because the prostate surrounds the upper part of the urethra (see Figure 4.2), enlargement constricts the urethra. Resultant symptoms include urinary frequency (including increased frequency of nocturnal urination), urinary urgency, and difficulty starting the flow of urine. Several treatments are available to relieve the pressure on the urethra and increase the flow of urine. They include medicines such as finasteride and alpha-adrenergic-agonist drugs ("alpha blockers") and surgical removal of part of the prostate (Wasson, 1998).

CANCER OF THE PROSTATE A more serious and life-threatening problem is prostate cancer. About one man in eight in the United States will develop prostate cancer (Bazell, 2000; Chang, 2000). Prostate cancer is the second most common form of cancer among men, after skin cancer, and the second leading cause of cancer deaths in men, after lung cancer. There are about 180,000 new cases of prostate cancer each year in the United States, and nearly 37,000 men die from it annually.

Prostate cancer involves the growth of malignant prostate tumors that can metastasize to bones and lymph nodes if not detected and treated early. African American men are one third more likely than European American men to develop prostate cancer (National Cancer Institute, 2000). African American men have less access to health care than European American men do, so prostate cancer is diagnosed later among them and they

Benign prostatic hyperplasia Enlargement of the prostate gland due to hormonal changes associated with aging and characterized by symptoms such as urinary frequency, urinary urgency, and difficulty starting the flow of urine.

a CLOSER look

Lance Armstrong—Winner of Two Contests

France had not had a king for more than 200 years, and the last one came to a bad end. But Lance Armstrong received a welcome fit for royalty in Paris, in July 2000, as he cruised to victory in the Tour de France.

Armstrong had won the world's premier bicycling race for the second consecutive year, and that was not his only accomplishment. In part because of his near-miraculous recovery from cancer three years earlier and his determined return to cycling, Armstrong has helped restore credibility to a sport that was badly tainted by drug scandals in the 1998 Tour. The cheers of the crowds, French and foreign, that greeted the American champion were loud, unambiguous, and enthusiastic.

When Armstrong received his trophy, he lifted his baby son, Luke, who was born after the 1999 Tour, and held him high to show the spectators. The child was conceived in vitro with sperm frozen before Armstrong began chemotherapy for testicular cancer in 1996. "This is the first time I win as a father," he said. "It was a hard Tour de France, and I'm very glad it's finished."

Armstrong is only the second American ever to win the 87-year-old race; he now has two victories to Greg LeMond's three.

Armstrong is hailed as much for being a cancer survivor as for being a world-class athlete. He had been given a less-than-50% chance of survival when his testicular cancer was diagnosed in 1996. It also spread to his lungs and brain.

Months of chemotherapy and brain surgery saved him, though he did not decide to return to cycling for another year. When he did, he trained harder than ever.

After winning the 2000 Tour, Armstrong was not a survivor—he was a celebrity. He was feted at the American Embassy with a party attended by friend and cycling fan Robin Williams, among other notables. His commercial endorsements have made him rich. But he has not forgotten the past.

Tour de France Winner Lance Armstrong. Lance Armstrong came back to win the grueling Tour de France after being successfully treated for testicular cancer. His cancer had spread to his lungs and brain, and he had to be treated with chemotherapy and surgery.

"It's always a struggle," he told reporters. "My health is not guaranteed forever. The fact that I'm still here and cancer-free is the most important thing in my life."

Source: From Anne Swardson (2000, July 23) "Armstrong Basks in Glow of Victory." The Washington Post online. Copyright © 2000 by The Washington Post. Adapted by permission.

have a higher mortality rate from the disease (Merrill & Brawley, 2000). Researchers have identified intake of animal fat as a potential risk factor. Men whose diets are rich in animal fats, especially fats from red meat, have a substantially higher chance of developing advanced prostate cancer than do men with a low intake of animal fat. The incidence of prostate cancer also increases with age (Tarone et al., 2000). More than 80% of cases of prostate cancer are diagnosed in men aged 65 or above (National Cancer Institute, 2000). Genetic factors are also apparently involved (Lichtenstein et al., 2000). Moreover, testosterone spurs the development of prostate cancer (D'Amico et al., 2000).

The early symptoms of cancer of the prostate may mimic those of benign prostate enlargement: urinary frequency and difficulty in urinating. Later symptoms include blood in the urine, pain or burning on urination, and pain in the lower back, pelvis, or upper thighs (National Cancer Institute, 2000). Most cases occur without noticeable symptoms in the early stages.

The American Cancer Society (2000) recommends that men receive annual rectal examinations beginning at about age 40. The physician inserts a finger into the rectum and feels for abnormalities in the prostate gland. Unfortunately, many men are reluctant to undergo a rectal examination, even though it is only mildly uncomfortable and may save their lives. Some are embarrassed or reluctant to discuss urinary problems with their physicians. Others may resist the rectal examination because they associate rectal insertion with male-male sex. Still others fear that they may have cancer and choose to remain ignorant. Avoidance of, or ignorance of the need for, regular exams is a major contributor to the death rate from prostate cancer.

When a cancerous growth is suspected on the basis of a rectal examination or a PSA blood test, further testing is usually done via additional blood tests, ultrasound, or biopsy. PSA (prostrate-specific antigen) is a protein that helps transform a gel-like substance in the prostate gland into a liquid that transports sperm when it is ejaculated. In the diseased or enlarged prostate, PSA seeps into the blood at higher levels, giving higher test readings. Early detection is important because treatment is most effective before the cancer has spread. Fifty-eight percent of cases of prostate cancer are discovered while they are still localized. The survival rate drops dramatically if the cancer has metastasized. Still, the overall survival rate has improved since the 1960s from 50% to more than 80% (National Cancer Institute, 2000).

The most widely used treatment for prostate cancer is surgical removal of the prostate gland (Klein, E. A., 2000). However, surgical prostate removal may damage surrounding nerves, leading to problems in controlling the flow of urine or in erection or ejaculation (Stanford et al., 2000). Recently introduced surgical techniques tend to spare the surrounding nerves and to reduce, but not eliminate, the risk of complications. Other treatments include radiation, hormone treatment, and anticancer drugs (Klein, E. A., 2000). Hormone treatment in the form of androgen (testosterone) suppression therapy and anticancer drugs may shrink the tumor and relieve pain for long periods of time. One study found that the combination of radiation therapy and androgen suppressive therapy was more effective than radiation therapy alone (D'Amico et al., 2000). Men who have their prostate glands removed are more likely to experience urinary incontinence (loss of control over urination) and sexual dysfunction (trouble attaining erection) than men who use radiation (Potosky et al., 2000). But many physicians argue that surgery remains the better choice in terms of survival rates. Among older men with slow-growing prostate cancer, physicians often prefer "watchful waiting" to surgery. The men may die from causes other than cancer. The treatment choices are explored further in the nearby Closer Look.

In sum, recently developed methods of detecting prostate cancer earlier are increasing the survival rate for men who develop the disease (Merrill & Brawley, 2000). A blood test for PSA can detect evidence of prostate cancer even among men whose prostates feel normal upon physical examination (Tarone et al., 2000). According to Dr. Steven Skates (1998) of Massachusetts General Hospital, PSA screening permits detection of prostate cancer an average of 11 years before other methods do. The 5-year survival rate is above 90% for prostate cancer that has *not* metastasized (National Cancer Institute, 2000). By the time there are other symptoms or cancer is detected via rectal examination, the cancer may have already metastasized (Tarone et al., 2000).

Researchers are currently investigating even earlier and more reliable ways of detecting prostate cancer. For example, high levels of insulin growth factor-I may provide an early marker of prostate cancer (Pollak et al., 1998). A urine test may also detect prostate cancer earlier than the PSA blood test by seeking changes in the GSTPI gene. Such changes are found among 90% of people with prostate cancer (Cairns, 2000).

a CLOSER look

More Options, and Decisions, for Men with Prostate Cancer

Mayor Rudolph W. Giuliani of New York decided on a treatment for his prostate cancer, which was diagnosed in April 2000, after months of private deliberation and public speculation. In mid-September, the mayor, along with his doctors at Mount Sinai Medical Center, announced he had received radioactive seed implants during an out-patient procedure earlier that day. But Mr. Giuliani said that in deciding on a treatment, he had great difficulty weighing his choices. Doctors refer to the process as constructing a "decision tree" or "clinical pathway," as patients weigh each pro and con.

Prostate cancer is the most common cancer, after skin cancer, among American men. Some 180,000 new cases are diagnosed each year, and about 37,000 men die of the disease, according to the American Cancer Society. African Americans and those with relatives who had prostate cancer are at increased risk. But the disease progresses at a relatively slow pace, so men in the early stages can take some time to explore the expanding treatment options available to them.

Prostate cancer is usually detected and assessed via a combination of tests. These include a blood test for prostate-specific antigen (PSA), recommended annually for men 50 and over (earlier for those at higher risk). Results under 4 nanograms per milliliter are usually considered normal. Results over 10 are high, and results between 4 and 10 are considered borderline.

Next comes the digital–rectal exam, in which the doctor palpates the prostate gland with a finger to detect abnormalities. A transrectal ultrasound may also be used. If cancer is suspected, a needle biopsy is done of the gland and sometimes of nearby lymph nodes. If a tumor is detected, pathologists will "grade" two pieces of tissue and rate them from 1 to 5 on the Gleason scale, which reflects the cancer's aggressiveness. The two scores are then added together. Sums of 2 through 4 are considered low (slower growing), 5 and 6 intermediate, and 7 to 10 high, with the worst prognosis. Finally, doctors may do other tests to determine whether the cancer has spread. These include radionuclide bone scans, CAT scans (coaxial tomography) and MRI's (magnetic resonance imaging).

Deciding on treatment can be daunting, partly because the options today are far better than 10 years ago. They include improved surgery, with "nerve sparing" to help reduce incontinence and impotence; external-beam radiation, wherein standard X-ray beams are directed at the prostate area; highly targeted proton radiation, in which protons are "shaped" into beams to match the shape of the tumor and to eliminate damage to surrounding tissue; radioactive seed implants, in which rice-size pellets are injected into the prostate to kill cancer cells; new hormone treatments to block production of testosterone, which feeds cancer cells; combination therapies using two or more treatments at once; and "watchful waiting," in which patients seek no treatment but simply monitor themselves for problems (a common practice in older patients).

Mayor Rudolph W. Giuliani of New York. In 2000, New York's Mayor Rudy Giuliani decided to treat his prostate cancer with radioactive seed implants during an out-patient procedure. There are many treatments today for prostate cancer, and people with the disease must weigh the benefits and risks of each

New treatments continue to be developed in clinical trials around the country. Here are the stories of seven men and the options they selected.

Damon O. Harris, 50, Chose Hormones, for Spreading Cancer

Damon O. Harris spent much of the 1970s as a singer with the Motown group, the Temptations. But now Mr. Harris, 50, spends time educating African American men about their higher-than-average risk of prostate cancer. Mr. Harris had never been screened for prostate cancer. But in early 1998, he noticed a lingering pain for several days after playing with a friend's newborn. He had a digital examination, and his PSA test came back at over 13. Within days, he was facing a cancer diagnosis. Mr. Harris began what he called "my crash course in Cancer 101." "I had little time to do research," he said. As it turned out, the decision in his case was simple. "My cancer was metastatic," he said. "I opted for the best and only treatment available, which is hormone therapy."

Mr. Harris, who lives in Reno, Nevada, immediately began taking an oral medication, Casodex, to neutralize testosterone and other male hormones in his system, followed by weekly injections of Lupron, which shuts down new production of male hormones, such as testosterone, needed for prostate growth. "I will be on it for the rest of my life," Mr. Harris said.

But he remains philosophical. "This was God's designation, and I had to accept it to overcome it," Mr. Harris said. "I was sad but not bitter. I had just met someone, and I cared for her very much. I was concerned how this would affect the relationship."

The most severe side effects of hormone therapy are hot flashes, weight gain, and reduced sexual desire. Mr. Harris experienced all three. "I no longer have the hot flashes, for which I'm glad," he said. "But I was concerned because I thought possibly my treatment wasn't working anymore. But they only last until your system adjusts." He exercises regularly to keep the weight off, and though his libido has changed, "I am still a man," he said. "More importantly, I've learned things about myself that I didn't know when I was complete."

Mr. Harris is in "tentative remission," though frequent tests and scans "sadly remind me of my situation." He plans to begin intermittent treatment soon to fight osteoporosis. For now, promoting early detection and treatment through the Damon Harris Cancer Foundation has become his life's work, he said. "It's important to me that other men not go through what I've experienced, particularly African Americans."

Eugene Waterman, 71, Chose Proton Beam Radiation

As a doctor, Eugene Waterman, 71, was especially aware of the risks of prostate cancer and the means of detecting it. In 1998 Dr. Waterman, a semiretired psychiatrist, learned that his PSA had suddenly risen to 11, very high. A biopsy diagnosed cancer with an aggressive Gleason score of 8. Fearful of the risks of surgery, Dr. Waterman, of Hot Springs, Arkansas, began researching proton beam radiation therapy (PBRT.). Unlike X-rays, the standard forms of external radiation, PBRT is highly specific and individualized. Protons can be "shaped" into a beam to match the exact shape of the tumor, whereas the standard treatment—with photons and electrons, which are harder to direct—may affect healthy surrounding tissue.

Dr. Waterman received this sophisticated form of therapy at the Loma Linda University Medical Center in California. But he also agreed to have standard radiation treatment to the pelvic area because of signs that the cancer might have spread.

"The side effects of radiation were mild," he said. "There was some fatigue at the end of the week and some diarrhea." He had some mild anal soreness, too, which was treated with a cortisone salve. But the treatment also involved hormone therapy, a Zoladex injection every three months for a year. The resulting testosterone depletion was difficult, Dr. Waterman said. Because the condition can lead to bone loss, he asked for a bone scan, and some borderline osteoporosis was found. Now he takes Fosomax twice a week.

He also experienced hot flashes 10 to 15 times a day. He suffered a complete loss of libido and sore, enlarged breasts. "But I have been off Zoladex for over a year," he said. "Testosterone levels are back to normal levels and side effects have gone away." His sexual ability has returned with help from Viagra, Dr. Waterman said. "But desire is not as strong as it was."

Today his health is good, and his PSA has remained at 0.26 for the last year, although he realizes there is still a "50–50 chance" of eventual relapse.

Thomas Sellers, 50, Had His Prostate Removed

Thomas Sellers, a 50-year-old resident of Brookline, Massachusetts, is not only a prostate cancer survivor but also chief financial officer of the New England division of the American Cancer Society. As a cancer professional and an African American, Mr. Sellers knew about the need to begin annual PSA tests at 45. A test in late 1998 showed that his PSA had leaped to 8.5. A biopsy confirmed cancer.

"I was not surprised, but I sure was scared and angry," Mr. Sellers said. He and his wife searched the Internet, and met with others who had had cancer and with a "full range" of surgeons and oncologists. "The most important thing was, we took our time, had the pathology reports read by different doctors, got an MRI to ensure that the cancer was still in the prostate," he said. Mr. Sellers and his wife took four months to reach a decision.

Mr. Sellers's doctor, like many urologists, immediately recommended surgery. "I dropped him quickly," said Mr. Sellers, aware of the options and not ready for surgery. But Mr. Sellers's cancer, a 7 on the Gleason score, was fairly aggressive, so watchful waiting was not an option, and radioactive seeds were less practical, "though that would've been my preference based on my own research," he said.

(continued)

More Options, and Decisions, for Men with Prostate Cancer *(continued)*

Ultimately, Mr. Sellers chose surgery, a nerve-sparing radical prostatectomy. Impotence, he said, worried him greatly, but "The likelihood of a cure because of early detection and the fact the cancer was still within the prostate outweighed fear of side effects." After surgery, Mr. Sellers wore a catheter for two weeks to drain urine. He took an eight-week medical leave and did not recover his full strength for six months, he said. He endured short-term incontinence, which he called disconcerting. "It took six months or more before I was completely dry, and nine months before I stopped wearing a pad," he said.

As for sexual activity, Mr. Sellers has some problems despite the nerve sparing, and he is "experimenting with Viagra, with intermittent success." But his cancer is cured and his PSA is back to zero.

Edward Kuenzi, 64, Enrolled in Clinical Trial

Edward Kuenzi, a 64-year-old former farmer, truck driver, and schoolteacher, said he received a prostate cancer diagnosis "by coincidence only." In 1993, Mr. Kuenzi's trucking company sent him in for a routine exam to get a truck driver's license. The digital exam was abnormal and the PSA was high, 14.5. Because his health was good and the cancer did not appear to have spread, Mr. Kuenzi chose surgery. His PSA returned to normal.

But 18 months after the operation, his PSA began to rise. "I realized I hadn't been cured," Mr. Kuenzi said. "My cancer had come back." Indeed, the cancer had spread beyond the prostate gland. A biopsy, a bone scan, and an MRI found some cancerous cells on a small spot just below the bladder. Mr. Kuenzi's doctor recommended external-beam radiation, at a relatively low dose. He received 39 treatments, with few side effects, he said. The radiation lowered Mr. Kuenzi's PSA. Now, four years later, it has again begun to rise.

He took another battery of tests, though no cancer was found. His doctor recommended "watchful waiting" and continued PSA monitoring, but that was not enough for Mr. Kuenzi. He felt he needed to do more, and he began seeking out clinical trials of experimental new drugs. Mr. Kuenzi, who lives in central Oregon, found a clinical trial that was enrolling patients at Oregon Health Sciences University in Portland involving a drug called Calcitrol, a form of vitamin D thought to slow the cancer growth. He takes a medication orally four times a week, curtails his calcium intake, and has monthly examinations at the trial center, which is about 60 miles north of his home.

"It's too soon to tell if it is helping me," Mr. Kuenzi said. "But there are no side effects and no problems." He expects to remain in the trial for a year or two.

Meanwhile, Mr. Kuenzi said, his health is "excellent." But he has suffered severe sexual dysfunction from the surgery. "The impotence just got worse with time," he said. (Viagra was not yet available.) After trying injections, rather unsuccessfully, he had a penile implant, which he calls "not as effective as it should be." But he says he and his wife "can still be very intimate, even without intercourse."

Melvin S. Katz, 56, Received Radiation Implants

"I denied for eight months that anything was wrong with me," said Melvin S. Katz, a 56-year-old health care consultant from Forest Hills, New York. "I noticed symptoms in July 1993 and didn't do anything about it until the following March." By then, Mr. Katz's PSA had leapt to 91. A biopsy showed cancer with a Gleason score of 8. "I wasn't normal anymore," he said. "I had cancer. I would die! Soon!" Only 48 at the time, he said he felt too young to have this "old man's disease."

Mr. Katz immediately went on combination hormone therapy (Eulexin plus Lupron) to halt the production of testosterone, the hormone that feeds prostate cancer. Meanwhile, he said, he made "the most difficult and agonizing decision" of his life. He received conflicting advice from doctors, survivors, family, and friends, although most urged surgery. Mr. Katz chose high-dose external-beam radiation five days a week for eight weeks at Columbia-Presbyterian Hospital. "It afforded me a quality of life," he explained, adding that side effects were mild. Three months after radiation, Mr. Katz stopped taking the hormones.

"I wanted to see if I was cured," he said. He was not. His PSA went from 0 to 25 in just nine months, and he resumed hormone treatment. Then, in 1995, a prostate biopsy showed lingering cancer. "I couldn't take any more external radiation, and surgery was risky because the radiation increased the chance of bleeding," he said.

His doctor at Columbia-Presbyterian suggested palladium radioactive seed implants, rather uncommon in 1995, and he agreed. The procedure was done under local anesthetic and lasted about an hour, while doctors injected rice-size pellets into his prostate. "It was one night in the hospital," Mr. Katz said, followed by a few days of discomfort.

The radiation lasts for six to nine months, "and the only restriction is no little children or pregnant ladies on your lap," Mr. Katz said. Meanwhile, his PSA dropped from 25 to 0.9. Two years ago, he stopped the hormone therapy, replacing it with an herbal formula called PC-Spes, which Mr. Katz credited with lowering his PSA to 0.3, where it remains. But the implants "definitely had an impact," he said. "They killed most of the cancer, allowing me to stabilize and keep it under control with hormones and PC-Spes."

George Ripol, 58, Needed to Try Several Treatments

George Ripol had a very experienced urologist, Dr. Jim Sehn, the surgeon for John Bobbitt. But Dr. Sehn could not detect anything abnormal about Mr. Ripol's prostate, even though his PSA had skyrocketed to 23. A digital examination, a 9-sample biopsy, and then a 12-sample biopsy all revealed nothing. It was not until the Johns Hopkins Brady Urological Institute performed a 24-sample biopsy that cancer was found.

"It was in the transition zone, right next to the urethra, and hard to find," said Mr. Ripol, a 58-year-old resident of Manassas, Virginia. "I was shocked, and panicked," Mr. Ripol said. "But in a way I was relieved. I felt it had to mean something. And it did." Surgery was recommended because he was relatively young and because of the apparent containment of the tumor. "Surgery seemed like the silver bullet," he said.

The prostatectomy, with nerve sparing, was performed in December. But subsequent pathology reports showed the possibility that the cancer remained in the pelvic area. And his PSA, 4.5, was troubling. "So my doctors recommended a double dose," he said. The doctors administered external-beam radiation with hormone therapy, which started a month before the radiation treatment and will continue for a year or two.

In May Mr. Ripol began monthly injections of Lupron, and he took Casodex pills daily. In June he began external-beam radiation, five days a week, for a total of 33 sessions. Side effects from the surgery were minimal, he said, as were those from radiation. "Maybe a little fatigue," he said, "a little irritation of the skin and bowel areas, but very little." Hormone therapy, on the other hand, hit like "male menopause," Mr. Ripol said. "The hot flashes are very disconcerting and continue to wake me up at night. I've just tried to get used to them."

One thing that has not returned—and may never return, because the radiation may have killed some nerves—is his sex drive. "We'll see," Mr. Ripol said. "Frankly, at my age, it's not a big deal, compared to the alternatives." In August, after all the treatments, Mr. Ripol received his first clean bill of health: a PSA under 0.1 "I'm still on the hormone therapy and expect that will last a year or two," he said. "But the mental relief is unbelievable."

John Sosdian, 79, Decided to Watch and Wait

John Sosdian, a 79-year-old retired Army electrician from Tinton Falls, New Jersey, began annual screenings for prostate cancer in 1992. Though Mr. Sosdian's PSA was slightly elevated, at 3.9, nothing abnormal was noted in a digital examination. For several years, his PSA remained in the 5 to 7 range and all exams were normal. Doctors thought the higher PSA numbers were explained by Mr. Sosdian's age. Then, in September 1999, the PSA jumped to 9.4, and a biopsy led to a cancer diagnosis. The tumor scored a 6 on the Gleason scale, showing a moderate growth rate, even though the digital examination still revealed nothing. Given Mr. Sosdian's age, surgery seemed risky. Mr. Sosdian's doctor recommended hormone therapy combined with radiation seed implants. But he also mentioned the idea of taking no action, something known as "watchful waiting."

Mr. Sosdian carefully researched the side effects of each treatment and was discouraged by what he found. He discussed the treatment extensively with his wife, Lorette. "We agreed that watchful waiting was the best selection," he said, "because of my age and strong desire to maintain the relationship and lifestyle I have, which I treasure."

Mr. Sosdian's doctor was not convinced he was doing the right thing, and he urged his patient to see a radiation oncologist before making a final decision. "He said I was putting my life on the line," Mr. Sosdian said. But Mr. Sosdian had "complete comfort and peace" with his decision, adding that "the sole purpose" of speaking with another doctor "was to talk me into having aggressive treatment." And, he figured, he probably will die from something else.

Mr. Sosdian has an exam every three months (nothing abnormal so far) and a PSA test. Last April, the PSA was down to 6.6, and it has never risen above the 9.4 at diagnosis. He believes that nutritional supplements and a better diet play a role.

"I feel so good about what I am doing," Mr. Sosdian said. "I rarely have the thought that I have cancer." He said he had "no concern if the cancer should metastasize, resulting in my death." But, he added, "watchful waiting doesn't necessarily mean no treatment." Instead, he labeled it "delayed therapy."

Mr. Sosdian will have a PSA test and an exam every 3 months and an ultrasound exam every 12 months for two years, followed by two PSAs and one ultrasound each year after that. If he needs treatment, he said, he will opt for something called "Prost-R-cision," which uses irradiation to excise the prostate without cutting muscles and nerves, thus reducing the risks of incontinence and impotence.

Source: From David Kirby (2000, October 3) "More Options, and Decisions, for Men with Prostate Cancer." *New York Times online.*

There seems to be little doubt that during the next few years we will have yet earlier and more reliable ways of detecting prostate cancer. For the time being, however, the American Cancer Society and other professional groups recommend that men of age 50 and above have an annual PSA blood test along with a rectal examination (Morgan et al., 1996). One caveat in the generally good news about prostate cancer is that the death rate from the disease has declined one and a half times as rapidly among European American men as among African American men over the past decade (Chang, 2000). African American men remain more likely to develop and die from prostate cancer, so they are encouraged to begin regular screening earlier—at the age of 40 (Morgan et al., 1996).

PROSTATITIS Many infectious agents can inflame the prostate, causing **prostatitis.** The chief symptoms are painful ejaculation and an ache or pain between the scrotum and the anal opening. Prostatitis is usually treated with antibiotics. Although aspirin and ibuprofen may relieve the pain, men with these symptoms should consult a physician. Painful ejaculation may discourage masturbation or coitus, which is ironic, because regular flushing of the prostate through ejaculation may be helpful in the treatment of prostatitis.

MALE SEXUAL FUNCTIONS

The male sexual functions of erection and ejaculation provide the means for sperm to travel from the male's reproductive tract to the female's. There the sperm cell and ovum unite to conceive a new human being. Of course, the natural endowment of reproduction with sensations of pleasure helps ensure that it will take place with or without knowledge of these biological facts.

Erection

Erection is caused by the engorgement of the penis with blood, such that the penis grows in size and stiffens. The erect penis is an efficient conduit, or funnel, for depositing sperm deep within the vagina.

In mechanical terms, erection is a hydraulic event. The spongy, cavernous masses of the penis are equipped to hold blood. Filling them with blood causes them to enlarge, much as a sponge swells when it absorbs water. This simple description belies the fact that erection is a remarkable feat of biological engineering (there they go again, notes the third author) that involves the cooperation of the vascular (blood) system and the nervous system.

In a few moments—as quickly as 10 or 15 seconds—the penis can double in length, become firm, and shift from a funnel for passing urine to one that expels semen. Moreover, the bladder is closed off when the male becomes sexually aroused, decreasing the likelihood that semen and urine will mix.

Yes, the blood that fills the penis during sexual arousal causes erectile tissue to expand. But what accounts for the firmness of an erection? A sponge that fills with water expands but does not grow hard. It happens that the corpora cavernosa are surrounded by a tough, fibrous covering called the *tunica albuginea.* Just as the rubber of a balloon resists the pressure of pumped-in air, this housing resists expansion, causing the penis to become rigid. The corpus spongiosum, which contains the penile urethra, also engorges with blood during erection. It does not become hard, however, because it lacks the fibrous casing. The penile glans, which is formed by the crowning of the spongiosum at the tip of the penis, turns a dark purplish hue as it becomes engorged, but it too does not stiffen.

Despite the advanced state of biological knowledge, some mechanics of erection are not completely understood. It is not entirely clear, for example, whether penile cavities

Prostatitis Inflammation of the prostate gland.

Erection The enlargement and stiffening of the penis as a consequence of its engorgement with blood.

become engorged because the veins that carry blood away from the penis do not keep pace with the rapid flow of blood entering the penis, or whether the returning blood flow is reduced by compression of the veins at the base of the penis (as stepping lightly on a hose slows the movement of water).

We do know that erection is reversed when more blood flows out of the erectile tissue than flows in, restoring the pre-erectile circulatory balance and shrinking the erectile tissue or spongy masses. The erectile tissue thus exerts less pressure against the fibrous covering, resulting in a loss of rigidity. Loss of erection occurs when sexual stimulation ceases or when the body returns to a (sexual) resting state following orgasm. Loss of erection can also occur in response to anxiety or perceived threats. Loss of erection in response to threat can be abrupt, as when a man in the "throes of passion" suddenly hears a suspicious noise in the adjoining room, suggestive of an intruder. Yet the "threats" that induce loss of erection are more likely to be psychological than physical. In our culture, men often measure their manhood by their sexual performance. A man who fears that he will be unable to perform successfully may experience **performance anxiety** that can prevent him from achieving erection or lead to a loss of erection at penetration.

The male capacity for erection quite literally spans the life cycle. Erections are common in babies, even within minutes after birth. Evidence from ultrasound studies shows that male fetuses may even have erections months prior to birth. Men who are well into their 80s and 90s continue to experience erections and to engage in coitus.

Erections are not limited to the conscious state. Men have nocturnal erections every 90 minutes or so as they sleep. These generally occur during REM (rapid eye movement) sleep. REM sleep is associated with dreaming. It is so named because the sleeper's eyes dart about rapidly under the closed eyelids during this stage. Erections occur during most periods of REM sleep.

The mechanism of nocturnal erection appears to be physiologically based. That is, erections occur along with dreams that may or may not have erotic content. Morning erections are actually nocturnal erections. They occur when the man is awakened during REM sleep, as by an alarm clock. Men sometimes erroneously believe that a morning erection is caused by the need to urinate. When the man awakens with both an erection and the need to urinate, he may mistakenly assume that the erection was caused by the pressure of his bladder.

It is not true that morning erections reflect the need to urinate. Morning erections are actually a form of nocturnal erection. ■

Spinal Reflexes and Sexual Response

Men may become sexually aroused by a range of stimuli, including tactile stimulation provided by their partners, visual stimulation (such as from scanning photos of nude models in men's magazines), and even mental stimulation from engaging in sexual fantasies. Regardless of the source of stimulation, the man's sexual responses, erection and ejaculation, occur by **reflex.**

Erection and ejaculation are reflexes: automatic, unlearned responses to sexual stimulation. So too are vaginal lubrication and orgasm in women. We do not control sexual reflexes voluntarily, as we might control the lifting of a finger or an arm. We can set the stage for them to occur by ensuring the proper stimulation, but once the stage is set, the reflexes are governed by automatic processes, not by conscious effort. Efforts to control sexual responses consciously by "force of will" can backfire and make it more difficult to become aroused (for example, to attain erection or vaginal lubrication). We need not "try" to become aroused. We need only expose ourselves to effective sexual stimulation and allow our reflexes to do the job for us.

Performance anxiety Feelings of dread and foreboding experienced in connection with sexual activity (or any other activity that might be judged by another person).

Reflex A simple, unlearned response to a stimulus that is mediated by the spine rather than the brain.

Truth or Fiction? REVISITED

It is not true that men can will themselves to have erections. Men can only set the stage for erections by providing physical or cognitive sexual stimulation. ■

The reflexes governing erection and ejaculation are controlled at the level of the spinal cord. Thus, they are considered spinal reflexes. How does erection occur? Erections may occur in response to different types of stimulation. Some erections occur from direct stimulation of the genitals, as from stroking, licking, or fondling of the penis or scrotum. Erectile responses to such direct stimulation involve a simple spinal reflex that does not require the direct participation of the brain.

Erections can also be initiated by the brain, without the genitals being touched or fondled at all. Such erections may occur when a man has sexual fantasies, when he views erotic materials, or when he catches a glimpse of a woman in a bikini on a beach. In the case of the "no-hands" type of erection, stimulation from the brain travels to the spinal cord, where the erectile reflex is triggered. To better understand how this reflex works, we need first to explain the concept of the reflex arc.

THE REFLEX ARC When you withdraw your hand from a hot stove or blink in response to a puff of air, you do so before you have any time to think about it. These responses, like erection, are reflexes that involve sensory neurons and effector neurons (see Figure 4.9). In response to a stimulus like a touch or a change in temperature, sensory neurons or receptors in the skin "fire" and thereby send messages to the spinal cord. The message is then transmitted to effector neurons that begin in the spinal cord and cause muscles to contract or glands to secrete chemical substances. Thus, when you accidentally touch a hot stove, sensory neurons in your fingers or hand deliver a message to the spinal cord, which triggers effector neurons to contract muscles that pull your hand away from the stove. The brain does not control this spinal reflex arc. That is not to say that the brain fails to "get the message" shortly afterwards. Sensory messages usually rise from the spinal cord to the brain to make us aware of stimulation. (Awareness "dwells" within the nerve cells, or gray matter, of the brain.) The experience of pain occurs when a message travels from the site of the injury to the spinal cord and then to

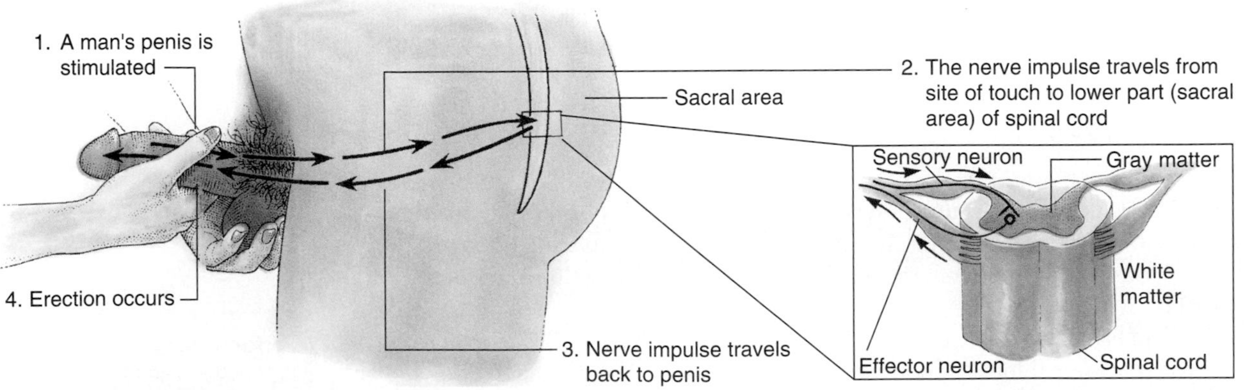
1. A man's penis is stimulated
2. The nerve impulse travels from site of touch to lower part (sacral area) of spinal cord
Sacral area
Sensory neuron
Gray matter
White matter
4. Erection occurs
3. Nerve impulse travels back to penis
Effector neuron
Spinal cord

FIGURE 4.9

Reflexes. Reflexes involve sensory neurons, effector neurons, and sometimes interneurons that connect the two in the spinal cord. Reflexes need not involve the brain, although messages to the brain may make us aware when reflexes are occurring. Reflexes are the product of "local government" in the spine.

receiving stations in the brain that "interpret" the message to produce the sensation of pain. The withdrawal of your hand from a harmful object begins before your brain even gets the message.

THE ROLE OF THE SPINAL CORD Let us look more closely at the spinal reflex that produces erection in response to tactile stimulation (touch). Tactile stimulation of the penis or nearby areas (lower abdomen, scrotum, or inner thighs) causes sensory neurons to transmit nerve messages (signals) to an erection center in the lower back, in an area of the spinal cord called the **sacrum.** The sacral erection center controls reflexive erections—that is, erections that occur in response to direct stimulation of the penis and nearby areas. When direct penile stimulation occurs, messages in the form of nerve impulses are received by this erection center, which in turn sends impulses to the genitalia via nerves that service the penis. These impulses cause arteries carrying blood to the corpora cavernosa and corpus spongiosum to dilate, so that more blood flows into these tissues; as these tissues expand, the penis becomes erect.

The existence of the sacral erection center makes it possible for men whose spinal cords have been injured or severed above the center to achieve erections (and ejaculate) in response to direct tactile stimulation of the penis. Erection occurs even though their injuries prevent nerve signals from reaching their brains. Because of the lack of communication between the genital organs and the brain, there are no sensations—no physical pleasure. Many spinal-injured men report that sex remains psychologically pleasurable and fulfilling nonetheless. They can observe the responses of their partners, and perhaps the brain fills in some sensations from memory or imagination.

THE ROLE OF THE BRAIN If direct penile stimulation triggers erection at the spinal level, what is the role of the brain? Although it may seem that the penis sometimes has a mind of its own, the brain plays an important role in regulating sexual responses.

Tactile (touch) stimulation of the penis may trigger the erection reflex through a simple reflex arc in the spinal cord. Penile sensations are then normally relayed to the brain, which generally results in sensations of pleasure and perhaps in a decision to focus on erotic stimulation. The sight of one's partner, erotic fantasies, memories, and so forth can result in messages being sent by the brain through the spinal cord to the arteries servicing the penis, helping to maintain the erection.

The brain can also originate messages that trigger the erectile reflex. The "no-hands" type of erection can occur while a man is recalling sexual memories, entertaining sexual fantasies, viewing erotic stimuli, or even asleep. In such cases, the brain plays a more direct role in the erectile response by transmitting nerve impulses to a second erection center located in the upper back in the lumbar region of the spinal cord. This higher spinal erection center serves as a "switchboard" between the brain and the penis, allowing perceptual, cognitive, and emotional responses to make their contributions. When the nerve pathways between the brain and the upper spinal cord are blocked or severed, men cannot achieve "psychogenic erections"—that is, erections in response to mental stimulation alone.

The brain can also stifle sexual response. A man who is highly anxious about his sexual abilities may be unable to achieve an erection even with the most direct, intense penile stimulation. Or a man who believes that sexual pleasure is sinful or dirty may be filled with anxiety and guilt and hence be unable to achieve erection when he is sexually stimulated by his partner.

In some males, especially adolescents, the erectile reflex is so easily tripped that incidental rubbing of the genitals against his own 'undergarments, the sight of an attractive passer-by, or a fleeting sexual fantasy produces erection. Spontaneous erections may occur under embarrassing circumstances, such as on a public beach, or just before classes

Sacrum The thick, triangular bone located near the bottom of the spinal column.

change in junior or senior high school. In an effort to distract himself from erotic fantasies and to allow an erection to subside, many a male adolescent in the classroom has desperately renewed his interest in his algebra or foreign language textbook before the bell has rung. (A well-placed towel may serve in a pinch on a public beach.)

As men mature, they require more penile stimulation to achieve full erection. Partners of men in their 30s and 40s need not feel that their attractiveness has waned if their lovers no longer have instant "no-hands" erections when they disrobe. It takes men longer to achieve erection as they age, and direct penile stimulation becomes a more critical source of arousal.

THE ROLE OF THE AUTONOMIC NERVOUS SYSTEM Although stimulation that brings about an erection can originate in the brain, this does not mean that erection is a voluntary response, like raising your arm. Whatever the original or dominant source of stimulation—direct penile stimulation or sexual fantasy—erection remains an unlearned, automatic reflex.

Automatic responses, such as erection, involve the division of the nervous system called the **autonomic nervous system** (ANS). *Autonomic* means "automatic." The ANS controls automatic bodily processes such as heartbeat, pupil dilation, respiration, and digestion. In contrast, voluntary movement (such as raising an arm) is under the control of the *somatic* division of the nervous system.

The ANS has two branches, the **sympathetic** and the **parasympathetic.** These branches have largely opposing effects; when they are activated at the same time, their effects become balanced out to some degree. In general, the sympathetic branch is in command during processes that involve a release of bodily energy from stored reserves, such as during running, performing some other athletic task, or being gripped by fear or anxiety. The sympathetic branch also governs the general mobilization of the body, such as by increasing the heart rate and respiration rate in response to threat.

The parasympathetic branch is most active during processes that restore reserves of energy, such as digestion. When we experience fear or anxiety, the sympathetic branch of the ANS quickens the heart rate. When we relax, the parasympathetic branch curbs the heart rate. The parasympathetic branch activates digestive processes, but the sympathetic branch inhibits digestive activity. Because the sympathetic branch is in command when we feel fear or anxiety, such stimuli can inhibit the activity of the parasympathetic system, thereby slowing down digestive process and possibly causing indigestion.

The divisions of the autonomic nervous system play different roles in sexual arousal and response. The nerves that cause penile arteries to dilate during erection belong to the parasympathetic branch of the autonomic nervous system. It is thus the parasympathetic system that largely governs erection. The nerves governing ejaculation belong to the sympathetic branch, however. One implication of this division of neural responsibility is that intense fear or anxiety, which involves sympathetic nervous system activity, may inhibit erection by counteracting the activity of the parasympathetic nervous system. Because sympathetic arousal is involved in triggering the ejaculatory reflex, anxiety or fear may also accelerate ejaculation, causing **premature ejaculation.** Intense emotions such as fear and anxiety can thus lead to problems in achieving or maintaining erection, as well as causing hasty ejaculation.

The connection between the emotions, sympathetic activity, and ejaculation can set up a vicious cycle. Anxiety in a sexual encounter may trigger premature ejaculation. During a subsequent sexual encounter, the man might fear a recurrence of premature ejaculation. This fear may act as a self-fulfilling prophecy. He may thus face further sexual encounters with even greater fear, possibly further hastening ejaculation—and possibly inhibiting erection itself. Methods for helping men with erectile dysfunction and pre-

Autonomic nervous system The division of the nervous system that regulates automatic bodily processes, such as heartbeat, pupil dilation, respiration, and digestion. Abbreviated *ANS.*

Sympathetic The branch of the ANS most active during emotional responses that draw on the body's reserves of energy, such as fear and anxiety. The sympathetic ANS largely controls ejaculation.

Parasympathetic The branch of the ANS most active during processes that restore the body's reserves of energy, such as digestion. The parasympathetic ANS largely controls erection.

Premature ejaculation A sexual dysfunction in which the male persistently ejaculates too early to afford the couple adequate sexual gratification. (Yes, what is on time in one relationship may be considered premature—or late—in another)

Peyronie's disease An abnormal condition characterized by an excessive curvature of the penis that can make erections painful.

Orgasm The climax of sexual excitement.

Paraplegic A person with sensory and motor paralysis of the lower half of the body.

mature ejaculation aim at reducing their levels of anxiety and thereby lessening sympathetic activity.

Because erections seem spontaneous at times, and because they often occur when the man would rather not have them, it may seem to men that the penis has a mind of its own. Despite this common folk belief, however, the penis possesses no guiding intelligence. It consists of spongy masses of erectile tissue, not the lovely dense gray matter that renders your thought processes so incisive.

Truth or Fiction? **REVISITED**

It is not true that the penis has a mind of its own. Although it may seem that the penis acts as a free agent at times, it is governed by the autonomic (automatic) nervous system. ■

ERECTILE ABNORMALITIES Some men find that their erect penises are slightly curved or bent. Some degree of curvature is perfectly normal, but men with **Peyronie's disease** have excessive curvature that can make erections painful or make it difficult to enjoy coitus. The condition is caused by a buildup of fibrous tissue in the penile shaft. Although some cases of Peyronie's disease appear to clear up on their own, most require medical attention.

Some men experience erections that persist for hours or days. This condition is called *priapism*, after Priapus of Greek myth, the son of Dionysus and Aphrodite who personified male procreative power. Priapism is often caused by leukemia, sickle-cell anemia, or diseases of the spinal cord, although in some cases the cause remains unknown. Priapism occurs when the mechanisms that drain the blood that makes the penis erect are damaged and so cannot return the blood to the circulatory system. The name of the disorder is truly a misnomer, because Priapus had a voracious sexual appetite. Men with priapism instead suffer from a painful condition that should receive medical attention. Priapism may become a medical emergency, because erection prolonged beyond six hours can starve penile tissues of oxygen, leading to tissue deterioration. Medical intervention in the form of drugs or surgery may be required to reverse the condition and allow blood to drain from the penis.

Ejaculation

Ejaculation, like erection, is a spinal reflex. It is triggered when sexual stimulation reaches a critical point or threshold. Ejaculation generally occurs together with **orgasm,** the sudden muscle contractions that occur at the peak of sexual excitement and result in abrupt release of the sexual tension that had built up during sexual arousal. Orgasm is accompanied by subjective sensations that are generally intensely pleasurable. Ejaculation, however, is simply the expulsion of semen from the tip of the penis. Orgasm and ejaculation are *not* synonymous, nor do they always occur simultaneously. For example, **paraplegics** can ejaculate if the area of the lower spinal cord that controls ejaculation is intact. They do not experience the subjective aspects of orgasm, however, because the sensations of orgasm do not reach the brain.

Truth or Fiction? **REVISITED**

It is true that men who are paralyzed below the waist can attain erection, engage in sexual intercourse, and ejaculate, if the spinal centers that control erection and ejaculation remain intact. ■

Conversely, prepubertal boys may experience orgasms even though they emit no ejaculate. Orgasms without ejaculate are termed "dry orgasms." Boys do not begin to produce seminal fluid (and sperm) until puberty. Mature men, too, can experience dry orgasms. They can take the form of "little orgasms" preceding a larger orgasm, or they can follow "wet orgasms" when sexual stimulation is continued but seminal fluids have not been replenished. These dry orgasms are perfectly normal. Dry orgasms can also be a result of retrograde ejaculation, however, as we shall see below.

Truth or Fiction? **REVISITED**

It is true that men can have orgasms without ejaculating. Such orgasms are termed dry orgasms. ■

Ejaculation occurs in two stages. The first phase, often called the **emission stage,** involves contractions of the prostate, the seminal vesicles, and the upper part of the vas deferens (the **ampulla**). The force of these contractions propels seminal fluid into the prostatic part of the urethral tract—a small tube called the **urethral bulb**—which balloons out as muscles close at either end, trapping the semen. It is at this point that the man perceives that orgasm is inevitable. Masters and Johnson term this feeling a sense of "ejaculatory inevitability" (Masters & Johnson, 1966). Men might colloquially describe the feeling as being about to "come." The man feels that a point of no return has been passed and that nothing can prevent ejaculation.

The second stage, which is often referred to as the **expulsion stage,** involves the propulsion of the seminal fluid through the urethra and out of the urethral opening at the tip of the penis. In this stage, muscles at the base of the penis and elsewhere contract rhythmically, forcefully expelling semen. The second stage is generally accompanied by the highly pleasurable sensations of orgasm.

In ejaculation, the seminal fluid is released from the urethral bulb and expelled by forceful contractions of the pelvic muscles that surround the urethral channel and the crura of the penis. The first few contractions are most intense and occur at 0.8-second intervals. Subsequent contractions lessen in intensity, and the interval between them gradually increases. Seminal fluid is expelled in spurts during the first few contractions. The contractions are so powerful that seminal fluid may be propelled as far as 12 to 24 inches, according to observations made by Masters and Johnson. Some men, however, report that semen travels but a few inches or just oozes from the penile opening. The force of the expulsion varies with the condition of the man's prostate, his general health, and his age. There is some correspondence between the force of the expulsion and the pleasure of orgasm. That is, more intense orgasms, psychologically speaking, often accompany more forceful ejaculations.

Like erection, ejaculation is regulated by two centers in the spinal cord: one in the sacral region and one in the higher lumbar region. When sexual arousal rises to a critical level—the point of ejaculatory inevitability—the lumbar ejaculatory center triggers the first stage of ejaculation, seminal emission. The lower, or sacral, ejaculatory center triggers the second stage of orgasm, the rhythmic muscle contractions that expel the ejaculate from the body.

Although ejaculation occurs by reflex, a man can delay ejaculation by maintaining the level of sexual stimulation below the critical threshold, or "point of no return." Men who suffer from premature ejaculation have been successfully treated in programs that train them to learn to recognize their "point of no return" and maintain sexual stimulation below it. (Issues concerning the definition and treatment of premature ejaculation are explored in Chapter 15 on sexual dysfunction.) Recognizing the point of no return and keeping stimulation beneath the critical level can also prolong coitus and enhance sexual pleasure for couples even when the man does not experience premature ejaculation.

RETROGRADE EJACULATION Some men experience **retrograde ejaculation,** in which the ejaculate empties into the bladder rather than being expelled from the body. During normal ejaculation an external sphincter opens, allowing seminal fluid to pass out of the body. Another sphincter, this one internal, closes off the opening to the bladder,

Emission phase The first phase of ejaculation, which involves contractions of the prostate gland, the seminal vesicles, and the upper part of the vas deferens.

Ampulla A sac or dilated part of a tube or canal.

Urethral bulb The small tube that makes up the prostatic part of the urethral tract and that balloons out as muscles close at either end, trapping semen prior to ejaculation.

Expulsion stage The second stage of ejaculation, during which muscles at the base of the penis and elsewhere contract rhythmically, forcefully expelling semen and providing pleasurable sensations.

Retrograde ejaculation Ejaculation in which the ejaculate empties into the bladder. (From the Latin *retrogradi,* which means "to go backward.")

preventing the seminal fluid from backing up into the bladder. In retrograde ejaculation, the actions of these sphincters are reversed. The external sphincter remains closed, preventing expulsion of the seminal fluid, while the internal sphincter opens, allowing the ejaculate to empty into the bladder. The result is a dry orgasm. No ejaculate is apparent because semen has backed up into the bladder. Retrograde ejaculation may be caused by prostate surgery (much less so now than in former years), by drugs such as tranquilizers, by certain illnesses, and by accidents. Retrograde ejaculation is usually harmless in itself, because the seminal fluid is later discharged with urine. Infertility can result, however, and there may be some changes in the sensations associated with orgasm. Persistent dry orgasms should be medically evaluated; their underlying cause may be a threat to health.

Male sexual functions, like female sexual functions, are complex. They involve the cooperation of the nervous system, the endocrine system, the cardiovascular system, and the musculoskeletal system. In Chapter 5 we learn more about how the female and male sex organs respond to sexual stimulation. In Chapter 6 we examine the similarities and differences between the genders with respect to sexual differentiation, behavior, and personality.

Human Sexuality ONLINE //

Web Sites Related to Male Sexual Anatomy and Physiology

The American Cancer Society: This official Web site of the American Cancer Society provides information on all types of cancer, cancer research, therapy, support groups, and community resources.
www.cancer.org/

The University of Michigan: The university's Web site for information related to prostate cancer, including diagnosis, treatment options, specialists, and current research. It is intended for both patients and health professionals.
www.cancer.med.umich.edu/prostcan/prostcan.html

The Journal of Urology®. This is the official journal of the American Urological Association. The Web site provides free access to tables of contents and abstracts of articles.
www.jurology.com/

The Circumcision Information and Resource Pages: This Internet resource that provides information about various aspects of circumcision. The Circumcision Information Pages within the Web site offer information written for laypeople.
www.cirp.org/

The Circumcision Resource Center: The Web site of this not-for-profit educational organization aims to inform the general public and professionals about circumcision.
www.circumcision.org/

summing up

EXTERNAL SEX ORGANS

The male external sex organs include the penis and the scrotum.

■ The Penis

Semen and urine pass out of the penis through the urethral opening. The penis contains cylinders that fill with blood and stiffen during sexual arousal. Circumcision—the surgical removal of the prepuce—has been carried out for religious and hygienic reasons. In our culture, the size of the penis is sometimes seen as a measure of a man's masculinity and his ability to please his sex partners, although there is little if any connection between the size of the penis and sexual performance.

The Scrotum

The scrotum is the pouch of loose skin that contains the testes. Each testicle is held in place by a spermatic cord, which contains the vas deferens and the cremaster muscle.

INTERNAL SEX ORGANS

The male internal sex organs consist of the testes, a system of tubes and ducts that conduct sperm, and organs that nourish and activate sperm.

The Testes

The testes serve two functions analogous to those of the ovaries. They secrete male sex hormones (androgens) and produce germ cells (sperm). The hypothalamus, pituitary gland, and testes keep blood testosterone at a more or less constant level through a hormonal negative-feedback loop. Testosterone is produced by interstitial cells. Sperm are produced by seminiferous tubules. Sperm are stored and mature in the epididymis.

The Vas Deferens

Each epididymis empties into a vas deferens that conducts sperm over the bladder.

The Seminal Vesicles

The seminal vesicles are glands that open into the ejaculatory ducts, where the fluids they secrete combine with and nourish sperm.

The Prostate Gland

The prostate gland secretes fluid that accounts for the texture and odor characteristic of semen.

Cowper's Glands

During sexual arousal, the Cowper's glands secrete a drop or so of clear, slippery fluid that appears at the urethral opening.

Semen

Sperm and the fluids contributed by the seminal vesicles, the prostate gland, and the Cowper's glands make up semen, the whitish fluid that is expelled through the tip of the penis during ejaculation.

DISEASES OF THE UROGENITAL SYSTEM

Urethritis

Men, like women, are subject to bladder and urethral inflammations, which are generally referred to as urethritis.

Cancer of the Testes

This is the most common form of solid-tumor cancer to strike young men between the ages of 20 and 34.

Disorders of the Prostate

The prostate gland generally becomes enlarged in men past the age of 50. Prostate cancer involves the growth of malignant prostate tumors that can metastasize to bones and lymph nodes. The chief symptoms of prostatitis are painful ejaculation and an ache or pain between the scrotum and anal opening.

MALE SEXUAL FUNCTIONS

Erection

Erection is the process by which the penis becomes engorged with blood, increases in size, and stiffens. Erection occurs in response to sexual stimulation but is also common during REM sleep.

Spinal Reflexes and Sexual Response

Erection and ejaculation occur by reflex. Reflexes involve sensory neurons and effector neurons, which meet in the spinal cord. There are two erection centers in the spinal cord. Although erection is a reflex, penile sensations are relayed to the brain, where they generally result in pleasure. Erection and ejaculation also involve the autonomic nervous system (ANS). The parasympathetic branch of the ANS largely governs erection, whereas the sympathetic branch largely controls ejaculation.

Ejaculation

Ejaculation, like erection, is a reflex. It is triggered when sexual stimulation reaches a critical threshold. Ejaculation usually (though not always) occurs with orgasm, but the terms are not synonymous. The emission phase of ejaculation involves contractions of the prostate, the seminal vesicles, and the upper part of the vas deferens. In the expulsion stage, semen is propelled through the urethra and out of the penis. In this stage, muscles at the base of the penis and elsewhere contract rhythmically, forcefully expelling semen. There are two ejaculation centers in the spinal cord. In retrograde ejaculation, the ejaculate empties into the bladder rather than being expelled from the body.

questions for critical thinking

1. Agree or disagree with the following statement, and support your answer: Men with bigger penises make more effective lovers.
2. Are you or the men in your family circumcised? Why or why not?
3. How are the testes analogous to the ovaries?
4. Is there such a thing as a "manopause"? What is the evidence?
5. Did you know how erection occurred before you took this course? Do the facts differ from your assumptions?
6. Given that erection and ejaculation are reflexes, how is it that people can consciously cause them to happen?
7. Agree or disagree with the following statement, and support your answer: The penis has a mind of its own.

chapter

5

Sexual Arousal and Response

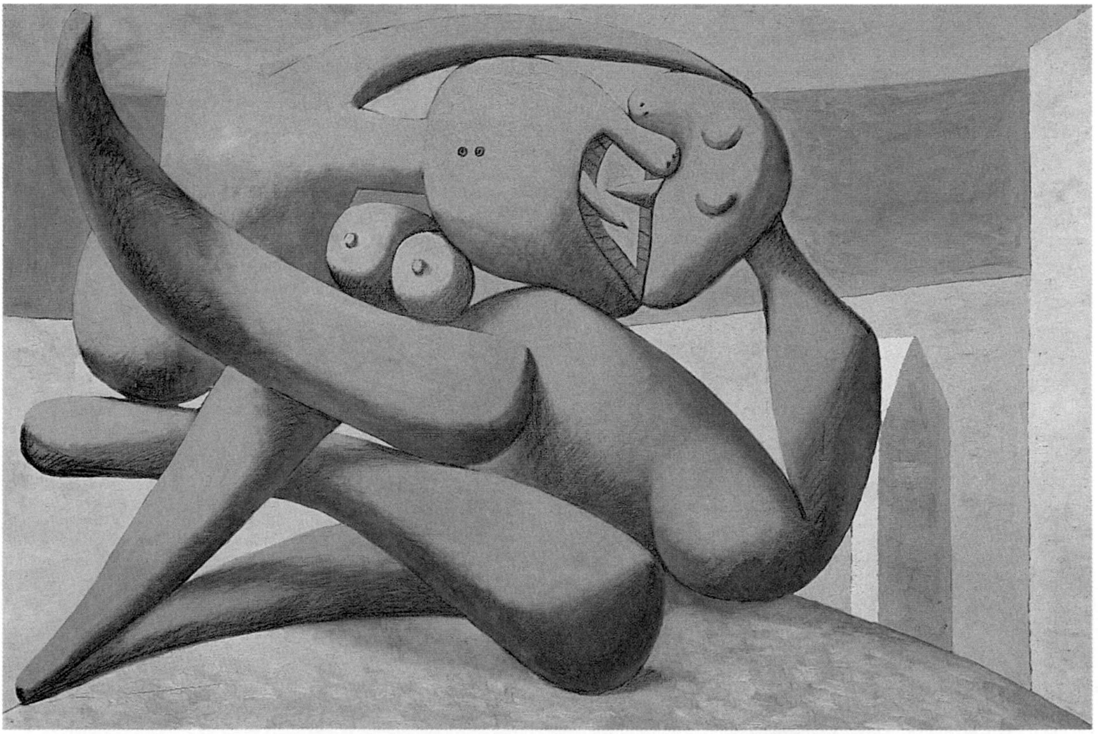

Pablo Picasso. *Figures at the Seaside*, January, 1931. © Copyright ARS, NY. Photo: R. G. Ojeda. Musée Picasso, Paris, France. Credit: Réunion des Musées Nationaux/Art Resource, NY. Credit: © 2002 Estate of Pablo Picasso/Artists Rights Society (ARS), New York.

outline

Truth or Fiction?

MAKING SENSE OF SEX: THE ROLE OF THE SENSES IN SEXUAL AROUSAL

Vision: The Better to See You With

Smell: Does the Nose Know Best?

A Closer Look The Buds and the Bees

The Skin Senses: Sex as a Touching Experience

Taste: On Savory Sex

Human Sexuality Online www.seduction.net

Hearing: The Better to Hear You With

APHRODISIACS: OF SPANISH FLIES AND RHINO HORNS

A Closer Look Sex Toys for Pandas?

Anaphrodisiacs

Psychoactive Drugs

SEXUAL RESPONSE AND THE BRAIN: CEREBRAL SEX?

The Geography of the Brain

Brain Mechanisms in Sexual Functioning

On Pushing the Right Buttons: Are There Pleasure Centers in the Brain?

SEX HORMONES: DO THEY "GOAD" US INTO SEX?

Sex Hormones and Sexual Behavior: Organizing and Activating Influences

Sex Hormones and Male Sexual Behavior

Sex Hormones and Female Sexual Behavior

THE SEXUAL RESPONSE CYCLE

Excitement Phase

Plateau Phase

Orgasmic Phase

Resolution Phase

Kaplan's Three Stages of Sexual Response: An Alternative Model

CONTROVERSIES ABOUT ORGASM

Multiple Orgasms

How Many Kinds of Orgasms Do Women Have? One, Two, or Three?

The G-Spot

Human Sexuality Online Web Sites Related to Sexual Arousal and Response

SUMMING UP

QUESTIONS FOR CRITICAL THINKING

Truth or Fiction?

___ The menstrual cycles of women who live together tend to become synchronized.

___ The primary erogenous zone is the brain.

___ "Spanish fly" will not turn your date on, but it may cure his or her warts.

___ Electrical stimulation of certain areas in the human brain can yield sensations similar to those of sexual pleasure and gratification.

___ Normal men produce estrogen, and normal women produce androgens.

___ Written descriptions of men's and women's experiences during orgasm cannot be differentiated.

___ Orgasms attained through sexual intercourse are more intense than those attained through masturbation.

What turns you on? What springs your heart into your mouth, tightens your throat, and opens the floodgates to your genitals? The sight of your lover undressing, a photo of Denzel Washington or Cindy Crawford, a sniff of some velvety perfume, a sip of wine?

Many factors contribute to sexual arousal. Some people are aroused by magazines with photographs of nude or seminude models that have been airbrushed to perfection. Some need only imagine Hollywood's latest sex symbol. Some become aroused by remembrances of past lovers. Some are stimulated by sexual fantasies of flings with strangers.

People vary greatly in the cues that excite them sexually and in the frequency with which they experience sexual thoughts and feelings. Some young people seem perpetually aroused or arousable. Some people rarely or never entertain sexual thoughts or fantasies.

In this chapter we look at factors that contribute to sexual arousal and the processes related to sexual response. Because our experience of the world is initiated by our senses, we begin the chapter by focusing on the role of the senses in sexual arousal.

MAKING SENSE OF SEX: THE ROLE OF THE SENSES IN SEXUAL AROUSAL

We come to apprehend the world around us through our senses—vision, hearing, smell, taste, and the skin senses, which include that all-important sense of touch. Each of the senses plays a role in our sexual experience, but some senses play larger roles than others.

Vision: The Better to See You With

It was the face of Helen of Troy, not her scent or her melodic voice, that "launched a thousand ships." Men's and women's magazines are filled with pictures of comely members of the other gender, not with "scratch and sniff" residues of their scents (but wait, a new marketing idea dawns!).

In matters of sexual attraction, people seem to have more in common with birds than with fellow mammals such as dogs and cats. Birds identify prospective mates within their species on the basis of their plumage and other visual markings. People also tend to be visually oriented when it comes to sexual attraction. By contrast, dogs and cats are attracted to each other more on the basis of scents that signal sexual receptivity.

Visual cues can be sexual turn-ons. We may be turned on by the sight of a lover in the nude, disrobing, or dressed in evening wear. Lingerie companies hope to convince customers that they will enhance their sex appeal by wearing strategically concealing and revealing nightwear. Some couples find it arousing to observe themselves making love in an overhead mirror or on videotape. Some people find sexually explicit movies arousing. Others are bored or offended by them. Though both genders can be sexually aroused by visually mediated erotica (a technical term for "porn flicks"), men are more interested in them.

Smell: Does the Nose Know Best?

Although the sense of smell plays a lesser role in governing sexual arousal in humans than in lower mammals, odors can be sexual turn-ons or turn-offs. Perfume companies, for example, bottle fragrances purported to be sexually arousing.

Most Westerners prefer their lovers to be clean and fresh smelling. People in our society learn to remove or mask odors by using soaps, deodorants, and perfumes or colognes. The ancient Egyptians invented scented bathing to rid themselves of offensive odors (Ramirez, 1990). The ancient Romans had such a passion for perfume that they would bathe in fragrances and even dab their horses and household pets (Ackerman, 1990).

Inclinations to find underarm or genital odors offensive may reflect cultural conditioning rather than biological predispositions. In some societies, genital secretions are considered **aphrodisiacs.** And the underarms? Check out the following section on pheromones.

PHEROMONES For centuries, people have searched for a love potion—a magical formula that could make other people fall in love with or be strongly attracted to the wearer. Some scientists suggest that such potions may already exist in the form of chemical secretions known as **pheromones.**

Pheromones are odorless chemicals that would be detected through a "sixth sense"—the *vomeronasal organ.* If people were to possess such an organ, it would be located in the nose, as it is in lower animals, and it would detect the chemicals and communicate information about them to the hypothalamus, where certain pheromones might affect sexual response (Cutler, 1999). People may use pheromones in many ways. Infants may use them to recognize their mothers, and adults might respond to them in seeking a mate. Lower animals use pheromones to stimulate sexual response, organize food gathering, maintain pecking orders, sound alarms, and mark territories (Cutler, 1999). Pheromones induce mating behavior in insects. Male rodents such as mice are extremely sensitive to several kinds of pheromones (Leinders-Zufall et al., 2000). Male rodents show less sexual arousal when their sense of smell is blocked, but the role of pheromones in sexual behavior becomes less vital as one moves upward through the ranks of the animal kingdom.

Only a few years ago, most researchers did not believe that pheromones played a role in human behavior. Today, however, it appears that people do possess vomeronasal organs (Bartoshuk & Beauchamp, 1994), and this field of research has attracted new interest.

In a typical study, Winnifred Cutler and her colleagues (1998) had heterosexual men wear a suspected male pheromone, whereas a control group wore a placebo. The men using the pheromone increased their frequency of sexual intercourse with their female partners but did not increase their frequency of masturbation. The researchers conclude that the substance increased the sexual attractiveness of the men to their partners, although they do not claim that it directly stimulated sexual behavior. In fact, it has not been conclusively shown that pheromones—or suspected pheromones—directly affect the behavior of people (Wysocki & Preti, 1998).

Even so, some other studies are also of interest. Consider a couple of double-blind studies that exposed men and women to certain steroids (androstadienone produced by males and estratetraenol produced by females) suspected of being pheromones. They found that both steroids enhanced the moods of women but not of men; the substances also apparently reduced feelings of nervousness and tension in women, but again, not in men (Grosser et al., 2000; Jacob & McClintock, 2000). The findings about estratetraenol are not terribly surprising. This substance is related to estrogen, and women tend to function best during the time of the month when estrogen levels are highest (Ross et al., 2000; Sourander, 1994). The fact that the women responded positively to the androstadienone is of somewhat greater interest. It suggests that women may generally feel somewhat better when they are around men, even if the chemical substances that may be connected with their moods have not been shown to have direct sexual effects. Of course, being in a good (or better) mood could indirectly contribute to a woman's interest in sexual intercourse. By the way, androstadienone is found on underarm skin and hair in men, so we appreciate the dedication of the humans who have participated in these studies. (Actually, we jest. The steroid itself is odorless.)

Aphrodisiac Any drug or other agent that is sexually arousing or increases sexual desire. (From *Aphrodite,* the Greek goddess of love and beauty).

Pheromones Chemical substances that are secreted externally by certain animals and that convey information to, or produce specific responses in, other members of the same species. (From the Greek *pherien,* which means "to bear [a message]" and *hormone.*)

a CLOSER look

The Buds and the Bees

It may be that we will have to cover our children's eyes when bees are at work in the yard. Researchers at the Australian National University (ANU) have learned that bees that land on certain flowers are after something other than pollen. It turns out that some plants emit chemical secretions that mimic the pheromones of female bees (Australia scientists, 2000). As a result, male bees try to mate with them. As a side effect, the bees transfer pollen from one plant to another, facilitating fertilization and the survival of these species of plants. No, there is no evidence that these plants are "trying" to dupe the bees (plants do not think); it just happens, evolutionarily speaking, that whatever genes contribute to the development of these chemicals are likely to be transmitted to the next generation.

The kinds of plants that are "in on" the pheromone scam are orchids. One type is found in Europe, and nine types are found in Australia, where the bees are apparently especially active. These orchids produce the same hydrocarbon compounds that are found in the pheromones of the female bee. Aren't they lovely?

The power of the pheromones apparently overrides bees' vision. Even so, one of the researchers noted that "The bees will . . . only try mating with the flowers a few times each" (cited in "Australia scientists," 2000). Better to learn late than never. (One imagines Dr. Evil suggesting, pinky in cheek, "Why don't the two of you get a room?")

A Pheromonal Scam. Australian researchers have discovered that certain orchids emit pheromones that mimic those of female bees. As a result, male bees attempt to mate with them (the flowers, that is). In the process, the bees pick up pollen and transfer it to other orchids, helping the orchids reproduce.

Of course we can conclude, if the current studies stand up to the scrutiny of replication and time, that substances that enhance the moods of women might make them more receptive to sexual advances. Still, the substances do not directly stimulate behavior, as pheromones do in lower animals. If pheromones with such effects on humans exist, they have not yet been isolated.

MENSTRUAL SYNCHRONY Research by several investigators suggests that exposure to other women's sweat can modify a woman's menstrual cycle. In one study, women exposed to underarm secretions from other women, which contained steroids that may function as pheromones, showed converging shifts in their menstrual cycles (Bartoshuk & Beauchamp, 1994; Preti et al., 1986). Similar synchronization of menstrual cycles has been observed among women who share dormitory rooms. In another study, 80% of the women who dabbed their upper lips with an extract of perspiration from other women began to menstruate in sync with the cycles of the donors after about three menstrual cycles (Cutler, 1999). A control group, who dabbed their lips with alcohol, showed no changes in their menstrual cycles. In yet another study from this research group, the length of the cycles of women with unusually short or long cycles began to normalize when they were exposed to an extract of *male* underarm perspiration (Preti et al., 1986).

The menstrual cycles of women who live together do tend to become synchronized, possibly in response to pheromones. ■

The Skin Senses: Sex as a Touching Experience

Our skin senses enable us to register pain, changes in temperature, and pressure (or touch). Whatever the roles of vision and smell in sexual attraction and arousal, the sense of touch has the most direct effects on sexual arousal and response. Any region of that sensitive layer we refer to as skin can become eroticized. The touch of your lover's hand upon your cheek, or your lover's gentle massage of your shoulders or back, can be sexually stimulating.

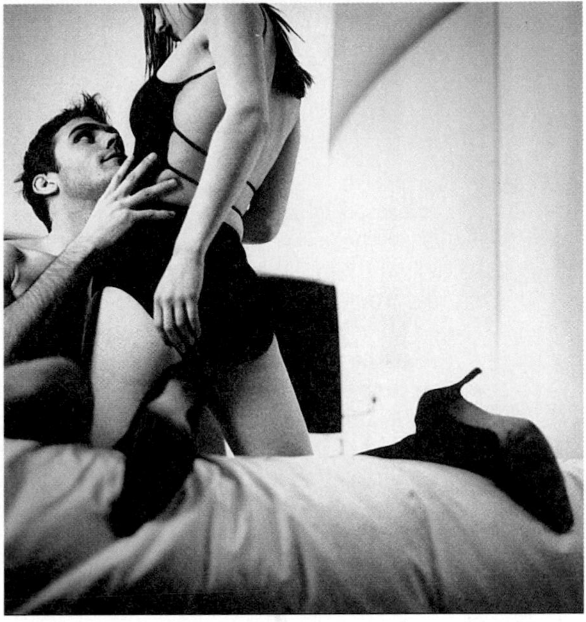

A Touching Experience. The sense of touch is intimately connected with sexual experience. The touch of a lover's hand on the cheek or a gentle massage can be sexually stimulating. Certain parts of the body—called erogenous zones—have special sexual significance because of their response to erotic stimulation.

EROGENOUS ZONES **Erogenous zones** are parts of the body that are especially sensitive to tactile sexual stimulation—to strokes and other caresses. **Primary erogenous zones** are erotically sensitive because they are richly endowed with nerve endings. **Secondary erogenous zones** are parts of the body that become erotically sensitized through experience.

Primary erogenous zones include the genitals; the inner thighs, perineum, buttocks, and anus; the breasts (especially the nipples); the ears (particularly the earlobes); the mouth, lips, and tongue; the neck; the navel; and, yes, the armpits. Preferences vary somewhat from person to person, reflecting possible biological, attitudinal, and experiential differences. Areas that are exquisitely sensitive for some people may produce virtually no reaction, or even discomfort, in others. Many women, for example, report little sensation when their breasts are stroked or kissed. Many men are uncomfortable when their nipples are caressed. On the other hand (or foot), many people find the areas between their toes sensitive to erotic stimulation and enjoy keeping a toehold on their partners during coitus.

Secondary erogenous zones become eroticized through association with sexual stimulation. For example, a woman might become sexually aroused when her lover gently caresses her shoulders, because such caresses have been incorporated as a regular feature of the couple's lovemaking. A few of the women observed by Masters and Johnson (1966) reached orgasm when the smalls of their backs were rubbed.

People are also highly responsive to images and fantasies. This is why the brain is sometimes referred to as the primary sexual organ or an erogenous zone. Some women report reaching orgasm through fantasy alone (Kinsey et al., 1953). Men regularly experience erection and nocturnal emissions ("wet dreams") without direct stimulation of the genitals.

However, the brain is not an erogenous zone. It is not stimulated directly by touch. (The brain processes tactile information from the skin, but it does not have sensory neurons to gather this information itself.) Nevertheless, the derivation of the term *erogenous* clearly applies to the brain. It *can* "give birth to erotic sensations" by producing fantasies, erotic memories, and other thoughts.

Truth or Fiction? **REVISITED**

It is not true that the primary erogenous zone is the brain. The brain is not an erogenous zone in the strict sense of the term. Erogenous zones are directly sensitive to touch. However, the brain processes information received through erogenous zones and can have such an impact on sexual arousal that it has been dubbed an erogenous zone in recognition of its importance.

Taste: On Savory Sex

Taste appears to play a minor role in sexual arousal and response, unless we digress into puns and note that taste in the form of a zesty meal or a delicate wine may contribute to arousal. In any event, some people are sexually aroused by the taste of genital secretions, such as vaginal secretions or seminal fluid. We do not know, however, whether these secretions are laced with chemicals that have biologically arousing effects or whether arousal reflects the meaning that these secretions have to the individual. That is, we may learn to become aroused by, or to seek out, flavors or odors that have been associated with sexual pleasure. (Others are turned off by the taste or odor of these secretions.)

Erogenous zones Parts of the body that are especially sensitive to tactile sexual stimulation. (*Erogenous* is derived from roots that mean "giving birth to erotic sensations.")

Primary erogenous zones Erogenous zones that are particularly sensitive because they are richly endowed with nerve endings.

Secondary erogenous zones Parts of the body that become erotically sensitized through experience.

Human Sexuality ONLINE //

www.seduction.net

The new millennium has seen some changes in Italy. Once upon a time, Latin Lovers would never have admitted to needing some help, but nowadays they can pay $192 for a seduction course that purports to teach them the skills of a Casanova.

The two-day course, which can be accessed via www.seduction.net, attracts students aged 19 to 60. Most are in their 30s. Lest the course be equated with male chauvinism, impresario Carlo della Torre notes that it is also open to women.

"Seduction is like an orchestra with all its musicians playing in sync. If one musician slips up, everything collapses," della Torre (cited in "Lackluster Latin Lovers," 2000) explains. "Seduction is based 60% on environment, 30% on the seducer and 10% [on] the object of seduction"—that is, the person being seduced. Sixty percent environment? That means we can let accoutrements like candlelight and a fine wine do part of the work.

Della Torre claims that his service screens applicants for the course, in order to deny admission to those who have "immoral motives." (Right.)

Della Torre also offers a course that sounds something like a U.S. TV show: "How to Meet and Marry a Millionaire." But della Torre warns that he does not guarantee the results. Pity.

The Age-Old Art of Seduction. The art of seduction is now being facilitated by Web sites such as www.seduction.net, an online "magazine" (e-zine) that offers advice and advertises courses (you've got to hand over cash) on the topic.

Hearing: The Better to Hear You With

The sense of hearing also provides an important medium for sexual arousal and response. Like visual and olfactory cues, sounds can be turn-ons or turn-offs. The sounds of one's lover, be they whispers, moans of pleasure, or animated sounds that may attend orgasm, may be arousing during the heat of passion. For some people, key words or vocal intonations may become as arousing as direct stimulation of an erogenous zone. Many people are aroused when their lovers "talk dirty." Spoken vulgarities spur their sexual arousal. Others find vulgar language offensive.

The sultry voices of screen sirens Lauren Bacall and Kathleen Turner have aroused the ardor of many a moviegoer. Teenagers may shriek at the music of groups such as Aerosmith or Pearl Jam, just as their parents (and grandparents) did when they heard the Beatles, Frank Sinatra, Johnny Mathis, or the King himself (Elvis, that is). Of course, the sex appeal of these performers encompasses much more than the sounds they produce.

Music itself can contribute to sexual arousal. Music can relax us and put us "in the mood" or evoke powerful associations ("They're playing our song!"). Many couples find background music "atmospheric"—a vital accoutrement of lovemaking.

Sounds can also be sexual turn-offs. Most of us would find funeral music a damper on sexual arousal. We may also be inhibited by scratchy, unnerving voices. Heavy-metal rock might be a sexual turn-off to many (are your authors showing their age?), but it could help set the right tone for others.

APHRODISIACS: OF SPANISH FLIES AND RHINO HORNS

> The only known aphrodisiac is variety.
> —Marc Connolly

An aphrodisiac is a substance that arouses or increases one's capacity for sexual pleasure or response. You may have heard of "Spanish fly," an alleged aphrodisiac once extracted from a Spanish beetle. (The beetle from which it was taken, *Lytta vesicatoria*, is near extinction.) A few drops in a date's drink was believed to make you irresistible. Spanish fly is but one of many purported aphrodisiacs. It is toxic, however, not sexually arousing. Spanish fly is now synthesized—but not as an aphrodisiac. The active ingredient, *cantharidin*, is a skin irritant that can burn off warts. If it can burn off warts, consider the damage it can do when taken internally. It irritates the urinary tract and can cause severe tissue damage or death. It inflames the urethra, producing a burning sensation in the penis that is sometimes misinterpreted as sexual feelings.

We should also be concerned about an expectancy effect, or placebo effect, when evaluating the effectiveness of a purported aphrodisiac (Brody, 1993). The belief that a substance has sexually stimulating effects may itself inspire sexual excitement. A person who tries a supposed aphrodisiac and feels sexually aroused may well attribute the turn-on to the effects of that substance, even if it had no direct effect on sex drive.

Truth or Fiction? REVISITED

It is true that "Spanish fly" may cure your date's warts but will not turn him or her on). Spanish fly contains cantharidin, a skin irritant that can in fact burn off warts. Cantharidin is not an aphrodisiac, but it can inflame the urethra, producing a burning sensation that could be misinterpreted as sexual feelings. ■

Foods that in some way resemble male genitals have now and then been considered aphrodisiacs. These include oysters, clams, bulls' testicles ("prairie oysters"), tomatoes, and "phallic" items such as celery stalks, bananas, and even ground-up reindeer antlers, elephant tusks, and rhinoceros horns (which is one derivation of the slang term *horny*).

Even potatoes have been held to be aphrodisiacs. Yes, potatoes. Shakespeare echoed this belief when he wrote, "Let the sky rain potatoes . . . ; let a tempest of provocation come." None of these foods or substances has been shown to be sexually stimulating, however—not even deep-fried potato skins with cheddar cheese and bacon. Sadly, myths about the sexually arousing properties of substances extracted from rhinoceroses and elephants are contributing to the rapidly diminishing numbers of these animals.

Other drugs and psychoactive substances may have certain effects on sexual arousal and response. The drug *yohimbine*, an extract from the African yohimbe tree, does stimulate blood flow to the genitals (Brody, 1993). However, its effects are limited and unreliable (Morales, 1993). Yohimbine also happens to be toxic (Brody, 1993).

Amyl nitrate (in the form of "snappers" or "poppers") has been used, mostly by gay men but also by some heterosexuals, in the belief that it heightens sensations of arousal and orgasm. Poppers dilate blood vessels in the brain and genitals, producing sensations of warmth in the pelvis and possibly facilitating erection and prolonging orgasm. Amyl nitrate does have some legitimate medical uses, such as helping reduce heart pain (angina) among cardiac patients. It is inhaled from ampules that "pop" open for rapid use when heart pain occurs. Poppers can cause dizziness, fainting, and migraine-type headaches, however. They should be taken only under a doctor's care for a legitimate medical need, not to intensify sexual sensations.

The drug Viagra was originally developed as a treatment for angina (heart pain) because it was thought that it would increase the blood flow to the heart. It does so, modestly. However, it is more effective at dilating blood vessels in the genital organs, thereby facilitating vasocongestion and erection in the male—and, according to some reports, sexual response in women as well. Viagra is marketed as a treatment for erectile dysfunction (also termed *impotence*). Is Viagra also an aphrodisiac? It is a matter of definition. Although Viagra facilitates erection, it still takes a sexual turn-on for erection to occur (Lewan, 1998). If an aphrodisiac must be directly sexually arousing, Viagra is not an aphrodisiac.

But certain drugs do appear to have aphrodisiac effects, apparently because they act on the brain mechanisms controlling the sex drive. For example, drugs that affect brain receptors for the neurotransmitter dopamine, such as the antidepressant drug *bupropion* (trade name Wellbutrin) and the drug L-dopa used in the treatment of Parkinson's disease, can increase the sex drive (Brody, 1993).

The most potent chemical "aphrodisiac" may be a naturally occurring substance in the body, the male sex hormone testosterone. It is the basic fuel of sexual desire in both genders (Brody, 1993; Williams, 1999).

The safest and perhaps the most effective method for increasing the sex drive may not be a drug or substance, but exercise. Regular exercise not only enhances general health. It also boosts energy and increases the sex drive in both genders (Brody, 1993). Perhaps the most potent aphrodisiac of all is novelty. Partners can invent new ways of sexually discovering one another. They can make love in novel places, experiment with different techniques, wear provocative clothing, share or enact fantasies, or whatever their imaginations inspire.

The nearby Closer Look raises the question of whether plastic balls and spruce branches are aphrodisiacs for pandas.

a CLOSER look

Sex Toys for Pandas?

The notoriously unsatisfactory sex lives of giant pandas may be boosted by giving the animals toys. The introduction of playthings into pens at China's Wolong Breeding Centre, scene of a panda baby boom, coincided with the best ever season for panda sex, said the article in *New Scientist* magazine. It was announced that six panda cubs had been born at Wolong, in the southwestern province of Sichuan, in four days.

Many male pandas raised in captivity develop repetitive behavior patterns called "stereotypies." They may pirouette around their enclosure or suck a paw for long periods and show little interest in females. Ronald Swaisgood, a behavioral biologist working at Wolong, decided to experiment with putting toys like stuffed sacks, plastic balls and spruce branches in the pandas' pens every few days.

After a year he found the pandas were displaying repetitive stereotypes only a third as often as before. Although the article said it had not been proved that male pandas mated more because of the toys, in the experimental year, four of the center's six males and eight of nine females mated—the highest number ever.

Swaisgood said he believed the toys had a deeper effect than simply giving the pandas more to occupy their time. "Since there was much less stereotypy even after I removed the items, I know they had some psychological effect," he said.

The giant panda, a Chinese national symbol, is an endangered species and famously difficult to breed in captivity. In the wild, pandas roam the mountainous areas of China's Sichuan, Gansu and Shaanxi provinces, but their habitat has been threatened by human encroachment and rampant logging. There are believed to be just 1,000 left in the wild.

Source: From "Toys May Boost Pandas' Sex Lives, Say Scientists" (2000, August 17). *Reuters News Agency Online.*

A Giant Panda. The survival of the giant panda may well depend on humans' ability to breed them in captivity. Pandas show very little interest in sex, unfortunately, but researchers have recently discovered that "toys" such as stuffed sacks, plastic balls, and spruce branches can give their sex lives a boost. These items are not aphrodisiacs but seem to provide a general source of stimulation that then spills over into the arena of sexual behavior.

Anaphrodisiacs

Some substances such as potassium nitrate (saltpeter) have been considered inhibitors of sexual response—**anaphrodisiacs.** Saltpeter, however, only indirectly dampens sexual arousal. As a diuretic that can increase the need to urinate, it may make the thought of sex unappealing. It does not directly dampen sexual response, however.

Other chemicals do dampen sexual arousal and response. Tranquilizers and central nervous system depressants, such as barbiturates, can reduce sexual desire and impair sexual performance. These drugs may paradoxically enhance sexual arousal in some people, however, by lessening sexual inhibitions or fear of possible repercussions from sexual activity. Antihypertensive drugs, which are used in the treatment of high blood pressure, may produce erectile and ejaculatory difficulties in men and may reduce sexual desire in both genders. Certain antidepressant drugs, such as fluoxetine (brand name: Prozac), amitriptyline (brand name: Elavil), and imipramine (brand name: Tofranil) appear to dampen sex drive (Brody, 1993; Meston & Gorzalka, 1992). Antidepressants may also impair erectile response and delay ejaculation in men and impair orgasmic responsiveness in women (Meston & Gorzalka, 1992). (Because they delay ejaculation, some of these drugs are used to treat premature ejaculation.)

Nicotine, the stimulant in tobacco smoke, constricts the blood vessels. Thus it can impede sexual arousal by reducing the capacity of the genitals to become engorged with blood. Chronic smoking can also reduce the blood levels of testosterone in men, which can in turn lessen sex drive or motivation.

Anti-androgen drugs may have anaphrodisiac effects. However, their effectiveness in modifying deviant behavior patterns such as sexual violence and sexual interest in children is questionable.

Psychoactive Drugs

Psychoactive drugs, such as alcohol and cocaine, are widely believed to have aphrodisiac effects. Yet their effects may reflect our expectations of them, or their effects on sexual inhibitions, rather than direct stimulation of sexual response.

ALCOHOL: THE "GREAT PROVOKER OF THREE THINGS" In Shakespeare's *Macbeth*, the following exchange takes place between Macduff and a porter:

PORTER: . . . drink, sir, is a great provoker of three things.
MACDUFF: What three things does drink especially provoke?
PORTER: Marry,[1] sir, nose-painting,[2] sleep, and urine. Lechery,[3] sir, it provokes, and unprovokes; it provokes the desire, but it takes away the performance.

Small amounts of alcohol are stimulating, but large amounts curb sexual response. This fact should not be surprising, because alcohol is a depressant. Alcohol reduces central nervous system activity. Large amounts of alcohol can severely impair sexual performance in both men and women.

People who drink moderate amounts of alcohol may feel more sexually aroused because of their expectations about alcohol, not because of its chemical properties (George et al., 2000). That is, people who expect alcohol to enhance sexual responsiveness may act the part. Expectations that alcohol serves as an aphrodisiac may lead men with problems achieving erection to turn to alcohol in hopes of finding a cure (Roehrich & Kinder,

[1]This contraction of the expression "By the Virgin Mary" was used by Elizabethans, such as Shakespeare, to speak emphatically yet avoid disrespect to the Virgin Mary.
[2]The reddening of the nose that occurs in many chronic alcoholics as a result of the bursting of small blood vessels.
[3]Excessive indulgence in sexual desire; related to the German lecken, which means "to lick."

Anaphrodisiacs Drugs or other agents whose effects are antagonistic to sexual arousal or sexual desire.

Anti-androgen A substance that decreases the levels of androgens in the bloodstream.

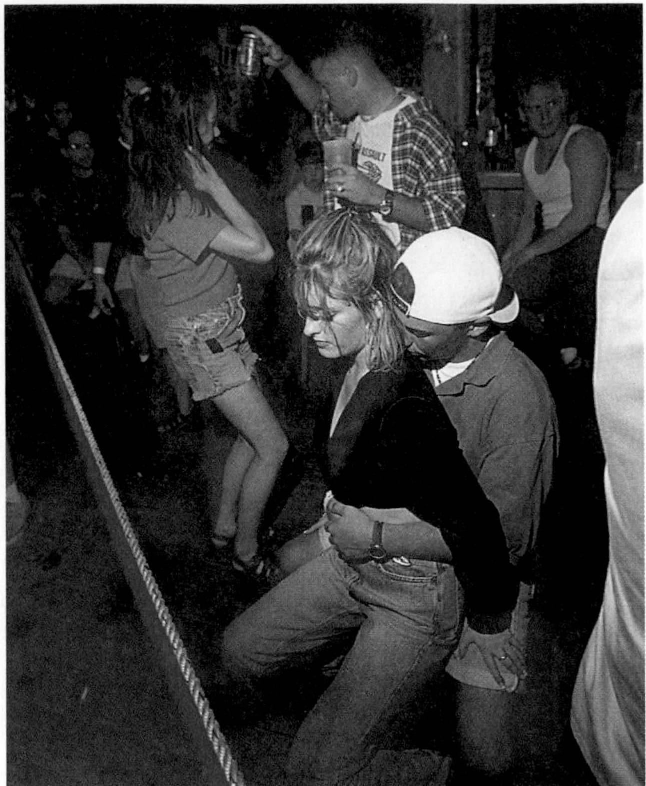

What Are the Effects of Alcohol on Sexual Behavior? Small doses of alcohol can be stimulating, can induce feelings of euphoria, and can lower inhibitions, all of which could be connected with sexual interest and facilitate social and sexual behavior. Furthermore, alcohol reduces fear of the consequences of engaging in risky behavior—sexual and otherwise. Alcohol also provides an excuse for engaging in otherwise unacceptable behavior, such as sexual intercourse on the first date (or upon a casual meeting). Alcohol is expected to be sexually liberating, and people often live up to social and cultural expectations. Yet, as a depressant drug, alcohol in large amounts will biochemically dampen sexual response and may make sexual response impossible.

1991). The fact is that alcohol is a depressant and can reduce sexual potency rather than restore it.

Alcohol may also lower sexual inhibitions, because it allows us to ascribe our behavior to the effects of the alcohol rather than to ourselves (Crowe & George, 1989; Lang, 1985). Alcohol is connected with a liberated social role and thus provides an excuse for dubious behavior. "It was the alcohol," people can say, "not me." When drinkng, people may express their sexual desires and do things that they would not do when sober. For example, a person who feels guilty about sex may become sexually active when drinking because he or she can later blame the alcohol.

Consider a survey of 1,100 undergraduates taken at the University of Virginia (Grossman, 1991). Forty-three percent of the respondents were classified as "heavy drinkers." That is, they had had at least five alcoholic beverages in a row on one occasion within the two weeks preceding the survey. *More than half* of the heavy drinkers reported that under the influence of alcohol, they had engaged in sexual activity with someone they would not ordinarily have become involved with. All in all, between 20% and 25% of the total sample had, when under the influence of alcohol, engaged in sexual activity they deemed unwise!

In laboratory studies, men who were misled into believing that they had drunk alcohol spent more time lingering over pornographic pictures as researchers looked on than did men who thought they had not drunk alcohol. They also reported a greater degree of sexual arousal (George et al., 2000). Did they believe that they would avert the censure of the onlookers because they could attribute their prurience to the alcohol? If so, perhaps alcohol provides an excuse for socially deviant behavior. The amount of time the men lingered over the pornographic material was unaffected by whether they had actually drunk alcohol (Lang et al., 1980; Lansky & Wilson, 1981).

Alcohol can also induce feelings of euphoria. Euphoric feelings may enhance sexual arousal and also sweep away qualms about expressing sexual desires. Alcohol also appears to impair the ability to weigh information ("information processing") that might otherwise inhibit sexual impulses (MacDonald et al., 2000; Steele & Josephs, 1990). When people drink, they may be less able to foresee the consequences of misconduct and less likely to ponder their standards of conduct.

HALLUCINOGENICS There is no evidence that marijuana and other hallucinogenic drugs directly stimulate sexual response. However, fairly to strongly intoxicated marijuana users claim to have more empathy with others, to be more aware of bodily sensations, and to experience time as passing more slowly. These sensations could heighten subjective feelings of sexual response. Some marijuana users report that the drug inhibits their sexual responsiveness, however (Wolman, 1985). The effects of the drug on sexual response may depend on the individual's prior experiences with the drug, on her or his attitudes toward the drug, and on the amount taken.

Other hallucinogenics, such as LSD and mescaline, have also been reported by some users to enhance sexual response. Again, these effects may reflect dosage level and user expectations, experiences, and attitudes toward the drugs, as well as altered perceptions.

STIMULANTS Stimulants such as amphetamines ("speed," "uppers," "bennies," "dexies") have been reputed to heighten arousal and sensations of orgasm. High doses can give rise to irritability, restlessness, hallucinations, paranoid delusions, insomnia, and loss of appetite. These drugs generally activate the central nervous system but are not known to have specific sexual effects. Nevertheless, arousing the nervous system can contribute to sexual arousal (Palace, 1995). The drugs can also elevate mood, and perhaps sexual pleasure is heightened by general elation.

Cocaine is a natural stimulant that is extracted from the leaves of the coca plant—the plant from which the soft drink Coca-Cola obtained its name. In fact, Coke (Coca-Cola), contained cocaine as part of its original formula. Cocaine was removed from the secret formula in 1906. Cocaine is ingested in various forms, snorted as a powder, smoked in hardened rock form ("crack" cocaine) or in a freebase form, or injected directly into the bloodstream in liquid form. Cocaine produces a euphoric rush, which tends to ebb quickly. Physically, cocaine constricts blood vessels (reducing the oxygen supply to the heart), elevates the blood pressure, and accelerates the heart rate.

Despite the popular belief that cocaine is an aphrodisiac, frequent use can lead to sexual dysfunctions, such as difficulty attaining erection and ejaculating among males, decreased vaginal lubrication in females, and sexual apathy in both men and women (Weiss & Mirin, 1987). Some people do report initial increased sexual pleasure with cocaine use; however, that increase may reflect cocaine's loosening of inhibitions. Over time, though, regular users may become dependent on cocaine for sexual arousal or lose the ability to enjoy sex (Weiss & Mirin, 1987).

SEXUAL RESPONSE AND THE BRAIN: CEREBRAL SEX?

The brain may not be an erogenous zone, but it plays a central role in sexual functioning (Fisher, 2000). Direct genital stimulation may trigger spinal reflexes that produce erection in the male and vaginal lubrication in the female without the direct involvement of the brain. The same reflexes may be triggered by sexual stimulation that originates in the brain in the form of erotic memories, fantasies, visual images, and thoughts. The brain may also inhibit sexual responsiveness, as when we experience guilt or anxiety in a sexual situation or when we suddenly realize, well into a sexual encounter, that we have left the car lights turned on. Let us explore the brain mechanisms involved in sexual functioning.

The Geography of the Brain

The brain consists of three major parts: the hindbrain, the midbrain, and the forebrain (see Figure 5.1). The lower part of the brain, called the hindbrain, consists of the **medulla,** the **pons,** and the **cerebellum.** The medulla helps regulate vital functions such as heart rate, respiration, and blood pressure. The pons relays information about body movement and plays a role in states or functions such as attention, sleep, and respiration. The cerebellum is involved in balance and coordination. Rising from the hindbrain into the forebrain is the **reticular activating system,** or RAS, which is involved in attention, arousal, and sleep.

Medulla An oblong area of the hindbrain involved in regulation of heart rate and respiration.

Pons A structure of the hindbrain that regulates respiration, attention, sleep, and dreaming.

Cerebellum A part of the hindbrain that governs muscle coordination and balance.

Reticular activating system A part of the brain that is active during attention, sleep, and arousal.

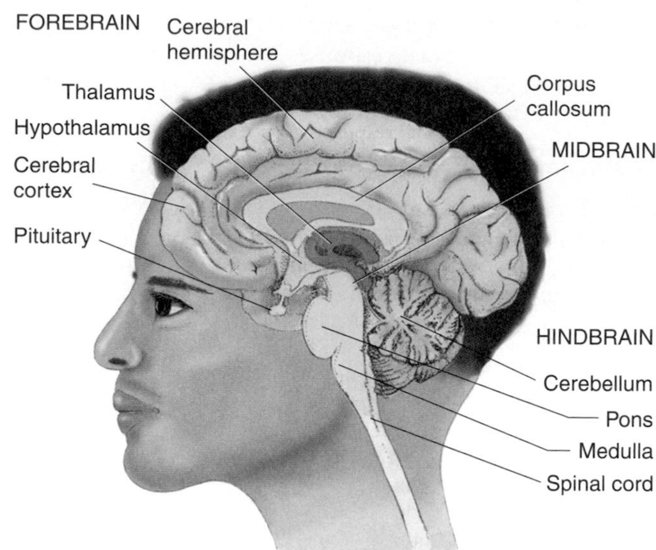

FIGURE 5.1

The Geography of the Brain. A view of the brain, split from top to bottom. Cells in the cerebral cortex transmit messages when we experience sexual thoughts and mental images. Cells in the cortex interpret sensory information as sexual turn-ons or turn-offs. The cerebral cortex may then transmit messages through the spinal cord that send blood coursing to the sex organs, leading to erection or vaginal lubrication.

Key areas in the forebrain, or frontal part of the brain, are the **thalamus, hypothalamus, limbic system,** and **cerebrum.** The thalamus, which lies in the center of the brain, plays a role in regulating sleep and attention and also relays sensory information to the cerebral cortex. Sensory information from the eyes, for example, is relayed by the thalamus to the visual areas of the cerebral cortex, where the information is processed. The hypothalamus lies between the thalamus and the pituitary gland and is involved in motivation and emotion; in body functions such as control of body temperature, concentration of fluids, and storage of nutrients; and in regulation of the menstrual cycle.

The limbic system consists of parts of the hypothalamus and contains other structures such as the hippocampus and the septum (see Figure 5.2). The limbic system is involved in memory and in regulation of hunger, aggression, and sexual behavior.

Your cerebrum is your crowning glory. The large mushroom-shaped outer surface of the cerebrum, convoluted with ridges and valleys, is called the **cerebral cortex.** It has two hemispheres and is involved in thinking, language, memory, and fantasy.

Thalamus An area near the center of the brain involved in the relay of sensory information to the cortex and in the functions of sleep and attention.

Hypothalamus A brain structure that lies below the thalamus and regulates body temperature, motivation, and emotion.

Limbic system A group of structures active in memory, motivation, and emotion; the structures that are part of this system form a fringe along the inner edge of the cerebrum.

Cerebrum The large mass of the forebrain; it consists of two hemispheres and is involved in thinking, memory, language, and fantasy.

Cerebral cortex The wrinkled surface area (gray matter) of the cerebrum.

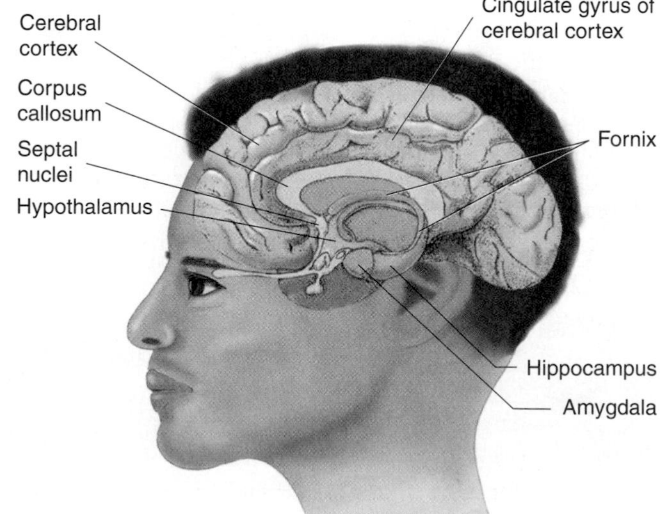

FIGURE 5.2

The Limbic System. The limbic system lies along the inner edge of the cerebrum. When part of a male rat's hypothalamus is electrically stimulated, the rat engages in its courting and mounting routine. Klüver and Bucy (1939) found that destruction of areas of the limbic system triggered continuous sexual behavior in monkeys. Electrical stimulation of the hippocampus and septal nuclei produces erections in monkeys.

Brain Mechanisms in Sexual Functioning

Parts of the brain—in particular the cerebral cortex and the limbic system—play key roles in sexual functioning. Cells in the cerebral cortex fire (transmit messages) when we experience sexual thoughts, images, wishes, fantasies, and the like. Cells in the cerebral cortex interpret sensory information as sexual turn-ons or turn-offs. The sight of your lover disrobing, the anticipation of a romantic kiss, a passing sexual fantasy, or the viewing of an erotic movie can trigger the firing of cortical cells. These cells, in turn, transmit messages through the spinal cord that send blood rushing to the genitals, causing erection or vaginal lubrication. The cortex also provides the conscious sense of self. The cortex judges sexual behavior to be proper or improper, moral or immoral, relaxing or anxiety- or guilt-provoking.

Areas of the brain below the cortex, especially the limbic system, also play roles in sexual processes (Everitt, 1990; Kimble, 1992). For example, when the rear part of a male rat's hypothalamus is stimulated by an electrical probe, the animal mechanically runs through its courting and mounting routine. It nibbles at the ears and the back of the neck of a female rat and mounts her when she responds. People, of course, are influenced by learning, fantasy, and values as well as by simple brain (or spinal) stimulation.

The importance of the limbic system in the sexual behavior in animals was demonstrated in experiments by Heinrich Klüver and Paul Bucy of the University of Chicago in 1939. Klüver and Bucy reported that destruction of areas of the limbic system triggered persistent sexual behaviors that included masturbation and male–female and male–male mounting attempts. The monkeys even tried to mount the experimenters. For obvious ethical reasons, researchers have not injured or destroyed parts of people's brains to observe the effects on humans.

Electrical stimulation of the hippocampus and septum of the limbic system has also been found to produce erections in laboratory monkeys (McLean, 1976). Electrical stimulation of a pathway in the thalamus, moreover, produced a seminal discharge in these monkeys—without erection. Stimulation of certain areas in the thalamus and hypothalamus may induce ejaculation. Still, the precise relationships among brain structures that regulate erection and ejaculation in animals or humans have not been fully mapped out.

On Pushing the Right Buttons: Are There Pleasure Centers in the Brain?

Research with electrical probes suggests that "pleasure centers" may exist in and near the hypothalamus in other animals and perhaps even in people. When electrodes are implanted in certain parts of the limbic system, investigators find that laboratory animals such as rats (Olds, 1956; Olds & Milner, 1954) will repeatedly press controls to receive bursts of electricity. Of course we cannot know what the rats experience, but people have reported that stimulation of these so-called pleasure centers led to feelings of sexual arousal and gratification.

Heath (1972) found that electrical stimulation of the septal region of the limbic system resulted in orgasmlike sensations in two people. Delgado (1969) reported that two female epileptic patients who received limbic stimulation as part of a diagnostic evaluation became sexually aroused by the stimulation:

> [One] reported a pleasant tingling sensation in the left side of her body "from my face down to the bottom of my legs." She started giggling and . . . [stated] that she enjoyed the sensation "very much." Repetition of these stimulations made the patient more communicative and flirtatious, and she ended by openly expressing her desire to marry the therapist. [The other patient reported] a pleasant sensation of relaxation and considerably increased her verbal output, which took on a more intimate character. [She] expressed her fondness for the therapist [whom she had just met], kissed his hands, and talked about her immense gratitude. (p. 145)

Truth or Fiction? **REVISITED**

It is true that electrical stimulation of certain areas in the human brain can yield sensations similar to those of sexual pleasure and gratification. Electrical stimulation of certain parts of the limbic system has yielded sensations similar to those of sexual gratification—at least in the relatively small number of people who have been studied.

We cannot say whether specific "pleasure centers" in the brain are responsible for sexual pleasure. We might wonder, however, whether we would bother to develop sexual relationships if pleasure centers could be accessed through direct stimulation. Before you run to the store for electrodes, you should note that very few researchers suggest that we may someday replace our lovers with battery-powered kits.

SEX HORMONES: DO THEY "GOAD" US INTO SEX?

In a TV situation comedy, a male adolescent was described as "a hormone with feet." Ask parents why teenagers act the way they do, and you are likely to hear a one-word answer: hormones! **Hormones** are chemical substances that are secreted by the ductless glands of the endocrine system and discharged directly into the bloodstream. The word *hormone* derives from the Greek *horman,* meaning "to stimulate" or "to goad." And we could say that they very much goad us into sexual activity. Hormones also regulate various bodily functions, including growth and resistance to stress as well as sexual functions.

Both men and women produce small amounts of the sex hormones of the other gender in their bodies. Testosterone, the major form of androgen, or male sex hormone, is secreted in small amounts by the adrenal glands (located above the kidneys) in both genders, but in much larger amounts by the testes. The ovaries produce small amounts of androgens but much larger amounts of the female sex hormones, estrogen and progesterone. The testes similarly produce small amounts of estrogen and progesterone.

Truth or Fiction? **REVISITED**

It is true that normal men produce estrogen and that normal women produce androgens. However, men do not produce as much estrogen as women do, and women do not produce as much of the male sex hormones as men do.

The hypothalamus and pituitary gland regulate gonadal secretion of sex hormones, specifically testosterone in males and estrogen and progesterone in females. At puberty, a surge of sex hormones causes the blossoming of reproductive maturation: the sperm-producing ability of the testes in males and the maturation of ova and ovulation in females. Sex hormones released at puberty also cause the flowering of **secondary sex characteristics.** In males, these include the lengthening of the vocal cords (and consequent lowering of the voice) and the growth of facial and pubic hair. In females, the breasts and hips become rounded with fatty tissue, and pubic hair grows.

Hormone A substance that is secreted by an endocrine gland and regulates various body functions. (From the Greek *horman,* which means "to stimulate" or "to goad.")

Secondary sex characteristics Physical traits that differentiate males from females but are not directly involved in reproduction.

Sex Hormones and Sexual Behavior: Organizing and Activating Influences

Sex hormones have organizing and activating effects on behavior. That is, they exert an influence on the type of behavior that is expressed (an *organizing* effect) and on the frequency or intensity of the drive that motivates the behavior and the ability to perform the behavior (*activating* effects). For example, sex hormones predispose lower animals and possibly people toward stereotypical masculine or feminine mating behaviors (an organizing effect). They also facilitate sexual response and influence sexual desire (activating effects).

Although sex hormones clearly determine the sexual "orientations" and drives of many lower animals, their roles in human sexual behavior may be relatively more subtle and are not so well understood. Much of our knowledge of the organizing and activating effects of sex hormones comes from research with other species in which hormone levels were manipulated by castration or injection. Ethical standards obviously prohibit such research with human infants.

The activating effects of testosterone can be clearly observed among male rats. For example, males who are castrated in adulthood and thus deprived of testosterone discontinue sexual behavior. If they are given injections of testosterone, however, they resume stereotypical male sexual behaviors, such as attempting to mount receptive females.

In rats, testosterone organizes or differentiates the brain in the masculine direction. As a result, adult male rats display stereotypical masculine behaviors upon activation by testosterone. Male fetuses and newborns normally have enough testosterone in their blood systems to organize their brains in the masculine direction. Female fetuses and newborns normally have lesser amounts of testosterone. Their brains thus become organized in a feminine direction. When female rats are prenatally exposed to large doses of testosterone, their sexual organs became somewhat masculinized, and they are predisposed toward masculine mating behaviors in adulthood (Kimble, 1992).

In rats and other rodents, sexual differentiation of the brain is not complete at birth. Female rodents who are given testosterone injections shortly before or shortly following birth (depending on the species) show typical masculine sexual patterns in adulthood; they try to mount other females and resisting mounting by males (Ellis & Ames, 1987).

Questions remain about the organizing effects of sex hormones on human sexual behavior. Prenatal sex hormones are known to play a role in the sexual differentiation of the genitalia and of the brain structures, such as the hypothalamus. Their role in patterning sexual behavior in adulthood remains unknown, however. Researchers have speculated that the brains of **transsexual** individuals may have been prenatally sexually differentiated in one direction, while their genitals were being differentiated in the other (Money, 1994). It has been speculated that prenatal sexual differentiation of the brain may also be connected with sexual orientation.

What of the activating effects of sex hormones on human sex drive and behavior? Although the countless attempts to extract or synthesize aphrodisiacs have failed to produce the real thing, men and women normally produce a genuine aphrodisiac—testosterone. Whatever the early organizing effects of sex hormones in humans, testosterone activates the sex drives of both men and women (Guzick & Hoeger, 2000).

Sex Hormones and Male Sexual Behavior

Evidence of the role of hormones in sex drive is found among men who have declines in testosterone levels as a result of chemical or surgical castration. Surgical castration (removal of the testes) is sometimes performed as a medical treatment for cancer of the prostate or other diseases of the male reproductive tract, such as genital tuberculosis. And some convicted sex offenders have voluntarily undergone castration as a condition of release.

Regardless of the reason for castration, men who are surgically or chemically castrated usually exhibit a gradual decrease in the incidence of sexual fantasies and loss of sexual desire (Bradford, 1998; Gijs & Gooren, 1996; Rösler & Witztum, 1998). They also gradually tend to lose the capacities to attain erection and to ejaculate—an indication that

Transsexual A person with a gender-identity disorder who feels that he or she is really a member of the other gender and is trapped in a body of the wrong gender.

testosterone is important in maintaining sexual functioning as well as drive, at least in males. Castrated men show great variation in their sexual interest and functioning, however. Some continue to experience sexual desires and are able to function sexually for years, even decades. Learning appears to play a large role in determining continued sexual response following castration. Males who were sexually experienced before castration show a more gradual decline in sexual activity. Those who were sexually inexperienced at the time show relatively little or no interest in sex. Male sexual motivation and functioning thus involve an interplay of hormonal influences and experience.

Further evidence of the relationship between hormonal levels and male sexuality is found in studies of men with **hypogonadism,** a condition marked by abnormally low levels of testosterone production. Hypogonadal men generally experience loss of sexual desire and a decline in sexual activity (Carani et al., 1990). Here again, hormones do not tell the whole story. Hypogonadal men are capable of erection, at least for a while, even though their sex drives may wane (Bancroft, 1984). The role of testosterone as an activator of sex drives in men is further supported by evidence of the effects of testosterone replacement in hypogonadal men. When such men obtain testosterone injections, their sex drives, fantasies, and activity are often restored to former levels (Cunningham et al., 1989; Goleman, 1988).

Though minimal levels of androgens are critical to male sexuality, there is no one-to-one correspondence between hormone levels and the sex drive or sexual performance in adults. In men who have ample supplies of testosterone, sexual interest and functioning depend more on learning, fantasies, attitudes, memories, and other psychosocial factors than on hormone levels. At puberty, however, hormonal variations may play a more direct role in stimulating sexual interest and activity in males. Udry and his colleagues (Udry et al., 1985; Udry et al., 1986; Udry & Billy, 1987) found, for example, that testosterone levels predicted sexual interest, masturbation rates, and the likelihood of engaging in sexual intercourse among teenage boys. A positive relationship also has been found between testosterone levels in adult men and frequency of sexual intercourse (Dabbs & Morris, 1990; Knussman et al., 1986). Moreover, drugs that reduce the levels of androgens in the blood system, called *anti-androgens,* lead to reductions in the sex drive and in sexual fantasies (Bradford, 1998).

Sex Hormones and Female Sexual Behavior

The female sex hormones estrogen and progesterone play prominent roles in promoting the changes that occur during puberty and in regulating the menstrual cycle. Female sex hormones do not appear to play a direct role in determining sexual motivation or response in human females, however.

In most mammals, females are sexually receptive only during estrus. Estrus is a brief period of fertility that corresponds to time of ovulation, and during estrus, females are said to be "in heat." Estrus occurs once a year in some species; in others, it occurs periodically during the year in so-called sexual or mating seasons. Estrogen peaks at time of ovulation, so there is a close relationship between fertility and sexual receptivity in most female mammals. Women's sexuality is not clearly linked to hormonal fluctuations, however. Unlike females of most other species of mammals, the human female is sexually responsive during all phases of the reproductive (menstrual) cycle—even during menstruation, when ovarian hormone levels are low—and after menopause.

There is some evidence, however, that sexual responsiveness in women is influenced by the presence of circulating androgens, or male sex hormones, in their bodies. The adrenal glands of women produce small amounts of androgens, just as they do in males (Guzick & Hoeger, 2000). The fact that women normally produce smaller amounts of

Hypogonadism An abnormal condition marked by abnormally low levels of testosterone production.

androgens than men does not mean that they necessarily have weaker sex drives. Rather, women appear to be more sensitive to smaller amounts of androgens. For women, it seems that less is more.

Women who receive **ovariectomies,** which are sometimes carried out when a hysterectomy is performed, no longer produce female sex hormones. Nevertheless, they continue to experience sex drives and interest as before. Loss of the ovarian hormone estradiol may cause vaginal dryness and make coitus painful, but it does not reduce sexual desire. (The dryness can be alleviated by a lubricating jelly or by estrogen-replacement therapy.) However, women whose adrenal glands *and* ovaries have been removed (so that they no longer produce androgens) gradually lose sexual desire. An active and enjoyable sexual history seems to ward off this loss, however, providing further evidence of the impact of cognitive and experiential factors on human sexual response.

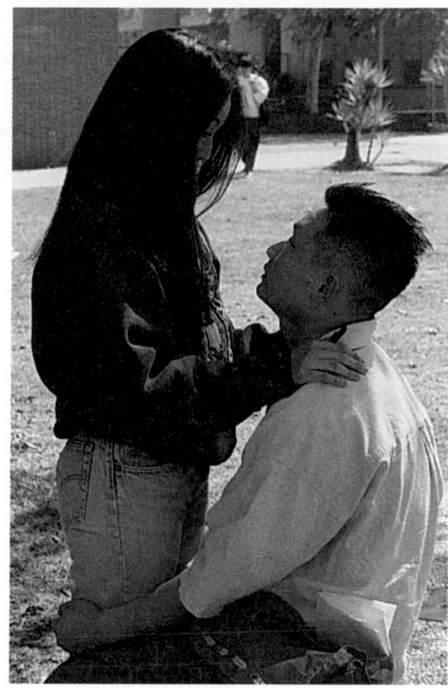

Are Adolescents "Hormones with Feet"? Research shows that levels of androgens are connected with sexual interest in both male and female adolescents. Hormone levels are more likely to predict sexual behavior in adolescent males, however, perhaps because society places greater restraints on female sexuality.

Research provides further evidence of the links between testosterone levels and women's sex drives (Williams, 1999). In the studies by Udry and his colleagues mentioned earlier, androgen levels were also found to predict sexual interest among teenage girls. In contrast to boys, however, girls' androgen levels were unrelated to the likelihood of coital experience. Androgens apparently affect sexual desire in both genders, but sexual interest may be more likely to be directly translated into sexual activity in men than in women (Bancroft, 1990). This gender difference may be explained by society's imposing greater restraints on adolescent female sexuality.

Other researchers report that women's sexual activity increases at points in the menstrual cycle when levels of androgens in the bloodstream are high (Morris et al., 1987). Another study was conducted with women whose ovaries had been surgically removed ("surgical menopause") as a way of treating disease. The ovaries supply major quantities of estrogen. Following surgery, the women in this study were treated with estrogen-replacement therapy (ERT), with ERT *plus* androgens, or with a placebo (an inert substance made to resemble an active drug) (Sherwin et al., 1985). This was a double-blind study. Neither the women nor their physicians knew which drug the women were receiving. The results showed that the combination of androgens and ERT heightened sexual desire and sexual fantasies more than ERT alone or the placebo. The combination also helps women maintain a sense of psychological well-being (Guzick & Hoeger, 2000).

Androgens thus play a more prominent role than ovarian hormones in activating and maintaining women's sex drives. As with men, however, women's sexuality is too complex to be explained fully by hormone levels. For example, an active and enjoyable sexual history seems to ward off the loss of sexual interest that generally follows the surgical removal of the adrenal glands and ovaries.

THE SEXUAL RESPONSE CYCLE

Although we may be culturally attuned to focus on gender differences rather than similarities, Masters and Johnson (1966) found that the physiological responses of men and women to sexual stimulation (whether from coitus, masturbation, or other sources) are quite alike. The sequence of changes in the body that takes place as men and women become progressively more aroused is referred to as the **sexual response cycle.** Masters and Johnson divided the cycle into four phases: *excitement, plateau, orgasm,* and *resolution.* Figure 5.3 suggests the levels of sexual arousal associated with each phase.

Ovariectomy Surgical removal of the ovaries.

Sexual response cycle Masters and Johnson's model of sexual response, which consists of four phases.

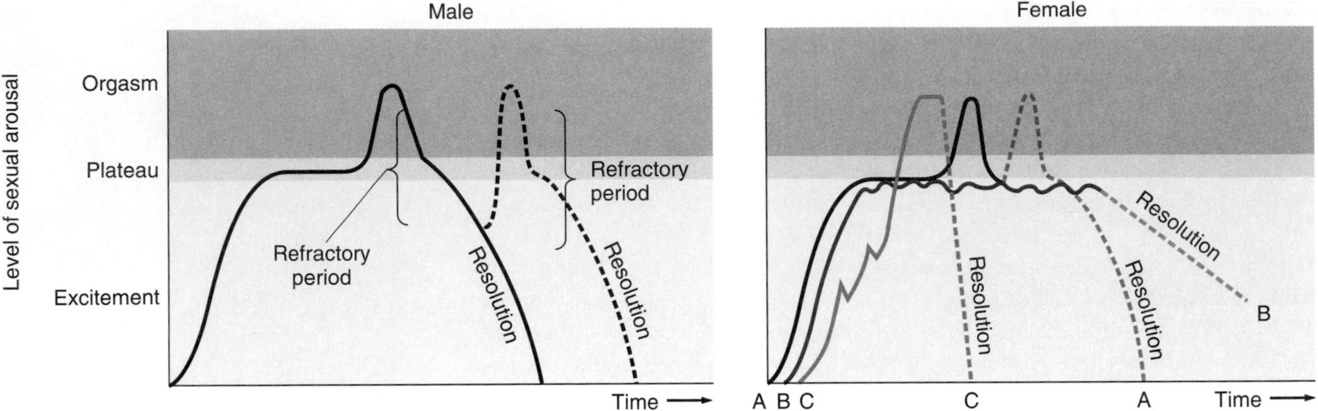

FIGURE 5.3

Levels of Sexual Arousal During the Phases of the Sexual Response Cycle. Masters and Johnson divide the sexual response cycle into four phases: excitement, plateau, orgasm, and resolution. During the resolution phase, the level of sexual arousal returns to the prearoused state. For men there is a refractory period following orgasm. As shown by the broken line, however, men can become rearoused to orgasm once the refractory period is past and their levels of sexual arousal have returned to preplateau levels. Pattern A for women shows a typical response cycle; the broken line indicates multiple orgasms, should they occur. Pattern B shows the cycle of a woman who reaches the plateau phase but for whom arousal is "resolved" without her experiencing orgasm. Pattern C shows the possibility of orgasm in a highly aroused woman who passes quickly through the plateau phase.

Both males and females experience **vasocongestion** and **myotonia** early in the response cycle. Vasocongestion is the swelling of the genital tissues with blood, which causes erection of the penis and engorgement of the area surrounding the vaginal opening. The testes, nipples, and even earlobes become engorged as blood vessels in these areas dilate.

Myotonia is muscle tension. It causes voluntary and involuntary muscle contractions, which produce facial grimaces, spasms in the hands and feet, and eventually, the spasms of orgasm. Let us follow these and the other bodily changes that constitute the sexual response cycle.

Vasocongestion The swelling of the genital tissues with blood, which causes erection of the penis and engorgement of the area surrounding the vaginal opening.

Myotonia Muscle tension.

Excitement phase The first phase of the sexual response cycle, which is characterized by erection in the male, by vaginal lubrication in the female, and by muscle tension and increases in heart rate in both males and females.

Sex flush A reddish rash that appears on the chest or breasts late in the excitement phase of the sexual response cycle.

Excitement Phase

In younger men, vasocongestion during the **excitement phase** produces penile erection as early as 3 to 8 seconds after stimulation begins. Erection may occur more slowly in older men, but the responses are essentially the same. Erection may subside and return as stimulation varies. The scrotal skin thickens, losing its baggy appearance. The testes increase in size. The testes and scrotum become elevated.

In the female, vaginal lubrication may start 10 to 30 seconds after stimulation begins. Vasocongestion swells the clitoris, flattens the labia majora and spreads them apart, and increases the size of the labia minora. The inner two thirds of the vagina expand. The vaginal walls thicken and, because of the inflow of blood, turn from their normal pink to a deeper hue. The uterus becomes engorged and elevated. The breasts enlarge, and blood vessels near the surface become more prominent.

Late in this phase, the skin may take on a rosy **sex flush,** which varies with intensity of arousal and is more pronounced in women. The nipples may become erect in both gen-

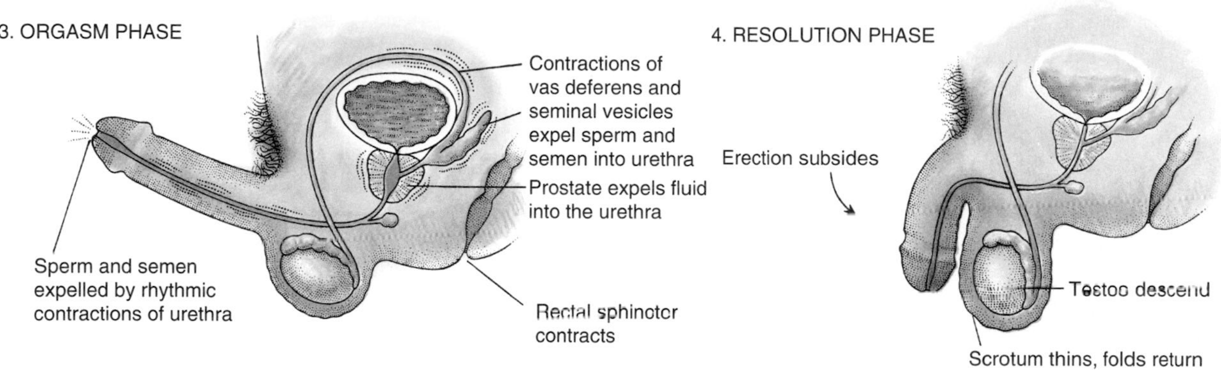
1. EXCITEMENT PHASE
Vasocongestion of penis results in erection
Meatus dilates
Testes begin elevation
Scrotal skin tenses, thickens
2. PLATEAU PHASE
The coronal ridge of the glans increases in diameter and turns a deeper reddish-purple
The Cowper's glands may release fluid
The testes become completely elevated and engorged when orgasm is imminent
Cowper's gland
3. ORGASM PHASE
Contractions of vas deferens and seminal vesicles expel sperm and semen into urethra
Prostate expels fluid into the urethra
Sperm and semen expelled by rhythmic contractions of urethra
Rectal sphincter contracts
4. RESOLUTION PHASE
Erection subsides
Testes descend
Scrotum thins, folds return

FIGURE 5.4

The Male Genitals During the Phases of the Sexual Response Cycle.

ders, especially in response to direct stimulation. Men and women show some increase in myotonia, heart rate, and blood pressure.

Plateau Phase

A plateau is a level region, and the level of arousal remains somewhat constant during the **plateau phase** of sexual response. Nevertheless, the plateau phase is an advanced state of arousal that precedes orgasm. Men in this phase show a slight increase in the circumference of the coronal ridge of the penis. The penile glans turns a purplish hue, a sign of vasocongestion. The testes are elevated further into position for ejaculation and may reach one and a half times their unaroused size. The Cowper's glands secrete a few droplets of fluid that are found at the tip of the penis (see Figure 5.4).

In women, vasocongestion swells the tissues of the outer third of the vagina, contracting the vaginal opening (thus preparing it to "grasp" the penis) and building the **orgasmic platform** (see Figure 5.5). The inner part of the vagina expands fully. The uterus becomes fully elevated. The clitoris withdraws beneath the clitoral hood and shortens. Thus a women (or her partner) may feel that the clitoris has become lost. This may be mistaken as a sign that the woman's sexual arousal is waning, when it is actually increasing.

Coloration of the labia minora, referred to as the **sex skin,** appears. The labia minora become a deep wine color in women who have borne children and bright red in women

Plateau phase The second phase of the sexual response cycle, which is characterized by increases in vasocongestion, muscle tension, heart rate, and blood pressure in preparation for orgasm.

Orgasmic platform The thickening of the walls of the outer third of the vagina, due to vasocongestion, that occurs during the plateau phase of the sexual response cycle.

Sex skin The reddening of the labia minora that occurs during the plateau phase.

FIGURE 5.5

The Female Genitals During the Phases of the Sexual Response Cycle.

1. EXCITEMENT PHASE

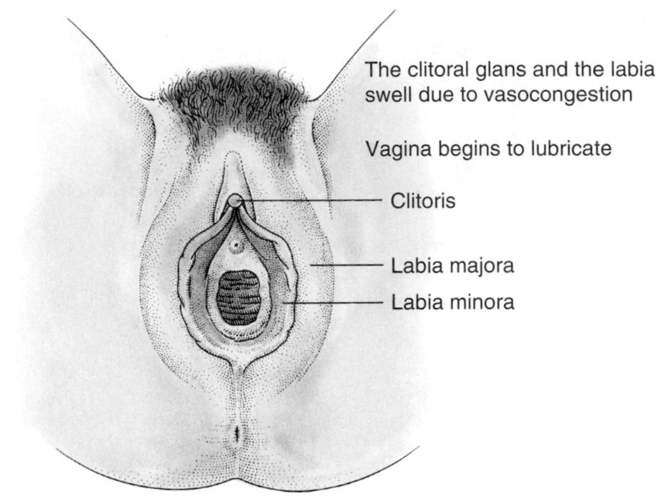
The clitoral glans and the labia swell due to vasocongestion
Vagina begins to lubricate
Clitoris
Labia majora
Labia minora

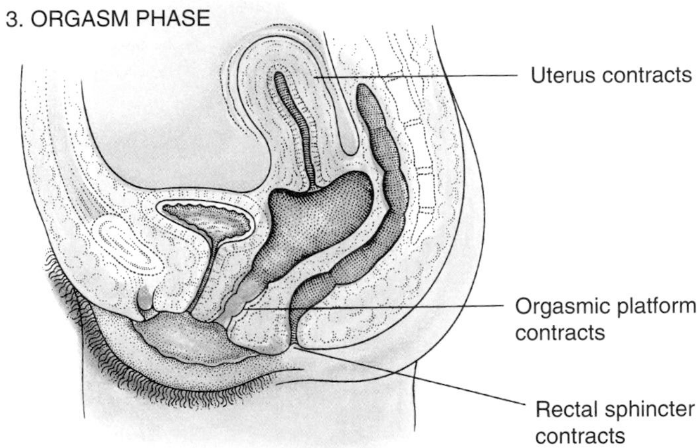
3. ORGASM PHASE
Uterus contracts
Orgasmic platform contracts
Rectal sphincter contracts

who have not. Further engorgement of the areolas of the breasts may make it seem that the nipples have lost part of their erection (see Figure 5.6). Bartholin's glands secrete a fluid that resembles mucus.

About one man in four, and about three women in four, show a sex flush, which often does not appear until the plateau phase. Myotonia may cause facial grimaces and spasmodic contractions in the hands and feet. Breathing becomes rapid, like panting, and the heart rate may increase to 100 to 160 beats per minute. Blood pressure continues to rise. The increase in heart rate is usually less dramatic with masturbation than during coitus.

Orgasmic Phase

The orgasmic phase in the male consists of two stages of muscular contractions. In the first stage, contractions of the vas deferens, the seminal vesicles, the ejaculatory duct, and the prostate gland cause seminal fluid to collect in the urethral bulb at the base of the penis (see Figure 5.4). The bulb expands to accommodate the fluid. The internal sphincter of the urinary bladder contracts, preventing seminal fluid from entering the bladder in a backward, retrograde ejaculation. The normal closing off of the bladder also serves to prevent urine from mixing with semen. The collection of semen in the urethral bulb produces feelings of ejaculatory inevitability—the sensation that nothing will stop the ejaculate from "coming." This sensation lasts for about 2 to 3 seconds.

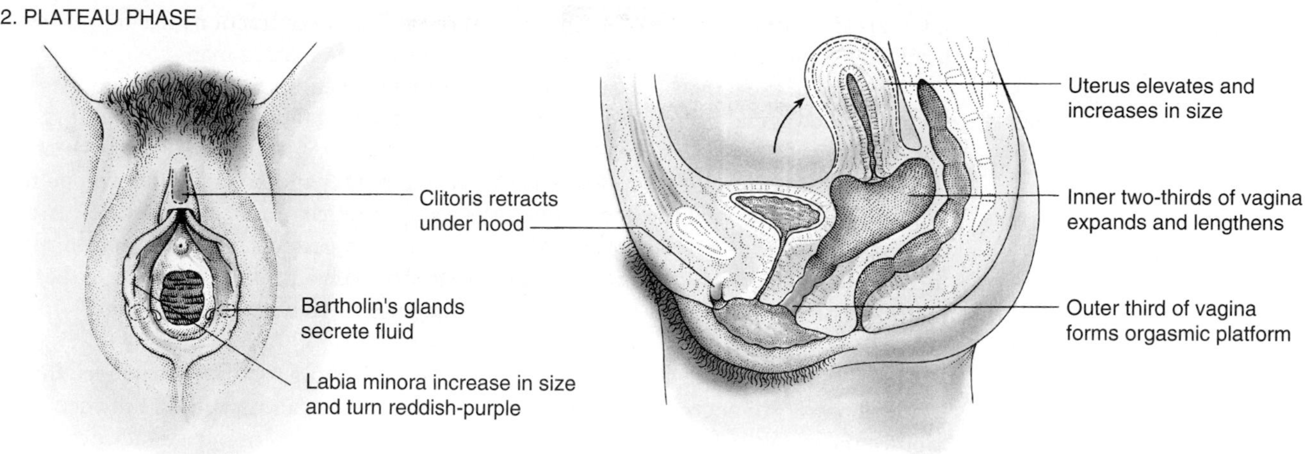
2. PLATEAU PHASE
Clitoris retracts under hood
Bartholin's glands secrete fluid
Labia minora increase in size and turn reddish-purple
Uterus elevates and increases in size
Inner two-thirds of vagina expands and lengthens
Outer third of vagina forms orgasmic platform

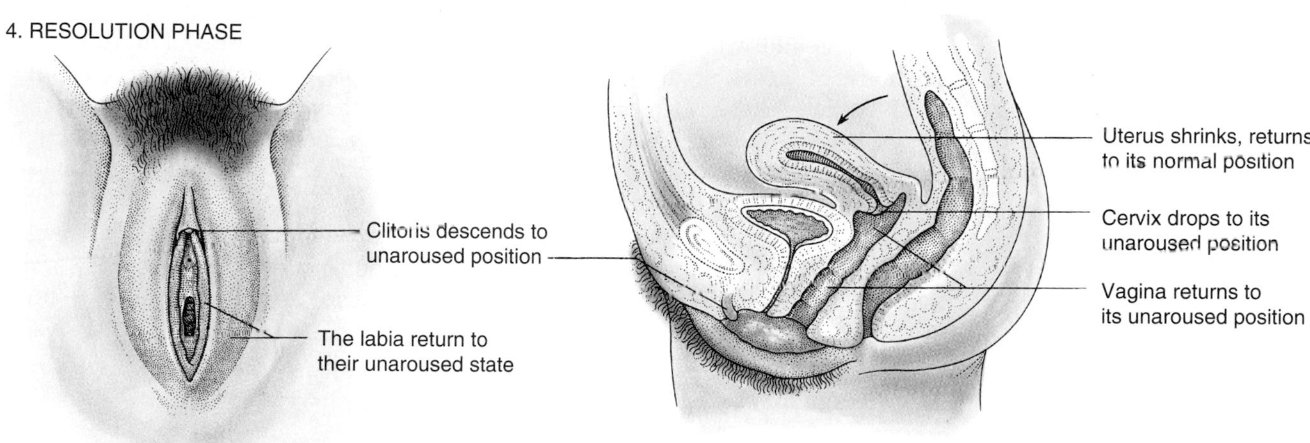
4. RESOLUTION PHASE
Clitoris descends to unaroused position
The labia return to their unaroused state
Uterus shrinks, returns to its normal position
Cervix drops to its unaroused position
Vagina returns to its unaroused position

In the second stage, the external sphincter of the bladder relaxes, allowing the passage of semen. Contractions of muscles surrounding the urethra and urethral bulb and the base of the penis propel the ejaculate through the urethra and out of the body. Sensations of pleasure tend to be related to the strength of the contractions and the amount of seminal fluid. The first 3 to 4 contractions are generally most intense and occur at 0.8-second

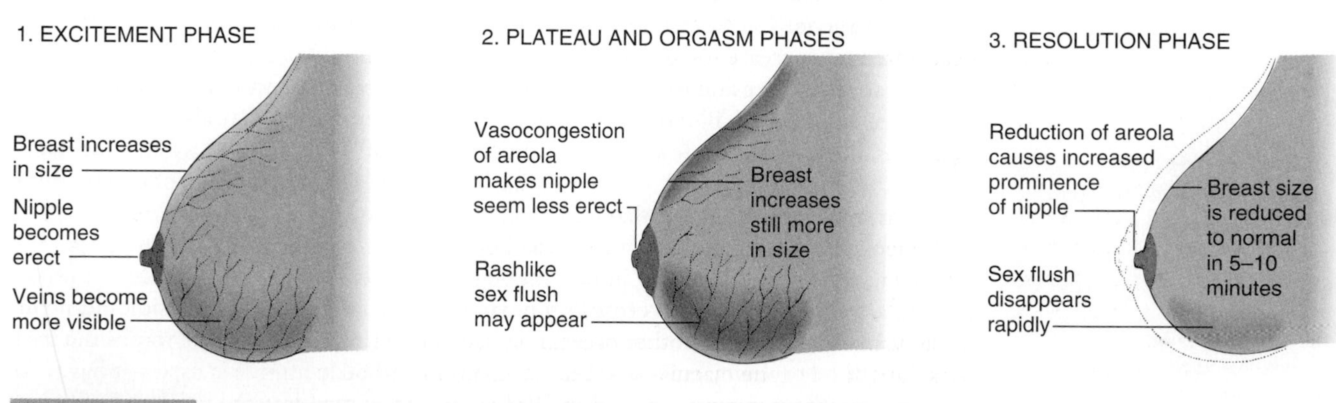
1. EXCITEMENT PHASE
Breast increases in size
Nipple becomes erect
Veins become more visible
2. PLATEAU AND ORGASM PHASES
Vasocongestion of areola makes nipple seem less erect
Breast increases still more in size
Rashlike sex flush may appear
3. RESOLUTION PHASE
Reduction of areola causes increased prominence of nipple
Breast size is reduced to normal in 5–10 minutes
Sex flush disappears rapidly

FIGURE 5.6

The Breasts During the Phases of the Sexual Response Cycle.

intervals (5 contractions every 4 seconds). Another 2 to 4 contractions occur at a somewhat slower pace. Rates and patterns vary somewhat from man to man.

Orgasm in the female is manifested by 3 to 15 contractions of the pelvic muscles that surround the vaginal barrel. The contractions first occur at 0.8-second intervals, producing, as in the male, a release of sexual tension. Another 3 to 6 weaker and slower contractions follow. The spacing of these contractions is generally more variable in women than in men. The uterus and the anal sphincter also contract rhythmically. Uterine contractions occur in waves from the top to the cervix. In both genders, muscles go into spasm throughout the body. Blood pressure and heart rate reach a peak, with the heart beating up to 180 times per minute. Respiration may increase to 40 breaths per minute.

SUBJECTIVE EXPERIENCE OF ORGASM The sensations of orgasm have challenged the descriptive powers of poets. Words like *rush*, *warmth*, *explosion*, and *release* do not adequately capture them. We may assume (rightly or wrongly) that others of our gender experience pretty much what we do, but can we understand the sensations of the other gender?

Several studies suggest that the orgasms of both genders may feel quite similar. In one, 48 men and women provided written descriptions of orgasms. The researchers (Proctor et al., 1974) modified the language (for example, changing *penis* to *genitals*) so that the authors' genders would not be apparent. They then asked 70 "experts" (psychologists, gynecologists, etc.) to indicate the gender of each author. The ratings were no more reliable than guesswork.

Truth or Fiction? **REVISITED**

It is true that written descriptions of men's and women's experiences during orgasm cannot be differentiated—at least when they are altered to exclude language that gives away exactly which anatomic features are involved. Thus, this Truth or Fiction item is only qualifiedly "true." (Hey, life is complex. Be tolerant.)

Resolution Phase

The period following orgasm, in which the body returns to its prearoused state, is called the **resolution phase.** Following ejaculation, the man loses his erection in two stages. The first occurs in about a minute. Half the volume of the erection is lost, as blood from the corpora cavernosa empties into the other parts of the body. The second stage occurs over a period of several minutes: The remaining tumescence subsides as the corpus spongiosum empties. The testes and scrotum return to normal size, and the scrotum regains its wrinkled appearance.

In women, orgasm also triggers release of blood from engorged areas. In the absence of continued stimulation, swelling of the areolas decreases; then the nipples return to normal size. The sex flush lightens rapidly. In about 5 to 10 seconds, the clitoris descends to its normal position. The clitoris, vaginal barrel, uterus, and labia gradually shrink to their prearoused sizes. The labia minora turn lighter (the "sex skin" disappears) in about 10 to 15 seconds.

Most muscle tension (myotonia) tends to dissipate within 5 minutes after orgasm in both men and women. Blood pressure, heart rate, and respiration may also return to their prearousal levels within a few minutes. About 30% to 40% of men and women find their palms, the soles of their feet, or their entire bodies covered with a sheen of perspiration. Both men and women may feel relaxed and satiated. However, . . .

Although the processes by which the body returns to its prearousal state are similar in men and women, there is an important gender difference during the resolution phase. Unlike women, men enter a **refractory period** during which they are physiologically incapable of experiencing another orgasm or ejaculation (in much the same way as the flash attachment to a camera cannot be set off again immediately after it is used—it has to be recharged). The refractory period of adolescent males may last only minutes, whereas that of men age 50 and above may last from several minutes (yes, "it could happen") to a day. Women do not undergo a refractory period and so can become quickly rearoused to

Resolution phase The fourth phase of the sexual response cycle, during which the body gradually returns to its prearoused state.

Refractory period A period of time following a response (e.g., orgasm) during which an individual is no longer responsive to stimulation (e.g., sexual stimulation).

the point of repeated (multiple) orgasm if they desire and receive continued sexual stimulation (see Figure 5.3).

Myotonia and vasocongestion may take an hour or more to dissipate in people who are aroused who do not reach orgasm. Persistent pelvic vasocongestion may cause "blue balls" in males—the slang term for a throbbing ache. Some men insist that their dates should consent to coitus, on the theory that it is unfair to decline after stimulating them to the point where they have this condition. This ache can be relieved through masturbation as well as coitus, however, or can be allowed to dissipate naturally. Although it may be uncomfortable, it is not dangerous and is no excuse to pressure or coerce another person into any sexual activity. "Blue" sensations are not limited to men. Women, too, may experience unpleasant pelvic throbbing if they have become highly aroused and do not find release (Barbach, 1975). Women, too, can relieve pelvic throbbing through masturbation.

Kaplan's Three Stages of Sexual Response: An Alternative Model

Helen Singer Kaplan was a prominent sex therapist and author of several professional books (1974, 1979, 1987) on sex therapy. Whereas Masters and Johnson had proposed a four-stage model of sexual response, Kaplan developed a three-stage model consisting of (1) desire, (2) excitement, and (3) orgasm. Kaplan's model is an outgrowth of her clinical experience in working with people with sexual dysfunctions. She believes that their problems can best be classified according to these three phases. Kaplan's model makes it convenient for clinicians to classify sexual dysfunctions involving desire (low or absent desire), excitement (such as problems with erection in the male or lubrication in the female), and orgasm (such as premature ejaculation in the male or orgasmic dysfunction in the female).

Masters and Johnson focus on physiological changes that occur during sexual stimulation. Kaplan's model includes two phases that are primarily physiological (*excitement*, consisting of initial vasocongestion of the genitals, resulting in erection in the male and vaginal lubrication in the female, and *orgasm*, marked by pelvic muscular contractions) and one phase that is primarily psychological (desire). Kaplan's stages of excitement and orgasm are clearly differentiated along physiological lines. (Masters and Johnson's excitement and plateau stages are less clearly differentiated in terms of differences in biological processes. Both primarily involve vasocongestion.)

Masters and Johnson view sexual response as composed of *successive* stages; the order is crucial and invariant. Kaplan treats her phases as relatively independent components of sexual response whose sequence is somewhat variable. For example, a person may experience sexual excitement and even orgasm, although sexual desire remains low. Excitement may also precede desire in some situations. For example, people with low sexual desire may find their sexual appetites sparked as their bodies respond to their partners' sexual stimulation. A person who lacks desire may not be motivated to seek sexual stimulation or may not be able to respond adequately to sexual stimuli, however.

Kaplan's model is noteworthy for designating desire as a separate phase of sexual response. Problems in lack of sexual interest or desire are among those most commonly brought to the attention of sex therapists.

CONTROVERSIES ABOUT ORGASM

Are women capable of experiencing multiple orgasms? Are men? Physiologically speaking, is there but one type of orgasm? Or are there different types of orgasms, depending on the site of stimulation? Do women ejaculate during orgasm? If so, what fluid do they emit?

Few other topics in human sexuality have aroused more controversies over the years than orgasm. We do not have all the answers, but some intriguing research findings have shed light on some of these continuing controversies.

Multiple Orgasms

Kinsey's report (Kinsey et al., 1953) that 14% of his female respondents regularly had **multiple orgasms** sent shock waves through the general community and even surprised his fellow scientists. Many people were aghast that women could have more than one orgasm in succession. There were comments (mostly by men, of course!) that the women in the Kinsey surveys must be "nymphomaniacs" who were incapable of being satisfied with the "normal" complement of one orgasm per occasion. However, only 13 years later, Masters and Johnson (1966) reported that most if not all women are capable of multiple orgasms. Though all women may have a biological capability for multiple orgasm, not all women report them. A recent survey of 720 nurses showed that only 43% reported experiencing multiple orgasms (Darling et al., 1991).

It is difficult to offer a precise definition of multiple orgasm. In Masters and Johnson's view, multiple orgasm involves the occurrence of one or more *additional* orgasms following the first, within a short period of time and before the body has returned to a preplateau level of arousal. By this definition, a person would not have experienced a multiple orgasm if he or she had two or more orgasms that were separated by a return to a prearoused state or a preplateau (excitement stage) level of arousal. (Note that the pattern shown by the broken line for the male in Figure 5.3 does *not* constitute a multiple orgasm, even if it occurs reasonably rapidly after the first orgasm, because he *does* return to a preplateau level of sexual arousal between orgasms.) The lines of demarcation between the excitement and plateau stages of arousal are not too obvious, however. Therefore, a person may experience two or more successive orgasms within a short time but not know whether these are, technically speaking, "multiple orgasms." Whether or not the orgasm fits the definition has no effect on the experience, but it does raise the question of whether both men and women are capable of multiple orgasm.

By Masters and Johnson's definition, men are not capable of achieving multiple orgasms, because they enter a refractory period following ejaculation during which they are physiologically incapable of achieving another orgasm or ejaculation. Put more simply, men who want more than one orgasm during one session may have to relax for a while and allow their sexual arousal to subside. Yet women can maintain a high level of arousal between multiple orgasms and have them in rapid succession.

Women do not enter a refractory period. Women can continue to have orgasms if they continue to receive effective stimulation (and, of course, are interested in continuing). Some men thus refrain from reaching orgasm until their partners have had the desired number. This differential capacity for multiple orgasms is one of the major gender differences in sexual response.

Some men have two or more orgasms without ejaculation ("dry orgasms") preceding a final ejaculatory orgasm. These men may not enter a refractory period following their initial dry orgasms and may therefore be able to maintain their level of stimulation at near-peak levels. Some men report multiple orgasms in which dry orgasms follow an ejaculatory orgasm, with little or no loss of erection between orgasms (Dunn & Trost, 1989). Some men report more varied patterns, with ejaculatory orgasms and dry orgasms preceding or following each other in different sequences.

Men who report multiple orgasms indicate that if they are highly aroused following an initial orgasm, and if sexual stimulation continues, they can achieve one or more subsequent orgasms before losing their erections (Dunn & Trost, 1989). Still, the evidence for multiple orgasms in men is largely anecdotal and limited to a relatively few case examples. Nor is it known whether such "multiple" orgasms meet the technical definition given by Masters and Johnson of two or more orgasms in rapid succession without a return to a *preplateau* stage of arousal in between. Nor do we know what percentage of men might be multiply orgasmic (Dunn & Trost, 1989).

Multiple orgasms One or more additional orgasms following the first, which occur within a short period of time and before the body has returned to a preplateau level of arousal.

Masters and Johnson found that some women experienced 20 or more orgasms by masturbating. Still, few women have multiple orgasms during most sexual encounters, and many are satisfied with just one per occasion. Some women who have read or heard about female orgasmic capacity wonder what is "wrong" with them if they are content with just one. Nothing is wrong with them, of course: A biological capacity does not create a behavioral requirement.

How Many Kinds of Orgasms Do Women Have? One, Two, or Three?

Until Masters and Johnson published their laboratory findings, many people believed that there were two types of female orgasms, as proposed by the psychoanalyst Sigmund Freud: the *clitoral orgasm* and the *vaginal orgasm.* Clitoral orgasms were achieved through direct clitoral stimulation, such as by masturbation. Clitoral orgasms were seen by psychoanalysts (mostly male psychoanalysts, naturally) as emblematic of a childhood fixation—a throwback to an erogenous pattern acquired during childhood masturbation.

The term *vaginal orgasm* referred to an orgasm achieved through deep penile thrusting during coitus and was theorized to be a sign of mature sexuality. Freud argued that women achieve sexual maturity when they forsake clitoral stimulation for vaginal stimulation. This view would be little more than an academic footnote except for the fact that some adult women who continue to require direct clitoral stimulation to reach orgasm, even during coitus, have been led by traditional (generally male) psychoanalysts to believe that they are sexually "fixated" at an immature stage or, at least, are sexually inadequate.

Despite Freudian theory, Masters and Johnson (1966) were able to find only one kind of orgasm, physiologically speaking, regardless of the source of stimulation (manual–clitoral or penile–vaginal). By monitoring physiological responses to sexual stimulation, they found that the female orgasm involves the same biological events whether it is reached through masturbation, petting, coitus, or just breast stimulation. All orgasms in volve spasmodic contractions of the pelvic muscles surrounding the vaginal barrel, leading to a release of sexual tension. Getrude Stein wrote, "A rose is a rose is a rose." Biologically speaking, the same principle can be applied to orgasm: "An orgasm is an orgasm is an orgasm." In men, it also matters not how orgasm is achieved—through masturbation, petting, oral sex, coitus, or fantasizing about a fellow student in chem lab. Orgasm still involves the same physiological processes: Involuntary contractions of the pelvic muscles at the base of the penis expel semen and release sexual tension. A woman or a man might prefer one source of orgasm to another (she or he might prefer achieving orgasm with a lover rather than by masturbation, or with one person rather than another), but the biological events that define orgasm remain the same.

Although orgasms attained through coitus or masturbation may be physiologically alike, there are certainly key psychological or subjective differences. (Were it not so, there would be fewer sexual relationships.) The coital experience, for example, is often accompanied by feelings of attachment, love, and connectedness to one's partner. Masturbation, by contrast, is more likely to be experienced solely as a sexual release.

Orgasms experienced through different means may also vary in physiological and subjective intensity. Masters and Johnson (1966) found that orgasms experienced during masturbation were generally more physiologically intense than those experienced during intercourse, perhaps because masturbation allows one to focus only on one's own pleasure and on ensuring that one receives effective stimulation to climax. This does not mean that orgasms during masturbation are more enjoyable or gratifying than those experienced through coitus, however. Given the emotional connectedness we may feel toward our lovers, we are unlikely to break off our relationships in favor of masturbation. Thus, "physiological intensity," as measured by laboratory instruments, does not translate directly into subjective pleasure or fulfillment.

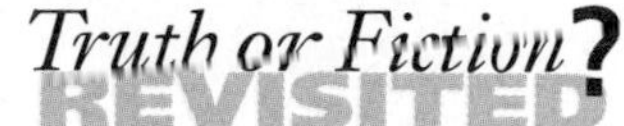

Actually, orgasms attained through masturbation appear to be more "intense" than those attained through coitus, at least in terms of physiological measurements. (This does not mean that they are more enjoyable.) ■

The purported distinction between clitoral and vaginal orgasms also rests on an assumption that the clitoris is not stimulated during coitus. Masters and Johnson showed this to be a *false* assumption. Penile coital thrusting draws the clitoral hood back and forth against the clitoris. Vaginal pressure also heightens blood flow in the clitoris, further setting the stage for orgasm (Lavoisier et al., 1995).

One might think that Masters and Johnson's research settled the question of whether there are different types of female orgasm. Other investigators, however, have proposed that there *are* distinct forms of female orgasm, but not those suggested by psychoanalytic theory. For example, Singer and Singer (1972) suggested that there are three types of female orgasm: *vulval, uterine*, and *blended.* According to the Singers, the vulval orgasm represents the type of orgasm described by Masters and Johnson (1966) that involves *vulval* contractions; that is, contractions of the vaginal barrel. Consistent with the findings of Masters and Johnson (1966), they note that the vulval orgasm remains the same regardless of the source of stimulation, clitoral or vaginal.

According to the Singers, the uterine orgasm does not involve vulval contractions. It occurs only in response to deep penile thrusting against the cervix. This thrusting slightly displaces the uterus and stimulates the tissues that cover the abdominal organs. The uterine orgasm is accompanied by a certain pattern of breathing: Gasping or gulping of air is followed by an involuntary holding of the breath as orgasm approaches. When orgasm is reached, the breath is explosively exhaled. The uterine orgasm is accompanied by deep feelings of relaxation and sexual satisfaction.

The third type, or blended orgasm, is described as combining features of the vulval and uterine orgasms. It involves both an involuntary breath-holding response and contractions of the pelvic muscles. The Singers note that the type of orgasm a woman experiences depends on factors such as the parts of the body that are stimulated and the duration of stimulation. Each produces its own kind of satisfaction, and no one type is necessarily better than or preferable to any other.

The Singers' hypothesis of three distinct forms of female orgasm remains controversial. Researchers initially scoffed at the idea that orgasms could arise from vaginal stimulation alone. The vagina, after all, especially the inner two thirds of the vaginal cavity, is relatively insensitive to stimulation (erotic or otherwise). Proponents of the Singers' model counter that the type of uterine orgasm described by the Singers is induced more by pressure resulting from deep pelvic thrusting than by touch.

The G-Spot

A part of the vagina, notably a bean-shaped area within the anterior wall, may have special erotic significance. This area is believed to lie about 1 to 2 inches from the vaginal entrance and to consist of a soft mass of tissue that swells from the size of a dime to that of a half-dollar when stimulated (Davidson et al., 1989). It has been called the **Grafenberg spot**—the "G-spot" for short (see Figure 5.7). This spot can be directly stimulated by the woman's or her partner's fingers or by penile thrusting in the rear-entry and the female-superior positions. Some researchers suggest that stimulation of the spot produces intense erotic sensations and that with prolonged stimulation, a distinct form of orgasm occurs. This orgasm is characterized by intense pleasure and, in some cases, by a biological event earlier thought to be exclusively male in nature: ejaculation (Perry & Whipple, 1981; Whipple & Komisaruk, 1988). These claims, like other claims of distinct forms of female orgasm, have been steeped in controversy.

The G-spot was named after a gynecologist, Ernest Grafenberg, who first suggested the erotic importance of this area. Grafenberg observed that orgasm in women could be induced by stimulating this area. He also claimed that such orgasms may be accompanied by the discharge of a milky fluid, or "ejaculate," from the urethra. In a laboratory experiment, Zaviacic and his colleagues (1988a, 1988b) found evidence of an ejaculate in 10 of 27 women studied. Some researchers believe that this fluid is urine that some women release involuntarily during orgasm (Alzate, 1985). Other researchers believe that it differs from urine (Zaviacic & Whipple, 1993). The nature of this fluid and its source remain controversial, but Zaviacic and Whipple (1993)suggest that the fluid may be released during sex by a "female prostate," a system of ducts and glands called *Skene's glands*, in much

Grafenberg spot A part of the anterior wall of the vagina, whose prolonged stimulation is theorized to cause particularly intense orgasms and a female ejaculation. Abbreviated *G-spot.*

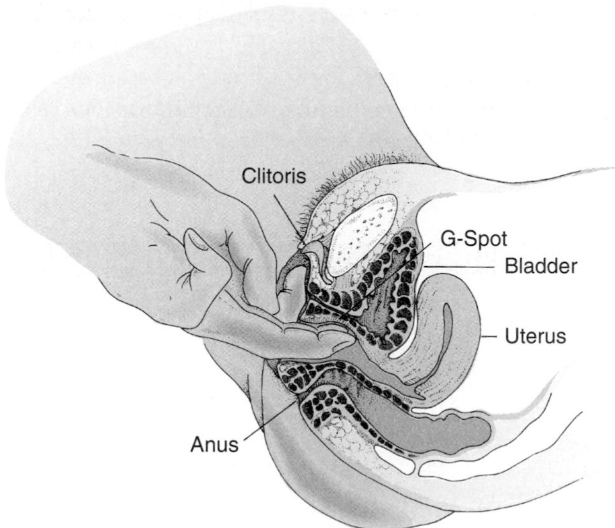

FIGURE 5.7

The Grafenberg Spot. It is theorized that the "G-spot" can be stimulated by fingers or by intercourse in the rear-entry or the female-superior positions. Does stimulation of the G-spot produce intense erotic sensations and a distinct form of orgasm?

the same way as semen is released by the prostate gland in men. Zaviacic and Whipple suggest that "many women who felt that they may be urinating during sex . . . [may be helped by] the knowledge that the fluid they expel may be different from urine and a normal phenomenon that occurs during sexual response" (1993, p. 149). Some women, however, may actually expel urine during sex, perhaps because of urinary stress incontinence (Zaviacic & Whipple, 1993). Zaviacic and Whipple also note that stimulation of the G-spot may be related to ejaculation in some women but not in others.

Even proponents of the existence of the G-spot recognize that it is difficult to locate, because it is not apparent to the eye (Ladas et al., 1982). Perry and Whipple (1981) suggest that women may try to locate the spot either by self-exploration or with the assistance of a partner. In either case, two fingers should be used to press deeply but gently into the front, or anterior, wall of the vagina to locate the spot, which may feel like a small lump within the anterior wall (see Figure 5.7). When the spot is stimulated by stroking, the woman may initially experience an urge to urinate, perhaps because the sensitive area lies close to the bladder and urethra. A few minutes of continued stimulation lead to strong sensations of sexual pleasure in some women, which is accompanied by vasocongestion that swells the area. More prolonged stimulation may lead to an intense orgasm; however, fear of loss of urinary control leads some women to avoid such prolonged stimulation (Ladas et al., 1982).

Perry and Whipple (1981) relate the G-spot to the Singers' model of three kinds of orgasms. They suggest that sustained stimulation of the G-spot can produce a uterine orgasm that is characterized by deeper sensations than those that occur during the vulval orgasm produced by clitoral stimulation. The connection between the G-spot and the Singers' model is controversial.

The very existence of the G-spot remains debatable. Ladas, Whipple, and Perry (1982) reported locating the G-spot in every one of more than 400 women they examined. Zaviacic and his colleagues (1988a&b) reported finding the spot in each of 27 women they examined. However, other researchers have been unable to find an area of heightened sensitivity corresponding to the G-spot (Alzate & Londono, 1984; Masters et al., 1989). Some researchers (e.g., Hock, 1983) deny the existence of the G-spot as a distinct anatomic structure. They argue that the entire anterior wall of the vagina, not just any one spot or area, is richly supplied with nerve endings and sensitive to erotic stimulation.

Although the existence of the G-spot continues to be debated among researchers, a recent survey of 1,289 professional women in the health and counseling professions revealed that a majority believe that the G-spot exists and that they have experienced sexual pleasure when it has been stimulated (Davidson et al., 1989). Still, there was considerable confusion among these women as to the precise location of this sensitive area. About three out of four women reported experiencing an orgasm from stimulation of this area, most frequently from manual stimulation.

More research is needed to determine the scientific basis of the claims for different kinds of orgasms in women and whether there are specific sites in the vagina, such as the G-spot, that may be especially sensitive to erotic stimulation.

Human Sexuality ONLINE //

Web Sites Related to Sexual Arousal and Response

This commercial Web site sells romantic gifts and lists of romantic "turn ons."
www.lovingyou.com/

Johan's Guide to Aphrodisiacs: Come to this Web site for interesting facts about who believes what, not for scientific confirmation. Everything you wanted to know, and more, about supposed aphrodisiacs from snakes' blood to animal genitals.
www.santesson.com/aphrodis/aphrhome.htm

The U.S. Food and Drug Administration: The Web site of the FDA offers "Facts About Aphrodisiacs"—or should we say "supposed aphrodisiacs?" A decent review of what is known about substances with supposed sexual powers.
www.fda.gov/fdac/features/196_love.html

This commercial Web site sells perfumes and other products laced with suspected pheromones. Be warned that the evidence is skewed toward encouraging purchases. You might as well look here to see the type of advertisements you will find. There is a list of references, but some are old, and they do not present an overall balanced view.
www.uneet.com/

Here is a Web site for foods—"A to Z"—that have supposed aphrodisiac effects. Lots of fun, but don't take it to heart (or to the supermarket). Entries under M include M&Ms and mangoes.
www.cookingcouple.com/atoz/

This Web site offers information on a testosterone gel that is rubbed into the skin to treat testosterone deficiency. Do *not* use such preparations without consulting a physician.
www.pharminfo.com/pubs/druginfoline/druginfo2_411p.html

This Web site has information about checking one's levels of estrogen and progesterone. Again, do *not* use such devices without consulting a physician.
store.yahoo.com/wdxcyberstore/eschec.html

summing up

MAKING SENSE OF SEX: THE ROLE OF THE SENSES IN SEXUAL AROUSAL

Each sense plays a role in sexual experience, but some play more of a role than others.

Vision: The Better to See You With
Visual information plays a major role in human sexual attraction. Visual cues can be sexual turn-ons or turn-offs.

Smell: Does the Nose Know Best?
Although the sense of smell plays a lesser role in governing sexual arousal in humans than in lower mammals, particular odors can be sexual turn-ons or turn-offs. Many organisms are sexually aroused by naturally produced chemicals called pheromones, but their role in human sexual behavior remains unclear.

The Skin Senses: Sex as a Touching Experience
The sense of touch has the most direct effects on sexual arousal and response. Erogenous zones are especially sensitive to tactile sexual stimulation.

Taste: On Savory Sex
Taste appears to play only a minor role in sexual arousal and response.

Hearing: The Better to Hear You With
Like visual and olfactory cues, sounds can be turn-ons or turn-offs.

APHRODISIACS: OF SPANISH FLIES AND RHINO HORNS

Alleged aphrodisiacs such as Spanish fly and foods that in some way resemble the genitals have not been shown to contribute to sexual arousal or response.

Psychoactive Drugs
The alleged aphrodisiac effects of psychoactive drugs, such as alcohol and cocaine, may reflect our expectations of them or their effects on sexual inhibitions, rather than direct stimulation of sexual response. Alcohol is also connected with a liberated social role and thus provides an external excuse for dubious behavior. Some people report increased sexual pleasure with an initial cocaine use, but frequent use can lead to sexual dysfunctions.

SEXUAL RESPONSE AND THE BRAIN: CEREBRAL SEX?

The brain plays a central role in sexual functioning.

■ **The Geography of the Brain**

The brain consists of three major parts: the hindbrain, the midbrain, and the forebrain.

■ **Brain Mechanisms in Sexual Functioning**

The cerebral cortex interprets sensory information as sexual turn-ons or turn-offs. The cortex transmits messages through the spinal cord that cause vasocongestion. Direct stimulation of parts of the limbic system may cause erection and ejaculation in male animals.

■ **On Pushing the Right Buttons: Are There Pleasure Centers in the Brain?**

Electrical stimulation of certain parts of the limbic system apparently yields sensations similar to those of sexual gratification.

SEX HORMONES: DO THEY "GOAD" US INTO SEX?

Sex hormones have organizing and activating effects on behavior. Both men and women produce one genuine aphrodisiac: testosterone. Female sex hormones do not appear to play a direct role in determining sexual motivation or response in human females. Yet levels of testosterone in the bloodstream have been associated with sexual interest in women.

THE SEXUAL RESPONSE CYCLE

Masters and Johnson found that the physiological responses of men and women to sexual stimulation are quite alike. Both experience vasocongestion and myotonia early in the response cycle.

■ **Excitement Phase**

Sexual excitement is characterized by erection in the male and vaginal lubrication in the female.

■ **Plateau Phase**

The plateau phase is an advanced state of arousal that precedes orgasm.

■ **Orgasmic Phase**

The third phase of the sexual response cycle is characterized by orgasmic contractions of the pelvic musculature. Orgasm in the male occurs in two stages of muscular contractions. Orgasm in the female is manifested by contractions of the pelvic muscles that surround the vaginal barrel.

■ **Resolution Phase**

During the resolution phase, the body returns to its prearoused state.

■ **Kaplan's Three Stages of Sexual Response: An Alternative Model**

Kaplan developed a three-stage model of sexual response consisting of desire, excitement, and orgasm. Kaplan's model makes it more convenient for clinicians to classify and treat sexual dysfunctions.

CONTROVERSIES ABOUT ORGASM

■ **Multiple Orgasms**

Multiple orgasm is the occurrence of one or more additional orgasms following the first, within a short period of time and before the body has returned to a preplateau level of arousal. Most women, but not most men, are capable of multiple orgasms.

■ **How Many Kinds of Orgasms Do Women Have? One, Two, or Three?**

Freud theorized the existence of two types of orgasms in women: clitoral and vaginal. Masters and Johnson found only one kind of orgasm among women. Singer and Singer suggested that there are three types of female orgasms: vulval, uterine, and blended.

■ **The G-Spot**

The G-spot—an allegedly distinct area of the vagina within the anterior wall—may have special erotic significance. Some researchers suggest that prolonged stimulation of this spot produces an orgasm that is characterized by intense pleasure and, in some women, by a type of ejaculation. The nature of this ejaculate remains in doubt.

questions for critical thinking

1. Do you find photos of nudes to be sexually arousing? Why or why not?
2. Do you find perfumes or colognes to enhance the attractiveness of other people? Why or why not?
3. What are the effects of alcohol on sexual arousal? Do the scientific facts contradict any beliefs you had held about alcohol?
4. Have you done anything under the influence of alcohol or other drugs that you otherwise would not have done? What role did the drug play?
5. If you could electrically stimulate a pleasure center in your brain and find as much pleasure as you do in sexual activity with another person, do you think you would bother to form romantic relationships? Why or why not?
6. Do you know people who have taken testosterone or estrogen for medical or other reasons? What were their reasons for using hormones? What were the effects?
7. Were you surprised to learn that the sensations of orgasm are apparently quite similar for women and men? Why or why not?

chapter

6

Gender Identity and Gender Roles

Pablo Picasso. *Harlequin at the Mirror.* 1923. Oil on canvas, 100 × 181 cm. © Copyright ARS, NY. Fundacion Coleccion Thyssen-Bornemisza, Madrid, Spain. Credit: Nimatallah/Art Resource, NY. Credit: © 2002 Estate of Pablo Picasso/Artists Rights Society (ARS), New York.

outline

Truth or Fiction?

PRENATAL SEXUAL DIFFERENTIATION

Genetic Factors in Sexual Differentiation

The Role of Sex Hormones in Sexual Differentiation

Descent of the Testes and the Ovaries

Sex-Chromosomal Abnormalities

Prenatal Sexual Differentiation of the Brain

GENDER IDENTITY

Nature and Nurture in Gender Identity

Hermaphroditism

Transsexualism

A Closer Look Boys Who Are Reared as Girls

A World of Diversity Using the Back Door in Iran: Can a Transsexual Change Her Mind?

A Closer Look Jayne Thomas Ph.D.—In Her Own Words

GENDER ROLES AND STEREOTYPES

SEXISM

GENDER DIFFERENCES: *Vive la Différence* or *Vive la Similarité*?

Differences in Cognitive Abilities

Differences in Personality

A World of Diversity Men and Women Iron Their Way to Domestic Equality

ON BECOMING A MAN OR A WOMAN: GENDER TYPING

Biological Perspectives

Cross-Cultural Perspectives

Psychological Perspectives

GENDER ROLES AND SEXUAL BEHAVIOR

Men as Sexually Aggressive, Women as Sexually Passive

Men as Overaroused, Women as Underaroused

PSYCHOLOGICAL ANDROGYNY: THE MORE TRAITS, THE MERRIER?

Psychological Androgyny, Psychological Well-Being, and Personal Development

Psychological Androgyny and Sexual Behavior

Human Sexuality Online Web Sites Related to Gender Identity and Gender Roles

SUMMING UP

QUESTIONS FOR CRITICAL THINKING

Truth or Fiction?

____ If male sex hormones were not present during critical stages of prenatal development, we would all develop female sexual organs.

____ The sex of a baby crocodile is determined by the temperature at which the egg develops.

____ Thousands of people have changed their genders through gender-reassignment surgery.

____ Men act more aggressively than women do.

____ A 2½-year-old child may know that he is a boy but think that he can grow up to be a mommy.

____ Adolescent girls who show a number of masculine traits are more popular than girls who thoroughly adopt the traditional feminine gender role.

We're halfway there. We've begun to raise our daughters more like sons—so now women are whole people. But fewer of us have the courage to raise our sons more like daughters. Yet until men raise children as much as women do—and are raised to raise children, whether or not they become fathers—they will have a far harder time developing in themselves those human qualities that are wrongly called "feminine," but are really those necessary to raise children: empathy, flexibility, patience, compassion, and the ability to let go.
—Gloria Steinem, graduation speech at Smith College, 1995

Whatever women do they must do twice as well as men to be thought half as good. Luckily, this is not difficult.
—Charlotte Whitton

I like men to behave like men—strong and childish.
—French author Françoise Sagan

These remarks from war correspondents in the battle of the genders signify key issues in the study of gender: gender roles, the actual differences between the genders, and the enduring problem of sexism. This chapter addresses the biological, psychological, and sociological aspects of gender. First we focus on sexual differentiation—the process by which males and females develop distinct reproductive anatomy. We then turn to gender roles—the complex behavior patterns that are deemed "masculine" or "feminine" in a particular culture. The chapter examines empirical findings on actual gender differences, which may challenge some of the preconceptions that many of us have about the differences between men and women. We next consider gender typing—the processes by which boys come to behave in line with what is expected of men (most of the time), and girls with what is expected of women (most of the time). We shall also explore the concept of psychological androgyny, which applies to people who display characteristics associated with both genders.

PRENATAL SEXUAL DIFFERENTIATION

Over the years many ideas have been proposed to account for **sexual differentiation.** Aristotle believed that the anatomical difference between males and females was due to the heat of semen at the time of sexual relations. Hot semen generated males, whereas cold semen made females[1] (National Center for Biotechnology Information, 2000). Others believed that sperm from the right testicle made females, but sperm from the left testicle made males. If this were so, males would indeed be sinister, because *sinister* means "left hand" or "unlucky side" in Latin.

According to the Bible, Adam was created first, and Eve issued forth from one of his ribs. From the standpoint of modern biological knowledge, however, it would be more accurate to say that "Adams" (that is, males) develop from "Eves" (females). Let us trace the development of sexual differentiation.

When a sperm cell fertilizes an ovum, 23 **chromosomes** from the male parent normally combine with 23 chromosomes from the female parent. The **zygote,** the beginning

[1]Just a wee bit of stereotyping here—the hot-blooded male and the frigid female—notes the third author. Hopefully, we have moved beyond.

Sexual differentiation The process by which males and females develop distinct reproductive anatomy.

Chromosome One of the rodlike structures, found in the nucleus of every living cell, that carry the genetic code in the form of genes.

Zygote A fertilized ovum (egg cell).

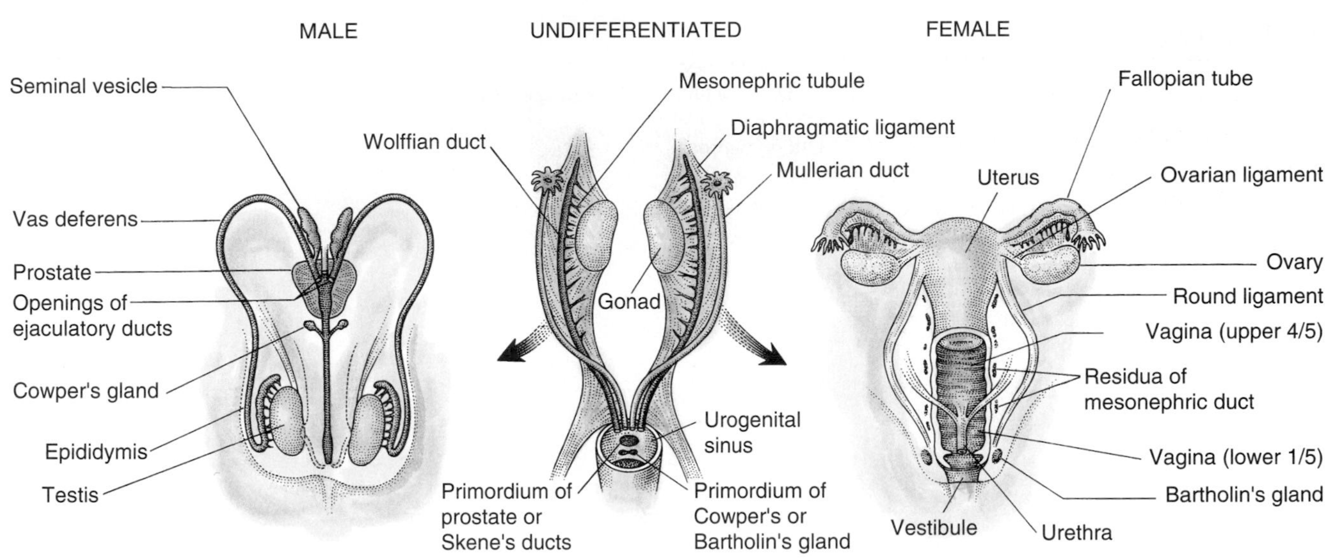

FIGURE 6.1

Development of the Internal Sex Organs from an Undifferentiated Stage at About 5 or 6 Weeks Following Conception.

of a new human being, is only 1/175 of an inch long. Yet, on this tiny stage, one's stamp as a unique individual has already been ensured—whether one will have black or blond hair, grow bald or develop a widow's peak, or become male or female.

The chromosomes from each parent combine to form 23 pairs. The 23rd pair are the sex chromosomes. An ovum carries an X sex chromosome, but a sperm carries either an X or a Y sex chromosome. If a sperm with an X sex chromosome fertilizes the ovum, the newly conceived person will normally develop as a female, with an XX sex chromosomal structure. If the sperm carries a Y sex chromosome, the child will normally develop as a male (XY).

After fertilization, the zygote divides repeatedly. After a few short weeks, one cell has become billions of cells. At about 3 weeks, a primitive heart begins to drive blood through the embryonic bloodstream. At about 5 to 6 weeks, when the **embryo** is only ¼ to ½ inch long, primitive gonads, ducts, and external genitals whose gender cannot be distinguished visually have formed (see Figures 6.1, 6.2). Each embryo possesses primitive external genitals, a pair of sexually undifferentiated gonads, and two sets of primitive duct structures, the Müllerian (female) ducts and the Wolffian (male) ducts.

During the first 6 weeks or so of prenatal development, embryonic structures of both genders develop along similar lines and resemble primitive female structures. At about the seventh week after conception, the genetic code (XX or XY) begins to assert itself, causing changes in the gonads, genital ducts, and external genitals. Genetic activity on the Y sex chromosome causes the testes to begin to differentiate (National Center for Biotechnology Information, 2000). Ovaries begin to differentiate if the Y chromosome is absent. Those rare individuals who have only one X sex chromosome instead of the typical XY or XX arrangement also become females, because they too lack the Y chromosome.

Thus, the basic blueprint of the human embryo is female. The genetic instructions in the Y sex chromosome cause the embryo to deviate from the female developmental course. "Adams" develop from embryos that otherwise would become "Eves."

By about the seventh week of prenatal development, strands of tissue begin to organize into seminiferous tubules. Female gonads begin to develop somewhat later than male gonads. The forerunners of follicles that will bear ova are not found until the fetal stage of development, about 10 weeks after conception. Ovaries begin to form at 11 or 12 weeks.

Embryo The stage of prenatal development that begins with implantation of a fertilized ovum in the uterus and concludes with development of the major organ systems at about two months after conception.

FIGURE 6.2

Development of the External Sex Organs from an Undifferentiated Stage at About 5 or 6 Weeks Following Conception.

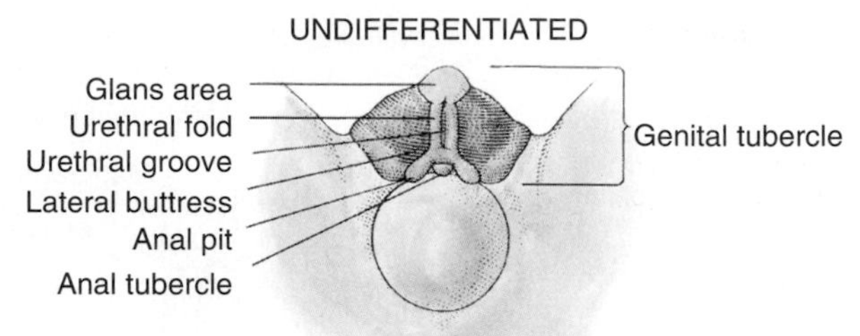

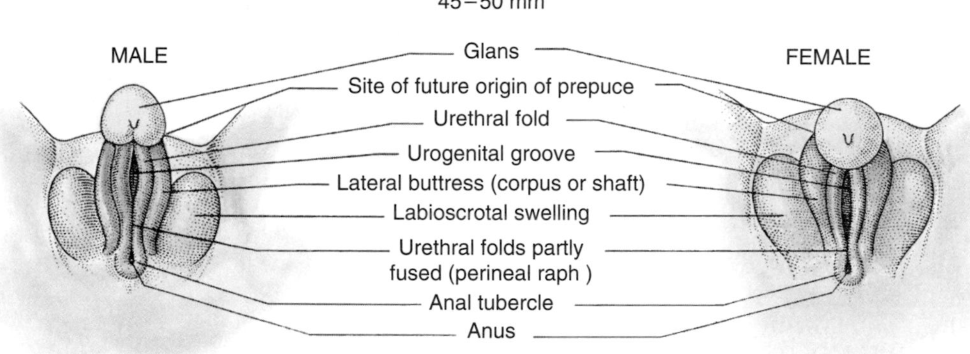

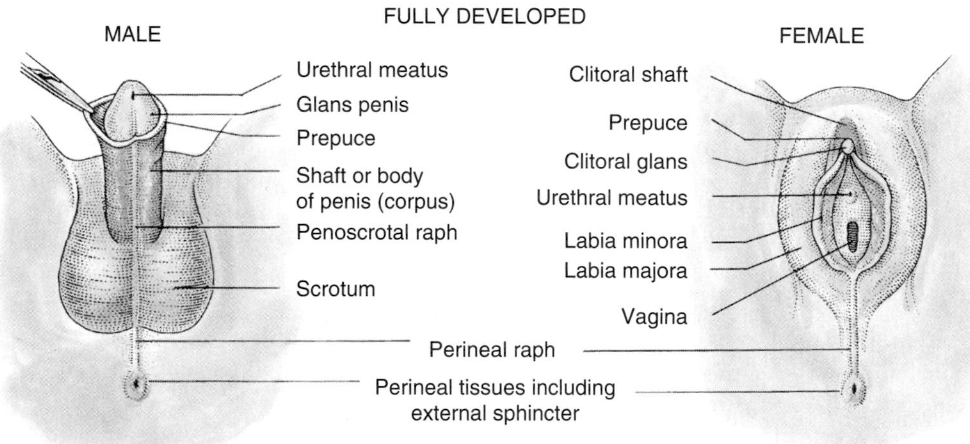

Genetic Factors in Sexual Differentiation

In recent years, we have learned much about the genetic determinants of sexual differentiation, and some of the findings are fascinating. For example, the interaction of a number of genes has led to the development of three different types of males in a crustacean, and quite complex mating strategies (Shuster & Sassaman, 1997). One sex-determining gene called "transformer" (tra) is needed in the development of female fruit flies. Chromosomal (XX) females with inactive tra attempt to mate with other females, but they are attractive to males because they still emit female pheromones (Arthur et al., 1998). Fruit fly researchers conclude that sex determination, sexual orientation, and sexual behavior patterns are all determined by the interactions of genes (O'Dell & Kaiser, 1997). In an article that could have been entitled "The Mouse That Roared," researcher Stephen Maxson (1998) reported that a number of genes that are involved in determining maleness in mice, including SRY (which stands for sex-determining region Y gene), are also connected with aggressiveness. All of these studies are suggestive of a role for genes in the determination of mating and other behavior patterns in humans. The studies with mice are of particular interest, because the National Center for Biotechnology Information (2000) notes that human SRY is similar to the SRY of mice. Nevertheless, as we see throughout this book, humans are also influenced by culture, values, experiences, and personal beliefs. In humans there is only rarely a direct line between genetics and behavior, especially social behaviors such as mating and aggression.

SRY is also one of the molecules involved in sex determination in humans (National Center for Biotechnology Information, 2000; Nordqvist & Lovell-Badge, 1994). It binds to DNA, which is a strand of genes—the basic building blocks of heredity—and distorts that DNA. The distortion alters the properties of the DNA and leads to the formation of the testes.[2] Another gene involved in sex determination has also been researched in mice: Sox 9. Sox 9 appears to regulate the expression of SRY (Overbeek, 1999). Females with XX sex chromosomal structure normally suppress the action of their own Sox 9, which in turn prevents the expression of SRY. However, when these XX mice are chemically prevented from turning off Sox 9, they develop as males, though albeit sterile males. Sperm cells, therefore, are likely to be created by other genes, not Sox 9 and SRY.

The Role of Sex Hormones in Sexual Differentiation

Without male sex hormones, or **androgens,** we would all develop into females in terms of anatomic structure (Federman, 1994). Once genes have done their work and testes develop in the embryo, they begin to produce androgens. The most important androgen, **testosterone,** spurs differentiation of the male (Wolffian) duct system (see Figure 6.1). Each Wolffian duct develops into an epididymis, vas deferens, and seminal vesicle. The external genitals, including the penis, begin to take shape at about the eighth week of development under the influence of another androgen, *dihydrotestosterone* (DHT). Yet another testicular hormone, one secreted during the fetal stage, prevents the Müllerian ducts from developing into the female duct system. It is appropriately termed the Müllerian inhibiting substance (MIS).

Small amounts of androgens are produced in female fetuses, but they are not normally sufficient to cause male sexual differentiation. In female fetuses, the relative absence of androgens causes degeneration of the Wolffian ducts and prompts development of female sexual organs. The Müllerian ducts evolve into fallopian tubes, the uterus, and the upper two thirds of the vagina. These developments occur even in the absence of female sex hormones. Although female sex hormones are crucial in puberty, they are not involved in fetal sexual differentiation. If a fetus with an XY sex chromosomal structure failed to produce testosterone, it would develop female sexual organs.

It is true that we would all develop female sexual organs if male sex hormones were not present during critical stages of prenatal development. Embryos develop female sexual organs in the absence of male sex hormones. ■

Descent of the Testes and the Ovaries

The testes and ovaries develop from slender structures high in the abdominal cavity. By about 10 weeks after conception, they have descended so that they are almost even with the upper edge of the pelvis. The ovaries remain there for the rest of the prenatal period. Later they rotate and descend farther to their adult position in the pelvis. About four months after conception, the testes normally descend into the scrotal sac through the **inguinal canal.** After their descent, this passageway is closed.

Androgens Male sex hormones.

Testosterone The male sex hormone that fosters the development of male sex characteristics and is connected with the sex drive.

Inguinal canal A fetal canal that connects the scrotum and the testes, allowing the latter to descend. (From the Latin *inguinus,* which means "near the groin.")

[2] The third author notes that this scientific finding seems to support the widely observed phenomenon that males are bent out of shape.

In a small percentage of males, one or both testes fail to descend. They remain in the abdomen at birth. The condition is termed **cryptorchidism.** In most cases of cryptorchidism, the testes migrate to the scrotum during infancy. In still other cases, the testes descend by puberty. Men with undescended testes are usually treated through surgery or hormonal therapy, because they are at higher risk for cancer of the testes. Sperm production is also impaired because the undescended testes are subjected to a higher-than-optimal body temperature, causing sterility.

Sex-Chromosomal Abnormalities

Abnormalities of the sex chromosomes can have profound effects on sexual characteristics, physical health, and psychological development. **Klinefelter syndrome,** a condition that affects about 1 in 500 males, is caused by an extra X sex chromosome, so the man has an XXY rather than an XY pattern. Men with this pattern fail to develop appropriate secondary sex characteristics. They have enlarged breasts and poor muscular development, and because they fail to produce sperm, they are infertile. They also tend to be mildly retarded.

Turner syndrome, found only in women, occurs in 1 in 2,000 to 5,000 girls. It is caused by the loss of some X chromosome material. These girls develop typical external genital organs, but they are short in stature and their ovaries do not develop or function normally. Girls with Turner syndrome have abnormal hairlines and tend to be involved in fewer social activities than other girls (Rovet & Ireland, 1994). However, they are not retarded (Saenger, 1996).

Prenatal Sexual Differentiation of the Brain

The brain, like the genital organs, undergoes prenatal sexual differentiation. Testosterone causes cells in the hypothalamus of male fetuses to become insensitive to the female sex hormone estrogen. In the absence of testosterone, as in female fetuses, the hypothalamus does develop sensitivity to estrogen.

Sensitivity to estrogen is important in the regulation of the menstrual cycle of women after puberty. The hypothalamus detects low levels of estrogen in the blood at the end of each cycle and initiates a new cycle by stimulating the pituitary gland to secrete FSH. FSH, in turn, stimulates estrogen production by the ovaries and the ripening of an immature follicle in an ovary. Sexual differentiation of the hypothalamus is most likely to occur during the second trimester of fetal development (Pillard & Weinrich, 1986).

GENDER IDENTITY

Our awareness of being male or being female—our **gender identity**—is one of the most obvious and important aspects of our self-concepts. Gender identity is not necessarily an automatic extension of our anatomic gender. Gender identity is psychological—a sense of being male or being female. **Gender assignment** reflects the child's anatomic gender and usually occurs at birth. Gender identity is so important to parents that they may want to know "Is it a boy or a girl?" before they begin to count fingers and toes.

Most children first become aware of their anatomic gender by about the age of 18 months. By 36 months, most children have acquired a firm sense of gender identity (Etaugh & Rathus, 1995).

Nature and Nurture in Gender Identity

What determines gender identity? Are our brains biologically programmed along masculine or feminine lines by prenatal sex hormones? Does the environment, in the form of postnatal learning experiences, shape our self-concepts as males or females? Or does gender identity reflect an intermingling of biological and environmental influences?

Cryptorchidism The condition defined by undescended testes. (From roots that mean "hidden testes.")

Klinefelter syndrome A sex-chromosomal disorder caused by an extra X sex chromosome.

Turner syndrome A sex-chromosomal disorder caused by loss of some X chromosome material.

Gender identity The psychological sense of being male or female.

Gender assignment The labeling of a newborn as a male or a female.

Gender identity is nearly always consistent with chromosomal gender. Such consistency does not certify that gender identity is biologically determined, however. We also tend to be reared as males or females, in accordance with our anatomic genders. How, then, can we sort out the roles of nature and nurture, of biology and the environment?

Clues may be found in the experiences of rare individuals, **pseudohermaphrodites,** who possess the gonads of one gender but external genitalia that are ambiguous or typical of the other gender. Pseudohermaphrodites are sometimes reared as members of the other gender (the gender other than their chromosomal gender). Researchers have wondered whether the gender identity of these children reflects their chromosomal and gonadal gender or the gender in accordance with which they were reared. Before going further with this, let us distinguish between true hermaphroditism and pseudohermaphroditism.

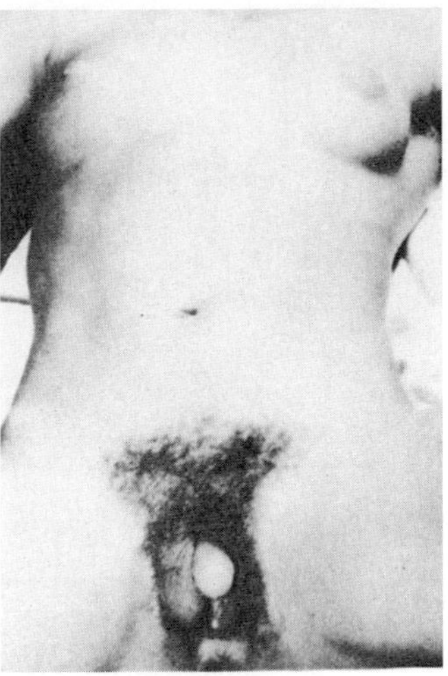

FIGURE 6.3

A True Hermaphrodite. This genetic (XX) female has one testicle and one ovary and the gender identity of a male.

Hermaphroditism

Hormonal errors during prenatal development produce various congenital defects. Some individuals are born with both ovarian and testicular tissue. They are called **hermaphrodites,** after the Greek myth of the son of Hermes and Aphrodite, whose body became united with that of a nymph while he was bathing. True hermaphrodites may have one gonad of each gender (a testicle and an ovary) or gonads that combine testicular and ovarian tissue.

Regardless of their genetic gender, hermaphrodites usually assume the gender identity and gender role of the gender assigned at birth. Figure 6.3 shows a genetic female (XX) with a right testicle and a left ovary. This person married and became a stepfather with a firm male identity. The roles of biology and environment remain tangled, however, because true hermaphrodites have gonadal tissue of both genders.

True hermaphroditism is extremely rare. More common is pseudohermaphroditism, which occurs in perhaps 1 infant in 1,000. The occurrence of pseudohermaphroditism has given scientists an opportunity to examine the roles of nature (biology) and nurture (environmental influences) in the shaping of gender identity.

PSEUDOHERMAPHRODITISM *Pseudohermaphrodites* ("false" hermaphrodites) have testes or ovaries, but not both. Unlike true hermaphrodites, their gonads (testes or ovaries) match their chromosomal gender. Because of prenatal hormonal errors, however, their external genitals and sometimes their internal reproductive anatomy are ambiguous or resemble those of the other gender.

The most common form of female pseudohermaphroditism is **androgenital syndrome,** in which a genetic (XX) female has female internal sexual structures (ovaries) but masculinized external genitals (see Figure 6.4). The clitoris is so enlarged that it may resemble a small penis. The syndrome occurs as a result of excessive levels of androgens. In some cases, the fetus's own adrenal glands produce excess androgen (the adrenal glands usually produce low levels of androgen). In other cases, mothers may have received synthetic androgens during their pregnancies. In the 1950s and 1960s, before these side effects were known, synthetic androgens were sometimes prescribed to help prevent miscarriages in women with histories of spontaneous abortions.

Another type of pseudohermaphroditism, **androgen-insensitivity syndrome,** describes genetic (XY) males who, as the result of a mutated gene, have lower-than-normal prenatal sensitivity to androgens (Adachi et al., 2000; Hughes, 2000). Consequently, their genitals do not become normally masculinized. At birth their external genitals are feminized, including a small vagina, and their testes are undescended. Because of insensitivity to androgens, the male duct system (epididymis, vas deferens, seminal vesicles, and ejaculatory ducts) fails to develop. Nevertheless, the fetal testes produce Müllerian inhibiting substance, preventing the development of a uterus or fallopian tubes. Genetic males with androgen-insensitivity syndrome usually have no or sparse pubic and axillary (underarm) hair, because the development of hair in these locations is dependent on androgens.

A third type of pseudohermaphroditism is named **Dominican Republic syndrome** because it was first documented in a group of 18 affected boys in two rural villages in that

Pseudohermaphrodites People who possess the gonads of one gender but external genitalia that are ambiguous or typical of the other gender.

Hermaphrodites People who possess both ovarian and testicular tissue. (From the names of the male and female Greek gods *Hermes* and *Aphrodite.*)

Androgenital syndrome A form of pseudohermaphroditism in which a genetic female has internal female sexual structures but masculinized external genitals.

Androgen-insensitivity syndrome A form of pseudohermaphroditism in which a genetic male is prenatally insensitive to androgens. As a result, his genitals do not become normally masculinized.

Dominican Republic syndrome A form of pseudohermaphroditism in which a genetic enzyme disorder prevents testosterone from masculinizing the external genitalia.

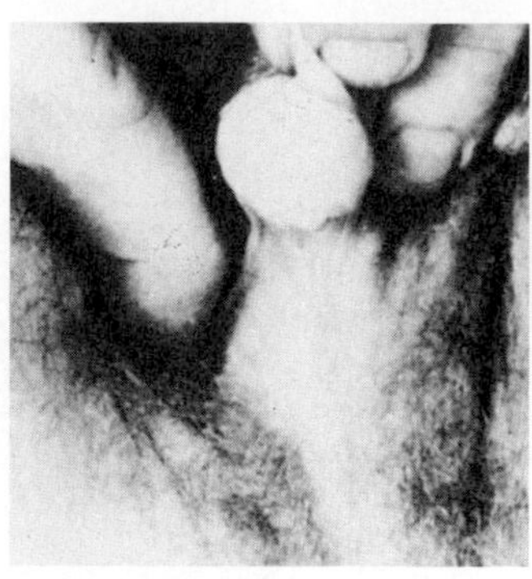

FIGURE 6.4

Pseudohermaphroditism. In androgenital syndrome, a genetic (XX) female has female internal sexual structures (ovaries) but masculinized external genitals.

nation (Imperato-McGinley et al., 1974). Dominican Republic syndrome is a genetic enzyme disorder that prevents testosterone from masculinizing the external genitalia. The boys were born with normal testes and internal male reproductive organs, but their external genitals were malformed. Their penises were stunted and resembled clitorises. Their scrotums were incompletely formed and resembled female labia. They also had partially formed vaginas.

PSEUDOHERMAPHRODITISM AND GENDER IDENTITY The experiences of pseudohermaphrodites provide insights into the origins of gender identity. The genitals of girls with androgenital syndrome are normally surgically feminized in infancy. The girls also receive hormone treatments to correct excessive adrenal output of androgens. As a result, they usually acquire a feminine gender identity and develop physically as normal females.

What if the syndrome is not identified early in life, however? Consider the cases of two children who were treated at Johns Hopkins University Hospital. The children both had androgenital syndrome, but their treatments and the outcomes were very different. Each child was genetically female (XX). Each had female internal sex organs. Because of prenatal exposure to synthetic male sex hormones, however, each developed masculinized external sex organs (Money & Ehrhardt, 1972).

The problem was identified in one child (let's call her Abby) in infancy. Her masculinized sex organs were removed surgically when she was 2 years old. Like many other girls, Abby was tomboyish during childhood, but she was always feminine in appearance and had a female gender identity. She began to develop breasts by the age of 12 but did not begin to menstruate until age 20. She dated boys, and her fantasy life revolved around marriage to a man.

The other child (let's call him James) was initially mistaken for a genetic male with stunted external sex organs. The error was discovered at the age of 3½. By then he had a firm male gender identity, so surgeons further masculinized his external sex organs rather than removing them. At puberty, hormone treatments stoked the development of body hair, male musculature, and other male secondary sex characteristics.

As an adolescent, James did poorly in school. Possibly in an effort to compensate for his poor grades, he joined a gang of "semidelinquents." He became one of "the boys." In contrast to Abby, James was sexually attracted to women.

Both children were pseudohermaphrodites. Both had internal female sexual organs and masculinized external organs, but they were treated and reared differently. In Abby's case, the newborn was designated female, surgically altered to remove the masculinized genitals, and reared as a girl. In James's case, the infant was labeled and reared as a boy. Each child acquired the gender identity of the assigned gender. Environmental influences may have played a critical role in shaping the gender identity of these children, although the issue is clouded by the fact that both individuals were exposed to large quantities of androgens in the uterus.

Further evidence of a role for psychosocial influences on gender identity is found in studies of genetic (XY) males with androgen-insensitivity syndrome. They possess testes but are born with feminine-appearing genitals and are typically reared as girls. They develop a female gender identity and stereotypical feminine interests. They show as much interest in dolls, dresses, and future roles as mothers and housewives as do genetic (XX) girls of the same age and social class (Brooks-Gunn & Matthews, 1979; Money & Ehrhardt, 1972).

The boys with Dominican Republic syndrome also resembled girls at birth and were reared as females. At puberty, however, their testes swung into normal testosterone production, causing startling changes: The testes descended, their voices deepened, their musculature filled out, and their "clitorises" expanded into penises. Of the 18 boys who were reared as girls, 17 shifted to a male gender identity. Sixteen of the 18 assumed a stereotypical masculine gender role. Of the remaining 2, 1 adopted a male gender identity but continued to maintain a feminine gender role, including wearing dresses. The other maintained a female gender identity and later sought gender-reassignment surgery to "correct" the pubertal masculinization.

The Dominican transformations would appear to suggest that gender identity is malleable. What of the roles of nature and nurture in the formation of gender identity, however? If environmental forces (nurture) were predominant, gender identity would be based on the gender in which the person is reared, regardless of biological abnormalities. With the Dominicans, however, pubertal biological changes led to changes in both gender identity and gender roles. Is nature (biology) then the primary determinant of gender identity? Unfortunately, the Dominican study does not allow clear separation of the effects of nature and nurture. It is possible that the pubertal surges of testosterone activated brain structures that were masculinized during prenatal development. Prenatal testosterone levels in these boys were presumably normal and could have affected the sexual differentiation of brain tissue, even though the genetic defect prevented the hormone from masculinizing the external genitalia.

What, then, can we conclude from studies of pseudohermaphrodites? Do they suggest that gender identity and the assumption of gender roles are influenced by psychosocial factors? Can pseudohermaphrodites acquire the gender identity of the other chromosomal gender when they are reared as members of that gender? The genetic (XX) females—Abby and James—were reared as members of different genders and acquired the gender identity in which they were reared. Yet both were treated with sex hormones appropriate to their assigned gender. Thus, sex hormones may have influenced their gender identity. The Dominican study also suggests that gender identity may not be fixed by early learning influences but may, rather, be subject to subsequent biological and/or psychosocial effects.

We should also recognize that the experiences of people affected by these hormonal errors may not generalize to others. Perhaps the brains of these children had not been clearly gender-typed prenatally and were thus capable of an unusual degree of postnatal flexibility in the assumption of gender identity. Perhaps in normal people the brain is more clearly differentiated prenatally, and hence gender identity is not so readily influenced by experience.

Most scientists conclude that gender identity is influenced by complex interactions between biological and psychosocial factors. Some place relatively greater emphasis on psychosocial factors (Money & Ehrhardt, 1972). Others emphasize the role of biological factors (Collaer & Hines, 1995; Diamond, 1996; Legato, 2000), even though they allow that nurture plays a role in gender identity. The debate over the relative contributions of nature and nurture is likely to continue.

In case you have had enough discussion of the complex issues surrounding the origins of gender in human beings, consider the crocodile. Crocodile eggs do not carry sex chromosomes. The baby's sex is determined, instead, by the temperature at which the eggs develop (Ackerman, 1991). Some (males) like it hot (at least in the mid-90s Fahrenheit), and some (females) like it not cold, perhaps, but under the mid-80s Fahrenheit.

It is true that the sex of a baby crocodile is determined by the temperature at which the egg develops. ■

Transsexualism

In 1953 an ex-GI who journeyed to Denmark for a "sex-change operation" made headlines. She became known as Christine (formerly George) Jorgensen. Since then, thousands of **transsexuals** (also called *transgendered* people) have undergone gender-reassignment surgery. Among the better known is the tennis player Dr. Renée Richards, formerly Dr. Richard Raskin.

Transsexuals People who have a gender-identity disorder in which they feel trapped in the body of the wrong gender.

Gender-reassignment surgery cannot implant the internal reproductive organs of the other gender. Instead, it generates the likeness of external genitals typical of the other

a CLOSER look

Boys Who Are Reared as Girls

Are children "psychosexually neutral" at birth? Can you surgically reassign a boy to the female gender, rear him as a girl, and have him feel that he is truly a girl as the years go on? Will cosmetic surgery, female sex hormone treatment, and laces and ribbons do it? Or will he be maladjusted and his male gender sort of "break through"? No one has sought to answer these questions by randomly selecting male babies and reassigning their genders. Evidence on the matter derives from studies of children who have lost their penises or failed to develop them through accidents or unusual medical conditions.

Getting Down to Cases

For example, in 1967, one of a pair of male twins lost his penis as a result of a circumcision accident. As this case study is related by Colapinto (2000), the parents wondered what to do. Johns Hopkins sexologist John Money believed that gender identity was sufficiently malleable that the boy could undergo gender-reassignment surgery (have his testes removed and an artificial vagina constructed) and female hormone treatments and be successfully reared as a girl. Money's view was based on experience with intersexual patients, who possess the genitals of both males and females. In most cases, their gender identity seemed to be shaped more by rearing than by sex chromosomes, gonads, or sex hormones.

For a number of years, the case seemed to supply evidence for the view that children may be psychosexually neutral at birth. The gender-reassigned twin, unlike his brother, seemed to develop like a "real girl," although exhibiting a number of "tomboyish" traits. But at the age of 14, when "she" was informed about the circumcision accident and the process of gender reassignment, she immediately decided to pursue life as a male. As an adult, he recalled that he had never felt quite comfortable as a girl—a view confirmed by the recollections of his mother. At the age 25, he married a woman and adopted her children. He said that he was sexually attracted to women only. This outcome would appear to support the view, shared by researchers such as Milton Diamond (1996), that gender identity is determined to a considerable extent in the uterus, as the fetal brain is being exposed to androgens.

Another Fascinating Case Study

Susan Bradley and her colleagues (1998) report on the development of another boy who suffered a circumcision accident in infancy. Again, John Money recommended gender reassignment, and the surgery was carried out at the age of 21 months. In this case, as Money found out in a follow-up at the age of 9, the individual was also tomboyish in behavior and personality traits but considered herself a girl.

She was interviewed subsequently at the ages of 16 and 26, and her situation was quite a bit more complex. She considered herself to be bisexual and had sexual relationships with both men and women. However, when last interviewed, she had begun living with a woman in what the authors label a "lesbian" relationship. Of course, if one remembers that the individual was born a male, the relationship with the woman is not with a person of the same gender at all. On the other hand, the individual did have the self-concept of being female.

A Larger Study

This celebrated case is far from the only one. On May 12, 2000, researchers from the Johns Hopkins University Hospital, including William G. Reiner, a psychiatrist and urologist, presented a paper on the subject to the Lawson Wilkins Pediatric Endocrine Society meeting in Boston. They recounted the development of 27 children who had been born without penises because of a rare condition called *cloacal exstrophy*. However, the children had normal testicles, male sex chromosomal structure, and male sex hormones.

Nevertheless, 25 of the 27 children were gender-reassigned shortly after birth. They were surgically castrated and reared as girls by their parents. As the years went on, all 25—now 5 to 16 years old—showed the rough-and-tumble play considered stereotypical of males. Of the 25, 14 declared themselves to be males. Reiner (2000) suggests that "with time and age, children may well know what their gender is, regardless of any and all information and child-rearing to the contrary," he said. "They seem to be quite capable of telling us who they are." Reiner (2000) also noted that the 2 of the 27 children who were not gender-reassigned fit in with male peers and appeared to be better adjusted than the children who were reassigned.

Marianne J. Legato (2000), a professor of medicine at Columbia University, believes that gender identity tends to be formed during the first trimester of pregnancy, even if children cannot verbally express their identities until a few years afterward. As Legato describes it, "When the brain has been masculinized by exposure to testosterone, it is kind of useless to say to this individual, 'You're a girl.' It is this impact of testosterone that gives males the feelings that they are men."

It should be noted that not all researchers, including Bradley and her colleagues (1998) agree on the prominence of prenatal male sex hormones in the formation of gender identity. However, the view that newborns are psychosexually neutral and that gender identity depends mainly on nurture has had rough sledding in recent years.

gender. This can be done more precisely with male-to-female than with female-to-male transsexuals. After such operations, people can participate in sexual activity and even attain orgasm, but they cannot conceive or bear children.

Truth or Fiction? REVISITED

It is not true that people have changed their genders through gender-reassignment surgery. Gender-reassignment surgery cannot implant the internal reproductive organs of the other gender. It only generates the appearance of the external genitals typical of the other gender.

Transsexuals experience **gender dysphoria.** That is, according to John Money (1994), they have the subjective experience of incongruity between their genital anatomy and their gender identity or role. They have the anatomic sex of one gender but feel that they are members of the other gender. As a result of this discrepancy, they wish to be rid of their own primary sex characteristics (their external genitals and internal sex organs) and to live as members of the other gender. A male transsexual perceives himself to be a female who, through some quirk of fate, was born with the wrong genital equipment. A female transsexual perceives herself as a man trapped in a woman's body. Although the prevalence of transsexualism remains unknown, it is thought to be rare. The number of transsexuals in the United States is estimated to be about 25,000. Perhaps 6,000 to 11,000 of them have undergone gender-reassignment surgery (Selvin, 1993).

Patterns of sexual attraction do not appear to be of great importance. Some transsexuals report never having had strong sexual feelings. Others are attracted to members of their own (anatomic) gender. They are unlikely to regard themselves as gay or lesbian, however. From their perspective, their lovers are members of the other gender. Still others are attracted to members of the other anatomic gender. Nonetheless, they all want to be rid of their own sex organs and to live as members of the other gender.

Transsexualism is not to be confused with a gay male or lesbian sexual orientation. Gay males and lesbians are erotically attracted to members of their own gender. A gay man may desire another man as a lover; a lesbian may sexually desire another woman. Gay men and lesbians perceive their gender identities to be consistent with their anatomic gender, however. They would no more want to be rid of their own genitals than would heterosexuals. As one gay man put it, "Just because I'm turned on by other men doesn't make me feel less like a man."

Transsexuals usually show cross-gender preferences in play and dress in early childhood. Many report that they have felt that they belong to the other gender for as long as they can remember. Only a few were unaware of their transsexual feelings until adolescence. Male transsexuals generally recall that as children, they preferred playing with dolls, enjoyed wearing frilly dresses, and disliked rough-and-tumble play. They were often perceived by their peers as "sissy boys." Female transsexuals usually report that as children they disliked dresses and acted much like "tomboys." They also preferred playing "boys' games" and doing so with boys. Female transsexuals appear to have an easier time adjusting than male transsexuals (Selvin, 1993). "Tomboys" generally find it easier to be accepted by their peers than "sissy boys." Even in adulthood, it may be easier for a female transsexual to don men's clothes and "pass" as a slightly built man than it is for a brawnier man to pass for a tall woman.

The transition to adolescence is particularly difficult for transsexuals. They find their bodies changing in ways that evoke their disgust. Female transsexuals abhor the onset of menstruation and the development of breasts. They may seek to disguise their budding breasts by binding them or wearing loose clothing. Some have mastectomies at the age of consent to remove the obvious reminder of what they perceive as "nature's mistake."

THEORETICAL PERSPECTIVES No clear understanding of the nature or causes of transsexualism has emerged (Money, 1994). Views on its origins somewhat parallel those on the origins of a gay male or lesbian sexual orientation, which is surprising, given the key differences that exist between gay people and transsexuals.

Gender dysphoria The subjective experience of incongruity between genital anatomy and gender identity or role.

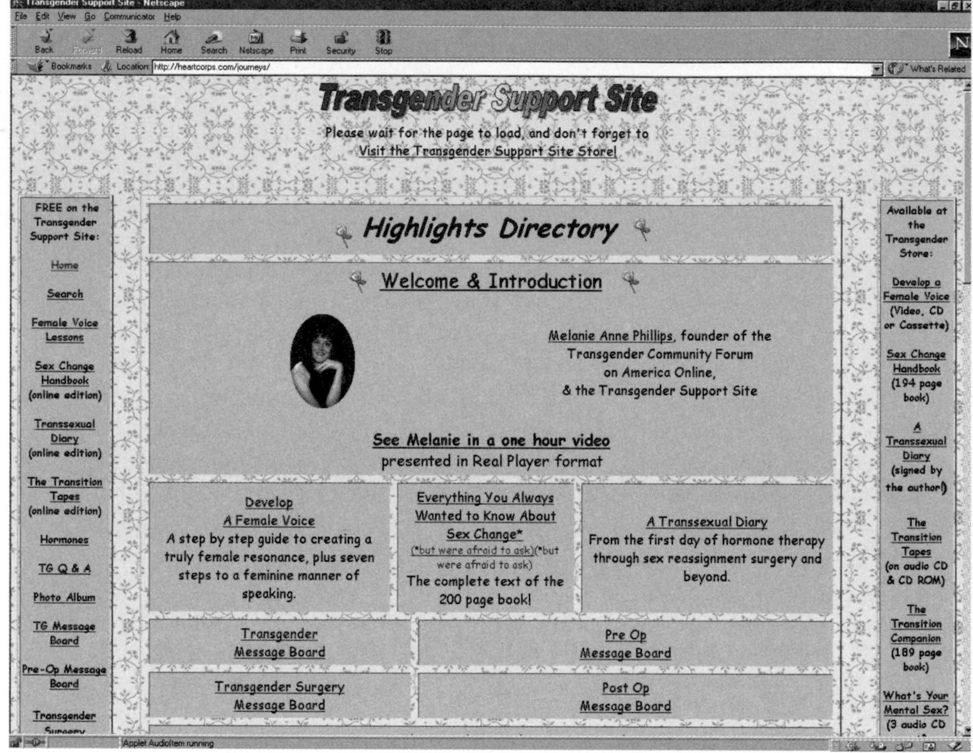

A Web Site That Helps Transgendered Individuals Adjust. Numerous Web sites are available to help transsexuals adjust. These Web sites frequently have links to local support groups and commercial aspects (online stores). (Reprinted by permission. www.heartcorps.com/journeys.)

Psychoanalytic theorists have focused on early parent–child relationships. Male-to-female transsexuals, in this view, may have had "close-binding mothers" (extremely close mother–son relationships) and "detached-hostile fathers" (fathers who were absent or uninterested) (Stoller, 1969). Such family circumstances may have fostered intense **identification** with the mother, to the point of an inversion of typical gender roles and identity. By the same token, girls with weak, ineffectual mothers and strong, masculine fathers may identify with their fathers, rejecting their own female identities.

There is some evidence that male-to-female transsexuals tend to have had unusually close relationships with their mothers during childhood. Female-to-male transsexuals tend to have identified more with their fathers and to have perceived their mothers as cold and rejecting (Pauly, 1974). Yet one problem with the psychoanalytic view is that the roles of cause and effect may be reversed. It could be that as children, transsexuals gravitate toward the parent of the other gender and reject the efforts of the parent of the same gender to reach out to them and engage them in gender-typed activities. These views also do not account for the many transsexuals whose family backgrounds fail to match the proposed patterns. Moreover, these views lack predictive power. Most children—in fact, the vast majority—who emerge from such family backgrounds do *not* become transsexuals (or gay). The early onset of transsexual feelings suggests that critical early learning experiences, if they occur, do so in the preschool years.

Transsexuals may also be influenced by prenatal hormonal imbalances. The brain is in some ways "masculinized" or "feminized" by sex hormones during prenatal development. Researchers in the Netherlands have discovered that a small region of the hypothalamus is about 50% larger in men than in women, and nearly 60% larger in men than in male-to-female transsexuals (Angier, 1995). It is possible that the brain could be influenced in one direction, even as the genitals are being differentiated in the other.

Identification In psychoanalytic theory, the process of incorporating within ourselves our perceptions of the behaviors, thoughts, and feelings of others.

GENDER REASSIGNMENT Gender reassignment for transsexuals has been controversial since its inception. Yet psychotherapy is not considered a reasonable alternative, because it has been generally unsuccessful in helping transsexuals accept their anatomic genders (Roberto, 1983; Tollison & Adams, 1979).

Using the Back Door in Iran: Can a Transsexual Change Her Mind?

An Iranian man who recently had a sex change to become a woman wants to reverse the operation because she finds life as a woman insufferable in Iran. The 25-year-old Maryam, formerly Mehran, underwent a sex change despite strong parental opposition. But she soon regretted the decision, finding it difficult to cope with "restrictions" surrounding a woman's life in the conservative Islamic society.

"I can't go on living with the new identity, after years of living as a man with no restrictions," she told the daily paper in Iran. "First I thought I would get used to it, but life has become painful and intolerable. So I want a new sex change."

Sex-change operations are legal in Iran, but there are no provisions for would-be transsexuals to test out their new identity first.

The social reforms of President Mohammad Khatami, who was elected with the overwhelming support of women, eased the lot of women somewhat. But women still struggle under the burdens of a legal code and a value system that severely limit their freedom of action and subordinate them to husbands, brothers, and fathers. Iran has a mandatory dress code for women, requiring them to cover their hair and body. Whereas men get on public transport through the front door, women must use the back door. Courts give a woman's testimony only half the weight of that of a man, and the inheritance, divorce, and child-custody laws overwhelmingly favor men. Official statistics show suicide rates among women far outstrip those of men—the opposite of Western societies.

Source: From "Iranian Transsexual Unhappy with Experience as Woman" (2000, June 19). *Reuters News Agency online.* Copyright © 2000 by Reuters Limited. Adapted by permission.

Surgery is one element of gender reassignment. Because the surgery is irreversible, health professionals conduct careful evaluations to determine that people seeking reassignment are competent to make such decisions and have thought through the consequences. They usually require that the transsexual live openly as a member of the other gender for a trial period of at least a year before surgery.

Once the decision is reached, a lifetime of hormone treatments is begun. Male-to-female transsexuals receive estrogen, which fosters the development of female secondary sex characteristics. It causes fatty deposits to develop in the breasts and hips, softens the skin, and inhibits growth of the beard. Female-to-male transsexuals receive androgens, which promote male secondary sex characteristics. The voice deepens, hair becomes distributed according to the male pattern, muscles enlarge, and the fatty deposits in the breasts and hips are lost. The clitoris may also grow more prominent. In the case of male-to-female transsexuals, "phonosurgery" can be done to raise the pitch of the voice (Brown et al., 2000).

Gender-reassignment surgery is largely cosmetic. Medical science cannot construct internal genital organs or gonads. Male-to-female surgery is generally more successful. The penis and testicles are first removed. Tissue from the penis is placed in an artificial vagina so that sensitive nerve endings will provide sexual sensations. A penis-shaped form of plastic or balsa wood is used to keep the vagina distended during healing.

In female-to-male transsexuals, the internal sex organs (ovaries, fallopian tubes, and uterus) are removed, along with the remaining fatty tissue in the breasts. The nipples are moved to keep them at the proper height on the torso. Either the urethra is rerouted through the enlarged clitoris, or an artificial penis and scrotum are constructed of tissue from the abdomen, the labia, and the perineum through a series of operations. In either case, the patient can urinate while standing, which appears to provide psychological gratification. Although the artificial penis does not stiffen and become erect naturally, a variety of methods, including implants, can be used to allow the artificial penis to approximate erection.

Some transsexuals hesitate to undertake surgery because they are repulsed by the prospect of such extreme medical intervention. Others forgo surgery so as not to jeopardize high-status careers or family relationships (Kockott & Fahrner, 1987). Such people continue to think of themselves as members of the other gender, however, even without surgery.

a CLOSER look

Jayne Thomas, Ph.D.— In Her Own Words

The "Glass Ceiling," male bashing, domestic violence, nagging, PMS, Viagra—these are but a few of the important issues examined in the Human Sexuality classes I instruct. As a participant–observer in my field, I see many of these topics aligning themselves as masculine/feminine or male/female. Ironically, I can both see and not see such distinctions. Certainly women have bumped up against, smudged (and in some cases even polished) this metaphorical limitation of women's advancement in the workplace (i.e., that "Glass Ceiling"). And most assuredly men have often found themselves "bashed" by angry women intent upon extracting a pound of flesh for centuries of felt unjust treatment. As previously mentioned, these distinctions between masculine and feminine, for me often become blurred; I must add that, having lived my life in both the roles of man and woman, I offer a rather unique perspective on masculinity and femininity.

Jayne Thomas, Ph.D. Jayne Thomas is chairperson of the Psychology Department of Los Angeles Mission College in Sylmar, California. She offers a comprehensive perspective on gender roles because she has lived as both a man and a woman.

Gender Identity Disorder

Gender Identity Disorder (GID) is defined by the American Psychiatric Association (2000) as a "strong and persistent cross-gender identification [accompanied by] a persistent discomfort with his or her sex or sense of appropriateness in the gender role of that sex" (p. 537). All of my life, I harbored the strongest conviction that I was inappropriately assigned to the wrong gender—that of a man—when inside I knew myself to be a woman. Even so (and like so many other GIDs) I continued a life-long struggle with this deeply felt mistake; I was successful in school, became a national swimming champion, received my college degrees, married twice (fathering children in both marriages) and was respected as a competent and good man in the workplace. However, the persistently unrelenting wrongfulness of my life continued. Not until my fourth decade was I truly able to address my gender issue.

Jay Thomas, Ph.D., underwent gender reassignment and officially became Jayne Thomas, Ph.D., in November of 1985, and what has transpired in the ensuing years has been the most enlightening of glimpses into the plight of humankind. As teachers we are constantly being taught by those we purport to instruct. My students, knowing my background (I share who I am when it is appropriate to do so), find me accessible in ways that many professors are not. Granted, I am continually asked the titillating questions that one watching *Geraldo* might ask and we do have fun with the answers (several years ago I even appeared on a few of the *Geraldo* shows). My students, however, are able to take our discussions beyond the sensational and superficial, and we enter into meaningful dialogue regarding gender differences in society and the workplace, sexual harassment, power and control issues in relationships, and what it really means to be a man or a woman.

Challenging Both the Masculine and the Feminine

Iconoclastically, I try to challenge both the masculine and feminine. "I know something none of you women know or

OUTCOMES OF GENDER-REASSIGNMENT SURGERY Following the introduction of gender-reassignment surgery in the United States in the 1960s, most reports of postoperative adjustment were positive (Pauly & Edgerton, 1986). An influential study conducted in the 1970s at the Gender Identity Clinic at Johns Hopkins University was quite negative, however (Meyer & Reter, 1979). The study included a control group of transsexuals who did will ever know in your lifetime," I can provocatively address the females in my audiences as Jayne. "I once lived as a man and have been treated as an equal. You never have nor will you experience such equality." Or, when a male student once came to my assistance in a classroom, fixing an errant video playback device and then strutting peacock-like back to his seat as only a satisfied male can, I teasingly commented to a nearby female student, "I used to be able to do that."

Having once lived as a man and now as a woman, I can honestly state that I see profound differences in our social/psychological/biological beings as man and woman. I have now experienced many of the ways in which women are treated as less than men. Jay worked as a consultant to a large banking firm in Los Angeles and continued in that capacity as a woman following her gender shift. Amazingly the world presented itself in a different perspective. [Jay's] technical presentations to management had generally been received in a positive manner and credit for my work fully acknowledged. Jayne now found management less accessible, [found] credit for her efforts less forthcoming and, in general, found herself working harder to be well prepared for each meeting than she ever had as a male. As a man, her forceful and impassioned presentations were an asset; as a woman they definitely seemed a liability. On one occasion, as Jayne, when I passionately asserted my position regarding what I felt to be an important issue, my emotion and disappointment in not getting my point across (my voice showed my frustration) were met with a nearby colleague (a man) reaching to touch my arm with words of reassurance, "There, there, take it easy, it will be all right." Believe me; that never happened to Jay. There was also an occasion when I had worked most diligently on a presentation to management only to find the company vice president more interested in the fragrance of my cologne than my technical agenda.

Certainly there are significant difference in the treatment of men and women, and yet I continue to be impressed with how similar we two genders really are. Although I have made this seemingly enormous change in lifestyle (and it is immense in so many ways), I continue as the same human being, perceiving the same world through these same sensory neurons. The difference—I now find myself a more comfortable and serene being than the paradoxical woman in a man's body, with anatomy and gender [that] have attained congruence.

Adjusting the Shifting Gender Roles

Does the shifting of gender role create difficulties in the GID's life? Most assuredly it does. Family and intimate relationships rank highest among those issues most problematic for the transitioning individual to resolve. When one shifts gender role, the effects of such a change are global; as ripples in a pond, the transformation radiates outward impacting all that have significantly touched the GID's life. My parents had never realized that their eldest son was dealing with such a life-long problem. Have they accepted or do they fully understand the magnitude of my issue? I fear not. After almost fifteen years of having lived as a female, my father continues to call me by my male name. I do not doubt my parents' or children's love for me, but so uninformed are we of the true significance of gender identity that a clear understanding seems light years away. Often I see my clients losing jobs, closeness with family members, [and] visitation rights with children and generally becoming relegated to the role of societal outcast. Someone once stated that "Everybody is born unique, but most of us die copies."—a great price my clients often pay for personal honesty and not living their lives as a version of how society deems they should.

Having lived as man and woman in the same lifetime, one personal truth seems clear. Rather than each gender attempting to change and convert the other to their own side, as I often see couples undertaking to accomplish (women need be more logical and men more sharing of their emotions), we might more productively come together in our relationships by building upon our gender uniqueness and strengths. Men and women have different perspectives, which can be used successfully to address life's issues.

Dr. Thomas teaches and lectures at colleges and universities on the West Coast, sharing her views of masculine and feminine with her students, conducting workshops, and continuing her experiment or *experience* of life.

not receive gender-reassignment surgery. Psychological adjustment was more positive among transsexuals in the control group than among those who had undergone surgery.

Other reviews report more positive outcomes for gender-reassignment surgery (Kockott & Fahrner, 1987; Lundstrom et al., 1984; Pauly & Edgerton, 1986). One study of 42 postoperative male-to-female transsexuals found that all but one would repeat the surgery and that the great majority found sexual activity more pleasurable as a "woman" (Bentler, 1976).

Reviewers of the international literature reported in 1984 that about 90% of transsexuals who undergo gender-reassignment surgery experience positive results (Lundstrom et al., 1984). In Canada, a follow-up study of 116 transsexuals (female-to-male and male-to-female) at least one year after surgery found that most of them were content with the results and were reasonably well adjusted (Blanchard et al., 1985). Positive results for surgery were also reported in a study of 141 Dutch transsexuals (Kuiper & Cohen-Kettenis, 1988). Nearly 9 out of 10 male-to-female and female-to-male transsexuals in a study of 23 transsexuals reported they were very pleased with the results of their gender-reassignment surgery (Lief & Hubschman, 1993). Still another study (Abramowitz, 1986) reported that about 2 out of 3 cases showed at least some postoperative improvement in psychological adjustment. These favorable results do not mean that postoperative transsexuals were ecstatic about their lives. In many cases it meant that they were less unhappy. Most transsexuals are socially maladjusted prior to gender reassignment, and many remain lonely and isolated afterward. Moreover, about half incur postoperative medical complications (Lindermalm et al., 1986).

Male-to-female transsexuals whose surgery permitted them to pass as members of the other gender showed better adjustment than those whose surgery left telltale signs (such as breast scarring and leftover erectile tissue) that they were not "real" women (Ross & Need, 1989). Social and family support also contributed to postsurgical adjustment (Ross & Need, 1989).

Male-to-female transsexuals outnumber female-to-males, but postoperative adjustment is apparently more favorable for female-to-males. Nearly 10% of male-to-female cases, as compared to 4% to 5% of female-to-males, have had disturbing outcomes, such as severe psychological disorders, hospitalization, requests for reversal surgery, and even suicide (Abramowitz, 1986). One reason for the relatively better postoperative adjustment of the female-to-male transsexuals may be society's more accepting attitudes toward women who desire to become men (Abramowitz, 1986). Female-to-male transsexuals tend to be better adjusted socially before surgery as well (Kockott & Fahrner, 1988; Pauly, 1974), so their superior postoperative adjustment may be nothing more than a selection factor.

Programs across the country help transsexuals come to terms with themselves and adjust to living in a society in which they rarely feel welcome (Selvin, 1993). One example is the Gender Identity Project in New York City's Greenwich Village, which sponsors meetings where transsexuals get together and share common concerns. Such programs help create a sense of community for a group of people who feel alienated from the larger society.

GENDER ROLES AND STEREOTYPES

"Why can't a woman be more like a man?" You may recall this lyric from the song that Professor Henry Higgins sings in the musical *My Fair Lady.* In the song the professor laments that women are emotional and fickle, whereas men are logical and dependable. The "emotional woman" is a **stereotype**—a fixed, oversimplified, and sometimes distorted idea about a group of people. The "logical man" is also a stereotype, albeit a more generous one. Even emotions are stereotyped. People assume that women are more likely to experience feelings of fear, sadness, and sympathy, whereas men are more likely to experience anger and pride (Plant et al., 2000). Gender roles are stereotypes in that they evoke fixed, conventional expectations of men and women.

Stereotype A fixed, oversimplified, conventional idea about a group of people.

Our gender identities—our identification of ourselves according to our concepts of masculinity and femininity—do not determine the roles or behaviors that are deemed

An Elementary School Teacher. If you think there is something wrong with this picture, it is because you have fallen prey to traditional gender-role stereotypes. Tradition has prevented many women from seeking jobs in "male" preserves such as construction work, the military, and various professions. Tradition has also prevented many men from obtaining work in "female" domains such as secretarial work, nursing, and teaching at the elementary school level.

masculine or feminine in our culture. Cultures have broad expectations of men and women that are termed **gender roles.**

Some might think that people "play" gender roles in the same way that many actors play roles. That is, they sense that they are pretending, and they conform to what the director wants. (The "director" here may be society at large.) John Money, however, writes that gender roles are very much a part of the person:

> In the language of the theater, a gender role is not a script handed to an actor but a role incorporated into the actor, who, [changed] by it, manifests it in person. An actor does not simply learn a role, he or she assimilates and lives it. So also with gender role: a child assimilates and lives it, is inhabited by it, has it as a belonging, and manifests it to others . . . in word and deed. There is no one cause of a gender role. It develops under the influence of multiple factors, sequentially over time, from prenatal life onwards. Nature alone is not responsible, nor is nurture alone. They work together, hand in glove. (Money, 1994, p. 166)

The stereotypical female exhibits traits such as gentleness, dependency, kindness, helpfulness, patience, and submissiveness. The masculine gender-role stereotype is one of toughness, gentlemanliness, and protectiveness. Females are generally seen as warm and emotional, males as independent, assertive, and competitive. The times are a-changing—somewhat. Women, as well as men, now bring home the bacon, but women are still more often expected to fry it in the pan and bear the primary responsibility for child rearing (Deaux & Lewis, 1983). A survey of 30 countries confirmed that these gender role stereotypes are widespread (Williams & Best, 1994; see Table 6.1). In some cultures, however, women are reared to be the hunters and food gatherers while men stay close to home and tend the children.

Gender roles Complex clusters of ways in which males and females are expected to behave.

TABLE 6.1

Gender-role stereotypes in 30 nations from around the world

Stereotypes of Males		Stereotypes of Females	
Active	Opinionated	Affectionate	Nervous
Adventurous	Pleasure-seeking	Appreciative	Patient
Aggressive	Precise	Cautious	Pleasant
Arrogant	Quick	Changeable	Prudish
Autocratic	Rational	Charming	Self-pitying
Capable	Realistic	Complaining	Sensitive
Coarse	Reckless	Complicated	Sentimental
Conceited	Resourceful	Confused	Sexy
Confident	Rigid	Dependent	Shy
Courageous	Robust	Dreamy	Softhearted
Cruel	Sharp-witted	Emotional	Sophisticated
Determined	Show-off	Excitable	Submissive
Disorderly	Steady	Fault-finding	Suggestible
Enterprising	Stern	Fearful	Superstitious
Hardheaded	Stingy	Fickle	Talkative
Individualistic	Stolid	Foolish	Timid
Inventive	Tough	Forgiving	Touchy
Loud	Unscrupulous	Frivolous	Unambitious
Obnoxious		Fussy	Understanding
		Gentle	Unstable
		Imaginative	Warm
		Kind	Weak
		Mild	Worrying
		Modest	

Source: From J. E. Williams and D. L. Best, "Cross-Cultural Views of Women and Men" in *Psychology and Culture*, p. 193, Table 1. Ed. by W. J. Lonner and R. Malpass. Copyright 1994 by Allyn & Bacon. Reprinted by permission.

Psychologists John Williams and Deborah Best (1994) found that people in 30 countries largely agreed on what constituted masculine and feminine gender-role stereotypes.

SEXISM

We have all encountered the effects of **sexism**—the prejudgment that because of gender, a person will possess certain negative traits. These negative traits are assumed to disqualify the person for certain vocations or to prevent him or her from performing adequately in these jobs or in some social situations.

Sexism may even lead us to interpret the same behavior in different ways, depending on whether it is performed by women or by men. We may see the man as "self-assertive" but the woman as "pushy." We may look upon *him* as flexible but brand *her* fickle and indecisive. *He* may be rational, whereas *she* is cold. *He* is tough when necessary, but *she* is bitchy. When the businesswoman engages in stereotypical masculine behaviors, the sexist reacts negatively by branding her abnormal or unhealthy.

Sexism may make it difficult for men to act in ways that are stereotyped as feminine. A sensitive woman is simply sensitive, but a sensitive man may be seen as a "sissy." A woman may be perceived as polite, whereas a man who exhibits the same behavior seems passive or weak. Only recently have men begun to enter occupational domains previously restricted largely to women, such as secretarial work, nursing, and teaching in the primary grades. Only recently have the gates opened to admit women into traditionally masculine professions such as engineering, law, and medicine.

Sexism The prejudgment that because of gender, a person will possess certain negative traits.

Although children of both genders have about the same general learning ability, stereotypes limit their horizons. Children tend to demonstrate preferences for gender-typed activities and toys by as early as 2 or 3 years of age. If they should stray from these expected preferences, their peers are sure to show them the "error of their ways." How many little girls are discouraged from considering professions like architecture and engineering because they are handed dolls, not blocks and fire trucks? How many little boys are discouraged from pursuing child-care and nursing professions because of the "funny" looks they get from others when they reach for dolls?

Children not only develop stereotyped attitudes about play activities; they also develop stereotypes about the differences between "man's work" and "woman's work." Women have been historically excluded from "male occupations," and stereotypical expectations concerning "men's work" and "women's work" filter down to the primary grades. For example, according to traditional stereotypes, women are *not expected* to excel in math. Exposure to such negative expectations may discourage women from pursuing careers in science and technology. Even when they choose a career in science or technology, women are often subject to discrimination in hiring, promotions, allocation of facilities for research, and funds to conduct research (Loder, 2000).

GENDER DIFFERENCES: *VIVE LA DIFFÉRENCE* OR *VIVE LA SIMILARITÉ?*

If the genders were not anatomically different, this book would never have been written. How do the genders differ in cognitive abilities and personality, however?

Differences in Cognitive Abilities

Assessments of intelligence do not show overall gender differences in cognitive abilities (Halpern & LaMay, 2000). However, reviews of the research suggest that girls are somewhat superior to boys in verbal abilities, such as verbal fluency, ability to generate words that are similar in meaning to other words, spelling, knowledge of foreign languages, and pronunciation (Halpern, 1997). Far more boys than girls have reading problems, ranging from reading below grade level to severe disabilities.

Males generally exceed females in visual–spatial abilities (Grön et al., 2000; Halpern & LaMay, 2000). Visual–spatial skills include the ability to follow a map when traveling to an unfamiliar location, to construct a puzzle or assemble a piece of equipment, and to perceive relationships between figures in space. (See Figure 6.5.)

Studies in the United States and elsewhere find that males generally obtain higher scores on math tests than females (Beller & Gafni, 2000; Gallagher et al., 2000; Halpern & LaMay, 2000). Yet differences in math are small and are narrowing at all ages (Hyde et al.,

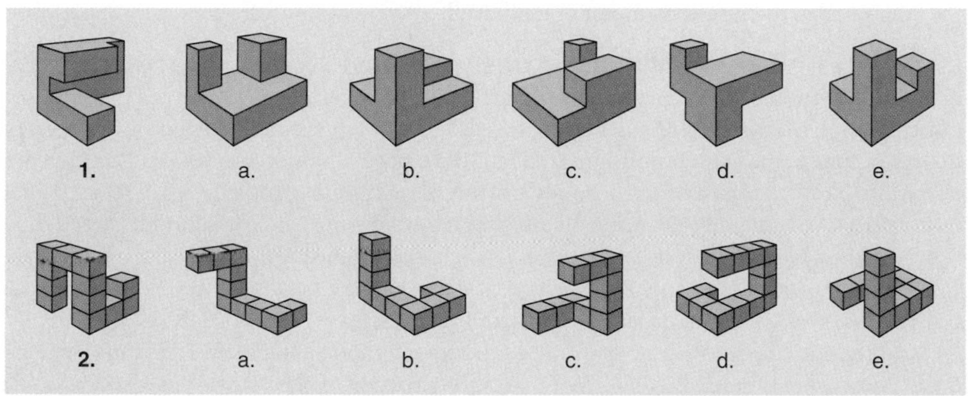

FIGURE 6.5

Rotating Geometric Figures in Space. Visual–spatial skills (for example, the ability to rotate geometric figures in space) are considered part of the male gender-role stereotype. But such gender differences are small and can be modified by training.

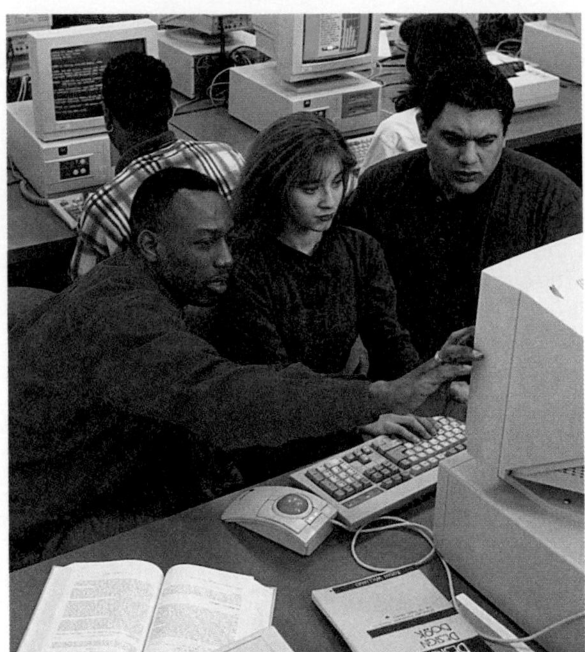

Are There Gender Differences in Cognitive Abilities? The physical differences between females and males are well established—and well celebrated! But are there cognitive differences between the genders? If so, what are they. How large are they? Are they the result of nature (heredity) or nurture (environmental influences such as educational experiences and cultural expectations)?

1990). Females excel in computational ability in elementary school. However, males excel in mathematical problem solving in high school and in college (Hyde et al., 1990). Differences in problem solving are reflected on the mathematics test of the Scholastic Aptitude Test (SAT). The mean score is 500, and about two thirds of the test takers receive scores between 400 and 600. Twice as many boys as girls attain math scores over 500. According to Byrnes and Takahira (1993), boys' superiority in math does not reflect gender per se. Instead, boys do well because of prior knowledge of math and their strategies for approaching math problems.

Three factors should caution us not to attach too much importance to these gender differences, however.

1. In most cases, the differences are small (Hyde & Plant, 1995). Differences in verbal, mathematical, and spatial abilities are also getting smaller (Hyde et al., 1990; Maccoby, 1990; Voyer et al., 1995).
2. These gender differences are *group* differences. Variation in ability on tests of verbal or math skills is larger *within each gender* than between the genders (Maccoby, 1990). Despite differences between groups of boys and girls, millions of boys exceed the "average" girl in writing and spelling skills. Likewise, millions of girls outperform the "average" boy in problem-solving and spatial tasks. The male gender has produced its Shakespeares and the female gender its Madame Curies.
3. The small differences that may exist may largely reflect environmental influences and cultural expectations. Spatial and math skills are stereotyped in our culture as masculine, whereas reading skills are stereotyped as feminine. Women who are given just a few hours of training in spatial skills—for example, rotating geometric figures or studying floor plans—perform at least as well men on tests of these skills (Baenninger & Elenteny, 1997; Lawton & Morrin, 1999).

Differences in Personality

There are also many gender differences in personality. According to a meta-analysis of the research literature, females exceed males in extraversion, anxiety, trust, and nurturance (Feingold, 1994). Males exceed females in assertiveness, tough-mindedness, and self-esteem. The assertiveness is connected with aggressiveness, as we see below. The tough-mindedness has unfortunate implications for men's health, as we also see. Your third author notes, with some displeasure, two factors that may largely account for the relatively lower self-esteem of females:

- Parents, on the average, prefer to have boys.
- Society has created an unlevel playing field in which females have to perform better than males to be seen as doing equally well.

DIFFERENCES IN COMMUNICATION STYLES: "HE'S JUST AN OLD CHATTERBOX" We have been inundated with cartoons of suburban housewives gossiping across the fence or pouring endless cups of coffee when "the girls" drop by for a chat. Research has shown, however, that in many situations, men spend more time talking than women do. Men are also more likely to introduce new topics and interrupt others (Brooks, 1982; Deaux, 1985; Hall, 1984). Girls tend to be more talkative than boys during early childhood. By the time they enter school, however, boys dominate classroom discussions (Sadker & Sadker, 1994). As girls mature, it appears that they learn to "take a back seat" to boys and let the boys do most of the talking when they are in mixed-gender groups (Hall, 1984).

Women are more willing than men to disclose their feelings and personal experiences, however (Dindia & Allen, 1992). The stereotype of the "strong and silent" male

Men and Women Iron Their Way to Domestic Equality

British men and women will iron out traditional imbalances over who does the housework, but not for another 15 years, a survey showed. The report by the Future Foundation think tank for the Abbey National Bank said gender roles have been breaking down as more women work outside the home.

In 1960, men spent about 10 minutes a day on cooking and cleaning, compared to 110 minutes for working women. These days, husbands and boyfriends do about 50 minutes of chores, about half the time put in by their employed partners. But the survey of 950 people showed that current trends pointed to hope of domestic equality by 2015, when both men and women will spend about an hour a day on housework.

"In the future, the most important thing will be who has the time or the inclination to do housework, and not whether they are a man or woman," Future Foundation founder Michael Willmott told the *Daily Telegraph* newspaper.

The survey revealed that two-thirds of men claimed to be taking on more chores than their fathers, with the same proportion of women saying they did less than their mothers.

Source: From "Men, Women Iron Way to Domestic Equality" (2000, October 2). *Reuters News Agency online.* Copyright © 2000 by Reuters Limited. Reprinted by permission.

may not discourage men from hogging the conversation, but it may inhibit them from expressing their personal feelings.

DIFFERENCES IN AGGRESSIVENESS In nearly all cultures (Ford & Beach, 1951; Mead, 1935), it is the males who march off to war and who battle for fame, glory, and shaving-cream-commercial contracts in stadiums and arenas. In most psychological studies on aggression, males have been found to behave more aggressively than females.

Truth or Fiction? **REVISITED**

It is true that men act more aggressively, on the whole, than women do. The question remains, *why?*

DIFFERENCES IN WILLINGNESS TO SEEK HEALTH CARE Men's life expectancies are seven years shorter, on the average, than women's. Part of the difference, according to surveys of physicians and of the general population, is due to women's greater willingness to seek health care (Courtenay, 2000). Men often let symptoms go until a problem that could have been prevented or readily treated becomes serious or life-threatening. Women, for example, are much more likely to check themselves for breast cancer than men are to even recognize the symptoms of prostate cancer. Many men have a "bullet-proof mentality." They are too strong to see the doctor in their 20s, too busy in their 30s, and too frightened later on.

ON BECOMING A MAN OR A WOMAN: GENDER TYPING

We have chronicled the biological process of sexual differentiation, and we have explored some gender differences in cognitive abilities and behavior. In this section, we consider various explanations of **gender typing.**

Biological Perspectives

Biological views on gender typing tend to focus on the roles of genetics and prenatal influences in predisposing men and women to gender-linked behavior patterns. Biological perspectives have also focused on the possible role of hormones in sculpting the brain during prenatal development.

THE EVOLUTIONARY PERSPECTIVE: IT'S ONLY NATURAL From the evolutionary perspective, the story of the survival of our ancient ancestors is etched in our genes. Those genes that bestow attributes that increase an organism's chances of surviving to produce viable offspring are most likely to be transmitted to future generations. We thus possess the genetic remnants of traits that helped our ancestors survive and reproduce (Bjorklund

Gender typing The process by which children acquire behavior that is deemed appropriate to their gender.

& Kipp, 1996; Fisher, 2000). This heritage influences our social and sexual behavior as well as our anatomic features.

According to the evolutionary perspective, men's traditional roles as hunters and warriors, and women's roles as caregivers and gatherers of fruits and vegetables, are bequeathed to us in our genes. Men are better suited to war and to the hunt because of physical attributes passed along since ancestral times. Upper-body strength, for example, would have enabled them to throw spears and overpower adversaries. Men also possess perceptual–cognitive advantages, such as superior visual–motor skills that favor aggression. Visual–motor skills would have enabled men to aim spears or bows and arrows effectively.

Women, it is argued, are genetically predisposed to be empathic and nurturant because these traits enabled ancestral women to respond to children's needs and enhance the likelihood that their children would flourish and eventually reproduce, thereby transmitting their own genetic legacy to future generations. Prehistoric women thus tended to stay close to home, care for the children, and gather edible plants, whereas men ventured from home to hunt and raid their neighbors' storehouses.

The evolutionary perspective is steeped in controversy. Although scientists do not dispute the importance of evolution in determining physical attributes, many are reluctant to attribute complex social behaviors, such as aggression and gender roles, to heredity. The evolutionary perspective implies that stereotypical gender roles—men as breadwinners and women as homemakers, for example—reflect the natural order of things. Critics contend that biology is not destiny, that our behavior is not dictated by our genes.

PRENATAL BRAIN ORGANIZATION Researchers have sought the origins of gender-typed behavior in the organization of the brain. Is it possible that the cornerstone of gender-typed behavior is laid in the brain before the first breath is taken?

The hemispheres of the brain are specialized to carry out certain functions (Levy, 1985). In most people, the right hemisphere ("right brain") appears to be specialized to perform visual–spatial tasks. The "left brain" appears to be more critical to verbal functions, such as speech, in most people.

We know that sex hormones are responsible for prenatal sexual differentiation of the genitals and for the gender-related structural differences in the hypothalamus of the developing prenatal brain. Sexual differentiation of the brain may also partly explain men's (slight!) superiority at spatial-relations tasks, such as interpreting road maps and visualizing objects in space. Testosterone in the brains of male fetuses spurs greater growth of the right hemisphere and slows the rate of growth of the left hemisphere. This difference may be connected with the ability to accomplish spatial-relations tasks.

Might boys' inclinations toward aggression and rough-and-tumble play also be prenatally imprinted in the brain? Some theorists argue that prenatal sex hormones may masculinize or feminize the brain by creating predispositions that are consistent with gender-role stereotypes, such as rough-and-tumble play and aggressive behavior in males (Collaer & Hines, 1995). Money (1994) allows a role for prenatal dispositions but argues that social learning plays a stronger role in gender typing. He claims that social learning is even potent enough to counteract prenatal predispositions.

Cross-Cultural Perspectives

The evolutionary perspective does not account for differences in gender roles that exist across cultures, especially neighboring cultures. The anthropologist Margaret Mead (1935) lived among several tribes on the South Pacific island of New Guinea and found that gender roles in these tribes differed not only from those of Western culture, but also from one another.

Among the Mundugumor, a tribe of headhunters and cannibals, both men and women were warlike and aggressive. The women disdained bearing and rearing children because it interrupted participation in warring parties against neighboring villages. The

Cross-Cultural Perspective on Gender Typing. Anthropologists have found that gender roles in various tribes differ not only from those found in Western culture but also from one another. People can be racially similar and live in similar physical environments and yet be subject to quite different cultural influences.

men and women of the Arapesh tribe were gentle and peaceful, by contrast. Both genders nurtured the children. The Tchambuli were even more unusual in terms of what we consider stereotypical behavior in our society. The men spent most of their time caring for children, gossiping, bickering, primping and applying makeup, and haggling over prices. Fish was the staple diet of the Tchambuli, and women brought home the daily catch. The women kept their heads shaven, disdained ornaments, and were more highly sexed and aggressive than the men.

Whatever the influence of biology on behavior, biological factors alone do not make men aggressive or independent, or women passive or submissive (Harris & Johnson, 2000; Whitten, 2001). Cultural expectations and learning play a large role.

GENDER ROLES AS A CULTURAL ADAPTATION The Mundugumor, Arapesh, and Tchambuli peoples of New Guinea and the members of modern industrialized societies all share the same biological makeup. Within each gender, the same sex hormones pulse through the arteries of the peoples of New Guinea as through the arteries of stockbrokers on Wall Street. (Yes, even stockbrokers are warm-blooded.) Yet despite this common biological makeup, wide cultural variations exist with respect to gender roles.

Anthropologists believe that cultural differences in gender roles can be explained in terms of the adaptations that cultures make to their social and natural environments (Werner & Cohen, 1990). Consider the differences in gender roles between the Sambian people of New Guinea and the !Kung people of Africa. The Sambians have rigidly defined gender roles. Boys are socialized to become warriors. Women tend the children and keep a discreet distance from men (Herdt, 1987). Among the !Kung people, however, women play a more active role in tribal affairs and enjoy more autonomy (Drapers, 1975).

How might such differences arise? Until recently, the Sambians were subject to repeated attacks from enemies and could survive only by rearing their sons as warriors (Werner & Cohen, 1990). Sambians' rigid gender roles may be seen as an adaptation to these onslaughts. The !Kung live in small, scattered groups and forage for their food. Both genders make substantial contributions to the food supply. In this egalitarian society, gender roles are more flexible. Both men and women have considerable autonomy and influence.

Psychological Perspectives

Children acquire awareness of gender-role stereotypes by the tender ages of 2½ to 3½ (Etaugh & Rathus, 1995). Both boys and girls generally agree, when asked to describe the differences between the genders, that boys build things, play with transportation toys such as cars and fire trucks, enjoy helping their fathers, and hit other children. Both boys and girls also agree that girls enjoy playing with dolls and helping their mothers cook and clean and that they are talkative, dependent on others for help, and nonviolent. They perceive the label "cruel" to be a masculine trait, whereas "cries a lot" is perceived as a feminine trait. By the time they are age 3, most children have become aware of the different stereotypical ways in which men and women dress and of the types of occupations that are considered appropriate for each gender (Etaugh & Rathus, 1995). Psychologists have applied psychodynamic, social-learning, and cognitive-developmental theories in an effort to explain how children acquire such knowledge and adopt stereotypical behavior patterns.

PSYCHODYNAMIC THEORY Sigmund Freud explained gender typing in terms of identification. Appropriate gender typing, in Freud's view, requires that boys come to identify with their fathers and girls with their mothers. Identification is completed, in Freud's view, as children resolve the **Oedipus complex** (which is sometimes called the Electra complex in girls).

According to Freud, the Oedipus complex occurs during the phallic period of psychosexual development, from the ages of 3 to 5. During this period, the child develops incestuous wishes for the parent of the other gender and comes to perceive the parent of the same gender as a rival. The complex is resolved by the child's forsaking incestuous wishes for the parent of the other gender and identifying with the parent of the same gender. Through identification with the same-gender parent, the child comes to develop gender-typed behaviors that are typically associated with that gender. Children display stereotypical gender-typed behaviors earlier than Freud would have predicted, however. Even during the first year, boys are more independent than girls. Girls are more quiet and restrained. Girls show preferences for dolls and soft toys, and boys for hard transportation toys, by the ages of 1½ to 3.

SOCIAL-LEARNING THEORY Social-learning theorists explain the development of gender-typed behavior in terms of processes such as observational learning, identification, and socialization. Children can learn what is deemed masculine or feminine by observational learning, as suggested by the results of an experiment by Perry and Bussey (1979). In this study, 8- and 9-year-old boys and girls watched adult role models indicate their preferences on each of 16 pairs of items—pairs such as toy cows versus toy horses and oranges versus apples. What the children didn't know was that the expressed preferences were made arbitrarily. The children then were asked to indicate their own preferences for the items represented in the pairs. The boys' choices agreed with the adult men's an average of 14 out of 16 times. Girls chose the pair item selected by the men an average of only 3 out of 16 times.

In social-learning theory, identification is viewed as a continuing and broadly based learning process in which rewards and punishments influence children to imitate adult models of the same gender—especially the parent of the same gender (Storms, 1979). Identification is more than imitation, however. In identification, the child not only imitates the behavior of the model but also tries to become like the model in broad terms.

Socialization also plays a role in gender typing. Almost from the moment a baby comes into the world, it is treated according to its gender. Parents tend to talk more to baby girls, and fathers especially engage in more roughhousing with boys (Jacklin et al., 1984). When children are old enough to speak, parents and other adults—even other children—begin to instruct children in how they are expected to behave. Parents may reward children for behavior they consider gender-appropriate and punish (or fail to reinforce)

Oedipus complex A conflict of the phallic stage in which the boy wishes to possess his mother sexually and perceives his father as a rival in love.

Socialization The process of guiding people into socially acceptable behavior patterns by means of information, rewards, and punishments.

them for behavior they consider inappropriate for their gender. Girls are encouraged to practice caretaking behaviors, which are intended to prepare them for traditional feminine adult roles. Boys are handed Legos or doctor sets to help prepare them for traditional masculine adult roles.

Fathers generally encourage their sons to develop assertive, instrumental behavior (that is, behavior that gets things done or accomplishes something) and their daughters to develop nurturant, cooperative behavior. Fathers are likely to cuddle their daughters gently. They are likely to carry their sons like footballs or toss them into the air. Fathers also tend to use heartier and harsher language with their sons, such as "How're yuh doin', Tiger?" and "Hey you, get your keester over here" (Jacklin et al., 1984). Being a nontraditionalist, your first author made sure to toss his young daughters into the air, which raised immediate objections from the relatives, who chastised him for being too rough. This, of course, led him to modify his behavior: He learned to toss his daughters into the air when the relatives were not around.

Generally speaking, from an early age, boys are more likely to receive toy cars and guns and athletic equipment and to be encouraged to compete aggressively. Even relatively sophisticated college students are likely to select traditionally masculine toys as gifts for boys and traditionally feminine toys for girls (Fisher-Thompson, 1990). Girls are spoken to more often, whereas boys are handled more frequently and more roughly. Whatever the biological determinants of gender differences in aggressiveness and verbal skills, early socialization experiences clearly contribute to gender typing.

Parental roles in gender typing are apparently changing. With more mothers working outside the home, daughters today are exposed to more women who represent career-minded role models than was the case in earlier generations. More parents today are encouraging their daughters to become career-minded and to engage in strenuous physical activities, such as organized sports. Many boys today are exposed to fathers who take a larger role than men used to in child care and household responsibilities.

Schools are also important socialization influences. Schools have been slow to adapt to recent changes in gender roles (Sadker & Sadker, 1994). They may be exposing children to masculine and feminine images that are even more rigid and polarized than those currently held in society at large. Teachers often expect girls to perform better than boys in reading and language arts and have higher expectations of boys in math and science. These expectations may be conveyed to children, patterning their choices of careers.

Sexism in America's schools is widespread. Many tests remain biased against girls, which reduces their chances of obtaining scholarships and gaining admission to more competitive colleges (Chira, 1992). Many science teachers and some math teachers tend to ignore girls in favor of boys. Such biases may discourage young women with aptitude in math and science from pursuing careers in these areas. Yet the report highlighted some promising developments. The traditional gender gap in math scores is narrowing; girls have made significant gains in catching up to boys. Moreover, special programs in math and science held for girls after school and in the summer have helped bolster the girls' confidence and interest in these subjects.

The popular media—books, magazines, radio, film, and especially television—also convey gender stereotypes (Remafedi, 1990). The media by and large portray men and women in traditional roles (Signorielli, 1990). Men more often play doctors, attorneys, and police officers. Women more often play nurses, secretaries, paralegals, and teachers. Even when women portray attorneys or police officers, they are more likely than men to handle family disputes. The male police officer is more likely to be shown in action roles; the male holds the court spellbound with a probing cross-examination. Working women are also more likely than men to be portrayed as undergoing role conflict—being pulled in opposite directions by job and family. One study reported that despite current awareness of sexism, "Women are often still depicted on television as half-clad and half-witted, and needing to be rescued by quick-thinking, fully clothed men" (Adelson, 1990). Ageism buttresses sexism in that female characters age 40 and above are only rarely depicted in roles other than mothers and grandmothers.

Social-learning theorists believe that aggression is largely influenced by learning. Boys are permitted—even encouraged—to engage in more aggressive behavior than girls. Nonetheless, females are likely to act aggressively under certain conditions. Ann Frodi and her colleagues (1977) reviewed 72 studies that examined gender differences in aggression. All in all, females acted as aggressively as men when they were given the physical means to do so and believed that aggression was justified. In an influential review article, Maccoby and Jacklin commented on the socialization influences that discourage aggression in girls:

> Aggression in general is less acceptable for girls, and is more actively discouraged in them, by either direct punishment, withdrawal of affection, or simply cognitive training that "that isn't the way girls act." Girls then build up greater anxieties about aggression, and greater inhibitions against displaying it. (1974, p. 234)

Social-learning theorists have made important contributions to our understanding of how rewards, punishments, and modeling influences foster gender-typed behavior patterns. How do children integrate gender-role expectations within their self-concepts? And how do their concepts concerning gender influence their development of gender-typed behavior? Let us consider two cognitive approaches to gender typing that shed light on these matters: cognitive-developmental theory and gender schema theory.

COGNITIVE-DEVELOPMENTAL THEORY Psychologist Lawrence Kohlberg (1966) proposed a cognitive-developmental view of gender typing. From this perspective, gender typing is not the product of environmental influences that mechanically "stamp in" gender-appropriate behavior. Rather, children themselves play an active role. They form con-

Gender Typing Through Observational Learning. According to social-learning theory, people learn about the gender roles that are available to them—and expected of them—at an early age. Gender schema theory adds that once children have learned the expected gender roles (i.e., the gender schema of their culture), they blend these roles with their self-concepts. Their self-esteem comes to be dependent on their adherence to the expected gender roles.

cepts, or **schemas,** about gender and then exhibit behavior that conforms to their gender concepts. These developments occur in stages and are entwined with general cognitive development.

According to Kohlberg, gender typing entails the emergence of three concepts: *gender identity, gender stability,* and *gender constancy.* Gender identity is usually acquired by the age of 3. By the age of 4 or 5, most children develop a concept of **gender stability**—the recognition that people retain their genders for a lifetime. Prior to this age, boys may think that they will become mommies when they grow up, and girls may think they will be daddies.

It is true that a 2½-year-old child may know that he is a boy but think that he can grow up to be a mommy. Children at this age have not yet developed gender stability. ■

The more sophisticated concept of **gender constancy** develops in most children by the age of 7 or 8. They recognize that gender does not change even when people alter their dress or behavior. Hence gender remains constant even when appearances change. A woman who wears her hair short (or shaves it off) remains a woman. A man who dons an apron and cooks dinner remains a man.

According to cognitive-developmental theory, children are motivated to behave in gender-appropriate ways once they have established the concepts of gender stability and gender constancy. They then make an active effort to learn which behavior patterns are considered "masculine" and which "feminine" (Perry & Bussey, 1979). Once they obtain this information, they imitate the "gender-appropriate" pattern. Thus, boys and girls who come to recognize that their genders will remain a fixed part of their identity will show preferences for "masculine" and "feminine" activities, respectively. Researchers find, for instance, that boys who had achieved gender constancy played with an uninteresting gender-typed toy for a longer period of time than did boys who hadn't yet achieved gender constancy (Frey & Ruble, 1992). Both groups of boys played with an interesting gender-typed toy for about the same time.

Cross-cultural studies of the United States, Samoa, Nepal, Belize, and Kenya find that the concepts of gender identity, gender stability, and gender constancy emerge in the order predicted by Kohlberg. However, gender-typed play often emerges at an earlier age than would be predicted by the cognitive-developmental theory. Many children make gender-typed choices of toys by the age of 2. Children as young as 18 months are likely to have developed a sense of gender identity, but gender stability and constancy are some years off. Gender identity alone thus seems sufficient to prompt children to assume gender-typed behavior patterns. Psychologist Sandra Bem (1983) also notes that Kohlberg's theory does not explain why the concept of gender plays such a prominent role in children's classification of people and behavior. Another cognitive view, gender schema theory, attempts to address these concerns.

GENDER SCHEMA THEORY: AN INFORMATION-PROCESSING APPROACH Gender schema theory proposes that children develop a **gender schema** as a means of organizing their perceptions of the world (Bem, 1981, 1985; Martin & Halverson, 1981). A gender schema is a cluster of mental representations about male and female physical qualities, behaviors, and personality traits. Gender gains prominence as a schema for organizing experience because of society's emphasis on it. Even young children start to mentally group people of the same gender in accordance with the traits that represent that gender.

Children's gender schemas determine how important gender-typed traits are to them. Consider the dimension of *strength–weakness.* Children may learn that strength is connected with maleness and weakness with femaleness. (Other dimensions, such as *light–dark,* are not gender-typed and thus may fall outside children's gender schemas.) Children also gather that some dimensions, such as strong–weak, are more important to one gender (in this case, male) than to the other.

Schema Concept; way of interpreting experience or processing information.

Gender stability The concept that people retain their genders for a lifetime.

Gender constancy The concept that people's genders do not change, even if they alter their dress or behavior.

Gender schema A cluster of mental representations about male and female physical qualities, behaviors, and personality traits.

Once children acquire a gender schema, they begin to judge themselves in accordance with traits considered appropriate to their genders. In doing so, they blend their developing self-concepts with the prominent gender schema of their culture. The gender schema furnishes standards for comparison. Children with self-concepts that are consistent with the prominent gender schema of their culture are likely to develop higher self-esteem than children whose self-concepts are inconsistent. Jack learns that muscle strength is a characteristic associated with "manliness." Thereafter, he is likely to think more highly of himself if he perceives himself as embodying this attribute than if he does not. Barbara is likely to discover that the dimension of kindness–cruelty is more crucial than strength–weakness to the way women are perceived in society.

According to gender schema theory, gender identity itself is sufficient to inspire gender-appropriate behavior. Once children develop a concept of gender identity, they begin to seek information concerning gender-typed traits and strive to live up to them. Jack will retaliate when provoked, because boys are expected to do so. Barbara will be "sugary and sweet" if such is expected of little girls. Thus, gender-typed behavior emerges earlier than cognitive-developmental theory predicts. Jack's and Barbara's self-esteem depends in part on how they measure up to the gender schema.

Research suggests that children do process information according to a gender schema (Levy & Carter, 1989; Stangor & Ruble, 1989). Objects and activities pertinent to a child's own gender are better retained in memory. Boys, for example, do a better job of remembering transportation toys they have been shown, whereas girls are better at recalling dolls and other "feminine" objects (Bradbard & Endsley, 1984). In another study, Martin and Halverson (1983) showed elementary school children pictures of children involved in "gender-consistent" or "gender-inconsistent" activities. Gender-consistent pictures showed boys doing things like sawing wood and playing with trains. Girls were shown doing things like cooking and cleaning. Gender-inconsistent pictures showed models of the other gender involved in gender-typed endeavors. A week later, the children were asked whether boys or girls had engaged in each activity. Boys and girls both made errors in their efforts to recall the genders of the models shown engaging in "gender-inconsistent" behavior.

Once established, gender schemas resist change, even when broad social changes occur. For this reason, many parents cannot accept their sons' wearing ("feminine") earrings. A couple of generations ago, parents had difficulty accepting daughters in jeans or sons in long hair.

GENDER ROLES AND SEXUAL BEHAVIOR

Gender roles have had a profound influence on dating practices and sexual behavior. Children learn at an early age that men usually make dates and initiate sexual interactions, whereas women usually serve as the "gatekeepers" in romantic relationships (Bailey et al., 2000). In their traditional role as gatekeepers, women are expected to wait to be asked out and to screen suitors. Men are expected to make the first (sexual) move and women to determine how far advances will proceed. Regrettably, some men refuse to take no for an answer. They feel that they have the right to force their dates into sexual relations.

Men as Sexually Aggressive, Women as Sexually Passive

The cultural expectation that men are initiators and women are gatekeepers is embedded within the larger stereotype that men are sexually aggressive and women are sexually passive. Men are expected to have a higher number of sex partners than women do (Mikach & Bailey, 1999). Men not only initiate sexual encounters; they are also expected to dictate all the "moves" thereafter, just as they are expected to take the lead on the dance floor. People who adhere to the masculine gender-role stereotype, whether male or female, are more likely to engage in risky (unprotected) sexual behavior (Belgrave et al., 2000). According to the stereotype, women are supposed to let men determine the choice, timing, and sequence of sexual positions and techniques. Unfortunately, the stereotype favors men's sexual preferences, denying women the opportunity to give and receive their preferred kinds of stimulation.

For example, a woman may more easily reach orgasm in the **female-superior position,** but her partner may prefer the **male-superior position.** If the man is calling the shots, she may not have the opportunity to reach orgasm. Even expressing her preferences may be deemed "unladylike."

The stereotypical masculine role also imposes constraints on men. Men are expected to take the lead in bringing their partners to orgasm, but they should not ask their partners what *they* like because men are expected to be natural experts. ("Real men" not only don't eat quiche; they also need not ask women how to make love.)

Fortunately, more flexible attitudes are emerging. Women are becoming more sexually assertive, and men are becoming more receptive to expressing tenderness and gentleness. Still, the roots of traditional gender roles run deep.

Men as Overaroused, Women as Underaroused

According to another stereotype, men become sexually aroused at puberty and remain at the ready throughout adulthood. Women, the sterotype continues, do not share men's natural interest in sex, and a woman discovers her own sexuality only when a man ignites her sexual flame. Men must continue to stoke women's sexual embers, lest they die out. This stereotype denies that "normal" women have spontaneous sexual desires or are readily aroused.

It was widely believed in the Victorian period (even by so-called sex experts!) that women were naturally asexual and "unbothered" by sexual desires. The contemporary residues of this stereotype hold that women do not enjoy sex as much as men and that women who openly express their sexual desires are "whores" or "sluts." The stereotype that women are undersexed also supports the traditional double standard: It is natural for men to sow their wild oats, but women who are sexually active outside of committed relationships are sluts or *nymphomaniacs.*

Despite the stereotype, women are no less arousable than men. Nor do they wait for men to discover their sexuality. Long before they have intimate relationships, children of both genders routinely discover that touching their genitals produces pleasurable sensations.

PSYCHOLOGICAL ANDROGYNY: THE MORE TRAITS, THE MERRIER?

Most people think of masculinity and femininity as opposite ends of one continuum (Storms, 1980). People tend to assume that the more masculine a person is, the less feminine he or she must be, and vice versa. Thus, a man who exhibits stereotypical feminine traits of nurturance, tenderness, and emotionality is often considered less masculine than other men. Women who compete with men in business are perceived not only as more masculine than other women but also as less feminine.

Some behavioral scientists, such as Sandra Bem, argue that masculinity and femininity instead constitute separate personality dimensions. A person who is highly masculine, whether male or female, may also possess feminine traits—and vice versa. People who exhibit "masculine" assertiveness and instrumental skills (skills in the sciences and business, for example) along with "feminine" nurturance and cooperation fit both the masculine and the feminine gender-role stereotypes. They are said to show **psychological androgyny** (see Figure 6.6). People only high in assertiveness and instrumental skills fit the masculine stereotype. People high only in traits such as nurturance and cooperation fit the

Female-superior position
A coital position in which the woman is on top.

Male-superior position
A coital position in which the man is on top.

Psychological androgyny
A state characterized by possession of both stereotypical masculine traits and stereotypical feminine traits.

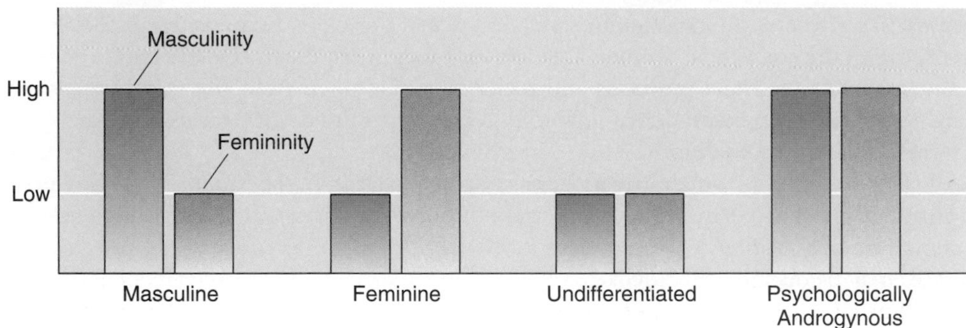

FIGURE 6.6

A Model of Psychological Androgyny. Some behavioral scientists argue that masculinity and femininity are independent personality dimensions. People who exhibit "masculine" assertiveness and instrumental skills along with "feminine" nurturance and cooperation are said to be psychologically androgynous. People high only in assertiveness and instrumental skills fit the masculine stereotype. People high only in traits such as nurturance and cooperation fit the feminine stereotype. People low in the stereotypical masculine *and* feminine patterns are considered undifferentiated.

feminine stereotype. People low in the stereotypical masculine *and* feminine patterns are considered "undifferentiated" in terms of gender-role stereotypes.

People who are psychologically androgynous may be capable of summoning up a wider range of masculine and feminine traits to meet the demands of various situations and to express their desires and talents. Researchers, for example, have found psychologically androgynous persons of both genders to show "masculine" independence under group pressures to conform and to show "feminine" nurturance in interactions with a kitten or baby (Bem, 1975; Bem et al., 1976). Androgynous men and women are more apt to share leadership responsibilities in mixed-gender groups (Porter et al., 1985). By contrast, "masculine" men and women tend to dominate such groups, whereas "feminine" men and women are likely to take a back seat.

Many people who oppose the constraints of traditional gender roles may perceive psychological androgyny as a desirable goal. Some feminist writers, however, criticize psychological androgyny on the basis that the concept is defined in terms of, and thereby perpetuates, belief in the existence of masculine and feminine gender roles (Lott, 1985).

Psychological Androgyny, Psychological Well-Being, and Personal Development

Psychologically androgynous people tend to have higher self-esteem and to be generally better adjusted psychologically than people who are feminine or undifferentiated. Yet it appears that these benefits are more strongly related to the presence of masculine traits than to the combination of masculine and feminine traits (Whitley, 1983; Williams & D'Alessandro, 1994). That is, masculine traits such as assertiveness and independence may be related to psychological well-being, whether or not they are combined with feminine traits such as warmth, nurturance, and cooperation.

There is also evidence that feminine traits, such as nurturance and sensitivity, appear to predict success in intimate relationships—in men as well as in women. Marital happiness as rated by husbands is positively related to femininity in the wives (Antill, 1983). More interestingly, perhaps, ratings of marital happiness from the wives were also positively correlated with their husbands' femininity. Androgynous men are more likely to express tender feelings of love toward their partners and to be more accepting of their partner's faults than are masculine-typed ("macho") men (Coleman & Ganong, 1985). It seems that both genders appreciate spouses who are sympathetic, able to express warmth and tenderness, and nurturant toward children.

Masculine and androgynous adolescents of both genders tend to be more popular and to have higher self-esteem than other adolescents (Lamke, 1982b). We might not be surprised, given the prevalence of sexism, that adolescent boys fare better if they possess stereotypical masculine traits. What is more surprising is that adolescent girls also fare better when they exhibit stereotypical masculine traits, such as assertiveness and independence. It seems that young women do not risk having others question their femininity if they exhibit masculine traits, an observation that provides more evidence that the constellations of traits we call masculinity and femininity are independent clusters.

It is true that adolescent girls who show a number of masculine traits are more popular than girls who thoroughly adopt the traditional feminine gender role. Their exhibiting masculine traits such as independence and assertiveness apparently does not compromise their friends' perceptions of their femininity.

Psychological Androgyny and Sexual Behavior

Some evidence shows psychologically androgynous men and women to be more comfortable with their sexuality than are masculine men and feminine women (Walfish & Mayerson, 1980). Perhaps they can draw upon a broader repertoire of sexual behaviors. They may be comfortable with cuddling and tender holding *and* with initiating and directing sexual interactions. Researchers also find that androgynous women experience orgasm more frequently (Radlove, 1983), and express greater sexual satisfaction (Kimlika et al., 1983), than feminine women do.

In this chapter, we have focused on the biology and psychology of gender. Our gender, both anatomically and psychologically, is a primary aspect of our sexuality. In the next chapter, we begin to explore how we express our sexuality through intimate relationships with others.

Human Sexuality ONLINE //

Web Sites Related to Gender Identity and Gender Roles

This is a support Web site for individuals who are transgendered (transsexual) or are wondering whether they are transgendered.
www.heartcorps.com/journeys/

A directory of transgender organizations in various states.
www.transgender.org/

Web site of the National Transgender Advocacy Coalition.
www.ntac.org/

Web site of a cosmetic (plastic) surgeon with some interesting information for male-to-female transsexuals on feminizing the face.
www.heathers.net/doctor-o.htm

An Australian support Web site for individuals who are transgendered.
host2.mbcomms.net.au/tg/satsg/

The Intersex Society of North America: This Web site is generally for individuals with ambiguous genitals, and the message is that people should feel free to be or remain what they are.
www.isna.org/

Femina: This Web site has links to generally responsible Web sites "for, by, and about women." Issues include gender roles and sexism.
femina.cybergrrl.com/

summing up

PRENATAL SEXUAL DIFFERENTIATION

During the first six weeks or so of prenatal development, embryonic structures of both genders develop along similar lines and resemble primitive female structures.

■ Genetic Factors in Sexual Differentiation

At about the seventh week after conception, the genetic code (XX or XY) begins to assert itself, causing changes in the gonads, genital ducts, and external genitals. Genetic factors, such as the SRY gene, are critical in sexual differentiation.

■ The Role of Sex Hormones in Sexual Differentiation

Testosterone spurs differentiation of the male (Wolffian) duct system. In the absence of testosterone, the Wolffian ducts degenerate, and female sex organs develop.

■ Descent of the Testes and the Ovaries

The testes and ovaries develop in the abdominal cavity. About four months after conception, the testes descend into the scrotal sac.

■ Sex-Chromosomal Abnormalities

Abnormalities of the sex chromosomes can have profound effects on sexual characteristics, physical health, and psychological development. Examples include Klinefelter syndrome and Turner syndrome.

■ Prenatal Sexual Differentiation of the Brain

Gender-specific changes occur in the hypothalamus during prenatal development. Testosterone causes cells in the hypothalamus of male fetuses to become insensitive to estrogen.

GENDER IDENTITY

One's gender identity is one's sense of being male or of being female.

■ Nature and Nurture in Gender Identity

Gender identity is nearly always consistent with anatomic gender.

■ Hermaphroditism

Some individuals—hermaphrodites—are born with both ovarian and testicular tissue. Hermaphrodites usually assume the gender identity and gender role of the gender assigned at birth. Pseudohermaphrodites can acquire the gender identity of the other chromosomal gender when they are reared as members of that gender.

■ Transsexualism

Transsexuals harbor a deep sense of discomfort about their anatomic gender. Hormone treatments and gender-reassignment surgery provide transsexuals with many of the characteristics of the other gender.

GENDER ROLES AND STEREOTYPES

Cultures have broad expectations of men and women that are termed *gender roles.* In our culture, the stereotypical female is seen as gentle, dependent, kind, helpful, patient, and submissive. The stereotypical male is tough, competitive, gentlemanly, and protective.

SEXISM

Sexism is the prejudgment that because of gender, a person will possess certain negative traits that disqualify him or her for certain vocations or prevent him or her from performing adequately in these jobs or in some social situations. Women have been historically excluded from "male occupations," and stereotypical expectations concerning "men's work" and "women's work" filter down to the primary grades.

GENDER DIFFERENCES: *VIVE LA DIFFÉRENCE* OR *VIVE LA SIMILARITÉ*?

■ Differences in Cognitive Abilities

Boys have historically been seen as excelling in math and spatial-relations skills, whereas girls have been viewed as excelling in language skills. Gender differences in these areas are small, however, and cultural expectations play a role in them.

■ Differences in Personality

Stereotypical gender preferences for toys and play activities are in evidence at an early age. Males are more aggressive than females, but the question is *why?*

ON BECOMING A MAN OR A WOMAN: GENDER TYPING

■ Biological Perspectives

Biological views on gender typing focus on the roles of genetics and prenatal influences in predisposing men and women to gender-linked behavior patterns. Testosterone in the brains of male fetuses spurs greater growth of the right hemisphere, which may be connected with the ability to manage spatial-relations tasks.

■ Cross-Cultural Perspectives

Anthropologists have found that differences in gender roles exist among preliterate cultures and even between neighboring cultures.

■ Psychological Perspectives

Psychologists have attempted to explain gender typing in terms of psychodynamic, social-learning, and cognitive-developmental theories. Freud explained gender typing in terms of identification with the parent of the same gender through resolution of the Oedipus complex.

Social-learning theorists explain the development of gender-typed behavior in terms of processes such as observational learning, identification, and socialization.

According to the cognitive-developmental view, children form concepts about gender and then make their behavior conform to their gender concepts. According to Kohlberg, gender typing entails the emergence of three concepts: gender identity, gender stability, and gender constancy.

Gender schema theory proposes that children develop a gender schema as a means of organizing their perceptions of the world. Once children acquire a gender schema, they begin to judge themselves in terms of traits considered appropriate to their genders. In doing so, they blend their developing self-concepts with the prominent gender schema of their culture.

GENDER ROLES AND SEXUAL BEHAVIOR

Stereotypical gender-role expectations affect dating practices and sexual behavior.

Men as Sexually Aggressive, Women as Sexually Passive

According to this stereotype, men are sexual initiators and women sexual gatekeepers. Men not only initiate sexual encounters; they are also expected to initiate all the "moves."

Men as Overaroused, Women as Underaroused

According to another stereotype, women do not share men's interest in sex and discover their own sexuality only when a man ignites their sexual flame.

PSYCHOLOGICAL ANDROGYNY: THE MORE TRAITS, THE MERRIER?

Masculinity and femininity may be two independent personality dimensions. People who combine stereotypical masculine and feminine behavior patterns are psychologically androgynous.

Psychological Androgyny, Psychological Well-Being, and Personal Development

Masculine and androgynous people of both genders tend to be better psychologically adjusted than people who are feminine or undifferentiated.

Psychological Androgyny and Sexual Behavior

Psychologically androgynous men and women are more comfortable with their sexuality than masculine men and feminine women.

questions for critical thinking

1. What is your gender identity? What is the relationship between your anatomic sex and your gender identity?
2. What is the difference between transsexualism and a gay male or lesbian sexual orientation?
3. How do most people from your sociocultural background feel about traditional gender roles? Explain.
4. Have you encountered sexism? Under what circumstances? What was its effect on you?
5. How would you explain gender differences in aggressive behavior? Do you know some men who are aggressive and some who are not? How would you explain the differences from person to person?
6. Do you consider yourself a highly masculine man/a highly feminine woman? Why or why not?

chapter

7

Attraction and Love

Pablo Picasso. *The Lovers,* 1919. Oil on canvas. © Copyright ARS, NY. Photo R. G. Ojeda. Musée Picasso, Paris, France. Credit: Réunion des Musées Nationaux/Art Resource, NY. Credit: © 2002 Estate of Pablo Picasso/Artists Rights Society (ARS), New York.

outline

Truth or Fiction?

ATTRACTION

Physical Attractiveness: How Important Is Looking Good?

Human Sexuality Online Hospital Holds 'Net Auction for Breast Reduction

A World of Diversity Wide-Eyed with . . . Beauty?

A World of Diversity Gender Differences in Preferences in Mates Across 37 Cultures

The Matching Hypothesis: Who Is "Right" for You?

Similarity in Attitudes: Do Opposites Attract?

Reciprocity: If You Like Me, You Must Have Excellent Judgment

LOVE

The Greek Heritage

Romantic Love in Contemporary Western Culture

Questionnaire Are You a Romantic or a Realist? The Love Attitudes Scale

Contemporary Models of Love: Dare Science Intrude?

Questionnaire Sternberg's Triangular Love Scale

Human Sexuality Online Web Sites Related to Attraction and Love

SUMMING UP

QUESTIONS FOR CRITICAL THINKING

Truth or Fiction?

___ Beauty is in the eye of the beholder.

___ College men would like women to be thinner than the women want to be.

___ People are regarded as more attractive when they are smiling.

___ Women who are randomly assigned names like Kathy and Jennifer are rated as more attractive than women assigned names like Harriet and Gertrude.

___ Physical appeal is the most important trait we seek in partners for long-term relationships.

___ "Opposites attract." That is, we are more apt to be attracted to people who disagree with our views and tastes than to people who share them.

___ It is possible to be in love with someone who is not also a friend.

Candy and Stretch. A new technique for controlling weight gain? No, these are the names of a couple who have just met at a camera club that doubles as a meeting place for singles.

Candy and Stretch stand above the crowd—literally. She is almost 6 feet tall, an attractive woman in her early 30s. He is more plain looking, but "wholesome." He is in his late 30s and 6 feet 5 inches tall. Stretch has been in the group for some time. Candy is a new member. Let us follow them as they meet during a coffee break. As you will see, there are some differences between what they say and what they think.[1]

	They Say	*They Think*
STRETCH:	Well, you're certainly a welcome addition to our group.	(Can't I ever say something clever?)
CANDY:	Thank you. It certainly is friendly and interesting.	(He's cute.)
STRETCH:	My friends call me Stretch. It's left over from my basketball days. Silly, but I'm used to it.	(It's safer than saying my name is David Stein.)
CANDY:	My name is Candy.	(At least my nickname is. He doesn't have to hear Hortense O'Brien.)
STRETCH:	What kind of camera is that?	(Why couldn't a girl named Candy be Jewish? It's only a nickname, isn't it?)
CANDY:	Just this old German one of my uncle's. I borrowed it from the office.	(He could be Irish. And that camera looks expensive.)
STRETCH:	May I? (He takes her camera, brushing her hand and then tingling with the touch.) Fine lens. You work for your uncle?	(Now I've done it. Brought up work.)
CANDY:	Ever since college.	(Okay, so what if I only went for a year?)
	It's more than being just a secretary. I get into sales, too.	(If he asks what I sell, I'll tell him anything except underwear.)
STRETCH:	Sales? That's funny. I'm in sales, too, but mainly as an executive. I run our department.	(Is there a nice way to say used cars? I'd better change the subject.)
	I started using cameras on trips. Last time I was in the Bahamas, I took—	(Great legs! And the way her hips move—)
CANDY:	Oh! Do you go to the Bahamas, too? I love those islands.	(So I went just once, and it was for the brassiere manufacturers' convention At least we're off the subject of jobs.)
STRETCH:		(She's probably been around. Well, at least we're off the subject of jobs.)
	I did a little underwater work there last summer. Fantastic colors. So rich in life.	(And lonelier than hell.)
CANDY:		(Look at that build. He must swim like a fish. I should learn.)
	I wish I'd had time when I was there. I love the water.	(Well, I do. At the beach, anyway, where I can wade in and not go too deep.)

[1]Bach and Deutsch (1970).

So begins a relationship. Candy and Stretch have a drink and talk, talk, talk—sharing their likes and dislikes. Amazingly, they seem to agree on everything, from clothing to cars to politics. The attraction they feel is very strong, and neither of them is willing to turn the other off by disagreeing.

They fall in love. Weeks later they still agree on everything, even though there is one topic they avoid scrupulously: religion. Their different backgrounds became apparent once they exchanged last names. That doesn't mean they have to talk about it, however.

They delay introductions to their parents. The O'Briens and the Steins are narrow-minded about religion. If the truth be known, so are Candy and Stretch. Candy errs by telling Stretch, "You're not like other Jews I know." Stretch also voices his feelings now and then. After Candy has nursed him through a cold, he remarks, "You know, you're very Jewish." Candy and Stretch play the games required to maintain the relationship. They tell themselves that their remarks were mistakes, and, after all, anyone can make mistakes. They were meant as compliments, weren't they? Yet each is becoming isolated from family and friends.

One of the topics they ignore is birth control. As a Catholic, Candy does not take oral contraceptives. (Stretch later claimed that he had assumed she did.) Candy becomes pregnant, and they get married. Only through professional help do they learn each other's genuine feelings. And on several occasions, the union comes close to dissolving.

How do we explain the goings-on in this tangled web of deception? Candy and Stretch felt strongly attracted to each other. What determines who is attractive? Why did Candy and Stretch pretend to agree on everything? Why did they put off introductions to their parents?

Two possible consequences of attraction are friendship and love. Candy and Stretch "fell in love." What *is* love? When the first author was a teenager, the answer was "Five feet of heaven in a ponytail." However, this answer may lack scientific merit. We see that there are different forms of love and that our concepts of love are far from universal.

ATTRACTION

Let us explore some of the factors that determine interpersonal attraction.

Physical Attractiveness: How Important Is Looking Good?

We might like to think of ourselves as so sophisticated that physical attractiveness does not move us. We might like to claim that sensitivity, warmth, and intelligence are more important. However, we may never learn about other people's personalities if they do not meet our minimal standards for physical attractiveness. Research shows that physical attractiveness is a major determinant of interpersonal and sexual attraction (Hensley, 1992). In fact, physical appearance is the key factor in consideration of partners for dates, sex, and marriage (Hatfield & Sprecher, 1986).

Beauty and Culture. Can you find Mr. or Ms. Right among these people? Are your judgments of physical beauty based on universal standards or on your cultural experiences?

IS BEAUTY IN THE EYE OF THE BEHOLDER? What determines physical attractiveness? Are our standards fully subjective, or is there broad agreement on what is attractive? Cross-cultural studies show that people universally want physically appealing partners (Ford & Beach, 1951). However, is that which appeals in one culture repulsive in others?

In certain African tribes, long necks and round, disklike lips are signs of feminine beauty. Women thus stretch their necks and lips to make themselves more appealing. Women of the Nama tribe persistently tug at their labia majora to make them "beautiful"—that is, prominent and elongated (Ford & Beach, 1951).

Truth or Fiction? **REVISITED**

Beauty may not be completely in the eye of the beholder. Although personal tastes may vary within and across cultures, there are cultural standards for physical attractiveness. ■

In our culture, women consider taller men to be more attractive (Hensley, 1994; Sheppard & Strathman, 1989). Undergraduate women prefer their dates to be about 6 inches taller than they are. Undergraduate men, on the average, prefer women who are about 4½ inches shorter (Gillis & Avis, 1980). Tall women are not viewed so positively.

Shortness, though, is perceived as a liability for both men and women (Jackson & Ervin, 1992).

Some women of Candy's stature find that shorter men are discouraged from asking them out. Some walk with a hunch, as though to minimize their height. A neighbor of the first and third authors refers to herself as 5 feet 13 inches tall.

Female plumpness is valued in many—perhaps most—preliterate societies (Anderson et al., 1992; Frayser, 1985). Wide hips and a broad pelvis are widely recognized as sexually appealing. In our culture, however, slenderness is in style. Some young women suffer from an eating disorder called **anorexia nervosa,** in which they literally starve themselves to conform to the contemporary ideal. Both genders find slenderness (though not anorexic thinness) attractive, especially for females (Fallon & Rozin, 1985; Franzoi & Herzog, 1987; Rozin & Fallon, 1988).

The hourglass figure is popular in the United States. In one study, 87 African American college undergraduates—both male and female—rated women of average weight with a waist-to-hip ratio of 0.7 to 0.8 as most attractive and desirable for long term relationships (Singh, 1994a). Neither very thin nor obese women were found to be as attractive, regardless of the waist-to-hip ratio. Findings were similar for a sample of 188 European American students (Singh, 1994b).

Do men idealize the *Penthouse* centerfold? What bust size do men prefer? Women's beliefs that men prefer large breasts may be somewhat exaggerated. The belief that men want women to have bursting bustlines leads many women to seek breast implants in the attempt to live up to an ideal that men themselves don't generally hold (Rosenthal, 1992). Researchers in one study showed young men and women (ages 17 to 25 years) a continuum of male and female figures that differed only in the size of the bust for the female figures and in that of the pectorals for the male figures (Thompson & Tantleff, 1992). The participants were asked to indicate the ideal size for their own gender and the size they believed the average man and woman would prefer.

The results show some support for the "big is better" stereotype—for both men and women. Women's conception of ideal bust size was greater than their actual average size. Men preferred women with still larger breasts, but not nearly so large as the breasts women *believed* that men prefer. Men believed that their male peers preferred women with much bustier figures than their peers themselves said they preferred. Ample breasts and chests may be preferred by the other gender, but people seem to have an exaggerated idea of the sizes that the other gender actually prefers.

And some women, of course, have too much of a good thing. Some of them go for breast-reduction surgery, as noted in the nearby Human Sexuality Online feature.

Both genders find obese people unattractive, but there are gender differences in impressions of the most pleasing body shape. On the average, college men think that their present physiques are close to ideal and appealing to women (Fallon & Rozin, 1985). College women generally see themselves as much heavier than the figure that is most alluring to men, and heavier still than the figure they perceive as the ideal feminine form. Both genders are wrong about the preferences of the other gender, however. Men actually prefer women to be somewhat heavier than the women imagine they would. Women prefer their men to be a bit leaner than the men would have expected.

It is not true that college men would like women to be thinner than the women want to be. College men actually prefer women who are heavier than the women expect, though still slender . On the other hand, college women prefer men who are thinner than the men imagine the women prefer.

Anorexia nervosa A potentially life-threatening eating disorder characterized by refusal to maintain a healthful body weight, intense fear of being overweight, a distorted body image, and, in females, lack of menstruation (amenorrhea.)

HOW TRAITS AND NAMES AFFECT PERCEPTIONS OF PHYSICAL ATTRACTIVENESS: ON THE IMPORTANCE OF *NOT* BEING ERNEST Both genders rate the attractiveness of faces higher when they are shown smiling in photographs than when they are shown in a nonsmiling pose (Mueser et al., 1984). (Photographers are not ignorant of this fact.) Thus,

Human Sexuality ONLINE //

Hospital Holds 'Net Auction for Breast Reduction

Some women have too much of a good thing and opt for breast-reduction surgery. To help meet the perceived need, a South African private hospital held an Internet auction offering a range of plastic surgery operations, from breast reductions to fat removal, with a starting price of one rand ($0.14).

"The bidding [was] open-ended for one operation per day during the week. The operation [went] to the highest bidder," said Jack Shevel, chief executive officer of Netcare, South Africa's biggest private hospital group. Shevel noted that the average price for breast reduction surgery was about 14,000 rand ($2,026).

Bidders who logged on to www.bidorbuy.co.za were offered a breast reduction operation Monday, laser skin resurfacing on Tuesday, fat removal Wednesday, facial hair removal on Thursday, and eyelid surgery on Friday.

"We are looking to attract uninsured patients. The target market for cosmetic surgery is the higher-income bracket. We will see what they are prepared to pay," Shevel said.

Source: From "Hospital Offers 'Net Auction for Breast Reduction" (2000, July 14). *Reuters News Agency online.*

The Surgical Search for Physical Perfection. In today's world, physical perfection—or at least, improvement—is sometimes just a scalpel away. But that doesn't mean the process is easy, free of side effects, or even always properly carried out. Let the buyer beware!

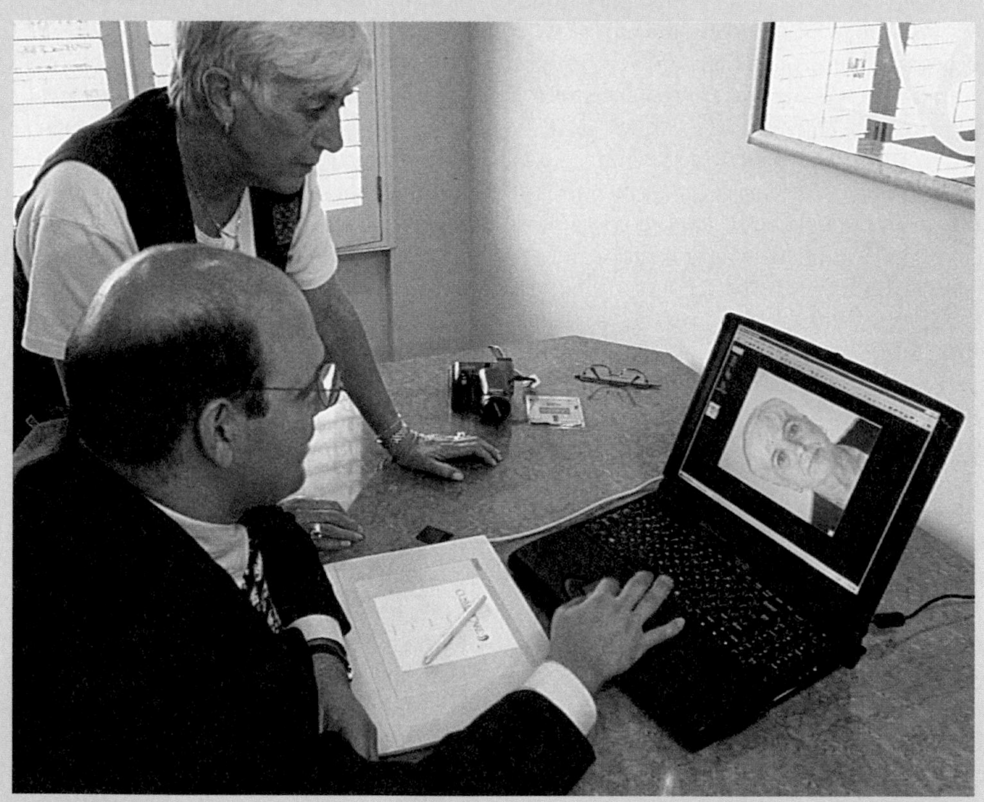

there is good reason to "put on a happy face" when you meet people socially or ask someone out on a date. Context is important, however. It may be more appropriate in a business context to maintain a more serious countenance. The effects of a smile may also be a greater determinant of attractiveness in women than in men (Deutsch et al., 1987).

It is true that people are regarded as more attractive when they are smiling. It may indeed make sense to "put on a happy face" when meeting people.

Gender-role expectations may affect perceptions of attractiveness. Women who viewed videos of prospective dates found men who acted outgoing and self-expressive

Wide-Eyed with . . . Beauty?

Some aspects of beauty seem to be largely cross-cultural. Research suggests that European Americans, African Americans, Asian Americans, and Latino and Latina Americans tend to agree on the facial features that they find to be attractive (Cunningham et al., 1995). They all prefer female faces with large eyes, greater distance between the eyes, small noses, narrower faces with smaller chins, high, expressive eyebrows, larger lower lips, and a well-groomed, full head of hair.

Consider the methodology of a study that compared the facial preferences of people in Japan and England. Perrett (1994) created computer composites of the faces of 60 women. Figure 7.1(a) is a composite of the 15 women who were rated the most attractive. He then used computer enhancement to exaggerate the differences between the composite of the 60—that is, the average face—and the composite of the 15 most attractive women. He found that both Japanese and British men deemed women with large eyes, high cheekbones, and narrow jaws to be the most attractive (Perret, 1994). Computer enhancement resulted in the image shown in Figure 7.1(b). The enhanced composite has even larger eyes, yet higher cheekbones, and a still narrower jaw. Figure 7.1(b) was then rated as the most attractive image. Similar results were found for the image of a Japanese woman.

Cunningham and his colleagues (1995) reported historical anecdotes that suggest that the facial preferences of people as diverse as Europeans, black Africans, Native Americans, Indians (in India, that is), and Chinese are quite consistent. They quoted from Charles Darwin's 1871 treatise *The Descent of Man, and Selection in Relation to Sex:*

> Mr. Winwood Reade . . . who has had ample opportunities for observation [with Black Africans] who have never associated with Europeans is convinced that their ideas of beauty are, on the whole, the same as ours; and Dr. Rohlfs writes to me [to] the same effect with respect to Borneo and the countries inhabited by the Pullo tribes. . . . Capt. Burton believes that a woman whom we consider beautiful is admired throughout the world.

Darwin believed that our physical preferences were largely inborn and related to survival of our species. What do you think? Do you believe that "their ideas of beauty are, on the whole, the same as ours"? Or do you think that research hasn't yet ferreted out significant cultural or ethnic differences that might exist? If there is ethnic consistency in these preferences, how would you explain them?

- Do you believe these preferences are coincidental?
- Do you think there has been more exchange of ideas among cultures than has been believed?
- Do you believe there is something instinctive about these preferences?

FIGURE 7.1

What Features Contribute to Facial Attractiveness? In both England and Japan, features such as large eyes, high cheekbones, and narrow jaws contribute to perceptions of the attractiveness of women. (a) A computer composite of the faces of 15 women rated as the most attractive of a group of 60. (b) A computer composite that exaggerates the features of these 15 women. That is, they are developed further in the direction that separates them from the average of the full group.

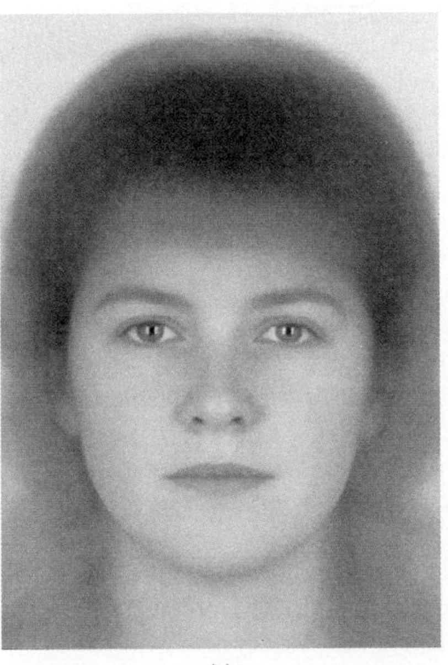
(a)

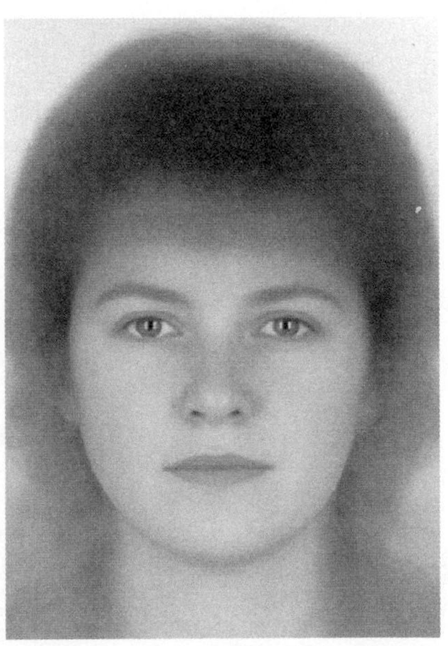
(b)

more appealing than men who were passive (Riggio & Woll, 1984). Another study found that highly feminine women are more likely to be attracted to dominant, "macho" men than less feminine women are (Maybach & Gold, 1994). Yet men who viewed videos in the Riggio and Woll (1984) study were put off by outgoing, self-expressive behavior in women. In another study, women rated videos of dominant college men (defined in this study as those who exerted social control over a troublesome interaction with an instructor) as more appealing than submissive men. Again, male viewers were put off by similarly dominant women (Sadalla et al., 1987). Despite recent changes in traditional gender-role stereotypes, many men in the United States still prefer demure women. This is not to suggest that dominant, self-expressive women should stifle themselves to attract traditional men; the relationships would probably rub them both the wrong way.

Names also may affect perceptions of physical appeal. In one study, women who were randomly assigned names like Kathy, Jennifer, and Christine were rated more attractive than women assigned the names Harriet, Gertrude, and Ethel (Garwood et al., 1980). Seems silly, does it not? After all, our parents name us, and there need be no relationship between our names and our physical appeal. On the other hand, we may choose whether to keep our names or to use nicknames. If you are unhappy with your name, why not assume a more popular nickname? Beginning college or taking a new job is an ideal time for doing so. Men, too, can scuttle their Sylvesters and Ernests if they prefer. If you have an unusual name and are content with it, be yourself, however.

Truth or Fiction? REVISITED

It is true that women who are randomly assigned names like Kathy and Jennifer are rated as more attractive than women assigned names like Harriet and Gertrude. Perhaps observers thought that the Harriets and Gertrudes of the world would use more contemporary nicknames if they were "with it." ■

WHAT DO YOU LOOK FOR IN A LONG-TERM, MEANINGFUL RELATIONSHIP? Your second author conducted a survey of college men and women in the early 1980s and found that psychological characteristics such as warmth, fidelity, honesty, and sensitivity were rated higher in importance than physical attractiveness as desirable qualities in a prospective partner for a meaningful, long-term relationship (Nevid, 1984). Physical attractiveness won out when students were asked to consider what qualities are most important in a partner for a sexual relationship. Overall, however, men placed greater emphasis on the physical characteristics of their partners for both types of relationships than did women. Women placed more value on qualities such as warmth, assertiveness, wit, and ambition. The single most highly desired quality that students wanted in long-term partners was honesty. Honestly.

Truth or Fiction? REVISITED

It is not true that physical appeal is the most important trait we seek in partners for long-term relationships. According to the results of one study, honesty was reported to be more vital in partners for long-term relationships. ■

Although personal qualities may assume more prominent roles in determining partner preferences in long-term relationships, physical appeal probably plays a "filtering" role. Unless a prospective date meets minimal physical standards, we might not look beneath the surface for "more meaningful" traits.

Nevid's results have been replicated in studies on initial attraction and on choice of mates. Women place relatively greater emphasis than men on such traits as vocational status, earning potential, expressiveness, kindness, consideration, dependability, and fondness for children. Men give relatively more consideration to youth, physical attractiveness, cooking ability (can't they switch on the microwave by themselves?), and frugality (Howard et al., 1987; Sprecher et al., 1994). When it comes to mate selection, females in a sample of students from Germany and the Netherlands also emphasized the financial prospects and status of a potential mate, whereas males emphasized the importance of physical attractiveness (de Raad & Doddema-Winsemius, 1992). A study of more than

TABLE 7.1

Gender differences in mate preferences

How willing would you be to marry someone who . . .	Men	Women
was not "good-looking"?	3.41	4.42**
was older than you by 5 or more years?	4.15	5.29**
was younger than you by 5 or more years?	4.54	2.80**
was not likely to hold a steady job?	2.73	1.62**
would earn much less than you?	4.60	3.76**
would earn much more than you?	5.19	5.93**
had more education than you?	5.22	5.82**
had less education than you?	4.67	4.08**
had been married before?	3.35	3.44
already had children?	2.84	3.11*
was of a different religion?	4.24	4.31
was of a different race?	3.08	2.84**

*Difference statistically significant at the .01 level of confidence.

**Difference statistically significant at the .001 level of confidence.

Source: Based on information in Susan Sprecher, Quintin Sullivan, and Elaine Hatfield (1994). "Mate Selection Preferences: Gender Differences Examined in a National Sample." *Journal of Personality and Social Psychology*, 66(6): 1074–1080. Copyright © 1994 by the American Psychological Association. Reprinted by permission.

200 Korean college students found that in mate selection, women placed more emphasis than men on education, jobs, and family of origin (Brown, 1994). Men placed more emphasis on physical attractiveness and affection. (Yes, men were more "romantic" and women more pragmatic.)

Susan Sprecher and her colleagues (1994) surveyed a national probability sample of 13,017 English- or Spanish-speaking people, age 19 or above, living in households in the United States. In one section of their questionnaire, they asked respondents how willing they would be to marry someone who was older, younger, of a different religion, not likely to hold a steady job, not good-looking, and so forth. Each item was followed by a 7-point scale in which 1 meant "not at all willing" and 7 meant "very willing." As shown in Table 7.1, women were more willing than men to marry someone who was not good-looking. On the other hand, women were less willing to marry someone not likely to hold a steady job.

ARE ATTRACTIVENESS PREFERENCES INHERITED? On the surface, gender differences in perceptions of attractiveness seem unbearably sexist, and perhaps they are. Yet some evolutionary psychologists believe that evolutionary forces favor the continuation of gender differences in preferences for mates because certain preferred traits offer reproductive advantages (Bjorklund & Kipp, 1996; Fisher, 2000). Some physical features, such as cleanliness, good complexion, clear eyes, good teeth, good hair, firm muscle tone, and a steady gait are universally appealing to both genders (Ford & Beach, 1951). Perhaps they are markers of reproductive potential (Symons, 1995). Youth and health may be relatively more important to a woman's appeal, because these characteristics tend to be associated with her reproductive capacity: The "biological clock" limits her reproductive potential. Physical characteristics associated with a woman's youthfulness, such as smooth skin, firm muscle tone, and lustrous hair, may thus have become more closely linked to a woman's appeal (Buss, 1994). A man's reproductive value, however, may depend more on how well

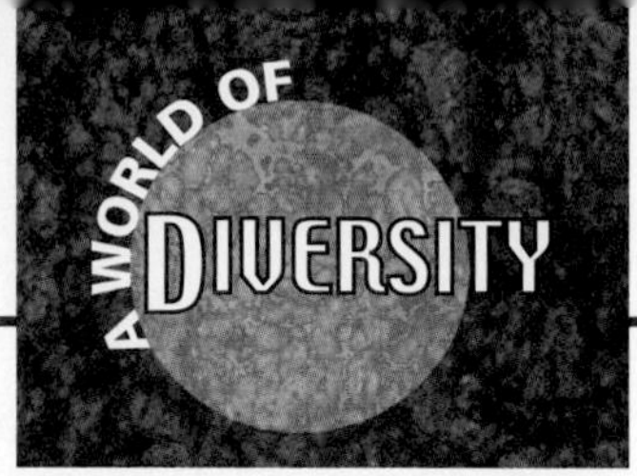

Gender Differences in Preferences in Mates Across 37 Cultures

What do men in Nigeria, Japan, Brazil, Canada, and the United States have in common? For one thing, men in these countries report that they prefer mates who are younger than themselves. Buss (1994) reviewed survey evidence on the preferred age difference between oneself and one's mate in 37 cultures (representing 33 countries) in Europe, Africa, Asia, Australia, New Zealand, and North and South America. In every culture, men preferred younger mates (the range was from 0.38 year to 6.45 years). Women, however, preferred older mates (the range was from 1.82 years to 5.1 years).

Gender differences in the preferred age of mates paralleled actual differences in age of men and women at the time of marriage. Men were between 2 and 5 years older, on the average, than their brides at the time of marriage. The smallest average age difference at marriage, 2.1 years, was found in Poland. The largest difference, 4.92 years, was found in Greece. Men in the mainland United States are 2.71 years older than women at the time of marriage. In Canada, men are 2.51 years older than their mates, on the average.

Buss finds that in all 37 cultures, men placed greater value on a prospective partner's "good looks" than did women. On the other hand, women in 36 of 37 cultures placed greater value on "good earning capacity" in prospective mates.

The consistency of Buss's findings lends credence to the notion that there are widespread gender differences in preferences with respect to age, physical characteristics, and financial status of prospective mates. Generally speaking, men place greater value on the physical attractiveness and relative youth of prospective mates. Women place relatively greater value on the earning capacity of prospective mates. Buss interprets women's preferences for relatively older mates as additional evidence that women appraise future mates on the basis of their ability to provide for a wife and family, because age and income tend to be linked among men.

Despite these gender differences in preferences for mates, Buss finds that both men and women placed greater weight on personal qualities than on the looks or income potential of prospective mates. In *all* 37 cultures, the characteristics "kind, understanding," and "intelligent" were rated higher than earning power or physical attractiveness.

he can provide for his family than on his age or physical appeal. The value of men as reproducers, therefore, is more intertwined with factors that contribute to a stable environment for child rearing, such as economic status and reliability. Evolutionary psychologists argue that these gender differences in mate preferences may have been passed down through the generations as part of our genetic heritage (Buss, 1994; Symons, 1995).

Men's interest in younger women occurs in both preliterate and industrialized societies (Buss, 1994). Female jealousy of younger women is another thread that spans cultures. Sexual competition, according to Margaret Mead, generally involves

> the struggle between stronger older men and weaker younger men or between more attractive younger women and more entrenched older ones. (Mead, 1967, p. 198)

The evolutionary view of gender differences in preferences for mates is largely speculative and not fully consistent with the evidence. Despite gender differences, both men and women report that they place greater weight on personal characteristics than on physical features in judging prospective mates (Buss, 1994). Many women, like men, still prefer physically appealing partners (Bixler, 1989). Women also tend to marry men similar to themselves in physical attractiveness as well as socioeconomic standing. Note also that older men are more likely than younger men to die from natural causes. From the standpoint of reproductive advantages, women would thus achieve greater success by marrying fit, younger males who are likely to survive during the child-rearing years than by marrying older, higher-status males. Moreover, similar cultural influences, rather than inherited dispositions, may explain commonalities across cultures in gender differences in mate preferences. For example, in societies in which women are economically dependent on men, a man's appeal may depend more on his financial resources than on his physical appeal.

The Matching Hypothesis: Who Is "Right" for You?

Who Is Right for You? Research shows that people tend to pair off with others who are similar in physical characteristics and personality traits. (The brand of eyeglasses and the specific content of the tattoo may not be crucial, however.)

Do not despair if you are less than exquisite in appearance, along with most of us mere mortals. You may be saved from permanently blending in with the wallpaper by the effects of the **matching hypothesis.** This concept holds that people tend to develop romantic relationships with people who are similar to themselves in physical attractiveness rather than with the local Will Smith or Britney Spears look-alike.

Researchers have found that people who are dating steadily, engaged, or married tend to be matched in physical attractiveness (Kalick, 1988). Young married couples even tend to be matched in weight (Schafer & Keith, 1990). The central motive for seeking "matches" seems to be fear of rejection by more appealing people (Bernstein et al., 1983).

There are exceptions to the matching hypothesis. Now and then we find a beautiful woman married to a plain or homely man (or vice versa). How do we explain it? What, after all, would *she* see in *him*? According to one study (Bar-Tal & Saxe, 1976), people judging "mismatched" pairs may tend to ascribe wealth, intelligence, or success to the man. We seek an unseen factor that will balance the physical attractiveness of one partner. For some mismatched couples, similarities in attitudes and personalities may balance out differences in physical attractiveness.

MORE THAN BEAUTY Matching applies not only to physical appeal. Our sex and marital partners tend to be like us in race and ethnicity, age, level of education, and religion. Consider some findings of the NHSLS study (Michael and others, 1994):

- The sex partners of nearly 94% of unmarried European American men are European American women. About 2% of single European American men are partnered with Latina American women, 2% with Asian American women, and less than 1% with African American women.
- The sex partners of nearly 82% of African American men are African American women. Nearly 8% of African American men are partnered with European American women. Under 5% are partnered with Latina American women.
- About 83% of the women and men in the study chose partners within 5 years of their own age and of the same or a similar religion.
- Of all the women in the study, *not one* with a graduate college degree had a partner who had not finished high school.
- Men with a college degree almost never had sexual relationships with women with much more or much less education than they had.

Similarity in Attitudes: Do Opposites Attract?

Why do the great majority of us have partners from our own backgrounds? One reason is that marriages are made in the neighborhood, not in heaven (Michael et al., 1994). That is, we tend to live among, and thus come into contact with, people who are similar to us in background. Another reason is that we are drawn to people whose attitudes are similar to ours. People similar in background are more likely to be similar in their attitudes. Similarity in attitudes and tastes is a key contributor to attraction, friendships, and love relationships (Cappella & Palmer, 1990; Griffin & Sparks, 1990; Laumann et al., 1994).

Matching hypothesis The concept that people tend to develop romantic relationships with people who are similar to themselves in attractiveness.

Let us also note a gender difference. Evidence shows that women place greater weight than men on attitude similarity as a determinant of attraction to a stranger of the other gender, whereas men place more value on physical attractiveness (Feingold, 1991).

We also tend to *assume* that people we find attractive share our attitudes (Dawes, 1989; Marks et al., 1981). The physical attraction between Candy and Stretch motivated them to pretend that their preferences, tastes, and opinions coincided. They entered into an unspoken agreement not to discuss their religious differences. When sexual attraction is strong, perhaps we want to think that we can iron out all the kinks in the relationship. Although similarity may be important in determining initial attraction, compatibility appears to be a stronger predictor of maintaining an intimate relationship (Vinacke et al., 1988).

Truth or Fiction? REVISITED

Actually, it is not true that opposites attract. We are actually *less* apt to be attracted to people who disagree with our views and tastes than to people who share them.

Reciprocity: If You Like Me, You Must Have Excellent Judgment

Has anyone told you that you are good-looking, brilliant, and emotionally mature to boot? That your taste is elegant? Ah, what superb judgment!

When we feel admired and complimented, we tend to return these feelings and behaviors. This is called **reciprocity**. Reciprocity is a potent determinant of attraction (Condon & Crano, 1988). We tend to be much more warm, helpful, and candid when we are with strangers who we believe like us (Clark et al., 1989; Curtis & Miller, 1986). We even tend to welcome positive comments from others when we know those remarks to be inaccurate (Swann et al., 1987).

Perhaps the power of reciprocity has enabled many couples to become happy with one another and reasonably well adjusted. By reciprocating positive words and actions, a person can perhaps stoke neutral or mild feelings into robust, affirmative feelings of attraction.

Attraction can lead to feelings of love. Let us now turn to that most fascinating topic.

LOVE

For thousands of years, poets have sought to capture love in words. A seventeenth-century poet wrote that his love was like "a red, red rose." In Sinclair Lewis's novel *Elmer Gantry*, love is "the morning and the evening star." Love is beautiful and elusive. It shines, brilliant and heavenly. Passion and romantic love are also earthy and sexy, brimming with sexual desire.

Romantic love is hardly unique to our culture. Researchers report finding evidence of romantic love in 147 of the 166 different cultures they studied in a recent cross-cultural comparison (Jankowiak & Fischer, 1992). Romantic love occurs even in most preliterate societies. The apparent absence of romantic love in the remaining 19 cultures, the investigators suspect, was most probably due to the limitations of their study methods (Gelman, 1993).

Our culture idealizes the concept of romantic love. Thus we readily identify with the plight of the "star-crossed" lovers in *Romeo and Juliet* and *West Side Story*, who sacrificed themselves for love. We learn that "love makes the world go round" and that "love is everything." Virtually all of the participants in the Janus and Janus nationwide survey (96% of the men and 98% of the women) reported that love is important to them (Janus & Janus, 1993). Like other aspects of sexual and social behavior among humans, the concept of love must be understood within a cultural context. Luckily (or miserably), we have such a context in Western culture. . . .

Reciprocity Mutual exchange.

The Greek Heritage

The concept of love can be traced back at least as far as the classical age of Greece. The Greeks distinguished four concepts related to the modern meanings of love: *storge*, *agape*, *philia*, and *eros*.

Storge is loving attachment, deep friendship, or nonsexual affection. It is the emotion that binds friends and parents and children. Some scholars believe that even romantic love is a form of attachment that is similar to the types of attachments infants have to their mothers (Carter, 1998; Stephan & Bachman, 1999; Tucker & Anders, 1999).

Agape is similar to generosity and charity. It implies the wish to share one's bounty and is epitomized by anonymous donations of money. In relationships, it is characterized by selfless giving. Agape, according to Lee's research, is the kind of love least frequently found between adults in committed relationships.

Philia is closest in meaning to friendship. It is based on liking and respect, rather than sexual desire. It involves the desire to do and enjoy things with the other person and to see him or her when one is lonely or bored.

Eros is closest in meaning to our concept of passion. Eros was a character in Greek mythology (transformed in Roman mythology into Cupido, now called Cupid) who would shoot unsuspecting people with his love arrows, causing them to fall madly in love with the person who was nearest to them at the time. Erotic love embraces sudden passionate desire: "love at first sight" and "falling head over heels in love." Younger college students are more likely to believe in love at first sight and that "love conquers all" than older (and wiser?) college students (Knox et al., 1999a). Passion can be so gripping that one is convinced one's life has been changed forever. This feeling of sudden transformation was captured by the Italian poet Dante Alighieri (1265–1321), who exclaimed upon first beholding his beloved Beatrice, "*Incipit vita nuova*," which can be translated as "My life begins anew." Romantic love can also be earthy and sexy. In fact, sexual arousal and desire may be the strongest component of passionate or romantic love. Romantic love begins with a powerful physical attraction or feelings of passion and is associated with strong physiological arousal.

Unlike the Greeks, we tend to use the word *love* to describe everything from feelings of affection toward another to romantic ardor to sexual intercourse ("making love"). Still, different types or styles of love are recognized in our own culture, as we shall see.

Romantic Love in Contemporary Western Culture

The experience of *romantic love*, as opposed to loving attachment or sexual arousal per se, occurs within a cultural context in which the concept is idealized. Western culture has a long tradition of idealizing the concept of romantic love, as represented, for instance, by romantic fairy tales that have been passed down through the generations. In fact, our exposure to the concept of romantic love may begin with hearing the fairy tales of Sleeping Beauty, Cinderella, and Snow White—along with their princes charming. Later perhaps, the concept of romantic love blossoms with exposure to romantic novels, television and film scripts, and the heady tales of friends and relatives.

During adolescence, strong sexual arousal, along with an idealized image of the object of our desires, leads us to label our feelings as love. We may learn to speak of love rather than lust, because sexual desire in the absence of a committed relationship might be viewed as primitive or animalistic. Being "in love" ennobles attraction and sexual arousal, not only to society but also to oneself. Unlike lust, love can be discussed at the dinner table. If others think we are too young to experience "the real thing"—which presumably includes knowledge of and respect for the other person's personality traits—our feelings may be called "puppy love" or a "crush."

Western society maintains much of the double standard toward sexuality. Thus, women are more often expected to justify sexual experiences as involving someone they

Storge (STORE-gay) Loving attachment and nonsexual affection; the type of emotion that binds parents to children.

Agape (AH-gah-pay) Selfless love; a kind of love that is similar to generosity and charity.

Philia (FEEL-yuh) Friendship love, which is based on liking and respect rather than sexual desire.

Eros The kind of love that is closest in meaning to the modern-day concept of passion.

Are You a Romantic or a Realist? The Love Attitudes Scale

David Knox of East Carolina University contrasts romantic love with *realistic love*—the kind of love that is maintained across the years. Partners who share a realistic love have the blinders off. They accept and cherish each other, warts and all.

Knox (1983) developed the Love Attitudes Scale to evaluate the degree to which people hold a romantic or a realistic view of love. How about you? Are you a realist or a romantic when it comes to matters of the heart? To find out, complete the scale, and then turn to the scoring key in the Appendix.

Directions: Circle the number that best represents your opinion on each item according to the following code. Add up your scores to arrive at a total score.

1 = Strongly agree (SA); 2 = Mildly agree (MA); 3 = Undecided (U); 4 = Mildly disagree (MD); 5 = Strongly disagree (SD)

	SA	MA	U	MD	SD
1. Love doesn't make sense. It just is.	1	2	3	4	5
2. When you fall "head over heels" in love, it's sure to be the real thing.	1	2	3	4	5
3. To be in love with someone you would like to marry but can't is a tragedy.	1	2	3	4	5
4. When love hits, you know it.	1	2	3	4	5
5. Common interests are really unimportant; as long as each of you is truly in love, you will adjust.	1	2	3	4	5
6. It doesn't matter if you marry after you have known your partner for only a short time as long as you know you are in love.	1	2	3	4	5
7. If you are going to love a person, you will "know" after a short time.	1	2	3	4	5
8. As long as two people love each other, the educational differences they have really do not matter.	1	2	3	4	5
9. You can love someone even though you do not like any of that person's friends.	1	2	3	4	5
10. When you are in love, you are usually in a daze.	1	2	3	4	5
11. Love "at first sight" is often the deepest and most enduring type of love.	1	2	3	4	5
12. When you are in love, it really does not matter what your partner does because you will love him or her anyway.	1	2	3	4	5

love. Young men usually need not attribute sexual urges to love, so men are more likely to deem love a "mushy" concept. The vast majority of people in the United States nonetheless believe that romantic love is a prerequisite to marriage. Romantic love is rated by young people as the single most important reason for marriage (Roper Organization, 1985).

Which is the more romantic gender? Although the question may well incite an argument in mixed company, the Janus and Janus (1993) nationwide survey of adult Americans found that a slightly greater percentage of the single men (82%) than of the single women (77%) perceived themselves as romantic. Yet among married people, the figures were reversed, with 79% of the women describing themselves as romantic, compared to

13. As long as you really love a person, you will be able to solve the problems you have with that person.	1	2	3	4	5
14. Usually you can really love and be happy with only one or two people in the world.	1	2	3	4	5
15. Regardless of other factors, if you truly love another person, that is a good enough reason to marry that person.	1	2	3	4	5
16. It is necessary to be in love with the one you marry to be happy.	1	2	3	4	5
17. Love is more of a feeling than a relationship.	1	2	3	4	5
18. People should not get married unless they are in love.	1	2	3	4	5
19. Most people truly love only once during their lives.	1	2	3	4	5
20. Somewhere there is an ideal mate for most people.	1	2	3	4	5
21. In most cases, you will "know it" when you meet the right partner.	1	2	3	4	5
22. Jealousy usually varies directly with love; that is, the more you are in love, the greater your tendency to become jealous will be.	1	2	3	4	5
23. When you are in love, you are motivated by what you feel rather than by what you think.	1	2	3	4	5
24. Love is best described as an exciting rather than a calm thing.	1	2	3	4	5
25. Most divorces probably result from falling out of love rather than failing to adjust.	1	2	3	4	5
26. When you are in love, your judgment is usually not too clear.	1	2	3	4	5
27. Love often comes only once in a lifetime.	1	2	3	4	5
28. Love is often a violent and uncontrollable emotion.	1	2	3	4	5
29. When selecting a marriage partner, differences in social class and religion are of small importance compared with love.	1	2	3	4	5
30. No matter what anyone says, love cannot be understood.	1	2	3	4	5

Total Score on the Love Attitudes Scale: ______

Source: From D. Knox. *The Love Attitudes Inventory*, rev. ed. Copyright © 1993 by David Knox. Family Life Publications. Permission granted by the author (see www.heartchoice.com).

72% of the men. Perhaps there is some truth to the stereotype that men are more romantic during the courtship stage of relationships than during marriage. Then again, maybe self-perceptions of being romantic don't quite jibe with the reality. In any event, you can explore *your* self-perceptions as a romantic or a realist when it comes to love by completing the accompanying Love Attitudes Scale.

When reciprocated, romantic love is usually a source of deep fulfillment and ecstasy (Hatfield, 1988). When love is unrequited, however, it can lead to emptiness, anxiety, or despair. Romantic love can thus teeter between states of ecstasy and misery (Hatfield, 1988). Perhaps no other feature of our lives can lift us up so high or plunge us so low as romantic love.

Infatuation or "True Love"? Infatuation is a state of intense absorption in another person. It is characterized by sexual longing and general excitement. Infatuation is often referred to as passion or as a crush. Infatuation is assumed to fade as relationships develop.

INFATUATION VERSUS "TRUE LOVE": WILL TIME TELL? Perhaps you first noticed each other when your eyes met across a crowded room, like the star-crossed lovers in *West Side Story.* Or perhaps you met when you were both assigned to the same Bunsen burner in chemistry lab—less romantic, but closer to the flame. However it happened, the meeting triggered such an electric charge through your body that you could not get him (or her) out of your mind. Were you truly in love, however, or was it merely a passing fancy? Was it infatuation or "the real thing"—a "true," lasting, and mutual love? How do you tell them apart?

Perhaps you don't, at least not at first. **Infatuation** is a state of intense absorption in or focus on another person. It is usually accompanied by sexual desire, elation, and general physiological arousal or excitement. Some refer to passion as infatuation. Others dub it a "crush." Both monikers suggest that it is a passing fancy. In infatuation, your heart may pound whenever the other person draws near or enters your fantasies.

For the first month or two, infatuation and the more enduring forms of romantic love are hard to differentiate. At first, both may be characterized by intense focusing or absorption. Infatuated people may become so absorbed that they cannot sleep, work, or carry out routine chores. Logic and reason are swept aside. Infatuated people hold idealized images of their love objects and overlook their faults. Caution may be cast to the winds. In some cases, couples in the throes of infatuation rush to get married, only to find, a few weeks or months later, that they are not well suited to each other.

As time goes on, signs that distinguish infatuation from a lasting romantic love begin to emerge. The partners begin to view each other more realistically and are better able to determine whether the relationship should continue. Although the tendency to idealize one's lover is strongest at the outset of a relationship, we should note that a so-called "positive illusion" tends to persist in relationships (Martz et al., 1998). That is, we maintain some tendency to differentiate our partners from the average and also to differentiate the value of our relationships from the average.

Infatuation has been likened to a state of passionate love (Sternberg, 1986) that is based on intense feelings of passion but not on the deeper feelings of attachment and caring that typify a more lasting mutual love. Although infatuation may be a passing fancy, it can be supplanted by the deeper feelings of attachment and caring that characterize more lasting love relationships.

Note, too, that infatuation is not a necessary first step on the path to a lasting mutual love. Some couples develop deep feelings of love without ever experiencing the fireworks of infatuation (Sternberg, 1986). And sometimes one partner is infatuated while the other manages to keep his or her head below the clouds.

Contemporary Models of Love: Dare Science Intrude?

Despite the importance of love, scientists have historically paid little attention to it. Some people believe that love cannot be analyzed scientifically. Love, they maintain, should be left to the poets, philosophers, and theologians. Yet researchers are now applying the scientific method to the study of love. They recognize that love is a complex concept, involving many areas of experience—emotional, cognitive, and motivational (Sternberg & Grajek, 1984). They have reinforced the Greek view that there are different kinds and styles of love. Let us consider some of the views of love that have emerged from modern theorists and researchers.

Infatuation A state of intense absorption in or focus on another person, which is usually accompanied by sexual desire, elation, and general physiological arousal or excitement; passion.

LOVE AS APPRAISAL OF AROUSAL Social psychologists Ellen Berscheid and Elaine Hatfield (Berscheid & Walster, 1978; Walster & Walster, 1978) define **romantic love** in terms of a state of intense physiological arousal and the cognitive appraisal of that arousal as love. The physiological arousal may be experienced as a pounding heart, sweaty palms, and butterflies in the stomach when one is in the presence of, or thinks about, one's love interest. Cognitive appraisal of the arousal means attributing it to some cause, such as fear or love. The perception that one has fallen in love is thus derived from several simultaneous events: (1) a state of intense physiological arousal that is connected with an appropriate love object (that is, a person, not an event like a rock concert), (2) a cultural setting that idealizes romantic love, and (3) the attribution of the arousal to feelings of love toward the person.

STYLES OF LOVE Some psychologists speak in terms of *styles* of love. Clyde and Susan Hendrick (1986) developed a love attitude scale that suggests the existence of six styles of love among college students. The following is a list of these styles. Each one is exemplified by statements similar to those on the original scale. As you can see, the styles owe a debt to the Greeks:

1. *Romantic love (eros):* "My lover fits my ideal." "My lover and I were attracted to one another immediately."
2. *Game-playing love (ludus):* "I keep my lover up in the air about my commitment." "I get over love affairs pretty easily."
3. *Friendship (storge, philia):* "The best love grows out of an enduring friendship."
4. *Logical love (pragma):* "I consider a lover's potential in life before committing myself." "I consider whether my lover will be a good parent."
5. *Possessive, excited love (mania):* "I get so excited about my love that I cannot sleep." "When my lover ignores me, I get sick all over."
6. *Selfless love (agape):* "I would do anything I can to help my lover." "My lover's needs and wishes are more important than my own."

Most people who are "in love" experience a number of these styles, but the Hendricks (1986) found some interesting gender differences in styles of love. College men are significantly more likely to develop game-playing and romantic love styles. College women are more apt to develop friendly, logical, and possessive love styles. (There were no gender differences in selfless love.) The Hendricks and their colleagues (1988) found that romantically involved couples tend to experience the same kinds of love styles. They also showed that couples with romantic and selfless styles of love are more likely to remain together. A game-playing love style leads to unhappiness, however, and is one reason that relationships come to an end.

STERNBERG'S TRIANGULAR THEORY OF LOVE Psychologist Robert Sternberg (1986, 1987, 1988) offers a triangular theory of love. In his view, there are three distinct components of love:

1. *Intimacy:* the experience of warmth toward another person that arises from feelings of closeness, bondedness, and connectedness to the other. Intimacy also involves the desire to give and receive emotional support and to share one's innermost thoughts with the other.
2. *Passion:* an intense romantic or sexual desire for another person, which is accompanied by physiological arousal.
3. *Commitment:* a component of love that involves *dedication* to maintaining the relationship through good times and bad.

According to Sternberg's model, love can be conceptualized in terms of a triangle in which each vertex represents one of these basic elements of love (see Figure 7.2). The way the components are balanced can be represented by the shape of the triangle. For

Romantic love A kind of love characterized by feelings of passion and intimacy.

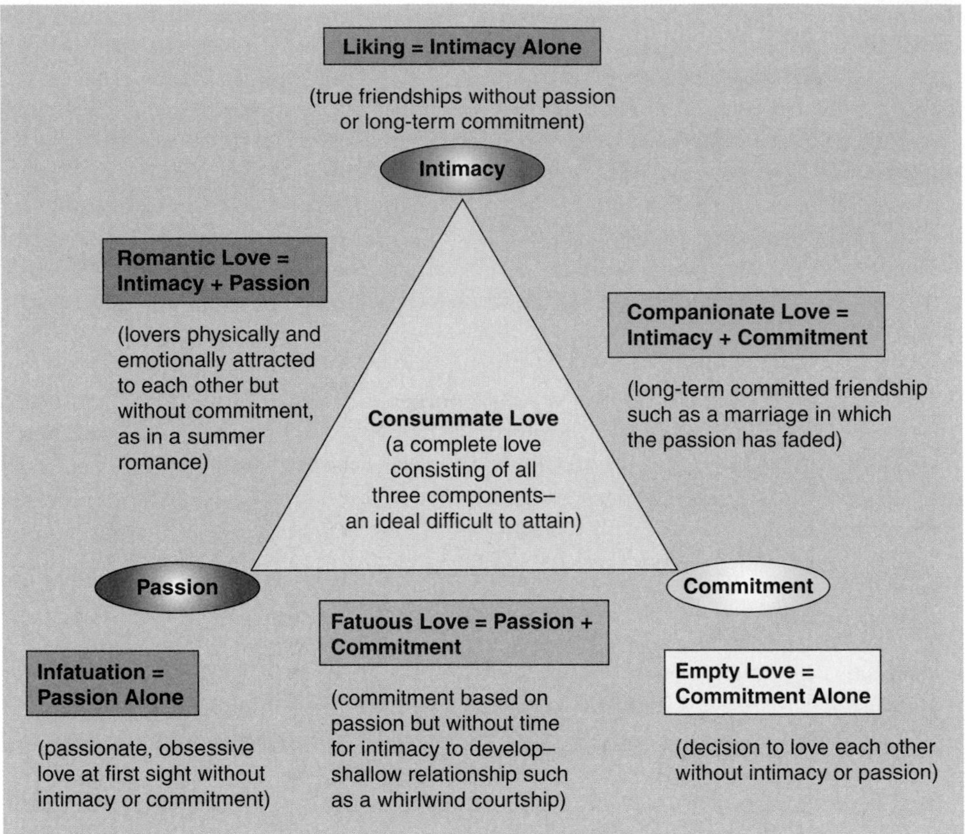

FIGURE 7.2

The Triangular Model of Love. According to psychologist Robert Sternberg, love consists of three components, represented by the vertices of this triangle. Various kinds of love consist of different combinations of these components. Romantic love, for example, consists of passion and intimacy. Consummate love—the cultural ideal—consists of all three.

example, a love in which all three components are equally balanced would be represented by an equilateral triangle, as in Figure 7.2.

Couples are apparently well matched if they possess corresponding levels of passion, intimacy, and commitment (Drigotas et al., 1999; Sternberg, 1988). Compatibility can be represented visually in terms of the congruence of the love triangles. Figure 7.3(a) shows a perfect match, in which the triangles are congruent. Figure 7.3(b) depicts a good match; the partners are similar in the three dimensions. Figure 7.3(c) shows a mismatch; major differences exist between the partners on all three components. Relationships may run aground when partners are mismatched. A relationship may fizzle, rather than sizzle, if one partner experiences more passion than the other, or if one wants a long-term relationship when the other's idea of commitment is to stay the whole night.

According to the Sternberg model, various combinations of the three elements of love characterize different types of love relationships (Sternberg, 1986, 1988) (see Figure 7.3 and Table 7.2). For example, *infatuation* (passionate love) is typified by strong sexual desire, but not by intimacy and commitment. The partners may each feel passionate love for the other, or, as in the case of Tom, such feelings may go unrequited:

> Tom sat behind Lisa in physics class. Tom hated physics, but he could not say the same for Lisa. One look at her was enough to change his life. He had fallen madly in love with her. Instead of listening to the teacher or looking at the blackboard, he would gaze at Lisa throughout the class. Lisa was aware of this and was not happy about it. She did not much care for Tom, and when he tried to start a conversation with her, she moved on as quickly as possible. Tom's staring and his awkwardness in talking to her made her feel uncomfortable. Tom, on the other hand, could think of little else besides Lisa, and his grades began to suffer as he spent the time he should have been devoting to his homework thinking about her. He was a man obsessed. The obsession might have gone on for quite some time had not both Tom and Lisa graduated that June and gone to different colleges. Tom never saw Lisa again, and after several unanswered love letters, he finally gave up on her. (Sternberg, 1988, p. 123)

TABLE 7.2

Types of love according to Sternberg's triangular model

Type	Description
1. Nonlove	A relationship in which all three components of love are absent. Most of our personal relationships are of this type—casual interactions or acquaintances that do not involve any elements of love.
2. Liking	A loving experience with another person or a friendship in which intimacy is present but passion and commitment are lacking.
3. Infatuation	A "kind of "love at first sight" in which one experiences passionate desire for another person in the absence of both the intimacy and the commitment components of love.
4. Empty love	A kind of love characterized by commitment (to maintain the relationship) in the absence of either passion or intimacy. Stagnant relationships that no longer involve the emotional intimacy or physical attraction that once characterized them are of this type.
5. Romantic love	A loving experience characterized by the combination of passion and intimacy, but without commitment.
6. Companionate love	A kind of love that derives from the combination of the intimacy and commitment components of love. This kind of love often occurs in marriages in which passionate attraction between the partners has died down and has been replaced by a kind of committed friendship.
7. Fatuous love	The type of love associated with whirlwind romances and "quicky marriages" in which the passion and commitment components of love are present, but intimacy is not.
8. Consummate love	The full or complete measure of love, involving the combination of passion, intimacy, and commitment. Many of us strive to attain this type of complete love in our romantic relationships. Maintaining it is often harder than achieving it.

Source: Adapted from Sternberg, 1988.

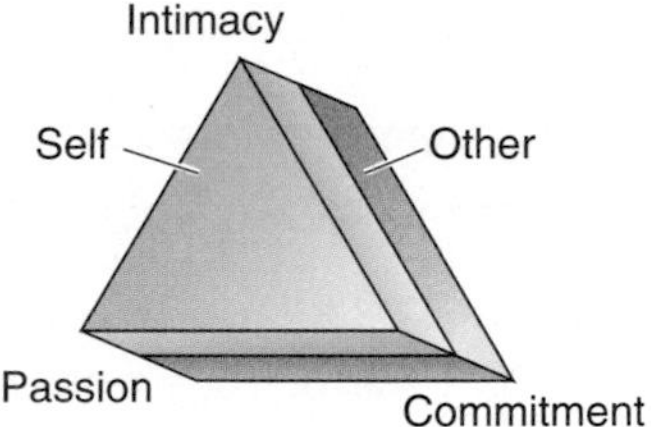

Perfectly matched involvements

(a)

Self
Other

Closely matched involvements

(b)

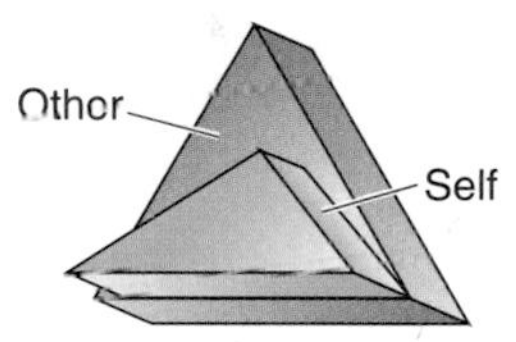

Severely mismatched involvements

(c)

FIGURE 7.3

Compatibility and Incompatibility, According to the Triangular Model of Love. Compatibility in terms of Sternberg's types of love can be represented as triangles. (a) A perfect match in which triangles are congruent. b) A good match; the partners are similar on the three dimensions. (c) A mismatch. Major differences exist between the partners on all three components.

Liking is a basis for friendship. It consists of feelings of closeness and emotional warmth without passion or commitment. Liking is not felt toward passing acquaintances. It is reserved for people to whom one feels close enough to share one's innermost feelings and thoughts. We sometimes develop these intimate relationships without making the commitment to maintaining a long-term relationship that typifies other types of love. Liking may develop into a passionate love, however, or into a more committed form of friendship (called *companionate love* in Sternberg's model).

Can lovers also be friends, or shall the twain never meet? Candy and Stretch's relationship lacked the quality most often associated with true friendship: the willingness to share confidences. Despite their physical intimacy, their relationship remained so superficial that they couldn't even share information about their religious backgrounds.

It is indeed possible to be in love with someone who is not also a friend. Being in love can refer to states of passion or infatuation, whereas friendship is usually based on shared interests, liking, and respect.

Candy and Stretch were "in love," although they were far from friends. Friendship and passionate love do not necessarily overlap. There is no reason why people in love

QUESTIONNAIRE

Sternberg's Triangular Love Scale

What are the strongest components of your love relationship? Intimacy? Passion? Commitment? All three components? Two of them?

To complete the following scale, fill in the blank spaces with the name of one person you love or care about deeply. Then rate your agreement with each of the items by using a 9-point scale in which 1 = "not at all," 5 = "moderately," and 9 = "extremely." Use points in between to indicate intermediate levels of agreement between these values. Then consult the scoring key in the Appendix.

Intimacy Component

_____ 1. I am actively supportive of _________'s well-being.

_____ 2. I have a warm relationship with _________.

_____ 3. I am able to count on _________ in times of need.

_____ 4. _________ is able to count on me in times of need.

_____ 5. I am willing to share myself and my possessions with _________.

_____ 6. I receive considerable emotional support from _________.

_____ 7. I give considerable emotional support to _________.

_____ 8. I communicate well with _________.

_____ 9. I value _________ greatly in my life.

_____ 10. I feel close to _________.

_____ 11. I have a comfortable relationship with _________.

_____ 12. I feel that I really understand _________.

_____ 13. I feel that _________ really understands me.

_____ 14. I feel that I can really trust _________.

_____ 15. I share deeply personal information about myself with _________.

Passion Component

_____ 16. Just seeing _________ excites me.

_____ 17. I find myself thinking about _________ frequently during the day.

_____ 18. My relationship with _________ is very romantic.

_____ 19. I find _________ to be very personally attractive.

_____ 20. I idealize _________.

should not become good friends, however—perhaps even the best of friends. Sternberg's model recognizes that the intimacy we find in true friendships and the passion we find in love are blended in two forms of love: romantic love and consummate love. These types of love differ along the dimension of commitment, however.

Romantic love has both passion and intimacy but lacks commitment. Romantic love may burn brightly and then flicker out. Or it may develop into a more complete love, called *consummate love,* in which all three components flower. Desire is accompanied by a deeper intimacy and commitment. The flames of passion can be stoked across the years, even if they do not burn quite so brightly as they once did. Consummate love is most special, and it certainly is an ideal toward which many Westerners strive.

In *empty love,* by contrast, there is nothing *but* commitment. Neither the warm, emotional embrace of intimacy nor the flame of passion exists. With empty love, one's lover is a person whom one tolerates and remains with because of a sense of duty. People often remain in an empty-love relationship because of either personal prescription or social prescription (Cox et al., 1997). *Personal prescription* is based on the belief that one should persist in a relationship. *Social prescription* is based on the belief that one's friends or family members believe that it is right to persist in a relationship.

Sometimes a love relationship has both passion and commitment but lacks intimacy. Sternberg calls this *fatuous (foolish) love.* Fatuous love is associated with whirlwind

21. I cannot imagine another person making me as happy as ________ does.
22. I would rather be with ________ than anyone else.
23. There is nothing more important to me than my relationship with ________.
24. I especially like physical contact with ________.
25. There is something almost "magical" about my relationship with ________.
26. I adore ________.
27. I cannot imagine life without ________.
28. My relationship with ________ is passionate.
29. When I see romantic movies and read romantic books, I think of ________.
30. I fantasize about ________.

Commitment Component

31. I know that I care about ________.
32. I am committed to maintaining my relationship with ________.
33. Because of my commitment to ________, I would not let other people come between us.
34. I have confidence in the stability of my relationship with ________.
35. I could not let anything get in the way of my commitment to ________.
36. I expect my love for ________ to last for the rest of my life.
37. I will always feel a strong responsibility for ________.
38. I view my commitment to ________ as a solid one.
39. I cannot imagine ending my relationship with ________.
40. I am certain of my love for ________.
41. I view my relationship with ________ as permanent.
42. I view my relationship with ________ as a good decision.
43. I feel a sense of responsibility toward ________.
44. I plan to continue my relationship with ________.
45. Even when ________ is hard to deal with, I remain committed to our relationship.

Source: From Robert J. Sternberg (1988). *The Triangle of Love: Intimacy, Passion, Commitment.*

courtships that burn brightly but briefly as the partners come to realize that they are not well matched. Intimacy can develop in such relationships, but couples who rush into marriage often find that the realities of marriage give the lie to their expectations:

> They expect a marriage made in heaven, but do not realize what they must do truly to maintain such a marriage. They base the relationship on passion and are disappointed when the passion starts to fade. They feel shortchanged—they have gotten much less than they bargained for. The problem, of course, is that they bargained for too much of one thing [passion] and not enough of another [intimacy]. (Sternberg, 1988, p. 128)

In *companionate love,* finally, intimacy and commitment are strong, but passion is lacking. This form of love typifies long-term (so-called Platonic) friendships and those marriages in which passion has ebbed but a deep and abiding friendship remains. Berscheid and Walster defined companionate love as "the affection we feel for those with whom our lives are deeply entwined" (1978, p. 9).

Although romantic love may become transformed into companionate love, the process by which this transformation takes place remains vague (Shaver et al., 1988). Companionate love need not be sexless or lacking in romance, however. Although passion may have ebbed, the giving and receiving of sexual pleasure can help strengthen bonds. Partners may feel that their sex lives have even become more deeply satisfying as they seek to please each other by practicing what they have learned about each other's sexual needs and wants.

The balance among Sternberg's three aspects of love is likely to shift through the course of a relationship. A healthful dose of all three components—found in consummate love—typifies, for many of us, an ideal marriage. At the outset of marriage, passions may be strong but intimacy weak. Couples may only just be getting to know each other's innermost thoughts and feelings. Time alone does not cause intimacy and commitment to grow, however. Some couples are able to peer into each other's deeper selves and form meaningful commitments at relatively early stages in their relationships. Other long-married couples may remain distant or waver in their commitment. Some couples experience only a faint flickering of passion early in the relationship. Then it becomes quickly extinguished. For some, the flames of passion burn ever bright. Yet many married couples find that passion tends to fade while intimacy and commitment grow stronger.

Knowing about these components of love may help couples avoid pitfalls. Couples who recognize that passion exerts a strong pull early in a relationship may be less likely to rush into marriage. Couples who recognize that it is normal for passions to fade may avoid assuming that their love is at an end when it may simply be changing into a deeper, more intimate and committed form of love. This knowledge may also encourage couples to focus on finding ways of rekindling the embers of romance, rather than looking to escape at the first sign that the embers have cooled.

Researchers have tested some facets of the triangular model. One study reported mixed results. As the model would predict, married adults reported higher levels of commitment to their relationships than did unmarried adults (Acker & Davis, 1992). Yet the expected decline in passion over time was found only for women. Critics contend that Sternberg's model does not account for all the nuances and complexities of love (Murstein, 1988). The model tells us little, for example, about the *goals* of love or the *sources* of love. In fairness, Sternberg's model is a major contribution to the scientific study of love, which has only recently begun. Poets, philosophers, and theologians, by comparison, have been writing about love for millennia.

In this chapter, we have discussed interpersonal attraction—the force that initiates social contact. In the next chapter, we follow the development of social contacts into intimate relationships.

Human Sexuality ONLINE //

Web Sites Related to Attraction and Love

The Web site of an online "megamagazine" with links to print magazines (*Marie Claire, Cosmopolitan, Good Housekeeping,* etc.), and with articles and advice on topics such as careers, cars, entertainment, family, fashion and beauty, fitness, food, health and wellness, home and garden, money, news and politics, pregnancy and babies, sex and romance, shopping, small business, technology and the Internet, travel, and weddings.
www.women.com

Columbia University's "Go Ask Alice." This Web site offers online advice about relationships in a question-and-answer format.
www.goaskalice.columbia.edu

e-How: This Web site offers advice about dating and forming romantic relationships.
www.ehow.com

MSNBC Sexploration: This Web site offers information and advice about romantic relationships.
www.msnbc.com/news/SEXPLORATIONH_Front.asp

Our weight affects our attractiveness, and weight control is a major concern for many. The following three Web sites offer some helpful and valid advice on this issue.

The American Obesity Association:
www.obesity.org/

The National Association for Health and Fitness:
www.physicalfitness.org/

The journal *Obesity Research:*
www.obesityresearch.org/

summing up

ATTRACTION

A number of factors determine interpersonal attraction.

■ Physical Attractiveness: How Important Is Looking Good?
Physical attractiveness is a major determinant of sexual attraction. In our culture, slenderness is in style. Both genders consider smiling faces more attractive. Socially dominant men, but not dominant women, are usually found attractive. Women place greater emphasis on such traits as vocational status and earning potential, whereas men give more consideration to physical attractiveness. Some evolutionary psychologists believe that evolutionary forces favor such gender differences in preferred traits because these traits offer reproductive advantages.

■ The Matching Hypothesis: Who Is Right for You?
According to the matching hypothesis, people tend to develop romantic relationships with people who are similar to themselves in attractiveness.

■ Similarity in Attitudes: Do Opposites Attract?
Similarity in attitudes and tastes is a strong contributor to attraction, friendships, and love relationships.

■ Reciprocity: If You Like Me, You Must Have Excellent Judgment
Through reciprocation of positive words and actions, neutral or mild feelings may be stoked into feelings of attraction.

LOVE

In our culture, we are brought up to idealize the concept of romantic love.

■ The Greek Heritage
The Greeks had four concepts related to the modern meanings of love: storge, agape, philia, and eros.

■ Romantic Love in Contemporary Western Culture
Western culture has a long tradition of idealizing the concept of romantic love. Most people in the United States see romantic love as prerequisite to marriage. Early in a relationship, infatuation and more enduring forms of romantic love may be indistinguishable.

■ Contemporary Models of Love: Dare Science Intrude?
Researchers are now applying the scientific method to the study of love.

Berscheid and Hatfield define romantic love in terms of intense physiological arousal and cognitive appraisal of that arousal as love.

Hendrick and Hendrick suggest that there are six styles of love among college students: romantic love, game-playing love, friendship, logical love, possessive love, and selfless love.

Sternberg suggests that there are three distinct components of love: intimacy, passion, and commitment. Various combinations of these components typify different kinds of love. Romantic love is characterized by the combination of passion and intimacy.

questions for critical thinking

1. How important is physical attractiveness to you in your evaluation of partners for dating, sex, and marriage? (Are you being truthful?)
2. What personality traits are most important to you in your evaluation of partners for dating, sex, and marriage?
3. Would you date someone of another race? Another religion? Why or why not? What would be the response of your parents or other members of your sociocultural group if you did?
4. Have you been in love? How did you know you were in love?

chapter

8

Relationships, Intimacy, and Communication

Pablo Picasso. *The Embrace*, 1903. Pastel. © ARS, NY. Photo: Arnaudet. Musée de l'Orangerie, Paris, France. Credit: Réunion des Musées Nationaux/Art Resource, NY. Credit: © 2002 Estate of Pablo Picasso/Artists Rights Society (ARS), New York.

outline

Truth or Fiction?

THE ABC(DE)'S OF ROMANTIC RELATIONSHIPS

A World of Diversity Don't Flirt (with Anyone!) in Qir

The A's—Attraction

Human Sexuality Online Modem Matchmaking

The B's—Building

A Closer Look How to Improve Date-Seeking Skills

Human Sexuality Online In Modern E-Mail Romance, "Trash" Is Just a Click Away

The C's—Continuation

The D's—Deterioration

Human Sexuality Online Long-Distance Romance, Web Enabled

The E's—Ending

Human Sexuality Online Advice for Young Women: How to Prepare for Married Life

LONELINESS: "ALL THE LONELY PEOPLE, WHERE DO THEY ALL COME FROM?"

Causes of Loneliness

Coping with Loneliness

INTIMACY

Knowing and Liking Yourself

Trusting and Caring

A Closer Look Heaven Sent: The Blind Date Who Is Your Destiny

Being Honest

Making a Commitment

Maintaining Individuality When the *I* Becomes *We*

The Michelangelo Phenomenon

Communicating

COMMUNICATION SKILLS FOR ENHANCING RELATIONSHIPS AND SEXUAL RELATIONS

Common Difficulties in Sexual Communication

Getting Started

Listening to the Other Side

Learning About Your Partner's Needs

Providing Information

Making Requests

Delivering Criticism

Receiving Criticism

When Communication Is Not Enough: Handling Impasses

Human Sexuality Online Web Sites Related to Relationships, Intimacy, and Communication

SUMMING UP

QUESTIONS FOR CRITICAL THINKING

Truth or Fiction?

F Making small talk is an insincere way to begin a relationship.

F Only phonies practice opening lines.

F Swift self-disclosure of intimate information is the best way to deepen a new relationship.

T Many people remain lonely because they fear being rejected by others.

T People can have intimate relationships without being sexually intimate.

F "Love is all you need." That is, when partners truly love one another, they instinctively know how to satisfy each other sexually.

F If you are criticized, the best course is to retaliate.

T Relationships come to an end when the partners cannot resolve their differences.

F Disagreement is destructive to a relationship.

Will you, won't you, will you, won't you, will you join the dance?

—Lewis Carrol, *Alice in Wonderland*

No man is an island, entire of it self.

—John Donne

One, two. One, two." A great opening line? In the film *Play It Again, Sam*, Woody Allen plays the role of Allan Felix, a social klutz who has just been divorced. Diane Keaton plays his platonic friend Linda. At a bar one evening with Linda and her husband, Allan Felix spots a young woman on the dance floor who is so attractive that he wishes *he* could have *her* children.

The thing to do, Linda prompts him, is to begin dancing, then dance over to her and "say something." With a bit more prodding, Linda convinces Allan to dance. It's so simple, she tells him. He need only keep time—"One, two, one, two."

"One, two," repeats Allan. Linda shoves him off toward his dream woman.

Hesitantly, Allan dances up to her. Working up courage, he says, "One, two. One, two, one, two." He is ignored and finds his way back to Linda.

"Allan, try something more meaningful," Linda implores.

Once more, Allan dances nervously back toward the woman of his dreams. He stammers, "Three, four, three, four."

"*Speak* to her, Allan," Linda insists.

He dances up to her again and tries, "You interested in dancing at all?"

"Get lost, creep," she replies.

Allan dances rapidly back toward Linda. "What'd she say?" Linda asks.

"She'd rather not," he shrugs.

So much for "One, two, one, two," and, for that matter, "Three, four, three, four." Striking up a relationship requires some social skills, and the first few conversational steps can be big ones.

In this chapter we first define the stages that lead to intimacy in relationships. We define intimacy and see that not all relationships achieve this level of interrelatedness, even some supposedly deep and permanent relationships such as marriage. Moreover, we do not all have partners with whom we can develop intimate relationships; some of us remain alone and, perhaps, lonely. There are steps we can take to overcome loneliness, however, as we illustrate in the pages that follow. Finally, we discuss the ways in which communication contributes to relationships and sexual satisfaction, and we enumerate ways of enhancing communication skills.

THE ABC(DE)'S OF ROMANTIC RELATIONSHIPS

Romantic relationships, like people, undergo stages of development. According to **social-exchange theory,** this development reflects the unfolding of social exchanges, which involve the rewards and costs of maintaining the relationship, compared to the rewards and costs of dissolving it. During each stage, positive factors sway partners toward maintaining and enhancing their relationship. Negative factors incline them toward letting it deteriorate and end (Karney & Bradbury, 1995).

Social-exchange theory The view that the development of a relationship reflects the unfolding of social exchanges—that is, the rewards and costs of maintaining the relationship as opposed to those of ending it.

Don't Flirt (with Anyone!) in Qir

The arrest of a shopkeeper accused by a judge of flirting with his wife has triggered riots in the Iranian town of Qir. Newspapers said the unrest had broken out Sunday after a local judge ordered the arrest of the grocer, accusing him of flirting with his wife while the couple were shopping at the local bazaar.

The daily Bahar said the grocer had been beaten up during several hours of detention. Fellow shopkeepers closed down the bazaar in Qir, a small town in Fars province, in protest, and young people burned cars and set fire to the local court house, a police station and the district government's office.

Local judiciary chief Hossein-Ali Amiri told reporters that a "misunderstanding" between the judge and the shopkeeper had led to the tensions. Bahar said the grocer was cross-eyed, giving a false impression to the judge's wife that he was leering at her.

Like many provincial towns in Iran, Qir, 220 km (140 miles) south of the provincial capital Shiraz, has a conservative religious environment with strong family values. Amiri said the riots came after the district government ignored a complaint from the public about the grocer's arrest. There were no reports of any injuries.

Source: Text from "Riot Erupts after Grocer Arrested for 'Flirting'" (2000, July 31). *Reuters News Agency online.*

George Levinger (1980) proposes an **ABCDE model** to describe the stages of romantic relationships: *A*ttraction, *B*uilding, *C*ontinuation, *D*eterioration, and *E*nding.

The A's—Attraction

Attraction occurs when two people become aware of each other and find one another appealing or enticing. We may find ourselves attracted to an enchanting person "across a crowded room," in a nearby office, or in a new class. We may meet others through blind dates, introductions by mutual friends, computer match-ups, or by "accident." Initial feelings of attraction are largely based on visual impressions, though we may also form initial impressions by overhearing people speak or hearing others talk about them.

Being in a good mood apparently heightens feelings of attraction. George Levinger and his colleagues (Forgas et al., 1994) exposed 128 male and female moviegoers to either a happy or a sad film. Those who had been shown the happy film reported more positive feelings about their partners and their relationships. (Think twice about what you take your date to see!)

According to the NHSLS study (Michael et al., 1994), married people are most likely to have met their spouses through mutual friends (35%) or via self-introductions (32%) (see Figure 8.1). Other sources of introductions are family members (15%) and coworkers, classmates, or neighbors (13%). Mutual friends and self-introductions are also the most common ways of meeting for unmarried couples (Michael et al., 1994). As we see in the nearby Human Sexuality Online feature, more and more couples are also meeting online.

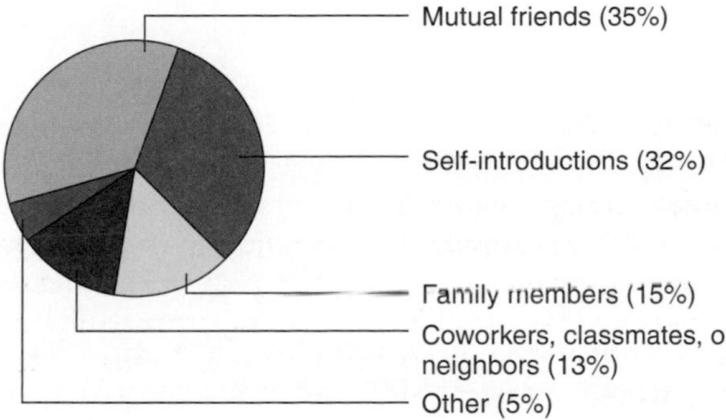

FIGURE 8.1

How Married People Met Their Partners. According to the NHSLS study (Michael et al., 1994), two thirds of married people met their partners through either mutual friends or self-introduction.

ABCDE model Levinger's view, which approaches romantic relationships in terms of five stages: Attraction, Building, Continuation, Deterioration, and Ending.

Human Sexuality ONLINE //

Modem Matchmaking

Maybe they're not matches made in heaven—but close.

Loveseekers are swarming into cyberspace.

Online has become the hot new place for finding sweethearts.

"The Net is what singles bars and mixers used to be," says Stanford University psychologist Al Cooper. "It's turning into the place for smart, eligible people to meet."

Bill Stanfield, 47, says he can't imagine marrying a woman after dating for only a few months. But he and his wife Jacqueline wed "and I've never regretted it. There were no surprises."

Of course, they didn't really date. They met online. "We spent hundreds and hundreds of hours communicating. There were no distractions. . . . I knew my first wife for three years before we got married, but I didn't know her as well as I knew Jackie."

The Stanfield's courtship, peculiar as it sounds, is hardly unique. Love at AOL, an area on America Online designed for searching singles, posts 30,000 "personal photo ads," with more than 700,000 visits to the area monthly. It's the service's most heavily used content area, says Anne Bentley of AOL.

Match.Com, largest independent dating service on the Web, offers 85,000 member "profiles," adding about 6,000 new faces weekly, corporate vice president Fran Maier reports.

Why such a crowd?

Surveys show Net traffic is dominated by affluent, educated 20- to 40-somethings, often overworked professionals for whom "dating isn't so great," Cooper says. Many are "solid people, good long-term prospects but not good 'daters.' The guys may be a little shy; they're not adept at small talk and don't have great pickup lines."

Finding someone online with similar interests and values can feel a lot easier than hitting bars or "meat market" clubs, Cooper says.

Net liaisons can be "female-friendly" too, he adds: Communication is queen; sex stays off the front burner (at least for a while); and guys don't snub women if they're not 10-pluses.

A feeling of intimacy often develops quickly. Online talk lacks nonverbal cues that can put the brakes on romantic encounters, says Santa Clara, California, psychologist Coralie Scherer. Disapproving glances, tone-of-voice changes, hesitance and breaks in eye contact aren't there. "All you know is what they tell you," says Scherer, so honesty can't make or break the deal.

The Stanfields, who married three months after meeting online, quickly realized they shared a passionate interest in hiking, camping and outdoor beauty. "We were really honest with each other about our personal likes and dislikes, what we want out of life," he says.

But honesty can also be elusive online. "Gender-bending" is so common that experts estimate that two out of three "women" in many chat rooms, particularly the sex-oriented ones, are men.

Changing or omitting other truths is common. San Diego psychologist Marlene Maheu tells about one client, "an intelligent, sophisticated professional woman," who recently flew hundreds of miles to meet a Net boyfriend and found "not only was he obese, but he looked considerably older than he claimed to be."

Women play that game, too. New York psychologist Judy Kuriansky, host of a nationally syndicated radio call-in show, took one frantic call from a lady about to meet her longtime online sweetheart. She'd neglected to mention she weighed 350 pounds and wondered what to do next.

"Omissions" can be more diabolical—even dangerous. Kuriansky recently heard a self-satisfied Michigan woman describe her "hobby": She's traveled coast to coast visiting cities she'd always wanted to see, at male expense, after feigning online interest in a string of men.

Maheu tells of one client recently stalked after giving her phone number to an e-mail friend who used it to find her address. "I've heard of similar situations quite often," Maheu reports.

Several "Netiquette" books advise how to bring cyberspace romantic encounters safely down to earth. *Looking for Love Online* by Richard Rogers suggests ways to sniff out phonies and crazies, and how to surround in-person meetings with safety nets and escape hatches.

Denise Beaupre, an Attleboro, Massachusetts, single mother, was pretty sure after three months of constant contact that her online friend, Scott Arena, was a wonderful guy. Still, she brought two friends to her first meeting (and one friend brought a taser gun).

"He was a sweetheart from the beginning though. . . . Look, I've been followed home from bars. Online is much safer, if you're careful about it."

Net romance may be most threatening to addicts, those online for many hours a week. And addicts may also behave most deceptively to would-be partners.

University of Pittsburgh psychologist Kimberly Young reports a link between Net addiction and clinical depression. Addicts may be "very vulnerable. They're quick to jump into relationships but sensitive to rejection."

The B's—Building

Building a relationship follows initial attraction. Factors that motivate us to try to build relationships include similarity in the level of physical attractiveness, similarity in attitudes, and mutual liking and positive evaluations. Factors that may deter us from trying to build relationships include lack of physical appeal, dissimilarity in attitudes, and negative mutual evaluations.

Surface contact A probing phase of building a relationship in which people seek common ground and check out feelings of attraction.

NOT SO SMALL TALK: AN AUDITION FOR BUILDING A RELATIONSHIP In the early stages of building a relationship, we tend to probe each other with **surface contact:** We typi-

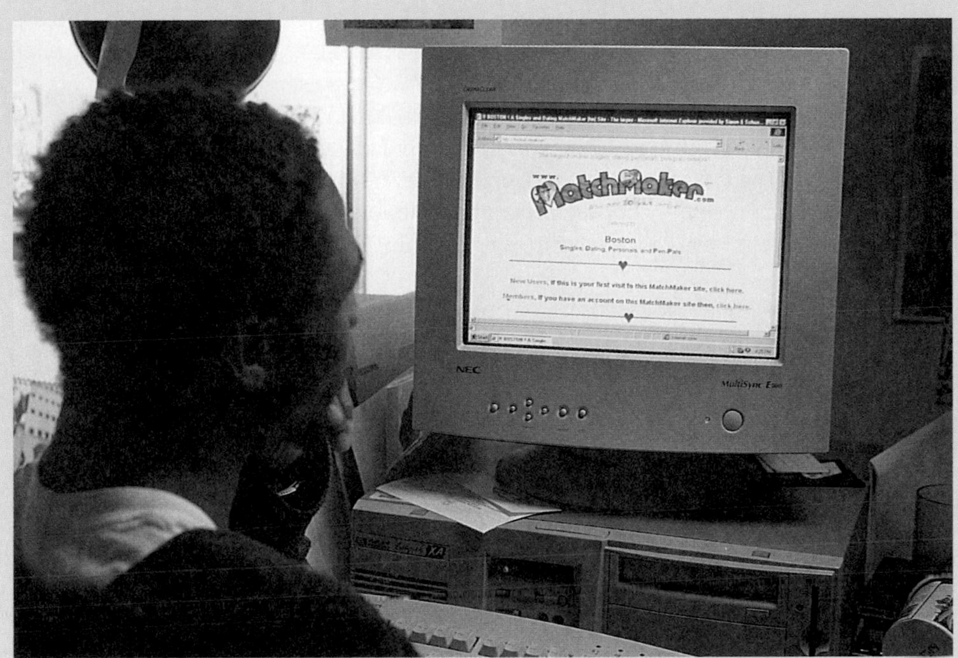

Matchmaking in the New Millennium? In the new millennium, matches are more likely to be made by modem than in heaven. For example, nearly one million people visit Love at AOL each month, an area on America Online that is designed to help searching singles find their matches. Many people click their way to Web sites like www.matchmaker.com. You can check out people online—including their photos and their interests—before taking the plunge and meeting them in person. E-mail can take the place of small talk. For many people, it's a lot less threatening than bars or "meat market" clubs.

Net addicts are also prone to donning online "personas" different from their everyday selves, her research shows.

"The chat rooms now are crowded with people who change their names daily. Who knows who they really are?" grumbles Joan Bounacos. She and husband George met online, married two years later and know a number of happily married couples matched years ago in cyberspace.

"Too many strange people are online now," Bounacos says, "and they want things to happen quick. . . . We were just good friends for a long time. What you bring to it is what you're going to get out of it."

That's exactly the point, argues psychologist Sherry Turkle, an MIT professor and author of *Life on the Screen.* She's studied several hundred adults who pursued relationships online.

"It's like a living Rorschach test," Turkle says. The absent nonverbal cues and isolation make Net encounters akin to traditional psychoanalysis in which patients "projected their greatest desires and fears" onto remote analysts. "They'll tell me, 'I was deceived,' and there is some deception. But when I look at their e-mail transcripts, I see less deception and more projection. These transcripts are often thin, people fill in the gaps as they wish."

How the gaps are filled in is the real tale. It's often a positive one. She's seen people use online encounters "as an occasion for self-reflection. They can learn a lot about themselves."

New York writer Sally Banks says she did just that. And Banks, alumna of several online liaisons, author of *Love Online,* is a bit embarrassed to admit she met her sweetheart, "the best relationship I've ever had," walking their dogs in Central Park. Her romantic forays on the Net harvested the self-revealing fodder for a real-world love.

"All those online relationships made me realize what I needed, what I was really looking for. I'll tell you what happened—I met myself online."

Source: From Marilyn Elias, "Modem Matchmaking." *USA Today,* August 14, 1997, pp. 1D, 2D. Copyright © 1997 by USA Today. Reprinted with permission.

cally look for common ground in the form of overlapping attitudes and interests, and we check out our feelings of attraction. At this point, the determination of whether to strive to develop the relationship is often made, at least in part, on the basis of **small talk.** Small talk allows an exchange of information but stresses breadth of topic coverage rather than in-depth discussion. Engaging in small talk may seem "phony," but premature self-disclosure of intimate information may repel the other person, as we shall see.

Small talk A superficial kind of conversation that allows exchange of information but stresses breadth of topic coverage rather than in-depth discussion.

Small talk is a trial balloon for friendship. Successful small talk encourages a couple to venture beneath the surface. At a cocktail party, people may flit about from person to person, exchanging small talk, but now and then a couple finds common ground and pairs off.

It is not true that small talk is an insincere way to begin a relationship. Actually, small talk is a realistic way to open the channels of communication. It allows couples to search for common ground and test feelings of attraction.

THE "OPENING LINE": HOW DO YOU GET THINGS STARTED? One kind of small talk is the greeting, or opening line. We usually precede verbal greetings with eye contact and decide to begin talking if this eye contact is reciprocated. Avoidance of eye contact may mean that the person is shy, but it could also signify lack of interest. If you would like to progress from initial attraction to surface contact, try a smile and direct eye contact. If the eye contact is reciprocated, choose an opening line, or greeting. Because your opening line can be important, you may prefer to say something more meaningful than "One, two, one, two."

Here are some types of greetings, or opening lines:

Verbal salutes, such as "Good morning."

Personal inquiries, such as "How are you doing?"

Compliments, such as "I like your outfit."

References to your mutual surroundings, such as "What do you think of that painting?" or "This is a nice apartment house, isn't it?"

References to people or events outside the immediate setting, such as "How do you like this weather we've been having?" (Opening gambits about the weather may work best when accompanied by a self-deprecating grin to acknowledge how corny the remark might seem.)

References to the other person's behavior, such as "I couldn't help noticing you were sitting alone," or "I see you out on this track every Saturday morning."

References to your own behavior, or to yourself, such as "Hi, my name is Allan Felix" (feel free to use your own name, if you prefer).

The simple "Hi" or "Hello" is very useful. A friendly glance followed by a cheerful hello ought to give you some idea of whether the attraction is reciprocated. If the hello is returned with a friendly smile and inviting eye contact, follow it up with another greeting, such as a reference to your surroundings, the other person's behavior, or your name.

Truth or Fiction? **REVISITED**

It is not true that only phonies practice opening lines. Actually, it can be helpful for everyone to practice opening lines. People are social beings, and opening lines help us initiate social relationships—with new friends as well as dates.

EXCHANGING "NAME, RANK, AND SERIAL NUMBER" Early exchanges are likely to include name, occupation, marital status, and hometown. This has been likened to exchanging "name, rank, and serial number" with the other person. Each person seeks a sociological profile of the other to discover common ground that may provide a basis for pursuing the conversation. An unspoken rule seems to be at work: "If I provide you with some information about myself, you will reciprocate by giving me an equal amount of information about yourself. Or . . . 'I'll tell you my hometown if you tell me yours' " (Knapp, 1978, p. 114). If the other person is unresponsive, she or he may not be attracted to you, and you may wish to try someone else. But you may also be awkward in your approach or perhaps turn the other person off by disclosing too much about yourself at once. The nearby Closer Look suggests ways of improving date-seeking skills.

SELF-DISCLOSURE: YOU TELL ME AND I'LL TELL YOU . . . CAREFULLY Opening up, or **self-disclosure**, is central to building intimate relationships. But just what sort of information is it safe to disclose upon first meeting someone? If you refuse to go beyond name, rank, and serial number, you may look uninterested or as though you are trying to keep things under wraps. If, on the other hand, you blurt out the fact that you have a terrible rash on your thigh, it's likely that you have disclosed too much too soon.

Research suggests that we should refrain from disclosing certain types of information too rapidly if we want to make a good impression. In one study, confederates of the experimenters (Wortman et al., 1976) engaged in 10-minute conversations with study par-

Self-disclosure The revelation of personal—perhaps intimate—information.

a CLOSER look

How to Improve Date-Seeking Skills

All right, now you know that Mr. or Ms. Right exists. So what do you do? How do you go about making a date?

Psychologists have learned that we can enhance our social skills, including our date-seeking skills, through the technique of *successive approximations.* (The third author interjects: "See here, you two psychologists, if you *mean* building skills step by step, why not just *say* 'step by step'?") That means that we can practice a series of tasks that are *graded in difficulty.* (The third author surrenders!) We can hone our social skills and gain self-confidence at each step. We can try out some of our skills on friends. Friends can role-play the prospective date and offer honest feedback about our behavior.

Here is an example of a series of graduated (*step-by-step*)[1] tasks that may help you sharpen your own date-seeking skills:

Easy Practice Level

Select a person with whom you are friendly but whom you have no desire to date. Practice making small talk about the weather, new films, TV shows, concerts, museum shows, political events, and personal hobbies.

Select a person you might have some interest in dating. Smile when you pass this person at work, school, or elsewhere, and say "Hi." Engage in this activity with other people of both genders to increase your skill at greeting others.

Speak into your mirror, using behavior rehearsal and role playing. Pretend you are in the process of sitting next to the person you would like to date, say, at lunch or in the laundry room. Say "Hello" with a broad smile, and introduce yourself. Work on the smile until it looks inviting and genuine. Make some comment about the food or the setting—the cafeteria, the office, whatever. Use a family member or confidant(e) to obtain feedback about the effectiveness of the smile, your tone of voice, posture, and choice of words.

Medium Practice Level

Sit down next to the person you want to date, and engage in small talk. If you are in a classroom, talk about a homework assignment, the seating arrangement, or the instructor (be kind). If you are at work, talk about the building or some recent interesting event in the neighborhood. Ask your intended date how he or she feels about the situation. If you are at a group meeting, such as Parents Without Partners, tell the other person that you are there for the first time and ask for advice on how to relate to the group.

Engage in small talk about the weather and local events. Channel the conversation into an exchange of personal information. Give your "name, rank, and serial number"—who you are, your college major or your occupation, where you're from, why or how you came to the school or company. The other person is likely to reciprocate and provide equivalent information. Ask how he or she feels about the class, place of business, city, hometown, and so on.

Rehearse asking the person out before your mirror, a family member, or a confidant(e). You may wish to ask the person out for a cup of coffee or to a film. It is somewhat less threatening to ask someone out to a gathering at which "some of us will be getting together." Or you may rehearse asking the person to accompany you to a cultural event, such as an exhibition at a museum or a concert—it's "sort of" a date but is also less anxiety-inducing.

Target Behavior Level

Ask the person out on a date in a manner consistent with your behavior rehearsal. If the person says he or she has a previous engagement or can't make it, you may wish to say something like "That's too bad" or "I'm sorry you can't make it," and add something like "Perhaps another time." This way you may be able to get a feeling for whether the person you asked out was just making an excuse or has a genuine interest in you but really couldn't accept the specific invitation.

Before asking this individual out again, pay attention to his or her apparent comfort level when you return to small talk on a couple of occasions. If there is still a chance, the person should smile and return your eye contact. The other person may also offer *you* an invitation. In any event, if you are turned down twice, do not ask a third time. And don't make a catastrophe out of the refusal. Look up. Note that the roof hasn't fallen in. The birds are still chirping in the trees. You are still paying taxes. Then give someone else a chance to appreciate your fine qualities.

[1]Happy now?

ticipants. Some confederates were "early disclosers," who shared intimate information early. Others, "late disclosers," shared intimate information toward the end of the conversation only. In both cases, the information was identical. Study participants then rated the disclosers. Early disclosers were rated less mature, less secure, less well adjusted, and less genuine than the late disclosers. Study participants also preferred to continue relationships with the late disclosers. We may say we value "openness" and "honesty" in our relationships, but it may be a social mistake to open up too soon.

Human Sexuality ONLINE //

In Modern E-Mail Romance, "Trash" Is Just a Click Away

One summer night, Jason Kellogg, a 24-year-old publishing assistant, left a bar in Brooklyn after spending a good deal of time talking with a woman he had met that evening. When he left the bar, his friends demanded the requisite information: How old is she? Is she from the area? Does she have any good-looking friends? And most important, did you get a phone number?

Mr. Kellogg hesitated on the last question. "Well, not exactly," he said. "She gave me her e-mail."

That was a first for Mr. Kellogg. He sent her an e-mail message a few days later, although nothing came of their correspondence. "I suppose it was better than a flat-out rejection," he said. "She was probably turned off by my bad spelling."

E-mail, which has long been an indispensable tool in business, is beginning to influence more romantic relationships as well. With more and more Americans in front of computers at work and spending time on them at home, e-mail is changing the ways people meet, court and even break up.

"It's changed every aspect of dating," said Sherry Amatenstein, a dating expert for the iVillage online network (www.ivillage.com) and author of *The Q & A Dating Book* (Adams Media, 2000). "Our new technology is both boon and curse for the modern dater."

One reason for the change in etiquette from requesting a phone number to requesting an e-mail address is comfort, particularly for women. "It's safer than a phone number," said Brenda Ross, dating adviser for About.com and creator of the Dating Advice for Geeks Web site (www.geekcheck.com). "With a few e-mails you can get some details, find out what a guy is like and then decide if you want to go out with him. It takes a lot of pressure off."

Some women are now using e-mail as a way to avoid people they have no intention of ever going out with. In many instances, Ms. Ross said, a woman doesn't want to give out her phone number, but she also doesn't want to be pestered all night for it by some libidinous Neanderthal. Solution: Give him your e-mail address and never respond, or give an e-mail address that you never use, an updating of the time-honored dating tradition of giving a wrong phone number. Eventually he will get the idea, and trashing an e-mail message is easier than dodging a phone call.

Of course, that may seem a bit disingenuous, but one thing about dating hasn't changed: It's still war out there.

Many single people have found that e-mail can be a solution to another age-old dating problem—when to call? Ellen Lavery, a 25-year-old from Manhattan, found herself in that situation a few months ago, the morning after a party where she exchanged phone numbers and e-mail addresses with a man. "The next day, I would have absolutely considered a phone call too soon, too desperate," she said, "a severe violation of the three-day rule. So I shot him off a brief 'We should get together sometime' e-mail. It seemed more acceptable and less threatening, but not overbearing."

Leslie, 31, a publishing executive in Manhattan, has used e-mail as a follow-up to a date. (She and several other people interviewed for this article spoke on the condition that their last names not be used.) "It would be a little too awkward to call, seeing as how you just saw that person the night before, and a letter is too formal, so e-mail comes in handy."

And what about that first e-mail note? Traditionally, the first phone call can be more nerve-racking than transporting nuclear weapons across a rope bridge. You have to be on your toes: funny but not obnoxious, charming yet not hammy, deep but not psychotic, all on the spot. An e-mail message, however, can go through multiple drafts; wording and tone can painstakingly be thought out, reviewed, edited and, if it's not quite right, sent back to committee. "I always show my e-mails to friends before I send them out," said Drew Brooks, a 25-year-old in Manhattan. "Involving other people is fun."

Single people who follow up on chance encounters with e-mail are finding that the awkward first stages of a relationship can be made easier online.

One reason relationships can move so far so fast over the Internet is the solitary nature of e-mail. "When you're e-mailing,

Truth or Fiction? **REVISITED**

It is not true that swift self-disclosure of intimate information is the best way to deepen a new relationship. Actually, swift or premature self-disclosure can make one seem distraught or awkward and thus turn people off.

If the surface contact provided by small talk and initial self-disclosure has been mutually rewarding, partners in a relationship tend to develop deeper feelings of liking for each other (Collins & Miller, 1994). Self-disclosure may continue to build gradually through the course of a relationship as partners come to trust each other enough to share confidences and more intimate feelings.

GENDER DIFFERENCES IN SELF-DISCLOSURE A woman complains to a friend: "He never opens up to me. It's like living with a stone wall." Women commonly declare that men are

How Will She Get Rid of Him? Perhaps she will give him a phony e-mail address. Some women now use e-mail as a way to avoid seeing people they do not want to go out with. They may give out their actual e-mail addresses but never respond, or they may give out seldom used or erroneous e-mail addresses. It's like giving out the wrong phone number. (Bye bye.)

you can be at home, cozy, in your pajamas," Ms. Amatenstein said. "It's so psychologically inviting, people say things they normally would not say."

Ling, a 23-year-old student in Manhattan who is dating someone she met on the Asian Avenue Web site, found that to be the case. "We just started talking, and pretty soon we were writing each other every day," she said. "It's like a psychiatrist's office: you get in, and you feel like you can say anything. It was weird when we first talked to each other on the phone, because we already knew so much about each other."

Mr. Brooks said that he and many of his friends find it easier to be themselves behind the veil of e-mail, saying whatever they want without the risk of a raised eyebrow, a nervous laugh or a dropped jaw. "In e-mail, your needs and feelings are much more important and apparent," he said, "because you're basically exchanging a series of monologues."

E-mail is also enabling people to do a little cyberflirting at work off the radar screen of nosy co-workers. "Although you speak on the phone, having a little light communication throughout the day can be a reinforcement," said Shalu Jaisinghani, 25, a financial consultant from Hoboken, New Jersey. "You get that two-second rush, ego boost, when you see that new-message icon flash across the screen indicating the guy you dig is thinking about you in the middle of the day."

Not all e-mail from a romantic interest, however, is an ego booster. Increasingly, people are finding that the distance of e-mail simplifies one of the potentially messier tasks in a relationship: the breakup. "That's one great thing about e-mail," Mr. Brooks said. "There's no ramifications for you in just blowing someone off."

Not everyone sees it that way, though. Brenda Ross, whose former boyfriend broke up with her via e-mail, after they had talked about the possibility of marriage, said it was about the "lamest" thing a person could do. "E-mail is a great way to break up with someone," she said, "if you're a coward."

Source: From Ed Boland (2000, October 19). "In Modern E-Mail Romance, 'Trash' Is Just a Click Away." *The New York Times online.*

loath to express their feelings (Tannen, 1990). Researchers find that men tend to be less willing to disclose their feelings, perhaps in adherence to the traditional "strong and silent" male stereotype.

Yet gender differences in self-disclosure tend to be small. Overall, researchers find that women are only slightly more revealing about themselves than men (Dindia & Allen, 1992). We should thus be careful not to jump to the conclusion that men are always more "tight-lipped." The belief that there are large gender differences in self-disclosure appears to be something of a myth.

Stereotypes are also a-changing—somewhat. We now see depictions in the media of the "new" man as someone who is able to express feelings without compromising his masculinity. We shall have to wait to see whether this "new" sensitive image replaces the rugged, reticent stereotype or is a passing fancy.

The C's—Continuation

Once a relationship has been established, the couple embarks upon the stage of continuation. Factors that encourage continuation include seeking ways to introduce variety and maintain interest (such as trying out new sexual practices and social activities), showing evidence of caring and positive evaluation (such as sending birthday and Valentine's Day cards), showing lack of jealousy, perceiving fairness in the relationship, and experiencing mutual feelings of general satisfaction. One of the developments in a continuing relationship is that of mutuality.

MUTUALITY: WHEN THE "WE," NOT THE "I'S," HAVE IT When feelings of attraction and the establishment of common ground lead a couple to regard themselves as "we," not as two "I's" who happen to be in the same place at the same time, they have attained what Levinger terms **mutuality.** Mutuality favors continuation and further deepening of the relationship.

As commitment to the relationship grows, there is a tendency for a cognitive shift to occur in which the two "I's"—that is, two independent persons—come to perceive themselves as a "we"—that is, a couple (Agnew et al., 1998). There is now cognitive interdependence (Agnew et al., 1998). Planning for the future, in little ways (What will I do this weekend?) and big ways (What will I do about my education and my career?), comes to include consideration of the needs and desires of one's partner.

Factors in the continuation stage that can throw the relationship into a downward spiral include boredom (such as falling into a rut in leisure activities or sexual practices), displaying evidence of negative evaluation (such as bickering and either forgetting anniversaries and other important dates or pretending that they do not exist), perceiving a lack of fairness in the relationship (such as one partner's always deciding how the couple will spend their free time), or experiencing feelings of jealousy and general dissatisfaction.

JEALOUSY

> O! beware, my lord, of jealousy;
> It is the green-ey'd monster . . .
>
> —William Shakespeare, *Othello*

Thus was Othello, the Moor of Venice, warned of jealousy in the Shakespearean play that bears his name. Yet Othello could not control his feelings and killed his beloved wife, Desdemona. The English poet John Dryden labeled jealousy a "tyrant of the mind." Anthropologists find evidence of jealousy in all cultures, although it may vary in amount and intensity across and within cultures. It appears to be more common and intense among cultures with a stronger *machismo* tradition, in which men are expected to display their virility.

Sexual jealousy is aroused when we suspect that an intimate relationship is threatened by a rival. Lovers can become jealous when others show sexual interest in their partners or when their partners show an interest (even a casual or nonsexual interest) in another. Jealousy can lead to loss of feelings of affection, to feelings of insecurity and rejection, to anxiety and loss of self-esteem, and to mistrust of one's partner and potential rivals (Peretti & Pudowski, 1997). Jealousy is one of the most commonly mentioned reasons why relationships fail (Zusman & Knox, 1998).

Feelings of possessiveness, which are related to jealousy, can also subject a relationship to stress. In extreme cases, jealousy can cause depression or give rise to spouse abuse, suicide, or (as with Othello) murder. But milder forms of jealousy are not necessarily destructive to a relationship. They may even serve the positive function of revealing how much one cares for one's partner.

Mutuality A phase in building a relationship in which members of a couple come to regard themselves as "we," no longer as two "I's" who happen to be in the same place at the same time.

What causes jealousy? In some cases, people become mistrustful of their current partners because their former partners cheated. Jealousy may also derive from low self-esteem or a lack of self-confidence. People with low self-esteem may experience sexual jealousy because they become overly dependent on their partners. They may also fear that they will not be able to find another partner if their present lover leaves.

Researchers have found gender differences in jealousy. Males seem to be most upset by sexual infidelity, females by emotional infidelity (Pines & Friedman, 1998; Wiederman & Kendall, 1999). That is, males are made more insecure and angry when their partners have sexual relations with someone else. Females are made more insecure and angry when their partners become emotionally attached to someone else. Researchers tend to tie this gender difference to evolutionary theory (Harris, 2000; Wiederman & Kendall, 1999). It is hypothesized that males are more upset by sexual infidelity because it confuses the issue of whose children a woman is bearing. Women are more upset by emotional infidelity because it threatens to deprive them of the resources they need to rear their children. However, at least one study found that women can be as upset by sexual infidelity as men can (Harris, 2000).

Many lovers—including many college students—play jealousy games. They let their partners know that they are attracted to other people. They flirt openly or manufacture tales to make their partners pay more attention to them, to test the relationship, to inflict pain, or to take revenge for a partner's disloyalty.

The accompanying Human Sexuality Online feature details how the Internet has helped some couples continue their relationships long-distance—and not.

The D's—Deterioration

Deterioration is the fourth stage of a relationship. It is not necessarily a stage that we seek, and it is certainly not an inevitability. Positive factors that can deter or slow deterioration include putting time and energy into the relationship, striving to cultivate the relationship, and showing patience—for example, giving the relationship a reasonable opportunity to improve. Negative factors that foster deterioration include failure to invest time and energy in the relationship, deciding to put an end to it, and simply permitting deterioration to proceed unchecked.

A relationship begins to deteriorate when one or both partners deem the relationship to be less enticing or rewarding than it has been. Couples who work toward maintaining and enhancing their relationships, however, may find that these become stronger and more meaningful.

ACTIVE AND PASSIVE RESPONSES TO DETERIORATION When a couple perceives their relationship to be deteriorating, they can respond in active or passive ways. Active means of response include doing something that may enhance the relationship (such as working on improving communication skills, negotiating differences, or seeking professional help) or making a decision to end the relationship. Passive methods of responding are basically characterized by waiting for something to happen—by just doing nothing. People can sit back passively and wait for the relationship to improve on its own (once in a great while, it does) or for the relationship to deteriorate to the point where it ends. ("Don't look at me; these things happen.")

It is irrational (and damaging to a relationship) to assume that suitable relationships require no investment of time and effort. No two of us are matched perfectly. Unless one member of a couple does double duty as a doormat, frictions inevitably surface. When problems arise, it is better to work to resolve them than to act as though they don't exist and hope that they will disappear of their own accord.

Human Sexuality ONLINE //

Long-Distance Romance, Web Enabled

If you can't be with the one you love, get a Webcam.

At least, that's what Simon Wong, an engineer in California, and Chiunwei Shu, a student in Arizona, have done to make it easier to live in different states and still date each other. Mr. Wong and Ms. Shu, who met when Ms. Shu was visiting California, say they spend at least two hours every night with their Webcams trained on their bedrooms.

"It's cool," Mr. Wong said. "I'll be at home watching TV—and she's not here, so we can't hang out—and she'll be at home doing her thing, reading a book, and I can check out what she's doing."

Ms. Shu agreed. "If I'm studying, it's not bothering me," she said. "But we can both look up and smile at each other."

High-tech communication isn't just giving e-commerce a global outlook or letting adolescent gamers shoot at one another across town or across the country. Tools like instant messages, videoconferencing software, Web phones and wireless communications are making it easier to keep love alive in long-distance relationships. Mr. Wong and Ms. Shu use NetMeeting, Microsoft's videoconferencing program, to gaze adoringly at each other over the Internet while they talk. They also keep in touch via instant messages and daily e-mail notes, usually with photos. Ms. Shu said she took a picture of herself and sent it to Mr. Wong every day so he could keep up to date on whether she looked happy, sad or stressed. But aside from that analogue standby, the phone, videoconferencing is their favorite means of communication.

"If it weren't for the Internet," Mr. Wong said, "we wouldn't spend the same percentage of the day communicating—you simply can't talk on the phone all the time. But it's easy to spend two hours on NetMeeting. It actually feels like we hang out all the time." There is an added benefit, Mr. Wong said, laughing: "If the phone rings on her side, I can still play the jealous boyfriend," he said. "Like, 'Who's calling you?' Or if we're talking, I can see her expression—if she looks bored, I can say, 'Are you really interested in this conversation?' "

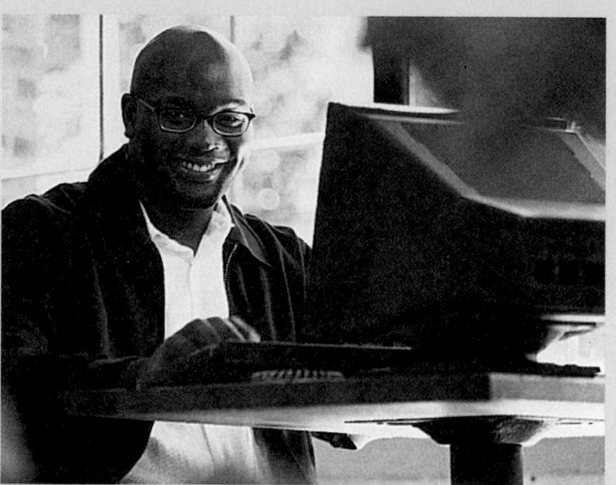

Keeping in Touch—Or at Least in Electronic Touch. When today's couples must be separated, many of them keep in touch by means of tools like instant messages, videoconferencing software, Web phones, and wireless communications. Some couple train Webcams on each other's bedrooms—not necessarily for sexual purposes, but to maintain a kind of intimacy.

California to Costa Rica to Beijing

Prentice Welch and Erika Lam, both 24, are old hands at using technology to keep in touch while they are apart. They have been sweethearts since high school, but in 1997 Ms. Lam moved to Costa Rica for a year of college while Mr. Welch stayed in Davis, California. "That was before we knew about instant messaging," Ms. Lam said.

The two kept in touch by sending e-mail almost every day. Ms. Lam used an online florist to send Mr. Welch flowers, and he sent her the Web address for his basketball league's site so she could keep up with his games. (But she didn't check. "It took too long," she explained.)

Two years later, Mr. Welch moved to Beijing for the school year while Ms. Lam stayed in San Francisco. This time, instant messaging was more common, and they started doing that right away. "That was huge to be able to chat live," said Mr. Welch, now a computer operator in San Francisco, where he and Ms. Lam share a home. "We'd set up dates, when we'd meet online every other day. And sometimes I'd be online, sitting in a cafe in Beijing, and we'd bump into each other. That was so cool."

Long-distance romance over the Web is not all sweetness and light. For one thing, annoying technical glitches sometimes crop

The E's—Ending

Ending is the fifth and final stage of a relationship. Like deterioration, it is not necessarily desirable and need not be inevitable. Various factors can prevent a deteriorating relationship from ending. For example, people who continue to find some sources of satisfaction, who are committed to maintaining the relationship, or who believe that they will eventually be able to overcome their problems are more likely to invest what they must to prevent the collapse.

up. Mr. Wong said that he and Ms. Shu sometimes spent an hour or two trying to get the software to work, to no avail. "Then it's like spending time on the relationship that ends up not being time spent together," Mr. Wong said. The technology can be distracting, he added. "Sometimes focusing on this little screen image distracts from the conversation," he said. "It's not as real."

Ms. Lam and Mr. Welch said that while instant messaging had helped them stay in touch, high-tech communication could make things harder.

Who Wins the Argument? The One Who Can Type Faster

Ms. Lam said that conflict over international data lines could be hard to manage. "An argument in cyberspace is the worst," she said. "You never get a feeling of resolution."

Mr. Welch agreed. "And when we'd get in fights—as much of a fight as you can have typing rapidly at each other—she'd always win because she can type faster," he said.

The limits of the technology could prove frustrating. "Sometimes it made me want to cry," Ms. Lam said. "It was like, I can't touch you through the screen."

Both agreed that even though they felt connected when they were online, they had become lonely as soon as they turned away from the computer.

"I'd be in some smoky Internet cafe," Mr. Welch said with melancholy in his voice.

"And I'd be alone in my dark room," Ms. Lam said with a shudder.

They resolved that problem when Ms. Lam moved to Beijing to be with Mr. Welch after a five-month separation. The couple are expecting their first child in October, and they say their days of Internet connections are behind them. "Next time we go somewhere," Mr. Welch said as he took Ms. Lam's hand, "we'll go together."

Not Maintaining a Relationship, But Ending One

For some people, the problem is not maintaining a relationship, but ending one. Ask Sun Kim, a 24-year-old research associate in a San Francisco cardiology laboratory. When her boyfriend moved to Korea last December for a consulting job, the two decided not to keep dating long distance but resolved to stay friends.

That, however, proved more difficult than it sounded. "While we were dating," Ms. Kim said, "he was really good at e-mailing. He worked in a dot-com and sat around in a cubicle and e-mailed me every day. He sent messages, stuff he found on the Internet, and he was the king of Egreetings." (Egreetings.com is an online greeting-card service.) When he moved to Korea, nothing changed.

"It was bad because we were supposed to be broken up," Ms. Kim said, "but he'd send flirty e-mails and use Net phones to call me from Korea—for free—all the time." She said her ex-boyfriend had also beamed a PocketSutra, a copy of the *Kama Sutra,* to her Handspring Visor.

"Geographically," Ms. Kim said, "the relationship changed, but most of it stayed the same through technology. It was like it was before, only I never saw him anymore. It was like, 'We're just friends, but I'm still going to be your virtual boyfriend.' "

Breaking Up Is Still Hard To Do

Frequent instant messaging, or IM, sessions made the relationship drag out for months, Ms. Kim said, even though they were continents apart. "About six months after he left," Ms. Kim said, "I was ready to date other people. I told him that, and it turned into this big 'define the relationship' discussion over IM" The time lag between replies opened up room for confusion. "I'd ask a question," Ms. Kim said, "and he wouldn't respond right away, and I'd wonder: Is he offended? Away from his desk? Working?"

Nonetheless, Ms. Kim said she thought that her former boyfriend preferred the medium—lag time or no—to face-to-face discussion. "At one point in this weird breakup discussion over instant messaging," she said, "I said, 'You're probably so relieved we're doing this over IM,' and he said, 'Yeah, I am, actually.'

"Finally, I told him: 'I think it would be a good idea if we don't communicate like this anymore. Call me on the phone or don't contact me.' There was no response, so I logged off. He called a few days later, and he hadn't gotten the message. His Yahoo Messenger had been logging him on and off so he never got my message. He thought everything was fine."

She added: "It's hard to convey emotions over IM. If you're mad, it's hard to convey that in your tone without coming right out and saying, 'I'm mad at you.' "

"In general," she said, "technology makes it hard to have an old-fashioned breakup, where you never see or speak to each other again. I kind of miss the old-fashioned breakup."

Source: From Sally McGrane (2000, August 31). "Long-Distance Romance, Web Enabled." *The New York Times online.*

According to social-exchange theory, relationships draw to a close when negative forces hold sway—when the partners find little satisfaction in the affiliation, when the barriers to leaving the relationship are low (that is, the social, religious, and financial constraints are manageable), and especially when alternative partners are available (Black et al., 1991; Karney & Bradbury, 1995). Problems in communication and jealousy are among the most common reasons for ending a relationship (Zusman & Knox, 1998). The availability of alternatives decreases one's commitment to and investment in a relationship

Human Sexuality ONLINE //

Advice for Young Women: How to Prepare for Married Life

Many humorous e-mails make their way around the Net, so that now you see them and . . . now you see them again. The fourth or fifth time we received this anonymous e-mail, we decided to share it with you. It compares advice from a (genuine!) home economics textbook of the 1950s with the somewhat different advice a young woman might profit from today. It goes to show how the times, they have a-changed. Much!

	From the Home Economics Text: How to Prepare for Married Life, 1950s Style	The Updated Version for the Young Woman of the New Millennium
1. Have dinner ready.	Plan ahead, even the night before, to have a delicious meal on time. This is a way of letting him know that you have been thinking about him, and are concerned about his needs. Most men are hungry when they come home and the prospects of a good meal are part of the warm welcome needed.	Make reservations ahead of time. If your day becomes too hectic, just leave him a voice mail message regarding where you'd like to eat and at what time. This lets him know that your day has been awful and gives him an opportunity to change your mood.
2. Prepare yourself.	Take 15 minutes to rest so you will be refreshed when he arrives. Touch up your make-up, put a ribbon in your hair and be fresh looking. He has just been with a lot of work-weary people. Be a little gay and a little more interesting. His boring day may need a lift.	A quick stop at the Clinique counter on your way home will do wonders for your outlook and will keep you from becoming irritated every time he opens his mouth. (Don't forget to use his credit card.)
3. Clear away clutter.	Make one last trip through the main part of the house just before your husband arrives, gathering up schoolbooks, toys, paper, etc. Then run a dust cloth over the tables. Your husband will feel he has reached a haven of rest and order, and it will give you a lift too.	Call the housekeeper and tell her (or him!) that any miscellaneous items left on the floor by the children can be placed in the Goodwill box in the garage.
4. Prepare the children.	Take a few minutes to wash the children's hands and faces if they are small, comb their hair, and if necessary, change their clothes. They are little treasures and he would like to see them playing the part.	Send the children to their rooms to watch television or play Nintendo. After all, both of them are from his previous marriage.
5. Minimize the noise.	At the time of his arrival, eliminate all noise of washer, dryer, or vacuum. Try to encourage the children to be quiet. Greet him with a warm smile and be glad to see him.	If you happen to be home when he arrives, be in the bathroom with the door locked.
6. Some DON'TS.	Don't greet him with problems or complaints. Don't complain if he's late for dinner. Count this as minor compared with what he might have gone through that day.	Don't greet him with problems and complaints. Let him speak first, and then your complaints will get more attention and remain fresh in his mind throughout dinner. Don't complain if he's late for dinner. Simply remind him that the leftovers are in the fridge and that you left the dishes for him to do.

	From the Home Economics Text: How to Prepare for Married Life, 1950s Style	The Updated Version for the Young Woman of the New Millennium
7. Make him comfortable.	Have him lean back in a comfortable chair or suggest he lie down in the bedroom. Have a cool or warm drink ready for him. Arrange his pillow and offer to take off his shoes. Speak in a low, soft, soothing and pleasant voice. Allow him to relax and unwind.	Tell him where he can find a blanket if he's cold. This will really show you care.
8. Listen to him.	You may have a dozen things to tell him, but the moment of his arrival is not the time. Let him talk first.	But don't ever let him get the last word.
9. Make the evening his.	Never complain if he does not take you out to dinner or to other places of entertainment. Instead, try to understand his world of strain and pressure and his need to be home and relax.	Never complain if he does not take you out to dinner or other places of entertainment; go with a friend or go shopping. (Use his credit card.) Familiarize him with the phrase Girls' Night Out.
10. The Goal.	Try to make your home a place of peace and order where your husband can relax.	Try to keep things amicable without reminding him that he thinks the world revolves around only him. Obviously, he's wrong.

What Advice Can We Give to This Young Couple? Shall we suggest that she plan ahead to have a delicious meal on the table when he comes home from his hard day of work? That she greet him with a ribbon in her hair to look as fresh as possible? That she do some picking up before he arrives so that their home will be his safe haven? (Perhaps not in the United States of the 2000s!)

(Knox et al., 1997; Rusbult et al., 1998). This fact has been widely recognized throughout the ages, which is one reason why patriarchal cultures like to keep their women locked up as much as possible. It also underlies the sexist advice that one should keep one's wife pregnant in summer and barefoot in winter.

About 6 out of 7 students at a large Southeastern university reported that they ended relationships by having frank discussions about them with their partners (Knox et al., 1998). Honesty helped them maintain friendly feelings once the romantic relationship had come to an end.

The swan song of a relationship—moving on—is not always a bad thing. When people are definitely incompatible, and when genuine attempts to preserve the relationship have faltered, ending the relationship can offer each partner a chance for happiness with someone else.

LONELINESS: "ALL THE LONELY PEOPLE, WHERE DO THEY ALL COME FROM?"

Many people start relationships because of loneliness. Loneliness and being alone are not synonymous. Loneliness is a state of painful isolation, of feeling cut off from others. Being alone, a state of solitude, can be quite desirable because it allows us to work, study, or reflect on the world around us. Solitude is usually a matter of choice; loneliness is not.

Lonely people tend to spend a lot of time by themselves, eat dinner alone, spend weekends alone, and participate in few social activities. They are unlikely to date. Some lonely people report having many friends, but a closer look suggests that these "friendships" are shallow. Lonely people are unlikely to share confidences. Loneliness tends to peak during adolescence. This is when peer relationships for most young people begin to supplant family ties. Loneliness is often connected with feelings of depression and of being "sick at heart."

Loneliness is even reported among some married people. In one study (Sadava & Matejcic, 1987), lonely wives tended to feel less liking and love for their partners and expressed less marital satisfaction. Lonely husbands reported less liking for their wives and less intimacy in their relationships.

Causes of Loneliness

The causes of loneliness are many and complex. Lonely people tend to have several of the following characteristics:

1. *Lack of social skills.* Lonely people often lack the interpersonal skills to make friends or cope with disagreements.
2. *Lack of interest in other people.*
3. *Lack of empathy.*
4. *Fear of rejection.* This fear is often connected with self-criticism of social skills and expectations of failure in relating to others (Schultz & Moore, 1984).
5. *Failure to disclose personal information to potential friends* (Solano et al., 1982).
6. *Cynicism about human nature* (for example, seeing people as out only for themselves).
7. *Demanding too much too soon.* The lonely perceive other people as cold and unfriendly in the early stages of a relationship.
8. *General pessimism.* When we expect the worst, we often get . . . you guessed it.
9. *An external locus of control.* That is, lonely people do not see themselves as capable of taking their lives into their own hands and achieving their goals.

It is true that many people remain lonely because they fear being rejected by others.

Coping with Loneliness

Psychologists have helped people cope with loneliness by fostering more adaptive ways of thinking and behaving. Lonely people often have distorted views of other people. They may have had one or two unfortunate experiences and may have jumped to the conclusion that people are generally selfish and not worth the effort of getting involved. Yes, some people *are* basically out for themselves, but believing that everyone is selfish can perpetuate loneliness by motivating us to avoid others.

What can you do to deal with loneliness in your own life? We are all different, and the methods that might help one person may not help another. But here is a list of suggestions compiled by Rathus and Fichner-Rathus (1994):

1. *Challenge your feelings of pessimism.* Adopt the attitude that things happen when you make them happen.
2. *Challenge your cynicism about human nature.* Yes, lots of people are selfish and not worth knowing, but if you assume that all people are like that, you can doom yourself to a lifetime of loneliness. Your task is to find people who possess the qualities that you value.
3. *Challenge the idea that failure in social relationships is awful and is thus a valid reason for giving up on them.* Sure, social rejection can be painful, but unless you happen to be Harrison Ford or Julia Roberts, you may not appeal to everyone. We must all learn to live with some rejection. But keep looking for the people who possess the qualities you value and who will find things of equal value in you.
4. *Follow the suggestions for improving your date-seeking skills spelled out in the Closer Look feature on page 235.* Sit down at a table with people in the cafeteria, not off in a corner by yourself. Smile and say hi to people who interest you. Practice opening lines for different occasions—and a few follow-up lines. Try them out in the mirror.
5. *Make numerous social contacts.* Join committees for student activities. Try intramural sports. Join social-action groups, such as environmental groups and community betterment groups. Join clubs, such as the photography club or the ski club. Get on the school yearbook or newspaper staff.
6. *Be assertive.* Express your genuine opinions.
7. *Become a good listener.* Ask people how they're doing. Ask them for their opinions about classes, politics, or the campus events of the day. Then actually *listen* to what they have to say. Tolerate diverse opinions; remember that no two of us are identical in our outlooks (not even your "perfectly" matched first and third authors). Maintain eye contact. Keep your face friendly. (No, you don't have to remain neutral and friendly if someone tells tasteless jokes or starts insulting a religious or ethnic group.)
8. *Give people the chance to know you.* Exchange opinions and talk about your interests. Yes, you'll turn some people off—who doesn't?—but how else will you learn whether you and another person share common ground?
9. *Fight fair.* Friends will inevitably disappoint you, and you'll want to tell them about it. Do so, but fairly. You can start by asking whether it's okay to be open about something. Then say, "I feel upset because you. . . ." You can ask your friend whether he or she realized that his or her behavior upset you. Try to work together to find a way to avoid recurrences. Finish by thanking your friend for helping you resolve the problem.
10. *Remember that you're worthy of friends.* It's true—warts and all. None of us is perfect. We're all unique, but you may connect with more people than you imagine. Give people a chance.
11. *Use your college counseling center.* Many thousands of students are lonely but don't know what to do about it. Others just can't find the courage to approach others. College counseling centers are very familiar with the problem of loneliness, and you should consider them a valuable resource. You might even ask whether there's a group at the center for students seeking to improve their dating or social skills.

INTIMACY

Intimacy involves feelings of emotional closeness and connectedness with another person and the desire to share each other's inmost thoughts and feelings. Intimate relationships are also characterized by attitudes of mutual trust, caring, and acceptance.

Sternberg's (1986) triangular theory of love (see Chapter 7) regards intimacy as a basic component of romantic love. But people can be intimate and *not* be in love, at least not in romantic love. Close friends and family members become emotionally intimate when they care deeply for each other and share their private feelings and experiences. It is not necessary for people to be *sexually* intimate to have an emotionally intimate relationship. Nor does sexual intimacy automatically produce emotional intimacy. People who are sexually involved may still fail to touch each other's lives in emotionally intimate ways. Even couples who fall in love may not be able to forge an intimate relationship because of unwillingness or inability to exchange inmost thoughts and feelings. Sometimes people are more emotionally intimate with friends than with spouses.

Truth or Fiction? **REVISITED**

It is true that people can have intimate relationships without being sexually intimate. Friends and relatives have nonsexual but intimate relationships. ■

Let us now consider some of the factors that are involved in intimate relationships.

Knowing and Liking Yourself

Some social scientists suggest that an initial step toward intimacy with others is getting to know and like *yourself*. By coming to know and value yourself, you identify your inmost feelings and needs and develop the security to share them.

Trusting and Caring

Two of the most important ingredients of an intimate relationship are trust and caring. When trust exists in a relationship, partners feel secure that disclosing intimate feelings will not lead to ridicule, rejection, or other kinds of harm. Trust usually builds gradually, as partners learn whether or not it is safe to share confidences.

Trusting and Caring. Trusting and caring are key ingredients of an intimate relationship. Trust gives partners the security they need to disclose intimate feelings. Caring is an emotional connection that allows intimacy to develop.

Intimacy Feelings of closeness and connectedness that are marked by the sharing of inmost thoughts and feelings.

a CLOSER look

Heaven Sent: The Blind Date Who Is Your Destiny

Michael Borer developed the LoveGety to help Japanese teenagers find romance. Using omnidirectional radio waves, the LoveGety senses anyone of the desired gender within a 50-foot range who is also carrying the device. Then it receives and displays data on how far that person might want to get involved. (The five modes are chat, dinner, dance, kiss, and love.)

Borer has an updated model for America. The Electronic Cupid will store an encrypted profile of its owner—based on background, hobbies, and morals—and when two carriers come within range, their devices will automatically swap profiles and evaluate their compatibility.

These gadgets, however, may be outdated soon after they hit the market. Online dating services will soon be able to stream personal data to personal digital assistants, beepers, and cellular phones. And because the data will be stored on servers rather than in the hardware itself, it can include the sort of exhaustive information, gleaned from lengthy questionnaires and personal essays, that traditional dating services pride themselves on supplying. Anyone who belongs to a service will be electronically notified when another member is in the vicinity and, in the amount of time it takes to call up a file, can download detailed information on that person and make some quick decisions about whether love is in the air.

Ted Selker of the MIT Media Lab says that this is just the beginning of a new kind of mating ritual. Selker is creating devices that monitor what he calls our "implicit" modes of communication, the "tells" in our computer-using patterns. One of Selker's text-based programs scans a person's e-mail, searching for common and quirky words, and then draws conclusions about what that language says about the individual's personality. Another project, the "interest tracker," is capable of scanning the eye movements of someone browsing the Web to determine what kind of information she or he is drawn to, down to the millisecond. It then uses that growing cache of data to constantly update the browser's personality profile. These profiles, which Selker says are infinitely more sophisticated—and accurate—than the paper-and-pencil profiles used by dating services, can then be cross-referenced with those stored and broadcast by others' machines, assessing much more than whether two people might hit it off for dinner and a movie.

When he is asked whether this doesn't take some of the mystery out of courtship, Selker's response is pragmatic. "You have to realize that a lifelong partner isn't just about combustible," he says. "Computers will help us understand who we really are as opposed to who we think we are."

Source: From Amanda Griscom (2000, June 11). "Heaven Sent: The Blind Date Who Is Your Destiny." *The New York Times Magazine,* p. 93. Copyright © 2000 by the New York Times Co. Adapted by permission.

Research shows that people come to trust their partners when they see that their partners have made sincere investments in the relationship, just as making sacrifices to be with one's partner (e.g., earning the disapproval of one's family, driving one's partner somewhere rather than studying, etc., etc.) (Wieselquist et al., 1999). Commitment and trust in a relationship can be seen as developing according to a model of **mutual cyclical growth.** According to this view,

- Feelings that one needs one's partner and the relationship promote a strong sense of commitment to and dependence on the relationship (Wieselquist et al., 1999).
- Commitment to the relationship encourages the individuals in the relationship to do things that are good for the relationship (to perform "pro-relationship acts).
- One's partner perceives the pro-relationship acts.
- Perception of the pro-relationship acts enhances the partner's trust in the other partner and in the relationship.
- Feelings of trust increase the willingness of the partners to increase their feelings that they need each other and the relationship.

And so the cycle again, and so commitment and trust grow.

Caring is an emotional bond that allows intimacy to develop. In caring relationships, partners seek to gratify each other's needs and interests. Caring also involves willingness to make sacrifices for the other person. Research shows that willingness to sacrifice is connected with strong commitment to the relationship, with a high level of satisfaction

Mutual cyclical growth The view that the need for one's partner promotes commitment; this commitment promotes acts that enhance the relationship; these acts build trust; and one's partner's commitment to the relationship increases.

in the relationship, and, interestingly, with a paucity of alternatives to the relationship (Van Lange et al., 1997). In other words, it may not be so easy to find another partner if one does not make the sacrifices required to remain in the relationship. Self-sacrifice can sometimes be self-serving!

Being Honest

Because intimacy involves the sharing of one's inmost thoughts and feelings, honesty is a core feature of intimacy. A person need not be an "open book" to develop and maintain intimacy, however. Some aspects of experience are better kept even from one's most intimate partners, for they may be too embarrassing or threatening to reveal (Finkenauer & Hazam, 2000). For example, we would not expect partners to disclose every passing sexual fantasy.

Intimate relationships thus usually involve balances in which some things are revealed and others are not. Total honesty could devastate a relationship (Finkenauer & Hazam, 2000). It would not be reasonable, for example, to expect intimate partners to divulge the details of past sexual experiences. The recipient may wonder, "Why is Kimball telling me this? Am I as good a lover as ______? Is Kimball still in love with ______? What else did Kimball do with ______?" Discretion thus also buttresses intimate relationships. As Gordon and Snyder (1989) put it, "Honesty means *saying what you mean*, not revealing every detail" (p. 24). Nor is intimacy established by frank but brutal criticism, even if it is honest.

Making a Commitment

Have you ever noticed that people may open up to strangers on airplanes or trains and yet find it hard to talk openly with people to whom they are closest? An intimate relationship involves more than the isolated act of baring one's soul to a stranger. Truly intimate relationships are marked by commitment or resolve to maintain the relationship through thick and thin (Drigotas et al., 1999a). By contrast, when we open up to strangers on a plane, we know we are unlikely to see them again.

This does not mean that intimate relationships require indefinite or lifelong commitments. A commitment, however, carries an obligation that the couple will work to overcome problems in the relationship rather than run for the exit at the first sign of trouble. Relationships tend to prosper when couples have a mutual level of commitment (Drigotas et al., 1999a). If one member of the couple is vowing undying love while the other has one foot out the door, the relationship doesn't have much of a future.

Maintaining Individuality When the *I* Becomes *We*

In committed relationships, a delicate balance exists between individuality and mutuality. In healthy unions, a strong sense of togetherness does not eradicate individuality. Partners in such relationships remain free to be themselves. Neither seeks to dominate or submerge himself or herself into the personality of the other. Each partner maintains individual interests, likes and dislikes, needs and goals.

The Michelangelo Phenomenon

Michelangelo was the Renaissance artist who breathed life into his statues by giving marble the semblance of flesh. For example, when you look closely at the arms of Michelangelo's statue of Moses, you feel that you can see veins straining with the contractions of muscle.

The **Michelangelo phenomenon** in relationships refers to a tendency for one's partner to help "sculpt" one in the direction of his or her ideal self. When our partners affirm our sense of what we wish to become, we are motivated to move in that direction, and our partners can make great contributions to our success (Drigotas et al., 1999b).

The Michelangelo phenomenon is also related to the issue of maintaining individuality in relationships. It can happen, of course, that our partners have their own sense of

Michelangelo phenomenon The tendency for one's partner to help "sculpt" one in the direction of his or her ideal self.

what they would like us to be—as, for example, when a man pretends he is "liberated" in order to earn the affection of a nontraditional woman but then attempts to shape her in more traditional directions as time goes on. In such a case, the Michelangelo phenomenon can have painful and ugly effects.

Communicating

Good communication is another hallmark of an intimate relationship. Partners are able to share their most personal thoughts and feelings clearly and honestly. Communication is a two-way street. It embraces sending and receiving messages. The good communicator is thus a skilled listener as well as a clear speaker (Tannen, 1990). We generally associate communication with *talk*, which involves the use of verbal messages to convey a thought or a feeling. Through talk, the speaker *encodes* a thought or a feeling into words. The listener *decodes* these words to extract the meaning of the message. Problems in verbal communication can arise at several levels:

1. *The speaker may use words differently than the listener, leading to misunderstandings or miscommunication.* For example, the speaker might say, "You look really cute tonight." The listener might object, saying, "You think I look *cute?* What's wrong with the way I look?" The speaker protests, "No, I didn't mean it that way. I meant you look good. Yes, good."
2. *The speaker's words may not match his or her tone of voice, facial expression, or body gestures.* For example, the speaker says, "Darling, I really don't mind if we visit your mother this weekend." But the speaker did not say, "I really want to visit your mother," so the words *don't mind* come across as a kind of snarl. In such cases, the listener is likely to place greater weight on *how* something is said than on *what* is said.
3. *The speaker may not be able to put into words what he or she truly means or feels.* Sometimes we grasp for words to express our feelings, but they do not come—or those that do come miss the mark.

NONVERBAL COMMUNICATION: THE BODY SPEAKS Although the spoken word is a primary form of communication, we often express our feelings through nonverbal channels as well, such as by tone of voice, gestures, body posture, and facial expressions (Tannen, 1990). People may place more weight on how words are said than on their denotative meaning. People also accentuate the meaning of their words through gestures, raising or lowering their voices, or using a sterner or softer tone of voice.

Nonverbal communication is used not only to accentuate the spoken word but also to express feelings directly. We sometimes are better able to convey our feelings through body language than by the use of words. A touch or a gaze into someone's eyes may express more about our feelings than words can. Parents express feelings of tenderness and caring toward infants by hugging, holding, and caressing them and by speaking in a gentle and soothing tone of voice, even though the meanings of the words cannot yet be grasped.

Let us examine some aspects of nonverbal communication.

ON "BEING UPTIGHT" AND "HANGING LOOSE" The ways that people carry themselves offer cues to their feelings and potential actions. People who are emotionally "uptight" often stand or sit rigidly and straight-backed. People who feel relaxed more literally "hang loose."

People who face us and lean toward us are in effect saying that they like us or care about what we are saying. If we overhear a conversation between a couple and see that the woman is leaning toward the man, but he is leaning back and playing with his hair, we tend to infer that he is not accepting what she is saying or has lost interest.

TOUCHING Touching is a powerful form of communication. Women are more apt than men to touch the people with whom they interact. When touching suggests too much intimacy, however, it can be annoying. Touching may also establish "property rights" over one's partner. People may touch their partners in public or hold their hands, not as a sign of affection, but as a signal to others that their partners are "taken" and that others should keep a respectful distance. Similarly, the wearing of an engagement or wedding ring signals unavailability.

GAZING AND STARING: THE LOOK OF LOVE? We gather information about the motives, feelings, and attitudes of others through eye contact. In Western culture, looking other people "right in the eye"—at least among men—is positively valued. People who do so appear self-assertive, direct, and candid. When people look away, they may be perceived as shy, deceitful, or depressed. In some Asian cultures, however, looking other people directly in the eye may seem an aggressive invasion.

Unaggressive gazing into another person's eyes—especially the eyes of a person whom one considers attractive—can create deep feelings of intimacy in our culture. In a laboratory study, couples who had just met were instructed to gaze into each other's eyes for two minutes. Afterwards, many reported experiencing feelings of passion (Kellerman et al., 1989). Is this what is meant by "the look of love"?

All in all, there are many ways in which we communicate with others through verbal and nonverbal channels. Let us now look at ways in which partners can learn to communicate better with each other, especially about sex. Many couples, even couples who are able to share their deepest thoughts and feelings, may flounder at communicating their sexual needs and preferences. Couples who have lived together for decades may know each other's tastes in food, music, and movies about as well as they know their own but may still hesitate to share their sexual likes and dislikes (Harris & Johnson, 2000; Whitten, 2001). They may also be reluctant, for fear of opening wounds in the relationship, to exchange their feelings about other aspects of their relationship, including each other's habits, appearance, and gender-stereotypical attitudes.

COMMUNICATION SKILLS FOR ENHANCING RELATIONSHIPS AND SEXUAL RELATIONS

Marital counselors and sex therapists might be as busy as the proverbial Maytag *repairperson*[2] if more couples communicated with each other about their sexual feelings. Unfortunately, when it comes to sex, *talk* may be the most overlooked four-letter word.

Many couples suffer for years because one or both partners are unwilling to speak up. Or problems arise when one partner misinterprets the other. One partner might interpret the other's groans or grimaces of pleasure as signs of pain and pull back during sex, leaving the other frustrated. Improved communication may be no panacea, but it helps. Clear communication can take the guesswork out of relationships, avert misunderstandings, relieve resentments and frustrations, and increase both sexual and general satisfaction with the relationship.

Common Difficulties in Sexual Communication

Why is it so difficult for couples to communicate about sex? Here are some possibilities:

ON "MAKING WHOOPIE"—IS SEX TALK VULGAR? Vulgarity, like beauty, is to some degree in the eye of the beholder. One couple's vulgarity may be another couple's pillow talk. Some people may maintain a Victorian belief that no talk about sex is fit for mixed company, even between intimate partners. Sex, that is, is something you may do, but not something to be talked about. Other couples may be willing in principle to talk about sex, but find the reality difficult because they lack an agreeable, common language.

How, for example, are they to refer to their genitals or to sexual activities? One partner may prefer to use coarse words to refer to them. (Because as the forbidden fruit is often the sweetest, some people feel sexually aroused when they or their partners "talk dirty.") The other might prefer more clinical terms. A partner who likes to use slang for

[2]"A bow in the direction of political correctness?" asks the third author.

the sex organs might be regarded by the other as vulgar or demeaning. One who uses clinical terms, such as *fellatio* or *coitus*, might be regarded as, well, clinical. Some couples compromise and try to use terms that are neither vulgar nor clinical. They might speak, for example, of "doing it" rather than "screwing" (and the like) at one extreme or of "engaging in sexual intercourse" at the other. (The title of the Eddie Cantor musical of the 1930s suggests that some people once spoke of "making whoopie.") Or they might speak of "kissing me down there" rather than of "eating me" or of practicing fellatio or cunnilingus.

ON IRRATIONAL BELIEFS Many couples also harbor irrational beliefs about relationships and sex, such as the notion that people should somehow *know* what their partners want without having to ask. The common misconception that people should know what pleases their partners undercuts communication. Men, in particular, seem burdened with the stereotype that they should have a natural expertise at sex. Women may feel it is "unladylike" to talk openly about their sexual needs and feelings. Both partners may hold the idealized romantic notion that "all you need is love" to achieve sexual happiness. But such knowledge does not arise from instinct or from love. It is learned—or it remains unknown.

A related irrational belief is that "one's partner will read one's mind." We may erroneously assume that if our partners truly loved us, they would somehow "read our minds" and know what types of sexual stimulation we desire. Unfortunately—or fortunately—others cannot read our minds. We must assume the responsibility for communicating our preferences.

Truth or Fiction? **REVISITED**

It is not true that "Love is all you need." Even when partners truly love each other, they do not instinctively know how to satisfy each other sexually. Communication is required. ■

Some people communicate more effectively than others, perhaps because they are more sensitive to others' needs or because their parents served as good models as communicators. But communication skills can be acquired at any time. Learning takes time and work, but the following guidelines should prove helpful if you want to enhance your communication skills. The skills can also improve communication in areas of intimacy other than the sexual.

Getting Started

How do you broach tough topics? Here are some ideas.

TALKING ABOUT TALKING You can start by talking about talking. You can inform your partner that it is difficult for you to talk about problems and conflicts: "You know, I've always found it awkward to find a way of bringing things up" or "You know, I think other people have an easier time than I do when it comes to talking about some things." You can allude to troublesome things that happened in the past when you attempted to resolve conflicts. This approach encourages your partner to invite you to proceed.

Broaching the subject of sex can be difficult. Even couples who gab endlessly about finances, children, and work may clam up about sex. Thus, it may be helpful first to agree to talk about talking about sex. You can admit that it is difficult to talk about sex. You can say that your sexual relationship is important to you and that you want to do everything you can to enhance it. Gently probe your partner's willingness to set aside time to talk about sex, preferably when you can dim the lights and avoid interruptions.

The "right time" may be when you are both relaxed, rested, and not pressed for time. The "right place" can be any place where you can enjoy privacy and talk undisturbed. Sex talk need not be limited to the bedroom. Couples may feel more comfortable talking about sex over dinner, when cuddling on the sofa, or when just relaxing together.

REQUESTING PERMISSION TO BRING UP A TOPIC Another possibility is to request permission to raise an issue. You can say something like this: "There's something on my mind. Do you have a few minutes? Is now a good time to tell you about it?" Or you can say, "There's something that we need to talk about, but I'm not sure how to bring it up. Can you help me with it?"

Active Listening. Effective communication requires listening to the other person's view of things. You can listen actively by maintaining eye contact and modifying your facial expression to show that you understand your partner's feelings and ideas. You can ask helpful questions, such as "Did I disappoint you when I . . . ?"

GIVING YOUR PARTNER PERMISSION TO SAY SOMETHING THAT MIGHT BE UPSETTING TO YOU You can tell your partner that it is okay to point out ways in which you can become a more effective lover. For example, you can say, "I know that you don't want to hurt my feelings, but I wonder if I'm doing anything that you'd rather I didn't do?"

Listening to the Other Side

Skilled listening involves such skills as active listening, paraphrasing, the use of reinforcement, and valuing your partner even when the two of you disagree.

LISTENING ACTIVELY To listen actively rather than passively, first adopt the attitude that you may actually learn something—or perceive things from another vantage point—by listening. Second, recognize that even though the other person is doing the talking, you shouldn't just sit there. In other words, it is not helpful to stare off into space while your partner is talking, or to offer a begrudging "mm-hmm" now and then to be polite. Instead, you can listen actively by maintaining eye contact and modifying your facial expression to show that you understand his or her feelings and ideas (Cole & Cole, 1999). For example, nod your head when appropriate.

Listening actively also involves asking helpful questions, such as "Would you please give me an example?"

An active listener does not simply hear what the other person is saying but also focuses attentively on the speaker's words and gestures to grasp the full meaning. Nonverbal cues may reveal more about the speaker's inner feelings than the spoken word. Good listeners do not interrupt, change the topic, or walk away when their partners are speaking.

PARAPHRASING Paraphrasing shows that you understand what your partner is trying to say. In paraphrasing, you recast or restate the speaker's words to confirm that you have understood correctly. For example, suppose your partner says, "You hardly ever say anything when we're making love. I don't want you to scream or make obligatory grunts or do something silly, but sometimes I wonder if I'm trying to make love to a brick wall." You can paraphrase this comment by saying something like: "So it's sort of hard to know if I'm really enjoying it."

REINFORCING THE OTHER PERSON'S WILLINGNESS TO COMMUNICATE Even when you disagree with what your partner is saying, you can maintain good relations and keep channels of communication open by saying something like "I really appreciate your taking the time to try to work this out with me" or "I hope you'll think it's okay if I don't see things entirely in the same way, but I'm glad that we had a chance to talk about it."

SHOWING THAT YOU VALUE YOUR PARTNER, EVEN WHEN THE TWO OF YOU DISAGREE When you disagree with your partner, do so in a way that shows that you still value your partner as a person. In other words, say something like "I love you very much, but it annoys me when you . . . " rather than "You're really contemptible for . . . " By so doing, you encourage your partner to disclose sensitive material without risk of attack or of losing your love or support.

Learning About Your Partner's Needs

Listening is basic to learning about another person's needs, but sometimes it helps to go a few steps further.

ASKING QUESTIONS TO DRAW THE OTHER PERSON OUT You can ask open-ended questions that allow for a broader exploration of issues, such as these:

"What do you like best about the way we make love?"
"Do you think that I do things to bug you?"
"Does it bother you that I go to bed later than you do?"
"Does anything disappoint you about our relationship?"
"Do you think that I do things that are inconsiderate when you're studying for a test?"

Closed-ended questions that call for a limited range of responses tend to be most useful when you're looking for a simple yes-or-no type of response. ("Would you rather make love with the stereo off?")

USING SELF-DISCLOSURE Self-disclosure is essential to developing intimacy. You can also use self-disclosure to learn more about your partner's needs, because communicating your own feelings and ideas invites reciprocation. For example, you might say, "There are times when I feel that I disappoint you when we make love. Should I be doing something differently?"

GRANTING PERMISSION FOR THE OTHER PERSON TO SAY SOMETHING THAT MIGHT UPSET YOU You can ask your partner to level with you about an irksome issue. You can say that you recognize that it may be awkward to discuss it but that you will try your best to listen conscientiously and not get too disturbed. You can also limit communication to one such difficult issue per conversation. If the entire emotional dam were to burst, the job of mopping up could be overwhelming.

Providing Information

There are many skillful ways of communicating information, including "accentuating the positive" and using verbal and nonverbal cues. When you want to get something across, remember that it is irrational to expect that your partner can read your mind. He or she can tell when you're wearing a grumpy face, but your expression does not provide much information about your specific feelings. When your partner asks, "What would you like me to do?" responding with "Well, I think you can figure out what I want" or "Just do whatever you think is best" is not very helpful. Only you know what pleases you. Your partner is not a mind reader.

ACCENTUATING THE POSITIVE Let your partner know when he or she is doing something right! Speak up or find another way to express your appreciation. Accentuating the positive is rewarding and also informs your partner about what pleases you. In other words, don't just wait around until your partner does something wrong and then seize the opportunity to complain!

USING VERBAL CUES Sexual activity provides an excellent opportunity for direct communication. You can say something like "Oh, that's great" or "Don't stop." Or you can ask for feedback, as in "How does this feel . . . ?"

Feedback provides direct guidance about what is pleasing. Partners can also make specific requests and suggestions.

USING NONVERBAL CUES Sexual communication also occurs without words. Couples learn to interpret each other's facial expressions as signs of pleasure, anxiety, boredom, even disgust. Our body language also communicates our likes and dislikes. Our partners may lean toward us or away from us when we touch them, or they may relax or tense up; in any case, they speak volumes in silence.

The following exercises may help couples use nonverbal cues to communicate their sexual likes and dislikes. Similar exercises are used by sex therapists to help couples with sexual dysfunctions.

1. *Taking turns petting*. Taking turns petting can help partners learn what turns one another on. Each partner takes turns caressing the other, stopping frequently enough to receive feedback by asking questions like "How does that feel?" The recipient is

responsible for giving feedback, which can be expressed either verbally ("Yes, that's it—yes, just like that" or "No, a little lighter than that") or nonverbally, such as by making certain appreciative or disapproving sounds. Verbal feedback is usually more direct and less prone to misinterpretation. The knowledge gained through this exercise can be incorporated into the couple's regular pattern of lovemaking.

2. *Directing your partner's hand.* Gently guiding your partner's hand—to show your partner where and how you like to be touched—is a most direct way of communicating sexual preferences. While taking turns petting, and during other acts of lovemaking, one partner can gently guide the other's fingers and hands through the most satisfying strokes and caresses. Women might show partners how to caress the breasts or clitoral shaft in this manner. Men might cup their partners' hands to show them how to stroke the penile shaft or caress the testes.
3. *Signaling.* Couples can use agreed-upon nonverbal cues to signal sexual pleasure. For example, one partner may rub the other in a certain way, or tap the other, to signal that something is being done right. The recipient of the signal takes mental notes and incorporates the pleasurable stimulation into the couple's lovemaking. This is a sort of "hit or miss" technique, but even near misses can be rewarding.

Making Requests

A basic part of improving relationships or lovemaking is asking partners to change their behavior—to do something differently or to stop doing something that hurts or is not gratifying. It is here that the skill of making requests comes to the fore.

BEING SPECIFIC Be specific in requesting changes. Telling your partner "I'd like you to be nicer to me" may accomplish little. Your partner may not know that his or her behavior is *not* nice and may not understand how to be "nicer." It is better to say something like "I would appreciate it if you would get coffee for yourself, or at least ask me in a more pleasant way." Or "I really have a hard time with the way you talk to me in front of your friends. It's as if you're trying to show them that you have control over me or something." Similarly, it may be less effective to say "I'd like you to be more loving" than to say "When we make love, I'd like you to kiss me more and tell me how you care about me."

Of course, you can precede your specific requests with openers such as "There's something on my mind. Is this a good time for me to bring it up with you?"

USING "*I*-Talk" Using the word *I* is an excellent way of expressing your feelings. Psychologists who help people become more assertive often encourage them to use the words *I*, *me*, and *my* in their speech, not just to express their feelings, but to buttress their sense of self-worth.

You are more likely to achieve desired results by framing requests in *I*-talk than by heaping criticisms on your partner. For example, "I would like it if we spent some time cuddling after sex" is superior to "You don't seem to care enough about me to want to hold me after we make love." Saying "I find it very painful when you use a harsh voice with me" is probably more effective than "Sometimes people's feelings get hurt when their boyfriends [girlfriends] speak to them harshly in front of their friends or families."

You may find it helpful to try out *I*-talk in front of a mirror or with a confidant(e) before using it with your partner. In this way, you can see whether your facial expression and tone of voice are consistent with what you are saying. Friends may also provide pointers on the content of what you are saying.

Delivering Criticism

Delivering criticism effectively is a skill. It requires focusing your partner's attention on the problem without causing resentment or reducing him or her to a trembling mass of guilt or fear.

EVALUATING YOUR MOTIVES First, weigh your goals forthrightly. Are you truly interested in gaining cooperation, or is your primary intention to punish your partner? If your goal is punishment, you may as well be coarse and disparaging, but this will invite reprisals. If your goal is to improve the relationship, however, a tactful approach may be in order.

PICKING THE RIGHT TIME AND PLACE Deliver criticism privately, not in front of friends or family. Your partner has a right to be upset when you make criticism public. Making private matters public induces indignation and cuts off communication.

BEING SPECIFIC Being specific may be even more important when delivering criticism than when making requests. By being specific about the *behavior* that disturbs you, you avoid the trap of disparaging your partner's personality or motives. For example, it may be more effective to say, "I could lose this job because you didn't write down the message" than to say, "You're completely irresponsible" or "You're a flake." Similarly, you will probably achieve better results by saying, "The bathroom looks and smells dirty when you throw your underwear on the floor" than by saying, "You're a filthy pig." It is more to the point (and less intimidating) to complain about specific, modifiable behavior than to try to overhaul another person's personality.

EXPRESSING DISPLEASURE IN TERMS OF YOUR OWN FEELINGS Your partner is likely to feel less threatened if you express displeasure in terms of your own feelings rather than directly attacking his or her personality. Attacks often arouse defensive behavior, and sometimes retaliation, and they seldom enhance relationships. When confronting your partner for failing to be sensitive to your sexual needs when making love, it may be more effective to say, "You know, it really *upsets* me that you don't seem to care about my feelings when we make love" than "You're so wrapped up in yourself that you never think about anyone else."

KEEPING CRITICISM AND COMPLAINTS TO THE PRESENT How many times have you been in an argument and heard things like "You never appreciated me!" or "Last summer you did the same thing!" Bringing up the past during conflicts muddles current issues and heightens resentment. When your partner forgets to jot down the details of the telephone message, it is more useful to note that "This was a vital phone call" than to dredge up the fact that "Three weeks ago you didn't tell me about the phone call from Chris, and as a result I missed out on seeing *Home Alone: The College Years.*" It's better to leave who did what to whom last year (or even last week) alone. Focus on the present.

EXPRESSING CRITICISM CONSTRUCTIVELY Be sensitive to your partner's needs by avoiding blunt criticisms or personal attacks and by suggesting constructive alternatives. Avoid saying things like "You're really a lousy lover." Say instead, "May I take your hand and show you what I'd like?" If you cannot criticize your partner constructively, it may be best not to criticize at all.

EXPRESSING CRITICISM POSITIVELY Whenever possible, express criticism positively, and combine it with a concrete request. When commenting on the lack of affection your partner displays during lovemaking, say, "I love it when you kiss me. Please kiss me more" rather than "You never kiss me when we're in bed and I'm sick of it."

Receiving Criticism

> Honest criticism is hard to take, particularly from a relative, a friend, an acquaintance, or a stranger.
>
> —Franklin P. Jones

Delivering criticism can be tricky, especially when you want to inspire cooperation. Receiving criticism can be even trickier. Nevertheless, the following suggestions offer some help.

CLARIFYING YOUR GOALS It is understandable if the hair on the backs of your arms does a headstand when you hear "It's time you did something about. . . ." After all, this is a blunt challenge. When we are confronted harshly, we are likely to become defensive and think of retaliating. But if your objective is to enhance the relationship, take a few moments to reflect. To resolve conflicts, we need to learn about the other person's concerns, keep lines of communication open, and find ways of changing problem behavior.

Thus, when your partner says, "It's about time you did something about the bathroom," stop and think before you summon up your most menacing voice and say, "Just what the hell is that supposed to mean?" Ask yourself what you want to find out.

ASKING CLARIFYING QUESTIONS Just as it's important to be specific when delivering criticism, it helps if you encourage the other person to be specific when you are on the receiving end of criticism. In the example of the complaint about the bathroom, you can help your partner be specific and, perhaps, avert the worst by asking clarifying questions, such as "Can you tell me exactly what you mean?" or "The bathroom?"

Consider a situation in which a lover says something like "You're one of the most irritating people I know." Rather than retaliate and further harm the relationship, you can say something like "How about forgoing the character assassination and telling me what I did that's bothering you?" This response requests an end to insults and asks your partner to be specific.

ACKNOWLEDGING THE CRITICISM Even when you disagree with a criticism, you can keep lines of communication open and show respect for your partner's feelings by acknowledging and paraphrasing the criticism.

On the other hand, if you are at fault, you can admit it. For example, you can say, "You're right. It was my day to clean the bathroom and it totally slipped my mind" or "I was so busy, I just couldn't get to it." Now the two of you should look for a way to work out the problem. When you acknowledge criticism, you cue your partner to back off and look for ways to improve the situation. What if your partner then becomes abusive and says something like "So you admit you blew it?" You might then try a little education in conflict resolution. You could say, "I admitted that I was at fault. If you're willing to work with me to find a way to handle it, great; but I'm not going to let you pound me into the ground over it."

REJECTING THE CRITICISM If you think that you were not at fault, express your feelings. Use *I*-talk and be specific. Don't seize the opportunity to point out angrily your partner's own shortcomings. By doing so, you may shut down lines of communication.

NEGOTIATING DIFFERENCES Negotiate your differences if you feel that there is merit on both sides of the argument. You may want to say something like "Would it help if I . . . ?" And if there's something about your obligation to clean the bathroom that seems totally out of place, perhaps you and your partner can work out an exchange—that is, you get relieved of cleaning the bathroom in exchange for tackling a chore that your partner finds equally odious.

If none of these approaches helps resolve the conflict, perhaps your partner is using the comment about the bathroom to express anger over other issues. You may be able to find out by saying something like "I've been trying to find a way to resolve this thing, but nothing I say seems to be helping. Is this really about the bathroom, or are there other things on your mind?"

And notice that we haven't suggested that you seize the opportunity to strike back by saying, "Who're you to complain about the bathroom? What about your breath and that pig sty you call your closet?" Retaliation is tempting and may make you feel good in the short run, but it can do a relationship more harm than good in the long run.

Truth or Fiction? **REVISITED**

It is not true that retaliation is the best course of action when you are criticized. Retaliation is an inferior way to handle criticism—that is, if your goal is to resolve conflict. ■

When Communication Is Not Enough: Handling Impasses

Communication helps build and maintain relationships, but sometimes partners have profound, substantial disagreements. In fact, it is normal to have disagreements from time to time. Even when their communication skills are superbly tuned, partners now and then reach an impasse.

Opening the lines of communication may also elicit hidden frustrations. These frustrations can lead partners to question seriously the value of continuing the relationship or to agree to consult a helping professional. Although some people feel it is best to "let sleeping dogs lie," the *skillful* airing of dissatisfactions can be healthful for a relationship. Couples who reach an impasse can follow several courses of action that may be helpful:

LOOKING AT THE SITUATION FROM THE OTHER PERSON'S PERSPECTIVE Research shows that taking one's partner's perspective (that is, looking at things from one's partner's point of view) during a dispute results in more positive feelings about the relationship and evokes greater effort to respond in a constructive manner (Arriaga & Rusbult, 1998). Taking the other person's perspective is also connected with less blaming of one's partner for the dispute and less inclination to respond in a destructive manner.

If there is an impasse, some of the conflict may be resolved by saying (so long as it is how you really feel) something like "I still disagree with you, but I can understand why you take your position." In this way, you recognize your partner's goodwill and, perhaps, lessen tensions.

SEEKING VALIDATING INFORMATION On the other hand, if you do not follow your partner's logic, you can say something like "Please believe me: I'm trying very hard to look at this from your point of view, but I can't follow your reasoning. Would you try to help me understand your point of view?"

TAKING A BREAK Sometimes when you reach a stalemate, it helps to allow the problem to "incubate" for a while. If you and your partner put the issue aside for a time, perhaps a resolution will dawn on one of you later on. If you wish, schedule a follow-up discussion so that the issue won't be swept under the rug.

TOLERATING DIFFERENTNESS Although we tend to form relationships with people who share similar attitudes, there is never a perfect overlap. A partner who pretends to be your clone will probably become a bore. Assuming that your relationship is generally rewarding and pleasurable, you may find it possible to tolerate certain differences between yourself and your partner. Respecting other people includes allowing them to be who they are. When we have a solid sense of who we are as individuals and of what we stand for, we are more apt to tolerate differentness in our partners.

Many relationships do come to an end when the partners cannot resolve their differences. However, it is possible for relationships to continue and improve even when we reach an impasse. We need to be able to tolerate differentness in other people if relationships are to improve. ■

Truth or Fiction? **REVISITED**

AGREEING TO DISAGREE When all else fails, you can agree to disagree on various issues. You can remain a solid, respected individual, and your partner can remain a worthwhile, effective person even if the two of you disagree from time to time.

Disagreement is not necessarily destructive to a relationship—unless you are convinced that it must be. Two people cannot see everything in the same way. Failure ever to disagree will have to leave at least one partner feeling frustrated now and then.

It is not true that disagreement is destructive to a relationship. The important thing is to try to resolve disagreements and, when they cannot be resolved, to handle the impasse productively. ■

Truth or Fiction? **REVISITED**

You can handle an impasse by focusing on the things that you and your partner have in common. Presumably there will be a number of them—some of them with little feet.

Human Sexuality ONLINE //

Web Sites Related to Relationships, Intimacy, and Communication

This commercial Web site serves as a dating service. According to the service, users of the site can read biographies of people who have signed up, see their photos, chat with them "in real time," and send them anonymous e-mail.
www.matchmaker.com

iVillage, "The Women's Network": At this Web site click on "Relationships" for quizzes (such as, "Rate Your Dates"), advice for singles, information about sex and romance, hot topics like domestic abuse, and "The Worst Date of the Week."
www.ivillage.com

"Brenda's Dating Advice for Geeks:" This Web site is intended for young people. Lighthearted, fun. Don't come here for serious psychological advice, but stop by for "horrendous" personal ads, mystery dates, "Dates from Hell," and—yes—Dating Tips.
www.geekcheck.com

Self-Help Magazine: This Web site features articles on issues in relationships, such as effective communication and finding a partner.
www.shpm.com

Smart Marriages: The Coalition for Marriage, Family, and Couples Education: This Web site answers questions about marriage and divorce.
www.smartmarriages.com

OutProud, The National Coalition for Gay, Lesbian & Bisexual Youth: This Web site offers information, resources, and support for gay, lesbian, and bisexual adults.
www.outproud.org

Duke University's Healthy Devil Online: This Web site has information about sexual communication and sexual choices.
healthydevil.stuaff.duke.edu/info/healthinfo.html

summing up

THE ABC(DE)'S OF ROMANTIC RELATIONSHIPS

Levinger proposes an ABCDE model of romantic relationships. The letters refer to five stages: attraction, building, continuation, deterioration, and ending.

■ The A's—Attraction

The major promoter of attraction is propinquity.

■ The B's—Building

Similarity in the level of physical attractiveness, similarity in attitudes, and liking motivate us to build relationships.

■ The C's—Continuation

Factors such as variety, caring, positive evaluations, lack of jealousy, perceived fairness in the relationship, and mutual feelings of satisfaction encourage us to continue relationships.

■ The D's—Deterioration

The factors that foster deterioration include failure to invest time and energy in the relationship, deciding to put an end to it, and simply permitting deterioration to proceed unchecked.

■ The E's—Ending

Relationships tend to end when the partners find little satisfaction in the affiliation, when alternative partners are available, when couples are not committed to preserving the relationship, and when they expect it to falter.

LONELINESS: "ALL THE LONELY PEOPLE, WHERE DO THEY ALL COME FROM?"

Loneliness is a state of painful isolation, a feeling that one is cut off from others.

■ Causes of Loneliness

The causes of loneliness include lack of social skills, lack of interest in other people, lack of empathy, fear of rejection, lack of self-disclosure, cynicism about human nature, demanding too much too soon, general pessimism, and an external locus of control.

■ Coping with Loneliness

People are helped to overcome loneliness by challenging self-defeating attitudes and social-skills training.

INTIMACY

Intimacy involves feelings of emotional closeness with another person and the desire to share each other's inmost thoughts and feelings.

■ Knowing and Liking Yourself

An initial step toward intimacy with others is getting to know and like yourself so that you can identify your inmost feelings and develop the security to share them.

■ Trusting and Caring

Intimate relationships require trust, caring, and tenderness.

■ **Being Honest**
Honesty is a core feature of intimacy.

■ **Making a Commitment**
Truly intimate relationships are marked by commitment—a resolve to maintain the relationship through thick and thin.

■ **Maintaining One's Individuality When the *I* Becomes *We***
In healthy unions, a strong sense of togetherness does not eradicate individuality.

■ **Communicating**
Communication is a two-way street. It embraces sending *and* receiving messages. We often express feelings through nonverbal channels such as tone of voice, gestures, body posture, and facial expressions.

COMMUNICATION SKILLS FOR ENHANCEING RELATIONSHIPS AND SEXUAL RELATIONS

■ **Common Difficulties in Sexual Communication**
Couples may find it difficult to talk about sex because of the lack of an agreeable common language. Many couples also harbor irrational beliefs about relationships and sex.

■ **Getting Started**
Ways of getting started in communicating include talking about talking, requesting permission to raise an issue, and granting one's partner permission to say things that might be upsetting.

■ **Listening to the Other Side**
Skilled listening involves elements such as active listening, paraphrasing, the use of reinforcement, and valuing your partner even when you disagree.

■ **Learning About Your Partner's Needs**
You can learn about your partner's needs by asking questions, using self-disclosure, and asking your partner to level with you about an irksome issue.

■ **Providing Information**
You can provide information by "accentuating the positive" and using verbal and nonverbal cues.

■ **Making Requests**
In making requests, it is helpful to take responsibility for what happens, to be specific, to be assertive, and to use "*I*-talk."

■ **Delivering Criticism**
In delivering criticism, it is helpful to evaluate your motives, to pick a good time and place, to be specific, to express displeasure in terms of your own feelings, to limit complaints to the present, and to express criticism constructively and positively.

■ **Receiving Criticism**
To receive criticism effectively, it is helpful to clarify goals, to ask clarifying questions, to acknowledge the criticism, to reject inappropriate criticisms, and to negotiate differences.

■ **When Communication Is Not Enough: Handling Impasses**
Partners can help manage impasses by trying to see things from the partner's perspective, seeking validating information, taking a break, tolerating differentness, and (when necessary) agreeing to disagree.

questions for critical thinking

1. Do you find it difficult to make small talk? Explain.
2. Have you ever had trouble deciding how much intimate information to disclose to a dating partner? What are the dangers in disclosing too much to a partner too soon? Are there some things that you feel that you should never share with a partner?
3. Have you ever had feelings of jealousy? How did they affect your relationship? Your self-esteem? How did you handle them?
4. What is the difference between loneliness and solitude?
5. What are the various meanings of *intimacy?* Are you now or have you been involved in an intimate relationship?
6. Do you have difficulty criticizing another person or accepting criticism? Why? How can you become better at delivering or receiving criticism?
7. Have you ever arrived at an impasse in a disagreement with a partner? What did you do about it? How did it work out?

chapter

9

Sexual Techniques and Behavior Patterns

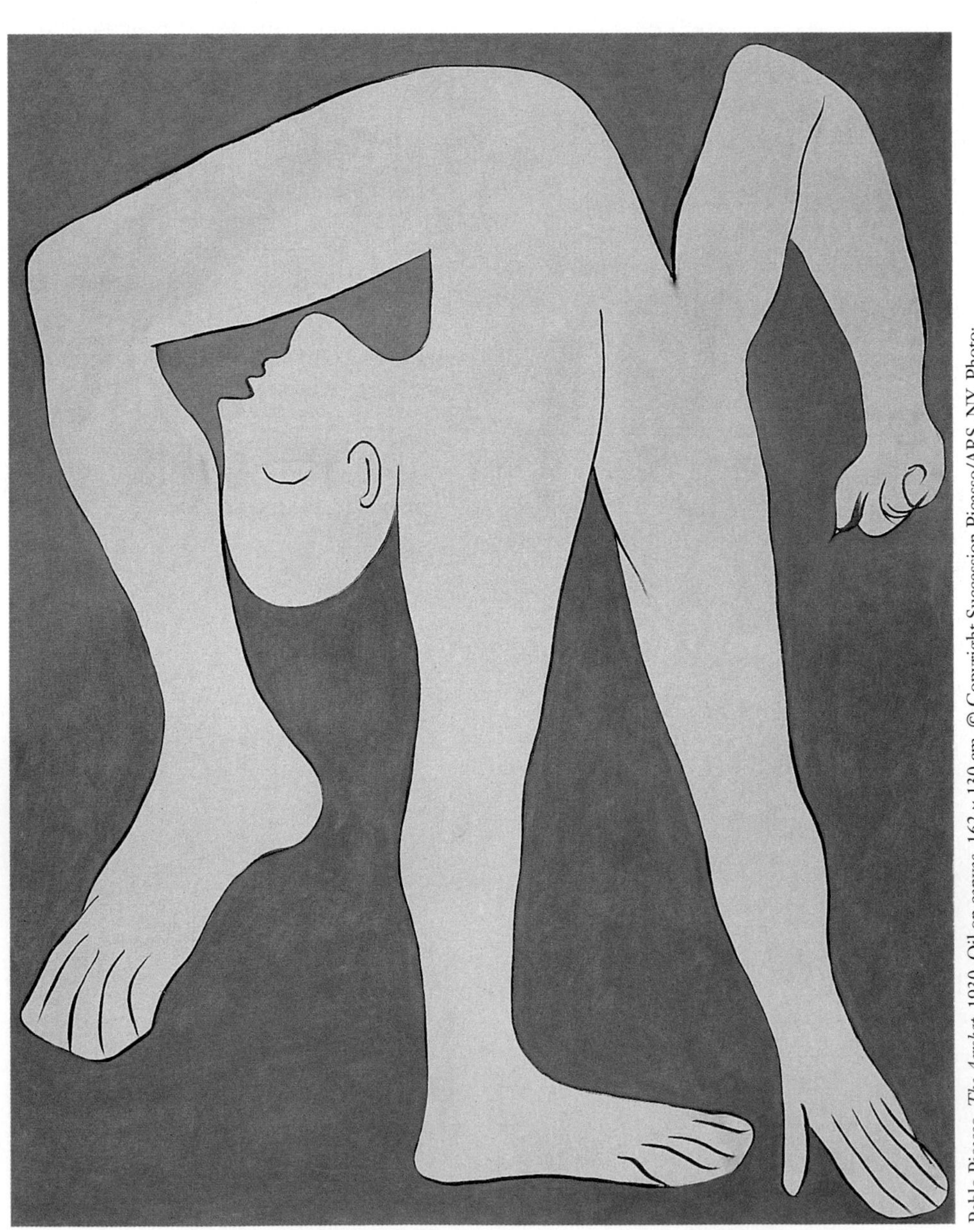

Pablo Picasso. *The Acrobat,* 1930. Oil on canvas. 162 × 130 cm. © Copyright Succession Picasso/ARS, NY. Photo: R. G. Ojeda. Musée Picasso, Paris, France. Credit: Réunion des Musées Nationaux/Art Resource, NY. Credit: © 2002 Estate of Pablo Picasso/Artists Rights Society (ARS), New York.

outline

Truth or Fiction?

SOLITARY SEXUAL BEHAVIOR

Masturbation

A Closer Look The Technology of Orgasm: "Hysteria," the Vibrator, and Women's Sexual Satisfaction

Use of Fantasy

SEXUAL BEHAVIOR WITH OTHERS

Foreplay

Kissing

Touching

Stimulation of the Breasts

Oral–Genital Stimulation

A World of Diversity Sociocultural Factors and Oral Sex

Sexual Intercourse: Positions and Techniques

A World of Diversity Are African Americans Sexually Permissive? Cultural Myths Versus Data

Human Sexuality Online Web Sites Related to Sexual Techniques and Behavior Patterns

SUMMING UP

QUESTIONS FOR CRITICAL THINKING

Truth or Fiction?

____ Married people rarely if ever masturbate.

____ European American men are more likely to masturbate than African American men are.

____ Women who masturbated during adolescence are less likely to find gratification in marital coitus than women who did not.

____ Women are more likely to reach orgasm through sexual intercourse than through masturbation.

____ Most women masturbate by inserting a finger or other object into the vagina.

____ Heterosexual people do not fantasize about sexual activity with people of their own gender.

____ Statistically speaking, oral sex is the norm for today's young married couples.

____ African Americans are more likely than European Americans to engage in oral sex.

____ When lovers fantasize about other people, the relationship is in trouble.

____ Anal sex is more common among less well-educated people.

This is the chapter that describes sexual techniques and statistical breakdowns of "who does what with whom." There is great variety in human sexual expression. Some of us practice few, if any, of the techniques in this chapter. Some of us practice most or all of them. Some of us practice some of them some of the time. Our knowledge of the prevalence of these techniques comes from sex surveys that began with Kinsey and have continued with the work of Hunt, the University of Chicago group, and others. Of course, surveys are plagued by problems such as nonrepresentative sampling, social desirability, and volunteer bias. Therefore, we must be cautious in generalizing on the basis of their results. Surveys provide our best "guesstimate" of the prevalence of sexual behaviors. They do not provide precise figures.

Readers of this textbook are as varied in their sexual values, preferences, and attitudes as is society in general. Some of the techniques discussed may thus strike some readers as indecent. Our aim is to provide information about the diversity of sexual expression. We are not seeking consensus on what is acceptable. Nor do we pass judgments or encourage readers to expand their sexual repertoires.

The human body is sensitive to many forms of sexual stimulation. Yet we reiterate the theme that biology is not destiny: A biological capacity does not impose a behavioral requirement. Cultural expectations, personal values, and individual experience—not only our biological capacities—determine our sexual behavior. What is right for you is right for you, but not necessarily for your neighbor.

We begin by reviewing the techniques that people practice by themselves to derive sexual pleasure: masturbation and sexual fantasy. We then consider techniques that involve a partner.

SOLITARY SEXUAL BEHAVIOR

Various forms of sexual expression do not require a partner or are not generally practiced in the presence of a partner. Masturbation is one of the principal forms of one-person sexual expression. Masturbation involves direct stimulation of the genitals. Other forms of individual sexual experience, such as sexual fantasy, may or may not be accompanied by genital stimulation.

Masturbation

> In solitude he pollutes himself, and with his own hand blights all his prospects for both this world and the next. Even after being solemnly warned, he will often continue this worse than beastly practice, deliberately forfeiting his right to health and happiness for a moment's mad sensuality.
>
> —J. H. Kellogg, M.D., *Plain Facts for Old and Young*, 1888

The word *masturbation* derives from the Latin *masturbari*, from the roots for "hand" and "to defile." The derivation provides clues to historical cultural attitudes toward the practice. **Masturbation** may be practiced by manual stimulation of the genitals, perhaps with the aid of artificial stimulation, such as a vibrator. It may employ an object, such as a pillow or a **dildo,** that touches the genitals. Even before we conceive of sexual experiences with others, we may learn early in childhood that touching our genitals can produce pleasure.

Pleasure is not the only reason why people masturbate. Table 9.1 lists reasons for masturbation, according to the findings of the NHSLS study.

Masturbation Sexual self-stimulation.

Dildo A penis-shaped object used in sexual activity.

TABLE 9.1

Reasons for masturbation (in percent of respondents who report reason), according to the NHSLS study

Reasons for Masturbation	Men	Women
To relax	26%	32%
To relieve sexual tension	73	63
Partners are unavailable	32	32
Partner does not want to engage in sexual activity	16	6
Boredom	11	5
To obtain physical pleasure	40	42
To help get to sleep	16	12
Fear of AIDS and other STIs	7	5
Other reasons	5	5

Source: From E. O. Laumann, J. H. Gagnon, R. T. Michael and S. Michaels (1994). *The Social Organization of Sexuality: Sexual Practices in the United States.* Chicago, IL: University of Chicago Press, Table 3.3, p. 86.

Within the Judeo-Christian tradition, masturbation—"onanism"—has been strongly condemned as sinful (Allen, 2000). Early Judeo-Christian attitudes toward masturbation reflected the censure that greeted nonprocreative sexual acts. In the Judeo-Christian tradition, masturbation has been referred to as onanism, a name that is derived from the biblical story of Onan. According to the Book of Genesis (38:9–11), Onan was the secondborn son of Judah. Judah's first son, Er, had died without an heir. Biblical law required that if a man died without leaving a male heir, his brother must take the widow as a wife (a union called a levirate marriage) and rear their first son as his brother's heir. Judah thus directed Onan to "Go in unto thy brother's wife, and perform the duty of a husband's brother unto her, and raise up seed to thy brother." But Onan "spilled [his seed] upon the ground" during relations with his deceased brother's wife and was struck down by God for his deed.

Although "onanism" has come to be associated with Judeo-Christian condemnation of masturbation, Onan's act was one of **coitus interruptus,** not masturbation. Both acts, however, involve nonprocreative sex—spilling the seed. Moreover, Onan's punishment seems to have had more to do with his failure to fulfill his lawful obligations than with his spilling his seed. Whatever its biblical origins, masturbation is prohibited under Jewish law. Historians suspect that Jews and Christians in ancient times condemned sexual practices that did not lead to pregnancy because of the need for an increase in their numbers. The need for progeny is also linked to the widespread view that coitus in marriage is the only morally acceptable avenue of sexual expression.

Until very recently, the history of cultural attitudes toward masturbation in Western society has been one of almost continual condemnation of the practice on moral and religious grounds—and even on medical grounds.

HISTORICAL MEDICAL VIEWS OF MASTURBATION Until recent years, masturbation was thought to be physically and mentally harmful, as well as degrading. The eighteenth-century physicians S. A. D. Tissot and Benjamin Rush (a signer of the Declaration of Independence) believed that masturbation caused tuberculosis, "nervous diseases," poor eyesight, memory loss, and epilepsy.

Coitus interruptus The practice of withdrawing the penis prior to ejaculation during sexual intercourse.

Many clergy and medical authorities of the nineteenth century were persuaded that certain foods had a stimulating effect on the sex organs. Hence, one form of advice to parents focused on modifying their children's diets to eliminate foods that were believed to excite the sexual organs, notably meat, coffee, tea, and chocolate, and to substitute "unstimulating" foods in their place, most notably grain products. In the 1830s, the Reverend Sylvester Graham developed a cracker, since called the graham cracker, to help people control their sexual impulses.

Yet another household name is that of a man who made his mark by introducing a bland diet that was also intended to help people, especially youngsters, control their sexual impulses. In the nineteenth-century United States, medical advice was largely disseminated to the general public through pamphlets and guides written by leading medical authorities. One of the more influential writers was the superintendent of the Battle Creek Sanatorium in Michigan, Dr. J. H. Kellogg (1852–1943), better known to you as the creator of the modern breakfast cereal. Kellogg identified 39 signs of masturbation, including acne, paleness, heart palpitations, rounded shoulders, weak backs, and convulsions. Kellogg, like Graham, believed that sexual desires could be controlled by sticking to a diet of simple foods, especially grains, including the corn flakes that have since borne his name. (We wonder how Kellogg would react to the energizing, sugar-coated cereals that bear his name now.)

Many nineteenth-century physicians also advised parents to take measures to prevent their children from masturbating. Kellogg suggested that parents bandage or cage their children's genitals or tie their hands. Some of the contraptions devised to prevent masturbation were barbarous (see Figure 9.1).

Several nineteenth-century scholars of sexuality joined the crusade against masturbation. Richard von Krafft-Ebing (in *Psychopathia Sexualis*, 1886) and Havelock Ellis (in *Studies in the Psychology of Sex*, 1900) condemned masturbation as psychologically dangerous. Krafft-Ebing linked masturbation to sexual orientation. Male masturbation, or so it was mistakenly believed, arrested the development of normal erotic instincts and led to **impotence** with women. Thus, it was thought to encourage male–male sexual activity.

MASTURBATION TODAY Despite this history, there is no scientific evidence that masturbation is harmful. Masturbation does not cause insanity, grow hair on the hands, or cause warts or any of the other psychological and physical ills once ascribed to it. Masturbation is physically harmless, save for rare injuries to the genitals from rough stimulation. Nor is masturbation in itself psychologically harmful, although it may be a sign of an adjustment problem if people use masturbation as an exclusive sexual outlet when opportunities for sexual relationships are available. Sex therapists have even found that masturbation has therapeutic benefits. It has emerged as a treatment for women who have difficulty reaching orgasm (Kay, 1992) (see Chapter 15).

Of course, people who consider masturbation wrong, harmful, or sinful may experience anxiety or guilt if they masturbate or wish to masturbate. These negative emotions are linked to their attitudes toward masturbation, not to masturbation per se (Michael et al., 1994).

Despite the widespread condemnation of masturbation in our society, surveys indicate that most people masturbate at some time. The incidence of masturbation is gener-

Impotence Recurrent difficulty in achieving or sustaining an erection sufficient to engage successfully in sexual intercourse. (The term has been replaced by the terms *male erectile disorder* and *erectile dysfunction,* as discussed in Chapter 15.)

FIGURE 9.1

Devices Designed to Curb Masturbation. Because of widespread beliefs that masturbation was harmful, various contraptions were introduced in the nineteenth century to prevent the practice in children. Some of the devices were barbarous.

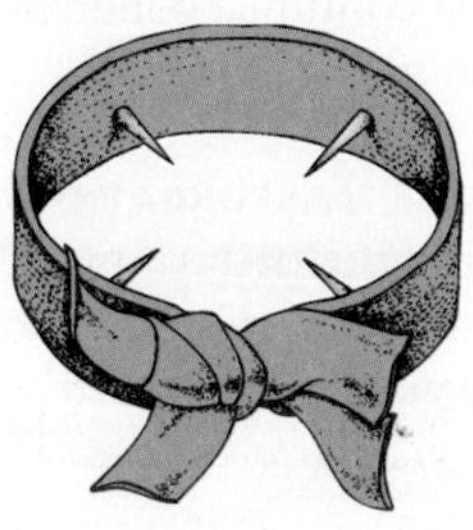

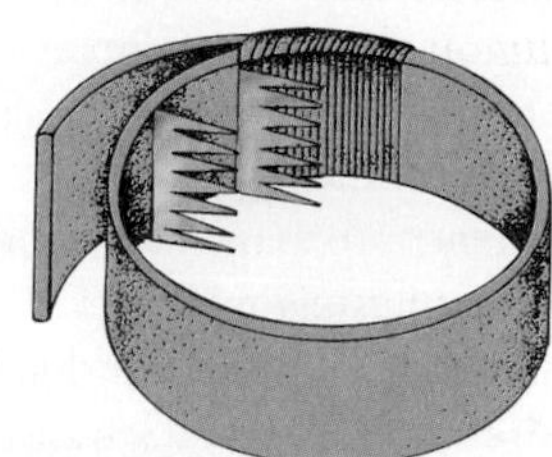

a CLOSER look

The Technology of Orgasm: "Hysteria," the Vibrator, and Women's Sexual Satisfaction

Rachel Maines was going to write a book about needlework in the late nineteenth and early twentieth centuries. In the course of her research, she noticed advertisements for vibrators. She turned her attention to the meaning and use of vibrators in U.S. history and wound up writing a book called *The Technology of Orgasm: "Hysteria," the Vibrator, and Women's Sexual Satisfaction* (1999).

It turns out that genital massage to orgasm—often using a vibrator—was a standard treatment for "hysteria," a health problem considered common in women. The treatment was usually carried out by a physician or a midwife. Genital massage would be used to bring the woman to "hysterical paroxysm" (orgasm, that is). The introduction of the vibrator in the 1880s made treatment more efficient.

Hysteria? What's that? In earlier centuries, the diagnosis of hysteria would be made on the basis of symptoms such as anxiety, irritability, nervousness, pelvic swelling, heaviness in the abdomen (bloating), and fainting. There were other symptoms as well, including sexual fantasies and vaginal lubrication. The word *hysteria* derives from the Greek word that means "uterus." The medical establishment believed that the uterus caused these symptoms by choking the patient because of sexual deprivation. Pregnancy would help; so would coitus. Single women were encouraged to get married, and married women were encouraged to get pregnant. Women without men might go horseback riding, try rocking chairs (yes, rocking chairs), or obtain genital massage. Maines found no evidence that physicians delighted in the task. Rather, they apparently relegated it to midwives when they could. Women, by the way, were not encouraged to masturbate as a way of achieving, uh, "hysterical paroxysm." Masturbation was seen as deviant and unhealthful. Use of the vibrator in the hands of the physician or midwife was seen as a medical treatment, not a sexual act (Heiman, 2000). An orgasm was a "hysterical crisis," not an orgasm.

One does not have to be a genius to recognize the "symptoms of hysteria" as related to menstruation. Today, of course, we realize that menstrual and premenstrual symptoms are related to the secretion of sex hormones. But it is fascinating to note that even as late as the mid-twentieth century, physicians and psychologists tended to attribute a wide variety of mental disorders to hysteria. Therefore, the behaviors connected with the disorders—such as the development of physical symptoms in response to stress—were expected in women and surprising when found in men.

Psychologist Julia Heiman (2000) believes that the content and the sources cited by Maines deserve further scrutiny by other scholars. It is surprising that this piece of history has remained relatively hidden until now. Yet the book does have a solid list of references, along with illustrations of the vibrators and treatment tables. It would appear that the practice described by Maines did in fact occur. The question remains how widespread it was.

ally greater among men than women. However, there are women who masturbate frequently and men who rarely if ever do so (Michael et al., 1994).

Nearly all of the adult men and about two thirds of the adult women in Kinsey's samples (Kinsey et al., 1948, 1953) reported that they had masturbated at some time. In a study of students in an urban university, 85% of the women and 95% of the men reported that they had masturbated (Person et al., 1989). Seventy-one percent of the women and 83% of the men reported masturbating during the previous three months.

Not all researchers find that the gender gap has narrowed to such an extent. A survey of students in a New England college found nearly twice as many men (81%) as women (45%) reporting some experience with masturbation (Leitenberg et al., 1993). About three in four men, as compared to only one in three women, reported masturbating during the previous year.

The NHSLS study also found a notable gender gap in reported frequencies of masturbation (Laumann et al., 1994). Table 9.2 shows how the sample group reported the frequency of masturbation during the past 12 months, broken down according to gender, age, marital status, level of education, religion, and race/ethnicity. Overall, 37% of the men and 58% of the women who were sampled reported that they had *not* masturbated during the past 12 months. Within every social category, men reported masturbating more frequently than women did. Despite the sexual revolution, women may still find

TABLE 9.2

Sociocultural factors and frequency of masturbation during past 12 months, as found in the NHSLS study

	FREQUENCY OF MASTURBATION (%)			
	NOT AT ALL		AT LEAST ONCE A WEEK	
Sociocultural Characteristics	**Men**	**Women**	**Men**	**Women**
Total population	36.7	58.3	26.7	7.6
Age				
18–24	41.2	64.4	29.2	9.4
25–29	28.9	58.3	32.7	9.9
30–34	27.6	51.1	34.6	8.6
35–39	38.5	52.3	20.8	6.6
40–44	34.5	49.8	28.7	8.7
45–49	35.2	55.6	27.2	8.6
50–54	52.5	71.8	13.9	2.3
55–59	51.7	77.6	10.3	2.4
Marital Status				
Never married (not cohabiting)	31.8	51.8	41.3	12.3
Married	42.6	62.9	16.5	4.7
Formerly married (not cohabiting)	30.2	52.7	34.9	9.6
Education				
Less than high school	54.8	75.1	19.2	7.6
High school graduate	45.1	68.4	20.0	5.6
Some college	33.2	51.3	30.8	6.9
College graduate	24.2	47.7	33.2	10.2
Advanced degree	18.6	41.2	33.6	13.7
Religion				
None	32.6	41.4	37.6	13.8
Liberal, moderate Protestant	28.9	55.1	28.2	7.4
Conservative Protestant	48.4	67.3	19.5	5.8
Catholic	34.0	57.3	24.9	6.6
Race/Ethnicity				
European American	33.4	55.7	28.3	7.3
African American	60.3	67.8	16.9	10.7
Latino and Latina American	33.1	65.5	24.4	4.7
Asian American	38.7	—*	31.3	—*
Native American	—*	—*	—*	—*

*Sample sizes too small to report findings.

Source: From E. O. Laumann, J. H. Gagnon, R. T. Michael, and S. Michaels (1994). *The Social Organization of Sexuality: Sexual Practices in the United States.* Chicago, IL: University of Chicago Press, Table 3.1, p. 82.

masturbation less pleasurable or acceptable than men do (Leitenberg et al., 1993). Women may still be subject to traditional socialization pressures that teach that sexual activity for pleasure's sake is more of a taboo for women than for men. Then, too, women are more likely to pursue sexual activity within the context of a relationship.

Married people are less likely to have masturbated during the past 12 months than never-married and formerly married people. Nevertheless, only 43% of the married men and 63% of the married women sampled said that they did not masturbate at all during the past year.

Truth or Fiction? **REVISITED**

It is not true that married people rarely if ever masturbate. The majority of married men and a sizable minority of married women in the United States report masturbating at least occasionally. ■

Education would appear to be a clearly liberating influence on masturbation. For both genders, people with more education reported more frequent masturbation. Perhaps better-educated people are less likely to believe the old horror stories about masturbation or to be subject to traditional social restrictions. Conservative religious beliefs appear to constrain masturbation. Conservative Protestants are apparently less likely to masturbate than liberal and moderate Protestants are. African Americans are notably less likely to report masturbating over a 12-month period than are other ethnic groups.

Truth or Fiction? **REVISITED**

It is true that European American men are more likely to masturbate than African American men are. Perhaps African American men are relatively more likely to adhere to traditional views concerning masturbation. ■

There appears to be a link between attitudes toward masturbation and orgasmic potential. A study of women revealed more negative attitudes toward masturbation among a group of 21- to 40-year-olds who had never achieved orgasm than among a comparison group of women who had (Kelly et al., 1990). Kinsey and his colleagues had reported links between prior masturbation and sexual satisfaction in marriage. Women who had masturbated during adolescence were more likely to find gratification in marital coitus than women who had not (Kinsey et al., 1953).

This evidence does not suggest that adolescents should be encouraged to masturbate to enhance the likelihood of sexual fulfillment as adults. A selection factor probably explains the link (see Figure 9.2). That is, people who masturbate early are probably generally more open to exploring their sexuality and learning about the types of stimulation that arouse them. These attitudes would carry over into marriage and increase the likelihood that women would seek the coital stimulation they need to achieve sexual gratification. Nevertheless, adolescent masturbation probably does also set the stage for marital sexual satisfaction by yielding information about the types of stimulation people need to obtain sexual gratification.

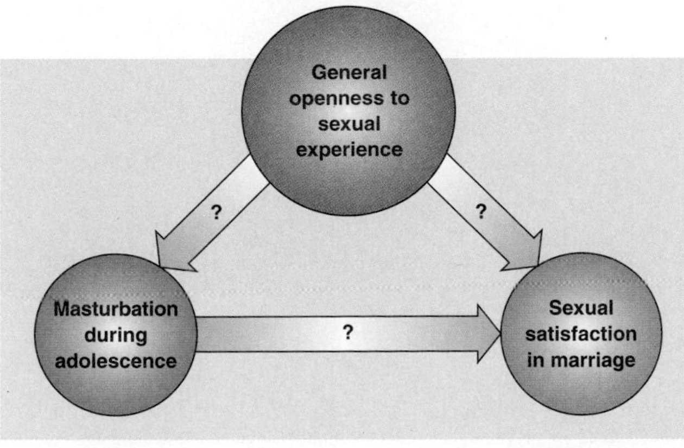

FIGURE 9.2

What Are the Connections Between Masturbation During Adolescence and Sexual Satisfaction in Marriage? There is a positive correlation between masturbation during adolescence and sexual satisfaction in marriage. What hypotheses can we make about the causal connections? Does experience with masturbation teach people about their sexual needs so that they are more likely to obtain adequate sexual stimulation in marriage? Are people who masturbate early generally more open to exploring their sexuality and learning about the types of stimulation that arouse them? Such attitudes might also increase the likelihood that people would seek the coital stimulation they need to achieve sexual gratification in marriage.

Truth or Fiction? **REVISITED**

Actually, women who masturbated during adolescence are *more* likely to find gratification in marital coitus than women who did not.

Other researchers find that women achieve orgasm more reliably through masturbation than through coitus. On the basis of her magazine survey, Hite (1977) reported that 92% of her sample could reliably achieve orgasm through masturbation. Only 30% regularly did so through coitus. Masters and Johnson (1966) also found that masturbation was a more reliable means for women to achieve orgasm than coitus, at least for women who accept masturbation as a sexual outlet.

Truth or Fiction? **REVISITED**

Women are actually *more* likely to reach orgasm through masturbation than through coitus.

In our efforts to correct misinformation about masturbation, we do not wish to leave the impression that there is anything wrong with people who choose *not* to masturbate. Nor do we wish to imply that people should masturbate. Although we may all be able to experience pleasure from self-stimulation, let us say once more that biology is not destiny. It is illogical to transform a biological capacity into a behavioral requirement.

MASTURBATION TECHNIQUES USED BY MALES

> Sex is like bridge—if you don't have a good partner, you'd better have a good hand.
> —Contemporary bathroom graffiti

Although masturbation techniques vary widely, most men report that they masturbate by manual manipulation of the penis (see Figure 9.3). Typically, they take one or two minutes to reach orgasm (Hite, 1981; Kinsey et al., 1948). Men tend to grip the penile shaft with one hand, jerking it up and down in a milking motion. Some men move the whole hand up and down the penis; while others use just two fingers, generally the thumb and index finger. Men usually shift from a gentler rubbing action during the flaccid or semierect state of arousal to a more vigorous milking motion once full erection takes place. Men are also likely to stroke the glans and frenulum lightly at the outset, but their grip tightens and their motions speed up as orgasm nears. At orgasm, the penile shaft may be gripped tightly, but the glans has become sensitive, and contact with it is usually avoided.

FIGURE 9.3

Male Masturbation. Masturbation techniques vary widely, but most men report that they masturbate by manual manipulation of the penis. They tend to grip the penile shaft with one hand and jerk it up and down in a milking motion.

(Likewise, women usually avoid stimulating the clitoris directly during orgasm because of increased sensitivity.)

Some men use soapsuds (which may become irritating) as a lubricant for masturbation during baths or showers. Other lubricants, such as petroleum jelly or K-Y jelly, may also be used to reduce friction and to simulate the moist conditions of coitus.

A few men prefer to masturbate by rubbing the penis and testicles against clothing or bedding (Kinsey et al., 1948). A very few men rub their genitals against inflatable dolls sold in sex shops. These dolls may come with artificial mouths or vaginas that can be filled with warm water to mimic the sensations of coitus. Artificial vaginas are also for sale.

Some men strap vibrators to the backs of their hands. Electrical vibrators save labor but do not simulate the type of up-and-down motions of the penis that men favor. Hence, they are not used very often. Most men rely heavily on fantasy or erotic photos or videos but do not use sex-shop devices.

MASTURBATION TECHNIQUES USED BY FEMALES Techniques of female masturbation also vary widely. In fact, Masters and Johnson reported never observing two women masturbate in precisely the same way. Even when the general technique was similar, women varied in the tempo and style of their self-caresses. But some general trends have been noted. Most women masturbate by massaging the mons, labia minora, and clitoral region with circular or back-and-forth motions (Hite, 1976; Kinsey et al., 1953). They may also straddle the clitoris with their fingers, stroking the shaft rather than the glans (see Figure 9.4). The glans may be lightly touched early during arousal, but because of its

FIGURE 9.4

Female Masturbation. Techniques of female masturbation vary so widely that Masters and Johnson reported never observing two women masturbating in precisely the same way. Most women masturbate by massaging the mons, labia minora, and clitoral region, however, either with circular or back-and-forth motions.

exquisite sensitivity, it is rarely stroked for any length of time during masturbation. Women typically achieve clitoral stimulation by rubbing or stroking the clitoral shaft or pulling or tugging on the vaginal lips. Some women also massage other sensitive areas, such as their breasts or nipples, with the free hand. Many women, like men, fantasize during masturbation (Leitenberg & Henning, 1995).

In contrast to the male myth (Kinsey and his colleagues [1953] describe it as a "male conceit") that women usually masturbate by simulating penile thrusting through the insertion of fingers or phallic objects into their vaginas, relatively few women actually do (Hite, 1976; Kinsey et al., 1953). Hite reported that only 1.5% of her respondents relied exclusively on vaginal insertion as a means of masturbation. Kinsey and his colleagues found that only one in five women had sometimes used vaginal insertions of objects during masturbation. Some women first experimented with the technique of vaginal insertions but then gave it up as they became more familiar with their sexual anatomy and capabilities. Others practiced the technique because their male partners found it sexually stimulating to watch them engage in this type of activity. Still, some women reported experiencing erotic pleasures from deep vaginal penetration.

Truth or Fiction? **REVISITED**

It is not true that most women masturbate by inserting a finger or other object into the vagina. Most women rely on clitoral stimulation.

Even when women do use insertion, they usually precede or combine it with clitoral stimulation. Sex shops sell dildos, which women can use to rub their vulvas or to insert vaginally. Penis-shaped vibrators may be used in a similar fashion. Many women masturbate during baths, some by spraying their genitals with water-massage shower heads.

Handheld electrical vibrators (see Figure 9.5) provide a constant massaging action against the genitals that can be erotic. Some women find this type of stimulation too intense, however, and favor vibrators that strap to the back of the hand and cause the fingers to vibrate during manual stimulation of the genitals. This type of vibration may numb the hand that is attached to the vibrator, however. Women who use vibrators often experiment with different models to find one with the shape and intensity of vibration that suits them.

Use of Fantasy

People may use sexual fantasies when they are alone or to heighten sexual excitement with a partner (Eggers, 2000; Leitenberg & Henning, 1995). Some couples find it sexually arousing to share fantasies or to enact them with their partners. Sexual fantasies may be experienced without sexual behavior, as in erotic dreams or daydreams. Masturbators of-

FIGURE 9.5

Electric Vibrators.

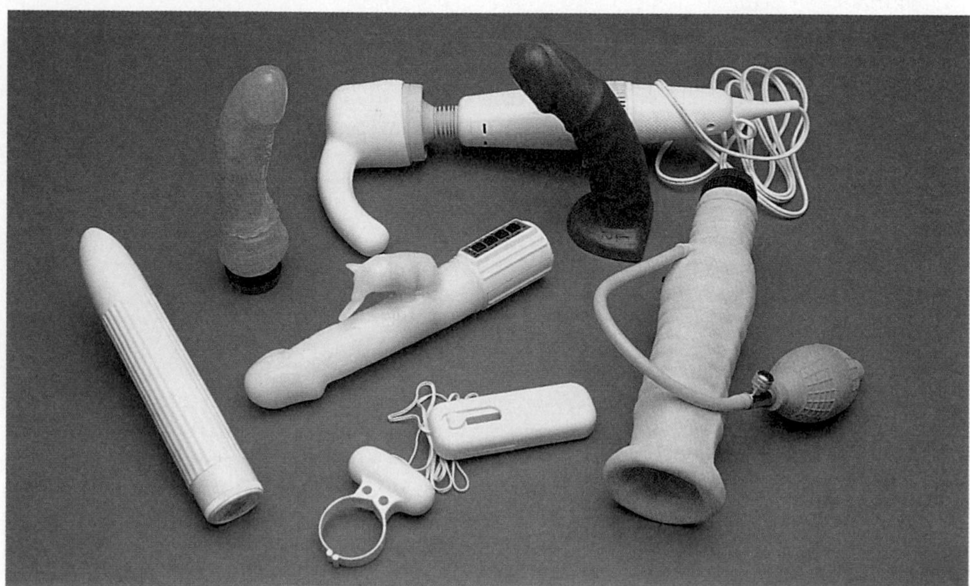

ten require some form of cognitive stimulation, such as indulging in a favorite fantasy or reading or viewing erotica, to increase their arousal to the point of orgasm.

How common are sexual fantasies? It is a popular belief that only people with deficient sex lives engage in sexual fantasies. However, evidence shows that the majority of men and women engage in sexual fantasies from time to time, especially during masturbation (Reinisch, 1990). Men fantasize more often than women do (Hsu et al., 1994). In their national survey, Laumann and his colleagues (1994) found that 54% of men and 19% of women said they thought about sex at least once a day.

A study of 212 married college women found that 88% of them reported having erotic fantasies. Most of them fantasized during both masturbation and coitus (Davidson & Hoffman, 1986). Another study showed that sexual daydreaming or fantasy in women was associated with a greater sex drive, with more sexual activity, and with a more positive attitude toward sex (Purifoy et al., 1992). We would hardly expect to find such a pattern among people with deficient sex lives.

Sexual fantasies may be reasonably realistic, such as imagining sexual activity with an attractive classmate. They may also involve flights of fancy, such as making love to a movie star. The most common fantasy theme reported among women involves someone they are or have been involved with (Leitenberg & Henning, 1995). In contrast to men's fantasies, women are more likely to focus on the partner's feelings, on touching, on their own responses to what is happening, and on the general tone of the sexual encounter (Leitenberg & Henning, 1995).

The most common masturbation fantasy reported by both genders in the *Playboy* sample was "having intercourse with a loved one" (Hunt, 1974). But note some interesting gender differences: Men are more likely to assume aggressive, dominant roles. Women are more apt to enact passive, submissive roles (Leitenberg & Henning, 1995; Person et al., 1989). Women are more likely to connect sexual activity to relationships and emotional involvement than men are (Barbach, 1995; Leitenberg & Henning, 1995). Perhaps for this reason, men are more likely than women—52% versus 40% in *The New York Times* poll—to think it's all right to fantasize about sex with someone other than their partner (Eggers, 2000).

Fantasy themes apparently parallel traditional gender stereotypes, but they may also reflect deeper currents. Sociobiologists might conjecture that women are relatively more likely to fantasize about the images of familiar lovers because female reproductive success in ancestral times was more likely to depend on a close, protective relationship with a stable partner (Symons, 1995). Women can bear and rear relatively few offspring. Thus, they would have had (and might still have) a relatively greater genetic investment in each reproductive opportunity.

Whether or not the wellspring of sexual fantasy themes lie in our genetic heritages, we should not confuse sexual fantasies with behavior. Fantasies are imaginary. Most people do not intend to act out their fantasies. They use them as a means of inducing or enhancing sexual pleasure (Reinisch, 1990). As one woman noted,

> My fantasies are so personal, and the pleasure I get from them derives so much, I think, from the fact that they are private and locked away in my imagination, that I wouldn't dream of trying to make them come true. . . . But act my fantasies out? Make them come true? No, absolutely not. My real life's not what they're about; I don't want those things to really happen to me, I simply want to imagine what it would be like. So that's where they'll stay. (Friday, 1973, p. 288)

What Friday says here of women's fantasies also applies to men:

> For many women, fantasy is a way of exploring, safely, all the ideas and actions which might frighten them in reality. In fantasy they can expand their reality, play out certain sexual variables and images in much the same way that children enter into fantasy as a form of play, of trying out desires, releasing energies for which they have no outlet in reality. Thinking about it, even getting excited over the image doesn't mean you want it as your reality . . . or else we all, night dreamers that we are, would be suppressed robbers, bisexuals, murderers, or even inanimate objects. (p. 41)

Fantasizing about forcing someone into sexual activity, or about being victimized, does not mean that one wants these events to occur (Leitenberg & Henning, 1995; Reinisch, 1990). Women who imagine themselves being sexually coerced remain in control of their fantasies. Real assault victims are not. Nor is it unusual for heterosexual people to have episodic fantasies about sexual activity with people of their own gender, or for gay males or lesbians to fantasize about sexual activity with people of the other gender. In neither case does the person necessarily intend to act out the fantasy.

Heterosexual people may in fact fantasize about sexual activity with people of their own gender. People do not necessarily express their fantasies in their behavior.

Why do people fantasize when they masturbate? Masturbation fantasies serve several functions. For one, they increase or facilitate sexual arousal. Sex therapists encourage clients to use sexual fantasies to enhance sexual arousal (e.g., Heiman & LoPiccolo, 1987). Sexual fantasies are highly arousing, in part because fantasizers can command the imagined sexual encounter. They may imagine that people who will not give them the time of day find them irresistible and are willing to fulfill their sexual desires. Or fantasizers may picture improbable or impossible arousing situations, such as sexual activity on a commercial airliner or while skydiving. Some masturbation fantasies may be arousing because they permit us to deviate from traditional gender roles. Women might fantasize about taking an aggressive role or forcing someone into sexual activity. Men, by contrast, may imagine being overtaken by a horde of sexually aggressive women. Other fantasies involve sexual transgressions or "forbidden" behaviors, such as exposing oneself, doing a striptease before strangers, engaging in sexual activity with strangers, or sadomasochistic sex (S&M).

Masturbation fantasies can also allow people to rehearse sexual encounters. We may envision the unfolding of an intended sexual encounter, from greeting a date at the door, through dinner and a movie, and finally to the bedroom. We can mentally rehearse what we would say and do as a way of preparing for the date. Finally, masturbation fantasies may fill in the missing love object when we are alone or our partners are away.

SEXUAL BEHAVIOR WITH OTHERS

Partners' feelings for one another, and the quality of their relationships, may be stronger determinants of their sexual arousal and response than the techniques that they employ. Partners are most likely to experience mutually enjoyable sexual interactions when they are sensitive to each other's sexual needs and incorporate techniques with which they are both comfortable. As with other aspects of sharing relationships, communication is the most important "sexual" technique.

Foreplay

Various forms of noncoital sex, such as cuddling, kissing, petting, and oral–genital contact, are used as **foreplay**. The pattern and duration of foreplay vary widely within and across cultures. Broude and Greene (1976) found that prolonged foreplay was the norm in about half of the societies in their cross-cultural sample. Foreplay was minimal in 1 out of 10 societies and virtually absent in about one third of them.

Within the United States, there is a gender difference in the amount of foreplay desired. A survey of college students revealed that women wanted longer periods of foreplay (and "afterplay") than men did (Denny et al., 1984). Because women usually require a longer period of stimulation during sex with a partner to reach orgasm, increasing the duration of foreplay may increase female coital responsiveness.

Foreplay is not limited to the human species. Virtually all species of mammals, from horses and sheep to dogs and chimpanzees, engage in some kind of foreplay. Depending

Foreplay Physical interactions that are sexually stimulating and set the stage for intercourse.

on the particular species, mating pairs may rub, lick, nuzzle, or playfully nip each other's genitals for minutes or hours preceding coitus (Geer et al., 1984).

Kissing, genital touching, and oral–genital contact may also be experienced as ends in themselves, not as preludes to coitus. Yet some people object to petting for petting's sake, equating it with masturbation as a form of sexual activity without a "product." Many people behave as though all sexual contact had to lead to coitus, perhaps because of the importance that our culture places on it.

Kissing

Kissing is almost universal in our culture, but it occurs less often among the world's cultures than manual or oral stimulation of the genitals (Frayser, 1985). Kissing is unknown in some cultures, including those of the Thonga of Africa and the Siriono of Bolivia. Variations in styles of kissing also exist across cultures (Ford & Beach, 1951). Kissing is now practiced in Japan because of the influence of Western culture, but it was previously unknown there. Instead of kissing, the Balinese of the South Pacific bring their faces close enough to each other to smell each other's perfume and feel the warmth of each other's skin. Europeans have wrongly dubbed this practice "rubbing noses." Among some preliterate societies, kissing consists of sucking the partner's lips and tongue and allowing saliva to pass from one mouth to the other.

Couples may kiss for its own enjoyment or as a prelude to intercourse, in which case it is a part of foreplay. In ***simple kissing,*** the partners keep their mouths closed. Simple kissing may develop into caresses of the lips with the tongue or into nibbling of the lower lip. In what Kinsey called ***deep kissing,*** which is also called French kissing or soul kissing, the partners part their lips and insert their tongues into each other's mouths. Some prefer the lips parted slightly. Others open their mouths widely.

Kissing may also be an affectionate gesture without erotic significance, as in kissing someone good night. Some people are accustomed to kissing relatives and close friends affectionately on the lips. Others limit kissing relatives to the cheek. Sustained kissing on the lips and deep kissing are nearly always erotic gestures.

Kissing is not limited to the partner's mouth. Kinsey found that more than 9 husbands out of 10 kissed their wives' breasts. Women usually prefer several minutes of body contact and gentle caresses before they want their partner to kiss their breasts or suck or lick their nipples. Women also usually do not prefer a hard sucking action unless they are highly aroused. Many women are reluctant to tell their partners that sucking hurts, because they do not want to interfere with their partner's pleasure.

Other parts of the body are also often kissed, including the hands and feet, the neck and earlobes, the insides of the thighs, and the genitals themselves.

Touching

Touching or caressing erogenous zones with the hands or other parts of the body can be highly arousing. Even simple hand-holding can be sexually stimulating for couples who are sexually attracted to one another. The hands are very rich in nerve endings.

Touching is a common form of foreplay. Both men and women generally prefer manual or oral stimulation of the genitals as a prelude to intercourse. Women generally prefer that direct caressing of the genitals be focused around the clitoris but not directly on the extremely sensitive clitoral glans. Men sometimes assume (often mistakenly) that their partners want them to insert their finger or fingers into the vagina as a form of foreplay. But not all women enjoy this form of stimulation. Some women go along with it because it's what their partners want or something they *think* their partners want. Ironically, men may do it because they assume that their *partners* want it. When in doubt, it would not hurt to ask. If you are not sure what to say, you can always blame us: "Listen, I read this thing in my human sexuality text, and I was wondering. . . ."

How Touching? Touching is a common form of foreplay. Most people like manual or oral stimulation of the genital organs as a prelude to sexual intercourse.

Masters and Johnson (1979) noted gender differences with respect to preferences in foreplay. Men typically prefer direct stroking of their genitals by their partner early in lovemaking. Women, however, tend to prefer that their partners caress their genitals after a period of general body contact that includes holding, hugging, and nongenital massage. This is not a hard and fast (or slow) rule, but it concurs with other observations that men tend to be more genitally oriented than women. Women are more likely to view sex within a broader framework of affection and love.

TECHNIQUES OF MANUAL STIMULATION OF THE GENITALS Here again, variability in technique is the rule, so partners need to communicate their preferences. The man's partner may use two hands to stimulate his genitals. One may be used to fondle the scrotum, by gently squeezing the skin between the fingers (taking care not to apply pressure to the testes themselves). The other hand may circle the coronal ridge and engage in gentle stroking of the penis, followed by more vigorous up and down movements as the man becomes more aroused.

The penis may also be gently rolled back and forth between the palms as though one were making a ball of clay into a sausage, increasing pressure as arousal progresses. Note that men who are highly aroused or who have just had an orgasm may find direct stimulation of the penile glans uncomfortable.

The woman may prefer that her partner approach genital stimulation gradually, following stimulation of other body parts. Genital stimulation may begin with light, stroking motions of the inner thighs and move on to the vaginal lips (labia) and the clitoral area. Women may enjoy pressure against the mons pubis from the heel of the hand, or they may like tactile stimulation of the labia, which are sensitive to stroking motions. Clitoral stimulation can focus on the clitoral shaft or the region surrounding the shaft, rather than on the clitoris itself, because of the extreme sensitivity of the clitoral glans to touch.

Moreover, the clitoris should not be stroked if it is dry, lest it become irritated. Because it produces no lubrication of its own, a finger may enter the outer portion of the vagina to apply some vaginal lubrication to the clitoral region.

Some, but not all, women enjoy having a finger inserted into the vagina, which can stroke the vaginal walls or simulate thrusting of the penis. Vaginal insertion is usually preferred, if at all, only after the woman has become highly aroused. Many women desire that their partners discontinue stroking motions while they are experiencing orgasm, but others want stimulation to continue. Men and women may physically guide their partners' hands or otherwise indicate what types of strokes they find most pleasurable.

If a finger is to be inserted into the vagina, it should be clean. Fingernails should be well trimmed. Inserting into the vagina fingers that have been in the anus is dangerous. The fingers may transfer microbes from the woman's digestive tract, where they do no harm, to the woman's reproductive tract, where they can cause serious infections.

Stimulation of the Breasts

Men are more likely to stimulate women's breasts than to enjoy having their own breasts fondled, even though the breasts (and especially the nipples) are erotically sensitive in both genders. Most, but not all, women enjoy stimulation of the breasts. Masters and Johnson report that some women are capable of achieving orgasm from stimulation of the breasts alone.

The hands and the mouth can be used to stimulate the breasts and the nipples. The desired type and intensity of stimulation of the breasts varies from person to person, so partners need to communicate their preferences.

Masters and Johnson (1979) found that gay men frequently stroked their partners' nipples before stimulating the penis itself. Although some heterosexual men enjoy having their breasts and nipples stimulated by their partners, many if not most do not. Many men are simply unaware that their breasts are erotically sensitive. Cultural conditioning may also play a major part in men's reluctance to having their breasts stimulated: Men may feel uncomfortable receiving a form of stimulation that they have learned to associate with the stereotypical feminine sexual role.

Oral–Genital Stimulation

Oral stimulation of the male genitals is called **fellatio.** Fellatio is referred to by slang terms such as *blow job, sucking, sucking off,* or *giving head.* Oral stimulation of the female genitals is called **cunnilingus,** which is referred to by slang expressions such as *eating* (a woman) or *going down* on her.

The popularity of oral–genital stimulation has increased dramatically since Kinsey's day, especially among young married couples. Kinsey and his colleagues (1948, 1953) found that at least 60% of married, *college-educated* couples had experienced oral–genital contact. Such experiences were reported by only about 20% of couples who had only a high school education and by 10% of couples who had only a grade school education.

The incidence of oral sex may have peaked during the sexual revolution. The *Playboy* survey in the early 1970s found that more than 90% of the married couples under 25 years of age—*across all educational levels*—reported that they had engaged in oral–genital sex (Hunt, 1974). But remember that 80% of the people approached by Hunt refused to participate in the survey. Therefore, there is likely to be a volunteer bias in his statistics. (The researchers who conducted the NHSLS study are less kind in discussing of the representativeness of Hunt's sample.)

According to the NHSLS study, the incidence of oral sex was somewhat lower in the 1990s (Laumann et al., 1994). About three of four men (77%) and two of three women (68%) reported playing the active role in oral sex during their lifetimes. Nearly four men in five (79%) and three of four women (73%) reported having been the recipients of oral–genital sex during their lifetimes. Among married couples, 80% of the men and 71% of the women had performed oral sex. Eighty percent of the men and 74% of the women had received oral sex. These are dramatic increases since Kinsey's day. Among young European American women (ages 18 to 36) in Kinsey's sample, 48% reported having engaged in fellatio. Fifty-one percent of their partners had engaged in cunnilingus.

It is true that oral sex is the norm for today's young married couples, statistically speaking. A majority of them report participating in this form of sexual expression. ■

Like touching, oral–genital stimulation can be used as a prelude to intercourse or as a sexual end in itself. If orgasm is reached through oral-genital stimulation, a woman may be concerned about tasting or swallowing a man's ejaculate. There is no scientific evidence that swallowing semen is harmful to one's health, unless the man is infected with a sexually transmitted disease in which semen can act as a conduit of infections (Reinisch, 1990). Note that oral–genital contact with the genitals of an infected partner, even without contact with semen, may transmit harmful organisms. Couples are thus advised to

Fellatio Oral stimulation of the male genitals.

Cunnilingus Oral stimulation of the female genitals.

Sociocultural Factors and Oral Sex

Table 9.3 shows the incidence of oral sex among men and women with various levels of education and from different racial/ethnic backgrounds in the NHSLS survey. As with masturbation, the incidence of oral sex is correlated with level of education. That is, more highly educated individuals are more likely to have practiced oral sex. Why? Perhaps education encourages experimentation. Perhaps education dispels myths that nontraditional behavior patterns are necessarily harmful. Note also that African American men and women are less likely to have engaged in oral sex than are people from other racial/ethnic backgrounds. African American men and women were also less likely than other ethnic groups to report masturbating during the past 12 months. African Americans may adhere more strictly to traditional ideas about what kinds of sexual behavior are proper and what kinds are not.

Findings from a national survey of more than 3,000 sexually active men between the ages of 20 and 39, conducted by the Battelle Human Affairs Research Center in Seattle, are consistent with the NHSLS findings. Seventy-five percent of the men reported performing oral sex. Seventy-nine percent reported receiving oral sex (Billy et al., 1993). Mirroring the racial differences observed by Laumann and his colleagues (1994), African American men in the Battelle survey were much less likely than their European American counterparts to have performed or received oral sex.

The results of a 1980s survey of college students are also consistent with the NHSLS data. Eighty-one percent of European American females reported that they had engaged in fellatio, as compared to 47% of African American females (Belcastro, 1985). Seventy-two percent of European American males had performed cunnilingus, compared to 50% of African American males. Although the prevalence of oral–genital sex since Kinsey's time has increased for both African Americans and European Americans, African Americans remain relatively less likely than European Americans to engage in oral sexual activity.

Truth or Fiction? REVISITED

It is not true that African Americans are more likely than European Americans to engage in oral sex. A review of the research suggests that the opposite is more likely to be the case.

TABLE 9.3

Percent of NHSLS study respondents who report experience with oral sex

	PERFORMED ORAL SEX		RECEIVED ORAL SEX	
	Men	**Women**	**Men**	**Women**
Education				
Less than high school	59.2%	42.1%	60.7%	49.6%
High school graduate	75.3	59.6	76.6	67.1
Some college	80.0	78.2	84.0	81.6
College graduate	83.7	78.9	84.6	83.1
Advanced college degree	80.5	79.0	81.4	81.9
Race/Ethnicity				
European American	81.4	75.3	81.4	78.9
African American	50.5	34.4	66.3	48.9
Latino and Latina American	70.7	59.7	73.2	63.7
Asian American	63.6	—*	72.7	—*

*There were not enough Asian American women in the sample to report meaningful findings.

Source: From E. O. Laumann, J. H. Gagnon, R. T. Michael, and S. Michaels (1994). *The Social Organization of Sexuality: Sexual Practices in the United States.* Chicago, IL: University of Chicago Press, Table 3.6, p. 98. Copyright © 1994 by University of Chicago Press. Reprinted by permission.

practice "safer sex" techniques (see Chapters 16 and 17) unless they know that they and their partners are free of sexually transmitted diseases (Reinisch, 1990).

TECHNIQUES OF FELLATIO Although the word *fellatio* is derived from a Latin root that means "to suck," a sucking action is generally not highly arousing. The up-and-down movements of the penis in the partner's mouth, and the licking of the penis, are generally the most stimulating. Gentle licking of the scrotum may also be highly arousing.

The mouth is stimulating to the penis because it contains warm, moist mucous membranes, as does the vagina. Muscles of the mouth and jaw can create varied pressure and movements. Erection may be stimulated by gently pulling the penis with the mouth (being careful never to touch the penis with the teeth) and simultaneously providing manual stimulation, as described earlier.

Higher levels of sexual arousal or orgasm can be promoted by moving the penis in and out of the mouth, simulating the motion of the penis in the vagina during intercourse. The speed of the motions can be varied, and manual stimulation near the base of the penis (firmly encircling the lower portion of the penis or providing pressure behind the scrotum) can also be stimulating.

Some people may gag during fellatio, a reflex that is triggered by pressure of the penis against the back of the tongue or against the throat. Gagging may be avoided if the man's partner grasps the shaft of the penis with one hand and controls the depth of penetration. Gagging is less likely to occur if the partner performing fellatio is on the top, rather than below, and if there is verbal communication about how deep the man should penetrate. Gagging may also be overcome by allowing gradually deeper penetration of the penis over successive occasions while keeping the throat muscles relaxed.

TECHNIQUES OF CUNNILINGUS Women can be highly aroused by their partner's tongue because it is soft, warm, and well lubricated. In contrast to a finger, the tongue can almost never be used too harshly. A woman may thus be more receptive to direct clitoral contact by a tongue. Cunnilingus provides such intense stimulation that many women find it to be the best means for achieving orgasm. Some women cannot reach orgasm in any other way (Hite, 1976).

In performing cunnilingus, the partner may begin by kissing and licking the woman's abdomen and inner thighs, gradually nearing the vulva. Gentle tugging at or sucking of the labia minora can be stimulating, but the partner should take care not to bite. Many women enjoy licking of the clitoral region, and others desire sucking of the clitoris itself. The tongue may also be inserted into the vagina, where it may imitate the thrusts of intercourse.

"69" The term *sixty-nine*, or *soixante-neuf* in French (pronounced swah-sahnt nuff), describes simultaneous oral–genital stimulation (see Figure 9.6). The numerals 6 and 9 are used because they resemble two partners who are upside-down and facing each other.

The "69" position has the psychologically positive feature of allowing couples to experience simultaneous stimulation, but it can be an awkward position if two people are not similar in size. Some couples avoid "69" because it deprives each partner of the opportunity to focus fully on receiving or providing sexual pleasure. A person who is receiving stimulation may find it distracting to have to focus on stimulating someone else at the same time.

The "69" technique may be practiced side by side or with one partner on top of the other. But here again there are no strict rules, and couples often alternate positions.

ABSTAINING FROM ORAL SEX Despite the popularity of oral sex among couples today, those who desire to abstain from it have no reason to consider themselves abnormal.

People offer various reasons for abstaining from oral sex. Although natural body odors may be arousing to some people, others are disturbed by the genital odors to which oral sex exposes them. Some people object on grounds of cleanliness. They view the genitals as "dirty" because of their proximity to the urinary and anal openings. Concerns about offensive odors or cleanliness may be relieved by thoroughly washing the genitals beforehand.

Some women prefer not to taste or swallow semen because they find the ejaculate "dirty," sinful, or repulsive. Others are put off by the taste or texture. Semen has a salty taste and a texture similar to that of an egg white. If couples are to engage in unprotected oral sex, open discussion of feelings can enhance pleasure and diminish anxiety. For example, the man can be encouraged to warn his partner or remove his penis from her mouth when he is nearing ejaculation.

FIGURE 9.6

Simultaneous Oral–Genital Contact. The "69" position allows partners to engage in simultaneous oral–genital stimulation.

Let us dispel a couple of myths about semen. For one thing, it is impossible to become pregnant by swallowing semen. For another, semen is not fattening. The average amount of semen expelled in the ejaculate contains only about 5 calories. On the other hand, it is not our intention to encourage swallowing of semen. The aesthetics of swallowing semen have little or nothing to do with concerns about pregnancy or weight. They involve the preferences of the individual.

Some of the objections expressed by people who are reluctant to engage in oral sex may be overcome. Others, such as beliefs that oral–genital contact is offensive or repulsive, are more deeply rooted. Shyness and embarrassment may also deter one from engaging in oral sex. A survey of college students found that shyness and embarrassment were the two reasons most frequently given for not engaging in oral sex (Gagnon & Simon, 1987). Oral sex *is* one of the most intimate types of lovemaking. After all, it provides a direct view of parts of the body we have been reared to keep private.

People may also object to oral sex because it is condemned by Judeo-Christian moral codes. In the Judeo-Christian tradition, any sexual contact that does not lead to procreation has been considered sinful. Couples may also be deterred from practicing oral sex because it is illegal in many states, even for married couples. Some people object to oral sex on grounds that it is unnatural, even though many other species practice some form of oral–genital contact (Ford & Beach, 1951).

Sexual Intercourse: Positions and Techniques

Sexual intercourse, or *coitus* (from the Latin *coire*, which means "to go together"), is sexual activity in which the penis is inserted into the vagina. Intercourse may take place in

FIGURE 9.7

The Male-Superior Coital Position. In this position the couple face one another. The man lies above the woman, perhaps supporting himself on his hands and knees rather than allowing his full weight to press against his partner. The position is also (somewhat disparagingly) referred to as the *missionary position.* (Primitive peoples have supposedly reported that this position never occurred to them until Western visitors described it.)

many different positions. Each position, however, must allow the genitals to be aligned so that the penis is contained by the vagina. In addition to varying positions, couples also vary the depth and rate of thrusting (in-and-out motions) and the sources of additional sexual stimulation.

Though the number of possible coital positions is virtually endless, we will focus on four of the most commonly used positions: the male-superior (man-on-top) position, the female-superior (woman-on-top) position, the lateral-entry (side-entry) position, and the rear-entry position. Although it does not technically fit the definition of sexual intercourse, we shall also discuss anal intercourse, a sexual technique used by both male–female and male–male couples.

THE MALE-SUPERIOR (MAN-ON-TOP) POSITION The male-superior position (this "superiority" simply reflects the couple's body positions, but it has sometimes been taken as a symbol of male domination) has also been called the **missionary position**. In this position the partners face one another. The man lies above the woman, perhaps supporting himself on his hands and knees rather than resting his full weight on his partner (see Figure 9.7). Even so, it is easier for the man to move than for the woman, which suggests that he is responsible for directing their activity.

Missionary position The coital position in which the man is on top. Also termed the *male-superior position*.

Many students of human sexuality suggest that it is preferable for the woman to guide the penis into the vagina, rather than having the man do so. The idea is that the woman can feel the location of the vaginal opening and determine the proper angle of entry. To accomplish this, the woman must feel comfortable "taking charge" of the couple's love-making. With the breaking down of the traditional stereotype of the female as passive, women are feeling more comfortable taking this role. On the other hand, if the couple prefers that the man guide his penis into his partner's vagina, the slight loss of efficiency need not trouble them, as long as he moves prudently to avoid hurting his partner.

The male-superior position has the advantage of permitting the couple to face one another so that kissing is easier. The woman may run her hands along her partner's body, stroking his buttocks and perhaps cupping a hand beneath his scrotum to increase stimulation as he reaches orgasm.

But the male-superior position makes it difficult for the man to caress his partner while simultaneously supporting himself with his hands. Therefore, the position may not be favored by women who enjoy having their partners provide manual clitoral stimulation during coitus. This position can be highly stimulating to the man, which can make it difficult for him to delay ejaculation. The position also limits the opportunity for the woman to control the angle, rate, and depth of penetration. It may thus be more difficult for her to attain the type of stimulation she may need to achieve orgasm, especially if she favors combining penile thrusting with manual clitoral stimulation. Finally, this position is not advisable during the late stages of pregnancy. At that time, the woman's distended abdomen would force the man to arch severely above her, lest he place undue pressure on her abdomen.

THE FEMALE-SUPERIOR (WOMAN-ON-TOP) POSITION In the female-superior position, the couple face one another with the woman on top. The woman straddles the male from above, controlling the angle of penile entry and the depth of thrusting (see Figure 9.8). Some women maintain a sitting position; others lie on top of their partners. Many women vary their position.

In the female-superior position, the woman is psychologically—and to some degree physically—in charge. She can move as rapidly or as slowly as she wishes with little effort, adjusting her body to vary the angle and depth of penetration. She can reach behind her to stroke her partner's scrotum or lean down to kiss him.

As in the male-superior position, kissing is relatively easy. This position has additional advantages. The man may readily reach the woman's buttocks or clitoris in order to provide manual stimulation. Assuming that the woman is shorter than he is, it is rather easy for him to stimulate her breasts orally (a pillow tucked behind his head may help). The woman can, in effect, guarantee that she receives adequate clitoral stimulation, either by the penis or manually by his hand or her own. This position thus facilitates orgasm in the woman. Because it tends to be less stimulating for the male than being on top, it may help him to control ejaculation. For these reasons, this position is commonly used by couples who are learning to overcome sexual difficulties.

THE LATERAL-ENTRY (SIDE-ENTRY) POSITION In the lateral-entry position, the man and woman lie side by side, facing one another (see Figure 9.9). This position has the advantages of allowing each partner relatively free movement and easy access to the other. The man and woman may kiss freely, and they can stroke each other's bodies with a free arm. The position is not physically taxing, because both partners are resting easily on the bedding. Thus it is an excellent position for prolonged coitus, for older couples, and when couples are somewhat fatigued.

Let us note some disadvantages to this position. First, inserting the penis into the vagina while lying side by side may be awkward. Many couples thus begin coitus in an-

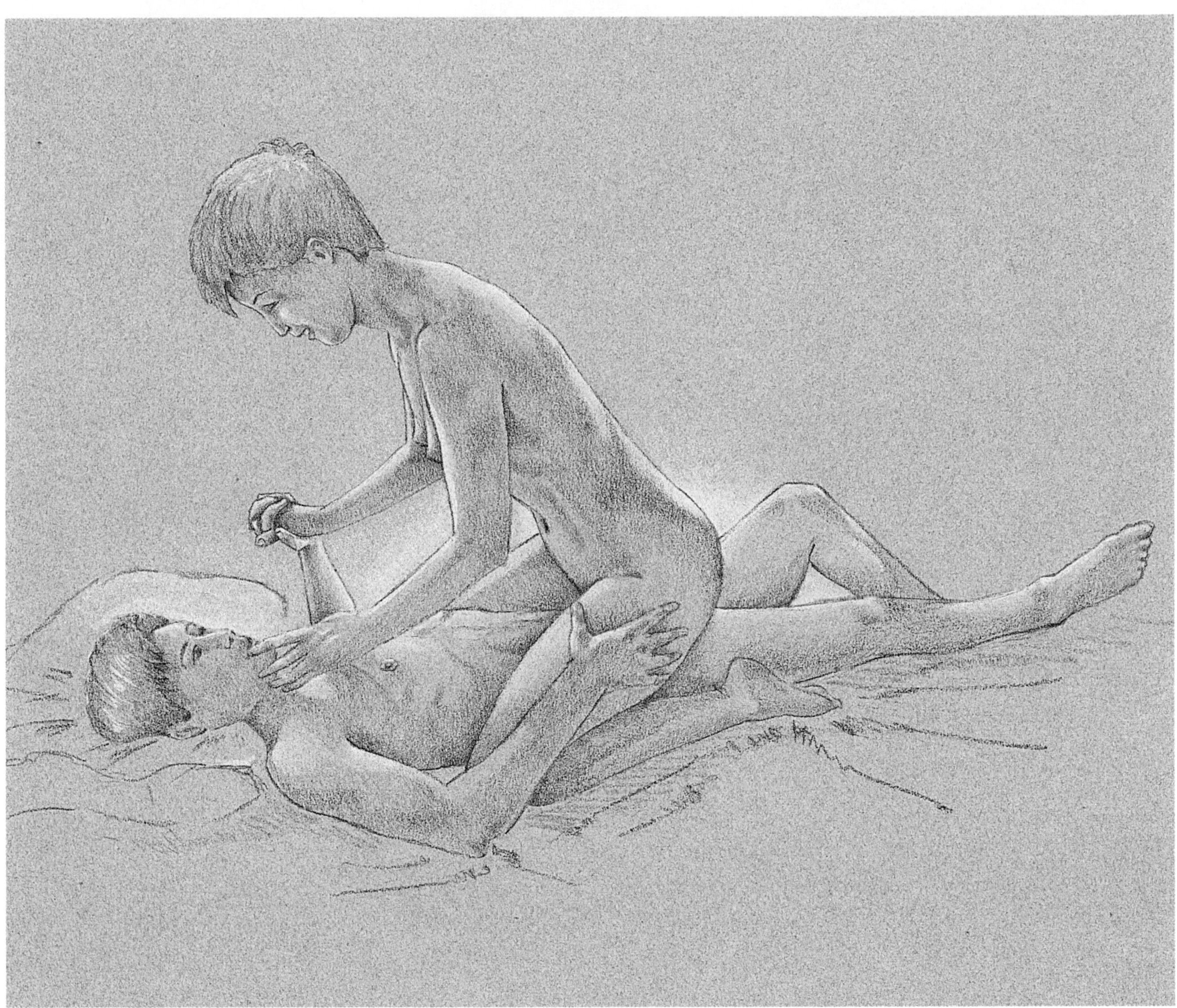

FIGURE 9.8

The Female-Superior Coital Position. The woman straddles the male from above, controlling the angle of penile entry and the depth of thrusting. The female-superior position puts the woman psychologically and physically in charge. The woman can ensure that she receives adequate clitoral stimulation from the penis or the hand. This position also tends to be less stimulating for the male and may thus help him to control ejaculation.

other position and then change into the lateral-entry position, often because they wish to prolong coitus. Second, one or both partners may have an arm lying beneath the other that will "fall asleep," or become numb because of the constricted blood supply. Third, women may not receive adequate clitoral stimulation from the penis in this position. Of course, such stimulation may be provided manually (by hand) or by switching to another position after a while. Fourth, it may be difficult to achieve deeper penetration of the penis. The lateral position is useful during pregnancy (at least until the final stages, when the distension of the woman's abdomen may make lateral entry difficult).

FIGURE 9.9

The Lateral-Entry Coital Position. In this position, the couple face each other side by side. Each partner has relatively free movement and easy access to the other. Because both partners rest easily on the bedding, it is an excellent position for prolonged coitus and for coitus when couples are fatigued.

THE REAR-ENTRY POSITION In the rear-entry position, the man faces the woman's rear. In one variation (see Figure 9.10), the woman supports herself on her hands and knees while the man supports himself on his knees, entering her from behind. In another, the couple lie alongside one another and the woman lifts one leg, draping it backward over her partner's thigh. The latter position is particularly useful during the later stages of pregnancy.

The rear-entry position may be highly stimulating for both partners. Men may enjoy viewing and pressing their abdomens against their partner's buttocks. The man can reach around or underneath to provide additional stimulation of the clitoris or breasts, and she may reach behind (if she is on her hands and knees) to stroke or grasp her partner's testicles.

Potential disadvantages to this position include the following: First, this position is the mating position used by most other mammals, which is why it is sometimes referred to as doggy style. Some couples may feel uncomfortable about using the position because of its association with animal mating patterns. The position is also impersonal in the sense that the partners do not face one another, which may create a sense of emotional distance. Because the man is at the woman's back, the couple may feel that he is very much in charge—he can see her, but she cannot readily see him. Physically, the penis does not provide adequate stimulation to the clitoris. The penis also tends to pop out of the vagina from time to time. Finally, air tends to enter the vagina during rear-entry coitus. When it is expelled, it can sound as though the woman has passed air through the anus—a possibly embarrassing though harmless occurrence.

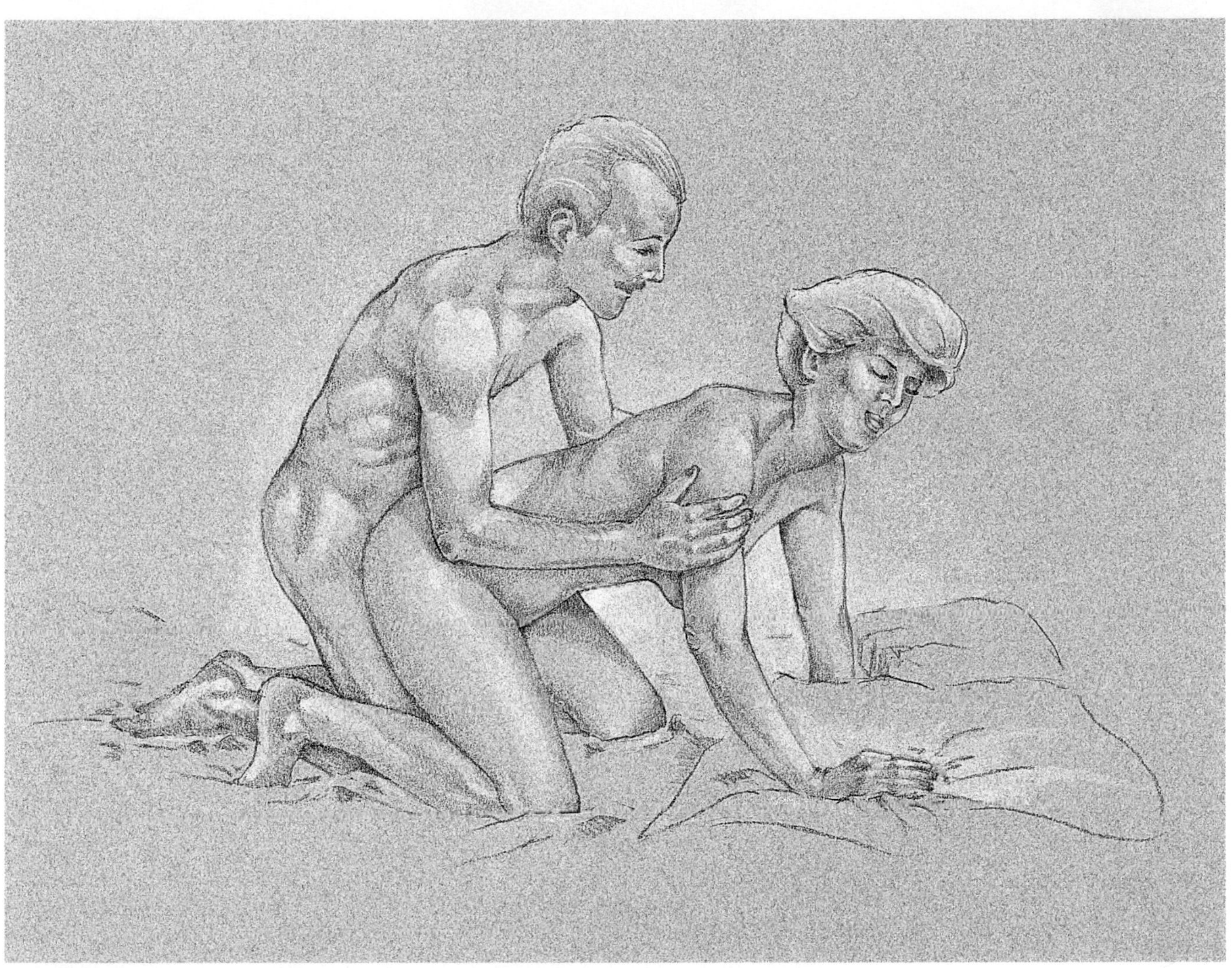

FIGURE 9.10

The Rear-Entry Coital Position. The rear-entry position is highly erotic for men who enjoy viewing and pressing their abdomens against their partners' buttocks. However, some couples feel uncomfortable about the position because of its association with animal mating patterns. The position is also impersonal in that the partners do not face one another. Moreover, some couples dislike the feeling that the man is psychologically in charge because he can see his partner but she cannot readily see him.

USE OF FANTASY DURING COITUS Like masturbation, mental excursions into fantasy during coitus may be used to enhance sexual arousal and response (Davidson & Hoffman, 1986). In a sense, coital fantasies allow couples to inject sexual variety and even offbeat sexual escapades into their sexual activity, without being unfaithful. Fantasies have historically been viewed as evil, however. People believed that fantasies, like dreams, were placed in the mind by agents of the devil. Despite this tradition, researchers find that most married people have engaged in coital fantasies (Davidson & Hoffman, 1986). In one study, 71% of the men and 72% of the women reported engaging in coital fantasies to enhance their sexual arousal (Zimmer et al., 1983). A more recent survey of a sample of 178 students, faculty, and staff members at a college in Vermont found that 84% reported fantasizing at least occasionally during intercourse (Cado & Leitenberg, 1990). Nor does there appear to be any connection between sexual dissatisfaction with one's relationship

Are African Americans Sexually Permissive? Cultural Myths Versus Data

African Americans have long been stereotyped as more sexually permissive than European Americans (Wyatt, 1989). The perpetuation of this stereotype is based more on cultural biases than on scientific evidence. Evidence from Kinsey's time through the 1980s has shown that African American teenagers begin intercourse at earlier ages, on the average, than European American teenagers (Wyatt, 1989). Such research has largely failed to take into account socioeconomic (social class) differences between the groups, however.

Researchers have begun to examine critically the existing stereotypes of African American sexuality. The 1988 National Survey of Family Growth, which surveyed nearly 8,500 American women ages 15 to 44, found that European American women were more likely to have had 10 or more sex partners than either African American or Latina American women (Lewin, 1992b).

Gail Wyatt (1989) used Kinsey-style interviews to examine various aspects of sexual behavior among 126 African American and 122 European American women, ages 18 to 36, from Los Angeles County. Wyatt's study is noteworthy for balancing European American and African American samples with respect to sociodemographic factors such as income level, education, and marital status. Wyatt reported no significant differences in age at first intercourse between the African American and European American women in her sample. The average age at first intercourse was 16.6 years for the total sample: 16.5 years for African American women and 16.7 years for European American women, respectively—an insignificant difference. Factors that predicted age at first intercourse were similar across groups. For both groups, perceptions that one's parents were more influential than one's friends during adolescence and the eventual attainment of higher educational levels were associated with a later age at first intercourse.

All in all, results of the Wyatt survey fail to support the stereotype of African American sexual permissiveness. The similarities between the races overshadowed their differences. We should caution, however, that neither sample was a national probability sample. The results may thus not generalize to all African Americans or all European Americans. Although more interracial research would be of interest, an even richer understanding of sexual behavior patterns might be gained by examining variability *within* racial groups according to factors such as socioeconomic status and family background.

and the use of coital fantasies (Davidson & Hoffman, 1986). Thus coital fantasies are not a form of compensation for an unrewarding sexual relationship.

Lest you think that coital fantasies arise only out of sexual monotony in marriage, consider that many if not most unmarried people also fantasize during sexual relations. In one study of sexually experienced, single undergraduates, Sue (1979) found that virtually the same percentages of men (58.6%) and women (59.4%) reported "sometimes" or "almost always" fantasizing during coitus.

Coital fantasies, like masturbation fantasies, run a gamut of themes. They include making love to another partner, group sex, orgies, images of past lovers or special erotic experiences, and making love in fantastic and wonderful places, among others.

Table 9.4 shows the kinds of fantasies reported to Hariton and Singer (1974) by a sample of married women from an affluent New York City suburb. The women included PTA members and regular churchgoers. Yet 65%—a strong majority—used fantasy. The use of fantasy was not a sign of marital difficulty. In fact, women who fantasized reported *better* sexual relations with their partners than those who did not. The content of these fantasies suggests that they serve to introduce novelty or variety into sexual relationships.

RELATIONSHIPS AND COITAL FANTASIES Studies of coital fantasies by Hariton and Singer (1974) and Sue (1979), among others, find that sexual fantasies during intercourse are common among couples with close relationships and are a means of facilitating sexual arousal. By facilitating sexual arousal, coital fantasies may help strengthen the intimate bond between the partners. Evidence does not show coital fantasies to be a sign of a troubled relationship.

Evidence fails to show that coital fantasies are a sign of a troubled relationship. ■

TABLE 9.4

Coital fantasies of married women

Fantasy	**Women Reporting Fantasy (%)**
Thoughts of an imaginary romantic lover enter my mind.	56
I relive a previous sexual experience.	52
I enjoy pretending that I am doing something forbidden.	50
I imagine that I am being overpowered or forced to surrender.	49
I am in a different place, like a car, motel, beach, woods, etc.	47
I imagine myself delighting many men.	43
I pretend that I struggle and resist before being aroused to surrender.	40
I imagine that I am observing myself or others having sex.	38
I pretend that I am another irresistibly sexy female.	38
I daydream that I am being made love to by more than one man at a time.	36
My thoughts center about feelings of weakness or helplessness.	33
I see myself as a striptease dancer, harem girl, or other performer.	28
I pretend that I am a whore or a prostitute.	25
I imagine that I am being forced to expose my body to a seducer.	19
My thoughts center around urination or defecation	2

Source: From E. B. Hariton and J. L. Singer (1974). "Women's Fantasies During Sexual Intercourse: Normative and Theoretical Implications," *Journal of Consulting and Clinical Psychology, 42*, Table 1, pp. 313–332.

Partners are often reluctant to share their coital fantasies, or even to admit having them. This is especially true when the fantasy is about someone other than the partner. The fantasizer might fear being accused of harboring extramarital desires. Or the fantasizer might fear that the partner will interpret fantasies as a sign of rejection: "What's the matter, don't I turn you on anymore?" Any perceived merits of self-disclosure are best weighed against one's partner's potential reactions to coital fantasies.

ANAL INTERCOURSE Anal intercourse can be practiced by male–female couples and male–male couples. It involves insertion of the penis into the rectum. The rectum is richly endowed with nerve endings and is thus highly sensitive to sexual stimulation. Anal intercourse is also referred to as "Greek culture," or lovemaking in the "Greek style," because of bisexuality in ancient Greece among males. It is also the major act that comes under the legal definition of sodomy. Both women and men may reach orgasm through receiving the penis in the rectum.

In anal intercourse, the penetrating male usually situates himself behind his partner. (He can also lie above or below his partner in a face-to-face position.) The receiving

Human Sexuality ONLINE //

Web Sites Related to Sexual Techniques and Behavior Patterns

Sexual Health InfoCenter: This Web site has information about sexual positions and techniques.
www.sexhealth.org/infocenter/infomain.htm

MSNBC Sexploration: This Web site offers information about enhancing loving relationships.
www.msnbc.com/news/SEXPLORATIONH_Front.asp

The Hearth Forums: These online forums discuss issues related to sexuality and other topics.
www.lynxcom.com/dearpaula/welcome.html

The Sex Coach at iVillage: This site offers advice about sexual techniques and sexual satisfaction.
www.ivillage.com/relationships

Women.com: This Web site has information about women's sexuality and sexual response.
women.com/sex/sex

partner can supplement anal stimulation with manual stimulation of the clitoral region or penis to reach orgasm.

Women often report wanting their partner's fingers in the anus at the height of passion or at the moment of orgasm. A finger in the rectum during orgasm can heighten sexual sensation because the anal sphincters contract during orgasm. Although some men also want a finger in the anus, many resist because they associate anal penetration with the female role or with male–male sexual activity. However, the desire to be entered by one's partner is not necessarily connected with sexual orientation. A gay male or lesbian sexual orientation refers to the eroticization of members of one's own gender, not to the desire to penetrate or be penetrated.

Many couples are repulsed by the idea of anal intercourse. They view it as unnatural, immoral, or risky. Yet others find anal sex to be an enjoyable sexual variation, though perhaps not a regular feature of their sexual diet.

The NHSLS survey found that 1 man in 4 (26%) and 1 woman in 5 (20%) reported having engaged in anal sex at some time during their lives (Laumann et al., 1994). Yet only about 1 person in 10 (10% of the men and 9% of the women) had engaged in anal sex during the past year. As with oral sex, there was a higher incidence of anal sex among more highly educated people in the NHSLS survey. For example, about 30% of the male college graduates had engaged in anal sex, as compared with 23% of male high school graduates. About 29% of the women with advanced college degrees had engaged in anal sex, as compared with about 17% of the women who had graduated only from high school (Laumann et al., 1994). Education appears to be a liberating experience in sexual experimentation. About 1 in 5 men in the Battelle sample reported having engaged in anal intercourse (Billy et al., 1993). As in the NHSLS survey, anal intercourse was more commonly reported among better-educated men.

Truth or Fiction? **REVISITED**

Anal sex actually turns out to be *less* common among less well-educated people. Perhaps education is a liberating influence on sexual experimentation.

Religion appears to be a restraining influence on anal sex. About 34% of the men and 36% of the women in the NHSLS survey who said that they had no religion reported engaging in anal sex during their lifetimes. Figures for male Christians ranged from the lower to upper 20s, and for female Christians, from the mid-teens to the lower 20s (Laumann et al., 1944, p. 99). There were too few Jews in the sample for the investigators to report meaningful figures for Jewish men and women.

Many couples kiss or lick the anus in their foreplay. This practice is called **anilingus.** Oral–anal sex carries a serious health risk, however, because microorganisms that cause intestinal diseases and various sexually transmitted diseases can be spread through oral–anal contact.

Many couples today hesitate to engage in anal intercourse because of the fear of AIDS and other sexually transmitted diseases (STIs). The AIDS virus and other mi-

Anilingus Oral stimulation of the anus.

croorganisms that cause STIs such as gonorrhea, syphilis, and hepatitis can be spread by anal intercourse, because small tears in the rectal tissues may allow the microbes to enter the recipient's blood system. Women also run a greater risk of contracting HIV, the virus that causes AIDS, from anal intercourse than from vaginal intercourse—just as receptive anal intercourse in gay men carries a high risk of infection (Voeller, 1991). However, partners who are both infection-free are at no risk of contracting STIs through any sexual act.

In this chapter, we have observed many of the variations in human sexual expression. No other species shows such diversity in sexual behavior. People show diversity not only in sexual behavior but also in sexual orientation, and that is the focus of the following chapter.

summing up

SOLITARY SEXUAL BEHAVIOR

Masturbation

Masturbation may be practiced by means of manual stimulation of the genitals, perhaps with the aid of an electric vibrator or an object that provides tactile stimulation. Within the Judeo-Christian tradition, masturbation has been condemned as sinful. Until recent years, masturbation was thought to be physically and mentally harmful, yet contemporary scholars see masturbation as harmless. Surveys indicate that most people have masturbated at some point in their lives.

Use of Fantasy

Sexual fantasies are often incorporated during masturbation or during sex with another person to heighten sexual response. Sexual fantasies range from the realistic to genuine flights of fancy. Many people fantasize about sexual activities that they would not actually engage in.

SEXUAL BEHAVIOR WITH OTHERS

Foreplay

The pattern and duration of foreplay vary widely within and across cultures. Women usually desire longer periods of foreplay than men do.

Kissing

Couples kiss for its own enjoyment or as a prelude to intercourse.

Touching

Touching or caressing erogenous zones with the hands or other parts of the body can be highly arousing. Men typically prefer direct stroking of their genitals by their partner early in lovemaking. Women, however, tend to prefer that their partners caress their genitals after a period of general body contact.

Stimulation of the Breasts

Most, but not all, women enjoy stimulation of the breasts. The hands and the mouth can be used to stimulate the breasts.

Oral–Genital Stimulation

The popularity of oral–genital stimulation has increased dramatically since Kinsey's day, but survey researchers find persistent differences in the frequency of these activities between African Americans and European Americans.

Sexual Intercourse: Positions and Techniques

Couples today use a greater variety of coital positions than in Kinsey's time. Four of the most commonly used coital positions are the male-superior position, the female-superior position, the lateral-entry position, and the rear-entry position.

questions for critical thinking

1. What are the beliefs of most people from your sociocultural group about masturbation? Do you share these beliefs? Why or why not?
2. What misinformation, if any, did you receive about masturbation as you were growing up? What was the source of the misinformation?
3. Do you have sexual fantasies? What do you fantasize about? Do you—or did you—wonder whether your sexual fantasies are "normal"? Explain.
4. How do their attitudes and values influence the kinds of foreplay and sexual techniques that people enjoy?
5. What is the connection between level of education and sexual practices such as oral sex? Do you believe that your college experience will have any effects on your own sexual behavior? Explain.

chapter

10

Sexual Orientation

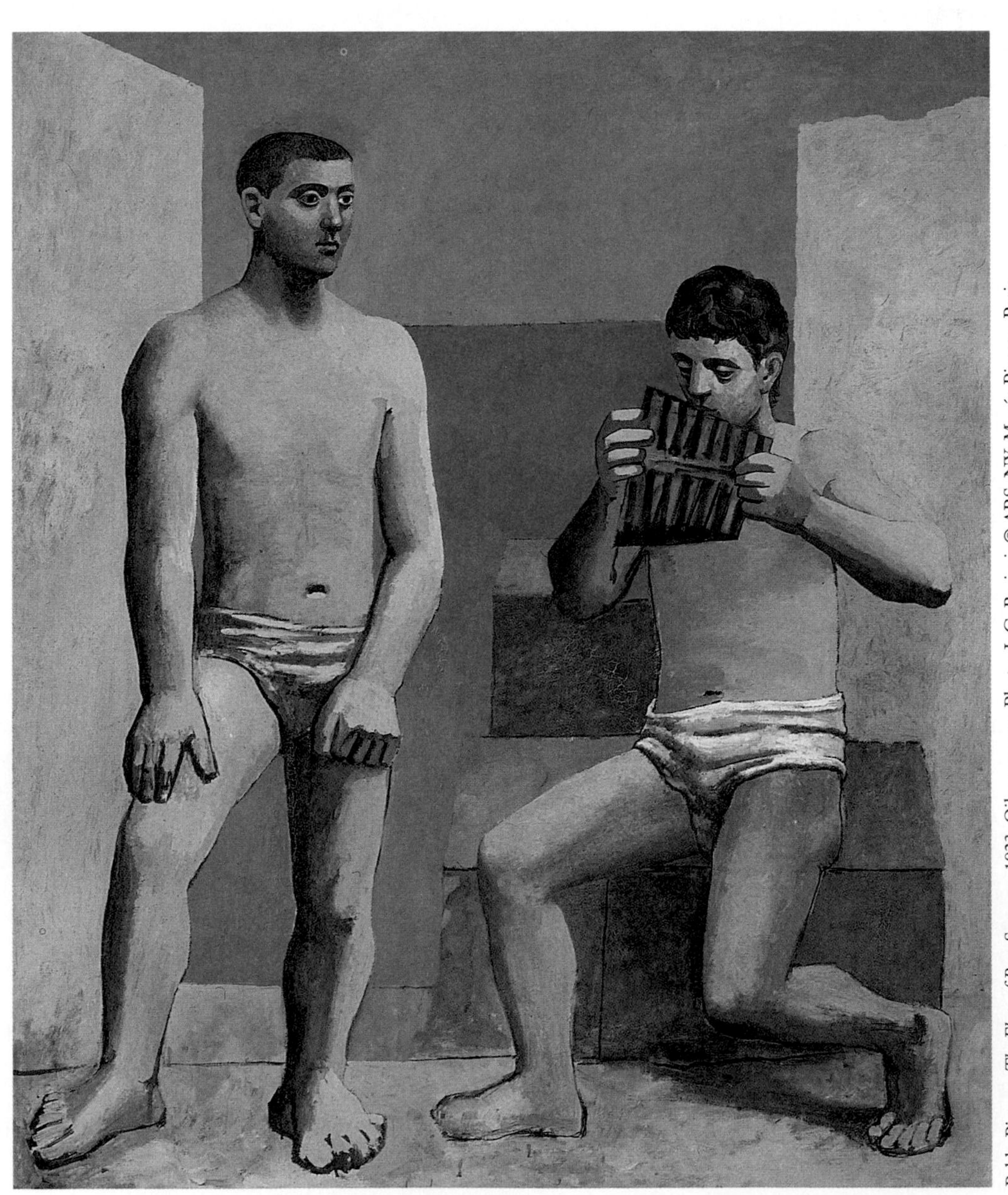

Pablo Picasso. *The Flute of Pan.* Summer 1923. Oil on canvas. Photo: J. G. Berizzi. © ARS, NY. Musée Picasso, Paris, France. Credit: Réunion des Musées Nationaux/Art Resource, NY. Credit: © 2002 Estate of Pablo Picasso/Artists Rights Society (ARS), New York.

outline

Truth or Fiction?

SEXUAL ORIENTATION

Coming to Terms with Terms

Sexual Orientation and Gender Identity

Classification of Sexual Orientation

Bisexuality

PERSPECTIVES ON GAY MALE AND LESBIAN SEXUAL ORIENTATIONS

Historical Perspectives

Cross-Cultural Perspectives

A World of Diversity Ethnicity and Sexual Orientation: A Matter of Belonging

Cross-Species Perspectives

Attitudes Toward Sexual Orientation in Contemporary Society

Biological Perspectives

Psychological Perspectives

Gender Nonconformity

ADJUSTMENT OF GAY MALES AND LESBIANS

Treatment of Gay Male and Lesbian Sexual Orientations

COMING OUT: COMING TO TERMS WITH BEING GAY

Coming Out to Oneself

Coming Out to Others

Human Sexuality Online Out Into the "Gay Global Village" of Cyberspace

PATTERNS OF SEXUAL ACTIVITY AMONG GAY MALES AND LESBIANS

Sexual Techniques

GAY LIFESTYLES

A Closer Look Don't Ask, Don't Tell—Far from Clear as a Bell

Lifestyle Differences Between Gay Males and Lesbians

Variations in Gay Lifestyles

Human Sexuality Online Web Sites Related to Sexual Orientation

SUMMING UP

QUESTIONS FOR CRITICAL THINKING

Truth or Fiction?

____ A lesbian lost custody of her daughter to the girl's father, even though he had served a prison term for murdering his first wife.

____ Gay males and lesbians would prefer to be members of the other gender.

____ Gay males and lesbians suffer from hormonal imbalances.

____ Gay males unconsciously fear women's genitals because they associate them with castration.

____ The American Psychiatric Association considers homosexuality a mental disorder.

____ Many gay couples have lifestyles similar to those of married heterosexual couples and are as well adjusted.

John, a nurse, was awarded custody of his 7-year-old son, Jacob, after a divorce. John's companion, Don, often picks Jacob up after school. "Who is that?" a teacher unfamiliar with the situation asked Jacob one day.

"That's my father's husband," Jacob replied matter-of-factly. (Gross, 1991)

Alyson Publications, a Boston publisher, added two titles to its children's list: "Heather Has Two Mommies" and "Daddy's Roommate." (Gross, 1991)

A Florida mother—a lesbian—lost custody of her 12-year-old daughter to the girl's father, even though he had served 8 years in prison for murdering his first wife. (Navarro, 1996)

Over a cup of coffee one blustery afternoon, Deidra Mack confessed to feeling the jitters about entering the stately brick church nearby, standing before a minister and making vows of lifetime commitment. But minutes later, there she was, in First Unitarian Universalist Society's austerely elegant sanctuary, facing her partner, Barbara Dolores. At the Rev. Gary Kowalski's invitation, the two women, both 28, spoke their vows and exchanged rings. "You have made for yourselves a marriage," the minister said. "Your lives are now joined." Family members and friends applauded. (Niebuhr, 1998)

This chapter is about **sexual orientation.** Sexual orientation concerns the *direction* of one's romantic interests and erotic attractions—toward members of the same gender, members of the other gender, or both.

As suggested in the news clips at the beginning of the chapter, many children are reared by gay male or lesbian couples. In most states, a gay male or lesbian sexual orientation is no longer grounds for parents to lose custody of their children (Dunlap, 1995). States have become generally more accepting of this living arrangement, because there is no evidence that children reared by gay male and lesbian parents are less well adjusted than other children (Allen & Burrell, 1996; Patterson, 1995). Still there are exceptions, as suggested by the Florida case in which custody of a girl was transferred from the girl's mother—a lesbian—to the father, even though the father had served a prison term for murdering his first wife. Nor is there evidence that being reared by a parent with a gay male or lesbian sexual orientation causes children to develop either that sexual orientation or confusion about their gender identity (Allen & Burrell, 1996; Patterson, 1995).

Truth or Fiction? **REVISITED**

It is true that a lesbian lost custody of her daughter to the girl's father, even though he had served a prison term for murdering his first wife. It happened in Florida in 1996. ■

In the chapter, we see that gay people, like heterosexual people, struggle to incorporate their sexuality within their personal identity, to find lovers, and to establish satisfying lifestyles. Civil union, perhaps a percursor of gay marriages, are now permitted in a number of states. In 1996, however, the United States Senate voted 85 to 14 that no state would have to recognize a gay marriage licensed in another state (Schmitt, 1996). Unlike heterosexual people, gay people in our culture face a backdrop of social intolerance, even if they commit themselves to long-term relationships.

Are These Women "Homosexuals," "Homophiles," or "Lesbians," or Do They Have a "Lesbian Sexual Orientation"? What terms shall we use to discuss sexual orientation? Many gay people object to the term *homosexual* because it focuses on sexual behavior rather than on relationships. Moreover, the term bears a social stigma. Therefore, they prefer terms such as *gay male* or *lesbian sexual orientation*, or even *homophile*, if terms must set them apart. The Greek root *philia* refers to love or friendship, not to sexual behavior.

SEXUAL ORIENTATION

Sexual orientation refers to one's erotic attractions toward, and interests in developing romantic relationships with, members of one's own or the other gender (American Psychological Association, 1998b). A **heterosexual orientation** refers to an erotic attraction to, and preference for developing romantic relationships with, members of the other gender. (Many gay people refer to heterosexual people as being *straight*, or as *straights*.) Note that we say *other gender*, not *opposite gender*. Social critics tell us that many problems that arise between men and women are based on the idea that they are polar opposites (Bem, 1993). Research and common sense suggest that men and women are more alike in personality and behavior than they are different, however.

A **homosexual orientation** refers to an erotic attraction to, and interest in forming romantic relationships with, members of one's own gender. The term *homosexuality* denotes sexual interest in members of one's own gender and applies to both men and women. Homosexual men are often referred to as **gay males.** Homosexual women are also called **lesbians** or *gay women*. Gay males and lesbians are also referred to collectively as gays or gay people. The term **bisexuality** refers to an orientation in which one is sexually attracted to, and interested in forming romantic relationships with, both males and females.

Coming to Terms with Terms

Now that we have defined the term *homosexuality*, let us note that we will use it only sparingly. Many gay people object to the term *homosexual*, because they feel that it draws too much attention to sexual behavior. Moreover, the term bears a social stigma. Many gays would prefer terms such as *gay male* or *lesbian sexual orientation*, or a term such as *homophile*, if a term must be used to set them apart. The Greek root *philia* suggests love and

Sexual orientation The direction of one's sexual interests—toward members of the same gender, members of the other gender, or both.

Heterosexual orientation Erotic attraction to, and preference for developing romantic relationships with, members of the other gender.

Homosexual orientation Erotic attraction to, and preference for developing romantic relationships with, members of one's own gender. (From the Greek *homos,* which means "same," not the Latin *homo,* which means "man").

Gay males Males who are erotically attracted to, and desire to form romantic relationships with other males.

Lesbians Females who are erotically attracted to, and desire to form romantic relationships with, other females. (After *Lesbos,* the Greek island on which, legend has it, female–female sexual activity was idealized).

Bisexuality Erotic attraction to, and interest in developing romantic relationships with, males and females.

friendship rather than sexual behavior. Thus, a homophile is a person who develops romantic love and emotional commitment to members of her or his own gender. Sexual activity is secondary.

We therefore use the terms *gay male* and *lesbian* instead of *homosexual* when we refer to sexual orientation. As noted by the American Psychological Association's (1991) Committee on Lesbian and Gay Concerns, the word *homosexual* has also been historically associated with concepts of deviance and mental illness. It perpetuates negative stereotypes of gay people. Also, the term is often used to refer to men only. It thus renders lesbians invisible.

Then, too, the word *homosexual* is ambiguous in meaning—that is, does it refer to sexual behavior or sexual orientation? In this book, your authors speak of male–female sexual behavior (not *heterosexual* behavior), of male–male sexual behavior, and of female–female sexual behavior.

Sexual Orientation and Gender Identity

Because gay people are attracted to members of their own gender, some people assume that they would prefer to be members of the other gender. Like heterosexual people, however, gay people have a gender identity that is consistent with their anatomic gender. Unlike transsexuals, gay people do not see themselves as being trapped in the body of the other gender.

It is not true that gay males and lesbians would prefer to be members of the other gender. Their gender identity is consistent with their anatomy.

Heterosexuals tend to focus almost exclusively on sexual aspects of male–male and female–female relationships. But the love relationships of gay people, like those of heterosexual people, involve more than sex. Gay people, like heterosexual people, spend only a small proportion of their time in sexual activity. More basic to a gay male or lesbian sexual orientation is the formation of romantic attachments with members of one's own gender. These attachments, like male-female attachments, provide a framework for love and intimacy. Although sex and love are common features of relationships, neither is a *necessary* prerequisite for a relationship. Sexual orientations are not defined by sexual activity per se but, rather, by the *direction* of one's romantic interests and erotic attractions.

Classification of Sexual Orientation

Determining a person's sexual orientation might seem to be a clear-cut task. Some people are exclusively gay and limit their sexual activities to partners of their own gender. Others are strictly heterosexual and limit their sexual activities to partners of the other gender. Many people fall somewhere in between, however. Where might we draw the line between a gay male and lesbian sexual orientation, on the one hand, and a heterosexual orientation, on the other? Where do we draw the line between these orientations and bisexuality?

It is possible—indeed not unusual—for heterosexual people to have had some sexual experiences with people of their own gender. Consider a survey of more than 7,000 male readers of *Playboy* magazine. Among those reporting sexual experiences with both men and women in adulthood, more than two out of three perceived themselves to be heterosexual rather than bisexual (Lever et al., 1992). For many of these men, sexual experiences with other men were limited to a brief period of their lives and did not alter their sexual orientations. Lacking heterosexual outlets, prison inmates may have sexual experiences with people of their own gender while they maintain their heterosexual identities. Such inmates would form sexual relationships with people of the other gender if they were available, and they return to male–female sexual behavior upon release from prison. Physical affection also helps some prisoners, male and female, cope with loneliness and isolation. Males who engage in prostitution with male clients may separate their sexual orientation from their "trade." Many fantasize about a female when permitting a client to fellate them. The behavior is male–male. The person's sexual *orientation* may be heterosexual.

Gay males and lesbians, too, may engage in male–female sexual activity while maintaining a gay sexual orientation. Some gay males and lesbians marry members of the other gender but continue to harbor unfulfilled desires for members of their own gender. Then, too, some people are bisexual but may not have acted upon their attraction to members of their own gender.

Sexual orientation is not necessarily expressed in sexual behavior. Many people come to perceive themselves as gay or heterosexual long before they ever engage in sex with members of their own gender. Some people, gay and heterosexual alike, adopt a celibate lifestyle for religious or ascetic reasons and abstain from sexual relationships. Some remain celibate not by choice but because of lack of partners.

People's erotic interests and fantasies may also shift over time. Gay males and lesbians may experience sporadic **heteroerotic** interests. Heterosexual people may have occasional **homoerotic** interests. Many heterosexual people report fantasies about sexual activity with people of their own gender. Many gay people have fantasies about sex with people of the other gender (Masters & Johnson, 1979). About 50% of one sample of lesbians reported that they are sometimes attracted to men (Bell & Weinberg, 1978).

Attraction to people of the other gender and attraction to people of one's own gender may thus not be mutually exclusive. People may have various degrees of sexual interest in, and sexual experience with, people of either gender. Kinsey and his colleagues recognized that the boundaries between gay male and lesbian sexual orientations, on the one hand, and a heterosexual orientation, on the other, are sometimes blurry. They thus proposed thinking in terms of a continuum of sexual orientation rather than two poles.

THE KINSEY CONTINUUM Before Kinsey, scientists generally viewed gay and heterosexual orientations as separate categories. People were viewed as either gay or heterosexual in their psychological makeup and erotic interests. Kinsey and his colleagues (1948, 1953), however, found evidence of degrees of gay and heterosexual orientations among people they surveyed, with bisexuality representing a midpoint between the two. As Kinsey and his colleagues noted,

> The world is not to be divided into sheep and goats. . . .Only the human mind invents categories and tries to force facts into separated pigeonholes. The living world is a continuum in each and every one of its aspects. (1948, p. 639)

Kinsey and his colleagues (1948, 1953) conceived a 7-point "heterosexual–homosexual continuum" (see Figure 10.1). People are located on the continuum according to their patterns of sexual attraction and behavior. People in category 0 are considered exclusively heterosexual. People in category 6 are considered exclusively gay.

What percentage of the population, then, is gay? The percentages depend on the standards one uses. Kinsey and his colleagues reported that about 4% of men and 1% to 3% of women in their samples were exclusively gay (6 on their scale). A larger percentage of people were considered predominantly gay (scale points 4 or 5) or predominantly heterosexual (1 or 2 on the scale). Some were classified as equally gay and heterosexual in

Heteroerotic Of an erotic nature and involving members of the other gender.

Homoerotic Of an erotic nature and involving members of one's own gender.

FIGURE 10.1

The Kinsey Continuum Kinsey and his colleagues conceived of a 7-point heterosexual–homosexual continuum that classifies people according to their homosexual behavior and the magnitude of their attraction to members of their own gender. People in category 0, who accounted for most of Kinsey's study participants, were considered exclusively heterosexual. People in category 6 were considered exclusively homosexual.

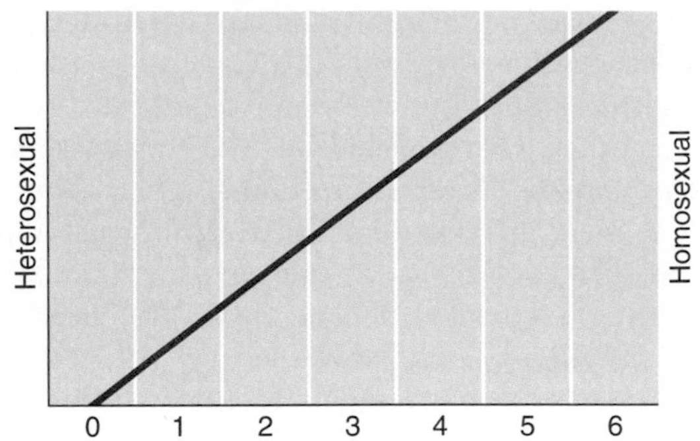

orientation and could be labeled bisexual (scale point 3). Most people were classified as exclusively heterosexual (scale point 0). Table 10.1 indicates the percentage of people who were classified according to the seven categories in Figure 10.1. This particular table is based on people 20 to 35 years of age.

Married people were more likely than single people to be classified as exclusively heterosexual. During Kinsey's day, single young people had fewer heterosexual outlets, which may have encouraged them to seek partners of their own gender. A gay male or lesbian sexual orientation may have influenced some to delay marriage, however. Men, overall, were more likely than women to report sexual experiences with people of their own gender.

Between the poles of the continuum lie people with various degrees of sexual interest in, and sexual experience with, people of their own gender. Thirty-seven percent of men and 13% of the women told interviewers that they had reached orgasm through sexual activity with someone of their own gender at some time after puberty. Fifty percent of men who remained single until age 35 reported they had reached orgasm through sexual activity with other males.

TABLE 10.1

Percentages of respondents aged 20 to 35 at each level of the Kinsey Continuum of Sexual Orientation

Rating Category		Females (%)	Males (%)
0:	Entirely heterosexual experience		
	Single	61–72	53–78
	Married	89–90	90–92
	Previously married	75–80	
1–6:	At least some sexual experience with people of one's own gender	11–20	18–42
2–6:	More than incidental sexual experience with people of one's own gender	6–14	13–38
3–6:	Homosexual as much or more than heterosexual	4–11	9–32
4–6:	Mostly homosexual	3–8	7–26
5–6:	Almost exclusively homosexual	2–6	5–22
6:	Exclusively homosexual	1–3	3–16

Source: From A. C. Kinsey, et al. (1953). *Sexual Behavior in the Human Female.* Philadelphia; PA: W. B. Saunders, p. 488. Copyright © 1953 by the authors. Reprinted by permission of the Kinsey Institute for Research in Sex, Gender, and Reproduction, Inc.

Many people in Kinsey's day were startled by these figures. They meant that sexual activity with people of one's own gender, especially among males, was more widespread than had been believed. But Kinsey's figures for male–male sexual activity may have been exaggerated. For instance, his finding that 37% of males had reached orgasm through sexual activity with other males was based on a sample that included a high proportion of former prisoners. Findings may also have been distorted by the researchers' efforts to recruit known gay people into their samples. Once these sources of possible bias are corrected, the incidence of male–male sexual activity drops from 37% to about 25%. (Even this percentage is much higher than people suspected at the time.) Because Kinsey's sample was not drawn at random, however, we cannot say whether it represented the general population.

A 1970 nationwide survey conducted by the Kinsey Institute obtained results (published in 1989) that were strikingly similar to Kinsey's early (adjusted) estimates. At least 20% of the randomly selected adult men in the United States had reached orgasm through male–male sexual activity at some point in their lives (Fay et al., 1989). Between 1% and 2% of men in the United States reported having male–male sexual experiences within the past year. A 1993 Louis Harris poll found that 4.4% of men and 3.6% of women reported engaging in sexual activity with a member of their own gender within the past 5 years (Barringer, 1993b). Kinsey Institute director June Reinisch (1990) examined the available evidence and estimated that more than 25% of men in the United States had had a male–male sexual experience in their teens or adult years.

Statistics concerning *past* sexual activity with a member of one's own gender can be misleading. They may represent a single episode or a brief period of adolescent experimentation. Half of the men who reported male–male sexual activity in Kinsey's sample limited it to the ages of 12 to 14. Another third had had male–male sexual experience by the age of 18, but not again.

Kinsey's research also showed that sexual behavior patterns can change, sometimes dramatically so (Sanders et al., 1990). Sexual experiences or feelings involving people of one's own gender are common, especially in adolescence, and do not mean that one will engage in sexual activity exclusively with people of one's own gender in adulthood (Bullough, 1990).

What did Kinsey find with respect to more enduring patterns of male–male and female–female sexual activity? Estimates based on an analysis of Kinsey's data (Gebhard, 1977) suggest that 13% of the men and 7% of the women, or 10% for both genders combined, were either predominantly or exclusively interested in, and sexually active with, people of their own gender for at least three years between the ages of 16 and 55.

The controversy over how many people are gay continues. Current estimates generally find lower percentages of gay people in the population than Kinsey did. Considering data drawn from studies conducted in the United States, Asia, and Pacific island countries, Milton Diamond (1993) estimates that only about 5% of men and 2% to 3% of women across different cultures have engaged in sexual activity with someone of their own gender on at least one occasion since adolescence. No cross-cultural studies show rates of sexual activity with people of the same gender as high as Kinsey's often-cited figure of 10% of predominant or exclusive sexual behavior with partners of one's own gender. Diamond also finds fewer people to have a bisexual orientation than a gay male or lesbian sexual orientation.

Research in the United States, Britain, France, and Denmark finds that about 3% of men surveyed *identify* themselves as gay (Hamer et al., 1993; Laumann et al., 1994). About 2% of the U.S. women surveyed *identify* themselves as lesbians (Janus & Janus, 1993; Laumann and others, 1994). Surveys in the United States, Britain, and France find that larger numbers of men (5% to 11%) and women (2% to 4%) report *engaging in sexual behavior* with members of their own gender within the past five years (Sell et al., 1995). Surveys show that still larger numbers of men (8% to 9%) and women (8% to 12%) report some *sexual attraction* to members of their own gender, but no sexual interaction since the age of 15 (Sell et al., 1995).

Sex surveys reveal the percentages only of people *willing to admit* to certain behaviors or sexual orientations (Cronin, 1993; Isay, 1993). "We can't count people who simply don't want to be counted" (Cronin, 1993). Surveys may omit gay people who hesitate to proclaim their sexual orientation because of social stigma or repression of their sexual feelings.

Keep in mind that the following factors affect survey results:

- The ways in which the questions are phrased (for example, do they look into sexual identity, sexual behavior, or sexual attraction—and over what period of time?)
- The social desirability of the professed behavior
- The gender of the interviewer
- The manner in which the survey was conducted, such as by means of personal interviews, phone calls, or written surveys
- The biases of respondents, such as volunteer bias

CHALLENGES TO THE KINSEY CONTINUUM Although the Kinsey continuum has been widely adopted by sex researchers, it is not universally accepted. Kinsey believed that exclusive heterosexual and gay sexual orientations lay at opposite poles of one continuum. Therefore, the more heterosexual a person is, the less gay that person is, and vice versa (Sanders et al., 1990).

Viewing gay and heterosexual orientations as opposite poles of one continuum is akin to the traditional view of masculinity and femininity as polar opposites, such that the more masculine one is, the less feminine, and vice versa. Viewing men and women as opposites has led to misunderstandings, and even hostility, between the genders (Bem, 1993). But we may also regard masculinity and femininity as independent personality dimensions. Similarly, the view of gay people and heterosexuals as opposites had also led to misunderstandings and hostility. Yet these sexual orientations may also be separate dimensions, rather than polar opposites.

Psychologist Michael Storms (1980) suggests that gay and heterosexual orientations are independent dimensions. Thus, one can be high or low on both dimensions simultaneously. Storms (1980) suggests that there are separate dimensions of responsiveness to male–female stimulation (heteroeroticism) and sexual stimulation that involves someone of the same gender (homoeroticism), as shown in Figure 10.2. According to this model, bisexuals are high in both dimensions, whereas people who are low in both are essentially asexual. According to Kinsey, bisexual individuals would be *less* responsive to stimulation

FIGURE 10.2

Heterosexuality and Homosexuality as Separate Dimensions According to this model, homosexuality and heterosexuality are independent dimensions. One can thus be high or low on both dimensions at the same time. Most people are high in one dimension. Bisexuals are high in both dimensions. People who are low in both are considered *asexual.*

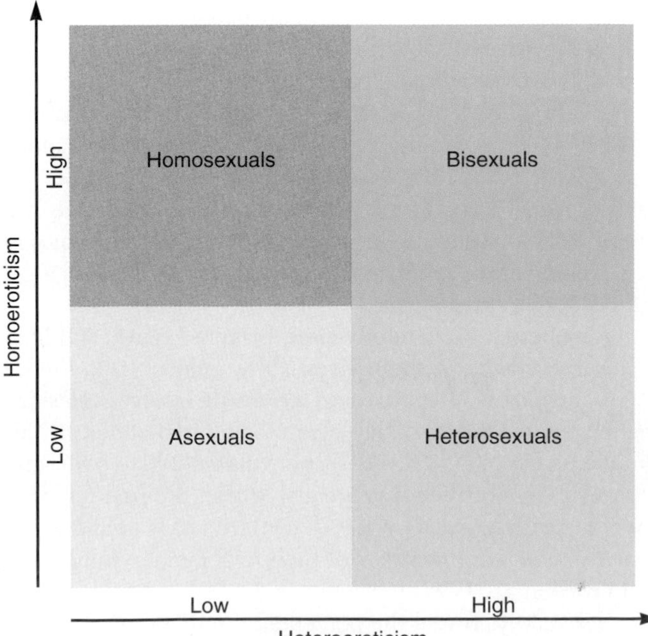

by people of the other gender than heterosexual people are, but *more* responsive than heterosexuals to stimulation by people of their own gender. According to the two-dimensional model, however, bisexual people may be just as responsive to stimulation by people of the other gender as heterosexual people are, and just as responsive to stimulation by people of their own gender as gay people are.

FANTASY AS THE MEASURE Kinsey had argued that the content of erotic fantasies was an excellent gauge of sexual orientation. To test this formulation, Storms (1980) investigated the erotic fantasies of heterosexual, gay, and bisexual people. Kinsey might have predicted that bisexual individuals would have *fewer* heteroerotic fantasies than heterosexual people and *more* homoerotic fantasies than heterosexual people. But Storms predicted that "bisexuals will report *as much* heteroerotic fantasy as heterosexual people and as much homoerotic fantasy as homosexuals" (p. 786).

Storms found that heterosexual students reported significantly more fantasies about the other gender than about their own gender. Gay students reported more frequent fantasies about their own gender. Bisexuals, as predicted, reported a high level of both kinds of fantasies. Gay students were also more likely to have fantasies about the other gender than heterosexual students were to have fantasies about their own gender. Given the stigma attached to homosexuality, it may be that heterosexual people are less likely to admit to fantasies about their own gender (even to themselves). It may also be that gay males and lesbians are influenced by the media and the culture at large, which, of course, continually portray socially desirable male–female interactions.

Whatever the reasons, Storms did find that many bisexual people show a high level of sexual interest, as measured by erotic fantasies, in both men and women. Thus, it may be that sexual interest in the other gender and sexual interest in one's own gender are independent dimensions rather than opposite poles of one continuum. Yet most sex researchers continue to use the Kinsey scale, or some derivative of it, for classifying sexual orientation (Sanders et al., 1990).

Bisexuality

> To me, I never felt like I had a stronger attraction to men or women. I didn't have a problem identifying myself as gay, but I knew that wasn't the whole picture.
>
> —A 29-year-old social worker from North Carolina
> who fell in love with a woman and then had a sexual
> relationship with his male roommate
>
> I don't limit myself to a guy or a girl. Whoever comes into my life, if we hit it off, great. That's happened with a lot of people I know. They'll say, "Guess what happened last night." It's very accepted.
>
> —A 22-year-old political science major
>
> She finally came out of the closet and said, "Mom, there's this girl and she's the most gorgeous girl you've ever seen." I said, "You're only for girls?" She said, "No, it's not just the girls. I can see a really good-looking guy, too."
>
> —The mother of a 19-year-old woman who uses a
> computer bulletin board to meet other bisexuals
>
> (Adapted from Gabriel, 1995b)

Bisexual people are attracted to both males and females. Many have a somewhat stronger attraction to one gender than to the other, yet they remain bisexual (Weinberg et al., 1994). Kinsey and his colleagues considered people rated as 3's to be bisexual (see Figure 10.1).

Depending on how one defines bisexuality, perhaps 1% to 4% of the population is bisexual. About 1% of the people (0.8% of the men and 0.9% of the women) surveyed in the NHSLS study (Laumann et al., 1994) reported having a bisexual *identity*. However, about 4% said they were sexually attracted to both women and men.

Bisexual people are sometimes said to "swing both ways," or to be "A/C-D/C" (as in "alternating current" and "direct current"). Some gay people (and some heterosexual people) believe that claims to bisexuality are a "cop-out" that people use to deny being gay. Perhaps they fear leaving their spouses or dread coming out (declaring their gay male or lesbian sexual orientation publicly). Others view bisexuality as a form of sexual experimentation with people of one's own gender by people who are mostly heterosexual. But many avowed bisexual people disagree. They report that they can maintain erotic interests in, and romantic relationships with, members of both genders. Some authors (e.g., Garber, 1995; Weinberg et al., 1994) insist that bisexuality is an authentic sexual orientation and not just a "cover" for a gay male or lesbian sexual orientation.

Some bisexual people follow lifestyles that permit them to satisfy their dual inclinations. Others feel pressured by heterosexual and gay people alike to commit themselves one way or the other (Garber, 1995; Weinberg et al., 1994). Some gay people also mask their sexual orientation by adopting a bisexual lifestyle. That is, they get married but also enter into clandestine sexual liaisons with members of their own gender.

PERSPECTIVES ON GAY MALE AND LESBIAN SEXUAL ORIENTATIONS

Gay male and lesbian sexual orientations have existed throughout history. Attitudes toward them have varied widely. They have been tolerated in some societies, openly encouraged in others, but condemned in most (Bullough, 1990). In this section we review historical and other perspectives on gay male and lesbian sexual orientations.

Historical Perspectives

In Western culture, few sexual practices have met with such widespread censure as sexual activities with members of one's own gender. Throughout much of history, male–male and female–female sexual behaviors were deemed sinful and criminal—an outrage against God and humanity (Lewis, 2000). Within the Judeo-Christian tradition, male–male sexual activity was regarded as a sin so vile that no one dared speak its name. Our legal system, grounded in this religious tradition, maintains criminal penalties for sexual practices commonly associated with male–male and female–female sex, such as anal and oral sex. Much of the criminalization of male–male and female–female sex has been directed against men (Bullough, 1990).

Jews and Christians have traditionally referred to male–male sexual activity as the sin of Sodom. Hence the origins of the term *sodomy*, which generally denotes anal intercourse (and sometimes oral–genital contact). According to the Book of Genesis, the city of Sodom was destroyed by God. Yet it is unclear what behavior incurred God's wrath. Pope Gregory III was not ambiguous, however, in his eighth-century account of the city's obliteration as a punishment for sexual activity with members of one's own gender. The Book of Leviticus was also clear in its condemnation:

> If a man lies with a man as with a woman, both of them have committed an abomination; they shall be put to death, their blood is upon them. (Leviticus 20:13)

Sexual activity with members of one's own gender was not the only sexual act that the early Christians considered sinful. Any nonprocreative sexual act was considered sinful, even within marriage. With the fall of the Roman Empire, the influence of Christianity spread across Western Europe. Christian beliefs were eventually encoded into secular law. By the late Middle Ages, most civil statutes throughout Western Europe contained penalties for nonprocreative sexual acts involving the discharge of semen, including oral or anal sex, masturbation, male–male sexual behavior, and bestiality (Boswell, 1990). Male–male and female–female sexual practices continue to be condemned by most Christian and Jewish denominations and by Islam.

The Roman Catholic Church, however, draws a distinction between a gay male or lesbian sexual orientation on the one hand, and male–male or female–female sexual behavior on the other, (Lewis, 2000). The behavior is the sin, not the orientation. The church sees sin as a temporary state from which a sinner, including a gay male or lesbian sinner, can be freed by behavioral change, contrition, and repentance.

Despite the history of opposition to gay male and lesbian sexual orientations, some churches today are performing marriages of gay couples—or, if marriage itself remains illegal, they are at least "blessing" these relationships. The St. Paul–Reformation Lutheran Church in St. Paul, Minnesota, has written a theological statement to explain why it blesses the relationships of gay couples. The church interprets the biblical creation story to mean that God did not intend for people to live alone and "that all human creatures are created for the sake of mutual companionship on this earth" (in Niebuhr, 1998). The Reverend Gary Kowalski, who "blesses" some 3 gay unions a year at the First Unitarian Universalist Society (along with performing about 20 traditional weddings), says, "The reason I perform same-sex ceremonies is I think churches and other social institutions should be in the business of encouraging long-term relationships" (Niebuhr, 1998).

Cross-Cultural Perspectives

Male–male sexual behavior has been practiced in many preliterate societies. In their review of the literature on 76 preliterate societies, Ford and Beach (1951) found that in 49 societies (64%), male–male sexual interactions were viewed as normal and deemed socially acceptable for some members of the group. The other 27 societies (36%) had sanctions against male–male sexual behavior. Nevertheless, male–male sexual activity persisted. In another cross-cultural analysis, Broude and Greene (1976) found that male–male sexual behavior was present but uncommon in 41% of a sample of 70 of the world's non-European societies. It was rare or absent in 59% of these societies. Broude and Greene also found evidence of societal disapproval and punishment of male–male sexual activity in 41% of a sample of 42 societies for which information was available.

Sexual behavior between adult males appears to be more common in societies that highly value female virginity before marriage and segregate young men and women (Davenport, 1977). Such factors may increase juvenile male–male experimentation, which may persist into adulthood.

Some societies permit or require some forms of male-male sexual activity. For example, sexual activities may be acceptable between older and younger males or between adolescents, but not between adult men. At the turn of the century, the Swans of North Africa expected all juvenile males to engage in sexual relations with older men. Fathers arranged for unmarried sons to be given to older men. Nearly all men were reported to have had such sexual relationships as boys. Later, between the ages of 16 and 20, they all married women.

Sexual activities between males are sometimes limited to rites that mark the young male's initiation into manhood. In some preliterate societies, semen is believed to boost strength and virility. Older males thus transmit semen to younger males through oral or anal sexual activities. Among the Sambian people of New Guinea, a tribe of warlike headhunters, 9- to 12-year-old males leave their parents' households and live in a "clubhouse" with other prepubertal and adolescent males. There they undergo sexual rites of passage. To acquire the fierce manhood of the headhunter, they perform fellatio on older males and drink "men's milk" (semen) (Herdt, 1981; Money, 1990; Stoller & Herdt, 1985). The initiate is enjoined to ingest as much semen as he can, "as if it were breast milk or food" (Herdt, 1981, p. 235). Ingestion of semen is believed to give rise to puberty. Following puberty, adolescents are fellated by younger males (Baldwin & Baldwin, 1989). By the age of 19, however, young men are expected to take brides and enter exclusively male–female sexual relationships.

These practices of Sambian culture might seem to suggest that the sexual orientations of males are fluid and malleable. The practices involve *behavior,* however, not *sexual orientation.* Male–male sexual behavior among Sambians takes place within a cultural context that bears little resemblance to consensual male–male sexual activity in Western society.

Ethnicity and Sexual Orientation: A Matter of Belonging

Lesbians and gay men frequently suffer the slings and arrows of an outraged society. Because of societal prejudices, it is difficult for many young people to come to terms with an emerging lesbian or gay male sexual orientation. You might assume that people who have been subjected to prejudice and discrimination—members of ethnic minority groups in the United States—would be more tolerant than others of people with a lesbian or gay male sexual orientation. However, according to psychologist Beverly Greene (1994) of St. John's University, such an assumption might not be warranted.

In an article that addresses the experiences of lesbians and gay men from ethnic minority groups, Greene (1994) notes that it is difficult to generalize about ethnic groups in the United States. For example, African Americans may find their cultural origins in the tribes of West Africa, but they have also been influenced by Christianity and by the local subcultures of their North American towns and cities. Native Americans represent hundreds of tribal groups, languages, and cultures. By and large, however, a lesbian or gay male sexual orientation is rejected by ethnic minority groups in the United States. Lesbians and gay males are pressured to keep their sexual orientations a secret or to move to communities where they can live openly without sanction.

Within traditional Latino and Latina American culture, the family is the primary social unit. Men are expected to support and defend the family, and women are expected to be submissive, respectable, and deferential to men (Morales, 1992). Because women are expected to remain virgins until marriage, men sometimes engage in male–male sexual behavior without considering themselves gay (Greene, 1994). Latino and Latina American culture frequently denies the sexuality of women. Thus, women who label themselves lesbians are doubly condemned—because they are lesbians and because they are confronting others with their sexuality. Because lesbians are independent of men, most Latino and Latina American heterosexual people view Latina American lesbians as threats to the tradition of male dominance (Trujillo, 1991).

Asian American cultures emphasize respect for elders, obedience to parents, and sharp distinctions in gender roles (Chan, 1992). The topic of sex is generally taboo within the family. Asian Americans, like Latino and Latina Americans, tend to assume that sex is unimportant to women. Women are also considered less important than men. Open admission of a lesbian or gay male sexual orientation is seen as rejection of traditional cultural roles and as a threat to the continuity of the family line (Chan, 1992; Garnets & Kimmel, 1991).

Because many African American men have had difficulty finding jobs, gender roles among African Americans have been more flexible than those found among European Americans and most other ethnic minority groups (Greene, 1994). Nevertheless, the African American community appears to strongly reject gay men and lesbians, pressuring them to remain secretive about their sexual orientations (Gomez & Smith, 1990; Poussaint, 1990). Greene (1994) hypothesizes that a number of factors influence African Americans to be hostile toward lesbians and gay men. One is strong allegiance to Christian beliefs

The prepubertal Sambian male does not *seek* sexual liaisons with other males. He is removed from his home and thrust into male–male sexual encounters by older males (Baldwin & Baldwin, 1989).

Little is known about female–female sexual activity in non-Western cultures. Evidence of female–female sexual behavior was found by Ford and Beach in only 17 of the 76 societies they studied. Perhaps it was more difficult to acquire data about female sexuality. Perhaps female sexual behavior in general, not just sexual activity with other females, was more likely to be repressed. Of course, it is also possible that women are less likely than men to develop sexual interests in, or romantic relationships with, members of their own gender. Whatever the reasons, this cross-cultural evidence is consistent with data from our own culture. Here, too, males are more likely than females to develop sexual interests in, or romantic relationships with, members of their own gender (Laumann et al., 1994).

Cross-Species Perspectives

Many of us have observed animals engaging in sexual behaviors that resemble male–male or female–female contacts among humans, such as mounting others of their own sex. But what do these behaviors signify?

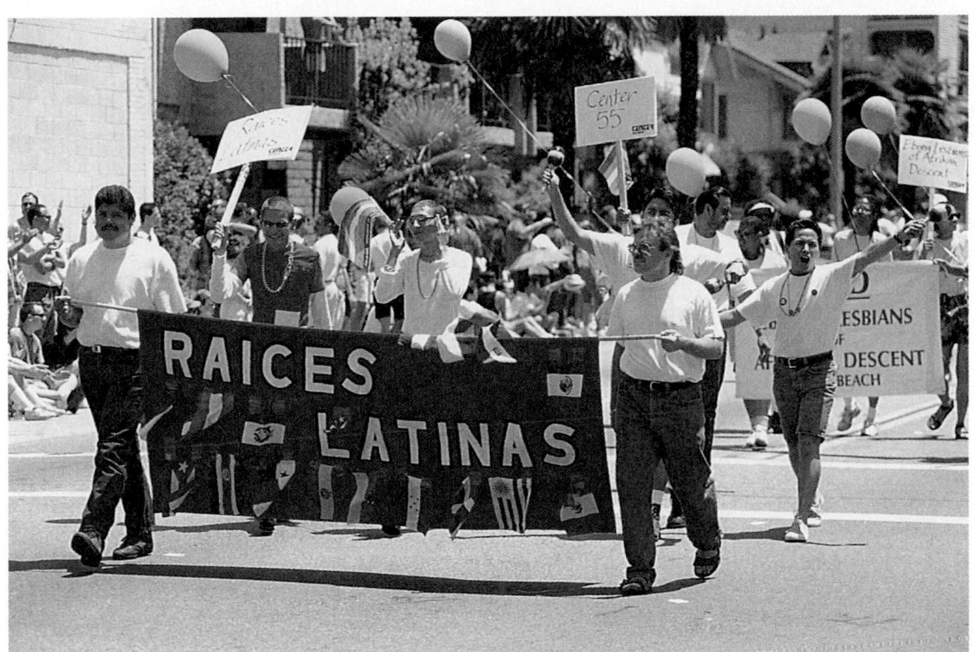

Ethnicity and Sexual Orientation. The experiences of lesbians and gay men can differ according to their ethnicity. For example, traditional Latino and Latina American culture supports strong differences in gender roles. Men are expected to support and defend the family. Women are expected to be submissive, respectable, and deferential to men. Latino and Latina American culture also often denies the sexuality of women. Thus, Latina American lesbians are multiply condemned—because they are lesbians, because they confront others with their sexuality, and because their independence from men is perceived as a threat to the tradition of male dominance.

and biblical scripture. Another is internalization of the dominant culture's stereotyping of African Americans as highly sexual beings. That is, many African Americans may feel a need to assert their sexual "normalcy."

Prior to the European conquest, sex may not have been discussed openly by Native Americans, but it was generally seen as a natural part of life. Individuals who incorporated both traditional feminine and masculine styles were generally accepted and even admired. The influence of the religions of the colonists led to greater rejection of lesbians and gay men, and pressure to move off the reservation to the big city (Greene, 1994). Native American lesbians and gay men, like Asian American lesbians and gay men, thus often feel doubly removed from their families.

If any generalization is possible, it may be that lesbians and gay men find more of a sense of belonging in the gay community than in their ethnic communities.

A male baboon may present his rear and allow himself to be mounted by another male. This behavior may resemble anal intercourse among gay men. Is the behavior sexually motivated, however? Mounting behavior among male baboons may represent a type of dominance ritual in which lower-ranking males adopt a submissive (feminine) posture to ward off attack from dominant males (Nadler, 1990). (Some male–male acts among people also involve themes of dominance, as in the case of a dominant male prisoner forcing a less dominant one to submit to anal intercourse.) In other cases, male baboons may be seeking favors or protection from more dominant males (Nadler, 1990). Among juvenile animals, male–male behaviors may also be a form of play. Females may also attempt to mount other females, but here too, the motives may not be the same as those of humans.

Sexual motivation appears to play a role in some, but not all, male–male and female–female sexual interactions among animals. Fellatio and anal intercourse to ejaculation among juvenile male orangutans may be a case in point, as may be thrusting by one adult female gorilla against another. Some male–male encounters among rhesus monkeys also appear to be sexually motivated. However, researchers have not found prolonged, exclusive male–male sexual activity among monkeys and apes when male–female opportunities are available.

Attitudes Toward Sexual Orientation in Contemporary Society

> Toni, a 20-year-old lesbian college student, was incredulous that her parents believed she had *chosen* her sexual orientation. They saw her lesbianism as linked to the general rebelliousness that had been stirred by attending a school known for its "radical" image. Toni was infuriated that they believed that she would "get over" her lesbian sexual orientation once she decided to accept adult responsibilities.
>
> —The authors' files

Historically speaking, negative attitudes toward gay people have pervaded our society. Male–male and female–female sexual behavior have at worst been viewed as a crime against God or against "nature." A national survey of males in their late teens (aged 15 to 19) showed that 9 out of 10 felt that sex between men was "disgusting"; 3 out of 5 could not even see themselves being friends with a gay man (Marsiglio, 1993b). Young males tend to be the harshest in their judgments of gay people, but a national poll conducted in June 1998 found that 59% of American adults believe that homosexuality is morally wrong, compared to 35% who say it is not wrong (Berke, 1998). (The remaining 6% had no answer or said they didn't know.)

Still, there are strong signs that Americans have grown more tolerant of gay people over the past few decades. For example, a poll conducted in June 1977 found that 56% of Americans said that gay people should have equal rights in terms of job opportunities (33% said they should not). In October 1989, 71% said gay people should have equal rights (compared to 18% who said they should not). By November 1996, 84% of Americans endorsed equal rights for gay people, and only 12% were opposed (Berke, 1998).

In 2000, Vermont became the first state to allow "civil unions" that grant gays the rights of married people (Wong, 2000). However, only about one American in three believes that such unions should be considered marriages. A poll by *The New York Times* found that only 28% of Americans agreed that "marriage between two people of the same sex should be legal" (Eggers, 2000). An Associated Press poll conducted in the same year found that 34% supported gay marriages (Lester, 2000). However, the direction of change is clearly toward becoming more liberal on the issue: Whereas only 13% of people older than 70 in the Eggers study (2000) reported that such unions should be legal, 42% of those under the age of 30 endorsed them. The Associated Press poll showed a similar age discrepancy: 54% of respondents aged 18 to 34 said gay people should be allowed to get married, compared with only 14% of those over 65 (Lester, 2000).

Interestingly, the word *marriage* seems to be a major barrier. The Associated Press poll of a national sample of 1,012 adults in the United States split the sample so that half were asked whether they approved of gay people getting married, whereas the other half were asked whether they approved of "domestic partnerships" that would give gay couples the same rights as married couples. As Table 10.2 shows, more people approved of domestic partnerships than marriage for gay couples. Despite the fact that a majority of those polled would not allow gay couples to form domestic partnerships that give the partners the same rights as married partners, a majority would allow domestic partners specific rights, as shown in Table 10.3. (Something of a "disconnect"? Ralph Waldo Emerson wrote, "A foolish consistency is the hobgoblin of little minds.") Perhaps the generalization is more difficult to handle than the specifics. That is, people may be more wary of giving gays carte blanche than health benefits.

Although most Americans believe that gay people should have equal access to jobs, some would bar them from teaching and similar activities because they believe that gay people, given the chance, will seduce children and recruit them into a gay lifestyle. Such beliefs have been used to prevent gay couples from becoming adoptive or foster parents and to deny them custody or visitation rights to their own children, following divorce. Some people who would bar gays from interacting with children fear that the children will be molested. Yet more than 90% of cases of child molestation involve heterosexual male

TABLE 10.2

The effect of wording in a poll on gay rights

Question	Percent Saying "Should"	Percent Saying "Should Not"	Percent Saying "Don't Know"	Percent Refusing to Answer
In general, do you think gays and lesbians should or should not be allowed to be legally married?	34	51	11	3
In general, do you think gays and lesbians should or should not be allowed to form a domestic partnership that would give the same-sex couple the same rights and benefits as opposite-sex marriage?	41	46	11	3

Source: Will Lester (2000, May 31). "Poll: Americans Back Some Gay Rights." *The Associated Press online.* Reprinted with permission of The Associated Press.

assailants (Gelles & Cornell, 1985; Gordon & Snyder, 1989). Nor are children who are reared or taught by gay men or lesbians more likely than others to become gay themselves. One study found that 36 of 37 children who were reared by lesbian or transsexual couples developed a heterosexual orientation (Green, 1978). Nor has it been shown that children reared by gay parents are more likely to encounter gender-identity conflicts (Allen & Burrell, 1996; Patterson, 1995). Nonetheless, 56% of respondents to a 1997 nationwide poll said it was a "bad thing" for gay couples to rear children (Berke, 1998). (Thirty-one percent believed that the sexual orientation of the parents made no difference in child rearing.)

We began this section with the case of Toni, whose parents assumed that she would change her lesbian sexual orientation once she "outgrew" her rebelliousness. Toni's parents assumed, as do many people, that Toni had *chosen* to be gay. As we will see in the section on biological perspectives, research evidence is accumulating to support the view that inborn biological factors play a strong role in the development of sexual orientation. Interestingly, people who believe that sexual orientation is biologically determined and inborn are more tolerant of gay people than are those who see sexual orientation as stemming from upbringing or other environmental factors. Over the years, more people have believed in environmental than in biological causes of sexual orientation (Berke, 1998), but the recent trend seems to favor the inborn, biological point of view.

TABLE 10.3

Percent of Americans who would approve of certain rights for gay people in domestic partnerships

Would Respondent Approve of—	Percent in Favor	Percent Opposed	Percent Who Say "Don't Know"	Percent Who Refuse to Answer
Providing health coverage to gay partners?	53	37	7	3
Providing Social Security benefits to gay partners?	50	41	6	3
Providing inheritance rights to gay partners?	56	32	9	3

Source: Will Lester (2000, May 31). "Poll: Americans Back Some Gay Rights." *The Associated Press online.* Reprinted with permission of The Associated Press.

For example, a nationwide poll conducted in June 1977 found that only 13% of Americans believed that gay people were "born with" their sexual orientation, compared to 56% who favored environmental causes. In October 1989, 18% of Americans said that gay people were born with their orientation, compared to 48% who favored an environmental explanation. In June 1998, 31% of Americans believed that gay people were born with their sexual orientations (Berke, 1998); 47% still favored the environmental approach. And an Associated Press poll conducted in 2000 found that 30% of Americans believe that gay people are born that way, compared to 46% who say that gays "choose" to be gay. The overall trend seems clear: More people are coming to look upon sexual orientation as something one is born with. Thus, fewer people are likely to believe that young people can be "seduced" into one sexual orientation or another.

HOMOPHOBIA

- **Homophobia** takes many forms, including
- Use of derogatory names (such as *queer, faggot,* and *dyke*).
- Telling disparaging "queer jokes."
- Barring gay people from housing, employment, or social opportunities.
- Taunting (verbal abuse).
- **Gay bashing** (physical abuse).

Homophobia derives from root words that mean "fear of homosexuals." Although homophobia is more common among heterosexuals, gay people can also be homophobic.

Although some psychologists link homophobia to fear of a gay male or lesbian sexual orientation within oneself, homophobic attitudes may also be embedded within a cluster of stereotypical gender-role attitudes toward family life (Cotten-Huston & Waite, 2000). These attitudes support male dominance and the belief that it is natural and appropriate for women to sacrifice for their husbands and children (Cotten-Huston & Waite, 2000; Marsiglio, 1993b). People who have a strong stake in maintaining stereotypical gender roles may feel more readily threatened by the existence of the gay male or lesbian sexual orientation, because gay people appear to confuse or reverse these roles. Men have more of a stake in maintaining the tradition of male dominance, so perhaps it is not surprising that college men are more intolerant of gay males than college women are (Schellenberg et al., 1999).

Homophobic attitudes are more common among males who identify with a traditional male gender role and those who hold a fundamentalist religious orientation (Cotten-Huston & Waite, 2000; Marsiglio, 1993b). Similarly, researchers find that college students who hold a conservative political orientation tend to be more accepting of negative attitudes toward gay people than are liberal students (Cotten-Huston & Waite, 2000). Other university samples find male students to be more homophobic than women (Kunkel & Temple, 1992; Schellenberg et al., 1999). In addition, business and science students at a Canadian University were more intolerant of gay people than students in the arts and social sciences (Schellenberg et al., 1999).

Generally speaking, heterosexual men are less tolerant of gay people than are heterosexual women (Kerns & Fine, 1994; Seltzer, 1992; Whitley & Kite, 1995). Perhaps some heterosexual men are threatened by the possibility of discovering male–male sexual impulses within themselves (Freiberg, 1995). In an outcome consistent with this view, Kite (1992) found that heterosexual males tend to hold more negative attitudes toward gay men than toward lesbians.

At least some homophobic men may have homoerotic impulses of which they are unaware. Denial of these impulses may be connected with their fear and disapproval of gay males. Henry Adams and his colleagues (1996) showed men sexually explicit videotapes of male–female, female–female, and male–male sexual activity and measured their sexual response by means of the penile plethysmograph. The plethysmograph measures penile circumference (size of erection). Participants were also asked to report how sexually aroused they felt in response to the videos. The men were also evaluated for their attitude toward gay males—homophobic or not homophobic. Men who were not homophobic

Homophobia A cluster of negative attitudes and feelings toward gay people, including intolerance, hatred, and fear. (From Greek roots that mean "fear" [of members of the] "same" [gender]).

Gay bashing Violence against homosexuals.

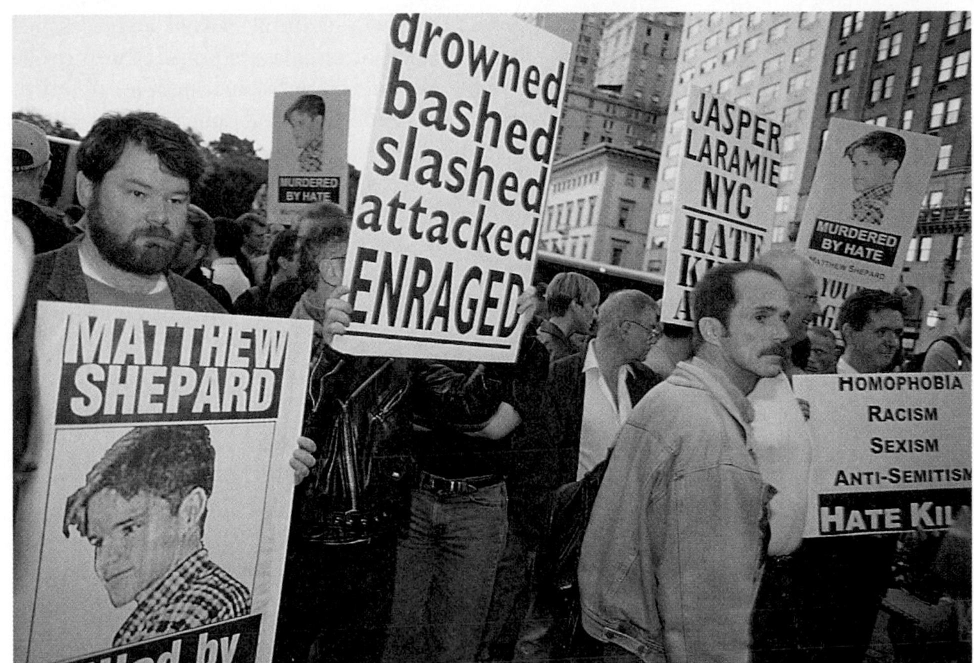

Protesting Homophobia. Gay men and lesbians are frequently assaulted because of their sexual orientation. Hatred of gays is termed *homophobia*, although the roots of the word mean "fear of gays." Why do so many people hate gay men and lesbians? Why was a college student, Matthew Shepard, killed by homophobes?

were sexually aroused, according to their penile circumference, only by videos of male–female and female–female sexual activity. The homophobic viewers were also aroused, in terms of penile circumference, by the video of male–male sexual activity. However, the homophobic men also reported that they did not *feel* aroused by the male–male sexual activity, a result that suggests that they were out of touch with their biological response.

GAY BASHING AND THE AIDS EPIDEMIC Although strides toward social acceptance of gay people have been made since Kinsey's day, the advent of AIDS has added fuel to the fire of hatred and prejudice. When AIDS first appeared, it struck primarily the gay male community. Some people in the larger society believed that the epidemic was a plague sent by God to punish gay people for sinful behavior.

The spread of the AIDS epidemic has been accompanied by a dramatic rise in the incidence of gay bashing (Freiberg, 1995). Perhaps the epidemic has served as a pretext for some people to attack gay males, whom they blame for spreading the disease.

Gay bashing also occurs on college campuses. A large percentage (77%) of 121 lesbian and gay male undergraduate students in one survey at Pennsylvania State University reported that they had been verbally insulted. Nearly 1 in 3 (31%) reported being chased or followed (D'Augelli, 1992a). Nearly 20% said they had been physically assaulted. Most of the victimizers were fellow students. Few of these incidents were reported to the authorities.

Students, too, tend to hold gay men more responsible for behaviors that put themselves at risk of contracting AIDS. One study found that students assigned a greater amount of blame to a hypothetical person who had contracted AIDS when the person was identified as a gay male than when the person was identified as a heterosexual male (Anderson, 1992).

Ironically, the percentage of gay men who account for new cases of AIDS has declined. By contrast, the percentages of persons who contract AIDS by sharing contaminated hypodermic syringes while injecting drugs, and via male–female sexual intercourse, have soared.

SEXUAL ORIENTATION AND THE LAW During the past generation, gay people have organized effective political groups to fight discrimination and to overturn the sodomy laws that have traditionally targeted them. Despite their success, sodomy laws are still on the books in about half of our states. Sodomy laws prohibit "unnatural" sexual acts, even between consenting adults. A gay male or lesbian sexual orientation is not illegal in itself. However, certain sexual acts that many gay people (and heterosexual people) practice, such as anal intercourse and oral–genital contact, fall under the legal definition of sodomy in many states. Sodomy laws also typically prohibit sexual contacts with animals. Although sodomy laws are usually intended to apply equally to all adults—married or unmarried, heterosexual or gay—the vast majority of prosecutions have been directed against gay people.

A 1986 Supreme Court decision (*Hardwick v. Bowers*) let stand a Georgia sodomy law that makes oral–genital sexual contact and anal–genital sexual contact crimes punishable by up to 20 years in prison, even when engaged in by consenting adults. The decision was a setback to gay rights organizations, which had looked to the Supreme Court to overturn state sodomy laws. Instead, the court held that states retain the right to enact sodomy laws and impose criminal penalties.

Many other countries, including Mexico, Holland, Italy, Spain, England, France, and the Scandinavian countries, have decriminalized male–male and female–female sexual activity (Carrera, 1981). In Canada, a consensual sexual act is prosecutable under sodomy laws only if one of the participants is under 21 years of age or the act is performed in public.

GAY ACTIVISM Nowhere in the United States have gay people been more politically effective than in San Francisco. They are well represented on the city police force and in other public agencies. The coming out of many gay people and their flocking to more tolerant urban centers have rendered them formidable political forces in these locales.

The Mattachine Society, named after gay medieval court jesters, was the first powerful gay rights organization. At first, membership was kept secret to protect members' social status in the community. Today members' names are published. Founded in Los Angeles in 1950, the Mattachine Society now has chapters in most metropolitan areas. The society publishes the *Mattachine Newsletter* and *Homosexual Citizen*. Another newsletter, the *Advocate*, is the nation's best-known gay newsletter.

The largest lesbian organization is the Daughters of Bilitis, named after the Greek courtesan who was loved by the lesbian poet Sappho. Founded in 1956, the group provides a forum for sharing social experiences and pursuing equal rights. The organization's newsletter is the *Ladder*.

The AIDS epidemic has had a profound effect on the political agenda of gay-rights organizations. These organizations combat the AIDS epidemic on several fronts:

1. They lobby for increased funding for AIDS research and treatment.
2. They educate the gay and wider communities on the dangers of high-risk sexual behavior.
3. They encourage gay men and others to adopt safer sex practices, including the use of condoms.
4. They protect the civil rights of people with AIDS and carriers of HIV (the virus that causes AIDS) with respect to employment, housing, and medical and dental treatment.
5. They provide counseling and support services for people with HIV/AIDS.

Some gay-rights organizations work within the mainstream political process. Others, such as the militant *Act-Up* organization, have taken a more strident and confrontational stance to secure more funding for AIDS research and treatment.

Some militant gay groups use the method of **outing** to combat discrimination. In this method, gay people unmask other gay people without their consent, forcing them "out of the closet." Many gay people protest this practice vehemently, terming it a "gay witch hunt."

Outing The revelation of the identities of gay people by other gay people. The method is intended to combat discrimination against gay people by forcing individuals out of the closet and into the fray.

STEREOTYPES AND SEXUAL BEHAVIOR Among heterosexual people, sexual aggressiveness is linked to the masculine gender role. Sexual passivity is linked to the feminine role. Some heterosexual people assume (often erroneously) that in gay male and lesbian relationships, one partner consistently assumes the masculine role in sexual relations, and the other the feminine role.

Many gay couples vary the active and passive roles, however. Among gay male couples, for example, roles in anal intercourse (*inserter* versus *insertee*) and in fellatio are often reversed. Contrary to popular assumptions, sexual behavior between lesbians seldom reflects distinct **butch–femme** gender roles. Most lesbians report both providing and receiving oral–genital stimulation. Typically, partners alternate roles or simultaneously perform and receive oral stimulation. Many gay people claim that the labels *masculine* and *feminine* only represent "the straight community's" efforts to pigeonhole them in terms that "straights" can understand.

Biological Perspectives

Is sexual orientation an *inborn* trait that is transmitted genetically, like eye color or height? Does it reflect hormonal influences? Biological perspectives focus on the possible roles of evolution, genetics, and hormonal influences in shaping sexual orientation.

THE EVOLUTIONARY PERSPECTIVE It might seem ironic that evolutionary theorists have endeavored to explain gay male and lesbian sexual orientations. After all, gay males and lesbians are not motivated to engage in sexual activity with members of the other gender. How, then, can these sexual orientations confer any evolutionary advantage?

To answer this question, we must look to the group or the species rather than the individual. Kirkpatrick (2000) suggests that male–male and female–female sexual behaviors derive from individual selection for reciprocal *altruism*. That is, strong male–male and female–female alliances have advantages for *group* survival in that they bind group members together emotionally. This hypothesis remains speculative.

GENETICS AND SEXUAL ORIENTATION Considerable evidence exists that gay male and lesbian sexual orientations run in families (Bailey et al., 1999; Bailey et al., 2000a; Dawood et al., 2000; Kendler et al., 2000). In one study, for example, 22% of the brothers of a sample of 51 predominantly gay men were either gay or bisexual themselves. This is nearly four times the proportion expected in the general population (Pillard & Weinrich, 1986). Although such evidence is consistent with a genetic explanation, families also share a common environment.

Twin studies also shed light on the possible role of heredity (Kendler et al., 2000). **Monozygotic (MZ) twins,** or identical twins, develop from a single fertilized ovum and share 100% of their heredity. **Dizygotic (DZ) twins,** or fraternal twins, develop from two fertilized ova. Like other brothers and sisters, DZ twins share only 50% of their heredity. Thus, if a gay male or lesbian sexual orientation were transmitted genetically, it should be found about twice as often among identical twins of gay people as among fraternal twins. Because MZ and DZ twins who are reared together share similar environmental influences, differences in the degree of **concordance** for a given trait between the types of twin pairs are further indicative of genetic origins.

Several studies have identified gay men who had either identical (MZ) or fraternal (DZ) twin brothers in order to examine the prevalence of a gay male sexual orientation in their twin brothers (Kendler et al., 2000). In an early study, Kallmann (1952) found 100% concordance for a gay male sexual orientation among the identical twin brothers of gay men, compared to 12% concordance for fraternal twin pairs in which one of the brothers was identified as gay. This seemed strong evidence indeed of genetic factors in sexual orientation. However, later studies have found much lower concordance rates among MZ twins (Bailey & Pillard, 1991; Kendler et al., 2000). These studies found many instances in which one of a pair of identical twins was gay and the other was not. Still, MZ twins do appear to have a higher concordance rate for a gay male sexual orientation than DZ twins. Bailey and Pillard (1991) reported a 52% concordance rate among MZ twin pairs versus a 22% rate among DZ twin pairs in their sample. In another study, researchers reported a concordance rate for a gay sexual orientation of 66% in MZ twins, compared to 30% among DZ twins (Whitam et al., 1993).

Butch A lesbian who assumes a traditional masculine gender role.

Femme A lesbian who assumes a traditional feminine gender role.

Monozygotic (MZ) twins Twins who develop from the same fertilized ovum; identical twins.

Dizygotic (DZ) twins Twins who develop from different fertilized ova; fraternal twins.

Concordance Agreement.

Bear in mind that MZ twins are more likely to be dressed alike and treated alike than DZ twins. Thus, their greater concordance for a gay sexual orientation may at least in part reflect environmental factors (Kendler et al., 2000). The greater concordance between MZ twins may signify only that MZ twins are likely to react more similarly than DZ twins to the same *environmental* influences (Bancroft, 1990).

Researchers have found evidence linking a region on the X sex chromosome to a gay male sexual orientation (Bailey et al., 1999). One group of researchers (Hamer et al., 1993) found that gay males in a sample of 114 gay men were more likely to have gay male relatives on their mothers' side of the family than would be expected on the basis of the prevalence of a gay male sexual orientation in the general population. Yet they did not have a greater than expected number of gay male relatives on the paternal side of the family. This pattern of inheritance is consistent with genetic traits, such as hemophilia, that are linked to the X sex chromosome, which men receive from their mothers.

The researchers then examined the X sex chromosome in 40 pairs of gay male, non-twin brothers. In 33 of the pairs, the brothers had identical DNA markers on the end tip of the X chromosome. For brothers overall in the general population, about half would be expected to have inherited this chromosomal structure. It is suspected, therefore, that this chromosomal region may hold a gene that predisposes men to a gay male sexual orientation.

The researchers cautioned that they had not found a particular gene linked to sexual orientation, just a general location where the gene may be found. Nor do scientists know how such a gene or combination of genes might account for sexual orientation. Perhaps a particular gene governs the development of proteins that sculpt parts of the brain in ways that favor the development of a gay male sexual orientation. On the other hand, a number of the gay brothers (7 of the 40 pairs) did *not* share the chromosomal marker.

Sexologist John Money (1994) agrees that genetic factors play a role in sexual orientation but believes that they do not fully govern it. Most researchers believe that sexual orientation is determined by the interaction of biological and experiential factors (Barlow & Durand, 1995).

HORMONAL INFLUENCES AND SEXUAL ORIENTATION Sex hormones strongly influence the mating behavior of other species (Crews, 1994). Researchers have thus looked into the possible role of hormonal factors in determining sexual orientation in humans.

Testosterone is essential to male sexual differentiation. Thus, levels of testosterone and its by-products in the blood and urine have been studied as possible influences on sexual orientation. Research has failed to connect sexual orientation in either gender with differences in the levels of either male or female sex hormones in adulthood (Friedman & Downey, 1994). In adulthood, testosterone appears to have **activating effects.** That is, it affects the intensity of sexual desire, but not the preference for partners of the same or the other gender (Whalen et al., 1990).

What of the possible *prenatal* effects of sex hormones? Pregnant rats in experiments were given anti-androgen drugs that block the effects of testosterone. When the drugs were given during critical periods in which the fetuses' brains were becoming sexually differentiated, male offspring were likely to show feminine mating patterns as adults (Ellis & Ames, 1987). The adult males became receptive to mounting attempts by other males and failed to mount females.

Do prenatal sex hormones play a similar role in determining sexual orientation in people? There is suggestive evidence. For example, Meyer-Bahlburg and his colleagues (1995) interviewed groups of women exposed prenatally to DES, a synthetic estrogen. They found that these women were more likely to be rated as lesbian or bisexual than women who were not exposed to DES. We do know that the genitals of gay people differentiate prenatally in accordance with their chromosomal gender (Whalen et al., 1990). It remains possible that imbalances in prenatal sex hormones cause brain tissue to be sexually differentiated in one direction, even though the genitals are differentiated in the other (Collaer & Hines, 1995).

Activating effects Those effects of sex hormones that influence the level of the sex drive, but not sexual orientation.

Truth or Fiction? **REVISITED**

It is not true that gay males and lesbians suffer from hormonal imbalances. Gay males and lesbians do not show imbalances of sex hormones in adulthood. However, the role of prenatal hormonal factors in the development of sexual orientation remains an open question. ■

THE STRUCTURE OF THE BRAIN Evidence suggests that there may be structural differences between the brains of heterosexual and gay men. In 1991, Simon LeVay, a neurobiologist at the Salk Institute in La Jolla, California, performed out autopsies on the brains of 35 AIDS victims—19 gay men and 16 presumably heterosexual men. He found that a segment of the hypothalamus—specifically, the *third interstitial nucleus of the anterior hypothalamus*—in the brains of the gay men was less than half the size of that same segment in the heterosexual men. The same brain segment was larger in the brain tissues of heterosexual men than in brain tissues obtained from a comparison group of 6 presumably heterosexual women. No significant differences in size were found between the brain tissues of the gay men and the women, however.

LeVay's findings are intriguing but preliminary. We do not know, for example, whether the structural differences he found are innate. Nor do the findings prove that biology is destiny. As Richard Nakamura, a scientist with the National Institute of Mental Health, commented, "This [LeVay's findings] shouldn't be taken to mean that you're automatically [gay] if you have a structure of one size versus a structure of another size" (Angier, 1991).

The belief that sexual orientation is innate, or inborn, has many adherents in both the scientific and general communities. Support for the possible influences of prenatal hormonal factors in "sculpting" the brain in a masculine or feminine direction is based largely on animal studies, however. Direct evidence with people is lacking. We must also be careful in generalizing results from other species to our own.

Psychological Perspectives

Do family relationships play a role in the origins of sexual orientation? What are the effects of childhood sexual experiences? Psychoanalytic theory and learning theory provide two of the major psychological approaches to understanding the origins of sexual orientation.

PSYCHOANALYTIC VIEWS Sigmund Freud, the originator of psychoanalytic theory, believed that children enter the world **polymorphous perverse.** That is, before they internalize social inhibitions, children are open to all forms of sexual stimulation. However, through proper resolution of the Oedipus complex, a boy will forsake his incestuous desires for his mother and come to identify with his father. As a result, his erotic attraction to his mother will eventually be transferred, or **displaced,** onto more appropriate *female* partners. A girl, through proper resolution of her Electra complex, will identify with her mother and seek erotic stimulation from men when she becomes sexually mature.

In Freud's view, a gay male or lesbian sexual orientation results from failure to resolve the Oedipus complex successfully by identifying with the parent of the same gender. In men, faulty resolution of the Oedipus complex is most likely to result from the so-called classic pattern of an emotionally "close-binding" mother and a "detached-hostile" father. A boy reared in such a family may come to identify with his mother and even to "transform himself into her" (Freud, 1922/1959, p. 40). He may thus become effeminate and develop sexual interests in men.

Freud believed that the mechanism of unresolved **castration anxiety** plays a role in a gay male sexual orientation. By the time the Oedipus complex takes effect, the boy will have learned from self-stimulation that he can obtain sexual pleasure from his penis. In his youthful fantasies, he associates this pleasure with mental images of his mother. Similarly, he is likely to have learned that females do not possess a penis. The psychoanalyst theorizes that somewhere along the line, the boy may also have been warned that his penis will be removed if he plays with himself. From all this, the boy may surmise that females—including his mother—once had penises but that they were removed.

Polymorphous perverse In psychoanalytic theory, being receptive to all forms of sexual stimulation.

Displacement In psychoanalytic theory, a defense mechanism that allows one to transfer unacceptable wishes or desires onto more appropriate or less threatening objects.

Castration anxiety In psychoanalytic theory, a man's fear that his genitals will be removed. Castration anxiety is an element of the Oedipus complex and is implicated in the directionality of erotic interests.

During the throes of the Oedipus complex, the boy unconsciously comes to fear that his father, his rival in love for the mother, will retaliate by removing the organ that the boy has come to associate with sexual pleasure. His fear causes him to repress his sexual desire for his mother and to identify with the potential aggressor—his father. The boy thus overcomes his castration anxiety and is headed along the path of adult heterosexuality.

If the Oedipus complex is not successfully resolved, castration anxiety may persist. When sexually mature, the man will not be able to tolerate sex with women. Their lack of a penis will arouse unconscious castration anxiety within himself. According to psychoanalytic theory, the boy may unconsciously associate the vagina with teeth, or other sharp instruments, and be unable to perform sexually with a woman.

The supposed Electra complex in little girls follows a somewhat different course. Freud believed that little girls become envious of boys' penises, because they lack their own. This concept of **penis envy** was one of Freud's most controversial beliefs. In Freud's view, jealousy leads little girls to resent their mothers, whom they blame for their anatomic "deficiency," and to turn from their mothers to their fathers as sexual objects. They now desire to possess the father, because the father's penis provides what they lack. But incestuous desires bring the girl into competition with her mother. Motivated by fear that her mother will withdraw her love if these desires persist, the girl normally represses them and identifies with her mother. She then develops traditional feminine interests and eventually seeks erotic stimulation from men. She supplants her childhood desire for a penis with a desire to marry a man and bear children. The baby, emitted from between her legs, serves as the ultimate penis substitute.

A girl who does not resolve her penis envy in childhood may "manifest homosexuality, . . . exhibit markedly masculine traits in the conduct of her later life, choose a masculine vocation, and so on" (Freud, 1922/1959, p. 50). The residue of this unresolved complex is continued penis envy, the striving to become a man by acting like a man and seeking sexual satisfaction with women: lesbianism.

In Freud's view, a gay male or lesbian sexual orientation is one result of assuming the gender role normally taken up by the other gender. The gay male is expected to be effeminate, the lesbian masculine. But, as noted in research into gender-typed behavior and sexual orientation, this stereotypical view of gay people is far from universal. Nor do the behaviors stereotypically associated with the other gender that are found in some adult gay men or women necessarily derive from Oedipal problems. Biological and other psychosocial factors may be involved.

A nagging problem is assessing the validity of Freudian theory is that many of its concepts, such as castration anxiety and penis envy, are believed to operate at an unconscious level. As such, they lie beyond the scope of scientific observation and measurement. We cannot learn whether boys experience castration anxiety by asking them, because the theory claims that **repression** will keep such anxieties out of awareness. Nor can we directly learn about penis envy by interviewing girls.

Truth or Fiction? **REVISITED**

Actually, the idea that male–female intercourse arouses "castration anxiety" in gay males has not been scientifically demonstrated.

RESEARCH ON PSYCHOANALYTIC THEORIES OF SEXUAL ORIENTATION Psychoanalysts—psychotherapists trained in the Freudian tradition—have researched sexual orientation largely through case studies. These studies have been carried out almost exclusively with gay males who were in therapy at the time.

In 1962, psychoanalyst Irving Bieber reported the results of questionnaires filled out by 77 psychiatrists on 106 gay male clients. In 1976, he reported the results of similar surveys of gay men in therapy.

Bieber claimed to find the "classic pattern" among gay males: Their orientation was determined by having a dominant, "smothering" mother and a passive, detached father. As the clients described them, the mothers were overprotective, seductive, and jealous of their sons. The fathers were aloof, unaffectionate, and hostile toward them. The father,

Penis envy In psychoanalytic theory, the girl's wish to have a penis.

Repression In psychoanalytic theory, the most basic defense mechanism through which threatening ideas and impulses are ejected from conscious awareness.

being detached, may have failed to buffer the close mother–son relationship. The mother's relationship with the father was typically disturbed, and the mother may have substituted a "close-binding" relationship with her son. This classic pattern was believed to result in the boy's developing a fear of sexual contacts with women. Though he would be unconscious of this dynamic, for him, any female partner would unconsciously represent his mother. Desire for her would stir unconscious fears of retaliation by the father—that is, castration anxiety.

Bieber's findings may be criticized on several grounds:

1. The study participants were all in analysis, and many wanted to become heterosexual. Thus we cannot generalize the results to well-adjusted gay men.
2. The analysts may have chosen cases that confirmed their theoretical perspectives. Or perhaps, over the course of analysis, clients were subtly encouraged to provide "evidence" that was consistent with psychoanalytic theory.
3. No clear evidence exists that gay males fear or are repulsed by female genitalia. To the contrary, many gay males have successfully, and repeatedly, engaged in coitus with women. They may simply find relationships with men to be more satisfying.
4. Many *heterosexual* men have family backgrounds that fit the classic pattern, and the families of many gay men do *not* fit the pattern. Researchers from the Kinsey Institute found that heterosexual males identified with their mothers as often as gay males did. A seductive mother–son relationship had little if any bearing on sexual orientation (Bell et al., 1981). Parent–child relations were also only weakly linked to lesbianism.
5. Retrospective accounts of childhood relationships with parents are subject to gaps in memory and to distortions.

More recent evidence has brought the issue of familial closeness between gay men and their parents into sharper perspective. Richard Pillard and his colleagues (Pillard & Weinrich, 1986; Pillard, 1990) found that gay males described themselves as more distant from their fathers during childhood than did either heterosexual controls or the gay men's own heterosexual brothers. The gay men in their sample also reported greater closeness to their mothers. Still, the father's psychological distance from the son may have reflected the son's alienation from him, not the reverse. That is, the son may have been so attached to his mother, or so uninterested in traditional masculine activities, that he rebuffed paternal attempts to engage him in conventional father–son activities.

In sum, family characteristics may play a role in the development of sexual orientation. However, there is great variation among the families of gay males and lesbians. No one pattern applies in all cases (Isay, 1990).

LEARNING THEORIES Learning theorists agree with Freud that early experiences play an important role in the development of sexual orientation. They focus on the role of reinforcement of early patterns of sexual behavior, however, rather than on the resolution of unconscious conflicts. People generally repeat pleasurable activities and discontinue painful ones. Thus, people may learn to engage in sexual activity with people of their own gender if childhood sexual experimentation with them is connected with sexual pleasure.

If sexual motivation is high, as it tends to be during adolescence, and the only outlets are with others of one's own gender, adolescents may experiment sexually with them. If these encounters are pleasurable, and heterosexual experiences are unpleasant, a firmer gay male or lesbian sexual orientation may develop (Gagnon & Simon, 1973). Conversely, pain, anxiety, or social disapproval may be connected with early contacts with people of one's own gender. In such cases, the child may learn to inhibit feelings of attraction to people of her or his own gender and develop a firmer heterosexual orientation.

Although learning may play a role in the development of a gay male or lesbian sexual orientation, learning theorists have not identified specific learning experiences that would lead to these orientations. Moreover, most adolescent encounters with people of the same gender, even if they are pleasurable, do not lead to an adult gay male or lesbian sexual orientation. Many heterosexual people have had adolescent encounters with members of their own gender without affecting their adult orientations. This is true even of people whose early sexual interactions with the other gender were fumbling and frustrating. Moreover, the overwhelming majority of gay males and lesbians were aware of sexual interest in people of their own gender *before* they had sexual encounters with them, pleasurable or otherwise (Bell et al., 1981).

Gender Nonconformity

On average, gay males tend to be somewhat feminine and lesbians to be somewhat masculine, but there is a good deal of variation within each group (Bailey et al., 1997; Dawood et al., 2000). Thus it seems that stereotypes of the effeminate gay male and the masculine lesbian are exaggerated. Gender nonconformity is rooted in childhood; gay males and lesbians are more likely than heterosexuals to report childhood behavior stereotypical of the other gender (Bailey & Zucker, 1995; Friedman & Downey, 1994). Many gay males and lesbians recall acting and feeling "different" from their childhood peers. Many gay males from a variety of groups, such as prisoners, psychiatric patients, and members of gay-rights organizations, report that they avoided participating in competitive sports as children and that they were more fearful of physical injury, and were more likely to avoid getting into fights, than heterosexual males (Dawood et al., 2000). Some gay males recall feeling different as early as the age 3 or 4 (Isay, 1990). Feelings of differentness were often related to behavior that is stereotypical of the other gender.

Gay males are also more likely to recall feeling more sensitive than their peers during childhood (Isay, 1990). They cried more easily. Their feelings were more readily hurt. They had more artistic interests. They had fewer male buddies but more female playmates. Gay males are also more likely than their heterosexual counterparts to have preferred "girls' toys." They preferred playing with girls to playing with trucks or guns or engaging in rough-and-tumble play (Dawood et al., 2000). Their preferences often led to their being called "sissies." Gay men also recall more cross-dressing during childhood. They preferred the company of older women to that of older men and engaged in childhood sex play with other boys rather than with girls.

There is also evidence of masculine-typed behavior among lesbians as children (Bailey & Zucker, 1995). Lesbians are more likely than heterosexual women to perceive themselves as having been "tomboys." They were more likely to prefer rough-and-tumble games to play with dolls and enjoyed wearing boy's clothing rather than "cutesy" dresses.

How might extreme childhood effeminacy lead to a gay male sexual orientation? Green (1987) speculates that the social detachment of these boys from male peers and role models (especially fathers) creates strong, unfulfilled cravings for male affection. This craving then leads them to seek males as partners in sex and love relationships in adolescence and adulthood.

Alan Bell of the Kinsey Institute believes that self-perceptions of differentness and social distance from other males during childhood lead these boys to develop erotic attractions that differ from those of other boys. These are erotic attractions toward members of their own gender.

Of course, there is another possibility, as suggested by research by J. Michael Bailey and his colleagues (Bailey et al., 2000; Dawood et al., 2000): Gender nonconformity appears to be somewhat heritable. Moreover, if a tendency toward homosexuality is inherited, gender nonconformity could well be that tendency.

All in all, the origins of a gay male or lesbian sexual orientation remain mysterious and complex—just as mysterious as the origins of heterosexuality. In reviewing theories and research, we are left with the impression that sexual orientation is unlikely to have a single cause. Sexual orientation appears to spring from multiple origins, including biological and psychosocial factors (Strickland, 1995). Genetic and biochemical factors (such as hormone levels) may affect the prenatal organization of the brain (Money, 1994). These factors may predispose people to a certain sexual orientation. But it may be that early socialization experiences are also necessary to give rise to a gay male, lesbian, heterosexual, or bisexual sexual orientation. The precise influences and interactions of these factors have so far eluded researchers.

ADJUSTMENT OF GAY MALES AND LESBIANS

Until 1973, the gay male and lesbian sexual orientations were in and of themselves considered to be mental illnesses by the American Psychiatric Association and were listed as such in their diagnostic manual. But in that year, the members of the association voted to drop a gay male or lesbian sexual orientation from its list of mental disorders,[1] although a diagnostic category remains for people with persistent and marked distress about their sexual orientation (American Psychiatric Association, 2000, p. 538).

Truth or Fiction? **REVISITED**

It is not true that the American Psychiatric Association considers homosexuality a mental disorder. The association has not considered a gay male or lesbian sexual orientation to be a mental disorder since 1973. ■

Many members of the organization objected to the vote, arguing that it was politically motived. Could the American Medical Association vote to drop cancer as a physical illness, they asked? Those who raised this kind of objection continued to believe that homosexuality is a disorder.

But for a number of years, research evidence seemed to suggest that gay males and lesbians were not any more likely than heterosexuals to suffer from mental disorders such as anxiety, depression, and schizophrenia (Reiss, 1980). Some researchers even pointed out that gay men and lesbians were, overall, more likely to be more highly educated than most Americans (Cronin, 1993). Yet some more recent, carefully controlled studies have found that gay males and lesbians are more likely than heterosexuals to experience feelings of anxiety and depression and that they are more prone to suicide (Bagley & D'Augelli, 2000; Ferguson et al., 1999; Herrell et al., 1999). Gay males, moreover, are more likely to have eating disorders (anorexia nervosa and bulimia nervosa) than heterosexual males (Ferguson et al., 1999).

Psychologist J. Michael Bailey (1999) of Northwestern University has carefully reviewed the issues surrounding the origins and correlates of homosexuality, and he wrote an interesting article about these recent studies in *Archives of General Psychiatry.* Bailey did not dispute the findings themselves. He did, however, have a great deal to say about their implications.

First, Bailey (1999) noted that many of the psychiatrists who had objected to the 1973 vote of the American Psychiatric Association were likely to claim vindication. Aha! They would say. "I told you so"—something, that is, was wrong with homosexuality. Bailey also noted that some social conservatives were likely to say that the troubles of homosexuals resulted from their "choice" of an immoral lifestyle. "If you make your bed, you must lie in it," they might argue. Needless to say, Bailey does not consider these interpretations to be warranted by the evidence. He proposes four possible interpretations:

1. Societal oppression causes the greater incidence of depression and suicidality that we find among gay males and lesbians. "Surely," writes Bailey, "it must be difficult for young people to come to grips with their homosexuality in a world where homosexual people are often scorned, mocked, mourned, and feared." Coming out is difficult at best, and gay males and lesbians, especially as adolescents, are likely to have few friends and to encounter a great deal of homophobia.

[1]We use the words *mental* and *psychological* interchangeably when we are talking about mental health (or psychological health) and mental disorders (or psychological disorders.)

2. Bailey acknowledges the possibility that homosexuality is a departure from the most common path of development and, as such, could be associated with other differences, some of which may lead to anxiety, depression, and other psychological health problems. Bailey does not "push for" this view but mentions it because it is a possibility. He does say that "Considerably more research would be necessary to validate the general hypothesis." Moreover, even if homosexuality is a departure from the more common path of development, this "departure" does not make the gay male or lesbian a "bad" person. As an example, consider the fact that nearsightedness is a departure from normal development but that we do not stigmatize nearsighted people.

3. A third possibility is connected with the view that sexual orientation may reflect atypical levels of sex hormones (particularly androgens) during prenatal development. In such a case, gay males might be prone to psychological disorders that more typically afflict women, and lesbians to disorders that more typically affect men. For example, women are more likely than men to encounter anxiety and depression, and gay males are more likely than heterosexual males to encounter these problems. Lesbians should then be more likely than other women to be diagnosed with antisocial personality disorder, a problem that is more often found in men. But so far, there is insufficient evidence for a general "reversal" of psychological health issues.

4. Bailey also suggests the possibility that psychological health problems among homosexual people reflects lifestyle differences that are connected with differences in sexual orientation. He notes that gay males are much more likely than heterosexual males to have eating disorders and also notes that "the gay male culture emphasizes physical attractiveness and thinness, just as the heterosexual culture emphasizes female physical attractiveness and thinness." (The incidence of eating disorders among gay males could also be connected with atypical levels of prenatal hormones, as discussed previously.) In this regard, we should note that about half of the professional male dancers in the United States are gay (Bailey & Oberschneider, 1997) and that eating disorders are especially common among dancers, who strive to remain thin at all costs.

We should reinforce the fact that Bailey insists that we must obtain more evidence before arriving at any judgments about why gay males and lesbians are more prone to mental or psychological health problems and suicidal ideas. Nevertheless, it is clear that gay males and lesbians do encounter stress from societal oppression and rejection. It is also clear that the adjustment of gays is connected with conflict over their sexual orientation (Simonsen et al., 2000).

There are also connections between lifestyle and health—physical and psychological—among gay males and lesbians, and among heterosexual people. Gay men and lesbians occupy all socioeconomic and vocational levels and follow a variety of lifestyles. In their classic research, Bell and Weinberg (1978) found variations in adjustment in the gay community that seem to mirror the variations in the heterosexual community. Gay people who lived with partners in stable, intimate relationships—so-called close couples—were about as well adjusted as married heterosexual couples. Older gay people who lived alone and had few sexual contacts were more poorly adjusted. So, too, are many heterosexual people who have similar lifestyles. All in all, Bell and Weinberg found that differences in adjustment were more likely to reflect the lifestyle of the individual than his or her sexual orientation.

Most gay males and lesbians who share close relationships with their partners are satisfied with the quality of their relationships. Researchers find that heterosexual and gay couples report similar levels of satisfaction with their relationships (Kurdek & Schmitt, 1986; Peplau & Cochran, 1990). Gay males and lesbians in enduring relationships generally report high levels of love, attachment, closeness, caring, and intimacy (Peplau & Cochran, 1990).

As with heterosexual people, not all of the relationships of gay people are satisfying. Among both groups, satisfaction is higher when both partners feel that the benefits they receive from the relationship outweigh the costs (Veniegas & Peplau, 1997). Like heterosexual people, gay men and lesbians are happier in relationships in which they share power and make joint decisions (Veniegas & Peplau, 1997).

TABLE 10.4

Percentages of gay people who wish they had received a "magic heterosexuality pill" at birth or could receive one today

	European American Males	African American Males	European American Females	African American Females
Desire magic pill at birth?	28	23	16	11
Desire magic pill today?	14	13	5	6

Source: Reprinted by permission of Simon & Schuster from A. P. Bell, and M. S. Weinberg (1978). *Homosexualities: A Study of Diversity Among Men and Women.* New York: Simon & Schuster, p. 339. Copyright © 1978 by Alan P. Bell and Martin S. Weinberg.

Treatment of Gay Male and Lesbian Sexual Orientations

Despite the judgment of the American Psychiatric Association, some people still view a gay male or a lesbian sexual orientation as an illness. If it is an illness, then it is something to be "cured," perhaps by medical or psychological means. However, the great majority of gay men and lesbians do not seek professional assistance to change their sexual orientations. Most see their sexual orientations as integral parts of their personal identities. Bell and Weinberg (1978) found only a few gay people who were interested in changing their sexual orientation—even if a "magic pill" were available to bring about the change (see Table 10.4). Only a minority of Bell and Weinberg's sample had *ever* considered discontinuing male–male or female–female sexual activity.

A few gay men and women do express an interest in changing their orientations, however. Helping professionals have tried to help them do so in a variety of ways. At one time, when it was commonly believed that hormonal imbalances influenced one's sexual orientation, hormone treatments were in vogue. There is no evidence that they were effective.

A few psychotherapists have reported changing the sexual orientations of some individuals. For example, Bieber (1976) claimed that about one in four clients changed his or her sexual orientation through psychoanalytic psychotherapy. But critics charge that these clients were highly motivated to change. Moreover, some of them began therapy as bisexuals. Their change in lifestyle may be attributable to initial motivation to change, not to the therapy itself.

Masters and Johnson (1979) employed methods used to treat sexual dysfunctions (see Chapter 15) to "reverse" patients' gay male or lesbian sexual orientations. For example, they involved gay males in a graded series of pleasurable activities with women, such as massage and genital stimulation. Masters and Johnson reported failure rates of 20% for the gay men and 23% for the lesbians they treated in their therapy program. At a five-year follow-up, more than 70% of the clients continued to engage in male–female sexual activity (Schwartz & Masters, 1984). For many reasons, however, these patients do not seem to represent the general gay population:

- Most of them were bisexuals. Only about one in five engaged exclusively in male–male or female–female sexual activities.
- More than half were married.
- All were motivated to switch their sexual orientations.

Regardless of these people's changes in sexual behavior, remember that sexual behavior is not the equivalent of sexual orientation. (It was thus inaccurate for Masters and Johnson to claim that they had reversed an individual's sexual *orientation.*) That is, many gay people engage in sexual activities with people of the other gender. Nevertheless, they still prefer to have such relationships with people of their own gender.

Critics of conversion therapy argue that gay men and women often enter therapy because of conflicts that arise from social prejudice. Such pressures create problems in self-acceptance among gay males and lesbians. Critics argue that the role of the therapist should be to help unburden the client of conflict and promote a more gratifying life as a gay person (Sleek, 1997).

COMING OUT: COMING TO TERMS WITH BEING GAY

> You must worship rugby and beer. And if you don't, God help you.
>
> —A Roman Catholic priest in New Zealand (cited in Shenon, 1995)

Because of the backdrop of social condemnation and discrimination, gay males and lesbians in our culture often struggle to come to terms with their sexual orientation. Gay men and lesbians usually speak of the process of accepting their sexual orientation as "coming out" or as "coming out of the closet." Coming out is a two-pronged process: coming out to oneself (recognizing one's gay male or lesbian sexual orientation) and coming out to others (declaring one's orientation to the world). Coming out can create a sense of pride in one's sexual orientation and foster the ability to form emotionally and sexually satisfying relationships with gay male or lesbian partners.

Coming Out to Oneself

Many gay people have a difficult time coming to recognize, let alone accept, their sexual orientation. Some have even considered or attempted suicide because of problems in self-acceptance (Bagley & D'Augelli, 2000):

> It (suicidal thinking) was because of my homosexuality. I was completely depressed that my homosexuality was leading me nowhere. My life seemed to be going around in a circle. I was still fighting against recognition of my homosexuality. I knew what I was but I didn't want to be.
>
> [I had] the feeling that homosexuality was a hopeless existence. I felt there was no future in it, that I was doomed to be alone. And from my Catholic background, I felt that homosexuality was very evil. (Bell & Weinberg, 1978, p. 202)

Consider the experience of a New Zealand gay adolescent who was subjected to merciless taunting until his parents pulled him out of school. "He threw a ball like a girl does," said a hot-line worker (cited in Shenon, 1995),"and the hassling just wouldn't stop. He was hassled nonstop from the day he got to that school until the day he left."

"You're judged constantly on your maleness," adds a Roman Catholic priest in New Zealand (cited in Shenon, 1995). "It can be a very intolerant country. You must worship rugby and beer. And if you don't, God help you. I think we've just been dreadful about this macho-ness."

Two researchers in human sexuality note that

> Sexual orientation emerges strongly during early adolescence. Youths with emerging identities that are gay, lesbian, or bisexual, living in generally hostile climates, face particular dilemmas. They are well aware that in many secondary schools the words "fag" and "dyke" are terms of denigration and that anyone who is openly gay, lesbian, or bisexual is open to social exclusion and psychological and physical persecution. Some of their families too will express negative feelings about people who are gay, lesbian, or bisexual; youths in such families may be victimized if they disclose that they are not heterosexual. (Bagley & D'Augelli, 2000)

For some people, coming to recognize and accept a gay male or lesbian sexual orientation involves gradually stripping away layers of denial. For others it may be a sudden awakening. Long-standing sexual interests in members of one's own gender may abruptly focus on a particular person, as happened with a graduate student named David:

> In college [David's] closest friend was gay. Although this friend had wanted to have sex with David and the attraction was mutual, David still could not associate this attraction with a sexuality that was not acceptable to him. In his first year of graduate school, when he was about 23, he fell in love and then suddenly and with a great sense of relief recognized and acknowledged to himself that he was homosexual. He then had sex for the first time and has subsequently been . . . open about his sexuality. (Isay, 1990, p. 295)

Recognition of a gay sexual orientation may be only the first step in a lifelong process of sexual identity formation. Acceptance of being gay becomes part of one's self-definition (Isay, 1990). The term *gay identity,* or *homosexual identity,* refers to the subjective sense of being gay.

Some gay men and lesbians who have not yet come to recognize or accept their gay sexual orientation get married. For some, marriage is a means of testing feelings toward people of the other gender. For others, marriage represents an attempt to conceal or overcome their sexual orientation (Buxton, 1994; Gabriel, 1995a). Perhaps 20% of gay males and a higher percentage of lesbians get married at least once (Bell & Weinberg, 1978). But such marriages tend to be unhappy and short-lived. They may be strained by one partner's concealing her or his gay sexual orientation. Alternatively, they may buckle under the open acknowledgment of being gay. Virtually all such marriages in Bell and Weinberg's study eventually ended in separation or divorce.

In the 1990s, many American couples included a gay or lesbian spouse. Thousands of gay men and lesbians who had conformed to social pressures by getting married were coming out of the closet and sometimes leaving home (Buxton, 1994; Gabriel, 1995a).

Coming Out to Others

There are different patterns of coming out to others. Coming out occasionally means making an open declaration to the world. Some individuals inform only one or a few select people. Others may tell friends but not family members.

Many gay men and lesbians remain reluctant to declare their sexual orientation, even to friends and family. Disclosure is fraught with the risk of loss of jobs, friendships, and social standing (Bagley & D'Augelli, 2000). A social worker who counsels lesbians described some fears that her clients have expressed about coming out to others:

> Will I lose my job, will I lose my house, will I lose my children, will I be attacked? I've seen people lose a job. Even if you don't lose it, you become marginal, excluded either actively or through uncomfortable vibes. (Cited in Barrett, 1990, p. 470)

Gay men and lesbians often anticipate that family members will have negative reactions, including denial, anger, and rejection (Bagley & D'Augelli, 2000). Family members and loved ones may refuse to hear or may be unwilling to accept reality, as Martha Barron Barrett notes in her book *Invisible Lives,* which chronicles the lives of a sample of lesbians in the United States.

> Parents, children, neighbors, and friends of lesbians deny, or compartmentalize, or struggle with their knowledge in the same way the women themselves do. "My parents know I've lived with my partner for six years. She goes home with me. We sleep in the same bed there. The word *lesbian* has never been mentioned. I told my mother and she said, 'Well, now that's over with. We don't need to mention it again.' She never has, and that was ten years ago. I don't know if she ever told my father." A husband may dismiss it as "just a phase," a boyfriend may interpret it as a sexual tease, a straight woman may believe "she's just saying that because she couldn't get a man."
>
> The strong message is, "Keep it quiet." Many lesbians do that by becoming invisible. . . .They (lesbians) leave their lesbian persona at home when they go to work on Monday morning. On Friday they don it again. That weekend at home, the flip side of the double life, is what most of heterosexual society never sees. (Barrett, 1990, p. 52)

Some families are more accepting. They may in fact have had suspicions and prepared themselves for such news. Then, too, many families are initially rejecting but often eventually come to at least grudging acceptance that a family member is gay.

Out Into the "Gay Global Village" of Cyberspace

The gay community has been active online for quite some time. An *Out* magazine survey shows that gay men and lesbians are more likely than the general population to use personal computers, modems, and online services. In an average month, 40,000 gay people spend more than 100,000 hours online with America Online's Gay and Lesbian Community Forum.

Planet Out (www.planetout.com), the electronic media company, has become a sort of "gay global village" in cyberspace. It has received the endorsement of virtually all the leading gay organizations. The Human Rights Campaign Fund, the National Gay and Lesbian Task Force, Parents and Friends of Lesbians and Gays, the Gay and Lesbian Victory Fund, Digital Queers, and the Gay and Lesbian Alliance Against Defamation all provide information on the service.

Planet Out is also a meeting place for millions of gay men, lesbians, bisexuals, and others who may be reluctant to associate with one another in public. A creative director of Netscape and the former head of design at Apple Computer said that chatting electronically with gay men and lesbians on America Online had given him the courage to discuss his gay sexual orientation openly. "It's something that would have been unthinkable for me even a year or two ago," he said. "If I had relied on more traditional ways of meeting people, like going to bars, or going to meetings of various organizations, or picking up gay publications, it never would have happened" (Lewis, 1995).

PATTERNS OF SEXUAL ACTIVITY AMONG GAY MALES AND LESBIANS

Heterosexual people are often confused about the sexual practices of gay couples. They may wonder, "Just what do they do?" Generally speaking, gay couples express themselves sexually through as wide a range of activities as heterosexual couples, with the exception of vaginal intercourse. But there are shades of difference between gay and heterosexual couples in sexual techniques.

Sexual Techniques

Gay male couples tend to engage in sexual activities such as kissing, hugging, petting, mutual masturbation, fellatio, and anal intercourse. Laboratory observations of sexual relations between gay males by Masters and Johnson (1979) showed that gay males spent a good deal of time caressing their partners' bodies before approaching the genitals. After hugging and kissing, 31 of 42 gay male couples observed by Masters and Johnson used oral or manual nipple stimulation.

Not all gay males enjoy or practice anal intercourse. Of those who do, most alternate between being the inserter and being the insertee. The frequency of anal intercourse among gay males is apparently declining in the face of the AIDS epidemic (Catania et al., 1991; Centers for Disease Control, 1990a; Sonenstein et al., 1989).

Risks of infection and of injury to the rectum or anus are associated with another sexual practice called fisting. Fisting is the insertion of the fist or hand into the rectum, usually after the bowels have been evacuated with an enema.

Sexual techniques practiced by lesbians vary. Lesbian couples report kissing, manual and oral breast stimulation, and manual and oral stimulation of the genitals (Kinsey et al., 1953). Manual genital stimulation is the most common and frequent sexual activity among lesbian couples (Bell & Weinberg, 1978). Most lesbian couples also engage in genital apposition. That is, they position themselves in such a way as to rub their genitals together rhythmically (Kinsey et al., 1953). Like gay males, lesbians spend a good deal of time holding, kissing, and caressing each other's bodies before they approach the breasts and genitals. By contrast, heterosexual males tend to move quickly to stimulate their partners' breasts or start directly with genital stimulation (Masters & Johnson, 1979).

Like heterosexual women, lesbians are less genitally oriented and less fixated on orgasm than men. Lesbians generally begin stimulating their partners with more general genital stimulation rather than direct clitoral stimulation, whereas heterosexual males often begin by stimulating the clitoris (Masters & Johnson, 1979). Nor do lesbian couples generally engage in deep penetration of the vagina with fingers. Rather, they may use more shallow vaginal penetration, focusing stimulation on the vaginal lips and entrance. Images of lesbians strapping on dildos for vaginal penetration exist more in the imagination of uninformed heterosexual people than in the sexual repertoire of lesbian couples (Masters & Johnson, 1979). The emotional components of lovemaking—gentle touching, cuddling, and hugging—are important elements of sexual sharing in lesbian relationships.

GAY LIFESTYLES

One of the mistakes that laypeople (and some researchers) make is to treat gay people as though they were all the same. According to Bell and Weinberg (1978), gay people do not adopt a single, stereotypical lifestyle. That is why the authors termed their report *Homosexualities: A Study of Diversity Among Men and Women.* Variations in sexual expression exist within and across sexual orientations. Descriptions of gay and heterosexual lifestyles must consider individual differences.

Gay men and lesbians in larger U.S. urban centers can usually look to gay communal structures to provide services and support. These include gay-rights organizations and gay-oriented newspapers, magazines, bookstores, housing cooperatives,

Sexual Orientation and Lifestyle. Gay people, like heterosexuals, have a variety of lifestyles. Some go from partner to partner. Others form stable relationships. Some avoid sexual intimacy and committed relationships altogether. As Alexandre Dumas said, "All generalizations are dangerous. Even this one."

medical services, and other support services (Gagnon, 1990). The gay community provides a sense of acceptance and belonging that gay people do not typically find in society at large. Gay people still encounter discrimination in the workplace, in housing, and in the military.

In the year 2000, Pentagon investigators surveyed a random sample of 71,570 officers and enlisted personnel at 38 military bases and aboard 11 warships in the United States and overseas—with the exception of personnel at Fort Campbell, Kentucky, where Pfc. Barry Winchell had been bludgeoned to death with a baseball bat the previous year (Myers, 2000a). Results of the survey are shown in Table 10.5.

The issue of restrictions on gay people serving in the armed services has continued to provoke controversy. In 1993, President Clinton encountered opposition when he sought to follow through on his campaign pledge to permit gay men and women to serve openly in the military. As a compromise measure, he instituted a "Don't ask, Don't tell" policy. It permits gay males and lesbians to serve so long as they do not publicly express or reveal their sexual orientation. Moreover, military officials are prohibited from inquiring about the sexual orientation of recruits. As noted in the accompanying Closer Look, the policy has required adjustment over the years.

Gay-rights organizations fight for rights for gay people to participate fully in society—to teach in public schools, to adopt children, to live together in sanctioned relationships, and to serve openly and proudly in the military. Cafés and social clubs provide places where gay men and lesbians can socialize and be open about their sexual orientations. Organizations such as New York's Gay Men's Health Crisis (GMHC) provide medical, social, and psychological assistance to gay males who have been afflicted by AIDS.

TABLE 10.5

Anti-gay bias in the military: Results of U.S. armed forces survey of 71,570 military personnel
80% of service members reported hearing anti-gay remarks within the past year.
85% of service members believed that anti-gay comments were tolerated within their units.
33% of service members reported hearing anti-gay remarks "often" or "very often."
5% of service members believed that harassment based on sexual orientation was tolerated by superiors in the chain of command.
10% said that harassment based on sexual orientation was tolerated by colleagues in their unit.
Of those who had witnessed an instance of harassment based on sexual orientation, 22% said that a more senior person had also witnessed the incident. In 73% of those cases, the senior person did nothing immediately to stop it.
78% of service members reported feeling free to report anti-gay harassment to their superiors without fear of retribution.
Only 16% of service members who reported witnessing or experiencing harassment based on sexual orientation did report it.

Source: As reported in Steven Lee Myers (2000, March 25). "Survey of Troops Finds Anti-gay Bias Common in Service." *The New York Times online.* Copyright © 2000 by the New York Times Company. Reprinted by permission.

a CLOSER look

"Don't Ask, Don't Tell" Far from Clear as a Bell

In 2000, the Pentagon issued a report criticizing the military's implementation of its policy on gays and vowed to roll out a massive training program aimed at reducing the level of anti-gay harassment in the armed services. As part of the new effort, military leaders will be instructed explicitly that they should never ask about a person's sexual orientation, no matter what the circumstance. "The question, 'Are you homosexual?' is never in order. Ever," Bernard Rostker, undersecretary of defense for personnel and readiness, said emphatically at a Pentagon news conference. "The days of the witch hunt, the days of stakeouts, are over."

The new policy was spurred by a survey conducted by the Defense Department's inspector general, which found that harassment of gay men and lesbians is commonplace and widely tolerated in the U.S. military. The survey found that 80% of those questioned had heard offensive comments about gays within the previous year.

The proposal to eliminate anti-gay behavior was the recommendation of a panel convened to review the military's "don't ask, don't tell" policy toward gays. The policy arose after President Clinton failed to persuade Congress and the Pentagon to allow gays to serve openly and holds that gays can serve in the military as long as they don't reveal their sexual orientation.

As part of the 13-point anti-harassment plan, the panel called for improving training to clear up misconceptions about the policy, for measuring the effectiveness of the new training, and for making commanders responsible for implementing the policy correctly. The panel's "Anti-Harassment Action Plan" stated that the Pentagon should make it clear that "commanders and leaders will be held accountable for failure to enforce this directive."

The Army released its own report on the command climate at Fort Campbell, Ky., where a gay soldier was beaten to death in 1999 by another soldier wielding a baseball bat and shouting anti-gay epithets. That murder had provoked questioning by many, including Clinton, of whether the "don't ask, don't tell" policy is really working. Overall, the Army report portrays the unit in which Pvt. Barry Winchell served—D Company, 2nd Battalion, 502nd Infantry Regiment, 101st Airborne Division—as demoralized, lacking in the required number of officers, with some soldiers drinking heavily and an abusive top sergeant. But the Army report clears the chain of command above that sergeant. When Winchell reported to the company commander that the sergeant had called him a "faggot," the report says, the commander counseled the sergeant. General Eric K. Shinseki, the Army chief of staff, said that the abusive sergeant ultimately was moved.

In detailing its program, the DiBattiste panel borrowed some key phrases used often by those opposed to permitting the openly gay to serve. For example, the panel stated as an "overarching principle" that treating individuals with dignity and respect is "essential to good order and discipline." Opponents often argue that permitting openly gay soldiers will undercut military discipline. Similarly, the panel turned another argument of the opponents on its head by stating that anti-gay harassment undercuts "unit cohesion."

Generally, the panel's strategy appears to have been to adopt much of the approach that the military has used to counter discrimination against black and female service personnel. Yet a major difference remains: Those programs deal with obvious, visible characteristics, whereas the program the Pentagon is about to embark on is intended to train the troops to respect people who can't identify themselves.

What Is the Sexual Orientation of These Soldiers? During Bill Clinton's presidency, a "Don't Ask, Don't Tell" policy was adopted toward gay people in the military. If they divulge their sexual orientation, they may be removed from the military, but superiors are not allowed to ask them about it.

Source: From T. E. Ricks (2000, July 22). "Pentagon Vows to Enforce 'Don't ask.'" *The Washington Post online*, p. A01.

Not all gay people feel that they are a part of the "gay community" or participate in gay rights organizations, however. For many, their sexual orientation is a part of their identity but not a dominant theme that governs their social and political activities.

Lifestyle Differences Between Gay Males and Lesbians

Much of our knowledge of lifestyle patterns among gay males and lesbians comes from research that predates the AIDS epidemic. Researchers have consistently found that gay males are more likely than lesbians to engage in casual sex with many partners. Lesbians more often confine their sexual activity to a committed, affectionate relationship (Bell & Weinberg, 1978; Peplau & Cochran, 1990). Bell and Weinberg reported that 84% of gay males, compared to about 7% of lesbian women, report having more than 50 partners in their lifetimes. Seventy-nine percent of gay males in their study, compared to only 6% of lesbians, reported that more than half of their partners were strangers. Even gay males within committed relationships have more permissive attitudes toward extracurricular sexual activity than lesbians do (Blumstein & Schwartz, 1990; Peplau & Cochran, 1990).

Traditionally, the gay bar was an arena for making sexual contacts (Bell & Weinberg, 1978). **Cruising** is the name gay people give to searching for a sex partner, principally for casual sex. One can "cruise," and one can "be cruised." Gay males were more likely than lesbians to cruise in public places, such as gay bars. Lesbians were more likely to find partners among friends, at work, and at informal social gatherings.

Today, with the threat of AIDS hanging over every casual sexual encounter, cruising has lost popularity. Many gay bathhouses, long a setting for casual sexual contacts, have closed down—or have been closed down by authorities—because of AIDS. These closings are but one sign of changes in gay communities that have occurred since the advent of AIDS. Many gay males have changed their behavior to prevent contracting or spreading AIDS. Evidence shows that as a group, gay males have become more likely to limit or avoid anal and oral sex, especially with new partners. They are now more likely to use latex condoms when they do practice these techniques, to limit their sexual contacts to partners they know, and to rely more on masturbation as a sexual outlet (Centers for Disease Control, 1990a, 1990b; Siegel et al., 1988). Despite the advent of AIDS, some gay men, like some heterosexual people, continue high-risk behavior. For example, they engage in unprotected anal intercourse and sexual activity with multiple partners.

Research also shows that extracurricular sexual activity is common among gay male couples. One study surveyed 943 gay males and 1,510 married heterosexual males who had been living with a partner for 2 to 10 years. Four of five (79%) of the gay males reported having sexual relations with another partner during the preceding year, compared to only 11% of the heterosexual males (Blumstein & Schwartz, 1990). Among couples who had been together for longer than 10 years, 94% of gay men reported extracurricular activity at some time during their primary relationships.

Variations in Gay Lifestyles

Bell and Weinberg (1978) found that about three out of four gay couples they studied could be classified as representing one of five lifestyles: *close couples, open couples, functionals, dysfunctionals,* and *asexuals.* **Close couples** strongly resembled married couples. They evidenced deep emotional commitment and few outside sexual relationships. Almost three times as many lesbians (28%) as gay males (10%) lived in such committed, intimate relationships. Gay people living in close relationships showed fewer social and psychological problems than those with any other lifestyle.

Cruising The name homosexuals give to searching for a sex partner.

Close couples Bell and Weinberg's term for gay couples whose relationships resemble marriage in their depth of commitment and exclusiveness.

Human Sexuality ONLINE //

Web Sites Related to Sexual Orientation

This is the Web site of the National Gay and Lesbian Task Force. NGLTF is a national organization that works for the civil rights of gay, lesbian, bisexual, and transgendered (GLBT) people. It is a national resource center that trains and assists gay males and lesbians to lobby for "pro-GLBT" legislation concerning families, employment, health care, hate crimes, and more. "ANGLTF celebrates diversity and builds bridges across race, sexual orientation, gender identity, religion, ethnicity, age, disability and income to fortify the gay, lesbian, bisexual and transgender civil rights movement."
www.ngltf.org/

Excite.com: This Web site has recent-history items about gay and lesbian life in the United States. Topics range from Kinsey's research to gays in the military.
home.excite.com/relationships/gay_and_lesbian/history/20th_century/

Boston Daughters of Bilitis (1151 Mass. Ave., Old Cambridge Baptist Church, Cambridge 02138): The Web site of the Boston chapter provides various social activities and support for lesbians who are just coming out and for newcomers to the Boston area. It is the surviving chapter of the first lesbian organization in the country, which was founded in San Francisco in 1955. Activities are open to women over the age of 18. You can also call (617) 661–3633 for information about events.
www.bostonphoenix.com/listings/1in10/resources/DAUGHTERS_OF_BILITIS.html

PlanetOut.com: This Web site contains information about politics, entertainment, money, and careers. You can also shop or meet people here. This is the major commercial Web site for gay males and lesbians.
www.planetout.com/pno/

The Advocate: The Web site for this magazine for gay males and lesbians contains up-to-date news, profiles of notable people (gay and "straight"), reviews of media events, and articles about the adjustment of gay males and lesbians.
www.advocate.com/

ACT UP—the AIDS Coalition to Unleash Power: ACT UP is a highly activist organization dedicated mainly to health issues among gay people. The main focus in these Web sites is on AIDS and on the promotion of research and the widening of access to effective treatment for HIV/AIDS.

San Francisco chapter:
www.actupgg.org/main.shtml

New York chapter:
www.actupny.org/

The Washington, DC chapter:
www.actupdc.org/

(There are others.)

Truth or Fiction? REVISITED

It is true that many gay couples have lifestyles similar to those of married heterosexual couples and are as well adjusted. They are referred to as *close couples* by Bell and Weinberg. ■

Partners in **open couples** lived together but engaged in clandestine affairs. Gay people in open couples were not as well adjusted as those in close couples. Nevertheless, their overall adjustment was similar to that of heterosexual people. Still other gay people lived alone and had sexual contacts with numerous partners—a kind of "swinging singles" gay lifestyle. Some of those who lived alone, **functionals,** appeared to have adapted well to their swinging lifestyle and were sociable and well adjusted. Others, called **dysfunctionals,** had sexual, social, or psychological problems. Dysfunctionals often were anxious, were unhappy, and found it difficult to form intimate relationships. **Asexuals** also lived alone but were distinguished by having few sexual contacts. Asexuals tended to be older than gay people in the other groups. They did not have the adjustment problems of dysfunctionals, but they too did not form intimate relationships. Although they were largely asexual in terms of their behavior, their sexual orientation was clearly gay.

The Bell and Weinberg study described diversity of lifestyles in the gay community in the 1970s, before the AIDS epidemic struck. Although such lifestyles continue today to a certain extent, the AIDS epidemic has inhibited promiscuous sex within the gay community. The specter of AIDS has had a more profound effect on the lifestyles and sexual practices of gay males than on any other group in society.

Open couples Bell and Weinberg's term for gay couples who live together but engage in secret affairs.

Functionals Bell and Weinberg's term for gay people who live alone, have adapted well to a swinging lifestyle, and are sociable and well adjusted.

Dysfunctionals Bell and Weinberg's term for gay people who live alone and have sexual, social, or psychological problems.

Asexuals Bell and Weinberg's term for gay people who live alone and have few sexual contacts.

summing up

SEXUAL ORIENTATION

Coming to Terms with Terms
Sexual orientation describes the directionality of one's sexual and romantic interests—toward members of the same gender, members of the other gender, or both. Gay male and lesbian sexual orientations denote sexual and romantic interest in members of one's own gender.

Sexual Orientation and Gender Identity
Gay males and lesbians have a gender identity that is consistent with their chromosomal and anatomic sex.

Classification of Sexual Orientation
Kinsey and his colleagues found evidence of degrees of homosexuality and heterosexuality, with bisexuality representing a midpoint between the two. Heterosexuality and homosexuality may be separate dimensions rather than polar opposites, however.

PERSPECTIVES ON GAY MALE AND LESBIAN SEXUAL ORIENTATIONS

Historical Perspectives
Throughout much of Western history, gay people have been deemed sinful and criminal.

Cross-Cultural Perspectives
Male–male sexual behavior is practiced by at least some members of many preliterate societies.

Cross-Species Perspectives
Many animals engage in behaviors that resemble male–male contacts among humans, but we must be cautious in ascribing motives to animals.

Attitudes Toward Sexual Orientation in Contemporary Society
The majority of people in our society view gay people negatively. During the past generation, gay males and lesbians have organized effective political groups to fight discrimination and overturn sodomy laws that have traditionally targeted them.

Biological Perspectives
Evidence of a genetic contribution to sexual orientation is accumulating. Research has failed to connect sexual orientation with differences in current (adult) levels of sex hormones. Prenatal sex hormones may play a role in determining sexual orientation in humans, however.

Psychological Perspectives
Psychoanalytic theory connects sexual orientation with unconscious castration anxiety and improper resolution of the Oedipus complex. Learning theorists focus on the role of reinforcement of early patterns of sexual behavior.

Gender Nonconformity
Although only a few gay people fit the stereotypes of "swishy" men and "butch" women, research finds that as children, homosexuals report a greater incidence of behavior stereotypical of the other gender than do heterosexual reference groups.

ADJUSTMENT OF GAY MALES AND LESBIANS

Evidence has failed to show that gay males, lesbians, and bisexuals are more emotionally unstable or more subject to psychiatric disorders than heterosexual people are.

COMING OUT: COMING TO TERMS WITH BEING GAY

Gay people in our culture struggle to come to terms with their sexual orientation against a backdrop of social condemnation and antagonism. Coming out is a two-pronged process: coming out to oneself and coming out to others. Recognizing and accepting one's gay sexual orientation may occur as a gradual process or as a sudden awakening.

PATTERNS OF SEXUAL ACTIVITY AMONG GAY MALES AND LESBIANS

Gay people generally express themselves sexually through as wide a range of activities as heterosexual people do, with the exception of vaginal intercourse.

GAY LIFESTYLES

Gay people do not adopt a single, stereotypical lifestyle.

Lifestyle Differences Between Gay Males and Lesbians
Gay males are more likely than lesbians to engage in casual sex with many partners. Lesbians more often confine sexual activity to a committed, affectionate relationship. Many gay males have changed their sexual behavior to prevent contracting or spreading AIDS.

Variations in Gay Lifestyles
The majority of gay people studied by Bell and Weinberg could be classified as representing one of five lifestyles: close couples, open couples, functionals, dysfunctionals, and asexuals.

questions for critical thinking

1. What are the attitudes of people from your sociocultural group toward gay males and lesbians? Do you share these attitudes? Why or why not?
2. Why do you think men tend to be more homophobic than women?
3. Why are people who believe the causes of sexual orientation are biological more tolerant than others of a gay male or lesbian sexual orientation?
4. Agree or disagree with the following statement, and support your answer: We should help gay males and lesbians change their sexual orientation whenever possible.
5. Are you a gay male or lesbian? If so, have you come out? If so, what was the experience like? If not, why not?
6. What did you learn about gay males and lesbians from this chapter? What misinformation has been corrected? What is completely new to you?
7. Do you believe that the information in this chapter has been presented in a straightforward or a biased manner? Explain.

chapter

11

Conception, Pregnancy, and Childbirth

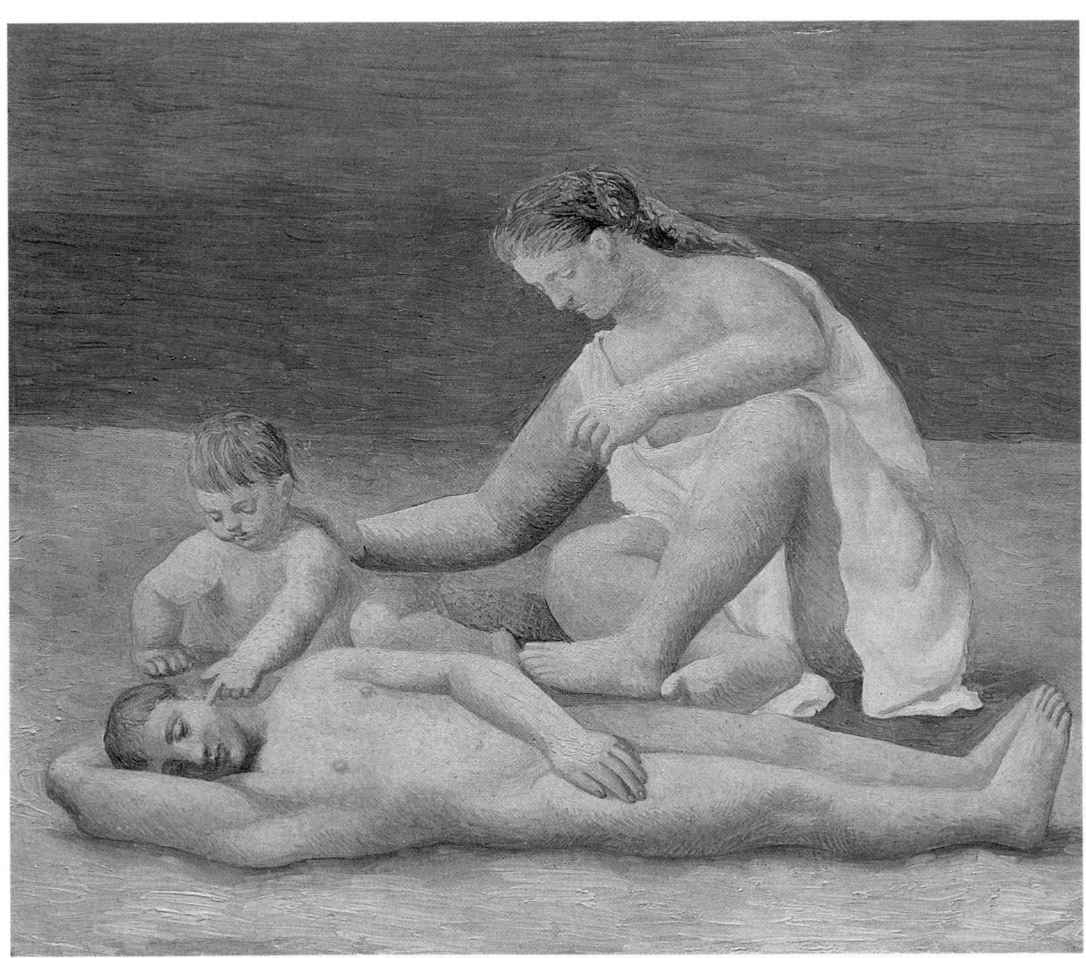

Pablo Picasso. *Family on the Seashore*, 1922. Oil on wood panel, 17.6 × 20.2 cm. Photo: R. G. Ojeda. © Copyright ARS, NY. Musée Picasso, Paris, France. Credit: Réunion des Musées Nationaux/Art Resource, NY. Credit. © 2002 Estate of Pablo Picasso/Artists Rights Society (ARS), New York.

outline

Truth or Fiction?

CONCEPTION: AGAINST ALL ODDS

Optimizing the Chances of Conception

A Closer Look Selecting the Gender of Your Child

INFERTILITY AND ALTERNATIVE WAYS OF BECOMING PARENTS

Male Fertility Problems

Female Fertility Problems

A Closer Look Sex and the Single Lizard

PREGNANCY

Questionnaire Should You Have a Child?

Early Signs of Pregnancy

Pregnancy Tests

Early Effects of Pregnancy

Miscarriage (Spontaneous Abortion)

Sex During Pregnancy

Psychological Changes During Pregnancy

PRENATAL DEVELOPMENT

The Germinal Stage

The Embryonic Stage

The Fetal Stage

Environmental Influences on Prenatal Development

A Closer Look Spacing Children the Goldilocks Way

Chromosomal and Genetic Abnormalities

CHILDBIRTH

The Stages of Childbirth

Methods of Childbirth

Laboring Through the Birthing Options: Where Should a Child Be Born?

Human Sexuality Online Childbirth Advice Is Just Keystrokes Away

A World of Diversity A Racial Gap in Infant Deaths, and a Search for Reasons

BIRTH PROBLEMS

Anoxia

Preterm and Low-Birth-Weight Children

THE POSTPARTUM PERIOD

Maternal Depression

Breast-Feeding Versus Bottle-Feeding

Human Sexuality Online Where to Get Help Breast-Feeding

Resumption of Ovulation and Menstruation

Resumption of Sexual Activity

Human Sexuality Online Web Sites Related to Conception, Pregnancy, and Childbirth

SUMMING UP

QUESTIONS FOR CRITICAL THINKING

Truth or Fiction?

T ____ Prolonged athletic activity may decrease fertility in the male.

F ____ A "test-tube baby" is grown in a large laboratory dish throughout the 9-month gestation period.

T ____ There is an all-female species of lizard that lays unfertilized eggs that develop into identical females generation after generation.

F ____ Morning sickness is limited to the first trimester.

T ____ For the first week after conception, a fertilized egg cell is not attached to its mother's body.

F ____ Pregnant women can have one or two alcoholic beverages a day without harming their babies.

T ____ A baby signals its mother when it is ready to be born.

T ____ In the United States, 1 birth in 5 is by cesarean section.

T ____ Couples should abstain from sexual activity for at least 6 weeks after childbirth.

On a balmy day in October, Elaine and her husband Dennis rush to catch the train to their jobs in the city. Elaine's workday is outwardly much the same as any other. Within her body, however, a drama is unfolding. Yesterday, hormones caused a follicle in her ovary to rupture, releasing its egg cell, or ovum. Like all women, Elaine possessed at birth all the ova she would ever have, each encased in a sac, or follicle. How this particular follicle was selected to ripen and release its ovum this month remains a mystery. For the next day or so, however, Elaine will be capable of conceiving.

When Elaine used her ovulation-timing kit the previous morning, it showed that she was about to ovulate. So later that night, Elaine and Dennis had made love, hoping that Elaine would conceive. Dennis ejaculated hundreds of millions of sperm within Elaine's vagina. Only a few thousand survived the journey through the cervix and uterus to the fallopian tube that contained the ovum, released just hours earlier. Of these, a few hundred remained to bombard the ovum. One succeeded in penetrating the ovum's covering, resulting in conception. From a single cell formed by the union of sperm and ovum, a new life began to form. The **zygote** is but 1/175 of an inch across—a tiny beginning.

Elaine is age 37. Four months into her pregnancy, Elaine obtains **amniocentesis** in order to check for the presence of chromosomal abnormalities, such as **Down syndrome.** (Down syndrome is more common among children born to women in their late 30s and older.) Amniocentesis also indicates the gender of the fetus. Although many parents prefer to know the gender of their baby before it is born, Elaine and Dennis ask their doctor not to inform them. "Why ruin the surprise?" Dennis tells his friends. So Elaine and Dennis are left to debate boys' *and* girls' names for the next few months.

CONCEPTION: AGAINST ALL ODDS

Conception is the union of a sperm cell and an ovum. On the one hand, conception is the beginning of a new human life. Conception is also the end of a fantastic voyage, however, in which a viable ovum—one of only several hundred that will mature and ripen during a woman's lifetime—unites with one of several hundred *million* sperm produced by the man in the average ejaculate.

Ova carry X sex chromosomes. Sperm carry either X or Y sex chromosomes. Girls are conceived from the union of an ovum and an X-bearing sperm, boys from the union of an ovum and a Y-bearing sperm. Sperm that bear Y sex chromosomes appear to be faster swimmers than those bearing X sex chromosomes. This is one of the reasons why between 120 and 150 boys are conceived for every 100 girls. What seem to be natural balancing factors favor the survival of female fetuses, however. Male fetuses are more likely to be lost in a **spontaneous abortion,** which often occurs during the first month of pregnancy. In many cases of early spontaneous abortion, the woman never realizes that she had been pregnant. Despite spontaneous abortions, boys still outnumber girls at birth. Boys suffer from a higher incidence of infant mortality, however. Thus the numbers of boys and girls in the population are further equalized by the time they mature to the point of pairing off.

Zygote A fertilized ovum.

Amniocentesis A procedure for drawing off and examining fetal cells in the amniotic fluid to determine whether various disorders are present in the fetus.

Down syndrome A chromosomal abnormality that leads to mental retardation and is caused by an extra chromosome on the 21st pair.

Spontaneous abortion The sudden, involuntary expulsion of the embryo or fetus from the uterus before it is capable of independent life.

The 200 to 400 million sperm in an average ejaculate may seem excessive, given that only 1 can fertilize an egg. Only 1 in 1,000 will ever arrive in the vicinity of an ovum, however. Millions deposited in the vagina simply flow out of the woman's body because of gravity, unless she remains prone for quite some time. Normal vaginal acidity kills many more. Many surviving sperm swim (against the current of fluid coming from the cervix) through the os and into the uterus. Surviving sperm may reach the fallopian tubes 60 to 90 minutes after ejaculation. About half the sperm end up in the wrong tube—that is, the one that does not contain the egg. Perhaps some 2,000 sperm find their way into the right tube. Fewer still manage to swim the final 2 inches against the currents generated by the cilia that line the tube.

The journey of sperm may not be random or blind. Fertile ova secrete a compound that appears to attract sperm cells (Angier, 1992). Sperm cells contain odor receptors that were once thought to be found only in the nasal cavity. It is thus conceivable (pardon the pun) that sperm cells are attracted to ova through a variation of the sense of smell.

Fertilization normally occurs in a fallopian tube. (Figure 11.1 shows sperm swarming around an egg in a fallopian tube.) Ova contain chromosomes, proteins, fats, and nutritious fluid and are surrounded by a gelatinous layer called the **zona pellucida.** This layer must be penetrated if fertilization is to occur. Sperm that have completed their journey secrete the enzyme **hyaluronidase,** which briefly thins the zona pellucida, enabling one sperm to penetrate. Once a sperm has entered, the zona pellucida thickens, locking other sperm out. The corresponding chromosomes in the sperm and ovum line up opposite each other. Conception occurs as the chromosomes from the sperm and ovum combine to form 23 new pairs, which carry a unique set of genetic instructions.

Optimizing the Chances of Conception

Some couples may wish to optimize their chances of conceiving during a particular month so that birth occurs at a desired time. Others may have difficulty conceiving and wish to

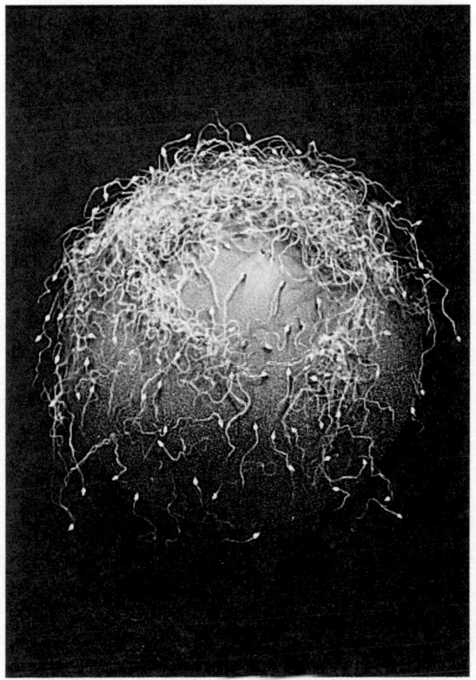

FIGURE 11.1

Human Sperm Swarming Around an Ovum in a Fallopian Tube. Fertilization normally occurs in a fallopian tube, not in the uterus.

Zona pellucida A gelatinous layer that surrounds an ovum. (From roots that mean "zone that light can shine through.")

Hyaluronidase An enzyme that briefly thins the zona pellucida, enabling one sperm to penetrate. (From roots that mean "substance that breaks down a glasslike fluid.")

a CLOSER look

Selecting the Gender of Your Child

Parents have wished that they could select the gender of their children for thousands of years. In many cultures, one gender—usually male—has been preferred over the other. In other cases, parents who already have girls (or boys) would like to be able to "balance" their family by having a boy (or girl). Then, too, there are a number of sex-linked diseases—diseases that show up only in sons, for example. In such cases, parents would feel more secure if they could elect to have only daughters (Bennett, 1998).

But how would one select the gender of one's children? Folklore is replete with methods. Some cultures have advised coitus under the full moon to conceive boys. The Greek philosopher Aristotle suggested that making love during a north wind would yield sons, a south wind daughters. Sour foods were once suggested for parents desirous of having boys. Those who wanted girls were advised to consume sweets. Husbands who yearned to have boys might be advised to wear their boots to bed. The thesis that the right testicle was responsible for seeding boys was popular at one time. Eighteenth-century French noblemen were advised to have their left testicles removed if they wanted to sire sons. It should go without saying that none of these methods worked.

These methods (or, should we say, non-methods?) supposedly led to the conception of children of the desired gender. However, methods applied after conception have also been used, such as the abortion of fetuses because of their gender. There are also many cultures in which infanticide has been practiced. Because boys have usually been considered more desirable than girls, female infanticide has been more common than male infanticide.

Today, more reliable ways to increase the odds of conceiving a child of one's desired gender seem to be in the offing.

Shettles's Approach

Sperm bearing the Y sex chromosome are smaller than those bearing the X sex chromosome and are faster swimmers. But sperm with the X sex chromosome are more durable. From these assumptions, Shettles (1982) derived a number of strategies for choosing the gender of one's children. For example, to increase the chances of having a boy, the couple should engage in coitus on the day of ovulation and the man should be penetrating deeply at the moment of ejaculation. To increase the chances of having a girl, the couple should engage in coitus two days or more before ovulation, and the man should ejaculate at a shallow depth of penetration. However, other researchers have not found these methods to be effective (Rawlins, 1998; Wilcox et al., 1995).

Sperm Separation Procedures

Several sperm separation procedures have also been in use. One is based on the relative swimming rates of Y- and X-bearing sperm. Another relies on the differences in electrical charges of the two types of sperm to separate them.

A more current and promising current sperm separation method relies on the fact that sperm carrying the Y sex chromosome have about 2.8% less genetic material (DNA) than sperm carrying the X sex chromosome (Kolata, 1998c). Edward Fugger (1998) and his colleagues have managed to pass the some 200 million human sperm in an ejaculation through a "DNA detector" and sort them on this basis. The cell sorter lined up the sperm individually in a stream, and then they were separated according to them amount of DNA they contained. Mothers were then artificially inseminated with the "proper" sperm. The results: The method may be able to create a female baby on about 85% of attempts, a male baby on about 65% of attempts. The method is more effective with cows and horses, because among these animals there is a larger difference (about 4%) in the amount of DNA be-

maximize their chances for a few months before consulting a fertility specialist. Some fairly simple procedures can dramatically increase the chances of conceiving for couples without serious fertility problems.

The ovum can be fertilized for about 4 to 20 hours after ovulation (Wilcox et al., 2000). Sperm are most active within 48 hours after ejaculation. So one way of optimizing the chances of conception is to engage in coitus within a few hours of ovulation. There are a number of ways to predict ovulation.

USING THE BASAL BODY TEMPERATURE CHART Few women have perfectly regular cycles, so they can only guess when they are ovulating (Wilcox et al., 2000). A basal body temperature (BBT) chart (Figure 11.2) may help provide a more reliable estimate.

Can You Select the Gender of Your Child? Would You Want to Do So? There are various approaches to selecting the gender of one's children, some of which are more reliable than others. If you could choose, would you prefer to have girls or boys? Do you believe that it is ethical to select the gender of your children? Why or why not?

tween male and female sex-chromosome-bearing sperm (Kolata, 1998c).

This method and others, should they prove reliable and affordable, raise many moral and ethical questions. Many individuals will wonder, for example, whether people have the right to select the gender of their children. In some cases, the word *right* will have a religious meaning. That is, some believe that only God should be able to determine the gender of a child. Others may think of the word *right* in terms of gender balancing within a culture. Because male children tend to be preferred, gender selection could lead quickly to an overabundance of males within a society. Some, like the Ethics Committee of the American Society of Reproductive Medicine (2000), consider gender selection for nonmedical reasons to be sexist. But many believe that selecting the gender of one's children is a purely personal matter. "By what authority," they challenge, "do some decide whether the rest of us have acceptable reasons for selecting the sex of our children?" (Holmes-Farley, 1998).

The new millennium is likely to bring yet more efficient and affordable methods for selecting the gender of infants. For that reason, we are very likely to have to cope with these moral and ethical questions—sooner rather than later.

As shown in the figure, body temperature is fairly even before ovulation, and early-morning body temperature is generally below 98.6 degrees Fahrenheit. But just before ovulation, basal temperature dips slightly. Then, on the day after ovulation, temperature tends to rise by about 0.4 to 0.8 degree and to remain higher until menstruation. In using the BBT method, a woman attempts to detect these temperature changes by tracking her temperature just after awakening each morning but before rising from bed. Thermometers that provide finely graded readings, such as electronic digital thermometers, are best suited for determining these minor changes. The couple record the woman's temperature and the day of the cycle (as well as the day of the month) and indicate whether they have engaged in coitus. With regular charting for six months, the woman may learn to predict the day of ovulation more accurately—assuming that her cycles are fairly regular.

FIGURE 11.2

A Basal Body Temperature (BBT) Chart. Because most women have somewhat irregular menstrual cycles, they may not be able to predict ovulation perfectly. The basal body temperature (BBT) chart helps them to do so. Body temperature is fairly even before ovulation but dips slightly just before ovulation. On the day following ovulation, temperature rises about 0.4°F to 0.8°F above the level before ovulation.

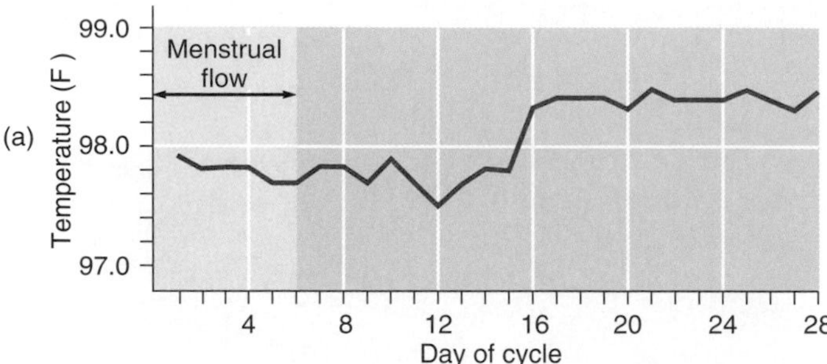

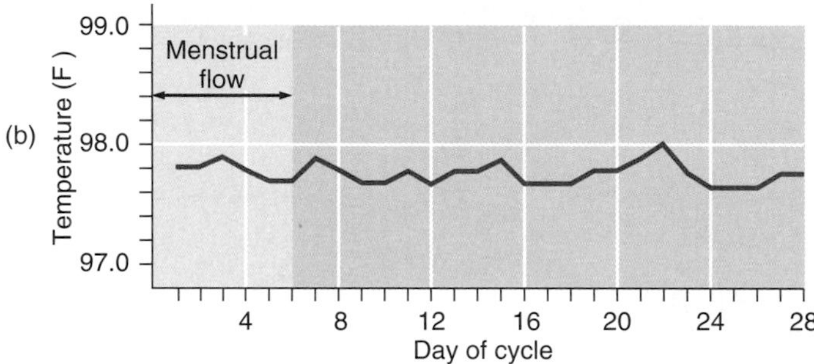

Opinion is divided as to whether it is better for couples to have coitus every 24 hours or every 36 to 48 hours for the several-day period during which ovulation is expected. More frequent coitus around the time of ovulation may increase the chances of conception. Less frequent coitus (that is, every 36 to 48 hours) leads to a higher sperm count during each ejaculation. Most fertility specialists recommend that couples seeking to conceive a baby have intercourse once every day or two during the week in which the woman expects to ovulate. Men with lower than normal sperm counts may be advised to wait 48 hours between ejaculations, however.

ANALYZING URINE OR SALIVA FOR LUTEINIZING HORMONE Over-the-counter kits are more accurate than the BBT method and predict ovulation by analyzing the woman's urine or saliva for the surge in luteinizing hormone (LH) that precedes ovulation by about 12 to 24 hours.

TRACKING VAGINAL MUCUS Women can track the thickness of their vaginal mucus during the phases of the menstrual cycle by rolling it between their fingers and noting changes in texture. The mucus is thick, white, and cloudy during most phases of the cycle. It becomes thin, slippery, and clear for a few days preceding ovulation. A day or so after ovulation, the mucus again thickens and becomes opaque.

ADDITIONAL CONSIDERATIONS Coitus in the male-superior position allows sperm to be deposited deeper in the vagina and minimizes leakage of sperm out of the vagina due to gravity. Women may improve their chances of conceiving by lying on their backs and drawing their knees close to their breasts after ejaculation. This position, perhaps aided by the use of a pillow beneath the buttocks, may prevent sperm from dripping out quickly and elevate the pool of semen in relation to the cervix. It thus makes gravity work for, rather than against, conception. Women may also avoid standing and may lie as still as possible for about 30 to 60 minutes after ejaculation to help sperm move toward the cervical opening.

Women with severely retroverted, or "tipped," uteruses may profit from supporting themselves on their elbows and knees and having their partners enter them from behind. Again, this position helps prevent semen from dripping out of the vagina.

The man should penetrate the woman as deeply as possible just before ejaculation, hold still during ejaculation, and then withdraw slowly in a straight line to avoid dispersing the pool of semen.

INFERTILITY AND ALTERNATIVE WAYS OF BECOMING PARENTS

For couples who want children, few problems are more frustrating than inability to conceive. Physicians often recommend that couples try to conceive on their own for six months before seeking medical assistance. The term **infertility** is usually not applied until the failure to conceive has persisted for more than a year.

Infertility concerns millions of Americans. Because the incidence of infertility increases with age, it is partially the result of a rise in couples who postpone childbearing until their 30s and 40s (Sheehy, 1995). All in all, about 15% of American couples have fertility problems (Howards, 1995). However, about half of them eventually succeed in conceiving a child (Jones & Toner, 1993). Many treatment options are available, ranging from drugs to stimulate ovulation to newer reproductive technologies, such as in vitro fertilization.

Male Fertility Problems

Although most concerns about fertility have traditionally centered on women, the problem lies with the man in about 30% of cases (Howards, 1995). In about 20% of cases, problems are found in both partners (Hatcher et al., 1998; Howards, 1995).

Fertility problems in the male reflect abnormalities such as:

1. Low sperm count
2. Irregularly shaped sperm—for example, malformed heads or tails
3. Low sperm **motility**
4. Chronic diseases such as diabetes, as well as infectious diseases such as sexually transmitted diseases
5. Injury to the testes
6. An **autoimmune response,** in which antibodies produced by the man deactivate his own sperm
7. A pituitary imbalance and/or thyroid disease

Problems in producing normal, abundant sperm may be caused by genetic factors, advanced age, hormonal problems, diabetes, injuries to the testes, varicose veins in the scrotum, drugs (alcohol, narcotics, marijuana, and/or tobacco), antihypertensive medications, environmental toxins, excess heat, and emotional stress. Sperm production gradually declines with age, but normal aging does not produce infertility. Men in late adulthood father children, even though conception may require more attempts.

Low sperm count (or the absence of sperm) is the most common problem. Sperm counts of 40 million to 150 million sperm per milliliter of semen are considered normal. A count of fewer than 20 million is generally regarded as low. Sperm production may be low among men with undescended testes that were not surgically corrected before puberty. Frequent ejaculation can reduce sperm counts. Sperm production may also be impaired in men whose testicles are consistently 1 or 2 degrees above the typical scrotal temperature of 94 to 95 degrees Fahrenheit (Leary, 1990). Frequent hot baths and tight-fitting underwear can also reduce sperm production, at least temporarily. Some men may encounter fertility problems from prolonged athletic activity, use of electric blankets, or even long, hot baths. In such cases the problem can be readily corrected. Male runners with fertility problems are often counseled to take time off to increase their sperm counts. Interestingly, the small amounts of estrogen that men produce may help keep sperm counts high ("Estrogen," 1997).

Sometimes the sperm count is adequate, but prostate, hormonal, or other factors deprive sperm of motility or deform them. Motility can also be hampered by scar tissue from infections. Scarring may prevent sperm from passing through parts of the

Infertility Inability to conceive a child.

Motility Self-propulsion. A measure of the viability of sperm cells.

Autoimmune response The production of antibodies that attack naturally occurring substances that are (incorrectly) recognized as being foreign or harmful.

male reproductive system, such as the vas deferens. To be considered normal, sperm must be able to swim for at least 2 hours after coitus and most (60% or more) must be normal in shape.

It is true that prolonged athletic activity may decrease fertility in the male. Such activity can raise the temperature of the scrotum, providing a less-than-optimal environment for sperm.

Sperm counts have been increased by surgical repair of the varicose veins in the scrotum. Microsurgery can also open blocked passageways that prevent the outflow of sperm (Schroeder-Printzen et al., 2000). Researchers are also investigating the effects on sperm production of special cooling undergarments. One device, a kind of athletic supporter that is kept slightly damp with distilled water, has been approved by the Federal Drug Administration. Seventy percent of the men whose infertility is due to higher-than-normal scrotal temperatures show increased sperm count and quality when they wear cooling undergarments (Leary, 1990; Silber, 1991).

ARTIFICIAL INSEMINATION The sperm of men with low sperm counts can be collected and quick-frozen. The sperm from multiple ejaculations can then be injected into a woman's uterus at the time of ovulation. This is one kind of **artificial insemination.** The sperm of a man with low sperm motility can also be injected into his partner's uterus, so that the sperm begin their journey closer to the fallopian tubes. Sperm from a donor can be used to artificially inseminate a woman whose partner is completely infertile or has an extremely low sperm count. The child then bears the genes of one of the parents, the mother. A donor can be chosen who resembles the man in physical traits and ethnic background.

A variation of artificial insemination has been used with some men with very low (or zero!) sperm counts in the semen, immature sperm, or immotile sperm. Immature sperm can be removed from a testicle by a thin needle and then directly injected into an egg in a laboratory dish (Brody, 1995b). The method has even been successful with a few men who have only tailless spermatids in the testes.

Female Fertility Problems

The major causes of infertility in women are:

1. Irregular ovulation, including failure to ovulate
2. Obstructions or malfunctions of the reproductive tract, which are often caused by infections or diseases involving the reproductive tract
3. Endometriosis
4. Declining hormone levels of estrogen and progesterone that occur with aging and may prevent the ovum from becoming fertilized or remaining implanted in the uterus

From 10% to 15% of female infertility problems stem from failure to ovulate. Many factors can play a role in failure to ovulate, including hormonal irregularities, malnutrition, genetic factors, stress, and chronic disease. Failure to ovulate may occur in response to low levels of body fat, as in the cases of women with eating disorders and athletes (Frisch, 1997).

Ovulation may often be induced by the use of fertility drugs such as *clomiphene* (Clomid). Clomiphene stimulates the pituitary gland to secrete FSH and LH, which in turn stimulates maturation of ova. Clomiphene leads to conception in the majority of cases of infertility that are due *solely* to irregular or absent ovulation (Reinisch, 1990). But because infertility can have multiple causes, only about half of women who use clomiphene become pregnant. Another infertility drug, Pergonal, contains a high concentration of FSH, which directly stimulates maturation of ovarian follicles. Like clomiphene, Pergonal has

Artificial insemination The introduction of sperm into the reproductive tract through means other than sexual intercourse.

Endometriosis An abnormal condition in which endometrial tissue is sloughed off into the abdominal cavity rather than out of the body during menstruation. The condition is characterized by abdominal pain and may cause infertility.

a CLOSER look

Sex and the Single Lizard

Some years back, a scandalous book emerged on the scene—*Sex and the Single Woman.* A somewhat different scandal has now emerged in the animal world. It could be dubbed "Sex and the Single Lizard."

National Geographic reports the discovery of a most unusual all-female species of lizard found in South America and the West Indies. At time for reproduction, no males need apply. With no male contact, the lizards lay unfertilized eggs. The hatchlings develop into identical females generation after generation (Cole, 1995). Why? They derive all of their hereditary material from their mothers. The scientific name of the species is *Gymnophthalmus underwoodi.* (We include the scientific name so that you will think that there is some value to this A Closer Look feature.)

The lizards are hybrids of two species of lizard, each of which reproduces normally—that is, by having male lizards fertilize the females' eggs.

It is true that there is an all-female species of lizard that lays unfertilized eggs that develop into identical females generation after generation.

The third author notes that she has three perfect daughters and no sons. The first author does not wish to discuss the matter further.

A Chip Off the Old . . . Mommy. There is a species of lizard that obtains all its genetic material from its mother. Yes, there is no generation gap.

high success rates with women whose infertility is due to lack of ovulation. Clomiphene and Pergonal have been linked to multiple births, including quadruplets and even quintuplets (Gleicher et al., 2000). The Centers for Disease Control and Prevention (2000c) estimate that 43% of the triplet and higher-order multiple births in the United States result from such fertility treatments. However, fewer than 10% of such pregnancies result in multiple births.

Local infections that scar the fallopian tubes and other organs impede the passage of sperm or ova. Such infections include pelvic inflammatory disease—an inflammation of the woman's internal reproductive tract that can be caused by various infectious agents, such as the bacteria responsible for gonorrhea and chlamydia (see Chapter 16).

In **endometriosis,** cells break away from the uterine lining (the endometrium) and become implanted and grow elsewhere. When they develop on the surface of the ovaries or fallopian tubes, they may block the passage of ova or impair conception. About one case in six of female sterility is believed to be due to endometriosis. Hormone treatments and surgery sometimes reduce the blockage to the point where the women can conceive. A physician may suspect endometriosis during a pelvic exam, but it is diagnosed with certainty by **laparoscopy.** A long, narrow tube is inserted through an incision in the navel, permitting the physician to inspect the organs in the pelvic cavity visually. The incision is practically undetectable.

Suspected blockage of the fallopian tubes may also be checked by a **Rubin test** or by a **hysterosalpingogram** (sometimes shortened to "hysterogram"). In a Rubin test, carbon dioxide gas is blown through the cervix. Its pressure is then monitored to determine whether it flows freely through the fallopian tubes into the abdomen or is trapped in the uterus. In the more common hysterosalpingogram, the movement of an injected dye is monitored by X-rays. This procedure may be uncomfortable.

Several methods help many couples with problems such as blocked fallopian tubes bear children.

Laparoscopy A medical procedure in which a long, narrow tube (laparoscope) is inserted through an incision in the navel, permitting the visual inspection of organs in the pelvic cavity. (From the Greek *lapara,* which means "flank.")

Rubin test A test in which carbon dioxide gas is blown through the cervix and its progress through the reproductive tract is tracked to determine whether the fallopian tubes are blocked.

Hysterosalpingogram A test in which a dye is injected into the reproductive tract and its progress is tracked by X-rays to determine whether the fallopian tubes are blocked. (From roots that mean "record of," "uterus," and "fallopian tubes.")

IN VITRO FERTILIZATION When Louise Brown was born in England in 1978 after being conceived by the method of **in vitro fertilization** (IVF), the event made headlines around the world. Louise was dubbed the world's first "test-tube baby." However, conception took place in a laboratory dish (not a test tube), and the embryo was implanted in the mother's uterus, where it developed to term. Before in vitro fertilization, fertility drugs stimulate ripening of ova. Ripe ova are then surgically removed from an ovary and placed in a laboratory dish along with the father's sperm. Fertilized ova are then injected into the mother's uterus to become implanted in the uterine wall.

Truth or Fiction? **REVISITED**

It is not true that a "test-tube baby" is grown in a large laboratory dish throughout the 9-month gestation period. A test-tube baby is actually conceived in a laboratory dish (which is similar to a test tube, perhaps), but the fertilized egg is then placed in the mother's uterus, where it must become implanted if it is to develop to term.

GIFT In **gamete intrafallopian transfer (GIFT),** sperm and ova are inserted together into a fallopian tube for fertilization. Conception occurs in a fallopian tube rather than a laboratory dish.

ZIFT **Zygote intrafallopian transfer (ZIFT)** involves a combination of IVF and GIFT. Sperm and ova are combined in a laboratory dish. After fertilization, the zygote is placed in the mother's fallopian tube to begin its journey to the uterus for implantation. ZIFT has an advantage over GIFT in that the fertility specialists can ascertain that fertilization has occurred before insertion is performed.

DONOR IVF "I tell her Mommy was having trouble with, I call them ovums, not eggs," says a 50-year-old female therapist in Los Angeles (cited in Stolberg, 1998a). "I say that I needed these to have a baby, and there was this wonderful woman and she was willing to give me some, and that was how she helped us. I want to be honest that we got pregnant in a special way."

That special way is termed **donor IVF,** which is a variation of the IVF procedure in which the ovum is taken from another woman, fertilized, and then injected into the uterus or fallopian tube of the intended mother. The procedure is used when the intended mother does not produce ova. The number of births effected by this method has been mushrooming in recent years (Stolberg, 1998a).

EMBRYONIC TRANSFER A similar method for women who do not produce ova of their own is **embryonic transfer.** In this method, a woman volunteer is artificially inseminated by the male partner of the infertile woman. Five days later the embryo is removed from the volunteer and inserted within the uterus of the mother-to-be, where it is hoped that it will become implanted.

In vitro and transfer methods are costly. Success with in vitro fertilization drops from nearly 30% in women in their mid-20s to 10% to 15% in women in their late 30s (Toner et al., 1991).

INTRACYTOPLASMIC SPERM INJECTION **Intracytoplasmic sperm injection** (ICSI) has been used when the man has too few sperm for IVF or when IVF fails. In this method, a single sperm is injected directly into an ovum. The method has enabled tens of thousands of men to conceive children ("Study raises concern," 2000). However, research with Belgian and Australian children who were conceived by this method suggests that it may be associated with an increase in the risk of birth defects. These defects include heart, stomach, kidney, and bladder problems, cleft palate, hernia ("New fertility treatment," 1997),

In vitro fertilization
A method of conception in which mature ova are surgically removed from an ovary and placed in a laboratory dish along with sperm.

Gamete intrafallopian transfer (GIFT) A method of conception in which sperm and ova are inserted into a fallopian tube to encourage conception.

Zygote intrafallopian transfer (ZIFT) A method of conception in which an ovum is fertilized in a laboratory dish and then placed in a fallopian tube.

Donor IVF A variation of in vitro fertilization in which the ovum is taken from one woman, fertilized, and then injected into the uterus or fallopian tube of another woman.

Embryonic transfer A method of conception in which a woman volunteer is artificially inseminated by the male partner of the intended mother, after which the embryo is removed from the volunteer and inserted within the uterus of the intended mother.

Intracytoplasmic sperm injection A method of conception in which a single sperm is injected directly into an ovum.

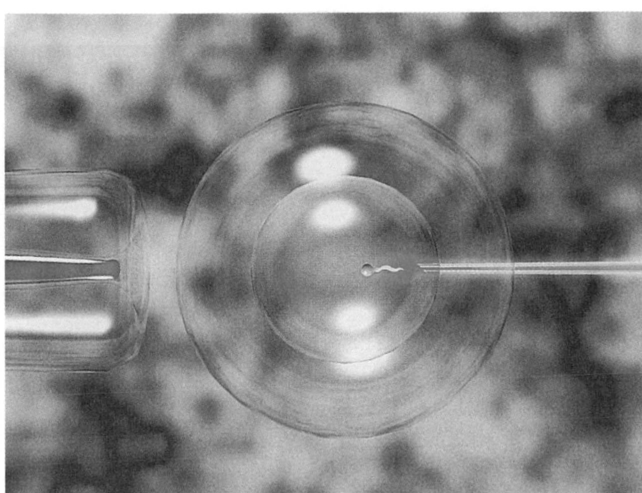

Intracytoplasmic Sperm Injection. This method of conception is sometimes used when the man has too few sperm for in vitro fertilization (IVF) or when IVF fails. As shown in the photograph, a thin (very thin!) needle injects a single sperm directly into an ovum.

and, in boys, malformation of the penis (Wennerholm et al., 2000). Although ICSI is connected with a higher fertilization rate than IVF, these concerns have led many experts to recommend that couples try IVF first (Fishel et al., 2000).

SURROGATE MOTHERHOOD A **surrogate mother** is artificially inseminated by the husband of the infertile woman and carries the baby to term. The surrogate signs a contract to turn the baby over to the infertile couple. Such contracts have been invalidated in some states, however, and surrogate mothers in these states cannot be compelled to hand over the babies.

ADOPTION Adoption is yet another way for people to obtain children. Despite the occasional conflicts in which adoptive parents are pitted against biological parents who change their minds about giving their children up for adoption, most adoptions result in the formation of loving new families. Many people in the United States find it easier to adopt infants from other countries, infants with special needs, or older children.

PREGNANCY

Women react to becoming pregnant in different ways. For those who are psychologically and economically prepared, pregnancy may be greeted with joyous celebration. Some women feel that pregnancy helps fulfill their sense of womanhood:

> Being pregnant meant I was a woman. I was enthralled with my belly growing. I went out right away and got maternity clothes.
>
> It gave me a sense that I was actually a woman. I had never felt sexy before . . . I felt very voluptuous. (Boston Women's Health Book Collective, 1992; Copyright © 1984, 1992 by The Boston Women's Health Book Collective. Reprinted with permission.)

On the other hand, an unwanted pregnancy may evoke feelings of fear and hopelessness, as occurs with many teenagers.

In this section we examine biological and psychological aspects of pregnancy: signs of pregnancy, prenatal development, complications, effects of drugs and sex, and the psychological experiences of pregnant women and fathers. The nearby questionnaire invites you to consider whether you should have a child.

Surrogate mother A woman who is impregnated with the sperm of a prospective father via artificial insemination, carries the embryo and fetus to term, and then gives the child to the prospective parents.

Early Effects of Pregnancy

Just a few days after conception, a woman may note tenderness of the breasts. Hormonal stimulation of the mammary glands may make the breasts more sensitive and cause sensations of tingling and fullness.

The term **morning sickness** refers to the nausea, food aversions, and vomiting that many women experience during pregnancy. Women carrying more than one child usually experience more nausea. Although called morning sickness, nausea and vomiting during pregnancy can occur any time during the day or night and is not a "sickness" at all but, rather, a perfectly normal part of pregnancy (Flaxman & Sherman, 2000). There is some evidence that the nausea and vomiting experienced by pregnant women reflect bodily changes that promote the development of the placenta (Huxley, 2000). Evolutionary biologists Samuel Flaxman and Paul Sherman (2000) reviewed records of some 20,000 pregnancies and reported in the *Quarterly Review of Biology* that morning sickness was associated with a healthy outcome, including lower incidences of miscarriage and stillbirth. The researchers suggest that the food aversions protect the mother and fetus from likely sources of toxins and disease-causing agents, which are most commonly found in spoiled meat. Although it may have provided an evolutionary advantage, the usefulness of morning sickness has waned over the millennia, especially with the advent of refrigeration to help safely store foods.

All of this is little consolation to pregnant women. In some cases, morning sickness is so severe that the woman cannot eat regularly and must be hospitalized to ensure that she and the fetus receive adequate nutrition. In milder cases, having small amounts of food in the stomach throughout the day is helpful. Many women find that eating a few crackers at bedtime and before getting out of bed in the morning is effective. Other women profit from medication. Many medications are available today, and women are advised to discuss the situation with their obstetricians rather than just assuming that they have to tough it out on their own. Morning sickness usually—but not always—subsides by about the twelfth week of pregnancy.

Although morning sickness tends to decrease as a pregnancy progresses, it is *not* true that it is limited to the first trimester. Nor is it limited to the morning. It can occur at any time.

Pregnant women may experience greater-than-normal fatigue during the early weeks, sleeping longer and falling asleep more readily than usual. Frequent urination, which may also be experienced, is caused by pressure from the swelling uterus on the bladder.

Miscarriage (Spontaneous Abortion)

Miscarriages have many causes, including chromosomal defects in the fetus and abnormalities of the placenta and uterus. Miscarriage is more prevalent among older mothers (Stein & Susser, 2000). About three in four miscarriages occur in the first 16 weeks of pregnancy, and the great majority of these occur in the first 7 weeks. Some miscarriages occur so early that the woman is not aware she was pregnant.

After a miscarriage, a couple may feel a deep sense of loss and undergo a period of mourning. Emotional support from friends and family often helps the couple cope with the loss. In most cases, women who miscarry can carry subsequent pregnancies to term.

Sex During Pregnancy

Most health professionals concur that coitus is safe throughout the course of pregnancy until the start of labor, provided that the pregnancy is developing normally and the woman has no history of miscarriages. Women who experience bleeding or cramps during pregnancy may be advised by their obstetricians not to engage in coitus.

Morning sickness Symptoms of pregnancy, including nausea, aversions to specific foods, and vomiting.

Miscarriage A spontaneous abortion.

Masters and Johnson (1966) reported an initial decline in sexual interest among pregnant women during the first trimester. There is increased interest during the second trimester and another decline in interest during the third. Many women show declines in sexual interest and activity during the first trimester because of fatigue, nausea, or misguided concerns that coitus will harm the embryo or fetus. Also during the first trimester, vasocongestion may cause tenderness of the breasts, discouraging fondling and sucking. One study found that 90% of 570 women were engaging in coitus at 5 months into their pregnancies (Byrd et al., 1998). Researchers in Israel reported a gradual decline in sexual interest and frequency of intercourse and orgasm during pregnancy among a sample of 219 women. The greatest decline occurred during the third trimester (Hart et al., 1991). Pain during intercourse is also commonly reported, especially in the third trimester.

As the woman's abdominal region swells, the popular male-superior position becomes unwieldy. The female-superior, lateral-entry, and rear-entry positions are common alternatives. Manual and oral sex can continue as usual.

Some women are concerned that the uterine contractions of orgasm may dislodge an embryo. Such concerns are usually unfounded, unless the woman has a history of miscarriage or is currently at risk of miscarriage. Women and their partners are advised to consult their obstetricians for the latest information.

Psychological Changes During Pregnancy

A woman's psychological response to pregnancy reflects her desire to be pregnant, her physical changes, and her attitudes toward these changes. Women with the financial, social, and psychological resources to meet the needs of pregnancy and child rearing may welcome pregnancy. Some describe it as the most wondrous experience of their lives. Other women may question their ability to handle their pregnancies and childbirth. Or they may fear that pregnancy will interfere with their careers or their mates' feelings about them. In general, women who want to have a baby and choose to become pregnant are better adjusted during their pregnancies.

The first trimester may be difficult for women who are ambivalent about pregnancy. At that stage symptoms like morning sickness are most pronounced, and women must come to terms with being pregnant. The second trimester is generally less tempestuous. Morning sickness and other symptoms have largely vanished. It is not yet difficult to move about, and the woman need not yet face the delivery. Women first note fetal movement during the second trimester, and for many the experience is stirring:

> I was lying on my stomach and felt—something, like someone lightly touching my deep insides. Then I just sat very still and . . . felt the hugeness of having something living growing in me. Then I said, No, it's not possible, it's too early yet, and then I started to cry. . . .That one moment was my first body awareness of another living thing inside me. (Boston Women's Health Book Collective, 1992; Copyright © 1984, 1992 by The Boston Women's Health Book Collective. Reprinted with permission.)

During the third trimester it is normal, especially for first-time mothers, to worry about the mechanics of delivery and whether the child will be normal. The woman becomes increasingly heavy and literally "bent out of shape." It may become difficult to get up from a chair or out of bed. She must sit farther from the steering wheel when driving. Muscle tension from supporting the extra weight in her abdomen may cause backaches. She may feel impatient in the days and weeks just before delivery.

Men, like women, respond to pregnancy according to the degree to which they want the child. Many men are proud and look forward to the child with great anticipation. In such cases, pregnancy may bring parents closer together. But fathers who are financially or emotionally unprepared may consider the pregnancy a "trap." Now and then an expectant father experiences some signs of pregnancy, including morning sickness and vomiting. This reaction is termed a **sympathetic pregnancy.**

Sympathetic pregnancy
The experiencing of a number of signs of pregnancy by the father.

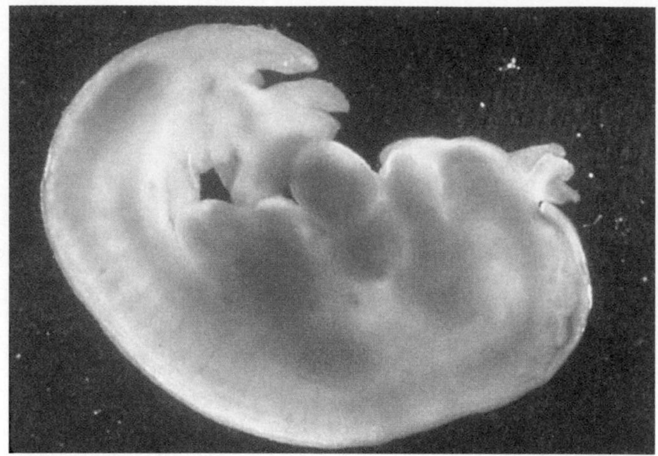

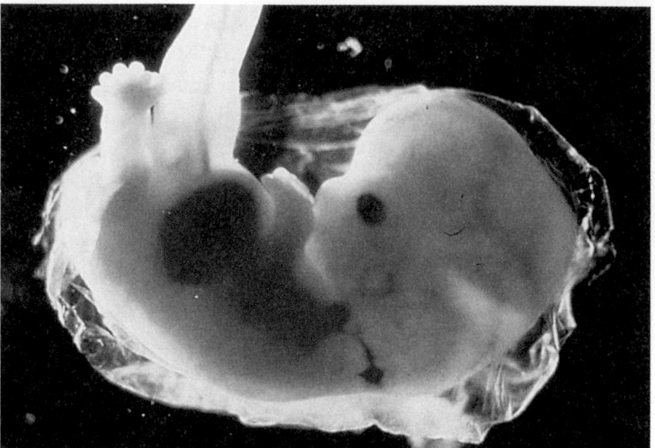

Prenatal Development. Developmental changes are most rapid and dramatic during prenatal development. Within a few months, a human embryo and then fetus advances from weighing a fraction of an ounce to several pounds, and from one cell to billions of cells.

PRENATAL DEVELOPMENT

We can date pregnancy from the onset of the last menstrual cycle before conception, which makes the normal gestation period 280 days. We can also date pregnancy from the date at which fertilization was assumed to have taken place, which normally corresponds to 2 weeks after the beginning of the woman's last menstrual cycle. In this case, the normal gestation period is 266 days.

Once pregnancy has been confirmed, the delivery date may be calculated by *Nagele's rule:*

- Jot down the date of the first day of the last menstrual period.
- Add 7 days.
- Subtract 3 months.
- Add 1 year.

For example, if the last period began on November 12, 2001, adding 7 days yields November 19, 2001. Then subtracting 3 months yields August 19, 2001. Adding 1 year gives a "due date" of August 19, 2002. Few babies are born exactly when they are due,[1] but the great majority are delivered during a 10-day period that spans the date.

[1]The first and third authors wish to boast, however, that their daughters Allyn and Jordan were born precisely on their due dates. At least one of them has been just as compulsive ever since. The second author adds that he and his wife Judy had their son Michael within 1 day of the due date. (Close, but no cigar, notes the first author.)

Shortly after conception, the single cell that results from the union of sperm and egg begins to multiply—becoming 2 cells, then 4, then 8, and so on. During the weeks and months that follow, tissues, structures, and organs begin to form, and the fetus gradually takes on the shape of a human being. By the time the fetus is born, it consists of hundreds of billions of cells—more cells than there are stars in the Milky Way galaxy. Prenatal development can be divided into three periods: the *germinal stage*, which corresponds to about the first 2 weeks, the *embryonic stage*, which coincides with the first 2 months, and the *fetal stage*. We also commonly speak of prenatal development in terms of three trimesters of 3 months each.

The Germinal Stage

Within 36 hours after conception, the zygote divides into 2 cells. It then divides repeatedly, becoming 32 cells within another 36 hours as it continues its journey to the uterus. It takes the zygote perhaps 3 or 4 days to reach the uterus. This mass of dividing cells then wanders about the uterus for perhaps another 3 or 4 days before it begins to become implanted in the uterine wall. Implantation takes about another week. This period from conception to implantation is termed the **germinal stage** or the **period of the ovum** (see Figure 11.3).

Truth or Fiction? REVISITED

It is true that for the first week after conception, a fertilized egg cell is not attached to its mother's body. Later it becomes implanted in the uterine wall. ■

Several days into the germinal stage, the cell mass takes the form of a fluid-filled ball of cells, which is called a **blastocyst.** Already some cell differentiation has begun. Cells begin to separate into groups that will eventually become different structures. Within a thickened mass of cells that is called the **embryonic disk,** two distinct inner layers of cells

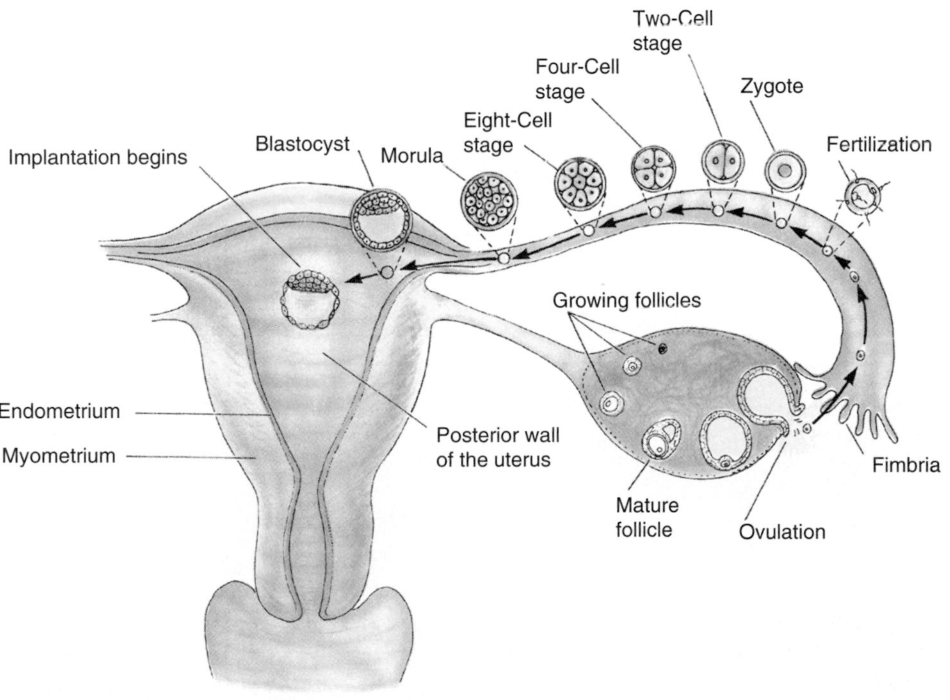

FIGURE 11.3

The Ovarian Cycle, Conception, and the Early Days of the Germinal Stage. The zygote first divides about 36 hours after conception. Continuing division creates the hollow sphere of cells termed the blastocyst. The blastocyst normally becomes implanted in the wall of the uterus.

Germinal stage The period of prenatal development before implantation in the uterus.

Period of the ovum Germinal stage.

Blastocyst A stage of embryonic development within the germinal stage of prenatal development, at which the embryo is a sphere of cells surrounding a cavity of fluid.

Embryonic disk The platelike inner part of the blastocyst, which differentiates into the ectoderm, mesoderm, and endoderm of the embryo.

are beginning to form. These cells will become the embryo and eventually the fetus. The outer part of the blastocyst, called the **trophoblast,** consists of several membranes from which the amniotic sac, placenta, and umbilical cord eventually develop.

Implantation may be accompanied by some bleeding, which results from the usual rupturing of some small blood vessels that line the uterus. Bleeding can also be a sign of a miscarriage—although most women who experience implantation bleeding do not miscarry but go on to have normal pregnancies and deliver healthy babies.

The Embryonic Stage

The period from implantation to about the eighth week of development is called the **embryonic stage.** The major organ systems of the body begin to differentiate during this stage.

Development of the embryo follows two trends, **cephalocaudal** and **proximodistal.** The apparently oversized heads depicted in Figure 11.4 represent embryos and fetuses at various stages of prenatal development. Growth of the head (the cephalic region) takes

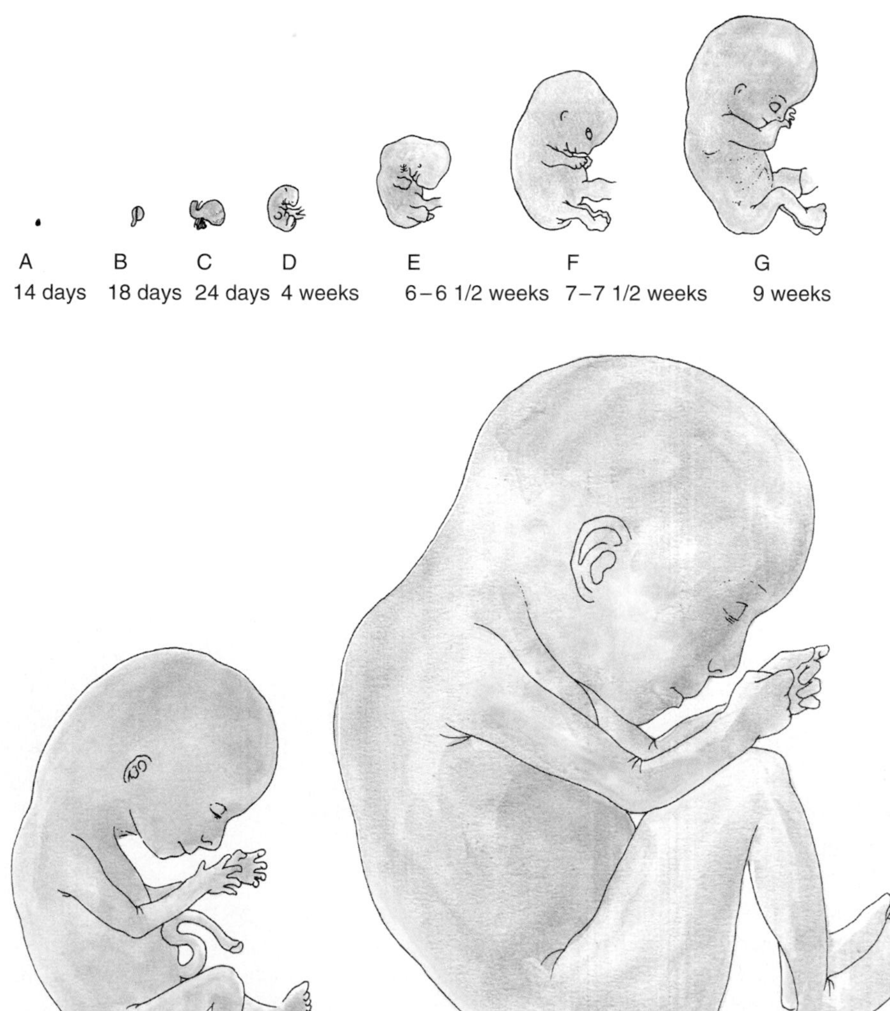

FIGURE 11.4

Human Embryos and Fetuses. Development is cephalocaudal and proximodistal. Growth of the head takes precedence over growth of the lower parts of the body.

Trophoblast The outer part of the blastocyst, from which the amniotic sac, placenta, and umbilical cord develop.

Embryonic stage The stage of prenatal development that lasts from implantation through the eighth week and is characterized by the differentiation of the major organ systems.

Cephalocaudal From the head downward. (From Latin roots that mean "head" and "tail.")

Proximodistal From the central axis of the body outward. (From Latin roots that mean "near" and "far.")

precedence over the growth of the lower parts of the body. You can also think of the body as containing a central axis that coincides with the spinal cord. The growth of the organ systems that lie close to this axis (that is, *proximal* to the axis) takes precedence over the growth of those that lie farther away toward the extremities (that is, *distal* to the axis). Relatively early maturation of the brain and organ systems that lie near the central axis allows these organs to facilitate further development of the embryo and fetus.

As the embryonic stage unfolds, the nervous system, sensory organs, hair, nails, and teeth and the outer layer of skin begin to develop from the outer layer of cells, or **ectoderm,** of the embryonic disk. By about 3 weeks after conception, two ridges appear in the embryo and fold together to form the **neural tube.** This tube develops into the nervous system. The inner layer of the embryonic disk is called the **endoderm.** From this layer develop the respiratory and digestive systems and organs such as the liver and the pancreas. A short time later in the embryonic stage, the middle layer of cells, or **mesoderm,** differentiates and develops into the reproductive, excretory, and circulatory systems, as well as the skeleton, the muscles, and the inner layer of the skin.

During the third week of development, the head and blood vessels begin to form. By the fourth week, a primitive heart begins to beat and pump blood in an embryo that measures but a fifth of an inch in length. The heart will normally continue to beat without rest for every minute of every day for the better part of a century. By the end of the first month of development, we can see the beginnings of the arms and legs—"arm buds" and "leg buds." The mouth, eyes, ears, and nose begin to take shape. The brain and other parts of the nervous system begin to develop.

The arms and legs develop in accordance with the proximodistal principle. First the upper arms and legs develop. Then the forearms and lower legs. Then the hands and feet form, followed by webbed fingers and toes by about 6 to 8 weeks into development. The webbing is gone by the end of the second month. By this time the head has become rounded, and the limbs have elongated and separated. Facial features are visible. All this has occurred in an embryo that is about 1 inch long and weighs 1/30 of an ounce. During the second month, nervous impulses also begin to travel through the developing nervous system.

THE AMNIOTIC SAC The embryo—and later the fetus—develop within a protective environment in the mother's uterus called the **amniotic sac,** which is surrounded by a clear membrane. The embryo and fetus are suspended within the sac in **amniotic fluid.** The amniotic fluid acts like a shock absorber. It cushions the embryo from damage that might result from the mother's movements. The fluid also helps maintain a steady temperature.

THE PLACENTA Nutrients and waste products are exchanged between mother and embryo (or fetus) through a mass of tissue called the **placenta.** The placenta is unique in origin. It develops from material supplied by both mother and embryo. Toward the end of the first trimester, it becomes a flattish, round organ about 7 inches in diameter and 1 inch thick—larger than the fetus itself. The fetus is connected to the placenta by the **umbilical cord.** The mother is connected to the placenta by the system of blood vessels in the uterine wall. The umbilical cord develops about 5 weeks after conception and reaches 20 inches in length. It contains two arteries through which maternal nutrients reach the embryo. A vein transports waste products back to the mother.

The circulatory systems of mother and embryo do not mix. A membrane in the placenta permits only certain substances to pass through, such as oxygen (from the mother to the fetus); carbon dioxide and other wastes (from the embryo or fetus to the mother, to be eliminated by the mother's lungs and kidneys); nutrients; some microscopic disease-causing organisms; and some drugs, including aspirin, narcotics, alcohol, and tranquilizers.

The placenta is also an endocrine gland. It secretes hormones that preserve the pregnancy, stimulate the uterine contractions that induce childbirth, and help prepare the breasts for breast-feeding. Some of these hormones may also cause the signs of pregnancy. HCG (human chorionic gonadotropin) stimulates the corpus luteum to continue to produce progesterone. The placenta itself secretes increasing amounts of estrogen and

Ectoderm The outermost cell layer of the newly formed embryo, from which the skin and nervous system develop.

Neural tube A hollow area in the blastocyst from which the nervous system will develop.

Endoderm The inner layer of the newly formed embryo, from which the lungs and digestive system develop.

Mesoderm The central layer of the embryo, from which the bones and muscles develop.

Amniotic sac The sac containing the fetus.

Amniotic fluid Fluid within the amniotic sac that suspends and protects the fetus.

Placenta An organ connected to the fetus by the umbilical cord. The placenta serves as a relay station between mother and fetus, allowing the exchange of nutrients and wastes.

Umbilical cord A tube that connects the fetus to the placenta.

progesterone. Ultimately, the placenta passes from the woman's body after delivery. For this reason it is also called the afterbirth.

The Fetal Stage

The fetal stage begins by the ninth week and continues until birth. By about the ninth or tenth week, the fetus begins to respond to the outside world by turning in the direction of external stimulation. By the end of the first trimester, the major organ systems, the fingers and toes, and the external genitals have been formed. The gender of the fetus can be determined visually. The eyes have become clearly distinguishable.

During the second trimester the fetus increases dramatically in size, and its organ systems continue to mature. The brain now contributes to the regulation of basic body functions. The fetus increases in weight from 1 *ounce* to 2 *pounds* and grows from about 4 to 14 inches in length. Soft, downy hair grows above the eyes and on the scalp. The skin turns ruddy because of blood vessels that show through the surface. (During the third trimester, layers of fat beneath the skin will give the red a pinkish hue.)

FETAL MOVEMENTS Usually by the middle of the fourth month the mother can feel the first fetal movements. By the end of the second trimester, the fetus moves its limbs so vigorously that the mother may complain of being kicked—often at 4:00 A.M. It opens and shuts its eyes, sucks its thumb, alternates between periods of wakefulness and sleep, and perceives lights and sounds. The fetus also does somersaults, which the mother will definitely feel. Fortunately, the umbilical cord will not break or strangle the fetus, no matter what acrobatic feats it performs.

Near the end of the second trimester the fetus approaches the **age of viability.** Still, only a minority of babies born at the end of the second trimester who weigh under 2 pounds will survive—even with intense medical efforts.

During the third trimester, the organ systems continue to mature and enlarge. The heart and lungs become increasingly capable of maintaining independent life. Typically, during the seventh month the fetus turns upside down in the uterus so that it will be head first, or in a **cephalic presentation,** for delivery. But some fetuses do not turn during this month. If such a fetus is born prematurely, it can have either a **breech presentation** (bottom first) or a shoulder-first presentation, which can complicate problems of prematurity. The closer to term (the full 9 months) the baby is born, the more likely it is that the presentation will be cephalic. If birth occurs at the end of the eighth month, the odds are overwhelmingly in favor of survival.

During the final months of pregnancy, the mother may become concerned that the fetus seems to be less active than before. Most of the time, the change in activity level is normal. The fetus has merely grown so large that it is cramped, and its movements are restricted.

Environmental Influences on Prenatal Development

Advances in scientific knowledge have made us more aware of the changes that take place during prenatal development. They have also heightened our awareness of the problems that can occur and what might be done to prevent them. We focus in this next section on the environmental factors that affect prenatal development. These include the mother's diet, maternal diseases and disorders, and the mother's use of drugs.

THE MOTHER'S DIET It is a common misconception that the fetus will take what it needs from its mother. Actually, malnutrition in the mother can adversely affect fetal development. Women who are too slender risk preterm deliveries and having babies who are low in birth weight (Cnattingius, 1998). Pregnant women who are adequately nourished are more likely to deliver babies of average or above-average size. Their infants are also less likely to develop colds and serious respiratory disorders. However, maternal *obesity* is linked with a higher risk of stillbirth (Cnattingius, 1998).

Age of viability The age at which a fetus can sustain independent life.

Cephalic presentation Emergence of the baby head first from the womb.

Breech presentation Emergence of the baby feet first from the womb.

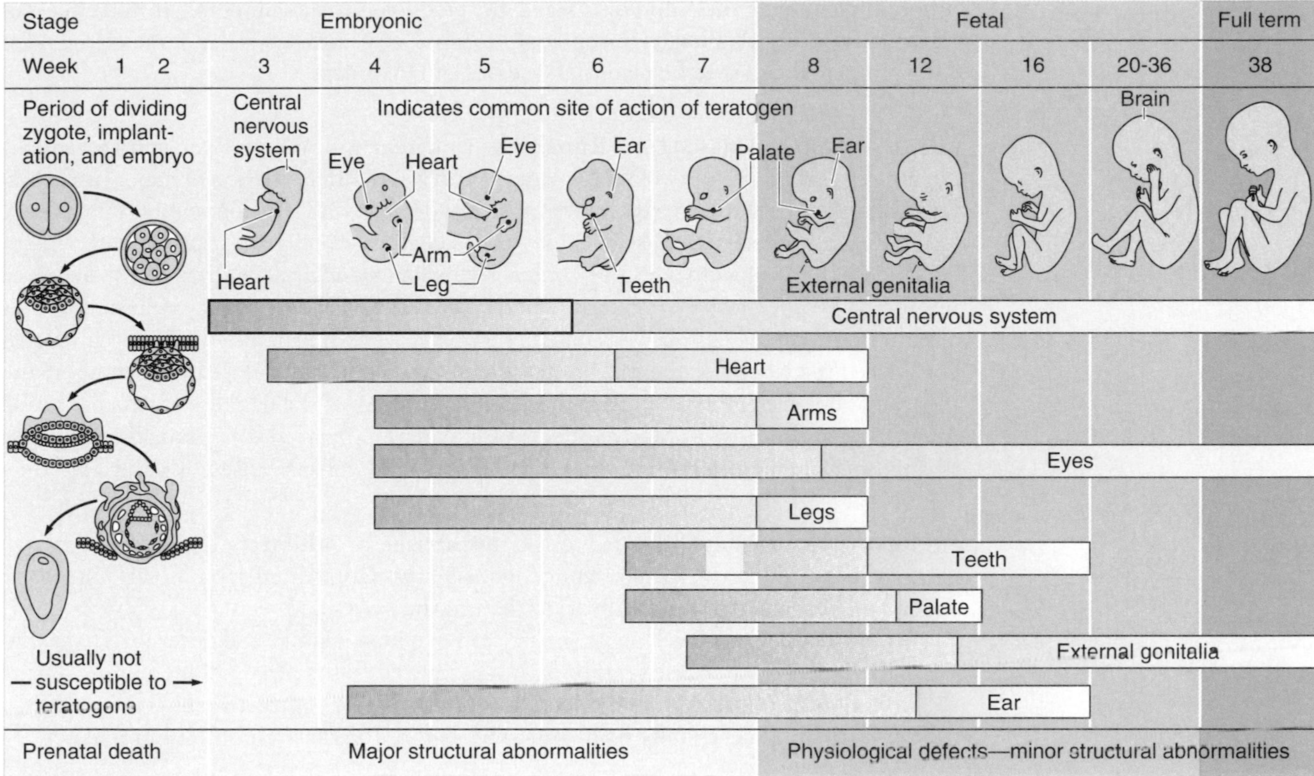

FIGURE 11.5

Critical Periods in Prenatal Development. The developing embryo is most vulnerable to teratogens when the organ systems are taking shape. The periods of greatest vulnerability of organ systems are shown in blue. Periods of lesser vulnerability are shown in yellow.

A woman can expect to gain at least 20 pounds during pregnancy because of the growth of the placenta, amniotic fluid, and the fetus itself. Most women will gain 25 pounds or so. Overweight women may gain less. Slender women may gain 30 pounds. Regular weight gains are most desirable—about ½ pound a week during the first half of pregnancy and about 1 pound a week during the second half.

MATERNAL DISEASES AND DISORDERS Environmental influences or agents that can harm the embryo or fetus are called **teratogens.** These include drugs taken by the mother, such as alcohol and even aspirin, as well as substances produced by the mother's body, such as Rh-positive antibodies. Other teratogens include the metals lead and mercury, radiation, and disease-causing organisms such as viruses and bacteria. Although many disease-causing organisms cannot pass through the placenta to infect the embryo or fetus, some extremely small organisms, such as those that cause syphilis, measles, mumps, and chicken pox, can. Some disorders, such as toxemia, are not transmitted to the embryo or fetus but can adversely affect the environment in which it develops.

CRITICAL PERIODS OF VULNERABILITY The times at which exposure to particular teratogens can cause the greatest harm are termed **critical periods of vulnerability.** Critical periods correspond to the times at which the structures most affected by the teratogens are developing (see Figure 11.5). The heart, for example, develops rapidly from the third to the fifth week after conception. It may be most vulnerable to certain teratogens at this time. The arms and legs, which develop later, are most vulnerable from the fourth through the eighth week of development. Because the major organ systems

Teratogens Environmental influences or agents that can damage an embryo or fetus. (From the Greek *teras,* which means "monster.")

Critical period of vulnerability A period of time during which an embryo or fetus is vulnerable to the effects of a teratogen.

differentiate during the embryonic stage, the embryo is most vulnerable to the effects of teratogens during this stage (Koren et al., 1998). Let us now consider some of the most damaging effects of specific maternal diseases and disorders.

RUBELLA (GERMAN MEASLES) **Rubella** is a viral infection. Women who contract rubella during the first month or two of pregnancy, when rapid differentiation of major organ systems is taking place, may bear children who are deaf or who develop mental retardation, heart disease, or cataracts. Risk of these defects declines as pregnancy progresses.

Nearly 85% of women in the United States had rubella as children and so acquired immunity. Women who do not know whether they have had rubella may be tested. If they are not immune, they can be vaccinated *before pregnancy.* Inoculation during pregnancy is considered risky because the vaccine causes a mild case of the disease in the mother, and this can affect the embryo or fetus. Increased awareness of the dangers of rubella during pregnancy, and of the preventive effects of inoculation, has led to a dramatic decline in the number of children born in the United States with defects caused by rubella.

SYPHILIS Maternal **syphilis** may cause miscarriage or **stillbirth,** or it may be passed along to the child in the form of congenital syphilis. Congenital syphilis can impair the vision and hearing, damage the liver, or deform the bones and teeth.

Routine blood tests early in pregnancy can diagnose syphilis and other problems. Because the bacteria that cause syphilis do not readily cross the placental membrane during the first months of pregnancy, the fetus will probably not contract syphilis if an infected mother is treated successfully with antibiotics before the fourth month of pregnancy.

ACQUIRED IMMUNODEFICIENCY SYNDROME (AIDS) **Acquired immunodeficiency syndrome** (AIDS) is caused by the *human immunodeficiency virus* (HIV). HIV is bloodborne and is sometimes transmitted through the placenta to infect the fetus. The rupturing of blood vessels in mother and baby during childbirth provides another opportunity for transmission of HIV. *However, the majority of babies born to mothers who are infected with HIV do not become infected themselves.* Using antiviral medication can also minimize the probability of transmission (see Chapter 16). HIV can also be transmitted to children by breast-feeding.

TOXEMIA **Toxemia,** a life-threatening condition characterized by high blood pressure, may afflict women late in the second trimester of pregnancy or early in the third. The first stage is termed *preeclampsia.* Its symptoms include protein in the urine, swelling from fluid retention, and high blood pressure, and it may be relatively mild. As preeclampsia worsens, the mother may have headaches and visual problems from the raised blood pressure, along with abdominal pain. If left untreated, the disease may progress to the final stage, termed *eclampsia.* Eclampsia can lead to maternal or fetal death. Babies born to women with toxemia are often undersized or premature.

Toxemia appears to be linked to malnutrition. Ironically, undernourished women may gain weight rapidly through fluid retention, but their swollen appearance may discourage them from eating. Pregnant women who gain weight rapidly but have not increased their food intake should consult their obstetricians.

ECTOPIC PREGNANCY In an **ectopic pregnancy,** the fertilized ovum implants itself somewhere other than in the uterus. Most ectopic pregnancies occur in a fallopian tube ("tubal pregnancies") when the ovum is prevented from moving into the uterus because of obstructions caused by infections. Ectopic pregnancies are more common among older women (Stein & Susser, 2000). If ectopic pregnancies do not abort spontaneously, they must be removed by surgery or use of medicines such as methotrexate, because the fetus cannot develop to term. Delay in removal may cause hemorrhaging and the death of the

Rubella A viral infection that can cause mental retardation and heart disease in an embryo. Also called *German measles.*

Syphilis A sexually transmitted disease caused by a bacterial infection.

Stillbirth The birth of a dead fetus.

Acquired immunodeficiency syndrome (AIDS) A sexually transmitted disease that destroys white blood cells in the immune system, leaving the body vulnerable to various "opportunistic" diseases.

Toxemia A life-threatening condition that is characterized by high blood pressure.

Ectopic pregnancy A pregnancy in which the fertilized ovum becomes implanted somewhere other than in the uterus.

a CLOSER look

Spacing Children the Goldilocks Way

No, the title of this feature does not refer to separating children in the back seat to prevent them from killing each other. It refers to how long women who are going to have more than one should wait between pregnancies. Some of the considerations are psychological: Is it good to have children close together so that they will be interested in doing the same things as they mature? Is it better to space them several years apart so that each can have more parental attention during the first years? Is it better to have them all as early as possible so that you (the parent) can be "child-free" while you are still young enough to enjoy it?

But then there are the health issues. That is, are there any risks attached to having children close together? Or far apart? What is too brief a period for spacing children? What is too long a period? What amount of time is just right?

Conde-Agudelo and Belizán (2000) studied the health records of more than 450,000 women in Uruguay. They specifically looked at women who had spaced their children at short intervals (becoming pregnant within 5 months of bearing the earlier child) and at long internals (becoming pregnant more than 59 months after bearing the earlier child). They compared both groups with women who had spaced their children at 18 to 23 months.

The researchers discovered that women who had spaced their children at short intervals had a great risk of maternal death, bleeding during the third trimester, premature rupture of membranes, and anemia. Women who spaced their children at more than 59 months had significantly greater risk of preeclampsia and eclampsia.

In terms of maternal health, 18 to 23 months seems to be about just right. Intervals of less than 5 months may be too short, and intervals of more than 59 months may be too long.

mother. A woman with a tubal pregnancy will not menstruate but may notice spotty bleeding and abdominal pain.

RH INCOMPATIBILITY In **Rh incompatibility,** antibodies produced by the mother are transmitted to a fetus or newborn infant. *Rh* is a blood protein found in some people's red blood cells. Rh incompatibility occurs when a woman who does not have this blood factor, and is thus *Rh-negative*, is carrying an *Rh-positive* fetus, which can happen if the father is Rh-positive. The negative–positive combination is found in about 10% of U.S. marriages. However, it becomes a problem only in a minority of the resulting pregnancies. In such cases the mother's antibodies attack the red blood cells of the fetus, which can cause brain damage or death. Rh incompatibility does not usually adversely affect a first child because the woman usually has not yet formed antibodies to the Rh factor.

Because mother and fetus have separate circulatory systems, it is unlikely that Rh-positive fetal red blood cells will enter the Rh-negative mother's body. The probability of an exchange of blood increases during childbirth, however, especially when the placenta becomes detached from the uterine wall. If an exchange of blood occurs, the mother will then produce antibodies to the baby's Rh-positive blood. The mother's antibodies may enter the fetal bloodstream and cause a condition called *fetal erythroblastosis*, which can result in anemia, mental deficiency, or even the death of the fetus or newborn infant.

Fortunately, blood-typing of pregnant women significantly decreases the threat of uncontrolled erythroblastosis. If an Rh-negative mother is injected with the vaccine Rhogan within 72 hours after delivery of an Rh-positive baby, she will not develop the dangerous antibodies and thus will not pass them on to the fetus in a subsequent pregnancy. A fetus or newborn child at risk for erythroblastosis may also receive a preventive blood transfusion, to remove the mother's Rh-positive antibodies from its blood.

Rh incompatibility A condition in which antibodies produced by a pregnant woman are transmitted to the fetus and may cause brain damage or death.

DRUGS TAKEN BY THE MOTHER (AND THE FATHER) Some widely used drugs, including nonprescription drugs, are linked with birth abnormalities (Koren et al., 1998). In the 1960s the drug thalidomide was marketed to pregnant women as a presumably safe treatment for nausea and insomnia. However, the drug caused birth deformities, including stunted or missing limbs (Koren et al., 1998). Maternal use of illegal drugs such as cocaine and marijuana may also place the fetus at risk.

Paternal use of certain drugs also may endanger the fetus. One question is whether drugs alter the genetic material in the father's sperm. The use of certain substances by those who come into contact with a pregnant woman can harm the fetus. For example, the mother's inhalation of second-hand tobacco or marijuana smoke can hurt the fetus.

Several antibiotics may harm a fetus, especially if they are taken during certain periods of fetal development. Tetracycline may yellow the teeth and deform the bones (Koren et al., 1998). Other antibiotics have been implicated in deafness and jaundice.

Acne drugs such as Accutane can cause physical and mental handicaps in the children of women who use them during pregnancy. Antihistamines, used commonly for allergies, may deform the fetus.

If you are pregnant or suspect that you are, it is advisable to consult your obstetrician before taking any and all drugs, not just prescription drugs. Your obstetrician can usually direct you to a safe and effective substitute for a drug that could harm a fetus.

HORMONES The hormones progestin and DES have sometimes been used to help women at risk of miscarriage maintain their pregnancies. When taken at about the time that sex organs differentiate, progestin—which is similar in composition to male sex hormones, can masculinize the external sex organs of embryos with female (XX) sex chromosomal structures. Progestin taken during the first trimester has also been linked to increased levels of aggressive behavior during childhood.

DES (short for *diethylstilbestrol*), a powerful estrogen, was given to many women at risk for miscarriage from the 1940s through the 1960s to help maintain their pregnancies. DES is suspected of causing cervical and testicular cancer in some of the children whose mothers used it when pregnant (Koren et al., 1998). Other problems have been reported as well. Daughters whose mothers used DES during their pregnancies have a higher-than-expected rate of miscarriages and premature deliveries. It was once suspected that men who were exposed prenatally to DES had higher than expected rates of infertility. However, research reveals no connection between in utero exposure to DES and male infertility (Wilcox et al., 1995). DES users themselves appear to be at high risk of some serious medical problems, such as breast cancer (Greenberg et al., 1984).

VITAMINS Many pregnant women are prescribed daily doses of multivitamins to maintain their own health and to promote the development of a healthy pregnancy. "Too much of a good thing" may be hazardous, however. High doses of vitamins such as A, B_6, D, and K have been linked to birth defects. Vitamin A excesses have been linked to cleft palate and eye damage, excesses of vitamin D to mental retardation.

NARCOTICS Narcotics such as heroin and methadone can readily pass from mother to fetus through the placental membrane. Narcotics are addictive. Fetuses of mothers who use them regularly during their pregnancies can become addicted in utero. At birth, such babies may undergo withdrawal and show muscle tension and agitation. Women who use narcotics are advised to notify their obstetricians so that measures can be taken to aid the infants before and after delivery.

TRANQUILIZERS AND SEDATIVES The tranquilizers Librium and Valium cross the placental membrane and may cause birth defects such as harelip. Sedatives, such as the barbiturate *phenobarbital*, are suspected of decreasing testosterone production and causing reproductive problems in the sons of women who use them during pregnancy.

HALLUCINOGENICS Use of hallucinogenic drugs such as marijuana and LSD during pregnancy has been linked to chromosomal damage in fetuses (National Academy of Sciences, 1982). The active ingredient in marijuana, THC, readily crosses the placenta, as does LSD. Use of marijuana can lead to decreased androgen production in male fetuses,

DES Diethylstilbestrol, an estrogen that was once given to women at risk for miscarriage to help maintain pregnancy.

which can interfere with the process of sexual differentiation. Research provides evidence that preschoolers whose mothers used marijuana during their pregnancies suffered more neurological and visual problems than did children of nonusers (Fried, 1986).

ALCOHOL Mothers who drink heavily during pregnancy expose the fetus to greater risk of birth defects, infant mortality, sensory and motor problems, and mental retardation (Barr et al., 1990; Coles, 1994). Nearly 40% of children whose mothers drank heavily during pregnancy develop **fetal alcohol syndrome** (FAS). FAS is a cluster of symptoms typified by developmental lags and characteristic facial features, such as an underdeveloped upper jaw, flattened nose, and widely spaced eyes. Infants with FAS are often smaller than average and have smaller-than-average brains. They may be mentally retarded, lack coordination, and have deformed limbs and heart problems.

Although research suggests that light drinking is unlikely to harm the fetus in most cases (Jacobson & Jacobson, 1994), FAS has been found even among the children of mothers who drank only 2 ounces of alcohol a day during the first trimester (Astley et al., 1992). Moreover, individual sensitivities to alcohol vary widely (Jacobson & Jacobson, 1994). The critical period for the development of the facial features associated with FAS seems to be the first 2 months of prenatal development, when the head is taking shape (Coles, 1994).

It is not true that pregnant women can have one or two alcoholic beverages a day without harming their babies. No safe minimal amount of drinking for pregnant women has been established. Therefore, pregnant women are advised to abstain from alcohol completely.

CIGARETTE SMOKING Cigarette smoke contains chemicals such as carbon monoxide and the stimulant nicotine that are transmitted to the fetus. It also lessens the amount of oxygen received by the fetus (Gruslin et al., 2000). Maternal smoking increases the risk of spontaneous abortion and complications during pregnancy such as premature rupturing of the amniotic sac, stillbirth, premature birth, low birth weight, and early infant mortality (English & Eskenazi, 1992; Floyd et al., 1993; USDHHS, 1992a). The health risks generally increase with the amount smoked.

Maternal smoking may also impair intellectual development. In one study, women who smoked during pregnancy were 50% more likely than women who did not to have children whose intelligence test scores placed them in the mentally retarded range (that is, beneath an IQ score of 70) when the children were 10 years old (Drews et al., 1996).

Low birth weight is the most common risk factor for infant disease and mortality (USDHHS, 1992a). Maternal smoking during pregnancy more than doubles the risk of low birth weight (Mayer et al., 1990). The combination of smoking and drinking alcohol places the child at greater risk of low birth weight than either practice alone (Day & Richardson, 1994). As many as 1 in 4 cases of low birth weight could be prevented if mothers-to-be stopped smoking during pregnancy (USDHHS, 1990). The earlier the pregnant smoker quits, the better for the baby (and for herself!). Simply cutting down on smoking during pregnancy may not offer much protection in preventing low birth weight, however (USDHHS, 1990).

Maternal smoking affects the fetal heart rate (Graca et al., 1991) and increases the risk of sudden infant death syndrome (SIDS) (Feng, 1993; Haglund & Cnattingius, 1990; Malloy et al., 1992; Schoendorf & Kiely, 1992; Zhang & Fried, 1992). Maternal smoking has also been linked to reduced lung function in newborns (Hanrahan et al., 1992) and to asthma in childhood (Martinez et al., 1992). Evidence also points to reduced attention spans, hyperactivity, and lower IQ and achievement test scores in children exposed to maternal smoking during and after pregnancy (Barr et al., 1990).

Smoking by the father (or other household members) may be dangerous to a fetus because secondary smoke (smoke exhaled by the smoker or emitted from the tip of a lit cigarette) may be absorbed by the mother and passed along to the fetus. Passive exposure to second-hand smoke during infancy is also linked to increased risk of SIDS (Schoendorf & Kiely, 1992).

Fetal alcohol syndrome
A cluster of symptoms caused by maternal drinking, in which the child shows developmental lags and characteristic facial features such as an underdeveloped upper jaw, flattened nose, and widely spaced eyes.

Don't Do It! Smoking cigarettes and pregnancy do not mix. Maternal smoking has been shown to increase the risk of spontaneous abortion and of complications during pregnancy. It is connected with premature rupturing of the amniotic sac, stillbirth, premature birth, low birth weight, early infant mortality, and even delayed intellectual development.

The majority of women in the United States of reproductive age drink alcohol, at least occasionally. More than 1 in 4 smoke cigarettes. Many of them do not suspend drug use until they learn that they are pregnant. Unfortunately, this knowledge may not be obtained until a woman is weeks into the pregnancy. Damage to the fetus may thus have already occurred. Many women are unwilling or unable to change their drug use habits even after learning they are pregnant. Among women who smoke, only 1 in 5 quits smoking when she becomes pregnant (Floyd et al., 1993).

Our clinical experience suggests that it may be easier for women to quit if they conceptualize their quitting as limited in time to the terms of their pregnancies, rather than as permanent. Then, of course, if they should remain abstinent after delivery, perhaps they will not be disappointed.

OTHER AGENTS X-rays increase the risk of malformed organs in the fetus, especially within a month and a half after conception. (Ultrasound has not been shown to harm the embryo or fetus.)

Chromosomal and Genetic Abnormalities

Not all of us have the normal complement of chromosomes. Some of us have genes that threaten our health or our existence (see Table 11.1).

DOWN SYNDROME Children with Down syndrome have characteristic round faces; wide, flat noses; and protruding tongues. They often suffer from respiratory problems and heart malformations, problems that tend to claim their lives by middle age—the "prime of life" for most of us. People with Down syndrome are also moderately mentally retarded, but they usually can learn to read and write. With a little help from family and social agencies, they may hold jobs and lead largely independent lives.

The risk of a child's having Down syndrome increases with the mother's age (see Table 11.2). Down syndrome is usually caused by an extra chromosome on the 21st pair. In about 95% of cases, Down syndrome is transmitted by the mother (Antonarakas et al., 1991). The inner corners of the eyes of people with the syndrome have a downward-sloping crease of skin that gives them a superficial resemblance to Asians. This is why the syndrome was once dubbed *mongolism*, a moniker that has since been rejected because of its racist overtones.

TABLE 11.1

Some chromosomal and genetic abnormalities

Health Problem	About . . .
Cystic fibrosis	A genetic disease in which the pancreas and lungs become clogged with mucus, which impairs the processes of respiration and digestion.
Down syndrome	A condition characterized by a 3rd chromosome on the 21st pair. The child with Down syndrome has a characteristic fold of skin over the eye and mental retardation. The risk of having a child with the syndrome increases as parents increase in age.
Hemophilia	A sex-linked disorder in which blood does not clot properly.
Huntington's chorea	A fatal neurological disorder whose onset occurs in middle adulthood.
Neural tube defects	Disorders of the brain or spine, such as *anencephaly,* in which part of the brain is missing, and *spina bifida,* in which part of the spine is exposed or missing. Anencephaly is fatal shortly after birth, but some spina bifida victims survive for a number of years, though with severe handicaps.
Phenylketonuria	A disorder in which children cannot metabolize phenylalanine, which builds up in the form of phenylpyruvic acid and causes mental retardation. The disorder can be diagnosed at birth and controlled by diet.
Retina blastoma	A form of blindness caused by a dominant gene.
Sickle-cell anemia	A blood disorder that mostly afflicts African Americans, in which deformed blood cells obstruct small blood vessels, decreasing their capacity to carry oxygen and heightening the risk of occasionally fatal infections.
Tay–Sachs disease	A fatal neurological disorder that primarily afflicts Jews of European origin.

SICKLE-CELL ANEMIA AND TAY–SACHS DISEASE Sickle-cell anemia and Tay–Sachs disease are genetic disorders that are most likely to afflict certain racial and ethnic groups. Sickle-cell anemia is most prevalent in the United States among African Americans. One of every 375 African Americans is affected by the disease, and 8% are carriers of the sickle-cell trait (Leary, 1993). In sickle-cell anemia, the red blood cells assume a sickle shape—hence the name—and form clumps that obstruct narrow blood vessels and

TABLE 11.2

Risk of giving birth to an infant with Down syndrome, according to age of the mother

Age of Mother	Probability of Down Syndrome	Age of Mother	Probability of Down Syndrome
30	1/885	40	1/109
31	1/826	41	1/85
32	1/725	42	1/67
33	1/592	43	1/53
34	1/465	44	1/41
35*	1/365	45	1/32
36	1/287	46	1/25
37	1/225	47	1/20
38	1/176	48	1/16
39	1/139	49	1/11

Source: From Samuels & Samuels (1986). "Risk of giving birth to a down-syndrome infant," in *The Well Pregnancy Book*, p. 240. Adapted with permission.

*Age at which testing for Down syndrome is usually first recommended.

diminish the supply of oxygen. As a result, victims can suffer problems ranging from swollen, painful joints to potentially lethal pneumonia and heart and kidney failure. Infections are a leading cause of death among those with the disease (Leary, 1993a).

Tay–Sachs disease is a fatal neurological disease of young children. Only 1 in 100,000 people in the United States is affected, but among Jews of Eastern European background the figure rises steeply to 1 in 3,600 (Hubbard & Wald, 1993). The disease is characterized by degeneration of the central nervous system and gives rise to retardation, loss of muscle control and paralysis, blindness, and deafness. Victims seldom live beyond the age of 5.

SEX-LINKED GENETIC ABNORMALITIES Some genetic defects, such as hemophilia, are sex-linked in that they are carried only on the X sex chromosome. They are transmitted from generation to generation as **recessive traits.** Females, each of whom has two X sex chromosomes, are less likely than males to be afflicted by sex-linked disorders, because the genes that carry the disorder would have to be present on both of their sex chromosomes for the disorder to be expressed. Sex-linked disorders are more likely to afflict sons of female carriers because they have only one X sex chromosome, which they inherit from their mothers. England's Queen Victoria was a hemophilia carrier and transmitted the condition to many of her children, who in turn carried it into several ruling families of Europe. For this reason hemophilia has been dubbed the "royal disease."

AVERTING CHROMOSOMAL AND GENETIC ABNORMALITIES On the basis of information about a couple's medical background and family history of genetic defects, genetic counselors help couples assess the risks of passing along genetic defects to their children. Some couples who face a high risk of passing along genetic defects to their children decide to adopt. Other couples decide to have an abortion if the fetus is determined to have certain abnormalities.

Various medical procedures are used to detect the presence of these disorders in the fetus. *Amniocentesis* is usually performed about 4 months into pregnancy but is sometimes done earlier. Fluid is drawn from the amniotic sac (or "bag of waters") with a syringe. Fetal cells in the fluid are grown in a culture and examined under a microscope for the presence of biochemical and chromosomal abnormalities. *Chorionic villus sampling* (CVS) is performed at about 10 weeks. A narrow tube is used to snip off material from the chorion, which is a membrane that contains the amniotic sac and fetus. The material is then analyzed. The risks of amniocentesis and CVS are comparable (Simpson, 2000). The tests detect Down syndrome, sickle-cell anemia, Tay–Sachs disease, spina bifida, muscular dystrophy, Rh incompatibility, and other conditions. The tests also identify the gender of the fetus.

In ultrasound, high-pitched sound waves are bounced off the fetus, like radar, revealing a picture of the fetus on a TV monitor and allowing the obstetrician to detect certain abnormalities. Obstetricians also use ultrasound to locate the fetus during amniocentesis in order to lower the probability of injuring it with the syringe.

An Ultrasound Image. A ultrasound image of the second author's son, Michael, at 12 weeks following conception. The head and upper torso (facing upward) can be seen by those of you willing to look for many hours in the upper middle section of the photo. Michael was handsome even then, his father points out. The first and third author note that they did not insist on including their own boring ultrasound photos in this text.

Recessive trait A trait that is not expressed when the gene or genes involved have been paired with dominant genes. Recessive traits are transmitted to future generations, however, and are expressed if they are paired with other recessive genes.

Parental blood tests can suggest the presence of problems such as sickle-cell anemia, Tay–Sachs disease, and neural tube defects. Still other tests examine fetal DNA and can indicate the presence of Huntington's chorea, cystic fibrosis, and other disorders. Blood tests also now allow detection of Down syndrome during the first trimester (Haddow et al., 1998).

CHILDBIRTH

Early in the ninth month of pregnancy, the fetus's head settles in the pelvis. This shift is called "dropping" or "lightening." The woman may actually feel lighter because of lessened pressure on the diaphragm. About a day or so before the beginning of labor, the woman may notice blood in her vaginal secretions because fetal pressure on the pelvis may rupture superficial blood vessels in the birth canal. Tissue that had plugged the cervix, possibly preventing entry of infectious agents from the vagina, becomes dislodged. There is a resultant discharge of bloody mucus. At about this time, 1 woman in 10 also has a rush of warm "water" from the vagina. The "water" is amniotic fluid, and it means that the amniotic sac has burst. Labor usually begins within a day after rupture of the amniotic sac. For most women the amniotic sac does not burst until the end of the first stage of childbirth. Other signs of impending labor include indigestion, diarrhea, abdominal cramps, and an ache in the small of the back. Labor begins with the onset of regular uterine contractions.

The first uterine contractions are relatively painless and are called **Braxton–Hicks contractions,** or false labor contractions. They are "false" because they do not widen the cervix or advance the baby through the birth canal. They tend to increase in frequency but are less regular than labor contractions. Real labor contractions, by contrast, become more intense when the woman moves around or walks.

The initiation of labor may involve the secretion of hormones by the fetal adrenal and pituitary glands that stimulate the placenta and the mother's uterus to secrete **prostaglandins.** Prostaglandins stimulate the uterine musculature to contract. It would make sense for the fetus to have a mechanism for signaling the mother that it is mature enough to sustain independent life. The mechanisms that initiate and maintain labor are not fully understood, however. Later in labor the pituitary gland releases **oxytocin,** a hormone that stimulates contractions strong enough to expel the baby.

It is probably true that a baby signals its mother (chemically) when it is ready to be born (that is, to sustain independent life).

The Stages of Childbirth

Childbirth begins with the onset of labor and has three stages.

THE FIRST STAGE In the first stage, uterine contractions **efface** and **dilate** the cervix to about 4 inches (10 cm) in diameter, so that the baby may pass. Stretching of the cervix causes most of the pain of childbirth. A woman may experience little or no pain if her cervix dilates easily and quickly. The first stage may last from a couple of hours to more than a day. Twelve to 24 hours of labor is considered about average for a first pregnancy. In later pregnancies labor takes about half this time.

The initial contractions are usually mild and spaced widely, at intervals of 10 to 20 minutes. They may last 20 to 40 seconds. As time passes, contractions become more frequent, long, strong, and regular.

Transition is the process that occurs when the cervix becomes almost fully dilated and the baby's head begins to move into the vagina, or birth canal. Contractions usually come quickly during transition. Transition usually lasts about 30 minutes or less and is often accompanied by feelings of nausea, chills, and intense pain.

Braxton–Hicks contractions So-called false labor contractions that are relatively painless.

Prostaglandins Uterine hormones that stimulate uterine contractions.

Oxytocin A pituitary hormone that stimulates uterine contractions.

Efface To become thin.

Dilate To open or widen.

Transition The process during which the cervix becomes almost fully dilated and the head of the fetus begins to move into the birth canal.

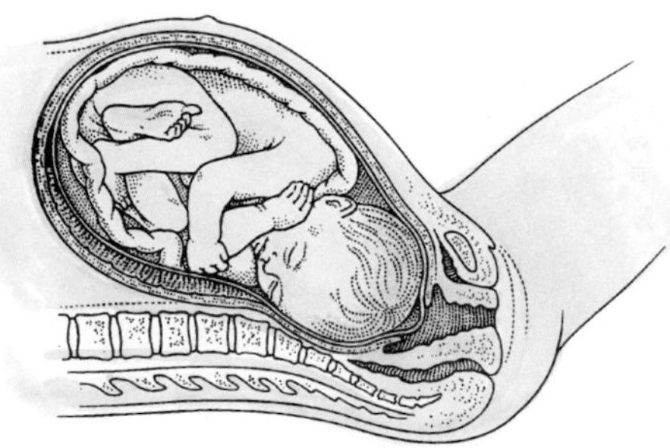

1. The second stage of labor begins

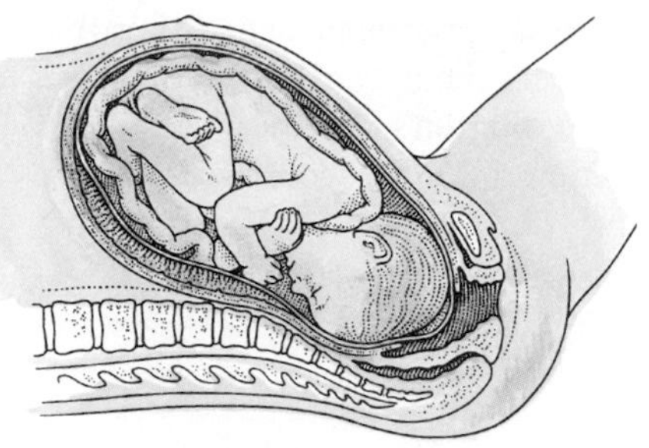

2. Further descent and rotation

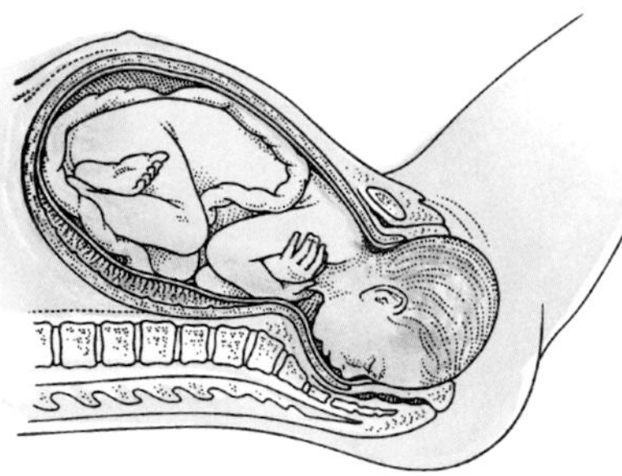

3. The crowning of the head

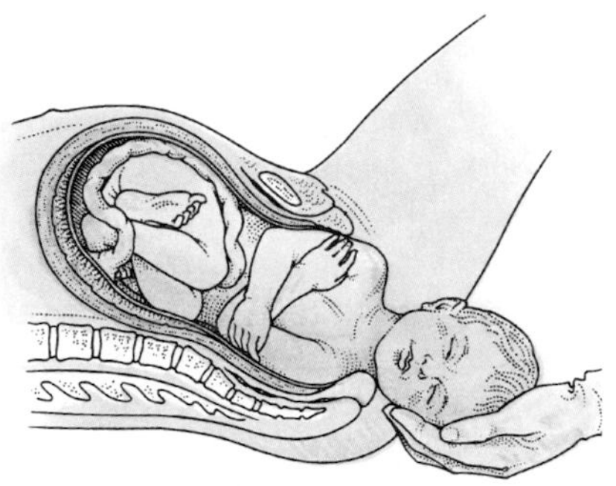

4. Anterior shoulder delivered

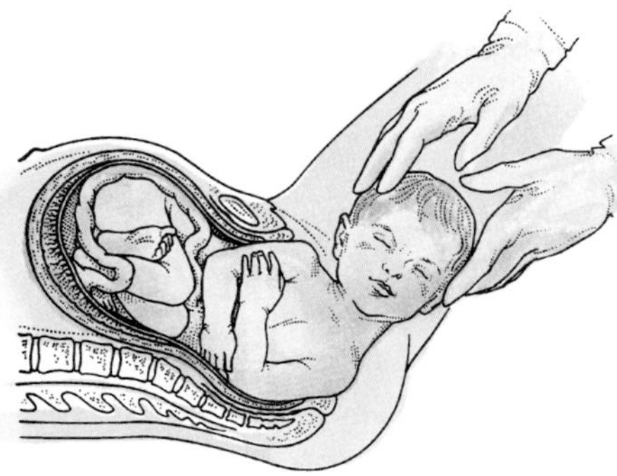

5. Posterior shoulder delivered

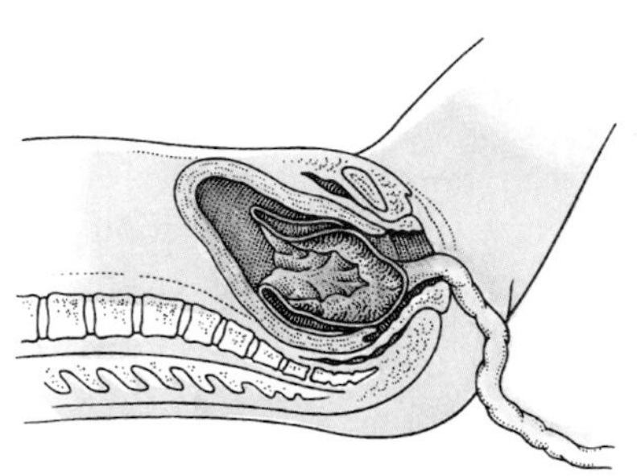

6. The third stage of labor begins with separation of the placenta from the uterine wall

FIGURE 11.6

The Stages of Childbirth. In the first stage, uterine contractions efface and dilate the cervix to about 4 inches so that the baby may pass through. The second stage begins with movement of the baby into the birth canal and ends with birth of the baby. During the third stage, the placenta separates from the uterine wall and is expelled through the birth canal.

THE SECOND STAGE The second stage of childbirth follows transition and begins when the cervix has become fully dilated and the baby begins to move into the vagina and first appears at the opening of the birth canal (see Figure 11.6). The woman may be taken to a delivery room for the second stage of childbirth. The second stage is shorter than the first stage. It lasts from a few minutes to a few hours and ends with the birth of the baby.

Each contraction of the second stage propels the baby farther along the birth canal (vagina). When the baby's head becomes visible at the vaginal opening, it is said to have *crowned.* The baby typically emerges fully a few minutes after crowning.

An **episiotomy** may be performed on the mother when the baby's head has crowned. The purpose is to prevent the random tearing of the **perineum** that can occur

Episiotomy A surgical incision in the perineum that widens the birth canal, preventing random tearing during childbirth.

Perineum The area between the vulva and the anus.

if it becomes extremely effaced. Episiotomies are controversial, however (Roberts, 2000). The incision can cause infection and pain and can create discomfort and itching as it heals. In some cases the discomfort interferes with coitus for months. For these and other reasons, a recent article in *Obstetrics & Gynecology* recommends that "Episiotomy should no longer be routine" (Eason & Feldman, 2000). In some cases, prenatal massage of the perineum can avert the necessity of episiotomy (Eason et al., 2000; Johanson, 2000). However, physicians generally agree that episiotomy should be performed if the baby's shoulders are too wide to emerge without causing tearing or if the baby's heartbeat drops for an extended period of time (Eason & Feldman, 2000). Having said all this, one study found that the strongest predictor of whether a physician will use episiotomy is his or her customary practice, not the condition of the woman in labor or of the baby (Robinson et al., 2000). (When interviewing an obstetrician, ask how often he or she performs an episiotomy!)

With or without an episiotomy, the baby's passageway to the external world is a tight fit. As a result, the baby may look as though it has been through a prizefight. Its head may be elongated, its nose flattened, and its ears bent. Although parents may be concerned about whether the baby's features will assume a more typical shape, they nearly always do.

THE THIRD STAGE The third, or placental, stage of childbirth may last from a few minutes to an hour or more. During this stage, the placenta is expelled. Detachment of the placenta from the uterine wall may cause some bleeding. The uterus begins the process of contracting to a smaller size. The attending physician sews up the episiotomy or any tears in the perineum.

IN THE NEW WORLD As the baby's head emerges, mucus is cleared from its mouth by means of suction aspiration to prevent the breathing passageway from being obstructed. Aspiration is often repeated once the baby is fully delivered. (Newly delivered babies are no longer routinely held upside down to help expel mucus. Nor is the baby slapped on the buttocks to stimulate breathing, as in old films.)

Once the baby is breathing adequately, the umbilical cord is clamped and severed about 3 inches from the baby's body. (After the birth of your first and third authors' third child, your first author was invited by the obstetrician to cut the umbilical cord—but your third author seized the scissors and cut the umbilical cord herself, squirting blood on the obstetrician's glasses. "Who gave the obstetrician the right to determine who would cut the umbilical cord?" she wanted to know.) The stump of the umbilical cord dries and falls off in its own time, usually in seven to ten days.

While the mother is in the third stage of labor, the nurse may perform procedures on the baby, such as placing drops of silver nitrate or an antibiotic ointment into the eyes. This procedure is required by most states to prevent bacterial infections in the newborn's eyes. Typically, the baby is also footprinted and (if the birth has taken place in a hospital) given an identification bracelet. Because neonates do not manufacture vitamin K on their own, the baby may also receive an injection of the vitamin to ensure that her or his blood will clot normally in case of bleeding.

Methods of Childbirth

Until this century, childbirth was usually an event that happened at home and involved the mother, a midwife, family, and friends. These days women in the United States and Canada typically give birth in hospitals attended by obstetricians who use surgical instruments and anesthetics to protect mothers and children from infection, complications, and pain. Medical procedures save lives but also make childbearing more impersonal. Social critics argue that these procedures have medicalized a natural process, usurping control over women's bodies and, through the use of drugs, denying many women the experience of giving birth.

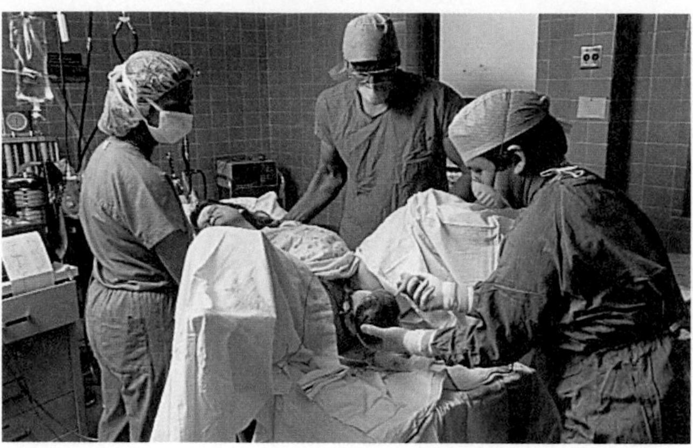

Coming into the World. Childbirth progresses through three stages. In the first stage, uterine contractions efface and dilate the cervix so that the baby can pass through. The second stage lasts from a few minutes to a few hours and ends with the birth of the baby. During the third stage, the placenta is expelled.

ANESTHETIZED CHILDBIRTH

> In sorrow thou shalt bring forth children.
> —Genesis 3:16

The Bible suggests that the ancients saw suffering as a woman's lot. But during the past two centuries, science and medicine have led to the expectation that women should experience minimal discomfort during childbirth. Today some anesthesia is used to minimize or eliminate pain in most U.S. deliveries.

General anesthesia first became popular when Queen Victoria of England delivered her eighth child under chloroform in 1853. General anesthesia, like the chloroform of old, induces unconsciousness. The drug sodium pentothal, a barbiturate, induces general anesthesia when it is injected into a vein. Barbiturates may also be taken orally to reduce anxiety while the woman remains awake. Women may also receive tranquilizers like Valium or narcotics like Demerol to help them relax and to blunt pain without inducing sleep.

Anesthetic drugs, as well as tranquilizers and narcotics, decrease the strength of uterine contractions during delivery. They may thus delay the process of cervical dilation and prolong labor. They also reduce the woman's ability to push the baby through the birth canal. And because they cross the placental membrane, they also lower the newborn's overall responsiveness.

Regional or **local anesthetics** block pain in parts of the body without generally depressing the mother's alertness or putting her to sleep. In a *pudendal block*, the external genitals are numbed by local injection. In an *epidural block* and a *spinal block*, an anesthetic is injected into the spinal canal, which temporarily numbs the mother's body below the waist. To prevent injury, the needles used for these injections do not come into contact with the spinal cord itself. Although local anesthesia decreases the responsiveness of the newborn baby, there is little evidence that medicated childbirth has serious, long-term consequences on children.

NATURAL CHILDBIRTH Partly as a reaction against the use of anesthetics, English obstetrician Grantly Dick-Read endorsed **natural childbirth** in his 1944 book *Childbirth Without Fear*. Dick-Read argued that women's labor pains were heightened by their fear of the unknown and resultant muscle tensions. Many of Dick-Read's contributions came to be regarded as accepted practice in modern childbirth procedures, such as the emphasis on informing women about the biological aspects of reproduction and childbirth, the encouragement of physical fitness, and the teaching of relaxation and breathing exercises.

PREPARED CHILDBIRTH: THE LAMAZE METHOD The French obstetrician Fernand Lamaze visited the Soviet Union in 1951 and found that many Russian women bore babies without anesthetics and without reporting a great deal of pain. Lamaze returned to Western Europe with some of the techniques the women used; they are now termed the **Lamaze method,** or *prepared childbirth*. Lamaze (1981) argued that women can learn to conserve energy during childbirth and reduce the pain of uterine contractions by associating the contractions with other responses, such as thinking of pleasant mental images such as beach scenes, or engaging in breathing and relaxation exercises.

A pregnant woman typically attends Lamaze classes with a "coach"—usually the father—who will aid her in the delivery room by timing contractions, offering emotional support, and coaching her in the breathing and relaxation exercises. The woman and her partner also receive more general information about childbirth. The father is integrated into the process, and many couples report that their marriages are strengthened as a result.

The Lamaze method is flexible about the use of anesthetics. Many women report some pain during delivery and obtain anesthetics. However, the Lamaze method appears to help women to gain a greater sense of control over the delivery process.

General anesthesia The use of drugs to put people to sleep and eliminate pain, as during childbirth.

Local anesthesia Anesthesia that eliminates pain in a specific area of the body, as during childbirth.

Natural childbirth A method of childbirth in which women use no anesthesia but are given other strategies for coping with discomfort and are educated about childbirth.

Lamaze method A childbirth method in which women learn about childbirth, learn to relax and to breathe in patterns that conserve energy and lessen pain, and have a coach (usually the father) present at childbirth. Also termed *prepared childbirth*.

CESAREAN SECTION In a **cesarean section,** the baby is delivered through surgery rather than naturally through the vagina. The term *section* is derived from the Latin for "to cut." Julius Caesar is said to have been delivered in this way, but health professionals believe this is unlikely. In a cesarean section (C-section for short) the woman is anesthetized, and incisions are made in the abdomen and uterus so that the surgeon can remove the baby. The incisions are then sewn up and the mother can begin walking, often on the same day, although generally with some discomfort for a while. Although most C-sections are without complications, some cause urinary tract infections, inflammation of the wall of the uterus, blood clots, or hemorrhaging ("After years of decline," 2000).

C-sections are most likely to be advised when normal delivery is difficult or threatening to the health of the mother or child. Vaginal deliveries can become difficult if the baby is large, the mother's pelvis is small or misshapen, or the mother is tired, weakened, or aging (Roberts, 2000). Herpes and HIV infections in the birth canal can be bypassed by C-section. C-sections are also likely to be performed if the baby presents for delivery in the breech position (feet downward) or the **transverse position** (lying crosswise) or if the baby is in distress. C-sections do appear to be less stressful to the baby, as assessed by sensitivity to pain shown during inoculations 8 weeks after delivery. Babies who have been delivered by C-section secrete less stress hormone (cortisol in the saliva) and cry less than babies who were delivered vaginally (Taylor et al., 2000).

Use of the C-section has mushroomed. More than 1 of every 5 births (22%) is now by C-section ("After years of decline," 2000). The peak year was 1988, when 1 birth in 4 was by C-section. Compare this figure to about 1 in 20 births in 1965. Much of the increase in the rate of C-sections reflects advances in medical technology (such as use of fetal monitors that allow doctors to detect fetal distress), fear of malpractice suits, financial incentives for hospitals and physicians, and, simply, current medical practice patterns (DiMatteo et al., 1996). Yet some women request C-sections to avoid the discomforts of vaginal delivery or to control the timing of the delivery. Critics claim that many C-sections are unnecessary, and the U.S. Department of Health and Human Services believes that a rate of 15 per 100 births would be more appropriate (Paul, 1996). But many obstetricians are concerned that a political push to lower the C-section rate could be dangerous ("After years of decline," 2000; Roberts, 2000). Despite the often justified reasons for doing C-sections, one of the reasons is clearly "doctors' habits" ("After years of decline," 2000).

Truth or Fiction? **REVISITED**

It is true that 1 U.S. birth in 5—actually 22%—is by cesarean section. The use of cesarean section mushroomed in the latter part of the twentieth century, to a peak of about 25% in 1988. ■

Some women who have C-sections experience negative emotional consequences. A meta-analysis of the results of studies on women who have C-sections reported that they are generally less satisfied with the birth process, that they are less likely to breast-feed, and that they interact somewhat less with their newborn babies (DiMatteo et al., 1996). Some studies (e.g., Durik et al., 2000) report differences for women who have planned as opposed to unplanned C-sections. Overall, however, it has not been shown that C-sections are connected with significant, enduring emotional consequences for mothers or their children.

Medical opinion formerly held that once a woman had a C-section, subsequent deliveries also had to be by C-section. Otherwise, uterine scars might rupture during labor. Research has shown that rupture is rare, however. In one study, only 10 of 3,249 women who chose to try vaginal delivery after a previous C-section had a uterine rupture (McMahon et al., 1996). Moreover, there were no maternal deaths. In any event, only about 1 woman in 4 (23.4%) among women who have previously had a C-section delivers subsequent babies vaginally ("After years of decline," 2000).

Consumer advocates advise pregnant women who would like to deliver vaginally, if possible, to ask about the rates of C-sections when they are choosing a physician and a hospital ("After years of decline," 2000). Women can try to choose obstetricians who have lower rates or who are open to a second opinion for elective surgery. But it should not be forgotten that there are excellent reasons for having C-sections. It makes no sense to avoid a C-section for "political" reasons if vaginal delivery might put the mother or the baby at risk.

Cesarean section A method of childbirth in which the fetus is delivered through a surgical incision in the abdomen.

Transverse position A crosswise birth position.

Laboring Through the Birthing Options: Where Should a Child Be Born?[2]

Want to deliver in a special suite? At home? In a pool? So many choices—which is right for you?

Women have never had so many choices in childbirth. They have the option to labor in a pool of warm water or at home in bed, in a cozy hospital "birthing suite" or in a traditional labor room. They can choose between an obstetrician or midwife—or both. How about some aromatherapy or acupuncture, yoga or Yanni to help ease the pain and discomfort? Whatever your desire, those in the baby-delivery business want to make sure your "birth experience" is all it can be.

"I think women definitely have a strong interest in getting back to natural, less-invasive childbirth," says Dr. Amy VanBlaricom, an obstetrician at the University of Washington in Seattle. She points to the increasing demand in many parts of the country for nurse-midwives—trained professionals, usually women, who stay with a woman throughout her labor, supporting her and working with techniques like massage to avoid surgery, forceps, and other interventions. There's also growing interest in doulas, lay women with minimal training who don't perform deliveries but offer support during childbirth.

More and more women also want a family atmosphere for their deliveries, often inviting their mothers or sisters, friends, and sometimes their other children to witness the blessed event. And unhappy with the days when obstetricians dictated every step of the way, today's mothers-to-be want control, many working through every detail of their "birth plan" with their providers.

The industry is eager to please. Hospitals or birthing centers with satisfied customers stand to gain much more than a one-time payment. Women who are happy with their care are likely to return to that institution for future deliveries or other medical services. And because women are the main health care decision-makers in the family, they are also likely to bring in their kids, their husbands, and their aging parents or grandparents. Women who feel as though the hospital somehow dampened one of life's greatest experiences may simply opt to take their future business elsewhere.

STATE-OF-THE-ART BIRTHING A few years ago, hospital administrators at Duke University Medical Center in Durham, North Carolina, recognized that their obstetric facilities, although offering high-quality care, weren't as cushy as competing hospitals in the area. Officials consulted with other medical institutions nationwide, conducted focus groups with area women, and began planning a new state-of-the-art birthing center.

"We realized that changes needed to be made if we were going to survive," says Dr. William Herbert, medical director of obstetrics at the hospital. "Our rooms were far inferior in terms of the amenities and expectations that people have now."

Herbert says women desire something more than the traditional, sterile hospital room. "They want a very home-like atmosphere that's family friendly, to celebrate delivery."

Duke opened its new birthing center within the hospital. Twenty-one private rooms serve as LDRPs—labor, delivery, recovery, and postpartum, all in one. Traditionally, labor and delivery take place in separate rooms or in the same room, after which a woman may be transferred to a recovery room and then a hospital room for the duration of her stay. Herbert says the LDRP concept is aimed at reducing the hassle for the mother as well as the medical staff.

"From the time the patient gets there until the time she leaves, she's in the same room," Herbert says. However, he notes that if a woman needed a caesarean section, she would be transferred to a nearby surgical operating room for the procedure, as would any woman with complications.

[2]This section is reprinted from Jacqueline Stenson (2000). Laboring through the birthing options. MSNBC online.

The rooms resemble high-class hotel suites more than drab hospital rooms. They're decorated in soft pastels, with hardwood floors, track lighting, armoires, televisions, refrigerators and private bathrooms with make-up mirrors and whirlpools (where women can labor to ease the pain). A day bed folds out for dads or other overnight guests. Artwork from area talent adorns the walls, and windows overlook a courtyard below. Bassinets allow the babies to stay in the same room with the new parents.

At the same time, the LDRPs are fully equipped with all the medical necessities for an uncomplicated birth, explains Herbert. Much of it is tucked into closets or behind wall hangings, yet all is within easy reach of the doctors and nurses. "We've merged safety, top-notch medical care and a family atmosphere," Herbert says.

There's no hard data on whether such amenities actually translate into improved birth outcomes, he notes, but they seem to make mothers feel more comfortable and perhaps reduce stress. And although Duke's emergency facilities are just down the hall, some doctors worry about the safety of delivering babies at free-standing birthing centers, where there are no surgical facilities should complications arise, and a woman would need to be transferred elsewhere.

"Birthing centers outside a hospital are a big problem for me," says Dr. Yvonne Thornton, a clinical professor of obstetrics and gynecology at the University of Medicine and Dentistry of New Jersey in Newark. Even if the nearest hospital is just a 10-minute drive away, Thornton says, too much time could pass while the woman is transferred onto the gurney and into the ambulance and while the ambulance fights traffic to get to the emergency room.

THE HOME BIRTH DEBATE The same goes for home births, she adds, where there are even fewer resources.

"People don't understand that women die, babies die. People keep forgetting that," she says. "Why are we going back to the dark ages? You don't have the necessary equipment should something go wrong."

Statistics show that each year in the United States, there are 7.5 maternal deaths for every 100,000 live births. In developing countries, where medical resources are scarce, there are 480 deaths per 100,000 births. Common causes of death are hemorrhage, pregnancy-induced hypertension, and infection.

Thornton says the dramatically reduced death rate in the United States is attributable to the fact that most women here have access to quality care. The overwhelming majority give birth in hospitals. But even in America, African American women die during childbirth at twice the rate of European Americans, she says, because many are poor and lack access to quality care.

VanBlaricom agrees that births outside of hospitals present a threat, particularly for women at high risk of complications. "Home births are not something I would recommend," she says. "The main reason is that the labor and delivery process is so unpredictable." Birthing at home probably poses less of a threat for women who've had uncomplicated pregnancies in the past, she notes, but there still is more risk than giving birth in a hospital. "With childbirth, you just never know ahead of time if there is going to be a problem," she adds.

But Marion McCartney, director of professional services at the American College of Nurse-Midwives in Washington, D.C., who has delivered babies in the home, says it can be done safely. "Our policy is that women have a right to choose where they give birth," McCartney says, adding that a certified nurse-midwife will carefully assess a woman's risk for complications and her proximity to emergency medical care before agreeing to assist with a home birth. "You'd like the woman to be able to have a C-section within 30 minutes," should the need arise, she says.But most nurse-midwives (who differ from lay midwives in terms of advanced training) practice in hospitals alongside obstetricians. The nurse-midwives typically handle the lower-risk births, with an obstetrician on hand to help should a problem arise.

Human Sexuality ONLINE //

Childbirth Advice Is Just Keystrokes Away

You're pregnant and it's midnight. You've just felt a couple of abdominal twinges and you're a little worried. You know your obstetrician's probably asleep and since you still have a couple of weeks to go before the baby's due, you don't want to rush out to the emergency room. Right now, some good, solid medical information would be ever so reassuring.

If you've got an Internet connection, help could be just a few keystrokes away. These days, the Web is teeming with birth-related sites. The Net-savvy mother-to-be can research labor issues, find a childbirth class, chat with other expectant women, and shop for everything from pregnancy vitamins to maternity fitness wear.

Of course, as with everything that appears on the Web, there's a certain amount of drivel and misinformation. So MSNBC has rounded up some experts to help you find the best places to dock your surfboard.

For trustworthy medical advice, you should probably stick to sites that are affiliated with either an academic institution or an established professional organization, suggests Dr. David Toub, director of quality improvement at Keystone Mercy Health Plan and a member of the department of obstetrics and gynecology at the Pennsylvania Hospital in Philadelphia.

One such site, Intelihealth, a joint venture between Johns Hopkins Medical Institutions and Aetna, offers its own experts and links to many reputable sources of information, says Dr. Pamela Yoder, medical director of women's health, obstetrics and gynecology, and maternal-fetal medicine at Provena Covenant Medical Center at the University of Illinois.

Another option is to seek sites that have input from recognized experts, such as **Obgyn.net**. This site also contains chats and forums that focus on such subjects as pregnancy, birth, and breast-feeding.

When she was pregnant, Dr. Kelly Shanahan often drifted over to **Obgyn.net**'s forums to talk with other mothers-to-be about such topics as swollen ankles. "Sometimes it's good to be able to vent to other people in the same situation," says Shanahan, a physician in private practice and chair of obstetrics and gynecology at the Barton Memorial Hospital in South Lake Tahoe, Calif.

And after the baby was born, Shanahan frequented the chat for new moms. "I may be an obstetrician, but I didn't know diddley about what to do with babies once they're out of the uterus," she says. "It was really helpful to get hints from other moms on how to cope with a newborn."

Another good site for conversation about pregnancy and baby-rearing is **iVillage.com**, Shanahan says. "This site also has an expert question and answer section," she adds.

When it comes to bulletin boards, women should remember that anyone can post anything, experts say.

"My impression is that UseNet has a lot of unmoderated free-for-alls that post incorrect, even libelous information," Toub says. "A recent troll on Sci.Med, for example, showed it contained postings that claimed that ACOG [the American College of Obstetricians and Gynecologists] was covering up the 'truth' about obstetricians crushing fetal skulls and creating shoulder dystocia during vaginal deliveries."

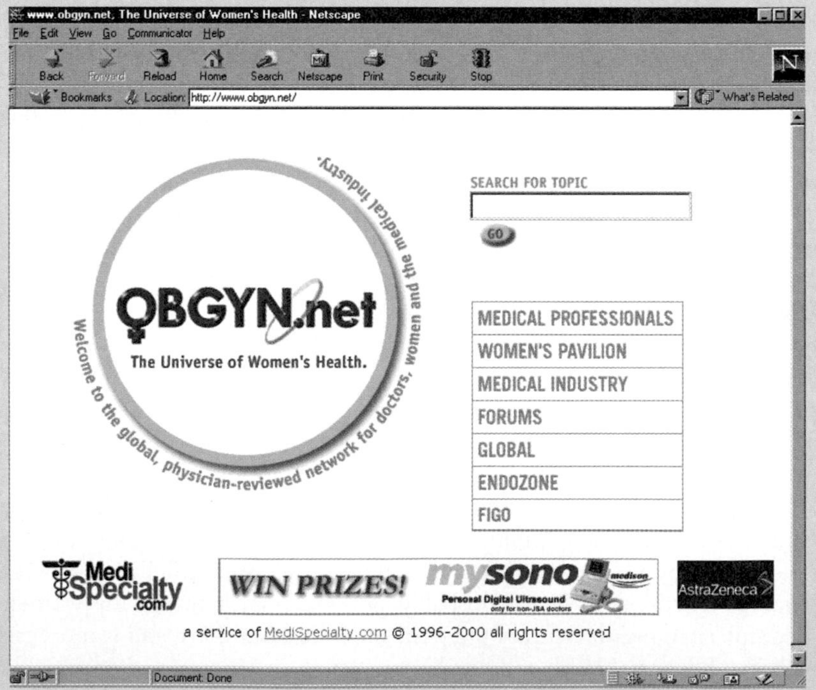

Surfing Online for Information About Pregnancy and Childbirth. Today's women can find online everything from advice from obstetricians to fitness wear for pregnant women.

Some studies show that midwife-assisted, low-risk births involve fewer C-sections, episiotomies (surgery performed to increase the size of the vaginal opening during childbirth), anesthesia, forceps, and other interventions than those supervised by an obstetrician. McCartney points to the extra support offered by midwives, who stay with the woman during the entire process, encouraging her to relax, to try different positions and techniques such as acupuncture or yoga to ease pain, and just to take things slowly. Midwives appeal to many women because of this openness to alternative techniques.

In other words, the medical oriented newsgroup has some not-so-correct advice mixed in with its information, he said.

Delivering at Home?

If you think you might want to deliver at home, consider checking out midwives listed on the Internet. But, "as with anything else on the Web, the key is to look carefully at the motivations and credentials of the person posting the information," cautions Cheri Van Hoover, a midwife who practices at Stanford University in Palo Alto, California. "Responsible midwives will disclose their background, credentials, and a way to contact them for more information."

A good starting place is the American College of Nurse-Midwives, Van Hoover suggests. Another good site is Childbirth.org, according to Dr. R. Daniel Braun, a clinical professor in the department of obstetrics and gynecology at the Indiana University School of Medicine in Indianapolis. "If you are going to have a home delivery this is an excellent site," Braun says. "The site is run by a doula. But remember that it is biased towards the 'natural' delivery."

Childbirth Classes

If you're looking for childbirth classes, there are plenty of listings on the Web, Van Hoover says. But again, women should be asking a few questions, she adds. For example, you might want to check, "Where did the instructors get their training?," Van Hoover suggests. "Which professional organizations do they belong to? Do they believe there is only one right way to give birth or are they flexible and individualized in their approach."

Ultimately, it could be the Web's easy shopping that you use the most.

One-Stop Shopping

For Shanahan, retail on the Web made all the difference. "I live in a rural community in the mountains," says Shanahan, who gave birth in December. "And getting out of here in the winter can sometimes be a problem."

One of Shanahan's favorite finds was a site marketing fitness wear for pregnant women. This site also contains information on exercise during pregnancy, including the recommendations of the American College of Obstetricians and Gynecologists (ACOG).

If you'd like to order vitamins or disposable diapers, for that matter, you might try Drugstore.com's pregnancy page, Shanahan suggests. The site also has accurate medical information, she says.

Another commercial site that provides expert information and advice is the Babycenter's birth and labor section. The site offers information on a variety of topics. You just select from the menu and click. Often sections are written by experts, whose biographies are readily available. And everything on the site is reviewed by a board that includes several ob-gyns.

Here you can also get answers to practical questions like "What should I take to the hospital?" or "When is it too late to change obstetricians?" The site includes discussions about the various types of childbirth classes and an interactive survey designed to help you figure out which kind of class would best fit your needs.

And if you were looking for insight on whether you could have contractions without being in labor, text under the heading "false labor" at the Babycenter site could put your mind at ease.

Some of the Best Web Sites

The American College of Obstetricians and Gynecologists
www.acog.org/

American College of Nurse Midwives
www.midwife.org/

Doulas of North America
www.dona.com/

Intelihealth
www.intelihealth.com/

Obgyn.net
www.obgyn.net/

iVillage.com
www.parentsplace.com/pregnancy/

Childbirth.org
www.childbirth.org/

Fitness Wear for Pregnant Women
www.fitmaternity.com/index.html

Drugstore.com
www.drugstore.com/

Birth and labor section of Babycenter.com
www.babycenter.com/birthandlabor/

Source: From Linda Carroll (2000). "Childbirth Advice Keystrokes Away: How to Find the Best Web Sites." MSNBC online. Reprinted by permission of MSNBC Interactive News.

DELIVERING WITH A DOULA Research also suggests that doulas can be a big help. "They're wonderful patient advocates, and they can be especially good for people who don't have a partner who can help," VanBlaricom says.

Another option that's popular in Europe, but much less so in America, is water birth. Although most U.S. experts say laboring in water in safe as long as the tub has been thoroughly disinfected, they're skeptical about underwater deliveries. "That's on the fringe of what we consider safe," VanBlaricom says. Thornton says she knows of two cases in which

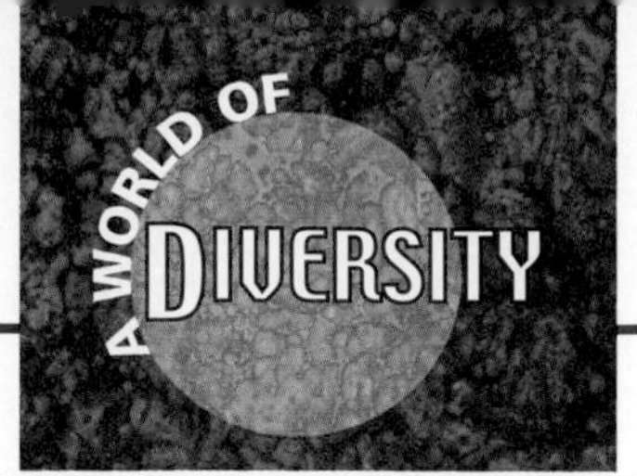

A Racial Gap in Infant Deaths, and a Search for Reasons

Ethelyn Bowers had a master's degree, an executive-level job, a husband who was a doctor, and access to some of the best medical care available. But her accomplishments and connections seemed to make little difference when she endured the premature births and subsequent deaths of three babies, two of whom were in a set of triplets.

A competitive businesswoman who is immersed in her career and the hectic schedule of her surviving children, now 9 and 11, Ms. Bowers is candid about her losses but does not dwell on their possible causes. "I just chalked it up to bad luck, mainly," said Ms. Bowers, a sales director for Lucent Technologies who lives with her family in Livingston, New Jersey.

But her husband is haunted by the notion that somehow, in a way experts have yet to fully grasp, the fact that he and his wife are African American was a factor in their children's deaths.

"At the time of the death, you don't really dissect out the reasons; all you really think about is the tremendous sorrow," said her husband, Dr. Charles H. Bowers, chief of obstetrics and gynecology at Kings County Hospital Center in Brooklyn. "In retrospect, I think you need to look at psychosocial causes—the euphemism I use for racism. If in fact my counterpart, a [European American] physician making six figures, had a wife the same age and her likelihood of losing their child was less than half of my wife's chances of losing my child, why should that be?"

It is a mystery that consumes not only Dr. Bowers but also a growing field of researchers struggling to explain a persistent racial gap in American infant mortality rates. For years, the number of babies who die before their first birthday has been a source of shame for public health advocates; in international comparisons of infant mortality, the World Health Organization has ranked the United States 25th, below Japan, Israel, and Western Europe. But even as infant mortality rates improve—to a record low national average of 7.2 deaths per 1,000 live births—the disparity between African Americans and European Americans has grown, from 2 to nearly 2.4 times the number of infant deaths.

Especially troubling is evidence that this is not simply because of poverty. Though college-educated African American women do better than impoverished women, they are still twice as likely to bury their babies as European American women. And immigrants have better pregnancy outcomes than assimilated minorities. What happens to their health once they move here? Why haven't education, better jobs, and health care made a bigger difference in closing the infant mortality gap?

"There's no single answer," said Dr. Solomon Iyasu, an epidemiologist at the Centers for Disease Control and Prevention in Atlanta. "Infant mortality is such a complex issue because it's driven both by medical issues [and] by social issues."

Because premature deliveries and low birth weights account for two-thirds of infant deaths, much recent research has focused on the causes of early labor. Medical complications from diabetes and high blood pressure, more prevalent illnesses in African Americans, can be blamed for some of the cases but do not account for the huge disparity. So researchers have begun exploring more subtle factors, like crime, pollution, and family support. Another hypothesis is that chronic stress caused by racial discrimination can elevate the hormones that set off premature labor.

"We're starting to think it's something about lifelong minority status," said Dr. James W. Collins, Jr., a neonatologist in Chicago who teaches pediatrics at Northwestern University Medical School.

Dr. Collins has compared the newborns of African Americans born in the United States with those of mothers who came directly from Africa and found that the immigrants' babies were bigger, with birth weights more compa-

the infants drowned. "I don't encourage water births," she says. "Why are we taking a chance here?"

With all the options available, experts encourage women to be fully informed of the risks and benefits of each before making a decision. "There's always something that someone can offer, but the bottom line is that you want a healthy baby," says Dr. Ruth Fretts, an assistant professor of obstetrics and gynecology at Beth Israel Deaconess Medical Center in Boston. "Make sure you don't miss the point."

BIRTH PROBLEMS

Most deliveries are uncomplicated, or "unremarkable" in the medical sense—although childbirth is the most remarkable experience of many parents' lives. Problems can and do occur, however. Some of the most common birth problems are anoxia and the birth of preterm and low-birth-weight babies.

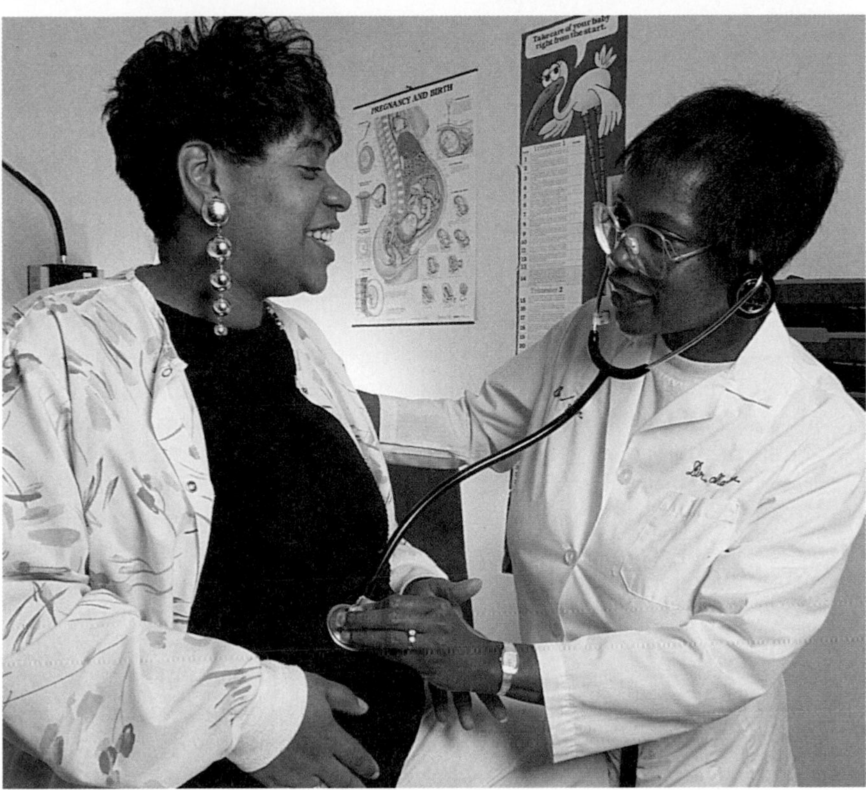

Why do African American Women Run Greater Risks of Infant and Maternal Mortality? It seems that more may be involved than the quality of prenatal care, but health care professionals are not sure exactly what. More research is needed in this area.

rable [to] European Americans than to African Americans. Another study involving interracial couples indicated that the mother's race was crucial; babies born to African American mothers and European American fathers had higher rates of low birth weight.

The intergenerational effects of poverty and discrimination are also being explored amid concerns that a woman's health and personal experiences—even before she becomes pregnant—can affect her children's health.

"It's not clear that if you have a woman who's first-generation college-educated that she's still not carrying the risks of generations of poverty," said Dr. Marie McCormick, a pediatrician and professor of child and maternal health at the Harvard School of Public Health. "There is also increasing evidence that having been born with a low birth weight yourself, you're more likely to have a low-birth-weight child."

Source: From Leslie Berger. (2000, June 25). "A Racial Gap in Infant Deaths, and a Search for Reasons." *The New York Times*, p. WH13.

Anoxia

Prenatal **anoxia** can cause various problems in the neonate and affect later development. It leads to complications such as brain damage and mental retardation. Prolonged anoxia during delivery can also result in cerebral palsy and possibly death.

The baby is supplied with oxygen through the umbilical cord. Passage through the birth canal squeezes the umbilical cord. Temporary squeezing, like holding one's breath for a moment, is unlikely to cause problems. (In fact, slight oxygen deprivation at birth is not unusual because the transition from receiving oxygen through the umbilical cord to breathing on its own may not happen immediately after the baby emerges.) Anoxia can result if constriction of the cord is prolonged, however. Prolonged constriction is more likely to occur with a breech presentation, because the baby's head presses the umbilical cord against the birth canal during delivery. Fetal monitoring can help detect anoxia early, however, before damage occurs. A C-section can be performed if the fetus appears to be in distress.

Anoxia Oxygen deprivation.

Preterm and Low-Birth-Weight Children

A neonate is considered to be premature, or **preterm,** if it is born before 37 weeks of gestation. The normal period of gestation is 40 weeks. Prematurity is generally linked with low birth weight, because the fetus normally makes dramatic gains in weight during the last weeks of pregnancy.

Regardless of the length of its gestation period, a newborn baby is considered to have a low birth weight if it weighs less than 5 pounds (about 2,500 grams). Preterm and low-birth-weight babies face a heightened risk of infant mortality from causes ranging from asphyxia and infections to sudden infant death syndrome (SIDS) (Berger, 2000; Kramer et al., 2000). Neurological and developmental problems are also common among preterm infants, especially those born at or prior to 25 weeks of gestation (Wood et al., 2000).

Twins and other multiple birth groupings are more likely to be of low birth weight than individual births (Blickstein et al., 2000). There is also a relationship between prematurity and the spacing of babies. Women who have babies less than 18 months or more than 59 months apart run the highest risk of having premature infants (Fuentes-Afflick & Hessol, 2000). On the other hand, it is common for women carrying more than one child to deliver prematurely, apparently because the babies are running out of space.

Preterm babies are relatively thin because they have not yet formed the layer of fat that accounts for the round, robust appearance of most full-term babies. Their muscles are immature, which weakens their sucking and breathing reflexes. Also, in the last weeks of pregnancy, fetuses secrete **surfactant** that prevent the walls of their airways from sticking together. Muscle weakness and incomplete lining of the airways with surfactant can cause a cluster of problems known as **respiratory distress syndrome,** which is responsible for many neonatal deaths. (Today, surfactant replacement therapy is extending the possibilities of survival in preterm infants [Cole, 2000].) Preterm babies may also suffer from underdeveloped immune systems, which leave them more vulnerable to infections.

Preterm infants usually remain in the hospital for a time. There they can be monitored and placed in incubators that provide a temperature-controlled environment and offer some protection from disease. If necessary, they may also receive oxygen. Although remarkable advances are being made in our ability to help preterm babies survive, the likelihood of developmental disabilities continues to increase dramatically for babies who are born at 25 weeks of gestation or earlier (Cole, 2000).

THE POSTPARTUM PERIOD

The weeks after delivery are called the **postpartum** period. The first few days postpartum are frequently happy ones. The long wait is over, as are the discomforts of childbirth. A sizable number of women experience feelings of depression, however, in the days and sometimes weeks and months after childbirth.

Preterm Born before 37 weeks of gestation.

Surfactant Substances that prevent the walls of the airways from sticking together.

Respiratory distress syndrome A cluster of breathing problems, including weak and irregular breathing, to which preterm babies are especially vulnerable.

Postpartum Following birth.

Maternal Depression

Many new mothers experience mood changes after childbirth. During the days or weeks following the delivery of their babies, as many as 80% of new mothers (Morris, 2000) experience periods of sadness, tearfulness, and irritability that are commonly called the "postpartum blues," the "maternity blues," or the "baby blues." This downswing in mood typically occurs around the third day after delivery (Samuels & Samuels, 1986). The baby blues usually last less than a week and are generally believed to be a normal response to hormonal and psychological changes that attend childbirth (Morris, 2000).

Some mothers experience more persistent and severe mood changes, called **postpartum depression** (PPD) (Morris, 2000). PPD may last a year or even longer. PPD can involve extreme sadness or despair, apathy, changes in appetite and sleep patterns, low self-esteem, and difficulty concentrating. A study of 1,033 married, middle-class, first-time mothers from the Pittsburgh area who had full-term, healthy infants found that 9.3% of them had experienced PPD (Campbell & Cohn, 1991). Some researchers (e.g., Gitlin & Pasnau, 1989) estimate that PPD affects up to 15% of new mothers.

Like the "maternity blues," PPD may reflect a combination of physiological and psychological factors. Hormonal changes may play a role in PPD, but women with PPD are more likely than those with the maternity blues to have been susceptible to depression before and during their pregnancies (O'Hara et al., 1984). Psychosocial factors such as stress, a troubled marriage, or the need to adjust to an unwanted or sick baby may all increase a woman's susceptibility to PPD (Gitlin & Pasnau, 1989; O'Hara et al., 1984, 1991). First-time mothers, single mothers, and mothers who lack social support from their partners or family members face the greatest risk of PPD (Gitlin & Pasnau, 1989).

New fathers may also have bouts of depression. New mothers are not the only ones who must adjust to the responsibilities of parenthood. Fathers too may feel overwhelmed or unable to cope. Perhaps more fathers would experience the "paternity blues" if mothers did not generally shoulder the lion's share of child-rearing chores.

Breast-Feeding Versus Bottle-Feeding

Only a minority of U.S. women breast-feed their children. One reason is that many women return to the work force shortly after childbirth. Some choose to share feeding chores with the father, who is equally equipped to prepare and hold a bottle, but not to breast-feed. Other women find breast-feeding inconvenient or unpleasant. Long-term comparisons of breast-fed and bottle-fed children show few, if any, significant differences. Breast-feeding does reduce the general risk of infections to the baby, however, by transmitting the mother's antibodies to the baby. Breast-feeding also reduces the incidence of allergies in babies, particularly in allergy-prone infants (Barisic, 1998). On the other hand, HIV (the AIDS virus) can be transmitted to infants via breast milk. According to UN estimates, one-third of the infants with HIV around the world were infected via breast milk (Meier, 1997). Moreover, when undernourished mothers in developing countries breast-feed their babies, the babies too can become malnourished (Crossette, 2000).

The hormones prolactin and oxytocin are involved in breast-feeding. **Prolactin** stimulates production of milk, or **lactation,** 2 to 3 days after delivery. Oxytocin causes the breasts to eject milk and is secreted in response to suckling. When an infant is weaned, secretion of prolactin and oxytocin is discontinued, and lactation comes to an end.

Uterine contractions that occur during breast-feeding help return the uterus to its typical size. Breast-feeding also delays resumption of normal menstrual cycles. Breast-feeding is not a perfectly reliable birth-control method, however. (But nursing women are advised not to use birth-control pills, because the hormone content of the pills is passed to the infant through the milk.)

Should a woman breast-feed her baby? The issue has become highly politicized (Law, 2000). Much of the vast literature on breast-feeding has little to do with the advantages of breast milk or formula, but with occupational and domestic arrangements, child day care, mother–infant bonding, and the politics of domestic decision making. Although breast-feeding has benefits for both mother and infant, each woman must weigh these benefits against the difficulties breast-feeding may pose for her. These include assuming the sole responsibility for nighttime feedings, the physical demands of producing and expelling milk, tendency for soreness in the breasts, and the inconvenience of being continually available to meet the infant's feeding needs. We suggest that women should breast-feed because they want to, not because they feel they must.

Postpartum depression Persistent and severe mood changes during the postpartum period, involving feelings of despair and apathy and characterized by changes in appetite and sleep, low self-esteem, and difficulty concentrating.

Prolactin A pituitary hormone that stimulates production of milk. (From roots that mean "for milk.")

Lactation Production of milk by the mammary glands.

Human Sexuality ONLINE //

Where to Get Help Breast-Feeding

During pregnancy, women tend to romanticize breast-feeding, assuming that this natural function will be pleasant and trouble-free. But many women encounter problems ranging from local irritation to difficulty maintaining the flow of milk. As noted by lactation consultant Corky Harvey (2000), breast-feeding can be painful, inconvenient, and stressful, especially at first.

Harvey suggests that women who are having problems breast-feeding call the hospital where they delivered and ask whether it has a lactation consultant or can refer them to one. Or call your pediatrician or obstetrician or a friend or relative who has successfully breast-fed. There is also (800) TELLYOU, the hotline of Medela, a supplier of lactation equipment. Medela can provide a list of nearby lactation centers that rent equipment and offer support. Another resource is a La Leche League chapter.

The Internet can also be helpful: **www.BreastfeedingTaskForLA.org** is the Web site of the Breastfeeding Task Force of Greater Los Angeles. Other Web sites:

www.breastfeeding.org

www.breastfeeding.com

www.lalecheleague.org

www.ILCA.org, the site for the International Lactation Consultants Association.

Resumption of Ovulation and Menstruation

For close to a month after delivery, women experience a reddish vaginal discharge called **lochia.** A non-nursing mother does not resume actual menstrual periods until 2 to 3 months postpartum. The first few cycles are likely to be irregular. Many women incorrectly assume that they will resume menstruating after childbirth by first having a menstrual period and then ovulating 2 weeks later. In most cases the opposite is true. Ovulation precedes the first menstrual period after childbirth. Thus a woman may become pregnant before the menstrual phase of her first postpartum cycle. Some women, but not all, who suffered premenstrual syndrome before their pregnancies find that their periods give them less discomfort after the birth of their children.

Resumption of Sexual Activity

The resumption of coitus depends on a couple's level of sexual interest, the healing of episiotomies or other injuries, fatigue, the recommendations of obstetricians, and, of course, tradition. Obstetricians usually advise a 6-week waiting period for safety and comfort. One study of 570 women found that they actually resumed sexual intercourse an average of 7 weeks after childbirth (Byrd et al., 1998).

Truth or Fiction? **REVISITED**

It is true that couples should abstain from coitus for at least 6 weeks following childbirth. However, many other kinds of sexual activities are safe and do not cause discomfort. (Check with your obstetrician.) ■

Women typically prefer to delay coitus until it becomes physically comfortable, generally when the episiotomy or other lacerations have healed and the lochia has ended. This may take several weeks. Women who breast-feed may also find that they have less vaginal lubrication, and the dryness can cause discomfort during coitus. K-Y jelly or other lubricants may help in such cases.

The return of sexual interest and resumption of sexual activity may take longer for some couples than for others. Sexual interest depends more on psychological than on physical factors. Many couples encounter declining sexual interest and activity in the first year following childbirth, generally because child care can sap energy and limit free time. Generally speaking, couples whose sexual relationships were satisfying before the baby arrived tend to show greater sexual interest and to resume sexual activity earlier than those who had less satisfying relationships beforehand. (No surprise.)

Lochia A reddish vaginal discharge that may persist for a month after delivery. (From the Greek *lochios,* which means "of childbirth.")

Human Sexuality ONLINE //

Web Sites Related to Conception, Pregnancy, and Childbirth

The following Web sites offer information about conception, pregnancy, and childbirth. Several other related Web sites are mentioned within the chapter.

The Web site of the American College of Obstetricians and Gynecologists. Information about how to find a physician, women's issues, legislation, and more:
www.acog.org/

The Web site of the American College of Nurse-Midwives:
www.midwife.org/

The Web site of Doulas of North America:
www.dona.com/

The Web site of "The Universe of Women's Health." Click on "Women's Pavilion" for information on topics such as pelvic pain, endometriosis, fetal monitoring, hysterectomy, infertility, laparoscopy, menopause, osteoporosis, pregnancy and birth, and ultrasound.
www.obgyn.net/

The Web site of Childbirth.org, which proclaims that "Birth is a natural process, not a medical procedure." There are a free online childbirth class, lists of books, and links to information about issues as diverse as episiotomy, postpartum depression, and fertility.
www.childbirth.org/

The Web site of La Leche League, an "international, nonprofit, nonsectarian organization dedicated to providing education, information, support, and encouragement to women who want to breast-feed." There are meetings for breast-feeding mothers, as well as future breast-feeding mothers, and information about how to obtain a consultant near you. The league also updates health care professionals on research on lactation management.
www.lalecheleague.org

summing up

CONCEPTION: AGAINST ALL ODDS

Conception is the union of a sperm cell and an ovum. Fertilization normally occurs in a fallopian tube.

Optimizing the Chances of Conception

Optimizing the chances of conception means engaging in coitus at the time of ovulation. Ovulation can be predicted by calculating the woman's basal body temperature, analyzing the woman's urine for luteinizing hormone, or tracking the thickness of vaginal mucus.

INFERTILITY AND ALTERNATIVE WAYS OF BECOMING PARENTS

Male Fertility Problems

Fertility problems in the male include low sperm count, irregularly shaped sperm, low sperm motility, certain chronic or infectious diseases, trauma to the testes, an autoimmune response to sperm, and pituitary imbalances and/or thyroid disease.

Female Fertility Problems

The major causes of infertility in women include irregular ovulation, obstructions or malfunctions of the reproductive tract, endometriosis, and the decline of hormone levels with age. Failure to ovulate may often be overcome by fertility drugs. Methods for overcoming other female fertility problems include in vitro fertilization, GIFT, ZIFT, donor IVF, embryonic transfer, and surrogate motherhood.

PREGNANCY

Early Signs of Pregnancy

Early signs include a missed period, presence of HCG in the blood or urine, and Hegar's sign.

Pregnancy Tests

Pregnancy tests detect the presence of human chorionic gonadotropin (HCG) in the woman's urine or blood.

Early Effects of Pregnancy

Early effects include tenderness in the breasts and morning sickness.

Miscarriage (Spontaneous Abortion)

Miscarriages have many causes, including chromosomal defects in the fetus and abnormalities of the placenta and uterus.

Sex During Pregnancy

Most health professionals concur that in most cases, coitus is safe until the start of labor.

Psychological Changes During Pregnancy

A woman's psychological response to pregnancy reflects her desire to be pregnant, her physical changes, and her attitudes toward these changes. Men, like women, respond to pregnancy according to the degree to which they want the child.

PRENATAL DEVELOPMENT

■ The Germinal Stage

The germinal stage is the period from conception to implantation.

■ The Embryonic Stage

The embryonic stage begins with implantation, extends to about the eighth week of development, and is characterized by differentiation of the major organ systems.

■ The Fetal Stage

The fetal stage begins by the ninth week, continues until the birth of the baby, and is characterized by continued maturation of the fetus's organ systems and dramatic increases in size.

■ Environmental Influences on Prenatal Development

Environmental factors that affect prenatal development include the mother's diet, maternal diseases and disorders, and drugs. Maternal malnutrition has been linked to low birth weight and infant mortality. Exposure to particular teratogens causes the greatest harm during critical periods of vulnerability.

■ Chromosomal and Genetic Abnormalities

Chromosomal and genetic abnormalities can lead to cystic fibrosis, Down syndrome, hemophilia, Huntington's chorea, neural tube defects, phenylketonuria, retina blastoma, sickle-cell anemia, and Tay–Sachs disease. Parental blood tests, amniocentesis, and ultrasound allow parents to learn whether the fetus has, or is at risk for, many such disorders.

CHILDBIRTH

■ The Stages of Childbirth

In the first stage, uterine contractions efface and dilate the cervix so that the baby may pass through. The first stage may last from a couple of hours to more than a day. The second stage lasts from a few minutes to a few hours and ends with the birth of the baby. During the third stage, the placenta is expelled.

■ Methods of Childbirth

Contemporary methods for facilitating childbirth include anesthetized childbirth, natural childbirth, the Lamaze method, and cesarean section.

■ Laboring Through the Birth Options: Where Should a Child Be Born?

In the United States most births occur in hospitals. Some parents seeking more intimate arrangements, however, opt for a birth center or home delivery.

BIRTH PROBLEMS

■ Anoxia

Prenatal anoxia can cause brain damage and mental retardation in the child.

■ Preterm and Low-Birth-Weight Children

Preterm and low-birth-weight babies have a heightened risk of infant mortality.

THE POSTPARTUM PERIOD

■ Maternal Depression

Many new mothers experience transient mood changes after childbirth. Women with postpartum depression experience lingering depressions following childbirth.

■ Breast-Feeding Versus Bottle-Feeding

Breast-feeding is connected with fewer infections and allergic reactions in the baby than bottle-feeding. Long-term studies show few differences between children whose parents used one or the other feeding method, however.

■ Resumption of Ovulation and Menstruation

The first few menstrual cycles following childbirth are likely to be irregular.

■ Resumption of Sexual Activity

Obstetricians usually advise waiting about 6 weeks after childbirth before resuming coitus. Couples need not wait this long to enjoy other forms of sexual activity.

questions for critical thinking

1. Agree or disagree with the following statement and support your answer. "Gender selection technology is sexist."

2. Today's reproductive technologies make it possible for women to have babies after menopause (by receiving donor eggs) or to become mothers to their own grandchildren (by using eggs donated by their daughters). It may even become possible for a woman to receive a transplanted ovary from an aborted fetus and later give birth to a child whose biological mother is the aborted fetus. These reproductive technologies have leapfrogged ahead of society's efforts to grapple with their ethical, moral, and legal implications. Where should the line be drawn in determining how far medical science should go in providing infertile people with reproductive alternatives?

3. Do you believe that it is appropriate to engage in sexual activity during pregnancy? Why, or why not?

4. Do you know anyone who drank alcohol or smoked during pregnancy? Would she talk about her drinking or smoking? What did she say?

5. Do you know of anyone who has been tested to learn whether her or his child is likely to have a genetic or chromosomal abnormality? Would the woman have terminated the pregnancy if an abnormality had been detected? Explain.

chapter

12

Contraception and Abortion

Pablo Picasso. *Portrait of Dora Maar.* 1936. Private Collection. Credit: Giraudon/Art Resource, NY. Credit: © 2002 Estate of Pablo Picasso/Artists Rights Society (ARS), New York.

outline

Truth or Fiction?

CONTRACEPTION

Contraception in the United States: The Legal Battle

A Closer Look Talking with Your Partner About Contraception

Selecting a Method of Contraception

METHODS OF CONTRACEPTION

Oral Contraceptives ("the Pill")

A World of Diversity African Americans and Birth Control—The Intersection of Suspicion and Hope

Norplant

A Closer Look Congratulations, Wheat Ridge, It's a . . . !

Intrauterine Devices (IUDs)

The Diaphragm

Spermicides

The Cervical Cap

Condoms

Douching

Withdrawal (Coitus Interruptus)

Fertility Awareness Methods (Rhythm Methods)

Sterilization

The Search Goes On

ABORTION

When Does Human Life Begin?

Historical and Legal Perspectives on Abortion

Questionnaire Pro-Choice or Pro-Life? Where Do You Stand?

Methods of Abortion

A Closer Look Partial-Birth Abortion

A Closer Look RU-486 in Europe

Psychological Consequences of Abortion

Human Sexuality Online Web Sites Related to Contraception and Abortion

SUMMING UP

QUESTIONS FOR CRITICAL THINKING

Truth or Fiction?

____ Ancient Egyptians used crocodile dung as a contraceptive substance.

____ Contraceptives not only prevent conception; they also prevent sexually transmitted infections.

____ There is an oral contraceptive that can be taken the morning after intercourse.

____ Sterilization operations can be surgically reversed.

____ Testosterone can be used as a male contraceptive device.

____ Abortions were legal in the newly founded United States.

____ The D&C is the most widely used abortion method in the United States.

It was a stifling day in July 1912. Margaret Sanger (1883–1966), a nurse practitioner, was summoned to the house of a woman near death from a botched self-induced abortion. Her husband had called a doctor, and the doctor sent for Sanger. Together, doctor and nurse worked feverishly through the days and nights that followed to stem an infection that had taken hold in the woman. Sanger later commented,

> Never had I worked so fast, never so concentratedly. The sultry days and nights were melted into a torpid inferno. It did not seem possible there could be such heat, and every bit of food, ice, and drugs had to be carried up three flights of stairs. . . .

After two interminable weeks, the woman began to recover. Her neighbors, who had feared the worst, came to express their joy. But the woman, who smiled wanly at those who came to see her, appeared more depressed and anxious than would be expected of someone who was recovering from a grave illness. By the end of the third week, when Sanger prepared to leave her patient, the woman, Mrs. Sachs, voiced the fear that was haunting her. Her face registered deep despair as she explained to Sanger that she dreaded becoming pregnant again and facing a choice between attempting another abortion, which she feared might kill her, and bearing a baby whose care it was beyond her means to support. She pleaded for information about contraception but Sanger could offer none. In 1912 it was a crime even for health professionals like Margaret Sanger to dispense information about contraceptives. Abortions, too, were illegal. Sanger tried to comfort her and promised to return to talk again.

Three months later she received another urgent call. It was Mr. Sachs. His wife was sick again—from the same cause. As Sanger recalled,

> For a wild moment I thought of sending someone else, but actually, of course, I hurried into my uniform, caught up my bag, and started out. All the way I longed for a subway wreck, an explosion, anything to keep me from having to enter that home again. But nothing happened, even to delay me. I turned into the dingy doorway and climbed the familiar stairs once more. The children were there, young little things.
>
> Mrs. Sachs was in a coma and died within ten minutes. I folded her still hands across her breast, remembering how they had pleaded with me, begging so humbly for the knowledge which was her right. I drew a sheet over her pallid face. Jake was sobbing, running his hands through his hair and pulling it out like an insane person. Over and over again he wailed, "My God! My God! My God!" (Sanger, 1938)

Today, partly because of the work of Margaret Sanger, who went on to become a key advocate for birth control, information about contraceptives is disseminated freely throughout the United States.

Methods of birth control include contraception and abortion. The term *contraception* refers to techniques that prevent conception. The term *abortion* refers to the termination of a pregnancy before the embryo or fetus is capable of surviving outside the womb.

CONTRACEPTION

People have been devising means of contraception since they became aware of the relationship between coitus and conception. Ironically, the safest and most effective method of contraception is also the least popular: abstinence. The Bible contains many references to contraceptive techniques, including vaginal sponges and contraceptive concoctions. It also refers to **coitus interruptus,** or withdrawal. The story of Onan, for example, implies knowledge of the withdrawal method.

Ancient Egyptian methods of birth control included douching with wine and garlic after coitus, and soaking crocodile dung in sour milk and stuffing the mixture deep within the vagina. The dung blocked the passage of many—if not all—sperm through the cervix and also soaked up sperm. The dung may also have done its job through a social mechanism. It may have discouraged all but the most ardent suitors!

Truth or Fiction? REVISITED

It is true that ancient Egyptians used crocodile dung as a contraceptive. ■

Greek and Roman women placed absorbent materials within the vagina to absorb semen. The use of sheaths or coverings for the penis has a long history. Sheaths worn over the penis as decorative covers can be traced to ancient Egypt (1350 B.C.). Sheaths of linen were first described in European writings in 1564 by the Italian anatomist Fallopius (from whom the name of the fallopian tube is derived). Linen sheaths were used, without success, as a barrier against syphilis. The term **condom** was not used to describe penile sheaths until the eighteenth century. At that time, sheaths made of animal intestines became popular as a means of preventing sexually transmitted infections and unwanted pregnancies. Among the early advocates of condoms as a method of contraception was the Italian adventurer and writer Giovanni Casanova (1725–1798). We now associate his name with men who are known for their amorous adventures. James Boswell (1740–1795), the biographer of Samuel Johnson, described his use of "armor," or condoms, in his graphic *London Journal.* On one occasion, however, he was so enamored with a street prostitute that he neglected to use his armor and contracted gonorrhea. Condoms made of rubber (hence the slang "rubbers") were introduced shortly after Charles Goodyear invented vulcanization of rubber in 1843. Many other forms of contraception were also used widely in the nineteenth century, including withdrawal, vaginal sponges, and douching.

Contraception in the United States: The Legal Battle

As methods of contraception grew more popular in the nineteenth century, opponents waged a battle to make contraception illegal. One powerful opponent of contraception was Anthony Comstock, who served for a time as the secretary of the New York Society for the Suppression of Vice. Comstock lobbied successfully for passage of a federal law in 1873—the Comstock law—that prohibited the dissemination of birth-control information through the mail on the grounds that it was "obscene and indecent." Many states passed even more restrictive laws. They outlawed passage of information from one person to another, even from physician to patient.

Consider the resistance that Margaret Sanger encountered when she challenged the laws restricting information about contraception. In 1914 she established the National Birth Control League, which published the magazine *The Woman Rebel. Rebel* did not publish birth-control information but challenged the view that it was obscene. Nevertheless, charges were brought against Sanger, and she fled to Europe before her trial. During her self-imposed exile, she visited birth-control clinics in the Netherlands. When the charges against her in the United States were dropped in 1916, Sanger returned and established a birth-control clinic in Brooklyn, New York. The clinic was closed by the police, and Sanger was arrested. Released on bail, she reopened the clinic and was thereupon sentenced to 30 days in jail. She successfully appealed the sentence. In 1918 the courts ruled that physicians must be allowed to disseminate information that might aid in the

Coitus interruptus A method of contraception in which the penis is withdrawn from the vagina prior to ejaculation. Also referred to as the *withdrawal method.*

Condom A sheath made of animal membrane or latex that covers the penis during coitus and serves as a barrier to sperm following ejaculation.

a CLOSER look

Talking with Your Partner About Contraception

When is the right time to discuss contraception? On a first date? When you are invited to meet your partner's family? When you are lost in amorous embraces? Broaching the topic can be awkward. *Not* broaching it can be disastrous. Often a man responds to the news that his partner is pregnant by saying something like: "But I thought you were *using* something!"

Technically speaking, the right time to discuss birth control is *anytime* that allows your contraceptive to become effective before you engage in coitus. That can mean weeks or months before, if you decide to use a prescription contraceptive such as the birth-control pill, the IUD, the diaphragm, or the cervical cap. Or it can mean a few moments before coitus, if you decide to use a condom and have one ready. Despite the obvious advantages of deciding upon contraception before coitus, the issue is often broached, if it is broached at all, only after the partners become sexually intimate.

Practically speaking, it is awkward—and perhaps presumptuous!—to discuss contraception when you meet or are on a first date. But at the very least, it is advisable to prepare oneself for the possibility of coitus. The man or woman may bring along a condom. The woman may already be on the pill, have an IUD in place, or a diaphragm.

Talking about contraception helps many couples make the transition from a casual relationship to an intimate one. Talking enables partners to share responsibility for their behavior. As a result, the woman is less likely to be resentful that the responsibility rests entirely on her.

Yes, it can be awkward or difficult to raise the topic. Couples may not feel that their relationship is secure enough. They might think, "We'll cross that bridge when we come to it." Not planning ahead, however, prevents the effective use of contraceptives that require advance planning.

Couples who choose to discuss contraception before engaging in coitus may benefit from these communication guidelines:

1. *Pick a strategic time and place.* Choose a time when the two of you are alone and are free of distractions. Pick a place that is comfortable and private.
2. *Couch your discussion in terms of your feelings about your partner and your relationship.* Talk about your general feelings toward your partner and your general relationship before you narrow in on contraception.
3. *Don't apologize for raising the topic.* You may feel embarrassed talking about sensitive topics like birth control. But you need not apologize for bringing up the subject. Apologizing suggests that you think you are doing something wrong.
4. *Raise the subject in a way that encourages candid discussion.* Use open-ended questions to explore your partner's attitudes. Say something like "I think our relationship has reached the point where we need to talk about contraception. I know we're not sleeping together yet, but some forms of contraception require advance planning. Have you been thinking about it?"
5. *Explore options.* Don't make demands of your partner. Don't say "Because we may start sleeping together, I think you should go on the pill." Rather, say something like "I know that many different types of contraceptives are available. Why don't we discuss which method might be best for us if we become intimate?" Use the opportunity to explore each other's views about birth control in general and specific techniques in particular.

cure and prevention of disease. Dismantling of the Comstock law had begun. With the financial support of a wealthy friend, Katherine Dexter McCormack, Sanger spurred research into the use of hormones as one approach to contraception. In 1960, only 6 years before Sanger's death, oral contraception—"the pill"—was finally marketed in the United States. In 1965 the Supreme Court struck down the last impediment to free use of contraception: a law preventing the sale of contraceptives in Connecticut (*Griswold v. Connecticut*, 1965). In 1973 abortion was in effect legalized by the Supreme Court in the case of *Roe v. Wade*, permitting women to terminate unwanted pregnancies.

Today, contraceptives are advertised in popular magazines and sold through vending machines in college dormitories. U.S. history is not a one-way road to unrestricted use of birth control, however. Recent Supreme Court decisions have set aside bits and pieces of *Roe v. Wade*, giving the states more discretion in the regulation of abortion and restricting access to abortions for minors. Use of **artificial contraception** continues to be opposed by many groups, including the Roman Catholic Church. Yet many individual Catholics, including many priests, hold liberal attitudes toward contraception.

Artificial contraception A method of contraception that applies a human-made device.

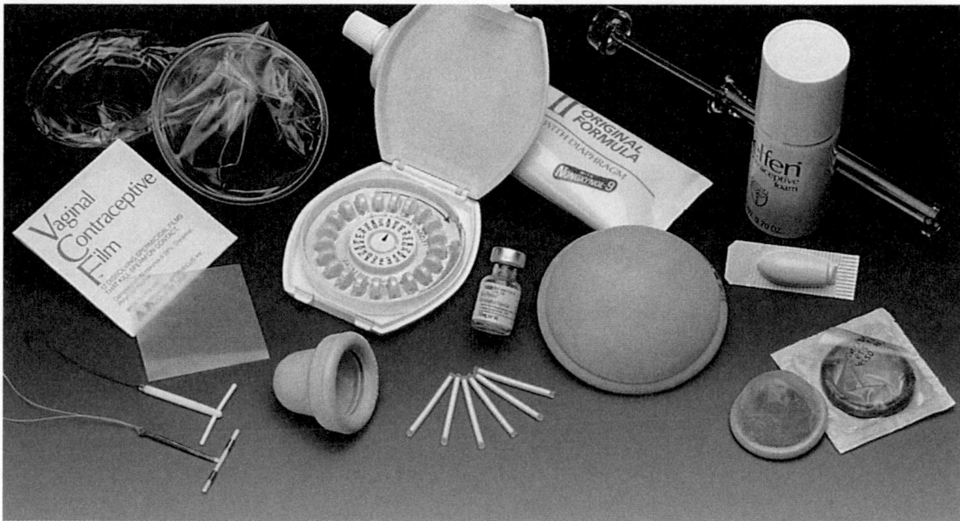

Selecting a Method of Contraception. Should you and your partner use contraception? If so, how can you determine which method is right for you? Issues you may want to consider include convenience, effectiveness, moral acceptability, safety, reversibility, and cost. Other issues include whether the method allows you and your partner to share the responsibility and whether it also affords protections from STIs.

Selecting a Method of Contraception

If you believe that you and your partner should use contraception, how will you determine which method is right for you? There is no simple answer. What is right for your friends may be wrong for you. You and your partner will make your own selections, but there are some issues you may want to consider:

1. *Convenience.* Is the method convenient? The convenience of a method depends on a number of factors. Does it require a device that must be purchased in advance? If so, can it be purchased over the counter as needed, or are a consultation with a doctor and a prescription required? Will the method work at a moment's notice, or, as with the birth-control pill, will it require time to reach maximum effectiveness? Some couples feel that few things dampen ardor and spontaneity more quickly than the need to pay attention to a contraceptive device in the heat of passion. Use of contraceptives like the condom and the diaphragm need not interrupt sexual activity, however. Both partners can share in applying the device. Some couples find that this becomes an erotic aspect of their lovemaking.
2. *Moral acceptability.* A method that is morally acceptable to one person may be objectionable to another. For example, some oral contraceptives prevent fertilization; others allow fertilization to occur but then prevent implantation of the fertilized ovum in the uterus. In the second case, the method of contraception may be considered to produce a form of early abortion, which is likely to concern people who object to abortion no matter how soon after conception it occurs. Yet the same people may have no moral objection to preventing fertilization.
3. *Cost.* Methods vary in cost. Some more costly methods involve devices (such as the diaphragm, the cervical cap, and the IUD) or hormones (pills or Norplant) that require medical visits in addition to the cost of the devices themselves. Other methods, such as rhythm methods, are essentially free.
4. *Sharing responsibility.* Most forms of birth control place the burden of responsibility largely, if not entirely, on the woman. The woman must consult with her doctor to obtain birth-control pills or other prescription devices such as diaphragms, cervical caps, Norplant, and IUDs. The woman must take birth-control pills reliably or check to see that her IUD remains in place.

TABLE 12.1

Approximate failure rates of various methods of birth control (in percentages of women using the method who become pregnant within the first year of use)

Method	% OF WOMEN EXPERIENCING AN ACCIDENTAL PREGNANCY WITHIN THE FIRST YEAR OF USE		% of Women Continuing Use at One Year[3]	Reversibility	Protection Against Sexually Transmitted Infections (STIs)
	Typical Use[1]	Perfect Use[2]			
Chance[4]	85	85		yes (unless fertility has been impaired by exposure to an STI)	no
Spermicides[5]	26	6	43	yes	no
Periodic abstinence	20		67	yes	no
Calendar		9			
Ovulation method		3			
Sympto-thermal[6]		2			
Post-ovulation		1			
Withdrawal	19	4		yes	no
Cervical cap[7]					
Parous women[9]	40	30	45	yes	some
Nulliparous[10] women	20	9	58	yes	some
Diaphragm[7]	20	6	58	yes	some

Sources: For failure rates and percentages of women discontinuing use, adapted from Hatcher et al. (1994, 1998). Information on reversibility and protection against STIs added.

[1]Among typical couples who initiate use of a method (not necessarily for the first time), the percentage who experience an accidental pregnancy during the first year if they do not stop use for any other reason.

[2]Among couples who initiate use of a method (not necessarily for the first time) and who use it perfectly (both consistently and correctly), the percentage who experience an accidental pregnancy during the first year if they do not stop use for any other reason.

[3]Among couples attempting to avoid pregnancy, the percentage who continue to use a method for one year.

[4]The percentages failing in columns (2) and (3) are based on data from populations where contraception is not used and from women who cease using contraception in order to become pregnant. Among such

Some couples prefer methods that allow for greater sharing of responsibility, such as alternating use of the condom and diaphragm. A man can also share the responsibility for the birth-control pill by accompanying his partner on her medical visits, sharing the expense, and helping her remember to take her pill.

5. *Safety.* How safe is the method? What are the side effects? What health risks are associated with its use? Can your partner's health or comfort be affected by its use?
6. *Reversibility.* Reversibility refers to the effects of a birth-control technique or device. In most cases the effects of birth-control methods can be fully reversed by discontinuing their use. In other cases reversibility may not occur immediately, as with oral contraceptives. One form of contraception, sterilization, should be considered irreversible, although many attempts at reversal have been successful.
7. *Protection against sexually transmitted infections (STIs).* Birth-control methods vary in the degree of protection they afford against STIs such as gonorrhea, chlamydia, and AIDS. This is especially important to people who are sexually active with one or more partners who are *not known* to be free of infectious diseases.

Method	% OF WOMEN EXPERIENCING AN ACCIDENTAL PREGNANCY WITHIN THE FIRST YEAR OF USE		% of Women Continuing Use at One Year[3]	Reversibility	Protection Against Sexually Transmitted Infections (STIs)
	Typical Use[1]	Perfect Use[2]			
Condom alone					
Female (Reality)	21	5	56	yes	yes
Male	14	3	63	yes	yes
Pill	3		72	yes	no, but may reduce the risk of PID[8]
Progestin only		0.5			
Combined		0.1			
IUD					
Progestasert	2.0	1.5	81	yes, except if fertility is impaired	no, and may increase the risk of PID
ParaGard Copper T 380A	0.8	0.6	78		
Depo-Provera	0.3	0.3	70	yes	no
Norplant (6 capsules)	0.05	0.05	85	yes	no
Female sterilization	0.5	0.5	100	questionable	no
Male sterilization	0.15	0.10	100	questionable	no

populations, about 89% become pregnant within one year. This estimate was lowered slightly (to 85%) to represent the percentage who would become pregnant within one year among women now relying on reversible methods of contraception if they abandoned contraception altogether.

[5]Foams, creams, gels, vaginal suppositories, and vaginal film.

[6]Cervical mucus (ovulation) method supplemented by calendar in the pre-ovulatory period and basal body temperature in the post-ovulatory period.

[7]With spermicidal cream or jelly.

[8]Pelvic inflammatory disease.

[9]Women who have borne children.

[10]Women who have not borne children.

Truth or Fiction? REVISITED

Some contraceptives prevent sexually transmitted infections as well as conception. Most methods, however, offer no protection against STIs. Therefore, this Truth or Fiction item is too broad to be true. ■

8. *Effectiveness.* Techniques and devices vary widely in their effectiveness in actual use. Despite the widespread availability of contraceptives, about two of three pregnancies in the United States are unplanned, and of these, about half result from contraceptive failures (Angier, 1993a). The failure rate for a particular method refers to the percentage of women who become pregnant when using the method for a given period of time, such as during the first year of use. Most contraceptive methods are not used correctly all or even much of the time. Thus it is instructive to compare the failure rate among people who use a particular method or device *perfectly* (consistently and correctly) with the failure rate among *typical* users. Failure rates among typical users are often considerably higher because of incorrect, unreliable, or inconsistent use. Table 12.1 shows the failure rates, continuation rates, reversibility, and degree of protection against STIs associated with various contraceptive methods.

METHODS OF CONTRACEPTION

There are many methods of contraception, including oral contraceptives (the pill), Norplant, intrauterine devices (IUDs), diaphragms, cervical caps, spermicides, condoms, douching, withdrawal (coitus interruptus), timing of ovulation (rhythm), and some devices under development.

Oral Contraceptives ("the Pill")

An **oral contraceptive** is commonly referred to as a birth-control pill or simply "the pill." However, there are many kinds of birth-control pills that vary in the type and dosages of hormones they contain. Birth-control pills fall into two major categories: combination pills and minipills.

Combination pills (such as Ortho-Novum, Ovcon, and Loestrin) contain a combination of synthetic forms of the hormones estrogen and progesterone (progestin). Most combination pills provide a steady dose of synthetic estrogen and progesterone. Other combination pills, called *multiphasic* pills, vary the dosage of these hormones across the menstrual cycle to reduce the overall dosages to which the woman is exposed and possible side effects. The **minipill** contains synthetic progesterone (progestin) only.

Available only by prescription, oral contraceptives are used by 28% of women in the United States who use reversible (nonsterilization) forms of contraception, or some 19 million women (Angier, 1993a). Birth-control pills are the most popular form of contraception among single women of reproductive age (Gilbert, 1996).

HOW THEY WORK Women cannot conceive when they are already pregnant because their bodies suppress maturation of egg follicles and ovulation. The combination pill fools the brain into acting as though the woman is already pregnant, so that no additional ova mature or are released. If ovulation does not take place, a woman cannot become pregnant.

In a normal menstrual cycle, low levels of estrogen during and just after the menstrual phase stimulate the pituitary gland to secrete FSH, which in turn stimulates the maturation of ovarian follicles. The estrogen in the combination pill inhibits FSH production, so follicles do not mature. The progesterone (progestin) inhibits the pituitary's secretion of LH, which would otherwise lead to ovulation. The woman continues to have menstrual periods, but there is no unfertilized ovum to be sloughed off in the menstrual flow.

The combination pill is taken for 21 days of the typical 28-day cycle. Then, for 7 days, the woman takes either no pill at all or an inert placebo pill to maintain the habit of taking a pill a day. The sudden drop in hormone levels causes the endometrium to disintegrate and menstruation to follow 3 or 4 days after the last pill has been taken. Then the cycle is repeated.

The progestin in the combination pill also increases the thickness and acidity of the cervical mucus. The mucus thus becomes a more resistant barrier to sperm and inhibits development of the endometrium. Therefore, even if an egg were somehow to mature and become fertilized in a fallopian tube, sperm would not be likely to survive the passage through the cervix. Even if sperm were somehow to succeed in fertilizing an egg, the failure of the endometrium to develop would mean that the fertilized ovum could not become implanted in the uterus. Progestin may also impede the progress of ova through the fallopian tubes and make it more difficult for sperm to penetrate ova.

The minipill contains progestin but no estrogen. Minipills are taken daily through the menstrual cycle, even during menstruation. They act in two ways. They thicken the cervical mucus to impede the passage of sperm through the cervix, and they render the inner lining of the uterus less receptive to a fertilized egg. Thus, even if the woman does conceive, the fertilized egg will pass from the body rather than becoming implanted in the uterine wall. It contains no estrogen, so the minipill does not usually prevent ovulation. The combination pill, by contrast, works directly to prevent ovulation. Since ovulation and fertilization may occur in women who use the minipill, some people see use of the minipill as an early abortion method. Others reserve the term *abortion* for methods of terminating pregnancy after successful implantation.

Oral contraceptive A contraceptive, consisting of sex hormones, which is taken by mouth.

Combination pill A birth-control pill that contains synthetic estrogen and progesterone.

Minipill A birth-control pill that contains synthetic progesterone but no estrogen.

EFFECTIVENESS OF BIRTH-CONTROL PILLS The failure rate of the birth-control pill associated with perfect use is very low: 0.5% or less, depending on the type of pill (see Table 12.1). The failure rate increases to 3% in typical use. Failures can occur when women forget to take the pill for 2 days or more, when they do not use backup methods when they first go on the pill, and when they switch from one brand to another. But forgetting to take the pill even for 1 day may alter the woman's hormonal balance, allowing ovulation—and fertilization.

REVERSIBILITY Use of oral contraceptives may temporarily reduce fertility after they are discontinued but is not associated with permanent infertility (Mishell, 1989). Nine of ten women begin ovulating regularly within 3 months of suspending use (Reinisch, 1990). Users of the pill who frequently start and stop usage may later incur fertility problems, however (Reinisch, 1990). When a woman appears not to be ovulating after going off the pill, a drug like clomiphene is often used to induce ovulation.

ADVANTAGES AND DISADVANTAGES The great advantage of oral contraception is that when used properly, it is nearly 100% effective. Unlike many other forms of contraception, such as the condom or diaphragm, its use does not interfere with sexual spontaneity or diminish sexual sensations. The sex act need not be interrupted, as it would be by use of a condom.

Birth-control pills may also have some *healthful* side effects. They appear to reduce the risk of pelvic inflammatory disease (PID), benign ovarian cysts, and fibrocystic (benign) breast growths (Gilbert, 1996). The pill regularizes menstrual cycles and reduces menstrual cramping and premenstrual discomfort. The pill may also be helpful in the treatment of iron-deficiency anemia and facial acne. The combination pill reduces the risks of ovarian and endometrial cancer, even for a number of years after the woman has stopped taking it (Gnagy et al., 2000; Hatcher & Guillebaud, 1998; Narod et al., 1998). The pill's protective effects against invasive ovarian cancer increase with the length of use (Gnagy et al., 2000).

The pill does have some disadvantages. It apparently heightens the risk of breast cancer somewhat among women who have a family history of the disease (Grabrick et al., 2000). It confers no protection against STIs. Moreover, it may reduce the effectiveness of antibiotics used to treat STIs. Going on the pill requires medical consultation, so a woman must plan to begin using the pill at least several weeks before becoming sexually active or before discontinuing the use of other contraceptives, and she must incur the expense of medical visits.

The main drawbacks of birth-control pills are potential side effects and possible health risks. Although a good deal of research suggests that the pill is safe for healthy women, in 2000 the American College of Obstetricians and Gynecologists released a bulletin suggesting caution in women with various preexisting medical conditions. These include hypertension, diabetes, migraine headaches, fibrocystic breast tissue, uterine fibroids, and elevated cholesterol level (Voelker, 2000).

The estrogen in combination pills may produce side effects such as nausea and vomiting, fluid retention (feeling bloated), weight gain, increased vaginal discharge, headaches, tenderness in the breasts, and dizziness. Many of these are temporary. When they persist, women may be switched from one pill to another, perhaps to one with lower doses of hormones. Pregnant women produce high estrogen levels in the corpus luteum and placenta. The combination pill artificially raises levels of estrogen, so it is not surprising that some women who use it have side effects that mimic the early signs of pregnancy, such as weight gain or nausea ("morning sickness"). Weight gain can result from estrogen (through fluid retention) or progestin (through increased appetite and development of muscle). Oral contraceptives may also increase blood pressure in some women, but clinically significant elevations are rare in women who use the low-dose pills available today (Hatcher et al., 1998). Still, it is wise for women who use the pill to have their blood pressure checked regularly. Women who encounter problems with high blood pressure from taking the pill are usually advised to switch to another form of contraception.

African Americans and Birth Control— The Intersection of Suspicion and Hope

Toni Cade wrote an essay, "The Pill: Genocide or Liberation?" about the gulf between men and women over the relationship of birth control to the African American liberation movement of the 1960s. She recalled a political meeting in which a tall African American man stood up and urged African American women to toss out the pill and hop onto the mattresses and breed revolutionaries as a way of combating European Americans' "genocidal program." Cade herself did not believe that producing more children would necessarily improve conditions for African Americans or, on the other hand, that the pill guaranteed women's liberation. She recognized that the pill gave women control over a key aspect of their lives—reproduction. But don't consider the African American man who viewed the pill as a genocidal tool to be a paranoid lunatic. According to law professor Dorothy Roberts (2000), his concerns about birth control as a kind of genocide arose from a history of efforts by European Americans to curtail reproduction among African Americans.

The meaning of birth control among African Americans reflects a history of racist denigration of African American childbearing, including specific campaigns to curtail African American fertility. African American women's attitudes toward the pill reflect racial injustice, gender inequality, and religious traditions. African American women recognize birth control as holding the potential for both liberation from unwanted pregnancy and denial of the opportunity to bear children.

The pill increased African American women's control over reproduction, but its use was controversial within the African American community. The pill was introduced during the 1960s, a time when scientists like Arthur Jensen and William Shockley were fostering genetic explanations for racial differences in IQ scores. During the 1960s and 1970s, many impoverished African American women were sterilized under federally funded programs. They were typically threatened with an end of welfare benefits or denial of medical care if they didn't agree to sterilization. Southern African American women were so commonly sterilized without their consent that the operation came to be called a "Mississippi appendectomy." Medical residents in the North performed medically unnecessary hysterectomies on African American women as practice. State legislators weighed sterilization bills aimed at the African Americans receiving welfare. The first author counseled an African American woman who had been sterilized following childbirth after being told, by the doctor, that the effect of the operation was temporary. She was shocked when she learned that she could not become pregnant again.

Thus it is understandable that many African Americans saw the pill as an-

Many women experience hormone withdrawal symptoms during periods when they do not take the pill (Sulak et al., 2000). These include headaches, pelvic pain, bloating, and breast tenderness.

Many women have avoided using the pill because of the risk of blood clots. The lower dosages of estrogen found in most types of birth-control pills today are associated with much lower risk of blood clots than was the case in the 1960s and 1970s, when higher dosages were used (Gilbert, 1996). Still, women who are at increased risk for blood clotting, such as women with a history of circulatory problems or stroke, are typically advised not to use the pill.

Women who are considering using the pill need to weigh the benefits and risks with their health care providers. For the great majority of young, healthy women in their 20s and early 30s, the pill is unlikely to cause blood clots or other cardiovascular problems (Hatcher & Guillebaud, 1998). Women who use the pill are no more likely than nonusers to develop cardiovascular problems later in life—even women who used the pill for more than 10 years (Stampfer et al., 1988).

Some women should not be on the pill at all (Calderone & Johnson, 1989; Hatcher et al., 1998; Reinisch, 1990). These include women who have had circulatory problems or blood clots and those who have suffered a heart attack or stroke or have a history of coronary disease, breast or uterine cancer, undiagnosed genital bleeding, liver tumors, or sickle-cell anemia (because of associated blood-clotting problems). Because of their increased risk of cardiovascular problems, caution should be exercised when the combination pill is used with women over 35 years of age who smoke (Hatcher et al., 1998). Nursing mothers should also avoid using the pill; the hormones may be passed to the baby in the mother's milk.

What Does Birth Control Mean for African Americans? We cannot really generalize about this issue, but many African Americans take into account a history of racial injustice, gender inequality, and religious traditions when they contemplate using birth control. Although birth control has liberated women of all races to pursue education, careers, and the lifestyle they choose, history reveals that African Americans have been subjected to multiple campaigns to control their numbers.

other tool in the effort to limit their population. Many African Americans viewed family planning programs as a means of racial genocide, especially when they involved sterilization. Many such programs were based on theories of the genetic inferiority and social degeneracy of African Americans. But some African American women did use birth control clinics to control the size of their families. The civil rights leader W. E. B. DuBois favored birth control as a way of improving African Americans' health and social status and argued that a high birthrate was not an efficient way of fighting discrimination. Birth control meant fewer mouths to feed, and it liberated women to become educated and join the work force. It is ironic that at the same time some African Americans advocated birth control as a way of bettering their lot in life, European American racists were promoting birth control as a means of preserving the oppressive social structure.

Roberts (2000) argues that we need to be aware of the history of the ways in which birth control has affected the lives of African Americans in order to understand why they approach it with both suspicion and hope. She notes that "Social justice requires both equal access to safe, user-controlled contraceptives and an end to the use of birth control as a means of population control."

Because the risks of cardiovascular complications generally increase with age, many women over the age of 35 have been encouraged by their gynecologists to use other forms of birth control. However, the American College of Obstetricians and Gynecologists believes that healthy nonsmokers can use oral contraceptives safely at least until the age of 45.

The pill may also have psychological effects. Some users report depression or irritability. Switching brands or altering doses may help. Evidence is lacking concerning the effects of lower-estrogen pills on sexual desire.

Progestin fosters male secondary sex characteristics, so women who take the minipill may develop acne, facial hair, thinning of scalp hair, reduction in breast size, vaginal dryness, and missed or shorter periods. Irregular bleeding, or so-called breakthrough bleeding, between menstrual periods is a common side effect of the minipill. Irregular bleeding should be brought to the attention of a health professional. Because they can produce vaginal dryness, minipills can hinder vaginal lubrication during intercourse, decreasing sexual sensations and rendering sex painful.

Researchers have also examined suspected links between the use of the pill and certain forms of cancer, especially breast cancer, which is sensitive to hormonal changes. Results from several large-scale studies show no overall increase in the rates of breast cancer among pill users, but it remains possible that some subgroups of women who use the pill are at increased risk (Hatcher et al., 1998). The evidence linking use of the pill to increased risk of cervical cancer is mixed, with some studies showing such a link and others showing none (Hatcher et al., 1998).

Women considering the pill are advised to have a thorough medical evaluation to rule out preexisting conditions that might make its use unsafe. The evaluation should include a detailed medical and family history and a physical exam, including a Pap smear, assessment of blood pressure, screening for STIs, urinalysis, breast and pelvic exams, and possibly an EKG (electrocardiogram). Women who begin to use the pill, regardless of their age or risk status, should pay attention to changes in their physical condition, have regular checkups, and promptly report any physical complaints or unusual symptoms to their physicians.

"MORNING-AFTER" PILLS The so-called morning-after pill, or postcoital contraceptive, actually comprises several types of pills that have high doses of estrogen and progestin. Because they are not taken regularly, they do not prevent ovulation from occurring. Instead, they stop fertilization from taking place or prevent the fertilized egg from implanting itself in the uterus. In this respect, they represent an early abortion technique. However, 65% of the respondents in a national poll taken by *The New York Times* said they would consider morning-after pills a form of birth control, and less than 20% considered them an abortion method (Goldberg & Elder, 1998).

Morning-after pills are most effective when taken within 72 hours after ovulation. Women who wait to see whether they have missed a period are no longer candidates for the morning-after pill. Depending on the brand, four, six, or eight pills are prescribed.

It is true that there are effective oral contraceptives that can be taken the morning after unprotected intercourse. They are termed "morning-after" pills. ■

Morning-after pills have a higher hormone content than most birth-control pills. For this reason, nausea is a common side effect, occurring in perhaps 70% of users. Nausea is usually mild and passes within a day or two after treatment, but it can be treated with antinausea medication (Hatcher et al., 1998). Vomiting should be brought to the attention of a physician, because the woman may need to take additional pills to make up for the ones possibly she may have lost in vomiting (Hatcher et al., 1998).

Because of the strength of the dosage, the morning-after pill is not recommended as a regular form of birth control. We also know little about possible long-term health complications. Morning-after pills are *one-time* forms of emergency protection (Hatcher et al., 1998), which may be most appropriate to use following rape or when regular contraceptive devices fail (for example, when a condom breaks or a diaphragm becomes dislodged). The morning-after pill is generally effective in preventing implantation of a fertilized ovum, but health professionals caution that when it fails, the fetus may be damaged by exposure to the hormones that the pill contains.

Norplant

The contraceptive implant *Norplant* consists of six matchstick-sized silicone tubes that contain progestin and are surgically embedded in a woman's upper arm. More than 1 million women in the United States have received Norplant since FDA approval in 1990 (Kolata, 1995).

HOW IT WORKS Norplant, like the pill, relies on female sex hormones to suppress fertility. But rather than the woman taking a pill once a day, tubes implanted in her body release a small, steady dose of progestin into her bloodstream, providing continuous contraceptive protection for as long as 5 years (Hatcher, 1998). The progestin in the Norplant system suppresses ovulation and thickens the cervical mucus so that sperm cannot pass. The contraceptive effect occurs within 24 hours of insertion. After 5 years the spent tubes are replaced. An alternative implant, Norplant-2, consists of two hormone-releasing tubes that provide at least 3 years of protection.

HOW IT IS USED Implantation takes a few minutes in a doctor's office and is carried out under local anesthesia.

EFFECTIVENESS Norplant is reported to have an extremely low failure rate of less than 1% per year across 5 years (see Table 12.1). The failure rate approximates that of surgical sterilization.

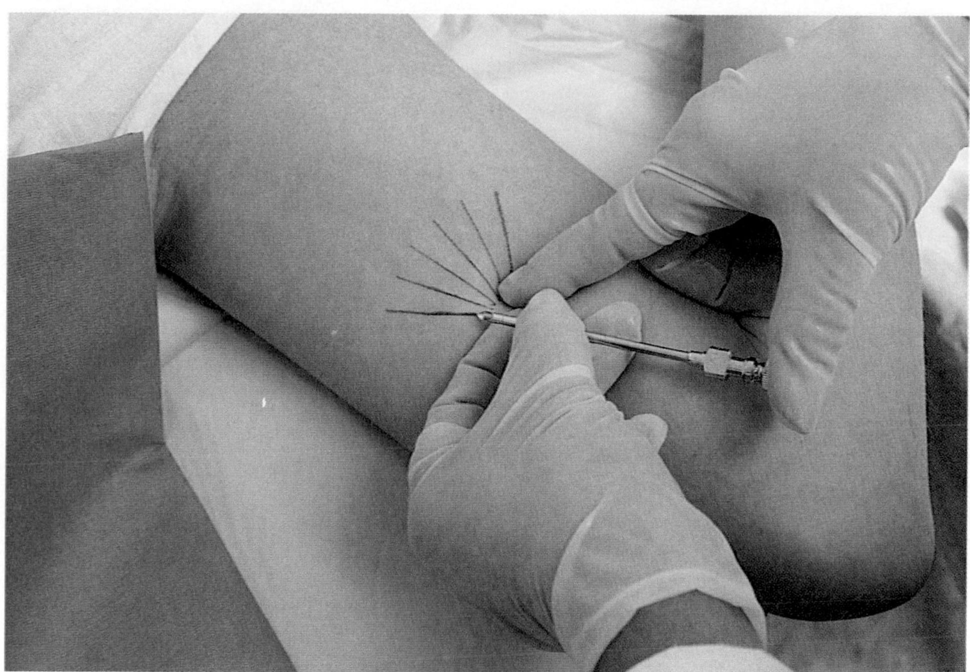

Norplant. This contraceptive implant consists of small silicone tubes that contain progestin and are surgically embedded in the woman's arm. Norplant uses female sex hormones to suppress fertility. The tubes release a small, steady dose of progestin into the bloodstream. Progestin suppresses ovulation and thickens the cervical mucus so that sperm cannot pass through it

REVERSIBILITY Although the failure rate of Norplant approximates that of sterilization, Norplant is fully reversible. Removal restores a normal likelihood of pregnancy.

ADVANTAGES AND DISADVANTAGES The key advantage of Norplant is its convenience. The hormone is dispensed automatically. The woman need not remember to take a pill each day, insert a device before coitus, or check that an IUD is in place. The most commonly reported side effect is abnormal menstrual bleeding (Hatcher, 1998; Mishell, 1989).

Many health professionals concerned about finding ways to reduce teenage pregnancy rates are enthusiastic about Norplant. School-based programs have been initiated to make Norplant available to young women, although this has fueled the debate over whether schools should be involved in distributing contraceptives to minors (see Chapter 13). The nearby Closer Look feature recounts another use of Norplant.

a CLOSER look

Congratulations, Wheat Ridge, It's a . . . !

The quality of mercy was seriously strained in a Denver suburb recently, when the town of Wheat Ridge had to bite the bullet and press the pause button on its nonviolent plan for curbing beaver proliferation. (Beavers, you see, have the horrendous habits of burrowing beneath walking paths [*flop!*] and turning trees into taste treats.)

According to Nick Fisher (1992) of the town's Animal and Park Enforcement Department, Wheat Ridge chewed over the situation and decided to defang its gnawing wildlife problem by biting into the mammals' reproductive patterns with Norplant. TV cameras were about to record the first implanting of the birth-control device on an anesthetized beaver when a veterinarian declared, at the last moment, "It's a male." Beavers, it happens, are "darn hard to sex," explained Dr. Robinson, because their sex organs are deep inside their bodies. (Is it possible that the good doctor meant to say "*dam* hard to sex" but was too genteel?) The program will have to await the identification of a toothsome tootsie.

Apparently Wheat Ridge had bitten off more than it could chew. By the time the town finds and identifies a female, we may all be long in the tooth. Frankly, dear readers, we wonder whether the beavers give a dam about all this.

FIGURE 12.1

Two IUDs: The Progestasert T and the Copper T 380A (ParaGard). The Progestasert T releases small quantities of progesterone (progestin) daily. The Copper T 380A is a T-shaped, copper-based device.

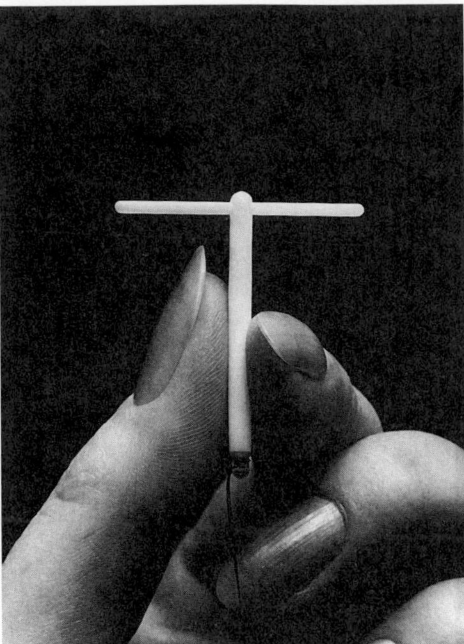

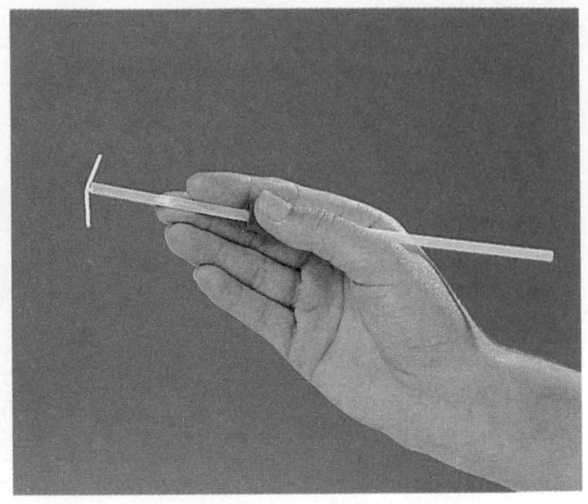

Intrauterine Devices (IUDs)

Camel drivers setting out on long desert journeys once placed round stones in the uteruses of female camels to prevent them from becoming pregnant and lost to service. The stones may have acted as primitive **intrauterine devices** (IUDs). IUDs are small objects of various shapes that are inserted into the uterus. IUDs have been used by humans since Greek times. Today, they are inserted into the uterus by a physician or nurse practitioner and are usually left in place for a year or more. Fine plastic threads or strings hang down from the IUD into the vagina, so that the woman can check to be sure it is still in place.

IUDs are used by about 1.5 million women in the United States and by more than 80 million women around the world (Hatcher et al., 1998). Nearly 60 million IUD users are in China, where nearly 1 in 3 married women uses an IUD during her childbearing years. By contrast, IUDs are used by only about 3% of married women of childbearing age in the United States. Married women in the United States are more than twice as likely as single women to use an IUD (Hatcher et al., 1998).

IUDs achieved their greatest popularity in the United States in the 1960s and 1970s. Then there was a sharp drop off in their use during the 1980s, after the use of a popular model, the Dalkon Shield, was linked to a high incidence of pelvic infections and tubal infertility (Hatcher et al., 1998).

Figure 12.1 shows two IUDs: the Progestasert T, which releases small quantities of progesterone (progestin) daily, and the Copper T 380A (ParaGard), a T-shaped, copper-based device. Because the Progestasert T must be replaced annually, and any insertion carries some risk of infection, health authorities recommend the use of the ParaGard device, which can be used for more than 8 years, unless the woman is allergic to copper (Stewart, 1998).

HOW THEY WORK We do not know exactly how IUDs work. A foreign body, such as the IUD, apparently irritates the uterine lining. This irritation gives rise to mild inflammation and to the production of antibodies that may be toxic to sperm or to fertilized ova and/or may prevent fertilized eggs from becoming implanted. Inflammation may also impair proliferation of the endometrium—another impediment to implantation. Progestin released by the Progestasert T also has effects like the progestin-only

Intrauterine device A small object that is inserted into the uterus and left in place to prevent conception. Abbreviated IUD.

minipill: It reduces the likelihood of fertilization and implantation. Action on fertilized ova may be considered to constitute an early abortion. Because IUDs may not prevent fertilization, people who oppose abortion, regardless of how soon it occurs after conception, also oppose the IUD.

EFFECTIVENESS The failure rate associated with typical use of the Progestasert T is about 2% (see Table 12.1). Most failures occur within 3 months of insertion, often because the device shifts position or is expelled. ParaGard is the most effective IUD. The first-year failure rate in typical use is 0.8%.

The IUD may irritate the muscular layer of the uterine wall, causing contractions that expel it through the vagina. The device is most likely to be expelled during menstruation, so users are advised to check their sanitary napkins or tampons before discarding them. Women who use IUDs are advised to check the string several times a month to ensure that the IUD is in place. Spontaneous expulsions occur in 2% to 10% of users within the first year of use (Stewart, 1998). Women who have not borne children have a higher expulsion rate (about 7%) than women who have (3%). Some family-planning clinics advise women to supplement their use of IUDs with other devices for the first 3 months, when the risks of a shift in position or expulsion are greatest.

Anti-inflammatory drugs, such as aspirin, and antibiotics may also decrease IUD effectiveness. Because many physicians recommend taking an aspirin every other day as a way of helping prevent certain kinds of cancer (Marcus, 1995), women are advised to discuss these issues with their gynecologists.

REVERSIBILITY IUDs may be removed readily by professionals. About 9 out of 10 former IUD users who wish to do so become pregnant within a year (Hatcher et al., 1998).

ADVANTAGES AND DISADVANTAGES The IUD has three major advantages: (1) It is highly effective; (2) it does not diminish sexual spontaneity or sexual sensations; and (3) once it is in place, the woman need not do anything more to prevent pregnancy (other than check that it remains in place). The small risk of failure is reduced in effect to zero if the couple also use an additional form of birth control, such as the diaphragm or condom.

The IUD also does not interfere with the woman's normal hormone production. Users continue to produce pituitary hormones that stimulate ovarian follicles to mature and rupture, thereby releasing mature ova and producing female sex hormones.

If IUDs are so effective and relatively "maintenance free," why are they not more popular? One reason is that insertion can be painful. Another reason is side effects. The most common side effects are excessive menstrual cramping, irregular bleeding (spotting) between periods, and heavier-than-usual menstrual bleeding (Hatcher et al., 1998). These generally occur shortly following insertion and are among the primary reasons why women ask to have the device removed. A more serious concern is the possible risk of pelvic inflammatory disease (PID), a serious disease that can become life-threatening if left untreated (Stewart, 1998). Women who use the IUD may have an increased risk of PID (Cates, 1998). The risk of infection is associated more with the insertion of the device (bacteria may enter the woman's reproductive tract during insertion) than with use of the device itself (Cates, 1998).

PID can produce scar tissue that blocks the fallopian tubes, causing infertility. Women with pelvic infections should not use an IUD (Stewart, 1998). Women who have risk factors for PID may also wish to consider the advisability of an IUD. Risk factors include a recent episode of gonorrhea or chlamydia, recurrent episodes of these STIs, sexual contact with multiple partners, and sexual contact with a partner who has had multiple sexual partners. All in all, the IUD may be best suited to women who have completed their families and are advised not to use oral contraceptives.

Another risk in using an IUD is that the device may perforate (tear) the uterine or cervical walls, which can cause bleeding, pain, and adhesions and become life threatening. Perforations are usually caused by improper insertion and occur in perhaps 1 case in

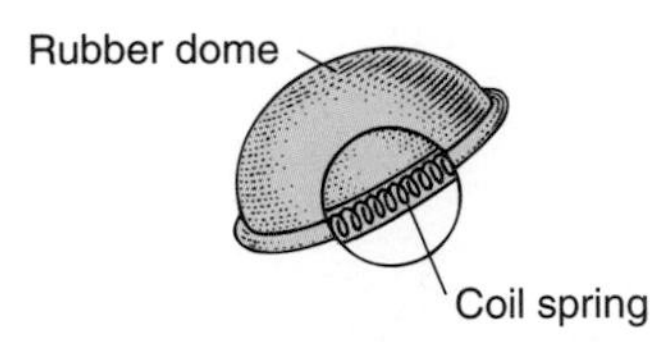

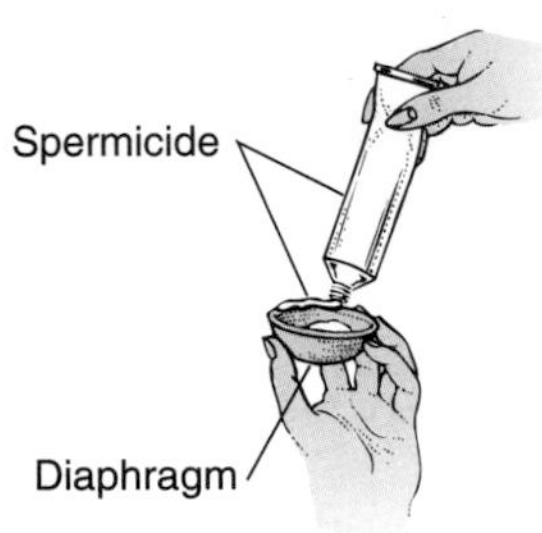

FIGURE 12.2

A Diaphragm. The diaphragm is a shallow cup or dome made of latex. Diaphragms must be fitted to the contour of the vagina by a health professional. The diaphragm forms a barrier to sperm but should be used in conjunction with a spermicidal cream or jelly.

1,000 (Reinisch, 1990). IUD users are also at greater risk for ectopic pregnancies, both during and after usage, and for miscarriage. Ectopic pregnancies occur in about 5% of women who become pregnant while using an IUD (Hatcher et al., 1998). IUD use is not recommended for women with a history of ectopic pregnancy. Women who become pregnant while using the IUD stand about a 50–50 chance of miscarriage (Stewart, 1998).

Despite the fact that the IUD irritates uterine tissues, there is no evidence that IUD users run a greater risk of cancer. Long-term data on the health effects of IUD use are limited, however.

Another drawback to the IUD is its cost. The typical cost of an IUD insertion in a family-planning clinic is a few hundred dollars. Potential expulsion of the IUD presents yet another disadvantage. Moreover, the IUD, like the pill, offers no protection against STIs. Finally, again like the pill, IUDs place the burden of contraception entirely on the woman.

The Diaphragm

Diaphragms were once used by about one third of U.S. couples who practiced birth control. When invented in 1882, they were a breakthrough. Their popularity declined only in the 1960s with the advent of the pill and the IUD. Today, fewer than 5% of married women use the diaphragm (Hatcher et al., 1998).

The diaphragm is a shallow cup or dome made of thin latex rubber (see Figure 12.2). The rim is a flexible metal ring covered with rubber. Diaphragms come in different sizes to allow a precise fit.

Diaphragms are available by prescription and must be fitted to the contour of the vagina by a health professional. Several sizes and types of diaphragms may be tried during a fitting. Women practice insertion in a health professional's office so that they can be guided as needed.

HOW IT WORKS The diaphragm is inserted and removed by the woman, much like a tampon. It is akin to a condom in that it forms a barrier against sperm when placed snugly over the cervical opening. Yet it is unreliable when use alone. Thus, the diaphragm should be used in conjunction with a spermicidal cream or jelly. The diaphragm's main function is to keep the spermicide in place.

HOW IT IS USED The diaphragm should be inserted no more than 2 hours before coitus, because the spermicides that are used may begin to lose effectiveness beyond this time. Some health professionals, however, suggest that the diaphragm may be inserted up to 6 hours preceding intercourse. (It seems reasonable to err on the side of caution and assume that there is a 2-hour time limit.) The woman or her partner places a tablespoonful of spermicidal cream or jelly on the inside of the cup and spreads it inside the rim. (Cream spread outside the rim might cause the diaphragm to slip.) The woman opens the inner lips of the vagina with one hand and folds the diaphragm with the other by squeezing the ring. She inserts the diaphragm against the cervix, with the inner side facing upward (see Figure 12.3). Her partner can help insert the diaphragm, but the woman is advised to check its placement. Some women prefer a plastic insertion device, but most find it easier to insert the diaphragm without it. The diaphragm should be left in place *at least 6 hours* after intercourse to allow the spermicide to kill any remaining sperm in the vagina (Hatcher et al., 1998). To guard against toxic shock syndrome (TSS), it should not be left in place longer than 24 hours.

After use, the diaphragm should be washed with mild soap and warm water and stored in a dry, cool place. When cared for properly, a diaphragm can last about 2 years. Women may need to be refitted after pregnancy or a change in weight of 10 pounds or more.

Diaphragm A shallow rubber cup or dome, fitted to the contour of a woman's vagina, that is coated with a spermicide and inserted prior to coitus to prevent conception.

EFFECTIVENESS If it is used consistently and correctly, the failure rate of the diaphragm is estimated to be 6% during the first year of use (see Table 12.1). In typical use, however, the failure rate is believed to be three times as high—18%. Some women become pregnant because they do not use the diaphragm during every coital experience. Others may

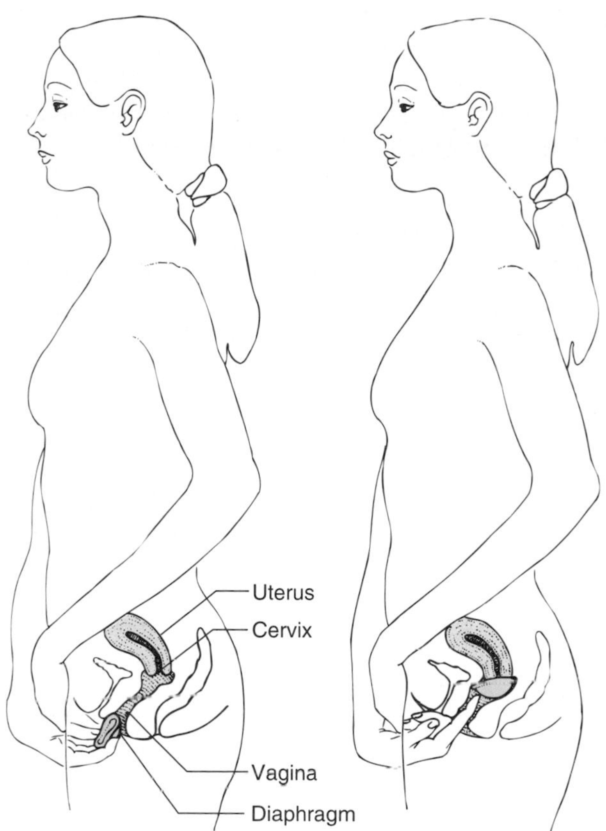

FIGURE 12.3

Insertion and Checking of the Diaphragm. Women are instructed in the insertion of the diaphragm by a health professional. In practice, a woman and her partner may find joint insertion an erotic experience.

insert it too early or not leave it in long enough. The diaphragm may not fit well, or it may slip—especially if the couple is acrobatic. A diaphragm may develop tiny holes or cracks. Women are advised to inspect the diaphragm for signs of wear and to consult their health professionals when in doubt. Effectiveness also is seriously compromised when the diaphragm is not used along with a correctly applied spermicide.

REVERSIBILITY The effects of the diaphragm are fully reversible. In order to become pregnant, the woman simply stops using it. The diaphragm has not been shown to influence subsequent fertility.

ADVANTAGES AND DISADVANTAGES The major advantage of the diaphragm is that when used correctly, it is a safe and effective means of birth control and does not alter the woman's hormone production or reproductive cycle. The diaphragm can be used as needed, whereas the pill must be used daily and the IUD remains in place whether or not the woman engages in coitus. Another advantage is the virtual absence of side effects. The few women who are allergic to the rubber in the diaphragm can switch to a plastic model.

The major disadvantage is the high pregnancy rate associated with typical use. Nearly 1 in 5 typical users (18%) of the diaphragm combined with spermicidal cream or jelly becomes pregnant during the first year of use (Hatcher et al., 1998). Another disadvantage is the need to insert the diaphragm prior to intercourse, which the couple may find disruptive. The woman's partner may find the taste of the spermicides used in conjunction with the diaphragm to be unpleasant during oral sex. The pressure exerted by the diaphragm against the vaginal and cervical walls may also irritate the urinary tract and cause urinary or even vaginal infections. Switching to a different-size diaphragm or one with a different type of rim may help alleviate this problem. About 1 woman or man in 20 may develop allergies to the particular spermicide that is used, which can lead to irritation of the genitals. This problem may also be alleviated by switching to another brand.

Spermicides

Spermicides are agents that kill sperm. They come in different forms, including jellies and creams, suppositories, aerosol foam, and a contraceptive film. Spermicides should be left in place in the vagina (no douching) for *at least 6 to 8 hours* after coitus (Cates & Raymond, 1998).

HOW THEY ARE USED Spermicidal jellies, creams, foam, and suppositories should be used no more than 60 minutes preceding coitus to provide maximum effectiveness (Cates & Raymond, 1998). Spermicidal jellies and creams come in tubes with plastic applicators that introduce the spermicide into the vagina (see Figure 12.4). Spermicidal foam is a fluffy white cream with the consistency of shaving cream. It is contained in a pressurized can and is introduced with a plastic applicator in much the same way as spermicidal jellies and creams.

Vaginal suppositories are inserted into the upper vagina, near the cervix, where they release spermicide as they dissolve. Unlike spermicidal jellies, creams, and foam, which become effective immediately when applied, suppositories must be inserted no less than 10 to 15 minutes before coitus so that they have sufficient time to dissolve (Cates & Raymond, 1998).

Spermicidal film consists of thin, 2-inch-square sheets saturated with spermicide. When placed in the vagina, they dissolve into a gel and release the spermicide. The spermicidal film should be inserted at least 5 minutes before intercourse to allow time for it to melt and for the spermicide to be dispersed. It remains effective for upwards of 1 hour. One disadvantage of the film that some users have noted is a tendency for it to adhere to the fingertips, which makes it difficult to insert correctly.

HOW THEY WORK Spermicides coat the cervical opening, blocking the passage of sperm and killing sperm by chemical action.

EFFECTIVENESS In typical use, the first-year failure rate of spermicides used alone is 21 pregnancies per year per 100 users (Cates & Raymond, 1998). When used correctly and consistently, the failure rate is estimated to drop to about 6 pregnancies per 100 users in the first year. All forms of spermicide are more effective when they are combined with other forms of contraception, such as the condom.

REVERSIBILITY Spermicides have not been linked with any changes in reproductive potential, so couples who wish to become pregnant simply stop using them.

ADVANTAGES AND DISADVANTAGES The major advantages of spermicides are that they do not alter the woman's natural biological processes and are applied only as needed. Unlike a diaphragm, they do not require a doctor's prescription or a fitting. They can be

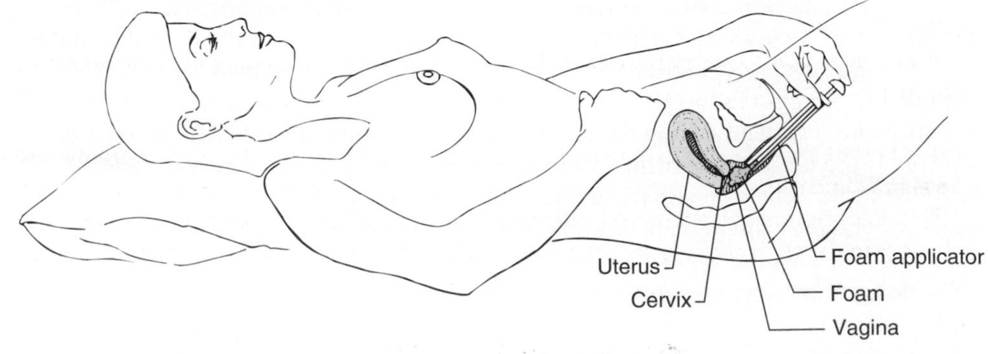

FIGURE 12.4

The Application of Spermicidal Foam. Spermicidal jellies and creams come in tubes with plastic applicators. Spermicidal foam comes in a pressurized can and is applied with a plastic applicator in much the same way as spermicidal jellies and creams.

bought in virtually any drugstore, and the average cost per use of the foam variety is modest—about 50 cents.

The major disadvantage is the high failure rate among typical users. Foam often fails when the can is not shaken enough, when too little is used, when it is not applied deeply enough within the vagina near the cervix, or when it is used after coitus has begun.

Spermicides are generally free of side effects but occasionally cause vaginal or penile irritation. Irritation is sometimes alleviated by changing brands. Some partners find the taste of spermicides unpleasant. (Couples can engage in oral sex before applying spermicides.) Spermicides may pose a danger to an embryo, so women who suspect they are pregnant are advised to suspend use until they find out for certain.

It was once thought that spermicides that contain nonoxynol-9 might provide some protection against STIs such as HIV/AIDS, genital herpes, trichomoniasis ("trich"), syphilis, and chlamydia (Reinisch, 1990). Yet a controlled experiment in Africa did not find that nonoxynol-9 afforded any protection against disease-causing agents (Roddy et al., 1998). Further research suggests that nonoxynol-9 is actually harmful when used as a **microbicide** (Perriëns, 2000). (Microbicides are chemical substances that kill viruses and bacteria. They are applied vaginally or rectally prior to sexual intercourse in the form of gels, creams, suppositories, or films. Any number of chemicals will kill viruses and bacteria; the challenge is to develop agents that will kill the microbes without harming the person using them.) UNAIDS oversaw a trial of the effectiveness of the spermicide as a means of preventing HIV infection among African and Thai prostitutes. It turned out that the group using nonoxynol-9 actually had a significantly higher rate of HIV infection (15%) than the group using the placebo (10%) (Stephenson, 2000b). Joseph Perriëns (2000), head of the UNAIDS microbicide effort, suggested that the local ulceration (irritation) caused by nonoxynol-9 might have made the vaginal tract an easier port of entry for HIV.

The Cervical Cap

The cervical cap, like the diaphragm, is a dome-shaped rubber cup. It comes in different sizes and must be fitted by a health professional. It is smaller than the diaphragm, however—about the size of a thimble—and is meant to fit snugly over the cervical opening.

HOW IT IS USED Like the diaphragm, the cap is intended to be used with a spermicide applied inside it (Hatcher et al., 1998). When inserting it, the woman (or her partner) fills the cap about a third full of spermicide. Then, squeezing the edges together, the woman inserts the cap high in the vagina, so that it presses firmly against the cervix. The woman can test the fit by running a finger around the cap to ensure that the cervical opening is covered. It should be left in place for at least 8 hours after intercourse. The cap provides continuous protection for upwards of 48 hours without the need for additional spermicide. To reduce the risk of toxic shock syndrome, the cap should not be left in place longer than 48 hours. Like the diaphragm, the cap should be cleaned after use and checked for wear and tear. When cared for properly, the cap should last for more than 3 years.

HOW IT WORKS Like the diaphragm, the cervical cap forms a barrier and also holds spermicide in place against the cervix. It prevents sperm from passing into the uterus and fallopian tubes and kills sperm by chemical action.

EFFECTIVENESS The failure rate in typical use is estimated to be high, ranging from 18% in women who have not borne children to 36% in women who have (Hatcher et al., 1998). Failures may be attributed, at least in part, to the cap's becoming dislodged and to changes in the cervix during the menstrual cycle, which can cause the cap to fit less snugly over the cervix.

REVERSIBILITY There is no evidence that the cervical cap affects fertility.

Microbicide A chemical substance that kills viruses and bacteria.

ADVANTAGES AND DISADVANTAGES Like the diaphragm, the cap is a mechanical device that does not affect the woman's hormonal production or reproductive cycle. The cap may be especially suited to women who cannot support a diaphragm because of lack of vaginal muscle tone. Because of concern that the cap may irritate cervical tissue, however, users are advised to have regular Pap tests.

Some women find the cap uncomfortable. The cap can also become dislodged during sexual activity or lose its fit as the cervix changes over the menstrual cycle. Side effects include urinary tract infections and allergic reactions or sensitivities to the rubber or spermicide. Other disadvantages include the expense and inconvenience of being fitted by a health professional. Moreover, some women are shaped so that the cap does not remain in place. For these reasons, and because they may be hard to obtain, cervical caps are not very popular in the United States.

Condoms

Condoms are also called "rubbers," "safes," **prophylactics** (because latex condoms protect against STIs), and "skins" (referring to those that are made from lamb intestines). Condoms lost popularity with the advent of the pill and the IUD. They are less effective than the pill or IUD, may disrupt sexual spontaneity, and can decrease coital sensations because they prevent the penis from actually touching the vaginal wall.

Condoms have been making a comeback, however, because those made of latex rubber can help prevent the spread of the AIDS virus (the *human immunodeficiency virus,* HIV) and other STIs and, to a lesser extent, because of concerns about side effects of the pill and the IUD. Largely because of concerns about AIDS and other STIs, the use of condoms among unmarried women has mushroomed.

Condom advertisements have appeared in mainstream media, including magazines such as *People* and *Cosmopolitan*, and on major TV networks. In a TV commercial, a man and a woman are shown hurriedly undressing. The man tells the woman that he forgot to bring a condom with him. She then tells him that he'd best forget it (making love, that is). Whether the media campaign will increase condom usage remains to be seen.

The renewed popularity of condoms has also been spurred by the increased assertiveness of contemporary women. They make the point (which should be obvious but all too often is not) that contraception is as much the man's responsibility as the woman's. Condoms alter the psychology of sexual relations. By using a condom, the man assumes much of the responsibility for contraception. Condoms are the only contraceptive device worn by men and the only readily reversible method of contraception that is available to men. Condoms are inexpensive and can be obtained without prescription from pharmacies, family-planning clinics, restrooms, and even vending machines in some college dormitories.

Some condoms are made of latex rubber. Thinner, more expensive condoms ("skins") are made from the intestinal membranes of lambs. The latter allow greater sexual sensation but do not protect so well against STIs. Only latex condoms are effective against the tiny AIDS virus. Condoms made of animal intestines have pores large enough to permit the AIDS virus and other viruses, such as the one that causes hepatitis B, to slip through (Hatcher et al., 1998). A few condoms are made from other materials, such as plastic (polyurethane). Questions remain about the effectiveness of polyurethane condoms. Some condoms have plain ends. Others have nipples or reservoirs (see Figure 12.5) that catch semen and may help prevent the condom from bursting during ejaculation.

HOW THEY WORK A condom is a cylindrical sheath that serves as a barrier, preventing the passage of sperm and disease-carrying microorganisms from the man to his partner. It also helps prevent infected vaginal fluids (and microorganisms) from entering the man's urethral opening or penetrating through small cracks in the skin of the penis.

Prophylactic An agent that protects against disease.

HOW THEY ARE USED The condom is rolled onto the penis once erection is achieved and before contact between the penis and the vagina (see Figure 12.6). If the condom is

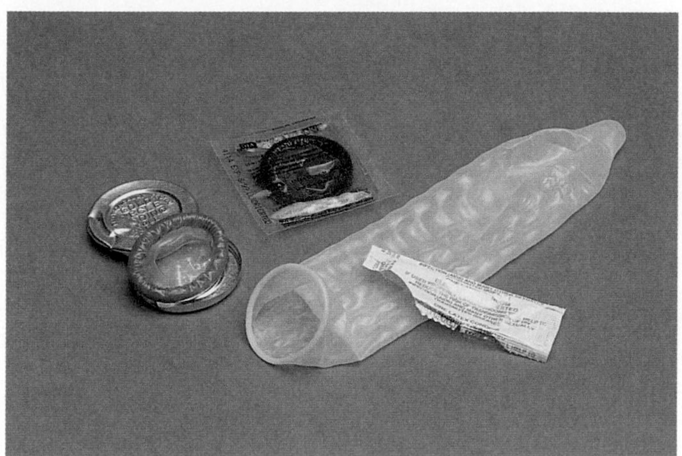

FIGURE 12.5

Condoms. Some condoms are plain-tipped, whereas others have nipples or reservoirs that catch semen and may help prevent the condom from bursting during ejaculation. Latex condoms form effective barriers to the tiny AIDS virus.

not used until moments before the point of ejaculation, sperm-carrying fluid from the Cowper's glands or from preorgasmic spasms may already have passed into the vagina. Nor does the condom afford protection against STIs if it is fitted after penetration.

Between 1% and 2% of condoms break or fall off during intercourse or when the penis is being withdrawn afterward (Warner & Hatcher, 1998). Condoms also sometimes slip down the shaft of the penis without falling off. To use a condom most effectively and to help prevent it from either breaking or falling off, a couple should observe the following guidelines:[1]

- Use a condom each and every time you have intercourse. Inexperienced users should also practice putting on a condom before they have occasion to use one with a partner.
- Handle the condom carefully, making sure not to damage it with your fingernails, teeth, or sharp objects.
- Place the condom on the erect penis before it touches the vulva.
- Uncircumcised men should pull back the foreskin before putting on the condom.
- If you use a spermicide, put some inside the tip of the condom before placing the condom on the penis. You may also wish to use additional spermicide applied by an applicator inside the vagina to provide extra protection, especially in the event that the condom breaks.
- Do not pull the condom tightly against the tip of the penis.
- For a condom without a reservoir tip, leave a small empty space—about half an inch—at the end of the condom to hold semen, yet do not allow any air to be trapped at the tip. Some condoms come equipped with a reservoir (nipple) tip that will hold semen.
- Unroll the condom all the way to the bottom of the penis.
- Ensure that adequate vaginal lubrication during intercourse is present, using lubricants if necessary. But use only water-based lubricants such as contraceptive jelly or K-Y jelly. Never use an oil-based lubricant that can weaken the latex material, such as petroleum jelly (Vaseline), cold cream, baby oil or lotion, mineral oil, massage oil, vegetable oil, Crisco, hand or body lotions, and most skin creams. Do not use saliva as a lubricant because it may contain infectious organisms, such as viruses.
- If the condom breaks during intercourse, withdraw the penis immediately, put on a new condom, and use more spermicide.
- After ejaculation, carefully withdraw the penis while it is still erect.
- Hold the rim of the condom firmly against the base of the penis as the penis is withdrawn, to prevent the condom from slipping off.

[1]Adapted from the Centers for Disease Control pamphlet *Condoms and Sexually Transmitted Diseases . . . Especially AIDs* (HHS Publication FDA 90-4329) and other sources.

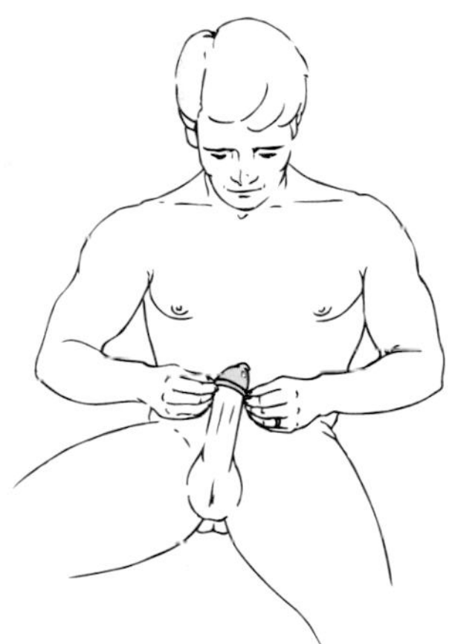

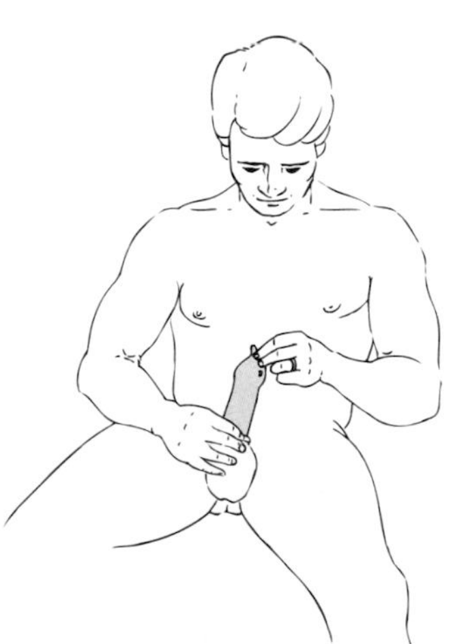

FIGURE 12.6

Applying a Condom. First the rolled-up condom is placed on the head of the penis, and then it is rolled down the shaft of the penis. If a condom without a reservoir tip is used, a ½-inch space should be left at the tip for the ejaculate to accumulate.

- Remove the condom carefully from the penis, making sure that semen doesn't leak out.
- Check the removed condom for tears or cracks. If any are found, immediately apply a spermicide containing nonoxynol-9 directly to the penis and within the vagina. Wrap the used condom in a tissue, and discard it in the garbage. (Condoms can be hard to flush down the toilet.) Wash your hands thoroughly with soap and water.

  Because condoms can be eroded by exposure to body heat or other sources of heat, they should not be kept for any length of time in a pocket or the glove compartment of a car. Nor should a condom be used more than once. Here are some other things you should *never* do with a condom:
- Never use teeth, scissors, or sharp fingernails to open a package of condoms. Open the condom package carefully to avoid tearing or puncturing the condom.
- Never test a condom by inflating it or stretching it.
- Never use a condom after its expiration date.
- Never use damaged condoms. Condoms that are sticky, gummy, discolored, brittle, or otherwise damaged and condoms that show signs of deterioration should be considered damaged.
- Never use a condom if the sealed packet containing the condom is damaged, cracked, or brittle; the condom itself may be damaged or defective.
- Do not open the sealed packet until you are ready to use the condom. A condom contained in a packet that has been opened can become dry and brittle within a few hours, causing it to tear more easily. The box that contains the condom packets, however, may be opened at any time.
- Never use the same condom twice. Use a new condom if you switch the site of intercourse, such as from the vagina to the anus or from the anus to the mouth, during a single sexual act.
- If you want to carry a condom with you, place it in a loose jacket pocket or purse, not in your pants pocket or in a wallet held in your pants pocket, where it might be exposed to body heat.
- Never buy condoms from vending machines that are exposed to extreme heat or placed in direct sunlight.

EFFECTIVENESS In typical use, the failure rate of the male condom is estimated at 12% (see Table 12.1). That is, 12 women out of 100 whose partners rely on condoms alone for contraception can expect to become pregnant during the first year of use. This rate drops dramatically if the condom is used correctly and combined with the use of a spermicide (Warner & Hatcher, 1998). The effectiveness of a condom and spermicide combined, when used correctly and consistently, rivals that of the birth-control pill.

REVERSIBILITY The condom is simply a mechanical barrier to sperm and does not compromise fertility. Therefore, a couple who wish to conceive a child simply discontinue its use.

ADVANTAGES AND DISADVANTAGES Condoms have the advantage of being readily available. They can be purchased without prescription. They require no fitting and can remain in sealed packages until needed. They are readily discarded after use. The combination of condoms and spermicides containing the ingredient nonoxynol-9 increases contraceptive effectiveness. Some condoms contain this spermicidal agent as a lubricant. When in doubt, ask a pharmacist.

Condoms do not affect production of hormones, ova, or sperm. Women whose partners use condoms ovulate normally. Men who use them produce sperm and ejaculate normally. With all these advantages, why are condoms not more popular?

One disadvantage of the condom is that it may render sex less spontaneous. The couple must interrupt lovemaking to apply the condom. Condoms may also lessen sexual sensations, especially for the man. Latex condoms do so more than animal membrane sheaths. Condoms also sometimes slip off or tear, allowing sperm to leak through.

On the other hand, condoms are almost entirely free of side effects. They offer protection against STIs that is unparalleled among contraceptive devices. They can also be used without prior medical consultation. Both partners can share putting on the condom, which makes it an erotic part of their lovemaking, not an intrusion. The use of textured or ultrathin condoms may increase sensitivity, especially for the male. Thus many couples find that their advantages outweigh their disadvantages. Sex in the age of AIDS has given condoms a new respectability, even a certain trendiness. Note, for example, the "designer colors" and styles on display at your local pharmacy. Advertisers now also target women in their ads, suggesting that women, like men, can come prepared with condoms.

It is tempting to claim that the condom has a perfect safety record and no side effects. Let us settle for "close to perfect." Some people have allergic reactions to the spermicides with which some lubricated condoms are coated or that the woman may apply. In such cases the couple may need to use a condom without a spermicidal lubricant or stop using supplemental spermicides. Some people are allergic to latex.

Women have an absolute right to insist that their male sex partners wear latex condoms if their partners are not latex-sensitive. STIs such as gonorrhea and chlamydia (see Chapter 16) do far more damage to a woman's reproductive tract than to a man's. Condoms can help protect women from vaginitis, pelvic inflammatory disease (PID), infections that can harm a fetus or cause infertility, and, most important, AIDS.

Douching

Many couples believe that if a woman **douches** shortly after coitus, she will not become pregnant. Women who douche for contraceptive purposes often use syringes to flush the vagina with water or a spermicidal agent. The water is intended to wash sperm out, the spermicides to kill them. Douching is ineffective, however, because large numbers of sperm move beyond the range of the douche seconds after ejaculation. In fact, squirting a liquid into the vagina may even propel sperm *toward* the uterus. Douching, at best, has a failure rate among typical users of 40% (Reinisch, 1990), too high to be considered reliable.

Regular douching may also alter the natural chemistry of the vagina, increasing the risk of vaginal infection. In short, douching is a "nonmethod" of contraception.

Withdrawal (Coitus Interruptus)

In **coitus interruptus** (withdrawal), the man removes his penis from the vagina before ejaculating. Withdrawal has a first-year failure rate among typical users of about 20% (Kowal, 1998). There are several reasons for these failures. The man may not withdraw in time. Even if the penis is withdrawn just before ejaculation, some ejaculate may still fall on the vaginal lips, and sperm may find their way to the fallopian tubes. Active sperm may also be present in the *pre*-ejaculatory secretions of fluid from the Cowper's glands, a discharge of which the man is usually unaware and cannot control. These sperm are capable of fertilizing an ovum even if the man withdraws before orgasm. Because of its unreliability and high failure rate, withdrawal, like douching, is also a nonmethod of contraception.

Fertility Awareness Methods (Rhythm Methods)

Fertility awareness methods, or *rhythm* methods, rely on awareness of the fertile segments of the woman's menstrual cycle. Terms such as *natural birth control* and *natural family planning* also refer to these methods. The essence of such methods is that coitus is avoided on days when conception is most likely. Fertility awareness methods are used by about 3% of married women aged 15 to 44 but by less than 1% of single women (U.S. Bureau of the Census, 1990b). Women aged 25 to 44 are more than twice as likely as 15- to 24-year-olds to use rhythm methods. Because the rhythm method does not employ artificial devices, it is acceptable to the Roman Catholic Church.

Douche To rinse or wash the vaginal canal by inserting a liquid and allowing it to drain out.

HOW THEY WORK A number of rhythm methods are used to predict the likelihood of conception. They are the mirror images of the methods that couples use to increase their chances of conceiving (see Chapter 11). Methods for enhancing the chances of conception seek to predict the time of ovulation so that the couple can arrange to have sperm present in the woman's reproductive tract at about that time. As methods of *birth control,* rhythm methods seek to predict ovulation so that the couple can *abstain* from coitus when the woman is fertile.

THE CALENDAR METHOD The **calendar method** assumes that ovulation occurs 14 days prior to menstruation. The couple abstains from intercourse during the period that begins 3 days prior to day 13 (because sperm are unlikely to survive for more than 72 hours in the female reproductive tract) and ends 2 days after day 15 (because an unfertilized ovum is unlikely to remain receptive to fertilization for longer than 48 hours). The period of abstention thus covers days 10 to 17 of the woman's cycle (Wilcox et al., 2000).

When a woman has regular 28-day cycles, predicting the period of abstention is relatively straightforward. Women with irregular cycles are generally advised to chart their cycles for 10 to 12 months to determine their shortest and longest cycles. The first day of menstruation counts as day 1 of the cycle. The last day of the cycle is the day preceding the onset of menstruation.

Consider a woman whose cycles vary from 23 to 33 days. In theory she will ovulate 14 days before menstruation begins. (To be safe, she should assume that ovulation will take place anywhere from 13 to 15 days before her period.) "Applying the rule of "3 days before" and "2 days after," she should avoid coitus from day 5 of her cycle, which corresponds to 3 days before her earliest expected ovulation (computed by subtracting 15 days from the 23 days of her shortest cycle and then subtracting 3 days), through day 22, which corresponds to 2 days after her latest expected ovulation (computed by subtracting 13 days from the 33 days of her longest cycle and then adding 2 days). Another way of determining this period of abstention would be to subtract 18 days from the woman's shortest cycle to determine the start of the "unsafe" period and 11 days from her longest cycle to determine the last "unsafe" day. The woman in the example has irregular cycles. She thus faces an 18-day abstention period each month—quite a burden for a sexually active couple.

Most women who follow the calendar method need to abstain from coitus for at least 10 days during the middle of each cycle. Moreover, the calendar method cannot ensure that the woman's longest or shortest menstrual cycles will occur during the 10- to 12-month period of baseline tracking. Some women, too, have such irregular cycles that the range of "unsafe" days cannot be predicted reliably even if baseline tracking is extended.

THE BASAL BODY TEMPERATURE (BBT) METHOD In the **basal body temperature (BBT) method,** the woman tracks her body temperature upon awakening each morning to detect the small changes that occur directly before and after ovulation. A woman's basal body temperature sometimes dips slightly just before ovulation and then tends to rise between 0.4 and 0.8 degree Fahrenheit just before, during, and after ovulation. It remains elevated until the onset of menstruation. (The rise in temperature is caused by the increased production of progesterone by the corpus luteum during the luteal phase of the cycle.) Thermometers that provide finely graded readings, such as electronic thermometers, are best suited for determining minor changes. A major problem with the BBT method is that it does not indicate the several *unsafe* preovulatory days during which sperm deposited in the vagina may remain viable. Rather, the BBT method indicates when a woman *has* ovulated. Thus many women use the calendar method to predict the number of "safe" days prior to ovulation and the BBT method to determine the number of "unsafe" days after. A woman would avoid coitus during the "unsafe preovulatory period (as determined by the calendar method) and then for 3 days when her temperature rises and remains elevated. A drawback of the BBT method is that changes in body temperature may also result from factors unrelated to ovulation, such as infections, sleepless-

Calendar method A fertility awareness (rhythm) method of contraception that relies on prediction of ovulation by tracking menstrual cycles, typically for a 10- to 12-month period, and assuming that ovulation occurs 14 days before menstruation.

Basal body temperature (BBT) method A fertility awareness method of contraception that relies on prediction of ovulation by tracking the woman's temperature during the course of the menstrual cycle.

ness, and stress. This is why some women triple-check themselves by also tracking their cervical mucus.

THE CERVICAL MUCUS (OVULATION) METHOD The **ovulation method** tracks changes in the **viscosity** of the cervical mucus. Following menstruation, the vagina feels rather dry. There is also little or no discharge from the cervix. These dry days are relatively safe. Then a mucous discharge appears in the vagina that is first thick, sticky, and white or cloudy in color. Coitus (or unprotected coitus) should be avoided at the first sign of any mucus. As the cycle progresses, the mucus discharge thins and clears, becoming slippery or stringy, like raw egg white. These are the **peak days.** This mucus discharge, called the *ovulatory mucus,* may be accompanied by a feeling of vaginal lubrication or wetness. Ovulation takes place about a day after the last peak day (about four days after this ovulatory mucus first appears). Then the mucus becomes cloudy and tacky once more. Intercourse may resume four days following the last peak day.

One problem with the mucus method is that some women have difficulty detecting changes in the mucus discharge. Such changes may also result from infections, certain medications, or contraceptive creams, jellies, or foam. Sexual arousal may also induce changes in viscosity.

OVULATION-PREDICTION KITS Predicting ovulation is more accurate with an ovulation-prediction kit. These kits enable women to test their urine daily for the presence of luteinizing hormone (LH). LH levels surge about 12 to 24 hours prior to ovulation. Ovulation-prediction kits are more accurate than the BBT method. Some couples use the kits to enhance their chances of conceiving a child by engaging in coitus when ovulation appears imminent. Others use them as a means of birth control to find out when to avoid coitus. When used correctly, ovulation-predicting kits are highly accurate.

Ovulation kits are expensive and require that the woman's urine be tested each morning. Nor do they reveal the full range of the unsafe *pre*ovulatory period during which sperm may remain viable in the vagina. A couple might thus choose to use the kits to determine the unsafe period following ovulation and use the calendar method to determine the unsafe period preceding ovulation.

EFFECTIVENESS The estimated first-year failure rate in typical use is 20%, which is high but no higher than that for the use of contraceptive devices such as the cervical cap or the female condom (see Table 12.1). Still, perhaps 1 in 5 typical users will become pregnant during the first year of use. (You may have heard the joke: "What do you call people who use the rhythm method? Parents!") Fewer failures occur when these methods are applied conscientiously, when a combination of rhythm methods is used, and when the woman's cycles are quite regular. Restricting coitus to the postovulatory period can reduce the pregnancy rate to 1% (Jennings et al., 1998). The trick is to determine when ovulation occurs. The pregnancy rate can be reduced to practically zero if rhythm methods are used with other forms of birth control, such as the condom or diaphragm.

ADVANTAGES AND DISADVANTAGES Because they are a natural form of birth control, rhythm methods appeal to many people who, for religious or other reasons, prefer not to use artificial means. No devices or chemicals are used, so there are no side effects. Nor do they cause any loss of sensation, as condoms do. Nor is there disruption of lovemaking, as with condoms, diaphragms, or foam, although lovemaking could be said to be quite "disrupted" during the period of abstention. Rhythm methods are inexpensive, except for ovulation-prediction kits. Both partners may share the responsibility for rhythm methods. The man, for example, can take his partner's temperature or assist with the charting. All rhythm methods are fully reversible.

A disadvantage is the fact that the reliability of rhythm methods is low. Rhythm methods may be unsuitable for women with irregular cycles. Women with irregular cycles who ovulate as early as a week after their menstrual flows can become pregnant even

Ovulation method A fertility awareness method of contraception that relies on prediction of ovulation by tracking the viscosity of the cervical mucus.

Viscosity Stickiness, consistency.

Peak days The days during the menstrual cycle when a woman is most likely to be fertile.

if they engage in unprotected intercourse only when they are menstruating, because some sperm remaining in a woman's reproductive tract may survive for up to 8 days and fertilize an ovum that is released at that time. Moreover, the rhythm method requires abstaining from coitus for several days, or perhaps weeks, each month. Rhythm methods also require that records of the menstrual cycle be kept for many months prior to implementation. Unlike diaphragms, condoms, or spermicides, rhythm methods cannot be used at a moment's notice. Finally, rhythm methods do not offer any protection against STIs.

Sterilization

Many people decide to be sterilized when they plan to have no children or no more children. With the exception of abstinence, sterilization is the most effective form of contraception. Yet the prospect of **sterilization** arouses strong feelings because a person is transformed all at once, and presumably permanently, from someone who might be capable of bearing children to someone who cannot. This transformation often involves a profound change in self-concept. These feelings are especially strong in men and women who link fertility to their sense of masculinity or femininity.

Still, more than a million sterilizations are performed in the United States each year. It is the most widely used form of birth control among married couples age 30 and above (Reinisch, 1990). Nineteen percent of the respondents in a 1992 national sample of nearly 7,000 women aged 15 to 50 reported being sterilized; 12% reported having partners who had undergone a vasectomy. Married women were far more likely to rely on a permanent method of contraception (tubal sterilization or vasectomy) than were single women (48% vs. 11%).

MALE STERILIZATION The male sterilization procedure used today is the **vasectomy.** About 500,000 vasectomies are performed each year in the United States. More than 15% of men in the United States have had vasectomies.

A vasectomy is usually carried out in a doctor's office, under local anesthesia, in 15 to 20 minutes. Small incisions are made in the scrotum. Each vas is cut, a small segment is removed, and the ends are tied off or cauterized (to prevent them from growing back together) (see Figure 12.7). Now sperm can no longer reach the urethra. Instead, they are harmlessly reabsorbed by the body.

The man can usually resume sexual relations within a few days. Because some sperm may be present in his reproductive tract for a few weeks, however, he is best advised to use an additional contraceptive method until his ejaculate shows a zero sperm count. Some health professionals recommend that the man have a follow-up sperm count a year after his vasectomy to ensure that the cut ends of the vas deferens have not grown together.

Vasectomy does not diminish sex drive or result in any change in sexual arousal, erectile or ejaculatory ability, or sensations of ejaculation. Male sex hormones and sperm are still produced by the testes. Without a passageway to the urethra, however, sperm are no longer expelled with the ejaculate. Sperm account for only about 1% of the ejaculate, so the volume of the ejaculate is not noticeably different.

Though there are no confirmed long-term health risks of vasectomy (Reinisch, 1990), two recent studies of more than 73,000 men who had undergone vasectomies raise concerns that the procedure may not be so risk-free as people generally believe. The studies showed that men who had had vasectomies more than 20 years earlier faced a slightly increased risk of prostate cancer (Altman, 1993a; Giovannucci et al., 1993a, 1993b). The studies found a correlation between vasectomies and the risk of prostate cancer but did not establish a causal connection. It is possible that other factors than the vasectomy itself explain the greater risk faced by vasectomized men. The results also conflict with earlier studies showing either no link between vasectomies and the risk of prostate cancer or even a *lower* risk among vasectomized men. Health professionals recommend that men with vasectomies have annual screenings for prostate cancer.

Sterilization Surgical procedures that render people incapable of reproduction without affecting sexual activity.

Vasectomy The surgical method of male sterilization in which sperm are prevented from reaching the urethra by cutting each vas deferens and tying it back or cauterizing it.

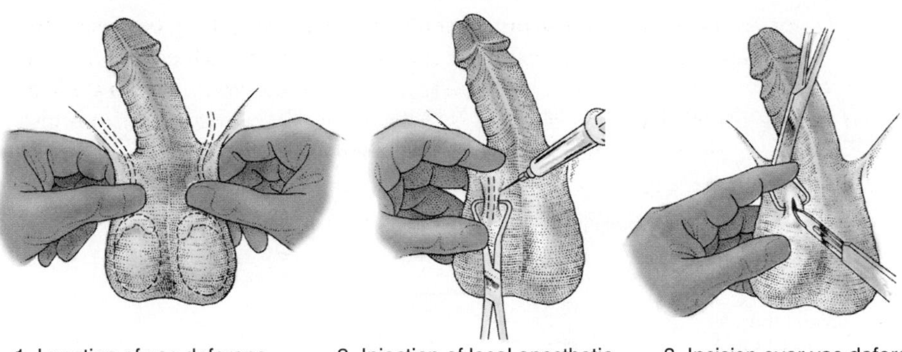

1. Location of vas deferens 2. Injection of local anesthetic 3. Incision over vas deferens

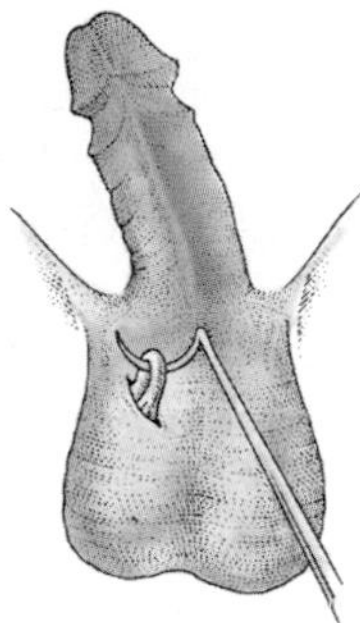

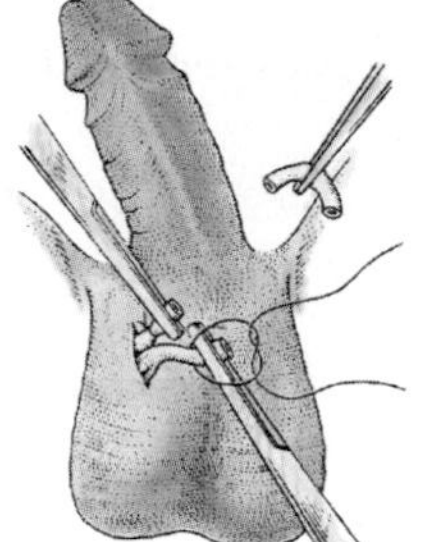

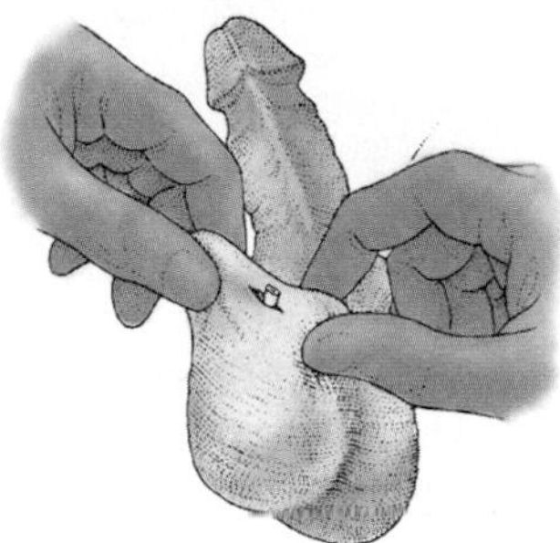

4. Isolation of vas from surrounding tissue 5. Removal of segment of vas; tying of ends 6. Return of vas to position; incision is closed and process is repeated on the other side

FIGURE 12.7

Vasectomy. The male sterilization procedure is usually carried out in a doctor's office, using local anesthesia. Small incisions are made in the scrotum. Each vas deferens is cut, and the ends are tied off or cauterized to prevent sperm from reaching the urethra. Sperm are harmlessly reabsorbed by the body after the operation.

The vasectomy is nearly 100% effective. Fewer than 2 pregnancies occur during the first year among 1,000 couples in which the man has undergone a vasectomy (see Table 12.1). The few failures stem from sperm remaining in the male's genital tract shortly after the operation or from the growing together of the segments of a vas deferens.

Reversibility is simple in concept but not in practice. Thus vasectomies should be considered permanent. In an operation to reverse a vasectomy, called a **vasovasotomy**, the ends of the vas deferens are sewn together, and in a few days they grow together. Estimates of success at reversal, as measured by subsequent pregnancies, range from 16% to 79% (Hatcher et al., 1998). Some vasectomized men develop antibodies that attack their own sperm. The production of antibodies does not appear to endanger the man's health (Hatcher et al., 1998), but it may contribute to infertility following reconnection.

Major studies reveal no deaths due to vasectomy in the United States (Reinisch, 1990). Few documented complications of vasectomies have been reported in the medical literature. Minor complications are reported in 4% or 5% of cases, however. They typically involve temporary local inflammation or swelling after the operation. Ice packs and anti-inflammatory drugs, such as aspirin, may help reduce swelling and discomfort. More serious but rarer medical complications include infection of the epididymis (Reinisch, 1990).

FEMALE STERILIZATION Nearly 4 in 10 (39%) married women under the age of 45 have been surgically sterilized (U.S. Bureau of the Census, 1990b). **Tubal sterilization,** also

Vasovasotomy The surgical method of reversing vasectomy in which the cut or cauterized ends of the vas deferens are sewn together.

Tubal sterilization The most common method of female sterilization, in which the fallopian tubes are surgically blocked to prevent the meeting of sperm and ova. Also called *tubal ligation*.

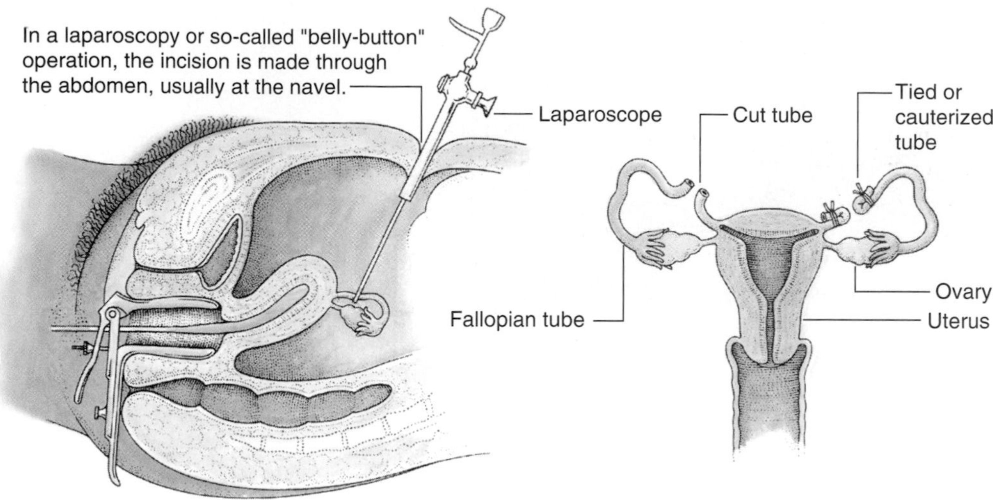

FIGURE 12.8

Laparoscopy. In this method of female sterilization, the surgeon approaches the fallopian tubes through a small incision in the abdomen just below the navel. A narrow instrument called a *laparoscope* is inserted through the incision, and a small section of each fallopian tube is cauterized, cut, or clamped to prevent ova from joining with sperm.

called *tubal ligation*, is the most common method of female sterilization. Tubal sterilization prevents ova and sperm from passing through the fallopian tubes. About 650,000 tubal sterilizations are performed each year in the United States (Altman, 1993c).

The two main surgical procedures for tubal sterilization are *minilaparotomy* and *laparoscopy*. In a **minilaparotomy,** a small incision is made in the abdomen, just above the pubic hairline, to provide access to the fallopian tubes. Each tube is cut and either tied back or clamped with a clip. In a **laparoscopy** (see Figure 12.8), sometimes called "belly-button surgery," the fallopian tubes are approached through a small incision in the abdomen just below the navel. The surgeon uses a narrow, lighted viewing instrument called a *laparoscope* to locate the tubes. A small section of each of the tubes is cauterized, cut, or clamped. The woman usually returns to her daily routine in a few days and can resume coitus when it becomes comfortable. In an alternative sterilization procedure, a **culpotomy,** the fallopian tubes are approached through an incision in the back wall of the vagina.

Minilaparotomy A kind of tubal sterilization in which a small incision is made in the abdomen to provide access to the fallopian tubes.

Laparoscopy Tubal sterilization by means of a *laparoscope,* which is inserted through a small incision just below the navel and used to cauterize, cut, or clamp the fallopian tubes. Sometimes referred to as "belly-button surgery."

Culpotomy A kind of tubal sterilization in which the fallopian tubes are approached through an incision in the back wall of the vagina.

Hysterectomy Surgical removal of the uterus. (*Not* appropriate as a method of sterilization.)

None of these methods disrupts sex drive or sexual response. Surgical sterilization does not induce premature menopause or alter the woman's production of sex hormones. The menstrual cycle is undisturbed. The unfertilized egg is simply reabsorbed by the body, rather than being sloughed off in the menstrual flow.

A **hysterectomy** also results in sterilization. A hysterectomy is a major operation that is commonly performed because of cancer or other diseases of the reproductive tract; it is inappropriate as a method of sterilization. Hysterectomy carries the risks of major surgery, and when the ovaries are removed along with the uterus, it induces a "surgical menopause" because the woman no longer produces female sex hormones.

Female sterilization is highly effective in preventing pregnancy, although slightly less effective than male sterilization. Overall, about 1 women in 200 (0.4%) is likely to become pregnant in the first year following a tubal sterilization (Hatcher et al., 1998); this is most likely to result from a failed surgical procedure or a pregnancy undetected at the time of the procedure. Like vasectomy, tubal ligation should be considered irreversible. Reversals are successful, as measured by subsequent pregnancies, in 43% to 88% of cases (Hatcher et al., 1998). Reversal is difficult and costly, however.

Truth or Fiction? REVISITED

Not all sterilization operations can be surgically reversed; therefore, the Truth or Fiction statement is too broad to be true. Sterilization is not advised for individuals who believe that they might later change their minds about becoming pregnant.

About 2% to 11% of women incur medical complications (Reinisch, 1990). The most common complications are abdominal infections, excessive bleeding, inadvertent punctures of nearby organs, and scarring. The use of general anesthesia (typical in laparoscopies and in some minilaparotomies) poses additional risks, as in any major operation. (Most of the deaths that are attributed to tubal sterilization actually result from the anesthesia [Reinisch, 1990]. The overall death rate is quite small, however—2 to 5 deaths per 100,000 operations.)

ADVANTAGES AND DISADVANTAGES OF STERILIZATION The major advantages of sterilization are effectiveness and permanence. Sterilization is nearly 100% effective. Following surgery, the couple need not do anything more to prevent conception. The permanence is also its major drawback, however. People sometimes change their minds about wanting to have children.

Sterilization procedures create varying risks of complications following surgery, and women generally incur greater risks than men. Another disadvantage of sterilization is that it affords no protection against STIs. People who are sterilized may still wish to use condoms and spermicides for protection against STIs.

The Search Goes On

The search for the perfect contraceptive continues. It is difficult to say precisely what that contraceptive will look like, but it will be safe and effective, and it will not interfere at all with sexual spontaneity or sexual pleasure.

Even now, in the third millennium, that contraceptive does not yet seem to be within our grasp. However, it does appear that we will be making new advances in mechanical- and chemical-barrier methods, systems for delivering hormones, intrauterine devices (IUDs), and systemic methods for men (such as a male pill) (Gabelnick, 1998).

MECHANICAL-BARRIER METHODS Mechanical barriers have the advantage of not introducing chemicals into people's circulatory systems. On the other hand, they have sometimes been cumbersome and, in the case of the male condom, reduced sexual sensations and spontaneity. Moreover, the male condom is "male controlled," and there is a desire for more female-controlled devices. One, the female condom, is already on the market.

THE FEMALE CONDOM The female condom consists of a polyurethane (plastic) sheath that is used to line the vagina during intercourse. It is held in place at each end by a flexible plastic ring. The female condom provides a secure but flexible shield that barricades against sperm but allows the penis to move freely within the vagina during coitus. It can be inserted as much as 8 hours before intercourse but should be removed immediately afterward (Hatcher et al., 1998). A new one must be used for each act of intercourse.

Like the male condom, the female condom may offer some protection against STIs. Dr. Mary E. Guinan (1992) of the Centers for Disease Control points out that women can use the female condom if their partners refuse to wear a male condom. Cynthia Pearson (1992) of the National Women's Health Network notes that the female condom "for the first time [gives] women control over exposure to sexually transmitted disease, including AIDS."

The female condom (brand name: Reality) carries a warning label that it appears to be less effective than the male latex condom in preventing pregnancies and transmission of STIs. During test trials, the pregnancy rate was estimated to range between 21% and 26%, but it is estimated to be as low as 5% among cautious users (Hatcher et al., 1998). Evidence concerning the effectiveness of the female condom in providing protection against STIs is scarce (Centers for Disease Control, 1993a).

Many women complain that the female condom is bulky and difficult to insert (Stewart, 1992). Still, it is the first barrier method of contraception that women control themselves. Naming the device Reality suggests that it may be used most widely by women faced with the reality of male partners who refuse to use condoms themselves or who fail to use them consistently or properly. The female condom costs several times as much as the male condom.

LEA'S SHIELD This diaphragm-like device fits all women. It has a valve that allows air to escape when it is put in place, creating suction that keeps it in close contact with the cervix. The valve permits uterine and cervical fluid to drain but prevents the passage of sperm. It is already marketed in Canada and Europe.

THE SILICONE DIAPHRAGM A diaphragm made of silicone is under development. It promises to be more user-friendly than earlier models—more comfortable and easy to use.

NONLATEX MALE CONDOMS The latex condom is often unused because it interferes with sexual spontaneity and pleasure. Condoms made of animal intestine provide more sexual sensations, but they permit the passage of some infectious organisms, including HIV. A polyurethane condom that allows more sexual sensations but prevents passage of HIV is currently on the market, and other nonlatex condoms are under development.

CHEMICAL-BARRIER METHODS New spermicides are under development—chemical barriers that will adhere better to the vaginal wall but produce less irritation. Such devices will take the form of contraceptive film, sponges, and so on. One sponge now available in Canada, Protectaid, uses low levels of active ingredients to avoid irritating the woman.

Let us note parenthetically that the search is on for chemical barriers that will kill disease organisms, such as HIV, but not harm sperm. To date, however, the search for a "microbicide" (agent that kills microorganisms) that does not damage sperm has been elusive.

HORMONAL METHODS Hormonal methods have excellent track records for women in terms of efficacy and safety. Therefore, there is not too much incentive to change them dramatically. Norplant and Depo-Provera have been quite successful with women who might find it burdensome to take a daily pill. New hormone preparations are also under development that take the form of vaginal rings, subdermal implants (like Norplant), suppositories, long-acting injections (like Depo-Provera), and hormone pills or injections for men.

THE VAGINAL RING The vaginal ring can be worn in the vagina for 3 months before replacement. Shaped like a diaphragm, the ring contains either a combination of estrogen and progestin or progestin only. The hormones are slowly released and pass into the bloodstream through the mucousal lining of the vagina. The vaginal ring is a convenient means of receiving a continuous dose of hormones without having to remember to take a pill. Research on the effectiveness of the vaginal ring is under way.

IMPROVED IMPLANTS Implanon is a single implant containing a progestin that would remain effective for 3 to 4 years. Improved versions of Norplant will necessitate fewer rods.

DEPO-PROVERA Depo-Provera (medroxyprogesterone acetate) is a long-acting, synthetic form of progesterone that works as a contraceptive by inhibiting ovulation. The progesterone signals the pituitary gland in the brain to stop producing hormones that would lead to the release of mature ova by the ovaries. Administered by injection once every 3 months, Depo-Provera is an effective form of contraception, with reported failure rates of less than 1 pregnancy per 100 women during the first year of use (Hatcher et al., 1998). A World Health Organization study of 12,000 women found no links between the drug and the risk of ovarian or cervical cancer (Walt, 1993). Yet the drug may produce such side effects as weight gain, menstrual irregularity, and spotting between periods

(Hatcher et al., 1998). Use of Depo-Provera has also been linked to *osteoporosis,* a condition involving bone loss that can cause bones to become brittle and fracture easily.

DESIGNER HORMONES Research is under way to find "designer" hormone preparations that act more specifically to prevent conception without acting on other bodily systems. Designer hormones would have fewer side effects and would carry less risk of fostering the growth of cancerous cells.

HORMONAL METHODS FOR MEN The male sex hormone testosterone has shown promise in reducing sperm production (Martin et al., 2000). The pituitary gland normally stimulates the testes to produce sperm. Testosterone suppresses the pituitary, in turn suppressing sperm production. Men who have received testosterone injections have shown declines in sperm production. Potential complications include an increased risk of prostate cancer. Testosterone also appears to increase cholesterol levels in the bloodstream, which may heighten the risk of cardiovascular disease.

Nevertheless, testosterone-delivery systems for men are under development. Eventually they will provide protection by reducing the sperm count for months or years. Current methods being tested in pilot studies involve drawbacks such as long induction periods that require supplementary contraception and also weekly injections—certainly not very user-friendly.

Truth or Fiction? **REVISITED**

It is true that testosterone can be used as a male contraceptive. However, side effects and other issues have prevented testosterone from being marketed for this use. ■

One "male pill" under development includes testosterone and another hormone called progestogen (Cohen, 1998). This pill is 95% effective in reducing the sperm counts to levels at which impregnation is highly unlikely, and it may be available within a few years. In a University of Edinburgh study, about 2 out of 3 men said that they would use such a pill if it were safe and effective (Cohen, 1998). Yet a survey of college men in the United States found that only 1 in 5 expressed willingness to take a male pill (Laird, 1994). The men believed that taking a male pill would be more of a bother than having one's partner take a female pill and more "contrary to nature"!

OTHER METHODS FOR MEN Another approach to male contraception was suggested when investigators in China found extremely low birth rates in communities in which cottonseed oil was used in cooking. They extracted from the cotton plant a drug, *gossypol,* that shows promise as a male contraceptive. Chinese studies reveal the drug to be nearly 100% effective in preventing pregnancies. The drug appears to nullify sperm production without affecting hormone levels or the sex drive. However, toxic effects of gossypol have limited its acceptability as a male contraceptive.

Some men are infertile because they produce antibodies that destroy their own sperm. It is also speculated that vasectomy causes some men's bodies to react to their own sperm as foreign substances and produce antibodies. This is an immunological response to sperm. Some researchers have suggested that it may be possible to develop ways to induce the body to produce such antibodies. Ideally, this procedure would be reversible.

Applying ultrasound waves to the testes has been shown to produce reversible sterility in laboratory rats, dogs, and monkeys. No one is quite certain how ultrasound works in inducing temporary sterilization. Moreover, its safety has not been demonstrated. A Chinese researcher has developed an electronic device that emits impulses that kill sperm and make a man sterile for up to a month ("China develops," 1998). The device is gaining some followers in China. It remains to be seen whether interest will spread to the West.

INTRAUTERINE DEVICES (IUDS) Use of existing IUDs has been connected with problems such as expulsion, cramping, and irritation or perforation of the uterus. Under development is a "frameless" IUD that would eliminate pressure against the uterus

(Gabelnick, 1998), thus minimizing cramping. The copper-releasing sleeves would be anchored by a polypropylene thread rather than by a frame.

Mirena, an IUD available in Europe, releases the chemical levonorgestrel. It is highly effective and proves up to a decade of protection against conception. Mirena is less likely than other IUDs to cause bleeding, and it may be helpful in avoiding PID. Its main disadvantage is that it is much more expensive than copper devices, but it may become available in the United States.

IMMUNOCONTRACEPTIVES Many chemicals are under development that will cause the body's immune system to prevent fertilization. For example, antigens that act on sperm or the zona pellucida of ova may be developed for use with women.

"Vaccines" for men are under development that shut down production of both sperm and testosterone or of sperm alone. A vaccine that shut down testosterone production would have to be used in conjunction with testosterone supplements, and therefore it would be better to shut down just sperm production. A vaccine based on follicle-stimulating hormone (FSH) might do the job.

THE ULTIMATE CONTRACEPTIVE DEVICE Wouldn't it be wonderful if people—female or male—could take a pill that would suspend fertility, with no side effects, until they took another pill that reversed the situation and made it possible again for them to conceive? It would be perfectly effective and perfectly safe. It would not interfere with sexual spontaneity. It would be quickly reversible if and when couples changed their minds about conceiving.

That pill may be a while off, but researchers are thinking about it.

ABORTION

"When I was young, I felt it myself, that it was wrong to kill a baby," said Maria (cited in Lewin, 1998a). Maria is a 17-year-old Latina American high school student who lives in Houston. She is waiting to be seen at an abortion clinic. "But when I got to be a teenager, and started having sex, things looked different. It's more complicated when it's your own life. I want to go to college, and I know that having a baby now, without a husband or anything, would make that very hard."

Maria is not all that unusual for a young person in the United States today. Abortion has little to do with politics to her. The *Roe v. Wade* decision was rendered 8 years before Maria was born. Like half the adults in the United States, Maria believes that abortion is murder (Goldberg & Elder, 1998). But like one third of the Americans who say abortion is murder, Maria believes that abortion is an acceptable solution to a very bad situation.

An **induced abortion** (in contrast to a spontaneous abortion, or miscarriage) is the purposeful termination of a pregnancy. Perhaps more than any other contemporary social issue, induced abortion (hereafter referred to simply as abortion) has divided neighbors and family members into opposing camps.

Thirty-seven million abortions worldwide are performed each year. In the United States, the abortion rate increased steadily from the early 1970s through 1980, leveled off somewhat in the 1980s, reached a peak in about 1990, when there were about 1.4 million abortions, and has declined to about 1.2 million abortions per year more recently (CDC, 2000g). The great majority of abortions in the United States—nearly 90%—occur during the first trimester. This is when they are safest for the woman and least costly.

About 43% of women in the United States have a voluntary abortion at some time (Davis, 2000). Some 80% of them are unmarried (Lewin, 1998a). Nearly half of them have had a previous abortion. Nearly half are mothers who already bear family responsibilities (Russo et al., 1992). They are young: Nearly three-quarters of them are under 30.

Induced abortion The purposeful termination of a pregnancy before the embryo or fetus is capable of sustaining independent life. (From the Latin *abortio,* which means "that which is miscarried.")

Sixty percent of them are European American. About 1 in 3 (35%) is African American. Nearly 1 in 6 is Latina American.

Abortion is practiced widely in Canada, Japan, Russia, and many European nations. It is less common in developing nations, largely because of sparse medical facilities. Abortion is rarely used as a primary means of birth control. It usually comes into play when other methods have failed.

The many reasons why women have abortions include psychological factors as well as external circumstances. Abortion is often motivated by a desire to reduce the risk of physical, economic, psychological, and social disadvantages that the woman perceives for herself and her present and future children should she take the pregnancy to term (Russo et al., 1992).

The national debate over abortion has been played out in recent years against a backdrop of demonstrations, marches, and occasional acts of violence, such as firebombing of abortion clinics and even murder. The right-to-life (pro-life) movement asserts that human life begins at conception and thus views abortion as the murder of an unborn child (Sagan & Dryan, 1990). Some in the pro-life movement brook no exception to their opposition to abortion. Others would permit abortion to save the mother's life or when a pregnancy results from rape or incest.

The pro-choice movement contends that abortion is a matter of personal choice and that the government has no right to interfere with a woman's right to terminate a pregnancy. Pro-choice advocates argue that women are free to control what happens within their bodies, including pregnancies.

When Does Human Life Begin?

Moral concerns about abortion often turn on the question of when human life begins. For some Christians, the matter revolves around when they believe the fetus obtains a soul.

In his thesis on *ensoulment*, the thirteenth-century Christian theologian Saint Thomas Aquinas wrote that a male fetus does not acquire a human soul until 40 days after conception. A female fetus does not acquire a soul until after 80 days. Scientists, too, have attempted to define when human life can be said to begin. Astronomer Carl Sagan, for example, wrote that fetal brain activity can be considered a scientific marker of human life (Sagan & Druyan, 1990). Brain activity is needed for thought, the quality that many consider most "human." Brain wave patterns typical of children do not begin until about the 30th week of pregnancy. Before then, the human fetus lacks the brain architecture to begin thinking (Sagan & Druyan, 1990). Of course, this line of reasoning raises the question of whether fetal brain wave activity can be equated to thought. (What would a fetus "think" about?) Moreover, some argue that a newly fertilized ovum carries the *potential* for human thought in the same way that the embryonic or fetal brain does. It could even be argued that sperm cells and ova are living things in that they carry out the biological processes characteristic of cellular life. All in all, the question of when *human* life begins is a matter of definition that is apparently unanswerable by science.

Historical and Legal Perspectives on Abortion

Societal attitudes toward abortion have varied across cultures and times in history. Abortion was permitted in ancient Greece and Rome, but women in ancient Assyria were impaled on stakes for attempting abortion. The Bible does not specifically prohibit abortion (Sagan & Dryan, 1990). For much of its history, the Roman Catholic Church held to Thomas Aquinas's belief that ensoulment of the fetus did not occur for at least 40 days after conception. In 1869, Pope Pius IX declared that human life begins at conception. Thus an abortion at any stage of pregnancy became murder in the eyes of the church and grounds for excommunication. The Roman Catholic Church has since opposed abortion during any stage of pregnancy.

QUESTIONNAIRE

Pro-Choice or Pro-Life? Where Do You Stand?

What does it mean to be "pro-life" on the abortion issue? What does it mean to be "pro-choice"? Which position is closer to your own views on abortion?

The *Reasoning About Abortion Questionnaire* (RAQ) (Parsons et al., 1990) assesses agreement with pro-life or pro-choice lines of reasoning about abortion. To find out which position is closer to your own, indicate your level of agreement or disagreement with each of the following items by circling the number of the statement that most closely represents your feelings. Then refer to the key in the Appendix to interpret your score.

5 = Strongly Agree (SA); 4 = Agree (A); 3 = Mixed Feelings (MF); 2 = Disagree (D); 1 = Strongly Disagree (SD)

	SA	A	MF	D	SD
1. Abortion is a matter of personal choice.	5	4	3	2	1
2. Abortion is a threat to our society.	5	4	3	2	1
3. A woman should have control over what is happening to her own body by having the option to choose abortion.	5	4	3	2	1
4. Only God, not people, can decide if a fetus should live.	5	4	3	2	1
5. Even if one believes that there may be some exceptions, abortion is still basically wrong.	5	4	3	2	1
6. Abortion violates an unborn person's fundamental right to life.	5	4	3	2	1
7. A woman should be able to exercise her rights to self-determination by choosing to have an abortion.	5	4	3	2	1
8. Outlawing abortion could take away a woman's sense of self and personal autonomy.	5	4	3	2	1
9. Outlawing abortion violates a woman's civil rights.	5	4	3	2	1
10. Abortion is morally unacceptable and unjustified.	5	4	3	2	1
11. In my reasoning, the notion that an unborn fetus may be a human life is not a deciding issue in considering abortion.	5	4	3	2	1
12. Abortion can be described as taking a life unjustly.	5	4	3	2	1
13. A woman should have the right to decide to have an abortion based on her own life circumstances.	5	4	3	2	1
14. If a woman feels that having a child might ruin her life, she should consider an abortion.	5	4	3	2	1
15. Abortion could destroy the sanctity of motherhood.	5	4	3	2	1
16. An unborn fetus is a viable human being with rights.	5	4	3	2	1
17. If a woman feels she can't care for a baby, she should be able to have an abortion.	5	4	3	2	1
18. Abortion is the destruction of one life for the convenience of another.	5	4	3	2	1
19. Abortion is the same as murder.	5	4	3	2	1
20. Even if one believes that there are times when abortion is immoral, it is still basically the woman's own choice.	5	4	3	2	1

Source: Questionnaire from N. K. Parsons, H. C. Richards, and G. O. P. Kanter (1990). "Validation of a Scale to Measure Reasoning About Abortion," *Journal of Consulting Psychology, 37.* Reprinted by permission of Nancy Parsons, Ph.D.

Abortion was legal in the United States from 1607 to 1828 (Hitt, 1998). Women were permitted to terminate a pregnancy until "quickening" occurred (the point at which the woman was first able to feel the fetus stirring within her) (Sagan & Dryan, 1990). More restrictive abortion laws emerged because of a national desire to increase the population and because of concerns voiced by physicians about protecting women from botched abortions (Hitt, 1998). By 1900 virtually all states in the union had enacted legislation banning abortion *at any point* during pregnancy, except when necessary to save the woman's life.

Truth or Fiction? **REVISITED**

It is true that abortions were legal in the United States prior to the Civil War—until the point in the pregnancy when the woman sensed fetal movements. ■

Abortion laws remained essentially unchanged until the late 1960s, when some states liberalized their abortion laws under mounting public pressure. Then, in 1973, the U.S. Supreme Court in effect legalized abortion nationwide in the landmark *Roe v. Wade* decision.

Roe v. Wade held that a woman's right to have an abortion was protected under the right to privacy guaranteed by the Constitution. The decision legalized abortions for any reason during the first trimester, leaving the decision to have an abortion entirely in the hands of the woman. In its ruling, the Court also noted that a fetus is not considered a person and is thus not entitled to constitutional protection. The Court ruled that states may regulate a woman's right to have an abortion during the second trimester to protect her health, such as by requiring her to obtain an abortion in a hospital rather than a doctor's office. The Court also held that when a fetus becomes viable, its rights override the mother's right to privacy. Because the fetus may become viable early in the third trimester, states may prohibit third-trimester abortions, except in cases in which an abortion is necessary to protect a woman's health or life.

In 1977, Congress enacted the Hyde amendment, which denies Medicaid funding for abortions except in cases in which the woman's life is endangered. In *Harris v. McRae* (1980), the U.S. Supreme Court essentially upheld the Hyde amendment (and similar state legislation) by ruling that federal and state governments are not required to pay for abortions for poor women who are receiving public assistance.

Since *Roe v. Wade*, most states have also enacted laws requiring parental consent or notification before a minor may have an abortion (Carlson, 1990). Sixty-nine percent of adults in the United States believe that parental permission should be required before a teenage girl has abortions (Carlson, 1990). Many pregnant teenage girls, however, especially those living in families with alcoholic or abusive parents, fear telling their parents that they are pregnant. In 1990 rulings involving state laws in Ohio and Minnesota, the U.S. Supreme Court upheld the rights of states to require that a minor seeking an abortion notify at least one parent and wait 48 hours before an abortion is performed. The Court provided an "escape clause," however: The minor girl may go before a judge instead. Carlson (1990) notes that many pregnant teenagers who are reluctant to reveal pregnancies to their parents may also hesitate to reveal them to authority figures such as judges. A 1993 U.S. Supreme Court ruling let stand a Mississippi law requiring minors to obtain approval either from both parents or from a judge.

Parental-consent laws have widespread popular support. Four out of five adults support such provisions, including many abortion rights supporters (Goldberg & Elder, 1998). Yet, with or without the rules, the majority of girls seeking abortions do consult their parents before going through with their plans. One of the arguments favoring parental approval is the belief that parents know what is best for their children. However, opponents say that parental-consent laws only serve to discourage girls from getting safer early abortions and force them to postpone abortions until the second trimester, when the risks are greater (Pliner & Yates, 1992).

FIGURE 12.9

Responses to the Question "Should a Woman Be Permitted or Forbidden to Have an Abortion During the (First, Second, or Third) Trimester of Pregnancy?" As you can see, respondents to *The New York Times* poll were less permissive of abortion during the later stages of pregnancy. [*Source:* C. Goldberg and J. Elder (1998, January 16). "Public Still Backs Abortion, But Wants Limits, Poll Says." *The New York Times,* pp. A1, A16. Copyright © 1998 by the New York Times Company. Reprinted by permission.]

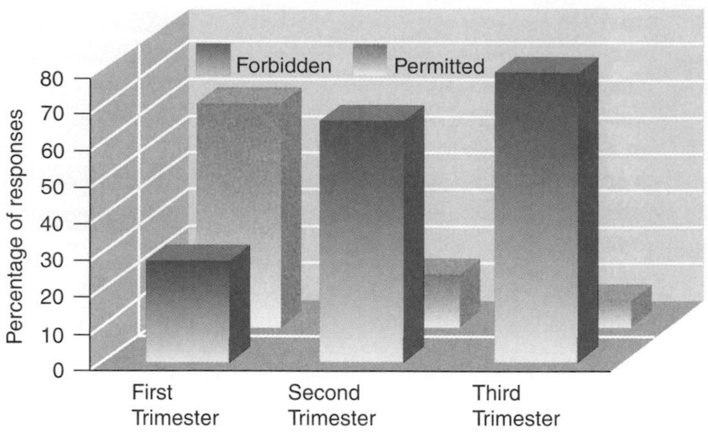

ATTITUDES TOWARD LEGALIZED ABORTION National public opinion polls taken since *Roe v. Wade* have consistently shown that a majority of people in the United States support the decision legalizing abortion (Goldberg & Elder, 1998). Catholics, Jews, and Protestants differ little in their support for abortion (Goldberg & Elder, 1998; Gordon & Snyder, 1989). On the other hand, African Americans are less likely than European Americans to support abortion. Researchers find that higher levels of education are associated with pro-choice attitudes. Reduced support for legal abortion is associated with religious commitment, conservative attitudes on premarital sex, and belief in having large families (Lynxwiler & Gay, 1994).

Most people in the United States favor legalized abortion, but not under all circumstances. A national poll by *The New York Times* found that 61% of Americans believe that abortion should be permitted during the first trimester (see Figure 12.9), but 70% said that abortion was unacceptable if the purpose was to prevent the child from interfering with the woman's career (Goldberg & Elder, 1998). Only 42% felt that abortion was justified to prevent interference with a teenage girl's education. But there was strong support for abortion when the woman was raped, her health was endangered, or there was the strong possibility of a deformity in the baby. Half (50%) of the respondents said that abortion was the same thing as murdering a child; 38% said that abortion was not murder because a fetus is not the same thing as a child. Not only are people in conflict with other people about abortion; they are also in conflict within themselves: One third (32%) of those who said that abortion was murder still agreed that abortion was sometimes the best course in a bad situation! Although some consider abortion a "women's issue," there were no differences between men's and women's attitudes among the respondents to *The New York Times* poll. Nor did the pollsters find that the age of the respondent was related to attitudes toward abortion. However, people with more years of education were more accepting of abortion (see Figure 12.10), and people who said that religion was extremely important to them were less tolerant of abortion than those who said that religion was not so important. Elizabeth Cook (1998), author of a book about abortion entitled *Between Two Absolutes,* suggests that a consensus about abortion may be emerging—that abortion should be allowed under some circumstances but is not to be taken lightly. Three quarters of the sample opposed a constitutional amendment that would ban abortion. Most Americans want the individual to have the right to an abortion, even if they disapprove of it.

The controversy over abortion has led to a decline in the numbers of abortion providers. Nearly 3 out of 5 (59%) doctors who perform abortions are 65 years of age or older (Hitt, 1998). The percentage of obstetricians who are willing to perform abortions dropped from 42% in 1983 to 33% in 1995 (Hitt, 1998). Only about 1 family practitioner

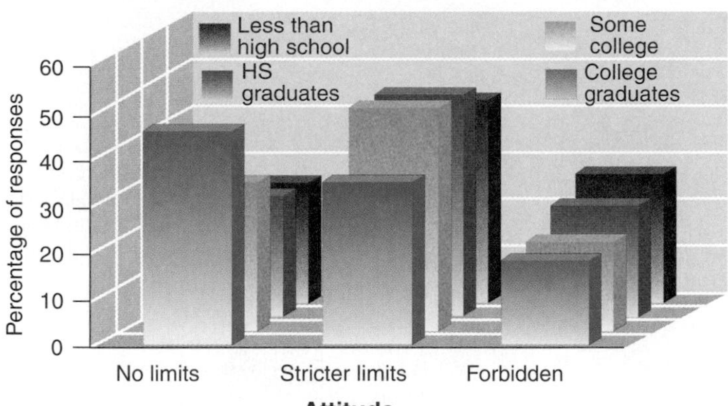

FIGURE 12.10

Attitudes Toward Abortion and Level of Education. Respondents to *The New York Times* poll with more years of education were more likely to say that abortion should be generally available (with no limits). People with fewer years of education were more likely to say that abortion should not be permitted (should be forbidden). [*Source:* C. Goldberg and J. Elder (1998, January 16). "Public Still Backs Abortion, But Wants Limits, Poll Says." *The New York Times,* pp. A1, A16. Copyright © 1998 by the New York Times Company. Reprinted by permission.]

in 6 has experience with any abortion method. Nationwide, 5 out of 6 counties have no abortion provider. Many women who seek abortions at abortion clinics must navigate past pro-life picketers to gain entrance. Many doctors who perform abortions feel that their lives are in jeopardy (Hitt, 1998). Consider the comments of one young doctor who was considering learning abortion methods: "But then I started to think, well, gosh," said the doctor, "I don't want to end up being known as the abortion person. Do you know what I mean?" (Hitt, 1998, p. 22). Yet doctors who perform abortions say that it eliminates the butchery that would otherwise go on in back alleys, provides poor women with the same access that affluent women have always managed to find, and raises general awareness of the need for contraception (Hitt, 1998).

Many people in the pro-choice movement argue that if abortions were to be made illegal again, thousands of women, especially poor women, would die or suffer serious physical consequences from botched or nonsterile abortions. People in the pro-life movement counter that alternatives to abortion such as adoption are available to pregnant women. Pro-choice advocates argue that the debate about abortion should be framed not only by notions of the mother's right to privacy but also by the issue of the quality of life of an unwanted child. They argue that minority and physically or mentally disabled children are often hard to place for adoption. These children often spend their childhoods being shuffled from one foster home to another. Pro-life advocates counter that killing a fetus eliminates any potential that it might have, despite hardships, of living a fruitful and meaningful life.

As noted by Tamar Lewin (1998a), *Roe v. Wade* is not in the personal memories of most individual women who choose to have abortions in the United States today. Many of them have a hard time recognizing that abortion was ever illegal and cannot imagine what it would mean if abortion were to be made illegal again.

Says a 19-year-old woman from Chicago who is waiting in an abortion clinic, "I never heard anything about *Roe v. Wade.* I know there is tension between different groups and that there are people who are really, really against abortion. My mom is totally against it. If she knew I was here, Wow! I can't even imagine what it was like when abortion was illegal. I just opened up the phone book. People definitely take it for granted today. I think of the movie 'Dirty Dancing.' I never understood why she was so sick, but it was because she got an illegal abortion" (Lewin, 1998a).

Says Sirena, a 24-year-old woman from Brooklyn who has two young children and has had three abortions: "I don't think there could ever be a time when abortion would be illegal. No one would let that happen. There would be too many girls throwing babies in the garbage or abusing their kids. But there's always going to be people who are against abortion, like my sister, and people who are pro-choice, like me. It's always been the same and it's always going to be" (Lewin, 1998a).

FIGURE 12.11

Vacuum Aspiration. This is the safest and most common method of abortion, but it can be performed only during the first trimester. An angled tube is inserted through the cervix into the uterus, and the uterine contents are then evacuated (emptied) by suction.

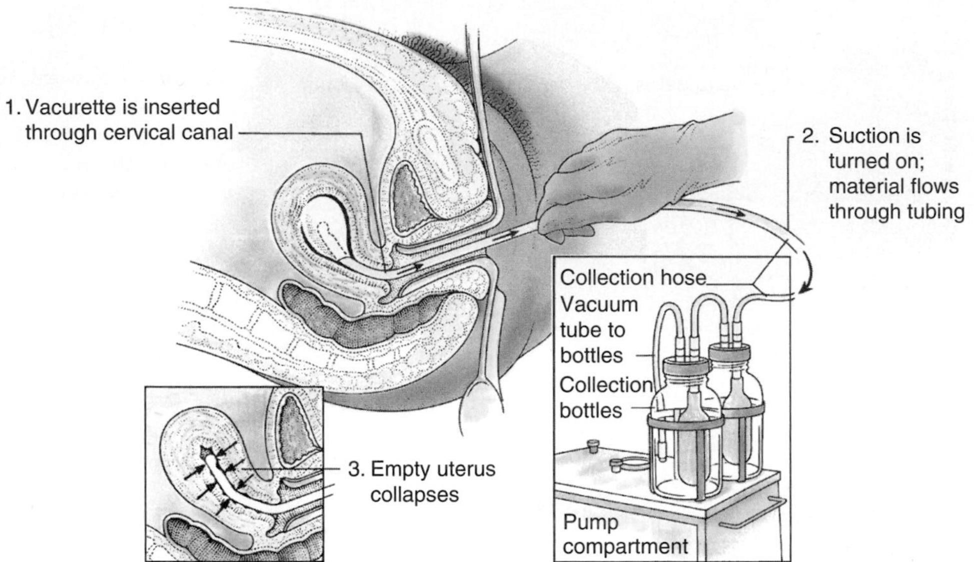
1. Vacurette is inserted through cervical canal
2. Suction is turned on; material flows through tubing
Collection hose
Vacuum tube to bottles
Collection bottles
3. Empty uterus collapses
Pump compartment

Methods of Abortion

Regardless of the moral, legal, and political issues that surround abortion, many abortion methods are in use today.

VACUUM ASPIRATION **Vacuum aspiration,** or suction curettage, is the safest and most common method of abortion. It accounts for more than 90% of abortions in the United States. It is relatively painless and inexpensive. It can be done with little or no anesthesia in a medical office or clinic, but only during the first trimester. Later, thinning of the uterine walls increases the risks of perforation and bleeding.

In the procedure, the cervix is usually dilated first by insertion of progressively larger curved metal rods, or "dilators," or by insertion, hours earlier, of a stick of seaweed called *Laminaria digitata. Laminaria* expands as it absorbs cervical moisture, providing a gentler means of opening the os. Then an angled tube connected to an aspirator (suction machine) is inserted through the cervix into the uterus. The uterine contents are then evacuated (emptied) by suction (see Figure 12.11). Possible complications include perforation of the uterus, infection, cervical lacerations, and hemorrhaging, but these are rare.

Truth or Fiction? **REVISITED**

It is not true that the D&C is the most widely used abortion method in the United States today. Vacuum aspiration is the most widely used abortion method.

Vacuum aspiration Removal of the uterine contents via suction. An abortion method used early in pregnancy. (From the Latin *aspirare,* which means "to breathe upon.")

D&C Abbreviation for *dilation and curettage,* an operation in which the cervix is dilated and the uterine contents are then gently scraped away.

D&E Abbreviation for *dilation and evacuation,* an abortion method in which the cervix is dilated prior to vacuum aspiration.

DILATION AND CURETTAGE (D&C) The **D&C** was once the customary method of performing abortions. It now accounts for only a small number of abortions in the United States. It is usually performed 8 to 20 weeks following the last menstrual period (LMP). Once the cervix has been dilated, the uterine contents are scraped from the uterine lining with a blunt scraping tool.

D&Cs are carried out in a hospital, usually under general anesthesia. The scraping increases the chances of hemorrhaging, infection, and perforation. Because of these risks, D&Cs have largely been replaced by the vacuum aspiration method. D&Cs are still used to treat various gynecological problems, however, such as abnormally heavy menstrual bleeding.

DILATION AND EVACUATION (D&E) The **D&E** is used most commonly during the second trimester, when vacuum aspiration alone would be too risky. The D&E combines suction and the D&C. First the cervix is dilated. The cervix must also be dilated more

a CLOSER look

Partial-Birth Abortion

Most Americans believe that early abortions should be permitted, although not without some limits (Goldberg & Elder, 1998). Moreover, the Supreme Court does not show signs of overturning the outcome of *Roe v. Wade*. Therefore, much of the recent battle over abortion has focused on more specific issues, such as whether parents should have the right to know or to approve if an underage daughter wants to have an abortion, or on the availability of specific methods, such as RU-486 and "partial-birth abortion." The latter is a late-term surgical abortion method that is referred to, medically, as an "intact dilation and extraction," or "intact D and X" (Hitt, 1998).

In a partial-birth abortion, the cervix is dilated, and then a fetus that may be 10 inches long is extracted from the mother through the birth canal. Brain tissue is destroyed to terminate quickly any life functions in the fetus.

Remember that support of abortion rights drops off rapidly as the stages of pregnancy progress (Figure 12.9). Therefore, most individuals who support early first-trimester abortion oppose the partial-birth abortion because of its timing. Others oppose it because of the nature of the disposal of the fetus. However, many pro-choice people argue in favor of partial-birth abortions for at least two reasons. One is that they are relatively rare and are usually performed when the health of the mother is at stake. Another is the "slippery-slope" argument—that is, the fear that surrendering a woman's right to have a partial-birth abortion could eventually lead to surrendering her right to have other kinds of abortions.

The general attitudes of the medical profession toward the partial-birth abortion may be expressed in the fact that the procedure is not to be found in medical books. Neither is the more technical term *intact dilation and extraction* (Hitt, 1998).

fully than with vacuum aspiration to allow for passage of the larger fetus. Then a suction tube is inserted to remove some of the contents of the uterus. But suction alone cannot safely remove all uterine contents, so the remaining contents are removed with forceps. A blunt scraper may also be used to scrape the uterine wall to make sure that the lining has been removed fully. Like the D&C, the D&E is usually performed in a hospital under general anesthesia. Most women recover quickly and relatively painlessly. In rare instances, however, complications arise. These include excessive bleeding, infection, and perforation of the uterine lining (Thompson, 1993).

INDUCING LABOR BY INTRA-AMNIOTIC INFUSION Second-trimester abortions are sometimes performed by chemically inducing premature labor and delivery. The procedure, which must be performed in a hospital, is called instillation, or **intra-amniotic infusion**. It is usually performed when fetal development has progressed beyond the point at which other methods are deemed safe. A saline (salt) solution or a solution of prostaglandins (hormones that stimulate uterine contractions during labor) is injected into the amniotic sac. Prostaglandins may also be administered by vaginal suppository. Uterine contractions (labor) begin within a few hours after infusion. The fetus and placenta are expelled from the uterus within the next 24 to 48 hours.

Intra-amniotic infusion accounts for only a small number of abortions. Medical complications, risks, and costs are greater with this procedure than with other methods of abortion. Overly rapid labor can tear the cervix, but previous dilation of the cervix with *Laminaria digitata* (see page 410) lessens the risk. Perforation, infection, and hemorrhaging are rare when prostaglandins are used, but about half the recipients experience nausea and vomiting, diarrhea, or headaches. Saline infusion can cause shock and even death if the solution is carelessly introduced into the bloodstream.

HYSTEROTOMY The **hysterotomy** is, in effect, a cesarean section. Incisions are made in the abdomen and uterus, and the fetus and uterine contents are removed. Hysterotomy may be performed during the late second trimester, between the 16th and 24th

Intra-amniotic infusion An abortion method in which a substance is injected into the amniotic sac to induce premature labor. Also called *instillation*.

Hysterotomy An abortion method in which the fetus is removed by cesarean section.

a CLOSER look

RU-486 in Europe

Even before she got to the Broussais Hospital clinic here last month, the 28-year-old woman knew that she wanted to have a drug-induced abortion. She hated the idea of a surgical procedure, of being touched by metal tools, of being put under.

"The other seemed more natural," she said. "It felt more private, too."

And it all went very smoothly, she said, sounding hardy for a moment. But then she added in a quieter voice that it still had not been easy. "The most difficult part was the pain in your head," she said 10 days later. "The decision to do it. Waiting for the drugs to work. And afterward I was crying all over the place."

While the United States approved the use of the abortion pill RU-486 in 2000, it had been in use in France for more than a decade. Across the country, government clinics routinely offer it as an alternative to surgery for women looking for abortions within 7 weeks of their last menstrual cycle. Officials estimate that about 30% of France's abortions are drug-induced.

But far from promoting abortion, as some anti-abortion activists had predicted, the pill here seems to have had little impact on the total number of abortions performed annually. The numbers have stayed about the same since 1990, despite a slow but steady increase in the use of RU-486, also known as mifepristone.

Nor has the procedure diminished the emotional pain that often accompanies an abortion. Many of those who work in France's abortion clinics say the drug-induced abortion can actually have more impact on a patient because the abortion takes place over several days and because a patient is awake and aware of the expulsion of the fetal tissue.

"The drugs are a way of de-medicalizing the procedure, and this is good for women" said Christine Der Andreassian, a nurse who has worked at Broussais since 1987 and has supervised thousands of abortions of both types. "But that is not to say that it trivializes the process. The pill is not magic. It does not negate the act of ending a pregnancy. It doesn't make the act disappear."

Restrictions

In the United States, the federal Food and Drug Administration requires that a woman be given written instructions on the pill's use and information about its side effects, which can include bleeding and cramping, headaches, vomiting, and diarrhea, and her doctor must sign a statement saying

weeks after the last menstrual period. It is performed very rarely, usually only when intra-amniotic infusion is not advised. A hysterotomy is major surgery that must be carried out under general anesthesia in a hospital. Hysterotomy involves risks of complications from the anesthesia and the surgery itself.

ABORTION DRUGS RU-486 (mifepristone) was approved by the U.S. Food and Drug Administration in 2000, 12 years after its debut in France. The chemical mifepristone induces early abortion by blocking the effects of progesterone. Progesterone is the hormone that stimulates proliferation of the endometrium, allowing implantation of the fertilized ovum and, subsequently, development of the placenta.

RU-486 can be used only within 49 days of the beginning of the woman's last menstrual period (FDA approves, 2000). The typical course is for the woman to take three mifepristone pills. Two days later, she is given a second oral drug, misoprostol, that causes uterine contractions to expel the embryo. There is also usually a follow-up visit within 2 weeks to make sure that the abortion is complete and the woman is well.

The pill was developed in France, and increasing numbers of women around the world are opting for early abortion without surgery (Davis, 2000). Nearly half of French women who seek abortion prefer RU-486 to surgical methods, including vacuum aspiration (Christin-Maitre et al., 2000). It is too early to tell what percentage of early abortions will be conducted by means of drugs in the United States.

Supporters of RU-486 argue that it offers a safe, noninvasive substitute for more costly and unpleasant abortion procedures (Christin-Maitre et al., 2000). Moreover, use she has read the instructions and will comply with them exactly. The woman must also agree to have a surgical abortion if the pills do not succeed.

In France, however, the procedure, which is virtually paid for by the government, must be performed under the watchful eye of the medical establishment. The abortion is a two-step process. A woman first takes mifepristone tablets, which block the action of progesterone, a hormone required to maintain a pregnancy, and then 36 to 48 hours later takes a second drug, misoprostol, which makes the uterus contract, expelling the fetal tissue.

Both sets of the pill must be taken in front of a doctor or a nurse in France, and after the second set of pills is swallowed, the woman is required to wait 3 or 4 hours at the doctor's office or in a health clinic until the tissue is expelled. (There is no such requirement in the United States.) The regulation is designed to make sure that a woman is not alone during those hours. As in the United States, she must return for a checkup within 2 weeks.

The abortion pill was approved for use in France in 1988. Three years later, Britain followed and in 1992 so did Sweden. More recently, the pill has won much wider approval. In 1999 Austria, Belgium, Finland, Greece, Israel, and Spain approved its use. The dosage and the protocol for use vary slightly from country to country. Some allow its use up to 9 weeks after a woman's last menstrual cycle.

Anne Weyman, the chief executive of the Family Planning Association of the United Kingdom, said that surveys show that women like "medical" abortions because they find them more natural and less frightening. However, some women prefer surgical abortions because they want to have it over quickly.

France has had only one death attributed to the pill, a 1991 case involving a woman, 31, who was a chain smoker and had already had 11 children. After that, the government restricted the pill's use for patients over 35 who are heavy smokers.

France this year decided to make a morning-after contraception pill called Norlevo available through pharmacies without prescriptions, hoping it would reduce the number of abortions, particularly among young girls. The French government tried this year to make such pills, which must be taken within 72 hours after sex, available from school nurses, too, but the way it issued its regulations was ruled illegal. The government is trying again to have the schools distribute the pills, but this time it will bring the measure before Parliament.

Source: From Suzanne Daley (2000, October 5). "Europe Finds Abortion Pill Is No Magic Cure-all. *The New York Times,* p. A3. Copyright © 2000 by the New York Times Co. Adapted by permission.

of RU-486 means that a woman need not run a gauntlet of demonstrators to visit an abortion clinic or hospital. Supporters also note that RU-486 may reduce the numbers of women who die each year from complications from self-induced abortions. Such women are usually too poor to avail themselves of legally sanctioned abortion facilities. Or they live in Third World countries that lack adequate medical services.

RU-486's introduction in the United States was delayed largely because of opposition by pro-life groups (Hausknecht, 1998). Opponents argued that RU-486 makes abortions more accessible and difficult to regulate. Pro-life groups consider abortion to be murder, whether it is induced by surgery or by a pill.

As the abortion debate continues, so does research into the use of other drugs (Christin-Maitre et al., 2000). A combination of the cancer drug methotrexate and misoprostol can also be used to terminate early pregnancy (Ngai et al., 2000). Methotrexate is toxic to the trophoblastic tissue of the embryo, and—as in combination with RU-486—misoprostol causes the uterus to expel the embryo.

Psychological Consequences of Abortion

The woman who faces an unwanted pregnancy may experience a range of negative emotions, including fear, anger directed inward ("How could I let this happen?"), guilt ("What would my parents think if they knew I were having an abortion?"), and ambivalence ("Will I regret it if I have an abortion? Will I regret it more if I don't?")

Consider the comments of Yardena, a 22-year-old woman who is the mother of a 3-year-old, had an abortion at the age of 16, and was waiting to be seen at the Planned Parenthood clinic when she learned that she was pregnant again: "I'm not pro-choice—I'm anti-abortion. I still have negative feelings about abortion, and I love children, but this is something I have to do at this point in my life" (Lewin, 1998a).

A 19-year-old college student says, "I don't like the idea of abortion being used as birth control. This is my first time [at the clinic] and my last. If I get pregnant again, which I won't, I'm having the child. That will mean I was stupid twice. Once is all right, but not the same mistake twice."

Whether to have an abortion is typically a painful decision—perhaps the most difficult decision a woman will ever make. Even women who apparently make the decision without hesitation may regret it later. Although the woman's partner is often overlooked in the research on abortion, he may encounter similar feelings.

Women's reactions depend on various factors, including the support they receive from others (or the lack thereof) and the strength of their relationships with their partners. Women with greater support from their male partners or parents tend to show a more positive emotional reaction following an abortion (Armsworth, 1991). Generally speaking, the sooner the abortion occurs, the less stressful it is. Women who have a difficult time reaching an abortion decision, who blame the pregnancy on their character, who have lower coping ability, and who have less social support experience more distress following abortion (Major & Cozzarelli, 1992).

Many men are very concerned and supportive of their partners. Others seek to detach themselves from the situation. And of course there are some cases in which the identity of the father is unknown. Some men consider pregnancy the woman's responsibility: "She's the one who let herself get pregnant." Some men reproach the woman for failing to take precautions. No wonder feminists insist that men share full responsibility for pregnancies.

All right. We know that abortion continues to be a political football. We know that people on both sides of the issue cite moral and social reasons why their views are correct. But what of the experiences of women who have abortions? How do they react? We reported a couple of anecdotes at the beginning of this section, but let's forget about anecdotes. We can all point to people who are well-adjusted following abortion and to others who are "a mess." What do the carefully conducted surveys tell us?

Frankly, their results are less than crystal clear. Consider one survey of several hundred women reported in *Archives of General Psychiatry* (Major et al., 2000) who showed up at one of three sites for a first-trimester abortion. More than 1,000 women were approached at random as they arrived at the clinic, and 882 (85%) agreed to be followed for 2 years so that their responses could be assessed at various times. Of these 882, 442 were actually followed for the 2 years. As you can see in Figure 12.12, the majority (72%) said they were satisfied with their decision to have the abortion. A majority said they would make the same decision if they had it to do over (69%) and that they had experienced more benefit than harm from having the abortion (72%). Moreover, 4 our of 5 women (80%) were *not* depressed.

Note that you can interpret these findings in any way that you like. Pro-choice advocates can say that the great majority of women appear to be psychologically well adjusted 2 years after having an abortion. But pro-life advocates can say that significant numbers of women are not. To see what we mean, assume that there are 1 million abortions per year. We'll keep the math straightforward. With this assumption in place, 720,000 women (72%) every year will say they are satisfied with their decision 2 years later. But 280,000 women will *not* say they are satisfied with their decision. Similarly, 310,000 women (31%) each year will *not* be able to say they would do it again, and 280,000 women each year will *not* be able to say that they found the effects of the abortion to be more beneficial than harmful.

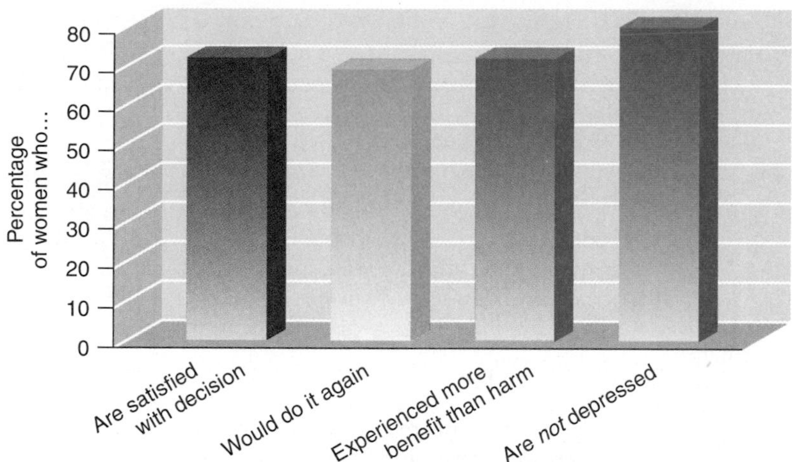

FIGURE 12.12

Women's Psychological Adjustment Two Years After Having Had an Abortion. Of several hundred respondents to an *Archives of General Psychiatry* survey, 72% of those who had had an abortion reported being satisfied with their decision 2 years after the fact. How do you interpret the finding? Do you focus on the finding that the great majority of women are satisfied with their choice or on the finding that significant numbers of women (28%) are not? [*Source:* Data from B. Major, C. Cozzarelli, M. L. Cooper, J. Zwbek, C. Richards, et al. (2000). "Psychological Responses of Women after First-Trimester Abortion." *Archives of General Psychiatry, 57*, 777–784.]

Our conclusion is that both sides are correct. The great majority of women appear to be psychologically well adjusted a couple of years after having an abortion. It is also true that hundreds of thousand of women per year cannot say that they are satisfied with their decision to have an abortion. In other words, numbers alone do not tell this tale. There is no simple answer to the question of the psychological effects of having an abortion.

But neither side is entitled to make overgeneralized, extreme claims. Pro-choice advocates cannot claim that women who have abortions suffer no psychological ill effects. (Some women do.) Nor can pro-life advocates argue that the psychological effects of having an abortion are devastating. (Most women who have abortions are well adjusted.)

Human Sexuality ONLINE //

Web Sites Related to Contraception and Abortion

This is the Web site of Planned Parenthood. "Founded in 1916, Planned Parenthood is the world's largest and oldest voluntary family-planning organization. Planned Parenthood is dedicated to the principles that every individual has a fundamental right to decide when or whether to have a child, and that every child should be wanted and loved." The Web site has links to information about contraception, press releases, legislative information, and local clinics.
www.plannedparenthood.org/

This is the Web site of the National Right to Life Organization. The National Right to Life Committee was founded in 1973 as a response to the United States Supreme Court decision released on January 22 of that year, which legalized abortion in all 50 states. The Web site contains the organization's views on when human life begins, the negative aspects of abortion, and so on.
www.nrlc.org/

This is the Web site of the Alan Guttmacher Institute (AGI). The mission of AGI is to "protect the reproductive choices of all women and men . . . to support their ability to obtain the information and services needed to . . . safeguard their health and exercise their individual responsibilities in regard to sexual behavior, reproduction and family formation. . . . AGI seeks to inform individual decision-making, encourage scientific inquiry and enlightened public debate, and promote the formation of sound public and private-sector programs and policies. . . . AGI recognizes the diversity and dignity of all individuals, and gives special attention to addressing the needs and concerns of those whose access to information, services or other societal benefits may be impeded by their age, by poverty or by virtue of gender, racial, ethnic, religious or cultural discrimination." AGI aims to promote the prevention of unintended pregnancies, guarantee women's freedom to terminate unwanted pregnancies, and help achieve healthy pregnancies and births.
www.agi-usa.org/

The following Web site contains a variety of information about abortion, including the text of court decisions, discussions of the moral issues involved in abortion, surveys (you can cast your vote) and the results of surveys on abortion, papal documents on abortion, and links to both pro-life and pro-choice organizations
ethics.acusd.edu/abortion.html

summing up

Birth-control methods include contraception and induced abortion.

CONTRACEPTION

Contraception in the United States: The Legal Battle
In the United States, Anthony Comstock lobbied successfully for passage of a federal law in 1873 that prohibited the dissemination of birth-control information through the mail on the grounds that it was obscene and indecent. In 1918 the courts ruled that physicians must be allowed to disseminate information that might aid in the cure and prevention of diseases. Dismantling of the Comstock law had begun.

Selecting a Method of Contraception
Issues surrounding the choice among methods of contraception involve their convenience, moral acceptability, cost, safety, reversibility, and effectiveness, the sharing of responsibility between the partners, and the protection the methods afford from STIs.

METHODS OF CONTRACEPTION

Oral Contraceptives ("the Pill")
Birth-control pills include combination pills and minipills. Combination pills contain estrogen and progestin and fool the brain into acting as though the woman is already pregnant, so that no additional ova mature or are released. Minipills contain progestin, thicken the cervical mucus to impede the passage of sperm through the cervix, and render the inner lining of the uterus less receptive to a fertilized egg. Oral contraception is nearly 100% effective. The main drawbacks are side effects and potential health risks. "Morning-after" pills prevent implantation of a fertilized ovum in the uterus.

Norplant
Norplant consists of tubes containing progestin that are surgically embedded under the skin of the woman's upper arm. Norplant provides continuous contraceptive protection for as long as 5 years.

Intrauterine Devices (IUDs)
The IUD apparently irritates the uterine lining, causing inflammation and the production of antibodies that may be toxic to sperm or fertilized ova and/or may prevent fertilized eggs from becoming implanted. The IUD is highly effective, but there are possible troublesome side effects and the potential for serious health complications.

The Diaphragm
The diaphragm covers the cervix and should be used with a spermicidal cream or jelly. It must be fitted by a health professional.

Spermicides
Spermicides block the passage of sperm and kill sperm. Their failure rate is high.

The Cervical Cap
Like the diaphragm, the cap covers the cervix and is most effective when used with a spermicide.

Condoms
Latex condoms afford protection against STIs. Condoms are the only contraceptive device worn by men and the only readily reversible method of contraception that is available to men.

Douching
Douching is ineffective as a contraceptive because large numbers of sperm may pass beyond the range of the douche within seconds after ejaculation.

Withdrawal (Coitus Interruptus)
Withdrawal requires no special equipment but has a high failure rate.

Fertility Awareness Methods (Rhythm Methods)
Rhythm methods rely on awareness of the fertile segments of the woman's menstrual cycle. Rhythm methods include the calendar method, the basal body temperature method, and the cervical mucus method. Their failure rate is high in typical use.

Sterilization
Sterilization methods should be considered permanent, although they can be reversed in many cases. The vasectomy is usually carried out under local anesthesia in 15 to 20 minutes. Female sterilization methods prevent ova and sperm from passing through the fallopian tubes.

Other Devices
The female condom is fitted over the vaginal opening and provides a shield that blocks sperm but allows the penis to move freely. The vaginal ring can be worn in the vagina for 3 months and delivers a continuous dose of hormones that relieves a woman of having to remember to take a pill. Depo-Provera is injected and supplies a continuous dosage of long-acting progesterone, which acts to inhibit ovulation.

ABORTION

Historical and Legal Perspectives on Abortion
In colonial times and through the mid-nineteenth century, women in the United States were permitted to terminate a pregnancy until "quickening" occurred. More restrictive abortion laws came into being after the Civil War. In 1973, the U.S. Supreme Court in effect legalized abortion nationwide in the landmark *Roe v. Wade* decision.

Methods of Abortion
Abortion methods in use today include vacuum aspiration, D&C, D&E, induction of labor by intra-amniotic infusion, hysterotomy, and drugs.

Psychological Consequences of Abortion
Research is mixed concerning the short-term and long-term psychological consequences of having an abortion.

questions for critical thinking

1. What is your personal reaction to the true story told by Margaret Sanger? Why?
2. What problems would you anticipate in trying to discuss contraception with a partner? Why?
3. Are some or all of the methods of contraception morally and ethically unacceptable to you? Explain.
4. Do you believe that your views on contraception are consistent with those of most people from your sociocultural background? Explain.
5. Do you know anyone who has had an abortion? What motivated the abortion? How did she react to the abortion? How did you feel at the time?
6. One of the issues concerning abortion is whether it is the taking of a human life. How do *you* define *human life*? When do *you* believe human life begins? At conception? When the embryo becomes implanted in the uterus? When the fetus begins to assume a human shape or develops human facial features? When the fetus is capable of sustaining independent life? Explain.
7. What are your own views on abortion? Are some abortion methods more acceptable to you than others? Explain.

chapter

13

Sexuality in Childhood and Adolescence

Pablo Picasso. *Figures.* 5 October, 1967. © Copyright ARS, NY. Coll. Picasso, Mougins, France. Credit: Scala/Art Resource, NY. Credit. © 2002 Estate of Pablo Picasso/Artists Rights Society (ARS), New York.

outline

Truth or Fiction?

INFANCY (0 TO 2 YEARS): THE SEARCH FOR THE ORIGINS OF HUMAN SEXUALITY

The Infant's Capacity for Sexual Response

A World of Diversity Cross-Cultural Perspectives on Childhood Sexuality

Masturbation

Genital Play

EARLY CHILDHOOD (3 TO 8 YEARS)

Masturbation

Male–Female Sexual Behavior

Male–Male and Female–Female Sexual Behavior

A Closer Look How Should Parents React to Children Who Masturbate?

PREADOLESCENCE (9 TO 13 YEARS)

Masturbation

Male–Female Sexual Behavior

Male–Male and Female–Female Sexual Behavior

Sex Education and Miseducation

A Closer Look Talking with Your Children About Sex

A Closer Look What Parents Want from Sex Education Courses

ADOLESCENCE

Puberty

A Closer Look Teenage Lingo: The Case of "Hooking Up"

Masturbation

Male–Female Sexual Behavior

A Closer Look "School Days, School Days," Third Millennium

A Closer Look The First Time

A World of Diversity Ethnicity, Teenagers, and HIV

Male–Male and Female–Female Sexual Behavior

TEENAGE PREGNANCY

Contraceptive Use Among Sexually Active Teens

TEENAGERS AND SEX: WHERE DO WE STAND?

Human Sexuality Online Web Sites Related to Sexuality in Childhood and Adolescence

SUMMING UP

QUESTIONS FOR CRITICAL THINKING

Truth or Fiction?

____ Many boys are born with erections.

____ Infants often engage in pelvic thrusting at 8 to 10 months of age.

____ Most children learn the facts of life from parents or from school sex education programs.

____ Sex education encourages sexual activity among children and adolescents.

____ Nocturnal emissions in boys accompany erotic dreams.

____ Petting is practically universal among adolescents in the United States.

____ About 800,000 adolescent girls in the United States become pregnant each year.

____ In some school districts, condoms are distributed to adolescents without parental consent.

My heart leaps up when I behold
A rainbow in the sky:
So was it when my life began;
So is it now I am a man;
So be it when I shall grow old,
Or let me die!
The Child is father of the Man;
And I could wish my days to be
Bound each to each by natural piety.
—William Wordsworth, "My Heart Leaps Up When I Behold"

What a life it would be, indeed, our hearts were to swell with the wonders of the world throughout our lives. In this chapter we begin our chronicle of sexual behavior across the life span, and we see the ways in which the child is father of the man. Within children's personal and social experiences lie the seeds of later sexual competence and self-esteem—or the seeds of incompetence, guilt, and shame. In the next chapter we shall see that our sexuality remains an integral part of our lives throughout our lives—one that has the potential to help our hearts leap up for all our days.

INFANCY (0 TO 2 YEARS): THE SEARCH FOR THE ORIGINS OF HUMAN SEXUALITY

Fetuses not only have erections, they also suck their fingers. The sucking reflex allows babies to gain nourishment, which is necessary for survival. But, as Sigmund Freud hypothesized, infants also seem to reap sensual pleasure from sucking fingers, pacifiers, nipples, or whatever else fits into the mouth. This is not surprising, given the sensitivity of the mouth's mucous lining.

Stimulation of the genitals in infancy may also produce sensations of pleasure. Parents who touch their infants' genitals while changing or washing them may discover the infants smiling or becoming excited. Infants discover the pleasure of self-stimulation (masturbation) for themselves when they gain the capacity to manipulate their genitals with their hands.

The Infant's Capacity for Sexual Response

Boys are often born with erections. Most have erections during the first few weeks.

It is true that many boys are born with erections. Ultrasound has revealed erections even in fetuses. ■

Signs of sexual arousal in infant girls, such as vaginal lubrication, are less readily detected. Yet evidence of lubrication and genital swelling has been reported (Martinson, 1976).

Do not interpret children's responses according to adult concepts of sexuality, however. Lubrication and erection are reflexes, not necessarily signals of "interest" in sex. Infants have the biological capacity for these reflexes, but we cannot say what, if anything, the reflexes mean to them.

PELVIC THRUSTING Pelvic thrusting is observed in infant monkeys, apes, and humans. These observations led ethologist John Bowlby (1969) to suggest that infantile sexual behavior may be the rule in mammals, not the exception. Thrusting has been observed in humans at 8 to 10 months of age and may be an expression of affection. Typically, the infant clings to the parent, nuzzles, and thrusts and rotates the pelvis for several seconds.

Cross-Cultural Perspectives on Childhood Sexuality

Although we are born with the capacity for sexual response, our expression of sexuality largely reflects the culture in which we are reared. Every culture encourages people to conform to socially acceptable behavior. Most people in a given culture develop similar behavior patterns and attitudes.

Cultures vary in their attitudes and practices concerning human sexuality, especially childhood and adolescent sexuality. Some may be characterized as sexually permissive, others as sexually restrictive. Broude and Greene (1976) analyzed attitudes toward, and frequency of, premarital sex among 114 of the world's societies. Most societies (55%) either disapproved of or disallowed premarital sex among females. About one in four societies (24%) permitted girls to engage in premarital sex. Another 21% tolerated discreet premarital sex for females. Despite societal prohibitions, premarital sex among females was common or universal in two thirds of the societies sampled. Female premarital sex was uncommon or absent in the other third. Premarital sex among males was even more common. Among 107 societies in Broude and Greene's cross-cultural sample, male premarital sex was universal or typical in 78%. It was atypical in the other 22%.

Sexually Permissive Societies
For and Beach (1951) noted that when masturbation was permitted in a society, children progressed from occasional genital touching to more purposeful masturbation by about 6 to 8 years of age. Sexually permissive cultures also allow sexual expression among peers. Among the Seniang people of Oceania, boys and girls publicly simulate coitus without fear of reproach by adults (Ford & Beach, 1951). The Lesu of Oceania believe that it is normal for children to imitate coital positions. The Chewu people of Africa believed that childhood sexual experimentation is necessary for adult fertility.

The Lecha people of the Himalayas believed that girls do not attain physical maturity unless they engage in early sexual intercourse. During early childhood, Lecha children engaged in mutual masturbation and attempted copulation. Girls began to engage in regular sexual intercourse by age 11 or 12. Among the Trobrianders of the South Pacific, girls were usually initiated into sexual intercourse by 6 or 8 years of age, boys by age 10 or 12. Among the Muria Gond of India, boys and girls as young as 10 lived together in dormitories and spent their evenings dancing, singing, and pairing off for sex.

Societies that permit sex play among children also tend to encourage open discussion of sex and to allow children to observe sexual behavior among adults. Among the Trukese of the South Pacific, children learned about sex by observing and asking adults. Lesu children, too, would observe adults, but there was at least one taboo: They were not to watch their own mothers.

Sexually Restrictive Societies
Ford and Beach (1951) found that only a minority of preliterate societies were sexually restrictive. The Apinaye people of South America warned their children not to masturbate and thrashed them if they were suspected of doing so. The Kwoma of New Guinea warned boys never to touch their genitals, even when urinating. They risked having their penises beaten with a stick if they touched them.

Sexually restrictive societies discourage masturbation and sex play among children. Children who disobey are punished. Such societies also tend not to talk about sex. Parents try to keep their children ignorant about reproduction. Premarital sex and watching adults engage in sex are also restricted.

Many societies hold to a sexual double standard by which boys are allowed greater sexual freedom than girls.

Truth or Fiction? **REVISITED**

It is true that infants often engage in pelvic thrusting at 8 to 10 months of age. However, there is no reason to believe that this thrusting means the same thing to infants as it does to adults. ■

ORGASM At least some infants seem capable of sexual responses that closely resemble orgasm. Kinsey and his colleagues (1953) noted that baby boys show behaviors that resemble adult orgasm by as early as 5 months, baby girls as early as 4 months. Orgasmic responses in boys are similar to those in men—but without ejaculation. Ejaculation occurs only after puberty.

Masturbation

Masturbation is typical for infants and young children and tends to start between 6 and 12 months. At early ages, children usually masturbate by rubbing the genitals against a soft object, such as a towel, bedding, or a doll. As the child matures and becomes capable of more coordinated hand movements, direct manual stimulation of the genitals is often preferred.

Masturbation to orgasm is rare until the second year, however (Reinisch, 1990). Some children begin masturbating to orgasm later. Some never do. All in all, however, an orgasmic response from masturbation is common among children, as it is among adults (Reinisch, 1990).

Genital Play

Children in the United States typically do not engage in genital play with others until about the age of 2. Then, as an expression of their curiosity about their environment and other people, they may investigate other children's genitals or hug, cuddle, kiss, or climb on top of them. None of this need cause concern. Spiro (1965) describes 2-year-olds at play in an Israeli kibbutz:

> Ofer [a boy] and Pnina [a girl] sit side by side on chamber pots. . . . Ofer puts his foot on Pnina's foot, she then does the same—this happens several times. . . . Finally, Pnina shifts her pot away, then moves back, then away . . . they laugh . . . Pnina stands up, lies on the table on her stomach, . . . Ofer pats her buttocks. . . . Ofer kicks Pnina gently, and they laugh . . . Pnina touches and caresses Ofer's leg with her foot . . . says "more more" . . . Ofer stands, then Pnina stands, both bounce up and down . . . both children are excited, bounce, laugh together . . . Pnina grabs Ofer's penis, and he pushes her away . . . she repeats, he pushes her away, and turns around . . . Pnina touches his buttocks. (p. 225)

There is no reason to infer that Ofer and Pnina were seeking sexual gratification. Rough-and-tumble play, including touching the genitals, is common among children.

EARLY CHILDHOOD (3 TO 8 YEARS)

> SUSAN: Once my younger sister and I were over at a girl friend's house playing in her bedroom. For some reason she pulled her pants down and exposed her rear to us. We were amazed to see she had an extra opening down there we didn't know about. My sister reciprocated by pulling her pants down so we could see if she had the same extra opening. We were amazed at our discovery, our mothers not having mentioned to us that we had a vagina!
>
> CHRISTOPHER: Nancy was a willing playmate, and we spent many hours together examining each other's bodies as doctor and nurse. We even once figured out a pact that we would continue these examinations and watch each other develop. That was before we had started school. (Morrison et al., 1980, p. 19)

These recollections of early childhood illustrate children's interest in sexual anatomy and sexual behavior. Children often show each other their bodies. The unwritten rule seems to be "I'll show you mine if you'll show me yours."

Masturbation

> KIM: I began to masturbate when I was 3 years old. My parents . . . tried long and hard to discourage me. They told me it wasn't nice for a young lady to have her hand between her legs.
>
> When I was five I remember my mother discovering that I masturbated with a rag doll I slept with. She was upset, but she didn't make a big deal about it. She just told me in a matter-of-fact way, "Do you know that what you're doing is called masturbating?" That didn't make much sense to me, except I got the impression she didn't want me to do it. (Morrison et al., 1980, pp. 4, 5)

Because of the difficulties in conducting such research, statistics concerning the incidence of masturbation in children are inconclusive. Parents may not wish to respond to questions about the sexual conduct of their children. Or if they do, they may have a tendency to present their children as little "gentlemen" and "ladies" and perhaps underreport their sexual activity. Their biases may also lead them not to perceive their children's genital touching as masturbation. Many parents will not even permit their adolescents, let alone younger chldren, to be interviewed about their sexual behavior

(Fisher & Hall, 1988). When we are asked to look back as adults, our memories may be less than accurate.

One recent study by psychologist William Friedrich (1998) of the Mayo Institute relied on interviews with the mothers of more than 1,100 children. The study designed was to establish what kinds of sexual behaviors can normally be expected in childhood. The goal was to help educators and other professionals determine when sexual behavior might be suggestive of childhood sexual abuse. The study did not provide data about masturbation per se, but as you can see in Table 13.1, it did offer some insight into how many children touch their "private parts." Friedrich suggests that behavior that occurs in at least 20% of children is normal from a statistical point of view.

At What Age Does Curiosity About Sex Develop? Children are naturally inquisitive about sexual anatomy and sexual behavior. Much curiosity is triggered when they become aware that males and females differ in anatomy.

Male–Female Sexual Behavior

> ALICIA: On my birthday when I was in the second grade, I remember a classmate, Tim, walked home with a friend and me. He kept chasing me to give me kisses all over my face, and I acted like I didn't want him to do it, yet I knew I liked it a lot; when he would stop, I thought he didn't like me anymore. (Morrison et al., 1980, p. 21)

Three- and 4-year-olds commonly express affection through kissing. Curiosity about the genitals may occur by this stage. Sex games like "show" and "playing doctor" may begin earlier, but they become common between the ages of 6 and 10 (Reinisch, 1990). Much of this sexual activity takes place in same-gender groups, although mixed-gender sex games are not uncommon. Children may show their genitals to each other, touch each other's genitals, or masturbate together.

Male–Male and Female–Female Sexual Behavior

> Arnold: When I was about 5, my cousin and I . . . went into the basement and dropped our pants. We touched each other's penises, and that was it. I guess I didn't realize the total significance of the secrecy in which we carried out this act. For later . . . my parents questioned me . . . and I told them exactly what we had done. They were horrified and told me that that was definitely forbidden. (Morrison et al., 1980, p. 24)

TABLE 13.1

Some sexual behaviors that occur commonly in childhood

Ages 2–5	**Boys**	**Girls**
Touches or tries to touch mother's or other women's breasts	42.4%	43.7%
Touches private parts when at home	60.2	43.8
Tries to look at people when they are nude or undressing	26.8	26.9
Ages 6–9		
Touches private parts when at home	39.8	20.7
Tries to look at people when they are nude or undressing	20.2	20.5
Ages 10–12		
Is very interested in the opposite sex	24.1	28.7

Source: William M. Friedrich. Cited in Susan Gilbert (1998, April 7). "New Light Shed on Normal Sexual Behavior in a Child." *The New York Times*, p. F7.

a CLOSER look

How Should Parents React to Children Who Masturbate?

Few parents today believe that children who masturbate set the stage for physical and mental maladies. Still, some parents react with concern, disgust, or shock when their children masturbate.

Parents who are unaware that masturbation is commonplace among children may erroneously assume that children who masturbate are oversexed or aberrant. The parent may pull a child's hands away and scold her or him. Some may even slap the child's hand. Once the child is capable of understanding speech, the parent may say things like "Don't touch down there! That's a bad thing to do. Stop doing that." Threats and punishments may be used. Or parents may fail to acknowledge the behavior openly but move the hands away from the genitals or pick up the child whenever he or she is discovered masturbating.

Sex educators Mary Calderone and Eric Johnson (1989) argue that punishment will not stop children from masturbating. It may cause them to become secretive and guilty about it, however. Sex guilt tends to persist and may impede sexual pleasure in marriage. June Reinisch (1990), director of the Kinsey Institute, notes that

> Parents who scowl, scold, or punish in response to a child's exploring his or her genitals may be teaching the child that this kind of pleasure is wrong and that the *child* is "bad" for engaging in this kind of behavior. This message may hinder the ability to give and receive erotic pleasures as an adult and ultimately interfere with the ability to establish a loving and intimate relationship. (p. 248)

Calderone, Johnson, and Reinisch concur that children need to learn that masturbation in public is not acceptable in our culture, however. Calderone and Johnson suggest that the child who masturbates in front of others can be told something like this:

> I'm glad you've found your body feels good, but when you want to touch your body that way, it's more private to be in your room by yourself. (p. 138)

Reinisch adds that parents are important shapers of their children's sexuality and, more broadly, of their self-esteem. Acknowledging the child's sexuality, rather than rejecting and discouraging it, can strengthen children's self-esteem, build a positive body image, and encourage competence and assertiveness.

Not all authorities—and certainly not all parents—endorse such views. Some object to masturbation on religious or moral grounds. Others feel uncomfortable or conflicted about masturbation themselves. Parents must decide for themselves how best to react when they discover their children masturbating.

Despite Arnold's parents' "horror," same-gender sexual play in childhood does not presage adult sexual orientation (Reinisch, 1990). It may, in fact, be more common than heterosexual play. It typically involves handling the other child's genitals, although it may include oral or anal contact. It may also include an outdoor variation of the game of "show" in which boys urinate together and see who can reach farthest or attain the highest trajectory.

We will end this section with the memories of a woman when she was on the edge of preadolescence.

> When I was 8 and had just learned about menstruation, I fashioned a small sanitary napkin for [my Barbie doll] out of neatly folded tissues. Rubber bands held it in place. "Look," said my bemused mother, "Barbie's got her little period. Now she can have a baby."[1] I was disappointed, but my girlfriends snickered in a way that satisfied me. You see, we all wanted Barbie to be, well. Dirty.
>
> Our Barbies had sex, at least our childish version of it. They hugged and kissed the few available boy dolls we had—clean-cut and oh-so-square Ken, the more relaxed and sexy Allan. Our Barbies also danced, pranced and strutted, but mostly they stripped. An adult friend tells me how she used to put her Barbie's low-backed bathing suit on backward, so the doll's breasts were exposed. I dressed mine in her candy-striped baby-sitter's apron—and nothing else. Girls respond intuitively to the doll's sexuality, and it lets them play out those roles in an endlessly compelling and yet ultimately safe manner. (McDonough, 1998, p. 70)

[1]Yes, yes, we know that the first several cycles of most girls are anovulatory (such that they cannot get pregnant), but we are recounting what someone *said.* So far as we know, it is difficult for Barbie dolls to get pregnant under the best of circumstances.

PREADOLESCENCE (9 TO 13 YEARS)

During preadolescence children typically form relationships with a close "best friend" that enable them to share secrets and confidences. The friends are usually peers of the same gender. Preadolescents also tend to socialize with larger networks of friends in gender-segregated groups. At this stage boys are likely to think that girls are "dorks." To girls at this stage, "dork" is too nice an epithet to apply to most boys.

Preadolescents grow increasingly preoccupied with, and self-conscious about, their bodies. Their peers pressure preadolescents to conform to dress codes, standards of "correct" slang, and group standards concerning sex and drugs. Peer disapproval can be an intense punishment.

Sexual urges are experienced by many preadolescents but may not emerge until adolescence. Sigmund Freud had theorized that sexual impulses are hidden ("latent") during preadolescence, but many preadolescents are quite active sexually.

Masturbation

> PAUL: When I was about 10, stories about masturbation got me worried. A friend and I went to a friend's older brother whom we respected and asked, "Is it really bad?" His reply stuck in my mind for years. "Well, it's like a bottle of olives—every time you take one out, there is one less in there." We were very worried because we thought we'd run out before we got to girls. (Morrison et al., 1980, p. 6)

Kinsey and his colleagues (1948, 1953) reported that masturbation is the primary means of achieving orgasm during preadolescence for both genders. They found that 45% of males and 15% of females masturbated by age 13.

Male–Female Sexual Behavior

Preadolescent sex play often involves mutual display of the genitals, with or without touching. Such sexual experiences are quite common and do not appear to impair future sexual adjustment (Leitenberg et al., 1989).

Although preadolescents tend to socialize in same-gender groups, interest in the other gender among heterosexuals tends to increase gradually as they approach puberty. Group dating and mixed-gender parties often provide preadolescents with their first exposure to heterosexual activities. Couples may not begin to pair off until early or middle adolescence.

Male–Male and Female–Female Sexual Behavior

Much preadolescent sexual behavior among members of the same gender is simply exploration. Some incidents reflect lack of availability of partners of the opposite gender. As with younger children, experiences with children of the same gender during preadolescence may be more common than heterosexual experiences (Leitenberg et al., 1989). These activities are usually limited to touching of each other's genitals or mutual masturbation. Because preadolescents generally socialize within their gender, it is not surprising that their sexual explorations are also often within their gender. Most same-gender sexual experiences involve single episodes or short-lived relationships and are not signs of a budding gay orientation.

Sex Education and Miseducation

> Imagine teaching driving the same way sex education is taught. You'd be told never to drive or ride as a passenger because you could be injured and go to the hospital. No one would ever take a car out. (Mark Miller, 1998)

Mark Miller is a graduate of Brown University's program in sexuality and society. He and other scholars of human sexuality lament the approaches of most sex education

a CLOSER look

Talking with Your Children About Sex

"Daddy, where do babies come from?"

"What are you asking me for? Go ask your mother."

Most children do not find it easy to talk to their parents about sex (Coles & Stokes, 1985). The parents may not find it any easier. Nearly half (47%) of the teens polled in a national survey said they would ask their friends, siblings, or sex partners if they wanted information about sex. Only about a third (36%) would turn to their parents (Coles & Stokes, 1985). Three out of four said that it is hard to talk about sex with their fathers. More than half (57%) found it difficult to approach their mothers. Regrettably, information received from peers is likely to be strewn with inaccuracies. Misinformed teenagers run a higher risk of unwanted pregnancies and STIs.

Yet most young children are curious about where babies come from, about what makes little girls different from little boys, and so on. Parents who avoid answering such questions convey their own uneasiness about sex and may teach children that sex is something to be ashamed of, not something they should discuss openly.

Some parents resist talking about sex with their children because they are insecure in their own knowledge. Reinisch (1990) argues that parents need not be sex experts to talk to their children about sex, however. Parents can turn to books in the local bookstore to fill in gaps in knowledge, or to books that are intended for parents to read to their children. They can also admit that they do not know the answer to a particular question. Reinisch suggests that children will respect parents who display such honesty.

In answering children's questions, parents need to think about what children can understand. The 4-year-old who wants to know where babies come from is probably not interested in detailed biological information. It may be enough to say, "From Mommy's uterus" and then point to the abdominal region. Why say "tummy"? "Tummy is wrong and confusing.

In their *Family Book About Sexuality*, Calderone and Johnson (1989) offer some pointers:

1. *Be willing to answer questions about sex.* Parents who respond to their children's questions about sex by saying, "Why do you want to know that?" squelch further questioning. The child is likely to interpret the parent's response as meaning "You shouldn't be interested in that."
2. *Use appropriate language.* As children develop awareness of their sexuality, they need to learn the names of their sex organs. They also need to learn that the "dirty words" that others use to refer to the sexual parts of the body are not acceptable in most situations, because they carry emotional connotations that can arouse negative feelings. Nor should parents use "silly words" to describe sexual organs:

> Another way parents send out negative messages about sexuality is by using silly words (or no words at all) to describe sexual anatomy. Whether they call genitals "pee-pee" or "privates" or nothing at all, parents are telling children that these body parts are significantly different, embarrassing, mysterious, or taboo compared to such other body parts as the eyes, nose, and knees, which have names openly used in conversation. (Reinisch, 1990, p. 248)

programs. Says Anke Ehrhardt (1998), a psychiatry professor at Columbia University, "In other countries, sex education is put in a positive context of loving relationships, but here we spread fear. And it hasn't worked. We have a much higher rate of teen pregnancy."

Yet fewer than 10% of American children receive comprehensive sex education in school (Bronner, 1998). Studies show that peers are the main source of sexual information. According to an ABC News *Nightline* poll (1995), 53% of American adults learned about sex from their friends; and 30% learned from their parents. When asked where teenagers today learn about sex, five of six (83%) said they learned from friends.

Truth or Fiction? **REVISITED**

It is not true that most children learn the facts of life from parents or from sex-education programs in school. Most children learn about sex from peers. Is the lamp on the street corner the key guiding light for U.S. youth?

Today, nearly all states mandate or recommend sex education, although its content and length vary widely. Most programs emphasize biological aspects of puberty and reproduction (Haffner, 1993). Few deal with abortion, masturbation, sexual orientation, or sexual pleasure.

3. *Give advice in the form of information that the child can use to make sound decisions, not as an imperial edict.* State your convictions, but label them as your own rather than something you are trying to impose on your child. Parents are not as likely to be effective by "laying down the law" as by providing information and encouraging discussion. Reinisch (1990) suggests combining information about sex with expressions of the parents' values and beliefs.

   Parents of teenage children often react to sexual experimentation with threats or punishments, which may cause adolescents to rebel or tune them out. Or the adolescent may learn to associate sex with fear and anger, which may persist even in adult relationships. Parents may find it more constructive to convey concern about the consequences of children's actions in a loving and nonthreatening way that invites an open response. Say, for example, "I'm worried about the way you are experimenting, and I'd like to give you some information that you may not have. Can we talk about it?" (Calderone & Johnson, 1989, p. 141).
4. *Share information in small doses.* Pick a time and place that feel natural for such discussions, such as when the child is preparing for bed or when you are riding in the car.
5. *Encourage the child to talk about sex.* Children may feel embarrassed about talking about sex, especially with family members. Make the child aware that you are always available to answer questions. Be "askable." But let the child postpone talking about a sensitive topic until the two of you are alone or the child feels comfortable. Books about sexuality may help a child open up. They can be left lying around or given to the child with a suggestion such as "This is a good book about sex, or at least I thought so. If you read it, then maybe we can talk about it" (Calderone & Johnson, 1989, p. 136).
6. *Respect privacy rights.* Most of us, parents and children alike, value our privacy at certain times. A parent who feels uncomfortable sharing a bathroom with a child can simply tell the child that Daddy or Mommy likes to be alone when using the bathroom. Or the parent might explain, "I like my privacy, so please knock and I'll tell you if it's okay to come in. I'll do the same for you" (Calderone & Johnson, 1989, p. 137). This can be said without a scolding or harsh tone. Privacy rights in the bedroom can be established by saying in a clear and unthreatening way, "Please knock when the door's closed and wait to be invited in" (Calderone & Johnson, 1989, p. 138). But it is just as important for the parent to respect the child's rights to privacy. The child is likely to feel grateful for the respect and to show respect in return.

Adolescents offer parents some additional advice on how to communicate with them (Pistella & Bonati, 1999):

1. Treat teenagers as equals.
2. Increase your knowledge about the lifestyles of today's teenagers and the peer pressures they experience.
3. *Listen*

Sex education in the schools, especially education about value-laden topics, remains a source of controversy. Some people argue that sex education ought to be left to parents and religious authorities. But the data suggest that the real alternatives to the schools are peers and the corner newsstand, which sells more copies of "adult" magazines than of textbooks. Many parents are concerned that teaching subjects such as sexual techniques and contraception encourages sexual experimentation, but research does not show that sex education increases sexual experimentation (Eisen & Zellman, 1987; Kolbe, 1998; Richardson, 1997).

Truth or Fiction? **REVISITED**

There is no evidence that sex education encourages sexual activity among children and adolescents. ■

Many school programs that offer information about contraception and other sensitive topics are limited to high school juniors and seniors. Sexual experimentation often begins earlier, however (Ehrhardt, 1998). Accurate information in preadolescence might prevent sexual mishaps. Many teens, for example, erroneously believe that a female cannot get pregnant from her first coital experience. Others believe that douching protects them from pregnancy and disease.

a CLOSER look

What Parents Want from Sex Education Courses

As a mother and librarian at a suburban high school outside Washington, Susan Madden holds few illusions about the sexual innocence of teenagers. She has overheard girls chattering in the corridors about condoms, shuddered at the scantily dressed stars on music videos and come across students' notes that read like intrepid dispatches from between the sheets.

"Now, it's everywhere, in your face," she said. None too pleased, Ms. Madden, a mother of three, turns to public schools for help. "There's so much bad information out there, and learning it from their friends is even worse."

While clashes over sex education often focus on parents who fear schools have been too permissive in teaching about sexuality, a recent survey suggests that the overwhelming majority of parents want schools to provide more, not less, sex education once children reach their teenage years. And they want discussions to cover a wide array of topics: abstinence, avoiding pregnancy, sexually transmitted infections, abortion, even sexual orientation.

The consensus found in the survey by the Henry J. Kaiser Family Foundation, a health research organization, appeared largely consistent throughout the country and cut across socioeconomic groups.

It also blurred the lines of a polarized debate between advocates of a conservative approach that frames marriage as the only acceptable venue for sexual relations and a more liberal one that says teenagers should receive comprehensive information about sex, delivered without value judgments.

Instead, parents said schools should borrow from both sides, discussing abstinence, but also advising students about how to use condoms, discuss birth control with a partner and get tested for AIDS.

"Parents want it all," said Steve Rabin, a senior vice president at the Kaiser Family Foundation.

The Kaiser report polled 1,501 sets of parents and students, as well as teachers and principals, from February to May 1999, with a margin of error of three percentage points for parents, teachers and students, and a margin of error of six percentage points for principals.

The survey uncovered a gap between what parents say they want and what schools deliver. Nearly two-thirds of parents said sex education should last half a semester or more, and 54% said boys and girls should be taught in separate classes. The typical class, though, includes boys and girls and consumes just one or two periods of a more general course in health education. Almost all classes teach children about the dangers of contracting AIDS and other sexually transmitted diseases, along with the basics of reproduction and some discussion of abstinence [See Table 13.2].

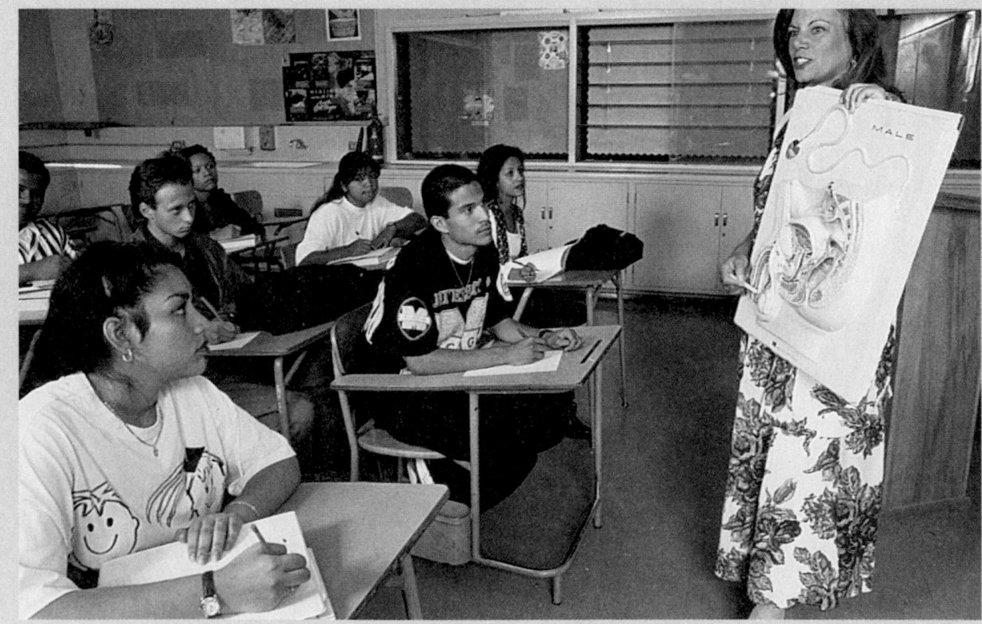

Sex Education. Despite the availability of sex education in most schools, more young people still learn about sex from their peers. Recent survey data reveal that most parents today want schools to provide more rather than less sex education for teenagers. They want sex education to cover abstinence, avoiding pregnancy, sexually transmitted infections, abortion, and even sexual orientation.

TABLE 13.2

What to teach about sex

Percentage of Parents Who Say Sex Education Should Cover . . .	
HIV/AIDS and other sexually transmitted infections	98%
Abstinence; what to do in cases of rape or sexual assault; how to talk with parents about sex	97
How to deal with pressure to have sex and the emotional consequences of sex	94
How to be tested for HIV and other sexually transmitted infections	92
The basics of reproduction and birth control	90
How to talk with a partner about birth control and sexually transmitted infections	88
How to use condoms	85
How to use and where to get other birth control	84
Abortion	79
Sexual orientation and homosexuality	76

Source: Kaiser Family Foundation; reported in Diana Jean Schemo. (2000, October 4). "Survey Finds Parents Favor More Detailed Sex Education." *The New York Times*, pp. A1, A27.

But most parents want sex education to cover much more. Some 84% want sex education to explain how to obtain and use birth control. Even more parents want schools to teach children how to be tested for HIV or AIDS [and] how to respond to pressure to have sex, discuss birth control with a partner and deal with the emotional consequences of sex. They want schools to tell students what to do if they are raped. Four out of five parents want teachers to discuss abortion with their children, and three out of four want their children to learn about homosexuality and sexual orientation in the classroom.

Benita Garcia, a 40-year-old homemaker and mother of three in suburban Maryland, said she believed teenagers needed to learn about sex in school. "They're living a crazy life," she said. "Sometimes they don't listen to what the parents have to say. They think we're antiquated." With a 13-year-old daughter, though, she is worried that talking about sex with students too early could "open their mentality when they're not ready for it."

Heather E. Cirmo, a spokeswoman for the Family Research Council, a conservative research group in Washington, said the report's findings would not change her organization's position favoring abstinence until marriage, with no instructions about obtaining birth control or using condoms to prevent diseases.

"We believe that the parents who are participating in the study have been duped into believing that comprehensive-based sex education is what's best for their children," Ms. Cirmo said. "If you have a standards-based approach to sex that says abstinence is what we expect from you, teens will live up to that."

The Kaiser foundation's findings confirmed earlier, if less thorough, polls tracking parental attitudes toward sex education. Various polls have found steadily increasing support since the 1970s for teaching about birth control in sex education courses. Now, as the generation that matured during the sexual revolution is watching its own children come of age, support for frank discussions of sexuality with them has become nearly universal. Many parents say the threat of AIDS lends particular urgency to such talks.

A 1981 survey done by the Gallup Organization, on behalf of the Phi Delta Kappa association of professional educators, showed 70% of Americans believed young people should learn about sex in high school. Of them, 84% believed high schools should teach about sexually transmitted diseases, a figure that was up to 98% of all parents in the Kaiser report. Some 54% of those favoring sex education in 1981 believed it should cover abortion, a sentiment now shared by 79% of all parents. And 45% of those who favored teaching about sex in high schools in 1981 wanted courses to include some discussion of homosexuality. Today, 76% of parents want teachers to discuss sexual orientation with their teenagers.

"Parents want lots of messages, protective messages, coming at their kids," said Mary McKay, an associate professor of sociology at the Columbia University School of Social Work. "And as kids get older, they want it coming from a variety of sources."

Ramon C. Cortines, chancellor of the New York City school system from 1993 to 1995, said the survey showed that schools should "teach what students need to know, not to please the politicians or parents who scream the loudest." He added, "More, not less, sex education should be taught in the classrooms."

(continued)

What Parents Want from Sex Education Courses *(continued)*

Educators around the country said they were not surprised by the broad support for ample information surrounding sex, though several described blocking of sex education by parents who view sex education as contributing to a breakdown in morality. In part, that is because parents who favor more thorough approaches to sex education nevertheless do not consider it a priority.

"I think the opponents are more vocal and more organized," said Arlene Zielke, a consultant to the Illinois Parent Teacher Association, of which she is a past president. "We're the largest parent organization in the state. Among our membership, we don't get the response back that this is the No. 1 priority, or 2, 3, 4 or 5."

Jan Wilkerson, chairwoman of the health concerns committee of the Texas Parent Teacher Association, served on a committee to update the health education curriculum of her state's public schools four years ago. But when it came to sex education, she found it impossible to overcome a well-organized opposition that insisted that only parents should discuss sex with their children.

"At that time in Texas, we had a 24-year-old grandfather," said Ms. Wilkerson, who lives in Diboll, in rural East Texas. "We had a 29-year-old grandmother in my town. I said, 'These people have families. But do those families have the ability to teach their children? Apparently not.' "

The Kaiser report found a disparity, however, between what teachers say they are teaching and what students say they are learning. A third of the teachers surveyed said they taught that "young people should only have sex when they are married," but only 18% of the students surveyed described that as the main message of their sex education.

The survey also suggested that the battle lines drawn by the camps that fight over sex education obscured, rather than described, the views of parents. Among the one-third of all parents who say schools should teach abstinence until marriage, for example, a substantial number also want schools to arm their children in case they do become sexually active, providing information about obtaining and using condoms, birth control and abortion.

Carol Callaway, a health and physical education teacher at Wakefield High School in Arlington, Virginia, said that her school district's policy was to teach teenagers to forswear sex until marriage, but that questions from students could take discussions elsewhere. "We might say until marriage, but what we're really trying to do is delay," she said.

Susan Soule, who teaches health education at Montgomery Blair High School in Silver Spring, Maryland, said teenagers from 76 nations and a variety of cultures filled her classroom. Many come from homes where their parents never married.

She could not insist that sex outside marriage is wrong, she said, adding, "I'm dealing with a population where marriage might not be the norm for them." Her school, more liberal than most, offers an elective course in sex education, and its roster of student organizations includes one devoted to gay pride.

Ms. Madden, an employee at Montgomery Blair whose children attended the school, said teachers often had more credibility than parents with adolescents, who seemed hardened these days by images of violence and nudity. "They're not allowed to be kids anymore," she said, and shook her head. "Teens have that invincible attitude, with drinking, driving, and sex, too."

Source: From Diana Jean Schemo. (2000, October 4). "Survey Finds Parents Favor More Detailed Sex Education. *The New York Times*, pp. A1, A27.

ADOLESCENCE

Adolescence is bounded by the advent of puberty at the earlier end and the capacity to take on adult responsibilities at the later end. In our society adolescents are "neither fish nor fowl," as the saying goes—neither children nor adults. Adolescents may be able to reproduce and are often taller than their parents, but they may not be allowed to get driver's licenses or attend R-rated films. They are prevented from working long hours and must usually stay in school until age 16. They cannot marry until they reach the "age of consent." The message is clear: Adults see adolescents as impulsive and as needing to be restricted for "their own good." Given these restrictions, a sex drive that is heightened by surges of sex hormones, and media inundation with sexual themes, it is not surprising that many adolescents are in conflict with their families about going around with certain friends, about sex, and about using the family car.

Adolescence. Adolescence begins with puberty. Adolescents in our society are "neither fish nor fowl"—neither children nor adults. They may be capable of reproduction and be larger than their parents, but they may not be allowed to get driver's licenses or attend R-rated films. Many adults see adolescents as impulsive—as needing to be controlled "for their own good." However, adolescents have a sex drive that is heightened by surges of sex hormones, and they are flooded with sexual themes in the media. Therefore, it is not surprising that many of them are in conflict with their families about issues of autonomy and sexual behavior.

Puberty

Puberty begins with the appearance of **secondary sex characteristics** and ends when the long bones make no further gains in length (see Table 13.3). The appearance of strands of pubic hair is often the first visible sign of puberty. Pubic hair tends to be light-colored, sparse, and straight at first. Then it spreads and grows darker, thicker, and coarser. Puberty also involves changes in **primary sex characteristics.** Once puberty begins, most major changes occur within three years in girls and within four years in boys (Etaugh & Rathus, 1995).

Toward the end of puberty, reproduction becomes possible. The two principal markers of reproductive potential are **menarche** in the girl and the first ejaculation in the boy. But these events do not generally herald immediate fertility.

Girls typically experience menarche between the ages of 10 and 18. In the mid-1800s, European girls typically achieved menarche by about age 17 (see Figure 13.1). Since then, the age of menarche has declined sharply among girls in Western nations, probably because of improved nutrition and health care. In the United States, the average age of menarche by the 1960s and 1970s had dropped to between 12½ and 13 (Etaugh & Rathus, 1995).

The **critical fat hypothesis** suggests that girls must reach a certain body weight (perhaps 103 to 109 pounds) to trigger pubertal changes such as menarche, and children today tend to achieve larger body sizes sooner. According to this hypothesis, body fat plays a crucial role because fat cells secrete leptin, a chemical that then signals the body to secrete a cascade of hormones that increases the levels of estrogen in the body. It is known that menarche comes later to girls who have a lower percentage of body fat, such as athletes (Frisch, 1997). Injections of leptin also cause laboratory animals to reach sexual maturity early (Angier, 1997). Whatever the exact triggering mechanism, the average age at which girls experience menarche has leveled off in recent years.

PUBERTAL CHANGES IN THE FEMALE First menstruation, or menarche, is the most obvious sign of puberty in girls. Yet other, less obvious changes have already set the stage for menstruation. Between 8 and 14 years of age, release of FSH by the pituitary gland causes the ovaries to begin to secrete estrogen. Estrogen has several major effects on pubertal development. For one, it stimulates the growth of breast tissue ("breast buds"), perhaps as early as age 8 or 9. The breasts usually begin to enlarge during the tenth year.

Puberty The stage of development during which reproduction first becomes possible. Puberty begins with the appearance of *secondary sex characteristics* and ends when the long bones make no further gains in length. (From the Latin *puber,* which means "of ripe age.")

Secondary sex characteristics Physical characteristics that differentiate males and females and that usually appear at puberty but are not directly involved in reproduction, such as the bodily distribution of hair and fat, the development of muscle mass, and deepening of the voice.

Primary sex characteristics Physical characteristics that differentiate males and females and are directly involved in reproduction, such as the sex organs.

Menarche The onset of menstruation; first menstruation. (From Greek roots that mean "month" [*men*] and "beginning" [*arche*].)

Critical fat hypothesis The view that girls must reach a certain body weight to trigger pubertal changes such as menarche.

TABLE 13.3

Stages of pubertal development

In Females	
Beginning sometime between ages 8 and 11	Pituitary hormones stimulate ovaries to increase production of estrogen. Internal reproductive organs begin to grow.
Beginning sometime between ages 9 and 15	First the areola (the darker area around the nipple) and then the breasts increase in size and become more rounded. Pubic hair becomes darker and coarser. Growth in height continues. Body fat continues to round body contours. A normal vaginal discharge becomes noticeable. Sweat and oil glands increase in activity, and acne may appear. Internal and external reproductive organs and genitals grow, making the vagina longer and the labia more pronounced.
Beginning sometime between ages 10 and 16	Areola and nipples grow, often forming a second mound sticking out from the rounded breast mound. Pubic hair begins to grow in a triangular shape and to cover the center of the mons. Underarm hair appears. Menarche occurs. Internal reproductive organs continue to develop. Ovaries may begin to release mature eggs capable of being fertilized. Growth in height slows.
Beginning sometime between ages 12 and 19	Breasts near adult size and shape. Public hair fully covers the mons and spreads to the top of the thighs. The voice may deepen slightly (but not as much as in males). Menstrual cycles gradually become more regular. Some further changes in body shape may occur into the young woman's early 20s.

This table is a general guideline. Changes may appear sooner or later than shown and do not always appear in the indicated sequence.

FIGURE 13.1

Age at Menarche. The age at menarche has been declining since the mid-1800s among girls in Western nations, apparently because of improved nutrition and health care. Menarche may be triggered by the accumulation of a critical percentage of body fat.

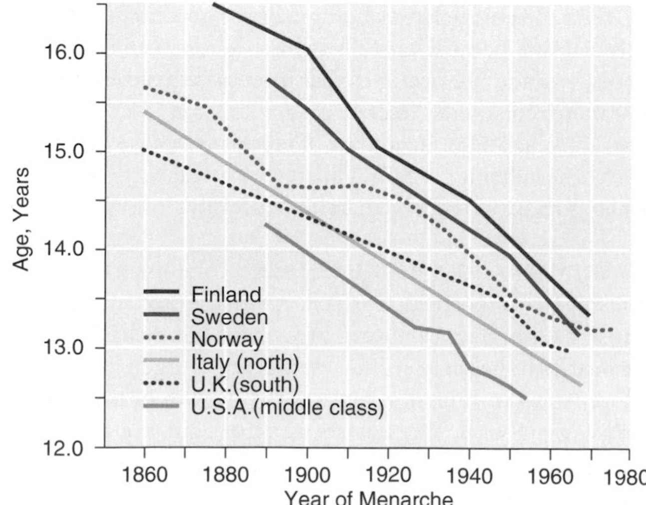

In Males

Beginning sometime between ages 9 and 15	The testicles begin to grow. The skin of the scrotum becomes redder and coarser. A few straight pubic hairs appear at the base of the penis. Muscle mass develops, and the boy begins to grow taller. The areola grows larger and darker.
Beginning sometime between ages 11 and 16	The penis begins to grow longer. The testicles and scrotum continue to grow. Pubic hair becomes coarser and more curled and spreads to cover the area between the legs. The body gains in height. The shoulders broaden. The hips narrow. The larynx enlarges, resulting in a deepening of the voice. Sparse facial and underarm hair appears.
Beginning sometime between ages 11 and 17	The penis begins to increase in circumference as well as in length (though more slowly). The testicles continue to increase in size. The texture of the pubic hair is more like an adult's. Growth of facial and underarm hair increases. Shaving may begin. First ejaculation occurs. In nearly half of all boys, gynecomastia (breast enlargement) occurs, which then decreases in a year or two. Increased skin oils may produce acne.
Beginning sometime between ages 14 and 18	The body nears final adult height, and the genitals achieve adult shape and size, with pubic hair spreading to the thighs and slightly upward toward the belly. Chest hair appears. Facial hair reaches full growth. Shaving becomes more frequent. For some young men, further increases in height, body hair, and muscle growth and strength continue into their early 20s.

Source: Copyright © 1990 by the Kinsey Institute for Research in Sex, Gender, and Reproduction. From The Kinsey Institute New Report on Sex.

Estrogen also promotes the growth of the uterus, thickening of the vaginal lining, and the growth of fatty and supporting tissue in the hips and buttocks. This tissue and the widening of the pelvis cause the hips to become rounded and permit childbearing. But growth of fatty deposits and connective tissue varies considerably. Some women may have pronounced breasts; others may have relatively large hips.

Small amounts of androgens produced by the female's adrenal glands, along with estrogen, stimulate development of pubic and underarm hair, beginning at about age 11. Excessive androgen production can darken or thicken facial hair.

Estrogen causes the labia to grow during puberty, but androgens cause the clitoris to develop. Estrogen stimulates growth of the vagina and uterus. Estrogen typically slows down the female growth spurt some years before that of the male. Girls deficient in estrogen during their late teens may grow quite tall, but most tall girls reach their heights because of normal genetically determined variations, not estrogen deficiency.

Estrogen production becomes cyclical in puberty and regulates the menstrual cycle. Following menarche, a girl's early menstrual cycles are typically **anovulatory**—without ovulation. Girls cannot become pregnant until ovulation occurs, and ovulation may lag behind menarche by as much as two years. And ovulation may not be reliable at first, so

Anovulatory Without ovulation.

a CLOSER look

Teenage Lingo: The Case of "Hooking Up"

"Only yesterday," notes the copywriter for the Farrar, Straus & Giroux fall catalog, "boys and girls spoke of embracing and kissing (necking) as getting to first base. Second base was deep kissing, plus groping and fondling this and that." No longer; first base is today deep kissing, also known as *tonsil hockey*. The writer then speeds up to date in orally touching second and rounding third base, which is now "going all the way," and slides home with a surprise twist of the old sex-as-baseball metaphor: "Home plate is being introduced by name."

The occasion for this recollection and updating of antediluvian teenage lingo is the promotion of a new book of essays and short fiction by Tom Wolfe titled *Hooking Up*. "How rarely our *hooked-up* boys and girls are introduced by name!" laments the promotion copy, which goes on to promise a chronicle of "everything from the sexual manners and mores of teenagers to fundamental changes in the way human beings now regard themselves, thanks to the new fields of genetics and neuroscience."

Wolfe has a sensitivity to *le mot juste*[2] in describing social phenomena. The title of his *The Right Stuff*, a book about the early astronauts, has now become part of the language, as is his popularization of the mathematicians' *pushing the envelope*. In selecting *Hooking Up* as his title, he is again on the cusp of usage.

When we hear a sultry seductress say to an aging Lothario, "We'll *hook up* one of these days," what does her promise mean? (A *Lothario* is a male deceiver, from a character in Nicholas Rowe's 1703 play *The Fair Penitent*. My need to point this out is what philologists call "coinage compulsion.")

The compound noun *hook-up* (which *The New York Times* no longer hyphenates) was born in a political context in 1903, as "a *hook-up* with the reform bunch," and meant a general linkage. In 1930, the term became specific, as "a national *hook-up*" came to denote a radio network.

As a verb, to *hook up* has for a century also meant "to marry," a synonym of "to get hitched," as a horse is to a wagon. But not until the 1980s did the meaning change to a less formal sexual involvement. It was first defined as "to pick someone up at a party" and then progressed to "become sexually involved with; to make out."

The swinging sense mainstreamed in 1995. "A few women insist," wrote *USA Today*, "they never go out with the intention of '*hooking up*' or having sex," while a CNN commentator noted, "The kids see shacking up and *hooking up* as the equivalent of marriage." In 1997, *The Cleveland Plain Dealer* quoted a Brown University student as saying, "In a normal Brown relationship, you meet, get drunk, *hook up* and then either avoid eye contact the next day or find yourself in a relationship." The scholarly reporter noted, "Depending on the context, a *hook-up* can mean anything from 20 minutes of strenuous kissing to spending the night together fully clothed to sexual intercourse."

To be *hooked*, taken from the fishing vocabulary, is to be addicted to drugs; however, with the addition of *up* to make the compound, the term has no sinister narcotics meaning. In current usage, which may not last long and is probably already fading, it most often means "have a sexual relationship." Nor is the "linking" verb limited to American English. An exasperated Liz Jones, editor of *Marie Claire*, wrote in *The Sunday Times of London* this year about men who are habitual sexual deceivers *(Lotharios)*, "Are all men like this or is it just the ones I *hook up* with?"

Linguae Engaged in Lubricious Osculation (Say That Again?)

Let's go back to first base. *Tonsil hockey*, as used at Farrar, Straus & Giroux to mean "deep kissing," is at least a decade

a girl may be relatively infertile. Some teenagers are highly fertile soon after menarche, however (Reinisch, 1990).

PUBERTAL CHANGES IN THE MALE At puberty the hypothalamus signals the pituitary to increase production of FSH and LH. These releasing hormones stimulate the testes to increase their output of testosterone. Testosterone prompts growth of the male genitals: the testes, scrotum, and penis. It fosters differentiation of male secondary sex characteristics: the growth of facial, body, and pubic hair and the deepening of the voice. Testicle growth, in turn, accelerates testosterone production and pubertal changes. The testes continue to grow, and the scrotal sac becomes larger and hangs loosely from the body. The penis widens and lengthens, and pubic hair appears.

"Hooking Up." In teenage slang, this kissing couple is "hooking up." Not so long ago, they might have been referred to as "getting to first base." Don't be surprised if you also hear their behavior referred to as "tonsil hockey."

old, having replaced *tonsil boxing.* A more recent variation is *tongue sushi,* which shows some metaphoric imagination: the Japanese *sushi*—cold rice rolled up with bits of raw fish and vegetables—is evoked to describe the mutual rolling-up of teenage linguae engaged in lubricious osculation.[3]

Tonsil hockey goalies have, in a spasm of good taste, rejected the phrase, popular in the 80s, "to suck face." That undeniably vivid but odious locution seems to have been replaced in some localities with the almost euphemistic *mess around.* Its variants include *mashing, macking* (from *smack,* the sound of a kiss) and *mugging,* the senses of which run the semantic gamut from "flirting" to "foreplay with no intention of intercourse." Those familiar with Old Slang would call it "taking a long lead off first base."

Though *hooking up* seems a mediumistic metaphor for what used to be euphemized as "sleeping together," it is more romantic than the phrase in current use on college campuses: *parallel parking.*

Source: From William Safire. (2000, June 18). "Hooking Up." *The New York Times Magazine online.*

[2]A French phrase that means "the right or proper word."

[3]Tongue kissing, that is.

By age 13 or 14, erections become frequent. Indeed, many junior high school boys dread being caught between classes with erections or being asked to stand before the class. Under the influence of testosterone, the prostate and seminal vesicles—the organs that produce semen—increase in size, and semen production begins. Boys typically experience their first ejaculation by age 13 or 14, most often through masturbation. There is much variation, however. First ejaculations may occur as early as age 8 or not until the early 20s (Reinisch, 1990). Mature sperm are not usually found in the ejaculate until about a year after the first ejaculation, at age 14 on the average (Kulin et al., 1989). But sperm may be present in the first ejaculate (Reinisch, 1990), so pubertal boys should not assume that they have an infertile "grace period" following their first ejaculation. About a year after their first ejaculation, boys may begin to experience **nocturnal emissions,**

Nocturnal emission Involuntary ejaculation of seminal fluid while asleep. Also referred to as a "wet dream," although the individual need not be dreaming about sex, or dreaming at all, at the time.

which are also called wet dreams because of the belief that nocturnal emissions accompany erotic dreams—which need not be so.

Despite the term *wet dreams*, nocturnal emissions in boys need not accompany erotic dreams. ■

Underarm hair appears at about age 15. Facial hair is at first a fuzz on the upper lip. A beard does not appear for another two or three years. Only half of U.S. boys shave (of necessity) by age 17. The beard and chest hair continue to develop past the age of 20. At age 14 or 15, the voice deepens because of the growth of the **larynx** and the lengthening of the vocal cords. Development is gradual, and the voices of adolescent boys sometimes crack embarrassingly.

Boys and girls undergo general growth spurts during puberty. Girls usually shoot up before boys. Individuals differ, however, and some boys spurt sooner than some girls.

Increases in muscle mass produce increases in weight. The shoulders and the circumference of the chest widen. At the age of 18 or so, men stop growing taller because estrogen prevents the long bones from making further gains in length (Smith et al., 1994). Males normally produce some estrogen in the adrenal glands and testes. Nearly one in two boys experiences temporary enlargement of the breasts, or **gynecomastia,** during puberty; this also is caused by estrogen.

Masturbation

Masturbation is a major sexual outlet during adolescence. About half of the adolescent boys (46%) in Coles and Stokes's (1985) national survey of 1,067 teenagers and about a quarter of the girls (24%) reported masturbating. The average age at which teenagers in the Coles and Stokes survey reported that they started to masturbate was 11 years 8 months.

A southern California survey of 641 teenagers showed that boys who masturbate do so two to three times a week, on the average, compared to about once a month for girls (Hass, 1979). Researchers find no links between adolescent masturbation or early sexual activity (for example, frequency of intercourse, number of different partners, or age at first intercourse) and sexual adjustment during young adulthood (Leitenberg et al., 1993).

Many teens still think of masturbation as shameful (Coles & Stokes, 1985). Only about one in three (31%) of the teens surveyed by Coles and Stokes reported feeling completely free of guilt over masturbation. One in five felt "a large amount" or "a great deal" of guilt. The others felt a "small" or "medium" amount of guilt.

Male–Female Sexual Behavior

Young people today start dating and "going steady" earlier than in past generations. These changes have implications for teenage pregnancy. Teens who date earlier (by age 14) are more likely to engage in coitus during high school (Miller et al., 1986). Teens who initiate sexual intercourse earlier are also less likely to use contraception and more likely to incur an unwanted pregnancy. If the young woman decides to keep her baby, she is also more likely to have to leave school and scuttle educational and vocational plans. Early dating does not always lead to early coitus, however. Nor does early coitus always lead to unwanted pregnancies. Still, some young women find their options in adulthood restricted by a chain of events that began in early adolescence.

PETTING Many adolescents use petting to express affection, satisfy their curiosities, heighten their sexual arousal, and reach orgasm while avoiding pregnancy and maintaining virginity.

An overwhelming majority (97%) of teenagers sampled in the Coles and Stokes (1985) survey had engaged in kissing (a form of light petting) by the age of 15. Girls tended to

Larynx A structure of muscle and cartilage that lies at the upper end of the trachea and contains the vocal cords; the voice box.

Gynecomastia Overdevelopment of a male's breasts. (From Greek roots that mean "woman" [*gyne*] and "breast" [*mastos*].)

engage in kissing earlier than boys, perhaps because girls tend to mature faster. By age 13, 73% of the girls and 66% of the boys had engaged in kissing.

Truth or Fiction? REVISITED

It is true that petting is practically universal among adolescents in the United States.

ORAL SEX

> The loose term "oral sex" is, I think, imprecise. Among the teenagers I talked to, there was widespread agreement that what oral sex consisted of, in middle school, was girls performing oral sex on boys. "I guess boys think it would be nasty" to perform oral sex on a girl, conjectured one 13-year-old girl. "It's different. It's just—the body parts—it's something different," said one 13-year-old boy, looking a little bit horrified. (Mundy, 2000; Copyright © 2000 by The Washington Post. Reprinted by permission.)

The incidence of premarital oral sex has increased twofold to threefold since Kinsey's time. About four in ten (41%) of the 17- and 18-year-old girls in the Coles and Stokes (1985) survey of the 1980s reported that they had performed fellatio. About a third of the boys reported performing cunnilingus. Many girls reported that they had engaged in fellatio for the partner's pleasure, not their own.

Table 13.4 highlights some of the results of a national survey reported by Gary Gates and Freya Sonenstein (2000). The incidence of oral sex increases with age. European American and Latino American males are more likely to have engaged in cunnilingus than African American males. Some adolescent couples use oral sex as a means of birth control. As one 17-year-old New York girl put it, "That's what we used to do before we could start having sex, because we didn't have protection and stuff" (Coles & Stokes, 1985, p. 60).

TABLE 13.4

Percentage of never-married males aged 15–19 who report ever having engaged in various sexual activities

	N	MASTURBATED BY A FEMALE	RECEIVED ORAL SEX FROM FEMALE	GAVE ORAL SEX TO FEMALE	HAD ANAL INTERCOURSE	HAD VAGINAL INTERCOURSE
			All Respondents			
Total	1297	53	49	49	11	55
			Age			
15–16	606	38	33	24	7	37
17–19	691	63	62	50	13	68
			Race/Ethnicity			
European American	474	53	51	42	9	50
African American	360	52	47	21	16	78
Latino American	426	48	44	37	16	58

Source: Reproduced with the permission of The Alan Guttmacher Institute from Gary J. Gates & Freya L. Sonenstein. (2000). "Heterosexual Genital Sexual Activity Among Adolescent Males: 1988 and 1995." *Family Planning Perspectives*, 32(6): 295–297, 304.

a CLOSER look

"School Days, School Days," Third Millennium

Morning brings the invitations. The casual ones. So routine are they that she hardly thinks about them, just waves them away like gnats. Today, for example, a boy came up to her in the hall and asked, "When are you going to let me hit that?" "That means, like, intercourse," the girl explains, with a sort of gum-popping matter-of-factness. She is 13.

She is an eighth-grader, fresh-faced, clear-eyed, with light brown hair and fluffy bangs and plucked eyebrows, her voice sweet and straightforward as, one morning in an unused classroom, she sits relating some of the other things guys say to her in the halls of her Montgomery County middle school, nestled in developed farmland in the central part of the county.

"They say, 'What's up with the dome?' " the girl continues, explaining that this is an invitation to perform oral sex, as is the more familiar: "When are you going to give me head?" She tells them never. She laughs. Whatever it takes to put them off. She has not done much more than kiss, though she and her female friends talk about sex a lot, especially oral sex. "They're like, 'It's not that bad once you do it. But it's scary the first time.' I guess they're nervous that they won't do it right. They said they didn't have any pleasure in it. They did it to make the boys happy, I guess."

She thinks that someday she will do it. She thinks that it will be gross.

The serious invitations come in the afternoon, after school, from two boys she knows well, boys who live within walking distance of her house, boys who call her up, or else she calls them, and they come over when her parents are still at work and the only other person in her house is a sibling. When the boys arrive they often say something like what one of them said just the other day: "Let's go to your room. You can give me some head and then we'll go downstairs." To which she replied: "No! You're nasty!"

It is a little complicated to explain who the two boys are. One of them—let's call him Boy A—used to be her boyfriend and is now just her friend. The two of them talk a lot—they're really close, they know each other's life story, he has told her everything about himself and his past, though she's not sure she believes all of it (how much past can an eighth-grader have?), and they've had conversations about oral sex. For example, the one in which she said to him, "If I do it to you and do it wrong, just tell me what I'm doing wrong so I can fix it."

The invitations also come from Boy B. One day, for example, back when she was going out with Boy A, she and Boy B were talking on the phone in the afternoon, and she invited the two of them over, and Boy B "was, like, 'Are you going to give him head?' And I was, like, 'No.' And then he asked about himself—he was, like, 'What about me?' And I was, like, 'No.' I was, like, 'Heck no!' and he was, like, 'Why?' And I was, like, 'Because I don't like you,' and he was, like, 'So? You can still do it!' "

"They always ask," she says. "Even if you say no 700 times, they'll always ask you."

What if the boys were to leave her alone and stop asking? "I would think they didn't like me or something," she says, "or that the other girls are prettier or, like, better than me."

What if she gives in? Just goes ahead and gets it over with? She has thought about this. For example, she knows that if she did it with Boy A, Boy A would tell Boy B, and likewise, if she did it with Boy B, Boy B would tell Boy A. So people would know. She doesn't think it would affect her reputation; you only get a bad reputation if you have sex with every boy who asks you. But the one thing she knows is that if she did it and people found out, her day would be one endless stream of requests. "They would ask me, and ask me, even more than they do now."

In other words, this girl—who asked, for obvious reasons, that her name not be used—is making complex moral calculations all day long, measuring popularity, fending off unwanted commentary, admitting to curiosity, assessing risk. At least until her mother gets home. "How was your day at school?" she usually asks.

Source: From L. Mundy (2000, July 16). "Sex and Sensibility." *The Washington Post online.*

PREMARITAL INTERCOURSE As we enter the new millennium, surveys reveal that about half of the high school students in the United States are sexually active (CDC, 2000e; Gates & Sonenstein, 2000). However, as noted in Table 13.4, African American males (78%) are much more likely than Latino American males (58%) and European American males (50%) to have engaged in sexual intercourse. Many young people feel as though they are caught betwixt and between. On the one hand, adults tell them to wait until they're older, to "just say no." On the other hand, the movies and television programs they see, and the stories they hear from their peers, reinforce the belief that everybody's "doing it."

The incidence of premarital intercourse for females has increased dramatically since Kinsey's day. In Kinsey's time, the sexual double standard held firm. Women were expected to remain virgins until marriage, but society looked the other way for men. By the age of 20, 77% of the single men but only 20% of the single women reported that they had engaged in premarital coitus. Of those still single by age 25, the figures rose to 83% for men but only 33% for women. The discrepancy between males and females is partly explained by the fact that men were often sexually initiated by prostitutes (Hunt, 1974). Rates of premarital coitus among young women should not be confused with promiscuity. Kinsey found that 53% of the females who had engaged in premarital coitus did so with one partner only (Kinsey et al., 1953).

MOTIVES FOR INTERCOURSE Premarital intercourse is motivated by a number of factors. Sex hormones, especially testosterone, activate sexual arousal. Thus the pubertal surge of hormones directly activates sexual arousal, at least among boys (Brooks-Gunn & Furstenberg, 1989). Hormonal changes may also have indirect effects on sexual experimentation (Brooks-Gunn & Furstenberg, 1989). About half of the men (51%) and one quarter of the women (24%) in the NHSLS study reported that their primary reason for the first coital experience was curiosity, or "readiness for sex" (Michael et al., 1994, p. 93).

Hormonal changes stoke the development of secondary sex characteristics. Adolescents whose secondary sex characteristics develop early may begin dating earlier, which may increase the likelihood of progressing toward sexual intercourse at an earlier age. Some early maturers are pressured into dating or sex, whether they are psychologically ready or not.

Many psychological motives are involved in sexual activity, including love, desire for pleasure, conformity to peer norms, seeking peer recognition, and the desire to dominate someone (Browning et al., 2000). Many adolescents engage in sexual intercourse because of feelings of love (Browning et al., 2000; Thompson, 1995). The NHSLS study found that affection for the partner was the primary reason for first intercourse among nearly half (48%) of the women and one quarter (25%) of the men sampled (Michael et al., 1994). Betsy believed that she was in love:

> I was seventeen when I had my first sexual experience. I had been going out with my boyfriend for about five months, during which time he had been continually pressuring me to have sex. He made it seem as though I had to comply or he would end the relationship. Because I was deeply in love with him (or so I thought), I allowed it to happen. (Copyright © 1991 by McIntyre, Formichella, Osterhout, and Gresh by arrangement with The Denise Marcil Literary Agency, Inc., p. 64)

Adolescents may consider intercourse a sign of maturity, a way for girls to reward a boyfriend for remaining loyal, or a means of punishing parents (Thompson, 1995). Some adolescents engage in coitus in response to peer pressure, especially from close friends (Dickson et al., 1998). Adolescents whose friends have engaged in sexual intercourse are more likely to engage in intercourse themselves. Coles and Stokes (1985) found that for 78% of the virgins in their national sample, few if any of their friends had engaged in intercourse. This was true of only 28% of the nonvirgins.

Sometimes the pressure comes from dating partners. About one quarter (24%) of the women sampled in the NHSLS study said that they went along with intercourse only for the sake of their partners (Michael et al., 1994):

> MEGAN (18, California): I have felt pressure before. My first boyfriend pressured me because he knew I loved him and that he could take advantage of my feelings. I was blinded by my feelings and I had sex with him. I hated it.
>
> AMY (18, Washington, D.C.): I was sexually pressured by my second boyfriend. He didn't love me, but he did want to have sex. I helped him sneak into my room in the middle of the night. Just before we were about to have sex, I realized that it wasn't something I wanted to do. I wanted my first time to be with someone I loved and who loved me. I stopped him, although he tried everything to get me to say yes. The next day we broke up, and I couldn't have been happier. (Copyright © 1991 by McIntyre, Formichella, Osterhout, and Gresh by arrangement with Avon Books, pp. 4–6)

a CLOSER look

The First Time

MARK: As we had no place to go, we went out into the woods with several blankets and made love. It was like something out of a Woody Allen movie. I couldn't get my pants off because I was shaking from nerves and from the cold. The nerves and cold made it all but impossible for me to get an erection and then after I had and we made love I couldn't find the car keys.

AMY: My first sexual experience occurred after the Junior Prom in high school in a car at the drive-in. We were both virgins, very uncertain, but very much in love. We had been going together since eighth grade. The experience was somewhat painful. I remember wondering if I would look different to my mother the next day. I guess I didn't because nothing was said. (Morrison et al., 1980, p. 608)

For those who do not remain celibate, there must be a first time. Given the inexperience and awkwardness of at least one member of the couple, and frequent feelings of guilt and fear, it is not surprising that most people, like Mark and Amy, don't get it quite right the first time. Adolescent boys and girls often report different concerns about first intercourse. The girl is more likely to be concerned about whether she is doing the right thing. The boy is more likely to be concerned about whether he is doing the thing right. Women are more likely than men to be physically and psychologically disappointed with the experience and to feel guilty afterwards (Darling et al., 1992; Sprecher et al., 1995). One study of 300 sexually experienced college students found that only 28% of the women considered their first encounter physically or psychologically satisfying. Yet 81% of college men were physically satisfied, and 67% were psychologically satisfied (Darling & Davidson, 1986).

Negative attitudes among women toward their first coital experience may have cultural roots. A comparison between American and Swedish women showed that American women reported more negative emotional reactions to their first premarital coitus (Schwartz, 1993). In general, the Swedish have more permissive attitudes about sex than do people in the United States. Negative emotional consequences of first intercourse reflect cultural norms or standards as well as the act itself. Of course, it is also possible that Swedish men are more responsive to their partners' needs.

Coles and Stokes (1985) found that most adolescent boys (60%) reported feeling "glad" after their first intercourse (see Table 13.5). Most adolescent girls (61%) expressed ambivalence. One in ten girls (11%) reported feeling "sorry," compared to only 1% of boys. Some females feel guilty about ending their virginity. Others find first intercourse painful or uncomfortable, in part because of the tearing of the hymen, in part because penetration may have been rushed or forced. The pain was more than one 15-year-old New York girl had anticipated:

[I] wasn't expecting it to hurt that much. It was like total pain. Even after the first minutes of pain, it's still like you're too sore to enjoy anything. I didn't expect that at all. (Coles & Stokes, 1985, p. 74)

About 8% of the men in the NHSLS study say they went along with intercourse for the sake of their partners (Michael et al., 1994). As one young man describes it,

MATT (18, New York): My girlfriend pressured me and I didn't handle it very well. I submitted so she wouldn't be mad or disappointed. (Copyright © 1991 by McIntyre, Formichella, Osterhout, and Gresh by arrangement with The Denise Marcil Literary Agency, Inc., p. 65)

FACTORS IN PREMARITAL INTERCOURSE Many young people abstain from premarital coitus for religious or moral reasons (Belgrave et al., 2000). Family influences are important determinants of adolescent sexual experience (White & DeBlassie, 1992). Other reasons include fear of being caught, of pregnancy, or of disease.

Studies of African American adolescent females have found that girls who are not sexually active, or who engage in less risky sexual activities, tend to be younger and more career-oriented, to live in two-parent households, to hold more conservative values about sexuality, and to be more influenced by family values and religion (Belgrave et al., 2000; Keith et al., 1991).

Teens who have higher educational goals and do better in school are less likely to engage in coitus than less academically oriented teens (Belgrave et al., 2000). The causal

TABLE 13.5

Feelings about first intercourse (percentage)

	Sorry	Ambivalent	Glad	No Feelings
Males	1	34	60	5
Females	11	61	23	4

Source: From *Sex and the American Teenager* by R. Coles and G. Stokes. Copyright © 1985 by Rolling Stone Press. Reprinted by permission of HarperCollins Publishers, Inc., and Wenner Media.

Yet for some, the quality of the relationship tempered pain:

> We were both so excited. We hadn't been able to sleep the night before. I can't remember that much leading up to it, but we had sex a few times—I guess about three times—that night. He really enjoyed it; I found it emotionally nice, but painful. It was like a *good* hurt, but still it hurt; it was uncomfortable. But it was something we both felt really good about. (Coles & Stokes, 1985, p. 74)

Young women are more likely to find their first coital experience satisfying when their partners are loving, gentle, and considerate (Weiss, 1983).

First intercourse is often awkward, even fumbling. The partners are still learning about their own sexual responses and how to please each other:

> KAREN (23, New York): I had sexual intercourse for the first time at age 18. My boyfriend and I had been going out for a year. For several months before we had intercourse we engaged in a lot of petting but not much genital contact. I was the more aggressive partner and I was the one who suggested we have intercourse. It was very awkward; the first time we tried, he couldn't get in. (Copyright © 1991 by McIntyre, Formichella, Osterhout, and Gresh by arrangement with Avon Books, pp. 50–51)

Let us note some gender differences in choice of first partners. For first-time intercourse, survey evidence shows that females are more likely than males to report having been in a committed relationship with their partners (Darling et al., 1992; Sprecher et al., 1995). In the study of some 1,600 college students by Susan Sprecher and her colleagues (1995), 60% of the women reported that their first partner was someone they were dating seriously, compared to 36% of the men. Men were more likely than women (23% versus 8%) to engage in first intercourse with someone they had been seeing for less than a week. Men experienced more pleasure, largely because they were more likely to reach orgasm. Women experienced more guilt. These gender differences are consistent with the traditional double standard that accords greater sexual freedom to men. Men are expected to "sow their wild oats" with casual partners. Women are expected to save themselves for a man with whom they share strong emotional ties and an enduring relationship.

connection between school performance and premarital sex is difficult to isolate, because adolescents who do well in school are also more likely to come from more stable families.

The likelihood of premarital intercourse is also linked to early dating, especially early steady dating (Belgrave et al., 2000). Adolescents who begin dating earlier may be more likely to progress through stages of petting to coitus.

The quality of the relationship between teens and their parents is important (Belgrave et al., 2000; Mundy, 2000). Teens who feel that they can talk to their parents are less likely to engage in coitus than those who describe communication with their parents as poor. Adolescents whose parents are permissive and impose few rules and restrictions are more likely to engage in premarital intercourse (Miller et al., 1986; Mundy, 2000). Parents who show interest in their children's behavior and communicate their concerns and expectations with understanding and respect may best influence their children to show sexual restraint.

TEENAGERS' VALUES AND THE SEXUAL DOUBLE STANDARD Do young people today subscribe to the traditional double standard that allows greater sexual freedom for men than for women? It may be premature to drive the nails into the coffin of the sexual double standard, but a common standard for judging acceptability of premarital sex may be emerging,

Ethnicity, Teenagers, and HIV

> "I was young and stupid. I wasn't scared of anything. It was just about having fun. I didn't listen to older people. I thought they didn't know what they were talking about."
>
> —Shernika, 16

> "I was scared. I was like, 'I shouldn't have done this.' But he didn't look like nothing was wrong with him."
>
> —George, 14

> "Everybody I've been with, I knew a long time, since childhood. I didn't think they'd have something like that."
>
> —Dicki, 16

They heard endless warnings about the dangers of intravenous drug use, about the wisdom of using condoms. Many even knew older relatives, neighbors, and friends who had died because of AIDS.

But somehow, it didn't add up. Not me, they said. Couldn't happen to me.

And so a 16-year-old girl became infected with HIV because she had unprotected sex with a friend. A 14-year-old boy never took his condom out of its wrapper, and now he is HIV-positive. A 16-year-old girl trusted her boyfriend, a drug user. She shouldn't have.

The experiences of African American teenagers are far from unique. African American youths have become the new face of HIV, making up about two thirds of the new cases among people under 25, according to recent studies by the CDC (2000b, 2000e).

"The disease is disappearing from the mainstream and becoming a disease of kids who are disenfranchised anyway," said Lawrence D'Angelo, who runs the Burgess Clinic for HIV-infected adolescents at Children's Hospital in Washington.

Risky sexual behavior is nothing new among young people. But medical professionals find it disturbing that the risk taking continues, despite extensive educational campaigns, and that African American youths are paying an especially high price. That is a major change from the early 1980s, when gay European American men made up the majority of young people infected with HIV.

"For many of our kids, HIV has become just one of the many problems in their lives, like are they going to get a good meal, who are they going to live with, are they going to school?" D'Angelo said.

D'Angelo and Ligia Peralta, who treats HIV-infected adolescents in Baltimore, began to notice the change in the 1990s when growing numbers of young African American women, most of them poor, began entering their programs.

"Theoretically, they do understand the risks," said Peralta, who runs Star Track, an HIV and AIDS clinic for adolescents at the University of Maryland Medical Center in Baltimore. "However, from there to recognizing that my partner, the person that I choose, may be infected with HIV, there's the major disconnect."

The disease is spread casually among young people, the doctors have found in the course of conducting a long-term HIV study. Once infected youths discover that they have HIV, many are unwilling to admit it to sexual partners.

at least in the university. A study of 666 undergraduate students in a Midwestern university showed no evidence of the double standard; students were not differently disposed toward the acceptability of premarital sex for one gender or the other (Sprecher, 1989).

On the other hand, this was a university sample. Younger people, less well educated, may be more prone to adhering to the traditional double standard. *The Washington Post* reporter Liza Mundy (2000) writes about middle schoolers in suburban Washington, DC, in the year 2000:

> There is still a strong double standard . . . that reduces the likelihood that sex, for a girl, is going to feel like an empowered act. Even if she feels empowered doing it, she's unlikely to feel that way afterward, when word gets around.
>
> "If a guy's having sex it's like, oh, who cares," said one eighth-grade girl. "But if a girl is, it's like, oh, she's a slut, she's a 'ho, she's nasty." (Copyright © 2000 by The Washington Post. Reprinted by permission.)

Male–Male and Female–Female Sexual Behavior

About 5% of the adolescents in the Coles and Stokes (1985) national survey reported sexual experiences with people of their own gender. More than nine out of ten experiences among adolescents of the same gender are between peers. Seduction of adolescents by gay

D'Angelo said that only about half tell their regular partners that they are HIV-positive, and almost none bothers to tell an occasional partner.

And with a stigma still attached to homosexuality in some communities, young African American men who have sex with other men tend to be less likely than their European American peers to identify themselves as gay, possibly causing them to ignore HIV-prevention messages aimed at gay men, said Helene Gayle, director of the CDC's National Center for HIV, STI and TB Prevention. "I think it means we have to make sure our prevention messages are keeping pace with the times and keeping pace with the population at risk," Gayle said.

Much to Think About. Adolescents need to think about the possibilities of unwanted pregnancy and of contracting HIV/AIDS and other STIs when they make decisions about sexual behavior.

The Personal Fable

Some say yet another factor—the sense that young people have that they can live forever—also is at work. "All youth—rich, poor, black, white—have this sense of invincibility, invulnerability," says Ronald King (2000), executive director of the HIV Community Coalition of Metropolitan Washington, explaining why many adolescents who know the risks still expose themselves to HIV infection

The developmental psychologist Jean Piaget noted that adolescents appear to believe in a **personal fable**—the belief that one's feelings and ideas are special, even unique, and that one is invulnerable. The personal fable is apparently connected with adolescent behavior patterns such as showing off and taking risks. Many adolescents have an "It can't happen to me" attitude; they assume they can smoke without risk of cancer or engage in sexual activity without risk of STIs or pregnancy.

Peralta and D'Angelo have found that adolescent girls in all groups are especially vulnerable to sexually transmitted infections, such as HIV, because the cervix at their age is not fully mature and is more susceptible to infection. Also, many of the young men and women in both programs were sexually abused as children and became sexually active at an early age. Most of the women also were infected by men 10 to 20 years older.

Although some national studies show that condom use has increased among young people in recent years and that they are delaying sex until their mid-teens, a recent survey of 4,500 patients at Children's Hospital showed that the average age when boys begin having sex is 12; for girls, the average age is 13.

Source: Adapted from Frazier, L. (2000, July 16). "The New Face of HIV is Young, Black. *The Washington Post*, p. C01; and Rathus, S. A. (2002). *Psychology in the New Millennium.* Fort Worth, TX: Harcourt.

male and lesbian adults is relatively rare. Most adolescent sexual encounters with people of the same gender are transitory. They most often include mutual masturbation, fondling, and genital display.

Many gay males and lesbians, of course, develop a firm sense of being gay during adolescence. Coming to terms with adolescence is often a difficult struggle, but it is frequently more intense for gay people (Baker, 1990) (see Chapter 10). Adolescents can be particularly cruel in their stigmatization, referring to gay peers as "homos," "queers," "faggots," and so on. Many adolescent gays therefore feel isolated and lonely and decide to cloak their sexual orientation. Many do not express their sexual orientation at all until after their high school years.

Adding to the strain of developing a gay identity in a largely hostile society is the threat of AIDS, which is all the more pressing a threat to young gay males because of the toll that AIDS has taken on the gay male community (Baker, 1990).

TEENAGE PREGNANCY

About 10% of American girls of ages 15 to 19 become pregnant each year (CDC, 2000e). This amounts to one in five sexually active girls and nearly 800,000 pregnancies a year, resulting in 500,000 births to young women who are generally incapable of

Personal fable The belief that one's feelings and ideas are special and that one is invulnerable.

providing children with the care and resources they need in today's world (CDC, 2000e, 2000f).

It is true that about 800,000 adolescent girls in the United States become pregnant each year. ■

Although these figures are cause for concern, the rate of births for teenagers actually dropped modestly during the 1990s (Centers for Disease Control, 2000e). Some pregnant teenagers plan their pregnancies, but the great majority do not. Nine in ten pregnancies among unmarried teenagers are unplanned (Centers for Disease Control, 2000e).

The consequences of unplanned teenage pregnancies can devastate young mothers, their children, and society at large. Even young people themselves perceive teenage parenthood to be disastrous (Moore & Stief, 1992). Teenage mothers are more likely to live in poverty and to receive welfare than their peers (Grogger & Bronars, 1993). Poverty, joblessness, and lack of hope for the future are recurrent themes in adolescent pregnancy (Desmond, 1994). Half of teenage mothers quit school and go on public assistance (Kantrowitz, 1990a). Few receive consistent emotional or financial help from the fathers, who generally cannot support themselves, much less a family. Working teenage mothers earn just half as much as those who give birth in their 20s (CDC, 2000e). Barely able to cope with one baby, many young mothers who give birth at age 15 or 16 have at least one more baby by the time they are 20. Among teenage girls who become pregnant, nearly one in five will become pregnant again within a year. More than 3 in 10 will have a repeat pregnancy within two years. Undereducated, unskilled, and overburdened, these young mothers face a constant uphill struggle.

Medical complications associated with teenage childbearing are highest of any group of fertile women except for those in their late 40s (Hess et al., 1993). In addition to high rates of miscarriage and stillbirths, children born to teenagers are at greater risk of prematurity, birth complications, and infant mortality (Fraser et al., 1995; Goldenberg & Klerman, 1995). These problems are often the result of inadequate prenatal care. However, even when prenatal care is adequate the children of teenagers are at greater risk than those of women in their 20s (Fraser et al., 1995; Goldenberg & Klerman, 1995).

The children of teenage mothers are at greater risk of physical, emotional, and intellectual problems in their preschool years, owing to poor nutrition and health care, family instability, and inadequate parenting (Furstenberg et al., 1989; Hechtman, 1989). They are more aggressive and impulsive as preschoolers than are children of older mothers (Furstenberg et al., 1989). They do more poorly in school. They are also more likely to suffer maternal abuse or neglect (Felsman et al., 1987; Kinard & Reinherz, 1987).

A number of factors have contributed to the increase in teenage pregnancy, including a loosening of traditional taboos on adolescent sexuality (Hechtman, 1989). Impaired family relationships, problems in school, emotional problems, misunderstandings about reproduction or contraception, and lack of contraception also play roles (Hechtman, 1989). Some adolescent girls believe that a baby will elicit a commitment from their partners or fill an emotional void. Some become pregnant as a way of rebelling against their parents. Some poor teenagers view early childbearing as the best of the severely limited options they perceive for their futures. But the largest number become pregnant because of misunderstandings about reproduction and contraception or miscalculations about the odds of conception. Even many teens who are relatively well informed about contraception fail to use it consistently (Hechtman, 1989).

More attention has been focused on teenage mothers, but young fathers bear an equal responsibility for teenage pregnancies. A survey based on a nationally representa-

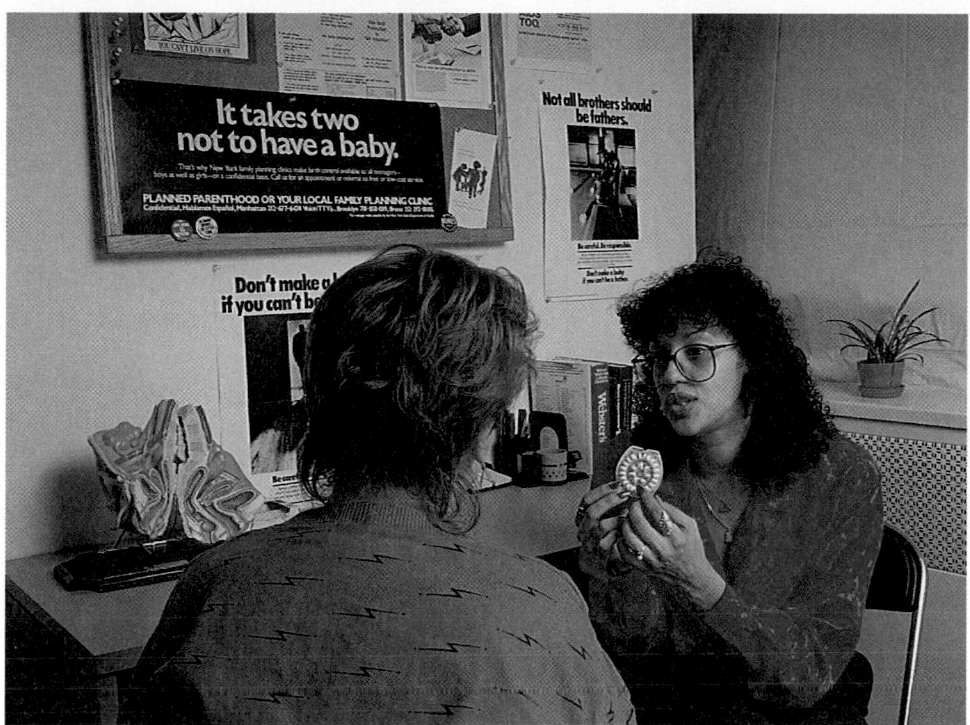

Teenage Pregnancy. Even with the recent decline in the rate of pregnancy among teenagers, nearly 10% of girls in the United States aged 15 to 19 become pregnant each year. Teenage mothers are more likely to live in poverty and to receive welfare than other girls their age. Half of them quit school and rely on public assistance. Few receive meaningful help from their babies' fathers.

tive sample of 1,880 young men aged 15 through 19 showed that socioeconomically disadvantaged young men in particular appeared to view paternity as a source of self-esteem and were consequently more likely than more affluent young men to say that fathering a child would make them feel like a real man and that they would be pleased—or at least not as upset—with an unplanned pregnancy (Marsiglio, 1993a). Thus poor young men were less likely to have used an effective contraceptive method during their most recent sexual experience.

Teenage fathers are more likely than their peers to have problems in school—both behavioral and academic—and to hold more pessimistic attitudes toward the future (Hanson et al., 1989). Steady dating also increased the risk of teenage fatherhood.

Contraceptive Use Among Sexually Active Teens

Sexually active teenagers use contraception inconsistently, if at all (Mundy, 2000). Contraception is most likely to be used by teens in stable, monogamous relationships (Baker et al., 1988). Even teens in monogamous relationships tend to use ineffective methods of contraception or to use effective methods inconsistently, however (Polit-O'Hara & Kahn, 1985).

Various factors determine use of contraceptives (Beck & Davies, 1987). Teenage girls who engage in more frequent intercourse are more likely to use contraception and to use more effective methods (DuRant & Sanders, 1989). Teens whose peers use contraceptives are more likely to do so themselves (Jorgensen et al., 1980). Older teenagers are more likely than younger ones to use contraception (Mosher & Bachrach, 1987). Younger teens who are sexually active may be less likely to use contraception because they lack information about contraception and because they do not always understand the repercussions of their actions (Handler, 1990). Younger teens may also have less access to contraceptives.

Teenage boys are more likely than teenage girls to know how to obtain and use condoms correctly, according to the results of a California survey of more than 1,000 high school students (Leland & Barth, 1992). Boys are also more likely to have used birth control during their first or most recent sexual experience. It appears that girls are more uncomfortable than boys in obtaining or using contraception, especially condoms (Leland & Barth, 1992).

Poor family relationships and communication with parents are associated with inconsistent contraceptive use (Brooks-Gunn & Furstenberg, 1989; Mundy, 2000). Poor performance in school and low educational ambitions predict irregular contraceptive use; they also predict early sexual initiation.

When asked to explain why they don't use contraceptives, sexually active teens often cite factors such not having intercourse often enough to use it and the fact that it gets in the way of sexual spontaneity (Mundy, 2000). Some teenagers get "carried away" and do not wish to disrupt sex by obtaining or applying contraceptives.

Myths also decrease the likelihood of using birth control. Some adolescents believe that they are too young to become pregnant. Others believe that pregnancy results only from repeated coitus or will not occur if they are standing up. Still other adolescents simply do not admit to themselves that they are engaging in coitus.

Teenagers who focus on the long-term consequences of their actions are more likely to use contraceptives. The quality of the relationship is also a factor. Satisfaction with the relationship is associated with more frequent intercourse *and* more consistent use of contraception (Jorgensen et al., 1980). More consistent contraceptive use is found in relationships in which the young woman takes the initiative in making decisions and resolving conflicts.

COMBATING TEENAGE PREGNANCY: A ROLE FOR THE SCHOOLS? Various means have been recommended to combat the problem of teenage pregnancy, including universal sex education, free contraceptive services for teenagers, open discussion of sex between parents and children, and dissemination through the media of information about responsible sex practices and contraception. Given the effects of sex education in other industrialized countries, many helping professionals believe that the rate of teenage pregnancy and the spread of sexually transmitted diseases in the United States could be curtailed through sex education about contraception and the provision of contraceptives.

Pregnancy prevention programs in the schools range from encouraging teens to delay sex ("saying no to early sex") through providing information about contraception to distributing condoms or referring students to contraceptive clinics (Furstenberg et al., 1989; Richardson, 1997). The vast majority of sex educators (86%) recommend abstinence to their students as the best way to prevent pregnancy and AIDS (Kantrowitz, 1990b). Fewer than half inform their students how to obtain contraceptives. Three out of four large school districts in the United States provide some instruction about methods of contraception and the use of condoms to prevent the spread of AIDS and other sexually transmitted diseases.

Evidence supports the effectiveness of programs that counsel abstinence, at least among younger teens. A school-based sex education program that focused on the development of skills needed to resist social and peer pressure to initiate sexual activity encouraged younger (junior high school) teens to postpone sexual involvement (Howard & McCabe, 1990). Program participants also had fewer pregnancies than students who did not participate in the program. However, programs that rely on the "just say no" model to encourage abstinence are less effective in persuading high school students to abstain from sexual activity (Wilson & Sanderson, 1988).

The Washington, DC Campaign to Prevent Teen Pregnancy is working to cut the rate of teenage pregnancy by encouraging parents and other adults to discuss the issues involved with teenagers (Donovan, 2000). The Web site of the Campaign has links to information and resources on teen pregnancy. For more information, go to www.teenpregnancydc.org.

TEENAGERS AND SEX: WHERE DO WE STAND?

> I've wondered what it must be like, for girls, and for boys, too, to be coming of age in an era shaped by both the '60s sexual revolution and the '80s AIDS onslaught; coming of age in the time of "Dawson's Creek" and "Sex in the City"; a time of HIV and Eminem; a time when Lorena and John Bobbitt and Monica Lewinsky and Bill Clinton have done more than anybody else to insinuate terms like "penis" and "oral sex" into family newspapers like this one. A time when the media—all media, even mainstream media—are more sexualized than they've ever been, and yet, at the same time, the consequences of sex are depicted so grimly, by cultural conservatives and liberals alike. What does it feel like—how does a kid respond—when the messages are so mixed and so insistent? (Mundy, 2000; Copyright © 2000 by The Washington Post. Reprinted by permission.)

There are times when the media inundations seems to suggest that everyone is sexually active, being infected with STIs, getting pregnant. Not so. A federal survey of more than 16,000 high school students found that in 1997, 48% students reported that they had become sexually active, as compared with 54% in 1991 (Kolbe, 1998). Sixteen percent of the respondents reported having sex with four or more partners, down from the 19% reported in 1991. Why the decline? Figure 13.2 shows ethnic differences

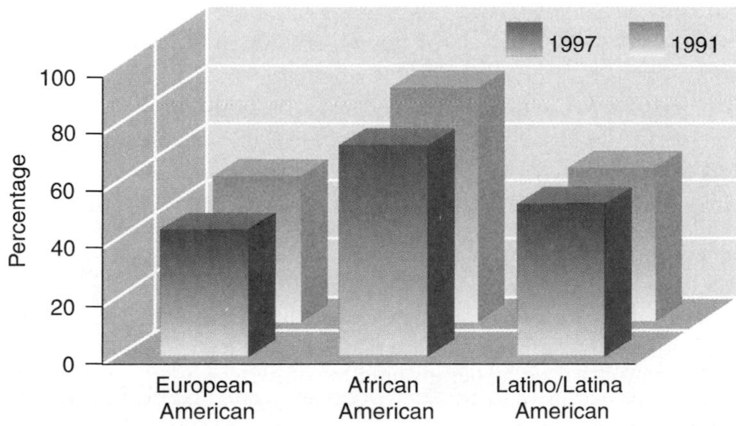

FIGURE 13.2

Percentage of U.S. High School Students Who Have Become Sexually Active, According to Ethnicity A poll by the Centers for Disease Control and Prevention found that fewer U.S. high school students were sexually active in 1997 than in 1991. African American students were most likely to be sexually active.

among students in 1991 and 1997. It reveals that the percentage of African American students who had become sexually active fell from 81% to 73% between 1991 and 1997. The percentage for European American students fell from 50% to 44%, and for Latino and Latina American students, the rate dropped from 53% to 52%. Lloyd Kolbe (1998), director of the CDC's Division of Adolescent and School Health (which released the report), speculated that efforts by school and parents to inform high school students about the risks of pregnancy and sexually transmitted infections had begun to pay off.

The 1990s saw a decline in the teenage pregnancy rate because of increased use of contraception and the leveling off of sexual activity among teenagers (Kolbe, 1998; Meckler, 1998). Still, the rate of teenage pregnancy remains much higher in the United States than in many other developed countries. The U.S. rate is twice as high as the comparable rate in England and Canada. It is nine times as high as the rate in Japan (Brody, 1998e).

Adolescent girls by and large obtain little advice at home or in school about how to resist sexual pressures. Nor do most of them have ready access to effective contraception. According to the Alan Guttmacher Institute (1998), fewer than half of the adolescents who are sexually active report using contraceptives consistently. As a result, about 800,000 teenage girls get pregnant each year, resulting in 500,000 births (CDC, 2000b, 2000e). This figure includes nearly 20% of the sexually active girls aged 15 through 19. Nearly 3 million teenagers in the United States also obtain a sexually transmitted infection each year. One-fifth of the Americans diagnosed with AIDS are in their 20s, and most of them were infected as adolescents.

According to a study published in the *British Medical Journal*, there are many reasons for the early age of first intercourse. For example, "Children are exposed to sexual images through the media. . . . Social and peer pressure may arise from the portrayal of sex as glamorous, pleasurable and adult, while negative consequences and the responsibilities involved in sexual relationships are seldom portrayed" (Dickson et al., 1998).

The New York Times columnist Jane Brody (1998e) asks, "Is this the message we want to convey to American youth—that sex is something you try, like roller-blading or water skiing, to see what it's like or to add to your roster of achievements?"

Remaining a virgin until married is no longer as important as it was early in the 20th century. Nor does becoming a single mother or having an abortion carry the stigma it once did. Nevertheless, there are problems. A study published in *Family Planning Perspectives* found that girls who had first intercourse at an early age, especially those who became pregnant and had children, tended to have low self-esteem and to be depressed (cited in Brody, 1998e).

Moreover, girls who become sexually active early are often not doing so voluntarily:

> It would be naive to think that increased sexual activity by young girls is part of some post-feminist equal opportunity movement. I sat in on one sex-ed class where a group of high school girls could not identify the female organ that is "the center of sexual excitement and pleasure" and another where a group of eighth-grade girls all agreed that it was normal for boys to masturbate, but not for girls. In three months of reporting, I heard very few girls make any bold expression of sensuality, apart from one who, during a discussion of risky behavior, timidly expressed the idea that "hugging feels—mmmmmmm!" whereupon she was roundly chastised by her group. One of her classmates pointed out that even hugging can be risky if the boy gets ambitious: "There are some people that be feeling on you when they're hugging!"
>
> Mostly, what I saw was girls being pressured to experiment, in some cases girls curious to experiment, and, invariably, girls condemned when they did. There is still a strong double standard, especially in this age group, that reduces the likelihood that sex, for a girl, is going to feel like an empowered act. Even if she feels empowered doing it, she's unlikely to feel that way afterward, when word gets around. (Mundy, 2000; Copyright © 2000 by The Washington Post. Reprinted by permission.)

Programs like those developed by Girls Inc., Growing Together, and Will Power/Won't Power seem to be helpful in delaying sexual activity among adolescent girls. Growing Together involves mother–daughter workshops that enhance communication about sex and other sensitive matters. Will Power/Won't Power is an assertiveness training program that helps teenagers resist sexual pressure without jeopardizing their relationships with their peers. Both programs provide information about the problems that adolescent girls encounter when they become pregnant. For information about the programs, contact Girls Inc. at (212) 509-2000 or see the Web site at www.girlsinc.org.

In this chapter we have chronicled sexuality in childhood and adolescence. In the next chapter we continue our journey through the life span.

Human Sexuality ONLINE //

Web Sites Related to Sexuality in Childhood and Adolescence

This is "ScarleTeen," A sex education Web site for teenagers. The people at ScarleTeen write that they provide "information which educates in ALL aspects of positive sexuality, including birth control, safe sex and sexually transmitted diseases, masturbation and self-pleasuring, anatomy, diverse sexual orientation and identification, sexual and romantic relationship and communication tools, and care and compassion in sexual technique and practice." They claim to have a "nonjudgmental and unbiased attitude of tolerance and understanding for teens, whether they choose to be sexually active or abstain." They are not into casual sex; they have "tools to encourage abstinence until readiness, such as information on masturbation and self-pleasuring, and to do so from a standpoint of positive sexuality, rather than anti-sexual behavior and sexual punishment." They encourage teenagers "to know as much as possible, and from an educated standpoint, make sound choices based on personal ethics and values gleaned from family, role models and the individual teen self." They have numerous bulletin boards, explain how to obtain condoms and so forth, and discuss STIs, etc.
www.scarleteen.com/

The Web site of the American Academy of Pediatrics (AAP). From the Web site: "the American Academy of Pediatrics (AAP) and its member pediatricians dedicate their efforts and resources . . . to attain optimal physical, mental and social health and well-being for all infants, children, adolescents and young adults." Check out the Press Release Archive and the Policy Statements.
www.aap.org

The Web site of SIECUS (The Sexuality Information and Education Council of the United States). Check out especially the "School Health Education Clearinghouse," the section "For Parents and Other Adults," and the section "For Teens."
www.siecus.org

A Web site administered by The Kaiser Family Foundation and Children Now. These organizations have launched a national campaign to encourage parents, and give them the tools, to talk to their kids early and often about tough issues such as sex, drugs, alcohol, AIDS, and violence.
www.talkingwithkids.org

The Web site of The National Parenting Center. This Web site contains many short articles written by child-rearing experts. Type in "teenager" or "adolescent" in the search area to find articles listed under individual categories such as "Talking About Sex," "Keeping the Lines of Communication Open," "Tips on Communicating with Teens and Adolescents," and "Privacy, Stress, and Setting Limits."
www.tnpc.com

The Web site of the National Campaign to Prevent Teen Pregnancy. The goal is to prevent teen pregnancy by supporting values and stimulating actions that are consistent with a pregnancy-free adolescence. The goal is to reduce the teen pregnancy rate by one third between 1996 and 2005. This site offers "Tips for Parents," "Tips for Parents from Teens," and "Tips for Teens from Teens."
www.teenpregnancy.org

The Web site of New York University's Child Study Center. The "Parenting" and "Is My Kid OK?" sections contain advice for parents on how to handle violence and concerns about a child's friends and friendships. The site also gives a rundown of different parenting styles and a section on "Matching Parenting Styles/Children's Temperaments."
www.aboutourkids.org

summing up

INFANCY (0 TO 2 YEARS): A SEARCH FOR THE ORIGINS OF HUMAN SEXUALITY

■ The Infant's Capacity for Sexual Response

Fetuses have been found to have erections and to suck their fingers. Stimulation of the genitals in infancy may produce sensations of pleasure. Pelvic thrusting has been observed in humans as early as 8 months of age. Masturbation may begin as early as 6 to 12 months. Some infants seem capable of sexual responses that closely resemble orgasm.

■ Masturbation

Self-stimulation for pleasure (masturbation) typically occurs among children as young as 6 to 12 months of age.

■ Genital Play

U.S. children typically do not engage in genital play with others until about the age of 2.

EARLY CHILDHOOD (3 TO 8 YEARS)

■ Masturbation

Statistics concerning the incidence of masturbation at ages 3 to 8 are inconclusive.

■ Male–Female Sexual Behavior

In early childhood, children show curiosity about the genitals and may play "doctor."

■ Male–Male and Female–Female Sexual Behavior

Same-gender sexual play may be more common than heterosexual play and does not presage adult sexual orientation.

PREADOLESCENCE (9 TO 13 YEARS)

Preadolescents tend to socialize with best friends and with large groups and to become self-conscious about their bodies.

■ Masturbation

Masturbation is apparently the primary means of achieving orgasm during preadolescence for both genders.

■ Male–Female Sexual Behavior

Preadolescent sex play often involves mutual display of the genitals, with or without touching. Group dating and mixed-gender parties often provide preadolescents with their first exposure to heterosexual activities.

■ Male–Male and Female–Female Sexual Behavior

Much preadolescent same-gender sexual behavior involves sexual exploration and is short-lived.

■ Sex Education and Miseducation

Despite the increased availability of sex education programs, peers apparently remain the major source of sexual information.

ADOLESCENCE

Adolescence is bounded by the advent of puberty at the earlier end and by the capacity to take on adult responsibilities at the later. The conflicts and distress experienced by many adolescents apparently reflect the cultural expectations to which they are exposed.

■ Puberty

Pubertal changes are ushered in by sex hormones. Puberty begins with the appearance of secondary sex characteristics and ends when the long bones make no further gains in length. Once puberty begins, most major changes in primary sex characteristics occur within three years in girls and within four years in boys.

■ Masturbation

Masturbation is a major sexual outlet during adolescence.

■ Male–Female Sexual Behavior

Adolescents today date and "go steady" earlier than in past generations, a change that has apparently increased the incidence of teenage pregnancy. Many adolescents use petting as a way of achieving sexual gratification without becoming pregnant or ending one's virginity. The incidence of premarital intercourse, especially for females, has increased dramatically since Kinsey's day.

■ Male–Male and Female–Female Sexual Behavior

Most adolescent same-gender sexual encounters are transitory. Coming to terms with adolescence is often more intense for gay males and lesbians, largely because gay male and lesbian sexual orientations are stigmatized in our society.

TEENAGE PREGNANCY

About 800,000 adolescents in the United States become pregnant each year. Rates of teenage pregnancy are connected with socioeconomic status.

■ Contraceptive Use Among Sexually Active Teen

Sexually active teenagers use contraception inconsistently, if at all. Thus more than 1 million teenage girls in the United States become pregnant each year.

questions for critical thinking

1. How did you feel when you saw that we were covering sexuality during infancy? Did it seem to you that we ought not be discussing sexuality during infancy (or childhood)? Explain.
2. Would you characterize your own sociocultural background as sexually permissive or restrictive? Why?
3. How old were you when you learned "the facts of life"? How did you learn them? Were there inaccuracies in your early sources of information? Explain.
4. Do you support sex education in the schools? If so, what topics should be covered? Which should not? Explain.
5. What are the attitudes of most people from your sociocultural background toward premarital sexuality? Do you share their attitudes? Why or why not?

chapter

14

Sexuality in Adulthood

Pablo Picasso. *The Embrace (L'Etreinte)*. Autumn 1900. Oil on cardboard, 20 7/8 × 22 in. (53 × 56 cm). © Copyright ARS, NY. Pushkin Museum of Fine Arts, Moscow, Russia. Credit: Scala/Art Resource, NY. Credit: © 2002 Estate of Pablo Picasso/Artists Rights Society (ARS), NY.

outline

Truth or Fiction?

SINGLEHOOD

COHABITATION: DARLING, WOULD YOU BE MY POSSLQ?

Some Facts About Cohabitation

Reasons for Cohabitation

Cohabitation First and Marriage Later: Benefit or Risk?

MARRIAGE

Historical Perspectives

A World of Diversity Russian Region Stops Men from Marrying Four Times

Why Do People Marry?

A World of Diversity Housewives Urged to Strike

Types of Marriage

Whom We Marry: Are Marriages Made in Heaven or in the Neighborhood?

MARITAL SEXUALITY

The Sexual Revolution Hits Home

Sexual Satisfaction

EXTRAMARITAL SEX

Patterns of Extramarital Sex

Attitudes Toward Extramarital Sex

Human Sexuality Online "Someone to Watch Over Me"? Yes—Snoopware

Effects of Extramarital Sex

Swinging

DOMESTIC VIOLENCE

DIVORCE

The Cost of Divorce

ALTERNATIVE LIFESTYLES

Open Marriage

Group Marriage

SEX IN THE LATER YEARS

Physical Changes

Patterns of Sexual Activity

SEX AND DISABILITY

Physical Disabilities

Psychological Disabilities

Human Sexuality Online Web Sites Related to Sexuality in Adulthood

SUMMING UP

QUESTIONS FOR CRITICAL THINKING

Truth or Fiction?

____ "Singlehood" (forgive the word) has become a more common U.S. lifestyle over the past few decades.

____ Divorced people are more likely than never-married people to cohabit.

____ In the ancient Hebrew and Greek civilizations, wives were viewed as their husbands' property.

____ Men are more romantic than women.

____ Most of today's sophisticated young people see nothing wrong with an occasional extramarital fling.

____ Men are more likely than women to commit acts of domestic violence.

____ Few women can reach orgasm after the age of 70.

____ People who are paralyzed as a result of spinal-cord injuries cannot become sexually aroused or engage in coitus.

Americans entering adulthood today face a wider range of sexual choices and lifestyles than earlier generations. The sexual revolution loosened traditional constraints on sexual choices, especially for women. Couples experiment with lifestyles that would have been unthinkable in earlier generations. An increasing number of young people choose to remain single as a way of life, not merely as a way station preceding the arrival of Mr. or Ms. Right.

In this chapter, we discuss diverse forms of adult sexuality in the United States today, including singlehood, marriage, and alternative lifestyles such as cohabitation, open marriage, and group marriage. Let us begin as people begin—with singlehood.

SINGLEHOOD

Recent years have seen a sharp increase in the numbers of single young people in our society. "Singlehood," not marriage, is now the most common lifestyle among people in their early 20s. Though marriages may be made in heaven, many Americans are saying heaven can wait. By the end of the millennium, one woman in four and three men in ten in the United States 15 years of age and older had never married (Table 14.1). Half a century earlier, in 1950, one woman in five and about one man in four aged 15 and above had never been married. The rate of marriages had also fallen off. Nearly 80% of men in the 20-to-24 age range were unmarried, up from 55% in 1970 (U.S. Bureau of the Census, 1999). The number of single women in this age group grew from about one in three in 1970 to more than three in five. The proportion of people who remain single into their late 20s and early 30s has more than doubled since 1970 (Edwards, 2000).

It is true that "singlehood" has become a more common U.S. lifestyle over the past few decades. More people are remaining single into their 20s and 30s than was the case a generation or two ago.

TABLE 14.1

Current marital status of the U.S. population aged 15 years and above

	MALES		FEMALES	
Year	Currently Married	Never Married	Currently Married	Never Married
1998	57.9%	31.2%	54.9%	24.7%
1990	60.7	29.9	56.9	22.8
1980	63.2	29.6	58.9	22.5
1970	66.8	28.1	61.9	22.1
1960	69.3	25.3	65.0	19.0
1950	67.5	26.4	65.8	20.0

Source: "Marital Status of the Population 15 Years Old and Over, by Sex and Race: 1950 to Present. U.S. Bureau of the Census. Internet release date: January 7, 1999.

The table does not include individuals who are currently divorced or widowed. The percentages of people who are currently married have declined more dramatically over the past half-century than the percentages of people who were never married have risen. This is due in part to the increase in the divorce rate. When we combine males and females, we find that 2.1% of adults aged 15 and above were currently divorced in 1950, compared with 9.3% in 1998. (About half of marriages end in divorce in recent years, but many divorced people get remarried.)

TABLE 14.2

Estimated median age at first marriage, by sex: 1890 to the present					
Year	Males	Females	Year	Males	Females
1998	26.7	25.0	1940	24.3	21.5
1990	26.1	23.9	1930	24.3	21.3
1980	24.7	22.0	1920	24.6	21.2
1970	23.2	20.8	1910	25.1	21.6
1960	22.8	20.3	1900	25.9	21.9
1950	22.8	20.3	1890	26.1	22.0

Source: Current Population Survey data. U.S. Bureau of the Census, "Marital Status and Living Arrangements: March 1998 (Update)," *Current Population Reports, Series P20–514* and earlier reports U.S. Bureau of the Census. Internet release date: January 7, 1999.

Several factors contribute to the increased proportion of singles. For one thing, more people are postponing marriage to pursue educational and career goals. Many young people are deciding to "live together" (cohabit), at least for a while, rather than get married. Also, as you can see in Table 14.2, people are getting married later. The typical man in the United States gets married at about 27 today, compared with 23 fifty years earlier (U.S. Bureau of the Census, 1999). The typical woman gets married today at about 25, compared with 20 fifty years earlier.

The increased prevalence of divorce also swells the ranks of single adults. When we combine males and females, we find that about 2% of adults aged 15 and above were currently divorced in 1950, compared to about 9% in 1998 (see Table 14.1).

Less social stigma is attached to remaining single today. Yet although single people are less likely today to be perceived as socially inadequate or as failures, some unmarried people still encounter stereotypes. Men who have never married may be suspected of being gay. Single women may feel that men perceive them as "loose." Nor are women over the age of 30 likely to be regarded as "spinsters" (Edwards, 2000). Some findings of a Time/CNN poll are shown in Figure 14.1.

Many single people do not choose to be single. Some remain single because they have not yet found Mr. or Ms. Right. However, many young people see singlehood as an

- In 1963, 83% of American women aged 25 to 55 were married, compared with about 65% today.
- When asked what they missed most because of being single, 75% of women said companionship, and only 4% said sex.
- Even as the birthrate has been falling among teenagers, it has been climbing among single adult women—up 15% since 1990 among women in their 30s.
- When single women were asked whether they would consider rearing a child on their own, 61% of those aged 18 to 49 said yes.
- Single women are an economic force. About a fifth of all home sales in 1999 were to unmarried women, up from 10% in 1985.
- Only 34% of single women said they would settle for less than a perfect mate if they couldn't find one, compared with 41% of men.

FIGURE 14.1

Some Results of a Time Magazine/CNN Poll, Year 2000. Single adult women in the United States are more self-confident and selective than they have ever been. They are assuming many of the social, economic, and sexual freedoms that had been reserved for single men. (*Source:* From "Single by Choice" by Tamala Edwards in *Time online*, August 28, 2000. Copyright © 2000 by Time, Inc. Reprinted by permission.)

Singles There is no single "singles scene." Although some singles meet in singles bars, many meet in more casual settings, such as the neighborhood laundromat. Some singles advertise online or in newspapers or magazines.

alternative, open-ended way of life, not just a temporary stage that precedes marriage. Now that career options for women have expanded, they are not as financially dependent on men as were their mothers and grandmothers. A number of career women, like young career-oriented men, choose to remain single (at least for a time) to focus on their careers.

Singlehood is not without its problems. Many single people are lonely. Some singles express concern about their lack of a steady, meaningful social relationship. Others, usually women, worry about their physical safety. Some people living alone find it difficult to satisfy their needs for intimacy, companionship, sex, and emotional support. Despite these concerns, most singles are well adjusted and content. Singles who have a greater number of friends and a supportive social network tend to be more satisfied with their lifestyles.

There is no single "singles scene." Single people differ in their sexual interests and lifestyles. Many achieve emotional and psychological security through a network of intimate relationships with friends. Most are sexually active and practice **serial monogamy.** Other singles have a primary sexual relationship with one steady partner but occasional brief flings. A few, even in this age of AIDS, are "swinging singles." That is, they pursue casual sexual encounters, or "one-night stands."

Some singles remain celibate, either by choice or for lack of opportunity. People choose **celibacy** for a number of reasons. Nuns and priests do so for religious reasons. Others believe that celibacy allows them to focus their energies and attention on work or to commit themselves to an important cause. They see celibacy as a temporary accommodation to other pursuits. Others remain celibate because they view sex outside of marriage as immoral. Still others remain celibate because they find the prospects of sexual activity aversive or unalluring, or because of fears of STIs.

COHABITATION: DARLING, WOULD YOU BE MY POSSLQ?

There is nothing I would not do
If you would be my POSSLQ
—Charles Osgood

POSSLQ? This unromantic abbreviation was introduced by the U.S. Bureau of the Census to refer to **cohabitation.** It stands for People of Opposite Sex Sharing Living Quarters and applies to unmarried couples who live together.

Some social scientists believe that cohabitation has become accepted within the social mainstream (Bumpass, 1995). Whether or not this is so, society in general has become more tolerant of it. We seldom hear cohabitation referred to as "living in sin" or "shacking up," as we once did. People today are more likely to refer to cohabitation with value-free expressions such as "living together."

Perhaps the current tolerance reflects society's adjustment to the increase in the numbers of cohabiting couples. Or perhaps the numbers of cohabiting couples have increased as a consequence of tolerance. The numbers of households that consist of an unmarried adult male and female couple living together in the United States quadrupled over the past 25 years (Smock, 2000). They grew from about 1.6 million couples in 1980 to 2.9 million couples in 1990 and nearly 5 million today (Smock, 2000).

Serial monogamy A pattern of becoming involved in one exclusive relationship after another, as opposed to engaging in multiple sexual relationships at the same time.

Celibacy Complete sexual abstinence. (Sometimes used to describe the state of being unmarried, especially in the case of people who take vows to remain single.)

Cohabitation Living together as though married but without legal sanction.

Some Facts About Cohabitation

Although much attention is focused on college students living together, cohabitation is actually more prevalent among less well-educated and less affluent people (Willis & Michael, 1994). The cohabitation rate is about twice as high among African American couples as among European American couples.

More than half (56%) the marriages that took place during the past decade were preceded by living together (Smock, 2000). It also turns out that nearly 55% of the couples who cohabit wind up getting married, which has led some social scientists to suggest that cohabitation, for many, is a new stage of engagement. Even so, about 40% of these couples get divorced later on, so "trial marriage" may not provide couples with the information they are seeking about each other.

We are reaching a time when we can say that half of the people living in the United States have cohabited at some time. Already, nearly half (48%) of women in their late 30s in the United States report having cohabited (Smock, 2000).

Children are common in cohabiting households. Nearly half of the divorced people who are cohabiting with new partners have children in the household (Smock, 2000). At least one out of three never-married cohabiting couples also have children living with them. Divorced people are more likely than people who have never been married to cohabit (Smock, 2000).

It is true that divorced people are more likely than never-married people to cohabit. The experience of divorce may make some people more willing to share their lives than their bank accounts—the second or third time around. ■

Willingness to cohabit is related to more liberal attitudes toward sexual behavior, less traditional views of marriage, and less traditional views of gender roles (Huffman et al., 1994; Knox et al., 1999b). Cohabitors are less likely than noncohabitors to attend church regularly (Laumann et al., 1994).

Reasons for Cohabitation

Why do people cohabit? Cohabitation, like marriage, is an alternative to the loneliness that can accompanying living alone. Romantic partners may have deep feelings for each other but not be ready to get married. Some couples prefer cohabitation because it provides a consistent relationship without the legal entanglements of marriage (Steinhauer, 1995).

Many cohabitors feel less commitment to their relationships than married people do (Nock, 1995). Ruth, an 84-year-old woman, has been living with her partner, age 85, for four years. "I'm a free spirit," she says. "I need my space. Sometimes we think of marriage, but then I think that I don't want to be tied down" (cited in Steinhauer, 1995, p. C7).

Ruth's comments are of interest because they fly in the face of stereotypes of women and older people. However, it is more often the man who is unwilling to make a marital commitment (Yorburg, 1995), as in the case of Mark. Mark, a 44-year-old computer consultant, lives with Nancy and their 7-year-old daughter Janet. Mark says, "We feel we are not primarily a couple but rather primarily individuals who happen to be in a couple. It allows me to be a little more at arm's length. Men don't like committing, so maybe this is just some sort of excuse" (cited in Steinhauer, 1995, p. C7).

Economic factors come into play as well. Emotionally committed couples may decide to cohabit because of the economic advantages of sharing household expenses. Cohabiting individuals who receive public assistance (social security or welfare checks) risk losing support if they get married (Steinhauer, 1995). Some older people live together rather than marry because of resistance from adult children (Yorburg, 1995). Some children

Cohabitation Cohabitation was once referred to as "living in sin," but it has become an increasingly common lifestyle. Some sociologists predict that cohabitation will replace marriage as the nation's most popular lifestyle sometime during the first century of the new millennium. Why do people cohabit? Cohabitation is an alternative to the loneliness of living by oneself. It also allows for the experience of an intimate relationship without some of the obligations of marriage. Men are more likely than women to want to cohabit rather than marry.

fear that a parent will be victimized by a needy senior citizen. Others may not want their inheritances to come into question or may not want to decide where to bury the remaining parent. Younger couples may cohabit secretly to maintain parental support that they might lose if they were to get married or to reveal their living arrangements.

Cohabitation First and Marriage Later: Benefit or Risk?

Cohabiting couples may believe that cohabitation will strengthen their eventual marriage by helping them iron out the kinks in their relationship. Yet cohabitors who later marry also run a serious risk of getting divorced. According to Pamela Smock's (2000) survey at the Institute for Social Research at the University of Michigan, 40% of couples who cohabited before tying the knot got divorced later on. Thus they may run a greater—not a lesser—risk of divorce than noncohabitors. Some studies suggest that the likelihood of divorce within 10 years of marriage is nearly twice as great among married couples who cohabited before marriage (Riche, 1988). A Swedish study found that the likelihood of marital dissolution was 80% greater among women who had cohabited before a first marriage than among women who had not (Bennett et al., 1988).

Why might cohabiting couples run a greater risk of divorce than couples who did not cohabit prior to marriage? Do not assume that cohabitation somehow causes divorce. We must be cautious about drawing causal conclusions from correlational data. Note that none of the couples in these studies were *randomly assigned* to cohabitation or noncohabitation. Therefore, *selection factors*—the factors that lead some couples to cohabit and others not to cohabit—may explain the results (see Figure 14.2). Cohabitors tend to be more

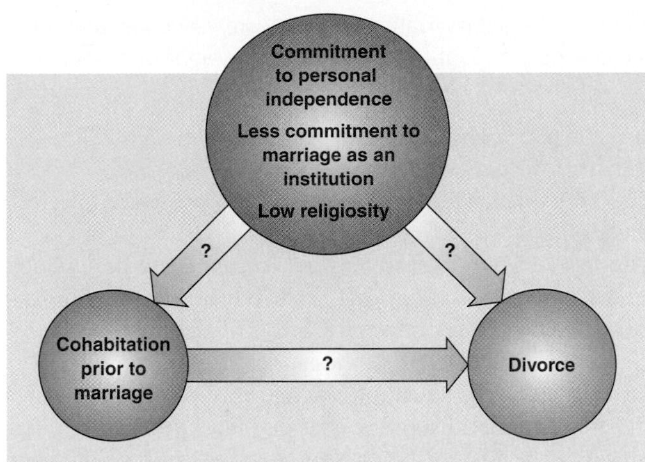

FIGURE 14.2

Does Cohabitation Prior to Marriage Increase the Risk of Eventual Divorce? There is a correlational relationship between cohabitation prior to marriage and the risk of divorce later on. Does cohabitation increase the risk of divorce, or do other factors—such as a commitment to personal independence—contribute both to the likelihood of cohabitation and to eventual divorce?

committed to personal independence than noncohabitors (Bumpass, 1995). They also tend to be less traditional and less religious. All in all, people who cohabit prior to marriage tend to be less committed to the values and interests traditionally associated with the institution of marriage. The attitudes of cohabitors, and not cohabitation itself, may thus account for their higher rates of marital dissolution.

MARRIAGE

Marriage is found in all human societies. Most people in every known society—sometimes nearly all—get married at least once.

Marriage is our most common lifestyle, and people see marriage as permanent. A recent *New York Times* poll asked, "If you got married today, would you expect to stay married for the rest of your life?" and 86% of respondents answered "Yes" (Eggers, 2000). Only 11% said no. In some cultures, such as among the Hindu of India, marriage is virtually universal, with more than 99% of the females eventually marrying. In the United States, most people still get married, but today about 28% of people aged 15 and above have never married. Table 14.1 shows a steady increase in the percentage of never-married adults over the past 40 years. However, cohabitation is becoming widespread in the United States as well. University of Wisconsin researcher Larry Bumpass has been following 10,000 people since the late 1980s. In 1995, he reported that 49% of people aged 35 to 39 were cohabiting, compared with 34% in the late 1980s. Bumpass estimates that nationwide, half of the adult population under age 40 is cohabiting. In 1995, he estimated that in 10 years, half of the adult population under the age of 50 would be cohabiting.

Historical Perspectives

Marriage in Western culture has a long and varied history. Among the ancient Hebrews, men dominated most aspects of life. This system was known as **patriarchy.** The man of the house had the right to choose wives for his sons and could himself take concubines or additional wives. He alone could institute divorce—a practice still found among Orthodox Jews. Although he could dally, his wife had to be scrupulous in the maintenance of her virtue. Failure to bear children was grounds for divorce. The wife was clearly considered to be part of the property of the man of the house: a **chattel.**

In classical Greece, women were also viewed as the property of men. Their central purposes were to care for the household and to bear children. Rarely were they viewed as suitable companions for men. During the golden age of Greece, men turned to high-class prostitutes for sensual sex and sophisticated conversation—not to their wives.

Patriarchy A form of social organization in which the father or eldest male runs the group or family; government, rule, or domination by men. (From the Greek *pater,* which means "father," and *archein,* which means "to rule."

Chattel A movable piece of personal property, such as furniture or livestock. (From the Old French word for "cattle.")

It is true that in the ancient Hebrew and Greek civilizations, wives were viewed as their husbands' property. The women's movement has fought to throw off the weight of millennia of such oppressive practices.

The Romans also had a powerful patriarchy. The oldest man directed family life and could, if he wished, sell his children into slavery and arrange their marriages and divorces. Marriages were typically arranged for financial or political gain. Women were in effect given by their fathers to their husbands.

The Christian tradition also has a strong patriarchal foundation. Male dominance was legitimized by biblical scripture, as we can see in this passage from the New Testament: "Wives, submit to your own husbands, as unto the Lord. For the husband is the head of the wife, as Christ is the head of the church" (Ephesians 5:23–24).

Patriarchal traditions in Western culture gradually weakened with time, and women came to be viewed as loving companions rather than mere chattels. They gradually gained more household responsibilities and were recognized as being capable of profiting from education. The notion that a married woman might rightfully seek personal fulfillment through a career unrelated to her husband's needs is a relatively recent development, however. The perception that married women have a right to sexual fulfillment is also new. As late as the nineteenth century, sex in marriage was seen largely as a means for bearing children and satisfying the husband's sexual needs. Although there were some caring and sensitive husbands, typical nineteenth-century sexual relations between husbands and wives were brief and brutal. Women derived little pleasure from them. Even among educated classes, sex was viewed as a husband's right and a wife's duty. Marriage manuals of the time held that a woman's role was to submit to her husband's advances in order to please him. Women were assumed to be motivated sexually only by the desire for children.

Marital roles in modern society have changed and are changing still. Some couples still adhere to traditional gender roles that ascribe breadwinning responsibilities to the husband and child care and homemaking roles to the wife. U.S. couples today are more likely to share or even reverse marital roles, however. Table 14.3 lists some of the discriminating features of traditional and so-called modern marriages.

TABLE 14.3

A comparison of traditional and modern marriages

Traditional Marriage	Modern Marriage
The emphasis is on ritual and traditional roles.	The emphasis is on companionship.
Couples do not live together before marriage.	Couples may live together before marriage.
The wife takes the husband's last name.	The wife may choose to keep her birth name.
The husband is dominant; the wife is submissive.	Neither spouse is dominant or submissive.
The roles for the husband and the wife are specific and rigid.	Both spouses have flexible roles.
There is one income (the husband's).	There may be two incomes; that is, the couple may share the breadwinning role. In many cases, the woman is the breadwinner.
The husband initiates sexual activity; the wife complies.	Either spouse may initiate (or refuse) sex.
The wife cares for the children.	The parents share child-rearing chores.
Education is considered important for the husband, not for the wife.	Education is considered equally important for both spouses.
The husband's career decides the location of the family residence.	The career of either spouse may determine the location of the family residence.

Source: From *Choices in Relationships: An Introduction to Marriage and the Family*, 3e by D. Knox and C. Schact. Copyright © 1991 by the authors. Reprinted with permission of Wadsworth, an imprint of the Wadsworth Group, a division of Thomson Learning.

Russian Region Stops Men from Marrying Four Times

Men in Russia's mainly Muslim Ingushetia region were told Friday they can no longer marry up to four women each, RIA news agency reported. Ingushetia's top court canceled two articles of law to end the traditional Islamic practice. RIA said the court took the step because the articles contradicted Russia's federal law. Russian President Vladimir Putin cracked down on such cases as part of a drive to rein in the region. Ingush leader Ruslan Aushev had enacted the articles last July and asked Russia's parliament to approve the changes, which were based on a traditional Islamic practice in the province on Russia's southern rim. RIA said 15 men had taken second or third wives since last year. Islam has been revived in Russia after years of being discouraged during officially atheist Soviet times.

Source: From "Russian Region Stops Men from Marrying Four Times" (2000, July 18). Reuters News Agency online. Copyright © 2000 by Reuters Limited. Adapted by permission.

Why Do People Marry?

Marriage meets personal and cultural needs. It legitimizes sexual relations and provides a legal sanction for deeply committed relationships. It permits the maintenance of a home life and provides an institution in which children can be supported and socialized into adopting the norms of the family and the culture at large. Marriage restricts sexual relations so that a man can be assured—or at least can assume—that his wife's children are his. Marriage also permits the orderly transmission of wealth from one family to another and from one generation to another. As late as the seventeenth and eighteenth centuries, most European marriages were arranged by the parents, generally on the basis of how the marriage would benefit the families.

Notions such as romantic love, equality, and the very radical concept that men as well as women would do well to aspire to the ideal of faithfulness are recent additions to the structure of marriage in Western society. Not until the nineteenth century did the notion of love as a basis for marriage become widespread in Western culture. In some preliterate societies, the very idea of being in love is considered a laughable concept—hardly a basis for marriage.

Today, because more people believe that premarital sex is acceptable between two people who feel affectionate toward each other, the desire to engage in sexual intercourse is less likely to motivate marriage. Marriage provides a sense of emotional and psychological security, however, and opportunities to share feelings, experiences, and ideas with someone with whom one forms a special attachment. Desires for companionship and intimacy are key goals in marriage today.

Broadly speaking, those people who want to get married do so because they believe that they will be happier if they get married. A Gallup poll conducted in the year 2000 suggests that their optimism may be well founded (Chambers, 2000). The poll revealed that married people are more likely than single people to report that they are "fairly happy" or "very happy." As Table 14.4 shows, married men and married women were more likely than their single counterparts to report that they were "very happy."

TABLE 14.4

Percent who report they are "very happy," according to marital status

	Married	Unmarried
Total	57%	36%
Men	53	35
Women	62	37

Source: From "Americans Are Overwhelmingly Happy and Optimistic about the Future of the U.S. Marital Status Strongly Affects Both Happiness and Optimism." By Chris Chambers in a Gallup News Service Survey conducted from October 6–9, 2000. Reprinted by permission of The Gallup Organization.

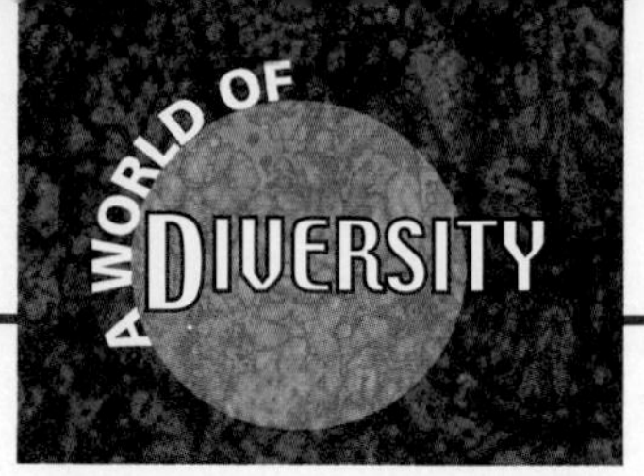

Housewives Urged to Strike

Authorities in Mexico City urged housewives to lay down their mops and cooking pans on a weekend in July 2000, to mark international domestic workers day. "We want the domestic work of women to be valued economically and socially," Gabriela Delgado of Mexico City's branch of the National Women's Institute said in calling on housewives to strike.

Although women make up 51.36% of Mexico's population of 98 million, they have traditionally taken a back seat in politics and business, and many Mexican men believe a women's place is in the home and kitchen. Mexican women were given the vote only in 1952. Delgado said international domestic workers day on Saturday was an opportunity to make women and men aware that household tasks should be equally divided among family members.

"Our problem is a patriarchal ideology that has forced women to take up the domestic work as something natural," she said.

Source: From "Housewives Urged to Strike." Reuters News Agency online, June 7, 2000. Copyright © 2000 by Reuters Limited. Reprinted by permission.

Types of Marriage

There are two major types of marriage: monogamy and polygamy. In **monogamy,** a husband and wife are wed only to each other. But let us not confuse monogamy, which is a form of matrimony, with sexual exclusivity. People who are monogamously wedded often do have extramarital affairs, as we shall see, but they are considered to be married to only one person at a time. In **polygamy,** a person has more than one spouse and is permitted sexual access to each of them.

Polygyny is by far the most prevalent form of polygamy among the world's preliterate societies (Ford & Beach, 1951; Frayser, 1985). **Polyandry** is practiced only rarely (see Chapter 1). In polygynous societies, men are permitted to have multiple wives if they can support them; more rarely, a man will have one wife and one or more concubines. The man's first wife typically has higher status than the others. Economic factors and the availability of prospective mates usually limit the opportunities for men to wed more than one woman at a time, however. In many cases, only wealthy men can afford to support multiple wives and the children of these unions. In addition, few if any societies have enough women to allow most men to have two or more wives (Harris & Johnson, 2000; Whitten, 2001). For these reasons, even in societies that prefer polygyny, fewer than half of the men at any given time actually have multiple mates (Ford & Beach, 1951).

Whom We Marry: Are Marriages Made in Heaven or in the Neighborhood?

Most preliterate societies regulate the selection of spouses in some way. The universal incest taboo proscribes matings between close relatives. Societal rules and customs also determine which people are desirable mates and which are not.

In Western cultures, mate selection is presumably free. Parents today seldom arrange marriages, although they may still encourage their child to date that wonderful son or daughter of the solid churchgoing couple who live down the street. Nevertheless, factors such as race, social class, and religion often determine the categories of people within which we seek mates (Laumann et al., 1994). People in our culture tend to marry others from the same geographical area and social class. Because neighborhoods are often made up of people from the same social class, storybook marriages like Cinderella's are the exception to the rule.

Because we make choices, we tend to marry people who attract us. These people are usually similar to us in physical attractiveness and attitudes, and even in minute details. We are more often than not similar to our mates in characteristics such as height, weight, personality traits, and intelligence (Buss, 1994; Lesnik-Oberstein & Cohen, 1984; Schafer & Keith, 1990). The people we marry also seem likely to meet our material, sexual, and psychological needs.

Monogamy Marriage to one person.

Polygamy Simultaneous marriage to more than one person.

Polygyny A form of marriage in which a man is married to more than one woman at the same time.

Polyandry A form of marriage in which a woman is married to more than one man at the same time.

The concept of "like marrying like" is termed **homogamy.** We usually marry people of the same racial/ethnic background, educational level, and religion. Interracial marriages account for fewer than 1 in every 100 marriages (U.S. Bureau of the Census, 1998). More than 9 of 10 marriages are between people of the same religion. Marriages between individuals who are alike may stand a better chance of survival, because the partners are more likely to share their values and attitudes (Michael et al., 1994). A study of undergraduates at a large southeastern university found that college women and men believe that homogamy in terms of background is connected with happy and lasting relationships (Knox et al., 1997b). Nevertheless, about 1 in 4 students on the same campus had dated interracially, and nearly half expressed a willingness to become involved in an interracial relationship (Knox et al., 2000). African Americans were somewhat more likely than European Americans to enter interracial relationships.

We also tend to follow *age homogamy* (Michael et al., 1994). Age homogamy—the selection of a partner who falls in one's own age range—may reflect the tendency to marry early in adulthood. Persons who marry late or who remarry tend not to select partners so close in age. Bridegrooms tend to be two to five years older than their wives, on the average, in European, North American, and South American countries (Buss, 1994).

Some marriages also show a **mating gradient.** The stereotype has been that an economically established older man would take an attractive, younger woman as his wife. But by and large, with boring predictability, we are attracted to and marry the boy or girl "next door." Most marriages seem to be made not in heaven, but in the neighborhood.

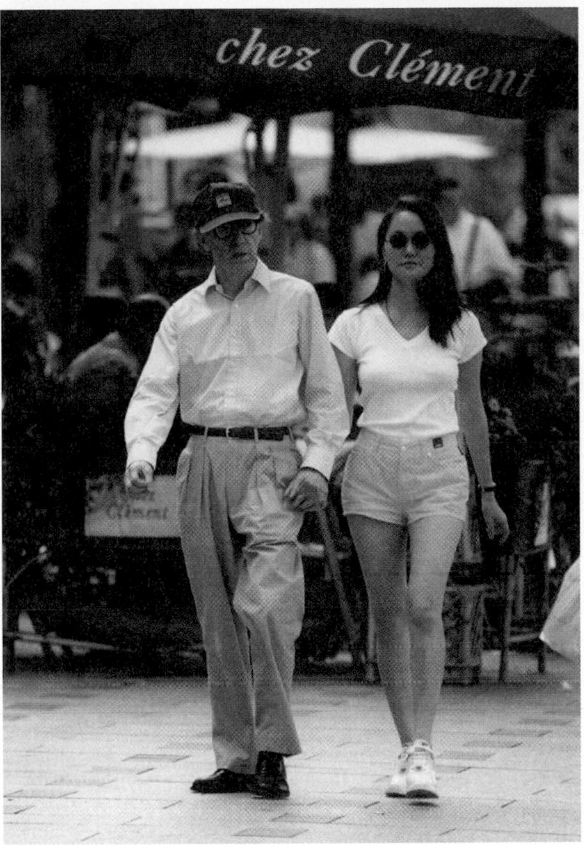

Whom Do We Marry? The marriage of Woody Allen and Soon Yi is typical in some ways, atypical in others. It is common for the man to be older than the woman. When there is a great disparity of ages, as there is here, the man is frequently well established and the woman is usually very attractive, as is the case here. But Allen and Soon Yi are of different races, which remains uncommon in the United States. Moreover, Soon Yi is the adopted daughter of Allen's ex-wife, Mia Farrow, which made the union somewhat scandalous at the time.

WHO HAS HIS HEAD IN THE CLOUDS? (*HINT:* THE QUESTION IS *NOT* PHRASED IN SEXIST LANGUAGE) When it comes to picking a mate, men tend to be the romantics, women the pragmatists. Men are more likely to believe that each person has one true love whom they are destined to find (Peplau & Gordon, 1985). Men are more likely to believe in love at first sight. Women, on the other hand, are more likely to value financial security as much as passion. Women are more likely to believe that they could form a loving relationship with many individuals. Women are also less likely to believe that love conquers all—especially economic problems.

Truth or Fiction? **REVISITED**

It is true that men are more romantic than women, when we define being *romantic* as believing in love at first sight or believing that there is just one person who is right for oneself. (Who is more "romantic" if we define *romantic* to mean that one wishes to confine sexual activity to a romantic relationship?) ■

Homogamy The practice of marrying someone who is similar to oneself in social background and standing. (From Greek roots that mean "same" [*homos*] and "marriage" [*gamos*].)

Mating gradient The tendency for women to "marry up" (in social or economic status) and for men to "marry down."

MARITAL SEXUALITY

Patterns of marital sexuality vary across cultures, yet anthropologists have noted some common threads (Harris & Johnson, 2000; Whitten, 2001). Privacy for sexual relations is valued in nearly all cultures. Most cultures also place restrictions on coitus during menstruation, during at least some stages of pregnancy, and for a time after childbirth.

Until the sexual revolution of the 1960s and 1970s, Western culture could have been characterized as sexually restrictive, even toward marital sex. Hunt noted that "Western civilization has long had the rare distinction of contaminating and restricting the sexual pleasure of married couples more severely than almost any other" (1974, p. 175).

The Sexual Revolution Hits Home

We usually think of the sexual revolution in terms of the changes in sexual behaviors and attitudes that occurred among young, unmarried people. It also ushered in profound changes in marital sexuality, however. Compared to Kinsey's "pre-revolution" samples from the late 1930s and 1940s, married couples today engage in coitus more frequently, with greater variety, and for longer durations of time. And they report higher levels of sexual satisfaction.

The sexual revolution also helped dislodge traditional male dominance in sexual behavior. In essence, male dominance is the view that sexual pleasure is meant for men but not women and that it is the duty of women to satisfy their husbands' sexual needs and serve as passive "receptacles." Assumptions of male dominance are found even in the writing of the (for his time) liberated Kinsey. Nowhere are they more evident than in his (and his colleagues') comment that men, who have the biological capacity to ejaculate a few moments after entry, need not feel bound to "wait" for women to reach orgasm (Kinsey et al., 1948, p. 580).

People caught up in social revolutions may not perceive themselves as revolutionaries. They may not even be aware that a "revolution" is taking place. The sexual revolution was not heralded by parades or massive demonstrations. There was no "storming of the Bastille" to proclaim sexual freedom. Organizations promoting "free love" did not attract large followings. Even the most famous slogan of the sexual revolution—"Make love, not war"—was more an indictment of the Vietnam War than a call for free love. Moreover, social movements are begun by a few but are spread by the perception that they define popular trends. Married people, as well as singles, thus made sexual decisions on the basis not only of their religious convictions and personal values, but also of their perceptions of the norm.

Liberalizing trends were communicated through popular media. Hundreds if not thousands of books and magazines with sexual content crowded bookstore shelves and supermarket display cases. During Kinsey's day, books like Henry Miller's *Tropic of Cancer* and D. H. Lawrence's *Lady Chatterley's Lover* were censored in the United States. Now they are readily available—and so tame, when compared to today's explicit books and magazines, that they barely raise an eyebrow. Similarly, since the late 1960s, pornographic films have been widely available. The VCR and the Internet have brought sexually explicit materials into middle-class suburban homes.

Scientific findings were also liberalizing influences. Kinsey's and Masters and Johnson's findings that normal women were capable not only of orgasm but also of multiple orgasms punctured traditional beliefs that sexual gratification was the birthright of men alone. TV shows, films, and radio talk shows began to portray women as sexual initiators who enjoy sex. All of these influences have encouraged U.S. women to forgo the traditional passive approach to sexuality.

The affluence of the post–World War II years encouraged more young people to pursue a college education and live away from home. College liberates not only through exposure to great books and scientific knowledge, but also through students' attitudes being challenged by bright, knowledgeable peers from different backgrounds. Prejudice and myth are often analyzed and discarded.

Americans have also become more mobile. Jobs carry young people thousands of miles from home. Their sexual attitudes are influenced by new acquaintances from many parts of the country and by people from other nations, not just by a few people from "the neighborhood."

The development of effective contraceptives also separated sex from reproduction. Motives for sexual pleasure became more open. The term *recreational sex* came into play. All these liberalizing forces have led to changes in the frequency of marital sex and in techniques of foreplay and coitus since Kinsey's day.

TABLE 14.5

Median weekly frequency of marital coitus, male and female estimates combined (Kinsey and Morton Hunt surveys)

KINSEY (1948, 1953)		MORTON HUNT (1974)	
Age	**Frequency**	**Age**	**Frequency**
16–25	2.45	18–24	3.25
26–35	1.95	25–34	2.55
36–45	1.40	35–44	2.00
46–55	.85	45–54	1.00
55–60	.50	55 and over	1.00

Source: From M. Hunt (1974). *Sexual Behavior in the 1970's.* New York: Playboy Press, p. 191. Reprinted with permission from Playboy Enterprises International, Inc. Copyright © 1974 by Morton Hunt.

CHANGES IN DURATION AND TECHNIQUES OF FOREPLAY Married women in Kinsey's sample reported an average (median) length of foreplay of about 12 minutes. This figure rose to nearly 15 minutes among the wives in the Morton Hunt survey (Hunt, 1974). Kinsey found that men at lower educational levels engaged in briefer periods of foreplay—generally only a minute or two—before penetration. The length of foreplay rose to 5 to 15 minutes among college-educated men. In a dramatic shift in sophistication since Kinsey's day, Hunt found that the typical duration of foreplay in the 1970s was 15 minutes for college-educated and noncollege males alike. Foreplay was relatively longer among younger married couples than among their elders, however.

Marital foreplay has also become more varied since Kinsey's day. Couples in more recent surveys report using a wider variety of foreplay techniques, including oral stimulation of the breasts and oral–genital contact (Blumstein & Schwartz, 1983; Hunt, 1974).

CHANGES IN FREQUENCY OF MARITAL COITUS How frequently do married couples engage in coitus? A comparison of the reports from Kinsey's couples and the Morton Hunt survey couples (Table 14.5) shows increases in coital frequency at every age level. However, keep in mind that because 80% of those contacted by the Morton Hunt survey refused to participate, the figures in the table may have a strong volunteer bias. Volunteers tend to be more open about their sexuality than people who refuse to divulge information about sex, so this bias would tend to inflate the statistical results (Laumann et al., 1994).

The data reported by the NHSLS study do not allow a direct comparison with the Kinsey and Morton Hunt figures. As shown in Table 14.6, however, the overwhelming

TABLE 14.6

Frequency of marital sexual relations during the past year, according to NHSLS study

Frequency of Sexual Relations	Men (%)	Women (%)
Not at all	1.3	3.0
A few times a year	12.8	11.9
A few times a month	42.5	46.5
Two to three times a week	36.1	31.9
Four or more times a week	7.3	6.6

Source: From E. O. Laumann, J. H. Gagnon, R. T. Michael, & S. Michaels (1994). *The Social Organization of Sexuality: Sexual Practices in the United States.* Chicago: University of Chicago Press, Table 3.4, pp. 88–89. Copyright © 1994 by University of Chicago Press. Adapted by permission.

majority of married men and women in the United States now report engaging in sexual relations either a few times per month or two to three times a week (Laumann et al., 1994). The average is seven times a month (Michael et al., 1994, p. 136). These figures are somewhat more in keeping with Kinsey's than with Morton Hunt's results, which may cast doubt on the Morton Hunt survey's statistics.

Kinsey and the University of Chicago group (who conducted the NHSLS study) did not find strong links between coital frequency and educational level. Studies consistently find that the frequency of sexual relations declines with age, however (Call et al., 1995; Laumann et al., 1994). At the ages of 50 to 59, for example, people reported engaging in sexual intercourse an average of four to five times per month (Laumann et al., 1994). Regardless of a couple's age, sexual frequency also appears to decline with years of marriage (Blumstein & Schwartz, 1990).

CHANGES IN TECHNIQUES AND DURATION OF COITUS In coitus, as in foreplay, the marital bed since Kinsey's day has become a stage on which the players act more varied roles. Today's couples use greater variety in coital positions.

Kinsey's study participants mainly limited coitus to the male-superior position. As many as 70% of Kinsey's males used the male-superior position exclusively (Kinsey et al., 1948). Perhaps three couples in ten used the female-superior position frequently. One in four or five used the lateral-entry position frequently, and about one in ten had used the rear-entry position. Younger and more highly educated men showed greater variety, however. Hunt (1974), by comparison, found that three quarters of the married couples in the Morton Hunt survey used the female-superior position at least occasionally. More than half had used the lateral-entry position, and about four in ten had used the rear-entry position.

An often overlooked but important difference between Kinsey's and current samples involves the length of intercourse. In Kinsey's time it was widely believed that the "virile" man ejaculated rapidly during intercourse. Kinsey estimated that most men reached orgasm within 2 minutes after penetration, many within 10 or 20 seconds. Kinsey recognized that women usually took longer to reach orgasm through coitus and that some clinicians were already asserting that a man's ejaculation was "premature" unless he delayed it until "the female (was) ready to reach orgasm" (1948, p. 580).

Even today's less highly educated couples appear to be more sophisticated than Kinsey's in their recognition of the need for sexual variety and in their focus on exchanging sexual pleasure rather than reaching orgasm rapidly (Michael et al., 1994). According to the NHSLS study, the "duration of the last sexual event" of three out of four married couples was 15 minutes to an hour. Eight percent to 9% of couples exceeded an hour (Michael et al., 1994). (About one unmarried, noncohabiting couple in three made love for an hour or more, suggesting that novelty and youth have their motivational aspects.)

Sexual Satisfaction

One index by which researchers measure sexual satisfaction is orgasmic consistency. Men tend to reach orgasm more consistently than women do. After 15 years of marriage, 45% of the wives in Kinsey's study reported reaching orgasm 90% to 100% of the time. After 15 years of marriage, 12% of the wives in Kinsey's study had not experienced orgasm.

Orgasmic consistency now seems higher than in Kinsey's day. The NHSLS study found that more than 90% of the men and about 70% of the women reported reaching orgasm "always" or "usually" with their primary partner during the 12 months prior to the survey (Laumann et al., 1994; Michael et al., 1994) (see Table 14.7). Three of four men (75%) and nearly three women in ten (28.6%) reported reaching orgasm on every occasion (not shown in Table 14.7). Only 2% of the married women reported never reaching orgasm with their husbands during the past year (not shown).

Women in their 40s were somewhat more likely to reach orgasm consistently than younger and older women. Perhaps women in their 40s have had more time to get in touch with their sexuality and may be more secure in their relationships than younger women are. Biological factors may contribute to the falloff for women and men in their

TABLE 14.7

Sociocultural factors and sexual satisfaction in primary relationship during last year

	ALWAYS OR USUALLY HAD AN ORGASM WITH PARTNER		HAS BEEN EXTREMELY PHYSICALLY SATISFIED WITH PARTNER		HAS BEEN EXTREMELY EMOTIONALLY SATISFIED WITH PARTNER	
Sociocultural Characteristics	**Men**	**Women**	**Men**	**Women**	**Men**	**Women**
Age						
18–24	92%	61%	44%	44%	41%	39%
25–29	94	71	50	39	46	40
30–39	97	70	45	41	39	38
40–49	97	78	44	42	38	42
50–59	91	73	53	32	52	32
Marital Status						
Noncohabiting	94	62	39	40	32	31
Cohabitating	95	68	44	46	35	44
Married	95	75	52	41	49	42
Race/Ethnicity*						
European American	96	70	47	40	43	38
African American	90	72	43	44	43	38
Latino and Latina American	96	68	51	39	43	39

*The numbers of Asian Americans and Native Americans were too small to generate reliable statistics.

Sources: E. O. Laumann, J. H. Gagnon, R. T. Michael, & S. Michaels (1994). *The Social Organization of Sexuality: Sexual Practices in the United States.* Chicago: University of Chicago Press, Table 3.7, pp. 116–117; and R. T. Michael, J. H. Gagnon, E. O. Laumann, & G. Kolata (1994). *Sex in America: A Definitive Survey.* Boston: Little, Brown, Table 9, pp. 128–129.

50s. The nature of the relationship is a factor for women. Married women were most likely to reach orgasm consistently, followed by cohabiting women and then noncohabiting women. Security in the relationship apparently promotes orgasmic consistency. There do not seem to be notable racial or ethnic differences.

Orgasm is not the only criterion for measuring pleasure or satisfaction in marital sex. The NHSLS study asked participants whether they had been extremely physically satisfied with their primary partners during the past year. It is apparent from Table 14.7 that men and women are comparable in general physical satisfaction—about 47% and 41%, respectively. Orgasm, then, is not a guarantee of satisfaction. And lack of orgasm is not necessarily a sign of dissatisfaction.

The emotional satisfaction in a marital relationship is also linked to sexual satisfaction. Table 14.7 shows that about 40% of men and women report being extremely emotionally satisfied with their primary partners. Closer relationships are connected with more consistent orgasm.

Other researchers have found that wives who talk openly to their husbands about their sexual feelings and needs report higher levels of sexual satisfaction (Banmen & Vogel, 1985; Tavris & Sadd, 1977). Women respondents to the *Redbook* survey who took an active role during sex were more satisfied with their sex lives than those who assumed the traditional passive female role (Tavris & Sadd, 1977).

EXTRAMARITAL SEX

> "Women seek soul mates; men seek playmates. Women believe that their affair is justified when it is for love; men, when it's *not* for love." (Janis Abrahms Spring, 1997)

It's next door and it's in the White House—Bill Clinton and Gennifer Flowers, Bill Clinton and Monica Lewinsky. Journalist Eric Alterman (1997) writes that adultery is everywhere today. When French President François Mitterand died, his wife, his mistress, and his illegitimate daughter were among the mourners. Alterman also mentions the examples of Kelly Flinn (who was forced to resign from the armed services), Frank Gifford (who retired from *Monday Night Football* in 1998), Bill Cosby (who admitted to an affair but denied that the woman's daughter was his child), and actor Eddie Murphy.

Why do people engage in extramarital sex? What does it mean? What are its effects on a marriage?

Some people engage in extramarital sex for variety (Lamanna & Riedmann, 1997). Some have affairs to break the routine of a confining marriage. Others enter affairs for reasons similar to the nonsexual reasons why adolescents often have sex: as a way of expressing hostility toward a spouse or retaliating for injustice. Husbands and wives who engage in affairs often report that they are not satisfied with or fulfilled by their marital relationships. Curiosity and desire for personal growth are often more prominent motives than marital dissatisfaction. Middle-aged people may have affairs to boost their self-esteem or to prove that they are still attractive.

Many times the sexual motive is less pressing than the desire for emotional closeness. Some women say they are seeking someone whom they can talk to or communicate with (Lamanna & Riedmann, 1997). There is a notable gender difference here. According to Janis Abrahms Spring, author of *After the Affair* (a self-help book designed to help people save their marriages after an affair), men may be seeking sex in affairs. Women, how-

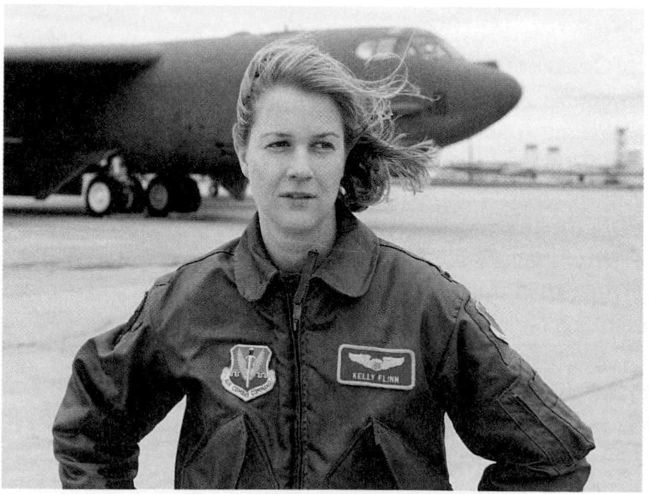

Convicted! Acquitted! Airwoman Kelly Flinn and many other officers in the military have been court marshaled for having adulterous affairs. Apparently the standards for people in the military differed from those that apply to their commander-in-chief—the President of the United States. Ex-President Bill Clinton remained in office despite numerous well-publicized extramarital affairs. Clinton's statements about his affair with White House intern Monica Lewinsky led him down the path to impeachment by the House of Representatives on charges of perjury and obstruction of justice. However, he was acquitted of these charges by the Senate—which soon thereafter included his wife Hillary Rodham Clinton, also here shown with him.

ever, are usually seeking "soul mates" (while men are seeking "playmates"). Spring (1997) notes that "Women believe that their affair is justified when it is for love; men, when it's *not* for love." (Seventy-seven percent of women who have had affairs, compared to 43% of the men, cite love as their justification [Townsend, 1995]). Men who have had affairs are more likely than women to cite a need for sexual excitement as a justification—75% versus 55% (Glass & Wright, 1992).

These data support the view, expressed repeatedly throughout this text, that women are less accepting of sex without emotional involvement (Townsend, 1995). Men are more likely than women to "separate sex and love; women appear to believe that love and sex go together and that falling in love justifies sexual involvement" (Glass & Wright, 1992, p. 361). Men (whether single, married, or cohabiting) are also generally more approving of extramarital affairs than women are (Glass & Wright, 1992). But note that these are all *group* differences. Many individual men are interested primarily in the extramarital relationship rather than the sex per se. Similarly, many women are out for the sex and not the relationship.

Patterns of Extramarital Sex

Let us begin with a few definitions. **Extramarital** sex (an "affair") is usually conducted without the spouse's knowledge or approval. Secret affairs are referred to as **conventional adultery,** infidelity, or simply "cheating." Conventional adultery runs the gamut from the "one-night stand" to the affair that persists for years. (Bill Clinton's affair with Gennifer Flowers was alleged to have continued for a dozen years.) In **consensual adultery,** extramarital relationships are conducted openly—that is, with the knowledge and consent of the partner. In what is called **swinging, comarital sex,** or mate swapping, the partner participates.

How many people "cheat" on their spouses? Viewers of TV talk shows may get the impression that everyone cheats, but surveys paint a different picture. In surveys conducted between 1988 and 1996 by the respected National Opinion Research Center, about one husband in four or five, and one wife in eight, admitted to marital infidelity (Alterman, 1997; "Cheating," 1993). Similarly, more than 90% of the married women and 75% of the married men in the NHSLS study reported remaining loyal to their spouses (Laumann et al., 1994). The vast majority of people who were cohabiting also reported that they were sexually faithful to their partners while they were living together (Laumann et al., 1994). Similarly, the overwhelming majority—86%—of respondents to a *New York Times* poll reported that they were "absolutely certain" that their partners were faithful to them (Eggers, 2000). What can we conclude? Perhaps two things: One is that men are about twice as likely as women to admit to affairs. The other is that only a minority of married people admit to affairs.

Those are the conclusions, but note that we said "*admit* to affairs." Having presented the percentages of reported extramarital sex, we point out that these reports cannot be verified. Even when they are assured of anonymity, people may be reluctant to reveal they have "cheated" on their spouses. There is likely to be an overall tendency to underreport the incidence of extramarital sex.

Attitudes Toward Extramarital Sex

The sexual revolution does not seem to have changed attitudes toward extramarital sex. About nine out of ten Americans say that affairs are "always wrong" or "almost always wrong" (Alterman, 1997). Three out of four Americans say that extramarital sex is "always wrong" (Berke, 1997). Another one in seven say it is "almost always wrong." Only about 1% say that extramarital sex is "not at all wrong." Most married couples embrace the value of monogamy as the cornerstone of their marital relationship (Blumstein & Schwartz, 1990).

Extramarital sex Sexual relations between a married person and someone other than his or her spouse.

Conventional adultery Extramarital sex that is kept hidden from one's spouse.

Consensual adultery Extramarital sex that is engaged in openly with the knowledge and consent of one's spouse.

Swinging A form of consensual adultery in which both spouses share extramarital sexual experiences. Also referred to as *mate swapping.*

Comarital sex Swinging; mate swapping.

Human Sexuality ONLINE //

"Someone to Watch Over Me"? Yes—Snoopware

We went to the SpectorSoft Web site (www.spectorsoft.com), and the following ad leapt at us from our monitor:

> Secretly record everything your spouse, children, and employees do online. Install Spector on your PC and it will record EVERYTHING anyone does on the Internet. Spector SECRETLY takes hundreds of snapshots every hour, very much like a surveillance camera. With Spector, you will be able to SEE what your kids and employees have been doing online and offline. Download now in less than 5 minutes. $49.95.

Spector is an example of "snoopware" (Lewis, 2000). Imagine that a phone-answering machine secretly recorded everything you said on it and then sent the audio clips to your spouse or parents or boss. It's as though a private investigator were standing behind you with a video camera and shooting, over your shoulder, a running record of everything that comes across your monitor—every Web site, every e-mail, every instant message, every file opened, every password, every credit card number, every sales transaction. It has been said that you can learn nearly everything you need to know about someone by going through his or her electronic garbage. Snoopware not only goes through your electronic garbage; it also makes a record of everything you say or save. Your privacy is a thing of the past. Snoopware is the most successful voyeur of them all.

Spy software was originally used primarily by law enforcement agencies and large corporations. But Spector, eBlaster, Cyber Snoop, the 007 Stealth Activity Recorder and Reporter, and similar spy programs are available to consumers today. Spector was originally designed to permit parents to track the Web sites and chat rooms visited by their children. The intention was to allow parents to be watchful but clandestine guardians. But the main customers today are suspicious spouses, mistrustful bosses, and private investigators.

Stealth Mode

The program's "stealth mode" is truly frightening. Only the person who installed it knows that it is peeping in the background. When the object of investigation is gone, the investigator keyboards a secret combination of keys (Crtl-Alt-Shift-S, all held at once) and a password. Everything that showed up on the user's monitor is then played back. Like the stealth bomber, the program is invisible to the computer user's personal radar. It isn't listed on any directory (don't bother instructing your computer to "find" a program with *Spector* or *snoop* in the title). Even expert teenage hackers won't locate it. (eBlaster is intended for people who do not have regular access to their target's computer. It hides in the background, like Spector, and then, every 15 minutes or so, sends a secret e-mail with the revealing information to whatever address is specified by the installer. The reports are sent when the target is connected to the Internet. It leaves no e-mail trail. It doesn't use the modem to dial the Internet itself; dialing could alert the target that there's more here than meets the eye and ear).

Spector and similar programs are not web filters like SurfWatch and Net Nanny, which prevent employees and children from accessing porno and other undesirable sites on the Net. Rather, they spy on people who erroneously assume that as long as no one made of flesh and blood is watching them, the images that flash across their monitor are known to them alone.

Testimonials

To those who are offended by the very existence of this software, SpectorSoft offers customer testimonials like the following:

- A Nashville woman used Spector to discover that her husband regularly visited porno sites and sex-related chat rooms. Spector replayed messages he sent women he met online, proposing liaisons and bragging about extramarital affairs throughout his marriage. The woman's divorce lawyer plopped dozens of pages of printouts in his lap, all time-stamped by the software. "Spector is no different than going out and hiring a private investigator," she said (cited in Lewis, 2000). "All it is is a more sophisticated way of doing something women and men have done for centuries."

It is not true that most of today's sophisticated young people see nothing wrong with an occasional extramarital fling. The sexual revolution never extended itself to extramarital affairs—at least not among the majority of married people.

CROSS-CULTURAL PERSPECTIVES ON EXTRAMARITAL SEX Most preliterate societies prohibit extramarital relationships for one or both spouses (Frayser, 1985). In cultures in which extramarital sex is permitted, husbands are typically allowed greater sexual freedom than wives. Frayser could find no societies that allowed extramarital sex for wives but not husbands.

Slightly more than half of the preliterate societies around the world permit men to have extramarital partners (Harris & Johnson, 2000; Whitten, 2001). Only about one in ten allows women the same opportunity. In some societies extramarital affairs are formalized within rules of social conduct. Among the Aleut people of Alaska's Aleutian Is-

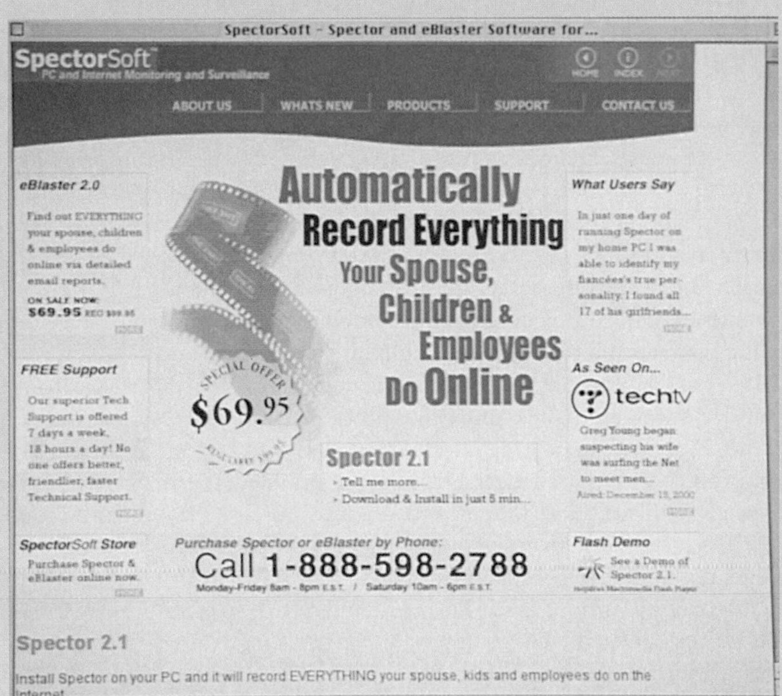

"Snoopware." It has been said that you can learn nearly everything you need to know about someone by going through his or her garbage. Snoopware goes through your electronic garbage; it makes a record of everything you say or save. What is most frightening is that someone could install snoopware on your computer and you would never know it.

- A Kansas man installed Spector on his wife's computer. The snoopware dutifully recorded e-mails between his wife and her boyfriend. They revealed not only the affair but also the boyfriend's wish to harm the husband physically. When confronted with the Spector record, his wife agreed to a divorce, moved out of town, and surrendered custody of the children without a fight.
- In yet another case, a woman hired a private investigator to check out her fiancé. He installed the software on the fiancé's computer and revealed that he was a philanderer. The marriage didn't happen.

Moral and Legal Questions

Spector raises many moral and legal questions. The legality of snoopware has not been adequately tested, but there are some obvious parallels between snoopware and tapping someone's phone line.

"Based on my reading of different state statutes, it is not clear if it is illegal or not," says Mike Godwin (cited in Lewis, 2000), the lawyer who wrote *Cyber Rights: Defending Free Speech in the Digital Age*. However, "If someone ever installed this software on my computers in my home without my permission or knowledge," he would take them to court.

lands, for example, men may offer visitors the opportunity to sleep with their wives as a gesture of hospitality. Among the Chukchee of Siberia, who often travel long distances from their homes, a married man is allowed to engage in sexual activity with his host's wife. The understanding is that he will reciprocate when the host visits him (Ford & Beach, 1951).

Kinship ties often determine sexual access to extramarital partners (Frayser, 1985). Among the people of the Marshall Islands in the Pacific, a woman is allowed to have a sexual relationship with her sister's husband. Among the Native American Comanches, a man is permitted to have intercourse with his brother's wife, if his brother consents. Customs sometimes limit extramarital intercourse to ceremonial occasions. The Fijians of Oceania, for example, engage in extramarital relationships only following the return of their men from warfare. In Western society, Mardi Gras and some out-of-town conventions represent ceremonial occasions for extramarital liaisons that might not be tolerated otherwise.

Effects of Extramarital Sex

The discovery of infidelity can evoke a range of emotional responses. The spouse may be filled with anger, jealousy, even shame. Feelings of inadequacy and doubts about one's attractiveness and desirability may surface. Infidelity may be seen by the betrayed spouse as a serious breach of trust and intimacy. Marriages that are not terminated in the wake of the disclosure may survive in a damaged condition (Charny & Parnass, 1995).

The harm an affair does to a marriage may reflect the meaning of the affair to the individual and his or her spouse. Deborah Lamberti, director of a counseling and psychotherapy center in New York City, points again to women's traditional intertwining of sex with relationships and argues that "Men don't view sex with another person as a reason to leave a primary relationship" (1997, pp. 131–132). Women may recognize this and be able to tell themselves that their husbands are sleeping with someone else just for physical reasons. But women are more concerned about remaining monogamous. Therefore, if a woman is sleeping with another man, she may already have a foot out the door, so to speak. Alterman (1997) also notes that a wife's affair may be an unforgivable blow to the husband's ego or pride. A woman may be more likely to see the transgression as a threat to the structure of her life.

If a person has an affair because the marriage is deeply troubled, the affair may be one more factor that speeds its dissolution. The effects on the marriage may depend on the nature of the affair. It may be easier to understand that a spouse has fallen prey to an isolated, unplanned encounter than to accept an extended affair (Charny & Parnass, 1995). In some cases the discovery of infidelity stimulates the couple to work to improve their relationship. If the extramarital activity continues, however, it may undermine the couple's efforts to restore their relationship.

Swinging

Swinging—also called "mate swapping" or **comarital sex**—is a form of consensual adultery in which both partners openly share sexual experiences with other people. Most swingers seek to avoid emotional entanglements with their swinging partners, but they may fail to separate their emotions from their sexual activity. Emotional intimacy between swinging partners can be even more threatening to the swingers' primary relationships than sexual intimacy. For various reasons, most swinging couples drop out after a short period of experimentation.

Swingers tend to be European American, to be fairly affluent and well-educated, and to have only nominal religious affiliations (Jenks, 1985). Solid statistics on the prevalence of swinging are lacking, however. Observers of sexual trends suggest that swinging, like other nontraditional forms of matrimony, appear to have largely vanished from the social scene (Harris & Johnson, 2000; Whitten, 2001). Concern over AIDS, coupled with the more conservative social climate of recent years, seems to have restricted swinging to a few devotees. Even during the heyday of the sexual revolution, only 2% of married males and fewer than 2% of married females in the Hunt survey had "swung," and many of these individuals had tried it just once (Hunt, 1974).

DOMESTIC VIOLENCE

Although the O. J. Simpson case may have brought the problem of domestic violence into the public eye, such battering is a national epidemic. At least one woman in eight is subjected to violence at the hands of her partner each year, and about 2,000 women die from it annually (Kyriacou et al., 1999; Zalar, 2000).

Women are more likely to be raped, injured, or killed by their current or former partners than by other assailants (Kyriacou et al., 1999; Zalar, 2000). Yet women are as likely as men to be violent (Magdol et al., 1997). In about half of the couples in whom domestic violence occurs, both partners are guilty of physical abuse. However, women are more likely than men to sustain physical injuries such as broken bones and other internal damage (Magdol et al., 1997).

Truth or Fiction? **REVISITED**

Actually, women are as likely as their male partners to engage in domestic violence. However, women are more likely to sustain serious injury.

It also appears that men are more likely to act, or go on the attack, whereas women are more likely to *react.* That is, male violence toward women often stems from factors that threaten their traditional dominance in relationships, such as unemployment and substance abuse. Women are often violent because of the stress of coping with an abusive partner (Magdol et al., 1997).

Domestic violence cuts across all strata of our society, but it is reported more commonly among people of lower socioeconomic levels. This difference may reflect the greater amount of stress experienced by people who are struggling financially. In many cases, a disparity in income between partners, such that the woman earns more than the man, not poverty per se, seems to contribute to domestic violence (McCloskey, 1996).

When socioeconomic level and other demographic factors are taken into account, ethnic or racial group differences are unrelated to the risk of domestic violence (Kyriacou et al., 1999; Zalar, 2000). Generally speaking, lack of opportunity and alienation from the larger society increase the potential for mental health problems, violence, and abuse within relationships (Kyriacou et al., 1999; Zalar, 2000). These factors tend to be found more often among members of ethnic minority groups.

Domestic violence often follows a triggering event such as criticism or rejection by one's partner or incidents that cause the man to feel trapped, insecure, or threatened (Kyriacou et al., 1999; Zalar, 2000). The use of alcohol or other drugs is also heavily connected with battering (Brookoff et al., 1997). Male batterers often have low self-esteem and a sense of personal inadequacy (Kyriacou et al., 1999; Zalar, 2000). They may become dependent on their partners for emotional support and feel threatened if they perceive their partners becoming more independent or developing separate interests.

Feminist theorists look upon domestic violence as a product of the power relationships that exist between men and women in our society. Men are socialized into dominant roles in which they expect women to be subordinate to their wishes (Wilson & Daly, 1996). Men learn that aggressive displays of masculine power are socially sanctioned and even glorified in some settings, such as in the athletic arena. These role expectations, together with a willingness to accept interpersonal violence as an appropriate means of resolving differences, create a context for domestic violence when the man perceives his partner to be threatening his sense of control or failing to meet his needs or respect his wishes. Men who batter may also have less power in their relationships and may be trying to make up for it through the use of force.

Feminist theorists argue further that our society supports domestic violence by appearing to condone it (Kyriacou et al., 1999; Zalar, 2000). The man who beats his wife may be taken aside and "talked to" by a police officer, rather than arrested on the spot. Even if he is arrested and convicted, his punishment is likely to be less severe (sometimes just a "slap on the wrist") than if he had assaulted a stranger.

Aftermath of Domestic Violence Former Boston Red Sox outfielder Wilfredo Cordero pleaded guilty to charges of beating his wife. He received a suspended sentence and was ordered to attend counseling for batterers. Were his arrest and sentencing a signal that the justice system takes domestic violence seriously, or was the fact that his sentence was suspended a sign that the courts still consider domestic violence a "family matter"? In any event, domestic violence usually worsens unless the victim leaves or reports the crime to the authorities.

DIVORCE

> My wife and I were considering a divorce, but after pricing lawyers we decided to buy a new car instead. —Henny Youngman
>
> Whenever I date a guy, I think, is this the man I want my children to spend their weekends with? —Rita Rudner

At least half of the marriages in the United States end in divorce (Carrère et al., 2000; Laumann et al., 1994). The divorce rate in the United States rose steadily through much of the twentieth century before leveling off in the 1980s. About one quarter (26%) of children below the age of 18 live in single-parent households (Barringer, 1991). Divorced women outnumber divorced men, in part because men are more likely to remarry following divorce (Saluter, 1992).

The relaxation of legal restrictions on divorce, especially the introduction of the so-called no-fault divorce, has made divorces easier to obtain. Until the mid-1960s, adultery was the only legal grounds for divorce in New York state. Other states were equally strict. But now no-fault divorce laws have been enacted in nearly every state, allowing a divorce to be granted without a finding of marital misconduct. The increased economic independence of women has also contributed to the rising divorce rate. More women today have the economic means of breaking away from a troubled marriage. Today, more people consider marriage an alterable condition than in prior generations.

People today hold higher expectations of marriage than did their parents or grandparents. They expect marriage to be personally fulfilling as well as to function as an institution for rearing children. Many demand the right to be happy in marriage. The most common reasons given for a divorce today are problems in communication and a lack of understanding. Key predictors of divorce today include a husband's criticism, defensiveness, contempt, and stonewalling—not lack of support (Carrère et al., 2000; Gottman et al., 1998).

Why do Americans believe that the divorce rate has risen so high? A *Time*/CNN poll asked a national sample, "Which is the main reason for the increase in the number of divorces?" Answers are shown in Table 14.8. Respondents were almost equally split in their answer to the question "Do you believe it should be harder than it is for married couples to get a divorce?" Half (50%) said yes, and 46% said no.

The Cost of Divorce

Divorce often causes financial and emotional problems. When a household splits, the resources often cannot maintain the earlier standard of living for each partner. Financially

TABLE 14.8

The main reason for the increase in the number of divorces in the United States, according to a *Time*/CNN poll

Reason	Percent of Respondents Citing Reason as Main Reason
Marriage is not taken seriously by the couples.	45%
Society has become more accepting of divorced people.	15
It is easier to get divorced than it used to be.	10
People who get divorced are selfish.	9
Changes in the earning power of women and men.	7
All of the above	2

Source: From "The Ties That Bind" by Walter Kirn, *Time*, August 18, 1997, pp. 48–50. Copyright © 1997 by Time, Inc. Reprinted by permission.

speaking, divorce hits women harder than men. According to the Population Reference Bureau report *Women, Work and Family in America,* a woman's household income drops by about 24% (Bianchi & Spain, 1997). A man's declines by about 6%. Women who have not pursued a career may have to struggle to compete with younger, more experienced workers. Divorced mothers often face the combined stress of being solely responsible for rearing their children and needing to increase their incomes to make ends meet. Divorced fathers may find it difficult to pay alimony and child support while attempting to establish a new lifestyle.

Divorce can also prompt feelings of failure as a spouse and parent, loneliness and uncertainty about the future, and depression. Married people appear to be better able to cope with the stresses and strains of life, perhaps because they can rely on each other for emotional support. Divorced and separated people have the highest rates of physical and mental illness in the population (Carrère et al., 2000; Gottman et al., 1998). Divorced people have higher rates of suicide than married people (Carrère et al., 2000; Gottman et al., 1998). On the other hand, divorce can be a time of personal growth and renewal. It can provide an opportunity for people to take stock of themselves and establish a new, more rewarding life.

Children are often the biggest losers when parents get a divorce (Ellis, 2000). On the other hand, chronic marital conflict or fighting is also connected with serious psychological distress in children and adolescents (Ellis, 2000; Erel & Burman, 1995; Harold et al., 1997). Boys have more trouble adjusting to conflict or divorce and may exhibit conduct problems at school and increased anxiety and dependence (Grych & Fincham, 1993; Holden & Ritchie, 1991).

Wallerstein and Blakeslee (1989) reported that about four out of ten children in their case studies showed problems such as anxiety, academic underachievement, decreased self-worth, and anger ten years after the divorce. A "sleeper effect" was also described. Apparently well-adjusted children of divorce developed problems in early adulthood, especially difficulties trusting that their partners in intimate relationships would make lasting commitments. Researchers attribute children's problems after divorce not only to the divorce itself but also to a subsequent decline in the quality of parenting. Children's adjustment is enhanced when parents maintain their parenting responsibilities and set aside their differences long enough to agree on child-rearing practices (Wallerstein & Blakeslee, 1989). Children of divorce also benefit when divorced parents encourage each other to continue to play important roles in their children's lives and avoid saying negative things about each other in their children's presence.

Despite the difficulties involved in adjusting to a divorce, most divorced people eventually bounce back. The majority remarry. Among older people, divorced men are more likely than divorced women to remarry—in part because men usually die earlier than women (and hence fewer prospective husbands are available), in part because older men tend to remarry younger women.

Remarriages are even more likely than first marriages to end in divorce (Lown & Dolan, 1988). One reason is the selection factor. That is, among people who get married, divorced people are a subgroup who are relatively less inclined than others to persist in a troubled marriage. Many divorced people who remarry are also encumbered with alimony and child-support payments that strain their new marriages. Many bring children from their earlier marriages to their new ones.

The stepfamily may soon be the most common family unit in the United States (CBS News, 1991). Six of ten stepfamilies eventually disband, often under the weight of the financial and emotional pressures of coping with the demands of a reconstituted family (CBS News, 1991). However, among remarriages that survive, the level of personal happiness in spouses is as high as among spouses in first marriages and much higher than the level among divorced people (Weingarten, 1985).

What of the long-term consequences of divorce? In one study, only one person in five reported, five years after the fact, that divorce had been a mistake (Wallerstein & Kelly, 1980). Most felt that the divorce had enhanced their lives. Most, however, also reported that they had underestimated the emotional pain they would endure.

ALTERNATIVE LIFESTYLES

Marriages are generally based on the expectation of sexual exclusivity. Alternative or nontraditional lifestyles, however, such as open marriages and group marriages, permit intimate relationships with people outside the marriage. Such alternative lifestyles attracted a flurry of attention during the heyday of the sexual revolution in the 1970s, but even then, they were more often talked about than practiced (Harris & Johnson, 2000; Whitten, 2001). Today, they attract still fewer adherents.

Open Marriage

Open marriage is based on the view that people's needs for intimacy are unlikely to be gratified through one relationship. Proponents argue that the core marriage can be enhanced if the partners have the opportunity to develop emotionally intimate relationships with others (O'Neill & O'Neill, 1972).

The prevalence of open marriages, like that of swinging, remains unknown. Nor are there much data on which to base conclusions about the success of open marriages. Rubin found no meaningful differences in marital adjustment between 130 couples with sexually open marriages and 130 couples who maintained sexual exclusivity. Still another study found no differences in longevity between sexually open and sexually exclusive marriages over a five-year period (Rubin & Adams, 1986).

Group Marriage

Group marriage also attracted some adherents during the sexual revolution of the 1960s and 1970s. In a group marriage, three or more people share an intimate relationship, although they are not legally married. Each member feels committed or married to at least two others. The major motive for group marriage is extension of intimacy beyond one spouse in order to increase personal fulfillment (Constantine & Constantine, 1973). Members of eight out of nine group marriages surveyed by the Constantines also admitted interest in a variety of sexual partners, however.

Group marriages differ from swinging in that participants share their innermost thoughts and feelings and expect their bonds to be permanent. Group marriages, therefore, are not perceived merely as vehicles for legitimizing mate swapping. Adherents expect children to profit by observing several adult role models and by escaping the "smothering" possessiveness of exclusive parent–child relationships.

Although they provide sexual variety, group marriages require adjustment to at least two other people—not just one. Moreover, legal and social problems can arise with respect to issues such as paternity and inheritance. Managing money may also be stressful, with arguments arising over joint accounts and who can spend how much for what. Sexual jealousies often arise as well. Given these problems, it is not surprising that group marriages are rare and have a high failure rate.

Group marriages are not unique to our culture. Although the practice has arisen rarely, other cultures have developed analogous marital customs (Werner & Cohen, 1990). Among the Chukchee people of Siberia, for example, group marriages sometimes involved as many as 10 couples. The men in such marriages, who considered themselves "companions-in-wives," had sexual rights to each wife.

All in all, the ideal of the traditional marriage remains strong in our culture. Men are somewhat more likely than women to participate or express an interest in lifestyles that permit greater sexual freedom and, perhaps, entail less personal responsibility (Knox, 1988). (Why is the third author not surprised?) But even most of those who have tried alternative lifestyles such as cohabitation, open marriage, or group marriage enter into traditional marriages at some time.

All in all, most adults in the United States still seem to feel about marriage the way Winston Churchill felt about democracy: It's flawed, laden with problems, and frustrating—but preferable to the alternatives.

Open marriage A marriage characterized by the personal privacy of the spouses and the agreed-upon liberty of each spouse to form intimate relationships, which may include sexually intimate relationships, with people other than the spouse.

Group marriage A social arrangement in which three or more people share an intimate relationship. Group marriages are illegal in the United States.

SEX IN THE LATER YEARS

Which is the fastest-growing segment of the U.S. population? People age 65 and above. More than 30 million people in the United States are senior citizens, and their number is growing fast. This "graying" of the United States may have a profound effect on our views of older people, especially concerning their sexuality. Many people in our culture see sexual activity as appropriate only for the young (Reiss, 1988). This belief falls within a constellation of unfounded cultural myths about older people, which includes the notions that older people are sexless, that older people with sexual urges are abnormal, and older males with sexual interests are "dirty old men."

Researchers find that sexual daydreaming, sex drive, and sexual activity tend to decline with age, whereas negative sexual attitudes tend to increase (Purifoy et al., 1992). However, research does not support the belief that people lose their sexuality as they age. Nearly all (95%) of the older people in one sample reported that they liked sex, and 75% reported that orgasm was essential to their sexual fulfillment (Starr & Weiner, 1981). People who are exposed to cultural views that sex among older people is deviant may renounce sex as they age, however. Those who remain sexually active may be bothered by guilt (Reiss, 1988).

Sexual activity among older people, as among other groups, is influenced not only by physical structures and changes but also by psychological well-being, feelings of intimacy (Shaw, 1994), and cultural expectations.

Physical Changes

Although many older people retain the capacity to respond sexually, physical changes do occur as the years pass (see Table 14.9). If we are aware of them, we will not view them as abnormal or find ourselves unprepared to cope with them. Many potential problems can be averted by adjusting our expectations or making some changes to accommodate the aging process.

CHANGES IN THE FEMALE Many of the physical changes in women stem from decline in the production of estrogen around the time of menopause. The vaginal walls lose much elasticity and the thick, corrugated texture of the childbearing years. They grow paler and

TABLE 14.9

Changes in sexual arousal often associated with aging

Changes in the Female	Changes in the Male
Reduced myotonia (muscle tension)	Longer time to erection and orgasm
Reduced vaginal lubrication	Need for more direct stimulation for erection and orgasm
Reduced elasticity of the vaginal walls	Less semen emitted during ejaculation
Smaller increases in breast size during sexual arousal	Erections may be less firm
Reduced intensity of muscle spasms at orgasm	Testicles may not elevate as high into the scrotum
	Less intense orgasmic contractions
	Lessened feeling of a need to ejaculate during sex
	Longer refractory period

Source: Copyright © 1990 by The Kinsey Institute for Research in Sex, Gender, and Reproduction. From *The Kinsey Institute New Report on Sex.* Reprinted with permission from St. Martin's Press, New York.

Sexuality in Late Adulthood. Are older people sexually active? If they are, are they abnormal or deviant? Although young people often find it difficult to imagine older people engaging in sexual activity, it is normal to retain sexual interest and activity for a lifetime.

thinner. Thus coitus may become irritating. The thinning of the walls may also exert greater pressure on the bladder and urethra during coitus, leading in some cases to urinary urgency and burning urination that may persist for days.

The vagina also shrinks in size. The labia majora lose much of their fatty deposits and become thin. The introitus becomes relatively constricted, and penile entry may become somewhat difficult. This "problem" has a positive aspect: Increased friction between the penis and vaginal walls may heighten sexual sensations. The uterus decreases in size after menopause and no longer becomes so congested during sexual arousal. Following menopause, women also produce less vaginal lubrication, and the lubrication that is produced may take minutes, rather than seconds, to appear. Lack of adequate lubrication is also a major reason for painful coitus.

Many of these changes may be slowed or reversed through estrogen-replacement therapy (see Chapter 3) or topical application of estrogen-containing cream. Natural lubrication may also be increased through more elaborate foreplay. The need for more foreplay may encourage the man to become a more considerate lover. (Older men too are likely to need more time to become aroused.) An artificial lubricant can also ease problems posed by difficult entry or painful thrusting.

Women's breasts show smaller increases in size with sexual arousal as they age, but the nipples still become erect. Because the muscle tone of the urethra and anal sphincters decreases, the spasms of orgasm become less powerful and fewer in number. Thus orgasms may feel less intense. The uterine contractions that occur during orgasm become discouragingly painful for some postmenopausal women. Despite these changes, women can retain their ability to achieve orgasm well into their advanced years. The subjective experience of orgasm also remains highly satisfying, despite the lessened intensity of muscular contractions.

Truth or Fiction? **REVISITED**

It is not true that few women reach orgasm after the age of 70. Actually, healthy women can retain the capacity for orgasm well into their advanced years.

CHANGES IN THE MALE Age-related changes tend to occur more gradually in men than in women and are not clearly connected with any one biological event, as they are with menopause in the woman. Male adolescents may achieve erection in a matter of seconds through sexual fantasy alone. After about age 50, men take progressively longer to achieve erection. Erections become less firm, perhaps because of lowered testosterone production. Older men may require minutes of direct stimulation of the penis to achieve an erection. Couples can adjust to these changes by extending the length and variety of foreplay.

Most men remain capable of erection throughout their lives. Erectile dysfunction is not inevitable with aging. Men generally require more time to reach orgasm as they age, however, which may also reflect lowered testosterone production. In the eyes of their sex partners, however, delayed ejaculation may make them better lovers.

The testes often decrease slightly in size and produce less testosterone with age. Testosterone production usually declines gradually from about age 40 to age 60 and then begins to level off. However, the decline is not inevitable and may be related to the man's general health. Sperm production tends to decline as the seminiferous tubules degenerate, but viable sperm may be produced quite late in life. Men in their 70s, 80s, and even 90s have fathered children.

Nocturnal erections tend to diminish in intensity, duration, and frequency as men age, but they do not normally disappear in healthy men (Reinisch, 1990; Schiavi et al., 1990). The refractory period tends to lengthen with age. An adolescent may require only a few minutes to regain erection and ejaculate again after a first orgasm, whereas a man in his 30s may require half an hour. Past age 50, the refractory period may increase to several hours.

Older men produce less ejaculate, and it may seep rather than shoot out. Though the contractions of orgasm still begin at 0.8-second intervals, they become weaker and fewer. Still, the number and strength of spasms do not translate precisely into subjective pleasure. An older male may enjoy orgasm as thoroughly as he did at a younger age. Attitudes and expectations can be as important as the contractions themselves.

An 82-year-old man commented as follows on his changing sexual abilities:

> I come maybe once in every three sexual encounters these days with my wife. My erection comes and goes, and it's not a big concern to us. I get as much pleasure from touching and thrusting as I do from an ejaculation. When I was younger it was inconceivable to me that I might enjoy sex without an orgasm, but I can see now that in those days I missed out on some pleasure by making orgasm such a focus. (Gordon & Snyder, 1989, p. 153)

Following orgasm, erection subsides more rapidly than it does in a younger man. A study of 65 healthy men aged 45 to 74 showed an age-related decline in sexual desire, arousal, and activity. Yet there were no differences between younger and older men in level of sexual satisfaction or enjoyment (Schiavi et al., 1990).

In sum, most physical changes do not bring a man's or a woman's sex life to a grinding halt. People's attitudes, sexual histories, and partners are usually more important factors in sexual behavior and enjoyment.

Patterns of Sexual Activity

Despite the decline in certain physical functions, older people can continue to lead a vibrant, fulfilling sex life. In fact, years of sexual experience may more than compensate for any diminution of physical responsiveness (Hodson & Skeen, 1994). A Roper Starch poll of 1,292 people in the United States aged 60 and above found that 74% of the men and 70% of the women who had remained sexually active said they were as satisfied with sex as they were in their 40s or that they were *more* satisfied (Leary, 1998). Unfortunately, people who overreact to expected changes in sexual response may conclude that their sex lives are over and give up on sexual activity or even on expressing any physical affection (Dunn, 1998).

In Kinsey's samples, 94% of the men and 84% of the women remained sexually active at the age of 60. Half of the 60- to 91-year-olds surveyed by Starr and Weiner (1981) reported having sexual relations on a regular basis—and half of these, at least once a week. A study of 200 healthy 80- to 102-year-olds found that 30% of the women and 62% of the men still engaged in intercourse (Bretschneider & McCoy, 1988). A study of 100 older men in England found that the key factor in whether they continued to engage in sexual activity was the availability of a partner, not physical condition (Jones et al., 1994). Sex therapist Helen Singer Kaplan (1990) concluded,

> The loss of sexuality is not an inevitable aspect of aging. . . . The results of these studies are remarkable in their consensus: Without exception, each investigator found that, providing they are in good health, the great majority of people remain sexually functional and active on a regular basis until virtually the end of life. Or, to put it more succinctly, 70% of healthy 70-year-olds remain sexually active, and are having sex at least once a week, and typically more often than that. . . .
>
> Although it is widely believed that sex no longer matters after middle age, the opposite is true, and sex often becomes *more* and not *less* important as a person grows older. Because sex is among the last pleasure-giving biological processes to deteriorate, it is potentially an enduring source of gratification at a time when these are becoming fewer and fewer, and a link to the joys of youth. These are important ingredients in the [older] person's emotional well-being. (pp. 185, 204)

Coital frequency tends to decline with age (Laumann et al., 1994). One study of people married for more than 50 years showed that nearly half (47%) had discontinued intercourse, and 92% reported a decline over the years (Ade-Ridder, 1985). Several factors played a role in declining activity, including physical problems, boredom, and cultural attitudes toward sex among the aging. Despite general trends, sexuality among older people is variable (Knox, 1988). Many older people engage in intercourse, oral sex, and masturbation at least as often as when they were younger; some become disgusted by sex; others simply lose interest. A six-year longitudinal study of older married couples found that one in five actually increased their coital frequency over time (Palmore, 1981).

Coital frequency, however, is not synonymous with sexual satisfaction. In a Canadian study of 215 married people who were middle-aged and older (51 to 81 years of age), those aged 65 or above showed lower coital frequency than younger respondents (Libman, 1989). No sizable differences emerged in the level of sexual satisfaction between older and younger groups, however.

The frequency of masturbation also generally declines with age for both men and women, although an increase may occur following a marital separation, a divorce, or the death of a spouse (Hegeler & Mortensen, 1977). Still, continued masturbation was reported by nearly half (46%) among a sample of people aged 60 to 91 (Starr & Weiner, 1981). This is a high level of acceptance among people who were reared during a time when masturbation was generally viewed as harmful (Kammeyer, 1990).

Couples may accommodate to the physical changes of aging by broadening their sexual repertoire to include more diverse forms of stimulation. Many respondents to a *Consumer Reports* survey reported using oral–genital stimulation, sexual fantasy, sexually explicit materials, anal stimulation, vibrators, and other sexual techniques to offset problems in achieving lubrication or erection (Brecher, 1984). Sexual satisfaction may be derived from manual or oral stimulation, cuddling, caressing, and tenderness, as well as from intercourse to orgasm.

The availability of a sexually interested and supportive partner may be the most important determinant of continued sexual activity. Many women discontinue sexual activity because of the death of their husbands. Women's life expectancy exceeds men's by an average of seven years.

SEX AND DISABILITY

Like older people, people with disabilities (especially those whose physical disabilities render them dependent on others) are often seen as sexless and childlike (Nosek et al., 1994). Such views are based on misconceptions about the sexual functioning of people with disabilities. Some of these myths and stereotypes may be eroding, however, in part because of the success of the civil and social rights movements of the disabled in the 1970s and the attention focused on the sexuality of people with disabilities in films such as *Coming Home*, *Born on the Fourth of July*, and *My Left Foot*.

A person may have been born with or acquire a bodily impairment or suffer a loss of function or a disfiguring change in appearance. Although the disability may require the person to make adjustments in order to perform sexually, most people with disabilities have the same sexual needs, feelings, and desires as people without disabilities. Their ability to express their sexual feelings and needs depends on the physical limitations imposed by their disabilities, their adjustment to their disabilities, and the availability of partners. The establishment of mature sexual relationships generally demands some distance from one's parents. Therefore, people with disabilities who are physically dependent on their parents may find it especially difficult to develop sexual relationships (Knight, 1989). Parents who acknowledge their children's sexual development can be helpful by facilitating dating. Far too often, parents become overprotective:

> Adolescent disabled girls have the same ideas, hopes, and dreams about sexuality as able-bodied girls. They will have learned the gender role expectations set for them by the media and others and may experience difficulty if they lack more substantive educational information about sexuality and sex function. In addition their expectations may come in conflict with the family which may have consistently protected or overindulged the child and not permitted her to "grow up." . . . In many cases the families are intensely con-

> cerned about the sexual and emotional vulnerability of the daughter and hope that "nothing bad" will happen to her. They may, therefore, encourage her to wear youthful clothing and to stay a safe little girl. The families can mistakenly assume there may be no sexual life ahead of her and protect her from this perceived bitter reality with youthful clothing and little-girlish ways. The result can be, of course, that the young emerging woman may become societally handicapped in learning how to conduct herself as a sexual woman. She will be infantilized. (Cole, 1988, pp. 282–283)

In such families, young people with disabilities get the message that sex is not for them. As they mature, they may need counseling to help them recognize the normalcy of their sexual feelings and to help them make responsible choices for exploring their sexuality.

Physical Disabilities

According to Margaret Nosek and her colleagues (1994), sexual wellness, even among the disabled, involves five factors:

- Positive sexual self-concept; seeing oneself as valuable sexually and as a person
- Knowledge about sexuality
- Positive, productive relationships
- Coping with barriers to sexuality (social, environmental, physical, and emotional)
- Maintaining the best possible general and sexual health, given one's limitations

This model applies to all of us, of course. Let us now consider aspects of specific physical disabilities and human sexuality.

CEREBRAL PALSY **Cerebral palsy** does not generally impair sexual interest, capacity for orgasm, or fertility (Reinisch, 1990). Depending on the nature and degree of muscle spasticity or lack of voluntary muscle control, however, afflicted people may be limited to certain types of sexual activities and coital positions.

The importance of sex education for disabled people is highlighted by a case of a man with cerebral palsy:

> A 40-year-old man with moderate cerebral palsy came to see me for sexual counseling. When asked why he came for counseling, he said, "I think I'm old enough to learn about sex." He was college educated, fully employed, and had been living by himself, away from his parents, for three years. In questioning him further, I found that the extent of his sexual knowledge was he knew he had a penis but knew nothing about the human sexual response or about female anatomy. When I asked him if any "white, sticky stuff" ever came out of his penis, he replied, "Yes, and doesn't that have something to do with my cerebral palsy?" (Knight, 1989, p. 186)

People with disabilities such as cerebral palsy often suffer social rejection during adolescence and perceive themselves as unfit for or unworthy of intimate sexual relationships, especially with people who are not disabled. They are often socialized into an asexual role. Sensitive counseling can help them understand and accept their sexuality, promote a more positive body image, and provide the social skills to establish intimate relationships (Edmonson, 1988).

SPINAL CORD INJURIES People who suffer physical disabilities as the result of traumatic injuries or physical illness must not only learn to cope with their physical limitations but also adjust to a world designed for nondisabled people (Trieschmann, 1989). Spinal cord injuries affect about 6,000 to 10,000 people annually in the United States (Seftel et al., 1991). The majority of persons who suffer disabling spinal cord injuries are young, active males. Automobile or pedestrian accidents account for about half of these cases. Other common causes include stabbing or bullet wounds, sports injuries, and falls. Depending on the location of the injury to the spinal cord, a loss of voluntary control (paralysis) can occur in either the legs (*paraplegia*) or all four limbs (*quadriplegia*). A loss of sensation may also occur in parts of the body that lie beneath the site of injury. Most people who suffer such injuries have relatively normal life spans, but the quality of their lives is profoundly affected.

Cerebral palsy A muscular disorder that is caused by damage to the central nervous system (usually prior to or during birth) and is characterized by spastic paralysis.

The effect of spinal cord injuries on sexual response depends on the site and severity of the injury. Men have two erection centers in the spinal cord: a higher center in the lumbar region that controls psychogenic erections and a lower one in the sacral region that controls reflexive erections. When damage occurs at or above the level of the lumbar center, men lose the capacity for psychogenic erections, the kinds of erections that occur in response to mental stimulation alone, such as when viewing erotic films or fantasizing. They may still be able to achieve reflexive erections from direct stimulation of the penis; these erections are controlled by the sacral erection center located in a lower portion of the spinal cord. However, they cannot feel any genital sensations because the nerve connections to the brain are severed. Men with damage to the sacral erection center lose the capacity for reflexive erections but can still achieve psychogenic erections so long as their upper spinal cords remains intact (Spark, 1991). Overall, researchers find that about three of four men with spinal cord injuries are able to achieve erections but that only about one in ten continues to ejaculate naturally (Geiger, 1981; Spark, 1991). Others can be helped to ejaculate with the aid of a vibrator (Szasz & Carpenter, 1989). Their brains may help to fill in some of the missing sensations associated with coitus and even orgasm. When direct stimulation does not cause erection, the woman can insert the limp penis into the vagina and gently thrust her hips, taking care not to dislodge the penis.

Although the frequency of sexual activity among men with spinal cord injuries tends to decline following the injury (Alexander et al., 1993), a study of almost 1,300 men with these injuries found that about one out of three (35%) continued to engage in sexual intercourse (Spark, 1991). Only about one in five of the men received any kind of sexual counseling to help them adjust sexually to their disability. The men typically reported increased interest in alternative sexual activities, especially those involving areas above the level of the spinal injury, such as the mouth, lips, neck and ears.

Retention of sexual response depends on the site and severity of the injury (Seftel et al., 1991). Women may lose the ability to experience genital sensations or to lubricate normally during sexual stimulation. However, breast sensations may remain intact, making this area even more erotogenic. Most women with spinal cord injuries can engage in coitus, become impregnated, and deliver vaginally. A survey of 27 women with spinal cord injuries showed that about half were able to experience orgasm (Kettl et al., 1991). Some also report "phantom orgasms" that provide intense psychological pleasure and are accompanied by nongenital sensations that are similar to those experienced by nondisabled women (Perduta-Fulginiti, 1992). Spinal-cord-injured women can heighten their sexual pleasure by learning to use fantasized orgasm, orgasmic imagery, and amplification of their physical sensations (Perduta-Fulginiti, 1992).

Actually, people who are paralyzed as a result of spinal cord injuries usually *can* become sexually aroused and engage in coitus. Most spinal-cord-injured women can become impregnated and bear healthy children. ■

Couples facing the challenge of spinal cord injury may expand their sexual repertoire to focus less on genital stimulation (except to attain the reflexes of erection and lubrication) and more on the parts of the body that retain sensation. Stimulation of some areas of the body, such as the ears, the neck, and the breasts (in both men and women) can yield pleasurable erotic sensations (Knight, 1989; Seftel et al., 1991).

SENSORY DISABILITIES Sensory disabilities, such as blindness and deafness, do not directly affect genital responsiveness. Still, sexuality may be affected in many ways. A person who has been blind since birth or early childhood may have difficulty understanding a partner's anatomy. Sex education curricula have been designed specifically to enable visually impaired people to learn about sexual anatomy via models. Anatomically correct dolls may be used to simulate positions of intercourse.

Deaf people, too, often lack knowledge about sex. Their ability to comprehend the social cues involved in forming and maintaining intimate relationships may also be impaired. Sex education programs based on sign language are helping many hearing-impaired people become more socially perceptive as well as knowledgeable about the physical aspects

Human Sexuality ONLINE //

Web Sites Related to Sexuality in Adulthood

This is the Web site of the Kinsey Institute at Indiana University. There is information about the institute, about graduate training, and about the Institute's archives, clinics, and so on. The archives contain information about sexual behavior in the United States.
www.indiana.edu/~kinsey/

The Web site of the Administration on Aging of the Department of Health and Human Services. Go to the site index and then to "Sexuality in Later Life." You will find a variety of sources of information on aging and sexuality.
www.aoa.dhhs.gov/

The Web site of the Sexual Health Network, a commercial Web site that provides access to information about sex, "education, mutual support, counseling, therapy, healthcare, products and other resources for people with disabilities, illness, or natural changes throughout the life cycle."
www.sexualhealth.com/

The Web site of the American Association for Marriage and Family Therapy. Statement of the AAMFT: "The AAMFT has been involved with the problems, needs and changing patterns of couples and family relationships. The association leads the way to increasing understanding, research and education in the field of marriage and family therapy, and ensuring that the public's needs are met by trained practitioners."
www.aamft.org/

The Web site of the Sexual Health InfoCenter, which debunks myths about aging and sexuality and provides information about age-related sexual changes.
www.sexhealth.org/infocenter/infomain.htm

The Web site of Rx.com, which contains information about sex in middle and late adulthood.
www.rx.com/reference/guides

The Web site of MSNBC Healthy Adam/Healthy Eve, which has news and advice on a range of topics related to health, aging, relationships, and sexuality.
www.msnbc.com/news/ADAMEVEHIDE_Front.asp

The Web site of the American Association of Retired Persons (AARP), which contains articles and information about sexuality in late adulthood.
www.aarp.org

of sex. People with visual and hearing impairments often lack self-esteem and self-confidence, problems that make it difficult for them to establish intimate relationships. Counseling may help them become more aware of their sexuality and develop social skills.

OTHER PHYSICAL DISABILITIES AND IMPAIRMENTS Specific disabilities pose particular challenges to, and limitations on, sexual functioning. **Arthritis** may make it difficult or painful for sufferers to bend their arms, knees, and hips during sexual activity. Coital positions that minimize discomfort and the application of moist heat to the joints before sexual relations may be helpful.

A male amputee may find that he is better balanced in the lateral-entry or female-superior position than in the male-superior position (Knight, 1989). A woman with limited hand function may find it difficult or impossible to insert a diaphragm and may need to request assistance from her partner or switch to another method of contraception (Cole, 1988). Sensitivity to each other's needs is as vital to couples in which one member has a disability as it is to nondisabled couples.

Psychological Disabilities

People with psychological disabilities, such as mental retardation, are often stereotyped as incapable of understanding their sexual impulses. Retarded people are sometimes assumed to maintain childlike innocence through their lives or to be devoid of sexuality. Some stereotype retarded people in the opposite direction: as having stronger-than-normal sex drives and being incapable of controlling them (Reinisch, 1990). Some mentally retarded people do act inappropriately—by masturbating publicly, for example. The stereotypes are exaggerated, however, and even many mentally retarded people who act inappropriately can be trained to follow social rules (Reinisch, 1990).

Arthritis A progressive disease characterized by inflammation or pain in the joints.

Parents and caretakers often discourage retarded people from learning about their sexuality or teach them to deny or suppress their sexual feelings. Although the physical changes of puberty may be delayed in mentally retarded people, most develop normal sexual needs (Edmonson, 1988). Most are capable of learning about their sexuality and can enter into rewarding and responsible intimate relationships.

One of the greatest impediments to sexual fulfillment among people with disabilities is finding a loving and supportive partner. Some people engage in sexual relations with people with disabilities out of sympathy. By and large, however, the partners are other people with disabilities or nondisabled people who have overcome stereotypes that portray disabled people as undesirable. Many partners have had some prior positive relationship, usually during childhood, with a person who had a disability (Knight, 1989). Experience facilitates acceptance of the idea that a disabled person can be desirable. Depending on the nature of the disability, the nondisabled partner may need to be open to assuming a more active sexual role to compensate for the limitations of the partner with the disability. Two partners with disabilities need to be sensitive to each other's needs and physical limitations. People with disabilities and their partners may also need to expand their sexual repertoires to incorporate ways of pleasuring each other that are not fixated on genital stimulation.

The message is simple: Sexuality can enrich the lives of nearly all adults at virtually any age.

summing up

SINGLEHOOD

Recent years have seen a sharp increase in the numbers of single young people in our society. Reasons include increased permissiveness toward premarital sex and, particularly for women, the desire to become established in a career. No one lifestyle characterizes single people.

COHABITATION: DARLING, WOULD YOU BE MY POSSLQ?

Some Facts About Cohabitation

Cohabitation is more prevalent among less well-educated and less affluent people.

Reasons for Cohabitation

Some couples prefer cohabitation because it provides a consistent intimate relationship without the legal and economic entanglements of marriage. Some emotionally committed couples cohabit because of the economic advantages of sharing household expenses.

Cohabitation First and Marriage Later: Benefit or Risk?

Cohabitors who later marry may run a greater risk of divorce than noncohabitors, perhaps because cohabitors are a more liberal group.

MARRIAGE

Marriage is found in all human societies and is our most common lifestyle.

Historical Perspectives

In historical patriarchies, the man of the house had the right to choose wives for his sons and could himself take concubines or additional wives.

Why Do People Marry?

Throughout Western history, marriages have legitimized sexual relations, sanctioned the permanence of a deeply committed relationship, provided for the orderly transmission of wealth, and established a setting for child rearing. In Western society today, romantic love is seen as an essential aspect of marriage.

Types of Marriage

The major types of marriage are monogamy and polygamy. Polygyny and polyandry are types of polygamy.

Whom We Marry: Are Marriages Made in Heaven or in the Neighborhood?

People in the United States tend to marry within their geographical area and social class. They tend to marry people similar to themselves in physical attractiveness, people whose attitudes are similar to their own, and people who seem likely to meet their material, sexual, and psychological needs.

MARITAL SEXUALITY

Until the sexual revolution, Western culture could be characterized as sexually restrictive, even in its attitudes toward marital sex.

The Sexual Revolution Hits Home

Married couples today engage in coitus more frequently and for longer durations of time than in Kinsey's day. They report higher levels of sexual satisfaction and engage in a greater variety of sexual activities.

Sexual Satisfaction

U.S. women are more likely to reach orgasm through marital sex today than they were in Kinsey's day. Wives take more active sexual roles than in Kinsey's day.

EXTRAMARITAL SEX

In most cases, extramarital relationships are conducted without the spouse's knowledge or approval. People may have affairs for sexual variety, to punish their spouses, to achieve emotional closeness, or to prove that they are attractive.

■ Patterns of Extramarital Sex
Precise figures on the prevalence of extramarital sex are lacking.

■ Attitudes Toward Extramarital Sex
Extramarital sex continues to be viewed negatively by the majority of married people in our society.

■ Effects of Extramarital Sex
The discovery of infidelity can evoke anger, jealousy, even shame. Affairs often, but not always, damage marriages.

■ Swinging
Only a small minority of married couples engage in swinging.

DOMESTIC VIOLENCE

Women and men are equally likely to engage in domestic violence, but women are more likely to sustain injury. Domestic violence is frequently connected with threats to men's dominance in relationships.

DIVORCE

About half the marriages in the United States end in divorce. Reasons include relaxed restrictions on divorce, greater financial independence among women, and wider acceptance of the idea that marriages should be happy.

■ The Cost of Divorce
Divorce is often associated with financial and emotional problems, however. Divorce can give rise to feelings of failure and depression and can make it difficult to rear children.

ALTERNATIVE LIFESTYLES

Alternative marital styles such as open marriages and group marriages permit intimate relationships with people outside the marriage.

■ Open Marriage
Open marriages allow each partner open companionship and personal privacy, which may include sexual intimacy with others.

■ Group Marriage
In a group marriage, three or more people share an intimate relationship, although they are not legally married.

SEX IN THE LATER YEARS

There are a number of unfounded cultural myths about sexuality among older people, including the stereotypes that older people are sexless and that older people with sexual urges are abnormal.

■ Physical Changes
Physical changes as the years pass can impair sexual activity. Many potential problems can be averted by altering our expectations and making changes to accommodate the aging process.

■ Patterns of Sexual Activity
Sexual activity tends to decline with age, but continued sexual activity can boost self-esteem and be an important source of gratification.

SEX AND DISABILITY

People with disabilities may suffer from prejudice that depicts them as sexless or as lacking the means to express their sexual needs or feelings.

■ Physical Disabilities
Cerebral palsy does not usually impair sexual interest, capacity for orgasm, or fertility, but afflicted people may be limited to certain types of sexual activities and coital positions. People with spinal cord injuries may be paralyzed and lose sensation below the waist. They often respond reflexively to direct genital stimulation. Sensory disabilities do not directly affect sexual response but may impair sexual knowledge and social skills.

■ Psychological Disabilities
Most mentally retarded people can learn the basics of their own sexuality and develop responsible intimate relationships.

questions for critical thinking

1. Is remaining single an option for you? Why or why not?
2. Do you know people who are living together without being married? What is their motivation for doing so? How do most people from their sociocultural background feel about cohabitation? Why?
3. Agree or disagree with the following statement, and support your answer: Cohabitation helps couples determine, before they get married, whether they are compatible.
4. Agree or disagree with the following statement, and support your answer: There is a right person for everyone, and they are destined to meet.
5. Agree or disagree with the following statement, and support your answer: Marriage is an old-fashioned, outdated lifestyle that has become irrelevant for today's sophisticated young people.
6. Why do you think that even the majority of "sexually liberated" people draw the line at extramarital sex?
7. How would you account for gender differences in the incidence of extramarital sex?
8. Do most people from your sociocultural background approve or disapprove of divorce as a remedy for an unhappy marriage? Do you agree with their beliefs? Explain.

chapter

15

Sexual Dysfunctions

Pablo Picasso. *The Sculptor*, 1931. Oil on plywood, 128.5 × 96 cm. Inv.: M. P. 135. Photo B. Hatala. © Copyright ARS, NY. Musée Picasso, Paris, France. Credit: Réunion des Musées Nationaux/Art Resource, NY. Credit: © 2002 Estate of Pablo Picasso/Artists Rights Society (ARS), NY.

outline

Truth or Fiction?

TYPES OF SEXUAL DYSFUNCTIONS

Sexual Desire Disorders

Sexual Arousal Disorders

Orgasmic Disorders

Sexual Pain Disorders

A World of Diversity Inis Beag and Mangaia—Worlds Apart

ORIGINS OF SEXUAL DYSFUNCTIONS

Organic Causes

Psychosocial Causes

A World of Diversity An Odd Couple: Koro and Dhat Syndromes

TREATMENT OF SEXUAL DYSFUNCTIONS

The Masters-and-Johnson Approach

The Helen Singer Kaplan Approach

Sexual Desire Disorders

A Closer Look "Sexercise" Credited with Boosting Performance

Sexual Arousal Disorders

Human Sexuality Online Buying Viagra and Other Drugs Online

Orgasmic Disorders

A Closer Look The "Orgasm Pill" for Women

Sexual Pain Disorders

Evaluation of Sex Therapy

HOW DO YOU FIND A QUALIFIED SEX THERAPIST?

Human Sexuality Online Web Sites Related to Sexual Dysfunctions

SUMMING UP

QUESTIONS FOR CRITICAL THINKING

Truth or Fiction?

____ Sexual dysfunctions are rare.

____ Only men can reach climax too early.

____ The most common cause of painful intercourse in women is vaginal infection.

____ During the 1980s, 2,000 people fell prey to the belief that their genitals were shrinking and retracting into their bodies.

____ Sex therapy teaches a man with erectile disorder how to will an erection.

____ A doctor once made a somewhat unusual presentation to a medical convention by dropping his pants to reveal an erection.

____ Many sex therapists recommend masturbation as the treatment for women who have never been able to reach orgasm.

____ A man can be prevented from ejaculating by squeezing his penis when he feels that he is about to ejaculate.

Derek Jones, 39, and his wife Pam, 37, had not attempted coitus for five years. Their sexual relations had been limited to fondling and caressing each other and to occasional oral–genital contact. They had given up at coitus because of Derek's persistent difficulty in attaining and sustaining erections. But recently they had begun trying again. Some nights Derek would have an erection enabling him to penetrate, only to find that he ejaculated too rapidly. Many nights he was unable to perform at all. Each failure was yet another blow to Derek's self-esteem. Pam kept secret her belief that he could not perform because he was no longer sexually attracted to her.

Terry, 24, has decided she is built differently from friends and for women she reads about. They all reach orgasm, it seems, at the drop of a hat. But she has never managed "one of those things." Her husband David, also 24, is considerate, but Terry knows that he too is frustrated and feels guilty with every ejaculation. Why should he enjoy sex if Terry cannot? Sex has become a chore rather than a source of pleasure, and David has been having some difficulty attaining erection. Terry wonders whether she should try to fake orgasm to hold on to him. But she is too embarrassed to ask a friend how to act.
—The Authors' Files

Derek and Terry have **sexual dysfunctions.** That is, they have difficulty in becoming sexually aroused or reaching orgasm. Many of us are troubled by sexual problems from time to time. Men occasionally have difficulty achieving an erection or ejaculate more rapidly than they would like. Most women occasionally have difficulty achieving orgasm or becoming sufficiently lubricated. People are not considered to have a sexual dysfunction unless the problem is persistent and causes distress, however.

Although there are different types of sexual dysfunctions, they share some features. People with sexual dysfunctions may avoid sexual opportunities for fear of failure. They may anticipate that sex will result in frustration or physical pain rather than pleasure and gratification. Because of the emphasis that our culture places on sexual competence, people with sexual dysfunctions may feel inadequate or incompetent, feelings that diminish their self-esteem. They may experience guilt, shame, frustration, depression, and anxiety.

Many people with sexual dysfunctions find it difficult to talk about them, even with their spouses or helping professionals. A woman who cannot reach orgasm with her husband may fake orgasms rather than "make a fuss." A man may find it difficult to admit erectile problems to his physician during a physical. Many physicians are also uncomfortable talking about sexual matters and may never inquire about sexual problems.

Because many people are reluctant to admit to sexual problems, we do not have precise figures on their frequencies. The best current information we have may be based on the National Health and Social Life Survey (Laumann et al., 1994) (see Table 15.1). The NHSLS group asked respondents for a yes or no answer to the question "During the last 12 months has there ever been a period of several months or more when you lacked interest in having sex; were unable to come to a climax; came to a climax too quickly; experienced physical pain during intercourse; did not find sex pleasurable; felt anxious about your ability to perform sexually; or (for men)

TABLE 15.1

Current sexual dysfunctions according to the NHSLS study (respondents reporting the problem within the past year)

	Men	Women
Pain during sex	3.0%	14.4%
Sex not pleasurable	8.1	21.2
Unable to reach orgasm	8.3	24.1
Lack of interest in sex	15.8	33.4
Anxiety about performance*	17.0	11.5
Reaching climax too early	28.5	10.3
Unable to keep an erection	10.4	—
Having trouble lubricating	—	18.8

*Anxiety about performance is not itself a sexual dysfunction. However, it figures prominently in sexual dysfunctions.

Source: E. O.Laumann, J. H. Gagnon, R. T. Michael, & S. Michaels (1994). *The Social Organization of Sexuality: Sexual Practices in the United States.* Chicago: University of Chicago Press, pp. 370–371. Copyright © 1994 by University of Chicago Press. Reprinted by permission.

had trouble achieving or maintaining an erection or (for women) had trouble lubricating?" Overall, women reported more problems than men in the areas of painful sex, lack of pleasure, inability to reach orgasm, and lack of interest in sex. Men were more likely than women to report reaching climax too early and being anxious about their performance. Difficulty keeping an erection (erectile disorder) increases with age from about 6% in the 18-to-24-year-old age group to about 20% in the 55-to-59-year-old age group. (But note that urologist Irwin Goldstein [1998] of the Boston University School of Medicine found that nearly *half* the men aged 40 to 70 in a Massachusetts survey reported problems in obtaining and maintaining erections!) The NHSLS figures represent persistent current problems. The incidences of occasional problems and of lifetime problems would be higher.

It is not true that sexual dysfunctions are rare. On the contrary, they are quite common.

TYPES OF SEXUAL DYSFUNCTIONS

The most widely used system of classification of sexual dysfunctions is based on the American Psychiatric Association's *Diagnostic and Statistical Manual of Mental Disorders* (the DSM) of 2000. The DSM, which is now in its fourth edition, groups sexual dysfunctions into four categories:

1. **Sexual desire disorders.** These involve dysfunctions in sexual desire, interest, or drive, in which the person experiences a lack of sexual desire or an aversion to genital sexual contact.
2. **Sexual arousal disorders.** Sexual arousal is principally characterized by erection in the male and by vaginal lubrication and swelling of the external genitalia in the female. In men, sexual arousal disorders involve recurrent difficulty in achieving or sustaining erections sufficient to engaging successfully in sexual intercourse. In women, they typically involve failure to become sufficiently lubricated.

Sexual dysfunctions Persistent or recurrent difficulties in becoming sexually aroused or reaching orgasm.

Sexual desire disorders Sexual dysfunctions in which people have persistent or recurrent lack of sexual desire or aversion to sexual contact.

Sexual arousal disorders Sexual dysfunctions in which people persistently or recurrently fail to become adequately sexually aroused to engage in or sustain sexual intercourse.

3. **Orgasmic disorders.** Men or women may encounter difficulties achieving orgasm or may reach orgasm more rapidly than they would like. Women are more likely to encounter difficulties reaching orgasm; men are more likely to experience overly rapid orgasm (premature ejaculation). But some men have trouble reaching orgasm during coitus, and a few women complain of overly rapid orgasms.
4. **Sexual pain disorders.** Both men and women may suffer from **dyspareunia** (painful intercourse). Women may experience **vaginismus,** which prevents penetration by the penis or renders penetration painful.

Sexual dysfunctions may also be classified as either lifelong or acquired. *Lifelong* dysfunctions have existed throughout the person's lifetime. *Acquired* dysfunctions develop following a period of normal functioning. They also may be classified as generalized or situational. *Generalized* dysfunctions affect a person's general sexual functioning. *Situational* dysfunctions affect sexual functioning only in some sexual situations (such as during coitus but not during masturbation) or occur with some partners but not with others. Consider, for example, a man who has never been able to achieve or maintain an erection during sexual relations with a partner but can do so during masturbation. His dysfunction would be classified as lifelong and situational.

Sexual Desire Disorders

Sexual desire disorders involve lack of sexual desire or interest and/or aversion to genital sexual activity. Although the prevalence of these disorders in the general population is not known, sex therapists report an increase in their frequency over the past generation (Schmidt, 1994).

LACK OF SEXUAL DESIRE People with little or no sexual interest or desire are said to have *hypoactive sexual desire disorder.* They also often report an absence of sexual fantasies. According to the NHSLS Study (Laumann et al., 1994), the problem is more common among women than men. Nevertheless, the belief that men are always eager and willing to engage in sexual activity is no more than a myth.

Lack of sexual desire does not imply that a person is unable to achieve erection, lubricate adequately, or reach orgasm. Some people with low sexual desire do have such problems. Others can become sexually aroused and reach orgasm when stimulated adequately. Many enjoy sexual activity, even if they are unlikely to initiate it. Many appreciate the affection and closeness of physical intimacy but have no interest in genital stimulation.

Hypoactive sexual desire is one of the most commonly diagnosed sexual dysfunctions (Letourneau & O'Donohue, 1993). Yet there is no clear consensus among clinicians and researchers concerning the definition of low sexual desire. How much sexual interest or desire is "normal"? There is no standard level of sexual desire—no 98.6 degree reading on the "sexual thermometer." Lack of desire is usually considered a problem when couples recognize that their level of sexual interest has gotten so low that little remains. Sometimes the lack of desire is limited to one partner. When one member of a couple is more interested in sex than the other, sex therapists often recommend that couples try to reach a compromise. They also attempt to uncover and resolve problems in the relationship that may dampen the sexual ardor of one or both partners.

Biological and psychosocial factors—hormonal deficiencies, depression, marital dissatisfaction, and so on—contribute to lack of desire. Among the medical conditions that diminish sexual desire are testosterone deficiencies, thyroid overactivity or underactivity, and temporal lobe epilepsy (Kresin, 1993). Sexual desire is stoked by testosterone, which is produced by men in the testes and by both genders in the adrenal glands (Tuiten et al., 2000). Women may experience less sexual desire when their adrenal glands are surgically removed. Low sexual interest, along with erectile difficulty, is also common among men with **hypogonadism.** (Hypogonadism is treated with testosterone [Brody, 1995c; Lue,

Orgasmic disorders Sexual dysfunctions in which people persistently or recurrently have difficulty reaching orgasm or reach orgasm more rapidly than they would like, despite attaining a level of sexual stimulation that would normally result in orgasm.

Sexual pain disorders Sexual dysfunctions in which people persistently or recurrently experience pain during coitus.

Dyspareunia A sexual dysfunction characterized by persistent or recurrent pain during sexual intercourse. (From roots that mean "badly paired.")

Vaginismus A sexual dysfunction characterized by involuntary contraction of the muscles surrounding the vaginal barrel, preventing penile penetration or rendering penetration painful.

Hypogonadism An endocrine disorder that reduces the output of testosterone.

2000].) The role of hormones in lack of desire among physically healthy men and women remains unclear.

Researchers find that men with hypoactive sexual desire disorder tend to be older than women with the disorder (Segraves & Segraves, 1991a). A gradual decline in sexual desire, at least among men, may be explained in part by the reduction in testosterone levels that occurs in middle and later life (Brody, 1995c). Abrupt changes in sexual desire, however, are more often explained by psychological and interpersonal factors such as depression, emotional stress, and problems in the relationship (Leiblum & Rosen, 1988; Schreiner-Engle & Schaivi, 1986).

Psychological problems can contribute to low sexual desire (Letourneau & O'Donohue, 1993). Anxiety is the most commonly reported factor. Various types of anxiety may be involved in dampening sexual desire, including performance anxiety (anxiety over being evaluated negatively), anxiety involving fears of pleasure or loss of control, and deeper sources of anxiety related to fears of castration or injury. Depression is a common cause of inhibited desire. A history of sexual assault has also been linked to low sexual desire.

Some medications, including those used to control anxiety or hypertension, may also reduce desire. Changing medications or doses may restore the person's previous level of desire.

SEXUAL AVERSION DISORDER People with low sexual desire may have little or no interest in sex, but they are not repelled by genital contact. Some people, however, find sex disgusting or aversive and avoid genital contact.

Sexual aversions are less common than lack of desire and remain poorly understood (Spark, 1991). Some researchers consider sexual aversion to be a *sexual phobia* or *sexual panic state* with intense, irrational fears of sexual contact and a pressing desire to avoid sexual situations (Kaplan, 1987). A history of erectile problems can cause sexual aversion in men. Men with such histories may be anxious in sexual situations because they trigger feelings of failure and shame. Their partners may also develop an aversion to sexual contact because of this anxiety and because of their own frustration. A history of sexual trauma, such as rape or childhood sexual abuse or incest, often figures prominently in cases of sexual aversion, especially in women.

Sexual Arousal Disorders

When we are sexually stimulated, our bodies normally respond with **vasocongestion,** which produces erection in the male and vaginal lubrication in the female. People with sexual arousal disorders, however, fail to achieve or sustain the lubrication or erection necessary to facilitate sexual activity. Or they lack the subjective feelings of sexual pleasure or excitement that normally accompany sexual arousal (American Psychiatric Association, 2000).

Problems of sexual arousal have sometimes been labeled *impotence* in the male and *frigidity* in the female. But these terms are pejorative, so many professionals prefer to use less threatening, more descriptive labels.

MALE ERECTILE DISORDER Sexual arousal disorder in the male is called **male erectile disorder** or *erectile dysfunction.* It is characterized by persistent difficulty in achieving or maintaining an erection sufficient to allow the completion of sexual activity. In most cases the failure is limited to sexual activity with partners, or with some partners and not others. It can thus be classified as *situational.* In rare cases the dysfunction is found during any sexual activity, including masturbation. In such cases, it is considered *generalized.* Some men with erectile disorder are unable to attain an erection with their partners. Others can achieve erection but not sustain it (or recover it) long enough for penetration and ejaculation.

Vasocongestion Engorgement of blood vessels with blood, which swells the genitals and breasts during sexual arousal.

Male erectile disorder Persistent difficulty achieving or maintaining an erection sufficient to allow the man to engage in or complete sexual intercourse. Also termed *erectile dysfunction.*

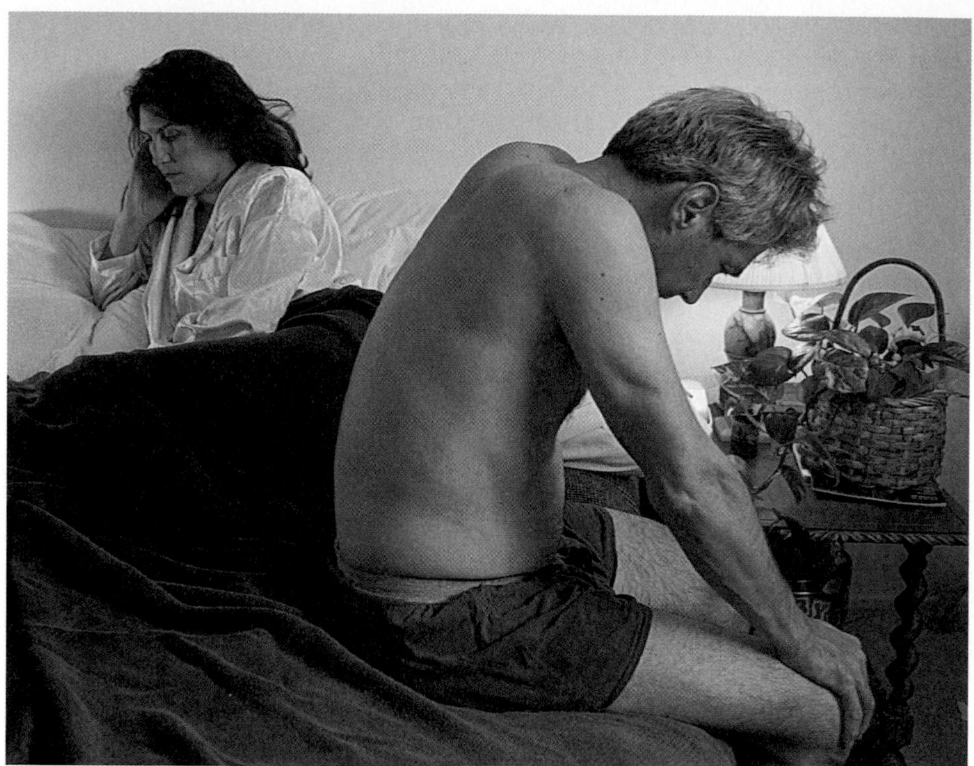

The Emotional Toll of Erectile Dysfunction. Male erectile disorder, or erectile dysfunction, is characterized by persistent difficulty in achieving or maintaining an erection sufficient to allow the completion of sexual activity. As many as 30 million men in the United States experience some degree of erectile dysfunction, and its incidence increases with age. Occasional erectile problems are common and may be caused by fatigue, alcohol, or anxiety about a new partner. But fear of recurrence can create a vicious cycle, in which anxiety leads to failure, and failure heightens anxiety.

As many as 30 million men in the United States experience some degree of erectile dysfunction (Goldstein, 1998). The incidence of erectile disorder increases with age and is believed to affect as many as one man in two between the ages of 40 and 70—at least intermittently (Feldman et al., 1994). According to the Massachusetts Male Aging Study, the incidence of complete erectile disorder triples from 5% at age 40 to 15% at age 70 (Feldman et al., 1994). Men with *lifelong* erectile disorder (formerly called *primary* erectile dysfunction) never had adequate erectile functioning. Erectile disorder more often develops after a period of normal functioning and is classified as *acquired erectile disorder* (formerly called *secondary* erectile dysfunction). Many such men engaged in years of successful coitus before the problem began.

Occasional problems in achieving or maintaining erection are quite common. Fatigue, alcohol, anxiety over impressing a new partner, and other factors may account for a transient episode. Even an isolated occurrence can lead to a persistent problem if the man fears recurrence, however. The more anxious and concerned the man becomes about his sexual ability, the more likely he is to suffer **performance anxiety.** This anxiety can contribute to repeated failure, and a vicious cycle of anxiety and failure may develop.

Performance anxiety Anxiety concerning one's ability to perform behaviors, especially behaviors that may be evaluated by other people.

A man with erectile problems may try to "will" an erection, which can compound the problem. Each failure may further demoralize and defeat him. He may ruminate about his sexual inadequacy, setting the stage for yet more anxiety. His partner may try to comfort and support him by saying things like "It can happen to anyone," "Don't worry about it," or "It will get better in time." But attempts at reassurance may be to no avail. As one client put it,

> I always felt inferior, like I was on probation, having to prove myself. I felt like I was up against the wall. You can't imagine how embarrassing this [erectile failure] was. It's like you walk out [naked] in front of an audience that you think is a nudist convention and it turns out to be a tuxedo convention.
>
> —The Authors' Files

The vicious cycle of anxiety and erectile failure may be interrupted if the man recognizes that occasional problems are normal and does not overreact. The emphasis in our culture on men's sexual prowess may spur them to view occasional failures as catastrophes rather than transient disappointments, however. Viewing occasional problems as an inconvenience, rather than a tragedy, may help avert the development of persistent erectile difficulties.

Performance anxiety is a prominent cause of erectile disorder. So are other psychological factors such as depression, lack of self-esteem, and problems in the relationship. Biological factors can also play a causal role, as we shall see.

FEMALE SEXUAL AROUSAL DISORDER Women may encounter persistent difficulties becoming sexually excited or sufficiently lubricated in response to sexual stimulation. In some cases these difficulties are lifelong. In others they develop after a period of normal functioning. In some cases difficulties are pervasive and occur during both masturbation and sex with a partner. More often they occur in specific situations. For example, they occur with some partners and not with others, or during coitus but not during oral–genital sex or masturbation.

Female sexual arousal disorder often accompanies other sexual disorders such as hypoactive sexual desire disorder and orgasmic disorder. Despite problems in becoming sexually aroused, women with sexual arousal disorder can often engage in coitus. Vaginal dryness may produce discomfort, however.

Female sexual arousal disorder, like its male counterpart, may have physical causes. A thorough evaluation by a medical specialist (a urologist in the case of a male, a gynecologist in the case of a female) is recommended. Any neurological, vascular, or hormonal problem that interferes with the lubrication or swelling response of the vagina to sexual stimulation may contribute to female sexual arousal disorder. For example, diabetes mellitus may lead to diminished sexual excitement in women because of the degeneration of the nerves servicing the clitoris and the blood vessel (vascular) damage it causes. Reduced estrogen production can also result in vaginal dryness.

Female sexual arousal disorder more commonly has psychological causes, however. In some cases, women harbor deep-seated anger and resentment toward their partners. They thus find it difficult to turn off these feelings when they go to bed. In other cases, sexual trauma is implicated. Survivors of sexual abuse often find it difficult to respond sexually to their partners. Childhood sexual abuse is especially prevalent in cases of female sexual arousal disorder (Morokoff, 1993). Feelings of helplessness, anger, or guilt—and or even flashbacks of the abuse—may surface when the woman begins sexual activity, undermining her ability to become aroused. Other psychosocial causes include anxiety or guilt about sex and ineffective stimulation by the partner (Morokoff, 1993).

Orgasmic Disorders

Three disorders can impair the orgasm phase of the sexual response cycle: (1) female orgasmic disorder, (2) male orgasmic disorder, and (3) premature ejaculation.

In female or male orgasmic disorder, the woman or man is persistently delayed in reaching orgasm or does not reach orgasm at all, despite achieving sexual stimulation that would normally be of sufficient intensity to result in orgasm. The problem is more common among women than men. In some cases, a person can reach orgasm without difficulty while engaging in sexual relations with one partner, but not with another.

MALE ORGASMIC DISORDER Male orgasmic disorder has also been termed *delayed ejaculation, retarded ejaculation*, and *ejaculatory incompetence.*[1] The problem may be lifelong or acquired, generalized or situational. There are very few cases of men who have never ejaculated. In most cases the disorder is limited to coitus. The man may be capable of ejaculating during masturbation or oral sex but may find it difficult, if not impossible—despite high levels of sexual excitement—to ejaculate during intercourse. There is a myth that men with male orgasmic disorder and their female partners enjoy his condition, because it enables him to "go on forever" (Dekker, 1993). Actually, the experience is frustrating for both partners.

Male orgasmic disorder is relatively rare in the general population and in clinical practice, where it is among the least frequently diagnosed disorders (Dekker, 1993). Research on the problem has been scarce (Dekker, 1993), and only a few individual or multiple case reports have appeared in the literature (e.g., Masters & Johnson, 1970; Rathus, 1978).

Male orgasmic disorder may be caused by physical problems, such as multiple sclerosis or neurological damage, that interfere with neural control of ejaculation. It may also be a side effect of certain drugs. Various psychological factors may also play a role, including performance anxiety, sexual guilt, and hostility toward one's partner. Helen Singer Kaplan (1974) suggests that some men with male orgasmic disorder may be unconsciously "holding back" their ejaculate from their partners because of underlying hostility or resentment. Masters and Johnson (1970) found that men with this problem frequently have strict religious backgrounds that may leave a residue of unresolved guilt about sex, which inhibits ejaculation. Emotional factors such as fear of pregnancy and anger toward one's partner can also play a role.

As with other sexual dysfunctions, men with male orgasmic disorder and their partners may "try harder." But trying harder often compounds rather than alleviates sexual problems. Sexual relations become a job to get done—a chore rather than an opportunity for pleasure and gratification.

FEMALE ORGASMIC DISORDER Women with female orgasmic disorder are unable to reach orgasm or have difficulty reaching orgasm following what would usually be an adequate amount of sexual stimulation. Women who have never achieved orgasm through any means are sometimes labeled **anorgasmic** or *preorgasmic.*

A woman who reaches orgasm through masturbation or oral sex may not necessarily reach orgasm dependably during coitus with her partner (Stock, 1993). Penile thrusting during coitus may not provide sufficient clitoral stimulation to facilitate orgasm. An orgasmic disorder may be diagnosed, however, if orgasm during coitus is impaired by factors such as sexual guilt or performance anxiety. Women who try to force an orgasm may also find themselves unable to do so. They may assume a **spectator role** and observe rather than fully participate in their sexual encounters. "Spectatoring" may further decrease the likelihood of orgasm.

Anorgasmic Never having reached orgasm. (Literally, "without orgasm.")

Spectator role A role, usually taken on because of performance anxiety, in which people observe their sexual encounters, rather than fully participating in them.

Premature ejaculation A sexual dysfunction in which ejaculation occurs with minimal sexual stimulation and before the man desires it.

PREMATURE EJACULATION A second type of male orgasmic disorder, premature ejaculation, was the male sexual dysfunction most commonly reported in the NHSLS study (see Table 15.1). Men with **premature ejaculation** ejaculate too rapidly to permit their

[1]Just as we find terms like *impotence* and *frigidity* unnecessarily pejorative, we prejer to use the more clinical-sounding but less offensive *male orgasmic disorder* or *delayed ejaculation* rather than "retarded" ejaculation or ejaculatory "incompetence."

partners or themselves to enjoy sexual relations fully. The degree of prematurity varies. Some men ejaculate during foreplay, even at the sight of their partner disrobing. But most ejaculate either just before or immediately after penetration or following a few coital thrusts (Kaplan, 1974).

Just what constitutes *prematurity?* Some definitions focus on a particular time period during which a man should be able to control ejaculation. Is ejaculation within 30 seconds of intromission premature? Within one minute? Ten minutes? There is no clear cutoff. Some scholars argue that the focus should be on whether the couple is satisfied with the duration of coitus rather than on a specific time period.

Helen Singer Kaplan (1974) suggested that the label *premature* should be applied to cases in which men persistently or recurrently lack voluntary control over their ejaculations. This may sound like a contradiction in terms, given that since ejaculation is a reflex, and reflexes need not involve thought or conscious control. Kaplan means that a man may control his ejaculation by learning to regulate the amount of sexual stimulation he experiences so that it remains below the threshold at which the ejaculation reflex is triggered.

RAPID FEMALE ORGASM: CAN WOMEN REACH ORGASM TOO QUICKLY? The female counterpart to premature ejaculation, *rapid orgasm*, is so rarely recognized as a problem that it is generally ignored by clinicians and is not classified as a sexual dysfunction in the DSM system. Still, some women experience orgasm rapidly and show little interest in continuing sexual activity so that their partners can achieve gratification. Many women who reach orgasm rapidly are open to continued sexual stimulation and capable of experiencing successive orgasms, however.

Sexual Pain Disorders

For most of us, coitus is a source of pleasure. For some of us, however, coitus gives rise to pain and discomfort.

DYSPAREUNIA Dyspareunia, or painful coitus, can afflict men or women. Dyspareunia is one of the most common sexual dysfunctions and is also a common complaint of women seeking gynecological services (Quevillon, 1993).

Pain is a sign that something is wrong—physically or psychologically. Dyspareunia may result from physical causes, emotional factors, or an interaction of the two (Meana & Binik, 1994). The most common cause of coital pain in women is inadequate lubrication. In such a case, additional foreplay or artificial lubrication may help. Vaginal infections and sexually transmitted diseases (STIs) may also produce coital pain, however. Allergic reactions to spermicides, even the latex material in condoms, can give rise to coital pain or irritation. Pain during deep thrusting may be caused by endometriosis or pelvic inflammatory disease (PID), by other diseases or structural disorders of the reproductive organs (Reid & Lininger, 1993), or by penile contact with the cervix.

The most common cause of painful intercourse in women is not vaginal infection but lack of adequate lubrication. ■

Psychological factors such as unresolved guilt or anxiety about sex or the lingering effects of sexual trauma may also be involved. These factors may inhibit lubrication and cause involuntary contractions of the vaginal musculature, making penetration painful or uncomfortable.

Painful intercourse is less common in men and is generally associated with genital infections that cause burning or painful ejaculation. Smegma under the penile foreskin of uncircumcised men may also irritate the penile glans during sexual contact.

Inis Beag and Mangaia—Worlds Apart

Let us invite you on a journey to two islands that are a world apart—sexually as well as geographically. The sexual attitudes and practices within these societies will shed some light on the role of cultural values in determining what is sexually normal and what is sexually dysfunctional.

Our first stop is the island of Inis Beag, which lies off the misty coast of Ireland. From the air Inis Beag is a green jewel, fertile and inviting. At ground level things do not appear quite so warm, however.

The residents of this Irish folk community do not believe that it is normal for women to experience orgasm. Anthropologist John Messenger (1971), who visited Inis Beag in the 1950s and 1960s, reported that any woman who finds pleasure in sex—especially the intense waves of pleasure that can accompany orgasm—is viewed as deviant. *Should women on Inis Beag, then, be diagnosed as orgasmically impaired?*

Premarital sex is all but unknown on Inis Beag. Prior to marriage, men and women socialize apart. Marriage comes relatively late—usually in the middle 30s for men and the middle 20s for women. Mothers teach their daughters that they will have to submit to their husbands' animal cravings in order to obey God's injunction to "be fruitful and multiply." *After this indoctrination, women show little interest in sex. Should they be diagnosed as having hypoactive sexual desire disorder?*

The women of Inis Beag need not be overly concerned about frequent sexual intercourse, however, because the men of the island believe that sexual activity will drain their strength. Consequently, men avoid sex on the eve of sporting activity or strenuous work. Because of taboos against nudity, married couples engage in intercourse with their underclothes on. Intercourse takes place in the dark—literally as well as figuratively.

During intercourse the man takes the male-superior position. The male is always the initiator. Foreplay is brief, rarely involving manual stimulation of the breasts and never including oral stimulation of the genitals. *Should people who have difficulty becoming sexually aroused under these circumstances be diagnosed as having sexual arousal disorders?* The man ejaculates as rapidly as he can, in the belief that he is the only partner with sexual needs and to spare his wife as best he can. Then he turns over and rapidly falls asleep. Once more the couple have done their duty. *Should the man be diagnosed as having premature ejaculation because he ejaculates rapidly?*

Our next stop is Mangaia. Mangaia is a Polynesian pearl of an island. It rises languidly out of the blue waters of the Pacific. It lies on the other side of the world from Inis Beag—in more ways than one.

From an early age, Mangaian boys and girls are encouraged to get in touch with their own sexuality through sexual play and masturbation (Marshall, 1971). At about the age of 13, Mangaian boys are initiated into manhood by adults who instruct them in sexual techniques. Mangaian males are taught the merit of bringing their female partners to multiple orgasms before ejaculating. *Are Mangaian males who ejaculate before their partners have multiple orgasms suffering from premature ejaculation?*

Boys practice their new techniques with girlfriends on secluded beaches or beneath the listing fronds of palms. They may visit girlfriends in the evening in the huts where they sleep with their families. Parents often listen for their

VAGINISMUS Vaginismus involves an involuntary contraction of the pelvic muscles that surround the outer third of the vaginal barrel. Vaginismus occurs reflexively during attempts at vaginal penetration, making entry by the penis painful or impossible. These reflexive contractions are accompanied by a deep-seated fear of penetration (Beck, 1993). Some women with vaginismus are unable to tolerate penetration by any object, including a finger, a tampon, or a physician's speculum. The prevalence of vaginismus is unknown.

The woman with vaginismus usually is not aware that she is contracting her vaginal muscles. In some cases, husbands of women with vaginismus develop erectile disorder after repeated failures at penetration (Masters & Johnson, 1970; Speckens et al., 1995).

Vaginismus is caused by a psychological fear of penetration rather than by a physical injury or defect (LoPiccolo & Stock, 1986). Women with vaginismus often have histories of sexual trauma, rape, or botched abortions that resulted in vaginal injuries. They may desire sexual relations. They may be capable of becoming sexually aroused and

Polynesia. Cultural expectations affect our judgments about what kinds of sexual behavior are functional and what kinds are dysfunctional. Some Polynesian cultures are sexually permissive. They encourage children to explore their sexuality. Men may be expected to bring their partners to orgasm several times before ejaculating. In such cultures, should men who ejaculate before their partners have multiple orgasms be diagnosed with premature ejaculation? (Paul Gauguin, *And the gold of their bodies.* Photo: B. Hatala. Musée d'Orsay, Paris, France. Courtesy of Réunion des Musées Nationaux/ Art Resource, NY.)

daughters to laugh and gasp so that they will know that they have reached orgasm with a visiting young man, called a "sleepcrawler." Usually they pretend to be asleep so as not to interfere with courtship and impede their daughters' chances of finding a suitable mate. Daughters may receive a nightly succession of sleepcrawlers.

Girls, too, learn techniques of coitus from their elders. Typically they are initiated by an experienced male relative. Mangaians look on virginity with disdain, because virgins do not know how to provide sexual pleasure. Thus the older relative makes his contribution to the family by initiating the girl.

Mangaians, by the way, expressed concern when they learned that many European and U.S. women do not regularly experience orgasm during coitus. Orgasm is apparently universal among Mangaian women. Therefore, Mangaians could only assume that Western women suffered from some abnormality of the sex organs. *Do they?*

All in all, the sharp contrasts between Inis Beag and Mangaia illustrate how concepts of normality are embedded within a cultural context. Behavior that is judged to be normal in one culture may be regarded as abnormal in another. How might our own cultural expectations influence our judgments about sexual dysfunction?

achieving orgasm. However, fear of penetration triggers an involuntary spasm of the vaginal musculature at the point of penile insertion. Vaginismus can also be a cause or an effect of dyspareunia. Women who experience painful coitus may develop a fear of penetration. Fear then leads to the development of involuntary vaginal contractions. Women with vaginismus may experience pain during coital attempts if the couple tries to force penetration. Vaginismus and dyspareunia may also give rise to, or result from, erectile disorder in men (Speckens et al., 1995). Feelings of failure and anxiety come to overwhelm both partners.

Table 15.2 shows differences in the incidences of current sexual dysfunctions and other problems between European Americans and African Americans, according to the NHSLS study (Laumann et al., 1994). African American men report a higher incidence than European American men of each of the sexual dysfunctions surveyed. African American women report a higher incidence of most sexual dysfunctions, with the exceptions of painful sex and trouble lubricating.

TABLE 15.2

European American and African American differences in the incidence of current sexual problems (respondents reporting the problem within the past year)

	European American Men (%)	African American Men (%)	European American Women (%)	African American Women (%)
Pain during sex	3.0	3.3	14.7	12.5
Sex not pleasurable	7.0	15.2	19.7	30.0
Unable to reach orgasm	7.4	9.9	23.2	29.2
Lack of interest in sex	14.7	20.0	30.9	44.5
Anxiety about performance	16.8	23.7	10.5	14.5
Reaching climax too early	27.7	33.8	7.5	20.4
Unable to keep an erection	9.9	14.5	—	—
Having trouble lubricating	—	—	20.7	13.0

Source: From E. O. Laumann, J. H. Gagnon, R. T. Michael, & S. Michaels (1994). *The Social Organization of Sexuality: Sexual Practices in the United States.* Chicago: University of Chicago Press, pp. 370–371. Copyright © 1994 by University of Chicago Press. Reprinted by permission.

ORIGINS OF SEXUAL DYSFUNCTIONS

Because sexual dysfunctions involve the sex organs, it was once assumed that they stemmed largely from organic or physical causes. Today it is widely believed that many cases reflect psychosocial factors such as sexual anxieties, lack of sexual knowledge, or marital dissatisfaction. Many cases also involve a combination of organic and psychological factors.

Organic Causes

Physical factors, such as fatigue and lowered testosterone levels, can dampen sexual desire and reduce responsiveness. Fatigue may lead to male erectile disorder and male orgasmic disorder, and to female orgasmic disorder and inadequate lubrication. But these will be isolated incidents unless the person attaches too much meaning to them and becomes concerned about future performance. Painful coitus, however, often reflects organic factors such as underlying infections. Orgasmic functioning in both men and women can be affected by various medical conditions, including coronary heart disease, diabetes mellitus, multiple sclerosis, spinal cord injuries; by complications from certain surgical procedures (such as removal of the prostate in men); by endocrinological (hormone) problems; and by the use of some medicines, such as drugs used to treat hypertension and psychiatric disorders. Generally speaking, Edward Laumann and his colleagues (1999) found that poor health can contribute to all kinds of sexual dysfunctions in men, but mostly to sexual pain in women.

It was once believed that the great majority of cases of erectile disorder had psychological causes (Masters & Johnson, 1970). It is now known that organic factors are involved in as many as 80% of cases (Brody, 1998c). But psychological factors such as anger and depression may prolong or worsen the problem even when there are organic factors (Feldman et al., 1994).

Organic causes of erectile disorder affect the flow of blood to and through the penis—a problem that becomes more common as men age—or damage the nerves involved in erection (Goldstein, 1998, 2000). Erectile problems can arise when clogged or narrow arteries leading to the penis deprive the penis of oxygen (Lipshultz, 1996). For example, erectile disorder is common among men with diabetes mellitus, a disease that can damage blood vessels and nerves. Eric Rimm (2000) of the Harvard School of Public Health studied 2,000 men and found that erectile dysfunction was connected with a large waist, physical inactivity, and drinking too much alcohol (or not having any alcohol!). The condition common to these men may be high cholesterol levels. Cholesterol can impede the flow of blood to the penis just as it impedes the flow of blood to the heart. Another study finds similar results: that erectile dysfunction is connected with heart disease and hypertension (Johannes et al., 2000). One or two drinks per day, exercise, and weight loss all help lower cholesterol levels.

We're not recommending that abstinent readers take up drinking to ward off or treat erectile problems. Try weight control and regular exercise. Examination of the Massachusetts Male Aging Study data base reveals that men who exercise regularly seem to ward off erectile dysfunction (Derby, 2000). Men who burned 200 calories or more a day in physical activity—an amount that be achieved by briskly walking two miles—cut their risk of erectile dysfunction about in half. Again, exercise seems to work by preventing the clogging of arteries, thereby keeping them clear for the flow of blood into the penis. There was no evidence in this study that men who had already developed erectile dysfunction could reduce the problem by taking up exercise. Young male readers are advised to consider the issue now.

Nerve damage resulting from prostate surgery may impair erectile response. Former senator and presidential candidate Bob Dole has spoken publicly about encountering erectile problems following removal of his prostate gland in 1991. Erectile disorder may also result from multiple sclerosis (MS), a disease in which nerve cells lose the protective coatings that facilitate transmission of neural messages. MS has also been implicated in male orgasmic disorder.

Syphilis, a sexually transmitted disease, can result in erectile failure if the bacteria that cause the disease invade the spinal cord and affect the cells that control erection. Chronic kidney disease, hypertension, cancer, emphysema, and coronary heart disease can all impair erectile response. So can endocrine disorders that impair testosterone production (Ralph & McNicholas, 2000).

Research with rats (Melis et al., 2000) and humans (Rajfer et al., 1992) suggests that many cases of erectile disorder involve failure of the body to produce enough nitric oxide. When nitric oxide comes into contact with the muscles encircling blood vessels in the penis, the muscles relax, allowing vasocongestion to occur and causing the penis to swell. Lack of nitric oxide allows blood to leak out of the penis and be reabsorbed by the body, thereby preventing the degree of engorgement necessary for erection.

Women also develop vascular or nervous disorders that impair genital blood flow, decreasing lubrication and sexual excitement, rendering intercourse painful, and reducing their ability to reach orgasm. As with men, these problems become more likely as women age.

People with sexual dysfunctions are generally advised to undergo a physical examination to determine whether their problems are biologically based. Men with erectile disorder may be evaluated in a sleep center to determine whether they attain erections while asleep. Healthy men usually have erections during rapid-eye-movement (REM) sleep, which occurs every 90 to 100 minutes. Men with organically based erectile disorder often do not have nocturnal erections. However, this technique, which is called nocturnal penile **tumescence** (NPT), may yield misleading results (Meisler & Carey, 1990).

Prescription drugs and illicit drugs account for many cases of erectile disorder. Antidepressant medication and antipsychotic drugs may impair erectile functioning and cause orgasmic disorders (Ashton et al., 2000; Michelson et al., 2000). Tranquilizers such as Valium and Xanax may cause orgasmic disorder in either gender. Antihypertensive drugs can lead to erectile failure (Ralph & McNicholas, 2000). Switching to hypertensive drugs that do not impair sexual response or adjusting the dose may help. Other drugs that can lead to erectile disorder include adrenergic blockers, diuretics, cholesterol-lowering drugs ("statins"), anticonvulsants, anti-Parkinson drugs, and dyspepsia and ulcer-healing drugs (Ralph & McNicholas, 2000).

Tumescence Swelling; erection. (From the Latin *tumere,* which means "to swell." *Tumor* has the same root.)

Central nervous system depressants such as alcohol, heroin, and morphine can reduce sexual desire and impair sexual arousal (Segraves & Segraves, 1993). Narcotics also depress testosterone production, which can further reduce sexual desire and lead to erectile failure (Spark, 1991). Marijuana use has been associated with reduced sexual desire and performance (Wilson et al., 2000).

It is commonly believed that cocaine is an aphrodisiac. However, regular use can cause erectile disorder or male orgasmic disorder and can reduce sexual desire in both genders (Weiss & Mirin, 1987). Some people report increased sexual pleasure from the initial use of cocaine, but repeated use can lead to dependency on the drug for sexual arousal. Long-term use may compromise the ability to experience sexual pleasure (Weiss & Mirin, 1987).

Despite the fact that alcohol can impair sexual arousal on a given occasion, Edward Laumann and his colleagues (1999) found no general relationship between alcohol consumption and the experiencing of sexual dysfunctions. However, problems can arise when people misattribute the sexually dampening effects of depressants such as alcohol to causes within themselves. In other words, if you are unable to perform sexually when you have had a few drinks, and you do not realize that alcohol can depress your performance, you may believe that something is wrong with you. This belief can create anxiety at your next sexual opportunity, and that anxiety can prevent normal functioning. A second failure may set off a vicious cycle in which self-doubt triggers more anxiety, and anxiety results in repeated failure, which heightens anxiety . . . and so on.

Psychosocial Causes

Psychosocial factors are connected with sexual dysfunctions. These include—but are not necessarily limited to—cultural influences, economic problems, psychosexual trauma, a gay sexual orientation, marital dissatisfaction, psychological conflict, lack of sexual skills, irrational beliefs, and performance anxiety (Laumann et al., 1999).

CULTURAL INFLUENCES Children reared in sexually repressive cultural or home environments may learn to respond to sex with feelings of anxiety and shame, rather than sexual arousal and pleasure. People whose parents instilled in them a sense of guilt over touching their genitals may find it difficult to accept their sex organs as sources of pleasure.

In Western cultures, sexual pleasure has traditionally been a male preserve. Young women may be reared to believe that sex is a duty to be performed for their husbands, not a source of personal pleasure. Although the traditional double standard may have diminished in the United States in recent years, girls may still be exposed to relatively repressive attitudes. Women are more likely than men in our culture to be taught to repress their sexual desires and even to fear their sexuality (Nichols, 1990a). Self-control and vigilance—not sexual awareness and acceptance—become identified as feminine virtues. Women reared with such attitudes may be less likely to learn about their sexual potentials or express their erotic preferences to their partners. Compared to women who readily reach orgasm, sexually active but anorgasmic women report more negative attitudes toward masturbation, greater sexual guilt, and greater discomfort talking with their partners about sexual activities that involve direct clitoral contact (cunnilingus and manual stimulation) (Kelly et al., 1990).

Many women who are exposed to negative attitudes about sex during childhood and adolescence find it difficult suddenly to view sex as a source of pleasure and satisfaction once they are married. The result of a lifetime of learning to turn themselves off sexually may lead to difficulties experiencing the full expression of sexual arousal and enjoyment when an acceptable opportunity arises (Morokoff, 1993).

PSYCHOSEXUAL TRAUMA Women and men who were sexually victimized in childhood are more likely to experience difficulty in becoming sexually aroused (Laumann et al., 1999). Learning theorists focus on the role of conditioned anxiety in explaining sexual dysfunctions. Sexual stimuli come to elicit anxiety when they have been paired with phys-

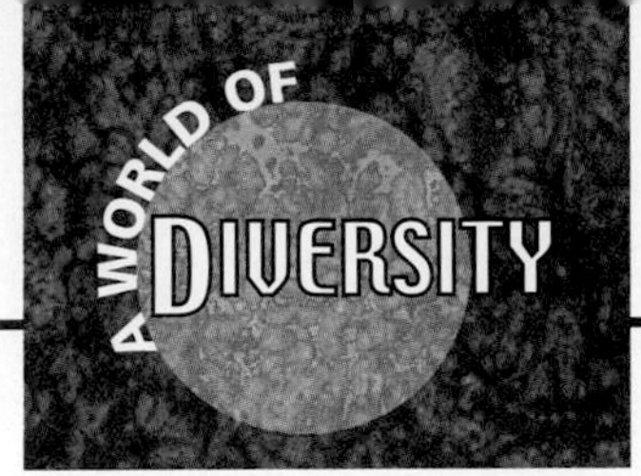

An Odd Couple: Koro and Dhat Syndromes

Many men in the United States are concerned that their penises are too small. However, they are not usually concerned that their genitals will retract into their bodies. Some people in the United States also still believe that men should avoid sex on the eve of an athletic event because sex saps the body of strength. This belief has a counterpart in the country of India.

Let us consider two Far Eastern sexual disorders: Koro syndrome and Dhat syndrome.

Koro Syndrome

Koro syndrome is found primarily in China and some other Far Eastern countries. People with Koro syndrome fear that their genitals are shrinking and retracting into the body, a condition that they believe will be fatal. Koro syndrome has been identified mainly in young men, although some cases have also been reported in women. People with Koro syndrome show signs of acute anxiety, including profuse sweating, breathlessness, and heart palpitations. Men with the disorder have been known to use mechanical devices, such as chopsticks, to try to prevent the penis from retracting into the body.

Koro syndrome has been traced as far back as 3000 B.C. Epidemics involving hundreds or thousands of people have been reported in such parts of Asia as China, Singapore, Thailand, and India (Tseng et al., 1992). In Guangdong Province in China, an epidemic involving more than 2,000 people occurred during the 1980s. Guangdong residents who did not fall victim to Koro were less superstitious and more intelligent than those who fell victim to the epidemic of fear (Tseng et al., 1992).

Reassurance by a health professional that fears that the genitals will retract into the body are unfounded often put an end to Koro episodes. In any event, Koro episodes tend to pass with time.

Truth or Fiction? **REVISITED**

It is true that in the 1980s, 2,000 people fell prey to the belief that their genitals were shrinking and retracting into their bodies. The syndrome is known as Koro syndrome and is found in the Far East. ■

Dhat Syndrome

Dhat syndrome is found among young Asian Indian males and involves excessive fears over the loss of seminal fluid during nocturnal emissions (Akhtar, 1988). Some men with Dhat syndrome also believe (incorrectly) that semen mixes with urine and is excreted by urinating.

There is a widespread belief within Indian culture (and other Near and Far Eastern cultures) that the loss of semen is harmful because it depletes the body of physical and mental energy (Chadda & Ahuja, 1990). Therefore, men with Dhat syndrome may visit physician after physician to find help in preventing nocturnal emissions or the imagined loss of semen mixed with urine. Dhat syndrome can best be understood within its cultural context:

> In India, attitudes toward semen and its loss constitute an organized, deep-seated belief system that can be traced back to the scriptures of the land. . . . [even as far back as the classic Indian sex manual, the *Kama Sutra*, which was believed to be written by the sage Vatsayana between the third and fifth centuries A.D.]. . . . Semen is considered to be the elixir of life, in both a physical and mystical sense. Its preservation is supposed to guarantee health and longevity. (Akhtar, 1988, p. 71)

It is a commonly held Hindu belief that it takes "40 meals to form one drop of blood, 40 drops of blood to fuse and form one drop of bone marrow, and 40 drops of this [to] produce one drop of semen" (Akhtar, 1988, p. 71). Some Indian males thus experience great anxiety over the involuntary loss of the fluid through nocturnal emissions (Akhtar, 1988). Dhat syndrome has also been associated with difficulty in achieving or maintaining erection, apparently because of excessive concern about loss of seminal fluid through ejaculation.

ically or psychologically painful experiences, such as rape, incest, or sexual molestation. Conditioned anxiety can stifle sexual arousal. Unresolved anger and misplaced guilt can also make it difficult for victims of rape and other sexual traumas to respond sexually, even years afterward. People who have been sexually victimized may harbor feelings of disgust and revulsion toward sex or deep-seated fears of sex that make it difficult for them to respond sexually, even with loving partners.

SEXUAL ORIENTATION Some gay males and lesbians test their sexual orientation by developing heterosexual relationships, even marrying and rearing children with partners of the other gender. Others may wish to maintain the appearance of heterosexuality to avoid the social stigma that society attaches to a gay sexual identity. In such cases, problems in arousal or performance with heterosexual partners can be due to a lack of heteroerotic interests (Laumann et al., 1999). A lack of heterosexual response is not a problem for people who are committed to a gay lifestyle, however. Sexual dysfunctions may also occur in gay relationships, as they do in heterosexual relationships.

INEFFECTIVE SEXUAL TECHNIQUES In some marriages, couples practice a narrow range of sexual techniques because they have fallen into a certain routine or because one partner controls the timing and sequence of sexual techniques. The woman who remains unknowledgeable about the erotic importance of her clitoris may be unlikely to seek direct clitoral stimulation. The man who responds to a temporary erectile failure by trying to force an erection may be unintentionally setting himself up for repeated failure. The couple who fail to communicate their sexual preferences or to experiment with altering their sexual techniques may find themselves losing interest. Brevity of foreplay and coitus may contribute to female orgasmic disorder.

EMOTIONAL FACTORS Orgasm involves a sudden loss of voluntary control. Fear of losing control, or "letting go," may block sexual arousal. Other emotional factors, especially depression, are often implicated in sexual dysfunctions (Ralph & McNicholas, 2000). Women with hypoactive sexual desire are more likely to report a history of depression (Schreiner-Engle & Schiavi, 1986). People who are depressed frequently report lessened sexual interest and may find it difficult to respond sexually. People who are exposed to a high level of emotional stress may also experience an ebbing of sexual interest and response.

PROBLEMS IN THE RELATIONSHIP Problems in the relationship are important and often pivotal factors in sexual dysfunctions (Catalan et al., 1990; Fish et al., 1994; Leiblum & Rosen, 1991). Problems in a relationship are not so easily left at the bedroom door. Couples usually find that their sexual relationships are no better than the other facets of their relationships. Couples who harbor resentment toward one another may make sex their arena of combat. They may fail to become aroused by their partners or "withhold" orgasm to make their partners feel guilty or inadequate.

Problems in communication may also play a role. Troubled relationships are usually characterized by poor communication. Partners who have difficulty communicating about other matters may be unlikely to communicate their sexual desires to each other.

The following case highlights how sexual dysfunctions can develop against the backdrop of a troubled relationship.

After living together for six months, Paul and Petula are contemplating marriage. But a problem has brought them to a sex therapy clinic. As Petula puts it, "For the last two months he hasn't been able to keep his erection after he enters me." Paul, 26, is a lawyer; Petula, 24, is a buyer for a large department store. They both grew up in middle-class, suburban families. They were introduced through mutual friends and began having intercourse, without difficulty, a few months into their relationship. At Petula's urging, Paul moved into her apartment, although he wasn't sure he was ready for such a step. A week later he began to have difficulty maintaining his erection during intercourse, although he felt strong desires for his partner. When his erection waned, he would try again but would lose his desire and be unable to achieve another erection. After a few times like this, Petula would become so angry that she began striking Paul in the chest and screaming at him. Paul, who at 200 pounds weighed more than twice as much as Petula, would just walk away, which angered Petula even more.

It became clear that sex was not the only trouble spot in their relationship. Petula complained that he would rather be with his friends and go to baseball games than spend time with her. When they were together at home, he would become absorbed in watching sports events on television and showed no interest in activities she enjoyed—attending the theater, visiting museums, etc. Because there was no evidence that the sexual difficulty was due to either organic problems or depression, a diagnosis of male erectile disorder was given. Neither Paul nor Petula was willing to discuss their nonsexual

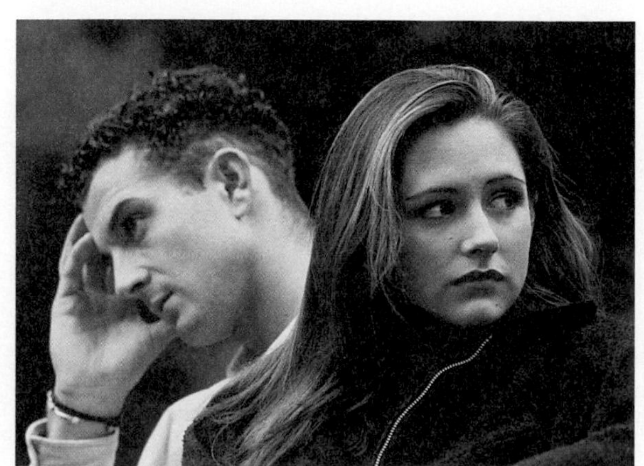

A Vicious Cycle? Marital Conflict and Sexual Desire. Marital conflicts many dampen sexual interest. Lack of sexual interest may then further increase marital strain and dissatisfaction.

problems with a therapist. Although the sexual problem was treated successfully with a form of sex therapy modeled after techniques developed by Masters and Johnson [see the discussion later in this chapter] and the couple later married, Paul's ambivalence continued well into their marriage, and there were future recurrences of sexual problems as well. (Adapted from Spitzer et al., 1989, pp. 149–150)

Although couples with strongly committed and supportive relationships can generally developing effective coping strategies for overcoming even the most severe sexual problems, couples in relationships with unresolved conflicts may derive little if any benefit from even the most advanced psychological, medical, or surgical treatment of sexual dysfunctions (Leiblum & Rosen, 1991).

PSYCHOLOGICAL CONFLICTS Within Freud's psychoanalytic theory, sexual dysfunctions are rooted in the failure to resolve successfully the Oedipus or Electra complex of early childhood. Sexual encounters in adulthood arouse unconscious anxieties and hostilities that are believed to reflect unresolved conflicts, thereby inhibiting sexual response.

A modern psychoanalytic theorist, Helen Singer Kaplan (1974), believes that sexual dysfunctions represent an interaction of *immediate* causes (such as poor techniques, marital conflict, performance anxiety, and lack of effective communication) and deep-seated or *remote* causes (such as unresolved childhood conflicts that predispose people to encounter sexual anxiety and hostility in adulthood). Kaplan believes there is value in combining direct behavioral techniques, which deal with the immediate causes of sexual dysfunctions, with psychoanalytic (insight-oriented) techniques that deal with the remote causes.

LACK OF SEXUAL SKILLS Sexual competency involves the acquisition of sexual knowledge and skills and is based largely on learning. We generally learn through trial and error, and by talking and reading about sex, what makes us and others feel good. Some people may not develop sexual competency because of a lack of opportunity to acquire knowledge and experience—even within marriage. People with sexual dysfunctions may have been reared in families in which discussions of sexuality were off limits and early sexual experimentation was harshly punished. Such early influences may have squelched the young person's sexual learning and experimentation or led her or him to associate anxiety or guilt with sex.

Helen Singer Kaplan (1974) notes that premature ejaculators may have failed to learn to recognize the level of sexual arousal that directly precedes their ejaculatory threshold—the level of stimulation that triggers the ejaculatory reflex. As a result they may be less able to employ self-control strategies to delay ejaculation, such as temporarily halting genital stimulation when they approach their "point of no return." Men with limited sexual experience are particularly unlikely to recognize their ejaculatory threshold and to regulate sexual stimulation so that it remains below this triggering point. Some men may try to delay ejaculation by keeping their minds blank or diverting their attention from sex (for example, by thinking about their forthcoming calculus exam). In so doing, they ignore their own sexual stimulation and do not learn "when to say when."

Although some men with premature ejaculation may have difficulty gauging their level of sexual arousal, research fails to show that premature ejaculators in general are less accurate than other men in making such judgments (Kinder & Curtiss, 1988; Strassberg et al., 1990). It may be that premature ejaculators are more physiologically sensitive to sexual stimulation, not that they are less capable of assessing their sexual arousal (Strassberg et al., 1990). Perhaps premature ejaculators simply require less stimulation to achieve orgasm.

IRRATIONAL BELIEFS Psychologist Albert Ellis (1962, 1977) points out that irrational beliefs and attitudes may contribute to sexual dysfunctions. Negative feelings such as anxiety and fear, Ellis submits, do not stem directly from the events we experience, but rather from our interpretations of these events. If a person encounters a certain event, like an erectile or orgasmic disorder, on a given day and then *believes* that the event is awful or catastrophic, he or she will exaggerate feelings of disappointment and set the stage for future problems.

PERFORMANCE ANXIETY Anxiety—especially performance anxiety—plays an important role in the development of sexual dysfunctions. Performance anxiety occurs when a person becomes overly concerned with how well he or she performs a certain act or task. Performance anxiety may place a dysfunctional individual in a role as spectator rather than performer. Rather than focusing on erotic sensations and allowing involuntary responses such as erection, lubrication, and orgasm to occur naturally, he or she focuses on self-doubts and fears and thinks, "Will I be able to do it this time? Will this be another failure?"

Performance anxiety can set the stage for a vicious cycle in which a sexual failure increases anxiety. Anxiety then leads to repeated failure, and so on. Sex therapists emphasize the need to break this vicious cycle by removing the need to perform.

In men, performance anxiety can inhibit erection while also triggering a premature ejaculation. (Erection, mediated by the parasympathetic nervous system, can be blocked by activation of the sympathetic nervous system in the form of anxiety. Because ejaculation, like anxiety, is mediated by the sympathetic nervous system, arousal of this system in the form of anxiety can increase the level of stimulation and thereby heighten the potential for premature ejaculation.)

In women, performance anxiety can reduce vaginal lubrication and contribute to orgasmic disorder. Women with performance anxieties may try to force an orgasm, only to find that the harder they try, the more elusive it becomes.

TREATMENT OF SEXUAL DYSFUNCTIONS

When Kinsey conducted his surveys in the 1930s and 1940s, there was no effective treatment for sexual dysfunctions. At the time, the predominant model of therapy for sexual dysfunctions was long-term psychoanalysis. Psychoanalysts believed that the sexual problem would abate only if the unconscious conflicts presumed to lie at the root of the problem were resolved through long-term therapy. Evidence of the effectiveness of psychoanalysis in treating sexual dysfunctions is still lacking, however.

Since that time, behavioral models of short-term treatment, collectively called **sex therapy**, have emerged. These models aim to modify the dysfunctional behavior as directly as possible. Sex therapists also recognize the roles of childhood conflicts, self-defeating attitudes, and the quality of the partners' relationship. Therefore, they draw upon various forms of therapy as needed (LoPiccolo, 1994; Rosen et al., 1994).

Although the particular approaches vary, sex therapies aim to

1. Change self-defeating beliefs and attitudes
2. Teach sexual skills
3. Enhance sexual knowledge
4. Improve sexual communication
5. Reduce performance anxiety

Sex therapy usually involves both partners, although individual therapy is preferred in some cases. Therapists find that granting people "permission" to experiment sexually or discuss negative attitudes about sex helps many people overcome sexual problems without the need for more intensive therapy.

Today biological treatments have also been emerging for various sexual dysfunctions. Since 1998, most public attention has been focused on Viagra, a drug that is helpful in most cases of erectile dysfunction. But biological treatments are also emerging for premature ejaculation, female orgasmic dysfunction, and lack of sexual desire.

In this section we will explore both psychological and behavioral approaches to the treatment of sexual dysfunctions. Let us begin with the groundbreaking work of Masters and Johnson.

The Masters-and-Johnson Approach

Masters and Johnson pioneered the use of direct behavioral approaches to treating sexual dysfunctions (Masters & Johnson, 1970). A female–male therapy team focuses on the couple as the unit of treatment during a two-week residential program. Masters and Johnson consider the couple, not the individual, dysfunctional. A couple may describe the hus-

Sex therapy A collective term for short-term behavioral models for treatment of sexual dysfunctions.

band's erectile disorder as the problem, but this problem is likely to have led to problems in the couple by the time they seek therapy. Similarly, a man whose wife has an orgasmic disorder is likely to be anxious about his ability to provide effective sexual stimulation.

Having a dual-therapist team permits each partner to discuss problems with someone of his or her own gender. It reduces the chance of therapist bias in favor of the female or male partner. It allows each partner to hear concerns expressed by another member of the other gender. Anxieties and resentments are aired, but the focus of treatment is behavioral change. Couples perform daily sexual homework assignments, such as **sensate focus exercises,** in the privacy of their own rooms.

Sensate focus sessions are carried out in the nude. Partners take turns giving and receiving stimulation in nongenital areas of the body. Without touching the breasts or genitals, the giver massages or fondles the receiving partner in order to provide pleasure under relaxing and nondemanding conditions. Because genital activity is restricted, there is no pressure to "perform." The giving partner is freed to engage in trial-and-error learning about the receiving partner's sensate preferences. The receiving partner is freed to enjoy the experience without feeling rushed to reciprocate or obliged to perform by becoming sexually aroused. The receiving partner's only responsibility is to direct the giving partner as needed. In addition to these general sensate focus exercises, Masters and Johnson used specific assignments designed to help couples overcome particular sexual dysfunctions.

Masters and Johnson were pioneers in the development of sex therapy. Yet many sex therapists have departed from the Masters-and-Johnson format. Many do not treat clients in an intensive residential program. Many question the necessity of female–male therapist teams. Researchers find that one therapist is about as effective as two, regardless of her or his gender (Libman et al., 1985). Nor does the therapeutic benefit seem to depend to any great extent on whether the sessions are conducted within a short period of time, as in the Masters-and-Johnson approach, or spaced over time (Libman et al., 1985). Some success has also been reported in minimal-contact programs in which participants are given written instructions rather than live therapy sessions (Mohr & Beutler, 1990). Other therapists have departed from the Masters-and-Johnson approach by working individually with preorgasmic women rather than the couple. Group treatment programs have also been used successfully in treating female orgasmic disorder (Killmann et al., 1987).

The Helen Singer Kaplan Approach

Kaplan (1974) calls her approach *psychosexual therapy.* Psychosexual therapy combines behavioral and psychoanalytic methods. Kaplan, as we have noted, believes that sexual dysfunctions have both *immediate* causes and *remote* causes (underlying intrapsychic conflicts that date from childhood). Kaplan begins therapy with the behavioral approach. She focuses on improving the couple's communication, eliminating performance anxiety, and fostering sexual skills and knowledge. She uses a brief form of insight-oriented therapy when it appears that remote causes impede response to the behavioral program. In so doing, she hopes to bring to awareness unconscious conflicts that are believed to have stifled the person's sexual desires or responsiveness. Although Kaplan reports a number of successful case studies, there are no controlled studies demonstrating that the combination of behavioral and insight-oriented, or psychoanalytic, techniques is more effective than the behavioral techniques alone.

Let us now consider some of the specific techniques that sex therapists have introduced in treating several of the major types of sexual dysfunctions.

Sexual Desire Disorders

Some sex therapists help kindle the sexual appetites of people with hypoactive sexual desire by prescribing self-stimulation exercises combined with erotic fantasies (LoPiccolo & Friedman, 1988). Sex therapists may also assist dysfunctional couples by prescribing sensate focus exercises, enhancing communication, and expanding the couple's repertoire of sexual skills. Sex therapists recognize that hypoactive sexual desire is often a complex

Sensate focus exercises Exercises in which sex partners take turns giving and receiving pleasurable stimulation in nongenital areas of the body.

a CLOSER look

"Sexercise" Credited with Boosting Performance

It's been locker room talk for years: Getting fit improves your love life.

Now research supports this "jock wisdom" about the connection between physical fitness and sexual ability, with new studies suggesting that some of the declines in male sexual functioning often attributed to age are actually the result of sedentary living (Goldstein, 2000).

"It's always amazed me that men come here so out of shape they can't walk up the stairs, and then they wonder why they're impotent," says LeRoy Nyberg, director of urology programs at the National Institute of Diabetes and Digestive and Kidney Disease (NIDDK). "Physical fitness is important to keep all bodily organs functioning at their peak—especially those that require a good blood supply."

"Erectile dysfunction isn't inevitable" with age, says Nyberg, who echoes other experts in encouraging patients to exercise for physical and sexual health. Two recent studies support this "sexercise" link, indicating that regular exercise boosts sexual performance, whereas inactivity increases men's risk of impotence.

"The only behavior we found that can reduce the problem of erectile dysfunction is regular physical activity," says John McKinlay, an epidemiologist at the New England Research Institutes in Watertown, Massachusetts, and co-author of several recent reports based on the Massachusetts Male Aging Study, the largest random-sample study ever done of the condition known as ED.

"The prevalence is higher than previously thought," says McKinlay, whose research indicated that 10% of men aged 40 to 70 had complete ED, and about half had at least mild ED. "We also found a very strong relationship between ED and coronary heart disease that is independent of all other risk factors, such as high cholesterol, hypertension, smoking, and diabetes."

Like heart disease, ED is "in some ways a plumbing problem, related to system-wide blockages in blood vessels," he says. "If there's going to be a blockage, it makes sense that it will turn up in the microvascular penile bed before it turns up in the large blood vessels of the heart." Massachusetts urologist Irwin Goldstein (2000) concurs.

For this reason, McKinlay calls ED "possibly the best marker we have of coronary heart disease, the biggest killer of aging men." Yet, he says, "most primary-care physicians routinely dismiss this very common problem" as an inevitable part of aging without considering the possibility that it may be a symptom of a life-threatening illness.

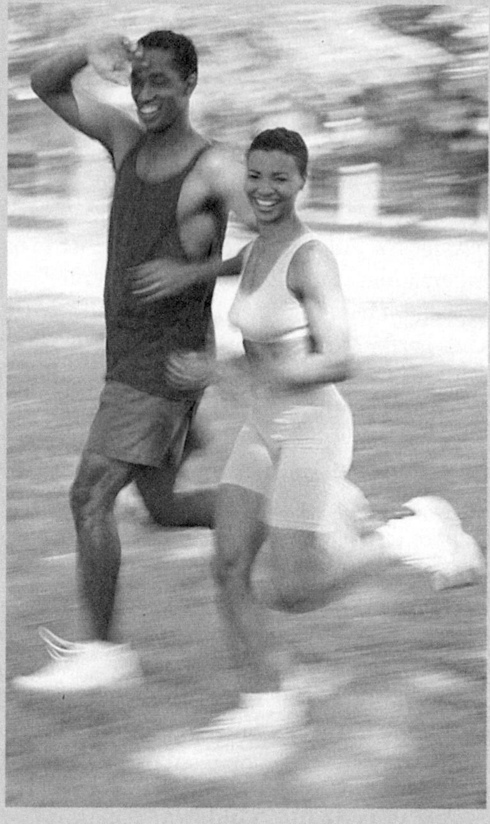

Are They Fit for Sex? Research now clearly shows that one's general health is an important factor in sexual functioning. In fact, males who maintain a healthful lifestyle are much less likely to develop erectile problems as they age.

problem that requires more intensive treatment than problems of the arousal or orgasm phases (Leiblum & Rosen, 1988). Helen Singer Kaplan (1987) argues that insight-oriented approaches are especially helpful in the treatment of hypoactive sexual desire and sexual aversion to uncover and resolve deep-seated psychological conflicts.

Some cases of hypoactive sexual desire in men involve hormonal deficiencies, especially deficiencies in testosterone. But testosterone replacement therapy works with only about half of men who have low testosterone levels (Rakic et al., 1997). Among women,

The Massachusetts Male Aging Study found a direct correlation between ED and physical activity, with sedentary men at highest risk for impotence and active men at lowest risk. The data suggest that sedentary men may be able to reduce their risk of developing ED by adopting regular physical activity equivalent to taking a brisk two-mile walk each day.

More evidence that exercise may improve men's sexual performance is offered by new research from the Harvard School of Public Health.

"Physical activity was strongly associated with better erectile functioning," concluded Eric B. Rimm and colleagues in a report presented at the recent annual meeting of the American Urological Association in Atlanta. His analysis of nearly 2,000 male health professionals, aged 51 to 87, revealed that men who exercised vigorously for approximately 20 to 30 minutes a day were about half as likely to have ED as men with the lowest level of physical activity. In addition, he found that as a man's waist size increased, so did his chance of ED.

These findings support previous research documenting the link between physical and sexual fitness for both men and women. Among the most intriguing:

- A University of California, San Diego, study of 78 sedentary but healthy middle-aged men who started exercising vigorously three to four days a week, for 60 minutes per session, found that new exercisers reported more frequent sexual activity and orgasms, more reliable function during sex, and a higher percentage of satisfying orgasms. "The degree of sexual enhancement correlated with the individual's fitness gain," reported study author James White.
- Vigorous exercise facilitates physiological sexual arousal in women, according to a study by Cindy Meston, an assistant professor of clinical psychology at the University of Texas at Austin. Her research on 35 women who watched a travel film and an erotic film on two separate occasions—once after cycling vigorously for 20 minutes—found that in the presence of a sexual stimulus, vigorous exercise appeared to have a "kind of priming effect," increasing blood flow into vaginal tissue.

Fitness appears to enhance sexual health through mechanisms that affect both body and mind. Physical boosts in endurance, muscle tone, body composition, blood flow, and flexibility may all improve sexual functioning. In addition, regular exercise reduces the risk of diseases associated with sexual dysfunction, such as diabetes and hypertension. Fit people also may be less likely to engage in behaviors linked with ED, such as smoking, or to take medications, such as drugs to treat high blood pressure, that may produce ED as a side effect.

Psychological benefits of regular exercise—including stress reduction, mood elevation, and heightened self-confidence—also may enrich a person's love life.

Because fitness improves physical and mental health and enhances appearance, it's not surprising that regular exercise is linked to a better sex life. In fact, now that drugs are available as a quick fix for ED, some health experts fear that they will undermine what has been a powerful incentive for getting sedentary men to "take steps" to improve their overall fitness.

"Viagra and other medications are not without side effects and risks," notes epidemiologist McKinlay. "Before taking these drugs, it's important to start healthy lifestyle changes."

And it's also important to remember that the benefits of exercise extend far beyond the bedroom, says NIDDK's Nyberg.

"Sedentary behaviors and their ill effects build up over a period of years," he notes. "So even though problems won't go away overnight with exercise, people who stick with their programs can get in shape and will boost their health."

Resources

Information on ED and other urologic conditions is available from

*The National Kidney and Urologic Diseases Information Clearinghouse, 3 Information Way, Bethesda, MD 20892-3580; on the Web at www.niddk.nih.gov/health/kidney/nkudic.htm.

*The Sexual Function Health Council of the American Foundation for Urologic Disease, 800-242-2383, www.afud.org.

Source: From Carol Krucoff. (2000, May 30). " 'Sexercise' Credited with Boosting Performance." *The Washington Post*, p. Z09.

as among men, lack of sexual desire can be connected with low levels of androgens, and testosterone shows promise in heightening desire (Tuiten et al., 2000).

When lack of desire is connected with depression, sexual interests may rebound when the depression lifts. Treatment in such cases may involve psychotherapy or chemotherapy, not sex therapy per se. When problems in the relationship are involved, marital or couples therapy may be indicated to improve the relationship. Once interpersonal problems are ironed out, sexual interest may return.

Treatment of sexual aversion disorder may involve a multifaceted approach, including biological treatments, such as the use of medications to reduce anxiety, and psychological treatments designed to help the individual overcome the underlying sexual phobia. Couples therapy may be used in cases in which sexual aversions arise from problems in relationships (Gold & Gold, 1993). Sensate focus exercises may be used to lessen generalized anxiety about sexual contact. But fears of specific aspects of the sexual act may need to be overcome through behavioral exercises in which the client learns to manage the stimuli that evoke fears of sexual contact:

> Bridget, 26, and Bryan, 30, had been married for four years but had never consummated their relationship because Bridget would panic whenever Bryan attempted coitus with her. Although she enjoyed foreplay and was capable of achieving orgasm with clitoral stimulation, her fears of sexual contact were triggered by Bryan's attempts at vaginal penetration. The therapist employed a program of gradual exposure to the feared stimuli to allow Bridget the opportunity to overcome her fears in small, graduated steps. First she was instructed to view her genitals in a mirror when she was alone—this in order to violate her long-standing prohibition against looking at and enjoying her body. This exercise initially made her feel anxious, but with repeated exposure she became comfortable performing it and then progressed to touching her genitals directly. When she became comfortable with this step, and reported experiencing pleasurable erotic sensations, she was instructed to insert a finger into the vagina. She encountered intense anxiety at this step and required daily practice for two weeks before she could tolerate inserting her finger into her vagina without discomfort. Her husband was then brought into the treatment process. The couple was instructed to have Bridget insert her own finger in her vagina while Bryan watched. When she was comfortable with this exercise, she then guided his finger into her vagina. Later he placed one and then two fingers into her vagina, while she controlled the depth, speed, and duration of penetration. When she felt ready, they proceeded to attempt penile penetration in the female-superior position, which allowed her to maintain control over penetration. Over time, Bridget became more comfortable with penetration to the point where the couple developed a normal sexual relationship. (Adapted from Kaplan, 1987, pp. 102–103)

Sexual Arousal Disorders

There are male and female sexual arousal disorders. Male sexual arousal disorder is also known as erectile disorder.

ERECTILE DISORDER Men with chronic erectile disorder may believe that they have "forgotten" how to have an erection. They may ask their therapists to "teach" them or "show them" how. Some of our clients have asked us to tell them what they should think or picture in their minds to achieve an erection, or how they should touch their partners or be touched. Erection is an involuntary reflex, however, not a skill. A man need not learn how to have an erection any more than he need learn how to breathe.

In sex therapy, women who have trouble becoming lubricated and men with erectile problems learn that they need not "do" anything to become sexually aroused. As long as their problems are psychologically and not organically based, they need only receive sexual stimulation under relaxed circumstances, so that anxiety does not inhibit their natural reflexes.

Truth or Fiction? **REVISITED**

It is not true that sex therapy teaches a man with erectile disorder how to will an erection. Men with erectile disorder are actually taught that it is not possible to will an erection. One can only set the stage for erection (or vaginal lubrication) to occur and then allow it to happen reflexively. ■

In order to reduce performance anxiety, the partners engage in nondemanding sexual contacts—contacts that do not demand lubrication or erection. They may start with nongenital sensate focus exercises in the style of Masters and Johnson. After a couple of sessions, sensate focus extends to the genitals. The position shown in Figure 15.1 allows

FIGURE 15.1

The Training Position Recommended by Masters and Johnson for Treatment of Erectile Disorder and Premature Ejaculation. By lying in front of her partner, who has his legs spread, the woman has ready access to his genitals. In one part of a program designed to overcome erectile disorder, she repeatedly "teases" him to erection and allows the erection to subside. Thus she avoids creating performance anxiety that could lead to loss of erection. Through repeated regaining of erection, the man loses his fear that loss of erection means it will not return.

the woman easy access to her partner's genitals. She repeatedly "teases" him to erection and allows the erection to subside. Thus she avoids creating performance anxiety that could lead to loss of erection. By repeatedly regaining his erection, the man loses the fear that loss of erection means it will not return. He learns also to focus on erotic sensations for their own sake. He experiences no demand to perform, because the couple is instructed to refrain from coitus.

Even when the dysfunctional partner can reliably achieve sexual excitement (denoted by erection in the male and lubrication in the female), the couple does not immediately attempt coitus; this might rekindle performance anxiety. Rather, the couple engages in a series of nondemanding, pleasurable sexual activities, which eventually culminate in coitus.

In Masters and Johnson's approach, the couple begin coitus after about 10 days of treatment. The woman teases the man to erection while she is sitting above him, straddling his thighs. When he is erect, *she* inserts the penis (to avoid fumbling attempts at entry) and moves slowly back and forth in a *nondemanding way.* Neither attempts to reach orgasm. If erection is lost, teasing and coitus are repeated. Once the couple become confident that erection can be retained—or reinstated if lost—they may increase coital thrusting gradually to reach orgasm.

BIOLOGICAL APPROACHES TO TREATMENT OF ERECTILE DISORDER The world's attention has recently been focused on biological approaches to treating erectile disorder. Perhaps the opening volley in the biological war against erectile disorder was fired in

1983, when a speaker made a somewhat unusual presentation to his audience at a medical convention. He dropped his trousers to reveal an erection. The erection was the result of not sexual stimulation or sexual fantasies, but of the injection of a drug called *alprostadil* directly into his penis. Alprostadil is a vasodilator; it relaxes the muscles surrounding the arteries in the penis, allowing more blood to flow in and thus increasing vasocongestion and causing erection. The speaker—a urologist—was giving a "live" demonstration of a biological method of treating erectile disorder.

Truth or Fiction? **REVISITED**

It is true that a doctor once dropped his pants to reveal an erection at a medical convention. He was demonstrating the effects of alprostadil. ■

Biological or biomedical approaches are helpful in treating erectile disorder, especially when organic factors are involved. Treatments include surgery, medication, and vacuum pumps (see Table 15.3).

SURGERY Two main types of surgery are used in treating erectile disorder: vascular surgery and the installation of penile implants. *Vascular surgery* can help in cases in which

TABLE 15.3

Biological treatments of erectile problems

	Surgery
Vascular surgery	Helps when blood vessels that supply the penis are blocked.
Penile implants	May be used when other treatments fail because of biological problems.
	Medication
Hormone therapy	Helps men and women with abnormally low levels of male sex hormones.
Injections	Muscle relaxants such as *alprostadil* and *phentolamine* are injected into the corpus cavernosum of the penis, relaxing the muscles that surround the arteries in the penis, allowing the vessels to dilate and blood to flow more freely.
Suppository	Alprostadil is inserted into the tip of the penis in gel form.
Oral medication	The oral form of sildenafil (Viagra) and the oral form of phentolamine (Vasomax) relax the muscles that surround the small blood vessels in the penis, allowing them to dilate so that blood can flow into them more freely. Apomorphine (Uprima) increases brain levels of the neurotransmitter dopamine. (Parkinson's disease is caused by the death of dopamine-producing cells and is often accompanied by erectile dysfunction. Men who use drugs that treat Parkinson's disease by increasing dopamine levels often have erections as side effects.)
	Vacuum Pump
Vacuum pump	A *vacuum constriction device* creates a vacuum when it is held over the penis. The vacuum induces erection by increasing the flow of blood into the penis. Rubber bands around the base of the penis maintain the erection.

the blood vessels that supply the penis are blocked or in which structural defects in the penis restrict blood flow (Cowley & Rogers, 1997). An arterial bypass operation reroutes vessels around the blockage.

A *penile implant* may be used when other treatments fail. Implants are either malleable (semirigid) or inflatable. The semirigid implant is made of rods of silicone rubber that remain in a *permanent* semirigid position. It is rigid enough for intercourse but permits the penis to hang reasonably close to the body at other times. The inflatable type requires more extensive surgery. Cylinders are implanted in the penis. A fluid reservoir is placed near the bladder, and a tiny pump is inserted in the scrotum. To attain erection, the man squeezes the pump, releasing fluid into the cylinders. When the erection is no longer needed, a release valve returns the fluid to the reservoir, deflating the penis. The inflatable implant more closely duplicates the normal processes of tumescence and detumescence. Some adverse side effects of penile implants have been reported, including destruction of erectile tissue, which impairs the man's ability to have normal erections. Penile implants do not affect sex drive, sexual sensations, or ejaculation.

Implant surgery is irreversible (Padma-Nathan et al., 1997). Therefore, the National Institutes of Health recommend that penile implants be used only when less invasive techniques, such as sex therapy and medication, are unsuccessful.

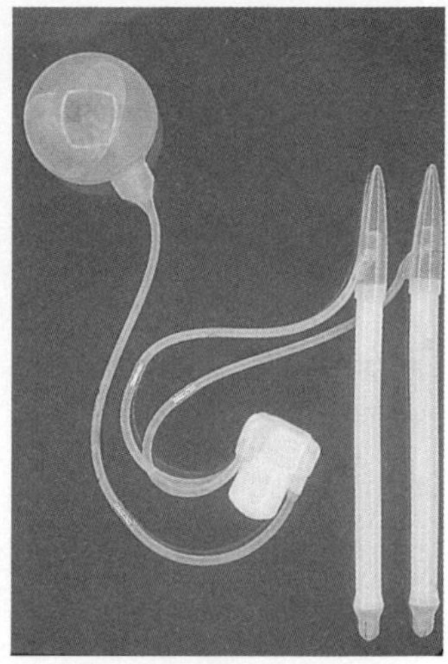

A Penile Implant. Penile implants provide erection when the man's cardiovascular system does not do the job. This implant consists of cylinders that are implanted in the penis. A fluid reservoir (top left) is placed near the bladder. A pump (lower middle) is typically inserted in the scrotum. Squeezing the pump forces fluid into the cylinders, inflating the penis. Tripping a release valve later returns the fluid to the reservoir, deflating the penis.

MEDICATION There are several ways in which medication can be used to help men with erectile problems. For example, *hormone (testosterone) treatments* help restore the sex drive and erectile ability in many men with abnormally low levels of testosterone (Lue, 2000; Rakic et al., 1997). There is no evidence that hormone therapy helps men who already have normal hormone levels. Testosterone does not help women who complain of lack of sexual desire, unless their problem follows a surgically induced menopause in which the testosterone-producing adrenal glands were removed along with the ovaries (Bagatell & Bremner, 1996).

The muscle relaxants *alprostadil* (brand names Caverject and Edex) and *phentolamine* (Invicorp) can be injected into the corpus cavernosum of the penis. These chemicals relax the muscles that surround the small blood vessels in the penis. The vessels dilate and allow blood to flow into the penis more freely. In the case of alprostadil, erections last for an hour or more and occur whether or not there is sexual stimulation. A physician teaches the man how to inject himself. Phentolamine is used along with the protein VIP, and erection occurs only when sexual stimulation is applied.

Injections are most effective for men with problems in the transmission of nerve signals that regulate erection (Altman, 1995c). They are less effective for men with vascular problems that restrict the flow of blood into the penis. Injections are also helpful with most men whose erectile problems are psychologically based (Linet & Ogrinc, 1996).

Penile injections may have side effects, including pain from the injection itself and prolonged, painful erections (*priapism*) (Ralph & McNicholas, 2000). Many men find the idea of penile injections distasteful and either reject them at the outset or drop out after a trial.

Alprostadil is also available as a suppository in gel form (brand name MUSE). It is then inserted into the tip of the penis via an applicator. The suppository helps men get around the "wince" factor that many experience with injections (Padma-Nathan et al., 1997). "Putting a needle in your penis is not everybody's idea of foreplay," notes Dr. John Seely (Kolata, 2000b), who is involved in the development of the oral medication Uprima (see Table 15.3).

Other medications are taken orally. The oral form of sildenafil is sold as Viagra, and the oral form of phentolamine will be marketed as Vasomax (Vasomax is currently under review by the FDA). The drug apomorphine (Uprima) heightens brain levels of dopamine, a neurotransmitter involved in erection. Researchers became aware of the

Human Sexuality ONLINE //

Buying Viagra and Other Drugs Online

Viagra is a prescription drug. Many men who might otherwise use Viagra are reluctant to discuss erectile dysfunction with their physicians. The anonymity of doing things on the Net is a lure—you don't have to admit your personal worries to your doctor face to face. Some men, unfortunately, do not even have a regular physician. What to do?

Many have discovered that by searching "Viagra" on the Net, they can find many Web sites where they can "consult" with online physicians, obtain a prescription, and order the drug for home delivery. Easy! A few questions and a fee and they've got it. But is it wise? Perhaps, perhaps not.

Prescriptions are needed for various drugs because physicians are better equipped than most laypeople to diagnose an individual's health problems, understand the chemical nature and side effects of the drugs that are available for treatment, and predict how the drugs will affect the individual patient. The physicians are also usually prepared to deal with the unexpected effects of the drugs, and there can be many.

Ask yourself what kind of physician will prescribe drugs online, without personally knowing the patient. Is it possible that some of them would have difficulty establishing private practices or getting jobs in hospitals? If you have a question about the drug once you use it, will you be able to get back to the prescriber easily for an answer, or will you wind up with an embarrassed call to your own physician or a trip to the emergency room?

You will find that many of the sites are online pharmacies that also advertise Propecia for treatment of hair loss, drugs that reverse the loss of pubic hair (yes, pubic hair), weight control drugs (such as Xenical) that prevent the absorption of some of the fat in food, herbal supplements, and so on. What you will tend to hit in your search are chemical armories of weapons that purport to arrest aging or even reverse the aging process. It may sound as though they will help you remain (or recover your status as) a studmuffin. But the fact is that medical science isn't there yet.

While surfing the Net, you will also come across sites that offer "natural" preparations, including a variety of herbs claimed to be as effective as Viagra, but without the side effects and without the need to get a prescription. Use some critical thinking: Are you convinced of the effectiveness and safety of these preparations? Because they are foods (sort of) rather than drugs, they escape the scrutiny of the Food and Drug Administration. That is, the government is not watching over their content or purity. Be warned.

And for Women . . .

Many sites of online drug stores also advertise the availability of Viagra for women with orgasmic dysfunction. But the benefits of Viagra for women remain somewhat in doubt. As with men, there are any number of online physicians who will prescribe Viagra to women after a brief screening process. But it is wiser to consult your gynecologist about the possible benefits of Viagra and other drugs aimed at treating sexual dysfunctions.

Legitimate Sites

There are some legitimate Web sites that are of interest. An example is www.viagra.com. This Web page is within the pharmaceutical company Pfizer's Web site. A disclaimer there reads, "The health information contained herein is for educational purposes only and is not intended to replace discussions with a health care provider. All decisions regarding patient care must be

potential benefits of dopamine-enhancing drugs through research with Parkinson's disease. Parkinson's is apparently caused by the death of dopamine-producing cells and is connected with loss of motor coordination and erectile dysfunction. L-dopa and other drugs that are used to treat Parkinson's raise dopamine levels and frequently have the "side effect" of erection (Kolata, 2000b). As of this writing, approval of Uprima by the FDA is in doubt as a consequence of side effects related mainly to nausea and low blood pressure (Twersky, 2000).

Viagra was hailed as a miracle drug when it hit the market in early 1998. It sold faster than any new drug had ever sold. Bob Dole announced, on the Larry King show, that he had participated in the clinical trials of Viagra a couple of years earlier and that he found it to be effective ("Cancer Survivor Bob Dole," 1998). At a press interview, Dole's wife Elizabeth Dole, who is president of the American Red Cross, echoed his view that Viagra is a "great drug" ("Mrs. Dole," 1998). The stock of Pfizer Company, which produces Viagra, shot up after the appearance of the drug. A study published in *The New England Journal of Medicine* tested the effects of Viagra on more than 800 men with erectile dysfunction due both to psychological and to a number of organic causes

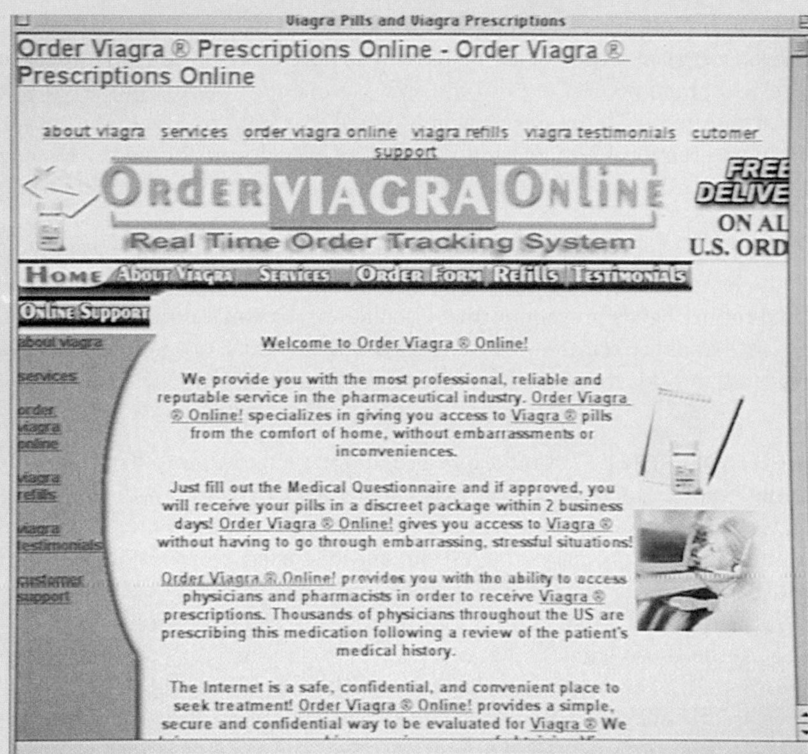

Should You Buy Viagra Online? Many Web sites enable men to consult with physicians and order Viagra online. Is it wise to purchase Viagra—or other prescription drugs—online? Prescriptions are needed for drugs when the person's diagnosis is in question and when the drugs have side effects. Physicians are better equipped than most laypeople to diagnose health problems, understand the chemical composition and effects of drugs, and predict how drugs will affect the individual.

made by a health care provider, who will consider the unique characteristics of the patient." You will find recent useful information about the drug there.

Another site of some interest is the Doctors Guide to the Internet: www.pslgroup.com. It contains information and links to news stories about treatments for erectile dysfunction and other sexual dysfunctions.

In sum, even if it is convenient to buy a drug online, you are well advised to get your prescription face to face—from a doctor you know and trust.

(Goldstein et al., 1998). Men with extreme physical problems, such as anatomical defects and spinal cord injury, were not included. In one phase of the study, 69% of attempts to engage in intercourse were successful for men taking Viagra, compared to 22% for men taking a placebo.

There are fascinating anecdotes about how Viagra has helped marriages—and how it has been a homewrecker. "I'd say that for 80% of men in a monogamous, stable relationship with a wife over many years, Viagra will bring relief and happiness," says Dr. Ira Sharlip, a San Francisco urologist (cited in Nordheimer, 1998). "As to what happens to the other 20%, nothing will surprise me," he added with a wry smile. For some men, Sharlip noted, restored sexual function can create issues of fidelity that previously had been moot. A man suddenly empowered sexually after years of inactivity may go flying out the door looking for someone else if his partner doesn't share his enthusiasm or rejects him.

A Greek study (Hatzichristou et al., 2000) found that Viagra is helpful to about three quarters of the men who have been using intracaverous injections for at least a year. Moreover, the great majority of men who have been using injections seem to be happy to switch to Viagra, although a few "stick" with the injections,[2] and some men switch back and forth.

VACUUM PUMPS Sounding like something from the "What will they think of next?" category, a *vacuum constriction device* (VCD) helps men achieve erections through vacuum pressure. The device (brand name ErecAid) consists of a cylinder that is connected to a hand-operated vacuum pump. It creates a vacuum when it is held over the limp penis. The vacuum induces erection by increasing the flow of blood into the penis. Rubber bands that are then applied around the base of the penis can maintain the erection for as long as 30 minutes.

The device has been used successfully by men with both organically and psychologically based erectile failure. However, side effects such as pain and black-and-blue marks are common. The rubber bands prevent normal ejaculation, so semen remains trapped in the urethra until the bands are released. The quality of the erections produced by the device is also considered inferior to spontaneous erections (Spark, 1991).

WHERE DO WE GO FROM HERE? It would appear that oral medications (pills) will be the most popular biological treatment of erectile problems. They are helpful with the majority of men treated and surmount the "wince factor." Viagra has side effects such as headaches, flushing, indigestion, nasal congestion, and distorted color vision, however (Goldstein et al., 1998). The likelihood of side effects increases with the dose. For example, about 20% of men who take the 50-mg pill have headaches, compared with about 30% of men who use the 100-mg pill (Goldstein et al., 1998). About 3% of users see things with a blue tinge for up to a few hours (Sponsel et al., 2000). (Given one of the slang meanings of the word *blue*, this side effect may not be inappropriate). Soon after Viagra was approved by the FDA, there were scattered reports of men with cardiovascular problems experiencing heart attacks. Because Viagra affects the flow of blood in the body, these reports created quite a scare. A carefully conducted study of the effects of Viagra on 14 older men with at least one severely constricted coronary artery suggests that Viagra by itself is not the problem (Herrmann et al., 2000). In this study, reported in the prestigious *New England Journal of Medicine* (www.nejm.com), Viagra was not shown to have adverse effects on the blood supply to the heart. It would appear that men who are using a class of medicines known as nitrates to help them manage cardiovascular disease run a greater risk from Viagra.

Vasomax promises to have fewer side effects than Viagra. Given the rapid development of these medications, within a few years most men with erectile problems will probably have effective oral medications available with few if any side effects.

FEMALE SEXUAL AROUSAL DISORDER Psychological treatments for female sexual arousal disorder parallel those for orgasmic disorder and are discussed in the following pages. Here let us briefly note that they involve sex education (labeling the parts, discussing their functions, and explaining how to arouse them), searching out and coping with possible cognitive interference (such as negative sexual attitudes), creating nondemanding situations in which sexual arousal may occur, and—when appropriate—working on problems in the relationship.

Yet many cases of female sexual arousal disorder reflect impaired blood flow to the genitals, just as in erectile disorder. Female sexual arousal involves vaginal lubrication, which permits sexual intercourse without a great deal of pain-causing friction. Lubrication is made possible by vasocongestion—the flow of blood into the genitals. Lack of lu-

[2]Yes, we confess to being guilty of a bad pun. Feel free to groan.

brication can reflect the physical effects of aging, menopause, or surgically induced menopause.

Sometimes all that is necessary to deal with lack of lubrication is an artificial lubricant such as K-Y Jelly. But lessened blood flow to the genitals can also sap sexual pleasure and, as a consequence, lessen a woman's desire for sex.

Just as biological treatments for erectile disorder are mushrooming, so are biological treatments for female sexual arousal disorder. Sadly—and certainly because of the lower amount of attention that has historically been paid to women's health problems—the development of treatments has lagged several paces behind the development of treatments for men. Ironically, the treatments that are emerging are highly similar to those that help men with erectile disorder.

For example, drugs identical or similar to those used for men are being investigated for use with women (Leland, 2000). Many trials have been undertaken with Viagra for women. Researchers are also developing alprostadil (the vasodilator) for use with women, largely in the form of creams that are inserted into the vagina to enhance the flow of blood and hence lubrication.

There is a perfect parallel in the area of lack of sexual desire for women who are low in sexual desire because of low levels of "male" sex hormones. (As an aside, we might begin to wonder out loud whether we should stop referring to estrogen as a female sex hormone and to testosterone as a male sex hormone. Both are produced by both women and men—although in different quantities—and both are intricately involved with women's and men's health, sexual functioning, and other kinds of behavior). Testosterone skin patches can be used by women who lack sexual desire because they lack adequate quantities of testosterone (Guzick & Hoeger, 2000; Shifren et al., 2000).

There is even a device—Eros—that is a parallel to the vacuum pump used by some men with erectile disorder. It is a clitoral device that was approved by the Food and Drug Administration in 2000 and is available by prescription. The clitoris swells during sexual arousal because of vasocongestion, and vasocongestion increases clitoral sexual sensations, thus moving somewhat in step with sexual interest and lubrication. The device creates "gentle" suction over the clitoris, increasing vasocongestion and sexual sensations (Leland, 2000; see Figure 15.2).

FIGURE 15.2

A Clitoral Device That Stimulates Genital Vasocongestion in Women by Creating (Gentle) Suction Over the Clitoris. This prescription device was approved by the FDA in 2000.

Orgasmic Disorders

Women who have never experienced orgasm often harbor negative sexual attitudes that cause anxiety and inhibit sexual response. Treatment in such cases may first address these attitudes.

Masters and Johnson use a couples-oriented approach in treating anorgasmic women. They begin with sensate focus exercises. Then, during genital massage and later during coitus, the woman guides her partner in the caresses and movements that she finds sexually exciting. Taking charge helps free the woman from the traditional stereotype of the passive, subordinate female role.

Masters and Johnson recommend a training position (see Figure 15.3) that gives the man access to his partner's breasts and genitals. She can guide his hands to show him the types of stimulation she enjoys. The genital play is *nondemanding.* The goal is to learn to provide and enjoy effective sexual stimulation, not to reach orgasm. The clitoris is not stimulated early, because doing so may produce a high level of stimulation before the woman is prepared.

After a number of occasions of genital play, the couple undertake coitus in the female-superior position (see Figure 15.4). This position allows the woman freedom of movement and control over her genital sensations. She is told to regard the penis as her "toy." The couple engage in several sessions of deliberately slow thrusting to sensitize the

FIGURE 15.3

The Training Position for Nondemanding Stimulation of the Female Genitals. This position gives the man access to his partner's breasts and genitals. She can guide his hands to show him what types of stimulation she enjoys.

woman to sensations produced by the penis and break the common counterproductive pattern of desperate, rapid thrusting.

Orgasm cannot be willed or forced. When a woman receives effective stimulation, feels free to focus on erotic sensations, and feels that nothing is being demanded of her, she will generally reach orgasm. Once the woman is able to attain orgasm in the female-superior position, the couple may extend their sexual repertoire to other positions.

Masters and Johnson prefer working with the couple in cases of anorgasmia, but other sex therapists prefer to begin working with the woman individually through masturbation (Barbach, 1975; Heiman & LoPiccolo, 1987). This approach assumes that the woman accepts masturbation as a therapy tool. Masturbation provides women with opportunities to learn about their own bodies at their own pace. It frees them of the need to rely on a partner or to please a partner. The sexual pleasure they experience helps counter

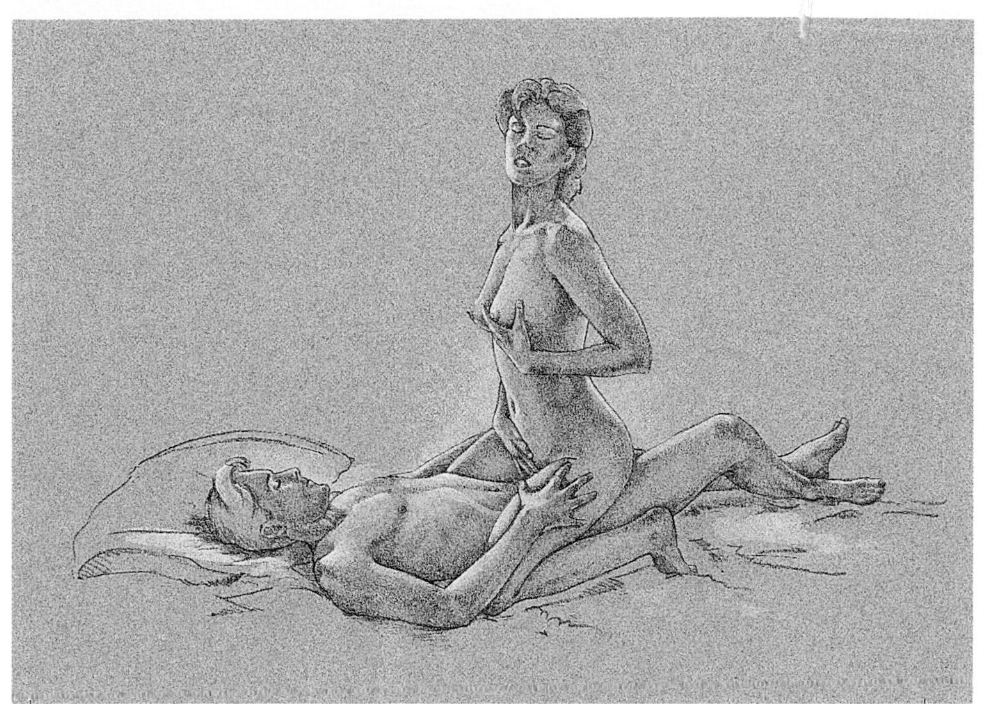

FIGURE 15.4

Coitus in the Female-Superior Position. In treatment of female orgasmic disorder, the couple undertake coitus in the female-superior position after a number of occasions of genital play. This position allows the woman freedom of movement and control over her genital sensations. She is told to regard the penis as her "toy." The couple engage in several sessions of deliberately slow thrusting to sensitize the woman to sensations produced by the penis and to break the common counterproductive pattern of desperate, rapid thrusting.

lingering sexual anxieties. Although there is some variation among therapists, the following elements are commonly found in directed masturbation programs:

1. *Education.* The woman and her sex partner (if she has one) are educated about female sexuality.
2. *Self-exploration.* Self-exploration is encouraged as a way of increasing the woman's sense of body awareness. She may hold a mirror between her legs to locate her sexual anatomic features. Kegel exercises may be prescribed to help tone and strengthen the pubococcygeus (PC) muscle that surrounds the vagina and to increase her awareness of genital sensations and her sense of control.
3. *Self-massage.* Once the woman feels comfortable about exploring her body, she creates a relaxing setting for self-massage. She chooses a time and place where she is free from distractions. She begins to explore the sensitivity of her body to touch, discovering and then repeating the caresses that she finds pleasurable. At first self-massage is not concentrated on the genitals. It encompasses other sensitive parts of the body. She may incorporate stimulation of the nipples and breasts and then direct genital stimulation, focusing on the clitoral area and experimenting with hand movements. Nonalcohol-based oils and lotions may be used to enhance the sensuous quality of the massage and to provide lubrication for the external genitalia. Kegel exercises may also be performed during self-stimulation to increase awareness of vaginal sensations and increase muscle tension. Some women use their dominant hand to stimulate their breasts while the other hand massages the genitals. No two women approach masturbation in quite the same way. In order to prevent performance anxiety, the woman does not attempt to reach orgasm during the first few occasions.
4. *Giving oneself permission.* The woman may be advised to practice assertive thoughts to dispel lingering guilt and anxiety about masturbation. For example, she might repeat to herself, "This is my body. I have a right to learn about my body and receive pleasure from it."

5. *Use of fantasy.* Arousal is heightened through the use of sexual images, fantasies, and fantasy aids, such as erotic written or visual materials.
6. *Allowing, not forcing, orgasm.* It may take weeks of masturbation to reach orgasm, especially for women who have never experienced orgasm. By focusing on her erotic sensations and fantasies, but not demanding orgasm, the woman lowers performance anxiety and creates the stimulating conditions needed to reach orgasm.
7. *Use of a vibrator.* A vibrator may be recommended to provide more intense stimulation, especially for women who find that manual stimulation is insufficient.
8. *Involvement of the partner.* Once the woman is capable of regularly achieving orgasm through masturbation, the focus may shift to the woman's sexual relationship with her partner. Nondemanding sensate focus exercises may be followed by nondemanding coitus. The female-superior position is often used. It enables the woman to control the depth, angle, and rate of thrusting. She thus ensures that she receives the kinds of stimulation she needs to reach orgasm.

Truth or Fiction? **REVISITED**

It is true that many sex therapists recommend masturbation as the treatment for women who have never been able to reach orgasm. Masturbation allows women (and men) to get in touch with their own sexual responses without relying on a partner. ■

Kaplan (1974) suggests a bridge maneuver to assist couples who are interested in making the transition from a combination of manual and coital stimulation to coital stimulation alone as a means for reaching orgasm. Manual stimulation during coitus is used until the woman senses that she is about to reach orgasm. Manual stimulation is then stopped and the woman thrusts with her pelvis to provide the stimulation necessary to reach orgasm. Over time the manual clitoral stimulation is discontinued earlier and earlier. Although some couples may prefer this "hands-off" approach to inducing orgasm, Kaplan points out that there is nothing wrong with combining manual stimulation and penile thrusting. There is no evidence that reliance on clitoral stimulation means that women are sexually immature. (Evidence has not borne out the theoretical psychoanalytic distinction between clitoral and vaginal orgasms.)

Our focus has been on sexual techniques, but it is worth noting that a combination of approaches that focusing on sexual techniques and underlying interpersonal problems may be more effective than focusing on sexual techniques alone, at least for couples whose relationships are troubled (Killmann et al., 1987; LoPiccolo & Stock, 1986).

MALE ORGASMIC DISORDER Treatment of male orgasmic disorder generally focuses on increasing sexual stimulation and reducing performance anxiety (LoPiccolo & Stock, 1986). Masters and Johnson instruct the couple to practice sensate focus exercises for several days, during which the man makes no attempt to ejaculate. The couple is then instructed to bring the man to orgasm in any way they can, usually by the woman's stroking his penis. Once the husband can ejaculate in the woman's presence, she brings him to the point at which he is about to ejaculate. Then, in the female-superior position, she inserts the penis and thrusts vigorously to bring him to orgasm. If he loses the feeling that he is about to ejaculate, the process is repeated. Even if ejaculation occurs at the point of penetration, it often helps break the pattern of inability to ejaculate within the vagina.

Squeeze technique A method for treating premature ejaculation whereby the tip of the penis is squeezed to prevent ejaculation temporarily.

PREMATURE EJACULATION In the Masters-and-Johnson approach, sensate focus exercises are followed by practice in the training position shown in Figure 15.1. The woman teases her partner to erection and uses the **squeeze technique** when he indicates that he

a CLOSER look

The "Orgasm Pill" for Women

Erectile disorder in men may be seen as a vascular problem or disease. As women get older, they too experience a reduced flow of blood to the genital region, which means that the clitoris becomes less engorged during sexual arousal and may be connected with feelings of lessened sexual arousal overall. Post-menopausal women also experience symptoms such as dryness of the vagina because of drop-off in secretion of estrogen. About 50% of adult women say that they have lost interest in sex or have difficulty becoming aroused (Kolata, 1998b).

Researchers (and many women) have asked whether Viagra or other drugs that enhance the flow of blood may also enhance the sexual experiences of women—when taken by the women, that is (Kolata, 1998a, 1998b). Some women report experiencing greater vaginal lubrication and stronger orgasms as a result of using Viagra (e.g., Berman, 2000). Table 15.4 shows the results of a study with 35 women who had received hysterectomies, as reported by Boston sex therapist Laura Berman.

Pilot studies with 500 European women ("Women might mark millennium," 1998) also found drugs that enhance the flow of blood to the genitals to be quite effective. It turns out that a number of pharmaceutical companies are developing Viagra-like drugs for women—drugs with names like VasoFem, Alista, and FemProx (Leland, 2000).

Yet another study of 577 women found that Viagra was no more effective than a placebo (sugar pill) in increasing sexual desire among women with sexual dysfunctions (Basson, 2000). How do we resolve this discrepancy? First, the women in the Basson study were aged 18 to 55. Those in the Berman study were generally middle-aged and had had hysterectomies. However, Berman also found sexual desire per se to be the *least changed* variable in her study. Prior to treatment, 52% of the women in her study reported low sexual desire, and this percentage fell only to 45% after use of Viagra. Sexual desire is a complex matter that involves the quality of relationships and other psychological factors as well as biological factors. Problems related to biological arousal, such as lack of vaginal lubrication and pain during sex, were reduced dramatically as a result of using Viagra (see Table 15.4). Perhaps drugs like Viagra are more effective at helping with the biological than with the psychological aspects of sexual relations and relationships.

TABLE 15.4

Impact of Viagra on the post-hysterectomy sexual complaints of women in the Berman (2000) Study

	Before Using Viagra	After Using Viagra
Low sexual sensations	100%	22%
Inability to reach orgasm	100	18
Little or no sexual desire	52	45
Little or no lubrication	67	40
Pain or discomfort during sex	68	33

Source: L. Berman (2000). Paper presented to the annual meeting of the American Urological Association, Atlanta, GA. Cited in "Women, Too, May Benefit from Viagra." (2000, May 1). Web posted by CNN.

is about to ejaculate. She squeezes the tip of the penis, which temporarily prevents ejaculation. This process is repeated three or four times in a 15- to 20-minute session before the man purposely ejaculates.

In the squeeze technique (which should be used only following personal instruction from a sex therapist), the woman holds the penis between the thumb and first two fingers of the same hand. The thumb presses against the frenulum. The fingers straddle the coronal ridge on the other side of the penis. Squeezing the thumb and forefingers together fairly hard for about 20 seconds (or until the man's urge to ejaculate passes) prevents ejaculation. The erect penis can withstand fairly strong pressure without discomfort, but erection may be partially lost.

It is true that a man can be prevented from ejaculating by squeezing his penis when he feels that he is about to ejaculate. This is the so-called squeeze technique.

After two or three days of these sessions, Masters and Johnson have the couple begin coitus in the female-superior position because it creates less pressure to ejaculate. The woman inserts the penis. At first she contains it without thrusting, allowing the man to get used to intravaginal sensations. If he signals that he is about to ejaculate, she lifts off and squeezes the penis. After some repetitions, she begins slowly to move backward and forward, lifting off and squeezing as needed. The man learns gradually to tolerate higher levels of sexual stimulation without ejaculating.

The alternating "stop-start" method for treating premature ejaculation was introduced by urologist James Semans (1956). The method can be applied to manual stimulation or coitus. For example, the woman can manually stimulate her partner until he is about to ejaculate. He then signals her to suspend sexual stimulation and allows his arousal to subside before stimulation is resumed. This process enables the man to recognize the cues that precede his point of ejaculatory inevitability, or "point of no return," and to tolerate longer periods of sexual stimulation. When the stop-start technique is applied to coitus, the couple begin with simple vaginal containment with no pelvic thrusting, preferably in the female-superior position. The man withdraws if he feels he is about to ejaculate. As the man's sense of control increases, thrusting can begin, along with variations in coital positions. The couple again stop when the man signals that he is approaching ejaculatory inevitability.

BIOLOGICAL APPROACHES TO TREATMENT OF PREMATURE EJACULATION Pilot studies have recently appeared in which drugs generally used for psychological problems have been helpful in treating premature ejaculation. One, clomipramine, is normally used to treat people with obsessive-compulsive disorder or schizophrenia. However, in a pilot study with 15 couples, low doses of clomipramine helped men engage in coitus five times longer than usual without ejaculating (Althof, 1994). So-called antidepressant drugs have also been helpful in treatment of premature ejaculation (Forster & King, 1994; Waldinger et al., 1994; Wise, 1994).

Why do drugs used to treat psychological problems help with premature ejaculation? The psychological problems are frequently connected with imbalances in body chemistry, such as **neurotransmitters**—the chemical messengers of the brain. Neurotransmitters are also involved in other bodily functions, including ejaculation. The antidepressant drugs (fluoxetine, paroxetine, and sertraline) all work by increasing the action of the neurotransmitter serotonin. Serotonin, in turn, may inhibit the ejaculatory reflex (Assalian, 1994; Shipko, 2000). It remains to be seen whether medications continue to show positive effects and to compare their effectiveness with psychological sex therapy techniques.

Sexual Pain Disorders

DYSPAREUNIA Dyspareunia, or painful intercourse, generally calls for medical intervention to indentify and treat any underlying physical problems, such as urinary tract genital infections, that might give rise to pain (Laumann et al., 1999). When dyspareunia is caused by vaginismus, treatment of vaginismus through a behavioral approach, described below, may eliminate pain.

Neurotransmitter A chemical that transmits messages from one brain cell to another.

VAGINISMUS Vaginismus is generally treated with behavioral exercises in which plastic vaginal dilators of increasing size are inserted to help relax the vaginal musculature. A gynecologist may first demonstrate insertion of the narrowest dilator. Later the woman herself practices insertion of wider dilators at home. The woman increases the size of the dilator as she becomes capable of tolerating insertion and containment (for 10 or 15 minutes) without discomfort or pain. The woman herself—not her partner or therapist—controls the pace of treatment (LoPiccolo & Stock, 1986). The woman's or her partner's fingers (first the littlest finger, then two fingers, and so on) may be used in place of the plastic dilators, with the woman controlling the speed and depth of penetration. When the woman is able to tolerate dilators (or fingers) equivalent in thickness to the penis, the couple may attempt coitus. Still, the woman should control insertion. Circumstances should be relaxed and nondemanding. The idea is to avoid resensitizing her to fears of penetration. Because vaginismus often occurs among women with a history of sexual trauma, such as rape or incest, treatment for the psychological effects of these experiences may also be in order (LoPiccolo & Stock, 1986).

Evaluation of Sex Therapy

Masters and Johnson (1970) reported an overall success rate of about 80% in treating sexual dysfunctions in their two-week intensive program. Some dysfunctions proved more difficult to treat than others. An analysis of treatment results from 1950 to 1977 showed success rates ranging from 67% for primary erectile disorder to 99% for vaginismus (Kolodny, 1981). A follow-up of 226 initial successes after a five-year period showed that 16 people, or 7%, experienced a "treatment reversal."

Masters and Johnson's critics note that they followed up on a disappointingly small percentage (29%) of their sample over a five-year period. Thus they may have seriously underestimated the actual number of treatment reversals (Adams, 1980; Zilbergeld & Evans, 1980). Zilbergeld and Evans (1980) also challenged their outcome measures. They argued that Masters and Johnson used a global measure of success that was not tied to specific criteria.

Masters and Johnson's sample may also have been biased in at least two ways. It consisted only of people who were willing (and could afford) to spend two weeks at their institute for full-time therapy. These people were better educated and more affluent than the general population. Masters and Johnson also denied treatment to a number of people whom they judged not "really interested" in changing. Thus their final sample may have been limited to people who were highly motivated, and some of the success of treatment may have been due to the clients' high level of motivation rather than to the treatment they received. Finally, the absence of a control group prevents us from knowing whether factors extraneous to the treatment itself may have been responsible for the apparent success.

Other researchers have reported more modest levels of success in treating erectile disorder than those reported at the Masters and Johnson Institute (Barlow, 1986). Nevertheless, long-term follow-up evaluations support the general effectiveness of sex therapy for erectile disorder (Everaerd, 1993). Yet problems do recur in some cases and are not always easily overcome (Everaerd, 1993). Some couples cope with recurring problems by using the techniques they learned during treatment, such as sensate focus exercises. The addition of biological treatments to the arsenal of treatments for erectile disorder has improved success rates to the point where the great majority of erectile problems can be successfully treated in one way or another (Lue, 2000).

We lack controlled studies of treatments of male orgasmic disorder (Dekker, 1993). Other than the original Masters-and-Johnson studies, results have been generally disappointing, with most people showing only modest improvement, if any (Dekker, 1993). New techniques in treating low sexual desire are also needed because the available techniques are often inadequate.

Sex therapy approaches to treating vaginismus and premature ejaculation have produced more consistent levels of success (Beck, 1993; O'Donohue et al., 1993). Reported success in treating vaginismus has ranged as high as 80% (Hawton & Catalan, 1990) to 100% in Masters and Johnson's (1970) original research. Treatment of premature ejaculation has resulted in success rates above 90% for the squeeze technique and the stop-start technique, but there are few data on the long-term results of treatment for premature ejaculation (LoPiccolo & Stock, 1986). Nor do we know why these techniques are effective (Kinder & Curtiss, 1988). Because the squeeze technique carries some risk of discomfort, many therapists prefer using the stop-start method.

LoPiccolo and Stock (1986) found that 95% of a sample of 150 previously anorgasmic women were able to achieve orgasm through a directed-masturbation program. About 85% of these women were able to reach orgasm through manual stimulation by their partners. Only about 40% were able to achieve orgasm during coitus, however. Christensen (1995) argues that prescribing masturbation in sex therapy can have the side effect of damaging the trust and openness that couples need for sexually rewarding relationships.

Generally speaking, couples-oriented treatment helps facilitate orgasm during coitus but is no guarantee that women will become orgasmic during coitus (Heiman & LoPiccolo, 1987; LoPiccolo & Stock, 1986). Nevertheless, the goal of achieving orgasm through some form of genital stimulation with a cooperative partner, such as through oral sex or by direct manual clitoral stimulation, is realistic for most women (LoPiccolo & Stock, 1986). Many couples who believe it is important for the woman to reach orgasm during coitus are able to accomplish this end by combining direct clitoral stimulation with coital stimulation (LoPiccolo & Stock, 1986).

Researchers have come to understand some of the factors that predict success in sex therapy. Couples are generally more likely to benefit from sex therapy if they have good relationships and are highly motivated (Killmann et al., 1987; McCabe & Delaney, 1992). It should come as no surprise that the success of treatment often depends on the quality of the relationship. Nor is it surprising that people are generally more successful when they are more motivated to take full advantage of the treatments they receive. It also appears that partners who acquire, from therapy, coping skills that they can use later, such as learning to respond to problems that arise by discussing them openly and by reinstating techniques learned in therapy, are generally better able to overcome recurrences of the problem. Sex therapy techniques may not be appropriate for people in whom profound personal problems or problems in the relationship underlie sexual dysfunctions (Pryde, 1989).

How Do You Find a Qualified Sex Therapist?

How would you locate a sex therapist if you had a sexual dysfunction? You might find advertisements for "sex therapists" in the Yellow Pages. But beware. Most states do not restrict usage of the term *sex therapist* to recognized professionals. In these states, anyone who wants to use the label may do so, including quacks and prostitutes.

Thus it is essential to determine that a sex therapist is a member of a recognized profession (such as psychology, social work, medicine, or marriage and family counseling) and has had training and supervision in applying sex therapy. Professionals are usually licensed or certified by their states. All states require licensing of psychologists and physicians, but some states do not license social workers or marriage counselors. If you have questions about the license laws in your state, contact your state's professional licensing board. The ethical standards of these professions prohibit practitioners from claiming expertise in sex therapy without suitable training.

If you are uncertain how to locate a qualified sex therapist in your area, you may obtain names of local practitioners from various sources, such as your university or college psychology department, health department, or counseling center; a medical or psychological association; a family physician; or your instructor. You may also seek services from a sex therapy clinic affiliated with a medical center or medical school in your area, many of which charge for services on a sliding scale, depending on your income level. You may also contact the American Association of Sex Educators, Counselors, and Therapists (AASECT), a professional organization that certifies sex therapists. They can provide you with the names of certified sex therapists in your area. They are located at 11 Dupont Circle, N.W., Washington, DC.

Ethical professionals are not annoyed or embarrassed if you ask them (1) what their profession is, (2) where they earned their advanced degree, and (3) whether they are licensed or certified by the state or if you inquire about (4) their fees, (5) their plans for treatment, and (6) the nature of their training in human sexuality and sex therapy. If the therapist hems and haws, asks why you are asking such questions, or fails to provide a direct answer, beware.

Professionals are also prohibited, by the ethical principles of their professions, from engaging in unethical practices, such as sexual relations with their clients. The nature of therapy creates an unequal power relationship between the therapist and the client. The therapist is perceived as an expert whose suggestions are likely to carry great authority. Clients may thus be vulnerable to exploitation by therapists who misuse their therapeutic authority. Let's be absolutely clear here: There is no therapeutic justification for a therapist to engage in sexual activity with a client. Any therapist who makes a sexual overture toward a client, or tries to persuade a client to engage in sexual relations, is acting unethically.

Human Sexuality ONLINE //

Web Sites Related to Sexual Dysfunctions

This is the Web site of WebMD, which contains news related to various sexual dysfunctions and links that enable one to find a sex therapist.
my.webmd.com/condition_center_hub/sex

The Web site of Dr. Koop, former surgeon general of the United States. It contains information about a range of health-related issues, including sexual dysfunction.
www.drkoop.com/index.asp

The Web site of the Sexual Health Info Center of the Encyclopaedia Britannica Online, which has information on sexual problems such as impotence, premature ejaculation, inability to achieve orgasm, and painful intercourse. Follow links for "romantic gifts."
www.sexhealth.org/infocenter/SexualDy/sexualdy.htm

Although it's a dot-org, this Web site is commercial, belonging to Inlet Medical, Inc., located in Minneapolis, Minnesota. This is a "medical device company that provides solutions to women's health problems through innovative instruments and kits for laparoscopic surgery." The site offers a good deal of information about dyspareunia and painful menstruation.
www.dyspareunia.org/

And don't forget these Web sites, even though they are cited elsewhere in the text:

The AASECT Web site: The American Association of Sex Educators, Counselors and Therapists. AASECT promotes sexual health through the development and advancement of the fields of sex therapy, counseling, and education. The association provides for the education and certification of sex educators, counselors, and therapists. AASECT also provides links that enable users to locate qualified sex therapists.
www.aasect.org/

The Web site of the Masters and Johnson Institute in St. Louis, which offers sex therapy and other kinds of help concerning problems with relationships.
www.mastersandjohnson.com

summing up

TYPES OF SEXUAL DYSFUNCTIONS

Sexual dysfunctions are difficulties in becoming sexually aroused or reaching orgasm.

■ **Sexual Desire Disorders**
These disorders involve dysfunctions in sexual desire, interest, or drive, in which the person experiences a lack of sexual desire or an aversion to genital sexual contact.

■ **Sexual Arousal Disorders**
In men, sexual arousal disorders involve recurrent difficulty in achieving or sustaining an erection sufficient to engaging successfully engage in sexual intercourse. In women, they typically involve failure to become sufficiently lubricated.

■ **Orgasmic Disorders**
Women are more likely to encounter difficulties reaching orgasm. Men are more likely to have premature ejaculation.

■ **Sexual Pain Disorders**
These disorders include dyspareunia and vaginismus.

ORIGINS OF SEXUAL DYSFUNCTIONS

Many sexual dysfunctions involve the interaction of organic and psychological factors.

■ **Organic Causes**
Fatigue may lead to erectile disorder in men and to orgasmic disorder and dyspareunia in women. Dyspareunia often reflects vaginal infections and STIs. Organic factors are believed to be involved in more than 50% of cases of erectile disorder. Medications and other drugs may also impair sexual functioning.

■ **Psychosocial Causes**
Psychosocial factors that are connected with sexual dysfunctions include cultural influences, psychosexual trauma, inclinations to a gay sexual orientation, marital dissatisfaction, psychological conflict, lack of sexual skills, irrational beliefs, and performance anxiety. Children reared in sexually repressive cultural or home environments may learn to respond to sex with feelings of anxiety and shame, rather than sexual arousal and pleasure. Many people do not acquire sexual competencies because they have had no opportunity to acquire knowledge and experience, even within marriage. Irrational beliefs and attitudes such as excessive needs for approval and perfectionism may also contribute to sexual problems. Performance anxiety may place a dysfunctional individual in a spectator rather than performer role.

TREATMENT OF SEXUAL DYSFUNCTIONS

Sex therapy aims to modify dysfunctional behavior directly by changing self-defeating beliefs and attitudes, fostering sexual skills and knowledge, enhancing sexual communication, and suggesting behavioral exercises to enhance sexual stimulation while reducing performance anxiety.

■ **The Masters-and-Johnson Approach**
Masters and Johnson pioneered the direct, behavioral approach to treating sexual dysfunctions. A male-and-female therapy team is employed during an in-residence program that focuses on the couple as the unit of treatment. Sensate focus exercises are used to enable the partners to give each other pleasure in a nondemanding situation.

■ **The Helen Singer Kaplan Approach**
Kaplan's *psychosexual therapy* combines behavioral and psychoanalytic methods.

■ **Sexual Desire Disorders**
Some sex therapists help kindle the sexual appetites of people with inhibited sexual desire by prescribing self-stimulation exercises combined with erotic fantasies.

■ **Sexual Arousal Disorders**
Men and women with impaired sexual arousal receive sexual stimulation from their partners under relaxed circumstances, so that anxiety does not inhibit their natural reflexes. Biological treatments such as the drug Viagra are also used with male erectile disorder.

■ **Orgasmic Disorders**
Masters and Johnson use a couples-oriented approach in treating anorgasmic women. Other sex therapists prefer a program of directed masturbation to enable women to learn about their own bodies at their own pace and free them of the need to rely on a partner or please a partner. Premature ejaculation is usually treated with the squeeze technique or the stop-start method. Biological treatment methods for female orgasmic disorder and premature ejaculation are under development.

■ **Sexual Pain Disorders**
Dyspareunia, or painful intercourse, is generally treated with medical intervention. Vaginismus is generally treated with plastic vaginal dilators of increasing size.

■ **Evaluation of Sex Therapy**
The success of sex therapy has varied with the type of sexual dysfunction treated.

questions for critical thinking

1. Why do you think so many people have difficulty talking about sexual problems or sexual dysfunctions? Why do you think men have more difficulty talking about them than women do?
2. There is no consensus among clinicians and researchers on the definition of low sexual desire. How much sexual interest or desire would seem to be normal to you? Why?
3. Are there any sexual attitudes common to people of your sociocultural background that can give rise to sexual problems or dysfunctions? What are these attitudes? Do you share them? Explain.
4. If you had a sexual dysfunction, do you think you would be willing to participate in sex therapy? Explain.
5. How would you go about finding a qualified sex therapist if you needed one?

chapter

16

Sexually Transmitted Infections

Pablo Picasso. *Women of Algiers (after Delacroix)*, 1955. Oil on canvas, 114 × 146 cm. © Copyright ARS, NY. Coll. Victor M. Ganz, New York, NY, USA. Credit: Scala/Art Resource, NY. Credit: © 2002 Estate of Pablo Picasso/Artists Rights Society (ARS), NY.

outline

Truth or Fiction?

AN EPIDEMIC

A Closer Look Talking with Your Partner about STIs

Questionnaire STI Attitude Scale

BACTERIAL INFECTIONS

Gonorrhea

Syphilis

Chlamydia

Other Bacterial Infections

VAGINAL INFECTIONS

Bacterial Vaginosis

Candidiasis

Trichomoniasis

VIRAL INFECTIONS

HIV Infection and AIDS

A World of Diversity HIV/AIDS: A Global Plague

A World of Diversity Gender Is a Crucial Issue in the Fight Against HIV/AIDS

Genital Herpes

Viral Hepatitis

Genital Warts

Molluscum Contagiosum

ECTOPARASITIC INFESTATIONS

Pediculosis

Scabies

PREVENTION OF STIs: MORE THAN SAFER SEX

Human Sexuality Online Web Sites Related to Sexually Transmitted Infections

Sources of Help

SUMMING UP

QUESTIONS FOR CRITICAL THINKING

Truth or Fiction?

____ Most women who contract gonorrhea do not develop symptoms.

____ Christopher Columbus brought more than beads, blankets, and tobacco back to Europe from the New World: He also brought syphilis.

____ Gonorrhea and syphilis can be contracted from toilet seats in public rest rooms.

____ If a syphilitic chancre (sore) goes away by itself, the infection does not require medical treatment.

____ Men too can develop vaginal infections.

____ As you are reading this page, you are engaged in search-and-destroy missions against foreign agents within your body.

____ Most people who are infected by HIV remain symptom-free and appear healthy for years.

____ Genital herpes can be transmitted only during flare-ups of the infection.

____ Pubic lice are of the same family of animals as crabs.

HAROLD and CARIN, both 20, have been dating for several months. They feel strong sexual attraction toward each other but have hesitated to become sexually intimate because of fears about AIDS. Harold believes that using condoms is no guarantee against infection and wants the two of them to be tested for HIV, the virus that causes AIDS. Carin has resisted undergoing an HIV test, partly because she feels insulted that Harold fears that she may be infected and, frankly, partly in fear of the test results. She has heard that symptoms may not develop for years after infection. She wonders whether she might have been infected by one of the men with whom she slept in the past.

KEISHA has genital herpes. A 19-year-old pre-law student, she has had no recurrences since the initial outbreak two years earlier. But she knows that herpes is a lifelong infection and may recur periodically from time to time. She also knows that she may inadvertently pass the herpes virus along to her sex partners, even to the man she eventually marries. She has begun thinking seriously about Steve, a man she has been dating for the past month. She would like to tell him that she has herpes before they become sexually intimate. Yet she fears that telling him might scare him off.

JOSÉ, 21, a math and computer science major, is planning a career in computer operations, hoping one day to run the computer systems for a large corporation. He lives off-campus with several of his buddies in a run-down house they've dubbed the "Nuclear Dumpsite." He has been dating Maria, a theater major, for several months. They have begun having sexual relations and have practiced "safer sex"—at least most of the time. During the past week he noticed a burning sensation while urinating. It seems to have passed now, so he figures that it was probably nothing to worry about. But he's not sure and wonders whether he should see a doctor.

Harold, Carin, Keisha, and José express some of the fears and concerns of a generation of young people who are becoming sexually active at a time when the threat of AIDS and other STIs (**sexually transmitted infections**) hangs over every sexual decision.

AIDS is indeed a scary thing, a very scary thing. But AIDS is only one of many STIs, though certainly the most deadly and frightening. Nearly every time you pick up a newspaper or turn on the radio or TV, you hear about AIDS, yet other STIs pose much wider threats (Stolberg, 1998b). In a study of more than 16,000 students on 19 U.S. college campuses, HIV (the virus that causes AIDS) was found in 30 blood samples (Gayle et al., 1990), or 0.2% of the students in the sample. *Chlamydia trachomatous* (the bacterium that causes chlamydia) was found in 1 sample in 10, or 10% of the college population! *Human papilloma virus* (HPV), the organism that causes genital warts, is estimated to be present in 20% of Americans over the age of 12 and in at least one-third of college women (Cannistra & Niloff, 1996).

College students have become reasonably well informed about AIDS. However, many are unaware that chlamydia can go undetected for years. Moreover, if it is left untreated, it can cause pelvic inflammation and infertility. Many—perhaps most—students are ignorant about HPV, which is linked to cervical cancer (Cannistra & Niloff,

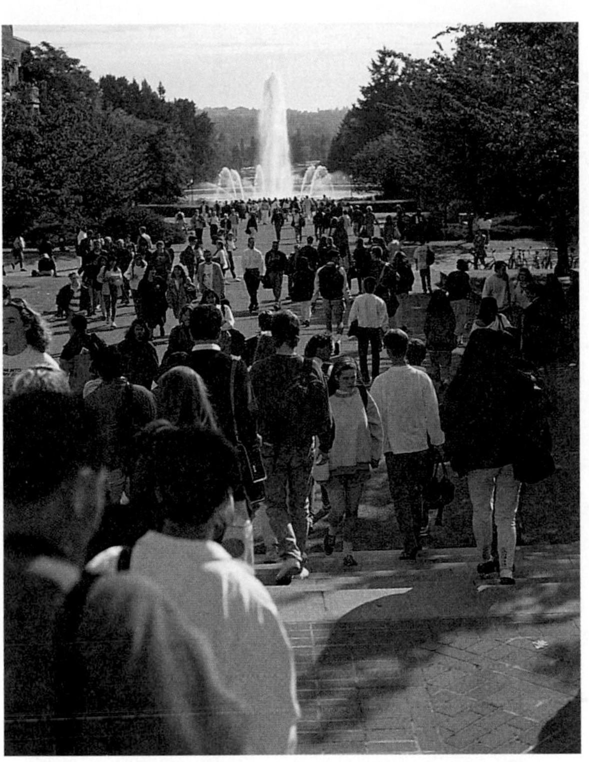

Who is at risk? Evidence suggests that young people are more sexually active than ever before. Thus it is as important as ever that they be aware of the risks involved and take responsibility for their sexual health.

1996; Josefsson et al., 2000). Yet the Federal Centers for Disease Control and Prevention (CDC) estimates that as many as 1 million new cases of HPV infection occur each year in the United States—more than syphilis, genital herpes, and AIDS combined (Penn, 1993). Whereas about 1 million Americans are thought to be infected with HIV, about 56 million are infected with other STI-causing viruses, such as those responsible for genital warts, herpes, and hepatitis (Barringer, 1993a). *None of this is intended to downplay the threat of AIDS. AIDS is lethal, and one case is one case too many.*

STIs are transmitted through sexual means, such as vaginal or anal intercourse or oral sex. They were formerly called *sexually transmitted diseases* (STDs) and before that *venereal diseases* (VDs)—after Venus, the Roman goddess of love.

Some STIs can be spread (and often are) through nonsexual contact as well. For example, AIDS and viral hepatitis may be spread by sharing contaminated needles. And yes, a few STIs (such as "crabs") may be picked up from bedding or other objects, such as moist towels, that harbor the infectious organisms that cause these STIs.

AN EPIDEMIC

STIs are rampant. The World Health Organization estimates that at least 333 million people around the world are stricken with curable STIs each year (WHO, 1995). The United States is believed to have the highest rate of infection by STIs in the industrialized world ("CDC tackles surge," 1998). In a recent year, for example, 150 cases of gonorrhea were reported per 100,000 people in the United States, compared with 18.6 per 100,000 in Canada and 3 per 100,000 in Sweden.

In the United States, 10 to 12 million new cases of STIs are reported to the disease control centers each year (Stolberg, 1998b), including 2.5 million adolescents (about one in six of these cases). At least one in four Americans is likely to contract an STI at some point in his or her life (Barringer, 1993a). Two cases in three occur in people under the age of 25. Yet many young people remain largely uninformed about the dangers of STIs.

Some people have STIs and do not realize it. Ignorance is not bliss, however. Some STIs may not produce noticeable symptoms, but they can be harmful if left untreated. STIs can also be painful and, in the cases of AIDS and advanced syphilis, lethal. An estimated 100,000 to 150,000 women become infertile each year because of STIs that are spreading through the reproductive system (Barringer, 1993a; Rosenberg & Gollub, 1992). Overall, STIs are believed to account for 15% to 30% of cases of infertility among women. In addition to their biological effects, STIs exact an emotional toll and strain relationships to the breaking point.

Why the surge in the incidence of STIs? One reason is the increased numbers of young people who engage in coitus. Many of them fail to use latex condoms consistently,

Sexually transmitted infections (STIs) Infections that are communicated through sexual contact. (Some, such as HIV/AIDS, can also be transmitted in other ways.)

a CLOSER look

Talking with Your Partner about STIs

Many people find it hard to talk about STIs with their partners. As one young woman explained,

> It's one thing to talk about being responsible . . . and a much harder thing to do it at the very moment. It's just plain hard to say to someone I am feeling very erotic with, "Oh, yes, before we go any further, can we have a conversation about STI?" It's hard to imagine murmuring into someone's ear at a time of passion, "Would you mind slipping on this condom or using this cream just in case one of us has STI?" Yet it seems awkward to bring it up any sooner if it's not clear between us that we want to make love. (Boston Women's Health Book Collective, 1992)

Because talking about STIs with sex partners can be awkward, many people wing it. They assume that their partners are free of STIs and hope for the best. Some people act as though not talking about AIDS and other STIs will cause these problems to go away. But the microbes that cause AIDS, herpes, chlamydia, genital warts, and other STIs will not disappear simply because we ignore them.

Imagine yourself in this situation: You've gone out with Chris a few times and you're keenly attracted. Chris is attractive, bright, and witty, shares some of your attitudes, and, all in all, is a powerful turn-on. Now the evening is winding down. You've been cuddling, and you think you know where things are heading.

Something clicks in your mind! You realize that as wonderful as this person is, you don't know every place Chris has "been." As healthy as Chris looks and acts, you don't know what's swimming around in Chris's bloodstream. Chris may not know either. In a moment of pent-up desire, Chris may also (how should we put this delicately?) lie about not being infected or about past sexual experiences.

What do you say now? How do you protect yourself without turning Chris off? Ah, the clumsiness! Asking about condoms or STIs is sort of making a verbal commitment to have sexual relations, and perhaps you're not sure that's what your partner intends. Even if it's clear that's where you're heading, will you seem too straightforward? Will you kill the romance? Undermine the spontaneity of the moment? Sure you might—life has its risks. But which is riskier: an awkward moment or being infected with a fatal illness? Let's put it another way: Are you *really* willing to die for sex? Given that few verbal responses are perfect, here are some things you can try:

1. You might say something like: "I've brought something and I'd like to use it . . . " (referring to a condom).
2. Or you can say, "I know this is a bit clumsy" [you are assertively expressing a feeling and asking permission to pursue a clumsy topic; Chris is likely to respond with "That's okay," or "Don't worry—what is it?"] "but the world isn't as safe as it used to be, and I think we should talk about what we're going to do."

The point is that your partner hasn't been living in a cave. Your partner is also aware of the dangers of STIs, especially of AIDS, and ought to be working with you to make things safe and unpressured. If your partner is pressing for unsafe sex and is inconsiderate of your feelings and concerns, you need to reassess whether you want to be with this person. We think you can do better.

if at all (Kamb et al., 1998). Some people do not use condoms because the woman is on the pill. Although birth-control pills are reliable methods of contraception, they do not prevent the spread of STIs. Another reason is that some infections, such as chlamydia, often have no symptoms. Therefore, some infected individuals unwittingly pass them along to others. Other risk factors include early sexual involvement and sex with multiple partners. Drug use is also associated with an increased risk of STIs. People who abuse drugs are more likely than others to engage in risky sexual practices (Lowry et al., 1994; Rotheram-Borus et al., 1994a). Moreover, certain forms of drug use, such as needle sharing, can directly transmit infectious organisms like HIV.

Education is critical to curtailing the epidemic among young people (Kamb et al., 1998). The goals are to promote responsible sexual decision making and to alter risky behavior. However, education may not be enough. Young people must also want to practice prevention and risk-reduction strategies if education is to change their behavior (Yarber & Parillo, 1992).

STI Attitude Scale

Do your attitudes toward sexually transmitted infections lead you to take risks that increase the chances of contracting one? The STI Attitude Scale (Yarber et al., 1989) was constructed to measure the attitudes of young adults toward STIs and to provide insight into these questions.

To complete the questionnaire, read each statement carefully. Record your reactions by circling the letters according to this code:

SA = Strongly Agree; A = Agree; U = Undecided; D = Disagree; SD = Strongly Disagree

Interpret your responses by referring to the scoring key in the Appendix.

1.	How one uses her or his sexuality has nothing to do with STI.	SA	A	U	D	SD
2.	It is easy to use the prevention methods that reduce one's chances of getting an STI.	SA	A	U	D	SD
3.	Responsible sex is one of the best ways of reducing the risk of STI.	SA	A	U	D	SD
4.	Getting early medical care is the main key to preventing harmful effects of STI.	SA	A	U	D	SD
5.	Choosing the right sex partner is important in reducing the risk of getting an STI.	SA	A	U	D	SD
6.	A high rate of STI should be a concern for all people.	SA	A	U	D	SD
7.	People with an STI have a duty to get their sex partners to medical care.	SA	A	U	D	SD
8.	The best way to get a sex partner to STI treatment is to take him or her to the doctor with you.	SA	A	U	D	SD
9.	Changing one's sex habits is necessary once the presence of an STI is known.	SA	A	U	D	SD
10.	I would dislike having to follow the medical steps for treating an STI.	SA	A	U	D	SD
11.	If I were sexually active, I would feel uneasy doing things before and after sex to prevent getting an STI.	SA	A	U	D	SD
12.	If I were sexually active, it would be insulting if a sex partner suggested we use a condom to avoid STI.	SA	A	U	D	SD
13.	I dislike talking about STI with my peers.	SA	A	U	D	SD
14.	I would be uncertain about going to the doctor unless I was sure I really had an STI.	SA	A	U	D	SD
15.	I would feel that I should take my sex partner with me to a clinic if I thought I had an STI.	SA	A	U	D	SD
16.	It would be embarrassing to discuss STI with one's partner if one were sexually active.	SA	A	U	D	SD
17.	If I were to have sex, the chance of getting an STI makes me uneasy about having sex with more than one person.	SA	A	U	D	SD
18.	I like the idea of sexual abstinence (not having sex) as the best way of avoiding STI.	SA	A	U	D	SD
19.	If I had an STI, I would cooperate with public health persons to find the sources of the STI.	SA	A	U	D	SD
20.	If I had an STI, I would avoid exposing others while I was being treated.	SA	A	U	D	SD
21.	I would have regular STI checkups if I were having sex with more than one partner.	SA	A	U	D	SD
22.	I intend to look for STI signs before deciding to have sex with anyone.	SA	A	U	D	SD
23.	I will limit my sex activity to just one partner because of the chances I might get an STI.	SA	A	U	D	SD
24.	I will avoid sexual contact any time I think there is even a slight chance of getting an STI.	SA	A	U	D	SD
25.	The chance of getting an STI would not stop me from having sex.	SA	A	U	D	SD
26.	If I had a chance, I would support community efforts toward controlling STI.	SA	A	U	D	SD
27.	I would be willing to work with others to make people aware of STI problems in my town.	SA	A	U	D	SD

Source: W. L. Yarber, M. R. Torabi, & C. H. Veenker (1989). "Development of a Three-Component Sexually Transmitted Diseases Attitude Scale." *Journal of Sex Education & Therapy*, 15: 36–49. Reprinted with permission.

BACTERIAL INFECTIONS

Without the one-celled microorganisms we call **bacteria,** there would be no wine. Bacteria are essential to fermentation. They also play vital roles in our digestive system. Unfortunately, bacteria also cause many diseases, such as pneumonia, tuberculosis, and meningitis—along with the common STIs gonorrhea, syphilis, and chlamydia.

Gonorrhea

Gonorrhea—also known as "the clap" or "the drip"—was once the most widespread STI in the United States, but it has been replaced by chlamydia. The rate of infection declined substantially from the mid-1980s through the mid-1990s, apparently because of safer sexual practices. However, the rate surged by about 9% in the late 1990s, possibly because of the advances in the treatment of HIV/AIDS. As HIV/AIDS becomes less frightening, people may engage in more spontaneous sexual behavior and increase the rate of STIs (CDC, 2000d). About 800,000 to 900,000 new cases of gonorrhea are now reported each year. Most new cases are contracted by people between the ages of 20 and 24.

Gonorrhea is caused by the gonococcus bacterium (see Table 16.1). A penile discharge that was probably gonorrhea is described in ancient Egyptian and Chinese writings and is mentioned in the Old Testament (Leviticus 15). Ancient Jews and Greeks assumed that the discharge was an involuntary loss of seminal fluid. In about 400 B.C., the Greek physician Hippocrates suggested that the loss stemmed from excessive sex, or "worship" of Aphrodite (whom the Romans would later rename *Venus*). The term *gonorrhea* is credited to the Greek physician Galen, who lived in the second century A.D. In 1879 Albert L. S. Neisser identified the gonococcus bacterium that bears his name: *Neisseria gonorrhoeae.*

Bacteria Plural of *bacterium,* a class of one-celled microorganisms that have no chlorophyll and can give rise to many illnesses. (From the Greek *baktron,* which means "stick," referring to the fact that many bacteria are rod-shaped.)

Gonorrhea An STI caused by the *Neisseria gonorrhoeae* bacterium and characterized by a discharge and burning urination. Left untreated, gonorrhea can give rise to pelvic inflammatory disease (PID) and infertility. (From the Greek *gonos,* which means "seed," and *rheein,* which means " to flow," referring to the fact that in ancient times, the penile discharge characteristic of the illness was erroneously interpreted as a loss of seminal fluid.)

Pharyngeal gonorrhea A gonorrheal infection of the pharynx (the cavity leading from the mouth and nasal passages to the larynx and esophagus) that is characterized by a sore throat.

Ophthalmia neonatorum A gonorrheal infection of the eyes of newborn children, who contract the disease by passing through an infected birth canal. (From the Greek *ophthalmos,* which means "eye.")

TRANSMISSION Gonococcal bacteria require a warm, moist environment like that found along the mucous membranes of the urinary tract in both genders and the cervix in women. Outside the body, they die in about a minute. There is no evidence that gonorrhea can be picked up from public toilet seats or by touching dry objects. In rare cases, gonorrhea is contracted by contact with a moist, warm towel or sheet used immediately beforehand by an infected person. Gonorrhea is nearly always transmitted by unprotected vaginal, oral, or anal sexual activity, or from mother to newborn during delivery.

A person who performs fellatio on an infected man may develop **pharyngeal gonorrhea,** which produces a throat infection. Mouth-to-mouth kissing and cunnilingus are less likely to spread gonorrhea. The eyes provide a good environment for the bacterium. Thus a person whose hands come into contact with infected genitals and who inadvertently touches his or her eyes afterward may infect them. A baby may contract gonorrhea of the eyes (**ophthalmia neonatorum**) when passing through the birth canal of an infected mother. This disorder may cause blindness, but it has become rare because the eyes of newborns are treated routinely with silver nitrate or penicillin ointment; these compounds are toxic to gonococcal bacteria.

A gonorrheal infection may be spread from the penis to the partner's rectum during anal intercourse. A cervical gonorrheal infection can be spread to the rectum if an infected woman and her partner follow vaginal intercourse with anal intercourse. Gonorrhea is less likely to be spread by vaginal discharge than by penile discharge.

Gonorrhea is highly contagious. Women stand nearly a 50% chance of contracting gonorrhea after one exposure. Men have a 25% risk of infection (Cates, 1998). The risks to women are apparently greater because women retain infected semen in the vagina. The risk of infection increases with repeated exposure.

TABLE 16.1

Causes, modes of transmission, symptoms, diagnosis, and treatment of major sexually transmitted infections (STIs)

STI AND PATHOGEN	MODES OF TRANSMISSION	SYMPTOMS	DIAGNOSIS	TREATMENT
		Bacterial Diseases		
Gonorrhea ("clap," "drip"): gonococcus bacterium *(Neisseria gonorrhoeae)*	Transmitted by vaginal, oral, or anal sexual activity or from mother to newborn during delivery	In men, yellowish, thick penile discharge, burning urination In women, increased vaginal discharge, burning urination, irregular menstrual bleeding (most women show no early symptoms)	Clinical inspection, culture of sample discharge	Antibiotics: ceftriaxone, ciprofloaxin, cefixime, ofloxacin
Syphilis: *Treponema pallidum*	Transmitted by vaginal, oral, or anal sexual activity or by touching an infectious chancre.	In primary stage, a hard, round, painless chancre or sore appears at site of infection within 2 to 4 weeks. May progress through secondary, latent, and tertiary stages, if left untreated	Primary-stage syphilis is diagnosed by clinical examination and by examination of fluid from a chancre in a dark-field test. Secondary-stage syphilis is diagnosed by blood test (the VDRL)	Penicillin, or doxycycline, tetracycline, or erythromycin for nonpregnant penicillin-allergic patients
Chlamydia and nongonococcal urethritis (NGU): *Chlamydia trachomatis* bacterium; NGU in men may also be caused by *Ureaplasma urealyticum* bacterium and other pathogens	Transmitted by vaginal, oral, or anal sexual activity; to the eye by touching one's eyes after touching the genitals of an infected partner; or to newborns passing through the birth canal of an infected mother	In women, frequent and painful urination, lower abdominal pain and inflammation, and vaginal discharge (but most women are symptom-free) In men, symptoms are similar to but milder than those of gonorrhea—burning or painful urination, slight penile discharge (most men are also symptom-free) Sore throat may indicate infection from oral–genital contact	The Abbott Testpack analyzes a cervical smear in women; in men, an extract of fluid from the penis is analyzed	Antibiotics: azithromycin, doxycycline, ofloxacin, amoxicillin
		Vaginitis		
Bacterial vaginosis: *Gardnerella vaginalis* bacterium and others	Can arise by overgrowth of organisms in vagina, allergic reactions, etc.; also transmitted by sexual contact	In women, thin, foul-smelling vaginal discharge, irritation of genitals and mild pain during urination In men, inflammation of penile foreskin and glans, urethritis, and cystitis May be symptom-free in both genders	Culture and examination of bacterium	Metronidazole, clindamycin

(continued)

TABLE 16.1 *(continued)*

Causes, modes of transmission, symptoms, diagnosis, and treatment of major sexually transmitted infections (STIs)

STI AND PATHOGEN	MODES OF TRANSMISSION	SYMPTOMS	DIAGNOSIS	TREATMENT
Candidiasis (moniliasis, thrush, "yeast infection"): *Candida albicans*—a yeast-like fungus	Can arise by overgrowth of fungus in vagina; may also be transmitted by sexual contact or by sharing a washcloth with an infected person	In women, vulval itching; white, cheesy, foul-smelling discharge; soreness or swelling of vaginal and vulval tissues In men, itching and burning on urination, or a reddening of the penis	Diagnosis usually made on basis of symptoms	Single dose of oral fluconazole, or suppositories of miconazole, clotrimazole, or butaconazole; modification of use of other medicines and chemical agents; keeping infected area dry
Trichomoniasis ("trich"): *Trichomonas vaginalis*—a protozoan (one-celled animal)	Almost always transmitted sexually	In women, foamy, yellowish, odorous vaginal discharge; itching or burning sensation in vulva. Many women are symptom-free In men, usually symptom-free, but mild urethritis is possible	Microscopic examination of a smear of vaginal secretions or of culture of the sample (latter method preferred)	Metronidazole
		Viral Diseases		
Oral herpes: *Herpes simplex virus–type 1 (HSV-1)*	Touching, kissing, sexual contact with sores or blisters; sharing cups, towels, toilet seats	Cold sores or fever blisters on the lips, mouth, or throat; herpetic sores on the genitals	Usually clinical inspection	Over-the-counter lip balms, cold-sore medications; check with your physician, however
Genital herpes: *Herpes simplex virus–type 2 (HSV-2)*	Almost always by means of vaginal, oral, or anal sexual activity; most contagious during active outbreaks of the disease	Painful, reddish bumps around the genitals, thighs, or buttocks; in women, may also be in the vagina or on the cervix. Bumps become blisters or sores that fill with pus and break, shedding viral particles. Other possible symptoms: burning urination, fever, aches and pains, swollen glands; in women, vaginal discharge	Clinical inspection of sores; culture and examination of fluid drawn from the base of a genital sore	There is no cure, but the antiviral drugs acyclovir, famciclovir, and valacyclovir may provide relief and prompt healing over; people with herpes often profit from counseling and group support as well
Viral hepatitis: hepatitis A, B, C, and D type viruses	Sexual contact, especially involving the anus (especially hepatitis A); contact with infected fecal matter; transfusion of contaminated blood (especially hepatitis B and C)	Ranges from being asymptomatic to mild flu-like symptoms and more severe symptoms, including fever, abdominal pain, vomiting, and "jaundiced" (yellowish) skin and eyes.	Examination of blood for hepatitis antibodies; liver biopsy	Treatment usually involves bed rest, intake of fluids, and (sometimes) antibiotics to ward off bacterial infections that might take hold because of lowered resistance. Alpha interferon is sometimes used in treating hepatitis C

STI AND PATHOGEN	MODES OF TRANSMISSION	SYMPTOMS	DIAGNOSIS	TREATMENT
Acquired immunodeficiency syndrome (AIDS): *Human immunodeficiency virus (HIV)*	HIV is transmitted by sexual contact; by infusion with contaminated blood; from mother to fetus during pregnancy; or through childbirth or breast-feeding	Infected people may initially have no symptoms or develop mild flu-like symptoms that may then disappear for many years prior to the development of "full-blown" AIDS. The symptoms of full-blown AIDS are fever, weight loss, fatigue, diarrhea, and opportunistic infections such as rare forms of cancer (Kaposi's sarcoma) and pneumonia (PCP)	Blood, saliva, or urine tests detect HIV antibodies. More expensive tests confirm the presence of the virus (HIV) itself. The diagnosis of AIDS is usually made on the basis of antibodies, a low count of CD4 cells, and/or the presence of indicator diseases	There is no cure for HIV infection or AIDS. Treatment is a combination of antiviral drugs including a protease inhibitor and nucleoside analogues such as Zidovudine
Genital warts (venereal warts): *Human papilloma virus (HPV)*	Transmission is by sexual and other forms of contact, such as with infected towels or clothing	Appearance of painless warts, often resembling cauliflowers, on the penis, foreskin, scrotum, or internal urethra in men and on the vulva, labia, wall of the vagina, or cervix in women. May occur around the anus and in the rectum of both genders	Clinical inspection	Methods include cryotherapy (freezing), podophyllin, trichloroacetic acid (TCA) or bichloroacetic acid (BCA), burning, and surgical removal
		Ectoparasitic Infestations		
Pediculosis ("crabs"): *Phthirus pubis (pubic lice)*	Transmission is by sexual contact or by contact with an infested towel, sheet, or toilet seat	Intense itching in pubic area and other hairy regions to which lice can attach	Clinical examination	Lindane (brand name: Kwell)—a prescription shampoo; nonprescription medications containing pyrethrins or piperonal butoxide (brand names: RID, Triple X)
Scabies: *Sarcoptes scabiei*	Transmission is by sexual contact or by contact with infested clothing or bed linen, towels, and other fabrics	Intense itching; reddish lines on skin where mites have burrowed in; welts and pus-filled blisters in affected areas	Clinical inspection	Lindane (Kwell)

Source: Table from the prepublication version of S. A. Rathus (2002). *Psychology in the New Millennium*, 8th ed. Harcourt, Inc.

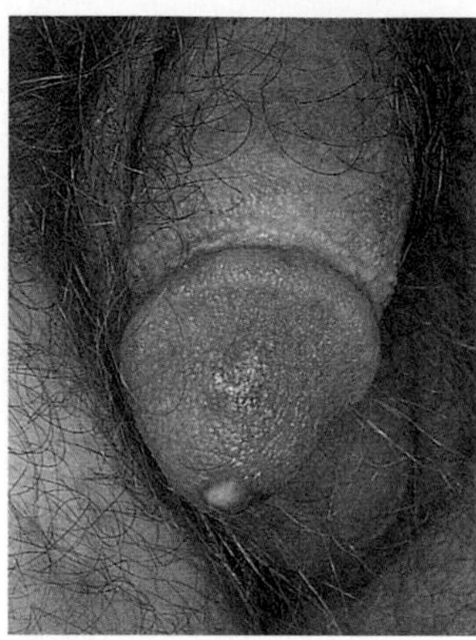

FIGURE 16.1

Gonorrheal Discharge. Gonorrhea in the male often causes a thick, yellowish, pus-like discharge from the penis.

SYMPTOMS Most men experience symptoms within two to five days after infection. Symptoms include a penile discharge that is clear at first (Figure 16.1). Within a day it turns yellow to yellow-green, thickens, and becomes pus-like. The urethra becomes inflamed, and urination is accompanied by a burning sensation. From 30% to 40% of males have swelling and tenderness in the lymph glands of the groin. Inflammation and other symptoms may become chronic if left untreated.

The initial symptoms of gonorrhea usually abate within a few weeks without treatment, leading people to think of gonorrhea as being no worse than a bad cold. However, the gonococcus bacterium usually continues to damage the body even though the early symptoms fade.

The primary site of infection in women is the cervix, where gonorrhea causes **cervicitis.** Cervicitis may cause a yellowish to yellow-green pus-like discharge that irritates the vulva. If the infection spreads to the urethra, women may also note burning urination. *About 80% of the women who contract gonorrhea have no symptoms during the early stages of the infection, however.* Because many infected women do not seek treatment until symptoms develop, they may innocently infect another sex partner.

When gonorrhea is not treated early, it may spread through the urogenital systems in both genders and strike the internal reproductive organs. In men, it can lead to **epididymitis,** which can cause fertility problems. Swelling and feelings of tenderness or pain in the scrotum are the principal symptoms of epididymitis. Fever may also be present. Occasionally the kidneys are affected.

In women, the bacterium can spread through the cervix to the uterus, fallopian tubes, ovaries, and other parts of the abdominal cavity, causing **pelvic inflammatory disease** (PID). Symptoms of PID include cramps, abdominal pain and tenderness, cervical tenderness and discharge, irregular menstrual cycles, coital pain, fever, nausea, and vomiting. PID may also occur without symptoms. Whether or not there are symptoms, PID can cause scarring that blocks the fallopian tubes, leading to infertility. PID is a serious illness that requires aggressive treatment with antibiotics. Surgery may be needed to remove infected tissue. Unfortunately, many women become aware of a gonococcal infection only when they develop PID. These consequences are all the more unfortunate because gonorrhea, when diagnosed and treated early, clears up rapidly in over 90% of cases.

Truth or Fiction? REVISITED

It is true that most women who contract gonorrhea do not develop symptoms. However, they may develop quite serious health problems later on.

Cervicitis Inflammation of the cervix.

Epididymitis Inflammation of the epididymis.

Pelvic inflammatory disease Inflammation of the pelvic region—possibly including the cervix, uterus, fallopian tubes, abdominal cavity, and ovaries—that can be caused by organisms such as *Neisseria gonorrhoeae.* Its symptoms are abdominal pain, tenderness, nausea, fever, and irregular menstrual cycles. The condition may lead to infertility. Abbreviated *PID.*

Syphilis An STI that is caused by the *Treponema pallidum* bacterium and may progress through several stages of development—often from a chancre to a skin rash to damage to the cardiovascular or central nervous system. (From the Greek *siphlos,* which means "maimed" or "crippled.")

DIAGNOSIS AND TREATMENT Diagnosis of gonorrhea involves clinical inspection of the genitals by a physician (a family practitioner, urologist, or gynecologist) and the culturing and examination of a sample of genital discharge.

Antibiotics are the standard treatment for gonorrhea. Penicillin was once the favored antibiotic, but the rise of penicillin-resistant strains of *Neisseria gonorrhoeae* has required that alternative antibiotics be used (Cates, 1998). An injection of the antibiotic ceftriaxone is often recommended. Other antibiotics that are used to treat gonorrhea include ciprofloaxin, cefixime, and ofloxacin. Because gonorrhea and chlamydia often occur together, people who are infected with gonorrhea are usually also treated for chlamydia through the use of another antibiotic (Cates, 1998). Sex partners of people with gonorrhea should also be examined.

Syphilis

No society wanted to be associated with the origins of **syphilis.** In Naples they called it "the French disease." In France it was "the Neapolitan disease." Many Italians called it "the Spanish disease," but in Spain they called it "the disease of Española" (modern Haiti).

In 1530 the Italian physician Girolamo Fracastoro wrote a poem about Syphilus, a shepherd boy. Syphilus was afflicted with the disease as retribution for insulting the sun god Apollo. In 1905 the German scientist Fritz Schaudinn isolated the bacterium that causes syphilis (see Figure 16.2). It is *Treponema pallidum* (*T. pallidum,* for short). The

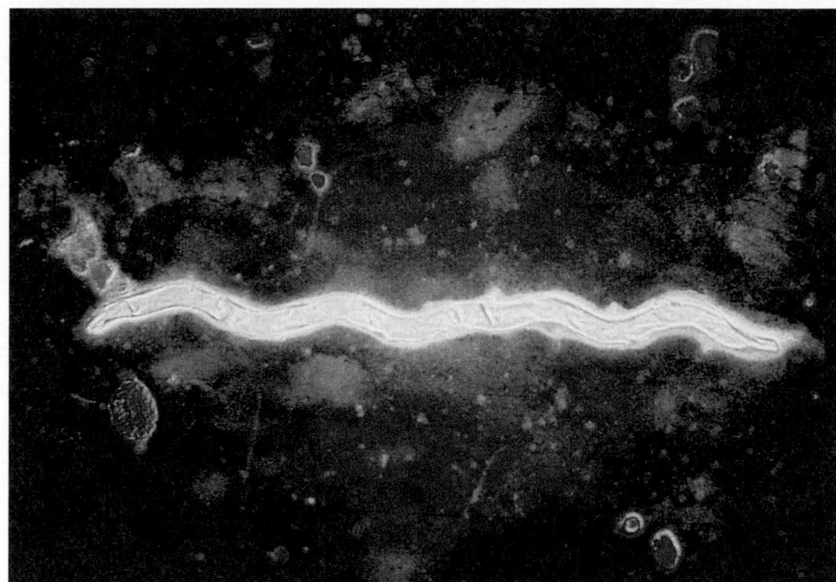

FIGURE 16.2

Treponema Pallidum. *Treponema pallidum* is the bacterium that causes syphilis. Because of its spiral shape, *T. pallidum* is also called a *spirochete*.

name is derived from Greek and Latin roots that mean a "faintly colored (pallid) turning thread"—a good description of the corkscrew-like shape of the microscopic organism. Because of the spiral shape, *T. pallidum* is also called a *spirochete*, from Greek roots that mean "spiral" and "hair."

THE ORIGINS OF SYPHILIS The origins of syphilis are controversial. The Columbian theory holds that Christopher Columbus returned to Spain from his first voyage to the West Indies (1492–1493) with more than beads, blankets, and tobacco. Then, from Spain, Spanish mercenaries may have carried the disease to Naples when they were hired to protect that city from French invaders. Then the French army may have contracted syphilis from prostitutes, who also practiced their profession with the Spaniards and Neapolitans. Sailors may have eventually spread syphilis to the East.

It is generally accepted that Columbus exhibited symptoms of advanced syphilis when he died in 1506. However, evidence reported in 1992 shows that syphilis existed in Europe prior to the voyages of Columbus ("Disease discovered Europe first?" 1992; Wilford, 1992).

It is not true that Christopher Columbus brought syphilis back to Europe from the New World. Evidence unearthed in the 1990s refutes this theory.

The incidence of syphilis decreased in the United States with the introduction of penicillin. But despite the availability of penicillin, there was a resurgence in syphilis during the 1980s (Rolfs & Nakashima, 1990); its incidence increased to about 50,000 a year by 1990 (Stolberg, 1998b). Then, during the 1990s, syphilis rates plunged to about 8,500 cases a year, the lowest on record, by 1997 (Bynum, 1998).

Researchers believe that the incidence of syphilis is linked to the use of cocaine (Minkoff et al., 1990; Rolfs et al., 1990). Cocaine users risk contracting syphilis through sex with multiple partners or with prostitutes, not through cocaine use per se (Rolfs et al., 1990). Syphilis and other STIs are often spread among drug users, prostitutes (many of whom abuse drugs themselves), and their sex partners (Farley et al., 1990).

Although syphilis is less widespread than it has been, its effects can be extremely harmful. They can include heart disease, blindness, gross confusion, and death. Syphilis killed the painter Paul Gauguin.

TRANSMISSION Syphilis, like gonorrhea, is most often transmitted by vaginal or anal intercourse or by oral–genital or oral–anal contact with an infected person. The spirochete is usually transmitted when open lesions on an infected person come into contact with the

FIGURE 16.3

Syphilis Chancre. The first stage, or primary stage, of a syphilis infection is marked by the appearance of a painless sore, or chancre, at the site of the infection.

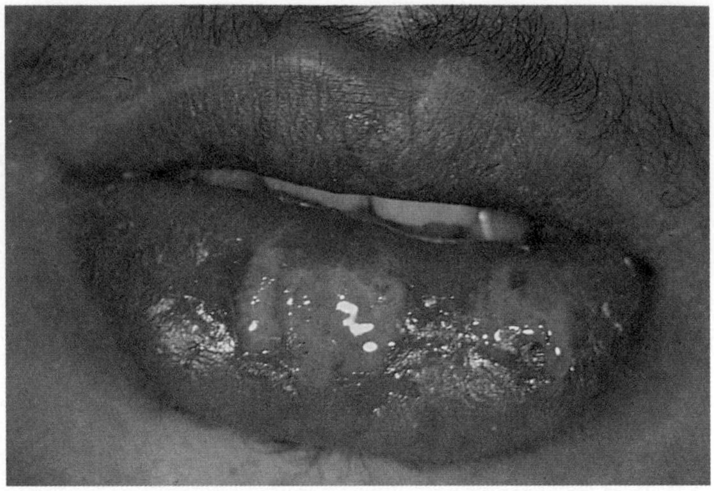

mucous membranes or skin abrasions of the partner's body during sexual activity. The chance of contracting syphilis from one sexual contact with an infected partner is estimated at 1 in 3 (Reinisch, 1990). Syphilis may also be contracted by touching an infectious **chancre,** but not by using the same toilet seat as an infected person.

Truth or Fiction? REVISITED

It is not true that gonorrhea and syphilis can be contracted from toilet seats in public rest rooms. Pubic lice can be contracted in this manner, however, as we will see later.

Pregnant women may transmit syphilis to their fetuses, because the spirochete can cross the placental membrane. Miscarriage, stillbirth, or **congenital syphilis** may result. Congenital syphilis may impair vision and hearing or deform bones and teeth. Blood tests are administered routinely during pregnancy to diagnose syphilis in the mother so that congenital problems in the baby may be averted. The fetus will probably not be harmed if an infected mother is treated before the fourth month of pregnancy.

SYMPTOMS AND COURSE OF ILLNESS Syphilis develops through several stages. In the first stage, or *primary stage,* of syphilis, a painless chancre (a hard, round, ulcer-like lesion with raised edges) appears at the site of infection 2 to 4 weeks after contact. When women are infected, the chancre usually forms on the vaginal walls or the cervix. It may also form on the external genitalia, most often on the labia. When men are infected, the chancre usually forms on the penile glans. It may also form on the scrotum or penile shaft. If the mode of transmission is oral sex, the chancre may appear on the lips or tongue (see Figure 16.3). If the infection is spread by anal sex, the rectum may be the site of the chancre. The chancre disappears within a few weeks, but if the infection remains untreated, syphilis will continue to work within the body.

The *secondary stage* begins a few weeks to a few months later. A skin rash develops, consisting of painless, reddish, raised bumps that darken after a while and burst, oozing a discharge. Other symptoms include sores in the mouth, painful swelling of joints, a sore throat, headaches, and fever. A person with syphilis may thus wrongly assume that he or she has the flu.

These symptoms also disappear. Syphilis then enters the *latent stage* and may lie dormant for 1 to 40 years. But spirochetes continue to multiply and burrow into the circulatory system, central nervous system (brain and spinal cord), and bones. The person may no longer be contagious to sex partners after several years in the latent stage, but a pregnant woman may transmit the infection to her newborn at any time (Wooldridge, 1991).

In many cases the disease eventually progresses to the late stage, or *tertiary stage.* A large ulcer may form on the skin, muscle tissue, digestive organs, lungs, liver, or other or-

Chancre A sore or ulcer.

Congenital syphilis A syphilis infection that is present at birth.

gans. This destructive ulcer can often be successfully treated, but still more serious damage can occur as the infection attacks the central nervous system or the cardiovascular system (the heart and the major blood vessels). Either outcome can be fatal. **Neurosyphilis** can cause brain damage, resulting in paralysis or the mental illness called **general paresis.**

The primary and secondary symptoms of syphilis inevitably disappear. Infected people may thus be tempted to believe that they are no longer at risk and fail to see a doctor. This is unfortunate, because failure to eradicate the infection through proper treatment may eventually lead to dire consequences.

Truth or Fiction? REVISITED

It is not true that the infection does not require medical treatment if a syphilitic chancre (sore) goes away by itself. The belief that medical treatment is unnecessary if the symptoms of an STI disappear by themselves is unfounded. Both gonorrhea and syphilis, for example, can damage the body even when their early symptoms have abated. ■

DIAGNOSIS AND TREATMENT Primary-stage syphilis is diagnosed by clinical examination. If a chancre is found, fluid drawn from it can be examined under a microscope. The spirochetes are usually quite visible. Blood tests are not definitive until the secondary stage begins. The most frequently used blood test is the **VDRL.** The VDRL tests for the presence of **antibodies** to *Treponema pallidum* in the blood.

Penicillin is the treatment of choice for syphilis, although for people allergic to penicillin, doxycycline and some other antibiotics can be used (Cates, 1998). Sex partners of persons infected with syphilis should also be evaluated by a physician.

Chlamydia

Chlamydia, another bacterial STI, is more common than gonorrhea and syphilis in the United States (Cates, 1998). Chlamydia infections are caused by the *Chlamydia trachomatis* bacterium, a parasitic organism that can survive only within cells. This bacterium can cause several different types of infection, including *nongonococcal urethritis (NGU)* in men and women, *epididymitis* (infection of the epididymis) in men, and *cervicitis* (infection of the cervix), *endometritis* (infection of the endometrium), and PID in women (Cates, 1998).

Nearly 4 million new chlamydia infections occur each year ("CDC tackles surge," 1998). The incidence of chlamydia infections is especially high among teenagers and college students (Burstein et al., 1998; Shafer et al., 1993). Researchers estimate that between 8% and 40% of teenage women become infected (Yarber & Parillo, 1992).

TRANSMISSION *Chlamydia trachomatis* is usually transmitted through sexual intercourse—vaginal or anal. *Chlamydia trachomatis* may also cause an eye infection if a person touches his or her eyes after handling the genitals of an infected partner. Oral sex with an infected partner can infect the throat. Newborns can acquire potentially serious chlamydia eye infections as they pass through the cervix of an infected mother during birth. Even newborns delivered by cesarean section may be infected if the amniotic sac breaks before delivery (Reinisch, 1990). Evidence from several studies suggests that between 2% and 26% of pregnant women in the United States carry the *Chlamydia trachomatis* bacterium in the cervix (Reinisch, 1990). Each year more than 100,000 infants are infected with the bacterium during birth (Graham & Blanco, 1990). Of these, about 75,000 develop eye infections and 30,000 develop pneumonia.

SYMPTOMS Chlamydia infections usually produce symptoms that are similar to, but milder than, those of gonorrhea. In men, *Chlamydia trachomatis* can lead to nongonococcal urethritis (NGU). *Urethritis* is an inflammation of the urethra. NGU refers to forms of urethritis that are not caused by the gonococcal bacterium. (NGU is generally diagnosed only in men. In women, an inflammation of the urethra caused by *Chlamydia trachomatis* is called a chlamydia infection or simply chlamydia.) NGU was formerly called nonspecific urethritis, or NSU. Many organisms can cause NGU. *Chlamydia trachomatis* accounts for about half of the cases among men (Cates, 1998).

Neurosyphilis Syphilitic infection of the central nervous system that can cause brain damage and death.

General paresis A progressive form of mental illness caused by neurosyphilis and characterized by gross confusion.

VDRL The test named after the Venereal Disease Research Laboratory of the U.S. Public Health Service that tests for the presence of antibodies to *Treponema pallidum* in the blood.

Antibodies Specialized proteins produced by the white blood cells of the immune system in response to disease organisms and other toxic substances. Antibodies recognize and attack the invading organisms or substances.

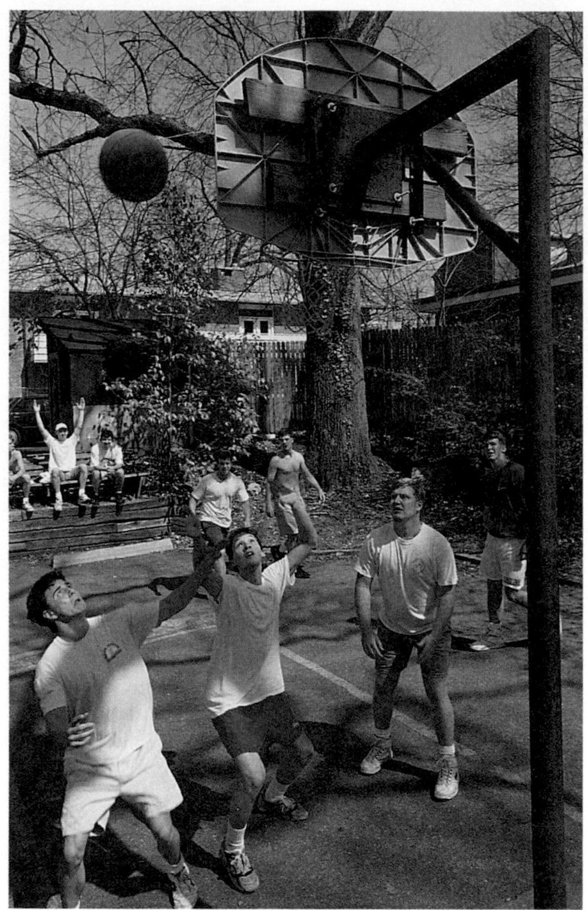

A silent disease. There are millions of new cases of chlamydia in the United States each year, and many—perhaps most—of them go undiagnosed and untreated. Because so many cases are symptom-free, infected individuals may unknowingly pass the infection to others and develop various health problems that could have been treated early.

NGU in men may give rise to a thin, whitish discharge from the penis and some burning or other pain during urination. These contrast with the yellow-green discharge and more intense pain produced by gonorrhea. There may be soreness in the scrotum and feelings of heaviness in the testes. NGU is about two to three times as prevalent among American men as gonorrhea (Cates, 1998). Men 20 to 24-years of age are most at risk of contracting gonorrhea and NGU, presumably because of their high levels of sexual activity (Bowie, 1990).

In women, chlamydial infections usually give rise to infections of the urethra or cervix. Women, like men, may experience burning when they urinate, genital irritation, and a mild (vaginal) discharge. Women are also likely to have pelvic pain and irregular menstrual cycles. The cervix may look swollen and inflamed.

Yet as many as 25% of men and 70% of women infected with chlamydia show no symptoms (Cates, 1998). For this reason, chlamydia has been dubbed "the silent disease." People without symptoms may go untreated and unknowingly pass along their infections to their partners. In women, an untreated chlamydial infection can spread throughout the reproductive system, leading to pelvic inflammatory disease (PID) and to scarring of the fallopian tubes, resulting in infertility (Garland et al., 1990; Hodgson et al., 1990). About half of the more than 1 million annual cases of PID are attributed to chlamydia (Cates, 1998). Women with a history of exposure to *Chlamydia trachomatis* also stand twice as much chance as others of incurring an ectopic (tubal) pregnancy (Sherman et al., 1990).

Untreated chlamydial infections can also damage the internal reproductive organs of men. About 50% of cases of epididymitis are caused by chlamydial infections (Cates, 1998). Yet only about 1% or 2% of men with untreated NGU caused by *Chlamydia trachomatis* go on to develop epididymitis. The long-term effects of untreated chlamydial infections in men remain undetermined.

Chlamydial infections frequently occur together with other STIs, most often gonorrhea. Nearly half of all cases of gonorrhea involve coexisting chlamydial infections (Cates, 1998).

DIAGNOSIS AND TREATMENT The Abbott Testpack permits physicians to verify a diagnosis of chlamydia in women. The test analyzes a cervical smear (like a Pap smear) and identifies 75% to 80% of infected cases. In men, a swab is inserted through the penile opening, and the extracted fluid is analyzed to detect the presence of *Chlamydia trachomatis*.

Antibiotics other than penicillin are highly effective in eradicating chlamydia infections. These include azithromycin, doxycycline, ofloxacin, and amoxicillin (Cates, 1998). (Penicillin, effective in treating gonorrhea, is ineffective against *Chlamydia trachomatis.*) Treatment of sex partners is considered critical regardless of whether the partner shows symptoms, to prevent the infection from bouncing back and forth (Cates, 1998). Moreover, a woman whose sex partner develops NGU should be examined for a chlamydial infection herself. At least 30% of these women will test positive for chlamydia even though they (and their partners) may be free of symptoms (Stamm & Holmes, 1990). Likewise, men whose sex partners develop urethral or cervical infections should be medically evaluated, whether or not they notice any symptoms. With chlamydia, both partners may be unaware that they are infected and oblivious to the internal damage the infection is causing. Because of the risks that untreated chlamydial infections pose, especially to women, and the high rate of symptom-free infections, many physicians screen young women for chlamydia during regular checkups.

Other Bacterial Diseases

Several other types of bacterial STIs occur less commonly in the United States and Canada. These include chancroid, shigellosis, granuloma inguinale, and lymphogranuloma venereum.

CHANCROID **Chancroid,** or "soft chancre," is caused by the bacterium *Hemophilus ducreyi.* It is more commonly found in the tropics and Eastern nations than in Western countries. The chancroid sore consists of a cluster of small bumps or pimples on the genitals, the perineum (the area of skin that lies between the genitals and the anus), or the anus itself. These lesions usually appear within 7 days of infection. Within a few days the lesion ruptures, producing an open sore or ulcer. Several ulcers may merge with other ulcers, forming giant ulcers (Ronald & Albritton, 1990). There is usually an accompanying swelling of a nearby lymph node. In contrast to the syphilis chancre, the chancroid ulcer has a soft rim (hence its name) and is painful in men. Women frequently do not experience any pain and may be unaware of being infected (Ronald & Albritton, 1990). The bacterium is typically transmitted through sexual or bodily contact with the lesion or its discharge. Diagnosis is usually confirmed by culturing the bacterium, which is found in pus from the sore, and examining it under a microscope. Antibiotics (azithromycin, ciprofloaxin, or erythromycin) are usually effective in treating the disease (Cates, 1998).

SHIGELLOSIS **Shigellosis** is caused by the *Shigella* bacterium and is characterized by fever and severe abdominal symptoms, including diarrhea and inflammation of the large intestine. About 25,000 cases of shigellosis are reported each year in the United States. Shigellosis can result from food poisoning, but it is also often contracted by oral contact with infected fecal material, which may stem from oral–anal sex. Shigellosis often resolves itself, but people with the disease may become severely dehydrated from the diarrhea. Severe cases are usually treated with antibiotics, such as tetracycline or ampicillin.

GRANULOMA INGUINALE Rare in the United States, **granuloma inguinale,** like chancroid, is more common in tropical regions. It is caused by the bacterium *Calymmatobacterium granulomatous* and is not as contagious as many other STIs. Primary symptoms are painless red bumps or sores in the groin area that ulcerate and spread. Like chancroid, it is usually spread by intimate bodily or sexual contact with a lesion or its discharge. Diagnosis is confirmed by microscopic examination of tissue of the rim of the sore. The antibiotics trimethoprim-sulfamethoxazole, doxycycline, ciprofloaxin, and erythromycin are effective in treating this disorder (Cates, 1998). If left untreated, however, the disease may lead to the development of fistulas (holes) in the rectum or bladder, destruction of the tissues or organs that underlie the infection, or scarring of skin tissue that results in **elephantiasis,** the condition that afflicted the so-called Elephant Man in the nineteenth century.

LYMPHOGRANULOMA VENEREUM (LGV) **Lymphogranuloma venereum** (LGV) is another tropical STI that occurs only rarely in the United States and Canada. Some U.S. soldiers returned home from Vietnam with cases of LGV. It is caused by several strains of the *Chlamydia trachomatis* bacterium. LGV usually enters the body through the penis, vulva, or cervix, where a small, painless sore may form. The sore may go unnoticed, but a nearby lymph gland in the groin swells and grows tender. Other symptoms mimic those of flu: chills, fever, and headache. Backache (especially in women) and arthritic complaints (painful joints) may also occur. If LGV is untreated, complications such as growths and fistulas in the genitals and elephantiasis of the legs and genitals may occur. Diagnosis is made by skin tests and blood tests. Antibiotics (doxycycline or erythromycin) are the usual treatment (Cates, 1998).

Chancroid An STI caused by the *Hemophilus ducreyi* bacterium. Also called *soft chancre.*

Shigellosis An STI caused by the *Shigella* bacterium.

Granuloma inguinale A tropical STI caused by the *Calymmatobacterium granulomatous* bacterium.

Elephantiasis A disease characterized by enlargement of parts of the body, especially the legs and genitals, and by hardening and ulceration of the surrounding skin. (From the Greek *elephas,* which means "elephant," referring to the resemblance of the affected skin areas to elephant hide.)

Lymphogranuloma venereum A tropical STI caused by the *Chlamydia trachomatis* bacterium.

VAGINAL INFECTIONS

Vaginitis is any kind of vaginal infection or inflammation. Women with vaginitis may encounter genital irritation or itching and burning during urination, but the most common symptom is an odorous discharge.

Most cases of vaginitis are caused by organisms that reside in the vagina or by sexually transmitted organisms. Organisms that reside in the vagina may overgrow and cause symptoms when the environmental balance of the vagina is upset by factors such as birth-control pills, antibiotics, dietary changes, excessive douching, or nylon underwear or pantyhose. (See Chapter 3 for suggestions on reducing the risk of vaginitis.) Still other cases are caused by sensitivities or allergic reactions to various chemicals.

The great majority of vaginal infections (Sobel, 1997) involve bacterial vaginosis (BV), candidiasis (commonly called a "yeast" infection), or trichomoniasis ("trich"). Bacterial vaginosis is the most common form of vaginitis, followed by candidiasis, then by trichomoniasis, but some cases involve combinations of the three.

The microbes that cause vaginal infections in women can also infect the man's urethral tract. A "vaginal infection" can be passed back and forth between sex partners (Sobel, 1997).

It is not literally true that men can develop vaginal infections. Only women have vaginas. However, the microbes that cause these infections in women may also cause problems for men.

Bacterial Vaginosis

Bacterial vaginosis (BV, formerly called *nonspecific vaginitis*) is most often caused by overgrowth of the bacterium *Gardnerella vaginalis* (Sobel, 1997). The bacterium is transmitted primarily through sexual contact. The most characteristic symptom in women is a thin, foul-smelling vaginal discharge, but infected women often have no symptoms. Diagnosis requires culturing the bacterium in the laboratory (Reinisch, 1990). Besides causing troublesome symptoms in some cases, BV may increase the risk of various gynecological problems, including infections of the reproductive tract (Hillier & Holmes, 1990). Oral treatment with metronidazole (brand name: Flagyl) is recommended and is effective in about 90% of cases (Sobel, 1997). Topical treatments with metronidazole or clindamycin are also effective. Recurrences are common, however.

Questions remain about whether the male partner should also be treated. The bacterium can usually be found in the male urethra but does not generally cause symptoms (Reinisch, 1990). Lacking symptoms, the male partner may unknowingly transmit the bacterium to others.

Candidiasis

Also known as *moniliasis*, *thrush*, or (most commonly) a yeast infection, **candidiasis** is caused by a yeast-like fungus, *Candida albicans*. Candidiasis commonly produces soreness, inflammation, and intense (sometimes maddening!) itching around the vulva that is accompanied by a thick, white, thick, curd-like vaginal discharge (see Figure 16.4). Yeast generally produces no symptoms when the vaginal environment is normal. Yeast infections can also occur in the mouth in both men and women and in the penis in men.

Infections most often arise from changes in the vaginal environment that allow the fungus to overgrow. Factors such as the use of antibiotics, birth-control pills, or intrauterine devices (IUDs), pregnancy, and diabetes may alter the vaginal balance, allowing the fungus that causes yeast infections to grow to infectious levels (Sobel, 1997). Wearing nylon underwear and tight, restrictive, poorly ventilated clothing may also set the stage for a yeast infection.

Vaginitis Any type of vaginal infection or inflammation.

Bacterial vaginosis A form of vaginitis usually caused by the *Gardnerella vaginalis* bacterium.

Candidiasis A form of vaginitis caused by a yeast-like fungus, *Candida albicans*.

Diet may play a role in recurrent yeast infections. Reducing one's intake of substances that produce excessive excretion of urinary sugars (such as dairy products, sugar, and artificial sweeteners) apparently reduces the frequency of recurrent yeast infections. The daily ingestion of one pint of yogurt containing active bacterial (*Lactobacillus acidophilus*) cultures helped reduce the rate of recurrent infections in one sample of women (Hilton et al., 1992).

Candidiasis can be passed back and forth between sex partners through vaginal intercourse. It can also be passed back and forth between the mouth and the genitals through oral–genital contact and can infect the anus through anal intercourse. However, most infections in women are believed to be caused by an overgrowth of "yeast" normally found in the vagina, not by sexual transmission. Still, it is advisable to evaluate both partners simultaneously. Whereas most men with *Candida* have no symptoms, some may develop NGU or a genital thrush that is accompanied by itching and burning during urination or reddening of the penis. Candidiasis may also be transmitted by nonsexual means, such as between women who share a washcloth.

About 75% of women will have at least one episode of candidiasis during their reproductive years (Sobel, 1997). About half of them will have recurrent infections. Recommended treatment is a single dose of oral fluconazole, or vaginal suppositories or creams containing miconazole, clotrimazole, or butaconazole (Cates, 1998; Sobel, 1997). Many of these treatments are sold over the counter. Ask the pharmacist which preparations contain these medicines (or read the labels). We also recommend that women with vaginal complaints consult their physicians before taking any of these medications, to ensure that they receive the proper diagnosis and treatment.

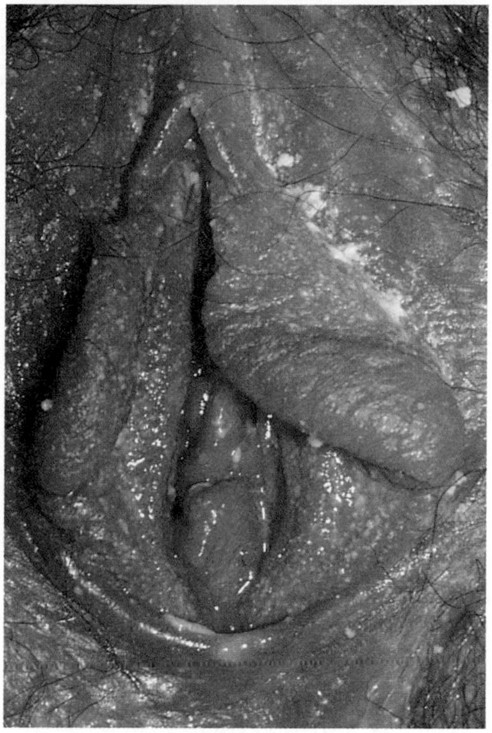

FIGURE 16.4

Candidiasis. A "yeast infection" is caused by the *Candida albicans* fungus and causes soreness, inflammation, and itching around the vulva that is accompanied by a thick, white vaginal discharge.

Trichomoniasis

Trichomoniasis ("trich") is caused by *Trichomonas vaginalis. Trichomonas vaginalis* is a one-celled parasite. Trichomoniasis is the most common parasitic STI. It accounts for 2 to 3 million cases a year among women in the United States (Sobel, 1997). Symptoms in women include burning or itching in the vulva, mild pain during urination or coitus, and an odorous, foamy whitish to yellowish-green discharge. Lower abdominal pain is reported by 5% to 12% of infected women (Rein & Muller, 1990). Many women notice that symptoms appear or worsen during, or just after, their menstrual periods. Trichomoniasis facilitates the transmission of HIV (Sobel, 1997) and is also linked to the development of tubal adhesions that can result in infertility (Grodstein et al., 1993). As with many other STIs, about half of infected women have no symptoms (Reinisch, 1990).

Candidiasis most often reflects an overgrowth of organisms normally found in the vagina. However, trichomoniasis is nearly always sexually transmitted. Because the parasite can survive for several hours on moist surfaces outside the body, trich can be communicated from contact with infected semen or vaginal discharges on towels, washcloths, and bedclothes. This parasite is one of the few disease agents that can be picked up from a toilet seat, but it would have to directly touch the penis or vulva (Reinisch, 1990).

Trichomonas vaginalis can cause NGU in the male, which can be symptom-free or can cause a slight penile discharge that is usually noticeable prior to first urination in the morning. There may be tingling, itching, and other irritating sensations in the urethral tract. Yet most infected men are symptom-free. Therefore, they can unwittingly transfer the organism to their sex partners. Perhaps three or four in ten male partners of infected women are found to harbor *Trichomonas vaginalis* themselves (Reinisch, 1990). Diagnosis is frequently made by microscopic examination of a smear of a woman's vaginal fluids in

Trichomoniasis A form of vaginitis caused by the protozoan *Trichomonas vaginalis.*

a physician's office. Diagnosis based on examination of cultures grown from the vaginal smear is considered more reliable, however.

Except during the first 3 months of pregnancy, trichomoniasis is usually treated in both genders with metronidazole (brand name: Flagyl). Both partners are treated, whether or not they report symptoms. When both partners are treated simultaneously, the success rate approaches 100% (Cates, 1998; Sobel, 1997).

VIRAL INFECTIONS

Viruses are tiny particles of DNA surrounded by a protein coating. They are incapable of reproducing on their own. When they invade a body cell, however, they can direct the cell's own reproductive machinery to spin off new viral particles that spread to other cells, causing infection. In this chapter we discuss several viral STIs: HIV/AIDS, herpes, viral hepatitis, genital warts, and molluscum contagiosum.

HIV Infection and AIDS

AIDS is the acronym for **acquired immunodeficiency syndrome.** AIDS is a fatal syndrome that is caused by the **human immunodeficiency virus (HIV).**[1] HIV attacks and disables the immune system, the body's natural line of defense, stripping it of its ability to fend off disease-causing organisms. HIV is believed to have originated in Africa in the 1930s as a virus that infected but was not lethal to chimpanzees (simian immunodeficiency virus, or SIV). SIV genetically converted to HIV either in the chimp or after it infected humans. However, it stayed in remote Africa until it was spread worldwide by the development of large cities, the global economy, and jet travel (Hillis, 2000).

Acquired immunodeficiency syndrome (AIDS) A condition caused by the human immunodeficiency virus (HIV) and characterized by destruction of the immune system so that the body is stripped of its ability to fend off life-threatening diseases.

Human immunodeficiency virus (HIV) A sexually transmitted virus that destroys white blood cells in the immune system, leaving the body vulnerable to life-threatening diseases.

Immune system A term for the body's complex of mechanisms for protecting itself from disease-causing agents such as pathogens.

Pathogen An agent, especially a microorganism, that can cause a disease. (From the Greek *pathos,* which means "suffering" or "disease," and *genic,* which means "forming" or "coming into being."

Leukocytes White blood cells that are essential to the body's defenses against infection. (From the Greek *leukos,* which means "white" and *kytos,* which means "a hollow" and is used in combination with other word forms to mean "cell.")

PREVALENCE OF HIV INFECTION AND AIDS Although Hillis targets the "birthdate" of HIV as 1931, AIDS was not described in the medical journals until 50 years later—in 1981 (Gottlieb, 1991). Early in the new millennium, more than 700,000 Americans are living with AIDS (CDC, 2000b). Nearly half a million Americans have died from it (CDC, 2000b). AIDS has become the leading killer of Americans aged 25 to 44. The incidence of AIDS is increasing most rapidly among women, people of color, people who share needles when they inject drugs, and people who engage in unprotected male–female sex. Nearly 34 million people worldwide are living with HIV/AIDS ("Global Plague of AIDS," 2000).

In the United States, AIDS predominantly affects men who engage in sexual activity with other men (about 50%) or share needles when injecting drugs (25% to 30%) (CDC, 2000b; see Table 16.2). Male–female sexual contact is the fasting-growing exposure category in the United States. Among women, male–female sexual contact now accounts for more than 40% of cases (CDC, 2000b; see Table 16.2).

THE IMMUNE SYSTEM AND AIDS AIDS is caused by a virus that attacks the body's **immune system**—the body's natural line of defense against disease-causing organisms. The immune system combats disease in a number of ways. It produces white blood cells that envelop and kill **pathogens** such as bacteria, viruses, and fungi; worn-out body cells; and cancer cells. White blood cells are referred to as **leukocytes.** Leukocytes engage in microscopic warfare. They undertake search-and-destroy missions. They identify and eradicate foreign agents and debilitated cells.

[1]Some people deny that AIDS is caused by HIV. It should be clearly understood that the scientific community is in agreement that HIV causes AIDS.

TABLE 16.2

Adolescents and adults with AIDS; number of cases by exposure category

Exposure Category	Male	Female	Total
Men who have sex with men	341,597	—	341,597
Injecting drug use	134,356	50,073	184,429
Men who have sex with men and inject drugs	46,582	—	46,582
Hemophilia/coagulation disorder	4,803	272	5,075
Heterosexual contact	26,530	47,946	74,477
Recipient of blood transfusion, blood components, or tissue	4,863	3,668	8,531
Risk not reported or identified	46,112	17,851	63,965
Totals	604,843	119,810	724,652

Source: Centers for Disease Control and Prevention (2000b). *HIV/AIDS Surveillance Report: U.S. HIV and AIDS Cases Reported Through December 1999*, 11(2).

It is true that as you read this page, you are engaged in search-and-destroy missions against foreign agents within your body. The white cells in your immune system continuously seek and destroy foreign pathogens within your body. ■

Leukocytes recognize foreign agents by their surface fragments. The surface fragments are termed **antigens** because the body reacts to their presence by developing specialized proteins, or **antibodies.** Antibodies attach themselves to the foreign bodies, inactivate them, and mark them for destruction. (Infection by HIV may be determined by examining the blood or saliva for the presence of antibodies to the virus.)

Rather than mark pathogens for destruction or war against them, special "memory lymphocytes" are held in reserve. Memory lymphocytes can remain in the bloodstream for years, and they form the basis for a quick immune response to an invader the second time around.[2]

Another function of the immune system is to promote **inflammation.** When you suffer an injury, blood vessels in the region initially contract to check bleeding. Then they dilate. Dilation expands blood flow to the injured region, causing the redness and warmth that identify inflammation. The elevated blood supply also brings in an army of leukocytes to combat invading microscopic life forms, such as bacteria, that might otherwise use the local injury to establish a beachhead into the body.

EFFECTS OF HIV ON THE IMMUNE SYSTEM Spikes (technically known as "gpl20" spikes) on the surface of HIV allow it to bind to sites on cells in the immune system (Sodroski et al., 1998). Like other viruses, HIV uses the cells it invades to spin off copies of itself. HIV uses the enzyme *reverse transcriptase* to cause the genes in the cells it attacks to make proteins that the virus needs in order to reproduce.

[2]Vaccination is the placement of a weakened form of an antigen in the body, which activates the creation of antibodies and memory lymphocytes. Smallpox has been annihilated by vaccination, and researchers are trying to develop a vaccine against the virus that causes AIDS.

Antigen A protein, toxin, or other substance to which the body reacts by producing antibodies. (Combined word formed from *anti*body *gen*erator.)

Antibodies Specialized proteins that develop in the body in response to antigens and that inactivate foreign bodies.

Inflammation Redness and warmth that develop at the site of an injury, reflecting the dilation of blood vessels that permits the expanded flow of leukocytes to the region.

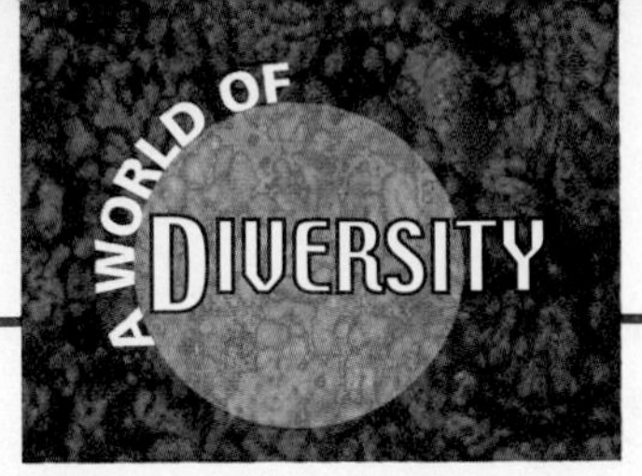

HIV/AIDS: A Global Plague

HIV infection and AIDS are a true global plague. According to the World Health Organization, nearly 34 million people around the world were living with HIV/AIDS as we entered the new millennium, although the United Nations estimates that about 90% of these individuals are unaware of their infections ("Global Plague of AIDS," 2000). More than 19 million people have already died of AIDS. Sub-Saharan Africa has been hardest hit by the epidemic, with more than 24 million people currently infected with HIV and another 14 million people already dead. Sub-Saharan Africa contains 10% of the world's population, scrapes by on 1% of the world's income, and bears the burden of two out of three people with HIV infection or AIDS (Benatar, 2000). Figure 16.5 summarizes tragic facts about HIV/AIDS in sub-Saharan Africa.

The incidences of new HIV infections are mushrooming in Central and Eastern Europe, India, China, Southeast Asia, Latin America, and the Caribbean.

Unlike many other infections, HIV/AIDS does not target older, infirm people. Because it is sexually transmitted, it afflicts the most productive sectors of the affected populations: young adults ("Global Plague of AIDS," 2000). In some parts of sub-Saharan Africa, one adult in four is infected, with the infection rates highest among skilled and unskilled workers, including medical professionals, engineers, teachers, and civil servants.

In many developing nations, there is little treatment for HIV/AIDS. For example, more than 90% of the people infected in sub-Saharan Africa received no treatment for the opportunistic diseases that hit people who are living with AIDS—diseases like pneumonia, tuberculosis, and brain infections ("Global Plague of AIDS," 2000; McNeil, 2000a). The combinations of drugs that are used to treat HIV/AIDS in industrialized nations can easily cost thousands of dollars per year per individual. Although many drug companies are cutting the costs of HIV/AIDS drugs for developing nations (McNeil, 2000b), the incomes of most sub-Saharan Africans can be measured in hundreds of dollar per year. Health care systems that could effectively distribute the drugs are also missing in many locations (McNeil, 2000b). As a result, 6 to 10 million of South Africa's 43 million people are expected to die of AIDS over the next 10 to 15 years (Josephson, 2000).

Sad to say, the governments of the world's wealthy nations seem to have largely given up on making expensive

- There were 5.6 million new HIV infections worldwide in 1999; 3.8 million of them were in Africa.
- There were 2.6 million dead of AIDS in 1999; 85% of them were in Africa.
- There were 13 million children orphaned by AIDS; 10 million of them were in sub-Saharan Africa.
- Between 2005 and 2010, life expectancy declined in sub-Saharan Africa from 59 to 45 years and in Zimbabwe, which is harder hit than most sub-Saharan African nations, from 61 to 33 years.
- There were more than 500,000 babies infected in 1999 by their mothers—most of them in sub-Saharan Africa.

FIGURE 16.5

HIV/AIDS in Sub-Saharan Africa at the End of the Millennium. Sub-Saharan Africa has been hardest hit by the HIV/AIDS epidemic. The losses in the region are staggering.

HIV directly attacks the immune system by invading and destroying a type of lymphocyte called the CD4 cell (see Figure 16.6).[3] The CD4 cell is the quarterback of the immune system. CD4 cells "recognize" invading pathogens and signal B-lymphocytes or B-cells—another kind of white blood cell—to produce antibodies that inactivate pathogens and mark them for annihilation. CD4 cells also signal another class of T-cells, called killer T-cells, to destroy infected cells. By attacking and destroying helper T-cells, HIV disables the very cells on which the body relies to fight off this and other diseases. As HIV cripples the body's defenses, the individual is exposed to infections that would not otherwise take hold. Cancer cells might also proliferate.

[3]CD4 cells are also known as T4 cells or helper T-cells.

drugs available to infected Africans and are focusing on prevention. One promising avenue of prevention is the relatively inexpensive short-term use of antiviral medicines by pregnant women. This measure would help prevent transmission of HIV from mothers to babies during childbirth and would markedly cut into the rate of new infections (Mofenson, 2000; Wood et al., 2000). Of course, this approach does nothing to help people who are already infected. Prevention is good—it is wonderful—but the fact is that making the most powerful drugs available to infected individuals in developing nations would cost but a fraction of the defense budgets of well-off nations.

There is some positive news from developing nations. The infection rates have been significantly cut in the African nations of Uganda and Senegal through sex education, testing for HIV infection, and distribution of condoms ("Global Plague of AIDS," 2000). Thailand has lowered its infection rate by imposing controls on prostitution, which was the country's greatest avenue of infection (MacAndrew, 2000).

TABLE 16.3

AIDS cases by race or ethnicity

Race or Ethnicity	Number of AIDS Cases	Percentage
European American	318,354	43.4%
African American	272,881	37.2
Latino and Latina American	133,703	18.2
Asian/Pacific Islander	5,347	0.7
American Indian/Alaska native	2,132	0.3
Race/ethnicity unknown	957	0.1

Source: Centers for Disease Control and Prevention (2000b). *HIV/AIDS Surveillance Report: U.S. HIV and AIDS Cases Reported Through December 1999,* 11(2).

In the United States

Disproportionately high numbers of African Americans and Latino and Latina Americans are living with HIV/AIDS in the United States (CDC, 2000b; see Table 16.3). Nearly half of the men and three-quarters of the women with AIDS are African American or Latino and Latina American (CDC, 2000b). Yet these ethnic groups make up only about one-quarter of the population. Death rates due to AIDS are much higher among African Americans and Latino and Latina Americans (especially Latino and Latina people of Puerto Rican origin) than among European Americans (CDC, 2000b), because African Americans and Latino and Latina Americans have less access to high-quality health care.

Ethnic differences in rates of transmission of HIV are connected with the practice of injecting illicit drugs. People who share needles with HIV-infected people when they inject drugs can become infected themselves. They can then also transmit the virus to their sex partners. People who share needles now account for one in four cases of people with AIDS. African Americans constitute more than half of the people who apparently became infected with HIV by injecting drugs (CDC, 2000b). Latino and Latina Americans account for another case in every five (CDC, 2000b). Drug abuse and the related problem of prostitution occur disproportionately in poor, urban communities with large populations of people of color. Thus it is not surprising that HIV infection and AIDS have affected these groups disproportionately.

The lessons from Uganda and Senegal pertain to the United States as well. Greater investment in sex education and use of condoms are also likely to cut the infection rates at home. And here, too, more needs to be done to help provide poor people with the drugs that can prolong their lives and help them remain productive members of the work force.

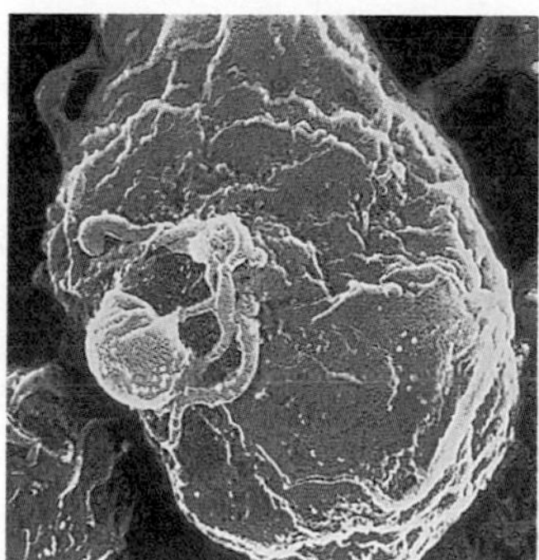

FIGURE 16.6

HIV (The AIDS Virus) Attacks a White Blood Cell. HIV progressively weakens the immune system, leaving the body vulnerable to infections and diseases that would otherwise be fended off.

Although the CD4 cells appear to be its main target, HIV also attacks other types of white blood cells.

The blood normally contains about 1,000 CD4 cells per cubic millimeter. The numbers of CD4 cells may remain at about this level for years following HIV infection. Many people show no symptoms and appear healthy while CD4 cells remain at this level. Then, for reasons that are not clearly understood, the levels of CD4 cells begin to drop off, although symptoms may not appear for a decade or more. As the numbers of CD4 cells decline, symptoms generally increase, and people fall prey to diseases that their weakened immune systems are unable to fight off. People become most vulnerable to opportunistic infections when the level of CD4 cells falls below 200 per cubic millimeter.

PROGRESSION OF HIV INFECTION AND AIDS HIV follows a complex course once it enters the body. Shortly after infection, the person may experience mild flu-like symptoms—fatigue, fever, headaches and muscle pain, lack of appetite, nausea, swollen glands, and possibly a rash. Such symptoms usually disappear within a few weeks. The infected person may dismiss them as a passing case of flu. People who enter this symptom-free or carrier state generally look and act well and do not realize that they are infectious. Thus they can unwittingly pass the virus along to others.

Most people who are infected with HIV remain symptom-free for years. Some enter a symptomatic state (previously labeled *AIDS-related complex*, or ARC) that is typically marked by symptoms such as chronically swollen lymph nodes and intermittent weight loss, fever, fatigue, and diarrhea. This symptomatic state does not constitute full-blown AIDS, but it shows that HIV is undermining the integrity of the immune system.

It is true that most people who are infected by HIV remain symptom-free and appear healthy for years. ■

Even during the years when HIV appears to be lying dormant in the body, billions of viral particles are being spun off. In a seesaw battle, the great majority of them are wiped out by the immune system. Eventually, however, in nearly all cases, the balance tips in favor of HIV. Then the virus's numbers swell. Perhaps a decade or more after the person is infected with HIV, and for reasons that remain unclear, the virus begins to overtake the immune system. It obliterates the cells that house it and spreads to other immune-system cells, eventually destroying or disabling the body's ability to defend itself from disease. About half of the people with HIV develop AIDS within 10 years of initial infection. For this reason, people who know that they are infected with HIV may feel that they are carrying time bombs within them. AIDS is classified as a *syndrome* because it is characterized by a variety of different symptoms. The beginnings of full-blown AIDS are often marked by such symptoms as swollen lymph nodes, fatigue, fever, night sweats, diarrhea, and weight loss that cannot be attributed to dieting or exercise.

AIDS is connected with the appearance of diseases such as pneumocystis carinii pneumonia (PCP), Kaposi's sarcoma (a form of cancer), toxoplasmosis of the brain (an infection of parasites), and *Herpes simplex* with chronic ulcers. These diseases are termed **opportunistic diseases** because they are not likely to emerge unless a disabled immune system provides the opportunity.

Opportunistic diseases Diseases that take hold only when the immune system is weakened and unable to fend them off. Kaposi's sarcoma and pneumocystis carinii pneumonia (PCP) are examples of opportunistic diseases found in people with AIDS.

About 10% of people with AIDS have a wasting syndrome. Wasting, the unintentional loss of more than 10% of a person's body weight, is connected with AIDS, some other infections, and cancer (Grunfeld, 1995). It appears that people with AIDS who waste away do so because they take in less energy, not because they burn more calories (Macallan et al., 1995).

As AIDS progresses, the individual grows thinner and more fatigued. He or she becomes unable to perform ordinary life functions and falls prey to opportunistic infections. If left untreated, AIDS nearly always results in death within a few years.

TRANSMISSION HIV can be transmitted by certain contaminated bodily fluids: blood, semen, vaginal secretions, and breast milk. The first three of these may enter the body through vaginal, anal, or oral–genital intercourse with an infected partner. An African study that followed seropositive mothers and their babies for 2 years found that the probability of transmission of HIV via breast milk was about 16.2% (1 in 6) (Nduati et al., 2000). Other avenues of infection include sharing a hypodermic needle with an infected person (as do many people who inject drugs), transfusion with contaminated blood, transplants of organs and tissues that have been infected with HIV, artificial insemination with infected semen, or being stuck by a needle used previously on an infected person. HIV may enter the body through tiny cuts or sores in the mucosal lining of the vagina, the rectum, and even the mouth. These cuts or sores can be so tiny that you are not aware of them.

Could This Couple Be Transmitting HIV? HIV is a blood-borne virus that is transmitted via various bodily fluids, including blood, semen, and vaginal fluids. However, the Centers for Disease Control and Prevention (CDC) have not found that HIV occurs in infectious quantities in saliva.

Transmission of HIV through kissing—even prolonged kissing or "French" kissing—is considered unlikely. When a person injects drugs, a small amount of his or her blood remains inside the needle and syringe. If the person is HIV-infected, the virus may be found in the blood remaining in the needle or syringe. Others who use the needle inject the infected blood into their bloodstream. HIV can also be spread by sharing needles used for other purposes, such as injecting steroids, ear piercing, or tattooing. If you are interested in having your ears pierced or in getting a tattoo, insist on seeing a qualified person who uses new or sterilized equipment. Ask questions about the safety measures that are followed before undergoing any such procedure.

HIV can also be transmitted from mother to fetus during pregnancy or from mother to child through childbirth or breast-feeding (Mofenson, 2000; Wood et al., 2000). Transmission is most likely to occur during childbirth.

Male-to-female transmission through vaginal intercourse is about twice as likely as female-to-male transmission (CDC, 2000b; see Table 16.3), partly because more of the virus is found in the ejaculate than in vaginal secretions. A man's ejaculate may also remain for many days in the vagina, providing greater opportunity for infection to occur. Male–female or male–male anal intercourse is especially risky, particularly to the recipient, because it often tears or abrades rectal tissue, facilitating entry of the virus into the bloodstream (Caceres & van-Griensven, 1994).

Male–female transmission via sexual intercourse is the primary route of HIV infection in Africa, Latin America, and Asia (Specter, 1998b). Worldwide, male–female sexual intercourse accounts for the majority of cases of HIV/AIDS. In the United States, many cases of male–female transmission occur within the community of people who inject drugs and their sex partners.

In the early years of the AIDS epidemic, HIV spread rapidly among hemophiliacs who had unknowingly been transfused with contaminated blood. Tennis great Arthur Ashe, who died of complications from AIDS in 1993, believed that he contracted HIV during a blood transfusion. But blood supplies are now routinely screened for HIV. Medical authorities now consider the risk of HIV transmission through transfusion of screened blood to be negligible if not entirely absent (Klein, 2000).

HIV may also be spread by donor semen, such as that used in artificial insemination. Many sperm banks do not test donors for STIs (Meyer, 1994). Cases have been reported of women who have become infected with hepatitis B, gonorrhea, trichomoniasis ("trich"), chlamydia, and even HIV via insemination with donor semen (Meyer, 1994).

FACTORS THAT AFFECT THE RISK OF SEXUAL TRANSMISSION Some people seem more likely than others to communicate HIV, and others seem to be especially vulnerable to HIV infection. Why, for instance, are some people infected by one sexual contact with an infected partner, whereas others are not infected during months or years of unprotected sex? Several factors appear to affect the risk of HIV infection and the development of AIDS:

- The probability of sexual transmission rises with the number of sexual contacts with an infected partner.
- The probability of transmission is also affected by the type of sexual activity. Anal intercourse, for example, provides a convenient port of entry for HIV because it often tears or abrades the rectal lining.

- The amount of virus in semen also affects the probability of infection. The quantity of HIV in the semen peaks shortly after initial infection and when full-blown AIDS develops.
- STIs such as genital warts, gonorrhea, trichomoniasis, and chlamydia inflame the genital region, which heightens the risk of sexual transmission of other STIs (Stolberg, 1998b). STIs that produce genital ulcers, such as syphilis and genital herpes, may heighten vulnerability to HIV infection by allowing the virus to enter the circulatory system through the ulcers.
- Circumcision lowers the risk of infection. HIV cannot accumulate under the folds of the foreskin in men who have been circumcised. Cells in the foreskin may also be particularly vulnerable to HIV infection (Cohen, 2000).
- Genetic factors may also be at work. About 1% of people of Western European descent have inherited a gene that prevents HIV from entering cells in the immune system from both parents and are apparently immune from HIV infection. Perhaps 20% of individuals of Western European descent have inherited the gene from one parent; HIV disease appears to progress more slowly in these people. Some prostitutes in Thailand and Africa, where HIV infection has been running rampant, also appear to be immune to HIV infection (Royce et al., 1997).

HOW HIV IS NOT *TRANSMITTED* There is much misinformation about the transmission of HIV. Let us consider some of the ways in which HIV is *not* transmitted:

1. *HIV is not transmitted from donating blood.* AIDS cannot be contracted by donating blood because needles are discarded after a single use. Unfortunately, many people have avoided donating blood because of unfounded fears of HIV transmission.
2. *HIV is not transmitted through casual, everyday contact.* There is no evidence of transmission of HIV by hugging someone, shaking hands, or bumping into strangers on buses and trains; by handling money, doorknobs, or other objects that have been touched by infected people; by sharing drinking fountains, public telephones, public toilets, or swimming pools; or by trying on clothing that has been worn by an infected person. Nor is HIV known to be transmitted by contact with urine, feces, sputum, sweat, tears, or nasal secretions, unless blood is clearly visible in these fluids. (Still, if you should have to clean up someone's urine, feces, nasal secretions, or especially blood, it would be wise to use rubber gloves and to wash your hands thoroughly immediately afterwards.)
3. *HIV is not transmitted by insect bites.* HIV is not transmitted by mosquito bites or the bites of other insects such as bedbugs, lice, or flies). Nor can you get HIV from contact with animals.
4. *HIV is not transmitted by airborne germs or contact with contaminated food.* People do not contract HIV from airborne germs, such as by sneezing or coughing, or by contact with contaminated food or by eating food prepared by a person infected with HIV.
5. *HIV is not transmitted through sharing work or home environments.* HIV has not been shown to be transmitted from infected people to family members, or others they live with, through any form of casual contact, such as hugging or touching, or through sharing bathrooms, food, or eating utensils, as long as there is no exchange of blood or genital secretions. There are some isolated reports of nonsexual transmission of HIV between children living in the same household. Investigators suspect that the route of transmission in these cases involved blood contact, such as the sharing of razor blades in one of the two reported cases and the sharing of a toothbrush in the other (the infected child in this case had bleeding gums) ("AIDS without needles or sex," 1993). No cases of HIV transmission have been documented to have occurred via nonsexual contact in schools or in the workplace.

DIAGNOSIS OF HIV INFECTION AND AIDS The most widely used test for HIV infection is the enzyme-linked immunosorbent assay (ELISA, for short), which may take several days to yield results. ELISA does not directly detect HIV in the blood. Instead, it reveals HIV antibodies. People may show an antibody response to HIV long before they develop symptoms of infection. A positive **(seropositive)** test result means that antibodies were found and usually[4] indicates that the person is infected with HIV. A negative **(seronegative)** outcome means that antibodies to HIV were not detected.

ELISA can be performed on samples of blood, saliva, or urine. A saliva test is not quite so accurate as a blood test, but it is less expensive and might encourage people who avoid blood tests to be tested. Although HIV antibodies can be detected in saliva, HIV itself is not found in measurable quantities. This is why kissing is not considered an avenue of transmission of HIV. The saliva is absorbed by a cotton pad on a stick that is placed between the lower gum and the cheek. The saliva in the cotton, like blood, undergoes ELISA in a laboratory.

The urine test is also not quite so accurate as the blood test. Yet administration of the test does not require specially trained health workers, nor does it use needles that can pose hazards to health care professionals. Luc Montagnier (1996), one of the discoverers of HIV, notes that the ease of use of the urine test could be of particular advantage in identifying HIV infection in people in developing countries.

When people receive positive ELISA results, the Western blot test or PCR (polymerase chain reaction) test can be performed to confirm the findings. The PCR test is expensive but detects the virus itself. It can take many months for people who have been exposed to HIV to develop antibodies. For this reason, repeated tests over a 6-month or even longer period of time from the date of possible exposure may be in order.

Tests that yield results more rapidly than ELISA are under development. The question is whether health professionals will be willing to sacrifice some accuracy for the sake of speedy results (FDA is hit on "rapid" AIDS tests, 2000).

TREATMENT OF HIV INFECTION AND AIDS For many years, researchers were frustrated by failure in the effort to develop effective vaccines and treatments for HIV infection and AIDS. There is still no safe, effective vaccine, but recent developments in drug therapy have raised hopes in the area of treatment.

VACCINES A number of experimental AIDS vaccines are being tested on people and animals (Barouch et al., 2000; Nabel & Sullivan, 2000). Several vaccines have proved from somewhat effective to strongly effective in animal trials (Barouch et al., 2000; Ho, 1995). One such vaccine has been developed by Norman Letvin and his colleagues (Barouch et al., 2000) and tested with monkeys. The Letvin team injected HIV-infected rhesus monkeys with DNA from outside of the virus and laced it with substances that boost immune system functioning (interleukin 2, an immune system protein, and an antibody called IgG). Monkeys who received the vaccine remained infected with HIV; however, they showed no symptoms of illness and kept the infection at low levels.

The ideal vaccine against AIDS would be safe and inexpensive and would confer lifetime protection against all strains of the disease with a single dose. Such a vaccine may take some time to develop; but researchers are becoming more hopeful.

DRUG THERAPY Today many drugs are used to combat HIV infection and AIDS. They offer a good deal of hope for people with HIV infection and AIDS, including pregnant women who are infected with HIV.

Zidovudine (previously referred to as AZT) has been the most widely used HIV/AIDS drug. Zidovudine is one of a number of so-called nucleoside analogues that inhibit replication (reproduction) of HIV by targeting the enzyme called reverse transcriptase. Other nucleoside analogues are ddI, ddC, d4T, and 3TC. Zidovudine in many

[4]But not always! Fetuses, for example, may receive antibodies from infected mothers, but not the virus itself. Some fetuses, however, do become infected with the virus.

Seropositive Having a pathogen or antibodies to that pathogen in the bloodstream.

Seronegative Lacking a pathogen or antibodies to that pathogen in the bloodstream.

cases delays the progression of the infection and increases the blood count of CD4 cells. HIV-infected pregnant women who use zidovudine reduce the rate of HIV infection in their newborns by two thirds (Connor et al., 1994). Zidovudine helps prevent transmission during childbirth by decreasing the amount of the virus in the mother's bloodstream (Meyer, 1998). Only 8% of the babies born to the zidovudine-treated women became infected with HIV, as compared to 25% of babies whose mothers were untreated. Zidovudine is generally used for 26 weeks prior to childbirth. However, using zidovudine for even the final few weeks cuts the transmission rate of HIV from mother to child by half (Meyer, 1998). Zidovudine is expensive, so briefer treatment may be of most help in developing nations or when the mother has not sought prenatal treatment early in pregnancy.

The results of the European Mode of Delivery Collaboration Trial on the efficacy of elective C-section versus vaginal delivery show that C-section further decreases the risk of maternal transmission of HIV to the baby (Ricci et al., 2000). The study enlisted about 400 seropositive mothers. All mothers received zidovudine during pregnancy. Half of the mothers were assigned at random to deliver vaginally, half by C-section. The HIV infection rate was 10.6% for babies delivered vaginally (not too different from the 8% reported by Meyer, 1998) and 1.7% for babies delivered by C-section. Thus the combination of zidovudine during pregnancy and delivery by C-section cuts the chance that a seropositive mother will transmit HIV to her baby to about 1 in 50.

A new generation of drugs that block the replication of HIV, called *protease inhibitors*, are more effective treatments for HIV/AIDS. Protease inhibitors block reproduction of HIV particles by targeting the protease enzyme. A combination, or "cocktail," of antiviral drugs including zidovudine, another nucleoside analogue, and a protease inhibitor has become the standard treatment of HIV/AIDS (Gallant, 2000). This combination—referred to as **HAART** (for "highly active antiretroviral therapy")—decreases the likelihood that HIV will develop resistance to treatment. It has reduced HIV below detectable levels in many infected people (Lederman & Valdez, 2000). It has created hope that AIDS will become increasingly manageable—a chronic health problem as opposed to a terminal illness. HAART is expensive, however, and many people who could benefit from it cannot afford it. A large number of pills must be taken at precise intervals, such that many people with HIV/AIDS find it difficult to adhere to the schedule. The side effects of these medicines can be unpleasant, including nausea and (in the case of protease inhibitors) unusual accumulations of fat, such as "buffalo humps" in the neck. But other drugs are coming into use that do not require the use of so many pills and may have milder side effects.

HAART has worked many wonders to date (Flexner, 1998; Lederman & Valdez, 2000). It cut the U.S. death rate from AIDS-related causes by 26% from 1995 to 1996 and by 44% from 1996 to 1997 (Altman, 1998). It cut the California and New York death rates from AIDS-related causes by 60% and 48%, respectively, from 1996 to 1997 (Altman, 1998). Yet even when HIV has been reduced to "undetectable levels" by ordinary means, scientists have been able to use methods of close scrutiny to locate it in "resting" (non-replicating) CD4 cells (Lederman & Valdez, 2000). Therefore, HAART does not appear to be a cure. The encouraging news is that after using HAART, some HIV-infected people can go on drug "vacations" without their HIV levels bouncing back. Perhaps HAART gave their immune systems the opportunity to learn to manage HIV. The not-so-encouraging news is the possibility that strains of HIV might arise that replicate rapidly even when they are bombarded by HAART (Lederman & Valdez, 2000).

Promising results are also reported in treating the opportunistic infections, such as PCP (Bozzette et al., 1995) and fungal infections (Powderly et al., 1995), that take hold in people with weakened immune systems (Clumeck, 1995).

Despite these improvements, more than 10,000 people in the United States die of AIDS each year (CDC, 2000b). Helene Gayle (2000), the head of the AIDS prevention effort of the Centers for Disease Control and Prevention, attributes these deaths to lack of early testing and treatment for some people, failure of some people with HIV/AIDS to

HAART (pronounced *HEART*) The acronym for "highly active antiretroviral therapy," which refers to the combination, or "cocktail," of drugs used to treat HIV/AIDS: a protease inhibitor in combination with a couple of other aniretroviral agents.

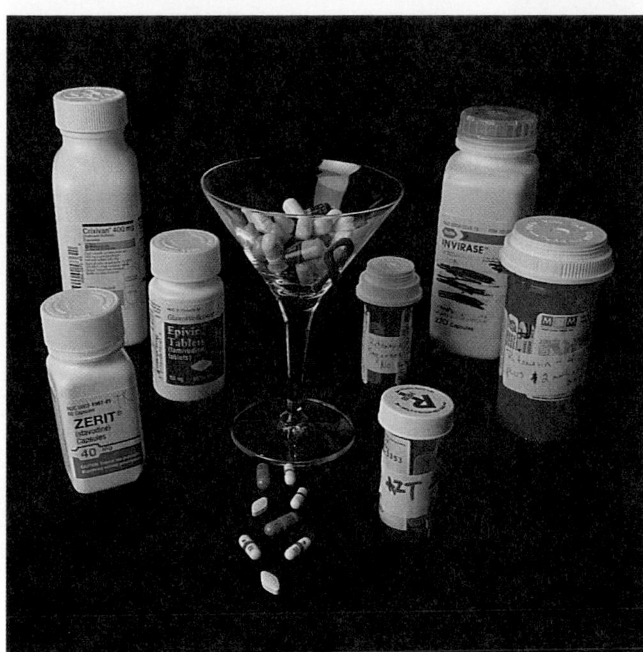

HAART. *HAART* is the acronym for "highly active antiretroviral therapy." It refers to a combination, or "cocktail," of antiviral drugs—including zidovudine, another nucleoside analogue, and a protease inhibitor—that has become the standard treatment of HIV/AIDS. The combination of drugs decreases the likelihood that HIV will develop resistance to drug therapy. HAART has reduced HIV in blood to below-detectable levels in many infected individuals. However, it is expensive, there are side effects, and it does not work for everyone.

follow HAART treatment regimens, and, in some cases, treatment failure. HAART does not help everyone with HIV/AIDS.

In sum, we finally have some good news about HIV/AIDS. Many people who had believed that they were doomed have been given a new lease on life. Only a few years earlier, infection with HIV was considered a death sentence. Almost everyone who was infected with HIV eventually developed symptoms, grew progressively weaker and sicker, and died. But HAART and better treatment of opportunistic diseases have made dramatic inroads into the death rate from AIDS in the United States. Yet the rate of *new* infections with HIV has remained more or less stable at about 40,000 cases per year (CDC, 2000b). The rate of HIV infection among gay men appears to have peaked in the late 1980s, but it remained high through the end of the millennium (Valleroy et al., 2000), possibly because the new drugs for HIV/AIDS allowed people to believe that HIV/AIDS was on the road to becoming a nuisance rather than a death sentence. On the other hand, the rate of transmission through male–female sexual activity has been increasing, especially in minority groups (CDC, 2000b).

Unfortunately, the advances in treatment in the United States do not extend to everyone in the world—or even to everyone in this country. Among women in the United States today, the risks of HIV/AIDS fall most heavily on poor women, mostly African American or Latina American, who live in urban areas. African American and Latina American women account for nearly three quarters of women with AIDS in the United States, although they make up only about one quarter of the female population (CDC, 2000b).

PREVENTION What can we do to curb the spread of HIV/AIDS? Given that we lack both vaccine and cure, prevention is our best hope. Our discussion of prevention will focus on sexual transmission, but other efforts have been made to prevent transmission of HIV from mother to child, through the injection of drugs with shared needles, and through blood transfusions. For example, HIV-infected women are advised to avoid breast-feeding. Zidovudine and other measures, such as C-section, decrease the probability of transmission through childbirth (Ricci et al., 2000). The screening of potential blood donors has rendered the probability of transmission via blood transfusion almost negligible. We have been less successful in reducing the risk of infection through unsafe

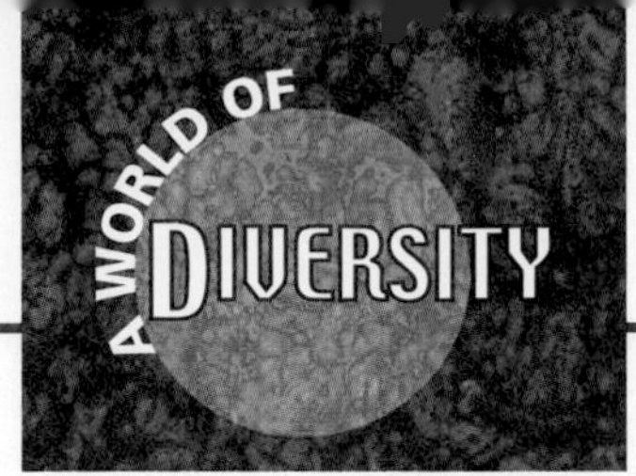

Gender Is a Crucial Issue in the Fight Against HIV/AIDS

Gender inequality is a fundamental driving force of the AIDS epidemic, a fact that we must address in responding to the epidemic, according to Peter Piot (2000), executive director of the Joint United Nations Program on HIV/AIDS (UNAIDS).

Biological and social factors make women and girls more vulnerable to AIDS than men, especially in adolescence and youth; in many places, HIV infection in young women has been found to be 3 to 5 times higher than among boys. Violence—or the threat of violence—against women increases their vulnerability to HIV and reduces their ability to protect themselves from infection.

"The slow, piecemeal reform we have seen in the past is not sufficient if women's rights and needs are to be taken seriously," said Dr. Piot. "Equity in all fields—health, education, environment, the economy—[is] essential if women are to act to protect themselves when it comes to HIV and AIDS." Dr. Piot was speaking at the UN General Assembly Special Session on the follow-up to the Fourth International Conference on Women.

Central among the issues of the Platform for Action elaborated at the Fourth International Conference on Women in Beijing in 1995 are those related to women and poverty, education and training, health, violence, the economy, and power and decision making. These issues also have a major impact on the spread of HIV/AIDS.

Today, women's increased vulnerability to HIV is becoming better understood. Women's economic dependence on men makes them less able to protect themselves, and social norms limit their access to information about sexual matters. At the same time, greater social acceptance of high-risk male sexual behavior can expose both men and their partners to infection.

"Slowing down the spread of HIV means important changes are needed in relationships between men and women," said Dr. Piot. "Men have a crucial role to play in bringing about this kind of radical change."

That change in men's behavior can influence the course of the AIDS epidemic was the theme of the year 2000 World AIDS Campaign to involve men more fully in preventing the spread of HIV, the virus that causes AIDS.

"We must stop seeing men as some kind of problem and begin seeing them as part of the solution," said Dr. Piot. "Working with men to change their behavior and attitudes has tremendous potential to slow down the epidemic. It will also improve the lives of men themselves, not to mention those of their families."

The World AIDS Campaign challenges harmful concepts of masculinity and contends that changing long-held beliefs and attitudes must be part of any effort to slow down the spread of AIDS. Efforts to raise awareness among men are focusing on how adult men look upon risk and sexuality and on how boys are socialized to become men.

Men and women should be encouraged to explore their own perceptions of gender roles and to recognize how stereotypes of both masculinity and femininity can be oppressive, Dr. Piot said. He added that AIDS prevention programs should give greater emphasis to the role of men in health care and that better ways should be found to give men and women the tools they need to communicate and take responsibility for issues of concern to them in preventing HIV/AIDS.

Source: Adapted from UNAIDS Press Release (2000, June 5). "Gender Is Crucial Issue in Fight Against AIDS, Says Head of UNAIDS." New York. Reproduced by kind permission of the Joint United Nations Programme on HIV/AIDS (UNAIDS).

Speaking Up. In the United States, males with HIV/AIDS outnumber females with this STI. However, biological and social factors make women and girls more vulnerable to HIV/AIDS than men, especially in adolescence and youth. Women around the world are more likely to be infected than men, and the numbers of infected women in the United States may be "catching up" with those of men. Violence—or the threat of violence—against women increases their vulnerability to HIV/AIDS and reduces their ability to protect themselves against infection.

negligible. We have been less successful in reducing the risk of infection through unsafe sexual contact.

Most prevention efforts focus on education. Sexually active people have been advised to alter their sexual behavior by practicing abstinence, by limiting their sexual experiences to a lifelong monogamous relationship, or by practicing "safe sex"—which, as we shall see, could more accurately be dubbed "safer sex."

The advent of HIV/AIDS presents the medical, mental health, and educational communities with an unprecedented challenge to develop programs to contain its spread and for compassionate treatment of people with HIV/AIDS. Although HIV/AIDS is frightening, it is preventable. For the latest information on HIV/AIDS, call the government AIDS hotline: 1-800-342-AIDS. The call is free and anonymous. We expand our discussion of prevention of HIV/AIDS in the section entitled "Prevention of STIs: More Than Safer Sex" on page 561.

Genital Herpes

The hysteria that surrounded the rapid spread of genital herpes in the 1970s and 1980s died down upon the advent of AIDS. Nevertheless, there are 200,000 to 500,000 new cases of genital herpes each year ("CDC tackles surge," 1998).

Once you get herpes, it's yours for life. After the initial attack, it remains an unwelcome guest in your body forever. It finds a cozy place to lie low until it stirs up trouble again. It causes recurrent outbreaks that often happen at the worst times, such as around final exams. This is not just bad luck. Stress can depress the functioning of the immune system and heighten the likelihood of outbreaks.

Not only are you stuck with the virus. You can also pass it along to sex partners for the rest of your life. Flare-ups may continue to recur, sometimes with annoying frequency. On the other hand, some people have no recurrences. Still others have mild, brief recurrences that become less frequent over time.

Different types of herpes that are caused by variants of the *Herpes simplex* virus. The most common type, ***Herpes simplex* virus type 1** (HSV-1) causes oral herpes. Oral herpes is characterized by cold sores or fever blisters on the lips or mouth. It can also be transferred to the genitals by the hands or by oral–genital contact (Mertz et al., 1992). **Genital herpes** is caused by a related but distinct virus, the ***Herpes simplex* virus type 2** (HSV-2). This virus produces painful, shallow sores and blisters on the genitals. HSV-2 can also be transferred to the mouth through oral–genital contact. Both types of herpes can be transmitted sexually. Another type of herpes virus (type 8) is also transmitted sexually and is connected with Kaposi's sarcoma, a kind of cancer found frequently in men with HIV/AIDS (Martin et al., 1998).

Physicians are not required to report cases of herpes to public health officials, so there are no precise statistics on its prevalence. It is estimated that more than 100 million people in the United States are infected with oral herpes and perhaps 30 million with genital herpes (Brody, 1993a).

TRANSMISSION Herpes can be transmitted through oral, anal, or vaginal sexual activity with an infected person (Martin et al., 1998; Mertz et al., 1992; Wald et al., 1995). The herpes viruses can also survive for several hours on toilet seats or other objects, where they can be picked up by direct contact. Oral herpes is easily contracted by drinking from the same cup as an infected person, by kissing, and even by sharing towels. But genital herpes is generally spread by coitus or by oral or anal sex.

Many people do not realize that they are infected; so they can unknowingly transmit the virus through sexual contact. And many of the people who do know they are infected don't realize that they can pass along the virus even when they have no noticeable outbreak (Mertz et al., 1992; Wald et al., 1995). Although genital herpes is most contagious during active flare-ups, it can also be transmitted when an infected partner has no symp-

***Herpes simplex* virus type 1** The virus that causes oral herpes, which is characterized by cold sores or fever blisters on the lips or mouth. Abbreviated *HSV-1*.

Genital herpes An STI caused by the *Herpes simplex* virus type 2 and characterized by painful; shallow sores and blisters on the genitals.

***Herpes simplex* virus type 2** The virus that causes genital herpes. Abbreviated *HSV-2*.

toms (genital sores or feelings of burning or itching in the genitals). Any intimate contact with an infected person carries some risk of transmission of the virus, even if the infected person never has another outbreak. People may also be infected with the virus and have *no* outbreaks, and yet pass the virus along to others.

Truth or Fiction? **REVISITED**

It is not true that genital herpes can be transmitted only during flare-ups of the infection. Genital herpes can be transmitted between flare-ups, although people are most contagious during outbreaks of the disease.

Herpes can also be spread from one part of the body to another by touching the infected area and then touching another body part. One potentially serious result is a herpes infection of the eye: **ocular herpes.** Thorough washing with soap and water after touching an infected area may reduce the risk of spreading the infection to other parts of the body. Still, it is best to avoid touching the infected area altogether, especially if there are active sores.

Women with genital herpes are more likely than the general population to have miscarriages. Passage through the birth canal of an infected mother can infect babies with genital herpes, damaging or killing them (Whitley et al., 1991). Obstetricians thus often perform cesarean sections if the mother has active lesions or **prodromal symptoms** at the time of delivery (Osborne & Adelson, 1990). Herpes can also place women at greater risk of genital cancers, such as cervical cancer. (All women, not just women with herpes, are advised to have regular pelvic examinations, including Pap tests for early detection of cervical cancer.)

SYMPTOMS Genital lesions or sores appear about 6 to 8 days after infection with genital herpes. At first they appear as reddish, painful bumps, or papules, along the penis or vulva (see Figure 16.7). They may also appear on the thighs or buttocks, in the vagina, or on the cervix. These papules turn into groups of small blisters that are filled with fluid containing infectious viral particles. The blisters are attacked by the body's immune system (white blood cells). They fill with pus, burst, and become extremely painful, shallow sores or ulcers surrounded by a red ring. People are especially infectious during such outbreaks, because the ulcers shed millions of viral particles. Other symptoms may include headaches and muscle aches, swollen lymph glands, fever, burning urination, and a vaginal discharge. The blisters crust over and heal in 1 to 3 weeks. Internal sores in the vagina or on the cervix may take 10 days longer than external (labial) sores to heal. Physicians thus advise infected women to avoid unprotected intercourse for at least 10 days after the healing of external sores.

Although the symptoms disappear, the disease does not. The virus remains in the body permanently, burrowing into nerve cells in the base of the spine, where it may lie dormant for years or for a lifetime. The infected person is least contagious during this dormant stage. For reasons that remain unclear, the virus becomes reactivated and gives rise to recurrences in most cases.

Recurrences may be related to infections (such as in a cold), stress, fatigue, depression, exposure to the sun, and hormonal changes such as those that occur during pregnancy or menstruation. Recurrences tend to occur within 3 to 12 months of the initial episode and to affect the same part of the body.

About 50% of people with recurrent herpes experience prodromal symptoms before an active outbreak (Reinisch, 1990). These warning signs may include feelings of burning, itching, pain, tingling, or tenderness in the affected area. These symptoms may be accompanied by sharp pains in the lower extremities, groin, or buttocks. People with herpes may be more infectious when prodromal symptoms appear. They are advised to avoid unprotected sex until the flare-up subsides.

The symptoms of oral herpes include sores or blisters on the lips, the inside of the mouth, the tongue, or the throat. Fever and feelings of sickness may occur. The gums may

Ocular herpes A herpes infection of the eye, usually caused by touching an infected area of the body and then touching the eye.

Prodromal symptoms Warning symptoms that signal the onset or flare-up of a disease. (From the Greek *prodromos,* which means "forerunner.")

swell and redden. The sores heal over in about 2 weeks, and the virus retreats into nerve cells at the base of the neck, where it lies dormant between flare-ups. About 90% of people with oral herpes experience recurrences. About half of them have five or more recurrences during the first 2 years after the initial outbreak (Reinisch, 1990).

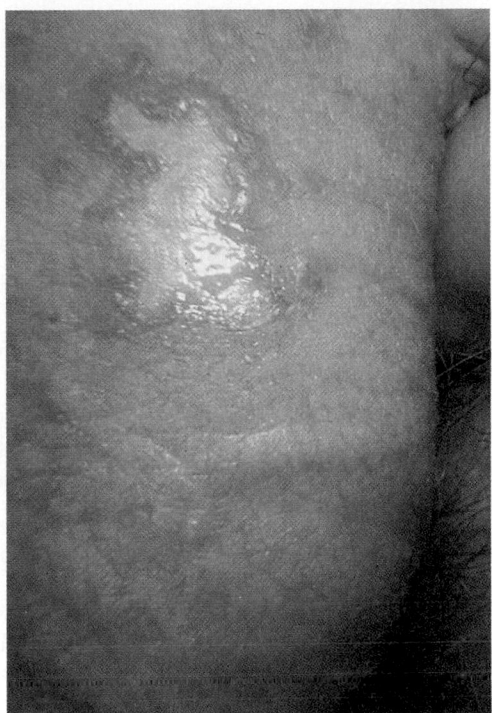

FIGURE 16.7

Herpes Lesion on the Male Genitals. Herpes lesions or sores can appear on the genitals in both men and women. In contrast to the syphilis chancre, they can be quite painful. Herpes is most likely to be transmitted during outbreaks of the disease (when the sores are present, that is), but it can be transmitted at other times as well. Stress increases the likelihood of outbreaks. Antiviral drugs tend to decrease the frequency, duration, and discomfort of outbreaks.

DIAGNOSIS AND TREATMENT Genital herpes is first diagnosed by clinical inspection of herpetic sores or ulcers in the mouth or on the genitals. A sample of fluid may be taken from the base of a genital sore and cultured in the laboratory to detect the growth of the virus.

There is encouraging news about the development of a vaccine against genital herpes called Simplirix (Laino, 2000). In pilot studies, the vaccine is reported to have prevented herpes outbreaks in more than 70% of women who had not previously had cold sores or genital herpes. The herpes virus that causes cold sores ironically confers some protection against genital herpes, and the vaccine does not help women who have already contracted genital herpes. It remains unclear why the vaccine has been ineffective with men.

Viruses, unlike the bacteria that cause gonorrhea or syphilis, do not respond to antibiotics. Antiviral drugs such as acyclovir (brand name: Zovirax), famciclovir (Diaz-Mitoma et al., 1998), and valacyclovir can relieve pain, speed healing, and reduce the duration of viral shedding (Cates, 1998). Acyclovir can be applied directly to the sores in ointment form, but it must be taken orally, in pill form, to help combat internal lesions in the vagina or on the cervix. Oral administration of antiviral drugs may reduce the severity of the initial episode and, if taken regularly, the frequency and duration of recurrent outbreaks (Cates, 1998). On the other hand, users may develop tolerance to these drugs, necessitating larger doses to maintain effectiveness (Drew, 2000).

Warm baths, loosely fitting clothing, aspirin, and cold, wet compresses may relieve pain during flare-ups. People with herpes are advised to maintain regular sleeping habits and to learn to manage stress.

COPING WITH GENITAL HERPES The psychological problems connected with herpes can be more distressing than the physical effects of the illness. The prospects of a lifetime of recurrences and concerns about infecting one's sex partners exacerbate the emotional impact of herpes. A "herpes syndrome" has been described that involves feelings of anger, depression, isolation, and shame—even thoughts of being tainted, ugly, or dangerous (Mirotznik et al., 1987).

An anonymous survey of Brooklyn College students suggests the emotional toll being exacted by the herpes epidemic. Among people with herpes, 55% reported strong emotional responses to it, and 28% reported moderate reactions. Only 17% reported mild reactions (Mirotznik et al., 1987). The majority reported feelings of fear and anger. Half indicated feeling "damaged." Most also believed that herpes had affected their sexual behavior; 75% said that they had avoided sex for "a long time" because of herpes. Although most had resumed sexual activity, many sought partners who were also infected so that they would not have to explain the disease, or risk transmitting it, to uninfected people. Given the consequences of herpes, you might think that students would be concerned about contracting the disease. Another survey at Brooklyn College revealed that most sexually active, uninfected students perceived herpes as having serious consequences. However, about three out of four were not fearful of contracting the disease, and nearly half had not changed their dating behavior because of it (Mirotznik, 1991).

People with herpes often feel angry, especially toward those who transmitted the disease to them. They may also feel anxious about making a long-term commitment or bearing children:

> After the first big episode of herpes, I felt distant from my body. When we began lovemaking again, I had a hard time having orgasms or trusting the rhythm of my responses. I shed some tears over that. I felt my body had been invaded. My body feels riddled with it; I'm somehow contaminated. And there is always that lingering anxiety: is my baby okay? It's unjust that the birth of my child may be affected. (Boston Women's Health Book Collective, 1992; Copyright © 1984, 1992 by The Boston Women's Health Book Collective. Reprinted with permission.)

Most people with herpes learn to cope. Some are helped by support groups that share ways of living with the disease. A caring and trusting partner is important. Joanne, a 26-year-old securities analyst, kept her herpes a secret from Jonathan during the first month they were dating. But when they approached the point of becoming sexually intimate, she felt obligated to tell him that she carried the virus:

> "I feared that telling him would scare him off. After all, who wants to have a relationship with someone who can give them herpes? After the first few dates I felt that this was the person I could spend the rest of my life with. I knew he also felt the same way. I had to tell him before things became too intense between us. Believe me, it wasn't easy blurting it out. He wasn't shocked or anything, although he did ask me all kinds of questions about it. I remember telling him that I got recurrences about once a year or so for about a week at a time. I told him that there was always the potential that I could infect him but that we would play it safe and avoid having sex whenever I had an outbreak. I also told him that even at other times I couldn't guarantee that it would be perfectly safe. He said at first that he needed some time to think about it. But later, that very night in fact, he called to tell me that he didn't want this to come between us and that we should try to make our relationship work."
>
> Joanne and Jonathan were married about six months later. A year after that their daughter Andrea was born. Jonathan remains uninfected. Joanne's occasional outbreaks are treated with acyclovir ointment and pass within a week or so.
>
> —The Authors' Files

The attitudes of people with herpes also play a role in their success in adjusting to it. People who view herpes as a manageable illness or problem, not as a medical disaster or a character deficit, seem to find it easier to cope.

Viral Hepatitis

Hepatitis is an inflammation of the liver that may be caused by such factors as chronic alcoholism and exposure to toxic materials. Viral hepatitis comprises several different types of hepatitis caused by related, but distinct, viruses. The major types are *hepatitis A* (formerly called infectious hepatitis), *hepatitis B* (formerly called serum hepatitis), *hepatitis C* (formerly called hepatitis non-A, non-B), and *hepatitis D.*

Most people with acute hepatitis have no symptoms. When symptoms do appear, they often include **jaundice,** feelings of weakness and nausea, loss of appetite, abdominal discomfort, whitish bowel movements, and brownish or tea-colored urine. The symptoms of hepatitis B tend to be more severe and long-lasting than those of hepatitis A or C. In about 10% of cases, hepatitis B leads to chronic liver disease. Hepatitis C tends to have milder symptoms but often leads to chronic liver disease such as cirrhosis or cancer of the liver. Hepatitis D—also called *delta hepatitis* or type D hepatitis—occurs only in the presence of hepatitis B. Hepatitis D, which has symptoms similar to those of hepatitis B, can produce severe liver damage and often leads to death.

Hepatitis An inflammation of the liver. (From the Greek *hepar,* which means "liver.")

Jaundice A yellowish discoloration of the skin and the whites of the eyes. (From the Old French *jaune,* which means "yellow.")

The hepatitis A virus is transmitted through contact with infected fecal matter found in contaminated food or water, and by oral contact with fecal matter, such as through oral–anal sexual activity (licking or mouthing the partner's anus). (It is largely because of the risk of hepatitis A that restaurant employees are required to wash their hands after using the toilet.) Ingesting uncooked infested shellfish is also a frequent means of transmission of hepatitis A (Lemon & Newbold, 1990).

Hepatitis B can be transmitted sexually through anal, vaginal, or oral intercourse with an infected partner; through transfusion with contaminated blood supplies; by the sharing of contaminated needles or syringes; and by contact with contaminated saliva, menstrual blood, nasal mucus, or semen (Lemon & Newbold, 1990). Sharing razors, toothbrushes, or other personal articles with an infected person can also transmit hepatitis B. Hepatitis C and hepatitis D can be transmitted sexually or through contact with contaminated blood. A person can transmit the viruses that cause hepatitis even if he or she is unaware of having any symptoms of the disease.

Hepatitis is usually diagnosed by testing blood samples for the presence of hepatitis antigens and antibodies. There is no cure for viral hepatitis. Bed rest and fluids are usually recommended until the acute stage of the infection subsides, generally in a few weeks. Full recovery may take months. A vaccine provides protection against hepatitis B and also against hepatitis D, because hepatitis D can occur only if hepatitis B is present (Cates, 1998).

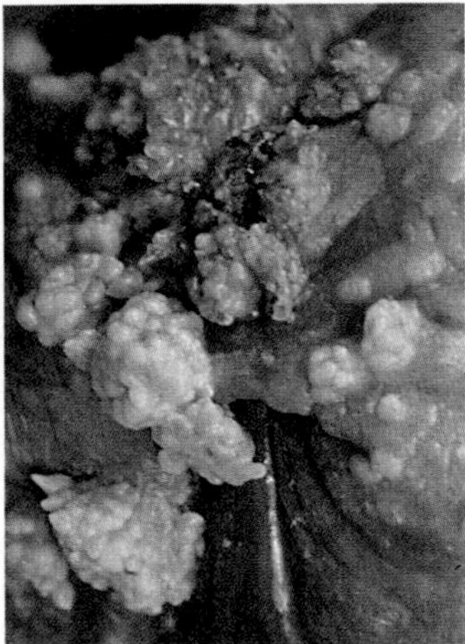

FIGURE 16.8

Genital Warts. Genital warts are caused by the human papilloma virus (HPV) and may have a cauliflower-like appearance. Many—perhaps most—cases occur where they can go visually undetected. HPV is implicated in cervical cancer, and women should be checked regularly for genital warts and other possibly "silent" STIs.

Genital Warts

The *human papilloma virus* (HPV) causes **genital warts** (formerly termed *venereal warts*). HPV, with signs of infection found in nearly half of the adult women in some countries, is the world's most common sexually transmitted infection (Ho et al., 1998).There are 800,000 newly diagnosed infections each year in the United States. It also appears that more than half of the sexually active college women in the United States are infected with HPV. Although the warts may appear in visible areas of the skin, in perhaps 7 out of 10 cases they appear in areas that cannot be seen, such as on the cervix in women or in the urethra in men (Reinisch, 1990). They occur most commonly among people 20 to 24 years old (Reinisch, 1990). Within a few months following infection, the warts are usually found in the genital and anal regions. Women are more susceptible to HPV infection because cells in the cervix divide swiftly, facilitating the multiplication of HPV (Blakeslee, 1992). Women who initiate coitus prior to the age of 18 and who have many sex partners are particularly susceptible to infection (Blakeslee, 1992). Similarly, it is estimated that nearly half of the sexually active teenage women in some U.S. cities are infected with HPV (Blakeslee, 1992).

Genital warts are similar to common plantar warts—itchy bumps that vary in size and shape. Genital warts are hard and yellow-gray when they form on dry skin. They take on pink, soft, cauliflower shapes in moist areas such as the lower vagina (see Figure 16.8). In men they appear on the penis, foreskin, and scrotum and in the urethra. They appear on the vulva, along the vaginal wall, and on the cervix in women. They can also occur outside the genital area—for example, in the mouth; on the lips, eyelids, or nipples; around the anus; or in the rectum.

Genital warts may not cause any symptoms, but those that form on the urethra can cause bleeding or painful discharges. HPV has been implicated in cancers of the genital organs, particularly cervical cancer and penile cancer (Koutsky et al., 1992). Ninety percent of cases of cervical cancer are linked to certain strains of HPV (Cannistra & Niloff, 1996; Reaney, 1998). Researchers in one study found that women who had a history of HPV infection were 11 times more likely than other women to develop cervical cancer during the 2-year study period (Koutsky et al., 1992). Moreover, women whose husbands visit prostitutes or have many sex partners are 5 to 11 times as likely as other married women to develop cervical cancer ("Roving mates," 1996). All in all, it would seem wise for women to safeguard themselves from HPV-related cervical cancer by limiting their number of sex partners (to reduce their risk of exposure to HPV), thinking about the extracurricular sexual activities of their mates, and having regular Pap smears.

Genital warts An STI that is caused by the human papilloma virus and takes the form of warts that appear around the genitals and anus.

HPV can be transmitted sexually through skin-to-skin contact during vaginal, anal, and oral sex. It can also be transmitted by other forms of contact, such as touching infected towels or clothing. The incubation period may vary from a few weeks to a couple of years.

Freezing the wart (*cryotherapy*) with liquid nitrogen is a preferred treatment. One alternative treatment involves painting or coating the warts over several days with podofilox solution or gel, imiquimod cream, trichloroacetic acid (TCA), or bichloroacetic acid (BCA) (Cates, 1998). An alcohol-based podophyllin solution causes the warts to dry up and fall off. Unfortunately, although the warts themselves may be removed, treatment does not rid the body of the virus (Cates, 1998). There may thus be recurrences. Podophyllin is not recommended for use with pregnant women or for treatment of warts that form on the cervix. The warts can also be treated by a doctor with electrodes (burning) or surgery (by laser or surgical removal).

Researchers are attempting to develop a vaccine for HPV, an effort based on the discovery of molecules that inform the immune system that that virus is present (Reaney, 1998). For the time being, however, prevention means using latex condoms, which help reduce the risk of contracting HPV. They do not eliminate the risk entirely, because the virus can be transmitted from areas of the skin not protected by condoms, such as the scrotum (Ochs, 1994). People with active warts should probably avoid sexual contact until the warts are removed and the area heals completely.

Molluscum Contagiosum

Molluscum contagiosum is caused by a pox virus that can be spread sexually. The virus causes painless, raised lesions to appear on the genitals, buttocks, thighs, or lower abdomen. Pinkish in appearance with a waxy or pearly top, they usually appear within 2 or 3 months of infection. Most infected people have between 10 and 20 lesions, although the number of lesions can range from 1 to perhaps 100 or more (Douglas, 1990). The lesions are generally not associated with serious complications and often disappear on their own within 6 months. They can also be treated by squeezing (like "popping" a blackhead) and thus causing them to exude the whitish center plug. Solutions of podophyllin, trichloroacetic acid (TCA), or silver nitrate are also used. Freezing with liquid nitrogen (*cryotherapy*) can be used to remove the lesions. However, do not try to treat any lesions on your own. See your doctor.

ECTOPARASITIC INFESTATIONS

Ectoparasites, as opposed to *endoparasites,* live on the outer surfaces of animals (*ecto* means "outer"). *Trichomonas vaginalis,* which causes trichomoniasis, is an endoparasite (*endo* means "inner"). Ectoparasites are larger than the agents that cause other STIs. In this section we consider two types of STIs caused by ectoparasites: pediculosis and scabies.

Molluscum contagiosum An STI is caused by a pox virus that causes painless, raised lesions to appear on the genitals, buttocks, thighs, or lower abdomen.

Ectoparasites Parasites that live on the outside of the host's body—in contrast to *endo*parasites, which live within the body. (From the Greek *ektos,* which means "outside.")

Pediculosis A parasitic infestation by pubic lice (*Pthirus pubis*) that causes itching.

Pediculosis

Pediculosis is the name given to an infestation of a parasite whose proper Latin name, *Pthirus pubis* (pubic lice), sounds rather too dignified for these bothersome (dare we say ugly?) creatures that are better known as crabs. Pubic lice are commonly called crabs because, under the microscope, they are somewhat similar in appearance to crabs (see Figure 16.9). They belong to a family of insects called biting lice. Another member of the family, the human head louse, is an annoying insect that clings to hair on the scalp and often spreads among schoolchildren.

It is not true that pubic lice are of the same family of animals as crabs. Viewed under a microscope, however, they look somewhat like crabs. ■

Truth or Fiction? **REVISITED**

In the adult stage, pubic lice are large enough to be seen with the naked eye. They are spread sexually but can also be transmitted via contact with an infested towel, sheet, or—yes—toilet seat. They can survive for only about 24 hours without a human host, but they may deposit eggs that can take up to 7 days to hatch in bedding or towels (Reinisch, 1990). Therefore, all bedding, towels, and clothes that have been used by an infested person must be either dry-cleaned or washed in hot water and dried on the hot cycle to ensure that they are safe. Fingers may also transmit the lice from the genitals to other hair-covered parts of the body, including the scalp and armpits. Sexual contact should be avoided until the infestation is eradicated.

Itching, ranging from the mildly irritating to the intolerable, is the most prominent symptom of a pubic lice infestation. The itching is caused by the "crabs" attaching themselves to the pubic hair and piercing the skin to feed on the blood of their hosts. (Yecch!) The life span of these insects is only about a month, but they are prolific egg-layers and may spawn several generations before they die. An infestation can be treated effectively with a prescription medication, a 1% solution of lindane (brand name: Kwell), which is available as a cream, lotion, or shampoo. Nonprescription medications containing pyrethrins or piperonyl butoxide (brand names: RID, Triple X, and others) will also do the job (Reinisch, 1990). Kwell is not recommended for use by pregnant or lactating women. A careful reexamination of the body is necessary after 4 to 7 days of treatment to ensure that all lice and eggs were killed (Reinisch, 1990).

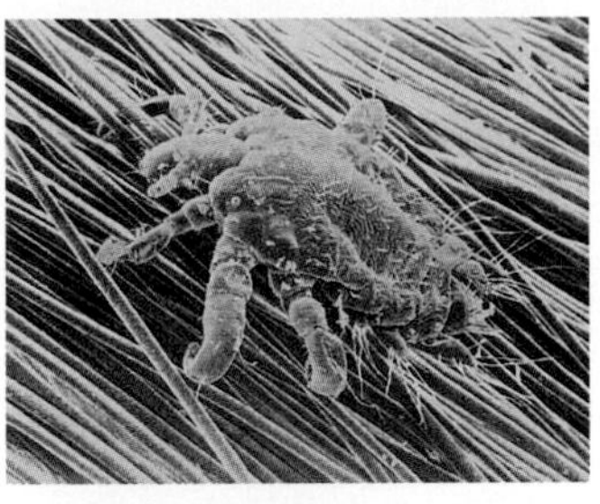

FIGURE 16.9

Pubic Lice. Pediculosis is an infestation by pubic lice (*Pthirus pubis.*) Pubic lice are commonly called "crabs" because of their appearance under a microscope.

Scabies

Scabies (short for *Sarcoptes scabiei*) is a parasitic infestation caused by a tiny mite that may be transmitted through sexual contact or contact with infested clothing, bed linen, towels, and other fabrics. The mites attach themselves to the base of pubic hair and burrow into the skin, where they lay eggs and subsist for the duration of their 30-day life span. Like pubic lice, scabies are often found in the genital region and cause itching and discomfort. They are also responsible for reddish lines (created by burrowing) and for sores, welts, or blisters on the skin. Unlike lice, they are too tiny to be seen by the naked eye. Diagnosis is made by detecting the mite or its by-products via microscopic examination of scrapings from suspicious-looking areas of skin (Levine, 1991). Scabies are most often found on the hands and wrists, but they may also appear on the genitals, buttocks, armpits, and feet (Reinisch, 1990). They do not appear above the neck—thankfully!

Scabies, like pubic lice, may be treated effectively with 1% lindane (Kwell). The entire body from the neck down must be coated with a thin layer of the medication, which should not be washed off for 8 hours. But lindane should not be used by women who are pregnant or lactating. To avoid reinfection, sex partners and others in close bodily contact with infected individuals should also be treated. Clothing and bed linen that the infected person has used must be washed and dried on the hot cycle or dry-cleaned. "s with "crabs," sexual contact should be avoided until the infestation is eliminated.

PREVENTION OF STIs: MORE THAN SAFER SEX

Prevention is the best way to control the spread of STIs, especially those for which there is no cure or vaccine. Prevention of even one case of an STI can prevent its spread to others—perhaps eventually to you.

Scabies A parasitic infestation caused by a tiny mite (*Sarcoptes scabiei*) that causes itching.

There are many things that you can do to lower your risk of contracting STIs. Safer sex is only one aspect of prevention.

CONSIDER ABSTINENCE OR MONOGAMY

> I don't even masturbate anymore. I'm so afraid I'll give myself something. I just want to be friends with myself.
>
> —Richard Lewis

The only fully effective strategies to prevent the sexual transmission of STIs are abstinence and maintaining a monogamous sexual relationship with an uninfected partner. If you are celibate, or if you and your sex partner are not infected and neither of you engages in sexual activity with anyone else, you have little to be concerned about. Many people are sexually active and have not committed themselves to a monogamous relationship, however. Even for those who seek monogamous relationships, there must always be that "first time."

BE KNOWLEDGEABLE ABOUT THE RISKS Be aware of the risks of STIs. Many of us try to put the dangers of STIs out of our minds—especially in moments of passion. Make a pact with yourself to refuse to play the dangerous game of pretending that the dangers of STIs do not exist or that you are somehow immune.

Prevention programs are apparently raising public awareness of HIV/AIDS. Seventy-nine percent of respondents to a 1993 *New York Times*/CBS News national poll correctly recognized that one cannot be infected with HIV by sharing a drinking glass with a person with AIDS, up from 34% in 1985. Eighty-nine percent recognized that you cannot be infected by a toilet seat, up from 49% in 1985. Nearly everyone knew that HIV can be transmitted by male–female sex (96%) and blood transfusions (98%). More people today are also personally acquainted with a person with AIDS. One third of those polled know someone who has died of AIDS or has either HIV infection or AIDS.

Knowledge about HIV transmission does not always translate into behavioral change, such as increased use of latex condoms (which can block the transmission of HIV) and other safer-sex practices. In fact, knowledge about transmission of HIV and other STIs is not clearly related to condom use (Geringer et al., 1993; Klepinger et al., 1993; Rotheram-Borus & Koopman, 1991a, 1991b). Despite widespread efforts to educate the public about the dangers of unprotected sex, negative attitudes toward using condoms persist, especially among males (Fisher et al., 1995). Consider the litany of complaints we come across: "They reduce sexual pleasure. . . . They're a nuisance to put on. . . . They cost too much. . . . They interrupt sex," and so on.

Virtually all young people are similarly aware of the sexual transmission of HIV (Wulfert & Wan, 1993). Why, then, do many of them continue to engage in risky sexual behavior?

For one thing, teenagers do not perceive risky behavior to be as dangerous as adults do. This generalization extends to drinking, smoking, failure to use seat belts, and drag racing, as well as to sexual behavior (Cohn et al., 1995). Young people also profit from being taught specific skills to protect themselves from engaging in risky behaviors, such as communication skills for discussing safer-sex options with their partners, assertiveness skills for ensuring that their partners respect their needs and interests, social skills to resist peer pressure, and correct use of condoms (Kamb et al., 1998). Greater efforts are needed to reverse peer norms that discourage condom use. For example, it may be advisable for adolescents themselves to teach condom use to their peers (Hodges et al., 1992).

Researchers have identified several factors that underlie risky sexual behavior among young people:

1. *Perceived low risk of infection.* One of the major stumbling blocks in promoting safer-sex practices is that many young heterosexuals perceive a low risk of contracting HIV (Nadeau et al., 1993; Oswalt & Matsen, 1993). People who perceive themselves as being at low risk are less likely to alter their behavior.

   Given the current low rate of known infections among heterosexuals who do not inject drugs, many heterosexuals may perceive risky sexual practices to be a reasonable gamble (Pinkerton & Abramson, 1992). Heterosexuals who have never had a friend or relative with HIV/AIDS may dismiss it as a problem that affects other types of people. Even gay men may operate under the "I'm not the type" fallacy and underestimate their personal risks. According to Pinkerton & Abramson (1992),

   > The frightening picture that emerges is one in which it is only the *other guy* (or gal) who gets AIDS: To the heterosexual who does not inject drugs, it's just gays and "druggies" that get AIDS; to the "average" gay man, it's those gay men who are overly promiscuous; and to the bath house participant, it's those who aren't "careful." (pp. 564–565)

2. *Negative attitudes toward condom use.* Many factors discourage condom use. Some people feel embarrassed to buy them. For others, the risk of being infected with HIV seems to fly out of their minds whenever the opportunity for sex arises. Some claim that a condom dampens romantic ardor in moments of passion by requiring an interruption of the sexual act to apply one. Some people just regard them as too much of a fuss. Many men say that condoms deprive them of sexual pleasure. Unless such obstacles to using condoms are overcome, efforts to stem the tide of HIV infection may be thwarted.

3. *Myth of personal invulnerability.* Some people subscribe to a myth of personal invulnerability and believe that they are somehow immune to AIDS and other diseases. Even students who are generally well informed about STIs may think of themselves as personally immune. One junior at the University of Miami (Ohio) explained to an interviewer why she did not insist that her partners use a condom: "I have an attitude—it may be wrong—that any guy I would sleep with would not have AIDS" (Johnson, 1990, p. A18). The adventurous spirit that we often associate with youth may confer on young people a dangerous sense of immortality and a greater willingness to take risks (CDC, 2000h; King, 2000). Perceptions of personal invulnerability help to explain why AIDS education does not always translate into behavioral change.

Even people who do not believe themselves to be immune may underestimate their risk of being infected with HIV (van der Velde et al., 1994). This finding holds true for poor inner-city residents (Hobfoll et al., 1993) as well as college students (Goldman & Harlow, 1993). People tend to view themselves as luckier than the norm. (After all, don't you believe that you alone hold the winning lottery ticket?) Moreover, because the transmission of HIV through casual male–female sexual encounters is rare, people who regularly engage in risky sexual practices and so far have remained uninfected may be lulled into a false sense of security.

REMAIN SOBER Alcohol and other drugs increase the likelihood of engaging in risky sexual behavior (Centers for Disease Control and Prevention, 2000a; MacDonald et al., 2000).

INSPECT YOURSELF AND YOUR PARTNER Inspect yourself for any discharge, bumps, rashes, warts, blisters, chancres, sores, lice, or foul odors. Do not expect to find telltale signs of an HIV infection, but infected people often have other STIs. Check out any unusual feature with a physician before you engage in sexual activity.

You may be able to work an inspection of your partner into foreplay—a good reason for making love the first time with the lights on. In particular, a woman may hold her partner's penis firmly, pulling the loose skin up and down, as though "milking" it. Then she can check for a discharge at the penile opening. The man may use his fingers to detect any sign of a vaginal discharge. Other visible features of STIs include herpes blisters, genital warts, syphilitic chancres or rashes, and pubic lice.

If you find anything that doesn't look, feel, or smell right, bring it to your partner's attention. Treat any unpleasant odor as a warning sign. Your partner may not be aware of the symptom and may be carrying an infection. If you notice any suspicious signs, refrain from further sexual contact until your partner has the chance to seek a medical evaluation. It is advisable to be informed about the common signs and symptoms of STIs, but you need not become a medical expert. Even if your concerns prove groundless, you can resume sexual relations without the uncertainty that you would have had if you had ignored them.

Of course, your partner may become defensive or hostile if you express a concern that he or she may be carrying an STI. Try to be empathic. The social stigma attached to people with STIs makes it difficult to accept the possibility of infection. It may be appropriate to point out that STIs are quite common (among college students, bank officers, military personnel, or . . . fill in the blanks) and that many people are unaware that they carry them.

But for your sake as well as your partner's, if you are not sure that sex is safe, stop. Think carefully about the risks, and seek expert advice.

USE LATEX CONDOMS Latex condoms are effective in blocking nearly all sexually transmissible organisms. In regular use, latex condoms reduce the rate of STI infections by about 50%, on the average (Hatcher et al., 1998). Improper use or inconsistent use is a common reason for failures in using condoms to prevent STI transmission (Hatcher et al., 1998). Yet even when used properly, condoms may be of limited or no value against disease-causing organisms that are transmitted externally, such as those that cause herpes, genital warts, and ectoparasitic infestations.

Condoms made from animal membranes ("skins") are less effective as barriers against STI-causing organisms (they contain pores that allow tiny microbes, including HIV, to penetrate). Even latex condoms are not 100% effective in preventing the transmission of HIV and other STIs. Condoms (and the people who use them) are fallible. Condoms can break or slip off.

USE BARRIER DEVICES WHEN PRACTICING ORAL SEX (FELLATIO OR CUNNILINGUS) If you decide to practice oral sex, use a condom before practicing fellatio and a dental dam (a square piece of latex rubber used by dentists during oral surgery) to cover the vagina before engaging in cunnilingus.

AVOID HIGH-RISK SEXUAL BEHAVIORS Avoid unprotected vaginal intercourse (intercourse without the use of a latex condom). Unprotected anal intercourse is one of riskiest practices. Other high-risk behaviors include unprotected oral–genital activity, oral–anal activity, insertion of a hand or fist ("fisting") into someone's rectum or vagina, or any activity in which you or your partner would come into contact with the other's blood, semen, or vaginal secretions. If you do engage in anal–genital sex and are uncertain whether you or your partner is infected, use a latex condom and spermicide. Oral–anal sex, or anilingus (sometimes called rimming), should be avoided because of the potential of transmitting microbes between the mouth and the anus.

Also avoid sexual contact people who have STIs, people who practice high-risk sexual behaviors, people who inject drugs, and prostitutes or people who frequent prostitutes.

WASH THE GENITALS BEFORE AND AFTER SEX Washing the genitals before and after sex removes a quantity of potentially harmful agents. Washing together may be incorporated into erotic foreplay. Right after intercourse, washing with soap and water may help reduce the risk of infection. Do not, however, deceive yourself into believing that washing your genitals is an effective substitute for safer sex. Most STIs are transmitted internally. Washing is useless against them.

Douching right after coitus might have some limited benefits for women. But frequent douching should be avoided because it may change the vaginal flora and encourage the growth of infectious organisms. Nor is immediate douching possible for women who use a diaphragm and spermicide, which must remain in the vagina for at least 6 to 8 hours after intercourse. But such women may profit from washing the external genitals immediately after coitus.

HAVE REGULAR MEDICAL CHECKUPS A sexually active person should have health examinations regularly—at least once a year. Many community clinics and family-planning centers scale their charges to ability to pay. Checkups are a small investment to make in one's own health. Many people are symptomless carriers of STIs, especially of chlamydial infections. Medical checkups enable them to learn about and receive treatment for disorders that might otherwise go unnoticed. Many physicians advise routine testing of asymptomatic young women for chlamydial infections to prevent the hidden damage that can occur when the infection goes untreated.

DISCUSS WHETHER YOU AND YOUR PARTNER SHOULD UNDERGO TESTING BEFORE INITIATING SEXUAL RELATIONS Some couples reach a mutual agreement to be tested for HIV and other STIs before they initiate sexual relations. (Some people simply insist that their prospective partners be tested before they initiate sexual relations.) But many people resist testing or feel insulted when their partners raise the issue. People usually assume that they are free of STIs if they have no symptoms and have been reasonably selective in their choice of partners. But STIs happen to the "nicest people," and the absence of symptoms is no guarantee of freedom from infection. Unless you have been celibate or have been involved in a monogamous relationship with an uninfected partner, you should consider yourself at risk of carrying or contracting an infectious STI.

ENGAGE IN NONCOITAL SEXUAL ACTIVITIES Other forms of sexual expression, such as hugging, massage, caressing, mutual masturbation, and rubbing bodies together without vaginal, anal, or oral contact, are low-risk ways of finding sexual pleasure, so long as neither semen nor vaginal fluids come into contact with mucous membranes or breaks in the skin. Many sexologists refer to such activities as **outercourse** to distinguish them from sexual intercourse. Sharing sexual fantasies can be very titillating, as can taking a bath or shower together. Vibrators, dildos, and other "sex toys" may also be erotically stimulating and carry a low risk of infection if they are washed thoroughly with soap and water before use and between uses by two people.

Outercourse Forms of sexual expression, such as massage, hugging, caressing, mutual masturbation, and rubbing bodies together, that do not involve the exchange of body fluids. (Contrast with *intercourse.*)

CONSULT YOUR PHYSICIAN IF YOU SUSPECT THAT YOU HAVE BEEN EXPOSED TO AN STI If you think you may have been exposed to an STI, such as HIV, see a doctor as soon as possible. Many STIs are detectable in the early stages of development and can be successfully treated. If HIV is in your bloodstream, there is some chance that it may be eradicated before it infects cells in the immune system. If you are infected with HIV, early treatment may keep virus levels low and prevent you from developing AIDS. Early intervention may also prevent the damage of an STI spreading to vital body organs. Be sensitive to any physical changes that may be symptomatic of STIs. Talk to a health professional when in doubt.

GET TO KNOW YOUR PARTNER BEFORE ENGAGING IN SEXUAL RELATIONS Be selective in your choice of sex partners. Having sex with multiple partners—especially "one-night stands"—increases your risk of sexual contact with an infected person. It also lessens your opportunity to get to know your partner well enough to discover whether he or she has participated in high-risk sexual practices or has had sex partners in the past who practiced high-risk behaviors.

AVOID OTHER HIGH-RISK BEHAVIORS Avoid contact with bodily substances (blood, semen, vaginal secretions, and fecal matter) from other people. Do not share hypodermic needles, razors, cuticle scissors, or other implements that could carry another person's blood. Be careful when handling wet towels, bed linen, or other material that might contain bodily substances.

Human Sexuality ONLINE //

Web Sites Related to Sexually Transmitted Infections

This is the Web site, within the Centers for Disease Control and Prevention, of *The HIV/AIDS Surveillance Report.* The report is issued every 6 months and contains tabular and graphical information about U.S. AIDS and HIV case reports, including data by state, metropolitan statistical area, mode of exposure to HIV, sex, race/ethnicity, age group, vital status, and case definition category.
www.cdc.gov/hiv/stats/hasrlink.htm

The NPIN Web site is a CDC service for distributing information about HIV/AIDS, other STIs, and tuberculosis.
www.cdcnpin.org/

The HIV/AIDS Treatment Information Service (ATIS) is a central resource for federally approved treatment guidelines for HIV and AIDS. ATIS is staffed by multilingual health information specialists who answer questions on HIV treatment options using a broad network of federal, national, and community-based information resources. You can call the confidential, personalized service at 1-800-HIV-0440 (1-800-448-0440). Bilingual specialists are available Monday through Friday from 9 A.M. to 7 P.M. EST.
www.hivatis.org/

HIV InSite is a project of the University of California San Francisco (UCSF) Positive Health Program at San Francisco General Hospital Medical Center and the UCSF Center for AIDS Prevention Studies, which are programs of the UCSF AIDS Research Institute.
hivinsite.ucsf.edu/

REACH es un sitio específico para todos—pacientes, familias, médicos y otros atendientes, investigadores, estudiantes, etc.—con directorios de vínculos a otras páginas electronicas que ahorran tiempo, organizados bajo una extensa lista de categorias.
www.kc-reach.org/spanish.html

Incluye gran información tanto para legos como profesionales. Es parte de NOAH (Acceso a Salud de Nueva York) quien provee de mucha información sobre la salud en general.
www.noah.cuny.edu/spaids/spaids.html

The Web site of UNAIDS (United Nations AIDS). It contains press releases, information about conferences, speeches, a newsletter, and video/sound clips.
www.us.unaids.org/

Sources of Help

Do you have questions about the signs and symptoms of STIs? Do you need assistance in coping with an STI? A number of organizations have established telephone hotlines that provide anonymous callers with information. Some organizations publish newsletters and other material to help people with particular diseases cope more effectively.

NATIONAL TOLL-FREE HOTLINES FOR INFORMATION ABOUT AIDS AND OTHER STIs
These hotlines provide information about AIDS and other STIs, as well as referral sources. You needn't give your name or identify yourself to obtain information.

National AIDS Hotline, Centers for Disease Control AIDS Hotline: (800) 342-AIDS: Information and referral resources nationwide, 24 hours a day.

National STI Hotline: (800) 227-8922 (in California, 800-982-5883) (A hotline sponsored by the American Social Health Association, which dispenses information about STI symptoms and refers callers to local STI clinics that provide confidential treatment at minimum or no cost.)

Spanish AIDS/SIDA Hotline: (800) 344-7432

AIDS Hotline for Teens: (800) 234-TEEN

AIDS Hotline for the Hearing Impaired: (800) 243-7889

Canadian Toll-Free Hotline (Toll-free in Canada): AIDS Committee of Toronto: (800) 267-6600

WHERE TO OBTAIN HELP OR INFORMATION ABOUT HERPES

National Herpes Hotline: (919) 361-8488

The Helper is a newsletter published by HELP (Herpetics Engaged in Living Productively), an organization that helps people with herpes cope with the disease. For copies of the newsletter, and for the address of the HELP chapter closest to you, either call the National STI Hotline (800-227-8922) or write to HELP, Herpes Resource Center, P.O. Box 13827, Research Triangle Park, NC 27709.

Herpes Resource Center, Box 100, Palo Alto, CA 94302

summing up

AN EPIDEMIC

More than 13 million people in the United States contract a sexually transmitted infection (STI) each year. Although public attention has been riveted on AIDS for a decade, other STIs, such as chlamydia and genital warts, pose wider threats.

BACTERIAL INFECTIONS

Bacteria are one-celled microorganisms that cause many illnesses.

■ **Gonorrhea**
Gonorrhea is caused by the gonococcus bacterium. For men, symptoms include a penile discharge and burning urination. Most women are asymptomatic. If left untreated, gonorrhea can attack the internal reproductive organs and lead to PID in women. Gonorrhea is treated with antibiotics.

■ **Syphilis**
Syphilis is caused by the *Treponema pallidum* bacterium. Syphilis undergoes several stages of development.

Although it can lie dormant for many years, it may also be lethal. Syphilis is treated with antibiotics.

■ Chlamydia

Chlamydia or chlamydial infections are caused by the *Chlamydia trachomatous* bacterium. The symptoms of chlamydial infections resemble those of gonorrhea but tend to be milder. Chlamydial infections also respond to antibiotics.

VAGINAL INFECTIONS

Vaginitis is usually characterized by a foul-smelling discharge, genital irritation, and burning during urination. Most cases involve bacterial vaginosis, candidiasis, or trichomoniasis.

■ Bacterial Vaginosis

Bacterial vaginosis is usually caused by the *Gardnerella vaginalis* bacterium.

■ Candidiasis

Candidiasis is caused by a yeast-like fungus, *Candida albicans*. Infections usually arise from changes in the vaginal environment that allow the fungus to overgrow.

■ Trichomoniasis

"Trich" is caused by a protozoan called *Trichomonas vaginalis*.

VIRAL INFECTIONS

Viruses are particles of DNA that reproduce by invading a body cell and directing the cell's own reproductive machinery to spin off new viral particles.

■ HIV Infection and AIDS

The number of people worldwide who are infected with HIV (human immunodeficiency virus) stands at about 40 million. AIDS is caused by HIV, which attacks the body's immune system. As HIV disables the body's natural defenses, the person becomes vulnerable to opportunistic diseases—such as serious infections and cancers—that are normally held in check. HIV is a blood-borne virus that is also found in semen, vaginal secretions, and breast milk. Common avenues of transmission include sexual intercourse, transfusion with contaminated blood, sharing a hypodermic needle with an infected person, childbirth, and breast-feeding. HIV infection is usually diagnosed through blood, saliva, and urine tests that detect HIV antibodies. There is no cure for AIDS, and an effective, safe vaccine has not yet been developed. The most effective form of treatment as this book goes to press is HAART, which includes a protease inhibitor along with other antiviral agents.

■ Gential Herpes

Oral herpes is caused by the *Herpes simplex* virus type 1 (HSV-1). Genital herpes is caused by the *Herpes simplex* virus type 2 (HSV-2), which produces painful, shallow sores and blisters on the genitals. A vaccine helps prevent women from contracting genital herpes but is ineffective with men. Antiviral drugs can relieve pain and speed healing during flare-ups.

■ Viral Hepatitis

There are several types of hepatitis caused by different hepatitis viruses. Most cases of hepatitis are transmitted sexually or by contact with contaminated blood or fecal matter.

■ Genital Warts

Genital warts are caused by the human papilloma virus (HPV). HPV has been linked to cancers of the genital tract. Freezing the wart is a preferred treatment for removal of the wart, but the virus remains in the body afterward.

■ Molluscum Contagiosum

This viral STI, caused by a pox virus, brings about an outbreak of painless, raised lesions on the genitals, buttocks, thighs, or lower abdomen. The lesions usually disappear on their own without serious complications within 6 months.

ECTOPARASITIC INFESTATIONS

■ Pediculosis

Pediculosis ("crabs") is caused by pubic lice (*Pthirus pubis*). Pubic lice attach themselves to pubic hair and feed on the blood of their hosts, which often causes itching. Infestations can be treated with the prescription medication lindane or with nonprescription medications that contain pyrethrins or piperonal butoxide.

■ Scabies

Scabies (*Sarcoptes scabiei*) is a parasitic infestation with a tiny mite that causes itching. Scabies, like pubic lice, is treated with lindane.

PREVENTION OF STIs: MORE THAN SAFER SEX

Strategies for preventing STIs include abstinence, monogamy, inspecting oneself and one's partner, using latex condoms, avoiding high-risk sex, washing the genitals before and after sex, having regular medical checkups, and getting to know one's partner before engaging in sexual activity.

questions for critical thinking

1. What problems would you anticipate in discussing concern over STIs with a dating partner? Would these concerns prevent you from talking about STIs? Explain.
2. If you have no symptoms of any STI, do you think it would be wise to have your physician check out your status anyway? Why or why not?
3. Does it seem to you that this textbook's presentation of information about STIs is straightforward? Or do you think it uses some scare tactics? Conversely, do you think it understates the need for caution? Explain.
4. Do you think that AIDS is something for you to be concerned about or something that affects "other people"? Explain.
5. Why do you think that knowledge about the deadliness of AIDS and about modes of transmission of HIV does not translate into people's ceasing risky behavior?
6. Do you favor distribution of condoms in the public schools as a means of attempting to prevent transmission of HIV/AIDS? Do you support the distribution of free hypodermic needles to people who inject drugs? Explain.
7. Will reading this chapter lead to any changes in your behavior?

chapter
17
Atypical Sexual Variations

Pablo Picasso. *Head on a red background*, 1930. 26 × 21 cm. Oil on panel. © Copyright Succession Picasso/ARS, NY. Photo R. G. Ojeda. Musée Picasso, Paris, France. Credit: Réunion des Musées Nationaux/Art Resource, NY. Credit: © 2002 Estate of Pablo Picasso/Artists Rights Society (ARS), NY.

outline

Truth or Fiction?

NORMAL VERSUS DEVIANT SEXUAL BEHAVIOR

THE PARAPHILIAS

Fetishism

Transvestism

A Closer Look A Case of Transvestism

Exhibitionism

A Closer Look How to Respond to an Exhibitionist

Obscene Telephone Calling

Voyeurism

A Closer Look Shooting Private Parts in Public Places

Sexual Masochism

Sexual Sadism

Frotteurism

Other Paraphilias

Human Sexuality Online "Cybersex Addiction"—A New Psychological Disorder?

A Closer Look Sexual Addiction and Compulsivity

THEORETICAL PERSPECTIVES

Biological Perspectives

Psychoanalytic Perspectives

Learning Perspectives

Sociological Perspectives

An Integrated Perspective: The "Lovemap"

TREATMENT OF THE PARAPHILIAS

Psychotherapy

Behavior Therapy

Biochemical Approaches

Human Sexuality Online Web Sites Related to Atypical Sexual Variations

SUMMING UP

QUESTIONS FOR CRITICAL THINKING

Truth or Fiction?

____ King Henry III of France insisted on being considered a woman and on being addressed as "Her Majesty."

____ Nude sunbathers are exhibitionists.

____ People who enjoy watching their mates undress are voyeurs.

____ Exhibitionists and voyeurs are never violent.

____ Some people cannot become sexually aroused unless they are bound, flogged, or humiliated by their sex partners.

____ It is considered normal to enjoy some mild forms of pain during sexual activity.

____ There is a subculture in the United States in which sexual sadists and sexual masochists form liaisons to inflict and receive pain and humiliation during sexual activity.

Aimee's life is mostly nothing but net. Aimee is a 29-year-old accountant who lives on the coast of North Carolina. She looks like the girl next door, with her red hair and freckles, but 5,000 or so strangers stop by for an intimate encounter every day. Not at her house, but at her Web site, www.acamgirl.com. Aimee is one of the Web's more familiar faces (Figure 17.1). Life in front of the four webcams in her house is her way of life. "I don't regret it for a second," she says (cited in Taylor, 2000).

Why is Aimee such a draw? She alerts the visitor to her Web site (or should we say "warns"?) that this is not an "adult site." There is no nudity, no sex. The cameras update the pictures twice a minute, and most of the pictures are downright boring. Her more exciting activities are rolling around the floor with her dogs, playing the guitar, and working at her computer. It is just as typical to find Aimee lying on her sofa.

But it isn't all watching; it's interactive. The visitor can chat with Aimee via microphone while he (or she) watches. The site also has "Aimee's own writings, including her diary ("Daily Scribbles"), and the e-mail comments of some of the visitors (the "Freak of the Week" feature). " sample commentary from a "freak":

> Aimee is the biggest male basher I have ever known. Her little web picture has her showing mega (and boy do I mean MEGA!!!!!) cleavage, when I asked her to show it, she got all bent out of shape. If you advertise something. . . .you better be able to back it up. . . . . . .got it!!! Do you see something wrong here? Really. . . . I'm not a hating person, I like to live and have fun, but when you have a person like Aimee always judging and putting people down . . . it makes you want to reach through the screen and choke the living shit out of her!!!!!

Why does Aimee do it? What motivates her? Why does she take the chance of arousing people who write this kind of e-mail? Living with the endless voyeurism is the price Aimee says she pays for the chance to have people read her writings. "You draw them in with the camera, and then they'll stay and read your stuff," she explains

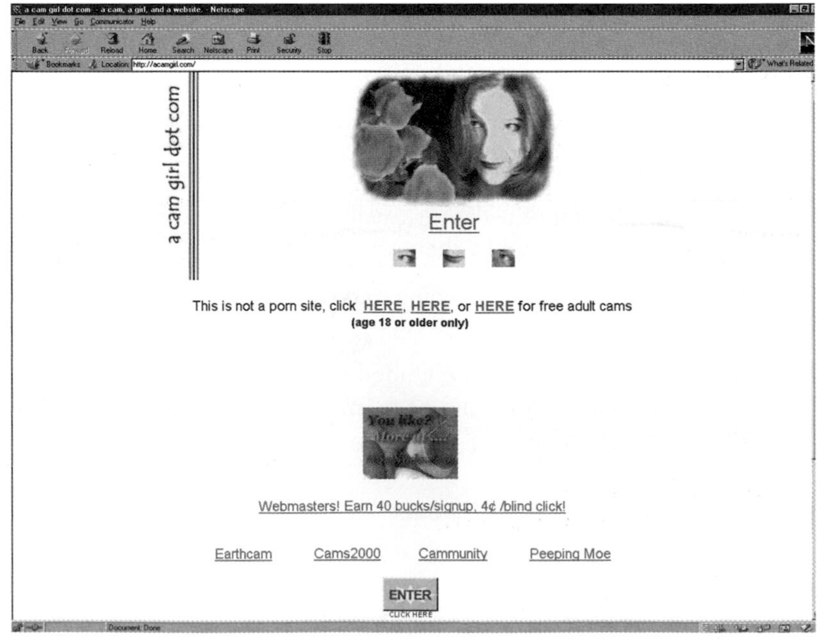

FIGURE 17.1

Aimee's Home Page, www.acamgirl.com. Aimee has some 5,000 visitors who drop in to watch her every day. Is she an *exhibitionist?* Are her electronic visitors *voyeurs?* What are the true definitions of these terms? What are the origins of exhibitionistic and voyeuristic behavior? (Reprinted by permission.)

TABLE 17.1

Percentage of adult Americans who say they would allow a TV show to film them . . .			
In pajamas	31%	Drunk	16%
Kissing	29%	Naked	8%
Crying	26%	Having sex	5%
Having an argument	25%		

Source: Yankelovich Partners telephone poll of 1,218 adult Americans for TIME/CNN. Reported in James Poniewozik et al. (2000, June 26). "We Like to Watch: Led by the Hit *Survivor*, Voyeurism Has Become TV's Hottest Genre. Why the Passion for Peeping?" *Time*, 155 (26); 56–62.

(Taylor, 2000). Of course, she also garners a great deal of attention, as in *Time* magazine and in this book.

Perhaps Aimee's motivation is, uh, literary, but there are thousands of Web sites where people—male, female, couples, trios, etc.—are all too happy to expose their bodies and engage in sexual acts while the visitor watches. Table 17.1 shows that there are a number of people in the United States who would not mind engaging in sexual activities in front of a camera.

Many webcam sites are adult pay sites, where visitors pay a fee and ask the person (usually female) in front of the webcam to engage in various kinds of sexual acts. The motivation of most of the people in front of the camera is simple: money.

The case of Aimee raises a number of questions. Aimee exposes aspects of her life to thousands of strangers. Could she be considered an *exhibitionist?* And what of those who stop by at her Web site and the Web sites of others with webcams? Are they *voyeurs?* Is what Aimee and her visitors are doing "normal," or is it abnormal or deviant? In this chapter we explore a number of sexual and sex-related behaviors that deviate from the norm in one sense or another. But first, let us consider in depth the question of how we define *normality*.

NORMAL VERSUS DEVIANT SEXUAL BEHAVIOR

One common approach to defining normality is based on a statistical norm. From this perspective, rare or unusual sexual behaviors are considered abnormal or deviant. The statistical approach may seem value-free, because the yardstick of normality is based on the frequency of behavior, not on any judgment of its social acceptability. Engaging in coitus while standing, or more than seven times a week, might be considered deviant by this standard. The choice of behaviors we subject to statistical comparison is not divorced from our underlying values, however. Sexual practices such as humming tunes from Rodgers and Hammerstein musicals while making love may be statistically infrequent (at least for people born after 1960) but would not be considered aberrant. *Statistical* infrequency, then, is not a sufficient criterion for classifying behavior as abnormal or deviant. We must also consider whether the sexual practice deviates from a *social* norm.

What is considered normal in one culture or at a particular time may be considered abnormal in other cultures and at other times. A gay male or lesbian sexual orientation was considered abnormal throughout most of Western history and was labeled a mental disorder by the American Psychiatric Association. But in 1973, a gay male or lesbian sexual orientation was dropped as a mental disorder from the association's official diagnostic manual. What is "normal" behavior for the female adolescent Trobriand islander (see Chapter 1) might be considered deviant—even *nymphomaniacal*—by the standards of Western culture.

In our own culture, sexual practices such as oral sex and masturbation were once considered deviant or abnormal. Today, however, they are practiced so widely in our society that few people would label them as deviant practices. Concepts of "normalcy" and "deviance," then, reflect the mores and customs of a particular culture at a given time.

Another way to determine sexual deviance is to classify sexual practices as deviant when they involve the persistent preference for nongenital sexual outlets (Seligman & Hardenburg, 2000). If a man prefers fondling a woman's panties to engaging in sexual relations with her, or prefers to masturbate against her foot rather than engaging in coitus, his behavior is likely to be labeled deviant.

Because of the confusing array of meanings of the terms *deviant* and *abnormal*, we prefer to speak about unusual patterns of sexual arousal or behavior as "atypical variations" in sexual behavior rather than as "sexual deviations." Atypical patterns of sexual arousal or behavior that become problematic in the eyes of the individual or of society are labeled *paraphilias* by the American Psychiatric Association (2000). Clinicians consider paraphilias to be mental disorders. But milder forms of these behaviors may be practiced by many people and fall within the normal spectrum of human sexuality (Brody, 1990a).

THE PARAPHILIAS

Paraphilias involve sexual arousal in response to unusual stimuli such as children or other nonconsenting persons (including unsuspecting people whom one watches or to whom one exposes one's genitals), nonhuman objects (such as shoes, leather, rubber, or undergarments), or pain or humiliation (Seligman & Hardenburg, 2000). The psychiatric diagnosis of paraphilia requires that the person has acted on the urges or is distinctly distressed by them.

People with paraphilias usually feel that their urges are insistent, demanding, or compulsory (Seligman & Hardenburg, 2000). They may describe themselves as overcome by these urges now and then. People with paraphilias tend to experience their urges as beyond their control, just as drug addicts and compulsive gamblers see themselves as helpless to avert irresistible urges. For these reasons, theorists have speculated that paraphilias may represent a type of sexual compulsion or an addiction.

Paraphilias vary in severity. In some cases the person can function sexually in the absence of the unusual stimuli and seldom if ever acts on his or her deviant urges. In other cases the person resorts to paraphilic behavior only in times of stress. In more extreme forms, the person repeatedly engages in paraphilic behavior and may become preoccupied with thoughts and fantasies about these experiences. In such cases the person may not be able to become sexually aroused without either fantasizing about the paraphilic stimulus or having it present. For some people, paraphilic behavior is the only means of attaining sexual gratification.

The person with a paraphilia typically replays the paraphilic act in sexual fantasies to stimulate arousal during masturbation or sexual relations. It is as though he or she is mentally replaying a videotape of the paraphilic scene. The scene grows stale after a while, however. According to sex researcher John Money, "the tape wears out and he has to perform another paraphilic act, in effect, to create a new movie" (quoted in Brody, 1990a, p. C12).

Some paraphilias are generally harmless and victimless, such as *fetishism* and cross-dressing to achieve sexual arousal (*transvestic fetishism*). Even being humiliated by one's partner may be relatively harmless if the partner consents. Other paraphilic behaviors, such as exposing oneself in public or enticing children into sexual relations, do have victims and may cause severe physical or psychological harm. They are also against the law. Sexual sadism, in which sexual arousal is connected to hurting or humiliating another person, can be a most harmful paraphilia when it is forced upon a nonconsenting person. Some brutal rapes involve sexual sadism.

Paraphilia A diagnostic category used by the American Psychiatric Association to describe atypical patterns of sexual arousal or behavior that become problematic in the eyes of the individual or society, such as fetishism and exhibitionism. The urges are recurrent and either are acted on or are distressing to the individual. (From Greek roots that means "to the side of" [*para-*] and "loving" [*philos*].)

Except in the case of sexual masochism, paraphilias are believed to occur almost exclusively among men (Seligman & Hardenburg, 2000). The prevalence of paraphilias in the general population remains unknown, because people are generally unwilling to talk about them. Much of what we have learned about paraphilias derives from the reported experiences of people who have been apprehended for performing illegal acts (such as exposing themselves in public) and the few who have voluntarily sought help. The characteristics of people who have not been identified or studied remain virtually unknown.

In the chapter, we discuss all the major types of paraphilia except *pedophilia.* In pedophilia, children become the objects of sexual arousal. Pedophilia often takes the form of sexual coercion of children, as in incest or sexual molestation. It is discussed in Chapter 18, as a form of sexual coercion.

Fetishism

The roots of the word *fetish* come from the French *fétiche,* which is thought to derive from the Portuguese *feitico,* meaning "magic charm." The "magic" in this case lies in an object's ability to arouse a person sexually. In **fetishism,** an inanimate object elicits sexual arousal. Articles of clothing (for example, women's panties, bras, lingerie, stockings, gloves, shoes, or boots) and materials made of rubber, leather, silk, or fur are among the more common fetishistic objects. Leather boots and high-heeled shoes are especially popular.

The fetishist may act on the urges to engage in fetishistic behavior, such as masturbating by stroking an object or while fantasizing about it, or he may be distressed about such urges or fantasies but not act on them. In a related paraphilia, **partialism,** people are excessively aroused by a particular body part, such as the feet, breasts, or buttocks.

Most fetishes and partialisms are harmless. Fetishistic practices are nearly always private and involve masturbation or are incorporated into coitus with a willing partner. Only rarely have fetishists coerced others into paraphilic activities. Yet some partialists have touched parts of women's bodies in public. And some fetishists have committed burglaries to acquire the fetishistic objects (Sargent, 1988).

Fetishism. In fetishism, inanimate objects such as leather shoes or boots, or parts of the body such as feet, elicit sexual arousal. Many fetishists cannot achieve sexual arousal without having contact with the desired objects or fantasizing about them. Women's undergarments and objects made of rubber, leather, silk, or fur are common fetishistic objects. Why is it that some men are more aroused by women's feet than by women's genital organs?

Fetishism A paraphilia in which an inanimate object such as an article of clothing or items made of rubber, leather, or silk elicit sexual arousal.

Partialism A paraphilia related to fetishism in which sexual arousal is exaggeratedly associated with a particular body part, such as feet, breasts, or buttocks.

Transvestism

Transvestism may be viewed as a type of fetish. Whereas other fetishists become sexually aroused by handling the fetishistic object while they masturbate, transvestites become excited by wearing articles of clothing—the fetishistic objects—of the other gender. A fetishist may find the object or sex involving the object to be erotically stimulating. For the transvestite, the object is sexually alluring only when it is worn. Transvestites are nearly always males. True transvestism has been described only among heterosexual males, although such men may occasionally engage in male–male sexual activity. Most are married and are otherwise masculine in behavior and dress.

Transvestism is often confused with transsexualism, but there are important differences between them. Transvestites cross-dress because they find it sexually arousing. Most transvestites have masculine gender identities and do not seek to change their anatomic sex. Some male transvestites show some aspects of a feminine gender identity, however (Doorn et al., 1994). Transsexuals, by contrast, cross-dress because they are uncomfortable with the attire associated with their anatomic sex.

As is true of fetishism in general, the origins of transvestism remain obscure. There is no evidence of biological abnormalities in transvestism (Buhrich et al., 1979). Family relationships appear to play a role, however. Transvestites are more likely than other people to be oldest children or only children (Schott, 1995). They also report closer relationships with their mothers than with their fathers (Schott, 1995). Some transvestites report a history of "petticoat punishment" during childhood. That is, they were humiliated by being dressed in girl's attire. Some authorities have speculated that the adult transvestite might be attempting psychologically to convert humiliation into mastery by achieving an erection and engaging in sexual activity despite being attired in female clothing (Geer et al., 1984). Transvestism can also be looked at as an attempt by males to escape the narrow confines of the masculine role (Bullough, 1991).

Cross-dressing is common in other cultures and has been reported in historical accounts of figures such as King Henry III of France. This 16th-century monarch wanted to be considered a woman and to be addressed as "Her Majesty" (Geer et al., 1984). But cross-dressing may occur in other cultures for reasons other than sexual arousal. In the case of Henry III, it appears that transsexualism, and not transvestism, was involved.

Transvestism. Transvestites cross-dress for purposes of obtaining sexual arousal and gratification. Are women who wear blue jeans engaging in transvestic activity?

Transvestism A paraphilia in which a person repeatedly cross-dresses to achieve sexual arousal or gratification or is troubled by persistent, recurring urges to cross-dress. (From the Latin roots *trans-*, which means "cross," and *vestis,* which means "garment.") Also known as *transvestic fetishism.*

a CLOSER look

A Case of Transvestism

Most transvestites are married and engage in sexual activity with their wives. Yet they seek additional sexual gratification through dressing as women, as in the case of Archie.

Archie was a 55-year-old plumber who had been cross-dressing for many years. There was a time when he would go out in public as a woman, but as his prominence in the community grew, he became more afraid of being discovered in public. His wife Myrna knew of his "peccadillo," especially because he borrowed many of her clothes. She urged him to stay at home, offering to help him with his "weirdness." For many years his paraphilia had been restricted to the home.

The couple came to the clinic at the urging of the wife. Myrna described how Archie had imposed his will on her for 20 years. Archie would wear her undergarments and masturbate while she told him how disgusting he was. (The couple also regularly engaged in "normal" sexual intercourse, which Myrna enjoyed.) The cross-dressing had come to a head because a teenage daughter had almost walked into the couple's bedroom while they were acting out Archie's fantasies.

With Myrna out of the consulting room, Archie explained how he grew up in a family with several older sisters. He described how underwear had been perpetually hanging to dry in the one bathroom. As an adolescent Archie experimented with rubbing against articles of underwear and then with trying them on. On one occasion a sister walked in while he was modeling panties before the mirror. She told him he was a "dredge to society," and he straightaway experienced unparalleled sexual excitement. He masturbated when she left the room, and his orgasm was the strongest of his young life.

Archie did not think that there was anything wrong with wearing women's undergarments and masturbating. He was not about to give it up, regardless of whether it destroyed his marriage. Myrna's main concern was finally separating herself from Archie's "sickness." She didn't care what he did any more, so long as he did it by himself. "Enough is enough," she said.

That was the compromise the couple worked out. Archie would engage in his fantasies by himself. He would do so when Myrna was not at home, and she would not be told of his activities. He would also be very, very careful to choose times when the children would not be around.

Six months later the couple were together and content. Archie had replaced Myrna's input into his fantasies with transvestic-sadomasochistic magazines. Myrna said, "I see no evil, hear no evil, smell no evil." They continued to have sexual intercourse. After a while, Myrna forgot to check to see which underwear had been used.

Truth or Fiction? REVISITED

It is true that King Henry III of France insisted on being considered a woman and on being addressed as "Her Majesty." The king appears to have been transsexual. ■

Some men cross-dress for reasons other than sexual arousal and so are not "true transvestites." Some men make a living by impersonating women like Marilyn Monroe and Madonna on stage and are not motivated by sexual arousal. Among some segments of the gay male community, it is fashionable to masquerade as women. Gay men do not usually cross-dress to become sexually stimulated, however.

Transvestism behaviors may range from wearing a single female garment when alone to sporting dresses, wigs, makeup, and feminine mannerisms at a transvestite club. Some transvestites become sexually aroused by masquerading as women and attracting the interest of unsuspecting males. They sometimes entice these men or string them along until they find some excuse to back out before their anatomic sex is revealed. The great majority of transvestites do not engage in antisocial or illegal behavior. Most practice their sexual predilection in private and would be horrified or embarrassed to be discovered by associates while dressed in female attire.

Although some transvestites persuade their female partners to permit them to wear feminine attire during their sexual activities, most keep their transvestic urges and activities to themselves. A survey of 504 transvestite men showed that most had kept their transvestism a secret from their wives-to-be, hoping not to be bothered by their urge to cross-dress once they were married (Weinberg & Bullough, 1986, 1988). The urges continued into their marriages, and the wives eventually discovered their husbands' secrets, however. Seventy of the wives were interviewed (Bullough & Weinberg, 1989). The wives tended to react with confusion, surprise, or shock to discovering their husbands' cross-dressing. Most tried to be understanding at first. Some assisted their husbands in their cross-dressing, such as by helping them apply makeup. Yet the longer the women were married, the more negative their attitudes tended to become toward their husbands' cross-dressing. Over time, wives generally learned to be tolerant, though not supportive, of their husbands' cross-dressing.

Exhibitionism

Exhibitionism ("flashing") entails persistent, powerful urges and sexual fantasies involving exposing one's genitals to unsuspecting strangers for the purpose of achieving sexual arousal or gratification. The urges are either acted on or are disturbing to the individual. Exhibitionists are nearly always males.

What we know of exhibitionists, as with most other people with paraphilias, is almost entirely derived from studies of men who have been apprehended or have been treated by mental health professionals. Such knowledge may yield a biased picture of exhibitionists. Although about one in three arrests for sexual offenses involves exhibitionism, relatively few reported incidents result in apprehension and conviction (Cox, 1988). Studies in England, Guatemala, the United States, and Hong Kong show that fewer than 20% of occurrences are reported to the police (Cox, 1988). The characteristics of most perpetrators may thus differ from those of people who have been available for study.

The prevalence of exhibitionism in the general population is unknown, but a survey of 846 college women at nine randomly selected U.S. universities found exposure to exhibitionism to be widespread. A third of the women reported that they had run into a "flasher" (Cox, 1988). A majority of the women had been approached for the first time (some had been approached more than once) by 16 years of age. Only 15 of the women had reported these incidents to the police. The clinical definition of exhibitionism involves exposure to a stranger, but about a third (36%) of the incidents among the college women were committed by acquaintances, relatives, or "good friends."

The archetypal exhibitionist is young, unhappily married, and sexually repressed. An exhibitionist may claim that marital coitus is reasonably satisfactory but that he also experiences the compulsion to expose himself to strangers. Many exhibitionists are single, however. They typically have difficulty relating to women and have been unable to establish meaningful heterosexual relationships.

Exhibitionism usually begins before age 18 (American Psychiatric Association, 2000). The urge to exhibit oneself, if not the actual act, usually begins in early adolescence, generally between the ages of 13 and 16 (Freund et al., 1988). The frequency of exhibitionism declines markedly after the age of 40 (American Psychiatric Association, 2000). The typical exhibitionist does not attempt further sexual contact with the victim. Thus he does not usually pose a physical threat (American Psychiatric Association, 2000).

The police may sometimes trivialize exhibitionism as a "nuisance crime," but the psychological consequences among victims, especially young children, indicate that exhibitionism is not victimless. Victims may feel violated and may be bothered by recurrent images or nightmares. They may harbor misplaced guilt that they had unwittingly enticed the exhibitionist. They may blame themselves for reacting excessively or for failing to apprehend the perpetrator. They may also develop fears of venturing out on their own.

Geer and his colleagues (1984) see exhibitionism as an indirect means of expressing hostility toward women. Exposing himself may be an attempt by the exhibitionist to strike back at women because of a belief that women have wronged him or damaged his self-esteem by failing to notice him or take him seriously. The direct expression of anger may be perceived as too risky, so the exhibitionist vents his rage by humiliating a defenseless stranger. The urge to expose oneself almost always follows a situation in which the exhibitionist feels that his masculinity has been insulted. Some evidence suggests that exhibitionists may be attempting to assert their masculinity by evoking a response from

Exhibitionism A paraphilia characterized by persistent, powerful urges and sexual fantasies that involve exposing one's genitals to unsuspecting strangers for the purpose of achieving sexual arousal or gratification.

their victims. A number of exhibitionists have reported that they hoped that the women would enjoy the experience and be impressed with the size of their penises (Langevin et al., 1979).

Other studies show exhibitionists to be shy, dependent, passive, lacking in sexual and social skills, and even inhibited (Dwyer, 1988). They tend to be self-critical, to have doubts about their masculinity, and to suffer from feelings of inadequacy and inferiority (Blair & Lanyon, 1981; Dwyer, 1988). Many have had poor relationships with their fathers and have overprotective mothers (Dwyer, 1988). Exhibitionists who are socially shy or inadequate may be using exhibitionism as a substitute for the intimate relationships they cannot develop.

The preferred victims are typically girls or young women (Freund & Blanchard, 1986). The typical exhibitionist drives up to, or walks in front of, a stranger and exposes his penis. In one sample of 130 exhibitionists, about 50% reported that they always or nearly always had erections when they exposed themselves (Langevin et al., 1979). After his victim has registered fear, disgust, confusion, or surprise, an exhibitionist typically covers himself and flees. He usually masturbates, either while exposing himself or shortly afterward while thinking about the act and the victim's response (American Psychiatric Association, 2000; Blair & Lanyon, 1981). Some exhibitionists ejaculate during the act. Most of the 238 exhibitionists in a Canadian study reported masturbating to orgasm while exposing themselves or afterward while fantasizing about it (Freund et al., 1988).

Exhibitionists may also need to risk being caught to experience a heightened erotic response (Stoller, 1977). The exhibitionist may even situate himself in such a way as to increase the risk. For example, he may expose himself in the same location or while sitting in his own, easily identifiable car.

Courts nowadays tend to be hard on exhibitionists, partly because of evidence that shows that some exhibitionists progress to more serious crimes of sexual aggression. In one sample, about 10% of rapists and child molesters had begun their "sexual careers" by exposing themselves to strangers (Abel et al., 1984). This does not mean that exhibitionists inevitably become rapists and child molesters. Most do not.

Definitions of exhibitionism also bring into focus the boundaries between normal and abnormal behavior. Are exotic dancers (stripteasers) or nude sunbathers exhibitionists? After all, aren't they also exposing themselves to strangers? But exotic dancers—male or female—remove their clothes to sexually excite or entertain an audience that is paying to watch them. Their motive is (usually) to earn a living. Sunbathers in their "birthday

a CLOSER look

How to Respond to an Exhibitionist

It is understandable that an unsuspecting woman who is exposed to an exhibitionist may react with shock, surprise, or fear. Unfortunately, her display of shock or fear may reinforce the flasher's tendencies to expose himself. She may fear that the flasher, who has already broken at least one social code, is likely to assault her physically as well. Fortunately, most exhibitionists do not seek actual sexual contact with their victims and run away before they can be apprehended by the police or passers-by.

Some women respond with anger, insults, and even arguments that the offender should feel ashamed. A display of anger may reinforce exhibitionism. We do not recommend that the victim insult the flasher, lest this provoke a violent response. Although most exhibitionists are nonviolent, about 1 in 10 has considered or attempted rape (Gebhard et al., 1965).

When possible, showing no reaction or simply continuing on one's way may be the best response. If women do desire to respond to the flasher, they might calmly say something like "You really need professional help. You should see a professional to help you with this problem." All should promptly report the incident to police, so that authorities can apprehend the offender.

suits" may also seek to sexually arouse others, not themselves. Of course, they may also be seeking an all-over tan or trying to avoid feeling encumbered by clothing. In any case, stripteasers and sunbathers do not expose themselves to unsuspecting others. Thus these behaviors are not regarded as exhibitionistic.

It is not true that nude sunbathers are exhibitionists—at least not in terms of the clinical definition of the disorder. Exhibitionists seek to become sexually aroused by exposing themselves to unsuspecting victims, not to show off their physical attractiveness.

It is also normal to become sexually excited while stripping before one's sex partner. Such stripping is done to excite a willing partner, not to surprise or shock a stranger.

Obscene Telephone Calling

Like exhibitionists, obscene phone callers (nearly all of whom are male) seek to become sexually aroused by shocking their victims. Whereas an exhibitionist exposes his genitals to produce the desired response, the obscene phone caller exposes himself verbally by uttering obscenities and sexual provocations to a nonconsenting person. Because of such similarities, obscene telephone calling is sometimes considered a subtype of exhibitionism. The American Psychiatric Association (2000) labels this type of paraphilia **telephone scatologia** (lewdness).

Relatively few obscene callers are women (Matek, 1988). Women who are charged with such offenses are generally motivated by rage for some actual or fantasized rejection rather than by the desire for sexual arousal. They use the phone to hurl sexual invectives against men whom they feel have wronged them. By contrast, male obscene phone callers are generally motivated by a desire for sexual excitement and usually choose their victims randomly from the phone book or by chance dialing. They typically masturbate during the phone call or shortly afterward. Despite the offensiveness of their actions, most obscene phone callers are not dangerous. Nor do most make repeat calls to the same person (Reinisch, 1990).

There are many patterns of obscene phone calling (Matek, 1988). Some callers limit themselves to obscenities. Others make sexual overtures. Some just breathe heavily into the receiver. Others describe their masturbatory activity to their victims. Some profess to have previously met the victim at a social gathering or through a mutual acquaintance. Some even present themselves as "taking a sex survey" and ask a series of personally revealing questions.

The typical obscene phone caller is a socially inadequate heterosexual male who has had difficulty forming intimate relationships with women. The relative safety and anonymity of the telephone may shield him from the risk of rejection. A reaction of shock or fright from his victims may fill him with feelings of power and control that are lacking in his life, especially in his relationships with women. The obscenities may vent the rage that he holds against women who have rejected him.

Obscene phone calls are illegal, but it has been difficult for authorities to track down perpetrators (Matek, 1988). Call tracing can help police track obscene or offending phone callers. Call tracing works in different ways in different locales. *Caller ID*, which reveals the caller's telephone number, is available in many places. Though this service may deter some obscene callers, others may use public phones instead of their home phones. Callers may also be able to block the display of their telephone numbers electronically. Check with your local telephone company if you are interested in these services.

Telephone scatologia A paraphilia characterized by the making of obscene telephone calls. (From the Greek *skatos,* which means "excrement.")

What should a woman do if she receives an obscene phone call? Advice generally parallels that given to women who are victimized by exhibitionists. Above all, women are advised to remain calm and not to reveal shock or fright, because such reactions tend to reinforce the caller and increase the probability of repeat calls. Women may be best advised to say nothing at all and gently hang up the receiver. A woman might alternatively offer a brief response that alludes to the caller's problems before hanging up. She might say in a calm but strong voice, "It's unfortunate that you have this problem. I think you should seek professional help." If she should receive repeated calls, the woman might request an unlisted number or contact the police about tracing the calls. Many women list themselves only by their initials in the phone directory so as to disguise their gender. But this practice is so widespread that obscene callers may assume that people listed by initials are women living alone.

Voyeurism

Voyeurism involves strong, repetitive urges to observe unsuspecting strangers who are naked, disrobing, or engaged in sexual relations (American Psychiatric Association, 2000). The voyeur becomes sexually aroused by the act of watching and typically does not seek sexual relations with the observed person. Like fetishism and exhibitionism, voyeurism is found almost exclusively among males. It usually begins before the age of 15 (American Psychiatric Association, 2000).

The voyeur may masturbate while peeping or afterward while replaying the incident in his imagination or engaging in voyeuristic fantasies. The voyeur may fantasize about making love to the observed person but have no intention of actually doing so.

Are people voyeurs who become sexually aroused by the sight of their lovers undressing? What about people who enjoy watching pornographic films or stripteasers? No, no, and no. The people being observed are not unsuspecting strangers. The lover knows that his or her partner is watching. Porn actors and strippers know that others will be viewing them. They would not be performing if they did not expect or have an audience.

It is perfectly normal for men and women to be sexually stimulated by the sight of other people who are nude, undressing, or engaged in sexual relations. Voyeurism is characterized by urges to spy on *unsuspecting* strangers.

It is not true that people who enjoy watching their mates undress are voyeurs. In such cases the person who is disrobing is knowingly and willingly observed, and the watcher's enjoyment is completely normal. Voyeurs target unsuspecting victims.

Voyeurs are also known as "peepers" or *peeping Toms.* Why "peeping Toms"? According to English legend, Lady Godiva asked the townspeople not to look upon her while she rode horseback in the nude to protest the oppressive tax that her husband, a landowner, had imposed on them. A tailor named Tom of Coventry was the only townsperson not to honor her request.

Voyeurs often put themselves in risky situations in which they face the prospect of being discovered or apprehended. They may risk physical injury by perching in trees or otherwise assuming precarious positions to catch a preferred view of their target. They will occupy rooftops and fire escapes in brutal winter weather. Peepers can be exceedingly patient in their outings. They may wait hour after hour, night after night, for a furtive glimpse of an unsuspecting person. One 25-year-old, recently married man secreted himself in his mother-in-law's closet, waiting for her to disrobe. Part of the sexual excitement seems to stem from the risks voyeurs run. The need for elements of risk in their voyeurism may explain why voyeurs are not known to frequent nude beaches or nudist camps where it is acceptable to look (though not to stare) at others.

Although most voyeurs are nonviolent, some commit violent crimes such as assault and rape (Langevin et al., 1985). Voyeurs who break into and enter homes or buildings, or who tap at windows to gain the attention of victims, are among the more dangerous.

Voyeurism A paraphilia characterized by strong, repetitive urges and related sexual fantasies of observing unsuspecting strangers who are naked, disrobing, or engaged in sexual relations. (From the French *voir,* which means "to see.")

a CLOSER look

Shooting Private Parts in Public Places

Police have this warning for Washington, D.C., women: Beware of geeks bearing tiny cameras.

In what they describe as a growing trend, police are beginning to catch video voyeurs trying to shoot private parts in public places. These men are aiming the latest compact camcorders up women's skirts in crowded stores and shopping malls, parks, and fairs—and often posting the pictures on the Internet.

One weekend, Fairfax County, Virginia, police arrested a 21-year-old man who was holding a palm-sized video camera under a woman's dress at a Tower Records store. Two weeks earlier at a Hecht's department store, Alexandria, Virginia, police nabbed a 19-year-old man toting a bulky VHS video camera. He was angling for similar shots in the china department, they said. At last year's Fairfax Fair, a man was arrested for videotaping from a camera bag dangling on a long strap down to his ankles.

"I had one guy who was doing it on Metro trains," said Alexandria Detective Harold Duquette, who tracked the suspect's movements across the region by watching days' worth of his videotapes. "I mean, he was in D.C. (District of Columbia). He was in tunnels. He was sitting on benches. He had one lady reading the newspaper."

"Upskirt Sites"

What began as a small photo gallery on the Internet a couple of years ago has rapidly expanded into a multitude of "Upskirt" sites, including one devoted entirely to shots taken up skirts in Maryland, said Duquette.

Through the Internet, many of these video peepers learn about new techniques and exchange stories, Duquette said. The most popular method, he said, is concealing a small video camera in a shoulder bag, with the lens pointing out of the top. The bag is either dangled under a woman's skirt or set on the ground next to her.

A Lack of Laws

Because the phenomenon is so new, there are few laws on the books to deal with it. Officials in Washington, D.C., and Maryland say they do not have a law that specifically fits the crime, unless the person assaults the victims or stalks them. Peeping Tom statutes apply to peering into dwellings, not up skirts.

In Virginia, police are using a relatively new state law—"unlawful filming, videotaping or photographing of another"—to prosecute people caught putting their lenses where they're not wanted. The statute, a misdemeanor with a maximum penalty of 12 months in jail, was written in 1994 and applies to circumstances in which victims had a "reasonable expectation of privacy," according to prosecutors.

Police believe there will be more arrests for this type of offense as people become aware of the problem. Duquette says that security guards, who are on the lookout for shoplifters, are becoming adept at spotting the peeping Toms. At Landmark Mall in Alexandria, the man with the VHS video camera strapped to his shoulder at Hecht's caught a security guard's attention.

It is incorrect to say that exhibitionists and voyeurs are *never* violent. Exhibitionistic and voyeuristic activities per se do not involve outright violence, but some exhibitionists and voyeurs have been known to be violent, and if provoked or angered, they may react violently.

Compared to other types of sex offenders, voyeurs tend to be less sexually experienced and are less likely to be married (Gebhard et al., 1965). Like exhibitionists, voyeurs tend to harbor feelings of inadequacy and to lack social and sexual skills (Dwyer, 1988). They may thus have difficulty forming romantic relationships with women. For this shy and socially inadequate type of voyeur, "peeping" affords sexual gratification without the risk of rejection. Yet not all voyeurs are socially awkward and inept with women.

Sexual masochism A paraphilia characterized by the desire or need for pain or humiliation to enhance sexual arousal so that gratification may be attained. (From the name of Leopold von Sacher-Masoch.)

Sexual Masochism

Although pleasure and pain may seem like polar opposites, some people experience sexual pleasure through having pain or humiliation inflicted on them by their sex partners. People who associate the receipt of pain or humiliation with sexual arousal are called **sexual masochists.** A sexual masochist either acts on or is distressed by persistent urges

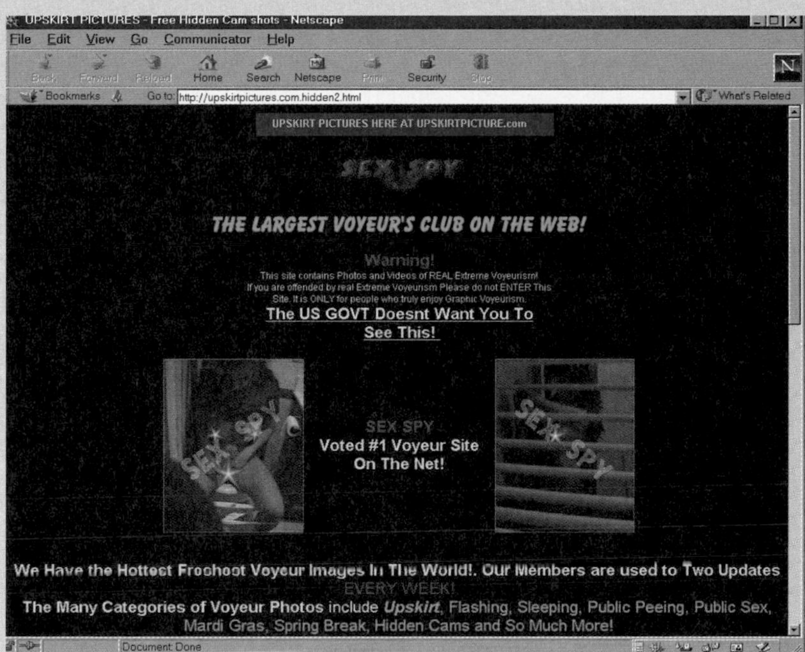

Shooting Private Parts in Public Places. In places like the malls in the suburbs surrounding Washington, D.C., police are beginning to catch people who are trying to shoot private parts in public places. They aim compact camcorders up women's skirts in crowded stores and shopping malls, parks, and fairs. Sometimes they post the pictures on the Internet. Often the pictures wind up for sale on sex sites.

"I observed the subject place the camera under a woman's dress," the guard, Charles Taylor, stated on the arrest warrant. "I asked the subject if he was videotaping underneath the woman's dress. The subject said yes."

Officials in many jurisdictions, including Prince William County, Virginia, say they haven't made any such arrests—and many weren't even aware of such behavior.

"Thanks for the warning," said Kim Chinn, a Prince William County police spokeswoman. "Next time I go to the mall, I'm wearing jeans."

Source: From Patricia Davis (1998, June 9). "Peeping Toms with Videocams Plague Malls." *The Washington Post online.* Copyright © 1998 by The Washington Post. Adapted by permission.

and sexual fantasies involving the desire to be bound, flogged, humiliated, or made to suffer in some way by a sexual partner so as to achieve sexual excitement. In extreme cases, the person is incapable of becoming sexually aroused unless pain or humiliation is incorporated into the sexual act.

It is true that some sexual masochists cannot become sexually aroused unless they are bound, flogged, or humiliated by their sex partners. ■

Sexual masochism is the only paraphilia that is found among women with some frequency (American Psychiatric Association, 2000). Even sexual masochism is much more prevalent among men than women, however. Male masochists may outnumber females by a margin of 20 to 1 (American Psychiatric Association, 2000).

The word *masochism* is derived from the name of the Austrian storyteller Leopold von Sacher-Masoch (1835–1895). He wrote tales of men who derived sexual satisfaction from having a female partner inflict pain on them, typically by flagellation (beating or whipping).

Sexual masochists may derive pleasure from various types of punishing experiences, including being restrained (a practice known as **bondage**), blindfolded (sensory bondage),

Bondage Ritual restraint, as by shackles, as practiced by many sexual masochists.

spanked, whipped, or made to perform humiliating acts, such as walking around on all fours and licking the boots or shoes of the sex partner or being subjected to vulgar insults. Some masochists have their partners humiliate them by urinating or defecating on them. Some masochists prefer a particular source of pain. Others seek an assortment. But we should not think that sexual masochists enjoy other types of pain that are not connected with their sexual practices. Sexual masochists are no more likely than anyone else to derive pleasure from the pain they experience when they accidentally stub their toes or touch a hot appliance. Pain has erotic value only within a sexual context. It must be part of an elaborate sexual ritual.

Sexual masochists and **sexual sadists** often form sexual relationships to meet each other's needs. Some sexual masochists enlist the services of prostitutes or obtain the cooperation of their regular sexual partners to enact their masochistic fantasies.

It may seem contradictory for pain to become connected with sexual pleasure. The association of sexual arousal with mildly painful stimuli is actually quite common, however. Kinsey and his colleagues (1953) reported that perhaps as many as one person in four has experienced erotic sensations from being bitten during lovemaking. The eroticization of mild forms of pain (love bites, hair pulls, minor scratches) may fall within the normal range of sexual variation. Pain from these sources increases overall bodily arousal, which may enhance sexual excitement. Some of us become sexually excited when our partners "talk dirty" to us or call us vulgar names. When the urge for pain for purposes of sexual arousal becomes so persistent or strong that it overshadows other sources of sexual stimulation, or when the masochistic experience causes physical or psychological harm, we may say that the boundary between normality and abnormality has been breached.

Truth or Fiction? **REVISITED**

It is in fact considered normal to enjoy some mild forms of pain during sexual activity. Love bites, hair pulls, and minor scratches are examples of sources of pain that are considered to fall within normal limits. ■

Baumeister (1988a) proposes that independent and responsible selfhood becomes burdensome or stressful at times. Sexual masochism provides a temporary reprieve from the responsibilities of independent selfhood. It is a blunting of one's ordinary level of self-awareness that is achieved by "focusing on immediate sensations (both painful and pleasant) and on being a sexual object" (Baumeister, 1988a, p. 54).

Sexual masochism can range from relatively benign to potentially lethal practices, such as **hypoxyphilia.** Hypoxyphiliacs put plastic bags over their heads, nooses around their necks, or pressure on their chests to deprive themselves of oxygen temporarily and enhance their sexual arousal. They usually fantasize that they are being strangled by a lover. They try to discontinue oxygen deprivation before they lose consciousness, but some miscalculations result in death by suffocation or strangulation (Blanchard & Hucker, 1991; Cosgray et al., 1991).

Sexual sadists People who become sexually aroused by inflicting pain or humiliation on others.

Hypoxyphilia A practice in which a person seeks to enhance sexual arousal, usually during masturbation, by becoming deprived of oxygen. (From the Greek root meaning "under" [*hypo*-].)

Sexual sadism A paraphilia characterized by the desire or need to inflict pain or humiliation on others to enhance sexual arousal so that gratification is attained. (From the name of the Marquis de Sade.)

Sexual Sadism

Sadism is named after the infamous Marquis de Sade (1774–1814), a Frenchman who wrote tales of becoming sexually aroused by inflicting pain or humiliation on others. The virtuous Justine, the heroine of his best-known novel of the same name, endures terrible suffering at the hands of fiendish men. She is at one time bound and spread-eagled so that bloodhounds can savage her. She then seeks refuge with a surgeon who tries to dismember her. Later she falls into the clutches of a saber-wielding mass murderer, but Nature saves her with a timely thunderbolt.

Sexual sadism is characterized by persistent, powerful urges and sexual fantasies involving the inflicting of pain and suffering on others to achieve sexual excitement or gratification. The urges are acted on or are disturbing enough to cause personal distress. Some sexual sadists cannot become sexually aroused unless they make their sex partners suffer. Others can become sexually excited without such acts.

Some sadists hurt or humiliate willing partners, such as prostitutes or sexual masochists. Others—a small minority—stalk and attack nonconsenting victims.

SADOMASOCHISM **Sadomasochism (S&M)** *mutually gratifying sexual interactions* between *consenting partners.* Occasional S&M is quite common among the general population. Couples may incorporate light forms of S&M in their lovemaking now and then, such as mild dominance and submission games or gentle physical restraint. It is also not uncommon for lovers to scratch or bite their partners to heighten their mutual arousal during coitus. They generally do not inflict severe pain or damage, however.

Twenty-two percent of the men and 12% of the women surveyed by Kinsey and his colleagues (1953) reported at least some sexual response to sadomasochistic stories. Although some milder forms of sadomasochism may fall within the boundaries of normal sexual variation, sadomasochism becomes pathological when such fantasies are acted on in ways that become destructive, dangerous, or distressing to oneself or others, as we find in the following case example:

Geoffrey Rush as the Marquis de Sade in the Film *Quills*. Sexual sadism is a fascinating topic in literature and among social and behavioral scientists. Why is it that some individuals do not obtain sexual gratification unless it is at the expense of others whom they subject to pain or humiliation? Why are sadists almost invariably male?

> A 25-year-old female graduate student described a range of masochistic experiences. She reported feelings of sexual excitement during arguments with her husband when he would scream at her or hit her in a rage. She would sometimes taunt him to make love to her in a brutal fashion, as though she were being raped. She found the brutality and sense of being punished to be sexually stimulating. She had also begun having sex with strange men and enjoyed being physically punished by them during sex more than any other type of sexual stimulus. Being beaten or whipped produced the most intense sexual experiences she had ever had. Although she recognized the dangers posed by her sexual behavior, and felt somewhat ashamed about it, she was not sure that she wanted treatment for "it" because of the pleasure that it provided her. (Adapted from Spitzer et al., 1989, pp. 87–88)

In one subculture, sadomasochism is the preferred or even the exclusive form of sexual gratification. People in this subculture seek one another out through mutual contacts, S&M social organizations, or personal ads in S&M magazines. The S&M subculture has spawned magazines and clubs catering to people who describe themselves as "into S&M," as well as sex shops that sell sadomasochistic paraphernalia. These include leather restraints and leather face masks that resemble the ancient masks of executioners.

It is true that there is a subculture in the United States in which sexual sadists and sexual masochists form liaisons to inflict and receive pain and humiliation during sexual activity. This *S&M* subculture is catered to by sex shops and magazines. ■

Participants in sadomasochism often engage in highly elaborate rituals involving dominance and submission. Rituals are staged as though they were scenes in a play (Weinberg et al., 1984). In the "master and slave" game, the sadist leads the masochist around by a leash. The masochist performs degrading or menial acts. In bondage and discipline (B&D), the dominant partner restrains the submissive partner and flagellates (spanks or whips) or sexually stimulates the submissive partner. The erotic appeal of bondage seems to be connected with controlling or being controlled.

Various types of stimulation may be used to administer pain during S&M encounters, but pain is not always employed. When it is, it is usually mild or moderate. Psychological pain, or humiliation, is perhaps as common as physical pain. Pain may also be used symbolically, as in the case of a sadist who uses a harmless, soft rubber paddle to spank the masochist. Thus the erotic appeal of pain for some S&M participants may derive from the ritual of control rather than from the pain itself (Weinberg, 1987).

Sadomasochism A mutually gratifying sexual interaction between consenting sex partners in which sexual arousal is associated with the infliction and receipt of pain or humiliation. Commonly known as *S&M*.

Extreme forms of pain, such as torture and severe beatings, are rarely reported by sadomasochists (Breslow et al., 1985). Masochists may seek pain, but they usually avoid serious injury and dangerous partners (Baumeister, 1988b).

S&M participants may be heterosexual, gay, or bisexual (Breslow et al., 1986). They may assume either the masochistic or the sadistic role, or they may alternate roles depending on the sexual script. People who seek sexual excitement by enacting both sadistic and masochistic roles are known as *sadomasochists.* In heterosexual relationships the partners may reverse traditional gender roles. The man may assume the submissive or masochistic role, and the woman may take the dominant or sadistic role (Reinisch, 1990).

A survey of S&M participants drawn from ads in S&M magazines found that about three out of four were male (Breslow et al., 1985). Most were married. Women respondents engaged in S&M more often and had more partners than men. (Apparently, a greater number of men than women seek partners for S&M.)

The causes of sexual masochism and sadism, as of other paraphilias, are unclear. Ford and Beach (1951) speculated that humans may possess a physiological capacity to experience heightened sexual arousal from the receipt or infliction of pain (which may explain the prevalence of love bites). Mild pain may heighten physiological arousal in both aggressor and victim, adding to the effects of sexual stimulation. Yet intense pain is likely to decrease rather than increase sexual arousal.

Pain may also have more direct biological links to pleasure. Natural chemicals called *endorphins,* similar to opiates, are released in the brain in response to pain and produce feelings of euphoria and general well-being. Perhaps, then, pleasure is derived from pain because of the release or augmentation of endorphins (Weinberg, 1987). This theory fails to explain the erotic appeal of sadomasochistic encounters that involve minimal or symbolic pain, however. Nor does it explain the erotic appeal to the sadist of inflicting pain.

Whatever their causes, the roots of sexual masochism and sadism apparently date to childhood. Sadomasochistic behavior commonly begins in early adulthood, but sadomasochistic fantasies are likely to have been present during childhood (American Psychiatric Association, 2000; Breslow et al., 1986).

Frotteurism

Frotteurism (also known as "mashing") is rubbing against or touching a nonconsenting person. As with other paraphilias, a diagnosis of frotteurism requires either acting on these urges or being distressed by them. Mashing has been reported exclusively among males (Spitzer et al., 1989).

Most mashing takes place in crowded places, such as buses, subway cars, or elevators. The man finds the rubbing or the touching, not the coercive nature of the act, to be sexually stimulating. While rubbing against a woman, he may fantasize a consensual, affectionate sexual relationship with her. Typically, the man incorporates images of his mashing within his masturbation fantasies. Mashing also incorporates a related practice, **toucherism:** the fondling of nonconsenting strangers.

Mashing may be so fleeting and furtive that the woman may not realize what has happened (Spitzer et al., 1989). Mashers thus stand little chance of being caught. Consider the case of a man who victimized a thousand or so women within a decade but was arrested only twice:

> Charles, 45, was seen by a psychiatrist following his second arrest for rubbing against a woman in the subway. He would select as his target a woman in her 20s as she entered the subway station. He would then position himself behind her on the platform and wait for the train to arrive. He would then follow her into the subway car and when the doors closed would begin bumping his penis against her buttocks, while fantasizing that they were enjoying having intercourse in a loving and consensual manner. About half of the time he would ejaculate into a plastic bag that he had wrapped around his penis to prevent staining his pants. He would then continue on his way to work. Sometimes when he hadn't ejaculated he would change trains and seek another victim. While he felt guilty for

Frotteurism A paraphilia characterized by recurrent, powerful sexual urges and related fantasies that involve rubbing against or touching a nonconsenting person. (From the French *frotter,* which means "to rub.")

Toucherism A practice related to frotteurism and characterized by the persistent urge to fondle nonconsenting strangers.

a time after each episode, he would soon become preoccupied with thoughts about his next encounter. He never gave any thought to the feelings his victims might have about what he had done to them. While he was married to the same woman for 25 years, he appears to be rather socially inept and unassertive, especially with women. (Adapted from Spitzer et al., 1989, pp. 106–107)

Although this masher was married, many mashers have difficulty forming relationships with women and are handicapped by fears of rejection. Mashing provides sexual contact in a relatively nonthreatening context.

Other Paraphilias

Let us consider some other, less common paraphilias.

ZOOPHILIA A person with **zoophilia** experiences repeated, intense urges and related fantasies involving sexual contact with animals. As with other paraphilias, the urges may be acted on or cause personal distress. A child or adolescent who shows some sexual response to an occasional episode of rough-and-tumble play with the family pet is not displaying zoophilia.

The term *bestiality* applies to actual sexual contact with an animal. Human sexual contact with animals, mythical and real, has a long history. Michelangelo's painting *Leda and the Swan* depicts the Greek god Zeus taking the form of a swan to mate with a woman, Leda. Zeus was also portrayed as taking the form of a bull or serpent to mate with humans. In the Old Testament, God is said to have put to death people who had sexual relations with animals. The Greek historian Herodotus notes that goats at the Egyptian temple at Mendes were trained to copulate with people.

Although the prevalence of zoophilia in the general population is unknown, Kinsey and his colleagues (1948, 1953) found that about 8% of the men and 3% to 4% of the women interviewed admitted to sexual contacts with animals. Men more often had sexual contact with farm animals, such as calves and sheep. Women more often reported sexual contacts with household pets. Men were more likely to masturbate or copulate with the animals. Women more often reported general body contact. People of both genders reported encouraging the animals to lick their genitals. A few women reported that they had trained a dog to engage in coitus with them. Urban–rural differences also emerged. Rates of bestiality were higher among boys reared on farms. Compared with only a few city boys, 17% of farm boys had reached orgasm at some time through sexual contact with dogs, cows, and goats. These contacts were generally restricted to adolescence, when human outlets were not available. Still, adults sometimes engage in sexual contacts with animals, generally because of curiosity or novelty or for a sexual release when human partners are unavailable. Whether or not such contacts constitute zoophilia depends on their frequency and intensity and on whether they cause the person distress. In most cases true zoophilia is associated with deep-seated psychological problems and difficulty developing intimate relationships with people.

NECROPHILIA In **necrophilia,** a rare paraphilia, a person desires sex with corpses. Three types of necrophilia have been identified (Rosman & Resnick, 1989). In *regular necrophilia*, the person has sex with a deceased person. In *necrophilic homicide*, the person commits murder to obtain a corpse for sexual purposes. In *necrophilic fantasy*, the person fantasizes about sex with a corpse but does not actually carry out necrophilic acts. Necrophiles often obtain jobs that provide them with access to corpses, such as working in cemeteries, morgues, or funeral homes. The primary motivation for necrophilia appears to be the desire to possess sexually a completely unresisting and nonrejecting partner (Rosman & Resnick, 1989). Many necrophiles are clearly mentally disturbed.

LESS COMMON PARAPHILIAS In **klismaphilia,** sexual arousal is derived from the use of enemas. Klismaphiles generally prefer the receiving role to the giving role. Klismaphiles may have derived sexual pleasure in childhood from the anal stimulation provided by parents giving them enemas.

Zoophilia A paraphilia involving persistent or repeated sexual urges and related fantasies that involve sexual contact with animals.

Necrophilia A paraphilia characterized by desire for sexual activity with corpses. (From the Greek *nekros,* which means "dead body.")

Klismaphilia A paraphilia in which sexual arousal is derived from the use of enemas.

Human Sexuality ONLINE //

"Cybersex Addiction"—A New Psychological Disorder?

Sex is the hottest topic among adult users of the Internet. Studies show that fully a third of all visits are directed to sexually oriented Web sites, chat rooms, and news groups.

For most people these forays into cybersex are relatively harmless recreational pursuits, but experts in the field say that the affordability, accessibility, and anonymity of the Internet are fueling a brand new psychological disorder—cybersex addiction—that appears to be spreading with astonishing rapidity and bringing turmoil to the lives of those affected.

Writing in the journal *Sexual Addiction and Compulsivity* (www.tandfdc.com/jnls/sac.htm), psychologist Al Cooper of Stanford University and his colleagues (2000) report that many of the men and women who now spend dozens of hours each week seeking sexual stimulation from their computers deny that they have a problem and refuse to seek help until their marriages and/or their jobs are in serious jeopardy.

For some people, the route to compulsive use of the Internet for sexual satisfaction is fast and short, said Dr. Mark Schwartz of Masters and Johnson in St. Louis. "Sex on the Net is like heroin," he said. "It grabs them and takes over their lives. And it's very difficult to treat because the people affected don't want to give it up."

Those most strongly hooked on Internet sex are likely to spend hours each day masturbating to pornographic images or having "mutual" online sex with someone contacted through a chat room. Occasionally, they progress to off-line affairs with sex partners they meet online.

Cooper and his colleagues (1999, 2000) conducted the largest and most detailed survey of online sex. Cooper calls the Net "the crack cocaine of sexual compulsivity." The survey, conducted online among 9,265 men and women who admitted surfing the Net for sexually oriented sites, indicated that at least 1% were seriously hooked on online sex. The survey found that as many as a third of Internet users visited some type of sexual site.

Extrapolated to the country as a whole, this would mean that a minimum of 200,000 men and women have become cybersex addicts in the last few years, Dr. Cooper said. And, he added, because the respondents were self-selected and because denial of the symptoms of sexual compulsivity is commonplace, the 1% figure is likely to be an underestimate.

"This is a hidden public health hazard exploding, in part, because very few are recognizing it as such or taking it seriously," Dr. Cooper wrote in the journal article.

Another author, Dr. Jennifer Schneider, a physician in Tucson, Arizona, who is associate editor of the journal, notes that even when cybersex addicts and their partners seek treatment, they often conceal their real problem, and therapists often fail to ask questions that would disclose it. As a result, the diagnosis of cybersex addiction is often missed.

Dr. Kimberly S. Young of the Center for Online Addiction in Bradford, Pennsylvania, wrote that "partially as a result of the general population and health care professionals not being attuned to the risks, seemingly harmless cyberromps can result in serious difficulties way beyond what was expected or intended."

According to Dr. Cooper, who works at the San Jose Marital and Sexuality Center in Santa Clara, California, cybersex compulsives are just like drug addicts. They "use the Internet as an important part of their sexual acting out, much like a drug addict who has a 'drug of choice,' " and often with serious harm to their home lives and livelihood. Especially vulnerable to becoming hooked on Internet sex, he wrote, are "those users whose sexuality may have

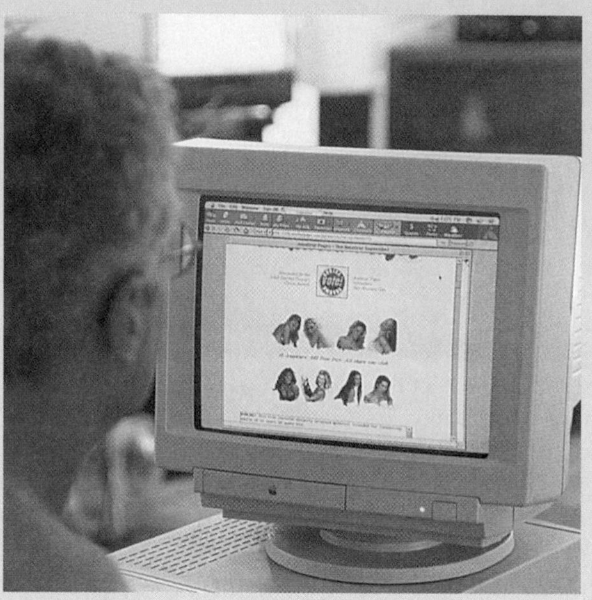

Do People Become Addicted to Cybersex? According to Dr. Mark Schwartz of the Masters and Johnson Institute, "Sex on the Net is like heroin," for some people. "It grabs them and takes over their lives." Some people spend hours each day masturbating to pornographic images they find online or engaging in "mutual" online sex with someone they contact through a chat room. Is becoming hooked on cybersex a safe kind of "addiction"? What do you think?

Coprophilia A paraphilia in which sexual arousal is attained in connection with feces. (From the Greek *copros,* which means "dung.")

In **coprophilia,** sexual arousal is connected with feces. The person may desire to be defecated on or to defecate on a sex partner. The association of feces with sexual arousal may also be a throwback to childhood. Many children appear to obtain anal sexual pleasure by holding in and then purposefully expelling feces. It may also be that the incidental connection between erections or sexual arousal and soiled diapers during infancy eroticizes feces.

been suppressed and limited all their lives [who] suddenly find an infinite supply of sexual opportunities" on the Internet.

A second survey conducted by Dr. Schneider among 94 family members affected by cybersex addiction reveals that the problem can arise even among people in loving marriages with ample sexual opportunities. "Sex on the Net is just so seductive and it's so easy to stumble upon it," she said. "People who are vulnerable can get hooked before they know it."

To those who say a behavioral compulsion is not a true addiction, Dr. Schneider responds with a definition of addiction that would clearly apply to cybersex abusers: "Loss of control, continuation of the behavior despite significant adverse consequences, and preoccupation or obsession with obtaining the drug or pursuing the behavior." Although behavioral addictions involve no external drugs, research suggests that they may cause changes in brain chemicals, like the release of endorphins, that help to perpetuate the behavior.

The sexual stimulation and release obtained through cybersex also contribute to the pursuit of the activity, Dr. Schwartz said. "Intense orgasms from the minimal investment of a few keystrokes are powerfully reinforcing," he noted, adding that "Cybersex affords easy, inexpensive access to a myriad of ritualized encounters with idealized partners."

Some cybersex addicts develop a conditioned response to the computer and become sexually aroused even before turning it on, Dr. Putnam said. This can exacerbate the problem for people whose jobs involve work on a computer. "Simply sitting down to work at the computer can start a sexual response that may facilitate online sexual activities," he wrote in the journal.

As with other addictions, tolerance to cybersex stimulation can develop, prompting the addict to take more and more risks to recapture the initial high, Dr. Schneider said. Online viewing that began as a harmless recreation can become an all-consuming activity and can even lead to real sexual encounters with people met online.

Cybersex compulsives can become so involved with their online activities that they ignore their partners and children and risk their jobs. In Dr. Cooper's survey, 20% of the men and 12% of the women reported they had used computers at work for some sexual pursuits. Many companies now monitor employees' online activities, and repeated visits to sexually oriented sites have cost people their jobs.

And some people, including two physicians, have landed in federal prison for two years because they downloaded child pornography when authorities were watching, Dr. Schwartz said.

Still, most who pursue cybersex consider it harmless and safe to do so. Although social and safety concerns and fear of discovery may prevent someone from visiting an adult bookstore or prostitute, there are no such constraints when pornography and sexual partners can be called up at any time of the day or night on a computer screen in one's home or office, Dr. Putnam said.

To those who say "What's the harm? They're not risking disease or death," Dr. Schneider, who has written extensively on sexual addiction, responds that the damage to a cybersex addict's life and family can be as devastating as that caused by compulsive gambling or addiction to alcohol or drugs. In her survey, 91 women and 3 men in committed relationships said they had experienced serious adverse consequences, including broken relationships, from their partners' cybersex addictions. Partners commonly reported feeling betrayed, devalued, deceived, ignored, abandoned, and unable to compete with a fantasy.

Among them was a 34-year-old woman married 14 years to a minister who she discovered was compulsively seeking sexual satisfaction by visiting pornographic sites on the Internet. "How can I compete with hundreds of anonymous others who are now in our bed, in his head?" the woman wrote. "Our bed is crowded with countless faceless strangers, where once we were intimate."

A 38-year-old woman married 18 years to a man who compulsively masturbates to images on the computer wrote that her husband had once had an extramarital affair and that "the online 'safe' cheating has just as dirty, filthy a feel to it as does the 'real-life' cheating."

Women who become cybersex addicts may face even greater risks than their male counterparts. Women, who tend to pursue relationships, are inclined to visit sexually oriented chat rooms rather than the pornographic Web sites that men prefer, Dr. Cooper said. As women become increasingly hooked on online sex, they are more likely to progress to off-line meetings, which can prove dangerous.

Children, too, often become victimized by cybersex addiction in a parent. As Dr. Schneider noted, children can stumble upon the pornographic material left on or near the computer or walk in on a parent masturbating at the computer. Several mothers in her survey were worried because their husbands surfed the Net while supposedly watching their children, who then viewed the pornography and sometimes the masturbation.

As Dr. Putnam put it, "Once people get hooked on cybersex, they tend to put themselves at risk and do things they wouldn't ordinarily do."

Source: From J. E. Brody (2000, May 16). "Cybersex Gives Birth to a Psychological Disorder." *The New York Times*, p. F7.

In **urophilia,** sexual arousal is associated with urine. As with coprophilia, the person may desire to be urinated on or to urinate on a sexual partner. Also like coprophilia, urophilia may have childhood origins. Stimulation of the urethral canal during urination may become associated with sexual pleasure. Or urine may have become eroticized by experiences in which erections occurred while the infant was clothed in a wet diaper.

Urophilia A paraphilia in which sexual arousal is associated with urine.

a CLOSER look

Sexual Addiction and Compulsivity

When it was revealed that President Bill Clinton had had a string of sexual contacts, culminating in his impeachment trial, a number of health professionals suggested that he had a sexual addiction. That is, he had lost control over his sexual behavior. What is a sexual addiction? What is sexual compulsivity?

What Is a Sexual Addiction?

A person with a sexual addiction

- Thinks about sex frequently
- Uses sexual behavior as a means of reducing anxiety
- Lacks control over his or her sexual impulses
- Typically experiences minimal satisfaction from sexual contacts
- Feels bad about his or her sexual contacts but engages in the behavior repeatedly
- Engages in illicit sexual behavior that endangers his or her own well-being and the well-being of his or her family
- Cannot resist sexual opportunities
- Continues sexual contacts that are nonintimate, dangerous, or undesirable

Characteristics of addiction include *tolerance* and *withdrawal symptoms.* As with other addictions, the sexually addicted person may experience tolerance—that is, may seek increasingly illicit sexual contacts or experiences. When the sexual activity is discontinued, the person may also experience withdrawal symptoms, such as anxiety and preoccupation with the craved activity.

What Is Sexual Compulsivity?

Sexual compulsivity differs from sexual addiction in that it is an obsessive-compulsive disorder. In an obsessive-compulsive disorder, an individual cannot eradicate certain thoughts or ideas from his or her mind and has extreme difficulty controlling his or her behavior. For example, some people have "checking" compulsions in which they check and recheck whether they have locked every door and window before they can leave home. (And then they may check some more and remain uneasy about the possibility of error.) Sexual compulsivity is manifested by means of sexual activity and may involve a specific paraphilia. Like sexual addiction, a sexual compulsion becomes the center of the person's life. As such, it may interfere with personal relationships, work, and health. There may also be legal consequences.

Sources: American Psychiatric Association, 2000; www.mastersandjohnson.com/, 2000.

THEORETICAL PERSPECTIVES

The paraphilias are among the most fascinating and perplexing variations in sexual behavior. We may find it difficult to understand how people can become sexually excited by fondling an article of clothing or by cross-dressing. It may also be hard to identify with people who feel compelled to exhibit their genitals or to rub their genitals against unsuspecting victims in crowded places. Perhaps we can recognize some voyeuristic tendencies in ourselves, but we cannot imagine peeping through binoculars while perched in a nearby tree or, for that matter, risking the social and legal consequences of being discovered in the act. Nor might we understand how people can become sexually turned on by inflicting or receiving pain.

Let us consider explanations that have been advanced from the major theoretical perspectives.

Biological Perspectives

Little is known about whether biological factors are involved in paraphilic behavior. Efforts to find concrete evidence of brain damage or abnormalities among people with paraphilias have so far failed (e.g., Langevin et al., 1989). Because testosterone is linked to sex drive, researchers have also focused on differences in testosterone levels between people with paraphilias and people without them. A recent study found evidence of some hormonal differences between a group of 16 male exhibitionists and controls (Lang et al., 1989). Although no differences in overall levels of testosterone were found, researchers reported that the exhibitionists evidenced significantly elevated levels of the measure of testosterone believed to be most closely linked to sex drive. This difference suggests that exhibitionists may have biologically elevated sex drives. The significance of this finding is limited by the fact that hormone levels among many of the people with paraphilias fell within the normal range.

Psychoanalytic Perspectives

Classical psychoanalytic theory suggests that paraphilias are psychological defenses, usually against unresolved castration anxiety dating back to the Oedipus complex (Fenichel, 1945). To the transvestic man, the sight of a woman's vagina threatens to arouse castration anxiety. It reminds him that women do not have penises and that he might suffer the same fate. Sequestering his penis beneath women's clothing symbolically asserts that women do have penises, which provides unconscious reassurance against his own fears of castration.

By exposing his genitals, the exhibitionist unconsciously seeks reassurance that his penis is secure. It is as though he were asserting, "Look! I have a penis!" Shock or surprise on the victim's face acknowledges that his penis still exists, temporarily relieving unconscious castration anxiety.

Masturbation with a fetishistic object (a shoe, for example) allows the fetishist to gratify his sexual desires while keeping a safe distance from the fantasized dangers that he unconsciously associates with sexual contact with women. Or the fetishistic object itself—the shoe—may unconsciously symbolize the penis. Stroking a woman's shoe during sexual relations, or fantasizing about one, may subconsciously provide reassurance that the man's own penis, symbolically represented by the fetishistic object, is secure.

In one psychoanalytic view of voyeurism, the man is unconsciously denying castration by searching for a penis among women victims. Other psychoanalytic views suggest that the voyeur is identifying with the man in the observed couple as he had identified with his own father during childhood observations of the parental coitus—the so-called *primal scene.* Perhaps he is trying to "master" the primal scene by compulsively reliving it.

Psychoanalytic explanations of sadism suggest that sadists are attempting to defend themselves against unconscious feelings of impotence and powerlessness by inflicting pain on others. The recipients' shouts of pain or confessions of unworthiness make sadists feel masculine and powerful.

Psychoanalytic theory suggests that masochism in the male may be the turning inward of aggressive impulses originally aimed toward the powerful, threatening father. The flagellation or bondage may also be unconsciously preferred to castration as a form of punishment for having unacceptable sexual feelings. Like the child who experiences relief when punishment is over, the sexual masochist willingly accepts immediate punishment in the form of flagellation or bondage in place of the future punishment of castration. Or a woman who witnessed her parents having coitus at an early age may have misperceived the father to be assaulting the mother. Her masochism then represents her unconscious reliving of her mother's (imagined) role with her father. Sexual masochists of either gender may also have so much guilt about sex that they can allow themselves to experience sexual pleasure only if they are adequately punished for it.

The paraphilias have provided a fertile ground for psychoanalytic theories. However, only clinical case studies support the role of such unconscious processes as unresolved castration anxiety. The basic shortcoming of psychoanalytic theory is that many of its key concepts involve unconscious mechanisms that cannot be directly observed or measured. Thus psychoanalytic theories remain interesting but speculative hypotheses about the origins of these unusual sexual behavior patterns.

Learning Perspectives

Learning theorists believe that fetishes and other paraphilias are learned behaviors that are acquired through experience. An object may acquire sexually arousing properties through association with sexual arousal or orgasm. Early proponents of the learning theory viewpoint were Alfred Kinsey and his colleagues, who wrote, in *Sexual Behavior in the Human Female* (1953),

> Even some of the most extremely variant types of human sexual behavior may need no more explanation than is provided by our understanding of the processes of learning and conditioning. Behavior which may appear bizarre, perverse, or unthinkably unacceptable to some persons, and even to most persons, may have significance for other individuals because of the way in which they have been conditioned. (pp. 645–646)

Like Pavlov's dogs, who learned to salivate in response to the ringing of a bell that had been repeatedly paired with food,

> An animal may become conditioned to respond not only to particular stimuli, but to objects and other phenomena which were associated with the original experience. . . . In the laboratory, male animals may respond to particular dishes, to particular boards, or to particular pieces of furniture with which some female has had contact. (Kinsey et al., 1953, p. 647)

According to the conditioning model, a boy who catches a glimpse of his mother's stockings hanging on the towel rack while he is masturbating may go on to develop a fetish for stockings (Breslow, 1989). Orgasm in the presence of the object would reinforce the erotic connection, especially if the experience occurs repeatedly.

In an early experimental test of the conditioning model, Rachman (1966) showed normal (nonfetishistic) males slides of nude women interspersed with slides of women's boots. After numerous repetitions, the men showed a sexual response to the boots alone. This "fetish" was weak and short-lived, however. It could still be argued that more persistent fetishes might be learned if pairings of such stimuli were to occur repeatedly during childhood. If fetishes were mechanically acquired by association, however, we might expect fetishists to be more attracted to objects that are inadvertently (and often repeatedly) associated with sexual activity, such as pillows, bedsheets, and even ceilings (Breslow, 1989). Yet we do not find this to be the case. The *meaning* of the object also seems to play a role.

Along these lines, Breslow (1989) proposed a learning theory explanation that describes the development of paraphilias in terms of the gradual acquisition of sexual arousal to an unusual object or activity through its incorporation in masturbatory fantasies. A transvestite, for example, may have achieved an erection while trying on his mother's panties in childhood. The paraphilic object or activity is then incorporated within masturbatory fantasies and is reinforced by orgasm. The paraphilic object or activity is then repeatedly used as a masturbatory aid, further strengthening the erotic bond.

We can also consider a role for parental approval as a reinforcing agent in the early development of transvestism. Parents who really wanted to have a girl have been known to dress their little boys in girls' clothing occasionally. The reinforcement of parental approval for cross-dressing may lead the boy to experiment with cross-dressing himself, which may then become eroticized if it is combined with masturbation or sexual fantasies and reinforced by sexual arousal or orgasm.

Fetishistic interests can often be traced to early childhood. Some rubber fetishists, for example, recall erotic interests in rubber objects since early childhood. Reinisch (1990) speculates that for some rubber fetishists, the earliest awareness of sexual arousal or response (such as erection) may have been associated with the presence of rubber pants, diapers, and so forth, such that a connection was formed between the two. Or perhaps the sexual attraction to objects associated with infancy derives from their association with feelings of being completely loved and cared for. The fetishistic act may represent an attempt to recapture sexual or loving feelings from early childhood.

McGuire and his colleagues (1965) report a case that provides some support for the role of learning in the development of exhibitionism. Two young males were surprised by an attractive woman while urinating. Although they were embarrassed at the time, their memories of the incident were sexually stimulating, and they masturbated repeatedly while fantasizing about it. The fantasies persisted, possibly reinforced by frequent orgasms. After a while, the young men purposely began to expose themselves to rekindle the high level of sexual excitement. Still, we should caution that only a very small percentage of men who have accidentally been discovered exposed have become exhibitionists.

Friedrich and Gerber (1994) studied five adolescent boys who practiced hypoxyphilia and found extensive early histories of choking in combination with physical or sexual abuse. The combination seems to have encouraged each of the boys to associate choking with sexual arousal.

Blair and Lanyon (1981) suggest that modeling plays a role in some cases. Parents may inadvertently model exhibitionistic behavior to young sons, which can lead the sons to eroticize the act of exposing themselves.

Learning explanations of sexual masochism focus on the pairing of sexual excitement with punishment. For example, a child may be punished when discovered masturbating. Or a boy may reflexively experience an erection if his penis accidentally rubs against the parent's body as he is being spanked. With repeated encounters like these, pain and pleasure may become linked in the person's sexual arousal system. Another learning explanation focuses on the association of pain with parental affection (Breslow, 1989; Sue et al., 1981). A child with cold and indifferent parents may be hugged only following a spanking. The pain or humiliation associated with the spanking becomes associated with the affection of the hug, which leads in later life to pain becoming a prerequisite for sexual pleasure. Yet these learning approaches fail to account for how sexual masochism develops within a larger and more complex sadomasochistic lifestyle (Breslow, 1989).

Learning theories of the origins of paraphilias also fail to consider the predisposing factors that may explain why some people who are exposed to early conditioning experiences develop paraphilic interests and others do not. Such predisposing factors may include poor self-esteem and difficulty forming intimate relationships. Many exhibitionists, voyeurs, frotteurs, and other people with paraphilias have few interpersonal skills in relating to women. They may avoid customary social interactions with women for fear of rejection. Their furtive, paraphilic behaviors may provide a sexual release with minimal risks of rejection or apprehension and may be maintained because they represent the only available source of sexual gratification or reinforcement. Some paraphilias, such as voyeurism, exhibitionism, and frotteurism, can also be conceptualized as *courtship disorders,* involving an exaggeration or distortion of the steps normally taken during courtship in identifying, approaching, and becoming more intimate with new sexual partners (Freund & Blanchard, 1986).

Sociological Perspectives

Most people indulge paraphilias privately. Sexual masochists and sadists require a partner, however, except for the few masochists who practice only autoerotic forms of masochism and the few sadists who stalk nonconsenting partners. Most sadomasochists also relate in one way or another to the sadomasochistic subculture. It is within the S&M subculture—the loosely connected network of S&M clubs, specialty shops, organizations, magazines, and so on—that S&M rituals are learned, sexual contacts made, sadomasochistic identities confirmed, and sexual paraphernalia acquired. But the S&M subculture exists in the context of the larger society, and the rituals it invents mirror the social and gender roles that exist in the larger society.

Martin Weinberg (1987) proposes a sociological model that focuses on the social context of sadomasochism. Noting that S&M rituals generally involve some form of dominance and submission, Weinberg attributes their erotic appeal to the opportunity to reverse the customary power relationships that exist between the genders and social classes in society at large. Within the confines of the carefully scripted S&M encounter, the meek can be powerful and the powerful meek (Geer et al., 1984). People from lower social classes or in menial jobs may be drawn to S&M by the opportunities it affords to enact a dominant role. They may have the opportunity to bark orders and commands that are followed unquestioningly. Those who customarily hold high-status positions that require them to be in control and responsible may be attracted by the opportunity to surrender control to another person. Dominance and submission games also offer opportunities to accentuate or reverse the gender stereotypes that identify masculinity with dominance and femininity with submissiveness.

Individual sadomasochistic interests may become institutionalized as an S&M subculture in societies (like ours) that have certain social characteristics: (1) Dominance–submission relationships are embedded within the culture, and aggression is socially valued. (2) There is an unequal distribution of power between people from different gender or social class categories. (3) There are enough affluent people to enable them to participate in such leisure-time activities. (4) Imagination and creativity, important elements in the development of S&M scripts and fantasies, are socially valued and encouraged (Weinberg, 1987).

An Integrated Perspective: The "Lovemap"

Like other sexual patterns, the paraphilias may have multiple biological, psychological, and sociocultural origins (Seligman & Hardenburg, 2000). Our understanding of them may thus be best approached from a theoretical framework that incorporates multiple perspectives, as found in the work of John Money and his colleagues.

Money and Lamacz (1989) trace the origins of paraphilias to childhood. They believe that childhood experiences etch a pattern in the brain, called a **lovemap,** that determines what types of stimuli and activities become sexually arousing to the individual. In the case of paraphilias, these lovemaps become distorted, or "vandalized," by early traumatic experiences, such as incest, overbearing antisexual upbringing, and physical abuse or neglect.

Research suggests that voyeurs and exhibitionists often were the victims of childhood sexual abuse (Dwyer, 1988). Not all children exposed to such influences develop paraphilic compulsions, however. For reasons that remain unknown, some children exposed to such influences appear to be more vulnerable than others to developing distorted lovemaps. A genetic predisposition, hormonal factors, brain abnormalities, or a combination of these and other factors may play a role in determining one's vulnerability to vandalized lovemaps (Brody, 1990a).

TREATMENT OF THE PARAPHILIAS

The treatment of these atypical patterns of sexual behavior raises a number of issues. First, people with paraphilias usually do not want or seek treatment, at least not voluntarily. They often deny that they are offenders, even after they are apprehended and convicted. They are generally seen by mental health workers only when they come into conflict with the law or at the urging of their family members or sexual partners who have discovered them performing the paraphilic behavior or found evidence of their paraphilic interests.

Paraphilic behavior is a source of pleasure, so many people are not motivated to give it up. The individual typically perceives his problems as stemming from society's intolerance, not from feelings of guilt or shame.

Second, helping professionals may encounter ethical problems when they are asked to contribute to a judicial process by trying to persuade a sex offender that he (virtually all are male) *ought* to change his behavior. Helping professionals traditionally help clients clarify or meet their own goals; it is not their role to impose societal goals on the individual. Many helping professionals believe that the criminal justice system, not they, ought to enforce social standards.

Lovemap A representation in the brain of the idealized lover and of idealized erotic activity with the lover.

The third issue is a treatment problem. Therapists realize that they are generally less successful with resistant or recalcitrant clients. Unless the motivation to change is present, therapeutic efforts are often wasted.

The fourth problem is the issue of perceived responsibility. Sex offenders almost invariably claim that they are unable to control their urges and impulses. Such claims of uncontrollability are often self-serving and may lead others to treat offenders with greater sympathy and understanding. Most therapies, however, are based on the belief that whatever causes may have led to the problem behavior, and however difficult it may be to resist these unusual sexual urges, accepting personal responsibility for one's actions is a prelude to change. Thus, if therapy is to be constructive, it is necessary to break through the client's personal mythology that he or she is powerless to control his or her behavior.

Despite these issues, many offenders are referred for treatment by the courts. A few seek therapy themselves because they have come to see how their behavior harms themselves or others. Let us consider some of the ways in which therapists treat people with these atypical sexual behavior patterns.

Psychotherapy

Psychoanalysis focuses on resolving the unconscious conflicts that are believed to originate in childhood and to give rise in adulthood to pathological problems such as paraphilias. The aim of therapy is to help bring unconscious conflicts, principally Oedipal conflicts, into conscious awareness so that they can be worked through in light of the individual's adult personality.

Although some favorable case results have been reported (e.g., Rosen, 1967), psychoanalytic therapy for the paraphilias has not been subjected to experimental analysis. We thus do not know whether successes are due to the psychoanalytic treatment itself or to other factors, such as spontaneous improvement or a client's willingness to change.

Behavior Therapy

Whereas traditional psychoanalysis tends to entail a lengthy process of exploration of the childhood origins of problem behaviors, **behavior therapy** is briefer and focuses directly on changing behavior. Behavior therapy has spawned a number of techniques to help eliminate paraphilic behaviors and strengthen appropriate sexual behaviors. These techniques include systematic desensitization, aversion therapy, social skills training, covert sensitization, and orgasmic reconditioning, to name a few.

Systematic desensitization attempts to break the link between the sexual stimulus (such as a fetishistic stimulus) and the inappropriate response (sexual arousal). The client is first taught to relax selected muscle groups in the body. Muscle relaxation is then paired repeatedly with each of a series of progressively more arousing paraphilic images or fantasies. Relaxation comes to replace sexual arousal in response to each of these stimuli, even the most provocative. In one case study, a fetishistic transvestite who had become attracted to his mother's lingerie at age 13 was taught to relax when presented with audiotaped scenes representing transvestite or fetishistic themes (Fensterheim & Kantor, 1980). He played such tapes daily while remaining relaxed. He later reported a complete absence of transvestite thoughts and activities.

In **aversion therapy,** the undesirable sexual behavior (for example, masturbation to fetishistic fantasies) is paired repeatedly with an aversive stimulus (such as a harmless but painful electric shock or a nausea-inducing chemical) in the hope that the client will develop a conditioned aversion to the paraphilic behavior.

Covert sensitization is a variation of aversion therapy in which paraphilic fantasies are paired with an aversive stimulus in imagination. In a broad-scale application, 38 **pedophiles** and 62 exhibitionists, more than half of whom were court-referred, were treated by pairing imagined aversive odors with fantasies of the problem behavior

Behavior therapy The systematic application of the principles of learning to help people modify problem behavior.

Systematic desensitization A method for terminating the connection between a stimulus (such as a fetishistic object) and an inappropriate response (such as sexual arousal to the paraphilic stimulus). Muscle relaxation is practiced in connection with each stimulus in a series of increasingly arousing stimuli, so that the person learns to remain relaxed (and not sexually aroused) in their presence.

Aversion therapy A method for terminating undesirable sexual behavior in which the behavior is repeatedly paired with an aversive stimulus such as electric shock so that a conditioned aversion develops.

Covert sensitization A form of aversion therapy in which thoughts of engaging in undesirable behavior are paired repeatedly with imagined aversive stimuli.

Pedophiles Persons with pedophilia, a paraphilia involving sexual interest in children.

Is Her Behavior Appropriate or Inappropriate? Most theorists suggest that early learning experiences contribute to the development of paraphilias. Is this woman's interaction with her young child of the sort that can lead to sexual problems as the child matures?

(Maletzky, 1980). Clients were instructed to fantasize pedophiliac or exhibitionistic scenes. Then,

> At a point . . . when sexual pleasure is aroused, aversive images are presented. . . . Examples might include a pedophiliac fellating a child, but discovering a festering sore on the boy's penis, an exhibitionist exposing to a woman but suddenly being discovered by his wife or the police, or a pedophiliac laying a young boy down in a field, only to lie next to him in a pile of dog feces. (Maletzky, 1980, p. 308)

Maletzky used this treatment weekly for 6 months and then followed it with booster sessions every 3 months over a 3-year period. The procedure resulted in at least a 75% reduction in the deviant activities and fantasies for over 80% of the study participants at follow-up periods of up to 36 months.

Social skills training focuses on helping the individual improve his ability to relate to the other gender. The therapist might first model a desired behavior, such as how to ask a woman out on a date or how to handle a rejection. The client might then role-play the behavior, with the therapist playing the part of the woman. Following the role-play enactment, the therapist would provide feedback and additional guidance and modeling to help the client improve his skills. This process would be repeated until the client mastered the skill.

Orgasmic reconditioning aims to increase sexual arousal to socially appropriate sexual stimuli by pairing culturally appropriate imagery with orgasmic pleasure. The person is instructed to become sexually aroused by masturbating to paraphilic images or fantasies. But as he approaches the point of orgasm, he switches to appropriate imagery and focuses on it during orgasm. In a case example, Davison (1977) reports reduction of sadistic fantasies in a 21-year-old college man. The client was instructed to attain an erection in any way he could, even through the use of the sadistic fantasies he wished to eliminate. But once erection was achieved, he was to masturbate while looking at photos of *Playboy* models. Orgasm was thus paired with nonsadistic images. These images and fantasies eventually acquire the capacity to elicit sexual arousal. Orgasmic reconditioning is often combined with other techniques, such as social skills training, so that more desirable social behaviors can be strengthened as well (Adams et al., 1981).

Although behavior therapy techniques tend to have higher reported success rates than most other methods, they too are limited by reliance on uncontrolled case studies. Without appropriate controls, we cannot isolate the effective elements of therapy or determine that the results were not due merely to the passage of time or other factors unrelated to the treatment. It is possible that clients who are highly motivated to change may succeed in doing so with *any* systematic approach.

Biochemical Approaches

There is no biological "cure" for the paraphilias. No drug or surgical technique eliminates paraphilic urges and behavior. Yet some progress has recently been reported in using antidepressants such as Prozac (fluoxetine hydrochloride) in treating exhibitionism, voyeurism, and fetishism (Lorefice, 1991; Miller, 1995; Roesler & Witztum, 2000). Why Prozac? In addition to treating depression, Prozac has been helpful in treating obsessive-compulsive disorder, a type of emotional disorder involving recurrent obsessions (intrusive ideas) and/or compulsions (urges to repeat a certain behavior or thought). Researchers speculate that paraphilias may be linked to obsessive-compulsive disorder (Kruesi et al., 1992). People with paraphilias often experience intrusive, repetitive thoughts or images of the paraphilic object or stimulus, such as mental images of young children. Many also report feeling compelled to carry out the paraphilic acts repeatedly. Paraphilias may belong to what researchers have dubbed an obsessive-compulsive spectrum of behaviors (Kruesi et al., 1992).

Social skills training Behavior therapy methods that rely on a therapist's coaching and practice to build social skills.

Orgasmic reconditioning A method for strengthening the connection between sexual arousal and appropriate sexual stimuli (such as fantasies about an adult of the other gender) by repeatedly pairing the desired stimuli with orgasm.

Human Sexuality ONLINE //

Web Sites Related to Atypical Sexual Variations

This is the Web site of the National Council on Sexual Addiction and Compulsivity. "NCSAC is a private, non-profit organization dedicated to the promotion of public and professional recognition, awareness and understanding of Sexual Addiction, Sexual Compulsivity and Sexual Offending. NCSAC provides access to education, information and referral resources encouraging wellness for all those we serve."
www.ncsac.org/

The Web site of Sexaholics Anonymous (SA), P.O. Box 111910, Nashville, TN 37222–1910. (615) 331–6230:
www.sa.org

The Web site of Sex and Love Addicts Anonymous (SLAA), P. O. Box 338, Norwood, MA 02062–0338. (781) 255–8825:
www.slaafws.org

fwsoffice@slaafws.com

The Web sites of Sex Addicts Anonymous (SAA), PO Box 70949, Houston, TX 77270. (713) 869–4902 or (800) 447–8191:
www.sexaa.org

info@www.saa-recovery.org

The Web site of Sexual Compulsives Anonymous (SCA), Old Chelsea Station, P.O. Box 1585, New York, NY 10013–0935. (212) 439–1123 or 800–977-HEAL:
www.sca-recovery.org

The Web site of Codependents (family members, partners, and friends) of Sex Addicts (COSA), 9337-B Katy Freeway, Suite 142, Houston, TX 77024. (612) 537–6904:
www.shore.net/~cosa

People who experience such intense urges that they are at risk of committing sexual offenses may be helped by **anti-androgen drugs,** which reduce the level of testosterone in the bloodstream (Roesler & Witztum, 2000). Testosterone is closely linked to sex drive and interest. *Medroxyprogesterone acetate* (MPA) (trade name: Depo-Provera), which is administered in weekly injections, is the anti-androgen that has been used most extensively in the treatment of sex offenders. In men, anti-androgens reduce testosterone to a level that is typical of a prepubertal boy (Bradford, 1998). They consequently reduce sexual desire and the frequency of erections and ejaculations (Bradford, 1998).

Depo-Provera suppresses the sexual appetite in men. It can lower the intensity of sex drive and erotic fantasies and urges so that the man may feel less compelled to act on them (Roesler & Witztum, 2000). Anti-androgens do not, however, eliminate paraphilic urges or behavior. As an analogy, consider the relationship between the accelerator pedal and the steering wheel of a car. The accelerator pedal controls the car's speed but not its direction. In much the same way in which releasing pressure on the accelerator pedal slows the car, the use of anti-androgens reduces the intensity of sex drives and desires. The types of stimuli that have erotic value are not affected by anti-androgens, however, any more than easing up on the accelerator alters the direction of the car.

The use of antiandrogens is sometimes incorrectly referred to as *chemical castration*. Surgical castration—the surgical removal of the testes—has sometimes been performed on convicted rapists and violent sex offenders (Roesler & Witztum, 2000). Surgical castration eliminates testicular sources of testosterone. Anti-androgens suppress, but do not eliminate, testicular production of testosterone. Also, unlike surgical castration, the effects of anti-androgens can be reversed when the treatment is terminated.

Evidence suggests that anti-androgens help some people when they are used in conjunction with psychological treatment (Roesler & Witztum, 2000). The value of anti-androgens has been limited by high refusal and dropout rates, however (Roesler & Witztum, 2000). Questions also remain concerning side effects.

Although we have amassed a great deal of research on atypical variations in sexual behavior, our understanding of them and our treatment approaches to them remain largely in their infancy.

Anti-androgen drug A chemical substance that reduces the sex drive by lowering the level of testosterone in the bloodstream.

summing up

NORMAL VERSUS DEVIANT SEXUAL BEHAVIOR

What is considered normal in one culture or at a particular time may be considered abnormal in other cultures or at other times. Atypical patterns of sexual arousal or behavior that become problematic in the eyes of the individual or society are labeled *paraphilias.*

THE PARAPHILIAS

Paraphilias involve sexual arousal in response to unusual stimuli such as children or other nonconsenting persons, certain objects, or pain or humiliation. The psychiatric diagnosis of paraphilia requires that the person has acted on these persistent urges or is distinctly distressed by them. Except in the case of sexual masochism, paraphilias are believed to occur almost exclusively among men.

■ Fetishism

In fetishism, an inanimate object comes to elicit sexual arousal. In partialism, people are inordinately aroused by a particular body part, such as the feet.

■ Transvestism

Whereas other fetishists become sexually aroused by handling the fetishistic object while they masturbate, transvestites become excited by wearing articles of clothing—the fetishistic objects—of the other gender.

■ Exhibitionism

An exhibitionist experiences the compulsion to expose himself to strangers. The typical exhibitionist does not attempt further sexual contact with the victim and so does not usually pose a physical threat.

■ Obscene Telephone Calling

The obscene phone caller is motivated to become sexually aroused by shocking his victim. Such callers typically masturbate during the phone call or shortly afterward.

■ Voyeurism

Voyeurs become sexually aroused by watching and do not seek sexual relations with the target. The voyeur may masturbate while peeping or afterward while engaging in voyeuristic fantasies. Like exhibitionists, voyeurs tend to harbor feelings of inadequacy and poor self-esteem and to lack social and sexual skills.

■ Sexual Masochism

Sexual masochists associate the receipt of pain or humiliation with sexual arousal. Sexual masochists and sexual sadists sometimes form liaisons to meet each other's needs.

■ Sexual Sadism

Sexual sadism is characterized by persistent, powerful urges and sexual fantasies involving the inflicting of pain and suffering on others to achieve sexual excitement or gratification. Sadomasochists enjoy playing both sadistic and masochistic roles.

■ Frotteurism

Most frotteuristic acts—rubbing against nonconsenting persons, also known as mashing—take place in crowded places, such as buses, subway cars, or elevators.

■ Other Paraphilias

Zoophiles desire to have sexual contact with animals. Necrophiles desire to have sexual contact with dead bodies.

THEORETICAL PERSPECTIVES

■ Biological Perspectives

The links between paraphilias and biological factors have yet to be fully explored.

■ Psychoanalytic Perspectives

Classical psychoanalytic theory suggests that paraphilias in males are psychological defenses against castration anxiety.

■ Learning Perspectives

Some learning theorists have argued that unusual stimuli may acquire sexually arousing properties through association with sexual arousal or orgasm. Another possibility is that unusual stimuli gradually acquire sexually arousing properties by being incorporated into masturbatory fantasies.

■ Sociological Perspectives

According to Weinberg's sociological model, the erotic appeal of S&M rituals may result from the opportunity to reverse the customary power relationships that exist between the gender and social classes in society at large.

■ An Integrated Perspective: The "Lovemap"

Money and Lamacz suggest that childhood experiences etch a pattern in the brain—a lovemap—that determines the types of stimuli and activities that become sexually arousing. In the case of paraphilias, these lovemaps become distorted by early traumatic experiences.

TREATMENT OF THE PARAPHILIAS

People with paraphilias may be motivated to seek help because of fears of exposure, criminal prosecution, or humiliation, but they seldom desire to surrender their sexual preferences.

■ Psychotherapy

Psychoanalysis aims to bring unconscious Oedipal conflicts into awareness so that they can be worked through in adulthood.

■ Behavior Therapy

Behavior therapy attempts to eliminate paraphilic behaviors through techniques such as systematic desensitization, aversion therapy, social skills training, covert sensitization, and orgasmic reconditioning.

■ Biochemical Approaches

The antidepressant Prozac has shown some promise in treating paraphilias. By reducing sex drives, anti-androgen drugs can help people who have difficulty combating paraphilic urges. They may be most helpful when used in conjunction with psychological treatment.

questions for critical thinking

1. Are the women on your campus who enjoy dressing in men's-style clothing to be considered transvestites? Why or why not?
2. Are lovers who become excited by seeing their partners undressing or in the nude to be considered voyeurs? Why or why not?
3. What problems do therapists encounter when they seek to treat people with paraphilias?
4. Do you have any thoughts on why nearly all people with paraphilias are males?

chapter

18

Sexual Coercion

Pablo Picasso. *Weeping Woman.* 1937. Tate Gallery, London, Great Britain. Credit: Tate Gallery, London/Art Resource, NY. Credit: © 2002 Estate of Pablo Picasso/Artists Rights Society (ARS), NY.

outline

Truth or Fiction?

RAPE

A World of Diversity Not All Sex Offenders Are Men

Incidence of Rape

Types of Rapes

A Closer Look Anatomy of a Date Rape: Ann and Jim

Social Attitudes and Myths That Encourage Rape

Sociocultural Factors in Rape

A Closer Look Feral Fowl Females Reject Rapists

Psychological Characteristics of Rapists

Adjustment of Rape Survivors

A World of Diversity Blaming the Victim to an Extreme

Treatment of Rape Survivors

Rape Prevention

A Closer Look Rape Prevention

VERBAL SEXUAL COERCION

A Closer Look Handling Sexual Pressure Lines

SEXUAL ABUSE OF CHILDREN

What Is Sexual Abuse of Children?

Patterns of Abuse

Pedophilia

Incest

Effects of Sexual Abuse of Children

A Closer Look New Methods for Getting to the Truth in Cases of Child Abuse

Prevention of Sexual Abuse of Children

Human Sexuality Online Online Registries of Sex Offenders: Boon or Boondoggle?

Treatment of Survivors of Sexual Abuse

TREATMENT OF RAPISTS AND CHILD MOLESTERS

SEXUAL HARASSMENT

Sexual Harassment in the Workplace

Sexual Harassment on Campus

Sexual Harassment in the Schools

How to Resist Sexual Harassment

Human Sexuality Online Web Sites Related to Sexual Coercion

SUMMING UP

QUESTIONS FOR CRITICAL THINKING

Truth or Fiction?

____ A woman is raped every 5 minutes in the United States.

____ The prevalence of rape is 20 times greater in the United States than in Japan.

____ The majority of rapes are committed by strangers in deserted neighborhoods or darkened alleyways.

____ Men who rape other men are gay.

____ Many women say no when they mean yes.

____ Most rapists are mentally ill.

____ Women who encounter a rapist should attempt to fight him off.

____ Father–daughter incest is the most common type of incest.

In recent years the airwaves have been flooded with incidents of sexual assault and sexual harassment involving celebrities, highly placed politicians, and members of the armed services. In the 1990s, media sharks went on a feeding frenzy in covering the rape trials of William Kennedy Smith and Mike Tyson, as well as the Senate confirmation hearings of Supreme Court nominee Clarence Thomas, who faced charges of sexual harassment leveled by a former assistant, Anita Hill. A few years later, the media focused on an affair between President Bill Clinton and a young White House intern named Monica Lewinsky. Many argued that their relationship constituted sexual harassment because of the disparity in power, even though Lewinsky was a willing partner.

Other high-profile cases included sexual abuse at an army training center in Maryland, the Tailhook incident, and sexual harassment by former U.S. senator Robert Packwood of Oregon ("Sex complaints," 1996). The Senate Ethics Committee voted unanimously to expel Packwood from the Senate in September 1995. Packwood resigned. In 2000, Lt. Gen. Claudia J. Kennedy, the highest-ranking woman in the Army, alleged that another general had "groped" her in her office (Myers, 2000b).

A team of professional writers could not have tickled the public's fancy more, but the plots and the characters in these events were very real. Many observers winced as they saw aspects of themselves—either as aggressors or as victims—being laid bare before the nation.

This chapter is about sexual coercion. Our topic includes rape and other forms of sexual pressure, including lying to seduce one's partner and sexual harassment. Sexual coercion also includes *any* sexual activity between an adult and a child. Even when children cooperate, sexual relations with children are coercive because children are below the legal age of consent.

RAPE

> I wanted to knock the woman off her pedestal, and I felt rape was the worst thing I could do to her.
>
> She wanted it, she was asking for it. She just said "no" so I wouldn't think she was easy. The only reason she yelled rape was she got home late and her husband knew she hadn't been out with her girlfriend.
>
> I found myself having sexual fantasies that would put women in precarious positions. I was thinking about this more and more, like devising a rack, perhaps, that would spread her legs as wide open as they could possibly be spread—something of this nature. I acted tough with them. The first one and the last pleaded for their virginity. I told them to do what they were told and they wouldn't get hurt. I said that if they didn't do what they were told, they would be sorry. I don't know if I actually threatened to kill them or not, but I very strongly feel that I never would have. The only thing is, perhaps if I continued on and hadn't been caught this time, seeing what happened from the first three times to the second three times—I just wonder—maybe in the next set of three somebody would have gotten hurt, you know, somebody would have really gotten hurt.
>
> (Groth & Birnbaum, 1979)

These statements by rapists show that rape is the subjugation of women by men by force or threat of force. Many social scientists view rape as an act of violence that is more connected with domination, anger, power, and sadism than with passion or sexual desire (Tedeschi & Felson, 1994).

For the first few thousand years of recorded history, the only rapes that were punished were those that defiled virgins. These rapes were considered crimes against prop-

Rape Sexual intercourse that takes place as a result of force or threats of force rather than consent. (The legal definition of rape varies from state to state.) See also *forcible rape* and *statutory rape.*

Not All Sex Offenders Are Men

She wore red coveralls—the uniform of the female prisoner. She looked exhausted and disheveled as she stood for sentencing. She is Mary Kay LeTourneau, a 35-year-old former grade school teacher in suburban Seattle. She is the woman who had sexual relations, and a baby, with a former student—a 13-year-old boy. LeTourneau wept as her lawyer pleaded for mercy and made no comment when the judge offered her a chance to speak. She was sentenced to 7 years and 5 months in prison and denounced by the judge who had been lenient some months earlier, in a case that has raised questions about whether female sex offenders can be held to different standards than men.

In November of 1997, LeTourneau had pleaded guilty to raping the boy and promised King County Superior Court Judge Linda Lau that LeTourneau would have no further contact with him. "I give you my word," LeTourneau had said. "It will not happen again." LeTourneau is the mother of four children by her ex-husband. She had no previous criminal record and was given a suspended prison sentence and ordered to undergo treatment. But in February of 1998, after hearing testimony that LeTourneau had again seen the boy, and possibly planned to flee with him, Judge Lau revoked the suspension and sent LeTourneau to prison. Prosecutors, state legislators, and professionals who treat sex offenders note that LeTourneau had escaped an earlier prison sentence largely because she was attractive and presented herself as being in love with an emotionally mature boy.

When LeTourneau escaped the prison sentence the previous November, critics pointed to a similar case involving a male teacher, Mark Blilie, who was 42-years-old. He also had no previous criminal history but was convicted of having sex with a 15-year-old former student. Blilie was imprisoned for at least 4 years. But women make up only a small fraction—less than 2%—of convicted sex offenders, which may have played a part in the judge's earlier leniency (Egan 1998).

The victim? The boy repeatedly protested to the media that he is *not* a victim—that he was is in love with his former teacher and is deeply saddened by her conviction. LeTourneau had claimed that the boy is emotionally mature beyond his years But people who treat sex offenders say they often justify sex with underage people by claiming that the victim is emotionally mature. "Exploiting a young teenager's desire to be older, to be liked, to be accepted is the currency of sex offenders who prey on children of that age," said Lucy Berliner, research director of the Seattle's Harborview Center for Sexual Assault (Egan, 1998).

The saga apparently continues. In April 2000 the boy, 16 at the time, filed law suits against the city of Des Moines, Washington, and his local school district for compensation for emotional suffering, lost income, and the cost of raising his two children, who were in the custody of his mother ("Boy to sue," 2000).

erty (virgins being the property of their fathers), not crimes against persons. In ancient Babylonia, rape laws applied to married women as well. Babylonian law required the assailant *and his victim* to be bound and thrown into a river. As the injured party (after all, *his* property had been damaged), the husband could choose to let his wife drown or save her. The unfair stigmatization of rape survivors is thus an age-old tradition. The ancient Hebrews stoned to death a married woman who was raped and her assailant. In the ancient Babylonian and Hebrew cultures, the wife was seen as guilty of adultery. Virgins who were raped within the city gates were also stoned by the Hebrews; it was thought that they could have maintained their purity by crying out.

The current definition of rape varies from state to state. **Forcible rape** is usually defined as sexual intercourse with a nonconsenting person by the use of force or the threat of force. **Statutory rape** refers to sexual intercourse with a person who is below the age of consent, even if the person cooperates.

Traditionally, a man could not be convicted of raping his wife, even though he might have forced her to submit to sexual activity by physical power or threats. This marital exclusion was derived from the English common law that held that a woman "gives herself over" to her husband when she becomes his wife and cannot then retract her consent. Today, however, most states have rape statutes that permit the prosecution of husbands who rape their wives. Many states have broadened the scope of rape laws to include forced sexual acts other than coitus, such as anal intercourse and oral–genital relations. Rape laws are now also applied to men who rape men and to women who coerce men into sexual activity or assist men in raping other women. Forcible rape is a form of **sexual assault**,

Forcible rape Sexual intercourse with a nonconsenting person obtained by the use of force or the threat of force.

Statutory rape Sexual intercourse with a person who is below the age of consent. Sexual intercourse under such conditions is considered statutory rape even though the person attacked may cooperate.

Sexual assault Any sexual activity that involves the use of force or the threat of force.

Even when a sexual attack does not meet the legal definition of rape, as in the case of forced penetration of the anus by an object such as a bottle or a broom handle, it can be prosecuted as sexual assault (Powell, 1996).

Incidence of Rape

The government's National Crime Victimization Survey (1995) estimates that 500,000 women are sexually assaulted each year. This figure includes 170,000 rapes and 140,000 attempted rapes. This means that a woman was raped about every 3 minutes on the average. The number of rapes has been outpacing population growth (FBI, 1991; U.S. Senate Committee on the Judiciary, 1991).

Historical studies have seriously underreported the incidence of rape (Schafran, 1995). They have largely relied on crime statistics, but the majority of rapes are neither reported to the police nor prosecuted (Gibbs, 1991). Many women choose not to report assaults because of concern that they will be humiliated by the criminal justice system. Others fear reprisal from their families or the rapist. Some simply assume that the offender will not be apprehended or prosecuted. Because they live in a culture in which women are often expected to "suffer in silence," Mexican American women are even more likely than European American women to remain quiet about rape and sexual abuse (Lira et al., 1999).

Truth or Fiction? **REVISITED**

It is true that in the United States a woman is raped at least every 5 minutes. Every 3 minutes is actually closer to the truth. ■

According to Schafran (1995), there are two reasons why even the National Crime Victimization Survey underestimates the incidence of rape in the United States. First, many women mistakenly believe that coercive sex is rape only when the rapist is a stranger. Second, many women mistakenly assume that only forced vaginal penetration is defined as rape. But many states define rape more broadly.

Other surveys report the following prevalence of rape or sexual assaults:

- 21% of a sample of more than 5,000 female members of a health maintenance plan reported being sexually assaulted (Koss, 1988).
- 18.5% of a sample of women clients of 257 psychotherapists in North Carolina reported being sexually assaulted (Dye & Roth, 1990).
- 15% of the 3,187 women sampled in a national survey of college students reported that they had been raped. An additional 12% reported that they had been victims of an attempted rape (Koss et al., 1987).
- Nearly 22% of the women in the NHSLS study reported being forced to do something sexual by a man (Laumann et al., 1994).

The weight of the evidence suggests that between 14% and 25% of women in the United States are raped at some point during their lives (Calhoun & Atkeson, 1991; Koss, 1993). The prevalence of reported rapes in the United States is 13 times greater than that in Great Britain and more than 20 times greater than that in Japan (*Newsweek*, 1990). Later we consider some of the cultural influences that make our society a breeding ground for rape.

It is true that the prevalence of rape is 20 times greater in the United States than in Japan. (That is, the prevalence of *reported* rape is more than 20 times greater in the United States.) ■

Women of all ages, races, and social classes are raped. Young women, however, are at greater risk than older women. Women of ages 16 to 24 are two to three times more likely to be raped than other women (National Crime Victimization Survey, 1995).

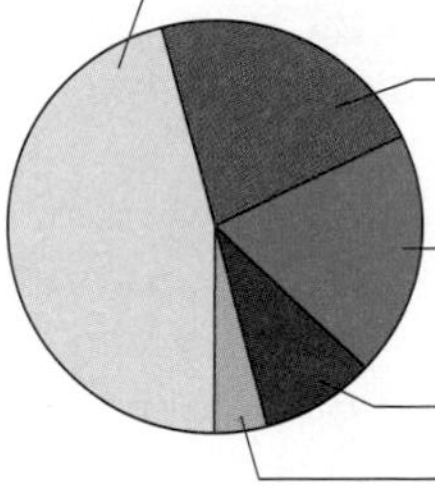

Someone with whom the respondent was in love (46%)
Someone that the respondent knew well (22%)
Acquaintance (19%)
Spouse (9%)
Stranger (4%)

FIGURE 18.1

Women's Relationships with Men Who Forced Them to Do Something Sexual That They Did Not Want to Do. Only 4% of the sexual assaults reported in the NHSLS study were perpetrated by strangers. (*Source:* Adapted from E. O. Laumann, J. H. Gagnon, R. T. Michael, & S. Michaels, [1994]. *The Social Organization of Sexuality: Sexual Practices in the United States.* Chicago: University of Chicago Press, Figure 9.3, p. 338. Copyright © 1994 by University of Chicago Press. Reprinted by permission.)

Types of Rapes

One of the central myths about rape in our culture is that most rapes are perpetrated by strangers lurking in dark alleyways or by intruders who climb through open windows in the middle of the night. Most women are raped by men they know, however—often by men they have come to trust. Figure 18.1 shows that only 4% of the women in the NHSLS study were "forced to do something sexual that they did not want to do" by a stranger. According to the National Crime Victimization Survey, 80% of rapes were committed by acquaintances of the victim (Schafran, 1995). The types of rape include stranger rape, acquaintance rape, marital rape, male rape, and rape by females.

It is not true that the majority of rapes are committed by strangers in deserted neighborhoods or darkened alleyways. Actually, most women are raped by men they know, not by strangers. ■

STRANGER RAPE **Stranger rape** refers to a rape that is committed by an assailant (or assailants) not previously known to the person attacked. The stranger rapist often selects targets who seem vulnerable—women who live alone, who are older or retarded, who are walking down deserted streets, or who are asleep or intoxicated. After choosing a target, the rapist may search for a safe time and place to commit the crime, such as a deserted, run-down part of town, a darkened street, a second-floor apartment without window bars or locks.

ACQUAINTANCE RAPE Women are more likely to be raped by men they know, such as classmates, fellow office workers, and even their brothers' friends, than by strangers (Schafran, 1995). **Acquaintance rapes** are much less likely than stranger rapes to be reported to the police (Schafran, 1995). One reason is that rape survivors may not perceive sexual assaults by acquaintances as rapes. Only 27% of the women in the national college survey who had been sexually assaulted saw themselves as rape victims (Koss et al., 1987). Despite increased public awareness of acquaintance rape, many still think of rapists as strangers lurking in shadows and believe that a woman should be able to resist a sexual advance unless the man uses a weapon (Calhoun & Atkeson, 1991). Acquaintance rapists tend to rationalize their behavior by subscribing to myths such as the traditional view that men are expected to assume a sexually aggressive role in dating and the belief that rapists are strangers. Even when acquaintance rapes are reported to police, they are often treated as "misunderstandings" or lovers' quarrels rather than as violent crimes.

DATE RAPE Date rape is a form of acquaintance rape. Studies of college women show a consistent trend: 10% to 20% of women report being forced into sexual intercourse by dates (Tang et al., 1995). These figures hold at the Chinese University of Hong Kong as well as in the United States (Tang et al., 1995). In one U.S. study, most reported date rapes were committed by men whom the women had known for nearly a year on the average (Muehlenhard & Linton, 1987). Rapes were more likely to occur when the couple had too much to drink and then parked in the man's car or went back to his residence. The

Stranger rape Rape that is committed by an assailant previously unknown to the person who is assaulted.

Acquaintance rape Rape by an acquaintance of the person who is assaulted.

man tended to perceive his partner's willingness to return home with him as a signal of sexual interest, even if she resisted his advances. In this study, most of the men ignored women's protests and overcame their resistance by force. None used a weapon. Only a few used threats of violence.

Men who commit date rape may believe that acceptance of a date indicates willingness to engage in coitus. They may think that women should reciprocate with coitus if they are taken to dinner. Other men assume that women who frequent places like singles bars are expressing tacit agreement to sex with men who show interest in them. Some date rapists believe that a woman who resists advances is just "protesting too much" so that she will not look "easy." They interpret resistance as coyness. It is a ploy in the cat-and-mouse game that typifies the "battle of the sexes" to them. They may believe that when a woman says no, she means maybe. When she says maybe, she means yes. They may thus not see themselves as committing rape. But they are.

The issue of consent lies at heart of whether a sexual act is rape. Unlike cases of stranger rape, date rape occurs within a context in which sexual relations could occur voluntarily. Thus the issue of consent can become murky. Juries and judges are often faced with a woman plaintiff who alleges that the male defendant, who may appear neatly dressed and looking like the boy next door, forced her into sexual relations against her will. As in the William Kennedy Smith and Mike Tyson trials, the defendant may concede that sexual intercourse took place but claim that it was consensual. Judges and juries face the task of discerning shadings in meaning regarding "consent." Attorneys on both sides vie to persuade them to see things their way.

Charges of date rape often come down to his word against hers. Her word often becomes less persuasive in the eyes of the jury if it was clear that she had consented to mutual activities beforehand, such as sharing dinner, attending the movies together, accompanying him to his home, sharing a drink alone, and perhaps kissing or petting. Let us state in no uncertain terms, however, that it does not matter whether the woman wore a "sexy" outfit, was "on the pill," or shared a passionate kiss or embrace with the man. If the encounter ended with the woman's being forcibly violated, then it is rape. When a woman says no, a man must take no for an answer.

The problem of date rape has been subjected to closer public scrutiny in recent years. "Take Back the Night" marches have become a common form of student protest on college campuses against the sexual misconduct of men (Gross, 1993). Many colleges have mandated date rape seminars and workshops.

THE GANG RAPE Groth and Birnbaum (1979) relate the story of Kurt, a 23-year-old European American, married father of three who was involved in a number of rapes with a friend, Pete.

> I always looked up to Pete and felt second-class to him. I felt I owed him and couldn't chicken out on the rapes. I worshiped him. He was the best fighter, lover, water-skier, motorcyclist I knew. Taking part in the sexual assaults made me feel equal to him. . . . I didn't have any friends and felt like a nobody. . . . He brought me into his bike club. He made me a somebody.
>
> I'd go to a shopping center and find a victim. I'd approach her with a knife or a gun and then bring her to him. He'd rape her first and then I would. . . . We raped about eight girls together over a four-month period. (p. 113)

By participating in a gang rape, the follower, like Kurt, is attempting to conform to the stereotype of the tough, competent, "masculine" he-man. Followers, however, appear to fortify the courage of the instigator of the act. One of Groth and Birnbaum's planners of such an assault remarks, "Having a partner is like having something to drink. I felt braver. I felt stronger. This gave me the courage to do something I might not have done on my own" (1979, p. 112).

Exercise of power appears to be the major motive behind gang rapes, although some attackers may also be expressing anger against women. Gang members often believe that

a CLOSER look

Anatomy of a Date Rape: Ann and Jim

Date rape is a pressing concern on college campuses, where thousands of women have been raped by men they knew or had dated, and where there is much controversy what exactly constitutes date rape (Gibbs, 1991). Consider the case of Ann (The College of New Jersey, 1991).

> I first met him at a party. He was really good looking and he had a great smile. I wanted to meet him but I wasn't sure how. I didn't want to appear too forward. Then he came over and introduced himself. We talked and found we had a lot in common. I really liked him. When he asked me over to his place for a drink, I thought it would be OK. He was such a good listener, and I wanted him to ask me out again.
>
> When we got to his room, the only place to sit was on the bed. I didn't want him to get the wrong idea, but what else could I do? We talked for a while and then he made his move. I was so startled. He started by kissing. I really liked him so the kissing was nice. But then he pushed me down on the bed. I tried to get up and I told him to stop. He was so much bigger and stronger. I got scared and I started to cry. I froze and he raped me.
>
> It took only a couple of minutes and it was terrible, he was so rough. When it was over he kept asking me what was wrong, like he didn't know. He had just forced himself on me and he thought that was OK. He drove me home and said he wanted to see me again. I'm so afraid to see him. I never thought it would happen to me.

College men on dates frequently perceive their dates' protests as part of an adversarial sex game. One male undergraduate said, "Hell, no" when asked whether a date had consented to sex. He added, "but she didn't say no, so she must have wanted it, too. . . . It's the way it works" (Celis, 1991). Consider the comments of Jim, the man who raped Ann (The College of New Jersey, 1991):

> I first met her at a party. She looked really hot, wearing a sexy dress that showed off her great body. We started talking right away. I knew that she liked me by the way she kept smiling and touching my arm while she was speaking. She seemed pretty relaxed so I asked her back to my place for a drink. . . . When she said yes, I knew that I was going to be lucky!
>
> When we got to my place, we sat on the bed kissing. At first, everything was great. Then, when I started to lay her down on the bed, she started twisting and saying she didn't want to. Most women don't like to appear too easy, so I knew that she was just going through the motions. When she stopped struggling, I knew that she would have to throw in some tears before we did it.
>
> She was still very upset afterwards, and I just don't understand it! If she didn't want to have sex, why did she come back to the room with me? You could tell by the way she dressed and acted that she was no virgin, so why she had to put up such a big struggle I don't know.

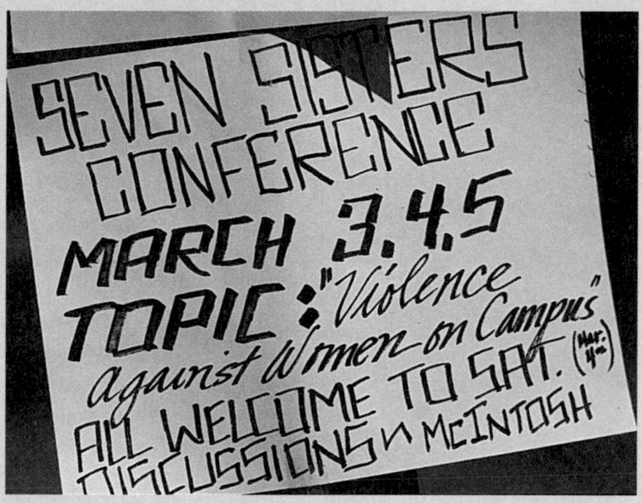

Combating Rape on Campus. Many colleges and universities have instituted rape-awareness programs to combat the problem of rape on campus. What is the prevalence of rape on your campus? What are you going to do about it?

Accepting a date is not the equivalent of consenting to coitus. Accompanying a man to his room or apartment is not the equivalent of consenting to coitus. Nor are kissing and petting the equivalent of consenting to coitus. Let us reiterate a point worth restating: When the woman says no, the man is obligated to take no for an answer.

once women engage in coitus, they are "whores." Thus each offending gang member may become more aggressive as he takes his turn.

The Koss college survey (Koss et al., 1987) showed that sexual assaults involving a group of assailants tend to be more vicious than individual assaults (Gidycz & Koss, 1990). Relatively few survivors of gang rape reported the attack to police or sought support from a crisis center.

MALE RAPE The prevalence of male rape is unknown because most assaults are not reported. Some estimates suggest that perhaps 1 in 10 rape survivors is a man (Gibbs, 1991). Most men who rape other men are heterosexual. Their motives tend to include domination and control, revenge and retaliation, sadism and degradation, and (when the rape is carried out by a gang member) status and affiliation (Groth & Birnbaum, 1979). Sexual motives are generally absent.

It is not true that men who rape other men are gay. Most men who rape other men are heterosexual. ■

Most male rapes occur in prison settings, but some occur outside prison walls. Male rape survivors tend to suffer greater physical injury than female survivors (Kaufman et al., 1980). Males are more often attacked by multiple assailants, are held captive longer, and are more often reluctant to report the assault (Gerrol & Resick, 1988; Groth & Burgess, 1980; Myers, 1989). After all, victimization does not fit the male stereotype of capacity for self-defense. Men are expected to be not only strong but also silent. Male rape survivors may suffer traumatic effects similar to those suffered by female rape survivors, however (Calhoun & Atkeson, 1991).

MARITAL RAPE

> The typical marital rapist is a man who still believes that husbands are supposed to "rule" their wives. This extends, he feels, to sexual matters. When he wants her, she should be glad, or at least willing. If she isn't, he has the right to force her. But in forcing her he gains far more than a few minutes of sexual pleasure; he humbles her and reasserts, in the most powerfully emotional way possible, that he is the ruler and she is the subject. (Hunt, 1979, p. 38)

When Hunt wrote these words, most states would not prosecute husbands for even the most violent of rapes. A number of recent legal changes have made it possible for sexually assaultive husbands to be convicted on rape charges, however. Nearly half the states have completely removed preferential treatment for assaultive husbands. In many other states, however, husbands still receive the benefit of the doubt unless there is evidence of gross brutality (Gibbs, 1991).

Marital rapes are probably more common than date rapes because a sexual relationship has already been established. A husband may believe that he is entitled to sexual access to his wife any time he desires it. He may be less willing to accept a rebuff. He may believe it is his wife's duty to satisfy his sexual needs even when she is uninterested. Although there are no precise statistics on marital rape, a committee of the U.S. Congress estimated that one wife in seven is likely to be raped by her husband (Gibbs, 1991).

Marital rape goes largely unreported and unrecognized by survivors as rape (Russell, 1982). Women may fail to report marital rape because of fear that no one will believe them.

Motives for marital rape vary. Some men use sex to dominate their wives. Others degrade their wives through sex, especially after arguments. Sexual coercion often occurs within a context of a pattern of marital violence, battering, and physical intimidation (Finkelhor & Yllo, 1982; Russell, 1982). In some cases, though, violence is limited to the sexual relationship (Finkelhor & Yllo, 1982). Some men see sex as the solution to all marital disputes. They think that if they can force their wives into coitus, "Everything will be OK."

Survivors of marital rape may be as fearful as survivors of stranger rape of serious injury or death (Kilpatrick et al., 1987). The long-term effects of marital rape on survivors are also similar to those experienced by survivors of stranger rape (Calhoun & Atkeson, 1991), including fear, depression, and sexual dysfunctions (Kilpatrick et al., 1987). Moreover, the woman who is raped by her husband usually continues to live with her assailant and may fear repeated attacks.

RAPE BY WOMEN Rape by women is rare. When it does occur, it often involves aiding or abetting men who are attacking another woman. Rape by women may occur in gang rape in which women follow male leaders to gain their approval. In such cases, a woman may be used to lure another woman to a reasonably safe place for the rape. Or the woman may hold the other woman down while she is assaulted.

But men have actually been raped by women (Sarrel & Masters, 1982; Struckman-Johnson, 1988). Sarrel and Masters reported 11 cases of men who were sexually assaulted by women, including one case of a 37-year-old man who was coerced into sexual intercourse by two women who accosted him at gunpoint. In another case, a 27-year-old man fell asleep in his hotel room with a woman he had just met in a bar and then awakened to find that he was bound to his bed, gagged, and blindfolded. He was then forced into sexual intercourse with four different women, who threatened him with castration if he did not perform satisfactorily. Rape of a man may not be recognized as rape in states that adhere to a legal definition that requires forced vaginal penetration of a woman by a man. In such cases, the assailant or assailants may be charged under other statutes governing physical or sexual assaults.

Social Attitudes and Myths That Encourage Rape

Many people believe a number of myths about rape, such as "Women say no when they mean yes," "All women like a man who is pushy and forceful," "The way women dress, they are just asking to be raped," and "Rapists are crazed by sexual desire" (Powell, 1996, p. 139). Yet another myth is that deep down, women want to be raped. Many women report rape fantasies, but they are in control of their fantasies. Rape is a complete lack of control, which no one wants (Powell, 1996). The belief that women desire to be overpowered and forced by men into sexual relations is a cowardly rationalization for violence (Powell, 1996).

Truth or Fiction? REVISITED

It is not true that women say no when they mean yes. This myth has the effect of encouraging rape. ■

Rape myths create a social climate that legitimizes rape. Though both men and women may subscribe to rape myths, researchers find that college men show greater acceptance of rape myths than do college women (Brady et al., 1991; Margolin et al., 1989). Men also cling more stubbornly to myths about date rape, even after taking date rape education classes designed to challenge these views (Lenihan et al., 1992). College men who endorse rape myths are more likely to see themselves as likely to commit rape (Malamuth, 1981). Such myths do not occur in a social vacuum. They are related to other social attitudes, including gender-role stereotyping, perception of sex as adversarial, and acceptance of violence in relationships.

Sociocultural Factors in Rape

Many observers contend that our society breeds rapists by socializing males into socially and sexually dominant roles (Lisak, 1991; Powell, 1996). Males are often reinforced from childhood for aggressive and competitive behavior (Lisak, 1991). Gender typing may also lead men to reject "feminine" traits such as tenderness and empathy that might restrain aggression (Lisak, 1991).

Research with college students supports the connection between stereotypical masculine identification and tendencies to rape or condone rape. Studies have compared students who believed strictly in traditional gender roles with student who held less rigid attitudes. The "traditionalists" express a greater likelihood of committing rape, are more accepting of violence against women, are more likely to blame rape survivors, and are more aroused by depictions of rape (Check & Malamuth, 1983; Raichle & Lambert, 2000). Other researchers found that college men who more closely identified with the traditional masculine gender role more often reported having engaged in verbal sexual coercion and forcible rape (Muehlenhard & Falcon, 1990).

a CLOSER look

Feral Fowl Females Reject Rapists

In most species, social dominance in the male is linked with privileges such as feeding and mating. We are familiar with male deer locking antlers in the battle for dominance and with male birds establishing territories before females will mate with them. Dominant walruses create herds of females and keep weaker males at bay. The gene pool thus remains "strong," because dominant males are more likely to get their genes into the next generation.

In some species of birds and other animals, females do not have the opportunity to choose mates. Copulations are forced (the typical mating pattern is one of rape), and paternity is determined by competition among the ejaculates of the various males. The fittest sperm wins.

Feral fowl females are also subject to forced copulation; however, it is not the fittest sperm that wins. Instead, female feral fowls eject the ejaculates of most males who copulate with them (Pizzari & Birkhead, 2000). The females prefer to copulate with socially dominant males, and when copulation is forced by subordinate males, they "retaliate" by ejecting their sperm.

Women may be socialized into assuming the helpless role. The feminine gender role includes such traits as submissiveness, passivity, cooperativeness, and even obedience to males. These traits may make it difficult for women to resist when faced with the threat of rape. A woman may lack aggressive skills and believe that physical resistance is inappropriate or that she cannot resist. Women are also taught that it is important to be sexually attractive to men. Women may thus unfairly blame themselves for the assault, believing that they somehow enticed the assailant.

Social influences may reinforce cultural themes that underlie rape, such as the belief that a truly masculine man should be sexually aggressive and overcome a woman's resistance until she "melts" in his arms (Stock, 1991). The popular belief that women fantasize about being overpowered sanctions coercive tactics to "awaken" a woman's sexual desires. Images from popular books and movies reinforce these themes. Remember Rhett Butler in *Gone with the Wind,* carrying a protesting Scarlett O'Hara up the stairs to her bedroom? Violent pornography, which fuses violence and erotic arousal, may also serve to legitimize rape (Stock, 1991).

Young men may come to view dates not as chances to get to know their partners but as opportunities for sexual conquest, in which the object is to overcome their partners' resistance. In one study, male college students expressed support for a man's right to kiss his female partner even if she resists (Margolin et al., 1989). Malamuth (1981) found that 35% of the college men in his sample said they would force a woman into sex if they knew they could get away with it.

The social philosopher Myriam Miedzian suggests that lessons learned in competitive sports may also predispose young Americans to sexual violence (Levy, 1991). Boys are often exposed to coaches who emphasize winning at all costs. They are taught to be dominant and to vanquish their opponents, even if winning means injuring or "taking out" the opposition. This philosophy, Miedzian argues, may be carried from the playing field into relationships with women. Some athletes distinguish between sports and dating relationship; others do not. Evidence shows that student athletes commit a disproportionate number of sexual assaults (Eskenazi, 1990). Many of these are gang rapes committed by groups of student athletes who live together and bond so strongly that they share even sexual experiences.

Sexual behavior and sports in our culture are linked through common idioms. A young man may be taunted by his friends after a date with a woman with such questions as "Did you score?" or, more bluntly, "Did you get in?" Consider, for example, the aggressive competitiveness with which this male college student views dating relationships between men and women:

> A man is supposed to view a date with a woman as a premeditated scheme for getting the most sex out of her. Everything he does, he judges in terms of one criterion—"getting laid." He's supposed to constantly pressure her to see how far he can get. She is his adversary, his opponent in a battle, and he begins to view her as a prize, an object, not a person. While she's dreaming about love, he's thinking about how to conquer her. (Powell, 1991, p. 55)

Psychological Characteristics of Rapists

Although sexual aggressiveness may be woven into our social fabric, not all men are equally vulnerable to such cultural influences (Burkhart & Fromuth, 1991). Personal factors are also involved. Are rapists mentally disturbed? Retarded? Driven by insatiable sexual urges?

Much of our knowledge of the psychological characteristics of rapists derives from studies of samples of incarcerated rapists (Harney & Muehlenhard, 1991). One conclusion that emerges is that rapists vary in their psychological characteristics, family backgrounds, mental health, and criminal histories (Prentky & Knight, 1991). Generally speaking, rapists are no less intelligent or more mentally ill than other people (Renzetti & Curran, 1989). Many rapists show no evidence of psychological disturbance (Dean & de Bruyn-Kops, 1982). This does not mean that their behavior is "normal." It means that most rapists are in control of their behavior and know it is illegal.

Truth or Fiction? **REVISITED**

It is not true that most rapists are mentally ill, even though their crimes might strike observers as being "sick." ■

Some rapists feel socially inadequate and report that they cannot find willing partners. Some lack social skills and avoid social interactions with women (Overholser & Beck, 1986). Others are no less skillful socially than nonrapists in the same socioeconomic group, however (Segal & Marshall, 1985). Some rapists are basically antisocial and have long histories of violent behavior (Knight et al., 1991). They tend to act on their impulses regardless of the cost to the person they attack. Some were sexually victimized or physically assaulted as children (Groth, 1978; Sack & Mason, 1980). As adults, they may be identifying with the aggressor role in interpersonal relationships. The use of alcohol may also dampen self-restraint and spur sexual aggressiveness.

For some rapists, violence and sexual arousal become enmeshed. Thus they seek to combine sex and violence to enhance their sexual arousal (Quinsey et al., 1984). Some rapists are more sexually aroused (as measured by the size of erections) by verbal descriptions, films, or audiotapes that portray themes of rape than are other people (Barbaree & Marshall, 1991). Other researchers, however, have failed to find deviant patterns of arousal in rapists (e.g., Baxter et al., 1986; Hall, 1989). These researchers find that, as a group, rapists, like most other people, are more aroused by stimuli depicting mutually consenting sexual activity than by rape stimuli.

Studies of incarcerated rapists can be criticized on the grounds that these samples may not represent the total population of rapists. It is estimated that fewer than 4% of rapists are caught and eventually imprisoned (Gibbs, 1991). Most rapes are committed by acquaintances. Acquaintance rapists are even less likely than stranger rapists to be arrested, convicted, and incarcerated.

To offset this methodological concern, researchers have turned to the survey method to study men who anonymously report that they have engaged in sexually coercive behaviors, including rape, but have not been identified by the criminal justice system.

Koss and her colleagues (1987) found that in their national college sample of nearly 3,000 college men, about 1 man in 13 (7.7%) admitted to committing or attempting rape.

Harney and Muehlenhard (1991) summarized research findings on self-identified sexually aggressive men. They are more likely than other men to

- Condone rape and violence against women
- Hold traditional gender-role attitudes
- Be sexually experienced
- Be hostile toward women
- Engage in sexual activity in order to express social dominance
- Be sexually aroused by depictions of rape
- Be irresponsible and lack a social conscience
- Have peer groups, such as fraternities, that pressure them into sexual activity

THE MOTIVES OF RAPISTS: THE SEARCH FOR TYPES Although sexual arousal is an obvious and important element in rape (Barbaree & Marshall, 1991), some researchers argue that sexual desire is not the basic motivation for rape (Gebhard et al., 1965; Groth & Birnbaum, 1979). Other researchers believe that sexual motivation plays a key role in at least some rapes (Hall & Hirschman, 1991). On the basis of their clinical work with more than 1,000 rapists, Groth and Birnbaum believe that there are three basic kinds of rape: anger rape, power rape, and sadistic rape.

- *Anger rape.* The **anger rape** is a vicious, unplanned attack that is triggered by anger and resentment toward women. Anger rapists usually employ more force than is needed to obtain compliance. The person they rape is often coerced into performing degrading and humiliating acts, fellatio, or anal intercourse. Typically, the anger rapist reports that he had suffered humiliations at the hands of women and used the rape as a means of revenge.
- *Power rape.* The man who commits a **power rape** is motivated by the desire to control and dominate the woman he rapes. Sexual gratification is secondary. The power rapist uses rape as an attempt "to resolve disturbing doubts about [his] masculine identity and worth, [or] to combat deep-seated feelings of insecurity and vulnerability" (Groth & Hobson, 1983, p. 165). Only enough force to subdue the woman is used.
- *Sadistic rape.* The **sadistic rape** is a ritualized, savage attack. Sadistic rapists often carefully plan their assaults and use a "con" or pretext to approach their targets, such as asking for directions or offering or requesting assistance (Dietz et al., 1990). Some sadists bind their victims and subject them to humiliating and degrading experiences and threats. Some torture or murder their victims (Dietz et al., 1990). Mutilation of the victim is common. Groth and Birnbaum (1979) suggest that sadistic rapists are often preoccupied with violent pornography but have little or no interest in nonviolent (consensual) pornography. Groth (1979) estimated that about 40% of rapes are anger rapes, 55% power rapes, and 5% sadistic rapes.

Adjustment of Rape Survivors

Many women who are raped fear for their lives during the attack (Calhoun & Atkeson, 1991). Whether or not weapons or threats are used, the experience of being dominated by an unpredictable and threatening assailant is terrifying. The woman does not know whether she will survive and may feel helpless to do anything about it. Afterwards, many survivors enter a **crisis.**

Many survivors are extremely distraught in the days and weeks following the rape (Koss, 1993; Valentiner et al., 1996). They have trouble sleeping and cry frequently. They tend to report eating problems, cystitis, headaches, irritability, mood changes, anxiety and depression, and menstrual irregularity. They may become withdrawn, sullen, and mistrustful (McArthur, 1990). People in the United States tend to believe that women who are raped are at least partly to blame for the assault (Bell et al., 1994). Therefore, some survivors experience feelings of guilt and shame (McArthur, 1990). Emotional distress tends to peak in severity by about three weeks following the assault and generally remains high for about a month before beginning to abate about a month or two later (Koss, 1993;

Anger rape A vicious, unplanned rape that is triggered by feelings of intense anger and resentment toward women.

Power rape Rape that is motivated by the desire to control and dominate the person assaulted.

Sadistic rape A highly ritualized, savage rape in which the person who is attacked is subjected to painful and humiliating experiences and threats.

Crisis A highly stressful situation that can involve shock, loss of self-esteem, and lessened capacity for making decisions.

What Motivates Rapists? Sure, rape has a good deal to do with sex—but is it sex for the sake of sex or is sex used as a weapon? Research suggests that many rapists use sex to express anger toward women, power over women, or sadistic impulses. In fact, some men do not become sexually aroused unless they hurt or humiliate women. Make no mistake about it: Rape is a crime of violence and should be treated as such.

Valentiner et al., 1996). About 1 survivor in 4 encounters emotional problems that linger beyond a year (Calhoun & Atkeson, 1991; Hanson, 1990).

According to a community survey of sexual assault survivors (both women and men) from Los Angeles, the most frequent emotional reactions to sexual assault were anger (59%), sadness (43%), and anxiety (40%) (see Table 18.1). A total of 447 persons from the sample of more than 3,000 adults reported that they had suffered a sexual assault, which

TABLE 18.1

Prevalence of emotional and behavioral reactions to sexual assault in Los Angeles community sample (in percents)

Reaction	Women	Men	Total
Fearful	45.5%	15.9*%	35.1%
Stopped doing things	31.8	15.2*	25.9
Fearful of sex	21.6	7.9*	16.8
Less sexual interest	32.5	6.9*	23.5
Less sexual pleasure	27.1	8.0*	20.4
Felt dishonored or spoiled	33.9	20.0*	29.1
Guilt	35.0	25.9*	31.8
Sad, blue, or depressed	50.6	28.8*	43.0
Anger	72.0	34.8*	59.0
Tense, nervous, or anxious	49.9	22.9*	40.4
Insomnia	25.2	11.6*	20.4
Loss/increase in appetite	15.8	8.4	13.2
Alcohol/drug use	5.1	5.9	5.4
Fearful of being alone	23.0	2.2*	15.7

*Indicates that the differences between men and women were statistically significant at the .05 level of significance; that is, there is less than a 5% chance that the differences between men and women were due to chance fluctuations.

Source: J. M. Siegel et al. (1990). "Reactions to Sexual Assault: A Community Study." *Journal of Interpersonal Violence*, 5: 229–246. Copyright © 1990 by authors. Reprinted by permission of Sage Publications, Inc.

included a range of coercive acts from fondling of the breasts or sex organs to sexual intercourse (Siegel et al., 1990). As shown in Table 18.1, women were significantly more likely to report 12 of the 14 reactions listed. Survivors who were physically threatened by their assailants reported greater fear and anxiety, depression, and sexual distress (reduced interest and pleasure, fear of sex). Attacks that resulted in forced intercourse were more distressing than those that did not.

Rape survivors may also suffer physical injuries and may contract sexually transmitted diseases, even AIDS, as a result of a sexual assault. In one study, concerns about contracting AIDS were reported by about 1 in 4 survivors (Baker et al., 1990).

Rape survivors may also be at risk for long-term health complications. In one study, survivors reported more current physical complaints, including gynecological problems, when they were assessed *two or more years* following the attack than did a reference group matched on demographic variables such as race and age (Waigandt et al., 1990). A study of women in the military who had survived rape and physical abuse found similar health-related problems a decade after the assault (Sadler et al., 2000).

Survivors may also encounter problems at work, such as trouble relating to coworkers or bosses or difficulty concentrating. Work adjustment, however, usually returns to normal within a year (Calhoun & Atkeson, 1991). Relationships with spouses or partners may also be impaired. Disturbances in sexual functioning are common and may last for years—or even a lifetime. Survivors often report a lack of sexual desire, fear of sex, and difficulty becoming sexually aroused (Becker et al., 1986). Some women never again experience the level of sexual pleasure they enjoyed before the assault (Calhoun & Atkeson, 1991).

SURVIVORS' REACTIONS TO THE ASSAULT: WHOM TO TELL? About two thirds (65%) of the 447 sexual assault survivors in the Los Angeles survey reported they had told someone about the assault (Golding et al., 1989). The person was usually a friend or a relative. Only 1 in 10 reported the assault to the police. Only 1 in 6 consulted a mental health professional. Survivors were more likely to inform police or physicians of attacks by strangers than of attacks by acquaintances.

Most women fail to report sexual assaults to police. Why? The reasons include fear of retaliation, the social stigma attached to the survivors of rape, doubts that others will believe them, feelings that it would be hopeless to try to bring charges against the perpetrator, concerns about negative publicity, and fears about the emotional distress to which they would be subjected if the case were to go to trial.

RAPE AND PSYCHOLOGICAL DISORDERS Rape survivors are at higher-than-average risk of developing anxiety disorders and depression, and of abusing alcohol and other substances (Koss, 1993). One anxiety disorder is **post-traumatic stress disorder** (PTSD) (Valentiner et al., 1996). PTSD is brought on by exposure to a traumatic event; for example, it is often seen in soldiers who have been in combat (American Psychiatric Association, 2000). People with PTSD may have flashbacks to the traumatic experience, disturbing dreams, emotional numbing, and nervousness. PTSD may persist for years. The person may also develop fears of situations connected with the traumatic event. For example, a woman who was raped on an elevator may develop a fear of riding in elevators by herself. Researchers also report that women who blame themselves for the rape tend to suffer more severe depression and adjustment problems, including sexual problems (Frazier, 1990; Wyatt et al., 1990).

Post-traumatic stress disorder A type of stress reaction brought on by a traumatic event and characterized by flashbacks of the experience in the form of disturbing dreams or intrusive recollections, a sense of emotional numbing or restricted range of feelings, and heightened body arousal. Abbreviated *PTSD.*

Rape trauma syndrome A two-phase reaction to rape that is characterized by disruption of the survivor's lifestyle (the acute phase) and reorganization of the survivor's life (the long-term phase).

RAPE TRAUMA SYNDROME Ann Burgess and Lynda Holmstrom (1974) identified some common response patterns in survivors of rape, which they labeled the **rape trauma**

syndrome. Through emergency room interviews with 92 women at Boston City Hospital, and telephone or in-person follow-up interviews, Burgess and Holmstrom found two phases in rape trauma syndrome:

The acute phase: disorganization. The acute phase typically lasts for several weeks following the attack. Many survivors are disorganized during this time and may benefit from rape trauma counseling. The woman may cry uncontrollably and experience feelings of anger, shame, fear, and nervousness. Some women present a calm composed face to the world, but inwardly have not yet come to terms with the traumatic experience. Calmness often gives way to venting of feelings later on.

The long-term process: reorganization. The long-term reorganization phase may last for years. The woman gradually comes to deal with her feelings and to reorganize her life. Lingering fears may lead rape survivors to move to safer surroundings. Women who informed the police may fear retaliation by the rapist. They may change their phone numbers, often to an unlisted number. Some take out-of-state trips, often visiting parents, although they do not necessarily tell them of the rape. Many survivors continue to be bothered by frightening dreams.

Burgess and Holmstrom noted that survivors of rape who do not disclose their attacks may have a silent rape reaction. Unfortunately, concealing the rape may prevent them from receiving social support. If survivors are children or adolescents when they are assaulted, their feelings may go unresolved for years.

IF YOU ARE RAPED . . . Elizabeth Powell (1996) offers the following suggestions to those who are raped:

1. Don't change anything about your body—don't wash, don't even comb your hair. Leave your clothes as they are. Otherwise you could destroy evidence.

2. Strongly consider reporting the incident to police. You may prevent another woman from being assaulted, and you will be taking charge, starting on the path from victim to survivor.

3. Ask a relative or friend to take you to a hospital if you can't get an ambulance or a police car. If you call the hospital, tell them why you're requesting an ambulance, in case they are able to send someone trained to deal with rape cases.

4. Seek help in an assertive way. Seek medical help. Injuries you are unaware of may be detected. Insist that a written or photographic record be made to document your condition. If you decide to file charges, the prosecutor may need this evidence to obtain a conviction.

5. Question health professionals. Ask about your biological risks. Ask what treatments are available. Ask for whatever will help make you comfortable. Call the shots. Demand confidentiality if that's what you want. Refuse what you don't want.

You may also wish to call a rape hotline or rape crisis center for advice, if one is available in your area. A rape crisis volunteer may be available to accompany you to the hospital and help see you through the medical evaluation and police investigation if you report the attack. It is not unusual for rape survivors to try to erase the details of the rape from their minds. However, trying to remember details clearly will enable you to provide an accurate description of the rapist to the police, including his clothing, type of car, and so on. This information may help police apprehend the rapist and may be of great value in the prosecution.

Blaming the Victim to an Extreme

This is a story about women . . . that begins with a man. His name is Ghazi al Marine. He had been married just three months—the happiest months of his life, he told us—when a killer came to their home and shot his wife.

"She was lying in that corner," he says, pointing. "The blood was everywhere."

That night, police arrested her brother. Apparently, he murdered his sister to preserve his family's honor—an honor that was stained when she was raped three years ago. From the time she was raped until her marriage, she had lived in a prison. Not because she had done anything wrong but because, like dozens of other women, she needed to be protected from her family.

The reputation of a family in certain countries rests on the reputation of its women. All it takes for violence to be justified is for a woman to be seen with a man she is not related to, for her to reject an unwanted suitor, or even [for her] to be the victim of a sexual crime.

In Jordan, more than 30 women a year are killed in the name of "honor," which means one-third of the nation's murders. In Pakistan, human rights organizations say, thousands of women and girls are stabbed, shot, maimed or burned each year.

In Bangladesh, a typical punishment is sulfuric acid thrown in the woman's face."Why did they do this to us?" asks Bina Akhtar in a *20/20* interview. "We didn't harm anyone." "It's the male feeling that they are responsible and authorized to control women's lives and bodies," says Asthma Khadar, a Jordanian human rights lawyer.

No Regrets

The men who commit these crimes know they will probably never be punished. The few who are arrested show very little remorse

"I'm proud that I killed her," says one man in prison. "I'm ashamed that she was my sister." These men will serve a few months in prison, at most, for the murder of a mother, a sister, a daughter, a wife.

"Most of the time," says Nazrin Huk, a Bangladeshi women's rights activist, "people who throw acid get away with it. Most people think, 'Oh, we can, you know, nothing is going to happen to us.' "

Protests in some of these countries have produced some signs of change. For the first time, Pakistan's President, Pervez Musharraf, has made this promise: "Pakistan will treat honor killings as murder."

In Jordan, the new king and queen have vowed to end the practice. The courts have begun in a few cases to issue longer sentences for honor crimes.

Ghazi al Marine has told his lawyer he will fight in the courts to make sure that his wife did not die in vain. But will he get justice?

"I hope so," he says. "He should be hanged . . . like everyone who kills a woman."

The honor crimes continue, but there is now at least the possibility for justice. And just a little less acceptance of violence in the name of honor.

Source: Reprinted with permission from Sheila MacVicar (2000, June 5). "Crimes in the Name of Honor." www.abcnews.com.

Treatment of Rape Survivors

Treatment of rape survivors typically involves a two-stage process of helping the victim through the crisis after the attack and then helping to foster long-term adjustment. Crisis intervention typically provides the survivor with support and information to help her express her feelings and develop strategies for coping with the trauma (Resick & Schnicke, 1990). Psychotherapy, involving group or individual approaches, can help the survivor cope with the emotional consequences of rape, avoid self-blame, improve self-esteem, validate the welter of feelings surrounding the experience, and help her establish or maintain loving relationships. Therapists also recognize the importance of helping the rape survivor mobilize social support (Ledray, 1990; Valentiner et al., 1996). Family, friends, religious leaders, and health care specialists are all potential sources of help. In major cities and many towns, concerned men and women have formed rape crisis centers and hotlines, peer counseling groups, and referral agencies geared to assessing and meeting survivors' needs after the assault. Some counselors are specially trained to mediate between survivors of rape and their loved ones—husbands, lovers, and so forth. These counselors help people to discuss and work through the often complex emotional legacy of rape. Phone numbers for these services can be obtained from feminist groups (for example, your local office

Taking Back the Night. Whose fault is it if a woman is raped when she goes out alone at night? Many women—and men who care about women—have marched to demonstrate their disgust for the men who might assault them if they were out walking by themselves and toward a society that too often blames the victim for what happens to her.

of the National Organization for Women, NOW), the police department, or the telephone directory.

Rape Prevention

The elimination of rape would probably require massive changes in cultural attitudes and socialization processes. Educational intervention on a smaller scale may reduce its incidence, however. A study of 276 undergraduates at Pitzer College in California showed that college men who were more knowledgeable about the trauma caused by rape were less likely to report that they might commit a rape (Hamilton & Yee, 1990). Many colleges and universities offer educational programs about date rape. The University of Washington, for example, offers students lectures and seminars on date rape (and also provides women with escorts to their homes or dorms after dark). Brown University requires all first-year students to attend orientation sessions on rape (Celis, 1991). The point of such programs is for men to learn that no means no, despite the widespread belief that some women like to be "talked into" sex.

Until the basic cultural attitudes that support rape change, however, "rape prevention" means that women must take a number of precautions. Why should women be advised to take measures to avoid rape? Is not the very listing of such measures a subtle way of blaming the woman if she should fall prey to an attacker? No! To provide the information

a CLOSER look

Rape Prevention

The New Our Bodies, Ourselves (Boston Women's Health Book Collective, 1992; Copyright © 1984, 1992 by The Boston Women's Health Book Collective. Reprinted with permission.) lists several suggestions that may help prevent rape:

- Establish a set of signals with other women in the building or neighborhood.
- List yourself in the phone directory and on the mailbox by your first initials only.
- Use dead-bolt locks.
- Lock all windows, and install iron grids on first-floor windows.
- Keep doorways and entries well lit.
- Keep your keys handy when approaching the car or the front door.
- Do not walk by yourself after dark.
- Avoid deserted areas.
- Do not allow strange men into your house or apartment without first checking their credentials.
- Keep your car doors locked and the windows up.
- Check out the back seat of your car before entering.
- Don't live in a risky building. (We realize that this suggestion may be of little use to poor women who have relatively little choice about where they live.)
- Don't give rides to hitchhikers (including women hitchhikers).
- Don't converse with strange men on the street.
- Shout "Fire!" not "Rape!" People are likely to flock to fires but to avoid scenes of violence.

Powell (1996) adds the following suggestions for avoiding date rape:

- Communicate your sexual limits to your date. Tell your partner how far you would like to go so that he will know what the limits are. For example, if your partner starts fondling you in ways that make you uncomfortable, you might say, "I'd prefer if you didn't touch me there. I really like you, but I prefer not getting so intimate at this point in our relationship."
- Meet new dates in public places, and avoid driving with a stranger or a group of people you've met. When meeting a new date, drive in your own car and meet your date at a public place. Don't drive with strangers or offer rides to strangers or groups of people. In some cases of date rape, the group disappears just prior to the assault.
- State your refusal definitively. Be firm in refusing a sexual overture. Look your partner straight in the eye. The more definite you are, the less likely that your partner will misinterpret your wishes.
- Become aware of your fears. Take notice of any fears of displeasing your partner that might stifle your assertiveness. If your partner is truly respectful of you, you need not fear an angry or demeaning response. But if your partner is not respectful, it is best to become aware of it and end the relationship right there.
- Pay attention to your "vibes." Trust your gut-level feelings. Many victims of acquaintance rape said afterward that they had had a strange feeling about the man but failed to pay attention to it.
- Be especially cautious if you are in a new environment, be it a college or a foreign country. You may be especially vulnerable to exploitation when you are becoming acquainted with a new environment, different people, and different customs.
- If you have broken off a relationship with someone you don't really like or feel good about, don't let him into your place. Many so-called date rapes are committed by ex-lovers and ex-boyfriends.

is not to blame the person who is attacked. The rapist is *always* responsible for the assault. But taking certain precautions, such as those discussed in the nearby A Closer Look feature, may lower a woman's risk of being assaulted.

CONFRONTING A RAPIST: SHOULD YOU FIGHT, FLEE, OR PLEAD? What if you are accosted by a rapist? Should you try to fight him off, flee, or try to plead with him to stop?

Some women have thwarted attacks by pleading or crying. Yet research has shown that less forceful forms of resistance, such as pleading, begging, or reasoning, can be dangerous strategies. They may not fend off the attack and may heighten the probability of injury (Bart & O'Brien, 1985). Screaming may be particularly effective in warding off some attacks (Byers & Lewis, 1988). Running away is sometimes an effective strategy for avoiding a rape (Bart & O'Brien, 1985), for example, but running may not be effective if the woman is outnumbered by a group of assailants (Gidycz & Koss, 1990). No single suggestion is likely to be helpful in all rape cases.

Self-defense training may help women become better prepared to fend off an assailant. Yet physical resistance may spur some rapists to become more aggressive (Powell, 1996). Federal statistics show that women who resist increase their chances of preventing the completion of a rape by 80%. However, resistance also increases by as much as threefold the odds of being physically injured (Brody, 1992c). A study of 116 rapes showed that women were more likely to resist physically if the attacker was a friend or relative, if the attacker made verbal threats, and if the attacker physically restrained her or injured her (Atkeson et al., 1989).

It is difficult, if not impossible, for people to think through their options clearly and calmly when they are suddenly attacked. Rape experts recommend that women rehearse alternative responses to a rape attack. The Boston Police Department recommends that whatever form of self-defense a woman intends to use, she should carefully think through how it is used and practice using it (Brody, 1992c). Thompson (1991) suggests that effective self-defense is built upon the use of multiple strategies, ranging from attempts to avoid potential rape situations (such as by installing home security systems or walking only in well-lit areas), to acquiescence when active resistance would seem too risky, to the use of more active verbal or physical forms of resistance in some low-risk situations.

Truth or Fiction? REVISITED

It is questionable whether women who encounter a rapist should attempt to fight off the assailant. Women must make their own decisions about whether to resist a rapist physically, on the basis of their assessment of the rapist, the situation, and their own ability to resist. ■

VERBAL SEXUAL COERCION

Verbal sexual coercion is persistent verbal pressure or the use of seduction "lines" to manipulate a person into sexual activity. A study of 194 male undergraduates in a southeastern university showed that 42% admitted to coercing a woman verbally into sex (Craig et al., 1989). In a survey of 325 college undergraduates from a northwestern state university, about 1 in 5 of the men reported having said things they didn't mean in order to induce women to engage in sexual intercourse (Lane & Gwartney-Gibbs, 1985). Women were more likely than men to have been pressured into sexual relations. One in 4 of the women respondents reported that they engaged in sexual intercourse with someone they would otherwise have rejected because they had "felt pressured by his continual arguments" (p. 56). About 1 in 15 of the *men* reported engaging in sexual intercourse unwillingly as a result of sexual pressure.

Verbal pressure and seduction "lines" are so commonly used in dating relationships that they are seldom recognized as coercion. Consider the man who deceives his partner into believing that he really loves her in order to persuade her to engage in sexual activity with him. His use of lies or deception can be considered a form of sexual coercion because it employs devious means to exploit his partner's emotional needs in order to curry sexual favor.

a CLOSER look

Handling Sexual Pressure Lines

Rape is an extreme form of sexual pressure that involves the use of threats or force to coerce an unwilling person into a sexual act. But sexual pressure may take a more subtle form, such as persistent verbal pressure or the use of seduction "lines" that aim, by deception or trickery, to manipulate another person into having sexual relations.

The following list of sexual pressure lines that men use with women was compiled by Powell (1996, pp. 71–73). Responses suggested by Powell are also shown.

Kind of Pressure Line	The Line	Possible Response
Lines that reassure you about the negative consequences	"Don't worry, I'm sterile."	I know you want to make me feel safer, but . . . well, I'm just not comfortable about sex without a condom. I've known a few people who were more fertile than they thought.
	"You can't get pregnant the first time."	Hey, where did you get your sex education? People can get pregnant any time they have intercourse, even if it's just for one second.
	"Don't worry—I'll pull out."	I know you want to reassure me, but people can get pregnant that way, even without ejaculation.
Lines that threaten you with rejection	"If you don't have sex, I'll find someone who will."	I can't believe you are making a threat like this. I'm furious that you would treat lovemaking like some kind of job, as if anyone will do.
Lines that attempt to put down the refuser	"You're such a bitch."	It's hard to believe you want to make love to me and think calling names will put me in the mood. I need to leave now.
	"Are you frigid?"	I resent being called names just because I tell you what I want to do with my body.
Lines that stress the beautiful experience being missed	"Our relationship will grow stronger."	I know you really would like to get more involved right now. But I need to wait. And lots of people have had their relationship grow stronger without intercourse.
Lines that might settle for less	"I don't want to do anything. I just want to lie next to you."	The way we're attracted to each other, I don't think that would be a good idea. As much as I care about you, I'd better not spend the night.
Lines to make you prove yourself	"If you loved me, you would."	You know I care a lot about you. But I feel very pressured when you try to get me to do something I'm not ready for. It's not fair to me. Please consider my feelings.
Lines that attempt to be logical but aren't	"You're my girlfriend—it's your obligation."	If you think sex is an obligation, we need to think about this relationship right now. (Watch out for any such talk—it is very common in abusers and rapists. At best, it's an irrational comment by an immature person.)
Lines that are totally transparent	"I'll say I love you after we do it."	Bye, now. (There is no way to deal with a person who would say such a thing.)

Although these sample responses can help you resist specific pressure lines, do not think that saying no to sexual pressure is a privilege that you earn by winning an argument. As Powell recognizes, "You don't *have* to explain." Your body is not debatable. You don't have to say anything except "I don't want to."

SEXUAL ABUSE OF CHILDREN

Many view sexual abuse of children as among the most heinous of crimes. Children who are sexually assaulted often suffer social and emotional problems that impair their development and persist into adulthood, affecting their self-esteem and their ability to form intimate relationships.

No one knows how many children are sexually abused. Although most sexually abused children are girls (Knudsen, 1991), one quarter to one third are boys (Finkelhor, 1990). A randomized national telephone survey of more than 2,600 adults showed that 9.5% of the men and 14.6% of the women reported having been sexually abused (a completed or attempted act of sexual intercourse) prior to age 19 (Finkelhor et al., 1990). These estimates may underrepresent the actual prevalence, given that people may fail to report such incidents because of faulty memories or of shame or embarrassment. In addition, about one in four people refused to participate in the survey, which in itself casts doubt on the sample's representativeness. Other researchers estimate that the prevalence of sexual abuse among boys ranges from 4% to 16% (Genuis et al., 1991; Janus & Janus, 1993; Kohn, 1987) and among girls exceeds 20% (Janus & Janus, 1993; Kohn, 1987). Whatever its actual prevalence, sexual abuse of children cuts across all racial, ethnic, and economic boundaries (Alter-Reid et al., 1986).

What Is Sexual Abuse of Children?

Sexual abuse of children ranges from exhibitionism, kissing, fondling, and sexual touching to oral sex and anal intercourse and, with girls, vaginal intercourse. Acts such as touching children's sexual organs while changing or bathing them, sleeping with children, or appearing nude before them are open to interpretation and are often innocent (Haugaard, 2000). Sexual contact between an adult and a child is abusive, even if the child is willing, because children are legally incapable of consenting to sexual activity. Although the age of consent varies among the states, sexual relations between adults and children under the age of consent is a criminal offense in every state.

Voluntary sexual activity *between children* of similar ages is not sexual abuse. Children often engage in consensual sex play with peers or siblings, as in "playing doctor" or in mutual masturbation. Although such experiences may be recalled in adulthood with feelings of shame or guilt, they are not typically as harmful as experiences with adults. When the experience involves coercion, or when one child is significantly older or in a position of power over the younger child, the sexual contact may be considered sexual abuse.

Patterns of Abuse

Children from stable, middle-class families appear to be generally at lower risk of encountering sexual abuse than children from poorer, less cohesive families (Finkelhor, 1984). In most cases, children who are sexually abused are not accosted by the proverbial stranger lurking in the school yard. In perhaps 75% to 80% of cases, the molesters are people who are close to them: relatives, steprelatives, family friends, and neighbors (Waterman & Lusk, 1986). Estimates of the percentage of sexually molested children who are abused by family members have ranged from 10% to 50% of cases (Waterman & Lusk, 1986). In a random survey of 521 Boston parents, 55% of the respondents who said they had been sexually abused as children reported that the perpetrator was a family member or acquaintance (Finkelhor, 1984).

Parents who discover that their child has been abused by a family member are often reluctant to notify authorities. Some may feel that such problems are "family matters" that are best kept private. Others may be reluctant to notify authorities for fear that it may shame the family or that they may be held accountable for failing to protect the child. The decision to report the abuse to the police depends largely on the relationship between the abuser and the person who discovers the abuse (Finkelhor, 1984). In the Boston community survey, none of the parents whose children were sexually abused by family members

So there really was a monster in her bedroom.

For many kids, there's a real reason to be afraid of the dark.

Last year in Indiana, there were 6,912 substantiated cases of sexual abuse. The trauma can be devastating for the child and for the family. So listen closely to the children around you.

If you hear something you don't want to believe, perhaps you should. For helpful information on child abuse prevention, contact the LaPorte County Child Abuse Prevention Council, 7451 Johnson Road, Michigan City, IN 46360. (219) 874-0007

LaPorte County Child Abuse Prevention Council

The monster in the bedroom. Not all monsters are make-believe. Some, like the perpetrators of incest, are members of the family.

notified the authorities. By contrast, 23% of the parents whose children had been abused by acquaintances notified the authorities, and 73% of the parents whose children were abused by strangers did so (Finkelhor, 1984).

Typically, the child initially trusts the abuser. Physical force is seldom needed to gain compliance, largely because of the child's helplessness, gullibility, and submission to adult authority. Whereas most sexually abused children are abused only once, those who are abused by family members are more likely to suffer repeated acts of abuse (Briere & Runtz, 1987; Dube & Hebert, 1988).

Genital fondling is the most common type of abuse (Knudsen, 1991). In one sample of women who had been molested in childhood, most of the contacts involved genital fondling (38% of cases) or exhibitionism (20% of cases). Intercourse occurred in only 4% of cases (Knudsen, 1991). Repeated abuse by a family member, however, commonly follows a pattern that begins with affectionate fondling during the preschool years, and progresses to oral sex or mutual masturbation during the early school years and then to sexual penetration (vaginal or anal intercourse) during preadolescence or adolescence (Waterman & Lusk, 1986).

Abused children rarely report the abuse, often because of fear of retaliation from the abuser or because they believe they will be blamed for it. Adults may suspect abuse if a child shows sudden personality changes or develops fears, problems in school, or difficulty eating or sleeping. A pediatrician may discover physical signs of abuse during a medical exam.

The average age at which most children are first sexually abused ranges from 6 to 12 years for girls and from 7 to 10 years for boys (Knudsen, 1991). Boys are relatively more likely to be abused in public places and by strangers and non-family members (Faller, 1989a; Knudsen, 1991). Boys are also more likely to be threatened and physically injured.

TYPES OF ABUSERS Researchers find that the overwhelming majority of people who sexually abuse children (both boys and girls) are males (Thomlison et al., 1991). Although most child abusers are adults, some are adolescents. Male adolescent sex offenders are more likely than other adolescents to have been molested themselves as boys (Becker et al., 1989; Muster, 1992). Some adolescent sex offenders may be imitating their own victimization. Adolescent child molesters also tend to feel socially inadequate and to be fearful of social interactions with age-mates of the other gender (Katz, 1990).

Although the great majority of sexual abusers are male, the number of female sexual abusers may be greater than has been generally believed (Banning, 1989). Many female sexual abusers may go undetected because society accords women a much freer range of physical contact with children than it does men. A woman who fondles a child might be seen as affectionate, or at worst seductive, whereas a man would be more likely to be perceived as a child molester (Banning, 1989).

What motivates a woman to sexually abuse children, even her own children? Some female abusers have histories of becoming dependent on, or rejected by, abusive males (Matthews et al., 1990). Some appear to have been manipulated by their husbands into engaging in sexual abuse. Others appear to have unmet emotional needs and low self-esteem and may have been seeking acceptance, closeness, and attention though sexual acts with children. Some, motivated by unresolved feelings of anger, revenge, powerlessness, or jealousy, may view their own and others' children as safe targets for venting these feelings. Some actually view their crimes as expressions of love.

The great majority of child molesters are male. Why? Banning (1989), arguing from a sociocultural framework, suggests that males in our culture are socialized into seeking partners who are younger and weaker, whom they can easily dominate. This pattern of socialization may take the extreme form of development of sexual interest in children and adolescent girls, who, because of their age, are more easily dominated than adult women.

Yet sexual interest in children may also be motivated by unusual patterns of sexual arousal in which children become the objects of sexual desire, sometimes to the exclusion of more appropriate (adult) stimuli. This brings us to pedophilia.

Pedophilia

The prevalence of **pedophilia** in the general population is unknown (Ames & Houston, 1990). Some pedophiles are so distressed by their urges that they never act on them. Many, however, molest young children and adolescents, often repeatedly. Some pedophiles are responsible for large numbers of sexual assaults on children. One study of 232 convicted pedophiles showed that they had molested an average of 76 children each (Abel et al., 1989). Incarcerated pedophiles have usually committed many more offenses than those for which they were convicted (Ames & Houston, 1990).

Although pedophiles are sometimes called child molesters, not all child molesters are pedophiles. Pedophilia involves persistent or recurrent sexual attraction to children. Some molesters, however, may seek sexual contacts with children only when they are under unusual stress or lack other sexual outlets. Thus they do not meet the clinical definition of pedophilia.

Pedophiles are almost exclusively male, although some isolated cases of female pedophiles have been reported (Cooper et al., 1990). Some pedophiles are sexually attracted only to children; others are sexually attracted to adults as well. Some pedophiles limit their sexual interest in children to incestuous relationships with family members; others abuse children to whom they are unrelated. Some pedophiles limit their sexual interest in children to looking at them or undressing them; others fondle them or masturbate in their presence. Some manipulate or coerce children into oral, anal, or vaginal intercourse.

Children tend not to be worldly wise. They can often be "taken in" by pedophiles who tell them that they want to "show them something," "teach them something," or do something with them that they will "like." Some pedophiles seek to gain the child's affection, and, later, to discourage the child from disclosing the sexual activity, by showering the child with attention and gifts. Others threaten the child or the child's family to prevent disclosure.

There is no consistent personality profile of the pedophile (Okami & Goldberg, 1992). Most pedophiles do not fit the common stereotype of the "dirty old man" in the trench coat who hangs around school yards. Most are otherwise law-abiding, well-respected citizens generally in their 30s and 40s. Many are married or divorced, with children of their own.

Research suggests that sexual attraction to children may be more common than is generally believed. Researchers in one study administered an anonymous survey to a sample of 193 college men (Briere & Runtz, 1989). A high percentage of the students—21%—admitted to having been sexually attracted to small children, 9% reported sexual fantasies involving young children, 5% reported masturbating to such fantasies, 7% reported that there was some likelihood that they would have sex with a young child if they knew they could avoid detection and punishment. Fortunately, most people with such erotic interests never act on them.

Pedophilia may have complex and varied origins. Some pedophiles who are lacking in social skills may turn to children after failing to establish gratifying relationships with adult women (Overholser & Beck, 1986; Tollison & Adams, 1979). Research generally supports the stereotype of the pedophile as weak, passive, and shy—a socially inept, isolated man who feels threatened by mature relationships (Ames & Houston, 1990; Wilson & Cox, 1983).

Pedophiles who engage in incestuous relationships with their own children present a somewhat different picture. They tend to fall on one or the other end of the dominance spectrum. Some are very dominant, others very passive. Few are found between these extremes (Ames & Houston, 1990).

Pedophilia A type of paraphilia that is defined by sexual attraction to unusual stimuli: children. (From the Greek *paidos,* which means "child," not the Latin *pedis,* which means "foot.")

Some pedophiles were sexually abused as children and may be attempting to establish feelings of mastery by reversing the situation (De Young, 1982). Cycles of abuse may be perpetuated from generation to generation if children who are sexually abused become victimizers or partners of victimizers as adults.

Incest

Incest involves people who are related by blood, or *consanguineous*. The law may also proscribe coitus between, say, a stepfather and stepdaughter. Although a few societies have permitted incestuous pairings among royalty, all known cultures have some sort of incest taboo.

PERSPECTIVES ON THE INCEST TABOO Speculations about the origin of incest taboos abound. One explanation holds that the incest taboo developed because it was adaptive for ancient humans to prevent the harmful effects of inbreeding that may result when genetic defects or diseases are carried within family bloodlines (Leavitt, 1990). Our ancient ancestors lacked knowledge of the mechanisms of genetics, but they may have observed that certain diseases or defects tend to run in families. Evidence does show that marriage between close relations is associated with an increased rate of genetic diseases, mental retardation, and other physical abnormalities (Ames & Houston, 1990). Inbreeding may also be counterproductive to survival because it decreases the amount of genetic variation in the gene pool. Therefore, it can reduce the ability of the population to adapt to changes in the environment.

Other theorists explain the incest taboo in terms of the role that it may play in maintaining stability in the family and establishing kinship ties within the larger social grouping (Harris & Johnson, 2000; Whitten, 2001). The anthropologist Bronislaw Malinowski (1927), for example, argued that the incest taboo serves to reduce sexual competition within the family. If left uncontrolled, competition would create rivalry and hostility such that the family might be unable to function as a social unit. Because the family unit fosters survival of a society, the incest taboo may have developed as a means of keeping the family intact.

Cooperation theory emphasizes the importance to the survival of the society of cooperative ties between family groups (Harris & Johnson, 2000; Whitten, 2001). Society is complex. Its survival requires the cooperation of large numbers of people. Marriage establishes kinship ties that lessen suspiciousness and hostility between family groups and foster cooperation. According to cooperation theory, the incest taboo was established to help ensure that people would marry outside their own families and thus create cohesive communities. Such theories are fascinating but remain speculative.

TYPES OF INCEST Most of our knowledge of incestuous relationships concerns father–daughter incest. Why? Most identified cases involve fathers who were eventually incarcerated.

About 1% of a sample of women in five American cities reported a sexual encounter with a father or stepfather (Cameron et al., 1986). Brother–sister incest, not parent–child incest, is the most common type of incest, however (Waterman & Lusk, 1986). Brother–sister incest is also believed to be greatly underreported, possibly because it tends to be transient and is apparently less harmful than parent–child incest. Finkelhor (1990) found that 21% of the college men in his sample, and 39% of the college women, reported incestuous relationships with a sibling of the other gender. Only 4% reported an incestuous relationship with their fathers. Incest between siblings of the same gender is rare (Waterman & Lusk, 1986). Mother–daughter incest is the rarest form of incest (Waterman & Lusk, 1986).

Incest Marriage or sexual relations between people who are so closely related (by "blood") that sexual relations are prohibited and punishable by law. (From the Latin *in-*, which means "not," and *castus,* which means "chaste.")

Truth or Fiction? **REVISITED**

It is not true that father–daughter incest is the most common type of incest. Father–daughter incest may be the most highly publicized variety of incest, but brother–sister incest is actually more common. ■

Let us further consider the two most common incest patterns, father–daughter incest and brother–sister incest.

FATHER–DAUGHTER INCEST Father–daughter incest often begins with affectionate cuddling or embraces and then progresses to teasing sexual play, lengthy caresses, hugs, kisses, and genital contact, even penetration. In some cases genital contact occurs more abruptly, usually when the father has been drinking or arguing with his wife. Force is not typically used to gain compliance, but daughters are sometimes physically overcome and injured by their fathers.

BROTHER–SISTER INCEST In sibling incest, the brother usually initiates sexual activity and assumes the dominant role (Meiselman, 1978). Some brothers and sisters may view their sexual activity as natural and not know that it is taboo (Knox, 1988).

Evidence on the effects of incest between brothers and sisters is mixed. In a study of college undergraduates, those who reported childhood incest with siblings did not reveal greater evidence of sexual adjustment problems than other undergraduates (Greenwald & Leitenberg, 1989). Sibling incest may be harmful for some children, however (Sorrenti-Little et al., 1984). Sibling incest is most likely to be harmful when it is recurrent or forced or when parental response is harsh (Knox, 1988; Laviola, 1989).

FAMILY FACTORS IN INCEST Incest frequently occurs within the context of general family disruption, as in families in which there is spouse abuse, a dysfunctional marriage, or alcoholic or physically abusive parents (Alter-Reid et al., 1986; Sirles & Franke, 1989; Waterman, 1986a). Stressful events in the father's life, such as the loss of a job or problems at work, often precede the initiation of incest (Waterman, 1986a).

Fathers who abuse older daughters tend to be domineering and authoritarian with their families (Waterman, 1986a). Fathers who abuse younger, preschool daughters are more likely to be passive, dependent, and low in self-esteem. As Waterman (1986a) notes,

> [The fathers] may need soothing and comforting, and may feel especially safe with preschool children: "I felt safe with her. . . . I didn't have to perform. She was so little that I knew she wouldn't and couldn't hurt me." (p. 215)

Marriages in incestuous families tend to be characterized by an uneven power relationship between the spouses. The abusive father is usually dominant. Another thread that frequently runs through incestuous families is a troubled sexual relationship between the spouses. The wife often rejects the husband sexually (Waterman, 1986a).

Gebhard and his colleagues (1965) found that many fathers who committed incest with their daughters were religiously devout, fundamentalist, and moralistic. Perhaps such men, when sexually frustrated, are less likely to seek extramarital and extrafamilial sexual outlets or to turn to masturbation as a sexual release. In many cases, the father is under stress but does not find adequate emotional and sexual support from his wife (Gagnon, 1977). He turns to a daughter as a wife surrogate, often when he has been drinking alcohol (Gebhard et al., 1965). The daughter may become, in her father's fantasies, the "woman of the house." This fantasy may become his justification for continuing the incestuous relationship. In some incestuous families, a role reversal occurs. The abused daughter assumes many of the mother's responsibilities for managing the household and caring for the younger children (Waterman, 1986a).

Incestuous abuse is often repeated from generation to generation. One study found that in 154 cases of children who were sexually abused within the family, more than a third of the male offenders and about half of the mothers had been either abused themselves or exposed to abuse as children (Faller, 1989b).

Sociocultural factors, such as poverty, overcrowded living conditions, and social or geographical isolation, may contribute to incest in some families (Waterman, 1986a). Sibling incest may be encouraged by the crowded living conditions and open sexuality that occur among some economically disadvantaged families (Waterman, 1986a).

Effects of Sexual Abuse of Children

The effects of sexual abuse are varied, and there is no single identifiable syndrome that emerges from sexual abuse (Saywitz et al., 2000). Nevertheless, sexual abuse, whether perpetrated by a family member, an acquaintance, or a stranger, often inflicts great psychological harm on the child. Children who are sexually abused may suffer from a litany of short- and long-term psychological complaints, including anger, depression, anxiety, eating disorders, inappropriate sexual behavior, aggressive behavior, self-destructive behavior, sexual promiscuity, drug abuse, suicide attempts, post-traumatic stress disorder, low self-esteem, sexual dysfunction, mistrust of others, and feelings of detachment (Meston & Heiman, 2000; Saywitz et al., 2000). Sexual abuse may also have physical effects such as genital injuries and may cause psychosomatic problems such as stomachaches and headaches.

Abused children commonly "act out." Younger children have tantrums or display aggressive or antisocial behavior. Older children turn to substance abuse (Finkelhor, 1990; Kendler et al., 2000). Some abused children become withdrawn and retreat into fantasy or refuse to leave the house. Regressive behaviors, such as thumb sucking, fear of the dark, and fear of strangers, are also common among sexually abused children. On the heels of the assault and in the ensuing years, many survivors of childhood sexual abuse—like many rape survivors—show signs of post-traumatic stress disorder. They suffer flashbacks, nightmares, numbing of emotions, and feelings of estrangement from others (Finkelhor, 1990).

The sexual development of abused children may also be adversely affected. For example, the survivor may become prematurely sexually active or promiscuous in adolescence and adulthood (Kendler et al., 2000; Tharinger, 1990). Researchers find that adolescent girls who are sexually abused tend to engage in consensual coitus at earlier ages than nonabused peers (Wyatt, 1988).

Researchers generally find more similarities than differences between the genders with respect to the effects of sexual abuse in childhood (Finkelhor, 1990). For example, both boys and girls tend to suffer fears and sleep disturbance. There are some gender differences, however. The most consistent gender difference appears to be that boys more often "externalize" their problems, perhaps by becoming more physically aggressive. Girls more often "internalize" their difficulties, such as by becoming depressed (Finkelhor, 1990; Gomes-Schwartz et al., 1990).

There has been comparatively little research comparing survivors of childhood sexual abuse from different ethnic groups. In one of the few reported studies, researchers compared Asian American children who had been sexually abused with random samples of African, European, and Latino and Latina American sexually abused children drawn from the same child-sex-abuse clinic (Rao et al., 1992). Compared to the other groups, Asian American children were less likely to exhibit anger and sexual acting out. They were also more likely to become suicidal.

The long-term consequences of sexual abuse in childhood tend to be greater for children who were abused by their fathers or stepfathers, who experienced penetration, who were forced, and who suffered more prolonged and severe abuse (Cheasty, 1998; Wyatt & Newcomb, 1990). Children who suffer incest often feel a deep sense of betrayal by the offender and, perhaps, by other family members—especially their mothers, whom they perceive as failing to protect them (Finkelhor, 1988). Incest survivors may feel powerless to control their bodies or their lives.

Late adolescence and early adulthood seem to pose especially difficult periods for survivors of childhood sexual abuse. Studies of women in these age groups reveal more psychological and social problems in abused women (Jackson et al., 1990; Kendler et al., 2000).

Effects of childhood sexual abuse are often long-lasting. In one study, researchers found evidence of greater psychological distress in a group of 54 adult women, ranging from 23 to 61 years of age, who had been sexually abused as children than in a matched group of nonabused women (Greenwald et al., 1990). A Dublin study found that female survivors of sexual abuse were more likely to feel depressed and to get divorces than the general population (Cheasty, 1998). Women who blame themselves for the abuse apparently have lower self-esteem and more depression than those who do not (Hoagwood, 1990).

a CLOSER look

New Methods for Getting to the Truth in Cases of Child Abuse

As social scientists continue to conduct research, what are the brightest prospects for determining the truth in cases of child abuse? The American Academy of Child and Adolescent Psychiatry (1998) points out that no particular pattern of behavior in children shows that they have been abused—or *who* has abused them. As many as one third of victims of sexual abuse show no symptoms. Microbiology and DNA testing are helpful only in cases where there is physical evidence, such as semen. There was a time when hopes were pinned on anatomically detailed dolls as vehicles for allowing children to describe incidences of sexual abuse. However, children's natural interest in the dolls' anatomy can be misinterpreted as a sign of abuse. Therefore, the American Professional Society on the Abuse of Children (1998) has taken the position that "Anatomical dolls should not be used as a diagnostic test for sexual abuse."

Asking children what has happened to them has led to many "false positives"—that is, cases in which innocent adults have been identified as guilty of abuse (Bruck et al., 2000). One problem is that children are highly suggestible. They pick up on interviewers' suggestions—made in the form of leading questions ("Did Mark touch your penis?")—and then often respond inaccurately. Suggestibility is also highlighted in the use of anatomically correct dolls. McGill University researcher Maggie Bruck (1998) notes that reports of inappropriate touching increase each time children are shown the dolls.

Yet it would appear that the best hope for unearthing childhood sexual abuse might simply be more careful interviewing. Guidelines for interviews that avoid suggesting abuse include the following (Goldberg, 1998):

- Interview the children in an open-ended way rather than describing specific events and asking whether they occurred. *Don't* say, "Did he touch your vagina?" (a leading question). Rather, say, "Tell me everything that happened in John's apartment."

How Do We Get at the Truth in Cases of Sexual Abuse of Children? Serious doubts have been raised about the use of anatomically correct dolls and about interviewing children. Children are quite likely to report remembering events suggested by the examiner, whether or not those events happened.

- Use interviewers who are unbiased—who have no preconceived ideas about whether abuse occurred or who might have been guilty of it.
- Keep the number of questions and the number of interviews to a minimum. Not only are interviews taxing to children, but exposure to many questions or interviews begins to plant suggestions in their minds.
- Children should not be asked to imagine sexual acts, because they may then become confused about what is real and what is fantasy.
- Children should not be pressured to respond, especially by bribing them or threatening them. Aside from the obvious inappropriateness of such courses of action, they may pressure children to invent abuses that did not occur.

Thus it may be that obtaining accurate information about childhood sexual abuse will not involve fancy electronic circuitry or microbiology. It may rely on careful questioning and common sense.

Prevention of Sexual Abuse of Children

Many of us were taught by our parents never to accept a ride or an offer of candy from a stranger. However, many instances of sexual abuse are perpetrated by familiar adults—often a family member or friend (Zielbauer, 2000). Prevention programs help children understand what sexual abuse is and how they can avoid it. A national survey showed that two out of three children in the United States have participated in school-based sex-abuse prevention programs (Goleman, 1993). In addition to learning to avoid strangers, children need to recognize the differences between acceptable touching, such as an affectionate embrace or pat on the head, and unacceptable or "bad" touching. Even children of elementary school age can learn the distinction between "good touching and bad touching" (Tutty, 1992). Good school-based programs are generally helpful in preparing children to handle an actual encounter with a potential molester (Goleman, 1993). Children who receive comprehensive training are more likely to use strategies such as running away, yelling, or saying no when they are threatened by an abuser. They are also more likely to report such incidents to adults.

Researchers recognize that children can easily be intimidated or overpowered by adults or older children. Children may be unable to say no in a sexually abusive situation, even though they want to and know it is the right thing to do (Waterman et al., 1986a). Although children may not always be able to prevent abuse, they can be encouraged to tell someone about it. Most prevention programs emphasize teaching children messages such as "It's not your fault," "Never keep a bad or scary secret," and

Human Sexuality ONLINE //

Online Registries of Sex Offenders: Boon or Boondoggle?

Congress began requiring states to maintain registries of sex offenders in 1996. States are now going further by posting the information, along with photos of the offenders and ugly details of their offenses, on Web pages. Anyone with a computer and modem can access them. "It's the wave," said Scott Matson (cited in Zielbauer, 2000), a researcher with the Center for Sex Offender Management. "Everyone's doing it now."

Supporters of the Web postings argue that bringing community notification to the Internet is an important and easy way to spread information on dangerous sex offenders. Putting information about sex offenders online "should make sex offenders think twice before offending again in Wisconsin, and that should certainly make this state a safer place," said Governor Tommy G. Thompson (Zielbauer, 2000) when he announced Wisconsin's intention to create an online registry.

Critics argue that these online registries can stigmatize and victimize marginal offenders and may prevent them from getting help. Thus the postings might eventually be responsible for the commission of more sex crimes than they prevent. Critics note that the sites do not explain how to protect children or how to evaluate the danger the offenders represent. David A. D'Amora, director of a treatment program for sexual abusers, believes that "Just throwing a bunch of things up on a screen is like throwing gasoline on a fire" (Zielbauer, 2000). For example, online registries have led to retributive violence against sex offenders. Ex-offenders have been harassed by neighbors, evicted by landlords, fired from their jobs, and even beaten by mobs.

- In 2000, two men beat a convicted child molester with a baseball bat in Florida. They threatened to kill him if he approached any children in the neighborhood.
- In 1999, four men in Dallas beat a 27-year-old retarded man whose address was erroneously listed on Texas's online registry.
- In 1998, a Linden, New Jersey, man, after receiving a police flier, fired bullets from a .45-caliber handgun into the house of a paroled rapist and was charged with aggravated assault.
- Also in 1998, residents of a trailer park in the state of Washington burned down a mobile home days before a released sex offender was to move into it.

The registration of information about sex offenders on the Internet is "evolutionary, . . . awkward and painful," says Roxanne Lieb, director of the Washington State Institute for Public Policy (Zielbauer, 2000). "I think it's important to remember that at least what we started with was not knowing, and that is worse."

"I think that when violent criminals are placed back in the community, the community has every right to know who they are and what they did," said Marc Klaas (Zielbauer, 2000), whose daughter was kidnaped and strangled by a sex offender. "You wouldn't want to hire a baby sitter, would you, that had spent years in prison for a violent crime. If you were a single mother, you certainly wouldn't want to start dating one of those characters. Community notification creates some kind of outer control on these individuals. If people know who they are, they're going to be more careful than if people don't know who they are."

"Always tell your parents about this, especially if someone says you shouldn't tell them" (Waterman et al., 1986a).

Children also need to be alerted to the types of threats they might receive for disclosing the abuse. They are more likely to resist threats if they are reassured that they will be believed if they disclose the abuse, that their parents will continue to love them, and that they and their families will be protected from the molester.

School-based prevention programs focus on protecting the child. In most states, teachers and helping professionals are required to report suspected abuse to authorities. Tighter controls and better screening are needed to monitor the hiring of day care employees. Administrators and teachers in preschool and day care facilities also need to be educated to recognize the signs of sexual abuse and to report suspected cases (Waterman et al., 1986). Treatment programs to help people who are sexually attracted to children *before* they commit abusive acts would also be of use.

Treatment of Survivors of Sexual Abuse

At least 75% of cases of sexual abuse of children go unreported. Psychotherapy in adulthood often becomes the first opportunity for survivors to confront residual feelings of pain, anger, and misplaced guilt (Alter-Reid et al., 1986; Ratican, 1992). Group or individual therapy can help improve survivors' self-esteem and their ability to develop intimate relationships. Confronting the trauma within a supportive therapeutic relationship may also help prevent the cycle of abuse from perpetuating itself from one generation to another (Alter-Reid et al., 1986).

Special programs that provide therapeutic services to abused children and adolescents have begun to appear. Most therapists recommend a multicomponent treatment approach, which may involve individual therapy for the child, mother, and father; group therapy for the adolescent or even preadolescent survivor; art therapy or play therapy for the younger child (e.g., using drawings or puppets to express feelings); marital counseling for the parents; and family therapy for the entire family (de Luca et al., 1992; Waterman, 1986b).

TREATMENT OF RAPISTS AND CHILD MOLESTERS

What does *treatment* mean? When a helping professional treats someone, the goal is usually to help that individual. When we speak of treating a sex offender, the goal is just as likely—or more likely—to be to help society by eliminating the problem behavior.

Rapists and child molesters are criminals, not patients. Most convicted rapists and child molesters are incarcerated as a form of punishment, not treatment. They may receive psychological treatment or rehabilitation in prison to help prepare them for release and reentry into society, however. The most common form of treatment is group therapy, which is based on the belief that although offenders may fool counselors, they do not easily fool one another (Kaplan, 1993). Yet the great majority of incarcerated sex offenders receive little or no treatment in prison (Goleman, 1992b). In California, for example, which has 15,000 incarcerated sex offenders, treatment has been provided in just one experimental program for only 46 rapists and child molesters (Goleman, 1992b).

The results of prison-based treatment programs are mixed at best. Consider a Canadian study of 54 rapists who participated in a treatment program. Following release from prison, 28% were later convicted of a sexual offense, and 43% were convicted of a violent offense (Rice et al., 1990). Treatment also failed to curb recidivism among a sample of 136 child molesters (Rice et al., 1991).

More promising findings resulted from innovative programs in prison facilities in California and Vermont (Goleman, 1992). In the Vermont program, the average rate of conviction for additional sex crimes following release was reduced by at least half in a group of sex offenders who received the treatment program, compared to a control group who did not. These innovative programs applied a variety of techniques. Empathy training was used to increase the offender's sensitivity to his victims. One empathy exercise had offenders write about their crimes from the perspective of the victim. The technique of covert sensitization was used to help offenders resist deviant sex fantasies, which often lead to deviant behavior. The offender would pair, in his imagination, scenes involving rape and molestation with aversive consequences. An exhibitionist, for example, might be asked to practice imagining that he is about to expose himself and is discovered in the act by his parents. A child molester might fantasize about sexually approaching a child, only to find himself confronted by police officers.

Another approach uses medical interventions such as castration to reduce testosterone levels and, consequently, offenders' sex drives. Experts do not agree on whether castration helps sex offenders control their sexual urges (Roesler & Witztum, 2000). Many castrated rapists report lowered sex drives, as might be expected from the reductions in testosterone production that result from removal of the testes. They may retain sexual interest and remain capable of erection, however. And some do repeat their crimes. Other researchers report lower recidivism rates among castrated offenders than among other offenders.

Surgical castration is an extreme measure. It raises ethical concerns because of its invasive character and irreversibility. Anti-androgen drugs such as Depo-Provera chemically reduce testosterone levels (Roesler & Witztum, 2000). Unlike surgical castration, such "chemical castration" is reversible (Roesler & Witztum, 2000).

SEXUAL HARASSMENT

Many Americans were spellbound by the Michael Douglas-Demi Moore film *Disclosure*. In the film, Demi Moore plays Michael Douglas's supervisor. She uses her power over him to attempt to harass him into sexual activity.

What *is* **sexual harassment?**

Sexual harassment can be difficult to define (Lewin, 1998b). For example, President Bill Clinton was accused of sexually touching or "groping" a resistant White House volunteer, Kathleen Willey, and of placing her hand on his penis (Lewin, 1998b). Such behavior would clearly constitute sexual harassment. But Clinton also engaged in fellatio with a young White House intern, Monica Lewinsky. Although Lewinsky participated voluntarily, some critics note that the White House *is* a workplace and that Clinton's power over the intern caused their interaction to constitute sexual harassment.

For legal purposes, sexual harassment in the workplace is usually defined as the "deliberate or repeated unsolicited verbal comments, gestures, or *physical contact*[1] of a sexual nature that is considered to be unwelcome by the recipient" (U.S. Merit Systems Protection Board, 1981, p. 2). Examples of sexual harassment can range from unwelcome sexual jokes, overtures, suggestive comments, and sexual innuendos to outright sexual assault. They may include behaviors such as the following (Powell 1996):

- Verbal harassment or abuse
- Subtle pressure for sexual activity
- Remarks about a person's clothing, body, or sexual activities
- Leering at or ogling a person's body
- Unwelcome touching, patting, or pinching
- Brushing against a person's body
- Demands for sexual favors accompanied by implied or overt threats concerning one's job or student status
- Physical assault. The idea here is that sexual assault of the kind of which President Clinton was accused would be sexual harassment, but it could also be much more

Sexual harassment Deliberate or repeated unsolicited verbal comments, gestures, or physical contact of a sexual nature that the recipient does not welcome.

[1]Our italics.

Both men and women can both commit, and be subjected to, sexual harassment. However, despite the plot of the film *Disclosure*, about 99% of harassers are men.

Charges of sexual harassment are often ignored or trivialized by co-workers and employers. The victim may hear, "Why make a big deal out of it? It's not like you were attacked in the street." Evidence shows, however, that persons subjected to sexual harassment do suffer from it. In one study, 75% of people who were sexually harassed reported reactions such as anxiety, irritability, lowered self-esteem, and anger (Gruber & Bjorn, 1986; Loy & Stewart, 1984). Some find harassment on the job so unbearable that they resign. College women have dropped courses, switched majors, or changed graduate programs or even colleges because they were unable to stop professors from sexually harassing them (Dziech & Weiner, 1984; Fitzgerald, 1993a, 1993b).

One reason why sexual harassment is so stressful is that, as with so many other forms of sexual exploitation or coercion, the blame tends to fall on the victim (Powell, 1996). Some harassers seem to believe that charges of harassment were exaggerated or that the victim "overreacted" or "took me too seriously." In our society, women are expected to be "nice"—to be passive and not "make a scene." The woman who assertively protects her rights may be seen as "strange" and disturbing or a "troublemaker." "Women are damned if they assert themselves and victimized if they don't" (Powell, 1991, p. 114).

Sexual harassment may have more to do with the abuse of power than with sexual desire (Goleman, 1991; Tedeschi & Felson, 1994). Relatively few cases of sexual harassment involve outright requests for sexual favors. Most involve the expression of power as a tactic to control or frighten someone, usually a woman. The harasser is usually in a dominant position and abuses that position by exploiting the victim's vulnerability. Sexual harassment may be used as a tactic of social control. It may be a means of keeping women "in their place." This is especially so in work settings that are traditional male preserves, such as the firehouse, the construction site, and the military academy. Sexual harassment expresses resentment and hostility toward women who venture beyond the boundaries of the traditional feminine role (Fitzgerald, 1993b).

Sexual harassment is not confined to the workplace or the university. It may also occur between patients and doctors and between therapists and clients. Therapists may use their power and influence to pressure clients into sexual relations. The harassment may be disguised, expressed in terms of the "therapeutic benefits" of sexual activity.

Two of the most common settings in which sexual harassment occurs are the workplace and the university.

Sexual Harassment in the Workplace

Harassers in the workplace can be employers, supervisors, co-workers, or clients of a company. In some cases clients make unwelcome sexual advances to employees that are ignored or approved of by the boss. If a worker asks a co-worker for a date and is refused, it is not sexual harassment. If the co-worker persists with unwelcome advances and does not take no for an answer, however, the behavior crosses the line and becomes harassment.

Perhaps the most severe form of sexual harassment, short of an outright assault, involves an employer or supervisor who demands sexual favors as a condition of employment or advancement. In 1980, the Equal Employment Opportunity Commission drafted a set of guidelines that expanded the definition of sexual harassment in the workplace to include any behavior of a sexual nature that interferes with an individual's work performance or creates a hostile, intimidating, or offensive work environment.

In 1986, the U.S. Supreme Court recognized sexual harassment as a form of sex discrimination under Title VII of the Civil Rights Act of 1964. It held that employers could be held accountable if such behavior was deemed to create a hostile or abusive work environment or to interfere with an employee's work performance. A 1993 Supreme Court ruling held that a person need not suffer psychological damage to sue an employer on grounds of sexual harassment ("Court, 9–0," 1993). Moreover, employers can be held responsible not only for their own actions but also for sexual harassment by their employees when they either knew or *should have known* that harassment was taking place and

Sexual Harassment. It is estimated that about half of the women in the workplace have experienced some kind of sexual harassment. Many victims keep such incidents to themselves for fear that they—the victims—will be blamed for the incidents or that their supervisors will fire them for complaining. In the case of sexual harassment, it is also possible for perpetrators to claim that they were misunderstood or that the victim is exaggerating.

failed to eliminate it promptly (McKinney & Maroules, 1991). To protect themselves, many companies and universities have developed programs to educate workers about sexual harassment, have established mechanisms for dealing with complaints, and have imposed sanctions against harassers.

Under the law, persons subjected to sexual harassment can obtain a court order to have the harassment stopped, have their jobs reinstated (when they have lost them by resisting sexual advances), receive back pay and lost benefits, and obtain monetary awards for the emotional strain imposed by the harassment. However, proving charges of sexual harassment is generally difficult, because there are usually no corroborating witnesses or evidence. As a result, relatively few persons who encounter sexual harassment in the workplace file formal complaints or seek legal remedies.

One survey found that most people who were sexually harassed handled the situation by ignoring the harasser (32%) or by saying something directly to the harasser (39%) (Loy & Stewart, 1984). Seventeen percent sought job transfers or quit their jobs. Relatively few, 2%, sought legal help. Only about 3% of women who have been sexually harassed file a formal complaint (Goleman, 1991). Like people subjected to other forms of sexual coercion, persons experiencing sexual harassment often do not report the offense for fear that they will not be believed or will be subjected to retaliation. Some fear that they will be branded as "troublemakers" or will lose their jobs (Goleman, 1991).

How common is sexual harassment in the workplace? Two thirds of the men interviewed by the *Harvard Business Review* said that reports of sexual harassment in the workplace were exaggerated (Castro, 1992). However, a survey by *Working Woman* magazine showed that more than 90% of the Fortune 500 companies had received complaints of sexual harassment from their employees. More than one third of the companies had been sued on charges of sexual harassment (Sandross, 1988). A survey of federal employees by the U.S. Merit System Protection Board found that 42% of females and 14% of males reported instances of sexual harassment (DeWitt, 1991). A 1991 *The New York Times*/CBS News poll found that 38% of the women sampled reported that they had been the object of sexual advances or remarks from supervisors or other men in positions of power (Kolbert, 1991). Overall, it may be that as many as one of every two women encounters some form of sexual harassment on the job or in college. Sexual harassment would thus be the most common form of sexual victimization (Fitzgerald, 1993b). Research overseas finds that about 70% of the women who work in Japan and 50% of those who work in Europe have encountered sexual harassment (Castro, 1992). Sexual harassment against women is more common in workplaces in which women have traditionally been underrepresented (Fitzgerald, 1993b), such as the construction site or the shipyard.

Sexual Harassment on Campus

Estimates of the frequency of sexual harassment of college students vary widely across studies, from 7% to 27% of men and from 12% to 65% of women (McKinney & Maroules, 1991). Overall, 25% to 30% of students report at least one incident of sexual harassment in college. The federal law prohibiting sex discrimination in academic institutions permits students to sue their schools for monetary damages for sexual harassment (Greenhouse, 1992).

Sexual harassment on campus usually involves the less severe forms of harassment, such as sexist comments and sexual remarks, as well as come-ons, suggestive looks, propositions, and light touching (McKinney & Maroules, 1991). Relatively few acts involve the use of direct pressure for sexual intercourse. Harassers are typically (but not always) male. Most students who encounter sexual harassment do not report the incident. If they do, it is usually to a confidante and not a person in authority.

Most forms of harassment involve unequal power relationships between the harasser and the person harassed. *Peer harassment* involves people who are equal in power, as in the cases of repeated sexual taunts from fellow employees, students, or colleagues. In some cases, the harasser may even have less formal power than the person harassed. For example, women professors have been sexually harassed by students (Grauerholz, 1989). Here, of course, the traditional social dominance of the male may override the academic position of the woman—at least in the mind of the offender.

The most common form of harassment by students, reported by nearly one third of the female professors polled in a recent survey, involved sexist remarks. Other common forms of harassment are obscene phone calls, undue attention, and sexual remarks. Outright sexual advances were reported by a few professors (Grauerholz, 1989).

Male faculty and staff members may also be subject to harassment. Of 235 male faculty members polled at one university, 6% reported being sexually harassed by a student. Nearly double that percentage (11%) reported that they had attempted to stroke, caress, or touch a student, however (Fitzgerald et al., 1988).

Sexual Harassment in the Schools

Playful sexual antics are common during adolescence. However, unwelcome sexual advances and lewd comments go beyond playfulness and have become a concern to many of America's teens. A nationwide survey of high school and junior high school students found that many boys and girls had encountered harassment in the form of others grabbing or groping them or subjecting them to sexually explicit putdowns when they were walking through school hallways (Henneberger, 1993).

The picture that emerges from a 1993 Louis Harris nationwide poll of teenagers in grades 8 through 11 indicates that sexual taunts and advances have become part of an unwelcome ritual for many students, especially girls, as they try to make their way through the hallways and stairwells of high schools (Barringer, 1993c; Henneberger, 1993). More than two out of three of the girls, and more than four out of ten of the boys, reported being touched, grabbed, or pinched at school. Two out of three boys and about one out of two girls reported harassing other students. Many of the harassers (41% of the boys and 31% of the girls) viewed their actions as "just part of school life" and "no big deal." Unwelcome sexual comments and advances had a negative impact on both boys and girls, but especially on girls. One out of three girls who experienced sexual harassment at school reported that it made them feel that they didn't want to go to school. Twenty-eight percent said that it made it more difficult to pay attention in class, and 20% said that it lowered their grades.

One problem is that teachers tend to tolerate sexually harassing behavior by male students against female classmates (Chira, 1992). Most of the harassment (about 80%)

reported in the Harris survey was committed by other students, but some students reported being harassed by teachers, coaches, custodians, and other adults.

How to Resist Sexual Harassment

What would you do if you were sexually harassed by an employer or a professor? How would you handle it? Would you try to ignore it and hope that it would stop? What actions might you take? We offer some suggestions, adapted from Powell (1996), that may be helpful. Recognize, however, that responsibility for sexual harassment always lies squarely with the perpetrator and with the organization that permits sexual harassment to take place, not with the person subjected to the harassment.

1. *Convey a professional attitude.* Harassment may be stopped cold by responding to the harasser with a businesslike, professional attitude.

2. *Discourage harassing behavior, and encourage appropriate behavior.* Harassment may also be stopped cold by shaping the harasser's behavior. Your reactions to the harasser may encourage businesslike behavior and discourage flirtatious or suggestive behavior. If a harassing professor suggests that you come back after school to review your term paper so that the two of you will be undisturbed, set limits assertively. Tell the professor that you'd feel more comfortable discussing the paper during regular office hours. Remain task-oriented. Stick to business. The harasser should quickly get the message that you insist on maintaining a strictly professional relationship. If the harasser persists, however, do not blame yourself. You are responsible only for your own actions. When the harasser persists, a more direct response may be appropriate: "Professor Jones, I'd like to keep our relationship on a purely professional basis, okay?"

3. *Avoid being alone with the harasser.* If you are being harassed by your professor but need some advice about preparing your term paper, approach him or her after class when other students are milling about, not privately during office hours. Or bring a friend to wait outside the office while you consult the professor.

4. *Maintain a record.* Keep a record of all incidents of harassment to use as documentation in the event that you decide to lodge an official complaint. The record should include the following: (1) where the incident took place; (2) the date and time; (3) what happened, including the exact words that were used, if you can recall them; (4) how you felt; and (5) the names of witnesses. Some people who have been subjected to sexual harassment have carried a hidden tape recorder during contacts with the harasser. Such recordings may not be admissible in a court of law, but they are persuasive in organizational grievance procedures. Use of a hidden tape recorder may be illegal in your state, however. It is thus advisable to check the law.

5. *Talk with the harasser.* It may be uncomfortable to address the issue directly with a harasser, but doing so puts the offender on notice that you are aware of the harassment and want it to stop. It may be helpful to frame your approach in terms of a description of the specific offending actions (e.g., "When we were alone in the office, you repeatedly attempted to touch me or brush up against me"); your feelings about the offending behavior ("It made me feel like my privacy was being violated. I'm very upset about this and haven't been sleeping well"); and what you would like the offender to do ("So I'd like you to agree never to attempt to touch me again, okay?"). Having a talk with the harasser may stop the harassment. If the harasser denies the accusations, it may be necessary to take further action.

6. *Write a letter to the harasser.* Set down on paper a record of the offending behavior, and put the harasser on notice that the harassment must stop. Your letter might (1) describe what happened ("Several times you have made sexist comments about my body"); (2) describe how you feel ("It made me feel like a sexual object when you talked to me that way"); and (3) describe what you would like the harasser to do ("I want you to stop making sexist comments to me").

7. *Seek support.* Support from people you trust can help you through the often trying process of resisting sexual harassment. Talking with others enables you to express your feelings and receive emotional support, encouragement, and advice. In addition, it may strengthen your case if you have the opportunity to identify and talk with other people who have been harassed by the offender.

8. *File a complaint.* Companies and organizations are required by law to respond reasonably to complaints of sexual harassment. In large organizations, a designated official (sometimes an ombudsman, affirmative action officer, or sexual harassment advisor) is usually charged with handling such complaints. Set up an appointment with this official to discuss your experiences. Ask about the grievance procedures in the organization and your right to confidentiality. Have available a record of the dates of the incidents, what happened, how you felt about it, and so on.

The two major government agencies that handle charges of sexual harassment are the Equal Employment Opportunity Commission (look under the government section of your phone book for the telephone number of the nearest office) and your state's Human Rights Commission (listed in your phone book under state or municipal government). These agencies may offer advice on how you can protect your legal rights and proceed with a formal complaint.

9. *Seek legal remedies.* Sexual harassment is illegal and actionable. If you are considering legal action, consult an attorney familiar with this area of law. You may be entitled to back pay (if you were fired for reasons arising from the sexual harassment), job reinstatement, and punitive damages.

In closing, we repeat: The question is not what persons who suffer rape, incest, and sexual harassment will do to redress the harm that has been done to them. The question is what all of us will do to reshape our society so that sex can no longer be used as an instrument of power, coercion, and violence.

Human Sexuality ONLINE //

Web Sites Related to Sexual Coercion

This is the Rape Recovery and Help Information Page, which offers information on topics of interest to rape survivors. Listed alphabetically, this is what we find for A–H: Advocates, Book Suggestions, Child Abuse, Christianity, Counseling Online, Date Rape, Date Rape Drug, Depression, Domestic Violence, Eating Disorders, Espanol, Online Crisis Centers, Flashbacks, If You're Raped, and Healing Tips.
www.geocities.com/HotSprings/2402/

The Web site of RAINN: The Rape, Abuse and Incest National Network (RAINN), a not-for-profit organization based in Washington, DC, that operates a toll-free hotline for victims of sexual assault. RAINN was founded by singer/songwriter Tori Amos (herself a rape survivor). RAINN's toll-free hotline is (800) 656-HOPE.
www.feminist.com/rainn.htm

A Web site of the American Medical Association: "Facts About Sexual Assault":
www.ama-assn.org/public/releases/assault/facts.htm

The Sexual Harassment Hotline Resource List of the Feminist Majority Foundation. The Web site offers advice to victims of harassment and links to state hotlines and news.
www.feminist.org/911/harass.html

A fact sheet about rape (sexual assault) at the Centers for Disease Control and Prevention:
www.cdc.gov/ncipc/factsheets/rape.htm

The Web site of the American Professional Society on the Abuse of Children:
www.apsac.org/

The Web site of Child Help USA. (You can also call, toll-free, 800-4-A-CHILD.)
www.childhelpusa.org/

The Web site of the National Clearinghouse on Child Abuse and Neglect Information, an agency of the U.S. Department of Health and Human Services. (Also call, toll-free, 800-FYI-3366.)
www.calib.com/nccanch/

The Web site of Survivors of Incest Anonymous, SIA World Service Office, P.O. Box 190, Benson, MD 21018. (410) 893-3322:
www.siawso.org

summing up

RAPE

Although sexual motivation plays a role in many rapes, the use of sex to express aggression, anger, and power is more fundamental to our understanding of rape. The definition of rape varies from state to state but usually refers to obtaining sexual intercourse with a nonconsenting person by the use of force or the threat of force.

Incidence of Rape

Surveys suggest that there are about 310,000 rapes and sexual assaults each year in the United States. Many rapes go unreported.

Types of Rapes

Types of rapes include stranger rape, acquaintance rape, marital rape, gang rape, male rape, and rape by females. Women are more likely to be raped by men they know than by strangers. Most male rapes occur in prison settings. Husbands who rape their wives can now be prosecuted under the rape laws in many states.

Social Attitudes and Myths That Encourage Rape

Social attitudes such as gender-role stereotyping, seeing sex as adversarial, and acceptance of violence in interpersonal relationships all help create a climate that encourages rape.

Sociocultural Factors in Rape

Many observers contend that our society breeds rapists by socializing males to be socially and sexually dominant.

Psychological Characteristics of Rapists

Incarcerated rapists vary in their psychological characteristics. Self-identified sexually aggressive men are more likely than other men to condone rape and violence against women, to exhibit traditional gender-role attitudes, to be hostile toward women, to engage in sexual activity to express social dominance, to be sexually aroused by rape, to lack a social conscience, and to have peer groups such as fraternities that pressure them into sexual activity. Groth and Birnbaum identified three basic kinds of rape: anger rape, power rape, and sadistic rape.

Adjustment of Rape Survivors

Rape survivors often experience post-traumatic stress disorder (PTSD). Burgess and Holmstrom identified some common response patterns in rape survivors that they labeled the rape trauma syndrome.

Treatment of Rape Survivors

Treatment of rape survivors typically involves helping them through the crisis period following the attack and then helping to foster long-term adjustment.

Rape Prevention

Rape prevention involves educating society at large and familiarizing women with a number of precautions that they can take. Whether or not women take precautions to prevent rape, however, the rapist is always the one responsible for the assault.

VERBAL SEXUAL COERCION

Verbal sexual coercion involves the use of verbal pressure or seduction lines to manipulate a person into having sexual relations.

SEXUAL ABUSE OF CHILDREN

Like rape, sexual abuse of children is greatly underreported.

What Is Sexual Abuse of Children?

Any form of sexual contact between an adult and a child is abusive, even if force or physical threat is not used, because children are legally incapable of consenting to sexual activity with adults.

Patterns of Abuse

Sexual abuse of children, like rape and other forms of sexual coercion, cuts across all socioeconomic classes. In most cases, the molesters are close to the children they abuse—relatives, step-relatives, family friends, and neighbors. Genital fondling is the most common type of abuse.

Pedophilia

Pedophilia is a type of paraphilia in which adults are sexually attracted to children. Pedophiles are almost exclusively male.

Incest

Incest is marriage or sexual relations between people who are so closely related that sex is prohibited and punished by virtue of the kinship tie. Several theories have been advanced to explain the development of the incest taboo, including theories based on the dangers of inbreeding, the role played by the taboo in maintaining stability in the family, and its role in maintaining a stable community cooperation theory). Father–daughter incest is more likely to be reported and prosecuted, but brother–sister incest is the most common type of incest. Incest frequently occurs within the context of general family disruption.

Effects of Sexual Abuse of Children

Children who are sexually abused often suffer social and emotional problems that impair their development and persist into adulthood, affecting their self-esteem and their formation of intimate relationships.

Prevention of Sexual Abuse of Children

In addition to learning to avoid strangers, children need to learn the difference between acceptable touching, such as an affectionate embrace or pat on the head, and unacceptable or "bad" touching. Children who may not be able to prevent abuse may nevertheless be encouraged to tell someone about the experience.

■ **Treatment of Survivors of Sexual Abuse**
Psychotherapy may help adult survivors of sexual abuse improve their self-esteem and ability to develop intimate relationships. Special programs provide therapeutic services to abused children and adolescents.

TREATMENT OF RAPISTS AND CHILD MOLESTERS

The effectiveness of prison-based rehabilitation programs and anti-androgen drugs in curbing repeat offenses requires further empirical support.

SEXUAL HARASSMENT

■ **Sexual Harassment in the Workplace**
Sexual harassment in the workplace involves "deliberate or repeated unsolicited verbal comments, gestures, or physical contact of a sexual nature that [are] unwelcome [to] the recipient." Sexual harassment may be used as a tactic to keep women "in their place," especially in work settings that are traditional male preserves.

■ **Sexual Harassment on Campus**
About 25% to 30% of students report at least one incident of sexual harassment in college. Professors may also be harassed by students.

■ **Sexual Harassment in the Schools**
Sexual harassment at school has become an unwelcome ritual that many junior and senior high school students are forced to endure.

■ **How to Resist Sexual Harassment**
There is no guaranteed way to put an end to sexual harassment, but some suggestions that have helped many individuals include conveying a professional attitude, avoiding being alone with the harasser, keeping a record of incidents, and seeking legal remedies. Whether or not these efforts are effective, responsibility for harassment lies with the harasser and with the institution that failed to prevent harassment of the student or employee.

questions for critical thinking

1. Agree or disagree with the following statement, and support your answer: "A woman who walks in a dangerous neighborhood or talks to a stranger deserves what she gets."
2. Agree or disagree with the following statement, and support your answer: "Our society breeds rapists by socializing males into socially and sexually dominant roles."
3. Would you lie to convince someone to engage in sexual activity? Would you lie to avoid sexual activity? Explain.
4. Why do you think that the great majority of child molesters are male?
5. Agree or disagree with the following statement, and support your answer: "We should punish sex offenders and not worry about 'treating' them."
6. Agree or disagree with the following statement, and support your answer: "Because of all the publicity, people who are intolerant of normal sexual advances, or who want to punish their supervisors, are now crying sexual harassment."
7. Imagine that you are sexually harassed and want to report it, but a friend advises, "Why make such a fuss? Is it really worth it? After all, complaining could backfire." How would you respond?

chapter

19

Commercial Sex

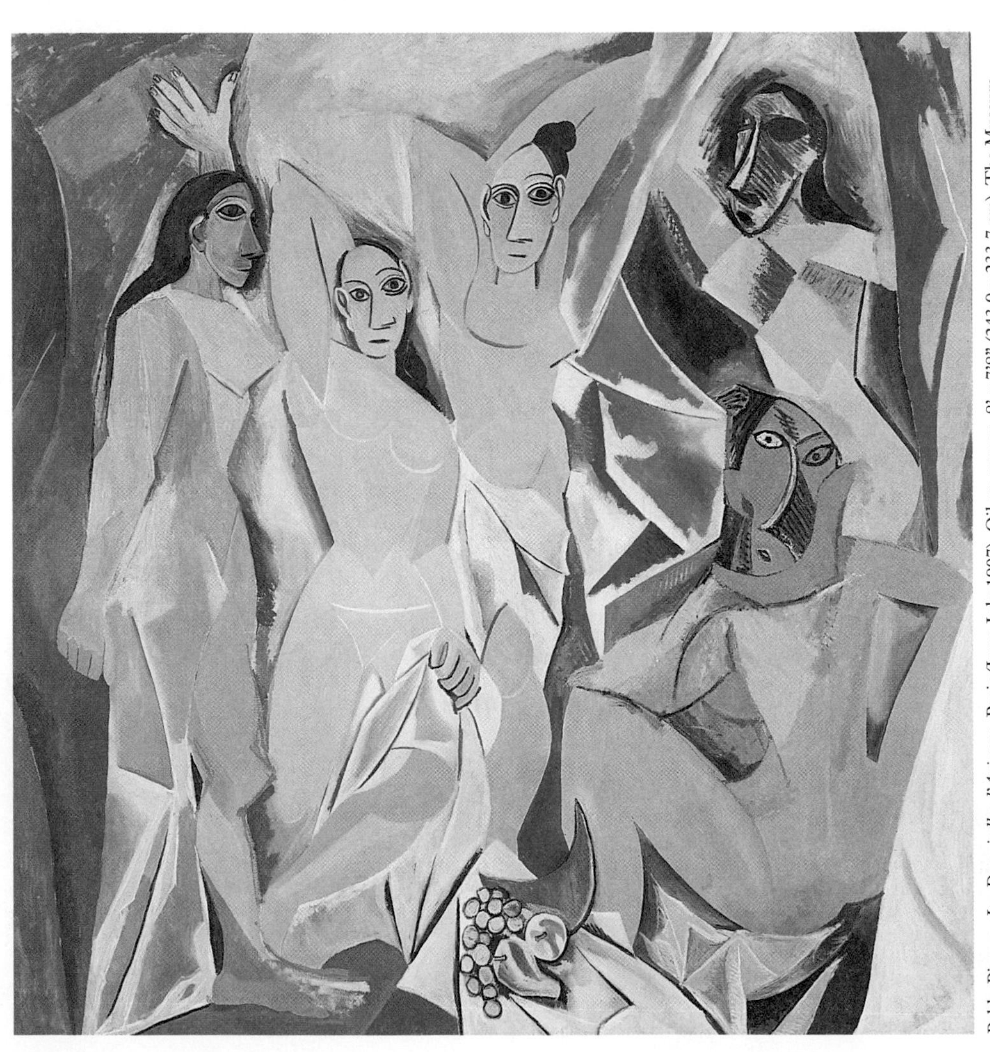

Pablo Picasso. *Les Demoiselles d'Avignon.* Paris (June-July 1907). Oil on canvas, 8' × 7'8" (243.9 × 233.7 cm). The Museum of Modern Art, New York. Acquired through the Lillie P. Bliss Bequest. Photograph © 2001 The Museum of Modern Art, New York. © 2002 Estate of Pablo Picasso/Artists Rights Society (ARS), New York.

outline

Truth or Fiction?

THE WORLD OF COMMERCIAL SEX: A DISNEYLAND FOR ADULTS

PROSTITUTION

Incidence of Prostitution in the United States Today

Types of Female Prostitutes

A World of Diversity The Joy of Amsterdam's Red-Light District A Police Guide

A World of Diversity From Russia, with . . . Sex?

Human Sexuality Online Online Sex Workers Have Rights Too

Characteristics of Female Prostitutes

A World of Diversity Sinking to a New Low in Romania

Customers of Female Prostitutes

A World of Diversity Schoolgirls as Sex Toys

Male Prostitution

HIV, AIDS, and Prostitution

PORNOGRAPHY AND OBSCENITY

What Is Pornographic?

Pornography and the Law

Human Sexuality Online Surfing for Sex

Prevalence and Use of Erotica and Pornography

Human Sexuality Online Filtering Sex Sites: Tough Work, But . . .

Pornography and Sexual Coercion

Human Sexuality Online Web Sites Related to Commercial Sex

SUMMING UP

QUESTIONS FOR CRITICAL THINKING

Truth or Fiction?

____ Prostitution is illegal throughout the United States.

____ The massage and escort services advertised in the Yellow Pages are fronts for prostitution.

____ Most female prostitutes were sexually abused as children.

____ Typical customers of prostitutes have difficulty forming sexual relationships with other women.

____ Most prostitutes take precautions to avoid becoming infected by or transmitting the AIDS virus.

____ Only males are sexually aroused by pornography.

A few years ago, England's floppy-haired actor Hugh Grant was caught with a prostitute in a BMW on Hollywood's Sunset Boulevard. Why? His girlfriend was supermodel Elizabeth Hurley, who at the time was promoting the perfume called "Pleasures." Grant exemplified good looks, charm, innocence, and success. Still, he sought the sexual services of a woman for hire.

Why, indeed. Why have millions of men (and some women) paid for sex throughout the ages? In the caverns of Wall Street, brokers and traders sell stocks and other financial instruments that drive the nation's commercial enterprises. On the floor of Chicago's Board of Trade, brokers trade commodities—soybeans, corn, wheat, and other goods. Prices go up, prices go down, responding to the law of supply and demand. On street corners a few short blocks from these financial institutions another sort of commerce takes place. In New York City and Chicago, as in Hollywood and the nation's smaller towns and villages, prostitutes exchange sex for money or for goods such as drugs.

THE WORLD OF COMMERCIAL SEX: A DISNEYLAND FOR ADULTS

Sex as commerce runs the gamut from adult movie theaters and bookshops to strip shows, sex toy shops, erotic hotels and motels, escort/outcall services, "massage parlors," "900" telephone services, cybersex (e.g., sex over the Internet), and the use of sex appeal in advertisements for legitimate products (Weizer, 1999). The "world of commercial sex is a kind of X-rated amusement park—Disneyland for Adults" (Edgley, 1989, p. 372).

PROSTITUTION

Hugh Grant ran afoul of the law because prostitution is illegal in California. **Prostitution** is illegal everywhere in the United States except for some rural counties in Nevada, where it is restricted to state-licensed brothels.

Truth or Fiction? REVISITED

It is not true that prostitution is illegal throughout the United States. It is legal in some counties in the state of Nevada, although it is restricted to regulated, state-licensed brothels.

Soliciting the services of a prostitute is also illegal in many states. Police rarely crack down on customers, or "johns," however. Hugh Grant came to their attention because he was engaging in a "lewd act" in a car parked on a public street (Beller, 1995). The few who are arrested are usually penalized with a small fine. On occasion, the names of convicted johns are published in local newspapers, which may deter men who fear publicity. On the other hand, some believe that Hugh Grant's indiscretion *boosted* receipts for his film *Nine Months*, which opened a few days later. Although prostitutes and their clients can be male or female, most prostitutes are female and virtually all customers are male.

Prostitution is often called "the world's oldest profession"—for good reason. It can be traced at least to ancient Mesopotamia, where temple prostitution flourished. The

Prostitution The sale of sexual activity for money or goods of value, such as drugs. (From the Latin *prostituere,* which means "to cause to stand in front of." The implication is that one is offering one's body for sale.)

Greek historian Herodotus noted that all women in the city were expected to put in some time at the temple. They would offer their bodies to passing strangers, who would make a "religious" donation.

Prostitution flourished in medieval Europe and during the sexually repressive Victorian period in the nineteenth century. Then, as now, the major motive for prostitution was economic. Many poor women were drawn to prostitution as a means of survival. In Victorian England, prostitution was widely regarded as a necessary outlet for men to satisfy their sexual appetites. It was widely held that women would not enjoy sex. Therefore, it was commonly believed that it was better for a man to visit a prostitute than to "soil" his wife with his carnal passions.

In the nineteenth-century United States, married and unmarried men frequented prostitutes regularly. Use of prostitution cut across all economic and social boundaries. Prostitution most often occurred within two contexts: sexual initiation for males and regular brothel visitation.

Incidence of Prostitution in the United States Today

No one knows how many prostitutes there are in the United States. The little we know of prostitution derives from sex surveys.

Almost two thirds of the European American males in Kinsey's sample (Kinsey et al., 1948) reported visiting a prostitute at least once. About 15% to 20% visited them regularly. The use of prostitutes varied with educational level, however. By the age of 20, about 50% of Kinsey's non-college-educated single males, but only 20% of his college-educated males, had visited a prostitute. By the age of 25, the figures swelled to about two thirds for non-college-educated males but only slightly above 25% for college men.

Kinsey's data foreshadowed a falling off of experience with prostitutes that seems linked to the decay of the sexual double standard. Kinsey found that 20% of his college-educated men had been sexually initiated by prostitutes. This figure was more than cut in half among generations who came of age in the 1960s and 1970s (Hunt, 1974). Less than 10% of both college-educated *and* non-college-educated males in the Morton Hunt survey (Hunt, 1974) of the 1970s reported being sexually initiated by a prostitute.

Why are young men less likely to visit prostitutes than they were in Kinsey's day? For one thing, young men in recent generations were more likely to become sexually initiated with their girlfriends. As Edgley (1989) put it, "An old and hallowed economic principle was at work; those who charge for a service cannot compete with those who give it away" (p. 392). Increased concern about sexually transmitted infections, especially AIDS, has also limited the use of prostitution. Nevertheless, prostitution continues to flourish.

Types of Female Prostitutes

Female prostitutes—commonly called *hookers, whores, working girls,* or *escorts*—are usually classified according to the settings in which they work. The major types of prostitutes today are streetwalkers; brothel or "house" prostitutes, many of whom work in massage parlors; and "escorts" or call girls (many prostitutes today have their customers "let their fingers do the walking through the Yellow Pages"). Traditional brothel prostitution is much less common today than before World War II.

STREETWALKERS Part of the mystery about Hugh Grant was why he would resort to the services of any prostitute. Another part of it was why he would seek out a **streetwalker.**

Streetwalkers Prostitutes who solicit customers on the streets.

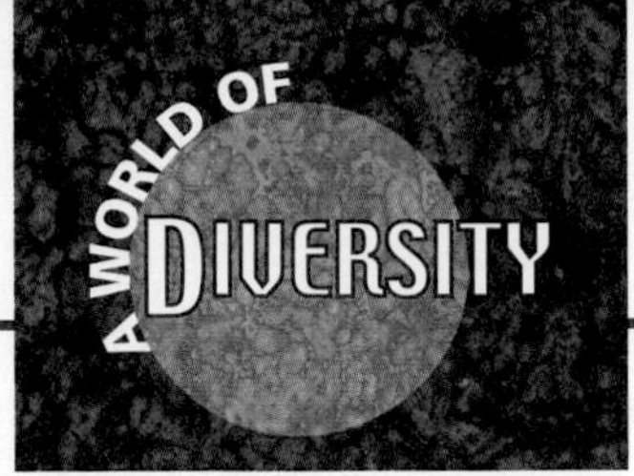

The Joy of Amsterdam's Red-Light District: A Police Guide

Rule No. 1, according to the new police guide for tourists to Amsterdam's red-light district, is: Have FUN, but act in a NORMAL manner.

Tolerant. Within limits. How very Dutch.

- Rule No. 6: PROSTITUTION. No pictures of the women.
- Rule No. 8: Soft drugs, NOT UNDER 18.
- And No. 10: Parking is NOT FREE.

"This brochure is NORMAL," insisted Superintendent Klaas Wilting, chief police spokesman, who must have repeated the word "normal" 30 times in an hour-long conversation about the brief brochure. He denied that the police felt defensive about publishing it. But he was distinctly irked that the FOREIGN press had taken such an interest. Here we go again, he implied: the old stereotype of Amsterdam as SIN CITY.

Actually, the guide is fairly subdued. Most of it is the kind of common-sense advice dished out by the police everywhere: watch out for pickpockets, don't play balletje (Amsterdam's three-card monte), resist those drunken impulses to take off all your clothes and jump into the canals.

Although it contains a map, it is not the usual guide, per se. It doesn't say which "coffee shops" sell the best marijuana, which live-sex shows are liveliest or which bordellos have the prettiest women. In fact, it is virtually impossible to find such information in print here. Most tourist guides to the Netherlands devote pages to the Vermeers in the Rijksmuseum, but scant sentences to the red-light district, even though it is unlikely the Vermeers get five million visitors a year.

So, Where Is It?

So what's the most common inquiry at the Amsterdam Tourist Service on Damrak Street?

"Where is the red-light district?" answered Djawehly Pelupessy, who works behind the counter. "They blush, we tell them, then they walk away fast. They're shy. They don't want to admit they want to see the women in the windows."

Toine Rodenburg, who manages the Banana Bar, the Erotic Museum, Theater Casa Rosso and other district landmarks, accuses the city of being "ambivalent or even hypocritical" toward the reality that sex lures far more visitors than culture.

"They take the taxes without saying thank you," he complained. "They put up signs everywhere saying 'Rijksmuseum' or 'Anne Frank House,' but there's only one for the red-light district, like it's something horrible. As if Amsterdam was all about tulips and wooden shoes. It's not. It's about freedom."

If it seems hard to tell whether the guide was written by a police officer or by a fed-up resident, it is because the answer to both questions is yes. The author, Officer Willem Schild, has patrolled this libertarian Arcadia for 12 years and, like Mr. Rodenburg, lives there. "That's me on the cover," he said proudly, pointing to the grinning policeman on a motorbike beside an Erotic Museum sign.

The guide reflects neighborhood thinking, repeated by everyone from Officer Schild to Erik Blom, a law student who sells tickets to the Moulin Rouge show, to Theodoor van Boven, founder of the Condomerie condom boutique. If the district has a problem, they say, it is not the law-abiding women sitting in pink windows wearing only fetching smiles and eye-popping underwear. Nor is it the mellowed-out hashish dealers who sit giggling with their coffee shop customers, debating which of Bob Marley's tunes was his chef-d'oeuvre.

Rather, if there is a problem in this district of 300-year-old gabled buildings and expensive real estate housing doctors, lawyers and plenty of families, it is the street drug dealers who sell heroin—or cough drops ground up to look like heroin—and those tourists, American spring break kids or English stag party louts who jump into the canals, howl their drunken lungs out at 4 A.M. and punch each other silly.

Blue-Rinse Grandmothers at the Theater Casa Rosso

Nobody really minds the middle-class tour groups that fill the sex shows in theaters that are cleaner and better lighted than many Off Broadway houses. Mr. Rodenburg seems to revel in the groups—often featuring blue-rinse grandmothers and other unexpected spectators—that flock to his Theater Casa Rosso. By prior arrangement with tour guides, he confided, the loudest-mouthed man in each group is the person selected by a leather-clad lady with whips to join her on stage—to his embarrassment and the general hilarity of his fellow tourists.

Superintendent Wilting's guide seems to be written in the same spirit of tolerance for human indulgence, while noting that despite popular notions about Amsterdam, the place does have rules.

All Drugs Are Not Equal

"Cocaine, heroin, LSD, ecstasy, etc. are strictly forbidden," the guide notes. "These drugs are always FAKE (washing powder, sugar, rat poison, vitamin C). On top of that, you may be forced to hand over your wallet." A small amount of marijuana for personal use can legally be consumed in a coffee shop, the leaflet says, adding, "When you feel sick after smoking, or eating spice cake, drink lots of water with sugar."

There is also a discreet hint about prostitutes: "If you visit one of the women, we would like to remind you

Amsterdam's Red-Light District. Prostitution is legal in Amsterdam's red-light district. Tourists wend their way through the narrow streets by the canals, "window-shopping." "Blue rinse" grandmothers join the audiences in the live sex shows. District landmarks include the Erotic Museum and the Theater Casa Rosso. Along with the Rembrandts, Rubens, and van Goghs, the district is such an ingrained part of Amsterdam's culture that the police have issued a guide to enjoying the district—safely.

they are not always women." But, here too, the guide demands decorum: "Don't shout or use bad language toward these women. SHOW SOME RESPECT."

That section was the most heavily edited by the department, Officer Schild said. He had advised tourists that the women charge $25 to $50 for 20 minutes, paid in advance. "So make a clear deal at the door and don't be too drunk to get things done," he had written. "Time is money for the ladies."

The department felt that was a bit TOO helpful, he said, although his precinct house's biggest headaches are frustrated drunks who get angry. All the women have alarm buttons, and the police then have to run or bicycle over to sort out the fight.

He also had a warning about "smart shops," which sell herbs that supposedly boost intelligence or virility, along with hallucinatory psilocybin mushrooms, which can mix badly with alcohol. "People think they're eagles and try to fly out of their hotel rooms," Officer Schild said. But the department is trying to ban smart shops, which also sell drug paraphernalia, so it did not want them included.

Officer Schild said he got to put in 95% of what he had hoped for, and takes the pamphlet around himself to bars and restaurants. "In the Netherlands," Superintendent Wilting noted, "we talk about the problems. We don't put them away."

Source: From D. G. McNeil, Jr. (2000, July 21). "The Joy of the Red-Light District: A Police Guide." *The New York Times online.*

From Russia, with . . . Sex?

Tamara, 19, has classic Slavic features, long blond hair, and deep green eyes. "I answered an ad to be a waitress," she told a reporter in a massage parlor near Tel Aviv's old Central Bus Station (Cited in Specter, 1998). "I'm not sure I would go back now if I could. What would I do there, stand on a bread line or work in a factory for no wages?"

"I didn't plan to do this," she added, surrounded by the red walls and leopard prints that decorate the brothel. "They took my passport, so I don't have much choice. But they do give me money. And believe me, it's better than anything I could ever get at home."

From Russia, with Love is the title of the James Bond novel by Ian Fleming, of course. But a new chapter is being written in the annals of prostitution, and it could very easily be titled *From Russia, with Sex.* Russia and its Slavic neighbors Ukraine and Belarus have largely replaced Thailand and the Philippines as the center of the worldwide business that trafficks in women. Economic problems in the former Soviet Union have opened a profitable market to the criminal gangs that have risen since the fall of communism (Smale, 2000). The routes along which women are transported run eastward to Thailand and Japan, where thousands of young Eastern European women now work against their will as prostitutes, and westward to the Adriatic Coast. Russian crime gangs based in Moscow arrange for security, connections with brothel owners in various countries, and false documents.

"It's no secret that the highest prices now go for the white women," says Marco Buffo (Specter, 1998), executive director of "On the Road," an organization that fights trafficking in women in Italy. "They are the novelty item now. It used to be Nigerians and Asians at the top of the market. Now it's [Eastern Europeans]."

Irina's is another case in point. Irina is from Ukraine, 21, and self-confident. Like many other Ukranian women, she wound up in Israel. She had answered a newspaper ad and crossed the Mediterranean on a tour boat for Haifa, intending to make her fortune dancing in the nude on tabletops. For her, Israel was a new world, filled with hope, until she was driven to a brothel and her boss burned her passport while she watched.

"I own you," he said (Specter, 1998). "You are my property, and you will work until you earn your way out. Don't try to leave. You have no papers and you don't speak Hebrew. You will be arrested and deported. Then we will get you and bring you back."

European women are in demand, and Ukraine has an apparently endless supply. Why? Because the economy of Ukraine is in turmoil and more than two thirds of the unemployed are women (Smale, 2000). Irina and others like her begin their hellish journeys seeking a better life. The average salary today in Ukraine is $30 a month, and young women are lured by local ads for lucrative jobs in foreign countries. Israel is a typical destination. Prostitution is legal here, although brothels are not. Brothel owners are sometimes fined, but they usually remain in business.

The women who are seeking their fortunes usually do not understand what is happening to them until it is too late. Once they reach their destination, their passports are confiscated, their cash is taken, and they become sexual slaves.

Although most prostitutes are streetwalkers, streetwalkers occupy the bottom rung in the hierarchy of prostitutes. They earn the lowest incomes and are usually the least desirable. They also incur the greatest risk of abuse by customers and **pimps.** Streetwalkers tend to come from poverty and to have had unhappy childhoods (Edgley, 1989). Perhaps as many as 80% are survivors of rape, sexual abuse, or incest (Gordon & Snyder, 1989). Many were teenage runaways who turned to prostitution to survive.

Streetwalkers operate in the open. They are thus more likely than other prostitutes to draw attention to themselves and risk arrest. To avoid arrest, streetwalkers may be indirect about their services. They may ask passers-by if they are interested in a "good time" or some "fun" rather than offering sex per se. In many cities, streetwalkers dress in revealing or protractive fashions.

There is the stereotype of the prostitute as a sexually unresponsive woman who feigns sexual arousal with johns while she keeps one eye glued to the clock. Most street prostitutes in a Philadelphia sample, however, reported that some forms of sex with customers were "very satisfying" (Savitz & Rosen, 1988). More than 60% of the prostitutes reported achieving orgasm with customers at least occasionally. Most prostitutes also reported that they had enjoyable sexual relationships in their private lives and were regularly orgasmic. Not surprisingly, they garnered more sexual enjoyment from their personal relationships than from their customers.

Pimps Men who serve as agents for prostitutes and live off their earnings. (From the Middle French *pimper,* which means "to dress smartly.")

A Streetwalker. It is estimated that several hundred thousand prostitutes work in the United States. Some prostitutes—called *streetwalkers*—ply their trade by walking the streets and hawking their wares to pedestrians and those who drive by. Others, who may be connected with "massage" or "escort" services, let their customers' fingers do the walking through the Yellow Pages. Still others advertise online.

In most locales, penalties for prostitution involve small fines or short jail terms. Many police departments, besieged by drug peddling and violent crimes, consider prostitution a "minor" or "nuisance" crime (Carvajal, 1995). Many prostitutes find the criminal justice system a revolving door. They pay the fine. They spend a night or two in jail. They return to the streets.

A bar prostitute is a variation of the streetwalker. She approaches men in a bar she frequents, rather than on the streets. Payoffs to bar owners or managers secure their cooperation, although the women are sometimes tolerated because they draw customers. Some streetwalkers work X-rated, or "adult," movie houses and may service their patrons in their seats with manual or oral sex. Payoffs may secure the cooperation of the management.

Many streetwalkers support a pimp. A pimp acts as lover–father–companion–master. He provides streetwalkers with protection, bail, and sometimes room and board, in exchange for a high percentage of their earnings, often more than 90%. As Chris, a streetwalker, explains, a prostitute cannot expect to survive long on the streets without a pimp.

> CHRIS: You can't really work the streets for yourself unless you got a man—not for a long length of time . . . 'cause the other pimps are not going to like it because you don't have anybody to represent you. They'll rob you, they'll hit you in the head if you don't have nobody to take up for you. Yeah, it happens. They give you a hassle. . . . [The men] will say, "Hey baby, what's your name? Where your man at? You got a man?"
>
> INTERVIEWER: So you can't hustle on your own?
>
> CHRIS: Not really, no. You can, you know, but not for long.
>
> (Romenesko & Miller, 1989, pp. 116–117)

Prostitutes are often physically abused by their pimps, who may use threats and beatings as a means of control. In a study of young streetwalkers in Boston, Virginia Ann Price (1989) observed that as the relationship progressed, the beatings became more vicious. Kim Romenesko and Eleanor Miller (1989), who studied streetwalkers in Milwaukee, commented, "Clearly, 'men' are the rulers of the underworld" (p. 117).

Streetwalkers do not stay in the business (sometimes referred to as "the life") very long (Edgley, 1989). Some make the transition to a more traditional life or get married.

Others die young from drug abuse, disease, suicide, or physical abuse from pimps or customers. Those who survive become less marketable with age.

Streetwalkers who work hotels and conventions generally hold a higher status than those who work the streets or bars. Clients are typically conventioneers or businessmen traveling away from home. The hotel prostitute must be skilled in conveying subtle messages to potential clients without attracting the attention of hotel management or security. They usually provide sexual services in the client's hotel room. Some hotel managers will tolerate known prostitutes (usually for a payoff under the table), so long as the woman conducts herself discreetly.

BROTHEL PROSTITUTION Many brothel prostitutes occupy a middle postition in the hierarchy of prostitutes, between streetwalkers on one side and call girls on the other (Edgley, 1989). They work in a brothel, or—more commonly today—in a massage parlor.

The life of the brothel (or "house") prostitute is usually neither as lucrative as that of the call girl nor as degrading as that of the streetwalker. Some prostitutes who work in massage parlors or for escort services may not consider themselves "real prostitutes" because they do not walk the streets and because they work for businesses that present a legitimate front (Edgley, 1989). *Cathouse, bordello, cat wagon, parlor house, whorehouse, joy house, sport house, house of ill repute*—these are but a handful of the names given to places in which prostitutes work. The heyday of the brothel is all but over in the United States. Formal brothels today are rare, except in Nevada, where they are legal but regulated.

Brothel prostitutes may try to maintain a traditional life apart from their prostitution activities. Ricker (1980) tells the story of one dual-world prostitute:

> Jan has been married 15 years and is the mother of three children—her earnings help keep them in a snooty Northern California boarding school. A relative newcomer [at the Chicken Ranch, a licensed Nevada brothel], she tried massage parlors, escort services, and other forms of "the business" [earlier]. Jan got into this line of work because she needed extra money. . . . Jan feels her home relationship has been improved by her job. "Before I went to work, my husband and I were always entertaining, trying to keep up that affluent suburban image. Now, when we're together we just enjoy each other and the children." She also feels her work has made her more interesting to her husband. "I've had to learn to relate to all kinds of different people. And I've learned how to play a variety of roles—I can be nurse, sophisticated companion, psychologist, temptress. . . ." No, Jan doesn't tell her children what she does. (p. 282)

Jan splits her fees with the management. In addition to her "split," she receives free room and board. Brothel prostitutes are on duty 3 weeks a month and on call 24 hours a day during that time. When a customer arrives, they step into the living room "lineup." After one has been chosen, they wait again, resting, reading, or watching television.

Jan doesn't see herself as having a bad life. Yet some brothel prostitutes lead lives of complete degradation (Jeffreys, 1998). Many poor Asian women have been recently lured to the United States by promises of the good life. Upon arrival, they have found themselves enslaved in brothels—working for tips and not allowed to leave. Such a brothel was recently closed by police at 208 Bowery in New York City. It housed more than 30 women smuggled from Thailand and came to the attention of police when one woman jumped from a window (Goldberg, 1995).

THE MASSAGE PARLOR Nature abhors a vacuum. "Massage parlors" have sprung up from coast to coast to fill the vacuum left by the departure of the brothels. Many massage parlors are legitimate establishments that provide massage—and only massage—to customers. Masseuses and masseurs are licensed in many states, and laws prohibit them from offering sexual services. Many localities require that the masseuse or masseur keep certain parts of her or his body clothed (some masseuses in suburban Detroit wear a scarf or a garter to comply) and not touch the client's genitals.

Many massage parlors serve as fronts for prostitution, however. They are often found in malls in middle-class suburbs, where there is ample parking (Carvajal, 1995). In these establishments, clients typically pay fees for a standard massage and then tip the workers for sexual extras.

Massage parlor prostitutes generally offer to perform manual stimulation of the penis ("a local"), oral sex, or less frequently, coitus ("full service"). Some massage parlor prostitutes are better educated than streetwalkers and brothel workers and would not work in those other venues.

ESCORT SERVICES If Hugh Grant had hired an "escort," we probably would never have learned of it. Conventioneers and businessmen are more likely to turn to the listings for "massage" and "escort services" in the telephone directory or under the personal ads in local newspapers than to seek hotel prostitutes. Services that provide "outcall" send masseuses (or masseurs) or escorts to the hotel room.

Escort services are typically (but not always) fronts for prostitution. Escort services are found in every major American city and present themselves as legitimate businesses providing escorts for men. Indeed, one will find female companionship for corporate functions and for unattached men traveling away from home under "escort services." Many escort services provide only prostitution, however, and clients of other escort services sometimes negotiate sexual services after formal escort duties are completed—or in their stead.

Not all of the massage and escort services advertised in the Yellow Pages are fronts for prostitution. Legitimate masseuses and masseurs often advertise that they are licensed by their states.

Prostitutes who work for escort services often come from middle-class backgrounds and are well educated—the better to hold their own in social conversation. Escort services may establish arrangements with legitimate companies to provide "escorts" for visiting customers or potential clients. Escort services also provide female escorts to "entertain" at conventions. Because of her high-society background, Sidney Biddle Barrow, the so-called Mayflower Madam, attracted a great deal of publicity when it was discovered that she ran an exclusive "escort service" in New York City in the 1980s.

CALL GIRLS **Call girls** occupy the highest rungs on the social ladder of female prostitution. Many of them overlap with escorts. Call girls tend to be the most attractive and well-educated prostitutes and tend to charge more for their services. Many come from middle-class backgrounds (Edgley, 1989). Unlike other types of prostitutes, call girls usually work on their own. Thus they need not split their income with a pimp, escort service, or massage parlor. Consequently, they can afford a luxurious lifestyle when business is good, living in expensive neighborhoods and wearing stylish clothes, and they can be more selective about the customers they will accept. Yet they incur expenses for answering services and laundry services, and for payoffs to landlords, doormen, and sometimes police to maintain their livelihood and avoid arrest.

Call girls often escort their clients to dinner and social functions and are expected to provide not only sex but also charming and gracious company and conversation (Edgley, 1989). Call girls often give clients the feeling that they are important and attractive. They may effectively simulate pleasure and orgasm and can create the illusion that time does not matter. It does, of course. To the call girl, as to other entrepreneurs, time is money.

Call girls may receive clients in their apartments ("incalls") or make "outcalls" to clients' homes and hotels. Some call girls trade or sell "black books" that list clients and their sexual preferences. To protect themselves from police and abusive clients, call girls may insist on reviewing a client's business card or learning his home telephone number before personal contact is made. They may investigate whether the customer is in fact the person he purports to be.

Call girls Prostitutes who arrange for their sexual contacts by telephone. *Call* refers both to telephone calls and to being "on call."

Human Sexuality ONLINE //

Online Sex Workers Have Rights Too

A German court ruled that people paid to talk dirty in the Internet's increasing numbers of sex chat rooms should enjoy the same rights as other workers, regardless of whether their job is "immoral." The court rejected claims by a north German firm offering live online sex chats that the immorality of the work done by its staff should exempt the company from having to make social security contributions for them.

A judge ruled that the morality of online sex services, which mostly employ women to meet a seemingly insatiable and largely male appetite for impersonal stimulation, was irrelevant and decided that staff should be treated as they would be in other jobs. Workers are to be paid according to the frequency and length of their one-to-one erotic conversations.

Even mainstream Internet portals in Germany, where topless women are a nightly fixture on national television, are awash with links to subscription-based Web sites promising such delights as "live chats with hundreds of the hottest girls." Because most firms who run these sites—along with countless premium-rate telephone sex lines offering similar services minus the accompanying porn pictures—employ staff on a similar freelance basis, the ruling could spawn new claims.

Source: From "Online Sex Workers Have Rights Too." (2000, August 10). *Reuters News Agency online.*

Sex Work Is Work. Sex workers have rights too, according to court rulings in various European cities. According to a German court, these rights include the right to social security pension programs.

Characteristics of Female Prostitutes

No single factor explains women's entry into prostitution. Yet poverty and sexual and/or physical abuse figure prominently in the backgrounds of many prostitutes. They often come from conflict-ridden or single-parent homes in poor urban areas or rural farming communities.

For impoverished young women with marginal skills, the life of the prostitute may seem alluring. It is an alternative to the menial and dismal work that is otherwise available. On the basis of their studies of streetwalkers in Milwaukee, Romensko and Miller (1989) concluded that

> Poverty, and the many concomitants of poverty manifest in American society, were the factors pushing these women into the world of illicit work. Conversely, the bright lights, money, and independence that the street seemingly offered—things that were largely absent from these women's lives prior to their entrance into street life—were enticements that drew women to street life. (p. 112)

Post-traumatic stress disorder A type of stress reaction brought on by a traumatic event and characterized by flashbacks of the experience in the form of disturbing dreams or intrusive recollections, a sense of emotional numbing or restricted range of feelings, and heightened body arousal. Abbreviated *PTSD.*

Researchers also find a high level of psychological disturbance among prostitutes. Melissa Farley and her colleagues (1998) interviewed nearly 500 prostitutes from the United States, Europe, Africa, and Asia and found that nearly two thirds of them could be diagnosed with **post-traumatic stress disorder** (PTSD). Only 5% of the general population have PTSD, and only 20% to 30% of combat veterans. The incidence of PTSD was the same for streetwalkers and brothel workers, although brothel prostitutes encountered less violence (Zuger, 1998). The prostitutes also developed high numbers of medical problems (other than sexually transmitted infections). Ninety percent said that they wanted to get out of "the life."

Cross-cultural evidence from a study of 41 prostitutes in Belgium showed similar patterns of psychopathology (de Schampheleire, 1990). Compared to other women (female flight attendants), streetwalkers in the Belgian study were more fearful, anxious, resentful, and depressed. They also had poorer relationships with their families and were in poorer health. In a study in the United States, teenage female prostitutes were more likely than normal female adolescents or delinquents who did not engage in prostitution to show signs of psychological disturbance and to have been placed in special-education classes in school (Gibson et al., 1988).

Poverty accounts for the entry of young women into prostitution in many countries. In some Third World nations, such as Thailand, many rural, impoverished parents in effect sell daughters to recruiters who place them in brothels in cities (Erlanger, 1991; Goldberg, 1995). Many of the women send home whatever money they can and also work hard to try to pay off the procurers and break free of their financial bonds.

In the United States and Canada, many initiates into prostitution are teenage runaways. The family backgrounds of teenage runaways vary in socioeconomic status (some come from middle-class or affluent homes, whereas others are reared in poverty). However, family discord and dysfunction frequently set the stage for their entry into street life and prostitution (Price, 1989). Many teenage runaways perceive life on the street to be the only possible escape from family strife and conflict, or from the physical, emotional, or sexual abuse they suffer at home. Despite its dangers, life on the streets appears more attractive than remaining in the troubled family environment. A study of teenage prostitutes in Boston found that a majority had come from broken homes and were reared by single parents or in reconstituted families consisting of half-siblings and stepsiblings (Price, 1989). Those who came from intact families reported high levels of family conflict.

Teenage runaways are at particularly high risk of unsafe sexual practices, such as unprotected sex with multiple partners. The average male teenage runaway in a New York City study reported having had 11 female sexual partners in his lifetime (Rotheram-Borus et al., 1992). Only 8% of the teenage boys reported using condoms consistently.

Teenage runaways with marginal skills and limited means of support may find few alternatives to prostitution. Adolescent prostitution results from the "necessities of street life—it is survival behavior more than it is sexual behavior" (Seng, 1989, p. 674). It is not long before the teenage runaway is approached by a pimp or a john. A study of 149 teenage runaways in Toronto found that 67% of the boys and 82% of the girls who had been away from home for more than a year had been offered money to engage in sexual activity with an adult (Hartman et al., 1987).

Studies of teenage prostitutes in the United States show that perhaps as many as one half to two thirds of female prostitutes had been sexually abused as children (Seng, 1989). One sample of 45 former prostitutes in Canada showed that 73% had been sexually abused in childhood (Bagley & Young, 1987).

It appears that the majority of female prostitutes were in fact sexually abused as children. ■

Some teenagers who have endured sexual abuse or incest learn how to detach themselves emotionally from sex to survive unwanted sexual experiences. The transition to prostitution may represent an extension of this unfortunate learning experience. Though teenage prostitutes may turn to prostitution primarily for money, Price (1989) recognized that survivors of sexual abuse or incest may also be attracted to prostitution because

Sinking to a New Low in Romania

There is a running debate about whether prostitution exploits women or provides another way to make a living. A recent story out of Bucharest, Romania, will set the issue straight forever: There is absolutely no doubt that prostitution exploits women.

Rarely in human history have prostitutes been compelled to sink so low as in Romania, but the fact is that after having sex with their clients, Romanian prostitutes are doing housework. Yes, housework. The sex trade was hit by economic recession, and the women had to provide added value to convince men to part with their money.

"We had to invent something," says a pimp (cited in "Prostitutes do housework too," 2000), "because people don't have money and clients are rare. After solving the [sexual] problem, the girls clean and cook, for free. All on the house. Men are happy because many of them live alone and the girls help them get rid of the three things which torment their lives: sex, cleaning and cooking."

But do they do windows?

they have learned that it is through their sexuality that they can obtain attention or love from adults:

> For a lonely adolescent, attention from tricks or a sugar daddy is not all that different from the attention he or she received at home. In fact, it may be preferable, as there may not be any physical abuse involved. (p. 84)

Not all sexually abused children become prostitutes, of course. Only 12% of one sample of predominantly female 16- to 18-year-olds who had been sexually abused became involved in prostitution (Seng, 1989). Abused children who run away from home are much more likely to become involved in prostitution than those who do not (Seng, 1989). Runaways are also more likely to become drug and alcohol abusers.

Nor do all teenage runaways become prostitutes. Though information on prevalence of prostitution among runaways is scarce, a Boston study of homeless street youth in the mid-1980s found that fewer than 20% had engaged in prostitution (Price, 1989). Though most had been approached for prostitution within a few days of living on the streets, the overwhelming majority refused.

In sum, evidence suggests that most female prostitutes are survivors of sexual abuse or incest. Running away from home appears to funnel many survivors of childhood sexual abuse into prostitution, yet not all survivors of sexual abuse—and not all teenage runaways—become prostitutes.

Customers of Female Prostitutes

Hugh Grant was a "john" or a "trick." That is how many prostitutes refer to their customers. Terms such as *patron*, *meatball*, *sucker*, and *beefbuyer* are also heard. Men who use female prostitutes come from all walks of life and represent all socioeconomic and racial groups. Many, perhaps most, are married men of middle-class background. A study of the address listings of customers (seized by the local police) of an "escort service" in a southern city showed that most of the clients lived in relatively affluent neighborhoods populated by a high percentage of well-educated, largely European American, married residents (Adams, 1987).

Most patrons are "occasional johns." Examples include traveling salesmen and military personnel who are stopping over in town without their regular sex partners. One study of 30 occasional johns showed that all had regular sex partners. They used prostitutes for novelty or sexual variety, not because they lacked other sexual outlets (Holzman & Pines, 1982).

"Habitual johns" use prostitutes as their major or exclusive sexual outlet. Some habitual johns have never established an intimate sexual relationship. Some wealthy men who wish to avoid intimate relationships habitually patronize call girls.

"Compulsive johns" feel driven to prostitutes to meet some psychological or sexual need. They may repeatedly resolve to stop using prostitutes but feel unable to control their compulsions. Some compulsive johns engage in acts of fetishism or transvestism with prostitutes but would not inform their wives or girlfriends of their variant interests. Some men who are compulsive users of prostitutes suffer from a **whore–Madonna complex.** They see women as either sinners or saints. They can permit themselves to enjoy sex only with prostitutes or would ask only prostitutes to engage in acts such as fellatio. They see marital coitus as a duty or an obligation.

MOTIVES FOR USING PROSTITUTES Though the reasons for using prostitutes vary, researchers have identified six of the most common motives (Edgley, 1989; Gagnon, 1977):

1. *Sex without negotiation.* Prostitution may be attractive to men who do not want to spend the time, effort, and money involved in dating and getting to know someone simply to obtain sex. As Edgley comments, the customer "gives [the prostitute] money for sex, she gives sex in return. The entire matter is simple, direct, and sure" (p. 393). Hugh Grant might have wanted to have sex, and his girlfriend was a continent and an ocean away. The streetwalker was willing to supply sex in the absence of a relationship.

2. *Sex without commitment.* Prostitutes require no commitment from the man other than payment for services rendered. The prostitute will not call him at home or expect to be called in return.

3. *Sex for eroticism and variety.* Many prostitutes offer "something extra" in the way of novel or kinky sex—for example, oral sex, use of costumes (such as leather attire), and S&M rituals (such as bondage and discipline or spanking). Men may desire such activity but not obtain it with their regular partners. They may even be afraid to mention the idea. Prostitutes may also be attractive to men who seek variety or novelty in sex partners.

4. *Prostitution as sociability.* In the nineteenth and early twentieth centuries, the brothel served not only as a place to obtain sex but also as a kind of "stopping off" place between home and work—referred to by some writers as a "third place" (Oldenberg & Brissett, 1980). The local tavern and pool hall are nonsexual "third places." At times, sex was secondary to the companionship and amiable conversation that men would find in brothels, especially in the days of the "bawdy houses" of the pioneer West.

5. *Sex away from home.* The greatest contemporary use of prostitution occurs among men who are away from home, such as businessmen at conventions and sports fans at out-of-town sporting events. In these typical all-male preserves, there is often intense peer pressure to engage in sexual adventures.

It is not true that typical customers of prostitutes have difficulty forming sexual relationships with other women. Many men often turn to prostitutes for sexual variety or because they seek sexual activity when they are away from home.

6. *Problematical sex.* People who have physical disabilities or disfiguring conditions sometimes seek the services of prostitutes because of difficulty attracting other partners or because of fears of rejection. (One prostitute at a Nevada brothel said that she was a favorite of the management because she accepted johns with cerebral palsy.) Although some prostitutes are selective about the clients they will accept, others will accept any client who is willing to pay. Men with sexual dysfunctions may also turn to prostitutes to help them overcome their problems. Some lonely men who lack sex partners may seek prostitutes as substitutes.

Whore–Madonna complex
A rigid stereotyping of women as either sinners or saints.

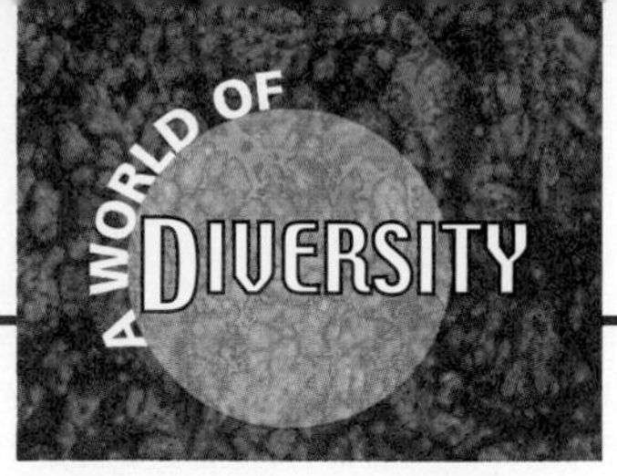

Schoolgirls as Sex Toys

A disturbing new national pastime has taken root in Japan: a male obsession with schoolgirls dressed in uniforms.

In Tokyo there are now several hundred "image clubs," where Japanese men pay $150 an hour to act out their fantasies with make-believe schoolgirls in make-believe classrooms, locker rooms or commuter trains. A customer may, for example, act the part of a teacher who walks into a classroom and tears the clothes off a schoolgirl. Or he may choose to fondle a schoolgirl on a crowded commuter train.

The clubs are but one part of a growing national obsession—and a growing national market—that the Japanese call "Loli-con," after Lolita. Hiroyuki Fukuda edits one Japanese magazine called *Anatomical Illustrations of Junior High School Girls* and another called *V-Club*, featuring pictures of naked elementary school girls. And a number of schoolgirls earn extra money as prostitutes.

Why? First because it's legal for men in Tokyo to have sex with children who are older than 12. And second, says Masao Miyamoto, a male psychiatrist, because many Japanese men feel threatened by the growing sophistication of older women. So they turn to schoolgirls.

Source: Reprinted from "Schoolgirls as Sex Toys" (1997, April 6). *The New York Times*, p. E2.

Japanese Schoolgirls. Sad to say, many Japanese men are obsessed with schoolgirls dressed in uniforms such as these. In Tokyo, it is legal for men to have sex with children who are older than 12. Many men are willing to pay for the privilege.

Male Prostitution

Male prostitution includes both male–male and male–female activities. Males prostitutes who service female clients—*gigolos*—are rare. Gigolos' clients are typically older, wealthy, unattached women. Gigolos may serve as escorts or as surrogate sons for the women, and they may or may not offer sexual services. Many gigolos are struggling actors or models.

The overwhelming majority of male prostitutes service gay men. Men who engage in male prostitution are called **hustlers.** Their patrons are typically called **scores.** Hustlers average 17 to 18 years of age and become initiated into prostitution at an average age of 14 or so (Coleman, 1989). They typically have less than eleventh-grade educations and few, if any, marketable skills. The majority come from working-class and lower-class backgrounds. Many male prostitutes, like many female prostitutes, come from families troubled by conflict, alcoholism, or abuse (Coleman, 1989). More than 4 out of 5 (83%) of a sample of 47 male prostitutes in Seattle, Washington, had been survivors of sexual

Hustlers Men who engage in prostitution with male customers.

Scores Customers of hustlers.

abuse (Boyer, 1989). Two out of 3 are survivors of rape or attempted rape. In another study, the factor that most clearly distinguished juvenile male prostitutes from male delinquent street youth was the absence of sexual victimization in the histories of the delinquents (Janus et al., 1984).

Hustlers may be gay, bisexual, or heterosexual in orientation. At least half of the male prostitutes surveyed are gay. Seventy percent of the male prostitutes in the Seattle study identified themselves as gay or bisexual. Only 30% identified themselves as heterosexual (Boyer, 1989).

The major motive for male prostitution, as for female prostitution, is money. In one study, 69% of male prostitutes cited money as their principal motive (Fisher et al., 1982). Running away from home typically serves as an entry point for male as well as female prostitution. In one study, 3 out of 4 male prostitutes had run away by an average age of 15 (Weisberg, 1985). Some had run away because of family problems, others because of a desire for adventure or independence (Coleman, 1989). Some gay male prostitutes are thrown out because their families cannot accept their sexual orientation (Coleman, 1989; Kruks, 1991). Researcher Debra Boyer (1989) comments:

> The responses of families to the [sexual orientation] of their sons ranged from strong condemnation to total rejection, for example "it's disgusting"; "we refuse to accept it"; "we don't want to talk about it"; "that's it, you are leaving." These young men suffered outright rejection from their families and were often literally thrown away as sullied human beings. (pp. 168–169)

For many of these gay "throwaways," prostitution may represent a way of obtaining adult acceptance, albeit within the context of exchanging sex for money. Another distinguishing characteristic of the gay prostitutes in the Seattle study was the lack of a network of gay friends. These young gay males became initiated into the gay subculture in the life of the street. Their only contacts with other gay men were through turning tricks or "hanging out" with other young hustlers. The ability to attract clients and the money they received inflated their self-esteem. As Boyer (1989) puts it, they were "no longer outcasts, but stars" (p. 177). Many hoped to find companionship or a meaningful relationship through prostitution. Some would fall in love with "tricks," only to be disappointed when the encounter or relationship fizzled.

Heterosexual hustlers may try to detach themselves psychologically from male clients by refusing to kiss or hug them or to perform fellatio. Gagnon (1977) writes, "As long as the [client's] head is below the [hustler's] belly button and contact is on the penis, it is the other person who is [gay]" (p. 264). To become aroused, heterosexual hustlers may fantasize about women while the "score" is fellating them. Many heterosexual male prostitutes maintain heterosexual relationships in their private lives while "turning tricks" with men to earn money.

Most hustlers are part-timers who continue some form of educational or vocational activity as they support themselves through prostitution. Drug dealing and drug use are also common among hustlers (Coleman, 1989). In one study, about 75% of male prostitutes used drugs while they were hustling (Fisher et al., 1982).

Unlike female prostitutes, hustlers typically are not attached to a pimp (Luckenbill, 1985). They generally make contacts with clients in gay bars and social clubs or by working the streets in areas frequented by gay men. They typically learn to hustle through their interactions with other hustlers and by watching other hustlers ply their trade (Luckenbill, 1985).

Coleman (1989) identifies several types of male prostitutes:

- *Kept boys* have relationships with older, economically secure men who keep them in an affluent lifestyle. The older male, or "sugar daddy," often assumes a parental role.
- *Call boys*, like call girls, may work on their own or through an agency or escort service.
- *Punks* are prison inmates who are used sexually by other inmates and rewarded with protection or goods such as cigarettes and drugs.

- *Drag prostitutes* are transvestites or presurgical male-to-female transsexuals who impersonate female prostitutes and have sex with men who are frequently unaware of their gender. Some drag prostitutes limits themselves to fellatio on their customers to conceal their genders. Others take the passive role in anal sex (Boles & Elifson, 1994).
- *Brothel prostitutes* are rarer than their female counterparts. Fewer houses of male prostitution exist.
- *Bar hustlers* and *street hustlers*, like their female counterparts, have the lowest status and ply their trade in gay bars or on streets frequented by gay passers-by. Street hustlers are the most common and typically the youngest subtype. They are also the most visible and consequently the ones most likely to draw the attention of the police.

Male prostitutes typically have shorter careers than their female counterparts (Price, 1989). By and large, male prostitution is an adolescent enterprise. The younger the hustler, the higher the fee he can command and the more tricks he can turn. By the time he reaches his mid-20s, he may be forced to engage in sexual activities he might have rejected when younger or to seek clients in sleazier places.

HIV, AIDS, and Prostitution

Concerns about the spread of sexually transmitted infections via prostitution is nothing new. In the 1960s, two of three prostitutes surveyed by Gebhard (1969) had contracted gonorrhea or syphilis. Though prostitutes are still exposed to a heightened risk of contracting or spreading these and other sexually transmitted infections, such as chlamydia, today the risk of AIDS poses a more deadly threat. The risk of HIV transmission has been linked to both male and female prostitution (Campbell, 1991; Van den Hoek et al., 1990; Yates et al., 1991).

Sex with prostitutes is the most important factor in the male–female transmission of HIV in Africa, where the infection is spread predominantly via male–female sexual intercourse (see Chapter 16). A Florida study showed that regular contact with female street prostitutes was a risk factor in the transmission of HIV in U.S. men (Castro et al., 1988). Prostitutes incur a greater risk of HIV transmission because they have sexual relations with many partners, often without protection (Quadagno et al., 1991). Moreover, many prostitutes and their clients and other sex partners inject drugs and share contaminated needles (Bloor et al., 1990; Freund et al., 1989). HIV may be spread by unprotected sex from prostitutes to customers and then to the customers' wives or lovers. Nationwide, about 12% of the prostitutes tested in the late 1980s were seropositive for HIV (Lambert, 1988).

Men who frequent male prostitutes may represent a vector, or conduit, for male–female sexual transmission of HIV. The rate of HIV infection is high among male prostitutes, and many if not most of their male customers describe themselves as either bisexual or heterosexual (Morse et al., 1992). One study found that the rate of HIV infection was 50% for gay male prostitutes, 36.5% for bisexual male prostitutes, and 18.5% for heterosexual male prostitutes (Boles & Elifson, 1994). Customers of such men may thus be exposing their wives and girlfriends to HIV.

Despite the obvious dangers of HIV transmission, many U.S. prostitutes have not altered either their sexual behavior or their patterns of drug use to any large extent. Only 30% of a sample of 20 street prostitutes in Camden, New Jersey, reported always using condoms (Freund et al., 1989). A study of 72 heroin-addicted female street prostitutes in southern California showed a high level of knowledge and fear about AIDS but a failure to change patterns of sexual behavior or drug use (Bellis, 1990). The prostitutes sampled did nothing to protect themselves or their clients from HIV. Their need for money led them to deny the risks to which they were exposing themselves. Surveys of adolescent prostitutes reveal a similar pattern of denial, as Price (1989) observes,

> Some [adolescent prostitutes] simply believe it will not happen to them, while others say they are willing to take the risk. Still others do not care whether they live or die, and expect their lives to end shortly anyway (p. 86).

Some changes in prostitution practices have taken place. Brothel owners in Nevada, for example, now require customers to wear condoms. But prostitutes who are bent on denial or are so filled with despair that they care nothing about their own or their clients' lives may be most resistant to change.

It is not true that most prostitutes take precautions to avoid becoming infected by or transmitting the AIDS virus. Most street prostitutes in the United States, and many brothel workers in foreign nations, do not take such precautions.

PORNOGRAPHY AND OBSCENITY

The production and distribution of sexually explicit materials is a boom industry (Lane, 2000). In the United States, 20 million "adult" magazines are sold each month. X-rated or adult movies (also called *porn* movies) have moved from sleazy adult theaters to the living rooms of middle America in the form of videocassette rentals and sales. Nobody knows how many people bring X-rated photos, videos, and even live sex from foreign nations into their homes via the Internet every day. However, it is estimated that there are 30,000 to 60,000 sex-oriented Web sites in use today (Lane, 2000).

Pornography is indeed popular, but it is also highly controversial. Many are opposed to pornography on moral grounds. Feminists oppose pornography because it portrays women in degrading and dehumanizing roles, as sex objects who are subservient to men's wishes, as sexually insatiable nymphomaniacs, or as sexual masochists who enjoy being raped and violated. Moreover, some feminists hold that depictions of women in sexually subordinate roles may encourage men to treat them as sex objects and increase the potential for rape (Scott & Schwalm, 1988).

Yet feminists are split on the issue of legal suppression of pornography. For example, the feminist writer Kate Millett argues that "We're better off hanging tight to the First Amendment so that we have freedom of speech" (Press et al., 1985).

What is pornography? How is it defined? Who uses it? Why? How does it affect users?

What Is Pornographic?

Webster's Deluxe Unabridged Dictionary defines **pornography** as "writing, pictures, etc., intended to arouse sexual desire." The inclusion of the word *intended* places the determination of what is pornographic in the mind of the person composing the work. Applying this definition makes it all but impossible to determine what is pornographic. If a film maker admits that he or she wanted to arouse the audience sexually, we may judge the work to be pornographic even if no naked bodies or explicit sex scenes are shown. On the other hand, explicit representations of people engaged in sexual activity would not be pornographic if the work was intended as an artistic expression, rather than created for its **prurient** value. Many works that were once prohibited in this country because of explicit sexual content, such as the novels *Tropic of Cancer* by Henry Miller, *Lady Chatterley's Lover* by D. H. Lawrence, and *Ulysses* by James Joyce, are now generally considered literary works rather than excursions into pornography. Even Mark Twain's *Huckleberry Finn*, John Steinbeck's *The Grapes of Wrath*, and Ernest Hemingway's *For Whom the Bell Tolls* have been banned in some places because local citizens found them offensive, obscene, or morally objectionable.

There is thus a subjective element in the definition of pornography. An erotic statue that sexually arouses viewers may not be considered pornographic if the sculptor's intent was artistic. A grainy photograph of a naked body that was intended to excite sexually may be pornographic. One alternative definition finds material pornographic when it is judged to be offensive by others. This definition, too, relies on subjective judgment—in this case, that of the person exposed to the material. In other words, one person's pornography is another person's work of art. Are the photos shown in Figure 19.1 pornographic? Why or why not?

Pornography Written, visual, or audiotaped material that is sexually explicit and produced for purposes of eliciting or enhancing sexual arousal. (From Greek roots that mean "to write about prostitutes.")

Prurient Tending to excite lust; lewd. (From the Latin *prurire*, which means "to itch" in the sense of "to long for.")

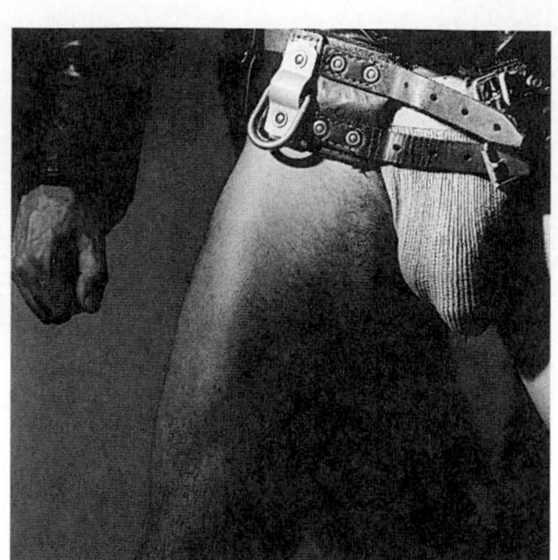

FIGURE 19.1

Art or Obscenity? The left-hand photo shows teenage sex icon Britney Spears in a live TV performance. Her minimalist outfit was flesh-colored. This seems to be an age in which pop stars are baring all or nearly all. The right-hand photo, *Patrice*, is from the *X Portfolio* of Robert Mapplethorpe. Complaints about an exhibition of the *X Portfolio* led authorities in Cincinnati, Ohio, to indict a gallery director for "pandering to obscenity." The director was subsequently acquitted.

Legislative bodies usually write laws about **obscenity** rather than pornography. Even the Supreme Court, however, has had a difficult time defining obscenity and determining where, if anywhere, laws against obscenity do not run afoul of the guarantee of free speech in the Bill of Rights. In the case of *Miller v. California* (1973), Supreme Court Justice William Brennan wrote that obscenity is "incapable of definition with sufficient clarity to withstand attack on vagueness grounds." Such vagueness did not deter Supreme Court Justice Potter Stewart from quipping that although he could not define obscenity in objective terms, he knew it when he saw it. Recall that *Huckleberry Finn* was once considered obscene, however.

Let us follow Mosher (1988) and define pornography as written, visual, or audiotaped material that is sexually explicit or graphic and produced for purposes of eliciting or enhancing sexual arousal. **Erotica** may be as sexually explicit as pornography is, but many writers use this term to refer to sexual materials that are artistically produced or motivated. Although some people find any erotic material offensive or obscene, it makes little sense to lump together artistic expressions of erotic themes and "hard-core" pornography.

Pornography is often classified as either *hard-core* (X-rated) or *soft-core* (R-rated). Hard-core pornography includes graphic and sexually explicit depictions of sex organs and sexual acts. Soft-core porn, as represented by R-rated films and *Playboy* photo spreads, features more stylized nude photos and suggested (or simulated) rather than explicit sexual acts. *Penthouse* has made the transition from soft-core to hard-core in many of its photo spreads.

Obscenity That which offends people's feelings or goes beyond prevailing standards of decency or modesty. (From the Latin *caenum*, which means "filth.")

Erotica Books, pictures, and so on that have to do with sexual love (from the Greek *eros*, which means "love"). Many contemporary writers use the term *erotica* to refer to sexual material that is artistically produced or motivated by artistic intent.

Pornography and the Law

Laws against obscenity provide the legal framework for outlawing the dissemination of pornography. Because the definition of obscenity relies on offending people or running afoul of community standards, that which is deemed obscene may vary from person to person and from culture to culture. Going topless on a public beach may be deemed obscene in some locales but not along beaches on the French Riviera or Rio de Janeiro, where this style of (un)dress is customary. The word *obscene* extends beyond sexual matters. One could judge TV violence or beer commercials obscene because they are personally offensive or are offensive to women, even if such depictions do not meet legal criteria for obscenity.

In the United States, legal prohibition of pornography as a form of obscenity dates to the nineteenth century. In 1873 Congress passed the Comstock Act, an antiobscenity bill. One effect of this bill was to outlaw the dissemination of information about birth control. The Comstock Act and similar laws made it a felony to mail obscene books, pamphlets, photographs, drawings, or letters. But what is obscene?

A landmark case in 1957 helped establish the legal basis of obscenity in the United States. In *Roth v. United States*, the U.S. Supreme Court ruled that portrayal of sexual activity was protected under the First Amendment to the Constitution unless its dominant theme dealt with "sex in a manner appealing to prurient interest" (*Roth v. United States*, 1957, p. 487). In a 1973 case, *Miller v. California*, the U.S. Supreme Court held that obscenity is based on a determination of

> (a) Whether the average person, applying contemporary community standards, would find that the work, taken as a whole, appeals to the prurient interest . . . ; (b) whether the work depicts or describes, in a patently offensive way, sexual conduct specifically defined by the applicable state law; and (c) whether the work, taken as a whole, lacks serious literary, artistic, political, or scientific value. (*Miller v. California*, 1973, p. 24)

Courts have since had to grapple with the *Miller* standard in judging whether material is obscene. *Miller* recognizes that judgments of obscenity may vary with "community standards." As a result, the same material may be considered obscene in one community but not in another. The *Miller* standard raises some obvious and unresolved questions. For example, who is the "average person" who can speak for a community? Many of us live in ethnically, racially, and religiously diverse communities. Can one viewpoint in *any* community truly represent the community? Even in relatively homogeneous communities, a diversity of opinion may exist on particular issues. Moreover, what is a *community*? Is it one's neighborhood, police precinct, municipality, or county? Or is it a larger political unit? What does "patently" offensive mean? Who judges "serious" literary, artistic, political, or scientific value? An attempt to clarify this last question was made in a 1987 case, *Pope v. Illinois*, in which the Supreme Court held that

> The proper inquiry is not whether an ordinary member of any given community would find serious literary, artistic, political, or scientific value in allegedly obscene material, but whether a reasonable person would find such value in the material, taken as a whole. (p. 445)

In this ruling, the court held that sexually explicit material could not be declared obscene even if many or most people in a particular community deemed it to lack serious literary, artistic, political, or scientific value, unless a "reasonable" person were to reach the same judgment. Whether the concept of a "reasonable" person will provide the courts with a clearer standard than that of an "average" or "ordinary" person for adjudicating obscenity cases remains to be seen.

Child pornography and violent, degrading, or dehumanizing pornography complicate matters further. People who do not find depictions, however explicit, of consensual sexual activity between adults to be obscene may regard child pornography or violent pornography to be obscene. Child pornography is clearly psychologically harmful to the juvenile actors (Silbert, 1989). Many people also object to sexually explicit material that portrays women as "sex objects" or in a position subordinate to men—gratifying men's sexual appetites.

Some footnotes: The U.S. Supreme Court ruled (in *Stanley v. Georgia*, 1969) that the possession of obscene material in one's home is not a criminal act (Sears, 1989). However, states may make it illegal to possess child pornography (Sears, 1989). Some communities (such as Indianapolis) have sought to ban pornography on the grounds that it discriminates against women. Pornography, that is, damages women's opportunities to enjoy equal rights by perpetuating stereotypes of women as subservient to men. The Indianapolis ordinance was overturned by a federal court on the grounds that it represented an unconstitutional infringement of free speech (Lewin, 1992). However, in a 1992 ruling, the Supreme Court of Canada redefined obscenity as involving sexually explicit material containing violence toward women or material that degrades or dehumanizes women (Lewin, 1992).

Human Sexuality ONLINE //

Surfing for Sex

It's everywhere. Singles do it, married people do it, even kids do it. State laws, electrical blackouts, even parental dictates can barely control it. What is it? Surfing for sex, of course. Sex is now available on the Internet. What was once found outside the arenas in ancient Rome ("fornicating"), in the brothels of Victorian England, in the bars that followed men out West, and in some adult bookstores and theaters now travels the world at the speed of light and enters homes (and offices!) through phone and cable lines. And millions of people around the world are clicking in to it.

Some are businessmen. Looking for companionship on their next trip, they may click into night life in San Francisco to find massage parlors, lap dancing, or particular "escorts." Escorts' photos and some of their sales pitch have been uploaded, and it takes little skill to save the data on one's hard drive. In fact, many men never leave home without it; check out the hard drive (no pun intended) in the laptop (no pun intended) of the businessman in the next seat in the airplane.

Some are adolescents looking for XXX-rated pictures. If their parents have placed parental controls on their Internet browsers, perhaps their friends' parents have not. They simply type key phrases like "sex pictures" or "Asian erotica" in a search engine, click "Find," and wait while computers come up with a list of likely hits. None of these "hits" quite does it? The user is prompted to modify the key phrase and try it again or simply to click "More." More and more and more.

What do they find? It varies. Sometimes they find something posted on the Web by an individual. Sometimes they find a particular vendor of hard-core photos and "live sex" that asks for credit card information to sign them up ("Only $9.95 per month"). The vendors often, but not always, ask to run "Adult checks" to verify that the user is over 18 and a resident of a state that allows users of that age access to pornographic materials. Sometimes users find electronic switchboards that invite them to click here or here or here for live sex, raw sex, amateur sex, she-male sex, the 1,000 top adult Web sites, the 5,000 top adult Web sites, college women, Japanese women, Japanese animated sex, live sex from Amsterdam, or live sex from your neighbor's basement (your neighbor may have created a Web site and be offering . . . whatever).

Some find other sites, other things. Consider Vancouver's multipurpose Consensual Sex Information Service (www.walnet.org/csis). This is about as traditional and honorable a Web site as one is likely to click across. Its home page is a bit raunchy, but compared to XXX-rated home pages, it looks like Mom and apple pie. CSIS does offer information about local and some not-so-local massage parlors and escort services, and it provides the electronic pathways that enable you to click your way to their electronic portals. But it also offers legal and medical information intended to be helpful to prostitutes and other sex workers (such as updates on efforts to have prostitution redefined as sex work). For example, sex workers would like prostitution legalized so that prostitutes would be protected by police (rather than harassed by police) and could obtain medical and retirement benefits.

In the new millennium, it seems that we have become less likely to find "dirty old men" leafing through adult magazines in the back of news shops. Rather, they—and younger men and women, many of whom are very freshly scrubbed—appear more and more likely to be spending their time surfing the Internet for sex.

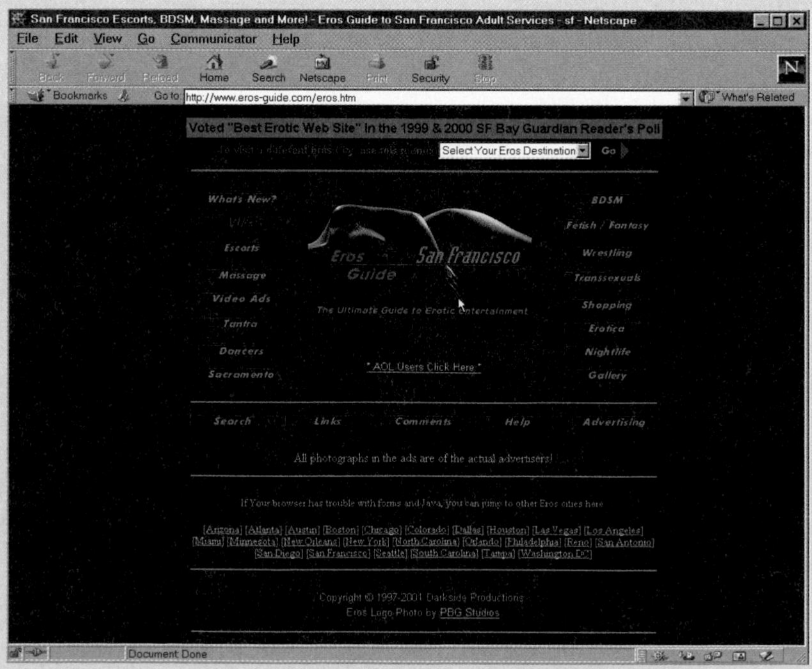

Eros Guide to San Francisco. Businessmen looking for sex on a trip to San Francisco can consult Web sites such as this to find escorts, masseuses, nightlife, and the like.

Prevalence and Use of Erotica and Pornography

Most of us have been exposed to sexually explicit materials, whether in the form of a novel, an article in *Playboy* or *Playgirl* (purchased, no doubt, for its literary value), or an X-rated film. The NHSLS study found that about one man in four (23%) and one woman in ten (11%) had bought an X-rated movie or video within the past year (Michael et al., 1994). Sixteen percent of the men and 4% of the women had bought sexually explicit magazines or books.

People in the United States are typically introduced to pornography by their high school years, often by peers (Bryant & Brown, 1989). Females are more likely to have been exposed to pornography by their boyfriends than the reverse (Bryant & Brown, 1989).

Pornography is typically used to elicit or enhance sexual arousal, often as a masturbation aid (Michael et al., 1994). Pornographic materials may also be used by couples to enhance sexual arousal during lovemaking. Couples may find that running a soft-core or hard-core video on the VCR, or reading an erotic story aloud, enlivens their sexual appetite or suggests novel sexual techniques.

Researchers have found that both men and women are physiologically sexually aroused by pornographic pictures, movies, and audiotaped passages (Goleman, 1995). That is, both men and women respond to pornographic stimuli with vasocongestion of the genitals and myotonia (muscle tension). However, there is a significant difference between physiological response and subjective feelings of arousal in women. Despite what is happening within their bodies, women tend to rate romantic scenes as more sexually arousing than sexually explicit scenes (Laan & Heiman, 1994).

Women are less accepting of sex without emotional involvement (Townsend, 1995). "Women like sex to come from an emotional connection," says marital and sex therapist Lonnie Barbach (1995). "For women there's a predisposition to allowing themselves to become turned on that romance allows. This wouldn't necessarily be measured by genital arousal."

Repeated exposure to the same pornographic materials progressively lessens the sexual response to them. People may become aroused by the familiar materials again if some time is allowed to go by. Novel materials are also likely to reactivate a sexual response (Meuwissen & Over, 1990; Zillmann, 1989).

Truth or Fiction? **REVISITED**

It is *not* true that only males are sexually aroused by pornography. Most men *and* women are sexually aroused, physiologically speaking, by erotica. As we see in the following section, however, their interest in, and subjective responses to, pornography may differ quite a bit. ■

GENDER DIFFERENCES IN RESPONSE TO PORNOGRAPHY Both genders can become physiologically aroused by erotic materials. Yet men and women do not necessarily share the same subjective response to them or level of interest in them. Visual pornography (sexually explicit pictures or films) is largely a male preserve (Symons, 1995; Winick, 1985). The majority of erotic visual materials are produced by men for men. Attempts to market visual materials to females have been largely unsuccessful (Symons, 1995). Women may read erotic romance novels, but they show little interest in acquiring erotic pictures, films, or videotapes (Lawrence & Herold, 1988). Women respondents to the Bryant (1985) survey were twice as likely to be disgusted by their initial exposure to X-rated material as to enjoy it. Men, by contrast, were twice as likely to report a positive response.

The reasons for these gender differences remain unclear (Bryant & Brown, 1989). Women may find erotica a "turn-off" when it portrays women in unflattering roles, as "whorish," as subservient to the sexual demands of men, and as sexually aroused by male domination and coercion. Symons (1979), an evolutionary biologist, believes that a basic evolutionary process is at work. He argues that ancestral men who were more sexually

Human Sexuality ONLINE //

Filtering Sex Sites: Tough Work, But . . .

Is the American Dream alive and well? Perhaps. Consider the travails of Yufeng Liang since he came to the United States from China in the mid-1990s.

Liang studied electrical engineering and computer imaging at Rutgers University, obtaining a doctorate. Although he had a job upon graduation, he also became a partner in an Internet startup company, to which he contributed his sweat equity.

Tough Work, But Someone's Got to Do It

Night after night, he sat at his computer keyboarding Internet addresses like babes.com and orgasmzone.com (Tierney, 2000). For hours on end, he studied thousands of pictures of naked women. His wife, Lufeng Leng, didn't discourage him. In fact, she urged him on.

"It was his job," she said. "Someone had to do it" (Tierney, 2000).

"At first, this job sounds like such a cool thing," Liang agreed, "but after a while you get fed up."

Dr. Liang looked tired of his Internet travels as he sat in his Manhattan office demonstrating the results of his labors. He keyboarded "sex" into the Yahoo search engine and waited for the hits to appear. Yahoo soon listed categories ranging from XXX movies to sexually transmitted infections. Dr. Liang chose "sex education" and then clicked on a site called "Life Begins." It turned out to be an innocent lesson for children. But then he clicked on a site titled "Mary's Place," which promised to provide "advice on sex, love, relationships, and more."

The "and more" was apparently the key, because Mary did not appear. In her place was a photo of a stop sign. Liang's filtering software had analyzed "Mary's Place" and decided it was about sex and not about education. The stop sign would prevent young minds from visiting the site.

Filtering software itself is not new. Other companies already protect children from surfing for sex by looking up a blacklist compiled by people who have patrolled the Internet and identified pornographic Web sites. Their software must be updated as new pornography sites arrive on the scene; otherwise, they are useless.

But Liang's Clicksafe program works not by means of a look-up table, but by actually analyzing the content of the Web sites that users click on to.

Fuzzy Logic

Clicksafe uses something called "fuzzy logic" to identify tip-off words and photos of naked bodies. "Mary's Place" triggers the stop sign because of the presence of telltale phrases such as *extremely personal* and *sexually explicit.*

Liang explains that his program uses an algorithm to evaluate phrases, not just single words. For example, the phrase *sex video* is more likely to trigger the stop sign than the words *sex* and *video* when they are encountered separately. The word *breasts* is judged to be okay so long as it's linked with the word *chicken.* Connect it with *huge,* and the stop sign pops up.

The Computer Knows Skin—European, African, Asian

If language alone doesn't provide the necessary information, the program "looks at" the photos on the Web page. It rapidly scans the pixels on the screen, searching out the peculiar tone that is connected with human skin, regardless of the race of the individual. Yes, a certain ratio of red, green, and blue in the pixel signals skin of people of every race. The software, which is available at www.clicksafe.com, really knows its skin.

The principal of St. Aloysius middle school in Harlem allowed his students to test the software. He believed that the software would make using the Internet a "safe, easy experience" for the schoolchildren (even if it stopped boys from surfing some sites that aroused their curiosity).

Programs such as Clicksafe also enable parents to select their preferred versions of censorship of the children's electronic travels. For example, they can screen out hate groups, cults, lingerie, even bikinis. (Sorry, guys, the electronic pathway to the *Sports Illustrated* swimsuit issue can be blocked with a stop sign.)

aroused by the sight of a passing female may have had reproductive advantages over their less arousable peers:

> The male's desire to look at female genitals, especially genitals he has not seen before, and to seek out opportunities to do so, is part of the motivational process that maximizes male reproductive opportunities. (p. 181)

Women, however, have fewer mating opportunities than men and must make the most of any reproductive opportunity by selecting the best possible mate and provider. To be sexually aroused by the sight of male genitalia might encourage random matings, which would undermine the woman's reproductive success.

Pornography and Sexual Coercion

Is pornography a harmless diversion or an inducement to commit sexual violence or other antisocial acts? Let us consider several sources of evidence in examining this highly charged issue, beginning with the findings of a 1970 government commission impaneled to review the evidence that was available at the time.

THE COMMISSION ON OBSCENITY AND PORNOGRAPHY In the 1960s, Congress created the Commission on Obscenity and Pornography to study the effects of pornography. Upon reviewing the existing research, the commission (Abelson et al., 1970) concluded that there was no evidence that pornography led to crimes of violence or to sexual offenses such as exhibitionism, voyeurism, or child molestation. Some people were sexually aroused by pornography and increased the frequency of their usual sexual activity, such as masturbation or coitus with regular partners, following exposure. They did not engage in antisocial behavior, however. These results have been replicated many times.

The commission found no causal link between pornography and delinquent behavior or sexual violence against women. The commission noted, for example, that when pornographic materials became widely available in Denmark following their legalization in late 1960s, there was no corresponding increase in the incidence of sex crimes. Finding pornography basically harmless (Edgley, 1989), the commission recommended that "federal, state, and local legislation should not seek to interfere with the right of adults who wish to read, obtain, or view explicit sexual materials" (Abelson et al., 1970, p. 58). Congress and then-president Richard Nixon rejected the commission's findings and recommendation, however, on moral and political—not scientific—grounds.

PORNOGRAPHY AND SEX OFFENDERS Another approach to examining the role of pornography in crimes of sexual violence involves comparing the experiences of sex offenders and nonoffenders with pornographic materials. A review of the research literature found little or no difference in the level of exposure to pornography between incarcerated sex offenders and comparison groups of felons who were incarcerated for nonsexual crimes (Marshall, 1989).

Yet evidence also shows that as many as one in three rapists and child molesters use pornography to become sexually aroused immediately before and during the commission of their crimes (Marshall, 1989). These findings suggest that pornography may stimulate sexually deviant urges in certain subgroups of men who are predisposed to commit crimes of sexual violence (Marshall, 1989).

THE MEESE COMMISSION REPORT In 1985, President Ronald Reagan appointed a committee headed by Attorney General Edwin Meese to reexamine the effects of pornography. In 1986, the U.S. Attorney General's Commission on Pornography, known as the Meese Commission, issued a report that reached very different conclusions from those of the 1970 commission. The Meese Commission claimed to find a causal link between sexual violence and exposure to violent pornography (U.S. Department of Justice, 1986). The commission asserted that a substantial increase in the proliferation of violent pornography had occurred since the earlier commission had been convened. Moreover, the report concluded that exposure to pornography that portrayed women in degrading or subservient roles increased the acceptability of rape in the minds of viewers. The commission found no evidence linking exposure to nonviolent, nondegrading pornography (consensual sexual activity between partners in equal roles) and sexual violence but noted that this type constituted only a small fraction of the pornographic materials on the market.

The commission issued 92 recommendations for tighter enforcement of obscenity laws and greater restrictions on the dissemination of pornography. Some recommendations focused on the enforcement of existing child pornography laws and the prosecution of offenders. Others encouraged states that had not already done so to make the knowing possession of child pornography a felony.

The commission's findings are controversial. Critics claim that although the Meese Commission did not blatantly falsify the data, its conclusions reflected an overgeneralization of laboratory-based findings (Wilcox, 1987). Two researchers in the area, Edward Donnerstein and Daniel Linz (1987), contended that the Meese Commission failed to distinguish between the effects of sexually explicit materials per se and the effects of violent materials. Evidence of links between exposure to sexually explicit materials (without violent content) and sexual aggression is lacking. Donnerstein and Linz concluded that violence, not sex, is the inducement to aggression. Nor do researchers confirm the Meese

Commission's belief in the rising incidence of violent pornography. Evidence does not show depictions of violence in media such as the magazines *Playboy* and *Hustler*, or in X-rated movies and videos, to have increased in recent years (Scott & Cuvelier, 1993). In fact, X-rated movies and videos contain less violence than general-release movies.

If nothing else, the Meese Commission raised awareness about different types of pornography and their potential effects on viewers' behaviors and attitudes. Let us take a look at the scientific evidence on the effects of violent and nonviolent pornography.

VIOLENT PORNOGRAPHY Laboratory-based studies have shown that men exposed to violent pornography are more likely to become aggressive against females and to show less sensitivity toward women who have been sexually assaulted. In one study (Donnerstein, 1980), 120 college men interacted with a male or female confederate (accomplice) of the experimenter, who treated them in either a neutral or a hostile manner. The study participants were then shown either neutral, nonviolent pornographic films or violent pornographic films. In the latter, a man forced himself into a woman's home and raped her. Participants were then given the opportunity to deliver electric shock to the male or female confederate, presumably to assist the confederate in learning a task. The measure of aggression was the intensity of the shock chosen. No shock was actually delivered, but participants did not know that the shock apparatus was fake. Unprovoked men who viewed violent pornographic films showed greater aggression toward the women than did unprovoked men who viewed nonviolent films, however. Provoked men who were shown violent pornography selected the highest shock levels of all. The film may have served as a model for retaliation.

In another study, aggression by males against female confederates following exposure to violent pornography (a rape scene) was increased by depictions of the woman being raped as either enjoying the experience or becoming sexually aroused during the rape (Donnerstein & Berkowitz, 1981). These findings suggest that depictions of women enjoying or becoming aroused by their victimization may legitimize violence against women in the viewer's mind, reinforcing the cultural myth that some women need to be dominated and are sexually aroused by an overpowering male.

Other research has shown that exposure to violent pornography leads men to become more accepting of rape, less sensitive to women survivors of rape, and more accepting of the use of violence in interpersonal relationships (Donnerstein & Linz, 1987; Malamuth, 1984).

Yet research suggests that it is the violence in violent pornography, not its sexual explicitness, that hardens men's attitudes toward rape survivors. In one study (Donnerstein & Linz, 1987), college men were exposed to films consisting of violent pornography, of nonviolent pornography (a couple having consensual intercourse), or of violence that was not sexually explicit. The violent pornographic and nonpornographic films both showed a woman being tied up and slapped at gunpoint, but the nonpornographic version contained no nudity or explicit sexual activity. The men had first been either angered or treated in a neutral manner by a female confederate of the experimenter. The results showed that, compared with nonviolent pornography, both violent pornographic films and violent *non*pornographic films produced greater acceptance of rape myths, greater reported willingness to force a woman into sexual activity, and greater reported likelihood of engaging in rape (if the man also knew that he could get away with it). These effects occurred regardless of whether or not the man was angered by the woman.

In another study, Linz and his colleagues (1988) assigned college men, at random, to watch five feature-length films over a two-week period. One group saw X-rated nonviolent films. A second group watched R-rated "slasher-type" violent films such as *Friday the 13th, Part 2*. ("Slasher films" show graphic violence, primarily directed at women, intermingled with mild erotic scenes.) A third group saw R-rated, nonviolent "teenage sex films" such as *Porky's*. Afterward, the men viewed a videotaped reenactment of a trial in which a woman accused a man of raping her. Men who had watched the R-rated slasher films showed less sensitivity toward the woman complainant than did men who had been exposed to the R- or the X-rated nonviolent films. Exposure to sexually explicit material *without violence*, whether soft-core (R-rated) or hard-core (X-rated), did not reduce sympathy toward the woman complainant.

On the basis of a review of the research literature, Linz (1989) concluded that short-term and prolonged exposure to sexual violence, whether sexually explicit or not, lessens sensitivity toward survivors of rape and increases acceptance of the use of force in sexual encounters. "Slasher" films are connected with the strongest antisocial effects.

Research on the effects of pornography should be interpreted with caution, however. Most of it has employed college students, whose behavior may or may not be typical of people in general or of people with propensities toward sexual violence. On the other hand, research discussed in Chapter 18 shows that many college men have engaged in rape, attempted rape, or some other form of sexual coercion. Another issue is that most studies in this area are laboratory-based experiments that involve simulated aggression or judgments of sympathy toward hypothetical women who have been portrayed as rape victims. None measured *actual* violence against women outside the lab. We still lack evidence that normal men have been, or would be, spurred to rape or sexually violate women because of exposure to violent pornography or other media depictions of violence.

NONVIOLENT PORNOGRAPHY Nonviolent pornography may not contain scenes of sexual violence, but it typically portrays women in degrading or dehumanizing roles—as sexually promiscuous, insatiable, and subservient. Might such portrayals of women reinforce traditional stereotypes of women as sex objects? Might they lead viewers to condone acts of rape by suggesting that women are essentially promiscuous? Might the depiction of women as readily sexually accessible inspire men to refuse to "take no for an answer" on dates?

Zillmann and Bryant (1984) exposed male and female study participants to six sessions of pornography over six consecutive weeks. Participants were exposed, during each weekly session, to a massive dose of pornography consisting of six nonviolent pornographic films ("Swedish Erotica"), or to an intermediate dose consisting of three pornographic and three neutral films, or to a no-dose control consisting of six nonsexual films. When later tested in a purportedly independent study, both males and females who received extended exposure to pornography, especially those who received the massive dose, gave more lenient punishments to a rapist who was depicted in a newspaper article. Moreover, males became more callous in their attitudes toward women.

Zillmann and Weaver (1989) argue that making women appear sexually permissive and promiscuous increases men's callousness toward women who have been sexually assaulted. Once men brand women as promiscuous, they lose respect for them. They see women as "public property" who have forfeited their rights to exercise choice in sex partners.

Not all researchers, however, find exposure to nonviolent pornography to increase callousness toward women. Some report finding that nonviolent pornography did not reduce the sensitivity of men (Linz et al., 1988) and women (Krafka, 1985) to female victims of sexual assault. Others report that nonviolent pornography did not increase men's aggression toward women in laboratory studies (Malamuth & Ceniti, 1986). Still others did not find that repeated exposure to nonviolent pornography made men adopt more callous attitudes toward women or increased their acceptance of rape myths (Padgett et al., 1989).

Given the inconsistencies in the research findings and the limited amount of research on nonviolent pornography, Linz (1989) concludes that

> The data, *overall*, do not support the contention that exposure to nonviolent pornography has significant adverse effects on attitudes toward rape as a crime or more general evaluations of rape victims. (p. 74)

Moreover, if nonviolent pornography can be connected with negative attitudes toward women, Linz (1989) suggests that such attitudes may result from the demeaning portrayals of women as sexual playthings valued only for their physical attributes and sexual availability, rather than from sexual explicitness per se.

Yet another concern is the possible effect of nonviolent pornography on the viewer's sexual and family values. Nonviolent pornography typically features impromptu sexual encounters between new acquaintances. Might repeated exposure to such material alter viewers' attitudes toward traditional sexual and family values? Zillmann (1989) reports intriguing evidence that repeated exposure to this type of nonviolent pornography loosens

Human Sexuality ONLINE //

Web Sites Related to Commercial Sex

This is the Web site of "Promise," an organization "For Women Escaping Prostitution." It offers information about prostitution and resources for women who choose to leave prostitution and are in need of support.
www.sirius.com/~promise/home.htm

The Web site of the ChildWatch International Research Network contains information about child prostitution.
www.childwatch.uio.no

This is the Web site of Serendipity, a libertarian-sponsored Web site that provides information about legal issues regarding commercial sex, including online sexual content.
serendipity.magnet.ch/index.html

The Web site of Child Safety Online, which has an overview of issues related to child safety online.
www.soc-um.org/netsum.html

A Canadian Web site, the Commercial Sex Information Service (CSIS) offers links to some massage parlors and other outlets for prostitution but also serves as a clearinghouse for information about laws, sexual health, and culture as these are related to prostitution (which the site refers to as "sex work").
www.walnet.org/csis

The Web site of ClickSafe enables the visitor to download software that will filter out sex sites so that they cannot be accessed by children, employees, and so on.
www.clicksafe.com

traditional sexual and family values. When compared to people who viewed nonsexual films, men and women who were exposed to weekly, hour-long sessions involving scenes of explicit sexual encounters between new acquaintances over a six-week period showed attitudinal changes including greater acceptance, compared to controls who viewed nonsexual films, of premarital and extramarital sex and of simultaneous sexual relationships with multiple partners. Men and women who viewed such pornography also reported desiring fewer children than people in the control groups and were relatively less committed to marriage as an "essential institution."

Zillmann (1989) argues that nonviolent pornography loosens traditional family values by projecting an image of sexual enjoyment without responsibility or obligations. Prolonged exposure to such pornography may also foster dissatisfaction with the physical appearance and sexual performance of one's intimate partners (Zillmann, 1989).

In sum, research on the effects of nonviolent pornography is so far inconclusive. The effects of nonviolent pornography may be more closely connected with whether women are presented in a dehumanizing manner than with sexual explicitness per se. No research has yet linked sexual explicitness itself with undesirable effects. It is of interest to note that only a minority of people in the United States support legislation making sexually explicit materials illegal, yet three out of four favor the criminalization of *violent* pornography (Harris, 1988).

summing up

THE WORLD OF COMMERCIAL SEX: A DISNEYLAND FOR ADULTS

Commercial sex runs the gamut from "adult" movie theaters and bookshops to strip shows, sex toy shops, brothels, escort services, massage parlors, and "900" telephone services.

PROSTITUTION

In the United States, prostitution is illegal everywhere except in certain rural counties in Nevada.

■ Incidence of Prostitution in the United States Today
Fewer young men in the United States use prostitutes than in Kinsey's day, apparently because of the liberalizing trends of the sexual revolution.

■ Types of Female Prostitutes
The major types of female prostitutes are streetwalkers, brothel prostitutes (many of whom work in massage parlors and for "escort services"), and call girls. Streetwalkers often support, and are abused by, pimps.

■ **Characteristics of Female Prostitutes**
No single factor explains entry into female prostitution, but poverty and sexual and/or physical abuse figure prominently in the backgrounds of many prostitutes. Teenage runaways with marginal skills and limited means of support may find few alternatives to prostitution.

■ **Customers of Female Prostitutes**
Those who patronize prostitutes are often referred to as "johns" or "tricks." Most patrons are "occasional johns" with regular sex partners. Some people use prostitutes habitually or compulsively, however.

■ **Male Prostitution**
Most male prostitutes are "hustlers" who service male clients. Hustlers typically begin selling sex in their teens and may be gay or heterosexual in orientation.

■ **HIV, AIDS, and Prostitution**
Prostitutes are at greater risk of contracting HIV infection because they have sexual relations with many partners, often without protection. Many prostitutes and their clients and other sex partners also inject drugs and share contaminated needles. HIV may be spread by unprotected sex from prostitutes to customers, then to the customers' wives or lovers. Despite the dangers of HIV transmission, many U.S. prostitutes have not altered either their sexual behavior or their patterns of drug use.

PORNOGRAPHY AND OBSCENITY

■ **What Is Pornographic?**
Pornography is "writing, pictures, etc., intended to arouse sexual desire." Erotica, by contrast, consists of sexual materials that are artistically produced or motivated. The judgment of what is pornographic or obscene varies from person to person and from culture to culture.

■ **Pornography and the Law**
In *Miller v. California*, the U.S. Supreme Court held that obscenity is based on a determination of "whether the average person, applying contemporary community standards, would find that the work . . . appeals to the prurient interest . . . ; whether the work depicts [sexual behavior] in a patently offensive way, [and] whether the work, taken as a whole, lacks serious literary, artistic, political, or scientific value." *Miller* recognizes that judgments of obscenity may vary with "community standards."

■ **Prevalence and Use of Erotica and Pornography**
People in the United States are typically introduced to pornography by their high school years. Although both genders can become physiologically aroused by erotic materials, men are relatively more interested in sexually explicit pictures and films.

■ **Pornography and Sexual Coercion**
A 1970 government commission found no harmful effects of pornographic material on normal people. A later government commission, the Meese Commission, issued a report in 1986 linking exposure to violent pornography with sexual aggression, and exposure to degrading but nonviolent pornography with increased acceptability of rape in the minds of viewers. Research evidence suggests that exposure to pornography may stimulate sexually deviant urges in some men who are predisposed to commit crimes of sexual violence. Research evidence generated in laboratory settings also suggests that violent pornography may stimulate college men to act more aggressively toward women. Some researchers argue that it is the violence in violent pornography, and not sexual explicitness per se, that promotes violence against women. The effects of nonviolent pornography on normal populations remain unclear.

questions for critical thinking

1. Agree or disagree with the following statement, and support your answer: "Prostitution should be legalized throughout the United States."
2. Why do you think that an attractive, popular actor like Hugh Grant would turn to a prostitute?
3. Agree or disagree with the following statement, and support your answer: "One person's pornography is another person's art."
4. Why do you think that men are more interested in pornography than women are?
5. In *Miller v. California*, the U.S. Supreme Court held that obscenity is based on "whether the average person, applying contemporary community standards, would find that the work, taken as a whole, appeals to the prurient interest." What is your "community"? What are its standards? What types of films, photographs, or written stories would an average person in your community consider obscene? Why?
6. Which do you consider to be the greater danger—the danger of having pornography available or the danger of censorship? Why?

chapter

20 Making Responsible Sexual Decisions—An Epilogue

Pablo Picasso. *The Two Saltimbanques or Harlequin and His Companion*, 1901. © Copyright ARS, NY. Pushkin Museum of Fine Arts, Moscow, Russia. Credit: Scala/Art Resource, NY. Credit: © 2002 Estate of Pablo Picasso/Artists Rights Society (ARS), New York.

outline

CHOICES, INFORMATION, AND DECISION MAKING

Conflict

Decisions, Decisions, Decisions . . .

VALUE SYSTEMS

Legalism

Human Sexuality Online What's Wrong with Virtual Adultery? A Sin Is a Sin Is a Sin

Situational Ethics

Ethical Relativism

Hedonism

Asceticism

Utilitarianism

Rationalism

THE BALANCE SHEET FOR DECISION MAKING

Human Sexuality Online Web Sites Related to Making Responsible Sexual Decisions

QUESTIONS FOR CRITICAL THINKING

Making choices is deeply intertwined with our sexual experience. Although sex is a natural function, the ways in which we express our sexuality are matters of personal choice. We choose how, where, and with whom to become sexually involved. We may face a wide array of sexual decisions: Whom should I date? When should my partner and I become sexually intimate? Should I initiate sexual relations or wait for my partner to approach me? Should my partner and I practice contraception? If so, which method? Should I use a condom to protect against sexually transmitted diseases (or insist that my partner does)? Should I be tested for HIV? Should I insist that my partner be tested for HIV before we engage in sexual relations?

CHOICES, INFORMATION, AND DECISION MAKING

We have presented you with information that can help you make responsible sexual decisions. Information alone cannot determine whether the options you consider are morally acceptable to you, however. Many issues raise moral concerns, such as premarital and extramarital sex, contraception, and abortion. Gathering information and weighing the scientific evidence will alert you to what is possible in the contemporary world, but only you can determine which of your options are compatible with your own moral values. We all have unique sets of moral values—as Americans, as members of one of America's hundreds of subcultures, as individuals. No single value system defines us all. Indeed, the world of diversity in which we live is a mosaic of different moral codes and cultural traditions and beliefs.

Conflict

The most difficult decisions involve choices between alternatives that each have positive and negative aspects or that each have compelling negative features.

Consider the issue of deciding whether to obtain screening for cancer. The pluses include freedom from anxiety (when results are negative) and early detection (when cancer is found). The negatives include the costs of screening, inconvenience, discomfort (in some techniques), and anxiety about hearing results one would rather not hear (if cancer is found) (White et al., 1994).

Consider contraceptive techniques. The pill is effective and does not interfere with sexual spontaneity, but it is costly and has side effects. The condom has no side effects (for the great majority of users), but some people find that it disrupts lovemaking and reduces sexual sensations. It is also not as effective as the pill.

Or consider the plight of Kim, a college student who becomes pregnant as a result of date rape: Should she have an abortion, which may be morally offensive to her, or bear an unwanted child? Then, too, should she report the crime and hope the criminal justice system will successfully prosecute her assailant, or should she try to swallow her bitter feelings and protect her anonymity? Or consider the college student who feels morally committed to remaining a virgin until marriage but fears enduring the disapproval of dating partners or having no dates at all.

When we are pulled in different directions at the same time, we experience psychological conflict. Conflict can be extremely stressful, especially if it is prolonged. People in conflict often feel "damned if they do and damned if they don't." They may also vacillate, taking tentative steps in one direction and then in the other. Kim, a survivor of date rape, takes a bus to an abortion clinic across town and gets off at the next stop. She checks the telephone number of the police but then hesitates to pick up the telephone. She is obsessed with the question of what is the right thing to do.

Part of Kim's conflict is over a moral issue. She would very much like to terminate her unwanted pregnancy by having an abortion, but she holds back because she firmly believes that the embryo within is a precious human life. Her conflict over whether to report her assailant is in part moral, in part practical. It is morally proper to report him, but what if she is not believed? What if the incident becomes widely known on campus and to her family? Moreover, she realizes that the longer she waits, the more it may appear that she is making up the story. And she knows that her neglecting to report the rape immediately has already compromised the authorities' efforts to collect evidence.

Decisions, Decisions, Decisions . . .

The way out of conflict is to make decisions. Making decisions involves choosing among various courses of action. The act of not making a formal decision may itself represent a tacit decision. For example, we may vacillate about whether to use a particular form of birth control but continue to engage in unprotected sex. Is this because we have not made a decision or because we have decided to let happen what will happen?

Gathering information helps us to predict the outcomes of the decisions we make. This textbook provides you with a broad data base concerning scientific developments and ways of relating to other people—including people who come from other cultures. By talking to people who have had similar questions or have wrestled with similar conflicts, you can also learn which decisions worked for them, which did not, and, perhaps, why. Or you may be able to learn what happens when people do not make decisions but simply hope for the best. You can also talk to your parents and friends, religious counselors, course instructors, psychologists, and other helping professionals. Regarding which contraceptive method to use, you may wish to discuss the information in this book with a physician, nurse, or other health counselor.

VALUE SYSTEMS

Our value systems provide another framework for judging the moral acceptability of sexual options. We often approach sexual decisions by determining whether the choices we face are compatible with our moral values. Our value systems do not necessarily determine the outcomes of all of our sexual decisions, however, any more than laws strictly determine people's behavior. Some of us adhere to stricter moral values. Some people make choices that bend or violate their values but try not to focus on the discrepancies too closely. Many of us also act on impulse, especially sexual impulses, without thinking through the consequences of our actions. Afterward, we may regret our decisions or feel guilty if our behavior was inconsistent with our values.

Our value systems—our sexual standards—have many sources: parents, peers, religious training, ethnic subcultures, the larger culture, and our appraisal of all these influences. Value systems that provide a guiding framework in our efforts to determine the moral acceptability of sexual choices include legalism, situational ethics, hedonism, asceticism, utilitarianism, and rationalism.

Legalism

The legalistic approach formulates ethical behavior on the basis of a code of moral laws derived from an external source, such as the creed of a particular religion (Knox, 1988). The Bible contains many examples of the moral code of the Jewish and Christian religions. In the Book of Leviticus (20:10–17) in the Old Testament, we find many of the biblical prohibitions against adultery, incest, sexual activity with people of one's own gender, and bestiality:

> And the man that committeth adultery with another man's wife, even he that committeth adultery with his neighbor's wife, both the adulterer and the adulteress shall surely be put to death. And the man that lieth with his father's wife . . . both of them shall surely be put to death; their blood shall be upon them. . . . And if a man lie with mankind, as with womankind, both of them have committed abomination: they shall surely be put to death; . . . And if a man lie with a beast, he shall surely be put to death; and ye shall slay the beast. . . . And if a man shall take his sister, . . . and see her nakedness, and she sees his nakedness: it is a shameful thing; and they shall be cut off in the sight of the children of their people. . . .

Leviticus also proscribes intercourse during menstruation and the forcing of daughters into prostitution.

Throughout the centuries, many Christians and Jews have adhered to religious laws that are intended to promote the solidarity of the family and of the community by minimizing interpersonal jealousies and frictions. Adultery, for example, could set neighbor against neighbor. Religious laws have also directed followers to be "fruitful and multiply" so as to ensure that an ample number of progeny would be available to meet the needs for labor and defense of the family and the community of religious followers.

Many religious followers today accept the moral codes of their religions as a matter of faith and commitment, not necessarily because they can logically or rationally derive these precepts from contemporary societal needs. Some people find it reassuring to be informed by religious authorities or scripture that a certain course of action is right or wrong. Others, however, have questioned how closely one must adhere to religious teachings in order to be ethical. A number of fundamentalist Protestants, conservative Catholics, and orthodox Jews would argue that the Bible is to be followed to the letter, not just "in spirit." They see the Bible, in all its literal details, as the revealed word of God.

Other Christians and Jews take a more liberal view. They say that the Bible was inspired by God but that it was written or transcribed by fallible humans and is subject to various interpretations. They may also assert that the Bible reflects the social setting of the time in which it was written, not just divine inspiration. At a time of burgeoning pop-

Human Sexuality ONLINE //

What's Wrong with Virtual Adultery? A Sin Is a Sin Is a Sin

Phone sex? Fugeddaboudit. That's Stone Age stuff. What has the editors of an Italian Catholic magazine worried in the new millennium is "adultery" over the Internet.

Adultery is adultery, even if it is virtual, according to *Famiglia Cristiana* (*Christian Family*), a magazine close to the Vatican. It is just as sinful as the real thing.

The question of the morality of flirting, falling in love, and perhaps betraying a spouse via the World Wide Web surfaced in the advice column of the latest issue of Italy's most widely circulated newsweekly.

A woman from the northern city of Varese wrote to the magazine, which is sold in Italian churches as well as news kiosks, asking for moral guidance as she surfed the Web.

"On the Internet you can fall in love, you can seek, you can think, you can truly desire. And you can commit adultery without leaving your home," the woman wrote in a long letter. "I ask myself what difference there is for the Church between a real extramarital affair and a virtual one. I ask myself how long this new (Internet) reality, which is so tangible, can be underestimated."

Father Antonio Sciortino, responding to the anonymous woman in the magazine's "Conversations with the Priest" column, had no hesitation. For the Church, there is no difference.

Committing Adultery in One's PC and in One's Heart

"Virtual reality can be just as much of a vice as reality made up of facts and actions," Sciortino, the magazine's editor-in-chief, wrote in his response. "Gospel morals attach a premium to what is inside a person and are just as concerned with bad thoughts as they are with bad actions," he explained. The priest recalled Christ's phrase in the Bible that if a man looks at a woman with lust he has already committed "adultery in his heart."

The story received wide play in Italy, where more than 97% of the population is nominally Roman Catholic.

"Cheating Online Is Adultery," ran the headline in *La Stampa* of Turin. "Famiglia Cristiana Gives Its Thumbs Down to Virtual Love." *La Stampa*'s article, which took up a third of a page, was replete with comments from respondents as diverse as a novelist and an actress known as a sex symbol.

Another newspaper, *Il Giornale,* dedicated half a page to the story and included a poll that said 40% of Italian women interviewed for it said they feared that their husbands might find a more fascinating woman while surfing the Web.

As Father Sciortino acknowledged in his response to the woman from Varese, the personal computer has changed many things in the world, even the way a marriage can fall apart.

Source: From P. Pullella, (2000, June 7). "Internet 'Adultery' a sin, Catholic magazine says." *Reuters News Agency online.*

ulation growth in many parts of the world, biblical injunctions to be fruitful and multiply may no longer be socially and environmentally sound. Prohibitions, such as that against coitus during menstruation, may have been based on prescientific perceptions of danger. Thus religious teachings may be viewed as a general framework for decision making rather than as a set of absolute rules.

Situational Ethics

Episcopal theologian Joseph Fletcher (1966, 1967) argued that ethical decision making should be guided by genuine love for others rather than by rigid moral rules. Fletcher advocated that sexual decision making should be based on the context of the particular situation that the person faces. For this reason, his view is termed *situational ethics.* According to Fletcher, a Roman Catholic woman will have been taught that abortion is the taking of a human life. Her situation, however—her love for her existing family and her recognition of her limited resources for providing for another child—might influence her to decide in favor of an abortion.

Fletcher argues that one's rules for conduct should not be inflexible. Rather, they should be general guidelines. "The situationist is prepared in any concrete case to suspend, ignore, or violate any principle if by doing so he can effect more good than by following it" (1966, p. 34).

Note that situational ethics is not an excuse for selfishness. The moral situation ethicist acts in a manner that he or she believes will lead to the greater good. Now and then, such an act may be frustrating, painful, or difficult. Situation ethicists also may find it difficult, as psychologist David Knox (1988) points out, to make sexual decisions on a case-by-case basis:

> "I don't know what's right anymore" reflects the uncertainty of a situational ethics view. Once a person decides that mutual love is the context justifying intercourse, how often and how soon should the person fall in love? Can love develop after two hours of conversation? How does one know that her or his own love feelings and those of a partner are genuine? The freedom that situational ethics brings to sexual decision making requires responsibility, maturity, and judgment. In some cases, individuals may deceive themselves by believing they are in love so they will not feel guilty about having intercourse. (Knox, 1988, p. 97; Copyright © 1991 by the authors. Reprinted with permission of Wadsworth, an imprint of the Wadsworth Group, a division of Thomson Leraning.)

Ethical Relativism

Ethical relativism assumes that diverse values are fundamental to human existence. Ethical relativists reject the idea that there is a single correct moral view. One person may believe that premarital sex is unacceptable under any circumstances, where as another may hold that "being in love" makes it acceptable. Still another person may believe that premarital sex is morally permissible without an emotional commitment between the partners.

The ethical relativist believes that there is no objective way of justifying one set of moral values over another. In this view, the essence of human morality is to derive one's own principles and apply them according to one's own conscience. From this perspective, whatever a person believes is right *is* right for him or her. Subscribing to ethical relativism does not free us from the need to develop moral beliefs, however. It challenges us, instead, to take responsibility for our decisions and to live up to them.

Opponents of ethical relativism believe that allowing people free rein to determine what is right or wrong may bring about social chaos and decay.

One form of ethical relativism is cultural relativism. From this perspective, what is right or wrong must be understood in terms of the cultural beliefs that affect sexual decision making. In some cultures, premarital sex is tolerated or even encouraged, whereas in others it is considered immoral. Cultural relativism, like ethical relativism, does not ascribe moral superiority to one cultural tradition over another.

Hedonism

The hedonist is guided by the pursuit of pleasure, not by whether a particular behavior is morally or situationally justified (Knox, 1988). "If it feels good, do it" expresses the hedonistic ethic. The hedonist believes that sexual desires, like hunger and thirst, do not invoke moral considerations.

Of course, one could make the case that there are two kinds of hedonism: short-sighted and long-term hedonism. Basing decisions on what feels good at the moment may fail to account for the long-term consequences of one's actions. The short-sighted hedonist may risk becoming so driven by the pursuit of pleasure that sex becomes a form of "addiction" in which sexual pleasure overshadows other important aspects of life.

On the other hand, taking a long-term hedonistic view, one can attempt to make responsible sexual decisions that can lead to a lifetime of sexual pleasure. Choosing to do what is necessary to maintain our sexual health, for instance, can help us to lead long lives and to enjoy our bodies throughout our span of years.

Asceticism

Religious celibates, such as Roman Catholic priests and nuns, choose asceticism (self-denial of material and sexual desires) in order to devote themselves to spiritual pursuits. Many ascetics in Eastern and Western religions seek to transcend physical and worldly desires.

Utilitarianism

Some people believe that it is necessary to adhere to specific religious principles or teachings if one is to be a moral or ethical person, but others believe that ethical guidelines can be based on principles other than religious ones. The English philosopher John Stuart Mill (1806–1873), for example, proposed an ethical system based on utilitarianism.

The core tenet of utilitarianism is that moral conduct is based on that which will bring about "the greatest good for the greatest number" (Mill, 1863). The utilitarian characterizes behavior as ethical when it does the greatest good and causes the least harm. This is not license. Utilitarians may come down hard in opposition to premarital sex and bearing children out of wedlock, for example, if they believe that these behavior patterns jeopardize a nation's health and social fabric. Mill's ethics generally require that we treat one another justly and honestly, because it serves the greater good for people to be true to their word and just in their dealings with others.

Rationalism

Rationalism involves the use of reason as the means of determining a course of action (Knox, 1988). The rationalist believes that decisions should be based on intellect and reasoning, rather than on emotions or strict obedience to a particular faith. The rationalist attempts to assess the facts in a sexual situation and then logically weigh the consequences of each course of action before making a decision. The rationalist shares with the utilitarian the belief that reasoning can lead to a course of ethical behavior. The rationalist is not bound, however, to the utilitarian code of making choices on the basis of the greatest good for the greatest number. The rationalist may contend that personal needs and desires outweigh the needs of the many.

The utilitarian may decide, for example, to prolong an unhappy marriage because she or he believes that the greater good (of the family and the community) is better served by maintaining an unhappy marriage than by dissolving it. The rationalist might decide that the personal consequences of continuing an unhappy marriage outweigh the consequences to the family or the community at large. The religious follower, adopting a legalistic view, might decide that divorce is or is not acceptable, depending on the tenets of the person's religion. The situation ethicist might interpret religious principles in light of how the greater good would be served in a particular situation.

These ethical systems represent general frameworks of moral reasoning or pathways for judging the moral acceptability of sexual and nonsexual behavior. Whereas some of us may adopt one or another of these systems in their purest forms, others adopt a system of moral reasoning that involves some combination or variation of these ethical systems. Some also shift from one ethical system to another from time to time, sometimes reasoning legalistically and sometimes adopting a more flexible, situationist approach. Despite these variations, these general guidelines of moral reasoning help us understand how people approach questions of the moral acceptability of sexual choices.

THE BALANCE SHEET FOR DECISION MAKING

You might use a balance sheet to help weigh the pluses and minuses of alternatives when making decisions. Sheri Oz (1994, 1995) suggests that it is particularly important to weigh the *costs* of deciding in one direction or another. Oz notes that in making decisions about sex, marriage, and divorce, we frequently weigh costs in self-esteem against costs in terms of the approval of other people—often our intimate partners.

Rosa, a Mexican American student at a California state college, has been involved in a sexual relationship for several weeks and is contemplating going on the pill. Rosa uses a balance sheet to list the projected gains and losses for herself and her partner according to five criteria:

Experiential: How will it feel? What is its effect on sexual relations? How does it affect sexual sensations and spontaneity?

Biological: How do contraceptive methods work? What are the side effects of the pill? What risks are involved? Is this birth control method effective in preventing STIs?

Financial: Can I afford it—literally? For example, how much do pills cost? How much do babies cost?

Ethical: Is what I am considering consistent with my religious and parental teachings? Is that important to me? Will I experience guilt, anxiety, or shame? Is this course of behavior consistent with my ideas concerning sex roles and who takes the responsibility for the decision? (For example, should the responsibility for birth control be placed on the woman's shoulders only?)

Legal: Am I contemplating something that is against the law? What could happen to my partner and me?

As shown in Table 20.1, Rosa filled in some of the spaces on the sheet to indicate the following pieces of information:

- Sex can be spontaneous.
- For all practical purposes, the pill can be considered 100% effective (I'm assuming I'll take them as directed).
- I'm 20 years old (young enough and healthy enough not to be too worried about the pill's potential effect on my health).
- The pill is kind of expensive.
- I have some moral concerns about using the pill.
- The pill is legal.

The remaining blank spaces prompted Rosa to consider other issues. For example, she realized that she had forgotten to think about the side effects of the pill and the fact that it is useless against STIs. The information prompted by using the balance sheet is placed in parentheses.

TABLE 20.1

Rosa's balance sheet for deciding whether to use birth-control pills

Issues	For Self or Partner	Positive Anticipations	Negative Anticipations
Experiential gains and losses	For self	Sex would be spontaneous	Possible side effects: nausea, bloating, etc.
	For partner	Sex would be spontaneous	None
Biological gains and losses	For self	High effectiveness	Probably no negative impact on health, but useless against STIs
	For partner	High effectiveness	None
Financial gains and losses	For self	Needn't worry about the cost of an unwanted child	Expensive!
	For partner	Needn't worry about the cost of an unwanted child	None!
Ethical gains and losses	For self	Would not risk bearing unwanted child or having to decide whether to have an abortion if became pregnant	Moderate concern about using artificial means of birth control; responsibility for birth control would be all mine
	For partner	Would not have to worry about causing pregnancy	None!
Legal issues	For self		None
	For partner		None

Rosa decided not to go on the pill for several reasons: (1) She was concerned about side effects. (2) She did not want to spend the money. (3) She felt that by taking the pill, she was assuming all responsibility for birth control and that she would rather share that responsibility with her partner. Her periods were regular, and she decided (1) to use the rhythm method (to abstain from intercourse for a week in the middle of her cycle) and (2) to have her partner use condoms (and spermicide containing nonoxynol-9) every time they made love. The condoms and spermicide would also afford protection against STIs.

After completing this balance sheet, Rosa might fill out a similar one for the alternative of using the condom before making a decision. The balance sheet does not automatically lead to one "correct" choice. You might, for example, reach a different decision than Rosa if you were faced with the same situation. Rather, the balance sheet is a means of organizing one's perceptions of the positive and negative features of each alternative for oneself and one's partner. Research has shown that users of balance sheets hold fewer regrets about "the road not taken" and are more likely to stick with their decisions.

We have come a long way together, and we wish you well. Perhaps we have helped you to phrase some very important questions and to find some answers. We also hope that we have encouraged you to try to understand other people's sexual beliefs and values in light of their cultural backgrounds. Understanding is an essential milestone on the pathway to respect, and respect is vital to resolving conflicts and establishing healthy relationships.

Human Sexuality ONLINE //

Web Sites Related to Making Responsible Sexual Decisions

Your authors take no positions on the controversies discussed in this book, except that readers should be well-informed about matters of sexual health and should make their own decisions. The following Web sites and pages offer information that may be helpful in such decision making.

A Web page in the SIECUS (The Sexuality Information and Education Council of the U.S.) Web site that links to articles on religion and sexuality. Traditionalists will see SIECUS as being liberal.
www.siecus.org/religion/

The Web site of the Society for Judgment and Decision Making, whose members include psychologists, economists, organizational researchers, decision analysts, and other researchers who study theories of decision making.
www.sjdm.org/

The Web site of "Ethics Updates," an academic site intended for college students and professors that explores topics such as relativism, utilitarianism, and pluralism as they are related to topics such as abortion, sexual orientation, and reproductive technologies.
ethics.acusd.edu/

The Web site of the World Wide Study Bible. Various translations of the Old and New Testaments, with copious commentary.
www.ccel.org/wwsb/

A papal document: "On the Value and Inviolability of Human Life," an Encyclical Letter by Pope John Paul II. The document espouses the Roman Catholic point of view.
listserv.american.edu/catholic/church/papal/jp.ii/jp2evanv.html

A profile of the philosopher John Stuart Mill.
www.utm.edu/research/iep/m/millјs.htm

A profile of Joseph Fletcher and situational ethics, hosted by Pepperdine University.
news.pepperdine.edu/gsep/class/ethics/Fletcher/default.html

The Web site of the Internet Encyclopedia of Philosophy. Look for a philosopher or a school of philosophy by clicking on the first letter of the name.
www.utm.edu/research/iep/

questions for critical thinking

1. Do any of the value systems discussed in this chapter seem similar to your own? What role does your sociocultural background play in your value system? Explain.
2. Agree or disagree with the following statement, and support your answer: "There is no objective way of justifying one set of moral values over another."
3. Agree or disagree with the following statement, and support your answer: "Sexual decisions should be based on intellect and reasoning, rather than on emotions or strict obedience to a particular faith."
4. What decisions are you facing in your own sex life? What sexual decisions are you likely to face in the future? As we return to our diverse cultural worlds, what value systems will you draw on in making these decisions?
5. What kinds of conflicts may emerge if your partner comes from another cultural background and holds different values? How can the information in this book be helpful to you?
6. Will you use the information in this textbook to examine stereotypes and folklore critically, or will you think "Oh, it's just another textbook" and toss it aside at the end of the term?

Scoring Keys for Self-Scoring Questionnaires

SCORING KEY FOR THE LOVE ATTITUDES SCALE (CHAPTER 7, PP. 218–219)

First add your scores on the 30 items of the scale to yield a total score.

Write down your total score here: ______ 107

Note that each of the items is keyed so that lower-numbered responses represent more romantic responses; higher-numbered responses represent more realistic responses. The lower your total score (30 is the lowest possible score), the more you identify with a romantic view of love. The higher your score (150 is the highest possible score), the more realistic you are in your love attitudes. A score of 90 places you at the midpoint on the romantic-realist dimension. You might wish to compare your score with your partner's score to see who is more the romantic or more the realist.

SCORING KEY FOR STERNBERG'S TRIANGULAR LOVE SCALE (CHAPTER 7, PP. 224–225)

First add your scores for the items on each of the three components—Intimacy, Passion, and Decision/Commitment—and divide each total by 15. This will yield an average rating for each subscale. An average rating of 5 on a particular subscale indicates a moderate level of the component represented by the subscale. A higher rating indicates a greater level. A lower rating indicates a lower level. Examining your ratings on these components will give you an idea of the degree to which you perceive your love relationship to be characterized by these three components of love. For example, you might find that passion is stronger than decision/commitment, a pattern that is common in the early stages of an intense romantic relationship. You might find it interesting to complete the questionnaire a few months or perhaps a year or so from now to see how your feelings about your relationship change over time. You might also ask your partner to complete the scale so that the two of you can compare your respective scores. Comparing your ratings for each component with those of your partner will give you an idea of the degree to which you and your partner see your relationship in a similar way.

Sternberg (1988) reports the results of administering the scale to a sample of 50 men and 51 women (average age of 31 years) from the New Haven (Connecticut) area who were either married or currently involved in a close relationship. Average scores for the three components were 7.39 for intimacy, 6.51 for passion, and 7.20 for commitment. High scores (scores representing approximately the top 15 percent of scores) were 8.6 for intimacy, 8.2 for passion, and 8.7 for commitment. Low scores, representing the bottom 15 percent of scores, were 6.2, 4.9, and 5.7 for the three components, respectively. Since romantic ardor might be more difficult to maintain over time, the lower average scores for passion might reflect the length of the relationships in which the people in the sample were involved, which averaged 6.3 years in length. Although you might want to compare your scores with those from this sample, we caution that the Sternberg sample was small and most likely does not accurately represent the general population.

SCORING KEY FOR REASONING ABOUT ABORTION SCALE (CHAPTER 12, P. 406)

First tally your scores for the following items: 1, 3, 7, 8, 9, 11, 13, 14, 17, 20. This score represents your support for a *pro-choice* point of view: __________.

Now tally your scores for the remaining items: 2, 4, 5, 6, 10, 12, 15, 16, 18, 19. This score represents your support for a *pro-life* point of view: __________.

Now subtract your *pro-choice* score from your *pro-life* score. Write the difference, including the sign, here: __________. A positive score indicates agreement with a pro-life philosophy. A negative score indicates agreement with a pro-choice philosophy. The higher your score, the more strongly you agree with the philosophy that you endorsed. Scores may range from –40 to +40.

One sample of 230 undergraduate students (115 of each gender) obtained a mean score of –7.48 and a median score of –13.33 (Parsons et al., 1990). This indicates that the students tended to be pro-choice in their attitudes. Another sample of 38 graduate students (31 women and 7 men) obtained mean scores of –11 to –12 and median scores of –17 to –18 on two separate occasions. Scores for other samples may vary.

SCORING KEY FOR STI ATTITUDE SCALE (CHAPTER 16, P. 531)

Scores on this scale are interpreted in terms of a predisposition to high-risk sexual behavior. The scale is composed of three subscales, which measure your predisposition to high-risk behavior on the basis of your *beliefs* about STIs (items 1–9), your *feelings* about STIs (items 10–18), and your *intentions to act* (items 19–27). Calculate a total for each subscale and for the total scale by using the point values below.

For items 1, 10–14, 16, 25:

Strongly Agree	=	5 points
Agree	=	4 points
Undecided	=	3 points
Disagree	=	2 points
Strongly Disagree	=	1 point

For items 2–9, 15, 17–24, 26, 27:

Strongly Agree	=	1 point
Agree	=	2 points
Undecided	=	3 points
Disagree	=	4 points
Strongly Disagree	=	5 points

Subtotal for *Beliefs* subscale (items 1–9): __________

Subtotal for *Feelings* subscale (items 10–18): __________

Subtotal for *Intentions to Act* subscale (items 19–27): __________

Total score (based on all items): __________

The higher your subscale and total scores, the greater the likelihood that your attitudes, feelings and intentions put you at risk of contracting an STI. The lower your scores, the less risk you are likely to incur. Although we have no norms at present, we suggest that subscale scores higher than 27 and total scores higher than 81 indicate that your responses are weighted more toward risky than safe behavior. We suggest that you weigh your particular risk by examining not only your scores but your responses to the individual scale items as well. Ask yourself, "How do my responses to these items increase or decrease my risk of contracting an STI?" Then ask yourself, "How might I change my attitudes to reduce my chances of contracting an STI?"

references

A

ABC News *Nightline* Poll. (1995, February 17). *Most favor sex ed in schools, believing it changes behavior.*

Abel, G. G., et al. (1984). Complications, consent, and cognitions in sex between children and adults. *International Journal of Law and Psychiatry, 7,* 89–103.

Abel, G. G., et al. (1989). The measurement of the cognitive distortions of child molesters. *Annals of Sex Research, 2,* 135–152.

Abelson, H., et al. (1970). Public attitudes toward and experience with erotic materials. In *Technical Reports of the Commission on Obscenity and Pornography,* Vol. 6. Washington, DC: U.S. Government Printing Office.

Abramowitz, S. (1986). Psychosocial outcomes of sex reassignment surgery. *Journal of Consulting and Clinical Psychology, 54,* 183–189.

Acker, M., & Davis, M. H. (1992). Intimacy, passion and commitment in adult romantic relationships: A test of the triangular theory of love. *Journal of Social and Personal Relationships, 9,* 21–50.

Ackerman, D. (1990). *A natural history of the senses.* New York: Random House.

Ackerman, D. (1991). *The moon by whale light.* New York: Random House.

Adachi, M., et al. (2000). Androgen-insensitivity syndrome as a possible coactivator disease. *The New England Journal of Medicine* online, 343(12).

Adams, H. E., et al. (1981). Behavior therapy with sexual deviations. In S. M. Turner, K. S. Calhoun, & H. E. Adams (eds.), *Handbook of clinical behavior therapy* (pp. 318–346). New York: Wiley.

Adams, H. E., Wright, L. W., Jr., & Lohr, B. A. (1996). Is homophobia associated with homosexual arousal? *Journal of Abnormal Psychology 105,* 440–445.

Adams, R. (1987). The role of prostitution in AIDS and other STDs. *Medical Aspects of Human Sexuality, 21,* 27–33.

Adams, R. A., et al. (1992). Components of a model adolescent AIDS/drug abuse prevention program: A delphi study. *Family Rations, 41,* 31–317.

Adams, V. (1980, August). Sex therapists in perspective. *Psychology Today,* pp. 35–36.

Ade-Ridder, L. (1985). Quality of marriage: A comparison between golden-wedding couples and couples married less than fifty years *Lifestyles, 7,* 224–237.

Adelson, A. (1990, November 19). Study attacks women's roles in TV. *The New York Times,* p. C18.

Adler, J. (1993, April 26). Sex in the snoring '90s. *Newsweek,* pp. 55, 57.

A flock with changing views. (1993, August 1). *The New York Times,* p. A26.

After years of decline, Caesareans on the rise again. (2000, August 29). The Associated Press online.

Agnew, C. R., Van Lange, P. A. M., Rusbult, C. E., & Langston, C. A. (1998). Cognitive interdependence: Commitment and the mental representation of close relationships. *Journal of Personality & Social Psychology, 74*(4), 939–954.

Ahmed, R. A. (1991). Women in Egypt and the Sudan. In L. L. Adler (ed.), *Women in cross-cultural perspective* (pp. 107–134). New York: Praeger.

AIDS without needles or sex. (1993, December 20). *Newsweek,* pp. 106–107.

Akhtar, S. (1988). Four culture-bound psychiatric syndromes in India. *The International Journal of Social Psychiatry, 34,* 70–74.

Alan Guttmacher Institute. (1998). *Sex and America's teenagers.* New York: Author.

Alexander, C. J., Sipski, M. L., & Findley, T. W. (1993). Sexual activities, desire, and satisfaction in males pre- and post-spinal cord injury. *Archives of Sexual Behavior, 22,* 217–228.

Allen, M., & Burrell, N. (1996). Comparing the impact of homosexual and heterosexual parents on children: Meta-analysis of existing research. *Journal of Homosexuality, 32*(2): 19–35.

Allen, P. L. (2000). *The wages of sin: Sex and disease, past and present.* Chicago: University of Chicago Press.

Alterman, E. (1997, November). Sex in the '90s. *Elle,* pp. 128–134.

Alter-Reid, K., et al. (1986). Sexual abuse of children: A review of the empirical findings. *Clinical Psychology Review, 6,* 249–266.

Althof, S. E. (1994). Paper presented to the annual meeting of the American Urological Association. San Francisco.

Altman, L. K. (1991, April 15). Study challenges federal research on risks of IUD's. *The New York Times,* p. A1.

Altman, L. K. (1992a, July 21). Women worldwide nearing higher rate for AIDS than men. *The New York Times,* p. C3.

Altman, L. K. (1992b, February 17). Two new studies link vasectomy to higher prostate cancer risk. *The New York Times,* p. C12.

Altman, L. K. (1993, February 21). New caution, and some reassurance, on vasectomy. *The New York Times,* Section 4, p. 2.

Altman, L. K. (1995a, February 2). Combination of drugs appears to slow AIDS virus, studies say. *The New York Times,* p. A20.

Altman, L. K. (1995b, February 7). Protein in saliva found to block AIDS virus in test tube study. *The New York Times,* p. C3.

Altman, L. K. (1995c, July 7). Drug for treating impotence is ready for sale, F. D. A. says. *The New York Times,* p. A13.

Altman, L. K. (1997, January 19). With AIDS advance, more disappointment. *The New York Times,* pp. A1, A14.

Altman, L. K. (1998a, February 3). Big drop in AIDS deaths attributed to drug therapies. *The New York Times* online.

Altman, L. K. (1998b, February 5). Doctors discuss unwanted side effects of new AIDS drugs. *The New York Times* online.

Altman, L. K. (1998c, April 21). Studies show second drug can prevent breast cancer. *The New York Times* online.

Alzate, H. (1985). Vaginal eroticism: A replication study. *Archives of Sexual Behavior, 14,* 529–537.

Alzate, H., & Londono, M. L. (1984). Vaginal erotic sensitivity. *Journal of Sex and Marital Therapy, 10,* 49–56.

American Academy of Child and Adolescent Psychiatry. (1998). Cited in Goldberg, C. (1998, September 8). Getting to the truth in child abuse cases: New methods. *The New York Times,* pp. F1, F5.

American Academy of Pediatrics. (1999, March). Circumcision policy statement. *Pediatrics, 103*(3), 686–693.

American Cancer Society. (2000). http://www.cancer.org/

American Professional Society on the Abuse of Children. (1998). Cited in Goldberg, C. (1998, September 8). Getting to the truth in child abuse cases: New methods. *The New York Times,* pp. F1, F5.

American Psychiatric Association. (2000). *Diagnostic and statistical manual of mental disorders,* 4th ed. Washington, DC: Author.

American Psychological Association, Committee on Gay and Lesbian Concerns. (1991). Avoiding heterosexual bias in language. *American Psychologist, 46,* 973–974.

American Psychological Association (1998a, March 16). Sexual harassment: Myths and realities. APA Public Information Home Page; www.apa.org.

American Psychological Association (1998b, October 4). Answers to your questions about sexual orientation and homosexuality. APA Public Information Home Page; www.apa.org.

Americans generally well-informed about AIDS, but many lack knowledge about preventive aspects. (1993). *Family Planning Perspectives, 25*, 139–140.

Ames, M. A., & Houston, D. A. (1990). Legal, social, and biological definitions of pedophilia. *Archives of Sexual Behavior, 19*, 333–342.

Anderson, J. L., et al. (1992). Was the Duchess of Windsor right? A cross-cultural review of the socioecology of ideals of female body shape. *Ethology and Sociobiology, 13*, 197–227.

Anderson, P. B., & Aymami, R. (1993). Reports of female initiating of sexual contact: Male and female differences. *Archives of Sexual Behavior, 22*, 335–343.

Anderson, V. N. (1992). For whom is this world just? Sexual orientation and AIDS. *Journal of Applied Social Psychology 22*, 248–259.

Angier, N. (1990, July 19). Scientists say gene on Y chromosome makes a man a man. *The New York Times*, pp. A1, 19.

Angier, N. (1991, August 30). Zone of brain linked to men's sexual orientation. *The New York Times*, A1, D18.

Angier, N. (1992, January 30). Odor receptors discovered in sperm cells. *The New York Times*, p. A19.

Angier, N. (1993a). Future of the pill may lie just over the counter. *The New York Times*, Section 4, p. 5.

Angier, N. (1993b). Reports suggest homosexuality is linked to genes. *The New York Times*, pp. A12, D21.

Angier, N. (1995, November 2). Study links brain to transsexuality. *The New York Times* online.

Angier, N. (1997). Chemical tied to fat control could help trigger puberty. *The New York Times*, pp. C1, C3.

Ansen, D., et al. (1993, January 18). A lost generation. *Newsweek*, pp. 16–23.

Antill, J. K. (1983). Sex role complementarity versus similarity in married couples. *Journal of Personality and Social Psychology, 52*, 260–267.

Antonarakas, S. E., et al. (1991). Prenatal origin of the extra chromosome in trisomy 21 as indicated by analysis of DNA polymorphisms. *The New England Journal of Medicine, 324*, 872–876.

Antoni, M. H., et al. (1990). Psychoneuroimmunology and HIV-1. *Journal of Consulting and Clinical Psychology, 58*, 38–49.

Antoni, M. H., et al. (1991). Cognitive-behavioral stress management intervention buffers distress responses and immunologic changes following notification of HIV-1 seropositivity. *Journal of Consulting and Clinical Psychology, 59*, 906–915.

Armstrong, K., Eisen, A., & Weber, B. (2000). Assessing the risk of breast cancer. *The New England Journal of Medicine* online, 342(8).

Armsworth, M. W. (1991). Psychological response to abortion. *Journal of Counseling and Development, 69*, 377–379.

Arriaga, X. B., & Rusbult, C. E. (1998). Standing in my partner's shoes: Partner perspective taking and reactions to accommodative dilemmas. *Personality & Social Psychology Bulletin, 24*(9), 927–948.

Arthur, B. I., Jr., et al. (1998). Sexual behaviour in *Drosophila* is irreversibly programmed during critical period. *Current Biology, 8*(21), 1187–1190.

Asbell, B. (1995). *The pill: A biography of the drug that changed the world.* New York: Random House.

Ashton, A. K., et al. (2000). Antidepressant-induced sexual dysfunction and ginkgo biloba. *American Journal of Psychiatry, 157*, 836–837.

Assalian, P. (1994). Premature ejaculation: Is it really psychogenic? *Journal of Sex Education and Therapy, 20*(1), 1–4.

Astley, S. J., et al. (1992). Analysis of facial shape in children gestationally exposed to marijuana, alcohol, and/or cocaine. *Pediatrics, 89*, 67–77.

Athletes rebuffed in nude photo Net case. (2000, June 22). Reuters News Agency.

Atkeson, B. M., Calhoun, K. S., & Morris, K. T. (1989). Victim resistance to rape: The relationship of previous victimization, demographics, and situational factors. *Archives of Sexual Behavior, 18*, 497–507.

Australia scientists find flowers dupe lonely bees. (2000, June 16). Reuters News Agency online.

Axinn, W. G. (1991). The influence of interviewer sex on responses to sensitive questions in Nepal. *Social Science Research, 20*, 303–318.

Azar, B. (1998). Communicating through pheromones. *APA Monitor, 29*(1), 1, 12.

B

Bach, G. R., & Deutsch, R. M. (1970). *Pairing.* New York: Peter H. Wyden.

Baenninger, M. A., & Elenteny, K. (1997). Cited in Azar, B. (1997). Environment can mitigate differences in spatial ability. *APA Monitor, 28*(6), 28.

Bagatell, C. J., & Bremner, W. J. (1996). Drug therapy: Androgens in men—Uses and abuses. *The New England Journal of Medicine, 334*, 707–714.

Bagley, C., & D'Augelli, A. R. (2000). Suicidal behaviour in gay, lesbian, and bisexual youth. *British Medical Journal, 320*, 1617–1618.

Bagley, C., & Young, L. (1987). Juvenile prostitution and child sexual abuse: A controlled study. *Canadian Journal of Community Mental Health, 6*, 5–26.

Bailey, J. M. (1999). Homosexuality and mental illness. *Archives of General Psychiatry, 56*(10), 883–884.

Bailey, J. M., Dunne, M. P., & Martin, N. G. (2000). Genetic and environmental influences on sexual orientation and its correlates in a Australian twin sample. *Journal of Personality & Social Psychology, 78*(3), 524–536.

Bailey, J. M., et al. (1999). A family history study of male sexual orientation using three independent samples. *Behavior Genetics, 29*(2), 79–86.

Bailey, J. M., Kim, P. Y., Hills, A., & Linsenmeier, J. A. W. (1997). Butch, femme, or straight acting? Partner preferences of gay men and lesbians. *Journal of Personality & Social Psychology, 73*(5), 960–973.

Bailey, J. M., Kirk, K. M., Zhu, G., Dunne, M. P., & Martin, N. G. (2000). Do individual differences in sociosexuality represent genetic or environmentally contingent strategies? Evidence from the Australian twin registry. *Journal of Personality & Social Psychology, 78*(3), 537–45.

Bailey, J. M., & Pillard, R. C. (1991). A genetic study of male sexual orientation. *Archives of General Psychiatry, 48*, 108–1096.

Bailey, J. M., Pillard, R. C., Neale, M. C., & Agyei, Y. (1993). Heritable factors influence sexual orientation in women. *Archives of General Psychiatry, 50*, 217–223.

Bailey, J. M., & Zucker, K. J. (1995). Childhood sex-typed behavior and sexual orientation: A conceptual analysis and quantitative review. *Developmental Psychology, 31*, 43–55.

Bailey, R. C. (2000). A study in rural Uganda of heterosexual transmission of human immunodeficiency virus. *The New England Journal of Medicine* online, 343(5).

Baker, J. N. (1990, Summer/Fall). Coming out. [Special Issue]. *Newsweek*, pp. 60–61.

Baker, S., Thalberg, S., & Morrison, D. (1988). Parents' behavioral norms as predictors of adolescent sexual activity and contraceptive use. *Adolescence, 23*, 278–281.

Baker, T. C., et al. (1990). Rape victims' concerns about possible exposure to HIV infection. *Journal of Interpersonal Violence, 5*, 49–60.

Baldwin, J. D., & Baldwin, J. I. (1989). The socialization of homosexuality and heterosexuality in a non-Western society. *Archives of Sexual Behavior, 18*, 13–29.

Bancroft, J. (1984). Hormones and human sexual behavior. *Journal of Sex and Marital Therapy, 10*, 3–21.

Bancroft, J. (1990). Commentary: Biological contributions to sexual orientation. In D. P. McWhirter, S. A. Sanders, & J. M. Reinisch (eds.), *Homosexuality/heterosexuality: Concepts of sexual orientation* (pp. 101–111). New York: Oxford University Press.

Bancroft, J., et al. (1983). Mood, sexuality, hormones, and the menstrual cycle. III: Sexuality and the role of androgens. *Psychosomatic Medicine, 45*, 509–516.

Banmen, J., & Vogel, N. (1985). The relationship between marital quality and interpersonal sexual communication. *Family Therapy, 12*, 45–58.

Banning, A. (1989). Mother–son incest: Confronting a prejudice. *Child Abuse and Neglect, 13*, 563–570.

Barbach, L. G. (1975). *For yourself: The fulfillment of female sexuality.* New York: Doubleday.

Barbach, L. G. (1995). Cited in Goleman, D. (1995, June 14). Sex fantasy research said to neglect women. *The New York Times*, p. C1.

Barbaree, H. E., & Marshall, W. L. (1991). The role of male sexual arousal in rape: Six models. *Journal of Consulting and Clinical Psychology, 59*, 621–630.

Barisic, S. (1998, February 18). Study: Breast milk is still best. The Associated Press online.

Barlow, D. H. (1986). Causes of sexual dysfunction: The role of anxiety and cognitive interference. *Journal of Consulting and Clinical Psychology, 54*, 140–148.

Barlow, D. H., & Durand, V. M. (1995). *Abnormal psychology.* Pacific Grove, CA: Brooks/Cole.

Barnard, N. D., Scialli, A. R., Hurlock, D., & Bertron, P. (2000). Diet and sex-hormone binding globulin, dysmenorrhea, and premenstrual symptoms. *Obstetrics & Gynecology, 95*, 245–250.

Barouch, D. H., et al. (2000). Control of viremia and prevention of clinical AIDS in Rhesus monkeys by cytokine-augmented DNA vaccination *Science, 290*(5491), 486–492.

Barr, H. M., Streissguth, A. P., Darby, B. L., & Sampson, P. D. (1990). Prenatal exposure to alcohol, caffeine, tobacco, and aspirin. *Developmental Psychology, 26*, 339–348.

Barrett, M. B. (1990). *Invisible lives: The truth about millions of women-loving women.* New York: Harper & Row (Perennial Library).

Barringer, F. (1993a, April 1). Viral sexual diseases are found in 1 of 5 in U.S. *The New York Times,* pp. A1, B9.

Barringer, F. (1993b, April 15). Sex survey of American men finds 1% are gay. *The New York Times,* p. A1.

Barringer, F. (1993c, June 2). School hallways as gauntlets of sexual taunts. *The New York Times,* p. B7.

Bart, P. B., & O'Brien, P. B. (1985). *Stopping rape: Successful survival strategies.* Elmsford, NY: Pergamon Press.

Bar-Tal, D., & Saxe, L. (1976). Perceptions of similarly and dissimilarly physically attractive couples and individuals. *Journal of Personality and Social Psychology, 33*, 772–781.

Bartel, G. D. (1970). Group sex among mid-Americans. *Journal of Sex Research, 6*, 113–131.

Bartlett, J. G. (1993). Zidovudine now or later? *The New England Journal of Medicine, 329*, pp. 351–352.

Bartoshuk, L. M., & Beauchamp, G. K. (1994). Chemical senses. *Annual Review of Psychology, 45*, 419–449.

Basson, R. (2000, May). Paper presented to the annual meeting of the American College of Obstetricians and Gynecologists, San Francisco.

Baumeister, R. F. (1988a). Gender differences in masochistic scripts. *Journal of Sex Research, 25*, 478–499.

Baumeister, R. F. (1988b). Masochism as escape from self. *Journal of Sex Research, 25*, 28–59.

Baxter, D. J., Barbaree, H. E., & Marshall, W. L. (1986). Sexual responses to consenting and forced sex in a large sample of rapists an nonrapists. *Behaviour Research and Therapy, 17*, 215–222.

Bazell, R. (2000, April 24). Battling prostate cancer. NBC News.

Beck, A. (1988). *Love is never enough.* New York: Harper & Row.

Beck, J. G. (1993). Vaginismus. In W. O'Donohue & J. H. Geer (eds.), *Handbook of sexual dysfunctions: Assessment and treatment* (pp. 381–397). Boston: Allyn & Bacon.

Becker, J. V., et al. (1986). Level of postassault sexual functioning in rape and incest victims. *Archives of Sexual Behavior, 15*, 37–49.

Becker, J. V., et al. (1989). Factors associated with erection in adolescent sex offenders. *Journal of Psychopathology and Behavioral Assessment, 11*, 353–362.

Belcastro, P. A. (1985). Sexual behavior differences between black and white students. *Journal of Sex Research, 21*, 56–67.

Belgrave, F. Z., van Oss Marian, B., & Chambers, D. B. (2000). Cultural, contextual, and intrapersonal predictors of risky sexual attitudes among urban African American girls in early adolescence. *Cultural Diversity and Ethnic Minority Psychology, 6*(3), 309–322.

Bell, A. P., & Weinberg, M. S. (1978). *Homosexualities: A study of diversity among men and women.* New York: Simon & Schuster.

Bell, A. P., Weinberg, M. S., & Hammersmith, S. K. (1981). *Sexual preference: Its development in men and women.* Bloomington, IN: University of Indiana Press.

Bell, S., Kuriloff, P. J., & Lottes, I. (1994) Understanding attributions of blame in stranger rape and date rape situations: In examination of gender, race, identification, and students' social perceptions of rape victims. *Journal of Applied Social Psychology, 24*(19), 1719–1734.

Beller, M., & Gafni, N. (2000). Can item format (multiple choice vs. open-ended) account for gender differences in mathematics achievement? *Sex Roles, 42*(1–2), 1–21.

Beller, T. (1995, July 13). Why? Don't ask. *The New York Times,* p. A23.

Bellis, D. J. (1990). Fear of AIDS and risk reduction among heroin–addicted female street prostitutes: Personal interviews with 72 Southern California subjects. *Journal of Alcohol and Drug Education, 35*, 26–37.

Bello, D. C., Pitts, R. E., & Etzel, M. J. (1983). The communications effects of controversial sexual content in television programs and commercials. *Journal of Advertising, 12*(3), 32–4

Bem, S. L. (1974). The measurement of psychological androgyny. *Journal of Consulting and Clinical Psychology, 42*, 151–16

Bem, S. L. (1975). Sex role adaptability: One consequence of psychological androgyny. *Journal of Personality and Social Psychology, 31*, 634–643.

Bem, S. L. (1981). Gender schema theory: A cognitive account of sex typing. *Psychological Review, 88*, 354–364.

Bem, S. L. (1983). Gender schema theory and its implications for child development: Raising gender-aschematic children in a gender-schematic society. *Signs, 8*, 598–616.

Bem, S. L. (1985). Androgyny and gender schema theory: A conceptual and empirical integration. In T. B. Sonderegger (ed.), *Nebraska symposium on motivation, 1984: Psychology and gender.* Lincoln: University of Nebraska Press.

Bem, S. L. (1993). *The lenses of gender.* New Haven, CT: Yale University Press.

Bem, S. L., Martyna, W., & Watson, C. (1976). Sex typing and androgyny: Further explorations of the expressive domain. *Journal of Personality and Social Psychology, 34*, 1016–1023.

Benatar, S. R. (2000). AIDS in the 21st century. *The New England Journal of Medicine, 342*(7).

Ben Hamida, S., Mineka, S., & Bailey, J. M. (1998). Sex differences in perceived controllability of mate value: An evolutionary perspective. *Journal of Personality & Social Psychology, 75*(4), 953–966.

Bennett, N. G. (1998, September 16). Who should decide the sex of a baby? *The New York Times,* p. A28.

Bennett, N. G., Blanc, A. K., & Bloom, D. E. (1988). Commitment and the modern union: Assessing the link between premarital cohabitation and subsequent marital stability. *American Sociological Review, 53*, 127–138.

Bentler, P. M. (1976). A typology of transsexualism: Gender identity theory and data. *Archives of Sexual Behavior, 5*, 567–584.

Berger, L. (2000, June 25). A racial gap in infant deaths, and a search for reasons. *The New York Times,* p. WH13.

Berke, R. L. (1997, June 15). Suddenly, the new politics of morality. *The New York Times,* p. E3.

Berke, R. L. (1998, August 2). Chasing the polls on gay rights. *The New York Times,* p. WK3.

Berliner, D. (1993). Cited in Blakeslee, S. (1993, September 7). Human nose may hold an additional organ for a real sixth sense. *The New York Times,* p. C3.

Berman, L. (2000). Paper presented to the annual meeting of the American Urological Association, Atlanta, GA. Cited in "Women, too, may benefit from Viagra." (2000, May 1). CNN.

Bernstein, S. J., et al. (1993). The appropriateness of hysterectomy: A comparison of care in seven health plans. *Journal of the American Medical Association, 269*, 2398–2402.

Bernstein, W. M., et al. (1983). Causal ambiguity and heterosexual affiliation. *Journal of Experimental Social Psychology, 19*, 78–92.

Berscheid, E. (1988). Some comments on love's anatomy: Or, whatever happened to old-fashioned lust? In R. J. Sternberg & M. L. Barnes (eds.), *The psychology of love* (pp. 359–374). New Haven, CT: Yale University Press.

Berscheid, E., & Walster, E. (1978). *Interpersonal attraction.* Reading, MA: Addison-Wesley.

Bianchi, S. M., & Spain, D. (1997). *Women, work and family in America.* Population Reference Bureau.

Bieber, I. (1976). A discussion of "Homosexuality: The ethical challenge." *Journal of Consulting and Clinical Psychology, 44*, 163–166.

Bieber, I., et al. (1962). *Homosexuality.* New York: Basic Books.

Biggar, R. J., & Melbye, M. (1992). Responses to anonymous questionnaires concerning sexual behavior: A method to examine potential biases. *American Journal of Public Health, 82*, 1506–1512.

Billingham, R. E., & Sack, A. Z. (1986). Gender differences in college students' willingness to participate in alternative marriage and family relationships. *Family Perspectives, 20*, 37–44.

Billy, J. O. G., et al. (1993). The sexual behavior of men in the United States. *Family Planning Perspectives, 25*, 52–60.

Binion, V. J. (1990). Psychological androgyny: A Black female perspective. *Sex Roles, 22*, 487–507.

Bixler, R. H. (1981). The incest controversy. *Psychological Reports, 49*, 267–283.

Bjorklund, D. F., & Kipp, K. (1996). Parental investment theory and gender differences in the evolution of inhibition mechanisms. *Psychological Bulletin, 120*, 163–188.

Black, L. E., Eastwood, M. M., Sprenkle, D. H., & Smith, E. (1991). An exploratory analysis of the construct of leavers versus left as it relates to Levinger's social exchange theory of attractions, barriers, and alternative attractions. *Journal of Divorce & Remarriage, 15*(1–2), 127–139.

Blair, C. D., & Lanyon, R. I. (1981). Exhibitionism: A critical review of the etiology and treatment. *Psychological Bulletin, 89* 439–43.

Blakeslee, S. (1992, January 22). An epidemic of genital warts raises concern but not alarm. *The New York Times,* p. C12.

Blanchard, R., & Hucker, S. J. (1991). Age, transvestism, bondage, and concurrent paraphilic activities in 117 fatal cases of autoerotic asphyxia. *British Journal of Psychiatry, 159*, 371–377.

Blanchard, R., Steiner, B. W., & Clemmensen, L. H. (1985). Gender dysphoria, gender reorientation, and the clinical management of transsexualism. *Journal of Consulting and Clinical Psychology 53*, 295–304.

Blickstein, I., Goldman, R. D., & Mazkereth, R. (2000). Risk for one or two very low birth weight twins: A population study. *Obstetrics Gynecology, 96*(3), 400–402.

Bloor, M., McKeganey, N., & Barnard, M. (1990). An ethnographic study of HIV–related risk practices among Glasgow rent boys and their clients: Report of a pilot study. *AIDS Care, 2*, 17–24.

Blumstein, P., & Schwartz, P. (1983). *American couples: Money, work, sex.* New York: William Morrow.

Blumstein, P. & Schwartz, P. (1990). Intimate relationships and the creation of sexuality. In D. P. McWhirter, S. A. Sanders, & J. M. Reinisch (eds.), *Homosexuality/heterosexuality: Concepts of sexual orientation* (pp. 307–320). New York: Oxford University Press.

Bohlen, C. (1995, April 4). Almost anything goes in birth science in Italy. *The New York Times*, p. A14.

Boles, J., & Elifson, K. W. (1994). Sexual identity and HIV: The male prostitute. *Journal of Sex Research, 31*, 39–46.

Boskind-White, M., & White, W. C. (1986). Bulmariexia: A historical-sociocultural perspective. In K. D. Brownell J. P. Foreyt (eds.), *Handbook of eating disorders: Physiology, psychology, and treatment of obesity, anorexia, and bulimia* (pp. 353–378). New York: Basic Books.

Bostick, R. M., et al. (1994). Predictors of cancer prevention attitudes and participation in cancer screening examinations. *Preventive Medicine: An International Journal Devoted to Practice and Theory, 23*(6), 816–826.

Boston Women's Health Book Collective. (1992). *The New Our Bodies, Ourselves.* New York: Simon & Schuster.

Boswell, J. (1990). Sexual and ethical categories in premodern Europe. In D. P. McWhirter, S. A. Sanders, & J. M. Reinisch (eds.), *Homosexuality/Heterosexuality: Concepts of sexual orientation* (pp. 15–31). New York: Oxford University Press.

Boulton, M., Hart, G., & Fitzpatrick, R. (1992). The sexual behaviour of bisexual men in relation to HIV transmission. *AIDS Care, 4*, 165–175.

Bowie, W. R. (1990). Approach to men with urethritis and urologic complications of sexually-transmitted diseases. *Medical Clinics of North America, 74*, 1543–1557.

Bowlby, J. (1969). *Attachment and loss*, Vol. 1. New York: Basic Books.

Boxer, S. (2000, July 22). Truth or lies? In sex surveys, you never know. *The New York Times* online.

Boyer, D. (1989). Male prostitution and homosexual identity. [Special Issue: Gay and lesbian youth: I]. *Journal of Homosexuality, 17*, 151–184.

Boy to sue city over tryst with teacher. (2000, April 14). Reuters News Agency.

Bozzette, S. A., et al. (1995). A randomized trial of three antipneumocystis agents in patients with advanced human immunodeficiency virus infection. *The New England Journal of Medicine, 332*, 693–699.

Bradbard, M. R., & Endsley, R. C. (1984). The effects of sex typed labeling on preschool children's information-seeking and retention. *Sex Roles, 9*, 247–261.

Bradford, J. M. W. (1998). Treatment of men with paraphilia. *The New England Journal of Medicine, 338*, 464–465.

Bradley, S. J., Oliver, G. D., Chernick, A. B., & Zucker, K. J. (1998). Experiment of nurture: Ablatio penis at 2 months, sex reassignment at 7 months, and a psychosexual follow-up in young adulthood. *Pediatrics* online, 102(1), p. e9.

Brady, E. C., et al. (1991). Date rape: Expectations, avoidance strategies, and attitudes toward victims. *Journal of Social Psychology, 131*, 427–429.

Breast fears fade. (1993, November). *Prevention*, pp. 19–20.

Brecher, E. M, and the Editors of Consumer Reports Books. (1984). *Love, sex, and aging.* Boston: Little, Brown.

Breslow, N. (1989). Sources of confusion in the study and treatment of sadomasochism. *Journal of Social Behavior and Personality, 4*, 263–274.

Breslow, N., Evans, L., & Langley, J. (1985). On the prevalence and roles of females in the sadomasochistic subculture: Report on an empirical study. *Archives of Sexual Behavior, 14*, 303–317

Breslow, N., Evans, L., & Langley, J. (1986). Comparisons among heterosexual, bisexual and homosexual male sadomasochists. *Journal of Homosexuality, 13*, 83–107.

Bretl, D. J., & Cantor, J. (1988). The portrayal of men and women in U.S. television commercials: A recent content analysis and trends over 15 years. *Sex Roles, 18*, 595–609.

Bretschneider, J., & McCoy, N. (1988). Sexual interest and behavior in healthy 80- to 102-year-olds. *Archives of Sexual Behavior, 17*, 109.

Brewer, J. J. (1982). A history of erotic art as illustrated in the collectors of the Institute for Sex Research (The "Kinsey Institute". In A. Hoch & H. I. Lief (eds.), *Sexology: Sexual biology, behavior and therapy* (pp. 318–321). Amsterdam, The Netherlands: Excerpta Medica.

Briere, J., & Runtz, M. (1987). Post sexual abuse trauma: Data and implications for clinical practice. *Journal of Interpersonal Violence, 2*, 367–379.

Briere, J., & Runtz, M. (1989). University males' sexual interest in children: Predicting potential indices of "pedophilia" in a nonforensic sample. *Child Abuse and Neglect, 13*, 65–75

Bright, S. (1995). *Susie Bright's sexwise.* Cleis.

Broder, M. S., Kanouse, D. E., Mittman, B. S., & Bernstein, S. J. (2000). The appropriateness of recommendations for hysterectomy. *Obstetrics & Gynecology, 95*, 199–206.

Brody, J. E. (1989, July 27). Research casts doubt on need for many Caesarean births as their rate soars. *The New York Times*, p. B5.

Brody, J. E. (1990a, January 23). Scientists trace aberrant sexuality. *The New York Times*, pp. C11, C12.

Brody, J. E. (1990b, August 2). In fight against breast cancer, mammograms are a crucial tool, but not foolproof. *The New York Times*, p. B5.

Brody, J. E. (1992a, April 29). How to outwit a rapist: Rehearse. *The New York Times*, p. C13.

Brody, J. E. (1992b, November 11). PMS is a worldwide phenomenon. *The New York Times*, p. C14.

Brody, J. E. (1993, August 4). A new look at an old quest for sexual stimulants. *The New York Times*, p. C12.

Brody, J. E. (1995a, May 3). Breast scans may indeed help women under 50. *The New York Times*, p. C11.

Brody, J. E. (1995b, July 19). Revolution in treating infertile men turns hopelessness to parenthood. *The New York Times*, p. C8.

Brody, J. E. (1995c, August 30). Hormone replacement therapy for men: When does it help? *The New York Times*, p. C8.

Brody, J. E. (1996, August 28). PMS need not be the worry it was just decades ago. *The New York Times*, p. C9.

Brody, J. E. (1997a, February 12). Abortion doesn't affect well-being, study says. *The New York Times* online.

Brody, J. E. (1997b, March 4). Drug researchers working to design customized estrogen. *The New York Times*, pp. C1, C10.

Brody, J. E. (1997c, August 20). Estrogen after menopause? A tough dilemma. *The New York Times*, p. C8.

Brody, J. E. (1998a, February 10). Genetic ties may be factor in violence in stepfamilies. *The New York Times*, pp. F1, F4.

Brody, J. E. (1998b, February 18). Studies confirm alcohol's link to breast cancer. *The New York Times* online.

Brody, J. E. (1998c). Sour note in the Viagra symphony. *The New York Times*, p. F7.

Brody, J. E. (1998d, September 15). Teen-agers and sex: Younger and more at risk. *The New York Times* online.

Brody, J. E. (1998e, September 8). Weighing the pros and cons of hormone therapy. *The New York Times*, p. F7.

Brody, J. E. (2000, May 16). Cybersex gives birth to a psychological disorder. *The New York Times*, pp. F7, F12.

Bronner, E. (1998, February 1). "Just say maybe. No sexology, please. We're Americans." *The New York Times*, p. WK6.

Brookoff, D., et al. (1997). Characteristics of participants in domestic violence: Assessment at the scene of domestic assault. *Journal of the American Medical Association, 277*, 1369–1373.

Brooks, V. R. (1982). Sex differences in student dominance behavior in female and male professors' classrooms. *Sex Roles, 8*, 683–690.

Brooks-Gunn, J., & Furstenberg, F. F. (1989). Adolescent sexual behavior. *American Psychologist, 44*, 249–257.

Brooks-Gunn, J., et al. (1986). Physical similarity of and disclosure of menarchal status to friends: Effects of grade and pubertal status. *Journal of Early Adolescence, 6*, 3–14.

Brooks-Gunn, J., & Matthews, W. S. (1979). *He and she: How children develop their sex-role identity.* Englewood Cliffs, NJ: Spectrum.

Broude, G. J., & Greene, S. J. (1976). Cross-cultural codes on twenty sexual attitudes and practices. *Ethnology, 15*, 409–4.

Brown, M., Perry, A., Cheesman, A. D., & Pring, T. (2000). Pitch change in male-to-female transsexuals: Has phonosurgery a role to play? *International Journal of Language & Communication Disorders, 35*(1), 129–136.

Brown, R. A. (1994). Romantic love and the spouse selection criteria of male and female Korean college students. *The Journal of Social Psychology, 134*(2), 183–189.

Browning, J. R., Hatfield, E., Kessler, D., & Levine, T. (2000). Sexual motives, gender, and sexual behavior. *Archives of Sexual Behavior, 29*(2), 135–153.

Bruck, M. (1998). Cited in Goldberg, C. (1998, September 8). Getting to the truth in child abuse cases: New methods. *The New York Times*, pp. F1, F5.

Bruck, M., Ceci, S. J., & Francoeur, E. (2000). Children's use of anatomically detailed dolls to report genital touching in a medical examination: Developmental and gender comparisons. *Journal of Experimental Psychology: Applied, 6*(1), 74–83.

Bryant, H., & Brasher, P. (1995). Breast implants and breast cancer—Reanalysis of a linkage study. *The New England Journal of Medicine, 332*, 1535–1539.

Bryant, J., & Brown, D. (1989). Uses of pornography. In D. Zillmann & J. Bryant (eds.), *Pornography: Research advances and policy considerations* (pp. 25–55). Hillsdale, NJ: Lawrence Erlbaum Associates.

Buhrich, N., et al. (1979). Plasma testosterone, serum FSH, and serum LH levels in transvestism. *Archives of Sexual Behavior, 8*, 49–54.

Bullough, V. (1990). The Kinsey Scale in historical perspective. In D. P. McWhirter, S. A. Sanders, & J. M. Reinisch (eds.), *Homosexuality/heterosexuality: Concepts of sexual orientation* (pp. 3–14). New York: Oxford University Press.

Bullough, V. L. (1991). Transvestism: A reexamination. *Journal of Psychology and Human Sexuality, 4*, 53–67.

Bullough, V. L., & Weinberg, T. S. (1989). Women married to transvestites: Problems and adjustments. *Journal of Psychology & Human Sexuality, 1*, 83–104.

Bumiller, E. (1990, October 25). Japan's abortion agony: In a country that prohibits the pill, reality collides with religion. *Washington Post.*

Bumpass, L. (1995, July 6). Cited in Steinhauer, J. No marriage, no apologies. *The New York Times*, pp. C1, C7.

Bumpass, L., Sweet, J., & Castro, T. (1988, August). *Changing patterns of remarriage.* Paper presented at the meeting of the American Sociological Association, Atlanta, Georgia.

Burgess, A. W., & Hartman, C. R. (1986). *Sexual exploitation of patients by health professionals.* New York: Praeger.

Burgess, A. W., & Holmstrom, L. L. (1974). Rape trauma syndrome. *American Journal of Psychiatry, 131*, 981–986.

Burkhart, B., & Fromuth, M. E. (1991). Individual psychological and social psychological understandings of sexual coercion. In E. Grauerholz & M. A. Koralewski (eds.), *Sexual coercion: A sourcebook on its nature, causes, and prevention* (pp. 7–89). Lexington, MA: Lexington Books.

Burstein, G. R., et al. (1998). Incident *Chlamydia trachomatis* infections among inner–city adolescent females. *Journal of the American Medical Association, 280*, 521–526.

Burt, M. R. (1980). Cultural myths and supports for rape. *Journal of Personality and Social Psychology, 38*, 217–230.

Buss, D. M. (1994). *The evolution of desire: Strategies of human mating.* New York: Basic Books.

Buve, A. (2000). Cited in Altman, L. K. (2000, July 11). Mystery factor is pondered at AIDS talk: Circumcision. *The New York Times* online.

Buxton, A. P. (1994). *The other side of the closet.* New York: Wiley.

Byers, E. S., & Lewis, K. (1988). Dating couples' disagreements over the desired level of sexual intimacy. *Journal of Sex Research, 24*, 15–29.

Bynum, R. (1998, June 26). U.S. syphilis rate hits record low. Associated Press online.

Byrd, J., Hyde, J. S., DeLamater, J. D., & Plant, E. A. (1998). Sexuality during pregnancy and the year postpartum. *Journal of Family Practice, 47*(4), 305–308.

Byrnes, J., & Takahira, S. (1993). Explaining gender differences on SAT-math items. *Developmental Psychology, 29*, 805–810.

C

Caceres, C. F., & van-Griensven, G. J. P. (1994). Male homosexual transmission of HIV-1. *AIDS, 8*(8), 1051–1061.

Cado, S., & Leitenberg, H. (1990). Guilt reactions to sexual fantasies during intercourse. *Archives of Sexual Behavior, 19*, 49–64.

Cairns, P. (2000). Cited in U.S. researchers seek earlier prostate cancer detection. (2000, April 3). Reuters News Agency online.

Calderone, M. S., & Johnson, E. W. (1989). *Family book about sexuality*, rev. ed. New York: Harper & Row.

Calhoun, K. S., & Atkeson, B. M. (1991). *Treatment of rape victims: Facilitating social adjustment.* New York: Pergamon Press.

Call, V., Sprecher, S., & Schwartz, P. (1995). The incidence and frequency of marital sex in a national sample. *Journal of Marriage and the Family, 57*, 639–652.

Cameron, P., et al. (1986). Child molestation and homosexuality. *Psychological Reports, 58*, 327–337.

Campbell, C. A. (1991). Prostitution, AIDS, and preventive health behavior. *Social Science and Medicine, 32*, 1367–1378.

Campbell, S. B., & Cohn, J. F. (1991). Prevalence and correlates of postpartum depression in first-time mothers. *Journal of Abnormal Psychology, 100*, 594–599.

Cancer survivor Bob Dole calls Viagra "great drug." (1998, May 7). Reuters News Agency online.

Cannistra, S. A., & Niloff, J. M. (1996). Cancer of the uterine cervix. *The New England Journal of Medicine, 334*, 1030–1038.

Cantalona, W. J. (1998). Cited in Story, P. (1998, June 2). Study: Prostate cancer deaths down. The Associated Press online.

Cappella, J. N., & Palmer, M. T. (1990). Attitude similarity, relational history, and attraction: The mediating effects of kinesic and vocal behaviors. *Communication Monographs, 5*, 161–183.

Carey, et al. (1992). Effectiveness of latex condoms as a barrier to human immunodeficiency virus-sized particles under conditions of simulated use. *Sexually Transmitted Diseases, 19*, 230–234.

Carlson, M. (1990, July 9). Abortion's hardest cases. *Time*, pp. 22–26.

Carrera, M. (1981). *Sex: The facts, the acts and your feelings.* New York: Crown.

Carrère, S., Buehlman, K. T., Gottman, J. M., Coan, J. A., & Ruckstuhl, L. (2000). Predicting marital stability and divorce in newlywed couples. *Journal of Family Psychology, 14*(1), 42–58.

Carrier, J. M. (1986). Childhood cross-gender behavior and adult homosexuality. *Archives of Sexual Behavior, 15*, 89–93.

Carter, C. S. (1998). Neuroendocrine perspectives on social attachment and love. *Psychoneuroendocrinology, 23*(8), 779–813.

Cartwright, R. D., et al. (1983). The traditional–liberated woman dimension: Social stereotype and self-concept. *Journal of Personality and Social Psychology, 44*, 581–588.

Carvajal, D. (1995). Oldest profession's newest home. *The New York Times*, pp. L29, L32.

Cash, T. F., & Duncan, N. C. (1984). Physical attractiveness stereotyping among black American college students. *Journal of Social Psychology, 122*, 71–77.

Castro, J. (1992, January 20). Sexual harassment: A guide. *Time*, p. 37.

Castro, K. G., et al. (1988). Transmission of HIV in Belle Glade, Florida: Lessons for other communities in the United States. *Science, 239*, 193–197.

Catalan, J., Hawton, K., & Day, A. (1990). Couples referred to a sexual dysfunction clinic: Psychological and physical morbidity. *British Journal of Psychiatry, 156*, 61–67.

Catania, J. A., et al. (1991). Changes in condom use among homosexual men in San Francisco. *Health Psychology, 10*, 190–199.

Catania, J. A., et al. (1992). Prevalence of AIDS-related risk factors and condom use in the United States. *Science, 258*, 1101–1106.

Cates, W., Jr. (1998). Reproductive tract infections. In Hatcher, R. A., et al. (1998). *Contraceptive technology*, 17th rev. ed (pp. 179–210). New York: Ardent Media.

Cates, W., Jr., & Raymond, E. G. (1998) Vaginal spermicides. In Hatcher, R. A., et al. (1998). *Contraceptive technology*, 17th rev. ed. (pp. 357–370). New York: Ardent Media.

Cates, W., Jr., & Wasserheit, J. N. (1991). Genital chlamydial infections: Epidemiology and reproductive sequelae. *American Journal of Obstetrics and Gynecology, 164*, 1171–1181.

CBS News. (1991, May 22). 48 hours: For better or worse.

CDC (See Centers for Disease Control and Prevention).

CDC tackles surge in sexual diseases. (1998, March 24). The Associated Press online.

Celis, W. (1991, January 2). Students trying to draw line between sex and an assault. *The New York Times*, pp. 1, B8.

Centers for Disease Control and Prevention. (2000a). Alcohol policy and sexually transmitted disease rates—United States, 1981–1995. *Morbidity and Mortality Weekly Report, 49*, 346–349.

Centers for Disease Control and Prevention. (2000b). *HIV/AIDS surveillance report: U.S. HIV and AIDS cases reported through December 1999, 11*(2).

Centers for Disease Control and Prevention. (2000c, June 23). Contribution of assisted reproductive technology and ovulation-inducing drugs to triplet and higher-order multiple births—United States, 1980–1997. *Morbidity and Mortality Weekly Report, 49*, 535–538.

Centers for Disease Control and Prevention. (2000d, June 23). Trends in gonorrhea rates—Selected states and United States, 1998. *Morbidity and Mortality Weekly Report.*

Centers for Disease Control and Prevention. (2000e, June 9). Youth risk behavior surveillance—United States, 1999. *Morbidity and Mortality Weekly Report, 49*(SS05); 1–96.

Centers for Disease Control and Prevention. (2000f). National and state-specific pregnancy rates among adolescents—United States, 1995B1997. *Morbidity and Mortality Weekly Report, 49*(27).

Centers for Disease Control and Prevention. (2000g). Abortion surveillance: Preliminary analysis—United States, 1997. *Morbidity and Mortality Weekly Report, 48*, 1171–1174, 1191.

Centers for Disease Control and Prevention. (2000h, June 9). Youth risk behavior surveillance—United States, 1999. *Morbidity and Mortality Weekly Report, 49*(SS05); 1–96.

The cervical cancer virus. (1995, September). *Discover*, pp. 24–26.

Chadda, R. K., & Ahuja, N. (1990). Dhat syndrome: A sex neurosis of the Indian subcontinent. *British Journal of Psychiatry, 156*, 577–579.

Chambers, C. (2000, October 13). Americans are overwhelmingly happy and optimistic about the future of the U.S. Marital status strongly affects both happiness and optimism. Princeton, NJ: Gallup News Service.

Chan, C. (1992). Cultural considerations in counseling Asian American lesbians and gay men. In S. Dworkin & F. Gutierrez (eds.), *Counseling gay men and lesbians: Journey to the end of the rainbow.* Alexandria, VA: American Association for Counseling and Development.

Charny, I. W., & Parnass, S. (1995). The impact of extramarital relationships on the continuation of marriages. *Journal of Sex and Marital Therapy, 21*, 100–115.

Cheasty, M. (1998). Cited in "Sexually abused women prone to divorce as adults." (1998, January 15). Reuters News Agency online.

Cheating going out of style but sex is popular as ever. (1993, October 19). *Newsday*, p. 2.

Check, J. V. P., & Malamuth, N. M. (1983). Sex-role stereotyping and reactions to depictions of stranger versus acquaintance rape. *Journal of Personality and Social Psychology, 45*, 344–356.

Chesney, M. A. (1996). Cited in Freiberg, P. (1996). New drugs give hope to AIDS patients. *APA Monitor, 27*(6), 28.

Chideya, F., et al. (1993, August 30). Endangered family. *Newsweek*, pp. 17–27.

China develops electronic contraceptive for men. (1998, March 15). Reuters News Agency online.

Chira, S. (1992, February 12). Bias against girls is found rife in schools, with lasting damage. *The New York Times*, pp. A1, 3.

Chivers, M. L., & Bailey, J. M. (2000). Sexual orientation of female-to-male transsexuals: A comparison of homosexual and nonhomosexual types. *Archives of Sexual Behavior, 29*(3), 259–278.

Chlebowski, R. T. (2000). Primary care: Reducing the risk of breast cancer. *The New England Journal of Medicine* online, 343(3).

Choi, P. Y. (1992). The psychological benefits of physical exercise: Implications for women and the menstrual cycle. [Special issue: The menstrual cycle]. *Journal of Reproductive and Infant Psychology, 10*, 111–115.

Christensen, C. (1995). Prescribed masturbation in sex therapy: A critique. *Journal of Sex and Marital Therapy, 21*(2), 87–99

Christin-Maitre, S., Bouchard, P., & Spitz, I. M. (2000). Drug therapy: Medical termination of pregnancy. *The New England Journal of Medicine* online, 342(13).

Clapper, R. L., & Lipsitt, L. P. (1991). A retrospective study of risk-taking and alcohol-mediated unprotected intercourse. *Journal of Substance Abuse, 3*, 91–96.

Clark, M. S., Mills, J. R., & Corcoran, D. M. (1989). Keeping track of needs and inputs of friends and strangers. *Personality and Social Psychology Bulletin, 15*, 533–542.

Clay, R. A. (1996a). Beating the "biological clock" with zest. *APA Monitor, 27*(2), 37.

Clay, R. A. (1996b). Older men are more involved fathers, studies show. *APA Monitor, 27*(2), 37.

Clumeck, N. (1995). Primary prophylaxis against opportunistic infections in patients with AIDS. *The New England Journal of Medicine, 332*, 739–740.

Cnattingius, S., Bergstrom, R., Lipworth, L., & Kramer, M. S. (1998). Prepregnancy weight and the risk of adverse pregnancy outcomes. *The New England Journal of Medicine, 338*, 147–152.

Cobb, J. (1996, January). The genes of 1995. *Discover*, p. 36.

Cochran, W. G., Mosteller, F., & Tukey, J. W. (1953). Statistical problems of the Kinsey Report. *Journal of the American Statistical Association, 48*, 673–716.

Cohen, E. (1998, March 26). Study: "Male pill" proves 95% effective. CNN.

Cohen, J. (1993). AIDS research: The mood is uncertain. *Science, 260*, 1254–1255.

Cohen, J. B. (1990, December 13). *A crosscutting perspective on the epidemiology of HIV infection in women.* Paper presented at the Women and AIDS Conference, Washington, DC (abstract).

Cohen, M. S. (2000). Preventing sexual transmission of HIV—New ideas from sub-Saharan Africa. *The New England Journal of Medicine, 342*(13), 970–973.

Cohen, R. (1996). Cited in Clay, R. A. (1996). Beating the "biological clock" with zest. *APA Monitor, 27*(2), 37.

Cohn, L. D., Macfarlane, S., Yanez, C., & Imai, W. K. (1995). Risk-perception: Differences between adolescents and adults. *Health Psychology, 14*, 217–222.

Colapinto, J. (2000). *As nature made him: The boy who was raised as a girl.* New York: HarperCollins.

Colditz, G. A., et al. (1993). Family history, age, and risk of breast cancer: Prospective data from the nurses' health study. *Journal of the American Medical Association, 270*, 338–343.

Cole, C. J. (1995, August). Cited in "Sex, the single lizard, and the missing parent." *National Geographic.*

Cole, C. L., & Cole, A. L. (1999). Marriage enrichment and prevention really works: Interpersonal competence training to maintain and enhance relationships. *Family Relations: Interdisciplinary Journal of Applied Family Studies, 48*(3), 273–275.

Cole, F. S. (2000). Extremely preterm birth—Defining the limits of hope. *The New England Journal of Medicine* online, 343(6).

Cole, H. M. (ed.). (1989). Intrauterine devices. *Journal of the American Medical Association, 261*, 2127–2130.

Cole, S. S. (1988). Women's sexuality, and disabilities. *Women and Therapy, 7*, 277–294.

Coleman, E. (1989). The development of male prostitution activity among gay and bisexual adolescents [Special Issue: Gay and lesbian youth]. *Journal of Homosexuality, 17*, 131–149.

Coleman, M., & Ganong, L. H. (1985). Love and sex role stereotypes: Do macho men and feminine women make better lovers? *Journal of Personality and Social Psychology, 49*, 170–176.

Coles, C. (1994). Critical periods for prenatal alcohol exposure: Evidence from animal and human studies. *Alcohol Health and Research World, 18*(1), 22–29.

Coles, R., & Stokes, G. (1985). *Sex and the American teenager.* New York: Harper & Row.

Collaer, M. L., & Hines, M. (1995). Human behavioral sex differences: A role for gonadal hormones during early development? *Psychological Bulletin, 118*, 55–107.

Collins, N. L., & Miller, L. C. (1994). Self-disclosure and liking: A meta-analytic review. *Psychological Bulletin, 116*, 457–475.

Conde-Agudelo, A., & Belizán, J. M. (2000). Maternal morbidity and mortality associated with interpregnancy interval: Cross sectional study. *British Medical Journal, 321*, 1255–1259.

Condon, J. W., & Crano, W. D. (1988). Inferred evaluation and the relation between attitude similarity and interpersonal attraction. *Journal of Personality and Social Psychology, 54*, 789–797.

Connor, E. M., et al. (1994). Reduction of maternal–infant transmission of human immunodeficiency virus type 1 with zidovudine treatment. *The New England Journal of Medicine, 331*, 1173–1180.

Constantine, L., & Constantine, J. (1973). *Group marriage.* New York: Macmillan.

Consumer's Union (1995, May). Consumer reports: How reliable are condoms? America Online.

Cooper, A., Delmonico, D. L., & Burg, R. (2000). Cybersex users, abusers, and compulsives: New findings and implications. *Sexual Addiction & Compulsivity*, 7(1–2), 5–29.

Cooper, A., Scherer, C. R., Boies, S. C., & Gordon, B. L. (1999). Sexuality on the Internet: From sexual exploration to pathological expression. *Professional Psychology: Research & Practice, 30*(2), 154–164.

Cooper, A. J., et al. (1990). A female sex offender with multiple paraphilias: A psychologic, physiologic (laboratory sexual arousal) and endocrine case study. *Canadian Journal of Psychiatry, 35*, 334–337.

Corey, L., & Holmes, K. K. (1996). Therapy for HIV infection—What have we learned? *The New England Journal of Medicine, 335*, 1142–14.

Cosgray, R. E., et al. (1991). Death from auto-erotic asphyxiation in a long-term psychiatric setting. *Perspectives in Psychiatric Care, 27*, 21–24.

Cotten-Huston, A. L., & Waite, B. M. (2000). Anti-homosexual attitudes in college students: Predictors and classroom interventions. *Journal of Homosexuality, 38*(3), 117–133.

Court, 9–0, makes sex harassment easier to prove. (1993, November 10). *The New York Times*, pp. A1, A22.

Courtenay, W. H. (2000). Engendering health: A social constructionist examination of men's health beliefs and behaviors. *Psychology of Men & Masculinity, 1*(1), 4–15.

Cowley, G., & Rogers, A. (1997, November 17). Rebuilding the male machine. *Newsweek*, pp. 66–67.

Cox, C. L., Wexler, M. O., Rusbult, C. E., & Gaines, S. O., Jr. (1997). Prescriptive support and commitment processes in close relationships. *Social Psychology Quarterly, 60*(1), 79–90.

Cox, D. J. (1988). Incidence and nature of male genital exposure behavior as reported by college women. *Journal of Sex Research, 24*, 227–234.

Craig, M. E., Kalichman, S. C., & Follingstad, D. R. (1989). Verbal coercive sexual behavior among college students. *Archives of Sexual Behavior, 18*, 421–434.

Cramer, D., et al. (1998, January). *The Lancet*. Cited in Associated Press (1998, January 9). Painkiller may lower cancer risk. America Online.

Crews, D. (1994). Animal sexuality. *Scientific American, 270*(1), 108–114.

Cronin, A. (1993, June 27). Two viewfinders, two views of Gay America. *The New York Times*, Section 4, p. 10.

Crossette, B. (1998, March 23). Mutilation seen as risk for the girls of immigrants. *The New York Times*, p. A3.

Crossette, B. (2000, August 29). Researchers raise fresh issues in breast-feeding debate. *The New York Times* online.

Crowe, L. C., & George, W. H. (1989). Alcohol and human sexuality: Review and integration. *Psychological Bulletin, 105*, 374–386.

Crum, C., & Ellner, P. (1985). Chlamydia infections: Making the diagnosis. *Contemporary Obstetrics and Gynecology, 25*, 153–159 ff.

Cummings, S. R., et al. (1999). The effect of raloxifene on risk of breast cancer in postmenopausal women: Results from the MORE randomized trial. *Journal of the American Medical Association, 281*, 2189–2197.

Cunningham, G., Cordero, E., & Thornby, J. (1989). Testosterone replacement with transdermal therapeutic systems. *Journal of the American Medical Association, 261*, 2525–2531.

Cunningham, M. R., et al. (1995). "Their ideas of beauty are, on the whole, the same as ours": Consistency and variability in the cross-cultural perception of female physical attractiveness. *Journal of Personality and Social Psychology, 68*(2) 261–279.

Curtis, R. C., & Miller, K. (1986). Believing another likes or dislikes you: Behavior making the beliefs come true. *Journal Personality and Social Psychology, 51*, 284–290.

Cutler, W. B. (1999). Human sex-attractant hormones: Discovery, research, development, and application in sex therapy. *Psychiatric Annals, 29*(1), 54–59.

Cutler, W. B., Friedmann, E., & McCoy, N. L. (1998). Pheromonal influences on sociosexual behavior in men. *Archives of Sexual Behavior, 27*(1), 1–13.

D

Dabbs, J. M., Jr., & Morris, R. (1990). Testosterone, social class, and antisocial behavior in a sample of 4,462 men. *Psychological Science, 1*, 1–3.

Daly, M., & Wilson, M. (1998). Cited in Brody, J. E. (1998, February 10). Genetic ties may be factor in violence in stepfamilies. *The New York Times*, pp. F1, F4.

D'Amico, A. V., et al. (2000). Biochemical outcome following external beam radiation therapy with or without androgen suppression therapy for clinically localized prostate cancer. *Journal of the American Medicaal Association, 284*, 1280–1283.

Danner, S. A., et al. (1995). A short-term study of the safety, pharmacokinetics, and efficacy of ritonavir, an inhibitor of HIV-1 protease. *The New England Journal of Medicine, 333*, 1528–1533.

Darling, C. A., & Davidson, J. K., Sr. (1986). Coitally active university students: Sexual behaviors, concerns, and challenges. *Adolescence, 21*, 403–419.12.

Darling, C. A., Davidson, J. K., & Jennings, D. A. (1991). The female sexual response revisited: Understanding the multiorgasmic experience in women. *Archives of Sexual Behavior, 20*, 527–540.

Darling, C. A., Davidson, J. K., & Passarello, L. C. (1992). The mystique of first intercourse among college youth: The role of partner contraceptive practices, and psychological reactions. *Journal of Youth and Adolescence, 21*, 97–117.

D'Augelli, A. R. (1992a). Lesbian and gay male undergraduates' experiences of harassment and fear on campus. *Journal of Interpersonal Violence, 7*, 383–395.

D'Augelli, A. R. (1992b). Sexual behavior patterns of gay university men: Implications for preventing HIV infection. *Journal of American College Health, 41*, 25–29.

Davenport, W. (1965). Sexual patterns and their regulation in a society of the Southwest Pacific. In F. A. Beach (ed.), *Sex and behavior.* New York: Wiley.

Davenport, W. (1977). Sex in cross-cultural perspective. In F. Beach (ed.), *Human sexuality in four perspectives.* (pp. 115–163). Baltimore, MD: Johns Hopkins University Press.

Davidson, J. K., & Hoffman, L. E. (1986). Sexual fantasies and sexual satisfaction: An empirical analysis of erotic thought. *Journal of Sex Research, 22*, 184–205.

Davidson, K. J., Darling, C., & Conway-Welch, C. (1989). The role of the Grafenberg spot and female ejaculation in the female orgasmic response: An empirical analysis. *Journal of Sex and Marital Therapy, 15*, 10–119.

Davidson, N. E. (1995). Hormone-replacement therapy–Breast versus heart versus bone. *The New England Journal of Medicine, 332*, 1638–1639.

Davis, A. (2000). Book review: A clinician's guide to medical and surgical abortion. *The New England Journal of Medicine* online, 343(3).

Davison, G. C. (1977). Elimination of a sadistic fantasy by a client-controlled counterconditioning technique. In J. Fischer & H. Gochios (eds.), *Handbook of behavior therapy with sexual problems.* New York: Pergamon Press.

Dawes, R. M. (1989). Statistical criteria for establishing a truly false consensus effect. *Journal of Experimental Social Psychology 25*, 1–17.

Dawood, K., Pillard, R. C., Horvath, C., Revelle, W., & Bailey, J. M. (2000). Familial aspects of male homosexuality. *Archives of Sexual Behavior, 29*(2), 155–163.

Day, N. L., & Richardson, G. A. (1994). Comparative teratogenicity of alcohol and other drugs. *Alcohol Health and Research World, 18* (1), 42–48.

Dean, C. W., & de Bruyn-Kops, E. (1982). *The crime and consequences of rape.* Springfield, IL: Charles C. Thomas.

DeAngelis, T. (1994). Educators reveal keys to success in classroom. *APA Monitor, 25*(1), 39–40.

Deaux, K. (1985). Sex and gender. *Annual Review of Psychology, 36*, 49–81.

Deaux, K., & Lewis, L. L. (1983). Assessment of gender stereotypes: Methodology and components. *Psychological Documents, 13*, 25 (Ms. No. 2583).

Deepening shame: A *Newsweek* investigation into the scandal that is rocking the navy. (1992, August 10). *Newsweek*, pp. 30–36.

Dekker, J. (1993). Inhibited male orgasm. In W. O'Donohue & J. H. Geer (eds.), *Handbook of sexual dysfunctions: Assessment and treatment* (pp. 279–301). Boston: Allyn & Bacon.

Delgado, J. (1969). Physical control of the mind. New York: Harper & Row.

Delmas, P. D., et al. (1997). Effects of raloxifene on bone mineral density, serum cholesterol concentrations, and uterine endometrium in postmenopausal women. *The New England Journal of Medicine, 337*, 1641–1648.

Del Priore, G., et al. (1995). Risk of ovarian cancer after treatment for infertility. *The New England Journal of Medicine, 332*, 1300.

de Luca, R. V., et al. (1992). Group treatment for child sexual abuse. [Special Issue: Violence and its aftermath]. *Canadian Psychology, 33*, 168–179.

Dennerstein, L., et al. (2000). A prospective population-based study of menopausal symptoms. *Obstetrics & Gynecology, 96*(3), 351–358.

Dennerstein, L., Dudley, E. C., Hopper, J. L., Guthrie, J. R., & Burger, H. G. (2000). A prospective population-based study of menopausal symptoms. *Obstetrics & Gynecology, 96*(3), 351-58.

Denny, N., Field, J., & Quadagno, D. (1984). Sex differences in sexual needs and desires. *Archives of Sexual Behavior, 13,* 233–245.

den Tonkelaar, I., & Oddens, B. J. (2000). Determinants of long-term hormone replacement therapy and reasons for early discontinuation. *Obstetrics & Gynecology, 95*(4), 507–512.

DeParle, J. (1993, July 14). Big rise in birth outside wedlock. *The New York Times,* pp. A1, A14.

de Raad, B., & Doddema–Winsemius, M. (1992). Factors in the assortment of human mates: Differential preferences in Germany and the Netherlands. *Personality and Individual Differences, 13,* 103–114.

Derby, C. A. (2000, October 2). Cited in "Study finds exercise reduces the risk of impotence." The Associated Press.

de Schampheleire, D. (1990). MMPI characteristics of professional prostitutes: A cross–cultural replication. *Journal of Personality Assessment, 54,* 343–350.

Desmond, A. M. (1994). Adolescent pregnancy in the United States: Not a minority issue. *Health Care for Women International, 15*(4), 325–331.

Deutsch, F. M., LeBaron, D., & Fryer, M. M. (1987). What is in a smile? *Psychology of Women Quarterly, 11,* 341–352.

DeWitt, K. (1991, October). The evolving concept of sexual harassment. *The New York Times.*

de Young, M. (1982). *The sexual victimization of children.* Jefferson, NC: McFarland & Company.

Diamond, M. (1993). Homosexuality and bisexuality in different populations. *Archives of Sexual Behavior, 22,* 291–310.

Diamond, M. (1996). Prenatal predisposition and the clinical management of some pediatric conditions. *Journal of Sex & Marital Therapy, 22*(3), 139–147.

Diaz–Mitoma, F., et al. (1998, September 9). Oral famciclovir for the suppression of recurrent genital herpes: A randomized controlled trial. *Journal of the American Medical Association* online.

Dickson, N., Paul, C., Herbison, P., & Silva, P. (1998). First sexual intercourse: Age, coercion, and later regrets reported by a birth cohort. *British Medical Journal, 316,* 29–33.

Dick–Read, G. (1944). *Childbirth without fear: The principles and practices of natural childbirth.* New York: Harper & Bros.

DiClemente, R. J. (1992). Epidemiology of AIDS, HIV prevalence, and HIV incidence among adolescents. *Journal of School Health, 62,* 325–330.

Dietz, P. E., Hazelwood, R. R., & Warren, J. (1990). The sexually sadistic criminal and his offenses. *Bulletin of the American Academy of Psychiatry and the Law, 18,* 163–178.

DiMatteo, M. R., et al. (1996). Cesarean childbirth and psychosocial outcomes: A meta-analysis. *Health Psychology, 15,* 303–314.

Dindia, K., & Allen, M. (1992). Sex differences in self-disclosure: A meta-analysis. *Psychological Bulletin, 112,* 106–124.

Disease discovered Europe first? (1992, November 24). *New York Newsday,* p. 49.

Dixon, J. (1991). Feminist reforms of sexual coercion. In E. Grauerholz & M. A. Koralewski (eds.), *Sexual coercion: A sourcebook on its nature, causes, and prevention* (pp. 161–171). Lexington, MA: Lexington Books.

Donnerstein, E. (1980). Aggressive erotica and violence against women. *Journal of Personality and Social Psychology, 39,* 269–277.

Donnerstein, E., & Berkowitz, L. (1981). Victim reactions in aggressive erotic films as a factor in violence against women. *Journal of Personality and Social Psychology, 41,* 710–724.

Donnerstein, E. I., & Linz, D. G. (1987). *The question of pornography.* New York: The Free Press.

Donoghue, D. (1995, August 20). The politics of homosexuality. *The New York Times Book Review,* pp. 3, 20.

Donovan, C. (2000, May 30). Confronting teen pregnancy. *The Washington Post,* p. Z17.

Doorn, C. D., Poortinga, J., & Verschoor, A. M. (1994). Cross-gender identity in transvestites and male transsexuals *Archives of Sexual Behavior, 23*(2), 185–201.

Douglas, J. M., Jr. (1990). *Molluscum contagiosum.* In K. K. Holmes, P. Mardh, P. F. Sparling, & P. J. Wiesner (eds.), *Sexually transmitted diseases,* 2nd ed. (pp. 443–448). New York: McGraw-Hill

Drapers, P. (1975). !Kung women: Contrasts in sexual egalitarianism in foraging and sedentary contexts. In R. R. Reiter (ed.), *Toward an anthropology of women* (pp. 77–109). New York: Monthly Review Press.

Drew, W. L. (2000). Ganciclovir resistance: A matter of time and titre. *Lancet, 356,* 609–610.

Drews, C. D., et al. (1996, April). *Pediatrics.* Cited in "Smokers more likely to bear retarded babies, study says." (1996, April 10). *The New York Times,* p. B7.

Dreyfuss, I. A. (1998. February 8.) Exercise may cut breast cancer risk. Associated Press online.

Drigotas, S. M., Rusbult, C. E., & Verette, J. (1999a). Level of commitment, mutuality of commitment, and couple well-being. *Personal Relationships, 6*(3) 389–409.

Drigotas, S. M., Rusbult, C. E., Wieselquist, J., & Whitton, S. W. (1999b). Close partner as sculptor of the ideal self: Behavioral affirmation and the Michelangelo phenomenon. *Journal of Personality & Social Psychology,* 77(2), 293–323.

Dube, R., & Hebert, M. (1988). Sexual abuse of children 12 years of age: A review of 511 cases. *Child Abuse and Neglect, 12,* 321–330.

Dugger, C. W. (1996, September 11). A refugee's body is intact but her family is torn. *The New York Times,* pp. A1, B6.

Dunlap, D. W. (1995, May 1). Support for gay adoptions seems to wane. *The New York Times,* p. A13.

Dunn, M. E. (1998). Cited in Leary, W. E. (1998, September 29). Older people enjoy sex, survey says. *The New York Times,* p. F8.

Dunn, M. E., & Trost, J. E. (1989). Male multiple orgasms: A descriptive study. *Archives of Sexual Behavior, 18,* 377–387.

Durfee, M. (1989). Prevention of child sexual abuse. *Psychiatric Clinics of North America, 12,* 445–453.

Durik, A. M., Hyde, J. S., & Clark, R. (2000). Sequelae of cesarean and vaginal deliveries: Psychosocial outcomes for mothers and infants. *Developmental Psychology, 36,* 2, 251–260.

Dwyer, M. (1988). Exhibitionism/voyeurism. *Journal of Social Work and Human Sexuality,* 7, 101–112.

Dziech, B. W., & Weiner, L. (1984). *The lecherous professor: Sexual harassment on campus.* Boston: Beacon Press.

E

Early prostate surgery is found very effective. (1996, August 28). *The New York Times,* p. C9.

Eason, E., & Feldman, P. (2000). Much ado about a little cut: Is episiotomy worthwhile? *Obstetrics & Gynecology, 95*(4), 616–618.

Eason, E., Labrecque, M., Wells, G., & Feldman, P. (2000). Preventing perineal trauma during childbirth: A systematic review. *Obstetrics Gynecology, 95,* 464–471.

Edgley, C. (1989). Commercial sex: Pornography, prostitution, and advertising. In K. McKinney & S. Sprecher (eds.), *Human sexuality: The societal and interpersonal context* (pp. 370–424). Norwood, NJ: Ablex.

Edlin, B. R., et al. (1994). Intersecting epidemics: Crack cocaine use and HIV infection among inner–city young adults. *The New England Journal of Medicine, 331,* 1422–1427.

Edmonson, B. (1988). Disability and sexual adjustment. In V. B. Van Hasselt, P. S. Strain, & M. Hersen (eds.), *Handbook of developmental and physical disabilities* (pp. 91–106). New York: Pergamon Press.

Edwards, T. M. (2000, August 28). Single by choice. *Time Magazine* online, 156 (9).

Eggers, D. (2000, May 7). Intimacies. *The New York Times Magazine,* pp. 76–77.

Ehrhardt, A. A. (1998). Cited in Bronner, E. (1998, February 1). Just say maybe. No sexology, please. We're Americans." *The New York Times,* p. WK6.

Eisen, M., & Zellman, G. (1987). Changes in incidence of sexual intercourse of unmarried teenagers following a community-based sex education program. *The Journal of Sex Research, 23,* 527–544

Ellertson, C. (2000). Nuisance or natural and healthy? Should monthly menstruation be optional for women? *The Lancet, 355,* 922.

Ellis, A. (1962). *Reason and emotion in psychotherapy.* New York: Lyle Stuart.

Ellis, A. (1977). The basic clinical theory of rational-emotive therapy. In A. Ellis & R. Grieger (eds.), *Handbook of rational-emotive therapy.* New York: Springer.

Ellis, E. M. (2000). *Divorce wars: Interventions with families in conflict.* Washington, DC: American Psychological Association.

Ellis, L., & Ames, M. A. (1987). Neurohormonal functioning and sexual orientation: A theory of homosexuality–heterosexuality. *Psychological Bulletin, 101,* 233–258.

Ellis, L., Burke, D., & Ames, M. A. (1987). Sexual orientation as a continuous variable: A comparison between the sexes. *Archives of Sexual Behavior, 16,* 523–529.

English, P. B., & Eskenazi, B. (1992). Reinterpreting the effects of maternal smoking on infant birthweight and perinatal mortality: A multivariate approach to birthweight standardization. *International Journal of Epidemiology, 21,* 1097–1105.

Erel, O., & Burman, B. (1995). Interrelatedness of marital relations and parent–child relations: A meta-analytic review. *Psychological Bulletin, 118,* 108–132.

Erlanger, S. (1991, July 14). A plague awaits. *The New York Times Magazine,* pp. 24ff.

Eron, J. J., et al. (1995). Treatment with lamivudine, zidovudine, or both in HIV-positive patients with 200 to 500 CD4+ cells per cubic millimeter. *The New England Journal of Medicine, 333,* 1662–1669.

Eskenazi, G. (1990, June 3). The male athlete and sexual assault. *The New York Times,* pp. L1, L4.

"Estrogen may aid fertility in men." (1997, December 16). *The New York Times,* p. F9.

Etaugh, C., & Rathus, S. A. (1995). *The world of children.* Ft. Worth, TX: Harcourt.

Ethics Committee of the American Society of Reproductive Medicine. (2000, June 7). Sex selection and preimplantation genetic diagnosis. *Webtrack.*

Everaerd, W. (1993). Male erectile disorder. In W. O'Donohue & J. H. Geer (eds.), *Handbook of sexual dysfunctions: Assessment and treatment* (pp. 201–224). Boston: Allyn & Bacon.

Everitt, B. J. (1990). Sexual motivation: A neural and behavioural analysis of the mechanisms underlying appetitive and copulatory responses of male rats. *Neuroscience and Biobehavioral Reviews, 14,* 217–232.

Exner, J. E., et al. (1977). Some psychological characteristics of prostitutes. *Journal of Personality Assessment, 41,* 474–485.

F

Faller, K. C. (1989a). The role relationship between victim and perpetrator as a predictor of characteristics of intrafamilial sexual abuse. *Child and Adolescent Social Work Journal, 6,* 217–229.

Faller, K. C. (1989b). Why sexual abuse? An exploration of the intergenerational hypothesis. *Child Abuse and Neglect, 13,* 543–548.

Fallon, A. E., & Rozin, P. (1985). Sex differences in perceptions of desirable body shape. *Journal of Abnormal Psychology, 94,* 102–105.

Farley, A. U., Hadler, J. L., & Gunn, R. A. (1990). The syphilis epidemic in Connecticut: Relationship to drug use and prostitution. *Sexually Transmitted Diseases, 17,* 163–168.

Fay, R. E., et al. (1989). Prevalence and patterns of same-gender sexual contact among men. *Science, 243,* 338–348.

FDA allows large-scale AIDS vaccine trial. (1998, June 3). The Associated Press; CNN.

FDA approves abortion pill. (2000, September 28). Associated Press online.

FDA approves an AIDS drug that is taken just once a day. (1998, September 19). *The New York Times* online.

FDA is hit on "rapid" AIDS tests. (2000, June 14). *Los Angeles Times,* p. A20.

Federal Bureau of Investigation. (1991). *Uniform crime reports.* Washington, DC: U.S. Department of Justice.

Federman, D. D. (1994). Life without estrogen. *The New England Journal of Medicine, 331,* 1088–1089.

Feingold, A. (1991). Sex differences in the effects of similarity and physical attractiveness on opposite–sex attraction. *Basic and Applied Social Psychology, 12,* 357–367.

Feingold, A. (1994). Gender differences in personality: A meta-analysis. *Psychological Bulletin, 116,* 429–456.

Feldman, H. A., Goldstein, I., Hatzichristou, D. G., Krane, R. J., McKinlay, J. B. (1994). Impotence and its medical and psychosocial correlates: Results of the Massachusetts Male Aging Study. *Journal of Urology, 151*(1), 54–61.

Felsman, D., Brannigan, G., & Yellin, P. (1987). Control theory in dealing with adolescent sexuality and pregnancy. *Journal of Sex Education and Therapy, 13,* 15–16.

Feng, T. (1993). Substance abuse in pregnancy. *Current Opinions in Obstetrics and Gynecology, 5,* 16–23.

Fenichel, O. (1945). *The psychoanalytic theory of neurosis.* New York: Norton.

Fensterheim, H., & Kantor, J. S. (1980). Behavioral approach to sexual disorders. In B. Wolman & J. Money (eds.), *Handbook of human sexuality.* Englewood Cliffs, NJ: Prentice-Hall.

Ferguson, D. M., Horwood, L. J., & Beautrais, A. L. (1999) Is sexual orientation related to mental health problems and suicidality in young people? *Archives of General Psychiatry, 56*(10), 876–880.

Ferrante, C. L., Haynes, A. M., & Kingsley, S. M. (1988). Image of women in television advertising. *Journal of Broadcasting and Electronic Media, 32,* 231–237.

Finkelhor, D. (1984). *Child sexual abuse: Theory and research.* New York: The Free Press.

Finkelhor, D. (1990). Early and long-term effects of child sexual abuse: An update. *Professional Psychology: Research and Practice, 21,* 325–330.

Finkelhor, D., & Hotaling, G. T. (1984). Sexual abuse in the National Incidence Study of Child Abuse and Neglect: An appraisal. *Child Abuse and Neglect, 8,* 22–33.

Finkelhor, D., & Russell, D. (1984). Women as perpetrators: Review of the evidence. In D. Finkelhor (ed.), *Child sexual abuse: Theory and research.* New York: The Free Press.

Finkelhor, D., & Yllo, K. (1982). Rape in marriage: A sociological view. In D. Finkelhor, R. J. Gelles, G. T. Hotaling, & M. A. Straus (eds.), *The dark side of families: Current family violence research* (pp. 119–130). Beverly Hills, CA: Sage.

Finkelhor, D., et al. (1990). Sexual abuse in a national survey of adult men and women: Prevalence, characteristics, and risk factors. *Child Abuse and Neglect, 14,* 19–28.

Finkenauer, C., & Hazam, H. (2000). Disclosure and secrecy in marriage: Do both contribute to marital satisfaction? *Journal of Social & Personal Relationships, 17*(2), 245–263.

Fish, L. S., Busby, D., & Killian, K. (1994). Structural couple therapy in the treatment of inhibited sexual drive. *American Journal of Family Therapy, 22*(2), 113–125.

Fishel, S., et al. (2000). Should ICSI be the treatment of choice for all cases of in-vitro conception? *Human Reproduction, 15*(6), 1278–1283.

Fisher, B., Weisberg, D. K., & Marotta, T. (1982). *Report on adolescent male prostitution.* San Francisco: Urban and Rural Systems Associates.

Fisher, H. E. (2000). Brains do it: Lust, attraction and attachment. *Cerebrum, 2,* 23–42.

Fisher, J. (1992, January 11). Cited in Associated Press, Boy beaver foils experts. *The New York Times,* p. 6.

Fisher, J. D., & Misovich, S. J. (1991). *1990 technical report on undergraduate students' AIDS-preventive behavior, AIDS-knowledge, and fear of AIDS.* Storrs, CT: University of Connecticut, Department of Psychology.

Fisher, T. D., & Hall, R. H. (1988). A scale for the comparison of the sexual attitudes of adolescents and their parents. *Journal Sex Research, 24,* 90–100.

Fisher, W. A., Fisher, J. D., & Rye, B. J. (1995). Understanding and promoting AIDS-preventive behavior: Insights from the theory of reasoned action. *Health Psychology, 14,* 255–264.

Fisher–Thompson, D. (1990). Adult sex typing of children's toys. *Sex Roles, 23,* 291–303.

Fitzgerald, L. F. (1993a). *Sexual harassment in higher education: Concepts and issues.* Washington, DC: National Education Association.

Fitzgerald, L. F. (1993b). Sexual harassment: Violence against women in the workplace. *American Psychologist, 48,* 1070–1076.

Fitzgerald, L. F., et al. (1988). Academic harassment: Sex and denial in scholarly garb. *Psychology of Women Quarterly, 12,* 329–340.

Flaxman, S. M., & Sherman, P. W. (2000). Morning sickness: A mechanism for protecting mother and embryo. *The Quarterly Review of Biology, 5*(2), 113–148.

Fletcher, J. (1966). *Situation ethics.* Philadelphia: Westminster Press.

Fletcher, J. (1967). *Moral responsibility: Situation ethics at work.* Philadelphia: Westminster Press.

Fletcher, J. M., et al. (1991). Neurobehavioral outcomes in diseases of childhood: Individual change models for pediatric human immunodeficiency viruses. *American Psychologist, 46,* 267–1277.

Flexner, C. (1998). Drug therapy: HIV-protease inhibitors. *The New England Journal of Medicine, 338,* 1281–1292.

Floyd, R. L., Rimer, B. K., Giovino, G. A., Mullen, P. D., & Sullivan, S. E. (1993). A review of smoking in pregnancy: Effects pregnancy outcomes and cessation efforts. *Annual Review of Public Health, 14,* 379–411.

Ford, C. S., & Beach, F. A. (1951). *Patterns of sexual behavior.* New York: Harper & Row.

Forgas, J. P., Levinger, G., & Moylan, S. J. (1994). Feeling good and feeling close: Affective influences on the perception of intimate relationships. *Personal Relationships, 1*(2), 165–184.

Forrest, J. D., & Fordyce, R. R. (1993). Women's contraceptive attitudes and use in 1992. *Family Planning Perspectives, 25,* 175–179.

Forster, P., & King, J. (1994). Fluoxetine for premature ejaculation. *American Journal of Psychiatry, 151*(10), 1523.

Frable, D. E. S., Johnson, A. E., & Kellman, H. (1997). Seeing masculine men, sexy women, and gender differences: Exposure to pornography and cognitive constructions of gender. *Journal Personality, 65*(2), 311–355.

Frank, E., & Wenger, N. K. (1997, December 15). Women doctors in menopause are twice as likely as other women to use hormone-replacement therapy. *Emory University's Annals of Internal Medicine.*

Franzoi, S. L., & Herzog, M. E. (1987). Judging physical attractiveness: What body aspects do we use? *Personality and Social Psychology Bulletin, 13,* 19–33.

Fraser, A. M., Brockert, J. E., & Ward, R. H. (1995). Association of young maternal age with adverse reproductive outcomes. *The New England Journal of Medicine, 332,* 1113–1117.

Frayser, S. (1985). *Varieties of sexual experience: An anthropological perspective on human sexuality.* New Haven, CT: Human Relations Area Files Press.

Frazier, P. A. (1990). Victim attributions and post–rape trauma. *Journal of Personality and Social Psychology, 59,* 298–304.

Freiberg, P. (1995). Psychologists examine attacks on homosexuals. *APA Monitor, 26*(6), 30–31.

Freiberg, P. (1996). New drugs give hope to AIDS patients. *APA Monitor, 27*(6), 28.

Freud, S. (1922/1959). Analysis of a phobia in a 5-year-old boy. In A. & J. Strachey (ed. & trans.), *Collected papers,* Vol. 3. New York: Basic Books. (Original work published 1909.)

Freund, K., & Blanchard, R. (1986). The concept of courtship disorder. *Journal of Sex and Marital Therapy, 12,* 79–92.

Freund, K., Watson, R., & Rienzo, D. (1988). The value of self-reports in the study of voyeurism and exhibitionism. *Annals of Sex Research 1,* 243–262.

Freund, M., Leonard, T. L., & Lee, N. (1989). Sexual behavior of resident street prostitutes with their clients in Camden, New Jersey. *Journal of Sex Research, 26,* 460–478.

Frey, K. S., & Ruble, D. N. (1992). Gender constancy and the "cost" of sex–typed behavior: A test of the conflict hypothesis. *Developmental Psychology, 28,* 714–721.

Friday, N. (1973). *My secret garden.* New York: Trident.

Fried, P. A. (1986). Marijuana use in pregnancy. In I. J. Chasnott (ed.), *Drug use in pregnancy: Mother and child.* Boston: MTP Press.

Friedman, L. C., et al. (1994). Dispositional optimism, self–efficacy, and health beliefs as predictors of breast self-examination. *American Journal of Preventive Medicine, 10*(3), 130–135.

Friedman, R. C., & Downey, J. I. (1994). Homosexuality. *The New England Journal of Medicine, 331,* 923–930.

Friedrich, W. N. (1998). Cited in Gilbert, S. (1998). New light shed on normal sexual behavior in a child. *The New York Times,* p. F7.

Friedrich, W. N., & Gerber, P. N. (1994). Autoerotic asphyxia: The development of a paraphilia. *Journal of the American Academy of Child and Adolescent Psychiatry, 33*(7), 970–974.

Frisch, R. (1997). Cited in Angier, N. (1997a). Chemical tied to fat control could help trigger puberty. *The New York Times,* pp. , C3.

Frodi, A. M., Macauley, J., & Thome, P. R. (1977). Are women always less aggressive than men? A review of the experimental literature. *Psychological Bulletin, 84,* 634–660.

Frost, M. H., et al. (2000). Long-term satisfaction and psychological and social function following bilateral prophylactic mastectomy. *Journal of the American Medical Association, 284,* 319–324

Fuchs, C. S., et al. (1995). Alcohol consumption and mortality among women. *The New England Journal of Medicine, 332,* 1245–1250.

Fuentes-Afflick, E., & Hessol, N. A. (2000). Interpregnancy interval and the risk of premature infants. *Obstetrics & Gynecology, 95,* 383–390.

Fugger, E. F., et al. (1998, September 9). *Human Reproduction.* Cited in Kolata, G. (1998, September 9). Researchers report success in method to pick baby's sex. *The New York Times* online.

Fukuyama, F. (1999, June 9). At last, Japan gets the pill. Is this good news? *The New York Times,* p. A29.

Fuleihan, G. E. (1997). Tissue specific estrogens—The promise for the future. *The New England Journal of Medicine, 337,* 1686–1687.

Furnham, A., & Voli, V. (1989). Gender stereotypes in Italian television advertisements. *Journal of Broadcasting and Electronic Media, 33,* 175–185.

Furstenberg, F. F., Jr., Brooks-Gunn, J., Chase-Lansdale, L. (1989). Teenaged pregnancy and childbearing. *American Psychologist, 44,* 313–320.

G

Gabelnick, H. L., (1998). Future methods. In Hatcher, R. A., et al. (1998). *Contraceptive technology* (17th rev. ed.) (pp. 615–622). New York: Ardent Media.

Gabriel, T. (1995a, April 23). When one spouse is gay and a marriage unravels. *The New York Times,* pp. A1, A22.

Gabriel, T. (1995b, June 12). A new generation seems ready to give bisexuality a place in the spectrum. *The New York Times,* p. A12.

Gaddis, A., & Brooks-Gunn, J. (1985). The male experience of pubertal change. *Journal of Youth and Adolescence, 14,* 61.

Gagnon, J. H. (1977). *Human sexualities.* Glenview, IL: Scott, Foresman.

Gagnon, J. H. (1990). Gender preferences in erotic relations: The Kinsey scale and sexual scripts. In D. P. McWhirter, S. A. Sanders, & J. M. Reinisch (eds.), *Homosexuality/heterosexuality: Concepts of sexual orientation* (pp. 177–207). New York: Oxford University Press.

Gagnon, J. H., & Simon, W. (1973). *Sexual conduct: The social origins of human sexuality.* Chicago: Aldine.

Gagnon, J. H., & Simon, W. (1987). The sexual scripting of oral–genital contacts. *Archives of Sexual Behavior, 16,* 1–25.

Gallagher, A. M., et al. (2000). Gender differences in advanced mathematical problem solving. *Journal of Experimental Child Psychology, 75*(3), 165–190.

Gallant, J. E. (2000). Strategies for long–term success in the treatment of HIV infection. *Journal of the American Medical Association, 283,* 1329–1334.

Garber, M. (1995). *Vice versa.* New York: Simon & Schuster.

Gardiner, H. W., & Gardiner, O. S. (1991). Women in Thailand. In L. L. Adler (ed.), *Women in cross-cultural perspective* (pp. 75–187). New York: Praeger.

Garland, S. M., Lees, M. I., & Skurrie, I. J. (1990). *Chlamydia trachomatis:* Role in tubal infertility. *Australian and New Zealand Journal of Obstetrics and Gynaecology, 30,* 83–86.

Garnets, L., & Kimmel, D. (1991). In J. D. Goodchilds (ed.), *Psychological perspectives on human diversity in America.* Washington, DC: American Psychological Association.

Garwood, S. G., et al. (1980). Beauty is only "name deep": The effect of first name in ratings of physical attraction. *Journal of Applied Social Psychology, 10,* 431–435.

Gates, G. J., & Sonenstein, F. L. (2000). Heterosexual genital sexual activity among adolescent males: 1988 and 1995. *Family Planning Perspectives, 32*(6), 295–297, 304.

Gay, P. (1984). *The bourgeois experience: Victoria to Freud.* New York: Oxford University Press.

Gayle, H. (2000). Cited in Carrns, A. (2000, July 13). Levels of HIV infection, AIDS fail to decline in the U.S., data show. *The Wall Street Journal Interactive Edition.*

Gayle, H., Selik, R., & Ccu, S. (1990). Surveillance for AIDS and HIV infection among Black and Hispanic children and women of childbearing age, 1981–1989. *Morbidity and Mortality Weekly Report: Progress in Chronic Disease Prevention, 39,* 23–29. Washington, DC: U.S. Department of Health and Human Services.

Gebhard, P. H. (1969). Misconceptions about female prostitutes. *Medical Aspects of Human Sexuality, 3,* 24–26.

Gebhard, P. H. (1976). The institute. In M. S. Weinberg (ed.), *Sex research: Studies from the Kinsey Institute.* New York: Oord University Press.

Gebhard, P. H. (1977). *Memorandum on the incidence of homosexuals in the United States.* Bloomington, IN: Indiana University Institute for Sex Research.

Gebhard, P. H., et al. (1965). *Sex offenders: An analysis of types.* New York: Harper & Row.

Geer, J., Heiman, J., & Leitenberg, H. (1984). *Human sexuality.* Englewood Cliffs, NJ: Prentice-Hall.

Geiger, R. (1981). Neurophysiology of sexual response in spinal cord injury. In D. Bullard & S. Knight (eds.), *Sexuality and physical disability: Personal perspectives.* St. Louis: Mosby.

Gelles, R. J., & Cornell, C. P. (1985). *Intimate violence in families.* Beverly Hills, CA: Sage.

Gelman, D., with P. Kandell. (1993, January 18). Isn't it romantic? *Newsweek,* pp. 60–61.

Genuis, M., Thomlison, B., & Bagley, C. (1991, Fall). Male victims of child sexual abuse: A brief overview of pertinent findings. [Special Issue: Child sexual abuse]. *Journal of Child and Youth–Care,* 1–6.

George, W. H., Stoner, S. A., Norris, J., Lopez, P. A., & Lehman, G. L. (2000). Alcohol expectancies and sexuality: A self-fulfilling prophecy analysis of dyadic perceptions and behavior. *Journal of Studies on Alcohol, 61*(1), 168–176.

Geringer, W. M., et al. (1993). Knowledge, attitudes, and behavior related to condom use and STDs in a high-risk population. *Journal of Sex Research, 30,* 75–83.

Gerrard, M. (1987). Sex, sex guilt, and contraceptive use revisited: The 1980s. *Journal of Personality and Social Psychology, 52,* 975–980.

Gerrol, R., & Resick, P. A. (1988, November). *Sex differences in social support and recovery from victimization.* Paper presented at the meeting of the Association for Advancement of Behavior Therapy, New York, NY.

Gibbs, N. (1991, June 3). When is it rape? *Time,* pp. 48–54.

Gibson, A. I., et al. (1988). Adolescent female prostitutes. *Archives of Sexual Behavior, 17,* 431–438.

Gidycz, C. A., & Koss, M. P. (1990). A comparison of group and individual sexual assault victims. *Psychology of Women Quarterly, 14,* 325–342.

Gijs, L., & Gooren, L. (1996). Hormonal and psychopharmacological interventions in the treatment of paraphilias: An update. *Journal of Sex Research, 33,* 273–290.

Gilbert, S. (1996, September 25). No long-term link is found between pill and breast cancer. *The New York Times,* p. C9

Gillis, J. S., & Avis, W. E. (1980). The male-taller norm in mate selection. *Personality and Social Psychology Bulletin, 6,* 396–401.

Gilmartin, B. G. (September 1975). That swing ing couple down the block. *Psychology Today, 54.*

Giovannucci, E., et al. (1993a). A prospective cohort study of vasectomy and prostate cancer in U.S. men. *Journal of the American Medical Association, 269,* 873–877.

Giovannucci, E., et al. (1993b). A retrospective cohort study of vasectomy and prostate cancer in U.S. men. *Journal of the American Medical Association, 269,* 878–882.

Gitlin, M. J., & Pasnau, R. O. (1989). Psychiatric syndromes linked to reproductive function in women: A review of current knowledge. *American Journal of Psychiatry, 146,* 1413–1422.

Glass, S. P., & Wright, T. L. (1992). Justifications of extramarital relationships: The association between attitudes, behaviors, and gender. *Journal of Sex Research, 29,* 361–387.

Gleicher, N., Oleske, D. M., Tur-Kaspa, I., Vidali, A., & Karande, V. (2000). Reducing the risk of high-order multiple pregnancy after ovarian stimulation with gonadotropins. *The New England Journal of Medicine, 343*(1), 2–7.

Global plague of AIDS. (2000, April 23). *The New York Times* online.

Gnagy, S., Ming, E. E., Devesa, S. S., Hartge, P., & Whittemore, A. S. (2000). Declining ovarian cancer rates in U.S. women in relation to parity and oral contraceptive use. *Epidemiology, 11*(2), 102–105.

Gold, E. B., et al. (2000). Relation of demographic and lifestyle factors to symptoms in a multiracial/ethnic population of women 40–55 years of age. *American Journal of Epidemiology, 152.*

Gold, R. S., & Skinner, M. J. (1992). Situational factor and thought processes associated with unprotected intercourse in young gay men. *AIDS, 6,* 1021–1030.

Gold, S. R., & Gold, R. G. (1993). Sexual aversions: A hidden disorder. In W. O'Donohue & J. H. Geer (eds.), *Handbook of sexual dysfunctions: Assessment and treatment* (pp. 83–102). Boston: Allyn & Bacon.

Goldberg, C. (1995, September 11). Sex slavery, Thailand to New York. *The New York Times,* pp. B1, B6.

Goldberg, C. (1998, September 8). Getting to the truth in child abuse cases: New methods. *The New York Times,* pp. F1, F5.

Goldberg, C., & Elder, J. (1998, January 16). Public still backs abortion, but wants limits, poll says. *The New York Times,* pp. A1, A16.

Goldberg, L. H., et al. (1993). Long-term suppression of recurrent genital herpes with acyclovir. *Archives of Dermatology, 129,* 582–587.

Goldenberg, R. L., & Klerman, L. V. (1995). Adolescent pregnancy—Another look. *The New England Journal of Medicine, 332,* 1161–1162.

Golding, J. M., et al. (1989, January). Social support sources following sexual assault. *Journal of Community Psychology, 17*(1), 92–107.

Goldman, J. A., & Harlow, L. L. (1993). Self-perception variables that mediate AIDS-preventive behavior in college students. *Health Psychology, 12,* 489–498.

Goldstein, A. M., & Clark, J. H. (1990). Treatment of uncomplicated gonococcal urethritis with single-dose ceftriaxone. *Sexually Transmitted Diseases, 17,* 181–183.

Goldstein, I. (1998). Cited in Kolata, G. (1998, April 4). Impotence pill: Would it also help women? *The New York Times,* pp. A1, A6.

Goldstein, I. (2000). Cited in Norton, A. (2000, September 1). Exercise helps men avoid impotence. *Reuters News Agency* online.

Goldstein, I., et al. (1998). Oral sildenafil in the treatment of erectile dysfunction. *The New England Journal of Medicine, 338,* 1397–1404.

Goleman, D. (1988, October 18). Chemistry of sexual desire yields its elusive secrets. *The New York Times,* pp. C1, C15.

Goleman, D. (1991, October 22). Sexual harassment: It's about power, not lust. *The New York Times,* pp. C1, C12.

Goleman, D. (1992). Therapies offer hope for sex offenders. *The New York Times,* pp. C1, C11.

Goleman, D. (1993, October 6). Abuse-prevention efforts aid children. *The New York Times,* p. C13.

Goleman, D. (1995, June 14). Sex fantasy research said to neglect women. *The New York Times,* p. C14.

Gomez, J., & Smith, B. (1990). Taking the home out of homophobia: Black lesbian health. In E. C. White (ed.), *The Black women's health book: Speaking for ourselves.* Seattle, WA: Seal Press.

Gomez-Schwartz, B., Horowitz, J., & Cardarelli, A. (1990). *Child sexual abuse: The initial effects.* Newbury Park, CA: Sage.

Goodwin, J. M., Cheeves, K., & Connell, V. (1990). Borderline and other severe symptoms in adult survivors of incestuous abuse. *Psychiatric Annals, 20,* 22–32.

Gordon, S., & Snyder, C. W. (1989). *Personal issues in human sexuality: A guidebook for better sexual health,* 2nd ed. Boston: Allyn & Bacon.

Gottlieb, M. S. (1991, June 5). AIDS—the second decade: Leadership is lacking. *The New York Times,* p. A29.

Gottman, J. M., Coan, J., Carrère, S., & Swanson, C. (1998). Predicting marital happiness and stability from newlywed interactions. *Journal of Marriage and the Family, 60,* 5–22.

Goulart, M., & Madover, S. (1991). An AIDS prevention program for homeless youth. *Journal of Adolescent Health, 12,* 573–575.

Graber, B. (1993). Medical aspects of sexual arousal disorders. In W. O'Donohue & J. H. Geer (eds.), *Handbook of sexual dysfunctions: Assessment and treatment* (pp. 103–156). Boston: Allyn & Bacon.

Grabrick, D. M., et al. (2000). Risk of breast cancer with oral contraceptive use in women with a family history of breast cancer. *Journal of the American Medical Association, 284*(14), 1791–1798.

Graca, L. M., Cardoso, C. G., Clode, N., & Calhaz-Jorge, C. (1991). Acute effects of maternal cigarette smoking on fetal heart rate and fetal body movements felt by the mother. *Journal of Perinatal Medicine, 19,* 385–390.

Graham, J. M., & Blanco, J. D. (1990). Chlamydial infections. *Primary Care: Clinics in Office Prac tice, 17,* 85–93.

Graham, S., et al. (1982). Sex patterns and herpes simplex virus type 2 in the epidemiology of cancer of the cervix. *American Journal of Epidemiology, 115,* 729–735.

Grauerholz, E. (1989). Sexual harassment of women professors by students: Exploring the dynamics of power, authority, and gender in a university setting. *Sex Roles, 21,* 789–301.

Green, R. (1978). Sexual identity of 37 children raised by homosexual or transsexual parents. *American Journal of Psychiatry, 135,* 692–697.

Green, R. (1987). *The "sissy boy syndrome" and the development of homosexuality.* New Haven, CT: Yale University Press.

Greenberg, E. R., et al. (1984). Breast cancer in mothers given diethylstilbestrol in pregnancy. *The New England Journal of Medicine, 311,* 1393–1398.

Greene, B. (1994). Ethnic-minority lesbians and gay men: Mental health and treatment issues. *Journal of Consulting and Clinical Psychology, 62,* 243–251.

Greenhouse, L. (1992, February 27). Court opens path for student suits in sex-bias cases. *The New York Times,* pp. A1, A16.

Greenwald, E., & Leitenberg, H. (1989). Long-term effects of sexual experiences with siblings and non-siblings during childhood. *Archives of Sexual Behavior, 18,* 389–399.

Griffin, E., & Sparks, G. G. (1990). Friends forever: A longitudinal exploration of intimacy in same-sex friends and platonic pairs. *Journal of Social and Personal Relationships, 7,* 29–46.

Griscom, A. (2000, June 11). Heaven sent: The blind date who is your destiny. *The New York Times Magazine,* p. 93.

Grodstein, F., Goldman, M. G., & Cramer, D. W. (1993). Relation of tubal infertility to history of sexually transmitted diseases. *American Journal of Epidemiology, 137,* 577–584.

Grodstein, F., et al. (1996). Postmenopausal estrogen and progestin use and the risk of cardiovascular disease. *New England Journal of Medicine, 335,* 453–461.

Grodstein, F., et al. (1997). Postmenopausal hormonal therapy and mortality. *The New England Journal of Medicine, 336*, 1769–1775.

Grogger, J., & Bronars, S. (1993). The socioeconomic consequences of teenage childbearing: Findings from a natural experiment. *Family Planning Perspectives, 25*, 156–161.

Grön, G., Wunderlich, A. P., Spitzer, M., Tomczak, R. & Riepe, M. W. (2000). Brain activation during human navigation: Gender-different neural networks as substrate of performance. *Nature Neuroscience, 3*(4), 404–408.

Gross, J. (1991, February 11). New challenge of youth: Growing up in gay home. *The New York Times*, pp. A1, B7.

Gross, J. (1992, July 13). Suffering in silence no more: Fighting sexual harassment. *The New York Times*, pp. A10, D10.

Gross, J. (1993, September 25). Combating rape on campus in a class on sexual consent. *The New York Times*, pp. 1, 9.

Grosser, B. I., Monti-Bloch, L., Jennings-White, C., & Berliner, D. L. (2000). Behavioral and electrophysiological effects of androstadienone, a human pheromone. *Psychoneuroendocrinology, 25*(3), 289–300.

Grossman, S. (1991, December 22). Cited in "Undergraduates drink heavily, survey discloses." *The New York Times*, p. 46.

Groth, A. N. (1978). Patterns of sexual assault against children and adolescents. In A. W. Burgess, A. N. Groth, L. L. Holmstrom, & S. M. Sgroi (eds.), *Sexual assault of children and adolescents.* Toronto, Canada: Lexington Books.

Groth, A. N., & Birnbaum, H. J. (1979). *Men who rape: The psychology of the offender.* New York: Plenum Press.

Groth, A. N., & Burgess, A. W. (1980). Male rape: Offenders and victims. *American Journal of Psychiatry, 137*, 806–810.

Groth, A., & Hobson, W. (1983). The dynamics of sexual assault. In L. Schlesinger & E. Revitch (eds.), *Sexual dynamics of antisocial behavior.* Springfield, IL: Thomas.

Gruber, J. E., & Bjorn, L. (1986). Women's responses to sexual harassment: An analysis of sociocultural, organizational, and personal resource models. *Social Science Quarterly, 67*, 814–826.

Grunfeld, C. (1995). What causes wasting in AIDS? *The New England Journal of Medicine, 333*, 123–124.

Gruslin, A., et al. (2000). Maternal smoking and fetal erythropoietin levels. *Obstetrics & Gynecology, 95*(4), 561–564.

Grych, J. H., & Fincham, F. D. (1993). Children's appraisals of marital conflict. *Child Development, 64*, 215–230.

Guinan, M. E. (1992, February 1). Cited in W. E. Leary, U.S. panel backs approval of first condom for women. *The New York Times*, p. 7.

Gupta, M. (1994). Sexuality in the Indian subcontinent. *Sexual and Marital Therapy, 9*(1), 57–69.

Guzick, D. S., & Hoeger, K. (2000). Sex, hormones, and hysterectomies. *The New England Journal of Medicine* online, 343 (10).

H

Haas-Hawkings, G., et al. (1985). A study of relatively immediate adjustment to widowhood in late life. *International Journal of Women's Studies, 8*, 158–165.

Haddow, J. E., et al. (1998). Screening of maternal serum for fetal Down's syndrome in the first trimester. *The New England Journal of Medicine, 338*, 955–961.

Haffner, D. (1993, August). *Sex education: Trends and issues.* Paper presented at the meeting of the American Psychological Association, Toronto, Canada.

Haglund, B., & Cnattingius, S. (1990). Cigarette smoking as a risk factor for sudden infant death syndrome: A population based study. *American Journal of Public Health, 80*, 29–32.

Hall, C. S. (1984). "A ubiquitous sex difference in dreams" revisited. *Journal of Personality and Social Psychology, 46*, 1109–1117.

Hall, G. C. N. (1989). Sexual arousal and arousability in a sexual offender population. *Journal of Abnormal Psychology, 98*, 145–149.

Hall, G. C. N., & Hirschman, R. (1991). Toward a theory of sexual aggression: A quadripartite model. *Journal of Consulting and Clinical Psychology, 59*, 662–669.

Hall, G. C. N., Sue, S., Narang, D. S., & Lilly, R. S. (2000). Culture-specific models of men's sexual aggression: Intra- and interpersonal determinants. *Cultural Diversity & Ethnic Minority Psychology, 6*(3), 252–267.

Halpern, D. F. (1997). Sex differences in intelligence: Implications for education. *American Psychologist, 52*, 1091–1102.

Halpern, D. F., & LaMay, M. L. (2000). The smarter sex: A critical review of sex differences in intelligence. *Educational Psychology Review, 12*(2), 229–246.

Halperin D. T., & Bailey, R. C. (1999). Male circumcision and HIV infection: 10 years and counting. *Lancet, 354*, 1813–1815.

Hamer, D. H., et al. (1993, July 16). A linkage between DNA markers on the X chromosome and male sexual orientation. *Science, 261*, 321–327.

Hamilton, M., & Yee, J. (1990). Rape knowledge and propensity to rape. *Journal of Research in Personality, 24*, 111–122.

Handler, A. (1990). The correlates of the initiation of sexual intercourse among young urban Black females. *Journal of Youth and Adolescence, 19*, 159–170.

Handsfield, H. (1984). Gonorrhea and uncomplicated gonococcal infection. In K. K. Holmes, et al. (eds.), *Sexually transmitted diseases* (pp. 205–220). New York: McGraw-Hill.

Haney, D. Q. (1997, September 30). AIDS virus resisting new drugs. *Daily Record*, pp. A1, A10.

Hanrahan, J. P., et al. (1992). The effect of maternal smoking during pregnancy on early infant lung function. *American Review of Respiratory Disease, 145*, 1129–1135.

Hanson, S. L., Morrison, D. R., & Ginsburg, A. L. (1989). The antecedents of teenage fatherhood. *Demography, 26*, 579–596.

Hanson, R. K. (1990). The psychological impact of sexual assault on women and children: A review. *Annals of Sex Research, 3*, 187–232.

Hariton, E. B., & Singer, J. L. (1974). Women's fantasies during sexual intercourse: Normative and theoretical implications. *Journal of Consulting and Clinical Psychology, 42*, 313–322.

Harney P. A., & Muehlenhard, C. L. (1991). Rape. In E. Grauerholz & M. A. Koralewski (eds.), *Sexual coercion: A sourcebook on its nature, causes, and prevention* (pp. 3–16). Lexington, MA: Lexington Books.

Harold, G. T., Fincham, F. D., Osborne, L. N., & Conger, R. D. (1997). Mom and Dad are at it again: Adolescent perceptions of marital conflict and adolescent psychological distress. *Developmental Psychology, 33*, 333–350.

Harris, C. R. (2000). Psychophysiological responses to imagined infidelity: The specific innate modular view of jealousy reconsidered. *Journal of Personality and Social Psychology, 78*(6), 1082–1091.

Harris, L. (1988). *Inside America.* New York: Vintage.

Harris, M., & Johnson, O. (2000). *Cultural anthropology*, 5th ed. Boston: Allyn & Bacon.

Hart, J., et al. (1991). Sexual behavior in pregnancy: A study of 219 women. *Journal of Sex Education and Therapy, 17*, 86–90.

Hartge, P. (1997). Abortion, breast cancer, and epidemiology. *The New England Journal of Medicine, 336*, 127–128.

Hartman, C. R., Burgess, A. W., & McCormack, A. (1987). Pathways and cycles of runaways: A model for understanding repetitive runaway behavior. *Hospital and Community Psychiatry, 38*, 292–299.

Harvey, C. (2000, September 4). Where to go for help. *The Los Angeles Times* online.

Harvey, S. (1987). Female sexual behavior: Fluctuations during the menstrual cycle. *Journal of Psychosomatic Research, 31*, 101–110.

Hass, A. (1979). *Teenage sexuality.* New York: Macmillan.

Hatcher, R. A. (1998). Depo-Provera, Norplant, and progestin-only pills (minipills). In Hatcher, R. A., et al. (1998). *Contraceptive technology*, 17th rev. ed. (pp. 467–510). New York: Ardent Media.

Hatcher, R. A., & Guillebaud, J. (1998). The pill: Combined oral contraceptives. In Hatcher, R. A., et al. (1998). *Contraceptive technology*, 17th rev. ed. (pp. 405–466). New York: Ardent Media.

Hatcher, R. A., et al. (1998). *Contraceptive technology*, 17th rev. ed. New York: Ardent Media.

Hatfield, E. (1988). Passionate and companionate love. In R. J. Sternberg & M. L. Barnes (eds.), *The psychology of love* (pp. 191–217). New Haven, CT: Yale University Press.

Hatfield, E., & Sprecher, S. (1986). Measuring passionate love in intimate relationships. *Journal of Adolescence, 9*, 383–410.

Hatzichristou, D. G., et al. (2000). Sildenafil versus intracavernous injection therapy: Efficacy and preference in patients on intracavernous injection for more than 1 year. *The Journal of Urology, 164*, 1197–1200.

Haugaard, J. J. (2000). The challenge of defining child sexual abuse. *American Psychologist, 55*(9), 1036–1039.

Hausknecht, R. U. (1998). Cited in Hitt, J. (1998, January 18). Who will do abortions here? *The New York Times Magazine*, p. 42.

Hawton, K., & Catalan, J. (1990). Sex therapy for vaginismus: Characteristics of couples and treatment outcomes. *Sexual and Marital Therapy, 5,* 39–48.

Hayes, C. D. (ed.). (1987). *Risking the future,* Vol. 1. Washington, DC: National Academy Press.

Heath, R. (1972). Pleasure and brain activity in man. *Journal of Nervous and Mental Disease, 154,* 3–18.

Hechtman, L. (1989). Teenage mothers and their children: Risks and problems: A review. *Canadian Journal of Psychiatry, 34,* 569–575.

Hegeler, S., & Mortensen, M. (1977). Sexual behavior in elderly Danish males. In R. Gemme & C. Wheeler (eds.), *Progress in sexology* (pp. 285–292). New York: Plenum Press.

Heiman, J. R. (2000). Book review: The technology of orgasm: "Hysteria," the vibrator, and women's sexual satisfaction. *The New England Journal of Medicine* online, 342 (25).

Heiman, J. R., & LoPiccolo, J. (1987). *Becoming orgasmic,* 2nd ed. Englewood Cliffs, NJ: Prentice-Hall.

Henderson, V. W., et al. (2000). Estrogen for Alzheimer's disease in women: Randomized, double-blind, placebo-controlled trial. *Neurology, 54,* 295–301.

Hendrick, C., & Hendrick, S. (1986). A theory and method of love. *Journal of Personality and Social Psychology, 50,* 392–402.

Henneberger, M. (with M. Marriott). (1993). For some, rituals of abuse replace youthful courtship. *The New York Times,* pp. A1, A33.

Henry, J. (1963). *Culture against man.* New York: Random House.

Hensley, W. E. (1992). Why does the best-looking person in the room always seem to be surrounded by admirers? *Psychological Reports, 70,* 457–458.

Hensley, W. E. (1994). Height as a basis for interpersonal attraction. *Adolescence,* 29, 469–474.

Herdt, G. H. (1981). *Guardians of the flutes: Idioms of masculinity.* New York: McGraw-Hill.

Herdt, G. H. (1987). *The Sambia: Ritual and gender in New Guinea.* New York: Holt.

Hernandez, J. T., & Smith, F. J. (1990). Inconsistencies and misperceptions putting college students at risk of HIV infection. *Journal of Adolescent Health Care, 11,* 295–297.

Herrell, R., et al. (1999). Sexual orientation and suicidality: A co-twin control study in adult men. *Archives of General Psychiatry, 56*(10), 867–874.

Herrington, D. M., et al. (2000). Effects of estrogen replacement on the progression of coronary-artery atherosclerosis. *The New England Journal of Medicine* online, 343(8).

Herrmann, H. C., Chang, G., Klugherz, B. D., Mahoney, P. D. (2000). Hemodynamic effects of sildenafil in men with severe coronary artery disease. *The New England Journal of Medicine, 342*(22), 1622–1626.

Herzog, L. (1989). Urinary tract infections and circumcision. *American Journal of Diseases of Children, 143,* 348–350.

Hess, B. B., Markson, E. W., & Stein, P. J. (1993). *Sociology,* 4th ed. New York: Macmillan.

Hillier, S., & Holmes, K. K. (1990). Bacterial vaginosis. In K. K. Holmes, P. Mardh, P. F. Sparling, & P. J. Wiesner (eds.), *Sexually transmitted diseases,* 2nd ed. (pp. 547–560). New York: McGraw-Hill.

Hillis, D. M. (2000, June 9). Origins of HIV. *Science, 288*(5472), 1757–1759.

Hilton, E., et al. (1992). Ingestion of yogurt containing *Lactobacillus acidophilus* as prophylaxis for candidal vaginitis. *Annals of Internal Medicine, 116,* 353–357.

Hingson, R. W., et al. (1990). Beliefs about AIDS, use of alcohol and drugs, and unprotected sex among Massachusetts adolescents. *American Journal of Public Health, 80,* 295–298.

Hite, S. (1976). *The Hite report.* New York: Macmillan.

Hite, S. (1977). *The Hite report: A nationwide study of female sexuality.* New York: Dell.

Hite, S. (1981). *The Hite report on male sexuality.* New York: Knopf.

Hitt, J. (1998, January 18). Who will do abortions here? *The New York Times Magazine,* pp. 20ff.

Ho, D. D. (1995). Time to hit HIV, early and hard. *The New England Journal of Medicine, 333,* 450–451.

Ho, G. Y. F., et al. (1998, February 12). *The New England Journal of Medicine, 338.* Cited in "Sexually active coeds at higher risk for HPV" (1998, February 11). Reuters News Agency online.

Hoagwood, K. (1990). Blame and adjustment among women sexually abused as children. *Women and Therapy, 9,* 89–110.

Hobfoll, S. E., Jackson, A. P., Lavin, J., Britton, P. J., & Shepherd, J. B. (1993). Safer sex knowledge, behavior, and attitudes of inner-city women. *Health Psychology, 12,* 481–488.

Hock, Z. (1983). The G Spot. *Journal of Sex and Marital Therapy, 9,* 166–167.

Hodges, B. C., et al. (1992). Gender-differences in adolescents' attitudes toward condom use. *Journal of School Health, 62,* 103–106.

Hodgson, R., et al. (1990). *Chlamydia trachomatis:* The prevalence, trend and importance in initial infertility management. *Australian and New Zealand Journal of Obstetrics and Gynaecology, 30,* 251–254.

Hodson, D. S., & Skeen, P. (1994). Sexuality and aging: The hammerlock of myths. *Journal of Applied Gerontology, 13*(3), 219–235.

Hoffman, C., & Hurst, N. (1990). Gender stereotypes: Perception or rationalization? *Journal of Personality and Social Psychology, 58,* 197–208.

Holden, G. W., & Ritchie, K. L. (1991). Linking extreme marital discord, child rearing, and child behavior problems. *Child Development, 62,* 311–327.

Holmes-Farley, S. R. (1998, September 16). Who should decide the sex of a baby? *The New York Times,* p. A28.

Holtzman, D., et al. (1992). HIV education and health education in the United States: A national survey of local school district policies and practices. *Journal of School Health, 62,* 421–427.

Holzman, H. R, & Pines, S. (1982). Buying sex: The phenomenology of being a john. *Deviant Behavior, 4,* 89–116.

Howard, J. A., Blumstein, P., & Schwartz, P. (1987). Social or evolutionary theories: Some observations on preferences in mate selection. *Journal of Personality and Social Psychology, 53,* 194–200.

Howard, M., & McCabe, J. B. (1990). Helping teenagers postpone sexual involvement. *Family Planning Perspectives, 22,* 21–26.

Howard, M. C. (1989). *Contemporary cultural anthropology,* 3rd ed. Glenview, IL: Scott, Foresman.

Howards, S. S. (1995). Current concepts: Treatment of male infertility. *The New England Journal of Medicine, 332,* 312–317.

Hsu, B., et al. (1994). Gender differences in sexual fantasy and behavior in a college population: A ten-year replication. *Journal of Sex and Marital Therapy, 20*(2), 103–118.

Hu, S. (1992, January 24). Cited in "Immune deficiency vaccine for monkeys succeeds." *The New York Times,* p. A17.

Hubbard, R., & Wald, E. (1993). *Exploding the gene myth.* Boston: Beacon Press.

Huffman, T., Chang, K., Rausch, P., & Schaffer, N. (1994). Gender differences and factors related to the disposition toward cohabitation. *Family Therapy, 21*(3), 171–184.

Hughes, I. A. (2000). A novel explanation for resistance to androgens. *The New England Journal of Medicine* online, 343(12).

Hunt, M. (1974). *Sexual behavior in the 1970's.* New York: Dell Books.

Hunt, M. (1979). Legal rape. *Family Circle,* January 9.

Hunter, D. J., et al. (1996). Cohort studies of fat intake and the risk of breast cancer—A pooled analysis. *The New England Journal of Medicine, 334,* 356–361.

Huxley, R. R. (2000). Nausea and vomiting in early pregnancy: Its role in placental development. *Obstetrics & Gynecology, 95,* 779–782.

Hyde, J. S., & Linn, M. C. (1988). Gender differences in verbal ability: A meta-analysis. *Psychological Bulletin, 104,* 53–69.

Hyde, J. S., Fennema, E., & Lamon, S. J. (1990). Gender differences in mathematics performance: A meta-analysis. *Psychological Bulletin, 107,* 139–155.

Hyde, J. S., & Plant, E. A. (1995). Magnitude of psychological gender differences: Another side to the story. *American Psychologist, 50,* 159–161.

I

Imperato-McGinley, J., et al. (1974). Steroid 5 reductase deficiency in man: An inherited form of male pseudohermaphroditism. *Science, 186,* 1213–1215.

Ingrassia, M. (1993, October 25). Abused and confused. *Newsweek,* pp. 57–58.

Iranian transsexual unhappy with experience as woman. (2000, June 19). Reuters News Agency online.

Isay, R. A. (1990). Psychoanalytic theory and the therapy of gay men. In D. P. McWhirter, S. A. Sanders, & J. M. Reinisch (eds.) *Homosexuality/heterosexuality: Concepts of sexual orientation* (pp. 283–303). New York: Oxford University Press.

Isay, R. A. (1993, April 23). Sex survey may say most about society's attitudes to gays. *The New York Times*, Section 4, p. 16 (letter).

Ison, C. A. (1990). Laboratory methods in genitourinary medicine: Methods of diagnosing gonorrhoea. *Genitourinary Medicine*, *66*, 453–459.

J

Jacklin, C. N, DiPietro, J. A., & Maccoby, E. E. (1984). Sex-typing behavior and sex-typing pressure in child–parent interaction. *Archives of Sexual Behavior*, *13*, 413–425.

Jackson, B. B., Taylor, J., & Pyngolil, M. (1991). How age conditions the relationship between climacteric status and health symptoms in African American women. *Research in Nursing and Health*, *14*, 1–9.

Jackson, J., et al. (1990). Young adult women who report childhood intrafamilial sexual abuse: Subsequent adjustment. *Archives of Sexual Behavior*, *19*, 211–221.

Jackson, L. A., & Ervin, K. S. (1992). Height stereotypes of women and men: The liabilities of shortness for both sexes. *Journal of Social Psychology*, *132*, 433–445.

Jacob, S., & McClintock, M. K. (2000). Psychological state and mood effects of steroidal chemosignals in women and men. *Hormones and Behavior*, *37*(1), 57–78.

Jacobsen, C. (1991). Redefining censorship: A feminist view. *Art Journal*, *50*(4), 42–55.

Jacobson, J. L., & Jacobson, S. W. (1994). Prenatal alcohol exposure and neurobehavioral development: Where is the threshold? *Alcohol Health and Research World*, *18*(1), 30–36.

Jamison, P. L., & Gebhard, P. H. (1988). Penis size increase between flaccid and erect states. An analysis of the Kinsey data. *Journal of Sex Research*, *24*, 177–183.

Jankowiak, W. R., & Fischer, E. F. (1992). A cross-cultural perspective on romantic love. *Ethnology*, *31*, 149–155.

Janus, M. D., Scanlon, B, & Prince, V. (1984). Youth prostitution. In A. W. Burgess (ed.), *Sex rings and child pornography*. Lexington, MA: Heath.

Janus, S. S., & Janus, C. L. (1993). *The Janus Report on Sexual Behavior*. New York: Wiley.

Jeffreys, S. (1998). *The idea of prostitution*. Spinifex.

Jencks, C., & Mayer, S. E. (1990). Residential segregation, job proximity, and Black job opportunities. In L. E. Lynn and M. McGeary (eds.), *Inner-city poverty in the United States*. Washington, DC: National Academy Press.

Jenks, R. (1985). Swinging: A replication and test of a theory. *The Journal of Sex Research*, *21*, 199–210.

Jennings, V. H., Lamprecht, V. M., & Kowal, D. (1998) Fertility awareness methods. In Hatcher, R. A., et al. (1998). *Contraceptive technology*, 17th rev. ed. (pp. 309–324). New York: Ardent Media.

Jetter, A. (2000, February 22). Breast cancer in Blacks spurs hunt for answers. *The New York Times*, p. D5.

Johannes, C. B., et al. (2000). Incidence of erectile dysfunction in men 40 to 69 years old: Longitudinal results from the Massachusetts male aging study. *The Journal of Urology*, *163*, 460.

Johanson, R. (2000). Perineal massage for prevention of perineal trauma in childbirth. *The Lancet*, *355*(9200), 250–251.

Johnson, D. (1990, March 8). AIDS clamor at colleges muffling older dangers. *The New York Times*, p. A18.

Johnson, K. A., & Williams, L. (1993). Risk of breast cancer in the nurses' health study: Applying the Gail model. *Journal of the American Medical Association*, *270*, 2925–2926.

Jones, A., et al. (1994). Erectile disorder and the elderly: An analysis of the case for funding. *Sexual and Marital Therapy*, *9*(1), 9–15.

Jones, E. F., et al. (1985). Teenage pregnancy in developed countries: Determinants and policy implications. *Family Planning Perspectives*, *17*, 53–62.

Jones, H. W., & Toner, J. P. (1993). The infertile couple. *The New England Journal of Medicine*, *329*, 1710–1715.

Jones, J. (1994). Embodied meaning: Menopause and the change of life. [Special Issue: Women's health and social work: Feminist perspectives.] *Social Work in Health Care*, *19*(3–4), 43–65.

Jorgensen, S. R., et al. (1980). Dyadic and social network influences on adolescent exposure to pregnancy risk. *Journal of Marriage and the Family*, *42*, 141–155.

Josefsson, A. M., et al. (2000). Viral load of human papilloma virus 16 as a determinant for development of cervical carcinoma in situ: A nested case-control study. *The Lancet*, *355*, 2189–2193.

Journal of the American Medical Association. (1993). Preventing HIV/AIDS among adolescents: Schools as agents of behavior change. *Journal of the American Medical Association*, *269*, 760–762 (editorial).

Judson, F. N. (1990). Gonorrhea. *Medical Clinics of North America*, *74*, 1353–1366.

K

Kagay, M. R. (1991, June 19). Poll finds AIDS causes single people to alter behavior. *The New York Times*, p. C3.

Kalick, S. M. (1988). Physical attractiveness as a status cue. *Journal of Experimental Social Psychology*, *24*, 469–489.

Kallmann, F. J. (1952). Comparative twin study on the genetic aspects of male homosexuality. *Journal of Nervous and Mental Disease*, *115*, 283–298.

Kamb, M. L., et al. (1998). Efficacy of risk-reduction counseling to prevent human immunodeficiency virus and sexually transmitted diseases: A randomized controlled trial. *Journal of the American Medical Association*, *280*, 1161–1167.

Kammeyer, K. C. W. (1990). *Marriage and family: A foundation for personal decisions*, 2nd ed. Boston: Allyn & Bacon, Inc.

Kammeyer, K. C. W., Ritzer, G., & Yetman, N. R. (1990). *Sociology: Experiencing changing societies*. Boston: Allyn & Bacon.

Kanin, E. J. (1985). Date rapists: Differential sexual socialization and relative deprivation. *Archives of Sexual Behavior*, *14*, 219–231.

Kantrowitz, B. (1990a, Summer/Fall Special Issue). High school homeroom. *Newsweek*, pp. 50–54.

Kantrowitz, B. (1990b, Summer/Fall Special Issue). The push for sex education. *Newsweek*, p. 52.

Kaplan, H. S. (1974). *The new sex therapy: Active treatment of sexual dysfunctions*. New York: Brunner/Mazel.

Kaplan, H. S. (1979). *Disorders of sexual desire*. New York: Simon & Schuster.

Kaplan, H. S. (1987). *Sexual aversion, sexual phobias, and panic disorder*. New York: Brunner/Mazel.

Kaplan, H. S. (1990). Sex, intimacy, and the aging process. *Journal of the American Academy of Psychoanalysis*, *18*, 185–205.

Karney, B. R., & Bradbury, T. N. (1995). The longitudinal course of marital quality and stability: A review of theory, method, and research. *Psychological Bulletin*, *118*, 3–34.

Kash, K. (1998). Cited in Zuger, A. (1998, January 6). Do breast self–exams save lives? Science still doesn't have answer. *The New York Times*.

Katz, M. H., & Gerberding, J. L. (1997). Postexposure treatment of people exposed to the human immunodeficiency virus through sexual contact or injection-drug use. *The New England Journal of Medicine*, *336*, 1097–1100.

Katz, R. C. (1990). Psychosocial adjustment in adolescent child molesters. *Child Abuse and Neglect*, *14*, 567–575.

Kaufman, A., et al. (1980). Male rape victims: Noninstitutionalized assault. *American Journal of Psychiatry*, *137*, 221–223.

Kavanagh, A. M., Mitchell, H., & Giles, G. G. (2000). Hormone replacement therapy and accuracy of mammographic screening. *The Lancet*, *355*, 270–274.

Kay, D. S. (1992). Masturbation and mental health: Uses and abuses. *Sexual and Marital Therapy*, *7*, 97–107.

Kegel, A. H. (1952). Sexual functions of the pubococcygeus muscle. *Western Journal of Surgery*, *60*, 521–524.

Keith, J. B., et al. (1991). Sexual activity and contraceptive use among low–income urban Black adolescent females. *Adolescence*, *26*, 769–785.

Kellerman, J., Lewis, J., & Laird, J. D. (1989). Looking and loving: The effects of mutual gaze on feelings of romantic love. *Journal of Research in Personality*, *23*, 145–161.

Kelly, M. P., Strassberg, D. S., & Kircher, J. R. (1990). Attitudinal and experiential correlates of anorgasmia. *Archives of Sexual Behavior*, *19*, 165–177.

Kendler, K. S., et al. (2000). Childhood sexual abuse and adult psychiatric and substance use disorders in women: An epidemiological and co-twin control analysis. *Archives of General Psychiatry*, *57*(10), 953–959.

Kendler, K. S., Thornton, L. M., Gilman, S. E., & Kessler, R. C. (2000). Sexual orientation in a U.S. national sample of twin and nontwin sibling pairs. *American Journal of Psychiatry*, *157*, 1843–1846.

Kenney, A. M., Guardad, S., & Brown, L. (1989). Sex education and AIDS education in the schools. *Family Planning Perspectives*, *21*, 56–64.

Kettl, P., et al. (1991). Female sexuality after spinal cord injury. *Sexuality and Disability*, *9*, 287–295.

Killmann, P. R., et al. (1987). The treatment of secondary orgasmic dysfunction II. *Journal of Sex and Marital Therapy*, *13*, 93–105.

Kilpatrick, D. G., et al. (1987, January). *Rape in marriage and dating relationships: How bad are they for mental health?* Paper presented at the meeting of the New York Academy of Science, New York, NY.

Kimble, D. P. (1992). *Biological psychology*, 2nd ed. Fort Worth, TX: Harcourt.

Kimlika, T., Cross, H., & Tarnai, J. (1983). A comparison of androgynous, feminine, masculine, and undifferentiated women on self-esteem, body satisfaction, and sexual satisfaction. *Psychology of Women Quarterly*, *1*, 291–294.

Kinard, E., & Reinherz, H. (1987). School aptitude and achievement in children of adolescent mothers. *Journal of Youth and Adolescence*, *16*, 69–78.

Kinder, B. N., & Curtiss, G. (1988). Specific components in the etiology, assessment, and treatment of male sexual dysfunctions: Controlled outcome studies. *Journal of Sex and Marital Therapy*, *14*, 40–48.

King, R. (2000). Cited in Frazier, L. (2000, July 16). The new face of HIV is young, black. *The Washington Post*, p. C01.

Kinsey, A. C., Pomeroy, W. B., & Martin, C. E. (1948). *Sexual behavior in the human male.* Philadelphia: W. B. Saunders.

Kinsey, A. C., Pomeroy, W. B., Martin, C. E., & Gebhard, P. H. (1953). *Sexual behavior in the human female.* Philadelphia: W. B. Saunders.

Kintsch, W. (1994). Text comprehension, memory, and learning. *American Psychologist*, *49*, 294–303.

Kirkpatrick, R. C. (2000). The evolution of human homosexual behavior. *Current Anthropology*, *41*(3), 385–413.

Kirn, W. (1997, August 18). The ties that bind. *Time*, pp. 48–50.

Kite, M. E. (1992). Individual differences in males' reactions to gay males and lesbians. *Journal of Applied Social Psychology*, *22*, 1222–1239.

Kjerulff, K. H., et al. (2000). Effectiveness of hysterectomy. *Obstetrics & Gynecology*, *95*, 319–326.

Klein, E. A. (2000). *Management of prostate cancer.* Totowa, NJ: Humana Press.

Klein, H. G. (2000). Will blood transfusion ever be safe enough? *Journal of the American Medical Association* online, 284(2).

Klepinger, D. H., et al. (1993). Perceptions of AIDS risk and severity and their association with risk-related behavior among U.S. men. *Family Planning Perspectives*, *25*, 74–82.

Klüver, H., & Bucy, P. C. (1939). Preliminary analysis of functions of the temporal lobes in monkeys. *Archives of Neurology and Psychiatry*, *42*, 979.

Knapp, M. L. (1978). *Social intercourse: A behavioral approach to counseling.* Champaign, IL: Research Press.

Knight, R. A., et al. (1991). *Antisocial personality disorder and Hare assessments of psychopathy among sexual offenders.* Manuscript in preparation.

Knight, S. E. (1989). Sexual concerns of the physically disabled. In B. W. Heller, L. M. Flohr, & L. S. Zegans (eds.), *Psychosocial interventions with physically disabled persons* (pp. 183–199). New Brunswick, NJ: Rutgers University Press.

Knox, D. (1983). *The love attitudes inventory*, rev. ed. Saluda, NC: Family Life Publications.

Knox, D. (1988). *Choices in relationships.* St. Paul, MN: West.

Knox, D., Gibson, L., Zusman, M., & Gallmeier, C. (1997). Why college students end relationships. *College Student Journal*, *31*(4), 449–452.

Knox, D., Schacht, C., & Zusman, M. E. (1999). Love relationships among college students. *College Student Journal*, *33*(1), 149–151.

Knox, D., Zusman, M. E., Buffington, C., & Hemphill, G. (2000). Interracial dating attitudes among college students. *College Student Journal*, *34*(1), 69–71.

Knox, D., Zusman, M. E., & Nieves, W. (1997). College students' homogamous preferences for a date and mate. *College Student Journal*, *31*(4), 445–448.

Knox, D., Zusman, M. E., & Nieves, W. (1998). Breaking away: How college students end love relationships. *College Student Journal*, *32*(4), 482–484.

Knox, D., Zusman, M. E., Snell, S., & Cooper, C. (1999). Characteristics of college students who cohabit. *College Student Journal*, *33*(4), 510–512.

Knudsen, D. D. (1991). Child sexual coercion. In E. Grauerholz & M. A. Koralewski (eds.), *Sexual coercion: A sourcebook on its nature, causes, and prevention* (pp. 17–28). Lexington, MA: Lexington Books.

Knussman, R., Christiansen, K., & Couwenbergs, C. (1986). Relations between sex hormone levels and sexual behavior in men. *Archives of Sexual Behavior*, *15*, 429–445.

Kockott, G., & Fahrner, E. (1988). Transsexuals who have not undergone surgery: A follow-up study. *Archives of Sexual Behavior*, *16*, 511–522.

Kohlberg, L. (1966). A cognitive-developmental analysis of children's sex-role concepts and attitudes. In E. E. Maccoby (ed.), *The development of sex differences.* Stanford, CA: Stanford University Press.

Kohn, A. (1987, February). Shattered innocence. *Psychology Today*, pp. 54–58.

Kolata, G. (1993a, Feburary 26). Studies say mammograms fail to help many women. *The New York Times*, pp. A1, A15.

Kolata, G. (1993b, December 14). Breast cancer screening under 50: Experts disagree if benefit exists: Statisticians find no proof that screening saves lives. *The New York Times*, pp. C1, C17.

Kolata, G. (1995, May 28). Will the lawyers kill off Norplant? *The New York Times*, pp. F1, F5.

Kolata, G. (1996a, February 28). Study reports small risk, if any, from breast implants. *The New York Times*, p. A12.

Kolata, G. (1996b, September 15). AIDS patients slipping through safety net. *The New York Times*, p. A24.

Kolata, G. (1998a, April 4). Impotence pill: Would it also help women? *The New York Times*, pp. A1, A6.

Kolata, G. (1998b, April 25). Doctors debate use of drug to help women's sex lives. *The New York Times* online.

Kolata, G. (1998c, September 9). Researchers report success in method to pick baby's sex. *The New York Times* online.

Kolata, G. (2000a, April 5). Estrogen tied to slight rise in heart attack. *The New York Times*, pp. A1, A20.

Kolata, G. (2000b, April 18). New name for impotence, and new drugs. *The New York Times*, pp. F6, F14.

Kolbe, L. (1998). Cited in "Poll shows decline in sex by high school students." (1998, September 18). *The New York Times*, p. A26.

Kolbert, E. (1991, October 11). Sexual harassment at work is pervasive, survey suggests. *The New York Times*, pp. A1, A17.

Kolodny, R. C. (1981). Evaluating sex therapy: Process and outcome at the Masters & Johnson Institute. *Journal of Sex Research*, *17*, 301–318.

Komisar, L. (1971). The image of women in advertising. In V. Gornick & B. Moran (eds.), *Women in sexist society.* New York: Basic Books.

Kon, I. S. (1995). *The sexual revolution in Russia.* New York: The Free Press.

Kontula, O., Rimpela, M., & Ojanlatva, A. (1992). Sexual knowledge, attitudes, fears and behaviors of adolescents in Finland (the KISS study). *Health Education Research*, *7*, 69–77.

Koop, C. E. (1988). *Understanding AIDS.* HHS Publication No. HHS-88-8404. Washington, DC: U.S. Government Printing Office.

Koren, G., Pastuszak, A., & Ito, S. (1998). Drug therapy: Drugs in pregnancy. *The New England Journal of Medicine*, *338*, 1128–1137.

Kornblum, W. J. (1994). *Sociology in a changing world*, 3rd ed. Ft. Worth, TX: Harcourt.

Koss, L. (1989). The Papanicolaou test for cervical cancer detection. *Journal of Sex Research*, *261*, 737.

Koss, M. P. (1993). Rape: Scope, impact, interventions, and public policy responses. *American Psychologist*, *48*, 1062–1069.

Koss, M. P., Gidycz, C. A., & Wisniewski, N. (1987). The scope of rape: Incidence and prevalence of sexual aggression and victimization in a national sample of higher education students. *Journal of Consulting and Clinical Psychology*, *55*, 162–170.

Koutsky, L. A., et al. (1992). A cohort study of the risk of cervical intraepithelial neoplasia Grade 2 or 3 in relation to Papillomavirus infection. *The New England Journal of Medicine*, *327*, 1272.

Kouzi, A. C., et al. (1992). Contraceptive behavior among intravenous drug users at risk for AIDS. *Psychology of Addictive Behaviors.*

Kowal, D. (1998). Coitus interruptus (withdrawal). In Hatcher, R. A., et al. (1998). *Contraceptive technology*, 17th rev. ed. (pp. 303–308). New York: Ardent Media.

Krafka, C. L. (1985). *Sexually explicit, sexually violent, and violent media: Effects of multiple naturalistic exposures and debriefing on female viewers.* Unpublished doctoral dissertation, University of Wisconsin-Madison.

Kramer, M. S., et al. (2000). The contribution of mild and moderate preterm birth to infant mortality. *Journal of the American Medical Association*, *284*, 843–849.

Kresin, D. (1993). Medical aspects of inhibited sexual desire disorder. In W. O'Donohue & J. H. Geer (eds.), *Handbook of sexual dysfunctions: Assessment and treatment* (pp. 15–52). Boston: Allyn & Bacon.

Krucoff, C. (2000 May 30). "Sexercise" credited with boosting performance. *The Washington Post*, p. Z09.

Kruesi, M. J. P., et al. (1992). Paraphilias: A double-blind cross-over comparison of clomipramine versus desipramine. *Archives of Sexual Behavior, 21,* 587–594.

Kruks, G. (1991). Gay and lesbian homeless/street youth: Special issues and concerns. [Special Issue: Homeless youth]. *Journal of Adolescent Health, 12,* 515–518.

Ku, L. C., Sonenstein, F. L., & Pleck, J. H. (1992). The association of AIDS education and sex education with sexual behavior and condom use among teenage men. *Family Planning Perspectives, 24,* 100–106.

Kuiper, B., & Cohen-Kettenis, P. (1988). Sex reassignment surgery: A study of 141 Dutch transsexuals. *Archives of Sexual Behavior, 17,* 439–457.

Kulin, H., et al. (1989). The onset of sperm production in pubertal boys. *American Journal of Diseases of Children, 143,* 190–193.

Kumar, U. (1991). Life stages in the development of the Hindu woman in India. In L. L. Adler (ed.), *Women in cross-cultural perspective* (pp. 143–158). New York: Praeger.

Kunkel, L. E., & Temple, L. L. (1992). Attitudes towards AIDS and homosexuals: Gender, marital status, and religion. *Journal of Applied Social Psychology, 22,* 1030–1040.

Kurdek, L. A, & Schmitt, J. P. (1986). Relationship quality of gay men in closed or open relationships. *Journal of Homosexuality, 12*(2), 85–99.

Kyriacou, D. N., et al. (1999). Risk factors for injury to women from domestic violence. *The New England Journal of Medicine* online, 341(25).

L

Laan, E., & Heiman, J. (1994). *Archives of Sexual Behavior.*

Lackluster Latin lovers surf for seduction (2000, June 1). Reuters News Agency online.

Ladas, A. K., Whipple, B., & Perry, J. D. (1982). *The G spot and other recent discoveries about human sexuality.* New York: Holt.

Laino, C. (2000, September 17). Herpes vaccine works well in women but fails to protect men, studies show. MSNBC online.

Laird, J. (1994). A male pill? Gender discrepancies in contraceptive commitment. *Feminism and Psychology, 4*(3), 458–468.

Lamanna, M. A., & Riedmann, A. (1997). *Marriages and families,* 6th ed. Belmont, CA: Wadsworth.

Lamaze, F. (1981). *Painless childbirth.* New York: Simon & Schuster.

Lambert, B. (1988, September 20). AIDS among prostitutes not as prevalent as believed, studies show. *The New York Times,* p. B1.

Lamberti, D. (1997). Cited in Alterman, E. (1997, November). Sex in the '90s. *Elle.*

Lamke, L. K. (1982a). Adjustment and sex-role orientation. *Journal of Youth and Adolescence, 11,* 247–259.

Lamke, L. K. (1982b). The impact of sex-role orientation on self-esteem in early adolescence. *Child Development, 53,* 1530–1535.

Lane, III, F. S. (2000). *Obscene profits: The entrepreneurs of pornography in the cyber age.* London: Routledge.

Lane, K. E., & Gwartney-Gibbs, P. A. (1985). Violence in the context of dating and sex. *Journal of Family Issues, 6,* 45–59.

Lang, A. R. (1985). The social psychology of drinking and human sexuality. *Journal of Drug Issues, 15,* 273–289.

Lang, A. R., et al. (1980). Expectancy, alcohol, and sex guilt as determinants of interest in and reaction to sexual stimuli. *Journal of Abnormal Psychology, 89,* 644–653.

Lang, R. A., et al. (1989). An examination of sex hormones in genital exhibitionists. *Annals of Sex Research, 2,* 67–75.

Langevin, R., Wright, P., & Handy, L. (1989). Characteristics of sex offenders who were sexually victimized as children. *Annals of Sex Research, 2,* 227–253.

Langevin, R., et al. (1979). Experimental studies of the etiology of genital exhibitionism. *Archives of Sexual Behavior, 8,* 307–332.

Langevin, R., et al. (1985). Sexual aggression: Constructing a predictive equation. In R. Langevin (ed.), *Erotic preference, gender identity, and aggression in men: New research studies.* (pp. 39–76). Hillsdale, NJ: Lawernce Erlbaum Associates.

Lansky, D., & Wilson, G. T. (1981). Alcohol, expectations, and sexual arousal in males: An information-processing analysis. *Journal of Abnormal Psychology, 90,* 35–45.

LaTour, M. S. (1990). Female nudity in print advertising: An analysis of gender differences in arousal and ad response. *Psychology and Marketing, 7,* 65–81.

Laumann, E. O. (1997). Cited in Gilbert, S. (1997, April 2). Study adds to doubts about benefits of circumcision. *The New York Times* online.

Laumann, E. O., Gagnon, J. H., Michael, R. T., & Michaels, S. (1994). *The social organization of sexuality: Sexual practices in the United States.* Chicago: University of Chicago Press.

Laumann, E. O., Masi, C. M., & Zuckerman, E. W., et al. (1997, April 2). Circumcision in the United States: Prevalence, prophylactic effects, and sexual practice. *The Journal of the American Medical Association,* 277, 1052–1057.

Laumann, E. O., Paik, A., & Rosen, R. C. (1999). Sexual dysfunction in the United States. Prevalence and predictors. *Journal of the American Medical Association, 281*(6), 537–544.

Laviola, M. (1989). Effects of older brother–younger sister incest: A review of four cases. *Journal of Family Violence, 4,* 259–274.

Lavoisier, P., et al. (1995). Clitoral blood flow increases following vaginal pressure stimulation. *Archives of Sexual Behavior, 24,* 37–45.

Law, J. (2000). The politics of breastfeeding: Assessing risk, dividing labor. *Signs, 25*(2), 407–450.

Lawrence, K., & Herold, E. S. (1988). Women's attitudes toward and experience with sexually explicit materials. *Journal of Sex Research, 24,* 161–169.

Lawson, C. (1993, August 5). Single but mothers by choice: "Who is my daddy?" can be answered in different ways. *The New York Times,* pp. C1, C9.

Lawton, C. A., & Morrin, K. A. (1999). Gender differences in pointing accuracy in computer-simulated 3D mazes. *Sex Roles, 40*(1–2), 73–92.

Leary, W. E. (1990, September 13). New focus on sperm brings fertility successes. *The New York Times,* p. B11.

Leary, W. E. (1992, December 10). Medical panel says most sexual impotence in men can be treated without surgery. *The New York Times,* p. D20.

Leary, W. E. (1993, April 28). Screening of all newborns urged for sickle cell disease. *The New York Times,* p. C11.

Leary, W. E. (1995, May 11). When it comes to giant sperm, this tiny fruit fly is a whale. *The New York Times,* p. A26.

Leary, W. E. (1998, September 29). Older people enjoy sex, survey says. *The New York Times,* p. F8.

Leavitt, G. C. (1990). Sociobiological explanations of incest avoidance: A critical review of evidential claims. *American Anthropologist, 92,* 971–993.

Lederman, M. M., & Valdez, H. (2000). Immune restoration with antiretroviral therapies: Implications for clinical management. *Journal of the American Medical Association, 284,* 223–228.

Ledray, L. E. (1990). Counseling rape victims: The nursing challenge. *Perspectives in Psychiatric Care, 26,* 21–27.

Lee, J. A. (1988). Love-styles. In R. J. Sternberg & M. L. Barnes (eds.), *The psychology of love* (pp. 38–67). New Haven, CT: Yale University Press.

Legato, M. J. (2000). Cited in "Study of children born without penises finds nature determines gender." (2000, May 12). The Associated Press online.

Leiblum, S. R., & Rosen, R. C. (1991). Couples therapy for erectile disorders: Conceptual and clinical considerations. [Special Issue: The treatment of male erectile disorders]. *Journal of Sex and Marital Therapy, 17,* 147–159.

Leigh, B. C., Morrison, D. M., Trocki, K., & Temple, M. T. (1994). Sexual behavior of American adolescents: Results from a U.S. national survey. *Journal of Adolescent Health, 15*(2), 117–125.

Leinders-Zufall, T., et al. (2000). Ultrasensitive pheromone detection by mammalian vomeronasal neurons. *Nature, 405,* 792–796.

Leitenberg, H., Detzer, M. J., & Srebnik, D. (1993). Gender differences in masturbation and the relation of masturbation experience in preadolescence and/or early adolescence to sexual behavior and sexual adjustment in young adulthood. *Archives of Sexual Behavior, 22,* 87–98.

Leitenberg, H., Greenwald, E., & Tarran, M. J. (1989). The relation between sexual activity among children during preadolescence and/or early adolescence and sexual behavior and sexual adjustment in young adulthood. *Archives of Sexual Behavior, 18,* 299–313.

Leitenberg, H., & Henning, K. (1995). Sexual fantasy. *Psychological Bulletin, 117,* 469–496.

Leland, J. (2000, May 29). The science of women and sex. Newsweek, pp. 48–54.

Leland, N. L., & Barth, R. P. (1992). Gender differences in knowledge, intentions, and behaviors concerning pregnancy and sexually transmitted disease prevention among adolescents. *Journal of Adolescent Health, 13,* 589–599.

Lemon, S. J., & Newbold, J. E. (1990). Viral hepatitis. In K. K. Holmes, P. Mardh, P. F. Sparling, & P. J. Wiesner (eds.), *Sexually transmitted diseases,* 2nd ed. (pp. 449–466). New York: McGraw-Hill.

Lenihan, G., Rawlins, M. E., Eberly, C. G., Buckley, B. & Masters, B. (1992). Gender differences in rape supportive attitudes before and after a date rape education intervention. *Journal of College Student Development, 33,* 331–338.

Lesnik-Oberstein, M., & Cohen, L. (1984). Cognitive style, sensation seeking, and assortive mating. *Journal of Personality and Social Psychology, 46,* 57–66.

Lester, W. (2000, May 31). Poll: Americans back some gay rights. The Associated Press online.

Letourneau, E., & O'Donohue, W. (1993). Sexual desire disorders. In W. O'Donohue & J. H. Geer (eds.), *Handbook of sexual dysfunctions: Assessment and treatment.* (pp. 53–81). Boston: Allyn & Bacon.

LeVay, S. (1991). A difference in hypothalamic structure between heterosexual and homosexual men. *Science, 253,* 1034–1037.

Lever, J., et al. (1992). Behavior patterns and sexual identity of bisexual males. *Journal of Sex Research, 29,* 141–167.

Levine, G. I. (1991). Sexually transmitted parasitic diseases. *Primary Care: Clinics in Office Practice, 18,* 101–128.

Levinger, G. (1980). Toward the analysis of close relationships. *Journal of Experimental Social Psychology, 16,* 510–544.

Levitt, E. E. (1988). Alternative life style and marital satisfaction: A brief report. *Annals of Sex Research, 1,* 455–461.

Levitt, E. E., & Mulcahy, J. J. (1995). The effect of intracavernosal injection of papaverine hydrochloride on orgasm latency. *Journal of Sex and Marital Therapy, 21,* 39–41.

Levy, D. S. (1991, September 16). Why Johnny might grow up violent and sexist. *Time,* pp. 16–19.

Levy, G. D., & Carter, D. B. (1989). Gender schema, gender constancy, and gender-role knowledge: The roles of cognitive factors in preschoolers' gender-role stereotype attributions. *Developmental Psychology, 25,* 444–449.

Levy, R. I. (1973). *The Tahitians.* Chicago: University of Chicago Press.

Lewan, T. (1998, May 2). Not all women thrilled with Viagra. The Associated Press online.

Lewin, T. (1991, February 8). Studies on teen-age sex cloud condom debate. *The New York Times,* p. A14.

Lewin, T. (1992a, February 28). Canada court says pornography harms women. *The New York Times,* p. B7.

Lewin, T. (1998a, January 17). Debate distant for many having abortions. *The New York Times,* pp. A1, A9.

Lewin, T. (1998b, March 23). Debate centers on definition of harassment. *The New York Times,* pp. A1, A28.

Lewis, D. K., & Watters, J. K. (1991). Sexual risk behavior among heterosexual intravenous drug users: Ethnic and gender variations. *AIDS, 5,* 77–83.

Lewis, P. H. (1995, August 21). Planet Out: "Gay global village" of cyberspace. *The New York Times,* p. D3.

Lewis, P. H. (2000, June 22). Snooping software enters the mainstream. *The New York Times* online.

Libman, E. (1989). Sociocultural and cognitive factors in aging and sexual expression: Conceptual and research issues. *Canadian Psychology, 30,* 560–567.

Libman, E., Fichten, C. S., & Brender, W. (1985). The role of therapeutic format in the treatment of sexual dysfunction: A review. *Clinical Psychology Review, 5,* 103–117.

Lichtenstein, P., et al. (2000). Environmental and heritable factors in the causation of cancer—Analyses of cohorts of twins from Sweden, Denmark, and Finland. *The New England Journal of Medicine, 343*(2), 78–85.

Lief, H. I, & Hubschman, L. (1993). Orgasm in the postoperative transsexual. *Archives of Sexual Behavior, 22* 145–155.

Lindermalm, G., Korlin, D., & Uddenberg, N. (1986). Long-term follow-up of "sex change" in 134 male to female transsexuals. *Archives of Sexual Behavior, 15,* 187–210.

Linet, O. I., & Ogrinc, F. G. (1996). Efficacy and safety of intracavernosal alprostadil in men with erectile dysfunction. *The New England Journal of Medicine, 334,* 873–877.

Linz, D. (1989). Exposure to sexually explicit materials and attitudes toward rape: A comparison of study results. *Journal of Sex Research, 26,* 50–84.

Linz, D., Donnerstein, E., & Penrod, S. (1988). The effects of long-term exposure to violent and sexually degrading depictions of women. *Journal of Personality and Social Psychology, 55,* 758–767.

Lipshultz, L. I. (1996). Injection therapy for erectile dysfunction. *The New England Journal of Medicine, 334,* 913–914.

Lira, L. R., Koss, M. P., & Russo, N. F. (1999). Mexican American women's definitions of rape and sexual abuse. *Hispanic Journal of Behavioral Sciences, 21*(3), 236–265.

Lisak, D. (1991). Sexual aggression, masculinity, and fathers. *Signs, 16,* 238–262.

Lisak, D., & Roth, S. (1990). Motives and psychodynamics of self-reported, unincarcerated rapists. *American Journal of Orthopsychiatry, 60,* 268–280.

Liskin, L., Wharton, C, & Blackburn, R. (1990). Condoms—now more than ever. *Population Reports, 8,* 1–36.

Loder, N. (2000). U.S. science shocked by revelations of sexual discrimination. *Nature, 405,* 713–714.

LoPiccolo, J. (1994) The evolution of sex therapy. *Sexual and Marital Therapy, 9*(1), 5–7.

LoPiccolo, J., & Friedman, J. (1988). Broad-spectrum treatment of low sexual desire: Integration of cognitive, behavioral, and systemic therapy. In S. Leiblum & R. Rosen (eds.), *Sexual desire disorders.* New York: Guilford Press.

LoPiccolo, J., & Stock, W. E. (1986). Treatment of sexual dysfunction. *Journal of Consulting and Clinical Psychology, 54,* 158–167.

Lorefice, L. S. (1991). Fluoxetine treatment of a fetish. *Journal of Clinical Psychiatry, 52.*

Lott, B. (1985). The potential enhancement of social/personality psychology through feminist research and vice versa. *American Psychologist, 40,* 155–164.

Lovdal, L. T. (1989). Sex role messages in television commercials: An update. *Sex Roles, 21,* 715–724.

Lown, J., & Dolan, E. (1988). Financial challenges in remarriage. *Lifestyles: Family and Economic Issues, 9,* 73–88.

Lowry, R., et al. (1994). Substance use and HIV–related sexual behaviors among U.S. high school students: Are they related? *American Journal of Public Health, 84*(7) 1116–1120.

Loy, P. H., & Stewart, L. P. (1984). The extent and effects of the sexual harassment of working women. *Sociological Focus, 17,* 31–43.

Lublin, J. (1995, September 28). Survey finds more Fortune 500 firms have at least two female directors. *Wall Street Journal,* p. B16.

Luckenbill, D. F. (1985). Entering male prostitution. *Urban Life, 14,* 131–153.

Lue, T. F. (2000). Drug therapy: Erectile dysfunction. *The New England Journal of Medicine* online, 342(24).

Lundstrom, B., Pauly, I., & Walinder, J. (1984). Outcome of sex reassignment surgery. *Acta Psychiatrica Scandinavica, 70,* 289–294.

Lynxwiler, J., & Gay, D. (1994). Reconsidering race differences in abortion attitudes. *Social Science Quarterly, 75*(1) 67–84.

Lyons, J. (1991). Artistic freedom and the university. *Art Journal, 50*(4), 77–83.

M

Macallan, D. C., et al. (1995). Energy expenditure and wasting in human immunodeficiency virus infection. *The New England Journal of Medicine, 333,* 83–88

MacAndrew, S. (2000). Sexual health through leadership and "sanuk" in Thailand. *British Medical Journal, 321*(7253), 114.

Maccoby, E. E. (1990). Gender and relationships: A developmental account. *American Psychologist, 45,* 513–520.

MacDonald, N. E., et al. (1990). High-risk STD/HIV behavior among college students. *Journal of the American Medical Association, 263,* 3155–3159.

MacDonald, T. K., MacDonald, G., Zanna, M. P., & Fong, G. T. (2000). Alcohol, sexual arousal, and intentions to use condoms in young men: Applying alcohol myopia theory to risky sexual behavior. *Health Psychology, 19,* 290–298.

MacFarquhar, N. (1996, August 8). Mutilation of Egyptian girls: Despite ban, it goes on. *The New York Times,* p. A3.

Macrae, C. N., & Shepherd, J. W. (1989). Sex differences in the perception of rape victims. *Journal of Interpersonal Violence, 4,* 278–288.

MacVicar, S. (2000, June 5). Crimes in the name of honor. www.abcnews.com.

Magdol, L., et al. (1997). Gender differences in partner violence in a birth cohort of 21-year-olds: Bridging the gap between clinical and epidemiological approaches. *Journal of Consulting and Clinical Psychology, 65,* 68–78.

Mahoney, E. R., et al. (1986). Sexual coercion and assault: Male socialization and female risk. *Sexual Coercion and Assault, 1,* 2–8.

Maines, R. P. (1999). *The technology of orgasm:* "Hysteria," the vibrator, and women's sexual satisfaction. Baltimore, MD: The Johns Hopkins University Press.

Major, B., & Cozzarelli, C. (1992). Psychosocial predictors of adjustment to abortion. *Journal of Social Issues, 48,* 121–142.

Major, B., Cozzarelli, C., Cooper, M. L., Zubek, J., Richards, C., et al. (2000). Psychological responses of women after first-trimester abortion. *Archives of General Psychiatry, 57,* 777–784.

Malamuth, N. M. (1981). Rape proclivity among males. *Journal of Social Issues, 37,* 138–157.

Malamuth, N. M. (1984). Aggression against women: Cultural and individual causes. In N. M. Malamuth & E. Donnerstein (eds.), *Pornography and sexual aggression* (pp. 19–52). Orlando, FL: Academic Press.

Malamuth, N. M., & Ceniti, J. (1986). Repeated exposure to violent and nonviolent pornography: Likelihood-of-raping ratings and laboratory aggression against women. *Aggressive Behavior, 12,* 129–137.

Maletzky, B. M. (1980). Self-referred vs. court-referred sexually deviant patients: Success with assisted covert sensitization. *Behavior Therapy, 11,* 306–314.

Malinowski, B. (1927). *Sex and repression in savage society.* London: Kegan Paul, Trench, Trubner & Co.

Malinowski, B. (1929). *The sexual life of savages in north-western Melanesia.* New York: Eugenics.

Malloy, M. H., Hoffman, H. J., Peterson, D. R. (1992). Sudden infant death syndrome and maternal smoking. *American Journal of Public Health, 82,* 1380–1382.

Mammograms on rise, federal study finds. (1992, April 29). *The New York Times,* p. C13.

Man pays victim's husband in fondling case. (2000, June 2). Reuters News Agency online.

Marchbanks, P. A., et al. (1988). Risk factors for ectopic pregnancy. *Journal of the American Medical Association, 259,* 1823–1827.

Marchbanks, P. A., et al. (2000). Cigarette smoking and epithelial ovarian cancer by histologic type. *Obstetrics & Gynecology, 95,* 255–260.

Marcus, A. J. (1995). Aspirin as a prophylaxis against colorectal cancer. *The New England Journal of Medicine, 333,* 656–658.

Margolin, L., Miller, M., & Moran, P. B. (1989). When a kiss is not just a kiss: Relating violations on consent in kissing to rape myth acceptance. *Sex Roles, 20,* 231–243.

Marks, G., Miller, N., & Maruyama, G. (1981). Effect of targets' physical attractiveness on assumption of similarity. *Journal of Personality and Social Psychology, 41,* 198–206.

Markowitz, M., et al. (1995). A preliminary study of ritonavir, an inhibitor of HIV-1 protease, to treat HIV-1 infection. *The New England Journal of Medicine, 333,* 1534–1539.

Marshall, D. (1971). Sexual behavior on Mangaia. In D. Marshall & R. Suggs (eds.), *Human sexual behavior: Variations in the ethnographic spectrum* (pp. 103–162). New York: Basic Books.

Marshall, W. L. (1989). Pornography and sex offenders. In D. Zillmann & J. Bryant (eds.), *Pornography: Research advances and policy considerations* (pp. 185–214). Hillsdale, NJ: Lawrence Erlbaum Associates.

Marsiglio, W. (1993a). Adolescent males' orientation toward paternity and contraception. *Family Planning Perspectives, 25,* 22–31.

Marsiglio, W. (1993b). Attitudes toward homosexual activity and gays as friends: A national survey of heterosexual 15- to 19-year-old males. *Journal of Sex Research, 30,* 12–17.

Martin, C. L., & Halverson, C. F., Jr. (1981). A schematic processing model of sex typing and stereotyping in children. *Child Development, 54,* 1119–1134.

Martin, C. L., & Halverson, C. F., Jr. (1983). The effects of sex-typing schemas on young children's memory. *Child Development, 54,* 563–574.

Martin, C. W., et al. (2000). Dose-finding study of oral desogestrel with testosterone pellets for suppression of the pituitary–testicular axis in normal men. *Human Reproduction, 15*(7), 1515–1524.

Martin, D. H. (1990). Chlamydial infections. *Medical Clinics of North America, 74,* 1367–1387.

Martin, J. N., et al. (1998). Sexual transmission and the natural history of human herpesvirus 8 infection. *The New England Journal of Medicine, 338,* 948–954.

Martinez, F. D., Cline, M., & Burrows, B. (1992). Increased incidence of asthma in children of smoking mothers. *Pediatrics, 89,* 21–26.

Martinson, F. M. (1976). Eroticism in infancy and childhood. *Journal of Sex Research, 2,* 251–262.

Martz, J. M., et al. (1998). Positive illusion in close relationships. *Personal Relationships, 5*(2) 159–181.

Marwick, C. (2000). Consensus panel considers osteoporosis. *Journal of the American Medical Association* online, 283(16).

Masters, W. H., & Johnson, V. E. (1966). *Human sexual response.* Boston: Little, Brown.

Masters, W. H., & Johnson, V. E. (1970). *Human sexual inadequacy.* Boston: Little, Brown.

Masters, W. H., & Johnson, V. E. (1979). *Homosexuality in perspective.* Boston: Little, Brown.

Masters, W. H., et al. (1989). *Human sexuality,* 4th ed. New York: HarperCollins.

Matek, O. (1988). Obscene phone callers. *Journal of Social Work and Human Sexuality,* 7, 113–130.

Matlin, M. W. (1996). *The psychology of women,* 3rd ed. Fort Worth, TX: Harcourt.

Matthews, K. A., et al. (1990). Influences of natural menopause on psychological characteristics and symptoms of middle-aged healthy women. *Journal of Consulting and Clinical Psychology, 58,* 345–351.

Matus, I. (1996.) Cited in Clay, R. A. (1996). Beating the "biological clock" with zest. *APA Monitor, 27*(2), 37.

Maxson, S. C. (1998). Homologous genes, aggression, and animal models. *Developmental Neuropsychology, 14*(1), 143–156.

Maybach, K. L., & Gold, S. R. (1994). Hyperfemininity and attraction to macho and non-macho men. *Journal of Sex Research, 31*(2), 91–98.

Mayer, J. P., Hawkins, B., & Todd, R. (1990). A randomized evaluation of smoking cessation interventions for pregnant women at a WIC [Women, Infants and Children] clinic. *American Journal of Public Health, 80,* 76–79.

McArthur, M. J. (1990). Reality therapy with rape victims. *Archives of Psychiatric Nursing, 4,* 360–365.

McCabe, M. P., & Delaney, S. M. (1992). An evaluation of therapeutic programs for the treatment of secondary inorgasmia in women. *Archives of Sexual Behavior, 21,* 69–89.

McCloskey, L. A. (1996). Socioeconomic and coercive power within the family. *Gender and Society, 10,* 449–463.

McConaghy, M. J. (1979). Gender performance and the genital basis of gender: Stages in the development of constancy of gender. *Child Development, 50,* 1223–1226.

McConaghy, N. (1987). Heterosexuality/homosexuality: Dichotomy or continuum. *Archives of Sexual Behavior, 16,* 411–424.

McCormack, W. M. (1990). Overview. *Sexually Transmitted Diseases, 57,* 187–191.

McCusker, J., et al. (1992). Maintenance of behavioral change in a cohort of homosexually active men. *AIDS, 6,* 861–868.

McDonough, Y. Z. (1998, January 24). What Barbie really taught me. *The New York Times Magazine,* p. 70.

McGuire, R. J., Carlisle, J. M., & Young, B. G. (1965). Sexual deviation as conditioned behavior: A hypothesis. *Behaviour Research and Therapy, 2,* 185–190.

McKeachie, W. (1994) Cited in DeAngelis, T. (1994). Educators reveal keys to success in classroom. *APA Monitor, 25*(1), 39–40.

McKinney, K. (1989). Social factors in contraceptive and abortion attitudes and behaviors. In K. McKinney & S. Sprecher (eds.), *Human sexuality: The societal and interpersonal context.* Norwood, NJ: Ablex.

McKinney, K., & Maroules, N. (1991). Sexual harassment. In E. Grauerholz & M. A. Koralewski (eds.), *Sexual coercion: A sourcebook on its nature, causes, and prevention* (pp. 29–44). Lexington, MA: Lexington Books.

McKirnan, D. J., Stokes, J. P., Doll, L., & Burzette, R. G. (1995). Bisexually active men: Social characteristics and sexual behavior. *Journal of Sex Research, 32,* 65–76.

McLean, P. M. (1976). Brain mechanisms of elemental sexual functions. In B. J. Sadock et al. (eds.), *The sexual experience.* Baltimore, MD: Williams & Wilkins.

McMahon, M. J., et al. (1996). Comparison of a trial of labor with an elective second cesarean section. *The New England Journal of Medicine, 335,* 689–695.

McNeil, D. G., Jr. (2000a, April 6). Simple antibiotic urged for Africans with HIV. *The New York Times* online.

McNeil, D. G., Jr. (2000b, May 12). Companies to cut costs of AIDS drugs for poor nations. *The New York Times* online.

Mead, M. (1935). *Sex and temperament in three primitive societies.* New York: Dell.

Mead, M. (1967). *Male and female: A study of the sexes in a changing world.* New York: William Morrow.

Meana, M., & Binik, Y. M. (1994). Painful coitus: A review of female dyspareunia. *Journal of Nervous and Mental Disease, 182*(5), 264–272.

Meier, B. (1997, June 8). In war against AIDS, battle over baby formula reignites. *The New York Times,* pp. A1, A16.

Meiselman, K. C. (1978). *Incest: A psychological study of causes and effects with treatment recommendations.* San Francisco: Jossey-Bass.

Meisler, A. W., & Carey, M. P. (1990). A critical reevaluation of nocturnal penile tumescence monitoring in the diagnosis of erectile dysfunction. *Journal of Nervous and Mental Disease, 178,* 78–89.

Melbye, M., et al. (1997). Induced abortion and the risk of breast cancer. *The New England Journal of Medicine, 336,* 81–85.

Melis, M. R., Succo, S., & Spano, M. S., & Argiolas, A. (2000). Effect of excitatory amino acid, dopamine, and oxytocin receptor antagonists on noncontact penile erections and paraventricular nitric oxide production in male rats. *Behavioral Neuroscience, 114*(4), 849–857.

Merrill, R. M., & Brawley, O. W. (2000). Prostate cancer incidence and mortality rates among White and Black men. *Epidemiology* online, 11(2).

Mertz, G. J., et al. (1992). Risk factors for the sexual transmission of genital herpes. *Annals of Internal Medicine, 116,* 197–202.

Messenger, J. C. (1971). Sex and repression in an Irish folk community. In D. S. Marshall and R. C. Suggs (eds.), *Human sexual behavior: Variations in the ethnographic spectrum* (pp. 3–37). New York: Basic Books.

Meston, C. M., & Gorzalka, B. B. (1992). Psychoactive drugs and human sexual behavior: The role of serotonergic activity. *Journal of Psychoactive Drugs, 24,* 1–40.

Meston, C. M., & Heiman, J. R. (2000). Sexual abuse and sexual function: An examination of sexually relevant cognitive processes. *Journal of Consulting and Clinical Psychology, 68*(3), 399–406.

Meuwissen, I., & Over, R. (1990). Habituation and dishabituation of female sexual arousal. *Behaviour Research and Therapy, 28,* 217–226.

Meyer, J. K., & Reter, D. J. (1979). Sex reassignment: Follow-up. *Archives of General Psychiatry, 36,* 1010–1015.

Meyer, T. (1998, February 18). AZT short treatment works. The Associated Press online.

Meyer-Bahlburg, H. F. L., et al. (1995). Prenatal estrogens and the development of homosexual orientation. *Developmental Psychology, 31*(1), 12–21.

Michael, R. T., Gagnon, J. H., Laumann, E. O., & Kolata, G. (1994). *Sex in America: A definitive survey.* Boston: Little, Brown.

Michelson, D., et al. (2000). Female sexual dysfunction associated with antidepressant administration: A randomized, placebo-controlled study of pharmacologic intervention. *American Journal of Psychiatry, 157,* 239–243.

Mikach, S. M., & Bailey, J. M. (1999). What distinguishes women with unusually high numbers of sex partners? *Evolution & Human Behavior, 20*(3), 141–150.

Mill, J. S. (1939). Utilitarianism. In E. A. Burtt (ed.), *The English philosophers.* New York: The Modern Library. (Original work published 1863.)

Miller, A. B., To, T., Baines, C. J., & Wall, C. (2000). Canadian National Breast Screening Study-2: 13-year results of a randomized trial in women aged 50–59 years. *Journal of the National Cancer Institute, 92,* 1490–1499.

Miller, B. C., McCoy, J. K., & Olson, T. D. (1986). Dating age and stage as correlates of adolescent sexual attitudes and behavior. *Journal of Adolescent Research, 1,* 361–371.

Miller, H. G., Turner, C. F., & Moses, L. E. (eds.) (1990). *AIDS: The second decade.* Washington, DC: National Academy.

Miller, M. (1998). Cited in Bronner, E. (1998, February 1). Just say maybe. "No sexology, please. We're Americans." *The New York Times,* p. WK6.

Miller, S. M., Shoda, Y., & Hurley, K. (1996). Applying cognitive-social theory to health-protective behavior: Breast self-examination in cancer screening. *Psychological Bulletin, 199,* 70–94.

Milliken, M.. (2000, June 15). Weddings brighten biggest Latin American prison. Reuters News Agency online.

Minai, N. (1981). *Women in Islam: Tradition and transition in the Middle East.* London: John Murray.

Minkoff, H. L, et al. (1990) The relationship of cocaine use to syphilis and human immunodeficiency virus infections among inner-city parturient women. *American Journal of Obstetrics and Gynecology, 163,* 521–526.

Mirotznik, J., et al. (1987). Genital herpes: An investigation of its attitudinal and behavioral correlates. *Journal of Sex Research, 23,* 266–272.

Mirotznik, J. (1991). Genital herpes: A survey of the attitudes, knowledge, and reported behaviors of college students at risk for infection. *Journal of Psychology and Human Sexuality, 4,* 73–99.

Mishell, D. R., Jr. (1989). Medical progress: Contraception. *The New England Journal of Medicine, 320,* 777–787.

Missailidis, K., & Gebre-Medhin, M. (2000). Female genital mutilation in eastern Ethiopia. *The Lancet, 356,* 137–138.

MMWR (1995). Update: AIDS among women—United States, 1994. *Morbidity and Mortality Weekly Report, 44,* 81–84.

Mofenson, L. M. (2000). Perinatal exposure to zidovudine—Benefits and risks. *The New England Journal of Medicine* online, 343(11).

Mohr, D. C., & Beutler, L. E. (1990). Erectile dysfunction: A review of diagnostic and treatment procedures. *Clinical Psychology Review,* 10, 123–150.

Money, J. (1990). Agenda and credenda of the Kinsey Scale. In D. P. McWhirter, S. A. Sanders, & J. M. Reinisch (eds.), *Homosexuality/heterosexuality: Concepts of sexual orientation* (pp. 41–60). New York: Oxford University Press.

Money, J. (1994). The concept of gender identity disorder in childhood and adolescence after 39 years. *Journal of Sex and Marital Therapy, 20*(3), 163–177.

Money, J., & Ehrhardt, A. (1972). *Man and woman, boy and girl.* Baltimore, MD: The Johns Hopkins University Press.

Money, J., & Lamacz, M. (1989). *Vandalized lovemaps.* Buffalo, NY: Prometheus Books.

Money, J., Lehne, G., & Pierre-Jerome, F. (1984). Micropenis: Adult follow-up and comparison of size against new norms. *Journal of Sex and Marital Therapy, 10,* 105–116.

Montagnier, L. (1996). Cited in Altman, L. K. (1996, August 7). Urine test for H. I. V. is approved. *The New York Times,* p. A14.

Moore, K. A., & Stief, T. M. (1992). Changes in marriage and fertility behavior: Behavior versus attitudes of young adults. *Youth and Society, 22,* 362–386.

Morales, A. (1993). Nonsurgical management options in impotence. *Hospital Practice, 28*(15–6), 19ff.

Morales, E. (1992). Latino gays and Latina lesbians. In S. Dworkin & F. Gutierrez (eds.), *Counseling gay men and lesbians: Journey to the end of the rainbow.* Alexandria, VA: American Association for Counseling and Development.

Moran, J. S., & Zenilman, J. M. (1990). Therapy for gonococcal infections: Options in 1989. *Reviews of Infectious Diseases* (Suppl. 6.), S633–S644.

More single mothers. (1993, July 26). *Time,* p. 16.

Morgan, T. O., et al. (1996). Age-specific reference ranges for serum prostate-specific antigen in black men. *The New England Journal of Medicine, 335,* 304–310.

Morokoff, P. J. (1993). Female sexual arousal disorder. In W. Donohue and J. H. Greer (eds.), *Handbook of sexual dysfunctions: Assessment and treatment* (pp. 157–199). Boston: Allyn & Bacon.

Morris, L. B. (2000, June 25). For the partum blues, a question of whether to medicate. *The New York Times* online.

Morris, N., et al. (1987). Marital sex frequency and midcycle female testosterone. *Archives of Sexual Behavior, 7,* 157–173.

Morrison, E. S., et al. (1980). *Growing up sexual.* New York: Van Nostrand Reinhold.

Morse, E., et al. (1992). Sexual behavior patterns of customers of male street prostitutes. *Archives of Sexual Behavior, 21,* 347.

Mortola, J. F. (1998). Premenstrual syndrome—Pathophysiologic considerations. *The New England Journal of Medicine, 338,* 256–257.

Mosher, D. L. (1988). Pornography defined: Sexual involvement theory, narrative context, and goodness–of–fit. *Journal of Psychology and Human Sexuality, 1,* 67–85.

Mosher, W. D., & Bachrach, C. A. (1987). First premarital contraceptive use. *Studies in Family Planning, 18,* 83.

Mrs. Dole calls Viagra "great drug." (1998, May 9). Associated Press online.

Muehlenhard, C. L., & Falcon, P. L. (1990). Men's heterosocial skill and attitudes toward women as predictors of verbal sexual coercion and forceful rape. *Sex Roles, 23,* 241–259.

Muehlenhard, C. L., & Linton, M. A. (1987). Date rape and sexual aggression in dating situations: Incidence and risk factors. *Journal of Counseling Psychology, 34,* 186–196.

Mueser, K. T., et al. (1984). You're only as pretty as you feel: Facial expression as a determinant of physical attractiveness. *Journal of Personality and Social Psychology, 46,* 469–478.

Mulnard, R. A., et al. (2000). Estrogen replacement therapy for treatment of mild to moderate Alzheimer disease. *Journal of the American Medical Association, 283,* 1007–1015.

Mulry, G., Kalichman, S. C., & Kelly, J. A. (1994). Substance use and unsafe sex among gay men: Global versus situational use of substances. *Journal of Sex Education and Therapy, 20*(3), 175–184.

Mundy, L. (2000, July 16). Sex and sensibility. *The Washington Post* online.

Murstein, B. I. (1988). A taxonomy of love. In R. J. Sternberg & M. L. Barnes (eds.), *The psychology of love.* (pp. 13–37). New Haven, CT: Yale University Press.

Murstein, B., Merighi, J. R., & Vyse, S. A. (1991). Love styles in the United States and France: A cross-cultural comparison. *Journal of Social and Clinical Psychology, 10*, pp. 37–46.

Muster, N. J., (1992). Treating the adolescent victim-turned-offender. *Adolescence, 27*, 441–450.

Myers, A. M., & Gonda, G. (1982). Utility of the masculinity–femininity construct: Comparison of traditional and androgyny approaches. *Journal of Personality and Social Psychology, 43*, 514–523.

Myers, M. F. (1989). Men sexually assaulted as adults and sexually abused as boys. *Archives of Sexual Behavior, 18*, 203–215.

Myers, S. L. (2000a, March 25). Survey of troops finds antigay bias common in service. *The New York Times* online.

Myers, S. L. (2000b, March 31). Female general in army alleges sex harassment. *The New York Times*, pp. A1, A22.

N

Nabel, G. J., Sullivan, N. J. (2000). Antibodies and resistance to natural HIV infection. *The New England Journal of Medicine* online, 343(17).

Nadeau, R., et al. (1993). Knowledge and beliefs regarding STDs and condoms among students. *Canadian Journal of Public Health, 84*, 181–185.

Nadler, R. D. (1990). Homosexual behavior in nonhuman primates. In D. P. McWhirter, S. A. Sanders, & J. M. Reinisch (eds.) *Homosexuality/heterosexuality: Concepts of sexual orientation* (pp. 138–170). New York: Oxford University Press.

Nakanishi, M. (1986). Perceptions of self-disclosure in initial interaction: A Japanese sample. *Human Communication Research, 13*, 167–190.

Narod, S. A., et al. (1998). Oral contraceptives and the risk of hereditary ovarian cancer. *The New England Journal of Medicine 339*, 424–428.

National Academy of Sciences, Institute of Medicine. (1982). *Marijuana and health.* Washington, DC: National Academic Press.

National Cancer Institute. (2000). http://www.nci.nih.gov/

National Center for Biotechnology Information (NCBI). (2000, March 30). National Institutes of Health. http://www.ncbi.nlm.nih.gov/disease/SRY.html

National Crime Victimization Survey (1995). U.S. Bureau of Justice Statistics. Washington, DC: U.S. Department of Justice.

Navarro, M. (1993, February 18). New York needle exchanges called surprisingly effective. *The New York Times*, pp. A.1, B4.

Navarro, M. (1996, August 31). Lesbian loses court appeal for custody of daughter. *The New York Times*, p. L7.

Nduati, R., et al. (2000). Effect of breastfeeding and formula feeding on transmission of HIV-1. *Journal of the American Medical Association, 283*, 1167–1174.

Nelson, R. (1988). Nonoperative management of impotence. *Journal of Urology, 139*, 2–5.

Nevid, J. S. (1984). Sex differences in factors of romantic attraction. *Sex Roles, 11*, 401–411.

Nevid, J. S., Rathus, S. A., & Greene, B. A. (2000). *Abnormal psychology in a changing world*, 4th ed. Upper Saddle River, NJ: Prentice-Hall.

New use for Prozac, A. (1995, June 19). *Newsweek*, p. 85.

Newcomb, P. A., & Storer, B. E. (1995). Postmenopausal hormone use and risk of large-bowel cancer. *Journal of the National Cancer Institute, 87*(14), 1067–1071.

New fertility treatment may increase birth defects. (1997, November 13). Reuters News Agency online.

Ngai, S. W., Tang, O. S., Chan, Y. M., & Ho, P. C. (2000). Vaginal misoprostol alone for medical abortion up to 9 weeks of gestation: Efficacy and acceptability. *Human Reproduction, 15*(5), 1159–1162.

Nichols, M. (1990). Lesbian relationships: Implications for the study of sexuality and gender. In D. P. McWhirter, S. A. Sanders, & J. M. Reinisch (eds.), *Homosexuality/heterosexuality: Concepts of sexual orientation* (pp. 350–364). New York: Oxford University Press.

Niebuhr, G. (1998, April 17). More clergy willing to perform same-sex marriages. *The New York Times* online.

Nieto, J. J., Cogswell, D., Jesinger, D., & Hardiman, P. (2000). Lipid effects of hormone replacement therapy with sequential transdermal 17-beta-estradiol and oral dydrogesterone. *Obstetrics & Gynecology*, 95, 111–114.

Nock, S. L. (1995). A comparison of marriages and cohabiting relationships. *Journal of Family Issues, 16*(1) 53–76.

Nordqvist, K., & Lovell-Badge, R. (1994). Setbacks on the road to sexual fulfillment. *Nature Genetics*, 7, 7–9.

Norman, J., & Harris, M. (1981). *The private life of the American teenager.* New York: Rawson Wade.

Nosek, M. A., et al. (1994). Wellness models and sexuality among women with physical disabilities. *Journal of Applied Rehabilitation Counseling, 25*(1), 50–58.

Nour, N. W. (2000). Cited in Dreifus, C. (2000, July 11). A conversation with Dr. Nawal M. Nour: A life devoted to stopping the suffering of mutilation. *The New York Times* online.

O

O'Connor, T. G., Caspi, A., DeFries, J. C., & Plomin, R. (2000). Are associations between parental divorce and children's adjustment genetically mediated? An adoption study. *Developmental Psychology, 36*(4), 429–437.

O'Dell, K. M. C., & Kaiser, K. (1997). Sexual behaviour: Secrets and flies. *Current Biology*, 7(6), R345–R347.

O'Donohue, W., Letourneau, E., & Geer, J. H. (1993). Premature ejaculation. In W. O'Donohue & J. H. Geer (eds.), *Handbook of sexual dysfunctions: Assessment and treatment.* (pp. 303–333). Boston: Allyn & Bacon.

O'Hara, M. W., Neunaber, D. J., & Zekoski, E. M. (1984). Prospective study of postpartum depression: Prevalence, course, and predictive factors. *Journal of Abnormal Psychology, 93*, 158–171.

O'Hara, M. W., et al. (1991). Prospective study of postpartum blues: Biological and psychosocial factors. *Archives of General Psychiatry, 48*, 801–806.

O'Neill, N., & O'Neill, G. (1972). *Open marriage.* New York: Evans.

Ochs, R. (1994, January 11). Cervical cancer comeback. *New York Newsday*, pp. 55, 57.

Okami, P., & Goldberg, A. (1992). Personality correlates of pedophilia: Are they reliable indicators? *Journal of Sex Research, 29*, 297–328.

Oldenburg, R., & Brissett, D. (1980, April). The essential hangout. *Psychology Today*, pp. 81–84.

Olds, J. (1956). Pleasure centers in the brain. *Scientific American, 193*, 105–116.

Olds, J., & Milner, P. (1954). Positive reinforcement produced by electrical stimulation of the septal area and other regions of the rat brain. *Journal of Comparative and Physiological Psychology, 47*, 419–427.

Orwoll, E., et al. (2000). Alendronate for the treatment of osteoporosis in men. *The New England Journal of Medicine, 343*(9), 604–610.

Osborne, N. G., & Adelson, M. D. (1990). Herpes simplex and human papillomavirus genital infections: Controversy over obstetric management. *Clinical Obstetrics and Gynecology, 33*, 801–811.

Osterhout, M. B., Formichella, A., & McIntyre, S. (1991). *Tell it like it is: Straight talk about sex.* New York: Avon Books.

Oswalt, R., & Matsen, K. (1993). Sex, AIDS, and the use of condoms: A survey of compliance in college students. *Psychological Reports, 72*, 764–766.

Overbeek, P. (1999). Cited in Philipkoski, K. (1999, October 28). Why men are that way. Wired Digital, Inc.

Overholser, J. C., & Beck, S. (1986). Multimethod assessment of rapists, child molesters, and three control groups on behavioral and psychological measures. *Journal of Consulting and Clinical Psychology, 54*, 682–687.

Oz, S. (1994). Decision making in divorce therapy: Cost–cost comparisons. *Journal of Marital and Family Therapy, 20*, 77–81.

Oz, S. (1995). A modified balance-sheet procedure for decision making in therapy: Cost–cost comparisons. *Professional Psychology: Research and Practice, 26*, 78–81.

P

Padgett, V. R., Brislin-Slütz, J. A., & Neal, J. A. (1989). Pornography, erotica, and attitudes toward women: The effects of repeated exposure. *Journal of Sex Research, 26*, 479–491.

Padma-Nathan, H., et al. (1997). Treatment of men with erectile dysfunction with transurethral alprostadil. *The New England Journal of Medicine, 336*, 1–7.

Palace, E. M. (1995). Modification of dysfunctional patterns of sexual arousal through autonomic arousal and false physiological feedback. *Journal of Consulting and Clinical Psychology, 63*, 604–615.

Palella, F. J., et al. (1998). Declining morbidity and mortality among patients with advanced human immunodeficiency virus infection. *The New England Journal of Medicine, 338*, 853–860.

Palmore, E. (1981). *Social patterns in normal aging: Findings from the Duke Longitudinal Study.* Durham, NC: Duke University Press.

Palson, C., & Palson, R. (1972). Swinging in wedlock. *Society, 9*(4), 28–37.

"Panel says Pap tests could almost end cervical cancer deaths." (April 4, 1996). *The New York Times*, p. A18.

Parsons, N. K., Richards, H. C., & Kanter, G. D. (1990). Validation of a scale to measure reasoning about abortion. *Journal of Counseling Psychology, 37*, 107–112.

Patterson, C. J. (1995). Special Issue: Sexual orientation and human development. *Developmental Psychology, 31*, 3–140.

Paul, R. H. (1996). Toward fewer cesarean sections—The role of a trial of labor. *The New England Journal of Medicine, 335*, 735–736.

Pauly, B., & Edgerton, M. (1986). The gender-identity movement. *Archives of Sexual Behavior, 15*, 315–329.

Pauly, I. B. (1974). Female transsexualism: Part 1. *Archives of Sexual Behavior, 3*, 487–508.

Pearson, C. A. (1992, February 1). Cited in Leary, W. E. (1992). U.S. panel backs approval of first condom for women. *The New York Times*, p. 7.

Penn, F. (1993, October 21). Cancer confusion: The risks and realities of human papilloma virus. *Manhattan Spirit*, pp. 14–15.

Peplau, L. A., & Cochran, S. D. (1990). A relationship perspective on homosexuality. In D. P. McWhirter, S. A. Sanders, & J. M. Reinisch (eds.), *Homosexuality/heterosexuality: Concepts of sexual orientation* (pp. 321–349). New York: Oxford University Press.

Peplau, L. A., Garnets, L. D., Spalding, L. R., Conley, T. D., & Veniegas, R. C. (1998). A critique of Bem's "Exotic Becomes Erotic" theory of sexual orientation. *Psychological Review, 105*(2), 387–394.

Peplau, L. A., & Gordon, S. L. (1985). Women and men in love: Sex differences in close heterosexual relationships. In V. O'Leary et al. (eds.), *Women, gender, and social psychology.* Hillsdale, NJ: Lawrence Erlbaum Associates.

Perduta-Fulginiti, P. S. (1992). Sexual functioning of women with complete spinal cord injury: Nursing implications. [Special Issue: Nursing roles and perspectives]. *Sexuality and Disability, 10*, 103–118.

Peretti, P. O., & Pudowski, B. C. (1997). Influence of jealousy on male and female college daters. *Social Behavior & Personality, 25*(2), 155–160.

Perrett, D. I. (1994). *Nature.* Cited in Brody, J. E. (1994, March 21). Notions of beauty transcend culture, new study suggests. *The New York Times*, p. A14.

Perriëns, J. (2000). Cited in "UNAIDS calls for continued commitment to microbicides." (2000, July 12). UNAIDS press release.

Perry, D. G., & Bussey, K. (1979). The social learning theory of sex differences: Imitation is alive and well. *Journal of Personality and Social Psychology, 37*, 1699–1712.

Perry, J. D., & Whipple, B. (1981). Pelvic muscle strength of female ejaculation: Evidence in support of a new theory of orgasm. *Journal of Sex Research, 17*, 22–39.

Perry, M. J., et al. (1994). High risk sexual behavior and alcohol consumption among bar-going gay men. *AIDS, 8*(9), 1321–1324.

Persky, H., et al. (1982). The relation of plasma androgen levels to sexual behaviors and attitudes of women. *Psychosomatic Medicine, 44*, 305–309.

Person, E. S., et al. (1989). Gender differences in sexual behaviors and fantasies in a college population. *Journal of Sex and Marital Therapy, 15*, 187–198.

Pillard, R. C. (1990). The Kinsey Scale: Is it familial? In D. P. McWhirter, S. A. Sanders, & J. M. Reinisch (eds.), *Homosexuality/heterosexuality: Concepts of sexual orientation* (pp. 88–100). New York: Oxford University Press.

Pillard, R. C., & Weinrich, J. D. (1986). Evidence of familial nature of male homosexuality. *Archives of Sexual Behavior, 43*, 808–812.

Pines, A. M., & Friedman, A. (1998). Gender differences in romantic jealousy. *Journal of Social Psychology, 138*(1), 54–71.

Pinkerton, S. D., & Abramson, P. R. (1992). Is risky sex rational? *Journal of Sex Research, 29*, 561–568.

Piot, P. Cited in UNAIDS Press Release (2000, June 5). Gender is crucial issue in fight against AIDS, says head of UNAIDS. New York.

Pistella, C. L., & Bonati, F. A. (1999). Adolescent women's recommendations for enhanced parent–adolescent communication about sexual behavior. *Child & Adolescent Social Work Journal, 16*(4), 305–315.

Pitnick, S. (1995, May 11). *Nature.*

Pizzari, T., & Birkhead, T. R. (2000). Female feral fowl eject sperm of subdominant males. *Nature, 405*, 787–789.

Plant, E. A., Hyde, J. S., Keltner, D., & Devine, P. G. (2000). The gender stereotyping of emotions. *Psychology of Women Quarterly, 24*(1), 81–92.

Platt, R., Rice, P., & McCormack, W. (1983). Risk of acquiring gonorrhea and prevalence of abnormal adrenal findings among women recently exposed to gonorrhea. *Journal of the American Medical Association, 250*, 3205–3209.

Pliner, A. J., & Yates, S. (1992). Psychological and legal issues in minors' rights to abortion. *Journal of Social Issues, 48*, 203–216.

Podolsky, D. (1991, April 15). Charting premenstrual woes. *U.S. News & World Report*, pp. 68–69.

Polit-O'Hara, D., & Kahn, J. (1985). Communication and adolescent contraceptive practices in adolescent couples. *Adolescence, 20*, 33–43.

Pollack, W. S. (1996). Cited in Clay, R. A. (1996). Older men are more involved fathers, studies show. *APA Monitor, 27*(2), 37.

Pollak, M., et al. (1998). *Science.* Cited in Weinraub, M. (1998, January 22). Study finds possible predictor of prostate cancer. Reuters.

Pomeroy, W. B. (1966). Normal vs. abnormal sex. *Sexology, 32*, 436–439.

Poniewozik, J., et al. (2000, June 26). We like to watch: Led by the hit *Survivor*, voyeurism has become TV's hottest genre. Why the passion for peeping? *Time, 155*(26), 56–62.

Population Council (1995). Cited in Brody, J. E. (1995, August 31). Abortion method using two drugs gains in a study. *The New York Times*, pp. A1, B12.

Porter, N., Geis, F. L., Cooper, E., & Newman, E. (1985). Androgyny and leadership in mixed-sex groups. *Journal of Personality and Social Psychology, 49*, 808–823.

Potosky, A. L., et al. (2000). Health outcomes after prostatectomy or radiotherapy for prostate cancer: Results from the Prostate Cancer Outcomes Study. *Journal of the National Cancer Institute, 92*, 1582–1592.

Poussaint, A. (1990, September). An honest look at Black gays and lesbians. *Ebony*, pp. 124ff.

Powderly, W. G., et al. (1995). A randomized trial comparing fluconazole with clotrimazole troches for the prevention of fungal infections in patients with advanced human immunodeficiency virus infection. *The New England Journal of Medicine, 332*, 700–705.

Powell, E. (1991). *Talking back to sexual pressure.* Minneapolis, MN: CompCare Publishers.

Powell, E. (1996). *Sex on your terms.* Boston: Allyn & Bacon.

Pratt, C., & Schmall, V. (1989). College students' attitudes toward elderly sexual behavior: Implications for family life education. *Family Relations, 38*, 137–141.

Prentky, R. A., & Knight, R. A. (1991). Identifying critical dimensions for discriminating among rapists. *Journal of Consulting and Clinical Psychology, 59*, 643–661.

Press, A., et al. (1985, March 18). The war against pornography. *Newsweek*, pp. 58–66.

Preti, G., Cutler, W. B., et al. (1986). Human axillary secretions influence women's menstrual cycles: The role of donor extract of females. *Hormones and Behavior, 20* 474–482.

Price, V. A. (1989). Characteristics and needs of Boston street youth: One agency's response [Special Issue: Runaway, homeless, and shut-out children and youth in Canada, Europe, and the United States]. *Children and Youth Services Review, 11*, 75–90.

Proctor, F., Wagner, N., & Butler, J. (1974). The differentiation of male and female orgasm: An experimental study. In N. Wagner (ed.), *Perspectives on human sexuality.* New York: Behavioral Publications.

Prostitutes do housework too. (2000, June 28). Reuters News Agency online.

Pryde, N. A. (1989). Sex therapy in context. *Sexual and Marital Therapy, 4*, 215–227.

Pullella, P. (2000, June 7). Internet "adultery" a sin, Catholic magazine says. Reuters News Agency online.

Purdy, M. (1995, November 6). A kind of sexual revolution: At some nursing homes, intimacy is a matter of policy. *The New York Times*, pp. B1, B6.

Purifoy, F. E., Grodsky, A., & Giambra, L. M. (1992). The relationship of sexual daydreaming to sexual activity, sexual drive, and sexual attitudes for women across the life-span. *Archives of Sexual Behavior, 21*, 369–375.

Q

Quadagno, D., et al. (1991). Women at risk for human immunodeficiency virus. *Journal of Psychology and Human Sexuality, 4*, 97–110.

Quam, J. K., & Whitford, G. S. (1992). Adaptation and age-related expectations of older gay and lesbian adults. *Gerontologist, 32,* 367–374.

Quevillon, R. P. (1993). Dyspareunia. In W. O'-Donohue & J. H. Geer (eds.), *Handbook of sexual dysfunctions: Assessment and treatment.* (pp. 367–380). Boston: Allyn & Bacon.

Quinn, T. C., et al. (2000). Viral load and heterosexual transmission of Human Immunodeficiency Virus Type 1. *The New England Journal of Medicine, 342*(13), 921–929.

Quinsey, V. L., Chaplin, T. C., & Upfold, D. (1984). Sexual arousal to nonsexual violence and sadomasochistic themes among rapists and non-sex-offenders. *Journal of Consulting and Clinical Psychology, 52,* 651–657.

R

Rachman, S. (1966). Sexual fetishism: An experimental analogue. *Psychological Record, 16,* 293–296.

Radlove, S. (1983). Sexual response and gender roles. In E. R. Allgeier & N. B. McCormick (eds.), *Changing boundaries: Gender roles and sexual behavior.* Palo Alto, CA: Mayfield.

Raichle, K., & Lambert, A. J. (2000). The role of political ideology in mediating judgments of blame in rape victims and their assailants: A test of the just world, personal responsibility, and legitimization hypotheses. *Personality & Social Psychology Bulletin, 26*(7), 853–863.

Rajfer, J., et al. (1992). Nitric oxide as a mediator of relaxation of the corpus cavernosum in response to nonadrenergic, noncholinergic neurotransmission. *The New England Journal of Medicine, 326,* 90–94.

Rakic, Z., Starcevic, V., Starcevic, V. P., & Marinkovic, J. (1997). Testosterone treatment in men with erectile disorder and low levels of total testosterone in serum. *Archives of Sexual Behavior, 26*(5), 495–504.

Ralph, D., & McNicholas, T. (2000). UK management guidelines for erectile dysfunction. *British Medical Journal, 321,* 499–503.

Ramirez, A. (1990, August 12). The success of sweet smell. *The New York Times,* p. 10F.

Rao, K., DiClemente, R. J., & Ponton, L. E. (1992). Child sexual abuse of Asians compared with other populations. *Journal of the American Academy of Child and Adolescent Psychiatry, 31,* 880–886.

Rathus, S. A. (1978). Treatment of recalcitrant ejaculatory incompetence. *Behavior Therapy, 9,* 962.

Rathus, S. A. (2002). *Psychology in the New Millenium,* 8th ed. Fort. Worth, TX: Harcourt.

Rathus, S. A., & Fichner-Rathus, L. (1997). *The Right Start.* New York: Addison Wesley Longman.

Ratican, K. L. (1992). Sexual abuse survivors: Identifying symptoms and special treatment considerations. *Journal of Counseling and Development, 71,* 33–38.

Rawlins, R. (1998). Cited in Kolata, G. (1998, September 9). Researchers report success in method to pick baby's sex. *The New York Times* online.

Raychaba, B. (1989). Canadian youth in care: Leaving care to be on our own with no direction from home [Special Issue: Runaway, homeless, and shut-out children and youth in Canada, Europe, and the United States]. *Children and Youth Services Review, 11,* 61–73.

Reaney, P. (1998, January 15). Discovery may lead to cervical cancer vaccine. Reuters News Agency online.

Reichart, C. A., et al. (1990). Evaluation of Abbott Testpack Chlamydia for detection of *Chlamydia trachomatis* in patients attending sexually transmitted diseases clinics. *Sexually Transmitted Diseases, 17,* 147–151.

Reid, T. R. (1990, December 24). Snug in their beds for Christmas Eve: In Japan, December 24th has become the hottest night of the year. *Washington Post.*

Rein, M. F., & Muller, M. (1990). *Trichomonas vaginalis* and trichomoniasis. In K. K. Holmes, P. Mardh, P. F. Sparling, & P. J. Wiesner (eds.), *Sexually transmitted diseases,* 2nd ed. (pp. 481–492). New York: McGraw-Hill.

Reiner, W. G. (2000). Cited in "Study of children born without penises finds nature determines gender." (2000, May 12). Associated Press online.

Reinisch, J. M. (1990). *The Kinsey Institute new report on sex: What you must know to be sexually literate.* New York: St. Martin's Press.

Reiss, B. F. (1988, Spring/Summer). The long-lived person and sexuality. *Dynamic Psychotherapy, 6,* 79–86.

Reiss, I. L. (1980). *Family systems in America.* New York: Holt, Rinehart & Winston.

Remafedi, G. (1990). Study group report on the impact of television portrayals of gender roles on youth. *Journal of Adolescent Health Care, 11*(1) , 59–61.

Resick, P. A., & Schnicke, M. K. (1990). Treating symptoms in adult victims of sexual assault. *Journal of Interpersonal Violence, 5,* 488–506.

Reuter (1996, September 18). The RU 486 abortion pill got a tentative seal of approval. America Online.

Ricci, E., Parazzini, F, & Pardi, G. (2000). Caesarean section and antiretroviral treatment. *The Lancet, 355*(9202), 496–502.

Rice, M. E., Harris, G. T., & Quinsey, V. L. (1990). A follow-up of rapists assessed in a maximum-security psychiatric facility. *Journal of Interpersonal Violence, 5,* 435–448.

Rice, M. E., Quinsey, V. L., & Harris, G. T. (1991). Sexual recidivism among child molesters released from a maximum security psychiatric institution. *Journal of Consulting and Clinical Psychology, 59,* 381–386.

Riche, M. (1988, November 23–26). Postmarital society. *American Demographics,* 60.

Ricker, A. L. (1980, November). Sex for sale in Los Vegas. *Cosmopolitan,* pp. 280–315.

Rickwood, A. M. K., Kenny, S. E., & Donnell, S. C. (2000). Towards evidence based circumcision of English boys: Survey of trends in practice. *British Medical Journal, 321,* 792–793.

Riggio, R. E., & Woll, S. B. (1984). The role of nonverbal cues and physical attractiveness in the selection of dating partners. *Journal of Social and Personal Relationships, 1,* 347–357.

Rimm, E. (2000). Lifestyle may play role in potential for impotence. Paper presented to the annual meeting of the American Urological Association, Atlanta, May.

Rindfuss, R. R. (1991, December 4). Cited in Pear, R. Larger number of new mothers are unmarried. *The New York Times,* p. A20.

Roberto, L. G. (1983). Issues in diagnosis and treatment of transsexualism. *Archives of Sexual Behavior, 12,* 445–473.

Roberts, D. (2000). Black women and the pill. *Family Planning Perspectives* online, 32 (2).

Roberts, J. M. (2000). Recent advances: Obstetrics. *British Medical Journal, 321*(7252), 33–35.

Robinson, J. N., Norwitz, E. R., Cohen, A. P., & Lieberman, E. (2000). Predictors of episiotomy use at first spontaneous vaginal delivery. *Obstetrics & Gynecology, 96*(2), 214–218.

Roddy, R. E., et al. (1998). A controlled trial of Nonoxynol 9 film to reduce male-to-female transmission of sexually transmitted diseases. *The New England Journal of Medicine, 339,* 504–510.

Rodriguez, I., Greer, C. A., Mok, M. Y., & Mombaerts, P. (2000). A putative pheromone receptor gene expressed in human olfactory mucosa. *Nature Genetics, 26*(1), 18–19.

Roehrich, L., & Kinder, B. N. (1991). Alcohol expectancies and male sexuality: Review and implications for sex therapy. *Journal of Sex & Marital Therapy, 17,* 45–54.

Roesler, A., & Witztum, E. (2000). Pharmacotherapy of paraphilias in the next millennium. *Behavioral Sciences & the Law, 18*(1) 43–56.

Rogan, H. (1984, October 30). Executive women find it difficult to balance demands of job, home. *The Wall Street Journal,* pp. 35, 55.

Roland, B., Zelkart, P., & Dubes, R. (1989). MMPI correlates of college women who reported experiencing child/adult sexual contact with father, stepfather, or with other persons. *Psychological Reports, 64,* 1159–1162.

Rolfs, R. T., Goldberg, M., & Sharrar, R. G. (1990). Risk factors for syphilis: Cocaine use and prostitution. *American Journal of Public Health, 80,* 853–857.

Rolfs, R. T., & Nakashima, A. K. (1990). Epidemiology of primary and secondary syphilis in the United States: 1981 through 1989. *Journal of the American Medical Association, 264,* 1432–1437.

Romenesko, K., & Miller, E. M. (1989). The second step in double jeopardy: Appropriating the labor of female street hustlers. *Crime and Delinquency, 35,* 109–135.

Ronald, A. R., & Albritton, W. (1990). Chancroid and *Haemophilus ducreyi.* In K. K. Holmes, P. Mardh, P. F. Sparling, & P. J. Wiesner (eds.), *Sexually transmitted diseases,* 2nd ed. (pp. 263–272). New York: McGraw-Hill.

Roper Organization. (1985). *The Virginia Slims American Women's* Poll. New York: Roper Organization.

Rose, P. G. (1996). Endometrial carcinoma. *The New England Journal of Medicine, 335,* 640–649.

Rose, R. J. (1995). Genes and human behavior. *Annual Review of Psychology, 46,* 625–654.

Rosen, I. (1967). *Pathology and treatment of sexual deviations.* London: Oxford University Press.

Rosen, R. C., & Beck, J. G. (1988). *Patterns of sexual arousal.* New York: Guilford.

Rosen, R. C., Leiblum, S. R., & Spector, I. P. (1994). Psychologically based treatment for male erectile disorder: A cognitive–interpersonal model. *Journal of Sex and Marital Therapy, 20*(2), 67–85.

Rosenthal, A. M. (1995, June 13). The possible dream. *The New York Times,* p. A25.

Rosenthal, E. (1992, July 22). Her image of his ideal, in a faulty mirror. *The New York Times,* p. C12.

Rösler, A., & Witztum, E. (1998). Treatment of men with paraphilia with a long-acting analogue of gonadotropin releasing hormone. *The New England Journal of Medicine, 338,* 416–422.

Rosman, J. P., & Resnick, P. J. (1989). Sexual attraction to corpses: A psychiatric review of necrophilia. *Bulletin of the American Academy of Psychiatry and the Law, 17,* 153–163.

Ross, J. L., Roeltgen, D., Feuillan, P., Kushner, H., & Cutler, W. B. (2000). Use of estrogen in young girls with Turner syndrome: Effects on memory. *Neurology, 54*(1), 164–170.

Ross, M., & Need, J. (1989). Effects of adequacy of gender reassignment surgery on psychological adjustment: A follow-up of fourteen male-to-female patients. *Archives of Sexual Behavior, 18,* 145–153.

Rossouw, J. E. (2000). Cited in Kolata, G. (2000, April 5). Estrogen tied to slight rise in heart attack. *The New York Times,* pp. A1, A20.

Rotheram-Borus, M. J., & Koopman, C. (1991a). HIV and adolescents. [Special Issue: Preventing the spread of the human immunodeficiency virus]. *Journal of Primary Prevention, 12,* 65–82.

Rotheram-Borus, M. J., & Koopman, C. (1991b). Sexual risk behaviors, AIDS knowledge, and beliefs about AIDS among runaways. *American Journal of Public Health, 81,* 209–211.

Rotheram-Borus, M. et al. (1992). Lifetime sexual behaviors among predominantly minority male runaways and gay/bisexual adolescents in New York City. *AIDS Education and Prevention, Fall Suppl.,* 34–42.

Rovet, J., & Ireland, L. (1994). Behavioral phenotype in children with Turner syndrome. *Journal of Pediatric Psychology, 19,* 779–790.

Roving mates called factor in cancer. (1996, August 7). *The New York Times,* p. A10.

Royce, R. A., Seña, A., Cates, W., Jr., & Cohen, M. S. (1997). Sexual transmission of HIV. *The New England Journal of Medicine, 336,* 1072–1078.

Rozin, P., & Fallon, A. (1988). Body image, attitudes to weight, and misperceptions of figure preferences of the opposite sex: A comparison of men and women in two generations. *Journal of Abnormal Psychology, 97,* 342–345.

Rubin, A., & Adams, J. (1986). Outcomes of sexually open marriages. *Journal of Sex Research, 22,* 311–319.

Rubinow, D. R., & Schmidt, P. J. (1995). The treatment of premenstrual syndrome—Forward into the past. *The New England Journal of Medicine, 332,* 1574–1575.

Ruefli, T., Yu, O., & Barton, J. (1992). Sexual risk taking in smaller cities: The case of Buffalo, New York. *Journal of Sex Research, 29,* 95–108.

Rusbult, C. E., Martz, J. M., & Agnew, C. R. (1998). The Investment Model Scale: Measuring commitment level, satisfaction level, quality of alternatives, and investment size. *Personal Relationships, 5*(4), 357–391.

Russell, D. (1982). *Rape in marriage.* New York: Macmillan.

Russo, N. F., Horn, J. D., & Schwartz, R. (1992). U.S. abortion in context: Selected characteristics and motivations of women seeking abortions. *Journal of Social Issues, 48,* 183–202.

Russo, N. F., & Marin, A. D. (1997, February). *Professional Psychology: Research and Practice.* Cited in Brody, J. E. (1997, February 12). Abortion doesn't affect well-being, study says. *The New York Times* online.

S

Sack, W. H., & Mason, R. (1980). Child abuse and conviction of sexual crimes: A preliminary finding. *Law and Human Behavior, 4,* 211–215.

Sadalla, E. K., Kenrick, D. T., & Vershure, B. (1987). Dominance and heterosexual attraction. *Journal of Personality and Social Psychology, 52,* 730–738.

Sadava, S. W., & Matejcic, C. (1987). Generalized and specific loneliness in early marriage. *Canadian Journal of Behavioural Science, 19,* 56–66.

Sadker, M., & Sadker, D. (1994). *How America's schools cheat girls.* New York: Scribner.

Sadler, A. G., Booth, B. M., Nielson, D., & Doebbeling, B. N. (2000). Health-related consequences of physical and sexual violence: Women in the military. *Obstetrics & Gynecology, 96*(3), 473–480.

Saenger, P. (1996). Turner's syndrome. *The New England Journal of Medicine, 335,* 1749–1754.

Safer, J. (1996, January 17). Childless by choice. *The New York Times,* p. A19.

Safire, W. (2000, June 18). Hooking up. *The New York Times Magazine* online.

Sagan, C., & Dryan, A. (1990, April 22). The question of abortion: A search for answers. *Parade Magazine,* pp. 4–8.

Saluter, A. F. (1992). Marital status and living arrangements: March 1992. *Current Population Reports,* Series P20–468.

Samuels, M., & Samuels, N. (1986). *The well pregnancy book.* New York: Simon & Schuster.

Sanchez-Guerrero, J., et al. (1995). Silicone breast implants and the risk of connective tissue diseases and symptoms, *The New England Journal of Medicine, 332,* 1666–1670.

Sanday, P. R. (1981). The socio-cultural context of rape: A cross-cultural study. *Journal of Social Issues, 37,* 5–27.

Sanders, S. A., Reinisch, J. M., & McWhirter, D. P. (1990). Homosexuality/heterosexuality: An overview. In D. P. McWhirter, S. A. Sanders, & J. M. Reinisch (eds.) *Homosexuality/heterosexuality: Concepts of sexual orientation* (pp. xix–xxvii). New York: Oxford University Press.

Sandross, R. (1988, December). Sexual harassment in the Fortune 500. *Working Woman,* p. 69.

Sanger, M. (1938). *Margaret Sanger: An autobiography.* New York: Norton.

Sargent, T. O. (1988). Fetishism. *Journal of Social Work and Human Sexuality, 7,* 27–42.

Sarrel, P., & Masters, W. (1982). Sexual molestation of men by women. *Archives of Sexual Behavior, 11,* 117–131.

Sauer, M. V., Paulson, R. J., & Lobo, R. A. (1990). A preliminary report on oocyte donation extending reproductive potential to women over 40. *The New England Journal of Medicine, 323,* 1157–1160.

Savitz, L., & Rosen, L. (1988). The sexuality of prostitutes: Sexual enjoyment reported by "streetwalkers." *Journal of Sex Research, 24,* 200–208.

Saywitz, K. J., Mannarino, A. P., Berliner, L., & Cohen, J. A. (2000). Treatment for sexually abused children and adolescents. *American Psychologist, 55*(9), 1040–1049.

Schachter, J. (1989). Why we need a program for the control of *Chlamydia trachomatis. The New England Journal of Medicine, 320,* 802–804.

Schafer, R. B., & Keith, P. M. (1990). Matching by weight in married couples: A life cycle perspective. *Journal of Social Psychology, 130,* 657–664.

Schafran, L. H. (1995, August 26). Rape is still underreported. *The New York Times,* p. A19.

Schairer, C., et al. (2000). Menopausal estrogen and estrogen-progestin replacement therapy and breast cancer risk. *Journal of the American Medical Association, 283,* 485–491.

Schellenberg, E. G., Hirt, J., & Sears, A. (1999). Attitudes toward homosexuals among students at a Canadian university. *Sex Roles, 40*(1–2), 139–152.

Schiavi, R. C., et al. (1990). Healthy aging and male sexual function. *American Journal of Psychiatry, 147,* 766–771.

Schillinger, L. (1995, June 10). More sex please, we're Russian. *The New York Times Book Review,* p. 49.

Schmidt, G. (1994). Sex therapy 1970–1994. *Nordisk–Sexologi, 12*(3), 178–183.

Schmidt, K. W., et al. (1992). Sexual behaviour related to psycho–social factors in a population of Danish homosexual and bisexual men. *Social Science and Medicine, 34,* 1119–1127.

Schmidt, P. J., et al. (1998). Differential behavioral effects of gonadal steroids in women with and in those without premenstrual syndrome. *The New England Journal of Medicine, 338,* 209–216.

Schmitt, E. (1996, September 11). Senators reject both job-bias ban and gay marriage. *The New York Times,* pp. A1, A16.

Schneemann, C. (1991). The obscene body/politic. *Art Journal, 50*(4), 28–35.

Schnell, D. J., O'Reilly, K. R. (1991). Patterns of sexual behavior change among homosexual/bisexual men—selected U.S. sites, 1987–1990. *Mortality and Morbidity Weekly Report, 40, 46,* 792–794.

Schoendorf, K. C., & Kiely, J. L. (1992). Relationship of sudden infant death syndrome to maternal during and after pregnancy. *Pediatrics, 90,* 905–908.

Schoolgirls as sex toys. (1997, April 6). *The New York Times,* p. E2.

Schork, K. (1990, August 19). The despair of Pakistan's women: Not even Benazir Bhutto could stop the repression. *Washington Post.*

Schott, R. L. (1995). The childhood and family dynamics of transvestites. *Archives of Sexual Behavior, 24,* 309–327.

Schreiner-Engle, P., & Schiavi, R. (1986). Lifetime psychopathology in individuals with low sexual desire. *Journal of Nervous and Mental Disease, 174*, 646–651.

Schroeder-Printzen, I., et al. (2000). Surgical therapy in infertile men with ejaculatory duct obstruction: technique and outcome of a standardized surgical approach. *Human Reproduction, 15*, 1364–1368.

Schultz, N. R., Jr., & Moore, D. W. (1984). Loneliness: Correlates, attributions, and coping among older adults. *Personality and Social Psychology Bulletin, 10*, 67–77.

Schwartz, I. M. (1993). Affective reactions of American and Swedish women to their first premarital coitus: A cross-cultural comparison. *Journal of Sex Research, 30*, 18–26.

Schwartz, M. F., & Masters, W. H. (1984). The Masters and Johnson treatment program for dissatisfied homosexual men. *American Journal of Psychiatry, 141*, 173–181.

Scott, J. E., & Cuvelier, S. J. (1993). Violence and sexual violence in pornography: Is it really increasing? *Archives of Sexual Behavior, 22*, 357–371.

Scott, J. E., & Schwalm, L. A. (1988). Rape rates and the circulation rates of adult magazines. *Journal of Sex Research, 24*, 241–250.

Seachrist, L. (1995, September 23). Amount of virus sets cancer risk. *Science News*, p. 197.

Sears, A. E. (1989). The legal case for restricting pornography. In D. Zillmann & J. Bryant (eds.), *Pornography: Research advances and policy considerations.* Hillsdale, NJ: Lawrence Erlbaum Associates.

Seattle-King County Department of Public Health (1991). The AIDS Prevention Project. The Seattle Star: A report to the communiity on what we're learning from the "Be a Star Study." Seattle: Author.

Seftel, A. D., Oates, R. D., & Krane, R. J. (1991). Disturbed sexual function in patients with spinal cord disease. *Neurologic Clinics, 9*, 757–778.

Segal, Z. V., & Marshall, W. L. (1985). Heterosexual social skills in a population of rapists and child molesters. *Journal of Consulting and Clinical Psychology, 53*, 55–63.

Segraves, R. T., & Segraves, K. B. (1993). Medical aspects of orgasm disorders. In W. O'Donohue & J. H. Geer (eds.), *Handbook of sexual dysfunctions: Assessment and treatment.* (pp. 225–252). Boston: Allyn & Bacon.

Seligman, L., & Hardenburg, S. A. (2000). Assessment and treatment of paraphilias. *Journal of Counseling & Development, 78*(1), 107–113.

Seligmann, J. (1993, July 26). Husbands no, babies yes. *Newsweek*, p. 53.

Sell, R. L., Wells, J. A., & Wypij, D. (1995). The prevalence of homosexual behavior and attraction in the United States, the United Kingdom, and France: Results of national, population-based samples. *Archives of Sexual Behavior, 24*, 235–248.

Seltzer, R. (1992). The social location of those holding antihomosexual attitudes. *Sex Roles, 26*, 391–398.

Selvin, B. W. (1993, June 1). Transsexuals are coming to terms with themselves and society. *New York Newsday*, pp. 55ff.

Semans, J. (1956). Premature ejaculation: A new approach. *Southern Medical Journal, 49*, 353–358.

Seng, M. J. (1989). Child sexual abuse and adolescent prostitution: A comparative analysis. *Adolescence, 24*, 665–675.

Severn, J., Belch, G. E., & Belch, M. A. (1990). The effects of sexual and non–sexual advertising appeals and information level on cognitive processing and communication effectiveness. *Journal of Advertising, 19*, 14–22.

Sex complaints flood phones set up by Army. (1996, November 12). *The New York Times*, p. A13.

Shafer, M. A., et al. (1993). Evaluation of urine-based screening strategies to detect *Chlamydia trachomatis* among sexually active asymptomatic young men. *Journal of the American Medical Association, 270*, 2065–2070.

Shapiro, J. P. (1992, July 13). The teen pregnancy boom. *U.S. News & World Report*, p. 38.

Shaver, P., Hazan, C., & Bradshaw, D. (1988). Love as attachment. In R. J. Sternberg & M. L. Barnes (eds.), *The psychology of love* (pp. 68–99). New Haven, CT: Yale University Press.

Sheehy, G. (1995). *New passages: Mapping your life across time.* New York: Random House.

Sheehy, G. (1997, November 17). Beyond virility, a new vision. *Newsweek*, p. 69.

Sheehy, G. (1998). *Understanding men*'s passages. New York: Random House.

Shenon, P. (1995, July 15). New Zealand seeks causes of suicides by young. *The New York Times*, p. A3.

Sheppard, J. A., & Strathman, A. J. (1989). Attractiveness and height: The role of stature in dating preference, frequency of dating, and perceptions of attractiveness. *Personality and Social Psychology Bulletin, 15*, 617–627.

Sherman, K. J., et al. (1990). Sexually transmitted diseases and tubal pregnancy. *Sexually Transmitted Diseases, 17*, 115–121.

Sherwin, B. B., Gelfand, M. M., & Brender, W. (1985). Androgen enhances sexual motivation in females: A prospective, crossover study of sex steroid administration in the surgical menopause. *Psychosomatic Medicine, 47*, 339–351.

Shettles, L. (1982, June). Predetermining children's sex. *Medical Aspects of Human Sexuality*, 172.

Shifren, J. L., et al. (2000). Transdermal testosterone treatment in women with impaired sexual function after oophorectomy. *The New England Journal of Medicine* online, 343(10).

Shipko, S. (2000, February 7). Antidepressants linked to sexual side effects. WebMD/Healtheon.

Shilts, R. (1987). *And the band played on: Politics, people, and the AIDS epidemic.* New York: Penguin Books.

Shlipak, M. G., et al. (2000). Estrogen and progestin, lipoprotein (a), and the risk of recurrent coronary heart disease events after menopause. *Journal of the American Medical Association, 283*, 1845–1852.

Shuster, S. M., & Sassaman, C. (1997). Genetic interaction between male mating strategy and sex ratio in a marine isopod. *Nature, 388*(6640), 373–377.

Siegel, J. M., et al. (1990). Reactions to sexual assault: A community study. *Journal of Interpersonal Violence, 5*, 229–246.

Siegel, K., et al. (1988). Patterns of change in sexual behavior among gay men in New York City. *Archives of Sexual Behavior, 17*, 481–497.

Signorielli, N. (1990). Children, television, and gender roles: Messages and impact. *Journal of Adolescent Health Care, 11*(1), 50–58.

Silber, S. J. (1991). *How to get pregnant with the new technology.* New York: Time Warner.

Silbert, M. H. (1989). The effects on juveniles of being used for pornography and prostitution. In D. Zillmann & J. Bryant (eds.), *Pornography: Research advances and policy considerations* (pp. 215–234). Hillsdale, NJ: Lawrence Erlbaum Associates.

Simonsen, G., Blazina, C., & Watkins, C. E., Jr. (2000). Gender role conflict and psychological well-being among gay men. *Journal of Counseling Psychology, 47*(1), 85–89.

Simpson, J. A., Campbell, B., & Berscheid, E. (1986). The association between romantic love and marriage: Kephart (1967) twice revisited. *Personality and Social Psychology Bulletin*, 363–372.

Simpson, J. L. (2000, June 1). Invasive diagnostic procedures for prenatal genetic diagnosis. *Journal Watch Women*'s Health.

Singer, J., & Singer, I. (1972). Types of female orgasm. *Journal of Sex Research, 8*, 255–267.

Singh, D. (1994a). Body fat distribution and perception of desirable female body shape by young Black men and women. *International Journal of Eating Disorders, 16*(3), 289–294.

Singh, D. (1994b). Is thin really beautiful and good? Relationship between waist–to–hip ratio (WHR) and female attractiveness. *Personality and Individual Differences, 16*(1), 123–132.

Sirles, E. A., & Franke, P. J. (1989). Factors influencing mothers' reactions to intrafamily sexual abuse. *Child Abuse & Neglect, 13*, 131–139.

Skates, S. (1998). Cited in Story, P. (1998, June 2). Study: Prostate cancer deaths down. The Associated Press online.

Sleek, S. (1997). Resolution raises concerns about conversion therapy. *APA Monitor, 27*(10), 15.

Slonim–Nevo, V. (1992). First premarital intercourse among Mexican–American and Anglo–American adolescent women: Interpreting ethnic differences. *Journal of Adolescent Research, 7*, 332–351.

Smale, A. (2000, June 11). After the fall, traffic in flesh, not dreams. *The New York Times* online.

Smart, C. R., et al. (1995, April). *Cancer.* Cited in Brody, J. E. (1995, May 3). Breast scans may indeed help women under 50. *The New York Times*, p. C11.

Smith, E. P., et al. (1994). Estrogen resistance caused by a mutation in the estrogen-receptor gene in a man. *The New England Journal of Medicine, 331*, 1056–1061.

Smock, P. J. (2000). *Annual Review of Sociology.* Cited in Nagourney, E. (2000, February 15). Study finds families bypassing marriage. *The New York Times*, p. F8.

Smolowe, J. (1993, June 14). New, improved and ready for battle. *Time*, pp. 48–51.

Sobel, J. D. (1997). Current concepts: Vaginitis. *The New England Journal of Medicine, 337*, 1896–1903.

Sodroski, J., et al. (1998). *Nature.* Cited in "Scientists uncover 'key' to AIDS virus." (1998, June 18). The Associated Press; CNN.

Solano, C. H., Batten, P. G., & Parish, E. A. (1982). Loneliness and patterns of self-disclosure. *Journal of Personality and Social Psychology, 43,* 524–531.

Sommerfeld, J. (2000, April 18). Lifting the curse: Should monthly periods be optional? MSNBC online.

Sonenstein, F. L., Pleck, J. H., & Ku, L. C. (1989). Sexual activity, condom use and AIDS awareness among adolescent males. *Family Planning Perspectives, 21,* 152–157.

Sorrenti-Little, I., Bagley, C., and Robertson, S. (1984). An operational definition of the long-term harmfulness of sexual relations with peers and adults by young children. *Canadian Child, 9,* 46–57.

Sourander, L. B. (1994). Geriatric aspects on estrogen effects and sexuality. *Gerontology, 40* (Suppl. 3), 14–17.

Southerland, D. (1990, May 27). Limited "sexual revolution" seen in China: Nationwide survey shows more liberal attitudes developing in conservative society. *Washington Post.*

Spark, R. F. (1991). *Male sexual health: A couple's* guide. Mount Vernon, NY: Consumer Reports Books.

Speckens, A. E. M., et al. (1995). Psychosexual functioning of partners of men with presumed non-organic erectile dysfunction: Cause or consequence of the disorder? *Archives of Sexual Behavior, 24,* 157–172.

Specter, M. (1998, January 11). Traffickers' new cargo: Naive Slavic women. *The New York Times.*

Spiro, M. E., (1965). *Children of the kibbutz.* New York: Schocken Books.

Spitzer, R. L., et al. (1989). *DSM-III-R casebook.* Washington, DC: American Psychiatric Press.

Sponsel, W. E., et al. (2000) Sildenafil and ocular perfusion. *The New England Journal of Medicine* online, 342(22).

Sprecher, S. (1989). Premarital sexual standards for different categories of individuals. *Journal of Sex Research, 26,* 232–248.

Sprecher, S., Barbee, A., & Schwartz, P. (1995). "Was it good for you, too?" Gender differences in first sexual intercourse experiences. *The Journal of Sex Research, 32,* 3–15.

Sprecher, S., Sullivan, Q., & Hatfield, E. (1994). Mate selection preferences: Gender differences examined in a national sample. *Journal of Personality and Social Psychology, 66*(6), 1074–1080.

Spring, J. A. (1997). Cited in Alterman, E. (1997, November). Sex in the '90s. *Elle,* p. 130.

Stamm, W. E., & Holmes, K. K. (1990). *Chlamydia trachomatis* infections of the adult. In K. K. Holmes, P. Mardh, P. F. Sparling, & P. J. Wiesner (eds.), *Sexually transmitted diseases,* 2nd ed. (pp. 181–194). New York: McGraw-Hill.

Stampfer, M. J., et al. (1988). A prospective study of past use of oral contraceptive agents and risk of cardiovascular diseases. *New England Journal of Medicine, 319,* 1313–1317.

Stanford, J. L., et al. (2000). Urinary and sexual function after radical prostatectomy for clinically localized prostate cancer: The Prostate Cancer Outcomes Study. *Journal of the American Medical Association, 283,* 354–360.

Stangor, C., & Ruble, D. N. (1989). Differential influences of gender schemata and gender constancy on children's information processing and behavior. *Social Cognition, 7,* 353–372.

Starr, B. D., & Weiner, M. B. (1981). *The Starr–Weiner report on sex and sexuality in the mature years.* New York: Stein and Day.

Steele, C. M., & Josephs, R. A. (1990). Alcohol myopia: Its prized and dangerous effects. *American Psychologist, 45,* 921–933.

Stein, Z., & Susser, M. (2000). The risks of having children in later life. *British Medical Journal, 320*(7251), 1681–1682.

Steiner, M., et al. (1995). Fluoxetine in the treatment of premenstrual dysphoria. *The New England Journal of Medicine, 332,* 1529–1534.

Steinhauer, J. (1995, July 6). No marriage, no apologies. *The New York Times,* pp. C1, C7.

Stephan, C. W., & Bachman, G. F. (1999). What's sex got to do with it? Attachment, love schemas, and sexuality. *Personal Relationships, 6*(1), 111–123.

Stephens, W. N. (1982). *The family in cross-cultural perspective.* Washington, DC: University Press of America.

Stephenson, J. (2000a). AIDS in South Africa takes center stage. *Journal of the American Medical Association* online, 284(2).

Stephenson, J. (2000b). Widely used spermicide may increase, not decrease, risk of HIV transmission. *Journal of the American Medical Association* online, 284(8).

Sternberg, R. J. (1986). A triangular theory of love. *Psychological Review, 93,* 119–135.

Sternberg, R. J. (1987). Liking versus loving: A comparative evaluation of theories. *Psychological Bulletin, 102,* 331–345.

Sternberg, R. J. (1988). *The triangle of love: Intimacy, passion, commitment.* New York: Basic Books.

Sternberg, R. J., & Grajek, S. (1984). The nature of love. *Journal of Personality and Social Psychology, 47,* 312–329.

Stewart, G. K. (1998). Intrauterine devices (IUDs). In Hatcher, R. A., et al. (1998). *Contraceptive technology,* 17th rev. ed. (pp. 511–544). New York: Ardent Media.

Stewart, F. H. (1992, February 1). Cited in Leary, W. E. (1992). U.S. panel backs approval of first condom for women. *The New York Times,* p. 7.

Still no pill for Japan. (1992, March 22). *The New York Times,* Section 4, p. 7.

Stock, W. E. (1991). Feminist explanations: Male power, hostility, and sexual coercion. In E. Grauerholz & M. A. Koralewski (eds.), *Sexual coercion: A sourcebook on its nature, causes, and prevention* (pp. 61–73). Lexington, MA: Lexington Books.

Stock, W. E. (1993). Inhibited female orgasm. In W. O'Donohue & J. H. Geer (eds.), *Handbook of sexual dysfunctions: Assessment and treatment.* (pp. 253–301). Boston: Allyn & Bacon.

Stolberg, S. G. (1998a, January 18). Quandary on donor eggs: What to tell the children. *The New York Times,* pp. 1, 20.

Stolberg, S. G. (1998b, March 9). U.S. awakes to epidemic of sexual diseases. *The New York Times,* pp. A1, A14.

Stoller, R. J. (1969). Parental influences in male transsexualism. In R. Green & J. Money (eds.), *Transsexualism and sex reassignment.* Baltimore, MD: The Johns Hopkins University Press.

Stoller, R. J. (1977). Sexual deviations. In R. Beach (ed.), *Human sexuality in four perspectives* (pp. 190–214). Baltimore, MD: The Johns Hopkins University Press.

Stoller, R. J., & Herdt, G. H. (1985). Theories of origins of male homosexuality. *Archives of General Psychiatry, 42,* 399–404.

Stone, S. (1989). Assessing oral contraceptive risks. *Medical Aspects of Human Sexuality,* 112–122.

Storms, M. D. (1980). Theories of sexual orientation. *Journal of Personality and Social Psychology, 38,* 783–792.

Strassberg, D. S., et al. (1990). The role of anxiety in premature ejaculation: A psychophysiological model. *Archives of Sexual Behavior, 19,* 251–257.

Strickland, B. R. (1995). Research on sexual orientation and human development: A commentary. *Developmental Psychology, 31,* 137–140.

Strom, S. (1993, April 18). Human pheromones. *The New York Times,* p. V12.

Struckman-Johnson, C. (1988). Forced sex on dates: It happens to men, too. *The Journal of Sex Research, 24,* 234–241.

Study raises concern about fertility technique. (2000, March 28). Reuters News Agency online.

Sue, D. (1979). Erotic fantasies of college students during coitus. *Journal of Sex Research, 15,* 299–305.

Sue, D., Sue, D. W., & Sue, S. (1981). *Understanding abnormal behavior.* Boston: Houghton Mifflin.

Sulak, P. J., et al., (2000). Hormone withdrawal symptoms in oral contraceptive users. *Obstetrics & Gynecology, 95,* 261–266.

Sullivan, A. (1995). *Virtually normal.* New York: Knopf.

Suppe, F. (1994). Explaining homosexuality: Philosophical issues, and who cares anyhow? *Journal of Homosexuality, 27*(3–4), 223–268.

Swann, W. B., Jr., et al. (1987). Cognitive-affective crossfire: When self-consistency meets self-enhancement. *Journal of Personality and Social Psychology, 52,* 881–889.

Symons, D. (1995). Cited in Goleman, D. (1995, June 14). Sex fantasy research said to neglect women. *The New York Times,* p. C14.

Szabo, R., & Short, R. V. (2000). How does male circumcision protect against HIV infection? *British Medical Journal, 320,* 1592–1594.

Szasz, G., & Carpenter, C. (1989). Clinical observations in vibratory stimulation of the penis of men with spinal cord injury. *Archives of Sexual Behavior, 18,* 461–474.

T

Tanfer, K., Grady, W. R., Klepinger, D. H., & Billy, J. O. G. (1993). Condom use among U.S. men, 1991. *Family Planning Perspectives, 25,* 61–66.

Tang, C. S., Critelli, J. W., & Porter, J. F. (1995). Sexual aggression and victimization in dating relationships among Chinese college students. *Archives of Sexual Behavior, 24,* 47–53.

Tang, M. C., Weiss, N. S., & Malone, K. E. (2000). Induced abortion in relation to breast cancer among parous women: A birth certificate registry study. *Epidemiology, 11*(2), 177–180.

Tannahill, R. (1980). *Sex in history.* Briarcliff Manor, NY: Stein and Day.

Tannen, D. (1990). *You just don't understand.* New York: Ballantine Books.

Tannenbaum, J. (1991). Robert Mapplethorpe: The Philadelphia story. *Art Journal, 50*(4), 71–76.

Tarone, R. E., Cho, K. C., & Brawley, O. W. (2000). Implications of stage-specific survival rates in assessing recent declines in prostate cancer mortality rates. *Epidemiology, 11*(2), 167–170.

Tavris, C., & Sadd, S. (1977). *The Redbook report on female sexuality.* New York: Delacorte.

Taylor, A., Fisk, N. M., & Glover, V. (2000). Mode of delivery and subsequent stress response. *The Lancet, 355,* 120.

Taylor, C. (2000, June 26). Looking online. *Time, 155*(26), 60–61.

Taylor, S. E., Klein, L. C., Lewis, B. P., Gurung, R. A. R., Gruenewald, T. L., & Updegraff, J. A. (In press). Biobehavioral responses to stress in females: Tend-and-befriend, not fight-or-flight. *Psychological Review.*

Tedeschi, J. T., & Felson, R. B. (1994). *Violence, aggression, and coercive actions.* Washington, DC: American Psychological Association.

Telushkin, J. (1991). *Jewish literacy.* New York: Morrow.

Teen-agers and AIDS: The risk worsens. (1992, April 14). *The New York Times,* p. C3.

Tharinger, D. (1990). Impact of child sexual abuse on developing sexuality. *Professional Psychology: Research and Practice, 21,* 331–337.

Thomason, J. L., & Gelbart, S. M. (1989). *Trichomonas vaginalis. Obstetrics & Gynecology, 74,* 536–541.

Thomlison, B., et al. (1991). Characteristics of Canadian male and female child sexual abuse victims. [Special Issue: Child sexual abuse]. *Journal of Child and Youth Care,* Fall, 65–76.

Thompson, J. K., & Tantleff, S. (1992). Female and male ratings of upper torso: Actual, ideal, and stereotypical conceptions. *Journal of Social Behavior and Personality,* 7, 345–354.

Thompson, M. E. (1991). Self-defense against sexual coercion: Theory, research, and practice. In E. Grauerholz & M. A. Koralewski (eds.), *Sexual coercion: A sourcebook on its nature, causes, and prevention* (pp. 111–121). Lexington, MA: Lexington Books.

Thompson, S. (1995). *Going all the way.* New York: Hill and Wang.

Thornhill, R., & Palmer, C. (2000). *A natural history of rape: Biological bases of sexual coercion.* Cambridge, MA: M.I.T. Press.

Tierney, J. (1994, January 9). Porn, the low-flung engine of progress. *The New York Times,* Section 2, pp. 1, 18.

Tierney, J. (2000, February 23). Paid to surf (and filter) sex sites. *The New York Times* online.

Ting, D., & Carter, J. H. (1990). Behavioral change through empowerment: Prevention of AIDS. *Journal of the National Medical Association, 84,* 225–228.

Tobias, S. (1982). Sexist equations. *Psychology Today, 16*(1), pp. 14–17.

Tollison, C. D., & Adams, H. E. (1979). *Sexual disorders: Treatment, theory, and research.* New York: Gardner Press.

Tomlinson, J. A. (1991). Burn it, hide it, flaunt it. *Art Journal, 50*(4), 59–64.

Toner, J. P., et al. (1991). Basal follicle-stimulating hormone level is a better predictor of in vitro fertilization performance than age. *Fertility and Sterility, 55,* 784–791.

Toomey, K. E., & Barnes, R. C. (1990). Treatment of *Chlamydia trachomatis* genital infection. *Reviews of Infectious Diseases* (Suppl. 6).

Townsend, J. M. (1995). Sex without emotional involvement: An evolutionary interpretation of sex differences. *Archives of Sexual Behavior, 24,* 173–206.

Trenton State College. (1991, Spring). Sexual Assault Victim Education and Support Unit (SAVES-U) Newsletter.

Trieschmann, R. (1989). Psychosocial adjustment to spinal cord injury. In B. W. Heller, L. M. Flohr, & L. S. Zegans (eds.), *Psychosocial interventions with physically disabled persons.* (pp. 117–136). New Brunswick, NJ: Rutgers University Press.

Trujillo, C. (ed.). (1991). *Chicana lesbians: The girls our mothers warned us about.* Berkeley, CA: Third Woman Press.

Trussell, J., Strickler, J., & Vaughan, B. (1993). Contraceptive efficacy of the diaphragm, the sponge, and the cervical cap. *Family Planning Perspectives, 25,* 100–105.

Tseng, W., et al. (1992). Koro epidemics in Guangdong, China: A questionnaire survey. *Journal of Nervous & Mental Disease, 180,* 117–123.

Tucker, J. S., & Anders, S. L. (1999). Attachment style, interpersonal perception accuracy, and relationship satisfaction in dating couples. *Personality & Social Psychology Bulletin, 25*(4), 403–412.

Tuiten, A., et al. (2000). Time course of effects of testosterone administration on sexual arousal in women. *Archives of General Psychiatry, 57,* 149–153.

Tutty, L. M. (1992). The ability of elementary school children to learn child sexual abuse prevention concepts. *Child Abuse and Neglect, 16,* 369–384.

Twersky, O. (2000, June 30). Drugmaker pulls impotence drug out of FDA approval process. Additional studies under way to address potential safety concerns. (my.webmd.com/condition_center_hub/sex)

Two viewfinders, two views of Gay America. (1993, June 27) *The New York Times,* Section 4, p. 10.

U

U.S. Bureau of the Census. (1990). Marital status and living arrangements: March 1990. *Current Population Reports,* Series P-20, No. 450. Washington, DC: U.S. Government Printing Office.

U.S. Bureau of the Census. (1996). *Statistical abstract of the United States,* 116th ed. Washington, DC: U.S. Government Printing Office.

U.S. Bureau of the Census. (1998). *Statistical abstract of the United States,* 118th ed. Washington, DC: U.S. Government Printing Office.

U.S. Bureau of the Census. (Internet release date: 1999, January 7). Marital status of the population 15 years old and over, by sex and race: 1950 to present.

U.S. Department of Health and Human Services. (USDHHS). (1990). *The health benefits of smoking cessation: A report of the Surgeon General* (DHHS Publication No. CDC 90–8416). Rockville, MD: Public Health Service, Centers for Disease Control, Center for Chronic Disease Prevention and Health Promotion, Office on Smoking and Health.

U.S. Department of Health and Human Services (USDHHS). (1991). *Strategies to control tobacco use in the United States: A blueprint for public health action in the 1990's* (NIH Publication No. 92–3316). Washington, DC: National Cancer Institute, Public Health Service, National Institutes of Health, National Cancer Institute.

U.S. Department of Health and Human Services. (USDHSS). (1992). *Smoking and health in the Americas* (DHHS Publication No. (CDC) 92–8419). Atlanta: Public Health Service, Center for Disease Control, National Center for Chronic Disease Prevention and Health Promotion. Office on Smoking and Health.

U.S. Department of Justice. (1986). *Attorney general's* commission on pornography: Final report. Washington, DC: U.S. Government Printing Office.

U.S. Merit Systems Protection Board. (1981). *Sexual harassment in the federal workplace: Is it a problem?* Washington, DC: Office of Merit Systems Review and Studies.

U.S. Senate Committee on the Judiciary. (1991). Violence against women: The increase of rape in America 1990. *Response to the Victimization of Women and Children, 14* (79, No. 2), 20–23.

Udry, J. R., & Billy, J. O. G. (1987). Initiation of coitus in early adolescence. *American Sociological Review, 52,* 841–855.

Udry, J. R., Talbert, L., & Morris, N. M. (1986). Biosocial foundations for adolescent female sexuality. *Demography, 23*(2), 217–230.

Udry, J., et al. (1985). Serum androgenic hormones motivate sexual behavior in adolescent boys. *Fertility and Sterility, 43,* 90–94.

V

Vaginal yeast infection can be an HIV warning. (1992, November 24). *New York Newsday,* p. 51.

Valentiner, D. P., Foa, E. B., Riggs, D. S., & Gershuny, B. S. (1996). Coping strategies and post-traumatic stress disorder in female victims of sexual and nonsexual assault. *Journal of Abnormal Psychology, 105,* 455–458.

Valian, V. (1998). *Why so slow? The advancement of women.* Cambridge, MA: M.I.T. Press.

Valleroy, L. A., et al. (2000). HIV prevalence and associated risks in young men who have sex with men. *Journal of the American Medical Association, 284,* 198–204.

Van den Hoek, A., Van Haastrecht, H. J., & Coutinho, R. A. (1990). Heterosexual behaviour of intravenous drug users in Amsterdam: Implications for the AIDS epidemic. *AIDS, 4,* 449–453.

Van der Velde, F. W., van der Pligt, J., & Hooykaas, C. (1994). Perceiving AIDS-related risk: Accuracy as a function of differences in actual risk. *Health Psychology, 13*, 25–33.

Van Lange, P. A. M., et al. (1997). Willingness to sacrifice in close relationships. *Journal of Personality & Social Psychology*, 72(6), 1373–1395.

Varas-Lorenzo, C., García-Rodríguez, L. A., Perez-Gutthann, & S., Duque-Oliart, A. (2000). Hormone replacement therapy and incidence of acute myocardial infarction: A population-based nested case-control study. *Circulation* online, 101.

Vazi, R., Best, D., Davis, S., & Kaiser, M. (1989). Evaluation of a testicular cancer curriculum for adolescents. *Journal of Pediatrics, 114*, 150–162.

Veniegas, R. C., & Peplau, L. A. (1997). Power and the quality of same-sex friendships. *Psychology of Women Quarterly, 21*(2), 279–297.

Vinacke, W., et al. (1988). Similarity and complementarity in intimate couples. *Genetic, Social, and General Psychology Monographs, 114*, 51–76.

Voelker, R. (2000). Advisory on contraceptives. *Journal of the American Medical Association* online, 284(8).

Voeller, B. (1991). AIDS and heterosexual anal intercourse. *Archives of Sexual Behavior, 20*, 233–276.

Von Krafft-Ebbing, R. (1978). *Psychopathia sexualis.* Philadelphia: F. A. Davis. (Original work published 1886.)

Voyer, D., Voyer, S., & Bryden, M. P. (1995). Magnitude of sex differences in spatial abilities: A meta-analysis and consideration of critical variables. *Psychological Bulletin, 117*, 250–270.

W

Waigandt, A., et al. (1990). The impact of sexual assault on physical health status. *Journal of Traumatic Stress, 3*, 93–102.

Wald, A., et al. (1995). Virologic characteristics of subclinical and symptomatic genital herpes infections. *The New England Journal of Medicine, 333*, 770–775.

Waldinger, M. D., Hengeveld, M. W., & Zwinderman, A. H. (1994). Paroxetine treatment of premature ejaculation: A double–blind, randomized, placebo–controlled study. *American Journal of Psychiatry, 151*(9), 1377–1379.

Walfish, S., & Mayerson, M. (1980). Sex role identity and attitudes toward sexuality. *Archives of Sexual Behavior, 9*, 199–204.

Wallace, H. M., & Vienonen, M. (1989). Teenage pregnancy in Sweden and Finland. *Journal of Adolescent Health Care, 10*, 231–236.

Wallerstein, J. S., & Blakeslee, S. (1989). *Second chances: Women and children a decade after divorce.* New York: Ticknor & Fields.

Wallerstein, J. S., & Kelly, J. B. (1980). *Surviving the breakup: How children and parents cope with divorce.* New York: Basic Books.

Walsh, P. C. (1996). Treatment of benign prostatic hyperplasia. *The New England Journal of Medicine, 335*, 586–587.

Walster, E., & Walster, G. W. (1978). *A new look at love.* Reading, MA: Addison-Wesley.

Walt, V. (1993, July 26). Some second thoughts on depo. *New York Newsday*, p. 13.

Walter, H. J., & Vaughan, R. D. (1993). AIDS risk reduction among a multiethnic sample of urban high school students. *Journal of the American Medical Association, 270*, 725–730.

Wardlaw, G. M., & Insel, P. M. (1990). *Perspectives in nutrition.* St. Louis: Times Mirror/Mosby College Publishing.

Warner, D. L., & Hatcher, R. A. (1998). Male condoms. In Hatcher, R. A., et al. (1998). *Contraceptive technology*, 17th rev. ed. (pp. 325–356). New York: Ardent Media.

Warren, C. W., et al. (1990). Assessing the reproductive behavior of on- and off-reservation American Indian females: Characteristics of two groups in Montana. *Social Biology, 37*, 69–83.

Wasson, J. H. (1998). Finasteride to prevent morbidity from benign prostate hyperplasia. *The New England Journal of Medicine, 338*, 612–613.

Waterman, J. (1986). Overview of treatment issues. In. K. MacFarlane, et al. (eds.), *Sexual abuse of young children: Evaluation and treatment.* (pp. 197–203). New York: Guilford.

Waterman, J., & Lusk, R. (1986). Scope of the problem. In K. MacFarlane, et al. (eds.), *Sexual abuse of young children: Evaluation and treatment* (pp. 3–14). New York: Guilford.

Waterman, J., et al. (1986). Challenges for the future. In K. MacFarlane, et al. (eds.), *Sexual abuse of young children: Evaluation and treatment* (pp. 315–332). New York: Guilford.

Weinberg, M. S., Williams, C. J., & Moser, C. (1984). The social constituents of sadomasochism. *Social Problems, 31*, 379–389.

Weinberg, M. S., et al. (1994). *Dual attraction.* New York: Oxford University Press.

Weinberg, T. S. (1987). Sadomasochism in the United States: A review of recent sociological literature. *Journal of Sex Research, 23*, 50–69.

Weinberg, T. S., & Bullough, V. L. (1986). *Women married to transvestites: Problems and adjustments.* Paper presented at the annual meeting of the Society for the Study of Social Problems, New York.

Weinberg, T. S., & Bullough, V. L. (1988). Alienation, self-image, and the importance of support groups for the wives of transvestites. *Journal of Sex Research, 24*, 262–268.

Weingarten, H. (1985). Marital status and well-being: A national study comparing first-married, currently divorced, and remarried adults. *Journal of Marriage and the Family, 47*, 653–662.

Weinstock, H. S., et al. (1993). Factors associated with condom use in a high-risk heterosexual population. *Sexually Transmitted Diseases, 20*, 14–20.

Weisberg, D. K. (1985). *Children of the night: A study of adolescent prostitution.* Lexington, MA: D. C. Heath.

Weiss, D. L. (1983). Affective reactions of women to their initial experince of coitus. *Journal of Sex Research, 19*, 209–237.

Weiss, R. D., & Mirin, S. M. (1987). *Cocaine.* Washington, D.C. American Psychiatric Press.

Weitzer, R. (1999). *Sex for sale: Prostitution, pornography, and the sex industry.* London: Routledge.

Wennerholm, U.-B., et al. (2000). Incidence of congenital malformations in children born after ICSI. *Human Reproduction, 15*, 944–948.

Werner, D., & Cohen, A. (1990). Instructor's edition. In C. R. Ember & M. Ember, *Anthropology*, 6th ed. (pp. I-1 to I-146). Englewood Cliffs, NJ: Prentice-Hall.

Westrom, L. V. (1990). *Chlamydia trachomatis*—Clinical significance and strategies of intervention. *Seminars in Dermatology, 9*, 117–125.

Whalen, R. E., Geary, D. C., & Johnson, F. (1990). Models of sexuality. In D. P. McWhirter, S. A. Sanders, & J. M. Reinisch (eds.), *Homosexuality/heterosexuality: Concepts of sexual orientation* (pp. 61–70). New York: Oxford University Press.

Wheeler, J., & Kilmann, P. R. (1983). Comarital sexual behavior: Individual and relationship variables. *Archives of Sexual Behavior, 12*, 295–306.

Whipple, B., & Komisaruk, B. R. (1988). Analgesia produced in women by genital self-stimulation. *Journal of Sex Research, 24*, 130–140.

Whitam, F. L. (1977). Childhood indicators of male homosexuality. *Archives of Sexual Behavior, 6*, 89–96.

Whitam, F. L., Diamond, M., & Martin, J. (1993). Homosexual orientation in twins: A report on 61 pairs and 3 triplet sets. *Archives of Sexual Behavior, 22*, 187–206.

White, S. D., & DeBlassie, R. R. (1992). Adolescent sexual behavior. *Adolescence, 27*, 183–191.

White, V. M., Wearing, A. J., & Hill, D. J. (1994). Is the conflict model of decision making applicable to the decision to be screened for cervical cancer? A field study. *Journal of Behavioral Decision Making*, 7(1), 57–72.

Whitley, B. E., Jr. (1983). Sex role orientation and self-esteem: A critical meta-analysis. *Journal of Personality and Social Psychology, 44*, 765–788.

Whitley, B. E., Jr., & Kite, M. E. (1995). Sex differences in attitudes toward homosexuality. *Psychological Bulletin, 117*, 146–154.

Whitley, R., et al. (1991). Predictors of morbidity and mortality in infants with herpes simplex virus infections. *The New England Journal of Medicine, 324*, 450–454.

Whittemore, A. S. (1994). The risk of ovarian cancer after treatment for infertility. *The New England Journal of Medicine, 331*, 805–806.

Whitten, P. (2001). *Anthropology: Contemporary perspectives*, 8th ed. Boston: Allyn & Bacon.

WHO (1995). Cited in "Rise in STDs concerns group." (1995, September 12). *Newsday*, p. B27.

Wiederman, M. W.. & Kendall, E. (1999). Evolution, sex, and jealousy: Investigation with a sample from Sweden. *Evolution & Human Behavior*, 20(2), 121–128.

Wieselquist, J., Rusbult, C. E., Foster, C. A., & Agnew, C. R. (1999). Commitment, pro-relationship behavior, and trust in close relationships. *Journal of Personality & Social Psychology*, 77(5), 942–966.

Wilcox, A. J. (1995). Fertility in men exposed prenatally to diethylstilbestrol. *The New England Journal of Medicine, 332*, 1411–1416.

Wilcox, A. J., et al. (1995). Timing of sexual intercourse in relation to ovulation: Effects on the probability of conception, survival of the pregnancy, and sex of the baby. *The New England Journal of Medicine, 333*, 1517–1521.

Wilcox, A. J., Dunson, D., & Baird, D. D. (2000). The timing of the "fertile window" in the menstrual cycle: Day-specific estimates from a prospective study. *British Medical Journal, 321*, 1259–1262.

Wilcox, B. L. (1987). Pornography, social science and politics: When research and ideology collide. *American Psychologist, 42,* 941.

Wilford, J. N. (1992, November 17). Clues etched in bone debunk theory of a plague's spread. *The New York Times,* pp. C1, C8.

Williams, D. E., & D'Alessandro, J. D. (1994). A comparison of three measures of androgyny and their relationship to psychological adjustment. *Journal of Social Behavior and Personality, 9*(3), 469–480.

Williams, J. E., & Best, D. L. (1994). Cross-cultural views of women and men. In W. J. Lonner & R. Malpass (eds.), *Psychology and culture.* Boston: Allyn & Bacon.

Williams, M. (1999, June 15). Study: Patch could restore sex drive. The Associated Press.

Williams, S. S., et al. (1992). College students use implicit personality theory instead of safer sex. *Journal of Applied Social Psychology, 22,* 921–933.

Willis, R. J., & Michael, R. T. (1994). Innovation in family formation: Evidence on cohabitation in the United States. In J. Eruisch & K. Ogawa (eds.), *The family, the market and the state in aging societies.* London: Oxford University Press.

Willoughby, T., Wood, E., & Khan, M. (1994). Isolating variables that impact on or detract from the effectiveness of elaboration strategies. *Journal of Educational Research, 86,* 279–289.

Wilson, G., & Cox, D. (1983). Personality of pedophile club members. *Personality and Individual Differences, 4,* 323–329.

Wilson, J. D., George, F., & Griffin, J. (1981). The hormonal control of sexual development. *Science, 211,* 1278–1284.

Wilson, M. I., & Daly, M. (1996). Male sexual proprietariness and violence against wives. *Current Directions in Psychological Science, 5,* 2–7.

Wilson, M. R., & Filsinger, E. E. (1986). Religiosity and marital adjustment: Multidimensional interrelationships. *Journal of Marriage and the Family, 48,* 147–151.

Wilson, S. N., & Sanderson, C. A. (1988). The sex report curriculum: Is "just say no" effective? *SIECUS Report, 17,* 10–11.

Wilson, W., et al. (2000). Brain morphological changes and early marijuana use: A magnetic resonance and positron emission tomography study. *Journal of Addictive Diseases, 19*(1), 1–22.

Winick, C. (1985). A content analysis of sexually explicit magazines sold in an adult bookstore. *Journal of Sex Research, 21,* 206–210.

Wise, T. N. (1994). Sertraline as a treatment for premature ejaculation. *Journal of Clinical Psychiatry, 55*(9), 417.

Wisell, T. E., et al. (1987). Declining frequency of circumcision: Implications for changes in the absolute incidence and male to female sex ratio of urinary tract infections in early infancy. *Pediatrics, 79,* 338–342.

Wolman, T. (1985). Drug addiction. In M. Farber (ed.), *Human sexuality* (pp. 277–285). New York: MacMillan.

Woloshyn, V. E., Paivio, A., & Pressley, M. (1994). Use of elaborative interrogation to help students acquire information consistent with prior knowledge and information inconsistent with prior knowledge. *Journal of Educational Psychology, 86,* 79–89.

Women might mark millennium with "orgasm pill." (1998, June 6). Reuters News Agency online.

Wong, E. (2000, April 19). Vermont Senate passes gay unions bill. *The New York Times* online.

Wood, E., et al. (2000). Extent to which low-level use of antiretroviral treatment could curb the AIDS epidemic in sub-Saharan Africa. *The Lancet, 355,* 2095–2100.

Wood, N. S., et al. (2000). Neurologic and developmental disability after extremely preterm birth. *The New England Journal of Medicine* online, 343(6).

Wooldridge, W. E. (1991). Syphilis: A new visit from an old enemy. *Postgraduate Medicine, 89,* 199–202.

Wortman, C. B., et al. (1976). Self-disclosure: An attributional perspective. *Journal of Personality and Social Psychology, 33,* 184–191.

WuDunn, S. (1995, July 9). Many Japanese women are resisting servility. *The New York Times,* p. 10.

WuDunn, S. (1996, January 25). In Japan, a ritual of mourning for abortions. *The New York Times,* A1, A8.

Wulfert, E., & Wan, C. K. (1993). Condom use: A self-efficacy model. *Health Psychology, 12,* 346–353.

Wyatt, G. E. (1985). The sexual abuse of Afro-American and white American women in childhood. *Child Abuse and Neglect, 9,* 507–519.

Wyatt, G. E. (1988). The relationship between child sexual abuse and adolescent sexual functioning in Afro-American and white American women. *Annals of the New York Academy of Sciences, 528,* 111–122.

Wyatt, G. E. (1989). Reexamining factors predicting Afro-American and white American women's age at first coitus. *Archives of Sexual Behavior, 18,* 271–298.

Wyatt, G. E., & Newcomb, M. (1990). Internal and external mediators of women's sexual abuse in childhood. *Journal of Consulting and Clinical Psychology, 58,* 758–767.

Wyatt, G. E., Notgrass, C. M., & Newcomb, M. (1990). Internal and external mediators of women's rape experiences. *Psychology of Women Quarterly, 14,* 153–176.

Wyatt, G. E., Peters, S. D., & Guthrie, D. (1988a). Kinsey revisited, Part I: Comparisons of the sexual socialization and sexual behavior of white women over 33 years. *Archives of Sexual Behavior, 17(3),* 201–209.

Wyatt, G. E., Peters, S. D., & Guthrie, D. (1988b). Kinsey revisited, Part II: Comparisons of the sexual socialization and sexual behavior of black women over 33 years. *Archives of Sexual Behavior, 17*(4), 289–332.

Wysocki, C. J., & Preti, G. (1998). Pheromonal influences. *Archives of Sexual Behavior, 27*(6), 627–629.

Y

Yarber, W. L., & Parillo, A. V. (1992). Adolescents and sexually transmitted diseases. *Journal of School Health, 62,* 331–338.

Yarber, W. L., Torabi, M. R., & Veenker, C. H. (1989). Development of a three-component sexually transmitted diseases attitude scale. *Journal of Sex Education & Therapy, 15,* 36–49.

Yates, G. L., et al. (1991). A risk profile comparison of homeless youth involved in prostitution and homeless youth not involved. [Special Issue: Homeless youth]. *Journal of Adolescent Health, 12,* 545–548.

Yorburg, B. (1995, July 9). Why couples choose to live together. *The New York Times,* p. 14.

Z

Zalar, R. W. (2000). Domestic violence. *The New England Journal of Medicine* online, 342(19).

Zaviacic, M., & Whipple, B. (1993). Update on the female prostate and the phenomenon of female ejaculation. *Journal of Sex Research, 30,* 148–151.

Zaviacic, M., et al. (1988a). Concentrations of fructose in female ejaculate and urine: A comparative biochemical study. *Journal of Sex Research, 24,* 319–325.

Zaviacic, M., et al. (1988b). Female urethral expulsions evoked by local digital stimulation of the G-spot: Differences in the response patterns. *Journal of Sex Research, 24,* 311–318.

Zenker, P. N., & Rolfs, R. T. (1990). Treatment of syphilis, 1989. *Reviews of Infectious Diseases* (Suppl. 6), S590-S609.

Zhang, J., & Fried, D. B. (1992). Relationship of maternal smoking during pregnancy to placenta previa. *American Journal of Preventative Medicine, 8,* 278–282.

Zielbauer, P. (2000, May 22). Sex offender listings on Web set off debate. *The New York Times* online.

Zilbergeld, B. (1978). *Male Sexuality.* Boston: Little, Brown.

Zilbergeld, B., & Evans, M. (1980). The inadequacy of Masters and Johnson. *Psychology Today, 14,* pp. 29ff.

Zillmann, D. (1989). Effects of prolonged consumption of pornography. In D. Zillmann & J. Bryant (eds.), *Pornography: Research advances and policy considerations* (pp. 127–157). Hillsdale, NJ: Lawrence Erlbaum Associates.

Zillmann, D., & Bryant, J. (1984). Effects of massive exposure to pornography. In N. M. Malamuth & E. Donnerstein (eds.), *Pornography and sexual aggression* (pp. 115–138). New York: Academic Press.

Zillmann, D., & Weaver, J. B. (1989). Pornography and men's sexual callousness toward women. In D. Zillmann & J. Bryant (eds.), *Pornography: Research advances and policy considerations* (pp. 95–125). Hillsdale, NJ: Lawrence Erlbaum Associates.

Zimmer, D., Borchardt, E., & Fischle, C. (1983). Sexual fantasies of sexually distressed and nondistressed men and women: An empirical investigation. *Journal of Sex and Marital Therapy, 9,* 38–50.

Zuger, A. (1998, January 6). Do breast self-exams save lives? Science still doesn't have answer. *The New York Times* online.

Zusman, M. E., & Knox, D. (1998). Relationship problems of casual and involved university students. *College Student Journal, 32*(4), 606–609.

name index

Page numbers followed by *t* and *f* indicate tables and figures, respectively.

Abel, G. G., 579, 623
Abelson, H., 661
Abramowitz, S., 186
Abramson, P. R., 563
Acker, M., 226
Ackerman, D., 143, 179
Acton, W., 16
Adachi, M., 177
Adams, H. E., 182, 304, 596, 623
Adams, J., 476
Adams, R., 650
Adams, V., 521
Ade-Ridder, L., 480
Adelson, A., 195
Adelson, M. D., 556
Adler, J., 49
Agnew, C. R., 238, 241–244, 247
Ahmed, R. A., 14
Ahuja, N., 501
Akhtar, S., 501
Albritton, W., 541
Alexander, C. J., 482
Alighieri, D., 217
Allen, M., 190, 237, 290, 303
Allen, P. L., 13, 263
Allen, W., 230
Alter-Reid, K., 621, 625, 629
Alterman, E., 468, 469, 472
Althof, S. E., 520
Altman, L. K., 79, 82, 85, 86, 88, 398, 400, 511, 552
Alzate, H., 166, 167
Amatenstein, S., 236–237
Ames, M. A., 155, 308, 623, 624
Anders, S. L., 217
Anderson, V. N., 209, 305
Angier, N., 119, 182, 309, 329, 379, 380, 431
Antill, J. K., 200
Antonarakas, S. E., 352
Argiolas, A., 499
Armstrong, K., 82, 83
Armstrong, L., 124
Armsworth, M. W., 414
Arriaga, X. B., 257
Arthur, B. I., Jr., 174
Asbell, B., 18
Assalian, P., 520
Astley, S. J., 351
Atkeson, B. M., 604, 605, 608, 612, 613, 614, 619
Augustine, Saint, 13
Avis, W. E., 208
Axinn, W. G., 49

Bacall, L., 146
Bach, G. R., 206
Bachman, G. F., 217
Bachrach, C. A., 446
Baenninger, M. A., 190
Bagatell, C. J., 511
Bagley, C., 313, 316, 317, 621, 625, 649
Bailey, J. M., 198, 307, 308, 312, 313–314
Bailey, R. C., 114
Baines, C. J., 84
Baird, D. D., 330, 396
Baker, J. N., 443, 614
Baker, S., 445
Baldwin, J. D., 299–300
Baldwin, J. I., 299–300
Bancroft, J., 156, 157, 308
Banmen, J., 467
Banning, A., 622
Barbach, L. G., 74, 163, 271, 516, 659
Barbaree, H. E., 611, 612
Barbee, A., 440, 441
Barisic, S., 367
Barlow, D. H., 308, 521
Barnard, M., 654
Barnard, N. D., 105
Barouch, D. H., 551
Barr, H. M., 351
Barrett, M. B., 317
Barringer, F., 295, 474, 529, 633
Bart, P. B., 619
Bar-Tal, D., 215
Bartel, G. D., 54
Barth, R. P., 446
Bartoshuk, L. M., 143, 144
Basson, R., 519
Batten, P. G., 244
Baumeister, R. F., 21, 584, 586
Baxter, D. J., 611
Bazell, R., 123
Beach, F. A., 25, 54, 190, 208, 213, 273, 278, 299, 421, 462, 471, 586
Beauchamp, G. K., 143, 144
Beautrais, A. L., 313
Beck, J. G., 55, 496, 522
Beck, S., 611, 623
Becker, J. V., 614, 622
Begley, S., 22–23
Belcastro, P. A., 275
Belgrave, F. Z., 198–199, 440, 441
Belizán, J. M., 349
Bell, A. P., 49, 293, 311–319, 322, 323
Bell, S., 612
Beller, T., 189, 640
Bellis, D. J., 654
Bem, S. L., 197, 200, 291, 296
Benatar, S. R., 546
Bennett, N. G., 330, 458
Bentler, P. M., 186
Berger, L., 364–365, 366
Bergstrom, R., 346
Berke, R. L., 302, 304, 469
Berkowitz, L., 662
Berliner, D. L., 143
Berliner, L., 626
Berman, L., 519
Bernstein, S. J., 78
Bernstein, W. M., 215
Berscheid, E., 221, 225, 227
Bertron, P., 105
Best, D., 122
Best, D. L., 187, 187*t*
Beutler, L. E., 505
Bianchi, S. M., 475
Bieber, I., 310–311, 315
Billy, J. O. G., 49, 156, 276, 286
Binik, Y. M., 495
Birkhead, T. R., 22, 610
Birnbaum, H. J., 602, 606, 608, 612
Bixler, R. H., 214
Bjorklund, D. F., 21, 191–192, 213
Bjorn, L., 631
Black, L. E., 241
Blair, C. D., 579, 593
Blakeslee, S., 475, 559
Blanchard, R., 186, 579, 584, 593
Blanco, J. D., 539
Blane, A. K., 458
Blazina, C., 314
Blickstein, I., 366
Blilie, M., 603
Bloom, D. E., 458
Bloor, M., 654
Blumstein, P., 52, 56, 212, 322, 465, 466, 469
Bohlen, C., 99
Boies, S. C., 588
Boland, E., 236–237
Boles, J., 654
Bonati, F. A., 427
Borchardt, E., 283
Borer, M., 247
Boswell, J., 298, 375
Bouchard, P., 412
Bowers, E., 364
Bowie, W. R., 540
Bowlby, J., 420
Boyer, D., 653
Bozzette, S. A., 552
Bradbard, M. R., 198
Bradbury, T. N., 230, 241
Bradford, J. M. W., 155, 156, 597
Bradley, S. J., 180
Bradshaw, D., 225
Brady, E. C., 609
Brannigan, G., 444
Brasher, P., 8585
Braun, R. D., 363
Brawley, O. W., 121, 125
Brecher, E. M., 480
Bremner, W. J., 511
Brender, W., 157, 505
Breslow, N., 586, 592, 593
Bretschneider, J., 479
Briere, J., 622, 623
Brislin-Slütz, J. A., 663
Brissett, D., 651
Britton, P. J., 563
Brockert, J. E., 444
Broder, M. S., 78
Brody, J. E., 83, 84, 86, 87, 97, 98, 102, 104*t*, 118, 147, 148, 149, 334, 448, 490, 491, 498, 555, 588–589, 594, 619, 674
Bronars, S., 444
Bronner, E., 47, 426
Brookoff, D., 473
Brooks, V. R., 190
Brooks-Gunn, J., 178, 439, 444, 446
Broude, G. J., 25, 272, 299, 421
Brown, D., 659
Brown, M., 183
Brown, R. A., 213
Browning, J. R., 439
Bruck, M., 627
Bryant, J., 85, 659, 663
Bryden, M. P., 190
Buckley, B., 609
Bucy, P. C., 152, 153
Buehlman, K. T., 474, 475
Buffington, C., 463
Buhrich, N., 576
Bullough, V. L., 295, 298, 576, 577–578
Bumiller, E., 7
Bumpass, L., 456, 459
Burg, R., 588
Burger, H. G., 95
Burgess, A. W., 608, 614–615, 649
Burkhart, B., 611
Burman, B., 475

Burrell, N., 290, 303
Burrows, B., 351
Burstein, G. R., 539
Busby, D., 502
Buss, D. M., 21, 23, 213, 214, 462, 463
Bussey, K., 194, 197
Butler, J., 162
Buve, A., 114
Buxton, A. P., 317
Byers, E. S., 619
Bynum, R., 537
Byrd, J., 341, 368
Byrnes, J., 189

Caceres, C. F., 549
Cade, T., 382
Cado, S., 283
Caesar, Julius, 13
Cairns, P., 125
Calderone, M. S., 382, 424, 426–427
Calhaz-Jorge, C., 351
Calhoun, K. S., 604, 605, 608, 612, 613, 614, 619
Caligula, 13
Call, V., 466
Callaway, C., 430
Calvin, J., 15
Cameron, P., 624
Campbell, C. A., 654
Campbell, S. B., 367
Cannistra, S. A., 76, 528, 559
Cappella, J. N., 215
Cardarelli, A., 626
Cardoso, C. G., 351
Carey, M. P., 499
Carlisle, J. M., 593
Carlson, M., 407
Carpenter, C., 482
Carrera, M., 306
Carrère, S., 474, 475
Carroll, L., 362–363
Carter, C. S., 217
Carter, D. B., 198
Carvajal, D., 645, 647
Casanova, G., 375
Castro, J., 632
Castro, K. G., 654
Catalan, J., 502, 522
Catania, J. A., 318
Cates, W., Jr., 387, 390, 532, 536, 539, 540, 541, 543, 544, 550, 557, 559, 560
Ccu, S., 528
Celis, W., 617
Ceniti, J., 663
Chadda, R. K., 501
Chambers, C., 461, 461*t*
Chambers, D. B., 198–199, 440, 441
Chan, C., 300
Chan, Y. M., 413
Chang, G., 514
Chang, K., 457
Chaplin, T. C., 611
Charny, I. W., 472
Chase-Lansdale, L., 444, 446
Cheasty, M. A., 626, 627
Check, E., 22–23
Check, J. V. P., 609
Cheesman, A. D., 183
Chernick, A. B., 180
Chira, S., 195, 633
Chlebowski, R. T., 85, 87, 88
Cho, K. C., 124, 125
Choi, P. Y., 105
Christensen, C., 522
Christiansen, K., 156
Christin-Maitre, S., 412, 413
Cirmo, H. E., 429
Clark, M. S., 216
Clark, R., 359
Clemmensen, L. H., 186
Cline, M., 351
Clinton, H. R., 468*f*
Clinton, W., 19, 53, 320, 321, 468*f*, 469, 590, 602, 630
Clode, N., 351
Clumeck, N., 552
Cnattingius, S., 346, 351
Coan, J. A., 474, 475
Cochran, S. D., 314, 322
Cochran, W. G., 46
Cogswell, D., 97
Cohen, A., 193, 476
Cohen, A. P., 357
Cohen, E., 403
Cohen, J. A., 626
Cohen, L., 462
Cohen, M. S., 114, 550
Cohen-Kettenis, P., 186
Cohn, J. F., 367
Cohn, L. D., 562
Colapinto, J., 180
Colditz, G. A., 83
Cole, A. L., 252
Cole, C. J., 335
Cole, C. L., 252
Cole, F. S., 366
Cole, S. S., 480–481, 483
Coleman, E., 652, 653–654
Coleman, M., 200–201
Coles, C., 351
Coles, R., 45, 426, 436–437, 439–440, 441, 442–443
Collaer, M. L., 179, 192, 308
Collins, J. W., Jr., 364–365
Collins, N. L., 236
Columbus, C., 537
Comstock, A., 375
Conde-Agudelo, A., 349
Condon, J. W., 216
Conger, R. D., 475
Connolly, M., 146
Connor, E. M., 552
Constantine, J., 476
Constantine, L., 476
Conway-Welch, C., 166, 167
Cook, E., 408
Cooper, A., 232, 588–589
Cooper, A. J., 623
Cooper, C., 457
Cooper, E., 200
Cooper, M. L., 414, 415*f*
Corcoran, D. M., 216
Cordero, E., 156
Cordero, W., 473*f*
Cornell, C. P., 303
Cortines, R. C., 429
Cosby, B., 468
Cosgray, R. E., 584
Cotton-Huston, A. L., 304
Courtenay, W. H., 191
Coutinho, E., 103
Coutinho, R. A., 654
Couwenbergs, C., 156
Cowley, G., 510–511
Cox, C. L., 224
Cox, D., 623
Cox, D. J., 578
Cozzarelli, C., 414, 415*f*
Craig, M. E., 619
Cramer, D., 79
Crano, W. D., 216
Crews, D., 308
Critelli, J. W., 605
Cronin, A., 295–296, 313
Cross, H., 201
Crossette, B., 70, 367
Crowe, L. C., 150
Cummings, S. R., 85, 88
Cunningham, G., 156
Cunningham, M. R., 211
Curtis, R. C., 216
Curtiss, G., 503, 522
Cutler, W. B., 78, 96, 143, 144
Cuvelier, S. J., 662

Dabbs, J. M., Jr., 156
D'Alessandro, J. D., 200
Daley, S., 412–413
Daly, M., 21, 473
D'Amico, A. V., 124, 125
D'Amora, D. A., 628
D'Angelo, L., 442–443
Darby, B. L., 351
Darling, C. A., 164, 166, 167, 440, 441
Darwin, C., 20, 20*f*, 211
D'Augelli, A. R., 305, 313, 316, 317
Davenport, W., 6, 25, 299
Davidson, J. K., 164, 271, 283–284, 440, 441
Davidson, K. J., 166
Davidson, K. M., 167
Davidson, N. E., 97
Davis, A., 404, 412
Davis, M. H., 226
Davis, P., 582–583
Davis, S., 122
Davison, G. C., 596
Dawes, R. M., 216
Dawood, K., 307, 312
Day, A., 502
Day, N. L., 351
de Bruyn-Kops, E., 611
de Luca, R. V., 629
de Raad, B., 212
de Schampheleire, D., 649
De Young, M., 624
Dean, C. W., 611
Deaux, K., 187, 190
DeBlassie, R. R., 440
Dekker, J., 494, 521
Del Priore, G., 79
DeLamater, J. D., 341, 368
Delaney, S. M., 522
Delgado, G., 462
Delgado, J., 153
della Torre, C., 146
Delmas, P. D., 96, 97
Delmonico, D. L, 588
den Tonkelaar, I., 96–97
Dennerstein, L., 95
Denny, N., 272
Der Andreassian, C., 412
Derby, C. A., 499
Desmond, A. M., 444
Detzer, M. J., 265, 267, 436
Deutsch, F. M., 210
Deutsch, R. M., 206
Devesa, S. S., 79, 381
DeWitt, K., 632
Diamond, M., 179, 180, 295, 307
Diaz-Mitoma, F., 557
Dick-Read, G., 358
Dickson, N., 439, 448
DiClemente, R. J., 626
Dietz, P. E., 612
DiMatteo, M. R., 359
Dindia, K., 190, 237
DiPietro, J. A., 194–195
Doddema-Winsemius, M., 212
Dolan, E., 475
Dole, B., 512–513
Donnell, S. C., 114
Donnerstein, E. I., 661–662, 663
Donovan, C., 447
Doorn, C. D., 576
Douglas, J. M., Jr., 560
Douglas, M., 630
Downey, J. I., 308, 312
Drapers, P., 193
Drew, W. L., 557
Drews, C. D., 351
Dreyfuss, I. A., 84
Drigotas, S. M., 222, 248
Dryan, A., 405, 407
Dryden, J., 238
Dube, R., 622
Dudley, E. C., 95
Dugger, C. W., 70
Dumas, A., 319*f*
Dunlap, D. W., 290
Dunn, M. E., 164, 479
Dunne, M. P., 198, 307, 312
Dunson, D., 330, 396
Duque-Oliart, A., 98
Duquette, H., 582
Durand, V. M., 308
Durik, 22
Durik, A. M., 359
Dwyer, M., 579, 582, 594
Dylan, B., 18
Dziech, B. W., 631

Eason, E., 357
Eastwood, M. M., 241
Eberly, C. G., 609
Edgerton, M., 186
Edgley, C., 640, 641, 644, 645–646, 647, 651, 661
Edmonson, B., 481, 484
Edwards, T. M., 454, 455
Eggers, D., 270, 271, 302, 459, 469
Ehrhardt, A., 178, 179, 426
Ehrhardt, A. A., 427
Eisen, A., 82, 83
Eisen, M., 427
Elder, J., 384, 404, 407, 408, 409*f*, 411
Elenteny, K., 190
Elias, M., 232–233
Elifson, K. W., 654
Ellertson, C., 102–103
Ellis, A., 503
Ellis, E. M., 475
Ellis, H., 17, 264

Ellis, L., 155, 308
Emerson, R. W., 302
Endsley, R. C., 198
English, P. B., 351
Erel, O., 475
Erlanger, S., 649
Ervin, K. S., 209
Eskenazi, B., 351
Eskenazi, G., 610
Etaugh, C., 176, 194, 431
Evans, L., 586
Evans, M., 521
Everaerd, W., 521
Everitt, B. J., 153

Fahrner, E., 183, 186
Falcon, P. L., 609
Faller, K. C., 622, 625
Fallon, A. E., 209
Farley, A. U., 537
Farley, M., 648–649
Federman, D. D., 175
Feingold, A., 190, 216
Feldman, H. A., 492, 498
Feldman, P., 357
Felsman, D., 444
Felson, R. B., 602, 631
Feng, T., 351
Fenichel, O., 591
Fennema, E., 189, 190
Ferguson, D. M., 313
Feuillan, P., 78, 96, 143
Fichner-Rathus, L., 10, 245
Fichten, C. S., 505
Field, J., 272
Filsinger, E. E., 56
Fincham, F. D, 475
Findley, T. W., 482
Finkelhor, D., 608, 621, 622, 626
Finkenauer, C., 248
Fischer, E. F., 216
Fischle, C., 283
Fish, L. S., 502
Fishel, S., 337
Fisher, B., 653
Fisher, H. E., 151, 191–192, 213
Fisher, J. D., 562
Fisher, N., 385
Fisher, T. D., 422–423
Fisher, W. A., 562
Fisher-Thompson, D., 195
Fisk, N. M., 359
Fitzgerald, L. F., 631, 632–633
Flaxman, S. M., 340
Fletcher, J., 671
Flexner, C., 552
Flinn, K., 468*f*
Flowers, G., 468, 469
Floyd, R. L., 351, 352
Foa, E. B., 612–613, 614, 616
Follingstad, D. R., 619
Fong, G. T., 150, 563
Ford, C. S., 25, 54, 190, 208, 213, 273, 278, 299, 421, 462, 471, 586
Forgas, J. P., 231
Forster, P., 520
Foster, C. A., 247
Fracastoro, G., 536
Frank, E., 97, 98, 100
Franke, P. J., 625
Franzoi, S. L., 209
Fraser, A. M., 444
Frayser, S., 25, 209, 273, 462, 470, 471
Frazier, L., 443
Frazier, P. A., 614
Freiberg, P., 304, 305
Fretts, R., 364
Freud, S., 17, 28–30, 35, 44, 194, 309–310, 420, 425, 503
Freund, K., 578, 579, 593
Freund, M., 654
Frey, K. S., 197
Friday, N., 271
Fried, D. B., 351
Fried, P. A., 351
Friedman, A., 239
Friedman, J., 505
Friedman, L. C., 87
Friedman, R. C., 308, 312
Friedrich, W. N., 423, 593
Frisch, R., 334, 431
Frodi, A., 196
Fromuth, M. E., 611
Frost, M. H., 87
Frustenberg, F. F., Jr., 444
Fryer, M. M., 210
Fuchs, C. S., 84
Fuentes-Afflick, E., 366
Fugger, E., 330330
Fuleihan, G. E., 98
Furstenberg, F. F., Jr., 439, 446

Gabelnick, H. L., 401, 403–404
Gabriel, T., 297, 317
Gafni, N., 189
Gagnon, J. H., 17, 31, 32*t*, 45, 47, 49, 50, 215, 231, 263*t*, 264–265, 266*t*, 271, 275, 276, 278, 286, 295, 297, 300, 311, 319–320, 439–440, 457, 462, 463, 465–466, 467*t*, 469, 474, 479, 488, 489*t*, 490, 497, 498*t*, 604, 605*f*, 625, 651, 653, 659
Gaines, S. O., Jr., 224
Gallagher, A. M., 189
Gallant, J. E., 552
Ganong, L. M., 200–201
Garber, M., 298
Garcia, B., 429
García-Rodríguez, L. A., 98
Garland, S. M., 540
Garnets, L., 300
Garwood, S. G., 212
Gates, G., 437, 438
Gathorne-Hardy, J., 47
Gauguin, P., 537
Gay, D., 408
Gay, P., 17
Gayle, H., 443, 528, 552–553
Geary, D. C., 308
Gebhard, P. H., 17, 115, 145, 265, 267, 269–270, 275, 293–297, 318, 421, 425, 439, 579, 582, 584, 585, 587, 592, 612, 625, 654
Gebre-Medhin, M., 70
Geer, J. H., 272–273, 522, 576, 578, 593–594
Geiger, R., 482
Geis, F. L., 200
Gelfand, M. M., 157
Gelles, R. J., 303
Gelman, D., 216
Genius, M., 621
George, W. H., 149, 150
Gerber, P. N., 593
Geringer, W. M., 562
Gerrol, R., 608
Gershuny, B. S., 612, 614
Giambra, L. M., 271, 477
Gibbs, N., 604, 607, 608, 611
Gibson, A. I., 649
Gidycz, C. A., 604, 605, 607, 611, 619
Gifford, F., 468
Gijs, L., 155
Gilbert, S., 84, 380, 381, 382, 423*t*
Giles, G. G., 84
Gillis, J. S., 208
Gilman, S. E., 307, 308, 626
Ginsburg, A. L., 445
Giovannucci, E., 398
Giovino, G. A., 351, 352
Gitlin, M. J., 367
Giuliani, R. W., 126
Glass, S. P., 469
Gleicher, N., 335
Glover, V., 359
Gnagy, S., 79, 381
Godiva, Lady, 581
Godwin, M., 471
Gold, E. P., 96–97
Gold, R. G., 508
Gold, S. R., 212, 508
Goldberg, A., 623
Goldberg, C., 384, 404, 407, 408, 409*f*, 411, 627, 646, 649
Goldberg, M., 537
Goldenberg, R. L., 444
Golding, J. M., 614
Goldman, J. A., 563
Goldman, R. D., 366
Goldstein, I., 489, 492, 498, 506, 512–513, 514
Goleman, D., 156, 628, 629, 631, 632, 659
Gomes-Schwartz, B., 626
Gomez, J., 300
Goodyear, C., 375
Gooren, L., 155
Gordon, B. L., 588
Gordon, S. L., 248, 303, 408, 463, 479, 644
Gorzalka, B. B., 149
Gottman, J. M., 474, 475
Grabrick, D. M., 84, 381
Graca, L. M., 351
Grady, W. R., 49
Grafenberg, E., 166–167
Graham, J. M., 539
Graham, S., 16, 264
Grajek, S., 220
Grant, H., 640, 647, 650
Grauerholz, E., 633
Green, R., 303, 312
Greenberg, E. R., 350
Greene, B., 300–301
Greene, S., 25
Greene, S. J., 272, 299, 421
Greenhouse, L., 633
Greenwald, E., 425, 625, 627
Greer, G., 80
Griffin, E., 215
Griscom, A., 247
Grodsky, A., 271, 477
Grodstein, F., 77, 543
Grogger, J., 444
Grön, G., 189
Gross, J., 290, 606
Grosser, B. I., 143
Grossman, S., 150
Groth, A. N., 602, 606, 608, 611, 612
Gruber, J. E., 631
Grunfeld, C., 548
Gruslin, A., 351
Grynch, J. H., 475
Guillebaud, J., 381, 382
Guinan, M. E., 401
Gunn, R. A., 537, 648–649
Gupta, M., 14
Guthrie, D., 51
Guthrie, J. R., 95
Guzick, D. S., 99, 155, 156, 157, 515
Gwartney-Gibbs, P. A., 619

Haddow, J. E., 355
Hadler, J. L., 537, 648–649
Haeri, S., 27
Haffner, D., 426
Haglund, B., 351
Hall, C. S., 190
Hall, G. C. N., 611, 612
Hall, R. H., 422–423
Halpern, D. F., 189
Halpern, D. T., 114
Halverson, C. F., Jr., 197, 198
Hamer, D. H., 295, 308
Hamilton, M., 617
Hammersmith, S. K., 49, 311, 312
Handler, A., 446
Handy, L., 590
Hanrahan, J. P., 351
Hanson, R. K., 613
Hanson, S. L., 445
Hardenburg, S. A., 575, 594, 674
Hardiman, P., 97
Hariton, E. B., 284, 285*t*
Harlow, L. L., 563
Harney, P. A., 611, 612
Harold, G. T., 475
Harris, C. R., 239
Harris, D. O., 126–127
Harris, G. T., 629
Harris, L., 664
Harris, M., 11, 25, 53, 193, 250, 462, 463, 470, 472, 476, 624
Hart, J., 341
Hartge, P., 79, 84, 381
Hartman, C. R., 649
Harvey, C., 368
Harvey, S., 95
Hass, A., 436
Hatcher, R. A., 333, 378*t*, 381–389, 391, 392, 393–394, 399, 400, 401, 402–403, 564
Hatfield, E., 207, 212, 213*t*, 218, 221, 227, 439
Hatzichristou, D. G., 492, 498, 513–514
Haugaard, J. J., 621
Hausknecht, R. U., 413
Hawkins, B., 351
Hawthorne, N., 16
Hawton, K., 502, 522
Hazam, H., 248
Hazan, C., 225
Hazelwood, R. R., 612
Heath, R., 153

Hebert, M., 622
Hechtman, L., 444
Hegeler, S., 480
Heiman, J. R., 265, 272–273, 516, 522, 576, 578, 593–594, 626, 659
Hemingway, E., 110, 655
Hemphill, G., 463
Henderson, V. W., 97
Hendrick, C., 221, 227
Hendrick, S., 221, 227
Hengeveld, M. W., 520
Henneberger, M., 633
Henning, K., 270, 271, 272
Henry III, King of France, 576
Hensley, W. E., 207, 208
Herbert, W., 360–361
Herbison, P., 439, 448
Herdt, G. H., 193, 299
Herodotus, 587
Herold, E. S., 659
Herrell, R., 313
Herrington, D. M., 97
Herrmann, H. C., 514
Hertzog, M. E., 209
Herzog, L., 114
Hess, B. B., 444
Hessol, N. A., 366
Hill, A., 602
Hill, D. J., 668
Hillier, S., 542
Hillis, D. M., 544
Hills, A., 312
Hilton, E., 543
Hines, M., 179, 192, 308
Hippocrates, 532
Hirschman, R., 612
Hirt, J., 304
Hite, S., 49, 268, 270, 277
Hitt, J., 407, 408–409, 411
Ho, D. D., 551
Ho, G. Y. E., 559
Ho, P. C., 413
Hoagwood, K., 627
Hobfall, S. E., 563
Hobson, W., 612
Hock, Z., 167
Hodges, B. C., 562
Hodgson, R., 540
Hodson, D. S., 479
Hoeger, K., 99, 155, 156, 157, 515
Hoffman, H. J., 351
Hoffman, L. E., 271, 283–284
Holden, G. W., 475
Holmes, K. K., 540, 542
Holmes-Farley, S. R., 331
Holmstrom, L., 614–615
Holzman, H. R., 650
Hooykaas, C., 563
Hopper, J. L., 95
Horn, J. D., 404–405
Horowitz, J., 626
Horvath, C., 307, 312
Horwood, L. J., 313
Houston, D. A., 623, 624
Howard, J. A., 212
Howard, M., 447
Howard, M. C., 25
Howards, S. S., 333
Hsu, B., 271
Hubbard, R., 354
Hubschman, L., 186
Hucker, S. J., 584
Huffman, T., 457
Hughes, I. A., 177
Huk, N., 616
Hunt, M., 43, 45, 47–48, 52, 53, 271, 275, 439, 463–464, 465–466, 472, 608, 641
Hunter, D. J., 84
Hurley, E., 640
Hurley, K., 82, 86
Hurlock, D., 105
Huxley, R. R., 340
Hyde, J. S., 22, 189, 190, 341, 359, 368

Imai, W. K., 562
Imperato-McGinley, J., 177–178
Ireland, L., 176
Isay, R. A., 295, 311, 312, 317
Ito, S., 347–348, 349–350
Iyasu, S., 364

Jacklin, C. N., 194–195, 196
Jackson, A. P., 563
Jackson, B. B., 99
Jackson, J., 626
Jackson, L. A., 209
Jacob, S., 143
Jacobson, J. L., 351
Jacobson, S. W., 351
Jamison, P. L., 115
Jankowiak, W. R., 216
Janus, C. L., 44, 48, 216, 218, 295, 621
Janus, M. D., 653
Janus, S. S., 44, 48, 216, 218, 295, 621
Jeffreys, S., 646
Jennings, D. A., 164
Jennings, V. H., 397
Jennings-White, C., 143
Jensen, A., 382
Jessinger, D., 97
Jetter, A., 83
Johannes, C. B., 499
Johnson, D., 563
Johnson, E., 424, 426–427
Johnson, E. W., 382
Johnson, F., 308
Johnson, K. A., 83
Johnson, O., 11, 25, 53, 193, 250, 462, 463, 470, 472, 476, 624
Johnson, S., 375
Johnson, V. E., 54–56, 101, 115, 136, 145, 157, 163, 164–166, 268, 274, 275, 293, 315, 318–319, 341, 464, 494, 496, 498, 504–505, 508–509, 518–522, 524
Jones, F. P., 255
Jones, H. W., 333
Jones, J., 99
Jones, L., 434
Jorgensen, C., 179
Jorgensen, S. R., 446
Josefsson, A. M., 528–529, 546
Josephs, R. A., 150
Joyce, J., 655

Kahn, J., 445
Kaiser, K., 174
Kaiser, M., 122
Kalichman, S. C., 619
Kalick, S. M., 215
Kallmann, F. J., 307
Kamb, M. L., 529–530, 562
Kammeyer, K. C. W., 6, 480
Kandell, P., 216
Kanouse, D. E., 78
Kanter, G. D., 406
Kantrowitz, B., 444
Kaplan, H. S., 163, 479, 491, 494, 495, 503, 505, 506, 508, 518, 629
Karande, V., 335
Karney, B. R., 230, 241
Kash, K., 87
Katz, M. S., 128–129
Katz, R. C., 622
Kaufman, A., 608
Kavanagh, A. M., 84
Kay, D. S., 264
Keaton, D., 230
Kegel, A., 74
Keith, J. B., 440
Keith, P. M., 215, 462
Kellerman, J., 250
Kellogg, J., 236
Kellogg, J. H., 262, 264
Kelly, J. B., 475
Kelly, M. P., 267, 500
Kendall, E., 239
Kendler, K. S., 307, 308, 626
Kennedy, C. J., 602
Kenny, S. E., 114
Kenrick, D. T., 212
Kessler, D., 439
Kessler, R. C., 307, 308, 626
Kettl, P., 482
Khadar, A., 616
Kiely, J. L., 351
Killian, K., 502
Killmann, P. R., 505, 518, 522
Kilpatrick, D. G., 608
Kim, P. Y., 312
Kimble, D. P., 153, 155
Kimlika, T., 201
Kimmel, D., 300
Kinard, E., 444
Kinder, B. N., 149, 503, 522
King, J., 520
King, R., 443, 563
Kinsey, A. C., 17–18, 45–47, 48, 49–52, 53, 145, 164, 265, 267, 268, 269–270, 273, 275, 293–297, 318, 421, 425, 439, 464, 465–466, 479, 504, 584, 585, 587, 592, 641
Kipp, K., 21, 191–192, 213
Kirby, D., 126–129
Kircher, J. R., 267, 500
Kirk, K. M., 198
Kirkpatrick, R. C., 307
Kite, M. E., 304
Kjerulff, K. H., 78
Klaas, M., 628
Klein, E. A., 125
Klein, H. G., 549
Klepinger, D. H., 49, 562
Klerman, L. V., 444
Klugherz, B. D., 514
Klüver, H., 152, 153
Knapp, M. L., 234
Knight, R. A., 611
Knight, S. E., 480, 481, 482, 483
Knox, D., 217, 218–219, 238, 241–244, 457, 460*t*, 463, 476, 480, 625, 669, 671, 672
Knudsen, D. D., 621, 622
Knussman, R., 156
Kockott, G., 183, 186
Kocoras, C., 116
Kohlberg, L., 196–197
Kohn, A., 621
Kolata, G., 47, 85, 98, 215, 231, 264–265, 330–331, 384, 439–440, 463, 466, 467*t*, 511, 512, 519, 659
Kolbe, L., 427, 447, 448
Kolbert, E., 632
Kolodny, R. C., 521
Komisaruk, B. R., 166
Kon, I. S., 7
Koopman, C., 562
Koren, G., 347–348, 349–350
Korlin, D., 186
Kosamu, B., 11
Koss, M. P., 604, 605, 607, 611, 612, 614, 619
Koutsky, L. A., 559
Kowal, D., 395, 397
Krafka, C. L., 663
Kramer, M. S., 346, 366
Krane, R. J., 481, 482, 492, 498
Kresin, D., 95, 490
Kruesi, M. J. P., 596
Kruff, C., 506–507
Kruks, G., 653
Ku, L. C., 318
Kuenzi, E., 128
Kuiper, B., 186
Kulin, H., 435
Kunkel, L. E., 304
Kurdek, L. A., 314
Kuriansky, J., 232
Kuriloff, P. J., 612
Kushner, H., 78, 96, 143
Kyriacou, D. N., 472, 473

Laan, E., 659
Ladas, A. K., 167
Laino, C., 557
Laird, J. D., 250, 403
Lamacz, M., 594, 598
Lamanna, M. A., 468
LaMay, M. L., 189
Lamaze, F., 358
Lambert, A. J., 609
Lambert, B., 654
Lamke, L. K., 201
Lamon, S. J., 189, 190
Lamprecht, V. M., 397
Landon, A., 43
Lane, III, F. S., 655
Lane, K. E., 619
Lang, A. R., 150
Lang, R. A., 591
Langevin, R., 579, 581, 590
Langley, J., 586
Langston, C. A., 238
Lansky, D., 150
Lanyon, R. I., 579, 593
Larijani, J., 27
Larijani, M. J., 27
Lau, L., 603
Laumann, E. O., 31, 32*t*, 45, 47, 49, 50, 113, 215, 231, 263*t*, 264–265, 266*t*, 271, 275, 276,

286, 295, 297, 300, 439–440, 457, 462, 463, 465–466, 467*t*, 469, 474, 479, 488, 489*t*, 490, 497, 498, 498*t*, 500–501, 520, 604, 605*f*, 659
Lavin, J., 563
Laviola, M., 625
Lavoisier, P., 165–166
Law, J., 367
Lawrence, D. H., 18, 464, 655
Lawrence, K., 659
Lawton, C. A., 190
Leary, W. E., 119, 120, 333, 334, 353, 354, 479
Leavitt, G. C., 624
LeBaron, D., 210
Lederman, M. M., 552
Ledray, L. E., 616
Lee, N., 654
Lees, M. I., 540
Legato, M. J., 179, 180
Lehman, G. L., 149, 150
Lehne, G., 115
Leiblum, S. R., 491, 502, 503, 504, 506
Leinders-Zufall, T., 143
Leitenberg, H., 265, 267, 270–273, 283, 425, 436, 576, 578, 593–594, 625
Leland, J., 515, 519
Leland, N. L., 446
Lemon, S. J., 559
Lenihan, G., 609
Leonard, T. L., 654
Lesnik-Oberstein, M., 462
Lester, W., 302, 303*t*
Letourneau, E., 490, 491, 522
LeTourneau, M. K., 603
Letvin, N., 551
LeVay, S., 309
Lever, J., 292
Levine, T., 439
Levinger, G., 231
Levy, D. S., 610
Levy, E., 103
Levy, G. D., 198
Levy, R. I., 192
Lewan, T., 147
Lewin, T., 33, 284, 404, 409, 414, 630, 657
Lewinsky, M., 19, 468*f*, 602, 630
Lewis, J., 250
Lewis, K., 619
Lewis, L. L., 187
Lewis, P. H., 298, 299, 318, 470, 471
Lewis, R., 562
Lewis, S., 216
Liang, Y., 660
Libman, E., 480, 505
Lichtenstein, P., 124
Lieb, R., 628
Lieberman, E., 357
Lief, H. I., 186
Lindermalm, G., 186
Linet, O. I., 511
Linsenmeier, J. A. W., 312
Linton, M. A., 605
Linz, D. G., 661–663
Lipshultz, L. I., 498
Lipworth, L., 346
Lira, L. R., 604
Lisak, D., 609
Liu, D., 50, 52
Lobo, R. A., 99
Loder, N., 189
Lohr, B. A., 304
Londono, M. L., 167
Lopez, P. A., 149, 150
LoPiccolo, J., 496, 504, 505, 516, 518, 521, 522
Lorefice, L. S., 596
Lott, B., 200
Lottes, I., 612
Lovell-Badge, R., 175
Lown, J., 475
Lowry, R., 530
Loy, P. H., 631, 632
Luckenbill, D. F., 653
Lue, T. F., 490, 511, 521
Lundstrom, B., 186
Lusk, R., 621, 622, 624
Luther, M., 15
Lynxwiler, J., 408

Macallan, D. C., 548
MacAndrew, S., 547
Macauley, J., 196
Maccoby, E. E., 190, 194–195, 196
MacDonald, G., 150, 563
MacDonald, T. K., 150, 563
Macfarlane, S., 562
MacFarquhar, N., 70
MacVicar, S., 616
Madden, S., 428
Magdol, L., 472–473
Maheu, M., 232
Mahoney, P. D., 514
Maines, R., 265
Major, B., 414, 415*f*
Malamuth, N. M., 609, 610, 662, 663
Maletzky, B. M., 596
Malinowski, B., 24, 54, 624
Malloy, M. H., 351
Malone, K. E., 84
Mannarino, A. P., 626
Mapplethorpe, R., 656*f*
Marchbanks, P. A., 78, 79
Marcus, A. J., 387
Margolin, L., 609, 610
Marinkovic, J., 506, 511
Marks, G., 216
Markson, E. W., 444
Marotta, T., 653
Maroules, N., 632, 633
Marshall, D., 496–497
Marshall, W. L., 611, 612, 661
Marsiglio, W., 302, 304, 445
Martin, C. E., 145, 164, 265, 267, 268, 269–270, 275, 293–297, 318, 421, 425, 439, 464, 466, 584, 585, 587, 592, 641
Martin, C. L., 197, 198
Martin, C. W., 403
Martin, J., 307
Martin, J. N., 555
Martin, N. G., 307, 312
Martinez, F. D., 351
Martinson, F. M., 420
Martyna, W., 200
Martz, J. M., 220, 241–244
Maruyama, G., 216
Marwick, C., 85, 88, 96
Masi, C. M., 113
Mason, R., 611
Masters, B., 609
Masters, W. H., 54–56, 101, 115, 136, 145, 157, 163, 164–166, 167, 268, 274, 275, 293, 315, 318–319, 341, 464, 494, 496, 498, 504–505, 508–509, 518–520, 521–522, 524, 609
Matejcic, C., 244
Matek, O., 580
Matsen, K., 563
Matson, S., 628
Matthews, K. A., 99, 622
Matthews, W. S., 178
Maxson, S., 174
Maybach, K. L., 212
Mayer, J. P., 351
Mazkereth, R., 366
McArthur, M. J., 612
McCabe, J. B., 447
McCabe, M. P., 522
McCartney, M., 361–362
McClintock, M. K., 143
McCloskey, L. A., 473
McCormack, A., 649
McCormack, K. D., 376
McCormack, W., 187
McCormick, M., 365
McCoy, J. K., 441
McCoy, N., 479
McDonough, Y. Z., 424
McGrane, S., 240–241
McGuire, R. J., 593
McKay, M., 429
McKeganey, N., 654
McKinlay, J. B., 492, 498, 506–507
McKinney, K., 632, 633
McLean, P. M., 153
McMahon, M. J., 359
McNeil, D. G., Jr., 546, 642–643
McNicholas, T., 499, 502, 511
McWhirter, D. P., 295, 296, 297
Mead, M., 24, 54, 190, 192, 214
Meana, M., 495
Meier, B., 367
Meiselman, K. C., 625
Meisler, A. W., 499
Melbye, M., 84
Melis, M. R., 499
Mendel, G., 20
Merrill, R. M., 124, 125
Mertz, G. J., 555
Messenger, J., 496
Meston, C. M., 149, 507, 626
Meuwissen, I., 659
Meyer, J. K., 186
Meyer, T., 552
Meyer-Bahlburg, H. F. L., 308
Michael, R. T., 31, 32*t*, 45, 47, 49, 50, 215, 231, 263*t*, 264–265, 266*t*, 271, 275, 276, 286, 295, 297, 300, 439–440, 457, 462, 463, 465–466, 467*t*, 469, 474, 479, 488, 489*t*, 490, 497, 498*t*, 604, 605*f*, 659
Michael, S., 467*t*
Michaels, S., 31, 32*t*, 45, 47, 49, 50, 215, 263*t*, 265, 266*t*, 271, 275, 276, 286, 295, 297, 300, 457, 462, 465–466, 469, 474, 479, 488, 489*t*, 490, 497, 498*t*, 604, 605*f*
Miedzian, M., 610
Mikach, S. M., 198
Mill, J. S., 672
Miller, A. B., 84
Miller, B. C., 441
Miller, E. M., 645, 648
Miller, H., 464, 655
Miller, K., 216
Miller, L. C., 236
Miller, M., 425–426, 609, 610
Miller, N., 216
Miller, S. M., 82, 86
Millet, K., 655
Mills, J. R., 216
Milner, P., 153
Minai, N., 14
Ming, E. E., 79, 381
Minkoff, H. L., 537
Mirin, S. M., 151, 500
Mirotznik, J., 557
Mishell, D. R., Jr., 381, 385
Missailidis, K., 70
Mitchell, H., 84
Mitterand, F., 468
Mittman, B. S., 78
Miyamoto, M., 652
Mofenson, L. M., 547, 549
Mohr, D. C., 505
Money, J., 115, 155, 178, 179, 180, 181, 187, 192, 299, 308, 312, 594, 598
Montagnier, L., 551
Monti-Bloch, L., 143
Moore, D., 630
Moore, D. W., 244
Moore, K. A., 444
Morales, A., 147
Morales, E., 300
Moran, P. B., 609, 610
Morgan, T. O., 130
Morokoff, P. J., 493, 500
Morrin, K. A., 190
Morris, K. T., 619
Morris, L. B., 366, 367
Morris, N. M., 156, 157
Morris, R., 156
Morrison, D., 445
Morrison, D. R., 445
Morrison, E. S., 72, 80–81, 422, 423, 425, 440
Morse, E., 654
Mortensen, M., 480
Mortola, J. F., 102, 103, 104, 105
Moser, C., 585
Mosher, C. D., 16–17
Mosher, D. L., 656
Mosher, W. D., 446
Mosteller, F., 46
Moylan, S. J., 231
Muehlenhard, C. L., 605, 609, 611, 612
Mueser, K. T., 209
Muhammad, 14, 26
Mullen, P. D., 351, 352
Muller, M., 543
Mulnard, R. A., 97
Mundy, L., 437, 438, 441, 442, 445, 446, 447, 448
Murphy, E., 468
Murstein, B. I., 226
Muster, N. J., 622
Myers, S. L., 320, 602
Myerson, M., 201

Nabel, G. J., 551
Nadeau, R., 563
Nadler, R. D., 301
Nakashima, A. K., 537
Narod, S. A., 381
Navarro, M., 290
Nduati, R., 549
Neal, J. A., 663
Need, J., 186
Neisser, A. L. S., 532
Neunaber, D. J., 367
Nevid, J. S., 212
Newbold, J. E., 559
Newcomb, M., 614, 626
Newcomb, P. A., 97
Newman, E., 200
Ngai, S. W., 413
Nichols, M., 500
Niebuhr, G., 290, 299
Nieto, J. J., 97
Nieves, W., 241–244, 463
Niloff, J. M., 76, 528, 559
Nock, S. L., 457
Nordqvist, K., 175
Norris, J., 149, 150
Norwitz, E. R., 357
Nosek, M. A., 480, 481
Notgrass, C. M., 614
Nour, N. W., 70–71
Nyberg, L., 506, 507

Oates, R. D., 481, 482
Oberschneider, 314
O'Brien, P. B., 619
Ochs, R., 560
Oddens, B. J., 96–97
O'Dell, K. M. C., 174
O'Donohue, W., 490, 491, 522
Ogrinc, F. G., 511
O'Hara, M. W., 367
Okami, P., 623
Oldenberg, R., 651
Olds, J., 153
Oleske, D. M., 335
Oliver, G. D., 180
Olson, T. D., 441
O'Neill, G., 476
O'Neill, N., 476
Orwoll, E., 98
Osborne, L. N., 475
Osborne, N. G., 556
Osgood, C., 456
Oswalt, R., 563
Over, R., 659
Overbeek, P., 175
Overholser, J. C., 611, 623
Oz, S., 673

Packwood, R., 602
Padgett, V. R., 663
Padma-Nathan, H., 511
Paik, A., 498, 500–501, 520
Palace, E. M., 151
Palmer, M. T., 215
Palmore, E., 480
Palson, C., 54
Palson, R., 54
Parazzini, F., 552, 553
Pardi, G., 552, 553
Parillo, A. V., 530, 531, 539
Parish, E. A., 244
Parnass, S., 472
Parsons, N. K., 406

Pasnau, R. O., 367
Passarello, L. C., 440, 441
Pastuszak, A., 347–348, 349–350
Patterson, C. J., 290, 303
Paul, C., 439, 448
Paul, R. H., 359
Paul, Saint, 13
Paulson, R. J., 99
Pauly, B., 186
Pauly, I. B., 182, 186
Pearson, C., 401
Penn, F., 529
Penrod, S., 662, 663
Peplau, L. A., 314, 322, 463
Peralta, L., 442, 443
Perduta-Fulginiti, P. S., 482
Peretti, P. O., 238
Perez-Gutthann, S., 98
Perrett, D. I., 211
Perriëns, J., 391
Perry, A., 183
Perry, D. G., 194, 197
Perry, J. D., 166, 167
Person, E. S., 265, 271
Peters, S. D., 51
Peterson, D. R., 351
Piaget, J., 443
Pierre-Jerome, F., 115
Pillard, R. C., 176, 307, 311, 312
Pines, A. M., 239
Pines, S., 650
Pinkerton, S. D., 563
Piot, P., 554
Pistella, C. L., 427
Pitnick, S., 120
Pizzari, T., 610
Plant, E. A., 187, 190, 341, 368
Pleck, J. H., 318
Pliner, A. J., 407
Pliny, 94
Polit-O'Hara, D., 445
Pollak, W. S., 125
Pomeroy, W. B., 47, 145, 164, 265, 267, 268, 269–270, 275, 293–297, 318, 421, 425, 439, 464, 466, 584, 585, 587, 592, 641
Poniewozsik, J., 673*t*
Ponton, L. E., 626
Poortinga, J., 576
Porter, J. F., 605
Porter, N., 200
Potosky, A. L., 125
Poussaint, A., 300
Powderly, W. G., 552
Powell, E., 604, 609, 611, 615, 618, 619, 620, 630, 631, 634
Prentky, R. A., 611
Press, A., 655
Preti, G., 143, 144
Price, V. A., 645, 649–650, 654–655
Prince, V., 653
Pring, T., 183
Proctor, F., 162
Pryde, N. A., 522
Pudowski, B. C., 238
Pullella, P., 670
Purifoy, F. E., 271, 477
Putin, V., 461
Pyngolil, M., 99

Quadagno, D., 272, 654
Quevillon, R. P., 495
Quinn, T. C., 114
Quinsey, V. L., 611, 629

Rabin, S., 428
Rachman, S., 592
Radlove, S., 201
Rafsanjani, A. A. H., 27
Raichle, K., 609
Rajfer, J., 499
Rakic, Z., 506, 511
Ralph, D., 499, 502, 511
Ramirez, A., 143
Rao, K., 626
Raskin, R., 179
Rathus, S. A., 29*t*, 176, 194, 245, 431, 494, 533–535*t*
Ratican, K. L., 629
Rausch, P., 457
Rawlins, M. E., 609
Rawlins, R., 330
Raymond, E. G., 390
Reaney, P., 559, 560
Reid, T. R., 7
Rein, M. F., 543
Reiner, W. G., 180
Reinherz, H., 444
Reinisch, J. M., 115, 123, 271, 272, 276, 295, 296, 297, 334, 381, 382, 387–388, 391, 395, 398, 399, 401, 422, 423, 424, 426–427, 434, 435, 478, 481, 483, 538, 539, 542, 556–557, 559, 561, 580, 586, 592
Reiss, B. F., 477
Reiss, I. L., 313
Remafedi, G., 195
Resick, P. A., 608, 616
Resnick, P. J., 587
Reter, D. J., 186
Revelle, W., 307, 312
Ricci, E., 552, 553
Rice, M. E., 629
Rice, P., 187
Richard, R., 179
Richards, C., 414, 415*f*
Richards, H. C., 406
Richardson, G. A., 351
Riche, M., 458
Ricker, A. L., 646
Ricks, T. E., 321
Rickwood, A. M. K., 114
Riedmann, A., 468
Rienzo, D., 578, 579
Riepe, M. W., 189
Riggio, R. E., 210–212
Riggs, D. S., 612–613, 614, 616
Rimer, B. K., 351, 352
Rimm, E., 499
Ripol, G., 129
Ritchie, K. L., 475
Ritzer, G., 6
Roberto, L. G., 182
Roberts, D., 356–357, 359, 382, 383
Robertson, S., 625
Robinson, J. N., 357
Roddy, R. E., 391
Rodenburg, T., 642
Roehrich, L., 149
Roeltgen, D., 78, 96, 143
Roesler, A., 596, 597, 630

Rogers, A., 510–511
Rolfs, R. T., 537
Romenesko, K., 645, 648
Ronald, A. R., 541
Roosevelt, F. D., 43
Rose, P. G., 77
Rosen, I., 595
Rosen, L., 644
Rosen, R. C., 55, 491, 498, 500–501, 502, 503, 504, 506, 520
Rosenthal, A. M., 71
Rosenthal, E., 209
Rösler, A., 155
Rosman, J. P., 587
Ross, B., 236, 237
Ross, J. L., 78, 96, 143
Ross, M., 186
Rossouw, J. E., 98
Rostker, B., 321
Rotheram-Borus, M. J., 530, 562, 649
Rovet, J., 176
Royce, R. A., 550
Rozin, P., 209
Rubin, A., 476
Rubinow, D. R., 103
Ruble, D. N., 197, 198
Ruckstuhl, L., 474, 475
Rudner, R., 474
Runtz, M., 622, 623
Rusbult, C. E., 222, 224, 238, 241–244, 247, 248, 257
Rush, B., 263
Rush, G., 585*f*
Russell, D., 608
Russo, N. F., 404–405, 604
Rye, B. J., 562

Sack, W. H., 611
Sadalla, E. K., 212
Sadava, S. W., 244
Sadd, S., 467
Sade, M. de, 584, 585*f*
Sadker, D., 190, 195
Sadker, M., 190, 195
Saenger, P., 176
Safire, W., 434–435
Saga, F., 172
Sagan, C., 405, 407
Saluter, A. F., 474
Sampson, P. D., 351
Samuels, M., 353*t*, 366
Samuels, N., 353*t*, 366
Sanchez-Guerrero, J., 8585
Sanders, S. A., 295, 296, 297, 445–446
Sanderson, C. A., 447
Sandross, R., 632
Sanger, M., 374, 375–376
Sargent, T. O., 575
Sarrel, P., 609
Sassaman, C., 174
Sauer, M. V., 99
Savitz, L., 644
Saxe, L., 215
Saywitz, K. J., 626
Scanlon, B., 653
Schacht, C., 217
Schafer, R. B., 215, 462
Schaffer, N., 457
Schafran, L. H., 604, 605
Schairer, C., 97

Schaudinn, F., 536
Schellenberg, E. G., 304
Schemo, D. J., 428–430
Scherer, C. R., 232, 588
Schiavi, R. C., 478, 479, 491, 502
Schild, W., 642
Schillinger, L., 7
Schmidt, G., 490
Schmidt, P. J., 103
Schmitt, E., 290
Schmitt, J. P., 314
Schneider, J., 588–589
Schnicke, M. K., 616
Schoendorf, K. C., 351
Schork, K., 7
Schott, R. L., 576
Schreiner-Engle, P., 491, 502
Schroeder-Printzen, I., 334
Schultz, N. R., Jr., 244
Schwalm, L. A., 655
Schwartz, I. M., 440
Schwartz, M., 588–589
Schwartz, M. F., 315
Schwartz, P., 52, 56, 212, 322, 440, 441, 465, 466, 469
Schwartz, R., 404–405
Scialli, A. R., 105
Sciolino, E., 26–27
Sciortino, A., 670
Scott, J. E., 655, 662
Sears, A. E., 304, 657
Seely, J., 511
Seftel, A. D., 481, 482
Segal, S., 102, 103
Segal, Z. V., 611
Segraves, K. B., 491, 500
Segraves, R. T., 491, 500
Seligman, L., 575, 594, 674
Selik, R., 528
Selker, T., 247
Sell, R. L., 295
Sellers, T., 127–128
Seltzer, R., 304
Selvin, B. W., 181, 186
Semans, J., 520
Seña, A., 550
Seng, M. J., 649, 650
Shafer, M. A., 539
Shakespeare, W., 147, 238
Shanahan, K., 362, 363
Sharrar, R. G., 537
Shaver, P., 225
Sheehy, G., 118, 333
Shenon, P., 316
Shepherd, J. B., 563
Sheppard, J. A., 208
Sherkat, S., 26
Sherman, K. J., 540
Sherman, P. W., 340
Sherwin, B. B., 157
Shettles, L., 330
Shevel, J., 210
Shifren, J. L., 515
Shild, W., 643
Shilibwa, A., 11
Shipko, S., 520
Shlipak, M. G., 97
Shockley, W., 382
Shoda, Y., 82, 86
Short, R. V., 114
Shuster, S. M., 174
Siegel, J. M., 613*t*, 614
Siegel, K., 322
Signorielli, N., 195
Silber, S. J., 334
Silbert, M. H., 657
Silva, P., 439, 448
Simon, W., 278, 311
Simonsen, G., 314
Simpson, J. L., 354
Simpson, O. J., 472
Singer, I., 166
Singer, J., 166
Singer, J. L., 284, 285*t*
Singh, D., 209
Sipski, M. L., 482
Sirles, E. A., 625
Skates, S., 125
Skeen, P., 479
Skinner, B. F., 30
Skurrie, I. J., 540
Sleek, S., 316
Smale, A., 644
Smart, C. R., 8585
Smith, B., 68, 300
Smith, E., 241
Smith, E. P., 436
Smith, W. K., 602, 606
Smock, P. J., 456, 457, 458
Snell, S., 457
Snyder, C. W., 248, 303, 408, 479, 644
Sobel, J. D., 542, 543, 544
Sodroski, J., 545
Solano, C. H., 244
Sommerfeld, J., 100, 102–103
Sonenstein, F. L., 318, 437, 438
Sorrenti-Little, I., 625
Soule, S., 430
Soules, M., 103
Sourander, L. B., 78, 143
Southerland, D., 52
Spain, D., 475
Spano, M. S., 499
Spark, R. F., 482, 491, 500, 514
Sparks, G. G., 215
Speckens, A. E. M., 496–497
Specter, M., 549, 644
Spector, I. P., 504
Spiro, M. E., 422
Spitz, I. M., 412
Spitzer, M., 189
Spitzer, R. L., 502–503, 585, 586–587
Sponsel, W. E., 514
Sprecher, S., 207, 212, 213*t*, 440, 441, 442, 466
Sprenkle, D. H., 241
Spring, J. A., 468–469
Srebnik, D., 265, 267, 436
Stamm, W. E., 540
Stampfer, M. J., 382
Stangor, C., 198
Starcevic, V. P., 506, 511
Starr, B. D., 477, 479, 480
Steele, C. M., 150
Stein, G., 165, 172
Stein, P. J., 444
Stein, Z., 340, 348
Steinbeck, J., 655
Steiner, B. W., 186
Steiner, M., 103
Steinhauer, J., 457
Stenson, J., 360–364
Stephan, C. W., 217
Stephens, W. N., 25, 40
Stephenson, J., 391
Sternberg, R., 221–222, 223–226, 227, 246
Sternberg, R. J., 220
Stewart, F. H., 402
Stewart, G. K., 386, 387, 388
Stewart, L. P., 631, 632
Stewart, P., 656
Stief, T. M., 444
Stock, W. E., 494, 496, 518, 521, 522, 610
Stokes, G., 45, 426, 436–437, 439–440, 441, 442–443
Stolberg, S. G., 336, 528, 529, 537, 550
Stoller, R. J., 182, 299, 579
Stoner, S. A., 149, 150
Storer, B. E., 97
Storms, M. D., 194, 199, 296–297
Strassberg, D. S., 267, 500, 503
Strathman, A. J., 208
Streissguth, A. P., 351
Strickland, B. R., 312
Struckman-Johnson, C., 609
Succo, S., 499
Sue, D., 284, 593
Sue, D. W., 593
Sue, S., 593
Sulak, P. J., 382
Sullivan, N. J., 551
Sullivan, Q., 212, 213*t*
Sullivan, S. E., 351, 352
Susser, M., 340, 348
Swaisgood, R., 148
Swann, W. B., Jr., 216
Swanson, C., 474, 475
Swardson, A., 124
Symons, D., 213, 214, 271, 659–660
Szabo, R., 114
Szasz, G., 482

Takahira, S., 189
Talbert, L., 156
Tanfer, K., 49
Tang, C. S., 605
Tang, M. C., 84
Tang, O. S., 413
Tannahill, R., 10, 11, 12, 15
Tannen, D., 236–237, 249
Tantleff, S., 209
Tarnai, J., 201
Tarone, R. E., 124, 125
Tarran, M. J., 425
Tavris, C., 467
Taylor, A., 359
Taylor, C., 672, 673
Taylor, J., 99
Tedeschi, J. T., 602, 631
Telushkin, J., 11
Temple, L. L., 304
Thabit, S. M., 70
Thalberg, S., 445
Tharinger, D., 626
Thomas, C., 602
Thomas, J., 184–185
Thome, P. R., 196
Thomlison, B., 621, 622
Thompson, D. S., 119, 120, 411
Thompson, J. K., 209
Thompson, M. E., 619
Thompson, S., 439
Thompson, T. G., 628
Thornby, J., 156
Thornton, L. M., 307, 308, 626
Thornton, Y., 361, 363–364
Tierney, J., 660
Tissot, S. A. D., 263
To, T., 84
Todd, R., 351
Tollison, C. D., 182, 623
Tomezak, R., 189
Toner, J. P., 333, 336
Torabi, M. R., 531
Townsend, J. M., 21, 469, 659
Trieschmann, R., 481
Trost, J. E., 164
Trujillo, C., 300
Tseng, W., 501
Tucker, J. S., 217
Tuiten, A., 490, 506–507
Tukey, J. W., 46
Tur-Kaspa, I., 335
Turkle, S., 233
Turner, K., 146
Tutty, L. M., 628
Twain, M., 111, 655
Twersky, O., 512
Tyson, M., 602, 606

Uddenberg, N., 186
Udry, J. R., 156
Upfold, D., 611

Valdez, H., 552
Valentiner, D. P., 612–613, 614, 616
Valleroy, L. A., 553
VanBlaricom, A., 360, 361, 363
Van den Hoek, A., 654
van der Plight, J., 563
van der Velde, F. W., 563
van-Griensven, G. J. P., 549
Van Haastrecht, H. J., 654
Van Hoover, C., 363
Van Lange, P. A. M., 238, 248
van Oss Marian, B., 198–199, 440, 441
Varas-Lorenzo, C., 98
Vatsyayana, 14
Vazi, R., 122
Veenker, C. H., 531
Veniegas, R. C., 314
Verette, J., 248
Verschoor, A. M., 576
Vershure, B., 212
Victoria, Queen, 16
Vidali, A., 335
Vinacke, W., 216
Voelker, R., 381
Voeller, B., 287
Vogel, N., 467
von Krafft-Ebing, R., 17, 264
von Sacher-Masoch, L., 583
Voyer, D., 190
Voyer, S., 190

Wagner, N., 162
Waigandt, A., 614
Waite, B. M., 304
Wald, A., 555
Wald, E., 354
Waldinger, M. D., 520
Walfish, S., 201
Walinder, J., 186
Walker, A., 71

Wall, C., 84
Wallerstein, J. S., 475
Walsh, P. C., 123
Walster, E., 221, 225
Walster, G. W., 221
Walt, V., 402
Walter, K., 474*t*
Wan, C. K., 562
Ward, R. H., 444
Warner, D. L., 393–394
Warren, C. W., 50
Warren, J., 612
Wasson, J. H., 123
Waterman, E., 127
Waterman, J., 621, 622, 624, 625–626, 628–629
Watkins, C. E., Jr., 314
Watson, C., 200
Watson, J. B., 30
Watson, R., 578, 579
Wearing, A. J., 668
Weaver, J. B., 663
Weber, B., 82, 83
Weinberg, M., 593–594
Weinberg, M. S., 49, 293, 297, 298, 311, 312, 314, 315, 316, 317, 318, 319, 322, 323, 585
Weinberg, T. S., 577–578, 585, 586
Weiner, L., 631
Weiner, M. B., 477, 479, 480
Weingarten, H., 475
Weinrich, J. D., 176, 307, 311
Weisberg, D. K., 653
Weiss, D. L., 441
Weiss, G., 103
Weiss, N. S., 84
Weiss, R. D., 151, 500
Weizer, R., 640
Wells, J. A., 295
Wenger, N. K., 97, 98, 100
Wennerholm, U.-B., 337
Werner, D., 193, 476
West, M., 111
Wexler, M. O., 224
Weyman, A., 413
Whalen, R. E., 308
Whipple, B., 166–167
Whitam, F. L., 307
White, S. D., 440
White, V. M., 668
Whitley, B. E., Jr., 200, 304
Whitley, R., 556
Whittemore, A. S., 79, 381
Whitten, P., 11, 25, 53, 193, 250, 462, 463, 470, 472, 476, 624
Whitton, C., 172
Whitton, S. W., 222, 248
Wiederman, M. W., 239
Wieselquist, J., 222, 247, 248
Wilcox, A. J., 330, 350, 396
Wilcox, B. L., 661
Wilford, J. N., 537
Wilkerson, J., 430
Willey, K., 630
Williams, C. J., 585
Williams, D. E., 200
Williams, J. E., 187, 187*t*
Williams, L., 83
Williams, M., 147, 157
Willis, R. J., 457
Willmott, M., 191
Wilson, G., 623
Wilson, G. T., 150
Wilson, M., 21
Wilson, M. I., 473
Wilson, M. R., 56
Wilson, S. N., 447
Wilson, W., 500
Wilting, K., 642
Winchell, B., 320
Winick, C., 659
Wise, T. N., 520
Wisniewski, N., 604, 605, 607, 611
Witztum, E., 155, 596, 597, 630
Wolfe, T., 434
Woll, S. B., 210–212
Wolman, T., 150
Wong, E., 302
Wood, E., 547, 549
Wood, N. S., 366
Wooldridge, W. E., 538
Wordsworth, W., 420
Wortman, C. B., 234
Wright, L. W., Jr., 304
Wright, P., 590
Wright, T. L., 469
Wulfert, E., 562
Wunderlich, A. P., 189
Wyatt, G., 50–51, 284
Wyatt, G. E., 614, 626
Wypij, D., 295
Wysocki, C. J., 143

Yaffe, 96
Yanez, C., 562
Yarber, W. L., 530, 539
Yates, G. L., 654
Yates, S., 407
Yee, J., 617
Yellin, P., 444
Yetman, N. R., 6
Yllo, K., 608
Yoder, P., 362
Yorburg, B., 457
Young, B., 25
Young, B. G., 593
Young, K., 232–233
Young, K. S., 588
Young, L., 649

Zalar, R. W., 472, 473
Zanna, M. P., 150, 563
Zaviacic, M., 166–167
Zekoski, E. M., 367
Zellman, G., 427
Zhang, J., 351
Zhu, G., 198
Zielbauer, P., 628
Zielke, A., 430
Zilbergeld, B., 115, 521
Zillmann, D., 659, 663–664
Zimmer, D., 283
Zubek, J., 414, 415*f*
Zucker, K. J., 180, 312
Zuckerman, E. W., 113
Zufer, A., 649
Zuger, A., 86
Zusman, M. E., 217, 238, 241–244, 457, 463
Zwinderman, A. H., 520

subject index

Page numbers followed by *t* and *f* indicate tables and figures, respectively.

ABCDE model of romantic relationships, 230–244
Abominations, early Christian, 13–14
Abortion, 374
  drugs causing, 412–413
  induced, 404–405
  premises of arguments on, 9
  spontaneous, 328, 340
Abortion providers, declining numbers of, 408–409
Abortion rights, 407
Abstinence
  periodic, failure rates of, 378*t*
  for STI prevention, 562
acamgirl.com, 572–573
Accutane, in fetal abnormalities, 350
Achilles-Patroclus relationship, 12
Acquaintance rape, 605
  psychological characteristics in, 611
Acquired immunodeficiency syndrome. *See* AIDS; HIV (human immunodeficiency virus) infection
Act-Up, 306
Activating effects, 308
Acyclovir (Zovirax), 534*t*, 557
Adam and Eve, original sin of, 13–14, 15
Adaptive traits, 21
Adolescence
  contraceptive use during, 445–447
  factors in premarital intercourse during, 440–441
  male-female sexual behavior during, 436–442
  male-male and female-female sexual behavior during, 442–443
  masturbation during, 436
  puberty in, 431–436
  sexual orientation during, 311–312, 316
  sexuality during, 430–443
  teenagers' values and sexual double standards in, 441–442
Adolescents
  pornography and, 659
  in prostitution, 649–650
Adoption, 337
Adrenal glands
  sex hormone production by, 154
  in sexual desire, 490
Adrenergic blockers, 499
Adultery
  in ancient Hebrew culture, 11
  consensual, 469
  conventional, 469
Adults
  alternative lifestyles among, 476
  cohabitation among, 456–459
  disabled, 480–484
  marriage among, 459–463
  sex in later years, 477–480
  single, 454–456
Advocate, 306
Affluence, post-World War II, 464
African Americans
  AIDS in, 553
  beauty standards of, 209
  number of sex partners among, 33
  premarital oral sex among, 437
  prostate cancer in, 123–124
  sexual activity among during adolescence, 447–448
  sexual dysfunctions among, 498*t*
Agape, 217, 221
Age
  homogamy, 463
  number of sex partners and, 31
Age of Enlightenment, 10
Ageism, 195
Aggressive behavior
  brain organization and, 192
  gender differences in, 191
  learning of, 196
  in men and women, 192–193
  pornography and, 661–663
  sexual, 198–199
Aging
  sexual activity and, 477–480
  sexual response and, 4
AIDS. *See also* HIV (human immunodeficiency virus) infection
  in African Americans, 553
  anal sex and, 286–287
  circumcision and, 113–114
  concerns about, 4, 528
  death rate from, 552–553
  diagnosis of, 535*t*, 550–551
  epidemic of, 529
  immune system and, 544–545
  incidence of, 529
  in Latina/Latino Americans, 553
  number of cases of by exposure category, 545*t*
  in prenatal development, 348
  prevention of, 553–555
  progression of, 548
  in prostitutes, 654–655
  public awareness of, 528–529, 562–563
  risk of in rape survivors, 614
  swinging and, 472
  transmission and symptoms of, 535*t*
  treatment of, 535*t*, 551–553
  vaccine for, 551
AIDS Hotlines, 555, 567
  for the Hearing Impaired, 567
  for Teens, 567
AIDS-related complex, 548
Alcohol
  aphrodisiac effects of, 149–150
  breast cancer and, 84
  in erectile dysfunction, 500
  fetal damage with, 351
  in rape, 611
  in STI transmission, 563
Aleutian Islanders, extramarital sex in, 470–471
Allergies, breast versus bottle feeding in, 367
Alpha-adrenergic-agonist drugs, 123
Alpha interferon, 534*t*
Alprostadil
  for erectile disorder, 510, 510*t*, 511
  for female sexual arousal disorder, 515
Alternative lifestyles, 476
Altruism, reciprocal, 307
Alzheimer's disease, 97
Amenorrhea, 101
American Association of Sex Educators, Counselors, and Therapists (AASECT), 523
American Cancer Society, breast exam recommendations of, 85
American Psychiatric Association, on homosexuality as mental disorder, 313
American Psychological Association, Committee on Lesbian and Gay Concerns, 292
Amitriptyline (Elavil), anaphrodisiac effects of, 149
Amniocentesis, 328
  for genetic disorders, 354
Amniotic fluid, 345
Amniotic sac, 345
Amoxicillin, 533*t*, 540
Ampicillin, 541
Ampulla, 78
Amputees, sexual activity in, 483
Amyl nitrate, 147
Anal-genital sex, 564
  laws against, 306
Anal intercourse, 285–287
  among adolescent males, 437*t*
  among gay males, 318
  in HIV transmission, 549
Analogous structures, 69
Analogues, 24
Anaphrodisiacs, 149
Androgen-insensitivity syndrome, 177, 178*f*
  gender identity and, 178
Androgen-suppression therapy, 125
Androgenital syndrome, 177, 178*f*
Androgens, 117, 154
  in sex drive, 155–156
  in sexual differentiation, 175
Androgyny, psychological, 199
Androstadienone, 143

Anesthesia, in childbirth, 358
Anger rape, 612, 613*f*
Anilingus, 286, 564
Animal fat-related prostate cancer, 124
Animals
  experimentation on, 58–59
  mating patterns of, 282, 283*f*
  promiscuity of female, 22–23
Anita Hill-Thomas Clarence case, 602
Anorexia nervosa
  amenorrhea and, 101
  beauty standards and, 209
Anorgasmic women, 494
  couples therapy for, 515–516
  directed masturbation program for, 516–518
Anovulatory menstrual cycles, 433–434
Anoxia, 365
  prenatal, 365
  prolonged during delivery, 365
Anthropological studies, 24–28
Anthropologists, 5–6
Anthropomorphism, 39
Anti-androgen drugs, 156
  anaphrodisiac effects of, 149
  for paraphilias, 597
  for rapists and child molesters, 630
Antibiotics
  in candidiasis, 542
  for chancroid, 541
  for chlamydia, 540
  in fetal abnormalities, 350
  for gonorrhea, 536
  for lymphogranuloma venereum, 541
  for STIs, 533*t*
  for viral hepatitis, 534*t*
Antibodies
  in immune system, 545
  to *Treponema pallidum*, 539
Anticancer drugs, 125
Anticonvulsants, 499
Antidepressants
  anaphrodisiac effects of, 149
  in erectile dysfunction, 499
  for paraphilias, 596
  for premature ejaculation, 520
Antigens, 545
Antihistamine-related fetal abnormalities, 350
Antihypertensives, 499
Antipsychotics, 499
Antiretroviral therapy, 552
Antiviral drugs, 557
Anxiety
  in dyspareunia, 495
  in erectile dysfunction, 492–493
  in female sexual arousal disorder, 493
  in lack of sexual desire, 491
  in male orgasmic disorder, 494
  in premature ejaculation, 134–135
  in sexual dysfunctions, 501, 503–504
  in vaginismus, 497
Aphrodisiacs, 12, 143, 146–151
Aphrodite, 12
Apollo, 12
Apomorphine (Uprima), 510*t*, 511–512
Arapesh society, 193
ARC. *See* AIDS-related complex
Archives of General Psychiatry, 414
Areola, 81
  engorgement of, 160
Arguments, premises of, 9
Art therapy, 629
Arthritis, 483
Artificial contraception, 376
Artificial insemination, 334
  in STI transmission, 549
Asceticism, 672
Asexuals, 323
Asian Americans, number of sex partners among, 33
Asian prostitutes, poverty of, 649
Athletes, sexual assaults by, 610
Attitudes, similarity of, 215–216
Attraction, 206–207
  factors determining, 207–216
  matching hypothesis of, 215
  reciprocity in, 216
  in romantic relationships, 230–231
  of similarity of attitudes, 215–216
Attractiveness. *See* Physical attractiveness
Autoimmune-related male infertility, 333, 334
Autonomic nervous system, 134–135
Aversion therapy, 595
Azithromycin
  for chancroid, 541
  for chlamydia, 533*t*, 540

Baboons, male-male sexual activity among, 301
Bacterial infections, 532
  sexually transmitted, 532–541
Bacterial vaginosis, 542
Bar hustlers, 654
Bar prostitute, 645
Barrier devices, 564
Bartholin's glands, 74
  fluid secretion of, 160
Basal body temperature method, 92, 396–397
  charting, 330–332
Battelle Memorial Institute of Seattle survey, 49
Battle Creek Sanatorium, 264
Beauty
  cultural differences in, 208*f*
  cultural standards of, 208–209
Behavior patterns
  sexual techniques and, 262
Behavioral therapy
  for paraphilias, 595–596
  for sexual aversion disorder, 508
  for sexual dysfunctions, 505
Behaviorists, 30
Beliefs
  irrational, 251
  in sexual attitudes, 7
  in sexual dysfunctions, 503
  supernatural, 10
Belly-button surgery, 400
Benign prostatic hyperplasia, 123
Bestiality, 13, 17, 587
  laws against, 306
Between Two Absolutes, 408
Bible, moral code of, 669–670
Bichloroacetic acid, 535*t*, 560
Biochemical therapy, 596–597
Biological clock, 213
Biological instincts, 20
Biological treatments
  for erectile disorder, 509–514
  for female sexual arousal disorder, 515
  for premature ejaculation, 520
  for sexual aversion disorder, 508
Biology, 19–20
  in gender typing, 191–192
  in sexual orientation, 307–309
Birth control. *See also* Abortion; Birth-control pills; Contraception
  ancient methods of, 375
  failure rates of various methods of, 378*t*
  natural. *See Coitus interruptus*
Birth-control pills, 18
  accidental pregnancy rates with, 379*t*
  advantages and disadvantages of, 381–384
  breast cancer and, 83–84
  in candidiasis, 542
  cardiovascular complications of, 383
  combination, 380
  contraindications to, 382
  decision making balance sheet for, 673–675
  effectiveness of, 381
  healthful side effects of, 381
  mechanism of, 380
  morning-after, 384
  reversibility of, 381
  types of, 380
Birth defects
  with drugs taken by parents, 349–350
  maternal rubella and, 348
Birth problems, 364–366
Birthing, options for, 360–364
Birthing center, state-of-the-art, 360–361
Bisexuality, 12, 291, 297–298
Bisexuals
  erotic fantasies of, 297
  responsiveness of to sexual stimulation, 296–297
Bladder, cystitis of, 70–71
Blastocyst, 343–344
Blended orgasm, 166
Blindness, 482–483
Blood supplies, HIV transmission via, 549
Blood tests
  for genetic screening, 355
  for lymphogranuloma venereum, 541
Blood transfusions, HIV transmission via, 549, 553–555
Blood-typing, pregnancy, 349
"Blue balls," 163
Body language
  in intimate communication, 249
  observation of, 54
Body shape, gender differences in standards of, 209
Bondage, 583–584
  psychoanalytic theory of, 591
Bottle feeding, 367–368
Brain
  damage of in paraphilias, 590
  in erection, 133–134
  geography of, 151–152
  pleasure centers in, 153–154
  prenatal organization of, 192
  prenatal sexual differentiation of, 176
  in sexual response, 151–154
  structure of and sexual orientation, 309
Braxton-Hicks contractions, 355
Breaking up, 241–244
Breast buds, 431
Breast feeding versus bottle feeding, 367–368
Breast implants, 85
Breast milk, HIV transmission via, 549
Breasts, 80, 88
  age-related changes in, 478
  cancer of, 82–88
  cultural views of, 81
  development of, 80–81
  lumps in, 84
  normal variations in, 82*f*
  in sexual excitement, 158
  in sexual response cycle phases, 161*f*

size of, 209
stimulation of, 275
Breech presentation, 346
Bridge maneuver, 518
Brothels, 646
condom use in, 655
enslavement in, 646
male prostitutes in, 654
poverty and, 649
Brother-sister incest, 625
Brother-sister marriage, 11
Bulbourethral glands, 121
Bupropion, aphrodisiac effects of, 147
Butaconazole, 534*t*, 543
Butch-femme gender roles, 307

Calcium supplements, 97
Calendar method, 396
Caligula, 13
Call boys, 653
Call girls, 647
Calymmatobacterium granulomatous, 541
Canadian Toll-Free Hotline: AIDS Committee of Toronto, 567
Cancer
breast, 82–88
cervical, screening for, 76–77, 80*f*
endometrial, 77
ovarian, 79
prostate, 123–130
of testes, 122–123
Candida albicans
transmission, symptoms, diagnosis, and treatment of, 534*t*
in vaginal candidiasis, 542–543
Candidiasis
factors in, 542
transmission, symptoms, diagnosis, and treatment of, 534*t*
vaginal, 542–543
Caring, 246–248
Case-study method, 44–45
limitations of, 44–45
Castration, 155–156
chemical
for rapists and child molesters, 630
Castration anxiety, 309
in paraphilias, 591
unconscious, 310
Casual sex, 18
among gay males versus lesbians, 322
Catholicism, number of sex partners and, 33
Cause-and-effect relationships, 57
in experimental methods, 59
Caverject. *See* Alprostadil
CD4 cells
in HIV infection, 546–548
zidovudine and, 550–551
Cefixime, 533*t*, 536
Ceftriaxone, 533*t*, 536
Celibacy, 456
among early Christians, 13
in Islamic society, 14
religious, 672
Central nervous system depressants, 500
Cephalic presentation, 346
Cephalocaudal development, 344–345
Cerebellum, 151
Cerebral cortex, 152
sensory processing in, 153
Cerebral palsy, 481
Cerebrum, 152
Cervical cancer
Pap test screen for, 80*f*
screening for, 76–77
Cervical cap, 391
advantages and disadvantages of, 392
effectiveness of, 391
failure rates of, 378*t*
insertion of, 391
mechanism of, 391
reversibility of, 391
Cervical mucus method, 397
Cervicitis
chlamydial, 539
with gonorrhea, 536
Cervix, 76–77
effacing and dilating, 355
Cesarean section, 359
in HIV transmission prevention, 552
in home birth, 361–362
Chancre, syphilitic, 538
Chancroid, 541
Chattel, 459
Chemical-barrier methods, 402
Chemotherapy, 507
Child molesters
of family members, 621–622
gay versus heterosexual, 302–303
gender of, 622–623
pornography and, 661
treatment of, 629–630
types of, 622–623
Child pornography, 657
Childbirth, 328. *See also* Delivery; Labor
first uterine contractions in, 355
location for, 360–364
methods of, 357
stages of, 356*f*, 358*f*
Childbirth Without Fear, 358
Childhood. *See also* Children
conflicts of in sexual dysfunction, 503
sexual behavior during, 420
Childlessness, 11
Children. *See also* Childhood
in cohabiting households, 457
in divorce, 475
as objects of sexual arousal, 574, 575
voluntary sexual activity between, 621
China
sexuality and religions of, 14
Chlamydia
diagnosis of, 533*t*, 540–541
incidence of, 539
symptoms of, 533*t*, 539–540
transmission of, 533*t*, 539
treatment of, 533*t*, 536
untreated, 540
Chlamydia trachomatis, 539
in lymphogranuloma venereum, 541
transmission of, 539
Cholesterol
in erectile dysfunction, 499
estrogen and, 97–98
Chorionic villus sampling, 354
Christianity
attitudes toward homosexuality in, 298–299
early, 13–14
marriage and, 460
moral code of, 669–671
Chromosomes, 20–21, 172
abnormalities of, 328, 352–355
hallucinogen-induced damage to, 350–351
sex, 173
Chukchee society
extramarital sex in, 471
men in, 25
Cigarette smoking
in fetal damage, 351–352
secondary smoke from, 351
Cilia, 120
Ciprofloaxin
for chancroid, 541
for gonorrhea, 533*t*, 536
Circulatory system, fetal, 345
Circumcision, 113–114
in HIV transmission risk, 550
Civil Rights Act of 1964, Title VII, 631–632
Clarifying questions, 256
Climacteric, 95
Clindamycin, 533*t*, 542
Clitoral device, 515*f*
Clitoral orgasm, 165
Clitoral stimulation
direct, 522
during masturbation, 269–270
Clitoris, 66, 68–69
engorgement of, 158
ignorance about erotic function of, 502
manual stimulation of, 274
oral stimulation of, 277
during plateau phase, 159
surgical removal of, 69
Cloaca, 112
Clomiphene (Clomid), 92, 334–335
Clomipramine, 520
Close-binding mother
in gay male sexual orientation, 310–311
in sexual orientation, 309
in transsexualism, 182
Close couples, gay, 322
Clotrimazole, 534*t*, 543
Cocaine
aphrodisiac effects of, 151
in birth defects, 350
in erectile dysfunction, 500
syphilis and, 537
Cognitive development
gender differences in, 189–190
in gender typing, 196–197
Cohabitation, 456
benefit and risk of, 458–459
facts about, 457
reasons for, 457–458
in Western culture, 16
Coital positions, 278–283
female-superior, 199, 280, 281*f*, 515–516, 517*f*, 520
lateral-entry, 280–281, 282*f*
male-superior, 279–280
during pregnancy, 341
rear-entry, 282, 283*f*
Coition machine, 55
Coitus
in adolescence, 438–439
anal, 285–287
changes in techniques and duration of, 466
clitoral stimulation during, 165–166
definition of, 5
delayed ejaculation during, 494
fantasy during, 283–284
female orgasmic disorder during, 494
first time, 440–441
frequency of
initial, 72–73
during menstruation, 95
orgasm through, 165

Coitus, *continued*
  painful, 490, 495–497
  positions in, 278–283
  during pregnancy, 340–341
  techniques in, 284–287
Coitus interruptus, 263, 375, 395
  failure rates of, 378*t*
College campus
  date rape on, 606
  sexual harassment on, 633
College students, body shape standards of, 209
Comanches, extramarital sex in, 471
Comarital sex, 469, 472
Coming out
  to oneself, 316–317
  to others, 317–318
Commercial sex, 640
  pornography and obscenity, 655–665
  prostitution, 640–655
Commission on Obscenity and Pornography, 661
Commitment
  in intimacy, 248
  in love, 221, 225
Communication
  getting started in, 251–252
  impasses in, 257
  intimacy and, 230, 249–250
  reinforcing partner's willingness for, 252
  styles of, 249
Communication skills
  delivering criticism, 254–255
  difficulties in sexual relations and, 250–251
  learning about partner's needs in, 252–253
  listening in, 252
  making requests, 254
  providing information, 253–254
  in receiving criticism, 255–256
  in romantic relationships, 251–252
Community standards, 657
Companionate love, 223*t*
  components of, 225–226
Compatibility, 222
  in triangular model of love, 223*f*
Comstock Act, 375, 656–657
  dismantling of, 376
Conception, 328–329
  basal body temperature charting for, 330–332
  definition of, 328
  factors influencing, 332
  luteinizing hormone levels and, 332
  optimizing chances of, 329–332
  vaginal mucus tracking for, 332
Conclusions, 39
  evidence and, 9
Concordance, 307
Concubines, 12
Conditioning model, 592
Condoms, 392
  accidental pregnancy rates with, 379*t*
  advantages and disadvantages of, 394–395
  allergic reactions to, 495
  applying, 393*f*
  in brothels, 655
  early advocates of, 375
  effectiveness of, 394
  excuses for not using, 562
  female, 401–402
  in gay sexual activity, 322
  guidelines for use of, 392–394
  latex, 394, 564
  mechanism of, 392
  negative attitudes toward, 563
  nonlatex, 402
  reversibility of, 394
  in teenage pregnancy prevention, 446
  types of, 393*f*
  use of by teenager boys, 446
Confidentiality, 46, 60
Conflict, 668–669
Consensual sex, 606
Constructs, 40
Consummate love, 223*t*
  components of, 224, 226
Contraception, 374, 375. *See also* Birth-control pills
  among sexually active teens, 445–447
  artificial, 376
  convenience of, 377
  cost of, 377
  effectiveness of, 379
  failure rates with, 379
  information about, 375–376
  legal battles over, 375–376
  methods of, 380–404
  moral acceptability of, 377
  responsible sexual decision about, 667
  reversibility of, 378
  safety of, 378
  search for perfect methods of, 401–404
  sharing responsibility for, 377–378
  STI protection with, 378–379
Contraceptive use, 380–384
  factors determining, 445–446
  oral. *See Birth-control pills*
  in sexual revolution, 464
  by teens, 445–448
Control, 40
  external locus of, 244
Control groups, 58
Convenience samples, 44
Conversion therapy, 315–316
Copper T 380A, 386
Coprophilia, 588
Copulation, 24
Corona, 112
Corpora cavernosa, 68, 73, 112
Corpus luteum, 88, 93
Corpus spongiosum, 112
Correlation, 56
Correlation coefficient, 56
Correlational method, 56
  limitations of, 56–57
Counselors, rape, 616–617
Couples therapy
  evaluation of, 522
  for hypoactive sexual desire, 507
  for orgasmic disorders, 515–516
  for sexual aversion disorder, 508
Courtesans, 12
Courtship disorders, 593
Covert sensitization
  for paraphilias, 595–596
  for rapists and child molesters, 630
Cowper's glands, 121
  during plateau phase, 159
Crabs. *See* Pediculosis
Cremaster muscle, 115
Crisis intervention, rape, 616
Critical fat hypothesis, 431
Critical periods, prenatal development, 347–348
Critical thinking
  about human sexuality, 8–9
  principles of, 8–9
  skills for, 7
Criticism
  acknowledging, 256
  clarifying goals in, 256
  constructive, 255
  delivery of, 254–255
  motives for, 255
  positive, 255
  rejection of, 256
  specific, 255
Crocodiles, sexual differentiation of, 179
Cross-cultural studies
  of beauty, 208–209
  of extramarital sex, 470–471
  of gay and lesbian sexual orientations, 299–300
  of gender identity, stability and constancy, 197
  of same-gender sexual activity, 295
Cross-cultural traditions
  gender roles in, 192–193
  kissing in, 273
  of sexual behaviors, 23, 24–25
Cross-dressing. *See also* Transvestism
  in gay males, 312
  sexual versus nonsexual reasons for, 576–577
Cross-gender preferences, 181
Cross-species sexual behavior, 24
Cruising, 322
Crura, 73
Crusaders, 15
Crush, 220
Cryotherapy
  for genital warts, 535*t*, 560
  for *molluscum contagiosum*, 560
Cryptorchidism, 122, 176
Culpotomy, 400
Cultural norms
  in negative attitudes among women toward first coital experience, 440
  in rape prevention, 617–618
Cultural relativism, 672
Cultural traditions
  beauty standards and, 208–209
  in defining sexual dysfunctions, 496–497
  gender roles and, 193
  in rape, 610
  sexual attitudes and, 7
  sexual behavior and, 24–28
  in sexual beliefs and behaviors, 6–7
  in sexual dysfunctions, 500
Cunnilingus, 13, 275
  during adolescence, 437
  barrier devices with, 564
  techniques of, 277
Cupid, 217
Cynicism, 244
  challenging feelings of, 245
Cystic fibrosis
  characteristics of, 353*t*
  screening for, 355
Cystitis, 70–71
Cysts, breast, 84

Dalkon Shield, 386
Dark Ages, 15
Dartos muscle, 115–116
Date rape, 605–606
  consent as issue in, 606
Dating
  during adolescence, 436
Dating relationships
  aggressive competition in, 610–611
  verbal coercion in, 619

Daughters of Bilitis, 306
ddC, 551–552
ddI, 551–552
Deafness, 482–483
Deception
  use of in research, 60
Decision making
  balance sheet for, 673–675
  conflict in, 668–669
  gathering information for, 669
  information for, 668
  value systems and, 669–673
Deep Throat, 18
Defense mechanisms, 28
Defloration, 73
Delayed ejaculation, 494
Delivery. *See also* Childbirth; Labor
  calculating date of, 342
  with doulas, 363–364
  location for, 360–364
  problems with, 364–366
  prolonged anoxia during, 365
Delta hepatitis. *See* Hepatitis D virus
Demographic variables, 40
Denial, 53
Dependent variables, 57–58
Depo-Provera, 402–403
  accidental pregnancy rates with, 379*t*
  for paraphilias, 597
  for rapists and child molesters, 630
Depression
  among gay males and lesbians, 313–314
  in erectile dysfunction, 493
  in hypoactive sexual desire, 507
  in lack of sexual desire, 490, 491
  postpartum, 366–367
  in sexual dysfunction, 502
Description, 39–40
Desensitization. *See* Systematic desensitization
Desire, 163
  in romantic love, 224
Detached-hostile father
  in gay male sexual orientation, 310–311
  in sexual orientation, 309
  in transsexualism, 182
Developmental stages, 29
Deviant sexual behavior
  definition of, 573–574
Diabetes
  in candidiasis, 542
  in erectile dysfunction, 498–499
  in male infertility, 334
Diagnostic and Statistical Manual of Mental Disorders, sexual dysfunction classifications of, 489
Diaphragm
  advantages and disadvantages of, 389
  checking of, 389*f*
  declining popularity of, 388
  effectiveness of, 388–389
  failure rates of, 378*t*
  insertion of, 388, 389*f*
  mechanism of, 388
  reversibility of, 389
  silicone, 402
Diet
  prenatal effect of, 346–347
  in recurrent candidiasis, 543
Diethylstilbestrol (DES)
  in cryptorchidism, 122
  in fetal abnormalities, 350
  prenatal influence of, 308
Differences, negotiation of, 256
Differentness, toleration of, 257
Dihydrotesterone (DHT), 175
Dilation, cervical, 355
Dilation and curettage (D&C), 410
Dilation and evacuation (D&E), 410–411
Dildo, 262
  in STI prevention, 565
Disability
  misconceptions about, 480
  parental overprotectiveness with, 480–481
  physical, 481–483
  psychological, 483–484
  sexual activity and, 480–484
Disagreement, 252
  agreement on, 257
Disfiguring conditions, prostitute services for customers with, 651
Displacement, 309
Diuretics, 499
Diversity
  sampling of, 42–44
  in sexual beliefs and behaviors, 6–7
Divorce
  conclusions about causes of, 9
  cost of, 474–475
  in early Christianity, 13
  incidence of, 474
  increasing rate of, 455
  long-term consequences of, 475
  predictors of, 474
  reasons for, 474*t*
  risk of among cohabiting couples who marry, 458–459
DNA, 20–21
Dolls, inflatable, 269
Domestic violence, 472–473
  aftermath of, 473*f*
  feminist theory on, 473
  gender differences in, 473
Dominican Republic syndrome, 177–178
  gender identity in, 178–179
Donor in vitro fertilization, 336
Dopamine
  deficiency of, 510*t*
  for erectile disorder, 512–513
Douche, 75
Douching, 395
  in STI prevention, 565
  for vaginal infection prevention, 76
Doulas, 363–364
Down syndrome, 328, 352
  characteristics of, 353*t*
  risk of by age of mother, 353*t*
Doxycycline
  for chlamydia, 533*t*, 540
  for lymphogranuloma venereum, 541
  for syphilis, 533*t*
Drag prostitutes, 654
Dream symbols, 29*t*
Drug users
  HIV infection in, 654
  syphilis transmission in, 537
D4T, 551–552
Dysfunctionals, 323
Dysmenorrhea, 100–101
  biological aspects of, 101
  primary and secondary, 100
Dyspareunia, 490, 495
  physical causes of, 495
  psychological factors in, 495
  treatment of, 520
  vaginismus and, 497
Eastern religions, sexuality in, 14–15
Eating disorders, 209
Eclampsia, 348
Economics, cohabitation and, 457–458
Ectoderm, 345
Ectoparasitic infestations, sexually transmitted, 560–561
Ectopic pregnancy, 78, 348–349
  with IUDs, 388
Edex. *See* Alprostadil
Education
  for rape prevention, 617
Efface, 355
Ego, 28
Egypt
  birth control and culture of, 375
  incest taboo in, 11
Ejaculation, 135–136, 161–162, 170
  declining production of, 479
  delayed, 136
  first, 435
  limbic system in, 152*f*, 153
  premature, 134–135
  as reflex, 131
  regulation of, 136
  retrograde, 136–137, 170
  stages of, 136
  with vasectomy, 398
Ejaculatory duct, 120
Ejaculatory incompetence, 494
Ejaculatory orgasm, 164
Electra complex, 194
  unresolved, 503
Elephantiasis, 541
Elmer Gantry, 216
Embryo, 173
  development of, 343, 344–345
  human, 342*f*
Embryonic disk, 343–344
Embryonic stage, 343, 344–346
Embryonic transfer, 336
Emotional closeness, desire for, 468–469
Emotional rigidity, 249
Emotional woman stereotype, 186–187
Emotions, in sexual dysfunctions, 502
Empathy
  lack of, 244
  training in, 630
Empirical approach, 38
Empty love, 223*t*
  components of, 224
Endocrine glands, 88
  major, 89*f*
  in menstrual cycle, 88
Endoderm, 345
Endometrial cancer, 77
Endometriosis, 77
  in dyspareunia, 495
  in infertility, 335
Endometritis, chlamydial, 539
Endometrium, 77
Endorphins, in sadomasochism, 586
Ensoulment, 405
Environmental influences, prenatal, 346–352
Enzyme-linked immunosorbent assay (ELISA), 550–551
Epididymis, 119
  development of, 175
Epididymitis
  chlamydial, 539
  with gonorrhea, 536
Epidural block, 358
Episiotomy, 73, 356–357

Equal Employment Opportunity Commission, sexual harassment guidelines of, 631
Erectile abnormalities, 135
Erectile disorder/dysfunction, 134–135, 491
  acquired, 492
  drugs causing, 499–500
  emotional toll of, 492*f*
  generalized, 491
  incidence and prevalence of, 492
  lifelong, 492
  masturbation and, 264
  occasional, 492
  organic causes of, 498–499
  performance anxiety in, 492–493
  in sexual aversion, 491
  situational, 491
  training position for, 509*f*
  treatment of, 508–509
Erection, 130–131
  age-related changes in, 478
  autonomic nervous system in, 134–135
  brain in, 133–134
  difficulty achieving or sustaining, 489
  first, 435
  in infant boys, 420
  nocturnal, 131
  psychogenic control of, 482
  as reflex, 131–132
  resolution of, 162
  during sexual response cycle, 158
  spinal cord in, 133
Erogenous zones, 29, 145
  primary, 145
  secondary, 145
Eros, 217
  definition of, 221
Eros device, 515
Erotic fantasy, 297
Erotic interests, 5
  shifting of, 293
Erotic love, 217
Erotic plasticity, 21–23
Erotica, 656
  prevalence and use of, 659–660
Eroticism, prostitutes for, 651
Erythromycin
  for chancroid, 541
  for lymphogranuloma venereum, 541
  for syphilis, 533*t*
Escort services, 647
  customers of, 650
Estimation, faulty, 52–53
Estratetraenol, 143
Estrogen, 78–79, 154
  in birth control pills, 380
  in breast cancer, 83–84
  deficit in, 95–96
  in female pubertal changes, 433–434
  in female sexual behavior, 156–157
  for menopause, 96–100
  placental secretion of, 345–346
  postmenopausal decline in, 477–478
  postmenopausal replacement of, 478
  in prenatal sexual differentiation of brain, 176
  in puberty, 431
  sexual functioning and, 157
Estrone, 97
Estrous cycle, 88
Estrus, 156
  versus menstruation, 88
Ethical relativism, 671–672
Ethics
  in sex research, 59–60
Ethnicity
  matching of, 215
  number of sex partners and, 33
  in teenage sexual activity, 447–448, 447*f*
Ethnocentrism, 28
Ethnographic-observation method, 54
Ethnography, 54
European Americans
  premarital oral sex among, 437
  sexual dysfunctions among, 498*t*
  swinging among, 472
European Mode of Delivery Collaboration Trial, on C-section for reduced HIV risk, 552
Eve, sin of, 15
Evidence
  drawing conclusions from, 9
  in scientific research, 38
Evolution, 20–23
  of attractiveness preferences, 213–214
  of gender typing, 191–192
  sexual orientation and, 307
Evolutionary psychology, 21–23
Exaggeration, 53
Excitement, 163
Exercise
  aphrodisiac effects of, 148
  breast cancer and, 84
  in erectile dysfunction, 499
  for menstrual discomfort, 105
Exhibitionism, 572–573
  boundaries of, 579–580
  definition of, 578
  hostility toward women and, 578–579
  learning theory of, 593
  origins of, 578
  prevalence of, 578
  progression to sexual aggression in, 579
  psychoanalytic theory of, 591
  risk of being caught in, 579
  victims of, 578
Exhibitionists
  archetypal, 578
  childhood sexual abuse of, 594
  covert sensitization of, 595–596
  personality of, 578–579
  preferred victims of, 579
  testosterone levels in, 591
Experiment, 57
Experimental groups, 58
Experimental method, 57
  aspects of, 57–59
  limitations of, 59
External genitals, primitive, 173
External sex organs
  development of, 174*f*
Extramarital sex, 468–469
  attitudes toward, 469–471
  cross-cultural perspectives on, 470–471
  effects of, 472
  patterns of, 469
  reasons for, 468–469

Facial traits, 209–212
Fairness, 245
Fallopian tubes, 78
  development of, 175
  fertilization in, 329
  scarring of, 335
Famciclovir, 534*t*, 557
Family
  acceptance of gay member, 317
  in Islamic society, 14
  prostitution and, 649
Family planning, natural. *See* Coitus interruptus
Family therapy, 629
Family values, 663–664
Fantasy
  with age, 480
  during coitus, 283–284
  masturbation, 270–272, 518
  in men, 271–272
  necrophilic, 587
  sexual orientation and, 293, 297
  sharing of, 565
  in solitary sexual behavior, 270–272
  themes of, 271–272
  in women, 271
Fat, breast cancer and, 84
Father-daughter incest, 625
Father-daughter marriages, 11
Fathers, teenage, 444–445
Fatigue, 498
Fatuous love, 223*t*
  components of, 224–225
Feces, sexual arousal connected with, 588
Federal Centers for Disease Control and Prevention (CDC), estimates on HPV infection, 529
Feedback loop, 117
Feelings, expressing displeasure in terms of, 255
Fellatio, 13
  barrier devices with, 564
  techniques of, 276–277
Female condom, 401–402
Female-female sexual behavior
  among adolescents, 442–443
  in early childhood, 423–424
  in non-Western cultures, 300
  in preadolescence, 425
  religious condemnation of, 298–299
Female figure, 209
Female idols, 10–11
Female impersonators, 577
Female orgasmic disorder, 494
Female play, nondemanding, 515, 516*f*
Female sexual anesthesia, 16–17
Female sexuality, liberation of, 18–19
Female-superior position, 199, 517*f*
  for anorgasmic women, 515–516
  for premature ejaculation, 520
Females. *See also* Women
  age-related sexual activity changes in, 477–478
  body weight triggering puberty in, 431
  first coital experience among, 440–441
  gender roles of
  plumpness of, 209
  pubertal changes in, 431–434
  pubertal development stages in, 432*t*
  rapid orgasm in, 495
  sex organs of, 173
  sexual arousal disorder of, 493
  sexual behavior of, 156–157
  sexual pleasure in, cultural influences on, 500
  sterilization of, 399–401
  teenage, sexual pressures on, 448–449
Feminine deodorant sprays, 75
Feminine role, 306–307
Feminine traits, 199–200
  intimacy and, 200–201

Femininity, 296
Feminists
  on domestic violence, 473
  pornography and, 655
  triggering events in, 473
Fertility
  female problems in, 334–337
  male problems in, 333–334
Fertility awareness methods, 395
  advantages and disadvantages of, 397–398
  basal body temperature, 396
  calendar, 396
  cervical mucus, 397
  effectiveness of, 397
  mechanism of, 396
  ovulation-prediction kits, 397
Fertility drugs, 334–335
Fertility symbols, 10–11
Fertilization, 119, 329
  in vitro, 336
  methods for, 20, 336–337
Fetal alcohol syndrome, 351
Fetal erythroblastosis, 349
Fetal movement, 341, 346
Fetal stage, 346
Fetishism, 575
  harmlessness of, 574
  learning theory of, 592
  psychoanalytic theory of, 591
  transvestism and, 576
Fetishists, 595
Fetus
  age of viability, 346
  development of, 344*f*
  prenatal development of, 342*f*, 343
  during second trimester, 346
  during third trimester, 346
Fibroadenomas, breast, 84
Fibrocystic breast growth, 381
Field study, 53–54
Fijians, extramarital sex in, 471
Film ratings, 656
Fimbriae, 78
Finasteride, 123
Fisting, 318
  avoidance of, 564
Fixation, 29
Flagellation, 583
  psychoanalytic theory of, 591
Flagyl (metronidazole), 544
Flashing. *See* Exhibitionism
Fluconazole, 534*t*, 543
Fluoxetine (Prozac)
  anaphrodisiac effects of, 149
  for paraphilias, 596
  for premature ejaculation, 520
Follicle, 79
Follicle-stimulating hormone (FSH), 90
  in contraceptive vaccine, 404
  in male pubertal changes, 434
  in men, 117
  in prenatal sexual differentiation of brain, 176
  in puberty, 431
  in secretory phase, 93
For Whom the Bell Tolls, 655
Foreplay, 5, 272–273
  brevity of in sexual dysfunction, 502
  changes in duration and techniques of, 465
  touching in, 273
Foreskin, 112–113
Fornication, 13
  Puritan attitudes toward, 16
Fragrance, 142–143
Framework, 40
Frenulum, 112
Frequency, 46–47
Freudian psychology, 17
  of Leonardo da Vinci, 44
  sexual theory in, 30
  theory of, 28–30
Friday the 13th, Part 2, 662
Friendship, 221
  basis of, 223
  passionate love and, 223–224
  small talk in, 232–234
Frigidity. *See* Female sexual arousal disorder
Frotteurism, 586–587
  learning theory of, 593
Fruit flies, sexual differentiation in, 174–175
FSH. *See* Follicle-stimulating hormone (FSH)
Functionals, 323
Fundamentalist Protestants, 670
Fundus, uterine, 77

G-spot, 166–168
Game-playing love, 221
Gamete intrafallopian transfer (GIFT), 336
Gamma-aminobutyric acid, 104
Gang rape, 606–607
  by athletes, 610
  by women, 608–609
Garden of Eden, 15
Gardnerella vaginalis, 542
  transmission, symptoms, diagnosis, and treatment of, 533*t*
Gatekeeper role
  in romantic relationships, 198
  of women, 33
Gay activism, 306
Gay bars, 322
Gay bashing, 304
  AIDS epidemic and, 305
Gay bathhouses, 322
Gay couples, types of, 322–323
Gay identity, acceptance of, 316–317
Gay lifestyles, 290, 319–322
  gender differences in, 322
  variations in, 322–323
Gay males, 291, 292
  adjustment of, 313–316
  during adolescence, 442–443
  biological perspectives on, 307–309
  close-binding mother and detached-hostile father in, 310–311
  contemporary social perspectives on, 302–307
  cross-cultural perspectives on, 299–300
  cross-dressing in, 577
  cross-species perspectives on, 300–301
  delayed marriage in, 294
  gender nonconformity in, 312–313
  historical perspectives on, 298–299
  lifestyles of, 314, 319–323
  male-female sexual behavior among, 293
  in prostitution, 653
  psychological perspectives on, 309–312
  sexual activity patterns among, 318–319
  sexual dysfunction in, 501–502
  versus transsexuals, 181
  treatment for sexual orientation of, 315–316
  X sex chromosome and, 308
Gay marriages, 299
Gay parents, 303
Gay people, 291. *See also* Gay males; Homosexuality; Lesbians; Sexual orientation
  bisexuality and, 298
  coming out, 316–318
  equal rights of, 302–303
  erotic fantasies of, 297
  fear of, 304–305
  Kinsey Institute Reports on, 49
  marriage among, 302
  in military, 320
  negative stereotypes of, 292
  openly, 295–296
  percentage of in population, 293–294
  social attitudes toward, 302–307
Gay rights
  AIDS epidemic and, 306
  effect of wording in poll on, 303*t*
  organizations for, 306, 320
  public opinion on, 302–303
Gay sexual orientation, 290
  defined as deviant or normal, 573–574
  percentage of in population, 295
  seeking treatment for, 315*t*
Gazing, in communication, 250
Gender
  constancy of, 197
  definition of, 5
  differences in
  number of sex partners and, 31
  selection of
  stability of, 197
  in survey bias, 49
Gender-appropriate behavior, 196–197
Gender assignment, 176
Gender dysphoria, 181
Gender identity, 5, 172, 176
  ambiguous, 177–179
  versus gender role, 187
  in gender typing, 197
  maleability of, 179
  nature and nurture in, 176–177
  pseudohermaphroditism and, 178–179
  psychological factors in, 30
  psychosocial influences on, 178
  sexual orientation and, 292
  transsexual, 179–186
Gender-inconsistent activities, 198
Gender nonconformity, 312–313
Gender reassignment surgery, 179–181
  in boys with Dominican Republic syndrome, 179
  outcomes of, 184
  for transsexuals, 182–186
Gender roles, 5
  acquisition of, 31*f*
  attractiveness and expectations of, 210–212
  break down of, 191
  as cultural adaptation, 193
  in gay relationships, 306–307
  homophobia and stereotypical attitudes toward, 304
  in masturbation fantasy, 272
  sexual behavior and, 198–199
  shifting, 185
  social learning of, 194–196
  stereotyped, 186–187, 188–189
  traditional, 33
Gender schema theory, 197–198
Gender typing, 191
  biological perspectives on, 191–192
  cognitive-developmental theory of, 196–197
  cross-cultural perspectives on, 192–193

Gender typing, *continued*
psychological perspectives on, 194–198
in rape, 609
social learning of, 194–196
through observational learning, 196*f*
General anesthesia, childbirth, 358
Generalizing, 42
Genes, 20
Genesis, on male-male sexual activity, 298
Genetic abnormalities, 352–355
prevention of, 354–355
Genetic evolution, 191–192
Genetics
in HIV transmission risk, 550
in sexual differentiation, 174–175
in sexual orientation, 307–308
in sexuality, 20–21
Genital herpes. *See also* Herpes simplex virus
coping with, 557–558
diagnosis and treatment of, 557
incidence of, 555
manifestations of, 555
prodromal symptoms of, 556
reporting cases of, 555
symptoms of, 556–557
transmission of, 555–556
types of, 555
Genital play
for anorgasmic women, 515
in infants, 422
Genital warts
causes of, 559
diagnosis of, 535*t*
risk factors in, 559
symptoms of, 535*t*, 559
transmission of, 535*t*, 560
treatment of, 535*t*, 560
vaccine for, 560
Genitals
external female, 66–74
internal female, 75–80
manual stimulation of, 274
negative attitudes toward, 66
vasocongestion in, 41
Germ cell, 117
German measles. *See* Rubella
Germinal stage, 343–344
Gigolos, 652
Goals
clarifying, 256
scientific, 39–40
Gonadotropin-releasing hormone, 90, 93
Gonadotropins, 90
Gone with the Wind, rape theme in, 610
Gonorrhea
asymptomatic, 536
causes of, 532
diagnosis of, 533*t*, 536
discharge in, 536*f*
incidence and prevalence of, 532
ophthalmia neonatorum with, 532
pharyngeal, 532
symptoms of, 533*t*, 536
transmission of, 532, 533*t*
treatment of, 533*t*, 536
Gossypol, 403
gp120 spikes, 545
Graafian follicle, 92
Granuloma inguinale, 541
The Grapes of Wrath, 655
Greece, ancient, 11–12
abortion in, 405
marriage in, 459
Greek culture
anal intercourse and, 285
birth control during, 375
love concepts in, 217
phallic symbols in, 110
Greek gods, sexuality of, 12
Griswold v. Connecticut, 376
Group marriage, 476
Group sampling, in Kinsey report, 46
Group survival, 307
Group therapy
for rapists and child molesters, 629
sex, 505
for sexual abuse survivors, 629
Growth spurts, 436
GSTPI gene, 125
Guilt, in dyspareunia, 495
Gynecologist, 70
Gynecomastia, 436

HAART, 552, 553*f*
Hallucinogenics
aphrodisiac effects of, 150–151
in chromosomal damage, 350–351
Hardwick v. Bowers, 306
Harris v. McRae, 407
Hawaii, incest taboo in, 11
HCG radioimmunoassay, beta subunit, 339
Headache, in dysmenorrhea, 101
Health, active role in, 7
Health care seeking, gender differences in, 191
Health education, 430
Hearing
impairment of, 482–483
in sexual arousal, 146
Heart, primitive, 345
Hebrews, ancient, 11
Hedonism, 672
Hegar's sign, 339
Helms, Jesse, 47
Helper T-cells, 546
Hemophilia, 354
characteristics of, 353*t*
Hemophiliacs, HIV transmission in, 549
Hemophilus ducreyi, 541
Hepatitis virus
diagnosis of, 559
hepatitis A, 558
hepatitis B, 558
hepatitis C, 558
hepatitis D, 558
symptoms, diagnosis, and treatment of, 534*t*
transmission of, 534*t*, 558–559
types of, 558
vaccine for, 559
Heracles (Hercules), 12
Herceptin, 85, 88
Hermaphroditism, 177
false, 177–179
Heroin
in erectile dysfunction, 500
in fetal abnormalities, 350
Herpes simplex virus
in AIDS, 548
type 1, 555
type 2, 555
Herpes syndrome, 557
Heteroerotic interests, in gay males and lesbians, 293
Heteroeroticism, responsiveness to, 296–297
Heterosexual-homosexual continuum, 293–296
challenges to, 296–297
Heterosexual orientation, 291
Heterosexuality, 290
as independent dimension, 296–297
Heterosexuals
erotic fantasies of, 297
exclusive, 294
High-risk behaviors, avoiding, 566
Highly active antiretroviral therapy. *See* HAART
Hinduism, sexuality and, 14–15
Hippocampus, 152
electrical stimulation of, 152*f*, 153
Historical perspectives, 9–19
The Hite Report, 49
on female masturbation, 268, 270
The Hite Report on Male Sexuality, 49
HIV (human immunodeficiency virus) infection, 544. *See also* AIDS
amount of virus in semen, 550
cesarean section to avert, 359
circumcision and, 113–114
concerns about, 528
condoms in protection against, 392
diagnosis of, 535*t*, 550–551
immune system effects of, 545–548
maternal-infant transmission of, 348
perceived low risk of, 563
prevalence of, 528, 544
prevention of, 553–555
progression of, 548
in prostitutes, 654–655
public awareness of, 562–563
responsible sexual decision about, 667
risk factors for, 549–550
sexual transmission of, 548–549
spermicidal protection against, 391
symptoms of, 535*t*
transmission of, 535*t*, 548–549
treatment of, 535*t*, 551
trichomoniasis and, 543
Home birth, 361–362
Home pregnancy tests, 339
Homoerotic interests, in heterosexual people, 293
Homoeroticism, responsiveness to, 296–297
Homogamy, 462–463
Homologous structures, 69
Homophile, 291–292
Homophobia, 304–305
Homosexual behavior, 11
Homosexual Citizen, 306
Homosexual identity, 316–317
Homosexual orientation, 291. *See also* Gay sexual orientation; Lesbian sexual orientation
Homosexualities: A Study of Diversity Among Men and Women, 319
Homosexualities (Bell, Weinberg), 49
Homosexuality. *See also* Gay sexual orientation; Lesbian sexual orientation
definition of, 291
as independent dimension, 296–297
oversimplification of factors in, 9
psychological origins of, 313
social condemnation of, 302
versus transsexualism, 181
in Western culture, 298–299
Homosexuality in Perspective (Master; Johnson), 55
Homosexuals
versus gay male and lesbian, 292
social stigma of, 291
Honesty, 248
Honeymoon cystitis, 70–71

Hopi Native Americans, 25
Hormonal contraceptives, 402
  for men, 403
Hormonal deficiencies
  in lack of sexual desire, 490–491
  in male hypoactive sexual desire, 506–507
Hormones. *See also* Estrogen; Sex hormones; Testosterone
  definition of, 154
  designer, 403
  in fetal abnormalities, 350
  in gender typing, 191–192
  in menstrual cycle regulation, 89
  in paraphilias, 590–591
  replacement of, 96–100
  in sexual behavior, 154–157
  in sexual orientation, 308–309
Hostility, in male orgasmic disorder, 494
Hourglass figure, 209
Housewives, strike of, 462
HPV. *See* Human papilloma virus (HPV)
Huckleberry Finn, obscenity in, 655, 656
Human behavior, control of, 40
Human chorionic gonadotropin (HCG)
  placental secretion of, 345
  in urine, 338–339
Human immunodeficiency virus. *See* AIDS; HIV (human immunodeficiency virus) infection
Human life, beginning of, 405
Human papilloma virus (HPV). *See also* Genital warts
  incidence of, 529
  prevalence of, 528
  public information about, 528–529
  symptoms of, 535*t*, 559
  transmission, diagnosis, and treatment of, 535*t*
  vaccine for, 560
Human Potential Movement, 18
Human Sexual Response (Masters; Johnson), 54–56
Human sexuality. *See* Sexuality
Humiliation, 583–584
  in sadomasochism, 585
Huntington's chorea
  characteristics of, 353*t*
  screening for, 355
Hustlers, 652
  bar, 654
  customers of, 652
  gay, 653
  heterosexual, 653
  street, 654
Hyaluronidase, 329
Hygiene, in STI prevention, 565
Hymen, 72–73
  cultural significance of, 73
  shapes of, 72*f*
Hypogonadism, 156
  in lack of sexual desire, 490–491
Hypothalamus, 152
  in male pubertal changes, 434
  in menstrual cycle regulation, 89–90
  in prenatal sexual differentiation of brain, 176
  sex hormone production by, 154
  third interstitial nucleus of anterior, 309
  in transsexualism, 182
Hypothesis, 38
  testing of, 39
Hypoxyphilia, 584
  learning theory of, 593
Hysterectomy, 77–78, 400
  complete versus partial, 78
Hysterosalpingogram, 335
Hysterotomy, 411–412

I-talk, 254
  in rejecting criticism, 256
Ibuprofen, 105
ICSH. *See* Interstitial-cell-stimulating hormone
Id, 28
Ideal self, "sculpting," 248–249
Identification, in transsexualism, 182
If You Are Raped . . ., 615
IgG antibody, 551
Imipramine (Tofranil), 149
Imiquimod cream, 560
Immune system
  AIDS/HIV and, 544
  CD4 cells in, 546–548
  HIV effects on, 545–548, 547*f*
  leukocytes and, 544–545
Immunocontraceptives, 404
Impasses, handling, 257
Implanon, 402
Implant surgery, penile, 510*t*, 511
Implantation, 344
Impotence. *See* Erectile disorder/dysfunction
Incas, incest taboo in, 11
Incest
  child's drawing of, 622*f*
  cross-cultural attitudes toward, 25
  definition of, 624
  family factors in, 625
  incidence of, 621–622
  in pedophiles, 623
  in sexual aversion, 491
  in sexual dysfunction, 501
  sociocultural factors in, 626
  taboo of, 11
  types of, 624–626
Incidence, 46–47
Incompatibility, in triangular model of love, 223*f*
Independent variables, 57
India, sexuality and religions of, 14–15
Individuality
  in intimacy, 248
  Michelangelo phenomenon and, 248–249
Infant mortality rate in teenage pregnancy, 444
Infants
  capacity for sexual response of, 420–421
  genital play in, 422
  masturbation in, 421–422
Infatuation, 222, 223*t*
  versus love, 220
Inferences, 39
Infertile couples, research helping, 20
Infertility, 333
  causes of, 333–337
  female fertility problems in, 334–337
  incidence of, 333
  male fertility problems in, 333–334
Infidelity. *See also* Extramarital sex
  discovery of, 472
  emotional, 239
  sexual, 239
Inflammation, 545
Information
  gathering, 668, 669
  processing of in gender typing, 197–198
  provision of, 253–254
Informed consent, 60
Infundibulum, 78
Inguinal canal, 175
Insect bites, HIV transmission and, 550
Insight-oriented therapy
  for hypoactive sexual desire, 506
  for sexual dysfunctions, 505
Insulin growth factor-I, 125
Intercourse. *See* Coitus
Interleukin 2, 551
Internal sex organs
  development of, 173*f*
  male, 116–122
International domestic workers day, 462
Interpersonal attraction, 207
Interpretations, alternative, 9
Interstitial cells, 117
Interstitial-cell-stimulating hormone, 117
Interviewers, gender of, 49
Interviews, 44–45
Intimacy, 221, 230, 246
  betrayal of, 472
  commitment in, 248
  communication in, 249–250
  honesty in, 248
  individuality and, 248
  in love, 224–225
  Michelangelo phenomenon and, 248–249
  psychological androgyny in success of, 200–201
  self-knowledge and, 246
  sexual versus emotional, 246
  trust and caring in, 246–248
Intra-amniotic infusion, 411
Intracytoplasmic sperm injection (ICSI), 336–337
Intrauterine devices (IUDs)
  accidental pregnancy rates with, 379*t*
  advantages and disadvantages of, 387–388
  in candidiasis, 542
  definition of, 386
  in ectopic pregnancy, 78–79
  effectiveness of, 387
  frameless, 403–404
  mechanisms of, 386–387
  primitive, 386
  reversibility of, 387
  sharing responsibility for, 377
  types of, 386*f*
Introitus, 72–73
Invisible Lives, 317
In vitro fertilization, 336
Irrational beliefs, 251
  in sexual dysfunctions, 501, 503
Islamic cultures, family and, 14
Israeli kibbutz, genital play among children in, 422
Issues, oversimplification of, 9
Isthmus, 78
Itaxol, 85

Janus Report on Sexual Behavior, 48
Japan
Jaundice, 558
Jealousy, 238–239
Johns, 650
  compulsive, 651
  habitual, 651
  occasional, 650
Jorgensen, Christine, sex change of, 179
Judaism
  moral code of, 669–671
  number of sex partners and, 33
Judeo-Christian tradition
  attitudes toward homosexuality in, 298–299

Judeo-Christian tradition, *continued*
marriage in, 459
masturbation in, 263
on oral sex, 278
Julius Caesar, bisexuality of, 13

K-Y Jelly, 515
Kama Sutra, 14–15
Kaplan approach, 505
Kaplan sexual response stages, 163
Kaposi's sarcoma, AIDS-related, 548
Karma, 15
Kegel exercises
for orgasmic disorders, 517
Kept boys, 653
Killer T-cells, 546
Kindness-cruelty dimension, 198
Kinsey continuum of sexual orientation, 293–296
challenges to, 296–297
Kinsey Institute
Reports on gay people, 49
research on male-male sexual activity, 294–295
Kinsey reports, 17–18, 45, 45*f*, 46–47
on duration of coitus, 466
on frequency of marital coitus, 465–466
on incidence of prostitution, 641
liberalizing effect of, 464
on premarital oral sex, 437
on premarital sex among adolescents, 439
on sexuality in later years, 479
Kinship ties, extramarital partners and, 471
Kissing, 273
in adolescents, 436–437
cross-cultural practices, 25
deep, 273
in HIV transmission, 549
simple, 273
Kissing gourami, 39
Klinefelter syndrome, 176
Klismaphilia, 587
!Kung people, 193
Kwell (lindane)
for pediculosis, 535*t*, 561
for scabies, 535*t*, 561

L-dopa, aphrodisiac effects of, 147
Labia majora, 66, 67–68
during sexual excitement, 158
Labia minora, 66, 68
coloration of during plateau phase, 159–160
engorgement of, 158
loss of coloration, 162
Labor. *See also* Childbirth; Delivery
false, 355
initiation of, 355
intra-amniotic infusion induced, 411
Laboratory-observation method, 54–56
Lactation, 367
Lactobacillus acidophilus, 543
Ladder, 306
Lady Chatterley's Lover, 18, 464, 655
Lamaze method, 358
Laminaria digitata, 410, 411
Landon-Roosevelt election, 43
Laparoscope, 400
Laparoscopy
for endometriosis, 335
in female sterilization, 400
Larynx, pubertal growth of in males, 436
Latina/Latino Americans
AIDS in, 553
number of sex partners among, 33
premarital oral sex among, 437
sexual activity during adolescence, 448
LDRP concept, 360–361
Learning theory, 30
of paraphilias, 592–593
of sexual orientation, 311–312
Lea's shield, 402
Legal issues
on abortion, 405–409
on contraception, 375–376
in pornography, 656–657
on sexual orientation, 305–306
Legalism, 669–671, 673
Lesbian sexual orientation, 290, 291
during adolescence, 442–443
defined as deviant or normal, 573–574
penis envy and, 310
versus transsexualism, 181
Lesbians, 291, 292
adjustment of, 313–316
biological perspectives on, 307–309
contemporary social perspectives on, 302–307
cross-cultural perspectives on, 299–300
cross-species perspectives on, 300–301
delayed marriage in, 294
gender nonconformity and, 312–313
historical perspectives on, 298–299
lifestyles of, 319–323
male-female sexual behavior among, 293
psychological perspectives on, 309–312
sexual activity patterns among, 318–319
sexual dysfunction in, 501–502
treatment for sexual orientation of, 315–316
Leukocytes, 544
in immune system, 545
Leviticus
on male-male sexual activity, 298
moral code of, 669–670
Lewd language, 633
Leydig's cells, 117
LH. *See* Luteinizing hormone
Liberalizing influences, 464
Librium, 350
Life expectancies, gender differences in, 191
Lifestyles
alternative, 476
gay, 319–323
health and, 314
Liking, 223
Limbic system, 152
in orgasm, 153–154
in sexual functioning, 153
Lindane (Kwell)
for pediculosis, 535*t*, 561
for scabies, 535*t*, 561
Liquid nitrogen
for genital warts, 560
for molluscum contagiosum, 560
Listening, 245
active, 252
in intimate communication, 249
Literary Digest poll, 43
Literature
obscene, 657
pornographic versus erotic, 655–656
Local anesthetics, childbirth, 358
Lochia, 368
Loestrin, 380
Logical love, 221
Loneliness, 244
causes of, 244
coping with, 245
Love, 216
as appraisal of arousal, 221
attraction and, 206–207
components of, 221, 225
contemporary models of, 220–226
in contemporary Western culture, 217–220
friendship and, 223
Greek concept of, 217
versus infatuation, 220
meanings of, 217
scientific study of, 220
Sternberg's triangular theory of, 221–226
styles of, 220, 221
unrequited, 219
Lovemap, 594
Low birth weight, 366
maternal smoking in, 351
LSD
aphrodisiac effects of, 151
in chromosomal damage, 350–351
Ludus, 221
Lumpectomy, 84
Lust, 13
Luteal phase, 93
Luteinizing hormone, 90
in male pubertal changes, 434
in men, 117
during ovulation, 397
in secretory phase, 93
urine or saliva analysis for, 332
Lymphocytes, memory, 545
Lymphogranuloma venereum, 541

Machismo, jealousy and, 238
Madonna, 15
Madonna-whore dichotomy, 15, 651
Magazine surveys, 48–49
Male contraceptive pills, 403
Male erectile disorder. *See* Erectile dysfunction/disorder
Male-female sexual behavior
in adolescence, 436–442
among gay males and lesbians, 293
in early childhood, 423
in ex-prison population, 293
in HIV transmission, 549
in preadolescence, 425
Male genitalia. *See* Male sex organs
Male-male anal intercourse, 549
Male-male sexual activity
among adolescents, 442–443
among nonhuman primates, 301
anal sex, 286
in ancient Greek culture, 12
cross-cultural perspectives on, 299–300
in early childhood, 423–424
Kinsey's incidence of, 294–295
in preadolescence, 425
religious condemnation of, 298–299
Male-male sexual impulses, sublimating, 44
Male orgasmic disorder, 494
treatment of, 518
Male prostitutes, types of, 653–654
Male prostitution, 652–654
motive for, 653
Male rape, 608
Male-superior position, 199
conception and, 332
Males
age-related sexual activity changes in, 477*t*, 478–479
contraceptive methods for, 403
extramarital sex among, 468–469

first coital experience among, 441
gender roles of
hormonal contraceptives for, 403
infertility problems in, 333–334
masturbation techniques used by, 268–269
psychological response to pregnancy in, 341
pubertal changes in, 434–436
pubertal development stages in, 433*t*
reproductive system of, 111*f*
as romantics, 463
sex organs of, 110–137
sexual behavior of, 155–156
sexual functions of, 130–137
sterilization of, 398–399
Mammals, sexual behavior among, 24
Mammary glands, 81. *See also* Breasts
milk production by, 367
Mammography, 82, 85–87
Mania, 221
Manual stimulation
of genitals, 274
during pregnancy, 341
Mardi Gras, 471
Marijuana
aphrodisiac effects of, 150
in birth defects, 350
in chromosomal damage, 350–351
Marital coitus, frequency of, 465–466
Marital conflict, 502*f*
Marital counseling, 629
Marital dissatisfaction, 490
Marital partners, 231*f*
Marital rape, 608
Marital relationships
legislated in ancient Hebrew culture, 11
sexual satisfaction and, 466–467
Marital roles, 460
Marital satisfaction, variables in, 56
Marital sexuality, 463–464
sexual revolution and, 464–466
sexual satisfaction and, 466–467
Marital status
happiness and, 461*t*
in masturbation frequency, 267
number of sex partners and, 31
of U.S. population aged 15 years and above, 454*t*
Marital therapy, 507
Marriage, 459
after cohabitation, 458–459
arranged, 462
cross-cultural forms of, 25
delayed with homosexual orientation, 294
early Christian attitude toward, 13
first, estimated median age at, 455*t*
foreplay in, 465
frequency of coitus in, 465–466
gay, 290, 299
group, 476
happiness and, 461
high expectations of, 474
historical perspectives on, 459–460
interracial, 463
in Islamic society, 14
open, 476
in Protestant Reformation, 15–16
reasons for, 461
selecting spouse in, 462–463
sexual satisfaction in, 466–467
as testing of gay identity, 317
traditional versus modern, 460*t*
trial, 457
types of, 462
in U.S. population aged 15 years and above, 454*t*
violence in, 608
Masculine gender role
in gay relationships, 307
rape and, 609
stereotypes of, 198–199
Masculine traits, 199–200
popularity and, 201
Masculinity, as personality dimension, 296
Mashing, 586–587
Masochism, sexual
definition of, 582–583
forms of, 583–584
learning theory of, 593
psychoanalytic theory of, 591
sexual sadism and, 584
in women and men, 583
Massachusetts Male Aging Study, on erectile dysfunction, 506–507
Massage, in sensate focus exercises, 505
Massage parlors, 646–647
Mastalgia, 101
Mastectomy, 84
Masters and Johnson sex therapy, 503
approach of, 504–505
for erectile disorder, 508–509
evaluation of, 521
for orgasmic disorders, 515–516
for premature ejaculation, 518–520
training position recommended by, 509*f*
Masters and Johnson studies
on erogenous zones, 145
liberalizing effect of, 464
Masturbation
among adolescence, 267–268, 436
among adolescent males, 437*t*
among women, 268
anxiety and guilt over, 264–265
in contemporary society, 264–268
cross-cultural attitudes toward, 25
cultural attitudes toward, 263
definition of, 5, 262
as deviant versus normal, 574
devices to curb, 264*f*
directed program of, 522
in early childhood, 422–423
early Christian attitude toward, 13
erectile dysfunction in, 491
for female orgasmic disorders, 516–518
female sexual arousal disorder and, 493
with fetishistic object, 575, 591
frequency of with age, 480
gender differences in, 265–267
guilt over, 436
historical medical views of, 263–264
in infants, 421–422
lubricants in, 269
multiple orgasms with, 164–165
orgasm through, 165
in preadolescence, 425
reasons for, 263*t*
to relieve pelvic throbbing, 163
rewards and punishments for, 30
sociocultural factors and frequency of, 266*t*
in Taoist tradition, 14
techniques of
Matching hypothesis, 215
Mate preferences, 212–213
gender differences in, 213*t*
Mate selection, 462–463
romantic, 463
Mate swapping, 18, 472
Maternal depression, postpartum, 366–367
Maternal diet, in prenatal development, 346–347
Mathematical abilities, gender differences in, 189–190
Mating behavior, 154–155
Mating gradient, 463
Mattachine Newsletter, 306
Mattachine Society, 306
Mayflower Madam, 647
"Me Decade," 18
Meaningful relationship, physical qualities in, 212–213
Mechanical-barrier methods, 401
Media, gender stereotypes in, 195
Medical interventions/traditions, 630
checkups in STI prevention, 565
on masturbation, 263–264
Medroxyprogesterone acetate (MPA; Depo-Provera), 402–403
for paraphilias, 597
Medulla, 151
Meese Commission Report, 661–662
Memory lymphocytes, 545
Menarche, 88, 431
age at, 432*t*
Menopause, 95–96
bleeding after, 77
hormone replacement therapy for, 96–100
Menstrual bleeding, 387
Menstrual cramping, 387
Menstrual cycle, 88
anovulatory, 433–434
calculating pregnancy onset from, 342
changes occurring during, 91*f*
phases of, 90–95
problems in, 100–105
regulation of, 89–90
sexual activity during, 157
Menstrual discomfort, 104–105
Menstrual phase, 93–95
Menstrual synchrony, 144
Menstruation
coitus during, 95
versus estrus, 88
postpartum resumption of, 368
Mental disorders, sexual orientation and, 313–316
Mental retardation
maternal rubella and, 348
sexual activity and, 483–484
stereotypes about, 483
Mescaline, aphrodisiac effects of, 151
Mesoderm, 345
Methadone, 350
Methotrexate, 413
Metronidazole
for bacterial vaginosis, 533*t*, 542
for trichomoniasis, 534*t*, 544
Michelangelo phenomenon, 248–249
Miconazole, 543
suppositories, 534*t*
Microbicides, 391, 402
Microsurgery, 334
Middle Ages, sexuality during, 15
Middle Eastern religions, 14
Mifepristone tablets, 413
Migraines, in dysmenorrhea, 101
Military
anti-gay bias in, 320*t*
gay people in, 320
Milk ducts, 81
Miller v. California, 656, 657

Minilaparotomy, 400
Minipill, 380
  progestin in, 383
Mirena, 404
MIS. *See* Müllerian inhibiting substance
Miscarriage. *See* Abortion, spontaneous
Misoprostol, 413
Missionary position, 279–280
Mittelschmerz, 92–93
Mobility, 464
Modeling, 30
Molluscum contagiosum, 560
Mongolism, 352
Moniliasis, 542–543
Monogamy, 11, 25, 462
  in marital relationship, 469
  serial, 456
  for STI prevention, 562
Mons veneris, 66, 67
Mood, in attraction, 230
Moral issues, conflict over, 668
Moral laws, 669–671
Moral reasoning, ethical frameworks for, 670–673
Morning-after pill, 384
Morning sickness, 340, 341
Morphine, 500
Mother, surrogate, 337
Mother-daughter incest, 624
Müllerian ducts, evolution of, 175
Müllerian inhibiting substance, 175
Multiple births, 366
Multiple orgasms, 164–165
Multiple sclerosis
  in erectile dysfunction, 499
  in male orgasmic disorder, 494
Mundugumor tribe, 192–193
MUSE suppository, 511
Music, 146
Mutations, 21
Mutual cyclical growth, 247
Mutual friends, in romantic relationships, 231
Mutuality, 238
  individuality and, 248
My Fair Lady, 186
Myometrium, 77
Myotonia, 158
  dissipation of, 163
  resolution of, 162
  during sexual response cycle, 160

Nagele's rule, 342
Nama tribe, beauty standards of, 208
Names, attractiveness and, 212
Narcotics
  in erectile dysfunction, 500
  in fetal abnormalities, 350
National AIDS Hotlines, CDC, 567
National Birth Control League, 375
National Cancer Institute, on breast cancer, 82, 85
National Crime Victimization Survey
  on acquaintance rape, 605
  on rapes, 604
National Health and Social Life Survey (NHSLS), 47
  on anal sex, 286
  of circumcision, 113–114
  on duration of coitus, 466
  on factors in number of sex partners, 32*t*
  on frequency of marital sexual relations, 465*t*
  on marital sexual satisfaction, 467
  on masturbation, 262, 263*t*
  on masturbation gender gap, 265–267
  of oral-genital stimulation, 275–276
  on premarital intercourse among adolescents, 439–440
  on sexual dysfunction, 488–489
Native Americans
  extramarital sex in, 471
  number of sex partners among, 33
Natural childbirth, 358
Natural selection, 20
Naturalistic-observation method, 53–54
Necrophilia, 17, 587
Necrophilic homicide, 587
Needle sharing, HIV transmission via, 530, 544, 549
Negative correlation, 56
Negotiating differences, 256
Neisseria gonorrhoeae, 532
  peniclin-resistant strains of, 536
  transmission, symptoms, diagnosis, and treatment of, 533*t*
Neonates
  causes of death in, 366
  delivery of, 357
Nerve damage, erectile dysfunction and, 499
Neural tube, 345
  defects of
Neurological damage, 494
Neurosyphilis, 539
Neurotransmitters
  in PMS, 103–104
  for premature ejaculation, 520
New Guinea
  cultural groups of, 193
  sexual orientation in, 299–300
New York Society for the Suppression of Vice, 375
New York's Gay Men's Health Crisis (GMHC), 320
NGU. *See* Urethritis, nongonococcal
Nicotine, anaphrodisiac effects of, 149
Nipples, 81
  male, 81
  during sexual excitement, 158–159
  stimulation of, 275
Nitric oxide, 499
Nocturnal emissions, 435–436
Nocturnal erections, 131
  age-related changes in, 478–479
Nocturnal penile tumescence, 499
Nonconformity, gender, 312–313
Nonlove, 223*t*
Nonoxynol-9, 391
Nonverbal communication
  in intimate communication, 249
  observation of, 54
Nonverbal cues, 253–254
Norlevo, 413
Normal sexual behavior, 573–574
Norplant, 384
  accidental pregnancy rates with, 379*t*
  advantages and disadvantages of, 385
  effectiveness of, 384
  implanting of, 385*f*
  improved versions of, 402
  reversibility of, 385
  sharing responsibility for, 377
  use and mechanisms of, 384
Nucleoside analogues
  for AIDS, 551–552
  for HIV infection, 553*f*
Nude sunbathing, 579–580
Nurturance, intimacy and, 200–201
Nutrition, menstrual discomfort and, 105

Oberlin, 16
Obesity
  acceptance of, 209
  stillbirth risk with, 346
Obscene telephone callers, 580
Obscene telephone calling, 580–581
  response to, 580–581
  tracing, 580
Obscenity. *See also* Erotica; Pornography
  versus art, 656*f*
  community standards for, 657
  definition of, 656
  enforcing laws against, 661
Observation methods, 44–56
Observational learning, 30
  gender typing through, 196*f*
Observational research, limitations of, 55–56
Observer effect, 55–56
Ocular herpes, 556
Odors, offensive, 143
Oedipus complex, 30, 194
  sexual orientation and, 309–310
Oedipus conflicts
  unresolved, in sexual dysfunction, 503
Ofloxacin
  for chlamydia, 533*t*, 540
  for gonorrhea, 533*t*, 536
Old Testament, moral code of, 669–670
Onan, 375
Onanism, 263
Open couples, gay, 323
Open marriage, 476
Opening lines, 234
Operational definitions, 40–42
Ophthalmia neonatorum, 532
Opportunistic infections, AIDS-related, 548
  treatment of, 552–553
Opposite gender, 291
Oral-anal sex, 564
Oral-anal stimulation, 286
Oral contraceptives. *See* Birth-control pills
Oral-genital sex, 47
Oral-genital stimulation, 275–278
  with age, 480
  among lesbians, 318
  laws against, 306
  simultaneous, 277, 278*f*
Oral herpes
  characteristics of, 555
  symptoms of, 556–557
  transmission of, 555
Oral sex, 19
  abstaining from, 277–278
  in adolescence, 437, 438
  among adolescent males, 437*t*
  barrier devices with, 564
  as deviant versus normal, 574
  during pregnancy, 341
Orangutans, male-male sexual activity among, 301
Organ transplant, HIV transmission via, 549
Orgasm, 163
  controversies about, 163–168
  difficulty reaching, 490
  dry, 135, 164
  female, 162
  G-spot and, 166–168
  in infants, 421
  limbic system stimulation in, 153–154
  male, 135–136

in marriage, 466–467
medical disorders and, 498
multiple, 164–165
phantom, 482
premature, 490
subjective experience of, 162
through coitus versus masturbation, 165
Orgasmic consistency, 466
Orgasmic disorders, 490
couples therapy for, 515–516
directed masturbation programs for, 516–518
female, 494
male, 494
premature ejaculation, 494–495
rapid female orgasm, 495
treatment of, 515–518
types of, 493
Orgasmic phase, 160–162
Orgasmic platform, 159
Orgasmic reconditioning, 596
Original sin, 13–14, 15
Ortho-Novum, 380
Orthodox Jewish beliefs, 670
Os, 76
Osteoporosis
calcium and, 97
hormone replacement and, 97–100
Othello, 238
Outercourse, 565
Outing, 306
Ovarian cycle, 343*f*
Ovarian follicle, 92*f*
Ovariectomy, 157
Ovaries, 78–79
birth-control pills preventing cysts in, 381
cancer of, 79
descent of, 175–176
development of, 173
hormone production by, 154
Ovcon, 380
Overgeneralization, 9
Oversimplifying, 9
Ovulation, 88
after menarche, 433–434
basal body temperature and, 396–397
induced, 334
irregular, 334
method of, 397
peak days in, 397
postpartum resumption of, 368
urine or saliva analysis for, 332
Ovulation prediction kits, 397
Ovulatory phase, 92–93
Ovum, 66
fertilization of, 329
magnified, 79*f*
period of, 343–344
X sex chromosomes in, 328
Oxytocin, 90
in breast feeding, 367
initiating labor, 355

Pain, 59–60
Pair bonding, 58–59
Pap test, 76–77, 79
in pelvic examination, 80
Paralysis, with spinal cord injury, 481
Paraphilias
APA diagnostic category of, 574
biological factors in, 590–591
definition of, 574
gender and, 575
integrated theory of, 594
learning theories of, 592–593
prevalence of, 575
psychoanalytic theories of, 591
sociological theories of, 593–594
treatment of, 594–595
types of, 575–589
victim versus victimless, 574
Paraphrasing, 252
Paraplegia, with spinal cord injury, 481
Paraplegics, ejaculation in, 135
Parasympathetic nervous system, 134
Parent-teen relationship, 441
Parent-child relationships, 182
Paresis, general, 539
Paroxetine, 520
Partialism, 575
Participant-observation method, 54
Partners
directing hands of, 254
disagreement with, 252
learning needs of, 252–253
listening to, 252
in masturbation for anorgasmic women, 518
NHSLS study on number of, 32*t*
number of, 31–33
perspective of, 257
sexual behavior with, 272–287
Passion
friendship and, 223–224
in love, 221, 224–225
versus love, 220
Passivity, sexual, 198–199
Paternity, knowledge of, 10–11
Pathogens
in AIDS, 544
destruction of, 545
Patriarchy
definition of, 459
marriage and, 460
women's status in, 462
Pavlovian conditioning, 592
Pederasty, 12
Pediculosis, 560
adult stage, 561
symptoms of, 535*t*, 561
transmission and diagnosis of, 535*t*
treatment of, 535*t*, 561
Pedophiles, 595
background of, 624
characteristics of, 623
covert sensitization of, 595–596
gender of, 623
incestuous, 623
Pedophilia, 574, 575, 623–624
Pelvic edema, in dysmenorrhea, 101
Pelvic examination
of female internal sex organs, 79–80
Pelvic inflammatory disease
birth control pills protection against, 381
chlamydial, 539, 540
condoms in protection against, 395
in dyspareunia, 495
in ectopic pregnancy, 78–79
with gonorrhea, 536
in infertility, 335
IUDs in, 387
with *Neisseria gonorrhoeae*, 532
Pelvic throbbing, 163
Pelvic thrusting, in infants, 420–421
Penetration, fear of, 497
Penicillin
for chlamydia, 540
for gonorrhea, 536
for syphilis, 533*t*, 539
Penile glans, 159
Penile implants, 510*t*, 511
Penile injections, 511
Penile strain gauge, 40
Penile thrusting, 166
Penis, 137
anatomy and physiology of, 111–113
artificial, 183
circumcision of, 113–114
clitoris and, 68–69
development of, 175
erection of, 130–131
manual stimulation of, 274
oral stimulation of, 276–277
during plateau phase, 159
root of, 112
during sexual arousal, 111*f*
shaft of, 112
size of, 114–115
Penis envy, 310
Pentagon, on gays in military, 320
Penthouse, 656
People of Opposite Sex Sharing Living Quarters (POSSLQ), 456
Performance anxiety, 131
in erectile dysfunction, 492–493
in lack of sexual desire, 491
in male orgasmic disorder, 494
sex therapy for, 508–509
in sexual dysfunctions, 503–504
treatment of for male orgasmic disorder, 518
Perfume, 142
Pergonal, 334–335
Perimetrium, 77
Perineum, 73
random tearing of, 356–357
Personal development, psychological androgyny and, 200–201
Personal experience, 4
Personal invulnerability myth, 563
Personal prescription, 224
Personality, gender differences in, 190–191
Pessimism
challenging feelings of, 245
in loneliness, 244
"Petticoat punishment," 576
Petting, 253–254
in adolescents, 436–437
Peyronie's disease, 135
Phallic symbols, 10, 110, 111
Phallic worship, 10
Pharyngeal gonorrhea, 532
Phenobarbital, 350
Phentolamine (Vasomax), 510*t*, 511
Phenylketonuria, 353*t*
Pheromones, 143–144
Philia, 217, 221
definition of, 291–292
Phimosis, 114
Photoplethysmography, vaginal, 40
Phthirus pubis, 535*t*
Physical attractiveness, 207
cultural standards of, 208–209
meaningful relationships and, 212–213
preferences for, 213–214
of similarity of attitudes, 215–216
traits affecting perceptions of, 209–212
Physical disabilities, 481
with cerebral palsy, 481
prostitute services for customers with, 651

Physical disabilities, *continued*
 sensory, 482–483
 with spinal cord injuries, 481–482
Pimps, 644, 645
Piperonyl butoxide, 535*t*, 561
Pituitary glands
 in menstrual cycle regulation, 89–90
 sex hormone production by, 154
Placenta, 345–346
 detachment of, 357
 expulsion of, 358*f*
Platonic love, 225
Play therapy, sexual abuse survivors, 629
Playboy, 656
 survey on sexual orientation, 292–293
Playboy Foundation Survey, 47–48
Pleasure centers, 153–154
Pluralistic society, values in, 6
PMS. *See* Premenstrual syndrome
Pneumocystis carinii pneumonia, 548
Podophyllin
 for genital warts, 535*t*, 560
 for molluscum contagiosum, 560
Polyandry, 25, 462
Polygamy, 11, 25, 462
Polygyny, 25, 462
Polymorphous perversion, 309
Pons, 151
Pop psychology movements, 18
Pope v. Illinois, 657
Poppers, 147
Populations, 42
 surveys of, 49
Porky's, 662
Pornography
 child, 657
 definition of, 655–656
 gender differences in response to, 659–660
 hard-core versus soft-core, 656
 Meese Commission Report on, 661–662
 nonviolent, 663–664
 prevalence and use of, 659–660
 sex offenders and, 661
 sexual and family values and, 663–664
 sexual coercion and, 660–664
 violent, 662–663
Positive, accentuating, 253
Positive correlation, 56
Possessive love, 221
Possessiveness, 238
Post-traumatic stress disorder (PTSD)
 in childhood sexual abuse victims, 626
 in prostitutes, 648–649
 in rape survivors, 614
Postpartum blues, 366–367
Postpartum period, 366
 breast-feeding versus bottle-feeding during, 367–368
 maternal depression in, 366–367
 resumption of ovulation and menstruation during, 368
 resumption of sexual activity during, 368
Potassium nitrate, 149
Power rape, 612
Pragma, 221
Preadolescence
 male-female sexual behavior during, 425
 male-male and female-female sexual behavior during, 425
 masturbation during, 425
 sex education during, 425–430
 sexuality during, 425–427
Prediction, 40
Preeclampsia, 348
Pregnancy, 328
 in candidiasis, 542
 dating of, 342
 drug use during, 351–352
 drugs taken by parents during, 349–350
 early effects of, 340
 early signs of, 338
 ectopic, 78, 348–349
 evaluating risks of, 4
 first trimester, 341
 HIV transmission during, 549
 obesity during, 346–347
 prevention programs for, 446–447
 psychological changes during, 341
 reaction to, 337
 rubella during, 348
 second trimester, 341
 sex during, 340–341
 smoking during, 351–352
 sympathetic, 341
 syphilis transmission during, 538
 teenage, 436, 443–445
 tests for, 338–339
 third trimester, 341
 unwanted, 337
 weight gain during, 347
Prehistoric sexuality, 10–11
Premarital sex, 18–19
 during adolescence, 438–439
 double standards for, 441–442
 factors in, 440–441
Premature ejaculation, 134–135, 490, 494–495
 definition of, 495
 degree of, 495
 lack of sexual skills in, 503
 Masters and Johnson approach for, 509*f*
 sex therapy for, 522
 squeeze technique for, 518–520
Prematurity
 shoulder-first presentation with, 346
 in teenage pregnancy, 444
Premenstrual syndrome, 101–104
 causes of, 103–104
 symptoms of, 104
 treatment of, 104
Prenatal development
 critical periods in, 347–348
 embryonic stage of, 344–346
 environmental influences on, 346–352
 fetal stage of, 346
 germinal stage of, 343–344
 of human embryo, 342*f*
 maternal disease and disorders in, 347
Preorgasmic women, 494
Prepuce, 68
Preterm infants, 366
Preventive programs
 for rape, 617–619
 for sexual abuse of children, 628–629
 for sexually transmitted infections, 562–565
Priapism, 135
Primal scene, voyeurism and, 591
Primary sex characteristics, 431
Prisons
 male-male sexual behavior in, 293
 male prostitutes in, 653
 male rape in, 608
 treatment programs for rapists and child molesters in, 629–630
Pro-choice movement, 405, 409
 emotional reaction to abortion and, 414–415
 premises of, 9
Pro-life movement, 405, 409
 emotional reaction to abortion and, 414–415
 premises of, 9
Probability sample, 43
Procreative goal, 13
Progestasert T, 386
Progesterone, 78–79, 154
 in female sexual behavior, 156–157
 placental secretion of, 345–346
 in Progestasert T, 386
Progesterone only pill, 380
Progestin
 in birth-control pills, 380, 383
 in Progestasert T, 386
Progestogen male contraceptive, 403
Prohibitions, religious, 670–671
Prolactin, 90
 in breast feeding, 367
Proliferative phase, 90–92
Promiscuity, evolutionary advantage of, 21–23
Prophylactic, 392
Prostaglandins
 in dysmenorrhea, 101
 initiating labor, 355
 in vaginal suppository, 411
Prostate gland, 121
 cancer of, 123–130
 disorders of, 123–130
 surgery on for erectile dysfunction, 499
Prostate-specific antigen (PSA) blood test, 125
Prostatitis, 130
Prostitutes
 in brothels, 646
 childhood sexual abuse of, 650
 dual-world, 646
 female
 hierarchy of, 644
 motives for using, 651
 responsiveness to sex of, 644
 risk of arrest, 644
 teenage, 649–650
Prostitution
 in ancient Greek culture, 12
 definition of, 640
 early Christian attitude toward, 13
 historic perspective on, 640–641
 HIV/AIDS and, 654–655
 incidence of, 664
 male, 652–654
 penalties for, 645
 poverty and, 648, 649
 psychological disorders and, 648–649
 as sociability, 651
 teenage runaways in, 649–650
 in Victorian era, 17
Protease inhibitors, 552, 553*f*
Protectaid, 402
Protestant Reformation, 15–16
Protestantism, number of sex partners and, 31–33
Proverbs, "good wife," 11
Proximodistal development, 344–345
Prozac (fluoxetine hydrochloride), 596
Prurient value, 655
PSA blood test, 130
Pseudohermaphrodites, 177
Pseudohermaphroditism, 177–178
 gender identity and, 178–179
Psychoactive drugs, aphrodisiac effects of, 147, 149–151
Psychoanalysis, 28
Psychoanalytic theory, 28–30
 of paraphilias, 591
 of sexual orientation, 309–311

Psychodynamic theory, 194
Psychogenic erections, 133
Psychological adjustment, in gay males and lesbians, 313–316
Psychological androgyny, 199–201
model of, 200*f*
personal development and, 200–201
sexual behavior and, 201
Psychological conflict, 668
in sexual dysfunctions, 503
Psychological disabilities, 483–484
Psychological disorders
among gay males and lesbians, 313–314
among prostitutes, 648–649
among rapists, 611–612
in female sexual arousal disorder, 493
Psychological well-being, 200–201
Psychologists, 5–6
Psychology
evolutionary, 21–23
in gender typing, 194–198
in sexual study, 17–18
Psychopathia Sexualis (Kraff-Ebing), 17
Psychosexual development, 29
Psychosexual therapy, 505
Psychosexual trauma, 500–501
Psychosocial factors, 500–501
Psychotherapy, 28–30
for female sexual arousal disorder, 515
for gay sexual orientation, 315
for hypoactive sexual desire, 507
for paraphilias, 595
for rape survivors, 616
for rapists and child molesters, 629
for sexual abuse survivors, 629
for vaginismus, 521
Pthirus pubis, 560–561
Puberty
definition of, 431
in mentally retarded people, 484
Pubic lice, 560–561
Public exposure. *See* Exhibitionism
Pubococcygeus muscle, 74
Pudendal block, 358
Pudendum, 66
Punishments, for sexual behavior, 30
Punks, 653
Puppy love, 217
Puritans, 15–16
Pyrethrins, 535*t*, 561

Quadriplegia, 481
Questionnaires, 45
Questions, drawing partner out, 253
Quickening, 407

Racial matching, 215
Radiotherapy
for cervical cancer, 76–77
for prostate cancer, 125
Raloxifene, 88
breast cancer risk and, 85
Random assignment, 58
Random sample, 43
Rape. *See also* Childhood sexual abuse; Incest; Sexual assault
as act of violence, 602
adjustment of survivors of, 612–615
attitudes toward, 602–603
definition of, 602
forcible, 603
incidence of, 604
motivations for, 613*f*
pornography and, 662
prevention of, 617–619
psychological characteristics of, 611–612
in sexual aversion, 491
in sexual dysfunction, 501
sexual sadism in, 574
social attitudes and myths encouraging, 609
sociocultural factors in, 609–611
statutory, 603
types of, 605–609
by women, 608–609
Rape crisis volunteers, 615
Rape hotlines, 615, 616
Rape survivors
adjustment of, 612–615
crisis of, 612
guilt and shame in, 612–613
health complications in, 614
male prostitutes as, 653
physical injuries in, 614
pornography and attitudes toward, 662–663
prevalence of emotional and behavioral reactions in, 613–614
psychological disorders in, 614
reporting rape by, 614
trauma syndromes in, 614–615
treatment of, 616–617
Rape trauma syndrome, 614–615
acute phase of, 615
long-term reorganization process in, 615
Rapists
confronting, 618–619
motives of, 612
pornography and, 661
psychological characteristics of, 611–612
treatment of, 629–630
Rapport, in survey, 45
Rationalism, 672–673
Recessive traits, 354
Reciprocal altruism, 307
Reciprocity, 216
Recreational sex, 464
Rectal examinations, 125
Red meat, 124
Reflex arc, 132–133
Reflexes
definition of, 131
sexual response and, 131–135
types of, 132*f*
Refractory period, 162
Regional anesthetics, childbirth, 358
Regressive behaviors, 626
Rehabilitation, 629
Reinforcement, 30
in sexual orientation, 311–312
Rejection, fear of, 244
Relationships, 230
coital fantasies and, 284–285
compatibility in, 222
deterioration of, 239
physical qualities in, 212–213
problems with
stages of, 230–244
Relativism
cultural, 672
ethical, 671–672
Relaxation exercises, Lamaze method, 358
Reliability, survey, 46
Religion
circumcision and, 113
masturbation and, 263
number of sex partners and, 31–33
Religious beliefs
on anal sex, 286
in masturbation frequency, 267
sexuality and, 10
Religious laws, 669–671
REM sleep, erections during, 131, 499
Remarriage, 475
Representative sample, 42
Repression, 28, 310
Reproduction
attractiveness preferences and, 213–214
mechanisms of, 19–20
Reproductive system
female, 69*f*
male, 111*f*
Repulsiveness, as grounds for divorce, 11
Requests, making, 254
Research
ethics in, 59–60
evidence in, 38
questions in
Respiratory distress syndrome, neonatal, 366
Respondents
exaggeration of, 53
inaccuracy of, 52–53
Response bias, social desirability, 53
Retardation, 483–484
Retarded ejaculation, 494
Reticular activating system, 151
Retina blastoma, 353*t*
Retrograde ejaculation, 136–137
Reverse transcriptase, 545
Rewards, for sexual behavior, 30
Rh incompatibility, 349
Rh-positive antibodies, 347, 349
Rhesus monkeys, male-male sexual activity among, 301
Rhogan, 349
Rhythm methods. *See* Fertility awareness methods
Richards, Dr. Renée, sex change of, 179
Rimming, 564
Roe v. Wade, 376, 401, 407
attitudes toward, 408–409
Roman Catholic Church
on abortion, 405
attitudes toward homosexuality of, 299
during Middle Ages, 15
on rhythm method, 395
Roman Empire, 13
abortion in, 405
attitudes toward homosexuality in, 298
birth control during, 375
early Christianity in, 13
marriage in, 460
phallic symbols in, 110
Romantic love, 216
characteristics of, 223*t*
components of, 224
in contemporary Western culture, 217–220
definition of, 221
transformed into companionate love, 225–226
Romantic relationships
attraction in, 231–232
building, 232–237
communication skills enhancing, 250–257
continuation of, 238–239
deterioration of, 239
early exchanges in, 234
ending of, 240–244
matching hypothesis of, 215
moving on from, 244
opening lines in, 234
self-disclosure in, 234–236
stages of, 230–244

Romantics, 463
Romeo and Juliet, 216
Roth v. United States, 657
RU-486, 411
  restrictions on, 412–413
  supporters of, 412–413
Rubella, 348
Rubin test, 335

Sacral erection center, 133
  in reflexive erection, 482
Sacrum, 133
  in ejaculation, 136
Sadism, sexual, 13
  characteristics of, 584
  definition of, 584
  sexual partners in, 585
Sadistic rape, 612
Sadists, orgasmic reconditioning of, 596
Sadomasochism, 17, 585–586
  definition of, 585
  gender and, 586
  incidence and prevalence of, 585
  learning theory of, 593
  participants in, 585
  social organizations for, 585
  sociological perspective on, 593–594
  types of stimulation in, 585–586
Safe sex, in HIV prevention, 555
Saline infusion, 411
Saliva analysis, 332
Saltpeter, 149
Sambian culture, 193
  sexual orientation in, 299–300
Same-gender sexual behavior
  among adolescents, 442–443
  in early childhood, 423–424
  in preadolescence, 425
Samples, 42
  of convenience, 44
  in magazine surveys, 48–49
Sampling
  group, 46
  methods for, 42–44
Sarcoptes scabiei, 561
  transmission, symptoms, diagnosis, and treatment of, 535*t*
Scabies, 561
  transmission, symptoms, diagnosis, and treatment of, 535*t*
The Scarlet Letter, 16
Scents, 142–143
Schema, 197
Scholastic Aptitude Test, gender differences in, 190
Schools
  child abuse prevention programs in, 629
  sexual harassment in, 633–634
  in teenage pregnancy prevention, 446–447
Science
  foundations of, 17–18
  rise of, 10
Scientific method, 38–39
  goals and, 39–40
  in human sexuality study, 38–42
Scores, 652
Scrotum, 115–116
  varicose veins in, 334
Secondary sex characteristics, 81
  appearance of, 431
  hormones in, 154
  male, 117
Secretory phase, 93
Sedative-related fetal abnormalities, 350
Sedentary lifestyle, 507
Seduction "lines," 619
Selection factor, 58
Self-concepts, in gender schema, 198
Self-control strategies, 503
Self-defense training, 619
Self-disclosure, 234–236
  in communication, 253
  early and late, 235
  failure of, 244
  gender differences in, 236–237
Self-esteem
  in erectile dysfunction, 493
  jealousy and, 239
  psychological androgyny and, 200–201
  with sensory disabilities, 482–483
  with sexual dysfunction, 488
Self-exploration, in directed masturbation, 517
Self-introductions, in romantic relationships, 231
Self-knowledge, 246
Self-massage, 517
Self-reporting, survey, 52
Selfless love, 221
Semen, 121–122
  collection of in urethral bulb, 170
  conserving, 16
  "hot" and "cold," 172
  myths about, 278
  passage of, 161
  taste and texture of, 277
Seminal fluid, ejaculation of, 136, 161–162
Seminal vesicles, 120
  development of, 175
  secretions of, 121
Seminiferous tubules, 118
Senate Ethics Committee, 602
Sensate focus exercises, 505
  for erectile disorder, 508–509
  for premature ejaculation, 518–520
  for sexual aversion disorder, 508
Senses, in sexual arousal, 142–146
Sensitivity, 200–201
Sensitization. *See* Covert sensitization
Sensory bondage, 583–584
Sensory disabilities, 482–483
Serial monogamy, 456
Seronegativity, 551
Seropositivity, 551
Serotonin
  in PMS, 103–104
  for premature ejaculation, 520
Sertraline, 520
Sex. *See also* Coitus; Commercial sex; Extramarital sex; Sexual activity
  away from home, 651
  broaching subject of, 251
  definitions of, 5
  for eroticism and variety, 651
  problematical, 651
  without commitment, 651
  without negotiation, 651
Sex and Temperament in Three Primitive Societies (Mead), 24
Sex-change operations. *See* Gender-reassignment surgery
Sex chromosomes
  abnormalities of, 176
  in ova and sperm, 328
Sex discrimination, Title VII, 631–632
Sex drive, 155–156
Sex education, 4
  controversy over, 427
  misinformation in, 426
  for preadolescents, 425–430
  in sign language, 482–483
  in teenage pregnancy prevention, 446–447
Sex flush, 158–159
  lightening of, 162
Sex games, early childhood, 423
Sex hormones, 154. *See also specific hormones*
  for female sexual arousal disorder, 515
  in female sexual behavior, 156–157
  human fetal exposure to, 59
  levels of in sexual orientation, 314
  in male sexual behavior, 155–156
  organizing and activating influences of, 154–155
  in premarital intercourse among adolescents, 439
  prenatal, 58, 155, 308
  in sexual differentiation, 175
  in sexual orientation, 308–309
Sex in America: A Definitive Survey, 47
Sex-linked genetic abnormalities, 354
Sex offenders
  issues related to treatment of, 594–595
  pornography and, 661
  treatment of, 629–630
Sex organs
  definition of, 5
  female
  male
Sex partners. *See* Partners
Sex research
  ethics in, 59–60
Sex skin, 159–160
Sex talk, 250–251
Sex the Measure of All Things: A Life of Alfred C. Kinsey (Gathorne-Hardy), 47
Sex therapists
  advertisements for, 522
  finding, 522–523
  male-female team, 504, 505
Sex therapy
  aims of, 504
  for erectile disorder, 508–509
  evaluation of, 521–522
  female-male teams in, 504
  finding qualified therapist for, 522–523
  group programs, 505
  Helen Singer Kaplan approach, 505
  for marital problems, 503
  Master and Johnson approach, 504–505
  for orgasmic disorders, 515–520
  for sexual arousal disorders, 508–515
  for sexual desire disorders, 505–508
  for sexual pain disorders, 520–521
Sex toys for STI prevention, 565
Sexism, 188–189
  socialization in, 195
Sexist remarks, 633
Sexologist, 17
Sexual abuse. *See* Childhood, sexual abuse during
Sexual activity. *See also* Coitus; Sexual behavior
  among gay males and lesbians, 318–319
  among teenagers, 447–449
  incidence and frequency of, 46–47
  in later years, 477–480
  noncoital, 565
  nonprocreative, 13–14
  patterns of with age, 479–480
  postpartum resumption of, 368
Sexual aggression, 198–199
Sexual anatomy, male, 110–137
Sexual anesthesia, 16

Sexual arousal, 142
  ability to assess, 503
  age-related changes in, 477*t*
  cognitive appraisal of, 221
  definition of, 40–41
  disorders of, 489, 491
  gender differences in, 199
  with homosexual and heterosexual videos, 304–305
  in infant girls, 420
  limbic system stimulation in, 153–154
  operational definition of, 41–42
  pain and, 584
  senses in, 142–146
Sexual assault, 603–604
  by athletes, 610
  emotional and behavioral reactions to, 613*t*
  by group of assailants, 606–607
  prevalence of, 604
  women's relationships with men committing, 605*f*
Sexual attraction, 207
Sexual aversion disorder, 491
  insight-oriented therapy for, 506
  treatment of, 508
Sexual behavior
  alcohol and, 149–150
  atypical, 472–599
  changing patterns of, 295
  in childhood and adolescence, 420
  definition of, 5
  during early childhood, 423–424
  evolutionary advantages of, 307
  gender roles and, 198–199
  high-risk, 322, 562
  normal versus deviant, 573–574
  with others, 272–287
  psychological androgyny and, 201
  rewards and punishments for, 30
  sexual orientation and, 292–293
  solitary, 262–272
  stereotypes and, 306–307
Sexual Behavior in the Human Female (Kinsey), 17
Sexual Behavior in the Human Male (Kinsey), 17
Sexual Behavior in the 1970's (Hunt), 47–48
Sexual beliefs, diversity of, 6–7
Sexual coercion, 602. *See also* Childhood sexual abuse; Rape; Sexual harassment
  pornography and, 660–664
  verbal, 619
Sexual communication
  common difficulties in, 250–251
  permission for, 251–252
Sexual decision making, 667
  balance sheet for, 673–675
  on case-by-case basis, 671
  conflict in, 668–669
  informed, 7, 668
  value systems and, 669–673
Sexual desire
  disorders of, 489, 490
  hypoactive, 490
  lack of, 490–491
  medications affecting, 491
Sexual deviations, 17
Sexual differentiation, 172
  of brain, 176
  causes of, 172
  genetic factors in, 174–175
  prenatal, 172–173
  sex hormones in, 175
  of testes and ovaries, 175–176
Sexual double standard, 441–442
Sexual dysfunction, 4, 488
  acquired, 490
  alcohol and, 500
  among European and African American men and women, 498*t*
  difficulty talking about, 488
  diminished self-esteem with, 488
  features of, 488
  immediate and remote causes of, 505
  innovative treatments of, 44
  life long, 490
  in men versus women, 489
  origins of, 498
  prevalence of, 488–489
  situational, 490
  treatment of, 504
  types of, 489–497
Sexual fantasy. *See* Fantasy
Sexual function, male, 130–137
Sexual guilt, 494
Sexual harassment
  behaviors of, 630–631
  on campus, 633
  definition of, 630
  documentation of, 634
  filing complaints against, 635
  incidence of, 631, 632
  legal remedies, 635
  motivations for, 631
  peer, 633
  proving charges of, 632
  resisting, 634–635
  in schools, 633–634
  in workplace, 631–633
Sexual intercourse. *See* Coitus
Sexual knowledge, 7
Sexual molestation, 501
Sexual orientation, 4, 290. *See also* Gay people; Heterosexuality; Homosexuality; Lesbians
  adolescent experience in, 311–312
  beliefs about, 8
  brain structure and, 309
  classification of, 292–297
  continuum of, 293–296
  cross-cultural perspectives on, 299–300
  definition of, 291
  early experience in, 310–311
  environmental influences in, 308
  environmental versus biological causes of, 303–304
  fantasy as measure of, 297
  gender identity and, 292
  genetics and, 307–308
  hormonal influences and, 308–309
  as independent dimensions, 296
  law and, 305–306
  learning theories of, 311–312
  lifestyle and, 319–323
  oversimplification of factors in, 9
  of parents, 303
  psychoanalytic theory of, 309–310
  psychological treatment for, 315–316
  in sadomasochism, 586
  sex hormone levels and, 314
  versus sexual behavior, 292
  in sexual dysfunctions, 501–502
  sexual experiences and, 292–293
  social attitudes toward, 302–307
  stereotypes and, 306–307
Sexual outlets, nongenital, 574
Sexual pain disorders, 490, 495–497
  treatment of, 520–521
Sexual panic state, 491
Sexual passivity, 306–307
Sexual performance
Sexual phobia, 491
Sexual pleasure
  double standards for, 500
  lack of subjective feelings of, 491
Sexual positions in *Kama Sutra*, 14–15
Sexual relations
  different definitions of, 53
  knowing partner in, 566
  physical attractiveness and, 212
Sexual response
  autonomic nervous system in, 134–135
  brain in, 133–134, 151–154
  cycle of, 157–158
  infant's capacity for, 420–421
  Kaplan's three stages of, 163
  preplateau stage of, 164
  with spinal cord injuries, 482
  spinal reflexes in, 131–135
Sexual responsive, 4
Sexual revolution, 18–19, 454, 464–466
  group marriage and, 476
Sexual sadism, 574
Sexual sadists, 584
Sexual satisfaction, 466–467
  with age, 480
  coitus frequency and, 56–57
  emotional factors in, 467
  sociocultural factors in, 467*t*
Sexual skills, 503
Sexual techniques, 262
  among gay males and lesbians, 318–319
  ineffective, in sexual dysfunctions, 502
Sexual trauma
  in dyspareunia, 495
  in female sexual arousal disorder, 493
  in sexual aversion, 491
  in vaginismus, 521
Sexual values, 663–664
Sexuality
  in adulthood, 454–485
  biological perspective on, 19–20
  critical thinking about, 8–9
  cross-cultural perspective on, 24–28
  cross-species perspective on, 24
  definition of, 5
  double standard toward, 217–218
  Eastern religions and, 14–15
  evolutionary perspective on, 20–23
  foundations of scientific study of, 17–18
  historical perspectives on, 9–19
  multiple perspectives on, 33
  psychological perspectives on, 28–30
  reasons for studying, 4
  scientific approach to, 38–42
  scientific goals and methods of, 39–40
  sociocultural perspectives on, 31–33
  study of, 5–6
  in twentieth century, 18
  values and, 6–7
Sexually abused children
  ethnic differences in, 626
  gender differences in, 626
  in late adolescence and early adulthood, 626
  long-lasting problems in, 626–627
  post-traumatic stress disorder in, 626
  regressive behaviors in, 626
Sexually aggressive men, 612
Sexually transmitted disease. *See* Sexually transmitted infections

Sexually transmitted infections. *See also* AIDS; HIV (human immunodeficiency virus); Human papilloma virus; Syphilis
- alcohol and, 563
- anal sex and, 286–287
- at-risk populations for, 529*f*
- bacterial, 532–541
- birth control pills and, 381
- causes of, 533–535*t*
- circumcision and, 113–114
- concerns about, 528
- condoms in protection against, 395
- consulting physician for, 565–566
- diagnosis of, 533–535*t*
- in dyspareunia, 495
- ectoparasitic infestations, 560–561
- epidemic of, 529–530
- in erectile dysfunction, 499
- female condoms in protection against, 401
- in HIV transmission, 550
- incidence of, 529
- inspection for signs of, 563–564
- prevalence of, 528
- prevention of, 561–566
- protection against, 378–379
- recognizing signs of, 4
- risks of, 562–563
- sources of help for, 567
- symptoms of, 533–535*t*
- testing for before sexual relations, 565
- transmission of, 529, 533–535*t*
- treatment of, 533–535*t*
- vaginal, 542–544
- viral, 544–560

Shame, 13
Shepard, Matthew, murder of, 305*f*
Shigellosis, 541
Sibling incest, 625, 626
Sickle-cell anemia, 353–354
- characteristics of, 353*t*
- screening for, 355

Signaling, 254
Sildenafil (Viagra), 510*t*, 511–514
Silicone diaphragm, 402
Silver nitrate, 560
Simplirix, 557
Singlehood, 454–456
Singles scene, 456
Situational ethics, 671
Sixty-nine, 277, 278*f*
Skene's glands, 166–167
Skepticism, 8
Skin senses, 144–145
Skin test, 541
Slasher films, 662–663
S&M. *See* Sadomasochism
Small talk, 232–234
Smell, in sexual arousal, 142–144
Smiling face, 209–210
Smoking, fetal abnormalities and, 350
Social attitudes, 609
Social change, in 1960s and 1970s, 18–19
Social contacts, 245
Social-desirability response bias, 53
Social-exchange theory, 230
- on ending of relationships, 241–244

Social learning, theory of, 30
Social norm, 573
The Social Organization of Sexuality: Sexual Practices in the United States, 47
Social prescription, 224
Social skills
- lack of, 244
- training in for paraphilias, 596

Socialization, 194–195
Sociocultural factors, 31–33
- in incest, 626
- in rape, 609–611

Sociocultural theorists, 5–6
Sociological theory, 593–594
Sodom, 298
Sodomy, 298
- laws prohibiting, 305–306

Solitary sexual behavior, 262
- fantasy in, 270–272
- masturbation, 262–270

Spanish AIDS/SIDA Hotline, 567
Spanish fly, 146–147
Spanking, 584
Spatial-relations skills, gender differences in, 192
Specific requests, 254
Spectator role, 494
Speculum, 80*f*
Sperm, 117
- activity of, 330
- in average ejaculate, 329
- declining production of, 478
- drugs affecting genetic material in, 350
- impaired production of, 176
- intracytoplasmic injection of, 336–337
- journey of, 329
- low count of, 333
- low motility of, 333–334
- mature, 435
- motility of, 120
- production of, 115–116, 118–119
- separation procedures for, 330–331
- sex chromosomes in, 328
- swarming around ovum, 329*f*

Sperm banks, 549
Spermatic cord, 115
Spermatids, 119
Spermatocytes, 119
Spermatogenesis, 118–119
Spermatozoa, 119
- passage of, 121*f*

Spermicides, 390
- advantages and disadvantages of, 390–391
- allergic reactions to, 495
- application of, 390
- with condom use, 394–395
- under development, 402
- with diaphragms, 388
- effectiveness of, 390
- failure rates of, 378*t*
- mechanism of, 390
- reversibility of, 390

Sphincters, underlying external female sex organs, 73
Spinal block, 358
Spinal cord
- in ejaculation, 136
- injuries to, 481–482
- in reflex arc, 132–133
- in sexual response, 133

Spinal reflexes, 131–135
Sports competition, 610–611
Squeeze technique, 518–520
- evaluation of, 522

SRY expression, 174–175
SSY chromosomes, 176
Stanley v. Georgia, 657
Staphylococcus aureus, 93–95
Staring, 250
Statins, 499
Statistical infrequency, 573
Statutory rape, 603
Stepfamily, 475
Stereotypes, 9
- gender-role, 186–187, 188–189
- sexual behavior and, 306–307

Sterilization, 398
- advantages and disadvantages of, 401
- female, 399–401
- male, 398–399

Sternberg's triangular love, 221–226
Steroids, pheromonal, 143
Stillbirth, 348
Stimulants, aphrodisiac effects of, 151
Stone Age art, 10
Stop-start method, 520
Storge, 217, 221
Straight people, 291
Stranger rape, 605
Stratified random sample, 43
Street hustlers, 654
Streetwalkers, 641–646
- clients of, 646
- motives for using, 651
- post-traumatic stress disorder in, 648–649
- risks to, 646

Strength-weakness dimension, 197–198
Stress, 59–60
Studies in the Psychology of Sex (Ellis), 17
Suction aspiration, newborn, 357
Suction curettage, 410
Sudden infant death syndrome, 366
- maternal smoking in, 351

Suicidality, among gay males and lesbians, 313–314, 316
Superego, 28
Support groups, genital herpes, 558
Suppository, 510*t*, 511
Surface contact, 232–233
Surfactant, fetal, 366
Surgery
- for cervical cancer, 76–77
- for erectile disorder, 510–511
- for gentile warts, 535*t*
- for prostate cancer, 125

Surrogate motherhood, 337
Surveys
- magazine, 48–49
- methods of, 45–53

Survival of fittest, 20
Swinging, 54, 469, 472, 476
"Swinging Sixties," 18
Symbolism, sexual, 28–29
Sympathetic nervous system, 134
Sympathetic pregnancy, 341
Syphilis
- congenital, 538
- diagnosis and treatment of, 533*t*, 539
- in erectile dysfunction, 499
- incidence of, 537
- latent stage of, 538
- origins of, 536–537
- in prenatal development, 348
- primary stage of, 538
- secondary stage of, 538
- symptoms and course of, 533*t*, 538–539
- tertiary stage of, 538–539
- transmission of, 533*t*, 537–538

Systematic desensitization, 595

T-cells, 546
"Take Back the Night" marches, 606
- against rape, 617*f*

Talking, about talking, 251
Tamoxifen, 85, 87–88
Tampons, in toxic shock syndrome, 93–95

Taoism, 14
Target populations, 42
Taste, in sexual arousal, 145
Taxol, 88
Tay-Sachs disease, 354
characteristics of, 353*t*
screening for, 355
3TC, 551–552
Tchambuli tribe, 193
Teenage fathers, 444–445
Teenage pregnancy, 407, 436
consequences of, 444
contraceptive use and, 445–447
declining rate of, 448
father's responsibility in, 444–445
incidence of, 443–444
medical complications of, 444
prevention of, 446–447
Teenage runaways
in male prostitution, 653
in prostitution, 649–650
Teenagers
sexual activity among, 447–449
teenpregnancydc.org, 447
Telephone scatologia, 580–581
Tension headaches, in dysmenorrhea, 101
Teratogens, 347
Terms
definition of, 8
different meanings of, 53
"Test-tube babies," 20, 336
Testes, 110, 111, 117–119
age-related changes in, 478
cancer of, 122–123
descent of, 175–176
hormonal control of, 117*f*
in menstrual cycle regulation, 89
self-examination of, 123
sex hormone production by, 154
undescended, 176
Testicles, 110, 111
undescended, 122
Testosterone, 116, 117, 154
activating effects of, 308
age-related changes in production of, 478
aphrodisiac effects of, 147
in male pubertal changes, 434
in menstrual cycle regulation, 89
organizing and differentiating effect of, 155
in paraphilias, 590–591
in premarital intercourse among adolescents, 439
in prenatal sexual differentiation of brain, 176
reduced levels of in sexual dysfunction, 498
reductions of for rapists and child molesters, 630
replacement therapy
in sexual desire, 490, 491
in sexual differentiation, 175
in sexual functioning, 155–156
skin patches, 515
vaccine with, 404
Testosterone-delivery contraceptive systems, 403
Tetracycline
in fetal abnormalities, 350
for shigellosis, 541
for syphilis, 533*t*
Thalamus, 152
Thalidomide, 349–350
Theory, frameworks of, 40
Thonga society, 25
Thrush, 542–543
Time Magazine/CNN poll
on divorce, 474
on single women, 455*f*
Tomboyishness, 312
Torture, 586
Toucherism, 586
Touching, 273–274
"bad," 628
in communication, 249–250
in sex therapy for erectile disorder, 508
in sexual arousal, 144–145
techniques of, 274
Toxemia, in prenatal development, 348
Toxic shock syndrome, 93–95
with diaphragms, 388
Toxoplasmosis, AIDS-related, 548
Tranquilizers
in erectile dysfunction, 499
in fetal abnormalities, 350
Transformer (tra) genes, 174
Transgender organization directory, 201
Transition, 355
Transplants, HIV transmission via, 549
Transsexualism, 179–186
gender reassignment and, 182–186
versus homosexuality, 181
theoretical perspectives on, 181–182
versus transvestism, 576
Transsexuals, 155
adolescence in, 181
as drag prostitutes, 654
female, 181
female-to-male, 182
gender dysphoria in, 181
gender role changes for, 185
male, 181
male-to-female, 182
Transvestic fetishism, 574
Transvestism
behaviors of, 577
definition of, 576
learning theory of, 592
versus nonsexual cross-dressing, 576–577
origins of, 576
secrecy of, 577–578
versus transsexualism, 576
Transvestites
as drag prostitutes, 654
married life of, 577–578
systematic desensitization for, 595
Treatment, 57
Treponema pallidum, 536–537
antibodies to, 539
symptoms, diagnosis, and treatment of, 533*t*
transmission of, 533*t*, 537–538
Triangular theory of love, 246
Trichloroacetic acid
for genital warts, 535*t*, 560
for molluscum contagiosum, 560
Trichomonas vaginalis, 543–544, 560
transmission, symptoms, diagnosis, and treatment of, 534*t*
Trichomoniasis
cause of, 543
diagnosis and treatment of, 534*t*, 543–544
transmission and symptoms of, 534*t*
Trobriand islanders, 24–25
childhood sexuality in, 30
ethnographic studies of, 54
normal sexual behavior for female adolescent, 574
Trophoblast, 344
Tropic of Cancer, 464, 655
Trust
betrayal of, 472
in intimacy, 246–248
Truth, meaning of, 8
Tubal ligation, 399–401
Tubal sterilization, 399–401
Tumescence, 499
Tunica albuginea, 130
Turner syndrome, 176
Twins
dizygotic, 307–308
low birth weight, 366
monozygotic, 307–308

Ulcers, chronic AIDS-related, 548
Ultrasound genetic screening, 354–355
Ulysses, 655
Umbilical cord, 345
severing of, 357
Unconscious mind, 28
United Nations AIDS (UNAIDS), microbicide effort of, 391
United States, contemporary society in, 9
Uprima (apomorphine), 510*t*, 511–512
Ureaplasma urealyticum, 533*t*
Urethra, female "ejaculate" from, 166–167
Urethral bulb, 136
during orgasmic phase, 170
Urethral meatus, 112
Urethral opening, female, 69–71
Urethritis, 122
nongonococcal, 539
Urinary control, 74
Urinary urgency, 70
Urination, eroticization of, 589
Urine
analysis of for ovulation, 332
released during female orgasm, 166–167
sexual arousal connected with, 589
Urogenital system disease, male, 122–130
Urologists, 122
Urophilia, 589
Uterine contractions
during breast feeding, 367
first, 355, 356*f*
in orgasm, 162
Uterine orgasm, 166
Uterus, 77
development of, 175
hysterectomy of, 77–78
layers of, 77
perforation of with IUDs, 387–388
retroverted, 332
Utilitarianism, 672

Vaccines, 545*n.2*
for AIDS, 551
contraceptive, 404
genital herpes, 557
hepatitis, 559
human papilloma virus, 560
Vacuum aspiration, 410
Vacuum constriction device (VCD), 511*t*, 514
Vacuum pump, 510*t*, 514
Vagina, 75–76
acidity of, 329
artificial, 269
development of, 175
involuntary contractions of, 497
manual stimulation of, 274
oral stimulation of, 277
during plateau phase, 159
postmenopausal changes in, 477–478

Vagina, *continued*
during sexual excitement, 158
touching of, 273
Vaginal dilators, 520–521
Vaginal infections
sexually transmitted, 542–544
Vaginal intercourse, 437*t*
Vaginal lining, "sweating" of, 74
Vaginal lubrication, 74
with condom use, 393
inadequate in dyspareunia, 495
in infant girls, 420
insufficient, 489
organic factors in, 499
postmenopausal changes in, 478
during sexual response cycle, 158
Vaginal mucus, tracking of, 332
Vaginal opening, 66, 72–73
parous and nonparous, 72*f*
during plateau phase, 159
Vaginal orgasm, 165
Vaginal photoplethysmography, 40
Vaginal ring, 402
Vaginal suppositories, 390
for candidiasis, 543
prostaglandins in, 411
Vaginal walls, 75
Vaginismus, 490, 496
incidence of, 497
psychological causes of, 496–497
sex therapy for, 522
treatment of, 520–521
Vaginitis, 75–76, 542
condoms in protection against, 395
prevention of, 76
Vaginosis
bacterial, 542
transmission, symptoms, diagnosis, and treatment of, 533*t*
Valacyclovir, 534*t*, 557
Validating information, 257
Validity, 46–47
Valium
in erectile dysfunction, 499
in fetal abnormalities, 350
Value systems, 669–673
Values
definition of, 6
sexual double standard and, 441–442
sexuality and, 6–7
Variables, 40
dependent, 57–58
independent, 57
Vas deferens, 115, 119–120
development of, 175
location and incision of in vasectomy, 399*f*
scarring of, 333–334
Vascular surgery, 510–511
Vasectomy, 119–120, 398–399
procedure of, 399*f*
reversibility of, 399
Vasocongestion, 41, 42, 158, 491
dissipation of, 163
persistent, 163
during plateau phase, 159
Vasodilators
for erectile disorder, 510
for female sexual arousal disorder, 515
Vasomax (phentolamine)
for erectile disorder, 510*t*, 511–512
side effects of, 514
Vasovasotomy, 399
Vatsyayana, 14
VDRL tests, 539
Venereal disease. *See* Sexually transmitted infections
Venus of Willendorf, 10*f*
Verbal abilities, gender differences in, 189–190
Verbal cues, 253
Vestibular bulbs, 74
Vestibule, 69
Viability, age of, 346
Viagra (sildenafil), 147, 504
benefits of, 513–514
for erectile disorder, 510*t*, 511–514
for female sexual arousal disorder, 515
impact of on post-hysterectomy sexual complaints, 519*t*
side effects of, 514
Vibrator
in masturbation, 518
in STI prevention, 565
types of, 270*f*
Victorian era, 16–17
attitudes of, 18, 199
Violence, 472–473. *See also* Domestic violence; Rape
pornography and, 661–663
Violent pornography, 610, 662–663
Viral infections, sexually transmitted, 544–560
Virgin Mary, cult of, 15
Virginity, early Christian attitude toward, 13
Viscosity, 397
Vision, in sexual arousal, 142
Visual impairment, 482–483
Visual-motor skills, 192
Visual-spatial abilities, 189
Vitamin A excess, 350
Vitamin B6, 105
Vitamin D excess, 350
Vitamin-related fetal abnormalities, 350
Voices, sultry, 146
Volunteer bias, 43–44, 52
Vomeronasal organ, 143
Voyeurism
definition of, 581
learning theory of, 593
psychoanalytic theory of, 591
risky behavior in, 581
Voyeurs
childhood sexual abuse of, 594
personality of, 582
violent behavior in, 581–582
Vulgarity, 250–251
Vulnerability, critical period of, 347–348
Vulva, 66
normal variations in, 67*f*
oral stimulation of, 277
Vulval orgasm, 166

Washington DC Campaign to Prevent Teen Pregnancy, 447
Wasting, AIDS-related, 548
Weight, erectile dysfunction and, 499
West Side Story, 216
Western culture
attitudes toward homosexuality in, 298–299
marriage in, 459–460, 461
mate selection in, 462
romantic love in, 217–220
sexual pleasure in, 500
sexually restrictive, 463–464
social change in, 16
White blood cells, HIV infection in, 547*f*
Whore-madonna complex, 15, 651
Wife, "good," 11
Will Power/Won't Power, 449
Withdrawal, 395. *See also* Coitus interruptus
Wolffian ducts, degeneration of, 175
The Woman Rebel, 375
Women
in ancient Greece, 12–13
anorgasmic, 494, 515–518
breasts, 80–88
conflicting concepts of, 15
dehumanized by pornography, 657, 663
external sex organs of, 66–74
extramarital sex among, 468–469
first coital experience among, 440–441
gatekeeping role of, 33
infertility problems in, 334–335
internal sex organs of, 75–80
in Islamic society, 14
masturbation and martial satisfaction among, 268
masturbation techniques used by, 269–270
menstrual cycle of, 88–100
menstrual problems in, 100–105
as property, 11
rape by, 608–609
response to pornography of, 659–660
sexually assertive, 199
sexually passive, 198
single, 455
Time Magazine/CNN poll on single life, 455*f*
underaroused, 199
in Victorian era, 16–17
violence against, 609–611
Woody Allen-Soon Yi marriage, 463*f*
Workplace, sexual harassment in, 631–633
World Health Organization (WHO), estimates of STIs, 529

X-ray-related fetal abnormalities, 352
X sex chromosomes, 119, 173, 328
in chromosomal abnormalities, 176
in gay male sexual orientation, 308
in genetic abnormalities, 354
Xanax, 499
XX sex chromosomal structure, 173
XY chromosomes, 176

Y chromosomes, 119, 328
Y gene, sex-determining region, 174–175
Y sex chromosome, 173
Yang, 14
Yeast infections
diet and, 543
factors contributing to, 542–543
vaginal, 542–543
Yin, 14
Yohimbine, 147

Zeus, 12
Zidovudine, 535*t*, 551–552, 553*f*
Zona pellucida, 329
Zoophilia, 587
Zovirax (acyclovir), 557
Zygote, 92, 172–173, 328
division of, 173
division of after conception, 343*f*
implantation of, 344
Zygote intrafallopian transfer (ZIFT), 336